NEW TRENDS IN DESIGN OF CONTROL SYSTEMS 1997

A Proceedings volume from the 2nd IFAC Workshop,
Smolenice, Slovak Republic, 7 - 10 September 1997

Edited by

Š. KOZÁK

Department of Automatic Control Systems,
Faculty of Electrical Engineering and Information Technology,
Slovak University of Technology, Bratislava, Slovak Republic

and

M. HUBA

Department of Automation and Control,
Faculty of Electrical Engineering and Information Technology,
Slovak University of Technology, Bratislava, Slovak Republic

Published for the

INTERNATIONAL FEDERATION OF AUTOMATIC CONTROL

by

PERGAMON

An Imprint of Elsevier Science

UK Elsevier Science Ltd, The Boulevard, Langford Lane, Kidlington, Oxford, OX5 1GB, UK
USA Elsevier Science Inc., 660 White Plains Road, Tarrytown, New York 10591-5153, USA
JAPAN Elsevier Science Japan, Tsunashima Building Annex, 3-20-12 Yushima, Bunkyo-ku, Tokyo 113, Japan

First edition 1998

Library of Congress Cataloging in Publication Data

A catalogue record for this book is available from the Library of Congress

British Library Cataloguing in Publication Data

A catalogue record for this book is available from the British Library

ISBN 0-08-042935 1

This volume was reproduced by means of the photo-offset process using the manuscripts supplied by the authors of the different papers. The manuscripts have been typed using different typewriters and typefaces. The lay-out, figures and tables of some papers did not agree completely with the standard requirements: consequently the reproduction does not display complete uniformity. To ensure rapid publication this discrepancy could not be changed: nor could the English be checked completely. Therefore, the readers are asked to excuse any deficiencies of this publication which may be due to the above mentioned reasons.

The Editors

Transferred to to Digital Printing 2009

IFAC WORKSHOP ON NEW TRENDS IN DESIGN OF CONTROL SYSTEMS

Sponsored by
International Federation of Automatic Control (IFAC)
Technical Committee on Control Design

Organized by
Slovak Society of Cybernetics and Informatics
(Slovak IFAC National Member Organization)

Supported by
Faculty of Electrical Engineering, Slovak University of Technology
Regotrans-Sauter, Slovak Republic
Microstep, Slovak Republic
Kameron, Slovak Republic
Termoreg, Slovak Republic

International Programme Committee (IPC)

Chairman: Š. Kozák , SK
Co-Chairmen: G. Hulkó, SK, J. Mikleš, SK, V. Veselý, SK

M. Alexík, SK
L. Basanez, E
G. Bekey, USA
Z. Binder, F
J. Bokor, H
M.A. Brdys, UK
P. Brunovský, SK
K. Burnham, UK
P. Colaneri, I
G. Conte, I
D.W. Clark, UK
F. Čapkovič, SK
K. Dobrovodský, SK
D. Donlagić, SLO
S. Engell, D
A.L. Fradkov, RUS
B. Frankovič, SK
E.D. Gilles, D
M. Günther, D
P. Hippe, D
H. Hoyer, D
V. Hrušák , CZ
B. Hrúz, SK
M. Huba, SK
H. Jörgl, A
L. Jurišica, SK
T. Kaczorek, PL
L. Keviczky, H
H. Koivo, FIN
F. Kraus, CH
V. Kučera, CZ
J.F. Lafay, F
C.P. Lewis, UK
S. Longhi, I
J. Lunze, D
R. Mesiar, SK
C. Moog, F
J. Murgaš, SK
A.J. Niemi, FIN
A. de Paor, IRL
D.Papadopoulos, GR
A. Perdon, I
R. Prokop, CZ
T.Ph. Proychev, BG
H. Puta, D
P.D. Roberts, UK
M. M'Saad, F
J. Sarnovský, SK
D.E. Seborg, USA
G. Schmidt, D
G. Sotirov, BG
R. Suarez, E
J. Suchý, SK
D. Siljak, USA
B. Tamm, EST
P. Tatjewski, PL
A. Tornambe, I
I. Vajk, H
P. Vavřín, CZ
I. Václavík, CH
J. Virkkunen, FIN
A. Víteček, CZ
E. Wagneur, F
Ch. Wurmthaler, D
A.E.Yakubovich, RUS
M. Žalman, SK
M. Zeitz, D
D. Zöbel, D

National Organizing Committee (NOC)

Chairman: M. Huba
Secretary: A. Kozáková

P. Bisták
M. Dúbravská
M. Budinský
B.Hrúz
M. Hypiusová
L. Juráčková
M. Karšai
P. Kuľha
E. Miklovičová
D. Mudrončík
I. Sekaj
Z. Skachová
N. Spurná
M. Zbihlej
K. Žáková

Preface

The IFAC Workshop "New Trends in Design of Control Systems" has its own history. It was in 1994 when we decided to organize this IFAC event for the first time and to invite scientists from the Slovak Republic and other countries to the Castle Smolenice in the Slovak Republic. Encouraged by the success of the 1st IFAC Workshop NTDCS '94 we decided to try to continue with the workshops every three years.

The 2nd IFAC Workshop "New Trends in Design of Control Systems" was organized by the Slovak Society of Cybernetics and Informatics (which is the Slovak IFAC National Member Organization) on 7-10 September, 1997 being sponsored by the IFAC Technical Committee on Control Design..

The aim of this Workshop was to bring together experts in various theoretical and applied areas of Control Theory to review recent advances and to anticipate major future developments of the discipline. The leading scientists participated in the International Program Committee. The Workshop attracted more than a hundred authors from which 92 presented their contributions in Smolenice.

The presented papers were organized within 6 sessions covering the following areas :

- Linear and Nonlinear Control System Design
- Large Scale Systems, Decentralized Control
- Intelligent Controllers, Neural, Fuzzy and Genetic Control Design
- Adaptive and Selftuning Control
- Control of Discrete Event Systems
- Control Design Problems in Robotics and Manufacturing

It is not an exaggeration to say that due to presentations of original papers of a high professional level both IFAC Workshops (held in 1994 and 1997) helped to outline new directions in Control Theory development and their applications in practice.

It is a pleasure for us to acknowledge the sponsorship of IFAC, especially of the TC on Control Design, as well as the continuous assistance from the IFAC Secretariat.

Bratislava, September 1997

Štefan Kozák
Mikuláš Huba
Workshop Editors

CONTENTS

LINEAR AND NONLINEAR CONTROL SYSTEM DESIGN

CONTROL DESIGN PROBLEMS IN ROBOTICS AND MANUFACTURING

CONTROL OF DISCRETE EVENT SYSTEMS

NEURAL, FUZZY AND GENETIC CONTROL SYSTEMS

LARGE-SCALE SYSTEMS

ADAPTIVE AND SELFTUNING CONTROL

2-D SINGULAR SOLUTIONS

NONCONVEX OPTIMAL CONTROL PROBLEMS

Vladimir A. Yakubovich

St.Petersburg State University,
Department of Mathematics and Mechanics,
Bibliotechnaja pl.2, Petrodvoretz, 198904, St.Petersburg, Russia.
E-mail: y@yakub.niimm.spb.su

Abstract: We present a short survey of some recent author's results concerning the method for solution of a class of global minimization problems. They may be nonconvex in general. The case of minimizing a quadratic functional in a Hilbert space subject to quadratic constraints is considered specially. Many applied optimal control problems can be formulated in this abstract form. For this case the method reduces the initial problem to the problem of minimization of related quadratic functional without quadratic constraint and to a finite-dimensional convex optimization problem. An example of application of this method, namely the deterministic infinite-horizon linear-quadratic problem of optimal control with quadratic constraints, is considered in details.

Keywords: Global minimization, linear quadratic constrained optimization problems, quadratic constraints.

1. INTRODUCTION

It is well known that many engineering problems are formulated naturally as global optimization problems with constraints given by some inequalities. These problems often are nonconvex. As well known, the nonconvexity is a reason of many difficulties in the optimization theory. To solve such problems, usually the Lagrangian multipliers method is used. But this method gives only necessary conditions. In many cases it gives a wide set of the solutions which contains the "superfluous" values. The problem to find the real solutions may be as difficult as the initial one. In Yakubovich (1992) we presented the method for the solution of a class of global minimization problems. (See also Yakubovich 1992, 1993). This method gives necessary and sufficient conditions. So it gives all the solutions of the considered constrained problem and only these solutions.

To characterize the class of problems which can be solved by this method efficiently and to explain the abstract settlement of the problem consider first two simple examples.

Example 1. It is required to minimize the functional

$$\mathcal{F} = \int_0^\infty y^2\,dt \to \min \tag{1}$$

subject to constraints

$$\frac{d^2y}{dt^2} = u, \quad y(0) = 1, \quad \frac{dy}{dt}(0) = 0, \tag{2}$$

$$|y| + |\frac{dy}{dt}| + |u| \in L_2(0,\infty), \tag{3}$$

$$\int_0^\infty u^2\,dt \le \alpha \int_0^\infty y^2\,dt. \tag{4}$$

(All values are real scalar, $\alpha > 0$ is a given number.)

Example 2. (Problem of optimal controller design for aircraft roll stabilization.) The motion equations are (see Gorbatenko et al.(1976), p.227)

$$\frac{dx_1}{dt} = x_2, \qquad \frac{dx_2}{dt} = \alpha x_2 + \beta u + \zeta(t), \qquad (5)$$

where $\alpha \neq 0$, $\beta \neq 0$ are given constants, x_1 is the roll angle, u is the angle of aileron deflection, $\zeta(t)$ is the white noise external disturbance:

$$\mathbf{E}\,\zeta(t) = 0, \qquad \mathbf{E}\,\zeta(t)\zeta(s) = \delta(t-s)\vartheta.$$

Here x_1, x_2 are random stationary processes. $\mathbf{E}$ denotes the mathematical expectation. It is required to minimize the energy of control

$$\mathbf{E}\,u^2 \to \min$$

subject to constraints (5) and

$$\mathbf{E}\,x_1^2 \leq \gamma_1, \qquad \mathbf{E}\,x_2^2 \leq \gamma_2.$$

(Here $\vartheta > 0$, $\gamma_1 > 0$, $\gamma_2 > 0$ are given numbers; $u(t)$ must not depend on future values of $x_1(t)$ and $x_2(t)$, the precise formulation is given later.)

Let us consider these examples from abstract point of view.

Take the first example. Let $\mathcal{H}$ be the three-dimensional L_2-space: $\mathcal{H} = L_2^{(3)}(0,\infty)$ and $\mathcal{Z} = \{z\}$ be the set of all triples $z = \begin{bmatrix} y(\cdot) \\ \dot{y}(\cdot) \\ u(\cdot) \end{bmatrix}$ satisfying (2) and (3). Obviously $\mathcal{Z}$ is a plane (an affine subspace) in the Hilbert space $\mathcal{Z}$. Let us introduce the quadratic functionals $\mathcal{F} = \int_0^\infty y^2\,dt$ and $\mathcal{G} = \int_0^\infty (\alpha y^2 - u^2)dt$ on $\mathcal{H}$. Then our problem rewrites as follows:

To minimize $\mathcal{F}(z)$ subject to $z \in \mathcal{Z}$, $\mathcal{G}(z) \geq 0$.

Consider the second example. Let $\mathcal{H}$ be the Hilbert space of three-dimensional random stationary processes $z(t) = \begin{bmatrix} z_1(t) \\ z_2(t) \\ z_3(t) \end{bmatrix}$. After the natural definition of admissible processes we will see that the equations (5) for $z = \begin{bmatrix} x_1(\cdot) \\ x_2(\cdot) \\ u(\cdot) \end{bmatrix}$ define certain plane $\mathcal{Z}$ in the Hilbert space $\mathcal{H}$. Let us introduce the quadratic functionals

$$\mathcal{F} = \mathbf{E}u^2, \quad \mathcal{G}_1 = \gamma_1 - \mathbf{E}x_1^2, \quad \mathcal{G}_2 = \gamma_2 - \mathbf{E}x_2^2.$$

Then our problem rewrites as follows:

To minimize $\mathcal{F}(z)$ subject to $z \in \mathcal{Z}$, $\mathcal{G}_1(z) \geq 0$, $\mathcal{G}_2(z) \geq 0$

It is clear that there is a lot of practically interesting concrete problems which can be rewritten in a similar form.

2. ABSTRACT GLOBAL CONSTRAINED OPTIMIZATION PROBLEM

Let $\mathcal{Z} = \{z\}$ be an arbitrary set and $\mathcal{F}(z)$, $\mathcal{G}_1(z), \ldots, \mathcal{G}_k(z), \mathcal{G}_{k+1}(z), \ldots, \mathcal{G}_{k+l}(z)$ be real functions. Let

$$\begin{aligned} \mathcal{N} = \{z \in \mathrm{Z} : \mathcal{G}_1(z) \geq 0, \ldots, \mathcal{G}_k(z) \geq 0, \\ \mathcal{G}_{k+1}(z) = 0, \ldots, \mathcal{G}_{k+l}(z) = 0\} \end{aligned} \qquad (5)$$

Consider the following abstract problem of global optimization:

$$\text{to minimize} \quad \mathcal{F}(z) \quad \text{subject to} \quad z \in \mathcal{N} \qquad (6)$$

The case when as in the previous examples $\mathcal{Z}$ is a plane in a Hilbert space and $\mathcal{F}$, $\mathcal{G}_j$ are quadratic functionals will be called the linear-quadratic case. In this case $\mathcal{F}$ or (and) $\mathcal{G}_j$ may be nonconvex, we have then a nonconvex problem of global minimization.

The method proposed in Yakubovich (1992a) (see also Yakubovich (1992b, 1993) gives a solution of the problem (6) under some assumptions. These assumptions will be formulated later. Now let us describe this method. Introduce the Lagrangian function

$$S(\tau, z) = \mathcal{F}(z) - \sum_{j=1}^{k+l} \tau_j \mathcal{G}_j(z) \qquad (7)$$

We shall write $\tau = [\tau_j]_{j=1}^{k+l} \geq 0$ iff $\tau_1 \geq 0, \ldots, \tau_k \geq 0$. The rule for solving the constrained problem (6) consists in the following four actions:

(I) For $\tau \geq 0$ solve the problem

$$\text{to minimize} \quad S(\tau, z) \quad \text{subject to} \quad z \in \mathrm{Z} \qquad (8)$$

More precisely, it is sufficient to determine

$$S_0(\tau) = \inf_{z \in \mathrm{Z}} S(\tau, z) \qquad (9)$$

(Remark that in linear-quadratic case this problem contains only linear constraints. We have efficient methods to solve it in many cases.)

(II) Find a solution τ^0 of the dual problem

$$\text{To maximize} \quad S_0(\tau) \quad \text{subject to} \quad \tau \geq 0 \qquad (10)$$

Suppose that τ^0 exists. (Remark that $S_0(\tau)$ is the concave function as an infimum of linear in τ functions. Therefore (10) is a finite dimensional convex problem. There is quite a lot methods to solve it.)

(III) For $\tau = \tau^0$ find all the solutions $z(\tau^0)$ of the problem (8). Suppose $z(\tau^0)$ exists.

(IV) Choose those $z^0 \in \{z(\tau^0)\}$ for which

$$\begin{aligned} \mathcal{G}_1(z^0) \geq 0, \ldots, \mathcal{G}_k(z^0) \geq 0, \\ \mathcal{G}_{k+1}(z^0) = 0, \ldots, \mathcal{G}_{k+l}(z^0) = 0 \end{aligned} \qquad (11)$$

$$\tau_j^0 \mathcal{G}_j(z^0) = 0 \qquad (j = 1, \ldots, k+l) \qquad (12)$$

We will say that *the rule* (I)–(IV) *is valid if the obtained set* $\{z^0\}$ *coincides with the set of all solutions of the initial constrained problem* (6).

The following theorem closes to some results (Yakubovich, 1992b; Matveev and Yakubovich, 1993).

Theorem 1. Let the dual problem (10) have a solution τ^0 and the set $\{z^0\}$ that results from execution of above actions (I)–(IV) be non-empty. Then this set coincides with the set of all solutions of the initial problem (6), so that the rule (I)–(IV) is valid. Moreover the following duality relation holds:

$$\inf_{z \in N} \mathcal{F}(z) = \max_{\tau \geq 0} S_0(\tau). \qquad (13)$$

Certainly we are interested to know beforehand without executing the actions (I)–(IV), in which cases our rule is valid. The following theorems 2, 3 give some answers.

Theorem 2. Assume that $\mathcal{Z}$ is linear normed space and the Frechet derivatives $\mathcal{F}'(z), \mathcal{G}'_j(z)$ exist and for all $\tau \geq 0$ close to $\tau^0 \geq 0$ the infimum

$$\inf_{z \in Z} S(\tau, z) = S[\tau, z(\tau)] \qquad (= S_0(\tau))$$

is attained in a certain point $z(\tau)$ and also $\left(\frac{\partial z(\tau)}{\partial \tau_j}\right)_{\tau^0}$ exist (one-sided if $\tau_j^0 = 0$). Then the rule (I)–(IV) is valid and $z^0 = z(\tau^0)$, that is we do not need to verify (11) and (12). (These conditions are fulfilled automatically.)

This theorem is a small generalization of Lemma 2 (Yakubovich, 1992b) and can be proved similarly to (Yakubovich, 1992b, p. 16].

3. LINEAR-QUADRATIC CONSTRAINED OPTIMIZATION PROBLEMS

Let us suppose for simplicity that $l = 0$ (there is no equality constraints) and that the inequalities constraints are regular:

$$\exists z_* : \quad \mathcal{G}_1(z_*) > 0, \ldots, \mathcal{G}_k(z_*) > 0.$$

Let H be a real Hilbert space, $M \subset H$ be a subspace and (as in above examples) $Z = M + c$ be a plane in H ($c \in H$). Let us denote $\mathcal{F}(z) = \mathcal{G}_0(z)$. Suppose that

$$\mathcal{G}_j(z) = \mathcal{G}_j^0(z) + (g_j, z) + \gamma_j \qquad (j = 0, \ldots, k)$$

are quadratic functionals, $\mathcal{G}_j^0(z) = (G_j z, z)$ where $G_j = G_j^*$ are bounded self-adjoint operators.

Theorem 3. If $k = 1$ then τ^0 exists and the rule (I)–(IV) is valid.

If $k \geq 2$ then τ^0 exists and the rule (I)–(IV) is valid under the following additional condition: there exist linear bounded operators ${}_q : \mathcal{H} \to \mathcal{H}$ such that

(i) $({}_q z_1, z_2) \to 0$ as $q \to \infty$ $(\forall z_1, z_2 \in H)$,

(ii) ${}_q M \subset M$,

(iii) $\mathcal{G}_j^0({}_q z) \to \mathcal{G}_j^0(z)$ as $q \to \infty$ $(\forall z \in M, \forall j = 0, \ldots, k)$.

This theorem follows from (Yakubovich, 1987) ($k = 1$) and from (Yakubovich 1992a, 1992b) ($k \geq 2$).

The next result requires some definitions. In our case Lagrangian function is

$$S(\tau, z) = (G(\tau)z, z) + 2(g(\tau), z) + \gamma(\tau), \qquad (14)$$

where

$$G(\tau) = G_0 - \sum_{j=1}^{k} \tau_j G_j,$$
$$g(\tau) = g_0 - \sum_{j=1}^{k} \tau_j g_j, \qquad (15)$$
$$\gamma(\tau) = \gamma_0 - \sum_{j=1}^{k} \tau_j \gamma_j$$

Let

$$D := \{\tau \geq 0 : \quad z^* G(\tau) z \geq 0, \quad \forall z \in M\}, \quad (16)$$

$$D^0 := \{\tau \geq 0 : \exists \varepsilon > 0, \ z^* G(\tau) z \geq \varepsilon |z|^2, \ \forall z \in M\} \qquad (17)$$

It is simple to see that $S_0(\tau) = -\infty$ for $\tau \notin D$. Therefore $\tau^0 \in D$. Let $\tau \in D^0$. Then $z(\tau)$ exists such that $S_0(\tau) = S[\tau, z(\tau)]$. Moreover $\frac{\partial z(\tau)}{\partial \tau_j}$ exist. (In fact $z(\tau) = -G(\tau)^{-1} g(\tau)$.) Therefore Theorem 1 implies

Theorem 4. If $\tau^0 \in D^0$ then the rule (I)–(IV) is valid and we do not need in the action (IV).

Thus we have three possibilities: 1) $\tau^0 \in D^0$, then the rule is valid; 2) $\tau^0 \in D \backslash D^0$, $\{z^0\} \neq \emptyset$, then the rule is valid also; 3) $\tau^0 \in D \backslash D^0$, $\{z^0\} = \emptyset$, then the rule may hold and may not hold. In this sense the rule is valid almost always. Remark that there are examples when the rule (I)–(IV) is not valid.

Let us pass to concrete constrained optimization problems.

4. INFINITE-HORIZON LINEAR-QUADRATIC DETERMINISTIC CONTROL PROBLEM WITH CONSTRAINTS

Consider the following optimization problem

$$\frac{dx}{dt} = Ax + Bu, \qquad x(0) = a, \qquad (18)$$

$$|x(\cdot)| \in L_2(0, \infty), \qquad |u(\cdot)| \in L_2(0, \infty), \quad (19)$$

$$\mathcal{G}_j = \gamma_j - \int_0^\infty G_j[x(t), u(t)]\, dt \geq 0 \quad (j = 1, \ldots, k), \qquad (20)$$

$$\mathcal{F} = \int_0^\infty F[x(t), u(t)]\, dt \to \min. \qquad (21)$$

Here $x(t) \in R^n$, $u(t) \in R^n$, A, B are constant matrices, the pair (A, B) is controllable, G_j and

F are quadratic forms in x and u. So the problem is to minimize the functional $\mathcal{F}$ subject to constraints (18)–(20). Obviously the Example 1 is a special case of this problem. In its turn, the problem (18)–(21) is a special case of the abstract problem (6) for $\mathcal{H} = L_2^{(n+m)}(0,\infty)$ (($n+m$) — dimensional $L_2(0,\infty)$-space), $z = [x(\cdot), u(\cdot)]$, $\mathcal{Z} = \{z : \text{(18) holds}\}$.

Let us apply Theorem 3 taking for operators $_q$ the shifts: $_q z(\cdot) = \hat{z}(\cdot)$ means that $\hat{z}(t) = z(t-q)$ for $t \geq q$, $\hat{z}(t) = 0$ for $0 \leq t < q$. Let us check the conditions (i), (ii), (iii).

(i) It follows from

$$(\mathrm{T}_q z_1, z_2) = \int_q^\infty z_2(t)^* z_1(t-q)\, dt$$

that

$$|({}_q z_1, z_2)|^2 \leq \int_q^\infty |z_2|^2\, dt \cdot \int_0^\infty |z_1|^2\, dt \to 0 \quad \text{as} \quad q \to +\infty.$$

Therefore (i) holds.

(ii) In our case M is a subspace

$$= \left\{ z = [x(\cdot), u(\cdot)] : \frac{dx}{dt} = Ax + Bu, \quad x(0) = 0 \right\}$$

It is obvious that $z(\cdot) \in \mathrm{M}$ implies ${}_q z(\cdot) \in \mathrm{M}$, that is (ii) holds.

(iii) In our case

$$\mathcal{G}_j^0(z) = -\int_0^\infty G_j[x(t), u(t)]\, dt,$$

therefore $\mathcal{G}_j^0(\mathrm{T}_q z) = \mathcal{G}_j^0(z)$, that is (iii) holds also.

By Theorem 3 the rule (I)–(IV) is valid for the problem (18)–(21). Our rule will reduce this constrained problem to a similar problem without quadratic constraints but containing parameters $\tau_1 \geq 0, \ldots, \tau_k \geq 0$ and to finite dimensional convex problem to determine these parameters.

Let us consider the use of our rule in more details.

In our case the Lagrangian function is

$$S(\tau, z) = \mathcal{K}(\tau, z) - \sum_{j=1}^k \tau_j \gamma_j,$$

where as before $z = [x(\cdot), u(\cdot)]$ and

$$\mathcal{K}(\tau, z) = \int_0^\infty K_\tau[x(t), u(t)] dt,$$

$$K_\tau(x, u) = F(x, u) + \sum_{j=1}^k \tau_j G_j[x, u].$$

The unconstrained problem is: for fixed $\tau \geq 0$

$$\text{to minimize} \quad \mathcal{K}(\tau, z) \quad \text{subject to} \quad (18), (19) \tag{22}$$

It is a usual linear-quadratic problem without quadratic constraints. Remark that we need a complete solution of this problem including the cases of indefinite form K_τ and the cases when $\inf_{z \in Z} \mathcal{K}(\tau, z)$ does not achieve. Such solution is given by Theorem 3 (Yakubovich, 1992b). Let us reformulate this result as follows.

Define the matrices $\hat{G}_\tau = \hat{G}_\tau^*$, g_τ, $\Gamma_\tau = \Gamma_\tau^*$ by

$$K_\tau(x, u) = \begin{bmatrix} x \\ u \end{bmatrix}^* \begin{bmatrix} \hat{G}_\tau & g_\tau \\ g_\tau^* & \Gamma_\tau \end{bmatrix} \begin{bmatrix} x \\ u \end{bmatrix} \tag{23}$$

In the case $\Gamma_\tau \geq 0$ form the following Lur'e equations for the $n \times n$ matrix $H_\tau = H_\tau^*$ and $n \times m$ matrix $\hat{h}_\tau$:

$$H_\tau A + A^* H_\tau + \hat{G}_\tau = \hat{h}_\tau \hat{h}_\tau^*, \quad H_\tau B + g_\tau + \hat{h}_\tau \Gamma_\tau^{1/2} = 0 \tag{24}$$

Consider also the condition

$$\det \begin{bmatrix} \lambda I_n - A & -B \\ \hat{h}_\tau^* & -\Gamma_\tau^{1/2} \end{bmatrix} \neq 0 \quad \text{for } \mathrm{Re}\,\lambda > 0 \tag{25}$$

If $\Gamma_\tau > 0$ equations (24) and condition (25) may be rewritten as follows:

$$H_\tau A + A^* H_\tau + \hat{G}_\tau = h_\tau \Gamma_\tau h_\tau^*, \quad H_\tau B + g_\tau + h_\tau \Gamma_\tau = 0 \tag{26}$$

$$\det[\lambda I_n - \hat{A}_\tau] \neq 0 \quad \text{as} \quad \mathrm{Re}\,\lambda \geq 0 \tag{27}$$

Here $h_\tau = \hat{h}_\tau \Gamma_\tau^{-1/2}$, $\hat{A}_\tau = A + B h_\tau$.

For the fixed $\tau \geq 0$ let us introduce "the week frequency domain condition" (WFC):

$$K_\tau(\tilde{x}, \tilde{u}) \geq 0 \text{ for all } \tilde{x} \in {}^n,\ \tilde{u} \in {}^m\ \omega \in {}^1 \quad \text{such that} \quad i\omega\tilde{x} = A\tilde{x} + B\tilde{u} \tag{28}$$

and "the strong frequency domain condition" (SFC):

$$\begin{aligned} &\exists\, \delta_\tau > 0 : \quad K_\tau(\tilde{x}, \tilde{u}) \geq \delta_0(|\tilde{x}|^2 + |\tilde{u}|^2) \\ &\text{for all} \quad \tilde{x} \in \mathrm{C}^n,\ \tilde{u} \in \mathrm{C}^m,\ \omega \in \mathrm{R}^1 \\ &\text{such that} \quad i\omega\tilde{x} = A\tilde{x} + B\tilde{u} \end{aligned} \tag{29}$$

It is well known (KYP-lemma) that WFC holds iff $\Gamma_\tau \geq 0$ and the Lur'e equations (24) have a solution $H_\tau = H_\tau^*$, $\hat{h}_\tau$ satisfying (25) and also that SFC holds iff $\Gamma_\tau > 0$ and the Lur'e equations (26) have solution $H_\tau = H_\tau^*$, h_τ satisfying (27).

Theorem 5. a) If WFC fails or, equivalently, Γ_τ is indefinite matrix or the Lur'e equations (24) have no solution then $S_0(\tau) := \inf_{z \in Z} S(\tau, z) = -\infty$.

b) If WFC holds then $\tau \in D$ and

$$S_0(\tau) = a^* H_\tau a - \sum_{j=1}^k \tau_j \gamma_j. \tag{30}$$

c) If SFC holds (it means that $\tau \in D^0$) then $\inf \mathcal{K}(\tau, z)$ is achieved and the optimal process

$z(\tau) = [x(\cdot), u(\cdot)]$ is realized by a stabilizing regulator $u = h_\tau^* x$, where h_τ (together with $H_\tau = H_\tau^*$) is a solution of Lur'e equations (26) satisfying the condition (27).

d) If WFC holds, SFC fails (it means that $\tau \in D \setminus D^0$), $\Gamma_\tau > 0$ and $a = x(0)$ belongs to the stable subspace of the matrix $\widehat{A}_\tau = A + Bh_\tau$ then the optimal process is performed by a nonstabilized regulator $u = h_\tau^* x$. If this regulator gives a solution $x(t) = e^{\widehat{A}_\tau t}a$ which does not belong to $L_2(0, \infty)$ then $\inf\limits_{z \in Z} S(\tau, z)$ is not achieved, there is no optimal process in this problem.

e) If WFC holds, SFC fails, $\det \Gamma_\tau = 0$ then $\inf\limits_{z \in Z} S(\tau, z)$ may possibly be not achieved.

So to solve the constrained optimization problem (18)–(21) we must

(I) Find D from WFC and determine $S_0(\tau)$ by (30),

(II) Determine $\tau^0 = \arg\max\limits_{\tau \geq 0} S_0(\tau) = \arg\max\limits_{\tau \in D} S_0(\tau)$. The value τ_0 does exist.

(III) Let $\tau^0 \in D^0$ (SFC holds). Determine h_{τ^0} by solving (26) for $\tau = \tau^0$ with the condition (27). The answer is $u = h_{\tau^0}^* x$.

(IV) Let $\tau^0 \in D \setminus D^0$, $\Gamma_{\tau^0} > 0$. To determine the solution h_{τ^0} of (26) for $\tau = \tau^0$ with the condition (25) or, equivalently, $\det[\lambda I_n - \widehat{A}_{\tau^0}] \neq 0$ as $\operatorname{Re}\lambda > 0$. If $x(0) = a$ belongs to the stable subspace of $\widehat{A}_{\tau^0}$ then the answer is $u = h_{\tau^0}^* x$. In the contrary case the problem (18)–(21) has no solutions.

(V) Let $\tau^0 \in D \setminus D^0$, $\det \Gamma_{\tau^0} = 0$. The problem (18)–(21) may have no solutions ($\inf \mathcal{F}$ possibly is not achieved).

Similarly may be solved the discrete infinite horizon linear-quadratic constrained problem.

Example 1 (The continuation). Let us consider again the constrained optimization problem (1)–(4). In this case $x = [y, \dot{y}]$, $z = [y(\cdot), \dot{y}(\cdot), u(\cdot)]$,

$$S(\tau, z) = \mathcal{K}(\tau, z) = \mathcal{F} - \tau \int_0^\infty (\alpha y^2 - u^2)dt, \quad \tau \geq 0,$$

$$K_\tau(y, \dot{y}, u) = \gamma y^2 + \tau u^2, \quad \gamma = 1 - \tau\alpha$$

Using Theorem 5, after some calculations we obtain

$$S_0(\tau) = -\infty \quad \text{if} \quad \tau > \alpha^{-1},$$
$$D^0 = [0, \alpha^{-1}), \quad D = [0, \alpha^{-1}],$$
$$S_0(\tau) = \sqrt{2}(1 - \tau\alpha)^{3/4}\tau^{1/4} \quad \textit{for } 0 \leq \tau \leq \alpha^{-1},$$
$$h(\tau) = \begin{bmatrix} h_1(\tau) \\ h_2(\tau) \end{bmatrix},$$
$$h_1(\tau) = -\sqrt{\gamma\tau^{-1}}, \quad h_2(\tau) = -\sqrt{2}\sqrt[4]{\gamma\tau^{-1}},$$
$$\tau^0 = \arg\max S_0(\tau) = (4\alpha)^{-1} \qquad (\in D^0)$$

The optimal control is realized by a regulator $u = h_1(\tau^0)y + h_2(\tau^0)\dot{y}$ that is by a regulator

$$u = -\sqrt{3\alpha}\, y - \sqrt{2}\sqrt[4]{3\alpha}\, \dot{y}$$

5. INFINITE-HORIZON LINEAR-QUADRATIC PERIODICAL CONTROL PROBLEM WITH CONSTRAINTS

Consider the following optimization problem

$$\frac{dx}{dt} = A(t)x + B(t)u, \qquad x(0) = a, \tag{31}$$

$$|x(\cdot)| \in L_2(0, \infty), \qquad |u(\cdot)| \in L_2(0, \infty), \tag{32}$$

$$\mathcal{G}_j = \gamma_j - \int_0^\infty G_j[t, x(t), u(t)]\, dt \geq 0 \quad (j = 1, \ldots, k) \tag{33}$$

$$\mathcal{F} = \int_0^\infty F[t, x(t), u(t)]\, dt \to \min \tag{34}$$

Here $A(t), B(t)$ are T-periodic matrix functions, $G_j(t, x, u)$, $F(t, x, u)$ are quadratic forms in x, u with T-periodic coefficients. All the coefficients are piecewise continuous functions.

By Theorem 3 our rule is valid for this problem. The operators T_q are defined as follows: $\mathrm{T}_q z(\cdot) = \widehat{z}(\cdot)$ means that $\widehat{z}(t) = z(t - qT)$ for $t \geq qT$, $\widehat{z}(t) = 0$ for $0 \leq t < qT$. Our rule reduces this problem to a similar one without constraints (33), but containing parameters $\tau_j \geq 0$. These parameters are determined from convex finite-dimensional problem $S_0(\tau) \to \max$. The solution of the corresponding problem without constraints (33) is given in (Yakubovich 1986, Gelfand and Vilenkin, 1961).

6. INFINITE-HORIZON LINEAR-QUADRATIC STOCHASTIC PROBLEM WITH CONSTRAINTS

Consider the optimization problem

$$\frac{dx}{dt} = Ax + Bu + \xi(t), \tag{35}$$

$$\mathbf{E}\xi(t) = 0, \qquad \mathbf{E}\xi(t)\xi(s)^* = \delta(t - s)Q, \tag{36}$$

$$\mathcal{G}_j = \gamma_j - \mathbf{E}\begin{bmatrix} x \\ u \end{bmatrix}^* G_j \begin{bmatrix} x \\ u \end{bmatrix} \geq 0 \;\; (j = 1, \ldots, k), \tag{37}$$

$$\mathcal{G}_0 = \mathbf{E}\begin{bmatrix} x \\ u \end{bmatrix}^* G_0 \begin{bmatrix} x \\ u \end{bmatrix} \to \min. \tag{38}$$

Here $\xi(t)$ is a vector white noise ((36) holds) or, precisely speaking, $\xi(t)$ is a generalized derivative $\xi(t) = d\eta(t)/dt$ of the random process $\eta(t)$ with noncorrelated increments, $x(t) \in \mathrm{R}^n$, $u(t) \in \mathrm{R}^m$ are stationary random processes, $\mathbf{E}$ denotes the mathematical expectation, A, B, G_j are constant matrices, the pair (A, B) is controllable, the equation (35) is understood as an equality of generalized random stationary processes. The admissible processes may be defined in different ways. We use the following definition, which is convenient from the theoretical point of view: the process $z(\cdot) = \begin{bmatrix} x(\cdot) \\ u(\cdot) \end{bmatrix}$ is admissible if it can be represented as a stochastic integral

$$z(t) = \int_{-\infty}^{t} Z(t-s)\, d\eta(s) \qquad (39)$$

where $Z(t)$ is $(n+m) \times n$ matrix function with entries in $L_2(0,\infty)$. Note that any stabilizing regulator $u = Cx$ (so that $A + BC$ is a Hurwitz matrix) defines an admissible process.

Using Theorem 3, we obtain that our rule is valid for this optimization problem. The operators T_q are defined as follows: $\mathrm{T}_q z(\cdot) = \hat{z}(\cdot)$ means $\hat{z}(t) = z(t-q)$. For more details see (Yakubovich, 1993). [1] There are some examples in (Yakubovich, 1993) concerning the problem of optimal controller design for aircraft roll stabilization.

7. CONCLUDING REMARKS

In the paper the method (Yakubovich 1992a,1992b and 1993) to solve some global optimization problems with the constraints is described. These problems may be nonconvex. The method reduces the initial problem to a similar one without constraints but containing parameters. These parameters must be determined from some convex finite-dimensional problem. Many problems important in practice may be solved by this method. Let me also note that some new important results extending this method to many new cases were obtained recently by Matveev A.S.

8. REFERENCES

Yakubovich V.A.(1992a). One method for the solution some special global minimization problems. *Vestnik of St.Petersburg University*, ser. 1, N 2, v. 8, 58-68 (in Russian).

Yakubovich V.A.(1992b). Nonconvex optimization problem: The infinite-horizon linear-quadratic control problem with quadratic constraints. *Systems & Control Letters*, **19**, 13-22.

Yakubovich V.A.(1993). Linear-quadratic optimization problems with quadratic constraints. *Proceedings of the 2nd European Control Conference*, **1** of 4, 346-349.

[1] Let us correct the misprints in (Yakubovich, 1993). The formula (16) must be written as $u(t) = h(\tau)^* x(t)$, in the third line after the formula (21) must be written $\mathbf{E}\zeta(t)\zeta(s)$, the definition of T_q must be as it is formulated above.

Gorbatenko S.A., E.M. Makashoff, Y.P. Polushkin and L.V. Sheftel (1976). Synthesis and analysis of motion of flighting apparatus. Moscow, *Mashinostroenie*, 352 (in Russian).

Matveev A.S.and V.A. Yakubovich V.A.(1993). Nonconvex problems of global optimization. St.Petersburg, *Math.J.*, **4**, 1217-1243. harvarditem[6] Yakubovich V.A.(1973). Minimization of quadratic functionals under quadratic constraints, and the necessity of the frequency condition in the quadratic criterion for absolute stability of nonlinear control systems. *Dokl. Acad. Nauk SSSR*, **209**, 1039-1042; English transl. in Soviet Math. Dokl., **14**,1973, 593-597.

Yakubovich V.A.(1987). Frequency theorem for periodic systems and the theory of analytical design of regulators, in "Method of Lyapunov Functions in Analysis of Systems Dynamics" (Matrosov V.M. and Anapolsky L.Yu., eds), Nauka: Novosibirsk, 281-290 (in Russian).

Yakubovich V.A.(1986). Linear-quadratic optimization problem and frequency theorem for periodic systems. *Sybirskii Mat.Zh.*, **27**, (No.2), 50-67 (in Russian).

Gelfand I.M., Vilenkin N.Ya.(1961). Generalized functions. *Moscow, Gos. Izd. Fiz. Mat. Literatury*, **4**, 471.

ROBUST EIGENVALUE ASSIGNMENT BY OUTPUT FEEDBACK

S. Longhi, G. Orlando and R. Zulli

Dipartimento di Elettronica e Automatica, Università di Ancona
v. Brecce Bianche, I-60131 Ancona, Italy

Abstract. An output periodic feedback is proposed for the robust eigenvalue assignment of linear, time-invariant, discrete-time systems. The condition numbers characterizing the eigenstructure of the closed-loop system are assumed as a robustness measure. The proposed periodic eigenvalue assignment algorithm has been tested on different examples, giving satisfactory results.

Keywords. Eigenvalue assignment; Output feedback; Periodic controllers; Discrete-time systems.

1. INTRODUCTION

The output feedback problem is a longstanding open question in control engineering. Reviews of the developed results in this area can be found in (Kimura, 1994; Syrmos et al., 1994). Given a linear time-invariant system Σ described by

$$x(k+1) = Ax(k) + Bu(k) \quad (1)$$
$$y(k) = Cx(k) \quad (2)$$

the problem consists of finding a static output feedback

$$u(k) = Ky(k) + u_0(k) \quad (3)$$

such that the poles of the closed-loop system Σ_c

$$x(k+1) = (A + BKC)x(k) + Bu_0(k) \quad (4)$$
$$y(k) = Cx(k) \quad (5)$$

coincide with a prescribed symmetric set of complex numbers $\mathcal{L}$ (a set of complex numbers subject to the requirement that non-real elements appear in conjugate pairs).

Recently, a periodic solution to this problem is introduced in (Aeyels and Willems, 1992). In this case, under a proper set of technical assumptions, it is possible to find an output feedback expressed by

$$u(k) = K(k)y(k) + u_0(k), \quad (6)$$

where $K(\cdot)$ is a periodic matrix of period ω, i.e. $K(k+\omega) = K(k)$, such that the poles of the ω-periodic closed-loop system coincide with a prescribed symmetric set of complex numbers $\mathcal{L}$. The poles of the ω-periodic closed-loop system are the eigenvalues of the monodromy matrix associated to the closed-loop system and expressed by

$$\Phi^K(k+\omega, k) := A^K(\omega - 1 + k)$$
$$A^K(\omega - 2 + k) \ldots A^K(k), \quad (7)$$

where $A^K(k) := A + BK(k)C$. The characteristic polynomial of the monodromy matrix $\Phi^K(k+\omega, k)$ is independent of k, nevertheless the eigenstructure of $\Phi^K(k+\omega, k)$ is generally dependent on time k.

The periodic solution to the state-feedback problem has been analyzed in (Longhi and Zulli, 1995). In this case the control law is expressed by

$$u(k) = F(k)x(k) + u_0(k), \quad (8)$$

and the periodic matrix $F(\cdot)$ is synthesized to assign a prescribed symmetric set of eigenvalue $\mathcal{L}$. The additional degrees of freedom made available in the periodic case are used to improve the robustness properties of the closed-loop system. The condition numbers characterizing the eigenstructure of the closed-loop system are assumed as a robustness measure.

The aim of this paper is to extend the results of (Longhi and Zulli, 1995) to the output feedback problem. In particular, an algorithm for the synthesis of a periodic output feedback able to assign a prescribed set of eigenvalues of $\Phi^K(k+\omega,k)$ and to improve the robustness properties of the closed-loop system is proposed. The condition numbers characterizing the eigenstructure of $\Phi^K(k+\omega,k)$ are assumed as a robustness measure.

The paper is organized as follows. Section 2 contains a set of preliminary results on the periodic solution to the problem of output eigenvalue assignment. A numerical algorithm for the solution of the robust output periodic eigenvalue assignment problem is introduced in Section 3 together with some numerical results.

2. OUTPUT PERIODIC EIGENVALUE ASSIGNMENT

The eigenstructure of the monodromy matrix $\Phi^K(k+\omega,k)$ is generally dependent on time k (Grasselli and Longhi, 1991). Therefore, the periodic eigenvalues assignment problem can be formally stated in the following way.

Problem 2.1 *Given a symmetric set of n distinct eigenvalues $\mathcal{L}$, an integer ω and an ω-periodic non singular matrix $V(\cdot)$, find, if it exists, an ω-periodic matrix $K(\cdot)$ for the output feedback control law* (6) *such that the following conditions are satisfied*

$$A^K(h)\,V(h) = V(h+1), \quad h = 0,1,...,\omega-2, \tag{9}$$

$$A^K(\omega-1)V(\omega-1) = V(0)\,\Lambda, \tag{10}$$

where $\Lambda := \text{diag}\{\lambda_i, \quad i=1,...,n,\}$, *with* $\lambda_i \in \mathcal{L}$.

The existence of a solution $K(\cdot)$ to the above stated problem implies that

$$\Phi^K(k+\omega,k)\,V(k) = V(k)\,\Lambda, \quad \forall k \in \mathbb{Z} \tag{11}$$

From this relation it is evident that the set of the eigenvalues of the closed-loop system coincides with the set $\mathcal{L}$ and the columns of $V(k)$ represent the linearly independent right eigenvectors of $\Phi^K(k+\omega,k)$.

Assume that the matrices B and C are full rank. Under this assumption the following decompositions are considered

$$B = [M_a \quad M_b]\begin{bmatrix} Z \\ 0 \end{bmatrix}, \tag{12}$$

$$C = [Y \quad 0]\begin{bmatrix} N_a \\ N_b \end{bmatrix}, \tag{13}$$

where $M = [M_a \quad M_b]$ and $N = [N_a' \quad N_b']'$ are orthogonal and Z and Y are non-singular. The QR factorization can be used to obtain these decompositions.

Now, for $h = 0,1,...,\omega-1$, consider the i-th column $v_i(h)$ of matrix $V(h)$ and define

$$\tilde{v}_i := [v_i(0)' \quad v_i(1)' \quad \dots \quad v_i(\omega-1)']' \in \mathbb{C}^{n\omega}, \quad i = 1,...,n. \tag{14}$$

The vector $\tilde{v}_i$ will be called the lifted representation of the i-th column of matrices $V(h), h = 0,1,...,\omega-1$. Note that $v_i(h)$ represents a right eigenvector of $\Phi^K(h+\omega,h)$ associated with the assigned eigenvalue $\lambda_i \in \mathcal{L}$, for $i = 1,...,n$. Analogously, for $h = 0,1,...,\omega-1$, consider the i-th row $w_i(h)'$ of the matrix $W(h) := V(h)^{-1}$ and define

$$\tilde{w}_i := [w_i(0)' \quad w_i(1)' \quad \cdots \quad w_i(\omega-1)']' \in \mathbb{C}^{n\omega}, \quad i = 1,...,n. \tag{15}$$

Moreover, define the following matrices

$$\mathcal{R}(\lambda) := \begin{bmatrix} 0 & I_{(\omega-1)n} \\ \lambda I_n & 0 \end{bmatrix}, \quad \lambda \in \mathbb{C}, \tag{16}$$

$$\mathcal{A} := \text{diag}\,\underbrace{\{A, A, ..., A\}}_{\omega \quad blocks}, \tag{17}$$

$$\mathcal{M}_b := \text{diag}\,\underbrace{\{M_b, M_b, ..., M_b\}}_{\omega \quad blocks}, \tag{18}$$

$$\mathcal{N}_b := \text{diag}\,\underbrace{\{N_b, N_b, ..., N_b\}}_{\omega \quad blocks}. \tag{19}$$

In order to introduce conditions for the solvability of the Problem 2.1, define:

$$\mathcal{P}_i := \ker \mathcal{M}_b'(\mathcal{R}(\lambda_i) - \mathcal{A}) \subset \mathbb{C}^{n\omega}, \quad i = 1,2,...,n, \tag{20}$$

$$\mathcal{Q}_i := \ker \mathcal{N}_b(\mathcal{R}'(\lambda_i) - \mathcal{A}') \subset \mathbb{C}^{n\omega}, \quad i = 1,2,...,n. \tag{21}$$

The dimensions of the subspaces $\mathcal{P}_i$ and $\mathcal{Q}_i$ are stated by the following lemma.

Lemma 2.1. *If system Σ is minimal:*

$$\dim(\mathcal{P}_i) = p\omega, \quad i = 1,2,...,n, \tag{22}$$

$$\dim(\mathcal{Q}_i) = q\omega, \quad i = 1,2,...,n. \tag{23}$$

The relation (22) is proved in (Longhi and Zulli, 1995). By duality, the relation (23) can be proved in a similar way (see Lemma 3.1 in Longhi and Zulli, 1995).

Theorem 2.1. *There exists a solution $K(\cdot)$ to the Problem 2.1 if and only if for $i = 1,2,...,n$ the lifted representations $\tilde{v}_i$ and $\tilde{w}_i$ belong, respectively, to the subspaces $\mathcal{P}_i$ and $\mathcal{Q}_i$.*

Proof. By the decompositions (12) and (13), the conditions (9) and (10) can be expressed in the following way:

$$\begin{bmatrix} Z \\ 0 \end{bmatrix} K(h)\,[Y \quad 0] = \begin{bmatrix} M_a' \\ M_b' \end{bmatrix} (V(h+1) - AV(h))W(h)\,[N_a' \quad N_b'], \qquad h = 1, 2, ..., \omega - 2, \quad (24)$$

$$\begin{bmatrix} Z \\ 0 \end{bmatrix} K(h)\,[Y \quad 0] = \begin{bmatrix} M_a' \\ M_b' \end{bmatrix} (V(0) - AV(\omega - 1))W(\omega - 1)\,[N_a' \quad N_b'], \quad (25)$$

which are satisfied if and only if the following relations hold:

$$M_b'(V(h+1) - AV(h)) = 0, \qquad h = 0, 1, \dots, \omega - 2, \quad (26)$$

$$M_b'(V(0)\Lambda - AV(\omega - 1)) = 0, \quad (27)$$

$$(W(h) - W(h+1)A)N_b' = 0, \qquad h = 0, 1, \dots, \omega - 2, \quad (28)$$

$$(\Lambda W(\omega - 1) - W(0)A)N_b' = 0, \quad (29)$$

where $W(h) = V(h)^{-1}$ for all integer k.

By the above introduced notations, the relations (26) – (29) are equivalent to the following ones:

$$\mathcal{M}_b'(\mathcal{R}(\lambda_i) - \mathcal{A})\tilde{v}_i = 0, \quad i = 1, 2, ..., n, \quad (30)$$

$$\mathcal{N}_b(\mathcal{R}(\lambda_i)' - \mathcal{A}')\tilde{w}_i = 0, \quad i = 1, 2, ..., n, \quad (31)$$

and this proves the theorem. △

If the conditions of the Theorem 2.1 are satisfied the output feedback control law which solves the Problem 2.1 can be calculated using the following equations:

$$K(h) = Z^{-1} M_a'(V(h+1)\,V^{-1}(h) - A)N_a'Y^{-1}, \qquad h = 0, 1, ..., \omega - 2, \quad (32)$$

$$K(\omega - 1) = Z^{-1} M_a'(V(0)\Lambda\, V^{-1}(\omega - 1) - A)N_a'Y^{-1}. \quad (33)$$

3. ROBUST OUTPUT PERIODIC EIGENVALUE ASSIGNMENT

If a periodic solution $K(\cdot)$ to Problem 2.1 exists, the matrix $V(h)$ is composed by the right eigenvectors associated to the monodromy matrix $\Phi^K(h+\omega, h)$. The eigenvalues of the monodromy matrix $\Phi^K(h+\omega, h)$ are independent of time h, while the eigenstructure is generally $\omega-$periodically time-varying (Grasselli and Longhi, 1991). Therefore the sensitivity of the eigenvalues to perturbations of the elements of $\Phi^K(h+\omega, h)$, measured by the condition number $\kappa_2(V(h)) := \|V(h)\|_2 \|V(h)^{-1}\|_2$, is generally $\omega-$periodically time-varying. Thus, it seems appropriate to consider as a time-independent robustness measure the following quantity (Longhi and Zulli, 1995):

$$\kappa_\omega := \sum_{h=0}^{\omega-1} \kappa_2(V(h)). \quad (34)$$

The minimization of κ_ω implies the simultaneous minimization of $\kappa_2(V(h))$, for $h = 0, 1, ..., \omega - 1$. Therefore, the robustness of the solution $K(\cdot)$ to the above stated problem depends on the assigned ω-periodic matrix $V(\cdot)$. This fact leads to the introduction of a different formulation of the periodic eigenvalue assignment problem. Consider the following notation:

$$\mathcal{I}(h) := \text{diag}\,\{\,\delta_{0,h}I_n,\ \ \delta_{1,h}I_n,\ \ \dots,\ \ \delta_{\omega-1,h}I_n\,\},$$

$$\delta_{j,h} := \begin{cases} 1, & h = j \\ 0, & h \neq j \end{cases}, j, h = 0, 1, ..., \omega - 1. \quad (35)$$

Problem 3.1. *Given a symmetric set of n distinct eigenvalues $\mathcal{L}$, find an integer ω and an ω-periodic non-singular matrix $V(\cdot)$ which minimize κ_ω, under the constraints that $\tilde{v}_i$ belongs to the subspace $\mathcal{P}_i$, $i = 1, ..., n$, and*

$$\tilde{w}_i{}'\mathcal{I}(h)\tilde{v}_i = 1, \quad h = 0, 1, ..., \omega, \quad i = 1, ..., n, \quad (36)$$

$$\tilde{w}_j{}'\mathcal{I}(h)\tilde{v}_i = 0, \quad h = 0, 1, ..., \omega, \quad j \neq i, \qquad j = 1, ..., n, i = 1, ..., n, \quad (37)$$

where $\tilde{w}_i$ belongs to the subspace $\mathcal{Q}_i$, $i = 1, ..., n$.

Once a solution $V(\cdot)$ of this problem has been found, an ω-periodic matrix $K(\cdot)$ for the output feedback control law is calculated using the relations (32) and (33).

In this way, the set of the eigenvalues of $\Phi^K(k+\omega, k)$ coincides with $\mathcal{L}$ and the robustness measure is minimized.

Now assume that system Σ is minimal and that the following condition is satisfied:

$$q + p \geq n + 1, \quad (38)$$

where q is the number of measured outputs, p is the number of control inputs and n is the number of state variables.

Condition (38) together with the minimality of Σ are the classical sufficient conditions for generic pole assignability by output feedback (Kimura, 1994; Syrmos et al., 1994).

Under these conditions, an algorithm for the computation of a ω-periodic matrix $V(\cdot)$ is introduced.

Algorithm 3.1

Step 0: Assign matrices A, B, C defining a system Σ and a symmetric set $\mathcal{L}$ of desired closed-loop eigenvalues. Verify the minimality of system Σ and the fulfillment of the condition (38). Set the period ω of the output feedback law (6) equal to one. Compute the matrices M_a, M_b, Z, N_a, N_b and Y, in order to obtain the decompositions of B and C expressed by (12) and (13). Sort the symmetric set $\mathcal{L}$ in such a way that the complex elements appear in consecutive conjugate pairs.

Step 1: Compute the matrices $\mathcal{A}$, $\mathcal{M}_b$ and $\mathcal{N}_b$ defined respectively by (17), (18) and (19).

Step 2: Set $i = 1$. While $i \leq n$, compute matrices $T(i) \in \mathbb{C}^{n\omega \times p\omega}$ and $S(i) \in \mathbb{C}^{n\omega \times q\omega}$ such that $\text{Im } T(i) = \mathcal{P}_i$ and $\text{Im } S(i) = \mathcal{Q}_i$, where $\mathcal{P}_i$ and $\mathcal{Q}_i$ are defined by (20) and (21). If $\lambda_i \in \mathbb{R}$ set $i = i + 1$, else compute $T(i+1) = T^*(i)$, $S(i+1) = S^*(i)$ and set $i = i + 2$.

Step 3: Denote with

$$[T_{1,j}(i)' \quad T_{2,j}(i)' \quad \ldots \quad T_{\omega,j}(i)']',$$

where $T_{\ell,j}(i) \in \mathbb{C}^{n\times 1}$, and with

$$[S_{1,j}(i)' \quad S_{2,j}(i)' \quad \ldots \quad S_{\omega,j}(i)']',$$

where $S_{\ell,j}(i) \in \mathbb{C}^{n\times 1}$, the block partitions of the j-th column of matrices $T(i)$ and $S(i)$ respectively, and solve the following minimization problem in the unknown scalar quantities $x_j(i)$, $j = 1, \ldots, p\omega$, $i = 1, \ldots, n$ and $y_j(i)$, $j = 1, \ldots, q\omega$, $i = 1, \ldots, n$:

$$\min \kappa_\omega = \min \sum_{h=1}^{\omega} \kappa_2 [\sum_{j=1}^{p\omega} [x_j(1) T_{h,j}(1) \quad x_j(2) T_{h,j}(2) \quad \ldots \quad x_j(n) T_{h,j}(n)]], \tag{39}$$

under the constrains:

$$\langle \sum_{\ell=1}^{p\omega} x_\ell(j) T_{h,\ell}(j), \sum_{\ell=1}^{q\omega} y_\ell(i) S_{h,\ell}(i) \rangle = 0,$$
$$j = 1, 2, \ldots, n, \quad i = 1, 2, \ldots, n, \quad j \neq i,$$
$$h = 0, 1, \ldots, \omega - 1, \tag{40}$$

$$\langle \sum_{\ell=1}^{p\omega} x_\ell(i) T_{h,\ell}(i), \sum_{\ell=1}^{q\omega} y_\ell(i) S_{h,\ell}(i) \rangle = 1,$$
$$i = 1, 2, \ldots, n, \quad h = 0, 1, \ldots, \omega - 1, \tag{41}$$

where $x_j(i) \in \mathbb{R}$ and $y_j(i) \in \mathbb{R}$ if λ_i is a real element, $x_j(i) \in \mathbb{C}$, $y_j(i) \in \mathbb{C}$ and $x_j(i+1) = x_j^*(i)$, $y_j(i+1) = y_j^*(i)$ if λ_i is a complex element.

Step 4: Compute the matrices $V(h)$, $h = 0, 1, \ldots, \omega - 1$, according to

$$V(h) = \sum_{j=1}^{p\omega} [\overline{x}_j(1) T_{h+1,j}(1) \quad \overline{x}_j(2) T_{h+1,j}(2) \quad \ldots \quad \overline{x}_j(n) T_{h+1,j}(n)],$$
$$h = 0, 1, \ldots, \omega - 1, \tag{42}$$

where $\overline{x}_j(i)$, $j = 1, \ldots, p\omega$, $i = 1, \ldots, n$ denote the solutions of the minimization problem of Step 3.

Step 5: If the value of κ_ω is satisfactory go to Step 6, else set $\omega = \omega + 1$ and go to Step 1.

Step 6: Compute the robust ω-periodic solution to Problem 2.1 on the basis of equations (32) and (33).

Note that the constraint that $\tilde{v}_i$ belongs to the subspace $\mathcal{P}_i$, $i = 1, \ldots, n$, has been implicitly considered in the objective function defined at Step 3. In fact, following of the notation introduced in the above algorithm, the vector $\tilde{v}_i \in \mathcal{P}_i \subset \mathbb{C}^{n\omega}$ can be written as

$$\tilde{v}_i = \sum_{j=1}^{p\omega} x_j(i) [T_{1,j}(i)' \quad T_{2,j}(i)' \quad \ldots \quad T_{\omega,j}(i)']',$$
$$i = 1, \ldots, n. \tag{43}$$

The minimization problem considered at Step 3 of the Algorithm 3.1 has been solved using the Nelder-Mead algorithm (Nemhauser et al., 1989) with a proper choice of the scalar quantities $x_j(i)$, $j = 1, \ldots, p\omega$, $i = 1, \ldots, n$. The main weakness of the algorithm is that it can be slow for a high number of variables. The use of more sophisticated minimization algorithms (Byers and Nash, 1989) can improve the overall performances of Algorithm 3.1.

Example 3.1. Consider a linear system Σ described by:

$$A = \begin{bmatrix} 2 & -1 & 0 \\ 0 & -1 & 2 \\ -2 & -2 & 2 \end{bmatrix},$$

$$B = \begin{bmatrix} 1 & \\ 0 & 0 \\ 0 & 1 \end{bmatrix},$$

$$C = \begin{bmatrix} 1 & 1 & 0 \\ 0 & 0 & 1 \end{bmatrix}.$$

The prescribed set of the closed-loop eigenvalues $\mathcal{L}$ is equal to $\{0.1, 0.2, 0.3\}$. In order to compare the proposed approach with classical time-invariant eigenvalue assignment algorithms, the symmetric set $\tilde{\mathcal{L}}$, composed by the ω-th roots of the elements of $\mathcal{L}$, is considered as desired set of closed-loop eigenvalues with time-invariant output feedback. In this example the algorithm proposed in (Misra and Patel, 1989) is able to assign the set of eigenvalues $\tilde{\mathcal{L}}$ and the resulting output feedback is given by:

$$K = \begin{bmatrix} 2.1193 & -3.7123 \\ -1.6315 & 1.2010 \end{bmatrix},$$

with a condition number $\kappa_2(V) = 532$. The proposed Algorithm 3.1 gives the following 2−periodic output feedback:

$$K(0) = \begin{bmatrix} -3.1493 & 3.5368 \\ 1.8968 & -1.3143 \end{bmatrix},$$

$$K(1) = \begin{bmatrix} -3.2112 & 5.6422 \\ 1.8627 & -1.5275 \end{bmatrix},$$

which assigns the prescribed set $\mathcal{L}$ of eigenvalues with a 2−periodic condition number $\kappa_2(V(0)) = 15$, $\kappa_2(V(1)) = 26$. The sensitivity of the closed-loop eigenvalues to the considered perturbations is shown in Figure 1. The results here reported point out an higher degree of robustness obtained by periodic output feedback.

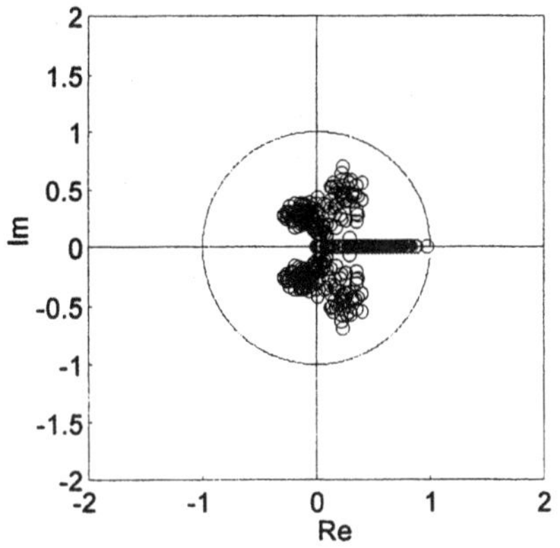

(a)

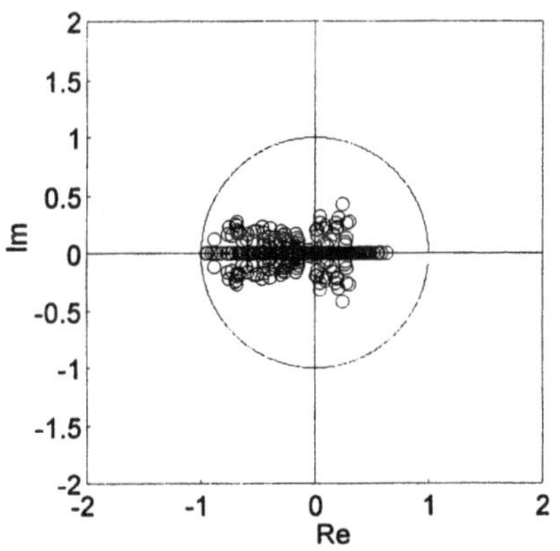

(b)

Fig. 1. Closed-loop eigenvalues of the Example 3.2: *(a)* time-invariant output feedback K and 4% random perturbations; *(b)* 2-periodic output feedback $K(\cdot)$ and 12% random perturbations.

Example 3.2. Consider the linearized model of an underwater vehicle reported in (Longhi and Rossolini, 1989). The matrices A and B of the discrete-time model obtained by sampling with a period of $0.5s$ are given by:

$$A = \begin{bmatrix} 0.995 & 0.485 & 0 & -0.004 & 0 & 0 \\ -0.017 & 0.942 & 0 & -0.018 & 0 & 0 \\ 0 & -0.003 & 0995 & 0485 & 0 & 0 \\ 0 & -0.014 & -0.017 & 0.942 & 0 & 0 \\ 0 & 0 & 0 & 0 & 1.0 & 0.5 \\ 0 & 0 & 0 & 0 & 0.0002 & 1.0 \end{bmatrix},$$

$$B = \begin{bmatrix} 0.0258 & -0.0002 & 0 \\ 0.1023 & -0.001 & 0 \\ -0.0001 & 0.0258 & 0 \\ -0.0008 & 0.1023 & 0 \\ 0 & 0 & 0.0055 \\ 0 & 0 & 0.0221 \end{bmatrix},$$

$$C = \begin{bmatrix} 1 & 0 & 0 & 0 & 0 & 0 \\ 0 & 0 & 1 & 0 & 0 & 0 \\ 0 & 0 & 0 & 0 & 1 & 0 \\ 0 & 0 & 0 & 0 & 0 & 1 \\ 0 & 1 & 0 & 0 & 0 & 0 \end{bmatrix}.$$

The prescribed set of closed-loop eigenvalues $\mathcal{L}$ is equal to $\{0.1, 0.15, 0.2, 0.25, 0.3, 0.4\}$.

In order to compare the proposed approach with classical time-invariant eigenvalue assignment algorithms, the symmetric set $\tilde{\mathcal{L}}$, composed by the ω-th roots of the elements of $\mathcal{L}$, is considered as desired set of closed-loop eigenvalues with time-invariant output feedback. In this example the algorithm proposed in (Misra and Patel, 1989) is able to assign the set of eigenvalues $\tilde{\mathcal{L}}$ and the resulting output feedback is given by:

$$K = \begin{bmatrix} 99.3629 & -207.6783 & 664.9939 & -305.9939 & 77.8306 \\ 95.2098 & -203.3400 & 664.1768 & -310.5233 & 78.7782 \\ 138.2669 & -248.3174 & 666.4442 & -263.5643 & 68.9542 \end{bmatrix}.$$

The proposed Algorithm 3.1 gives the following 2-periodic output feedback:

$$K(0) = \begin{bmatrix} -38.8282 & 0.5680 & 3.5556 & 0.6154 & -15.0988 \\ -162.1019 & -29.0630 & 80.2615 & 43.7207 & -18.1585 \\ -125.0455 & -11.4488 & -8.6405 & -42.3308 & -16.1172 \end{bmatrix},$$

$$K(1) = \begin{bmatrix} -48.7722 & 2.4851 & 6.8274 & 3.9642 & -22.7732 \\ 103.9492 & -16.8837 & -102.3648 & -91.7792 & 64.6128 \\ -163.8941 & 32.8859 & 19.6632 & -2.1848 & -61.4231 \end{bmatrix},$$

which leads to a closed-loop system with the prescribed set $\mathcal{L}$ of eigenvalues. The sensitivity of the closed-loop eigenvalues to the considered perturbations is shown in Figure 2. The results reported in this figure confirm better robustness properties obtained by periodic feedbacks.

4. CONCLUDING REMARKS

An algorithm for the synthesis of periodic output-feedback laws has been proposed. This algorithm produce output-feedback laws able to assign a prescribed set of eigenvalues and to minimize the condition numbers of the eigenvector matrices of the closed-loop system, representing the assumed robustness measure. The proposed algorithm has been tested on different examples showing better perfor-

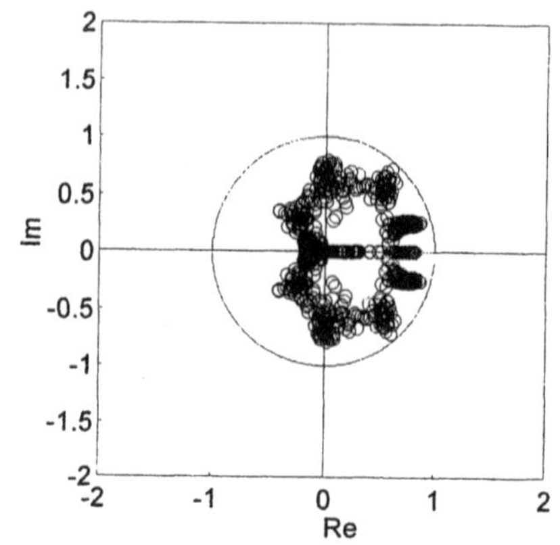

(a)

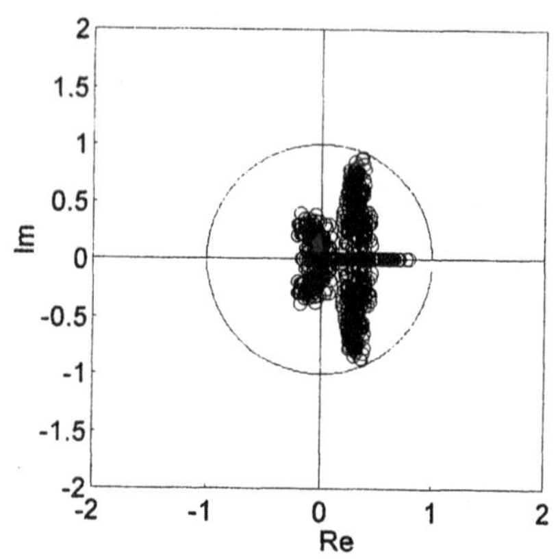

(b)

Fig. 2. Closed-loop eigenvalues of the Example 3.2: *(a)* time-invariant output feedback K and 0.4% random perturbations; *(b)* 2-periodic output feedback $K(\cdot)$ and 15% random perturbations.

mances than those provided by time-invariant feedbacks, as far as robustness is concerned.

Alternative robustness measures are under investigation.

5. REFERENCES

Aeyels, D. and J.L. Willems (1992). "Pole assignment for linear time-invariant systems by periodic memoryless output feedback", *Automatica*, **28**, pp. 1159-1168.

Byers, R. and S.G. Nash (1989). "Approaches to robust pole assignment", *International Journal of Control*, vol.49, no.1, pp. 97-117.

Grasselli, O.M. and S. Longhi (1991). '"The finite zero structure of linear periodic discrete-time systems", *Int. J. of System Science*, vol. 22, 1785-1806.

Kimura, H. (1994). "Pole assignment by output feedback: a longstanding open problem", *Proc. of the 33rd Conf. on Decision and Control*, Lake Buena Vista, Florida, pp. 2101-2105.

Longhi, S. and A. Rossolini (1989). "Adaptive control for an underwater vehicle: simulation studies and implementation details", *Proceedings of the IFAC Workshop on Expert Systems and Signal Processing in Marine Automation*, Lyngby, Denmark, pp.271-280.

Longhi, S. and R. Zulli (1995). "A robust periodic pole assignment algorithm", *IEEE Trans. Autom. Control*, **AC-40**, pp. 890-894.

Misra, P. and R. Patel (19989). "Numerical algorithms for eigenvalue assignment by constant and dynamic output feedback", *IEEE Trans. Autom. Control*, vol. 34, pp. 579-588.

Nemhauser, G.L., A.H.G. Rinnooy Kan and M.J. Todd (1989). *"Optimization"*, North Holland, Amsterdam, pp. 53-54.

Syrmos, V.L., C. Abdallah and P. Dorato (1994). "Static output feedback: a survey", *Proc. of the 33rd Conf. on Decision and Control*, Lake Buena Vista, Florida, pp. 837-842.

DESIGN OF PREDICTIVE CONTROLLER BASED UPON THE YOULA-KUČERA PARAMETRIZATION

M. Fikar * S. Engell ** J. Mikleš *

* *Department of Process Control, CHTF STU, Radlinského 9, SK-812 37 Bratislava, Slovakia, email: {fikar,mikles}@cvtstu.cvt.stuba.sk, Tel: + 42 7 39 52 69, Fax: + 42 7 39 31 98*
** *Process Control Group, Department of Chemical Engineering, University of Dortmund, D-44221 Dortmund, Germany, email: ast@ast.chemietechnik.uni-dortmund.de, Tel: +49 231 755 1, Fax: +49 231 755 51 29*

Abstract: A multivariable time varying controller is developed in the predictive control framework based upon the Youla-Kučera (YK) parametrization. This parametrization of the controller brings insight into closed-loop behaviour and stability can be guaranteed while the optimization is performed, as usually, over a finite horizon. Constraints on inputs and/or outputs can be handled.

Keywords: Stabilizing controllers, predictive control, constraints.

1. INTRODUCTION

Model Based Predictive Control (MBPC) has shown to be a successful control strategy and gained in popularity over the last two decades and is accepted for practical applications in process industry. This is mainly because the method provides high performance easily tuned controllers applicable to multivariable systems, process delays, and constraints. The best known representatives of such algorithms are Generalized Predictive Control (GPC) (Clarke *et al.* 1987) or (Q)DMC (Cutler and Ramaker 1980, Garcia and Morshedi 1986). A control sequence which minimizes a quadratic performance criterion is calculated and only its first element is applied. As the methods are open-loop and only a limited understanding of their stability properties exists, additional conditions on user defined parameters must be respected.

As a stable closed-loop behaviour is of course crucial several authors have addressed the issue of stability. As the basis of the proofs serve usually (Kwon and Pearson 1978, Keerthi and Gilbert 1988, Rossiter *et al.* 1995). The stable predictive constrol strategies implement zero terminal constraint requirement (Clarke and Scattolini 1991, Mosca and Zhang 1992) on state vector (or outputs).

In this contribution, a new type of predictive controller is presented. The algorithm can handle constraints and minimizes a quadratic objective function. The resulting predictive controller (PC) is piece-wise linear, has time-varying coefficients even in the non-adaptive case and its predictive part is a finite impulse filter (FIR). The predictive controller is based on the YK parametrization of all stabilizing controllers. This parametrization enables us to give a natural and efficient characterization of the necessary horizons for stability. Terminal constraints are inherently satisfied in the controller design and not artificially included afterwards to achieve stability. From the finiteness of the predictions follows the choice of the hori-

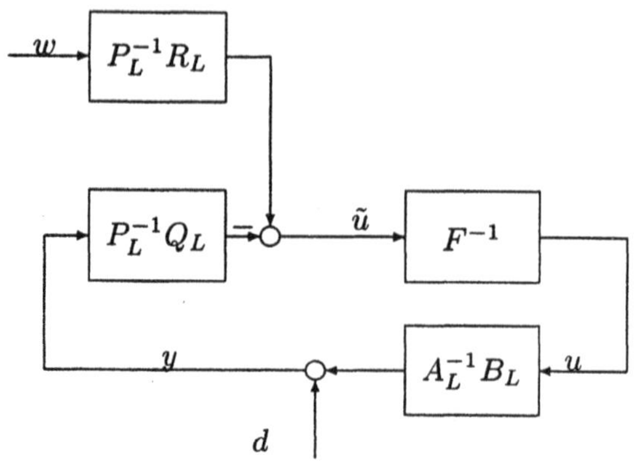

Fig. 1. Block diagram of the closed-loop system

zons and the fact that larger values of the horizons are superfluous. Hence the objective function can be reduced without loss of generality from an infinite to a finite horizon.

The idea behind the algorithm is to search for the controller in the family of all stabilizing controllers and to choose its parameters in such a way that the constraints are respected and a quadratic function is minimized.

2. PARAMETRIZATION OF THE CONTROLLER

We consider the closed-loop system illustrated in Fig. 1.

The controllable and observable process S_p with r inputs and r outputs is assumed to be modelled by the discrete-time linear time-invariant equation

$$A_L(z^{-1})y = B_L(z^{-1})u + H_y(z^{-1}) + d \quad (1)$$

where y, u, and d are process output, controller output, and measurable disturbance sequences of dimension r. $A_L, B_L \in \mathcal{R}_{rr}[z^{-1}], H_y \in \mathcal{R}_{r1}[z^{-1}]$ are polynomial matrices. The argument z^{-1} will be omitted in the following text for simplicity. $A_L(0)$ is assumed to be invertible, and $B_L(0) = 0$. H_y represents the influence of initial conditions. A_L and B_L are relatively left prime (r.l.p.) and describe the input - output properties of the plant.

We assume that the reference w and the disturbance d are constant r - vector sequences with step changes and are generated via

$$Fw = G_w, \quad Fd = G_d, \quad (2)$$

where $F = \text{diag}(1 - z^{-1}), G_w, G_d \in \mathcal{R}_{r1}[z^{-1}]$.

The controller is described by the equations

$$P_L\tilde{u} = R_Lw - Q_Ly + H_{\tilde{u}} \quad (3)$$
$$Fu = \tilde{u} + H_u \quad (4)$$

with $P_L, Q_L, R_L \in \mathcal{R}_{rr}[z^{-1}]$, $H_{\tilde{u}}$, $H_u \in \mathcal{R}_{r1}[z^{-1}]$. The triplet P_L, Q_L, R_L is r.l.p., $P_L(0)$ is invertible, and $H_{\tilde{u}}$ and H_u represent the initial conditions of the controller. The second equation assures that the controller has an integral part and achieves zero tracking error for the above class of reference vectors w and disturbances d.

We further define right matrix fraction descriptions of the process and of the controller

$$A_L^{-1}B_L = B_RA_R^{-1}, \quad P_L^{-1}Q_L = Q_RP_R^{-1}, \quad (5)$$

and assume that these descriptions are relatively right prime (r.r.p.).

Finally, to specify the closed loop poles we introduce a diagonal polynomial matrix $M \in \mathcal{R}_{rr}[z^{-1}]$, which is assumed to be stable and can be expressed in the form

$$M = mI \quad (6)$$

where m is a stable monic polynomial in the operator z^{-1}. We assume that the poles are the same for the each loop. However, this condition is not very restrictive as will be shown later.

Theorem 1. (Pole placement).
Internal stability of the closed-loop system with its poles defined by the matrix M can be achieved if and only if A_LF, B_L are r.l.p. The feedback part of the controller C_b results as a solution of the Diophantine equation:

$$\begin{aligned} P_LFA_R + Q_LB_R &= M \quad \text{or} \\ A_LFP_R + B_LQ_R &= M, \end{aligned} \quad (7)$$

and the feedforward part C_f is a solution of another Diophantine equation:

$$SF + B_RR_L = M, \quad (8)$$

where $S \in \mathcal{R}_{rr}[z^{-1}]$ is a free polynomial matrix.

Proof. See (Mikleš 1990, Vidyasagar 1985). □

There are infinitely many solutions of eqs. (7), (8). Usually the minimum degree solution is used, but it is as well possible to search among the general solutions to satisfy additional requirements. Then the minimum degree controller P_0, Q_0, R_0 only serves as a basis to find an expression for the set of all stabilizing controllers. In our case, all such controllers are given by the following theorem :

Theorem 2. A controller (P, Q, R) gives rise to the closed-loop denominator matrix M if and only if it can be expressed as

$$\begin{aligned} P_L &= P_{L0} + ZB_L, \quad P_R = P_{R0} + B_RZ \\ Q_L &= Q_{L0} - ZA_LF, \; Q_R = Q_{R0} - A_RFZ \end{aligned} \quad (9)$$

$$R_L = R_{L0} + FX \quad (10)$$

where X, Z are arbitrary proper stable transfer function matrices of appropriate dimensions and (P_0, Q_0, R_0) is a particular solution of eqs. (7),(8).

Proof. See (Middleton and Goodwin 1990). □

Remark X, Z will in the sequel be assumed to be polynomial matrices $X, Z \in \mathcal{R}_{rr}[z^{-1}]$ with constant degrees n_x, n_z respectively. As any polynomial matrix in $\mathcal{R}[z^{-1}]$ is proper stable transfer function matrix in $\mathcal{R}(z)$, this will assure the conditions needed in Theorem 2.

3. PREDICTIVE CONTROL

The objective function to be minimized is the standard finite horizon quadratic criterion

$$f(t) = \sum_{j=N_1}^{N} e(t+j)^T R_e e(t+j) + \sum_{j=0}^{N_u} \tilde{u}(t+j)^T R_u \tilde{u}(t+j) \quad (11)$$

where N_1, N, N_u are minimum, maximum, and control horizon respectively, $R_u = \mathrm{diag}(r_{ui})$, $R_e = \mathrm{diag}(r_{ei})$ are positive definite matrices used as weights on the control increments $\tilde{u}$ and the tracking error $e = w - y$, respectively.

To optimize (11), we need predictions of the signals $\tilde{u}$ and e. Their transforms can be obtained using equations (1)-(8) as

$$\tilde{u} = M^{-1}[A_R(R_L G_w + FH_{\tilde{u}}) - Q_R H_{yu}] \quad (12)$$

$$e = M^{-1}[SG_w - P_R H_{yu} - B_R H_{\tilde{u}}]. \quad (13)$$

where $H_{yu} = B_L H_u + FH_y + G_d$. For the purpose of constrained control, the signal u is also needed and can be constructed in a similar manner.

These predictions consist of two parts, *free* responses which depend only on past data and represent the behaviour of the basic minimum degree controller (P_0, Q_0, R_0), and forced responses depending on the matrices X, Z which can be manipulated to satisfy the control objectives. The key idea is to decompose the signals into these two parts and to rewrite the objective function (11) and the signal predictions (12), (13) to obtain a minimization problem which depends only on the coefficients of the matrices X, Z.

In our decomposition, the transformation of the original polynomial equations into mixed polynomial-matrix equations is used. The main idea is to rewrite the unknown polynomial matrices as column coefficient vectors.

Theorem 3. (Predictions).
The predictions of $\tilde{u} = \tilde{u}_0 + \tilde{u}_f$ and $e = e_0 + e_f$ are given as

$$\tilde{u}_0 = M^{-1}[A_R(R_{L0}G_w + FH_{\tilde{u}0}) - Q_{R0}H_{yu}] = M^{-1}\bar{U}_0 \quad (14)$$

$$e_0 = M^{-1}[S_0 G_w - P_{R0}H_{yu} - B_R H_{\tilde{u}0}] = M^{-1}\bar{E}_0 \quad (15)$$

$$\tilde{u}_f = M^{-1}FA_R(F_x x + F_z z) = M^{-1}\bar{U}_x x + M^{-1}\bar{U}_z z \quad (16)$$

$$e_f = -M^{-1}B_R(F_x x + F_z z) = M^{-1}\bar{E}_x x + M^{-1}\bar{E}_z z \quad (17)$$

where $x = \mathrm{coef}(\mathrm{vec}(X))$, $z = \mathrm{coef}(\mathrm{vec}(Z))$ (for the definitions of *coef*, *vec* see (Boekhoudt 1990)),

$$F_x = \sum_{j=0}^{n_x} (\Delta w(t-j)^T \otimes I_r)\langle P_x z^j \rangle_{cp} \quad (18)$$

$$F_z = \sum_{j=0}^{n_z} (\Delta d(t-j)^T \otimes I_r)\langle P_z z^j \rangle_{cp} \quad (19)$$

and P_x, P_z are shift polynomial matrices

$$P_x(z^{-1}) = \text{block diag}(1, z^{-1}, \ldots, z^{-n_x}) \quad (20)$$

$$P_z(z^{-1}) = \text{block diag}(1, z^{-1}, \ldots, z^{-n_z}). \quad (21)$$

Proof. (Fikar and Engell 1995). □

The free and the forced responses are then obtained via polynomial division $M^{-1}T, T \in \{\bar{U}_0, \ldots, \bar{E}_z\}$ as quotients with the degrees N_u and N respectively.

$$\tilde{u} = U_0 + U_x x + U_z z \qquad \deg(\tilde{u}) = N_u \quad (22)$$

$$e = E_0 + E_x x + E_z z \qquad \deg(e) = N. \quad (23)$$

The objective function can be transformed to the form

$$f(x,z) = c + 2g^T \begin{bmatrix} x \\ z \end{bmatrix} + [x^T, z^T] H \begin{bmatrix} x \\ z \end{bmatrix}. \quad (24)$$

The solution for x, z in the unconstrained case can be calculated analytically from (24) and is given as

$$\begin{bmatrix} x \\ z \end{bmatrix} = \arg\min \left(\left\| H \begin{bmatrix} x \\ z \end{bmatrix} + g \right\| \right). \quad (25)$$

If the constrained case is considered then the constraints follow from the equations (22), (23) with unknown vectors x, z and Quadratic Programming problem results.

4. STABILITY

At first we show the autoswitching property of the PC.

Theorem 4. (Autoswitching).
If the sequence $w(t)$ remains constant for $t \in [-k-1, 0]$, $0 < k \leq n_x$, then only the coefficients of the matrix $X(z^{-1})$ with powers z^{-l}, $k < l \leq n_x$, have an influence on the forced predictions. If $k > n_x$ then the X-part is disabled. The same follows for the sequence $d(t)$ and the matrix $Z(z^{-1})$.

Proof. (Fikar and Engell 1995) □

Thus if the input signals remain constant for n_x, n_z steps then the PC is switched off. However, this feature is not sufficient, as stability is required for arbitrary input sequences. For this additional conditions are needed. It is well known that no finite horizon control approach can be stable for all choices of horizons. This is the result of optimizing only over the next n steps. Our approach gives a transparent condition for the horizons given below.

Theorem 5. (Choice of horizons).
Let $n = \max(n_x, n_z)$, $N_1 = 1$ and let the horizons be equal or greater than

$$N = \deg(B_R) + n \tag{26}$$

$$N_u = \deg(A_R F) + n. \tag{27}$$

Further assume that the sequences w, d are bounded. Then PC is uniformly asymptotically stable.

Proof. (Fikar and Engell 1995) □

Corollary 1.
Increased values of N and N_u over the settings given by (26), (27) have no effect on the optimal values of X, Z.

Proof. Direct consequence of the previous theorem. □

The above theorems show that the PC is active only for a finite number of steps and that it influences the closed-loop behaviour for a finite time that is not a function of output and control horizons. Within this horizon, behaviour of the nominal controller is suppressed and an optimal finite-LQ trajectory is calculated.

Further, optimal values of X, Z remain unchanged if the horizons are increased over (26), (27). Hence the finite horizon can be made infinite but the number of optimized variables remains finite.

The approach which is closest to this is given by SGPC (Rossiter et al. 1995). It is shown there that the Youla-Kučera parameter is zero. This corresponds to the case when the nominal controller NC would be SGPC so that the predictive controller cannot improve the optimal trajectory. However, the condition that the controlled output should match the setpoint at the output horizon holds only if $M = I$. Otherwise the system is driven to the neighbourhood of the steady-state by PC and then the control is passed to NC. Moreover our strategy employs practically infinite output and control horizons that can be reduced without loss of generality. A further difference are the degrees of the polynomial matrices of the PC. As these are larger than minimal, the PC uses more information (compared to NC, SGPC) about the past closed-loop behaviour and optimizes it on line.

Further this approach makes it possible to extend the classical infinity horizon LQ controller to handle constraints. See (Fikar *et al.* 1996) for details. In this case if the constraints are active, a time varying controller results as in this article. However if the constraints are not active the controller reduces to classical time-invariant input-output LQ controller calculated via spectral factorization and Diophantine equations.

Remark If adaptive control is considered, the parameters of the plant model change at each sampling time and the sequence d must be estimated in each step from the plant description (1) as

$$d(t-i) = \hat{A}(t)y(t-i) - \hat{B}(t)u(t-i) \tag{28}$$

where $\hat{A}(t), \hat{B}(t)$ are the on-line estimates of A, B at time t.

4.1 *User Specifications*

The method can be tuned with the following user defined parameters:

- n_x, n_z: degrees of matrices X, Z,
- R_u, R_e: weighting matrices.

The degrees of the matrices X, Z. These are the most important parameters for the user. The higher the degrees, the more degrees of freedom exist and a smaller value of the objective function can be obtained. However, lower degrees result in fewer computations, so the values should be chosen carefully and depend on the speed of the computer as well as on the time constants of the process.

The degrees also determine the number of steps for which PC is switched on after a change in

w, d. Consider for example a step change of w. PC can modify n_x steps of the closed-loop behaviour and after these steps it is switched off. Therefore n_x should be chosen such, that after n_x steps the controlled output will be close to the desired setpoint.

As a dependency between N, N_u and n_x exists, one can use the general rule stated in GPC. The maximum horizon should be approximately equal to the rise-time of the controlled system. Hence n_x can be calculated for given N from Theorem 5 as

$$n_x = N - \max(\deg(B_R), \deg(A_R F)). \quad (29)$$

The same rules apply independently to n_z. A matrix Z shapes the response to disturbances, the values of which are usually small, smaller value of n_z is sufficient. Moreover, in the presence of measurement noise or undermodelled dynamics the disturbance is rarely constant and the feedback part of PC remains always active. Values of $n_z > 3$ in simulations showed no additional improvement of performance.

If constraints are present and active it is necessary to determine approximately for how many steps they will remain active for usual reference changes and disturbances. This is normally known in advance and depends mainly on how tight the process is controlled. The degrees n_x, n_z must then be increased accordingly.

The weighting matrices. The weighting matrices determine the activity of the controller. If the norm of R_u is large compared to the norm of R_e, more cautious control results. In the opposite situation, the control error vanishes faster, but the control actions can be excessive.

Other parameters. Unlike in standard GPC, the horizons are not directly set and can be manipulated indirectly by n_x, n_z. The minimum horizon N_1 can be set to 1. The influence of the nominal controller and its poles M is suppressed whenever a change of w, d occurs and the closed-loop behaviour is given by PC. Only after a finite number of steps after the last change of the input signals, PC is switched off and the system is controlled by the nominal closed-loop poles M.

4.2 *Feasibility of the solution*

There are two possibilities for the solution of the constrained QP problem to become infeasible. In one case there may be a large setpoint change or a large disturbance, for which the degrees n_x, n_z are too small. However, the sizes of these changes are usually known in advance and the degrees can be chosen sufficiently large to prevent such a situation. The second reason for infeasibility is inconsistency of the constraints with the model. This situation can happen if the controlled system is not stabilizable or it is caused by a wrong model in the adaptive control context. Thus infeasibility of the QP problem becomes a valuable tool to diagnose the status of the process. If the model of the process is perfect, then if it is not possible to reach given setpoints under the constraints the setpoints are infeasible and have to be changed or some of the constraints must be softened.

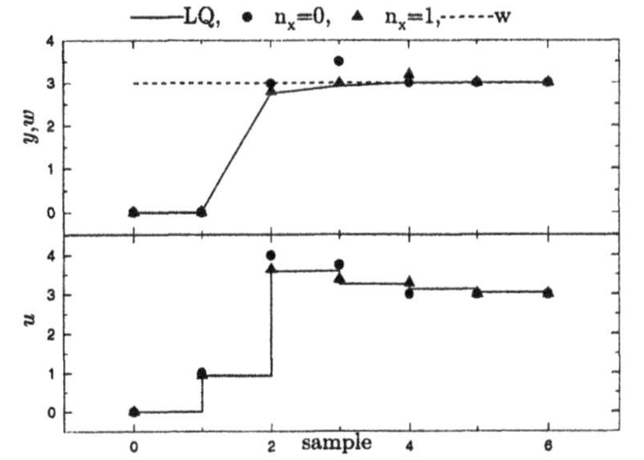

Fig. 2. Comparison of different finite and infinite horizon strategies

5. SIMULATIONS

The proposed algorithm was tested on the following unstable SISO system

$$S_p(z^{-1}) = \frac{3z^{-1} - z^{-2}}{1 + 2.5z^{-1} - 1.5z^{-2}}. \quad (30)$$

We assumed for simplicity no disturbance so that the free polynomial Z is equal to zero and only X polynomial is active.

In the first simulation, the choice of various degrees n_x is illustrated. The values of n_x were $0-4$. The values of other parameters were $M = 1$ (deat-beat nominal controller), $R_e = 10R_u$. The obtained results are shown in Figure 2. For comparison, the infinite horizon LQ controller minimizing the same cost function was also simulated (Fikar et al. 1996). It can be seen from the figure that the finite horizon optimal trajectory converges quickly for increasing values of n_x to the asymptotic trajectory. A further increase for $n_x > 1$ gave results practically identical to the infinite LQ-optimal trajectory.

The same controller was then used when a constraint $u \leq 3.2$ was considered. Note that the nominal deat-beat controller in this case gives an unstable response. The values of $n_x = 0, 1, 2$ resulted in infeasible quadratic programming problems. All higher values gave rise to stable constrained control with the same control and output

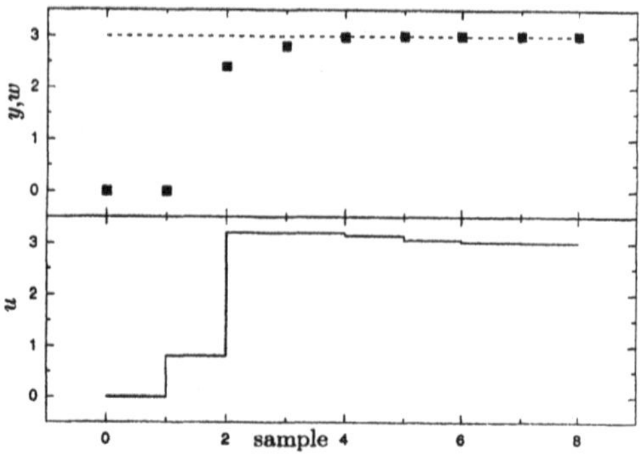

Fig. 3. Constrained control of the unstable system

trajectory. Figure 3 shows the results. The optimal polynomials X that minimized the cost function while satisfying the constraints were costant over the whole trajectory and given as

$$\begin{aligned}
n_x = 5 &: -0.234 - 0.099z^{-1} - 0.035z^{-2} \\
&\quad -0.012z^{-3} - 0.003z^{-4} - 0.001z^{-5} \\
n_x = 4 &: -0.234 - 0.099z^{-1} - 0.035z^{-2} \\
&\quad -0.011z^{-3} - 0.003dz^{-4} \\
n_x = 3 &: -0.234 - 0.100z^{-1} - 0.035z^{-2} \\
&\quad -0.010z^{-3}
\end{aligned}$$

It can be observed that increasing the value of n_x over the minimal feasible value $n_x = 3$ has only small influence on the lower order coefficients of X.

6. CONCLUSIONS

A new approach to the design of predictive controllers was presented. A quadratic long-range objective function is minimized subject to constraints on input/output signals in the subspace of all stabilizing controllers given by the Youla-Kučera parametrization. The derivation of the predictions is based on a matrix fraction description of all signals and systems utilizing transformations of the original polynomial equations to mixed polynomial-matrix equations. The stability of the time-varying closed-loop system was investigated and proved.

The Youla-Kučera parametrization of all stabilizing controllers applied to the problem of finite horizon receding control resulted in transparent properties. The time-varying part of the overparametrized controller is piece-wise constant and has the structure of a differentiating filter. Terminal constraints are inherently fulfilled as the important closed-loop signals are finite sequences.

It has been observed in simulations that the predictive controller can sometimes be time-invariant even in the constrained case. The study of conditions when this can happen is currently subject of the further research.

7. REFERENCES

Boekhoudt, P. (1990). Solution of polynomial equations in H_∞ optimal control. *Control Theory and Advanced Technology* **6**(3), 321 – 337.

Clarke, D. W. and R. Scattolini (1991). Constrained receding-horizon predictive control. *IEE Proc. D* **138**(4), 347 – 354.

Clarke, D. W., C. Mohtadi and P. S. Tuffs (1987). Generalized Predictive Control - part I. The basic algorithm. *Automatica* **23**(2), 137 – 148.

Cutler, C. R. and B. L. Ramaker (1980). Dynamic Matrix Control - A computer control algorithm. In: *Proc. of JACC, San Francisco.*

Fikar, M. and S. Engell (1995). Receding horizon predictive control based upon Youla-Kučera parametrization. *European Journal of Control.* (submitted for publication).

Fikar, M., P. Dostál and S. Engell (1996). An approach to design of predictive LQ-optimal controller. In: *Preprints of 2nd Conference Řízení Procesů ŘÍP '96, Horní Bečva, Czech Republic, June 3-6.* Vol. 1. pp. 70 – 75. (available from the first author).

Garcia, C. E. and A. M. Morshedi (1986). Quadratic programming solution of dynamic matrix control (QDMC). *Chem. Engng.Commun.* **46**, 73 – 87.

Keerthi, S. S. and E. G. Gilbert (1988). Optimal infinite-horizon feedback laws for a general class of constrained discrete-time systems: stability and moving horizon approximations. *Journal of Optimization Theory and Application* **57**, 265 – 293.

Kwon, W. H. and A. E. Pearson (1978). On feedback stabilization of time-varying discrete linear systems. *IEEE Tr. Aut. Control* **23**(3), 479 – 481.

Middleton, R. and G. C. Goodwin (1990). *Digital Control and Estimation. A Unified Approach.* Prentice Hall, Inc., Englewood Cliffs, N. J.

Mikleš, J. (1990). A multivariable self-tuning controller based on pole-placement design. *Automatica* **26**(2), 293 – 302.

Mosca, E. and J. Zhang (1992). Stable redesign of predictive control. *Automatica* **28**, 1229 – 1233.

Rossiter, J. A., B. Kouvaritakis and J. R. Gossner (1995). Feasibility and stability results for constrained stable predictive control. *Automatica* **31**(3), 863 – 877.

Vidyasagar, M. (1985). *Control System Synthesis: A Factorization Approach.* The MIT Press, Cambridge.

NONLINEAR CONTROL FOR DYNAMICAL SYSTEMS WITH CONSTRAINTS BASED ON THE BELLMAN-LYAPUNOV METHOD.

Valeriy Kudin, Sergey Lyashevskiy, Josef Kolacny.

Department of Electrical Drive and Automation, National Technical University "Kyiv Polytechnical Institute", pr. Pobedy, 37, 252056 Kyiv, Ukraine;
Department of Electrical Engineering Purdue University at Indianapolis 723 West Michigan Street, Room SL 160 B Indianapolis, Indiana 46202-5132, USA;
Technical University of Brno, Faculty of Electrical Engineering, Antoninska 1, 662 09 Brno, Chech Republic.

Abstract We treat the optimization issues for continuous-time dynamical systems whose state variables and control signals are bounded. The objective of this paper is to design of a nonlinear full state feedback control algorithm that satisfies the constraints and minimizes a nonlinear performance integral. The fundamental idea involves using a new class of the costs. The nonquadratic Lyapunov function is obtained by solving the Bellman-Lyapunov equation. The results are illustrated with an example.

Keywords: design, nonlinear systems, constraints, optimal control, Lyapunov functions.

1. INTRODUCTION.

The problem of extending nonlinear control algorithms to continuous-time systems has stimulated a number of fundamental research and new methods have given rise. The main motivation arises from the fact that classical variational theory suffers from the drawback for a class of high-order plants with constraints. The issue of designing nonlinear feedback controllers is a fundamental problem for control engineers. Methods for synthesis of control architectures are well established for linear systems with no constraints. However, control design based on a linear model might not work well in practice since the process variables and control signals usually are bounded. This has led to an interest in nonlinear problems and the control algorithm should be synthesized to achieve the desired performance when there are constraints in the process.

Much effort has been gone into designing of bounded plants and the problem of synthesizing dynamical systems of this type has attracted a considerable amount of interest (Bellman, 1957, Rekasius and Gibson, 1962, Hsu and Meyer, 1968, Letov, 1969). Primary among these investigations have been the pioneering studies (Hsu and Meyer, 1968, Letov, 1969, Pontryagin *et al.*, 1962, Sage, 1977). Some problems in the area of nonlinear optimization have been solved by authors (Kudin, 1976, Kudin and Lyashevskiy, 1990, Lyashevskiy, 1991). However, previous activities considered the optimization issues for systems subjected to hard bounds on control signals. Thus, constraints on the state vector were ignored. Our goal is to design a controller for a class of systems with constraints on the state variables and control signals. Therefore, we seek a control strategy subjected to additional constraints which depend on the solution of the system of differential equation forced by an admissible control.

2. THE PROCESS EQUATIONS.

Our model of a controllable time-invariant process is of the form

$$\dot{x} = Ax + Bu, |x_b| \le x_{b\,max}, |u| \le u_{max}, x(t_0) = x_0, \quad (1)$$

where $x \in R^c$ is the state vector; $x_b \in R^h$ is the vector of states which are bounded; $u \in R^m$ is the control vector; $A \in R^{c \times c}$ and $B \in R^{c \times m}$ are the constant plant matrices with rank(B)=m<c. Constraints on the control signals are expressed as $u \in U$, where $U \subset R^m$ is a prescribed constraint set. Piecewise continuous control function $u(\bullet):[t_o, t_f) \to R^m$ is admissible if the constraints are satisfied during the entire time interval $[t_o, t_f)$ in R^1. Let us assume that for each $t_o \in R$ and $x_o \in R^c$ there exists a solution, $x(\bullet):[t_o, t_f) \to R^c$, $t_o>0$, $t_o<t_f$, with $x(t_o) = x_o$ and, further, every solution can be extended over the interval $[t_o,\infty)$.

Thus obtained where the control $u(\bullet):[t_o, t_f) \to R^m$ satisfies $u \in U$, $U \subset R^m$, $\forall t \in [t_o, t_f)$, and generates the solution $x(\bullet):[t_o, t_f) \to R^c$, $x(t_o)=x_o$, such that

$$x \in X, X \subset R^c, \forall t \in [t_o, t_f) \quad (3)$$

As a mathematical model has been selected in the general form, the next step is define the description of the physical constraints on the state and control values.

Notations $|x_b| \le x_{bmax}$ and $|u_b| \le u_{bmax}$ are very general and do not indicate the type of nonlinear characteristics.

Nonlinearities of the state variables, which are bounded by $|x_b| \le x_{bmax}$, obviously can be represented as

$$x_b = \varphi(x), \quad (4)$$

where $\varphi(x)$ denotes a function vector which represents input / output nonlinear characteristics, $\varphi(\bullet):R^c \to R^h$. Our goal is to design analytically of the control algorithm. Bounded controller synthesis as a problem of finding the state feedbacks as a function of the state vector $f(x)$ subject to the constraints $|u| \le u_{max}$ will be considered now. Let us represent the control algorithm to be generated by a memoryless, state- feedback controller. One might write

$$u = \psi[f(x)], \quad (5)$$

where $\psi[f(x)]$ is a function vector which represents nonlinearities of the control signals, $\psi(\bullet):R^c \to R^m$.

Remark. Via the Bellman-Lyapunov concept it will be obtained in section 4 that

$$f(x) = -G^{-1}B^T(\partial V/\partial x), \quad (6)$$

where $V(x)$ is the Lyapunov function, $G \in R^{m \times m}$. Hence $u = \psi[f(x)] = -\psi[G^{-1}B^T(\partial V/\partial x)]$.■

Thus, the system (1)-(3) can be equivalently represented by the state equation of the form

$$\dot{x} = A \cdot x + B \cdot u, \; x_b = \varphi(x), \; u = \psi[f(x)], \; x(t_0) = x_0,$$
$$x \in X, \; X \subset R^c, \; u \in U, \; U \subset R^m, \; \forall t \in [t_0, t_f), \quad (7)$$

where $\varphi(\bullet):R^c \to R^h$ and $\psi(\bullet):R^c \to R^m$ are given functions. Several comments are in order here. Let the nonlinear characteristics of the state and control vectors may be described in the class of smooth functions. Hence, there exist C^1 functions $\varphi(\bullet):R^c \to R^h$ and $\psi(\bullet):R^c \to R^m$ such that $\varphi(\bullet)$, $\psi(\bullet)$, $\partial\varphi(\bullet)/\partial x$, $\partial\psi(\bullet)/\partial x$ and $\partial\psi(\bullet)/\partial u$ are definite and continuous for all $x \in R^c$ and $u \in R^m$ on $t \in [t_o, t_f)$. Thus, $\varphi(\bullet):R^c \to R^h$ and $\psi(\bullet):R^c \to R^m$ are continuously differentiable. Approximation of the nonlinear effects, such as saturations, relay characteristics, etc., by straight line segments only is sufficiently accurate in most cases.

Illustration. The responses of the ideal relay and saturation, which are described by a straight lines, are approximated with a high degree of accuracy by the following relationships:

$$x_{out} = \tanh(\eta x), \; x_{out} = \mathrm{erf}(\eta x), \text{ or } x_{out} = x^{1/(2m+1)},$$

where $\eta >> 1$ and m is a real positive number;

$$x_{out} = \tanh(\rho x), \; x_{out} = \mathrm{erf}(\rho x), \text{ or } x_{out} = \{1 - e^{-\sigma[\mathrm{tahn}(\eta x)]}\},$$

where $\rho > 0$ and $\sigma > 0$.

3. THE PERFORMANCE INDEX AND CONTROL OBJECTIVES.

In order to evaluate the performance of a dynamical process quantitatively one should select a cost integral. An optimal control is defined as one that minimizes/maximizes the cost. Therefore, having considered the modeling of the process let us discuss the performance index which should be used. The objective of this section is to provide physical motivation for the selection of an integral for the problem of constrained controller design. Some explanatory remarks about the performance index are in order before an attempt to design the control architecture for system (1)-(3) will be undertaken. Consider the nonquadratic integral

$$J[x(\bullet), u(\bullet)] = \int_{t_0}^{t_f} [W(x) + W(u)]dt, \quad (8)$$

where $W(x)$ and $W(u)$ are the integrand functions, $W(\bullet):R^c \to R$, $W(\bullet):R^m \to R$. Let us briefly discuss the performance criterion (8) and the problem of constructing of the integrand functions. Investigating properties of the functional will help us characterize the state trajectories and the control policy. Let us define the integrand functions. The requirements on the systems dynamics and nonlinear effects have to be taken into account. One should pose the procedure in the framework that the performance measure should serve as the error criterion in the process controller design whose purpose is to maintain plant (1)-(3) with regard the constraints and the specified control applications. Thus, as a pad of synthesis of bounded controllers for a class of plants with constraints let us specify integrands. The error criterion has to be positive definite and it is necessary to choose sufficiently smooth and positive definite **W(x)** and **W(u)** in such of matter that the following conditions hold

$$\phi_1(\|\mathbf{x}\|^d) \le \mathbf{W(x)} \le \phi_2(\|\mathbf{x}\|^d)\ \forall\ \mathbf{x} \in \mathbf{R}^c,\ \forall t\in[t_0, t_f),$$
$$\phi_3(\|\mathbf{u}\|^d) \le \mathbf{W(u)} \le \phi_4(\|\mathbf{u}\|^d),\ \forall\ \mathbf{u} \in \mathbf{R}^m,\ \forall t\in[t_0, t_f), \quad (9)$$

where $\phi_1(\bullet)$, $\phi_2(\bullet)$, $\phi_3(\bullet)$ and $\phi_4(\bullet)$ are nondecreasing functions, such that $\phi_1(0)=0$, $\phi_2(0)=0$, $\phi_3(0)=0$, $\phi_4(0)=0$, and $\phi_1(p_1)>0$, $\phi_2(p_2)>0$, $\phi_3(p_3)>0$, $\phi_4(p_4)>0$ whenever $p_1>0$, $p_2>0$, $p_3>0$, $p_4>0$; d denotes a positive constant (in general, d is not a real number, and one may put $d=2\alpha/(2\mu+1)$, where α and μ are real positive numbers).

Let $\mathbf{W}(\bullet):\mathbf{R}^c\to\mathbf{R}$, $\mathbf{W}(\bullet):\mathbf{R}^m\to\mathbf{R}$ be a given positive definite functions of class C^2. Since the smooth nonlinearities are considered it is natural to define integrand functions in the forms

$$\begin{aligned}\mathbf{W(x)} &= \mathbf{x}^T\mathbf{Qx} + \int\left[\varphi^{-1}(\mathbf{x}_b)\right]^T\mathbf{Q}_b d\mathbf{x}_b,\\ \mathbf{W(u)} &= \int\left[\psi^{-1}(\mathbf{u})\right]^T\mathbf{G}d\mathbf{u},\end{aligned} \quad (10)$$

where $\mathbf{Q} \in \mathbf{R}^{c\times c}$ is the positive semidefinite matrix; $\mathbf{Q}_b \in \mathbf{R}^{h\times h}$ and $\mathbf{G} \in \mathbf{R}^{m\times m}$ are positive definite matrices, and; further, matrices $\mathbf{Q}_b \in \mathbf{R}^{h\times h}$ and $\mathbf{G} \in \mathbf{R}^{m\times m}$ are analogous to the weighting matrices in the classical linear problem.; $\varphi^{-1}(\bullet):\mathbf{R}^h\to\mathbf{R}^h$ and $\psi^{-1}(\bullet):\mathbf{R}^m\to\mathbf{R}^m$ denote the function vectors.

The reason for using the positive definite nonlinear terms $\int\left[\varphi^{-1}(\mathbf{x}_b)\right]^T\mathbf{Q}_b d\mathbf{x}_b$ and $\int\left[\psi^{-1}(\mathbf{u})\right]^T\mathbf{G}d\mathbf{u}$ are as follows. The smooth nonlinearities $\mathbf{x}_b = \varphi(\mathbf{x})$ and $\mathbf{u} = \psi[f(\mathbf{x})]$ lie in the first and third quadrants interest (Bellman, 1957, Rekasius and Gibson, 1962, Hsu and Meyer, 1968, Letov, 1969). Hence, for all $\mathbf{x}_b \in \mathbf{R}^h$ and $\mathbf{u} \in \mathbf{R}^m$ the C^1 functions $\varphi^{-1}(\mathbf{x}_b)$ and $\psi^{-1}(\mathbf{u})$ are also lie in the first and third quadrants of the input-output plane. Moreover, the smooth functions $\varphi^{-1}(\bullet):\mathbf{R}^h\to\mathbf{R}^h$ and $\psi^{-1}(\bullet):\mathbf{R}^m\to\mathbf{R}^m$ are continuously differentiable. Thus, the positive definite integrands, given in (10), are smooth, continuously differentiable and satisfy conditions (9).

Illustration. Suppose that the input-output relation of the saturation nonlinearity is defined via the hyperbolic function as

$$\mathbf{x}_{bi} = \varphi_i(\mathbf{x}_i) = \tanh(\mathbf{x}_i),\ |\mathbf{x}_{bi}| \le 1, \quad (11)$$

where $\mathbf{x}_{bi}$ represents the output; $\mathbf{x}_i$ is the input of the nonlinearity. Using (10) for $q_{bii}=1$ one can obtain

$$\begin{aligned}&w_b(x_{bi}) = \int \mathrm{Artahn}(x_{bi})dx_{bi} =\\ &= \tfrac{1}{2}x_{bi}\ln[(1+x_{bi})/(1-x_{bi})] + \tfrac{1}{2}\ln(1-x_{bi}^2) =\\ &\ln[(1+x_{bi})/(1-x_{bi})]^{x_{bi}/2}(1-x_{bi}^2)^{1/2},\ |x_{bi}|<1,\end{aligned}$$

from which it can be concluded that $w_b(x_{bi}) > 0$, $\forall x_{bi} \ne 0$, and $w_b(x_{bi}) = 0$, $\forall x_{bi} = 0$. Thus, the $w_b(x_{bi})$ is positive definite. Moreover, this integrand reflects desired performance.■

Lemma 1. Consider a process described by (7) with $\varphi(\bullet)$ and $\psi(\bullet)$ continuous. The performance index (11) with (10) is a real valued and positive definite for all $\mathbf{x} \in \mathbf{R}^c$ and $\mathbf{u} \in \mathbf{R}^m$. Furthermore, the integrands (10) are continuously differentiable C^2 functions for which conditions (9) hold. ■

4. PROPOSED CONTROLLER.

In the case of designing of controllers for systems with bounded control signals most methods are based on solving the Hamilton-Jacobi-Bellman or Lyapunov's equations, as well as using the maximum principle. Consider a process described by (1). The problem is to find an admissible control which causes the system (1) to follow an admissible trajectories that minimizes/maximizes the performance index. The process defined by the equations (1)-(3), or in another notation (7), is to be controlled to minimize the cost (8) with (10). The Hamilton-Jacobi-Bellman equation is given by

$$\begin{aligned}-\partial V/\partial t = \min_{u\in U}\{&\mathbf{x}^T\mathbf{Qx} + \int\left[\varphi^{-1}(\mathbf{x}_b)\right]^T\mathbf{Q}_b d\mathbf{x}_b +\\ &+\int\left[\psi^{-1}(\mathbf{u})\right]^T\mathbf{G}d\mathbf{u} + (\partial V/\partial x)^T(\mathbf{Ax}+\mathbf{Bu})\}.\end{aligned} \quad (11)$$

The primary objective is to determine a control algorithm. To carry out the minimization process, let us differentiate both sides of equation (11) with respect to $\mathbf{u} \in \mathbf{R}^m$. Finally, the Hamilton- Jacobi-Bellman equation (11) gives the controller

$$\mathbf{u} = -\psi\left[\mathbf{G}^{-1}\mathbf{B}^T(\partial V/\partial x)\right]. \quad (12)$$

Lemma 2. The control law (12) guarantees at least a local minimum for the perfonnance index (8) with (10) for all $\mathbf{u} \in \mathbf{R}^m$. Moreover, if an optimal control exists, it is unique and represented by equality (12). ■

Thus, the bounded controller (12) was analytically determined as a nonlinear function of states; that is, $\mathbf{u} = \psi[f(\mathbf{x})]$, $\psi(\bullet):\mathbf{R}^c\to\mathbf{R}^m$, where $f(\mathbf{x})=\mathbf{G}^{-1}\mathbf{B}^T(\partial V/\partial x)$.

Hence, proposed control strategy connected with constraints of the input signals. Integration of a product of functions leads us to the formula

$$\int \Xi(\mathbf{u})[\Pi(\mathbf{u})] = \Xi(\mathbf{u})\Pi(\mathbf{u}) - \int [\Pi(\mathbf{u})]^T d[\Xi(\mathbf{u})]. \quad (13)$$

Then it follows that

$$\int [\{\psi^{-1}(\mathbf{u})\}]^T \mathbf{G} d\mathbf{u} = \int \{\psi^{-1}[\psi[\mathbf{G}^{-1}\mathbf{B}^T(\partial V/\partial x)]]\}^T \times$$
$$\times \mathbf{G} d\{\psi[\mathbf{G}^{-1}\mathbf{B}^T(\partial V/\partial x)]\} = (\partial V/\partial x)^T \mathbf{B}\psi[\mathbf{G}^{-1}\mathbf{B}^T(\partial V/\partial x)] \times$$
$$\times \int \{\psi[\mathbf{G}^{-1}\mathbf{B}^T(\partial V/\partial x)]\}^T d\{\mathbf{B}^T(\partial V/\partial x)\} \quad (14)$$

From equations (11) and (12), using above relationship, the Bellman-Lyapunov equation could be found at once in the form

$$-\partial V/\partial t = \mathbf{x}^T\mathbf{Q}\mathbf{x} + \int [\varphi^{-1}(\mathbf{x}_b)]^T \mathbf{Q}_b d\mathbf{x}_b + (\partial V/\partial x)^T \mathbf{A}\mathbf{x} -$$
$$- \int \{\psi[\mathbf{G}^{-1}\mathbf{B}^T(\partial V/\partial x)]\}^T d[\mathbf{B}^T(\partial V/\partial x)], \quad (15)$$

where $\int \{\psi[\mathbf{G}^{-1}\mathbf{B}^T(\partial \mathbf{V}/\partial \mathbf{x})]\}^T d[\mathbf{B}^T(\partial \mathbf{V}/\partial \mathbf{x})]$

and $\int [\varphi^{-1}(\mathbf{x}_b)]^T \mathbf{Q}_b d\mathbf{x}_b$ should be determined exactly or using the rational fraction expressions (Khovanskiy, 1956, Kudin, 1976, Lyashevskiy, 1991). This part of the section will be devoted to an examination of the minimization problem with a view toward determining and analysis necessary and sufficient conditions for optimality. The conditions which have to be satisfied if control $\mathbf{u}^o \in \mathbf{R}^m$ is an optimal one should be discussed. Let us define the Hamiltonian function $\mathbf{H}(\bullet):\mathbf{R}^c \times \mathbf{R}^m \times \mathbf{R}^c \to \mathbf{R}$ by (Hsu and Meyer, 1968, Pontryagin *et al.*, 1962, Sage and White, 1977, Tou, 1964)

$$\mathbf{H}(\mathbf{x},\mathbf{u},\mathbf{p}) = \mathbf{W}(\mathbf{x}) + \mathbf{W}(\mathbf{u}) + \mathbf{p}^T(\mathbf{A}\mathbf{x} + \mathbf{B}\mathbf{u}), \quad (16)$$

where $\mathbf{p} \in \mathbf{R}^c$ is a continuously differentiable function, $\mathbf{p}(\bullet):\mathbf{R}^c \to \mathbf{R}$. Using the definition of the Hamiltonian and the functional Hamilton-Jacobi-Bellman equation

$$-\partial V/\partial t = \min_{\mathbf{u}\in U}\{\mathbf{x}^T\mathbf{Q}\mathbf{x} + \int [\varphi^{-1}(\mathbf{x}_b)]^T \mathbf{Q}_b d\mathbf{x}_b +$$
$$\int [\psi^{-1}(\mathbf{u})]^T \mathbf{G} d\mathbf{u} + (\partial V/\partial x)^T(\mathbf{A}\mathbf{x} + \mathbf{B}\mathbf{u})\}. \quad (17)$$

one could observe a similarity. Moreover, the equations that constitute the maximum principle might be derived from the Hamilton- Jacobi-Bellman equation. The control issue is a minimization problem. Having mentioned the relationship between dynamic programming and maximum principle, let us discuss of necessary and sufficient conditions for optimality. The performance criterion, which is sufficiently smooth, has to satisfy the Hamilton-Jacobi-Bellman equation. This result represents necessary conditions for optimality. Sufficient conditions for a solution to the control problems have been given in (Hsu and Meyer, 1968).

If $\mathbf{J}[\mathbf{x}^o(\bullet),\mathbf{u}^o(\bullet)] \leq \mathbf{J}[\mathbf{x}(\bullet),\mathbf{u}(\bullet)]$ holds, then, $\mathbf{u}^o \in \mathbf{R}^m$ is said to be an absolute minimum. Here it is assumed that $\mathbf{u}^o(\bullet):[t_o, t_f) \to \mathbf{R}^m$ generates the solution $\mathbf{x}^o(\bullet):[t_o, t_f) \to \mathbf{R}^c$, $\mathbf{x}^o(t_o) = \mathbf{x}_o$, $\mathbf{x}^o \in \mathbf{X}$, $\mathbf{X} \subset \mathbf{R}^c$ .Hence $\mathbf{x}^o(\bullet):[t_o, t_f) \to \mathbf{R}^c$ denotes the corresponding optimal trajectory. Thus, if is an optimal control, then $(\mathbf{J}[\mathbf{x}(\bullet),\mathbf{u}(\bullet)] - \mathbf{J}[\mathbf{x}^o(\bullet),\mathbf{u}^o(\bullet)]) \geq 0$, $\forall\, \mathbf{x} \in \mathbf{X}$, $\mathbf{X} \subset \mathbf{R}^c$, $\mathbf{U} \subset \mathbf{R}^m$ $\forall t \in [t_o, t_f)$.

Lemma 3. For the control function to be a minimum of the performance cost, it is necessary that $\partial \mathbf{H}(\mathbf{x},\mathbf{u}^o,\mathbf{p})/\partial \mathbf{u} = 0$, a matrix $\partial^2 \mathbf{H}(\mathbf{x},\mathbf{u}^o,\mathbf{p})/\partial \mathbf{u}\cdot\partial \mathbf{u}^T \in \mathbf{R}^{m \times m}$ is positive definite for all $\mathbf{x} \in \mathbf{R}^c$, $\mathbf{u} \in \mathbf{R}^m$ on $[t_o, t_f)$, and, further, an inequality $(\mathbf{J}[\mathbf{x}(\bullet),\mathbf{u}(\bullet)] - \mathbf{J}[\mathbf{x}^o(\bullet),\mathbf{u}^o(\bullet)]) \geq 0$, $\forall\, \mathbf{x} \in \mathbf{X}$, $\mathbf{X} \subset \mathbf{R}^c$, $\mathbf{U} \subset \mathbf{R}^m$ $\forall t \in [t_o, t_f)$ has to be granted.▪

It will worth now discussing and interpreting the necessary and sufficient conditions for nonlinear control problem which is treating. The necessary condition $\partial \mathbf{H}(\mathbf{x},\mathbf{u}^o,\mathbf{p})/\partial \mathbf{u} = 0$, yields an extremum of the $\mathbf{H}(\mathbf{x},\mathbf{u},\mathbf{p})$ with respect to $\mathbf{u} \in \mathbf{R}^m$. As soon as the control law has been found as (12) from the Hamilton-Jacobi- Bellman equation (11), it must be proved that the second derivative of the $\mathbf{H}(\mathbf{x},\mathbf{u}^o,\mathbf{p})$ is positive definite, which represent a further necessary condition that equation (11) be a minimum. Therefore, the main question that arises is this: is the condition $\partial^2 \mathbf{H}(\mathbf{x},\mathbf{u}^o,\mathbf{p})/\partial \mathbf{u}\cdot\partial \mathbf{u}^T > 0$ has been satisfied? In order to answer to this question let us return to sections 2 and 3, and particularly the model of nonlinearities $|\mathbf{u}| \leq \mathbf{u}_{max}$ or $\mathbf{u} = \psi[f(\mathbf{x})]$, $\forall\, \mathbf{x} \in \mathbf{R}^c$, $\forall \mathbf{u} \in \mathbf{R}^m$ shall be considered. The characteristics of the nonlinear elements $\mathbf{u} = \psi[f(\mathbf{x})]$, $\forall\, \mathbf{x} \in \mathbf{R}^c$. $\forall \mathbf{u} \in \mathbf{R}^m$ lie in the first and third quadrants of the input-output plane. Functions $\psi^{-1}(u_i)$, i=1,2,...m, also are in the first and third quadrants, and, $\psi^{-1}(\mathbf{u})$, is an m-vector function of $\mathbf{u} \in \mathbf{R}^m$ which is continuously differentiable in its argument for all $\mathbf{u} \in \mathbf{R}^m$. Thus, $\phi_3(\|\mathbf{u}\|^d) \leq \int [\psi^{-1}(\mathbf{u})]^T \mathbf{G} d\mathbf{u} \leq \phi_4(\|\mathbf{u}\|^d)$ meets for all $\mathbf{u} \in \mathbf{R}^m$ on $t \in [t_o, t_f)$. Using above analysis is apparent that the matrix $\partial^2 \mathbf{H}(\mathbf{x},\mathbf{u}^o,\mathbf{p})/\partial \mathbf{u}\cdot\partial \mathbf{u}^T \in \mathbf{R}^{m \times m}$ is positive definite. Hence, requirement $\partial^2 \mathbf{H}(\mathbf{x},\mathbf{u}^o,\mathbf{p})/\partial \mathbf{u}\cdot\partial \mathbf{u}^T > 0$ holds for all admissible nonlinearities. Thus, the control does minimize the Hamiltonian, as well as the Hamiltonian-Jacobi-Bellman equation, globally. Further, the controller that minimizes the (8) is given by (12).

Necessary conditions allow to obtain of controllers which might be optimal. Thus, the control algorithm qualify only as a candidate for an optimal structure. Let us consider sufficient conditions which guarantee that a control law is an optimal one. The following Lemma gives sufficient conditions.

Lemma 4. If there exists a continuously differentiable positive definite C^1 function $\mathbf{V}(\bullet):\mathbf{R}^c \to \mathbf{R}$ such that

(h1) $v_1(\|x\|^d) \le \mathbf{V}(\mathbf{x}) \le v_2(\|x\|^d)\ \forall\ \mathbf{x} \in \mathbf{R}^c, \forall t \in [t_o, t_f)$,
(h2) $\mathbf{V}(\mathbf{x}) \le -v_3(\|x\|^d\ ),\ \forall\ \mathbf{x} \in \mathbf{R}^c, \forall t \in [t_o, t_f)$,
(h3) $\lim_{t \to t_f} \mathbf{V}(\mathbf{x}) \le \lim_{t \to t_f} \mathbf{V}(\mathbf{x}^o) = 0, \forall \mathbf{x} \in \mathbf{X}, \mathbf{X} \subset \mathbf{R}^c\ \forall t \in [t_o, t_f)$,
(h4) $\mathbf{W}(\mathbf{x}^o) + \mathbf{W}(\mathbf{u}^o) + \partial \mathbf{V}/\partial \mathbf{x}^o)^T(\mathbf{A}\mathbf{x}^o + \mathbf{B}\mathbf{u}^o) = 0$,
$\forall\ \mathbf{x} \in \mathbf{R}^c, \forall\ \mathbf{u} \in \mathbf{U}, \mathbf{U} \subset \mathbf{R}^m, \forall t \in [t_o, t_f)$,
(h5) $\mathbf{W}(\mathbf{x}) + \mathbf{W}(\mathbf{u}) + \partial \mathbf{V}/\partial \mathbf{x})^T(\mathbf{A}\mathbf{x} + \mathbf{B}\mathbf{u}) \ge 0$,
$\forall\ \mathbf{x} \in \mathbf{X}, \forall\ \mathbf{u} \in \mathbf{U}, \mathbf{U} \subset \mathbf{R}^m, \forall t \in [t_o, t_f)$,

then the control function $\mathbf{u}^o(\bullet):[t_o, t_f) \to \mathbf{R}^m$, $\mathbf{u}^o \in \mathbf{U}$, $\mathbf{U} \subset \mathbf{R}^m$, generating the solution of the system (1)-(3) $\mathbf{x}^o(\bullet):[t_o, t_f) \to \mathbf{R}^c$ with $\mathbf{x}^o(t_o) = \mathbf{x}_o$ is optimal with respect to the set of admissible state trajectories $\mathbf{x} \subset \mathbf{R}^c$ Here $v_1(\bullet)$, $v_2(\bullet)$, and $v_3(\bullet)$, are strictly increasing functions, such that $v_1(0)=0$, $v_2(0)=0$, $v_3(0)=0$, and $v_1(p_1) > 0$, $v_2(p_2) > 0$, $v_3(p_3) > 0$, whenever $p_1 > 0$, $p_2 > 0$, $p_3 > 0$.■

The problem of construction a suitable Lyapunov function shall be discussed now and some general remarks concerning the formulation and solution of the nonlinear control problem via the Bellman-Lyapunov formalism will be given. The nonquadratic integrands (10) were obtained by using models of the input-output nonlinear characteristics of the bounded state variables and the control signals. From (8) with (10) it follows at once that the minimum value of the functional is expressed in nonquadratic forms. Hence, it is reasonable to assume that the minimum value of the performance measure (8) with (10) is given by

$$\mathbf{I}_{\min} = \sum_{\alpha=1}^{\gamma} \mathbf{v}(0)^{2\alpha/(2\mu+1)}. \qquad (18)$$

Any scalar function satisfying the hypotheses of Lemma 4 can serve as a Lyapunov candidate for optimization problems. Assume that the Bellman-Lyapunov equation admits a Lyapunov function. Then, the solution of the partial differential equation (15) might be approximated by the nonquadratic C^1 function in the form (Kudin and Lyashevskiy, 1990)

$$\mathbf{V}(\mathbf{x}) = \sum_{\alpha=1}^{\gamma} \mathbf{v}(\mathbf{x})^{2\alpha/(2\mu+1)}; \qquad (19)$$

that is, the solution of the equation (15) is given by the continuously differentiable and positive definite Lyapunov candidate (19).

Remark. Function (19) leads to the particular Lyapunov candidates for the specified control problems. For example, in the case of the phenomenon of saturations the candidate $\mathbf{V}(\mathbf{x}) = \mathbf{v}^2(\mathbf{x}) + \mathbf{v}^4(\mathbf{x})$ should be used (see an example in section 5); for the minimum-fuel problem the Lyapunov candidate can be found as $\mathbf{V}(\mathbf{x}) = \mathbf{v}^{4/3}(\mathbf{x}) + \mathbf{v}^2(\mathbf{x})$, that is of class C^1 function (Kudin, 1976, Kudin and Lyashevskiy, 1990).■ Next a useful Definition will be presented

Definition If a Lyapunov candidate violates hypotheses (hl) - (h5) of Lemma 4, then this candidate should be eliminated as a suitable Lyapunov function. ■

For define unknown coefficients of the Lyapunov candidate (19) it is necessary to solve the equation (15). Relationships (15) and (19) represent a system of Γ nonlinear differential equations in Γ unknown coefficients of the function (19).. Once coefficients of the $\mathbf{V}(\mathbf{x})$ are found, the control law is obtained from relationship (19). The following theorem results.

Theorem. Given the continuous-lime system with constraints (1)-(3) and the real-valued positive definite cost integral (18) with (10). Let the plant is controllable. Assume that the minimum value of the performance index (18) with (10) is given by (18). The nonquadratic positive definite Lyapunov candidate of class C^1 (19) is the solution of the Bellman-Lyapunov equation (15) for all $\mathbf{x} \in \mathbf{R}^c$ and $\mathbf{u} \in \mathbf{R}^m$ on $t \in [t_0, t_f)$. The controller (12), generating the solution of the process (1)-(3), is optimal if and only if the hypotheses (h1)-(h5) are granted.

Proof. The proof of the Theorem is straightforward since the complete analysis has been established in sections 2, 3, 4. ■

5. ILLUSTRATIVE EXAMPLE: THE MINIMUM-ENERGY PROBLEM.

Let us now consider the minimum-energy problem in which the final time is not specified. Consider a process with the state equations

$$\dot{x}_1 = -x_1 + u,\ \dot{x}_2 = x_1,\ \dot{x}_3 = x_2. \qquad (20)$$

The control $u \in \mathbf{R}^1$ and the state variable $x_1 \in \mathbf{R}^1$, respectively, are bounded by $|u| \le 1$, $|x_1| \le 1$; that is,

$$\mathbf{U} = \{\mathbf{u} \in \mathbf{R}^1, |\, u\,| \le 1\}, \qquad (21)$$

$$\mathbf{X} = \{\mathbf{x} \in \mathbf{R}^3, |\, x_1 | \le 1, |\, x_2 | \le \infty, |\, x_3 | \le \infty\}. \qquad (22)$$

The system (20)-(22) is to be transferred from an arbitrary initial state $\mathbf{x}_0 \in \mathbf{R}^3$ to the origin. The state variable x_1, and the control signal u are subjected to the constraints, namely, the saturation effects. Let us approximate the nonlinear characteristics on the class of the hyperbolic C^1 functions. Examination of (8) and (10) leads us to the criterion

$$\mathbf{J}[\mathbf{x}(\bullet), \mathbf{u}(\bullet)] = \int_{t_0}^{t_f} [\int \mathrm{Artahn}(x_1)dx_1 + \int \mathrm{Artahn}(u)du]dt, \qquad (23)$$

For system (20)-(22) using index (23) one can obtain the following functional equation

$$-\frac{\partial V}{\partial t} = \min_{|u|\le 1}\{\int \mathrm{Artahn}(x_1)dx_1 + \int \mathrm{Artahn}(u)du + \left(\frac{\partial V}{\partial x}\right)^T \cdot [\mathbf{Ax} + \mathbf{Bu}]\}. \qquad (24)$$

Then, the nonlinear controller (12) has to be written as

$$u = -\tanh[(\partial V/\partial x_1,)]. \qquad (25)$$

From equations (24) and (25), the general method leads us to write the Bellman-Lyapunov equation (15) in the form

$$-\frac{\partial V}{\partial t} = \min_{|u|\le 1}\{\int \mathrm{Artahn}(x_1)dx_1 + \int \mathrm{Artahn}(u)du + \left(\frac{\partial V}{\partial x}\right)^T \mathbf{Ax} - \int \mathrm{tahn}\,(\partial V/\partial x_1)d(\partial V/\partial x_1)\}., \qquad (26)$$

where the integrands in the intervals $|x_1| < 1$, and $|\partial V/\partial \mathbf{x}_1| \le 2.5$ are approximated by

$$\begin{aligned}&\int \mathrm{Artahn}(\mathbf{x}_1)d\mathbf{x}_1 = \{\mathbf{x}_1 \ln[(1+\mathbf{x}_1)/(1-\mathbf{x}_1)] + \\ &+ \ln(1-\mathbf{x}_1^2)\}/2 = \mathbf{x}_1^2/2 + \mathbf{x}_1^4/12, \\ &\int \mathrm{tahn}(\partial V/\partial \mathbf{x}_1)d(\partial V/\partial \mathbf{x}_1) = \ln\cos(\partial V/\partial \mathbf{x}_1) \approx \\ &\approx (\partial V/\partial \mathbf{x}_1)^2/[3+0.05(\partial V/\partial \mathbf{x}_1)^2 \qquad (27)\end{aligned}$$

One can see that the positive definite form of the C^1 Lyapunov function, which meets the hypotheses (h1) and (h2) of Lemma 4,

$$\begin{aligned}V(x) = {}& 1/2(k_{11}x_1^2 + 2k_{12}x_1x_2 + 2k_{13}x_1x_3 + k_{22}x_2^2 \\ &+2k_{23}x_2x_3+k_{33}x_3^2) + 1/4(k_{41}x_1^4 + 4k_{42}x_1^3x + 6k_{43}x_1^2x_2^2 \\ &+4k_{44}x_1x_2^3 + 4k_{45}x_1^2x_2x_3 + 4k_{46}x_1x_2^2x_3 + 4k_{47}x_1x_2x_3^2 \\ &+4k_{48}x_1^3x_3+6k_{49}x_1^2x_3^2+4k_{410}x_1x_3^3+k_{411}x_2^4+4k_{412}x_2^3x_3 \\ &+ 6k_{413}x_2^2x_3^2 + 4k_{414}x_2x_3^3 + k_{415}x_3^4) \qquad (28)\end{aligned}$$

is a solution of equation (26). Using (28) and solving equation (26) the following nonlinear control structure could be obtained

$$\begin{aligned}u = {}& -\tanh(1.6x_1 + 2.5x_2 + x_3 + 0.52x_1^3+0.13x_1^2x_2 \\ &+ 0.07x_1x_2x_3 + 0.28x_1x_2^2+0.64x_1^2x_3+ 0.07x_1x_3^2 \\ &+0.63x_2^3+0.49x_2^2x_3+0.35x_2x_3^2+0.27x_3^3), \qquad (29)\end{aligned}$$

which is globally optimal in the class of an admissible control function constrained by $|u| \le 1$. Furthermore, the hypotheses (h3)-(h5) are granted. Thus, sufficient conditions for optimality are satisfied.

6. CONCLUSIONS

In this paper optimal control theory has been used to obtain an analytical solution of the nonlinear control problem for a class of continuous-time systems with state and control constraints. The presence of constraints complicates the application of variational methods. The central goal of the paper is further development of dynamic programming method as it applies to the nonlinear control issues. We have discussed the application of the Bellman- Lyapunov concept and the optimal control strategy has been obtained in the closed form. Performance objectives have been described and a nonquadratic cost integral, which assigns a unique positive real number to each trajectory of the plant, has discussed and treated. Our algorithm makes a comparison of the performance index value associated with an optimal control algorithm. An example of the minimum-energy system design was investigated in details.

REFERENCES.

Bellman R.E. (1957) *Dynamic Programming.* Princeton University Press,. Princeton

Hsu J.C. and A.U. Meyer (1968) *Modern Control Principles and Applications.*: McGraw-Hill Book Company,. New York.

Khovanskiy A.N. (1956) *Use of Continued Fractions and Their Generalizations in Approximate Analysis Problems.* Gostekhizdat Press, Moscow:.

Kudin V.F. and S.E. Lyashevskiy (1990) Extending the solution of a class of problem of analytical design of nonlinear controllers. *Automation and Remote Control*, vol. 51, pt. 1, no. 7, pp. 880-888,.

Kudin V.F. (1976) Choice of the functional to be minimized in nonlinear problem of analytical regulator design. *Engineering Cybernetics*, no. 5, pp. 192-199,.

Letov A.M (1969) *Flight Dynamics and Control.*: Nauka, Moscow.

Lyashevskiy S.E (1990) *Robust Stabilization of Continuous-Time and Discrete-Time Systems with Uncertain Parameters.*: The Academy of Sciences of the USSR Press, Kiev

Lyashevskiy S.E. (1991) *Methods of Synthesis of Robust Regulators for Servomechanism and Stabilization Problems.* The Academy of Sciences of the USSR Press, Kiev:

Pontryagin L.S., V. Boltyanskiy, R. Gamkrelidze and E. Mishchenko (1962) *The Mathematical Theory of Optimal Processes*, Addison-Wesley Publishing Company, Inc., Massachusetts.

Rekasius Z.V. and J. E. Gibson (1962) Stability analysis of nonlinear control systems by the second method of Lyapunov, *IRE Trans. Automat. Contr.*, vol. AC-7, no. 1, pp. 3-15,.

Sage A.P. and C. C. White (1977) *Optimum Systems Control.*: Prentice-Hall, Inc., Englewood Cliffs, New Jersey

Tou J.T.(1964) *Modern Control Theory.* McGraw-Hill Book Company, . New York.

DESIGN OF HYBRID CONTROL SYSTEMS BASED ON THE ROBUST POLE REGION ASSIGNMENT APPROACH

M. Budinský, Š. Kozák, M. Žalman

Department of Control and Automation, Faculty of Electrical Engineering and Information Technology, Slovak Technical University, Ilkovičova 3, 812 19 Bratislava, Slovak Republic. E-mail: budinsky@elf.stuba.sk, kozak@kasr.elf.stuba.sk, zalman@nov1.kar.elf.stuba.sk

Abstract: The paper deals with the robust synthesis and analysis of control systems with a hybrid structure which consists of a sampled continuous process and a discrete control algorithm. A method for the design of digital controllers for processes with uncertain physical plant parameters is presented. The Robust Pole Region Assignment and the Algebraic Approach to the Output Feedback Based Pole Placement in the discrete domain are applied. The paper proposes also a new robust stability/performance criterion for hybrid systems with interval parametric uncertainties. An application example is included.

Keywords: Robust Control, Digital Control, Algebraic Systems Theory, Diophantine equations, Robust Pole Region Assignment Approach, Hybrid Control Structure.

1.INTRODUCTION

Very frequent goal of practicable motivated feedback control system design is to preserve the stability and the performance of control in the face of uncertain or varying physical plant parameters when only upper and lower bounds are known for each uncertain parameter.

Furthermore, the accuracy of mathematical model used for controller design varies but is never perfect. Mainly, fast dynamic phenomena are usually neglected in the model. Therefore, at high frequencies, even the model order is unknown.

For these reasons it is desirable that the controller should be insensitive to parametric and dynamic uncertainties. In other words it should be robust.

Continuous technological processes are often controlled by discretely working computers. Hence it is important to develop robust analysis and synthesis methods for hybrid control systems consisting of a sampled continuous part and a discrete part with the zero order hold element An example of such system is shown in Fig.1.

In practice the performance requirements are often formulated in the following manner: The system responses should be composed of rapidly decaying and well damped (aperiodic) terms without overshoot. Trajectory tracking and disturbance rejection should be guaranteed for prescribed classes of reference and disturbance. No intersample ripples are allowed. Possible saturation effects (windup) must be regarded. These properties should be robust with respect to large variations in operating conditions.

Optimal synthesis of controllers to achieve robustness with respect to uncertain interval physical plant parameters is still scientifically an unsolved problem. There are, however, useful design tools that do not have inherent optimality characteristic, but provide a good engineering approximation to the optimal solution. They can be viewed as a feedback iterative process involving controller synthesis and robustness analysis as illustrated by Fig. 2.

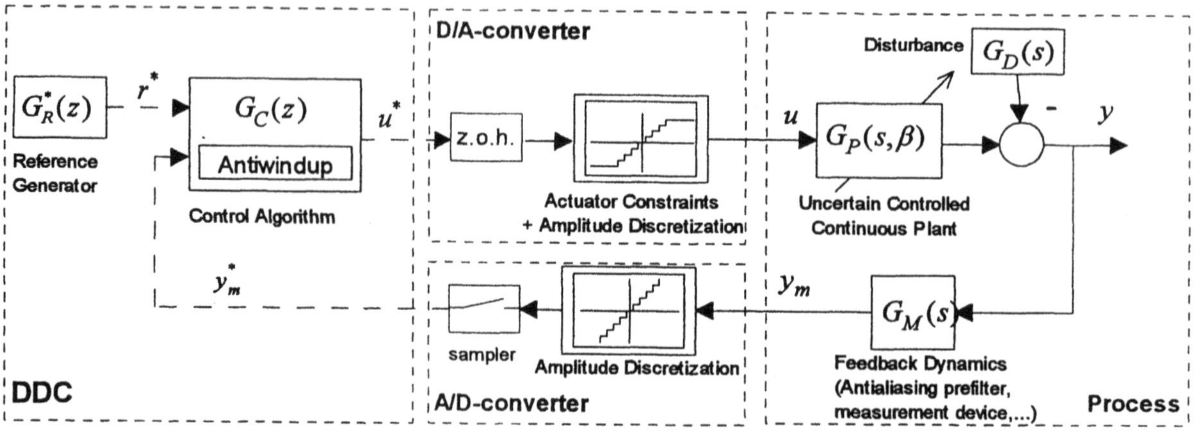

Fig. 1: Digital Control of Continuous Process

2.ROBUST POLE REGION ASSIGNMENT APPROACH

2.1 Physical Parameter Uncertainties

To deal with the parametric uncertainties the following family of plants is defined

$$G_P(s,\beta) = \frac{B(s,\beta)}{A(s,\beta)} = \frac{b_0(\beta)+\ldots+b_m(\beta)s^m}{a_0(\beta)+\ldots+a_n(\beta)s^n} \quad (1)$$

It is assumed that the polynomial coefficients are real and depend continuously on an unknown real parameter vector $\beta = [\beta_1, \beta_2, \ldots, \beta_l]^T$. Independent bounds are given for each parameter $\beta_i \in \langle \beta_i^-, \beta_i^+ \rangle$. As a result the set of possible parameter vectors is given by

$$\mathbf{B} = \{\beta = [\beta_1, \ldots, \beta_l]^T \mid \beta_i \in \langle \beta_i^-, \beta_i^+ \rangle, i = 1 \ldots l\} \quad (2)$$

A vertex is a parameter vector

$$\beta_v = [\beta_1 \ldots \beta_l]^T \text{ with } \beta_i \in \{\beta_i^-, \beta_i^+\} \quad (3)$$

It will also be assumed that $a_n(\beta) \neq 0$, $b_m(\beta) \neq 0$ for all $\beta \in \mathbf{B}$ and hence the degree of the polynomials do not vary in the operating domain **B**.

2.2 Generalised stability/performance criterion

The time responses of the control system should be composed of rapidly decaying and well damped terms for every $\beta \in \mathbf{B}$. Eigenvalues of linear control system indicate if this requirement is satisfied. Hence, this time domain requirement can be reformulated into the requirement that all closed loop poles should be placed in a prescribed region

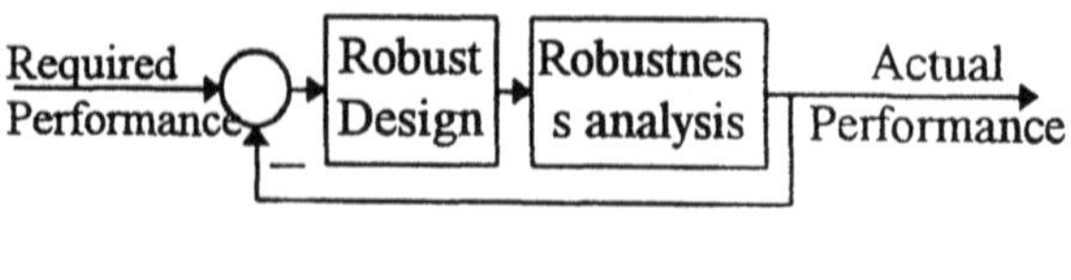

Fig.2. Iterative Procedure of Robust Design

γ in the complex plane and must remain in this region for all admissible parameter variations. The region is determined by the designer to comply for example with some damping and/or bandwidth requirements. This approach is called Robust Pole Region Assignment (Ackermann, 1993).

a) For purely continuous systems the region γ can be described by a boundary $\delta(\gamma)$ that consists of one or more contours in s-complex plane such that the membership $s_i \in \gamma$ is well defined. The boundary contours can be given as a value set of complex function $f(w)$, which is defined over real parameter $w \in W$ called generalised frequency. Set W is a subset of real numbers. Hence the boundary is defined by

$$\delta(\gamma) = \{s \mid s = f(w) \text{ for all } w \in W\} \quad (4)$$

b) For hybrid systems the region γ is defined in z-complex plane. Its boundary can be derived from (4) using the transform $z = e^{sT}$ (T - sampling period) under condition that the region in s-plane lies inside of the band given by imaginary axis limits $(-\pi/T, \pi/T)$. Then the boundary is described by:

$$\delta(\gamma) = \{z \mid z = e^{f(w)T} \text{ for all } w \in W\} \quad (5)$$

An example of such region is shown in Fig. 6.

Definition 1 (*Region Stability*): The system will be called γ -stable (or Region Stable), if it has all poles within the region γ .

Definition 2 (*Robust Region Stability*): The system is Robustly Region Stable if it is Region Stable for all admissible parameter variations.

2.3 Iterative approach to the controller design

An iterative method how to find a digital controller which is able to guarantee the Robust Region Stability of control system in Fig. 1 is based on the following steps:

STEP 1: Determine nominal model $G_{P0}(s) = G_P(s, \beta_0)$ and find corresponding Z-transform. Using equation (11) compute the minimal necessary number n of discrete closed loop poles $[z_1, \ldots z_n]$.

STEP 2: Select region γ.

STEP 3: Generate closed loop poles according to the following standard form

$$s_1 = -\frac{1}{T_m}, \quad T_m > 0$$
$$s_{i+1} = s_i b_i, \quad b_i > 1, \; i = 1,2,\ldots,n-1 \tag{6}$$

where $\mathbf{b} = [b_1, b_2, \ldots b_{n-1}]$ is the vector of tuning parameters (active damping coefficients) which enable the designer to form the closed loop root set. Constant T_m can be calculated in two ways:

a) if $b_1 = b_2 = \ldots = b_{n-1} = 1$ then $T_m = T_r \dfrac{(n-1)^{n-1}}{e^n (n-1)!}$ (7)

b) else $T_m = T_r A_m / e$. (8)

T_r represents a required settling time which should be prescribed by the designer. Equations (7) and (8) are derived according to the following definition:

<u>Definition 3</u> *(Settling Time)*: The settling time of a stable unit gain linear system (having only real poles and a zero order numerator) is equal to T_r if its impulse response amplitude is equal to eT_r.

An exact analytical formula similar to (7) cannot be derived for case (8). Hence, for given vector **b**, constant A_m in (8) has to be calculated numerically as the maximum of impulse response of the unit gain system which has a zero order numerator and poles generated using (6) for T_m=1.

STEP 4: Convert the prescribed poles s_i to the z-domain using equation $z_i = e^{s_i T}$. All poles z_i are distributed along the real axis within interval $\langle 0,1 \rangle$. Coefficients of active damping b_i determine a relative distance between every two consecutive poles. The role of parameter T_r is very similar to the role of the IMC filter time constant that is used for the unstructured uncertainty robustness tuning in the Internal Model Control (IMC) approach (Budinský and Kozák, 1994).

STEP 5: Use the procedure described in section 3 to find a digital controller that guaranties the pole placement prescribed in STEP 3.

STEP 6: For the designed controller plot the root set in the z-plane for uncertain parameters. Modify vector $\boldsymbol{b}$ of tuning parameters according to the following principle: The part of the root set that first tends to exceed the prescribed region γ determines which coefficient b_j must be increased. Then repeat steps 3, 4, 5, 6 until the complete root set remains in the prescribed region (if possible). The result should look like in Fig. 7.

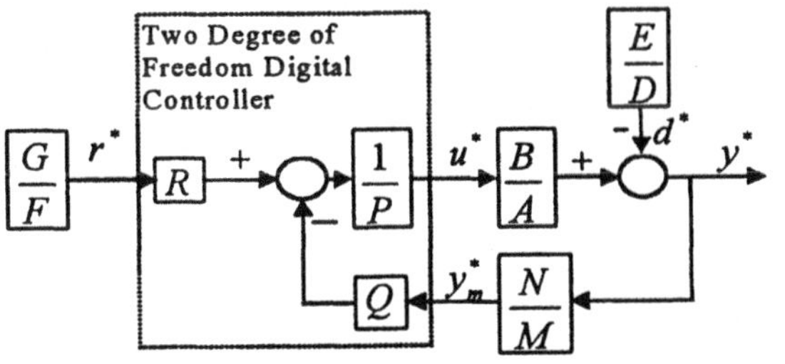

Fig.3. Linear Discrete Control System

STEP 7: Ensure that the coefficient values of controller polynomial Q (see Fig.3) are realistic (i.e. not extremely high). Otherwise the system will become very sensitive to a sensor noise, to the unstructured model uncertainties or to neglected nonlinearities. In such case it is necessary to reduce the parametric robustness requirements (uncertainty intervals must be shortened or region γ enlarged or settling time T_r prolonged).

STEP 8: Lemma 1 (in section 4) can be used to prove the Robust Region Stability.

In the described controller design procedure a new trade-off can be observed: *Robustness to physical real parameter uncertainties versus robustness to unmodelled dynamics* (i.e. unstructured complex uncertainties typical for H_∞-approach). To increase the robustness with respect to unstructured uncertainties it is necessary to reduce a control system bandwidth. On the other hand, to improve the robustness with respect to uncertain physical parameters it is necessary to augment the active damping. Keeping the same prescribed settling time it leads to the increase of the bandwidth.

3. FEEDFORWARD-FEEDBACK BASED ALGEBRAIC POLE-PLACEMENT

This section deals with a method which enables the designer to find such the digital controller that guarantees the nominal pole placement prescribed by section 2.3 (see STEP 5). The algebraic polynomial approach to the output feedback based pole placement in the discrete time domain is applied. To comply with the required asymptotic properties, the Internal Model Principle is included in the proposed design procedure. The two degree of freedom controller structure is utilized. It leads to the solution of two diophantine equations. The feed-forward feed-back structure favors the step responses with the zero or small over-shoot. No pole-zero cancellations between controller and process are used. Together with the involved Internal Model Principle it guarantees that the closed loop hybrid system behaves nicely without intersample ripples.

When all the continuous signals are observed only at the sampling intervals, then Fig.1 can be simplified to Fig.3. Nonlinearities are not considered.

Problem 1: Consider the discrete control system in Fig.3. Find the minimal degree polynomials $P(q)$, $Q(q)$ and $R(q)$ in delay operator $q = z^{-1}$ representing the two degree of freedom digital controller managed by equation

$$P(q)u^{*}(q) = R(q)r^{*}(q) - Q(q)y_{m}^{*}(q) \qquad (10)$$

The controller must guarantee that the discrete control system in Fig.3 satisfies the following two requirements:

1) Internal pole placement: The transfer functions from any point of the control system to any other point must have the same prescribed closed loop poles $z_1, \ldots, z_n$.

2) Asymptotic properties (Reference tracking and Disturbance rejection): The error sequence must vanish asymptotically for prescribed generic classes of reference and disturbance signals characterised by denominator polynomials $F(q)$ and $D(q)$. See Fig.3.

Theorem 1: (*Feedforward Feedback Based Pole Placement*). Polynomials P,Q,R solving problem 1 can be determined according to the following steps:

STEP 1: Find factor $J(q) = \mathrm{lcm}(A,F,D)/A$, where lcm( ) is the least common multiple of polynomials.

STEP 2: Compute number n which is the minimal degree ∂ of closed loop characteristic polynomial

$$n = \partial J(q) + \partial A(q) + \partial B(q) + \partial N(q) + \partial M(q) - 1 . \qquad (11)$$

STEP 3: Get characteristic polynomial $C(q)$ whose n roots are selected according to section 2.3.

STEP4: Find a minimal degree solution to diophantine equation $AJMX + BNQ = C$ where X and Q are unknown polynomials. Such solution is unique.

STEP 5: Find a solution to diophantine equation $FS+BMR=C$ where R and S are unknown polynomials. Minimise the degree of polynomial R only. Such the solution is unique.

STEP 6: $P(q) = J(q)X(q)$.

The proof and some necessary assumptions are omitted in this text bat they are available by the author. The control algorithm applies the simplest antiwindup principle: The controller uses the actual saturated process input u^{*} as the past values of the control signal.

4. ROBUST STABILITY/PERFORMANCE ANALYSIS

As it was mentioned in section 2.3 the root set plotting represents very important tool with respect to robust region stability analysis in the proposed approach. To improve the accuracy of root locus approximation, it is generally necessary to increase the number of grid points. In the case of more uncertain parameters, it can lead to more computations and time required. To save computational effort a generalised frequency domain based criterion of Robust Region Stability is proposed in this section. The criterion shows whether the uncertain systems in Fig.4-A or B have

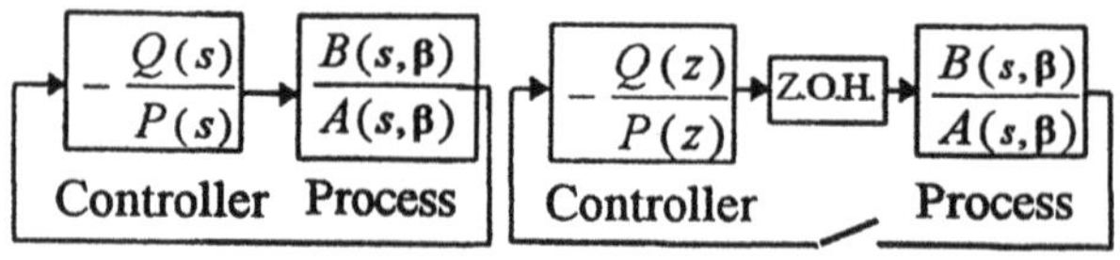

A B

Fig. 4. A-Continuous, B-Hybrid Control System

all their closed loop poles in the prescribed region γ for all $\boldsymbol{\beta} \in \mathbf{B}$. In the frame of this section it is allowed for the following two types of parametric uncertainties

a) Affine transfer functions

$$G_P(s,\underline{\beta}) = \frac{B(s,\underline{\beta})}{A(s,\underline{\beta})} = \frac{B_0(s) + \sum_{i=1}^{l} \beta_i B_i(s)}{A_0(s) + \sum_{i=1}^{l} \beta_i A_i(s)} \qquad (12)$$

b) Multilinear transfer functions

$$G_P(s,\underline{\beta}) = \frac{B(s,\underline{\beta})}{A(s,\underline{\beta})} = \frac{B_0(s) + \sum_{I_1} \beta_i B_i(s) +}{A_0(s) + \sum_{I_1} \beta_i A_i(s) +} \qquad (13)$$

$$\frac{+\sum_{I_2} \beta_i \beta_j B_{ij}(s) + \sum_{I_3} \beta_i \beta_j \beta_k B_{ijk}(s) + \ldots}{+\sum_{I_2} \beta_i \beta_j A_{ij}(s) + \sum_{I_3} \beta_i \beta_j \beta_k A_{ijk}(s) + \ldots}$$

where, e.g. $I_3 = \{\forall i,j,k \mid i \neq j \neq k,\ i = 1..l,\ j = 1..l, k = 1..l\}$

4.1 *Robust Region Stability of Continuous Control System*

Consider the continuous system in Fig. 4-A. Define a complex valued function

$$C(s,\boldsymbol{\beta}) = A(s,\boldsymbol{\beta})P(s) + B(s,\boldsymbol{\beta})Q(s) \qquad (14)$$

which corresponds to its characteristic polynomial. If s is substituted by function $s=f(w)$ defining the boundary of the considered region, then it can be rewritten in argument w of generalised frequency

$$C(w,\boldsymbol{\beta}) = A(w,\boldsymbol{\beta})P(w) + B(w,\boldsymbol{\beta})Q(w) \qquad (15)$$

Theorem 2: (*Condition of Robust Region Stability*) Consider the continuous control system in Fig. 4-A. Assume that the region γ and its boundary function $f(w)$ is given. Assume that there exists at least one nominal vector $\boldsymbol{\beta}_0 \in \mathbf{B}$ such that the closed loop system is γ-stable.

A) Necessary and sufficient condition for affine transfer function (12): The closed loop system is γ-stable for all vectors $\boldsymbol{\beta} \in \mathbf{B}$ if and only if the phase condition

$$\left|\angle\left(C(w,\boldsymbol{\beta}_v^-)\right) - \angle\left(C(w,\boldsymbol{\beta}_v^+)\right)\right| < 180^\circ, \quad \forall w \in W \quad (16)$$

is satisfied for all couples $\{\boldsymbol{\beta}_v^-, \boldsymbol{\beta}_v^+\}$ of such vertices which differs only in one parameter β_i. Their number is $l2^{l-1}$. (Symbol $\angle$ denotes phase of complex number)

B) Sufficient condition for multilinear transfer function (13): The system is γ-stable for all vectors $\boldsymbol{\beta} \in \mathbf{B}$ if the (16) is satisfied for all possible couples of vertices. Their number is equal to $N!/[2(N-2)!]$, where N=2^l.

The proof of this theorem is based on the results of The Boundary Crossing Theorem, The Edge Theorem and The Mapping Theorem. They can be found for example in the book of Ackermann (1993).

4.2 Region Stability of Hybrid Control System

This section brings an extension of theorem 2 for the hybrid control systems shown in Fig. 4-B.

Lemma 1: (*Condition of Robust Region-Stability for Hybrid Systems*) Theorem 2 is valid also for the hybrid control system (shown in Fig.4-B) with the following modification of the complex valued characteristic function C

$$C(z,s,\beta)=\frac{A(s,\beta)}{A_0(s)}A_0^*(z)P(z)+\frac{B(s,\beta)}{B_0(s)}B_0^*(z)Q(z) \qquad (17)$$

Using substitutions $z=e^{sT}$ and s=$f(w)$ the function (17) can be rewritten in common argument of the generalised frequency w

$$C(w,\beta)=\frac{A(w,\beta)}{A_0(w)}A_0^*(w)P(w)+\frac{B(w,\beta)}{B_0(w)}B_0^*(w)Q(w) \qquad (18)$$

Equations (17) and (18) represent modifications of equations (14) and (15). $B_0(s)$ and $A_0(s)$ are polynomials of the nominal transfer function for which the controller was designed. It is defined by equation $\frac{B_0(s)}{A_0(s)}=\frac{B(s,\beta_0)}{A(s,\beta_0)}$ where β_0 is a concrete nominal vector. Polynomials $B_0^*(z)$, $A_0^*(z)$ are polynomials of transfer function that is obtained by sampling of the nominal continuous transfer function considering the zero-order holder. It is assumed that the region γ was defined in the s-plane by the boundary defining function $f(w)$.

Function C now represents a generalisation of characteristic polynomial and will be called characteristic function of the closed loop of hybrid system. Till now the theoretical proof of Lemma 1 or the determination of its validity area is not finished yet. But its applicability is already proved by many concrete numerical examples reaching very exact results.

5. APPLICATION TO THE POSITION SERVOSYSTEM DESIGN

A simple example is given in this section which describes the application of proposed methods to the design of digital robust controller for a position servosystem with a DC motor and IRC position sensor (Fig.5). Varying inertial moment (J) of the system is considered as an uncertain parameter. Using the proposed approach the digital control

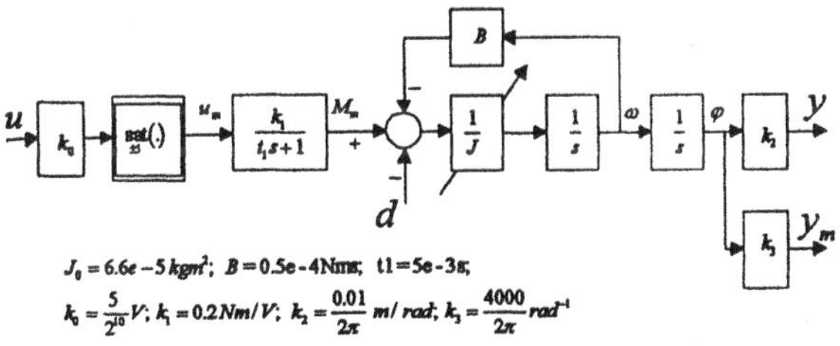

Fig. 5. Block diagram of the controlled process

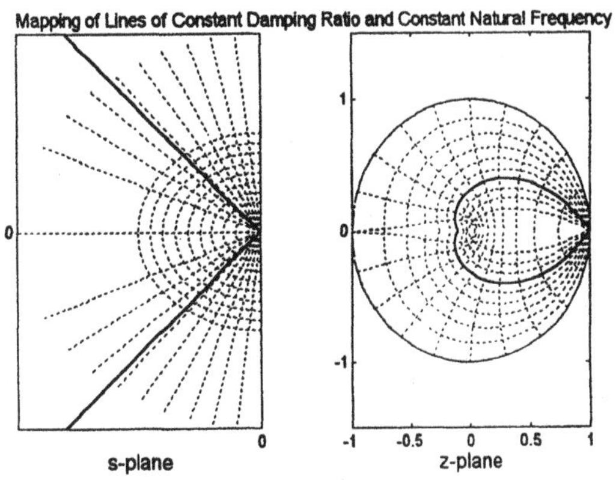

Fig. 6. Example of region γ in s-domain and z-domain

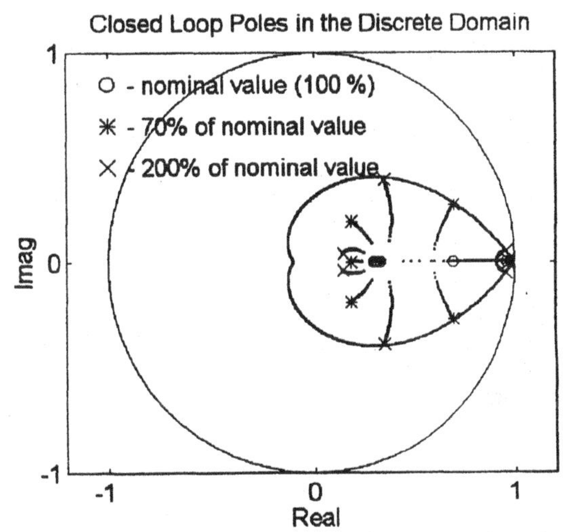

Fig.7. Root locus of robustly region-stable closed loop system for uncertain parameter J in the interval from 70% to 200% of nominal value.

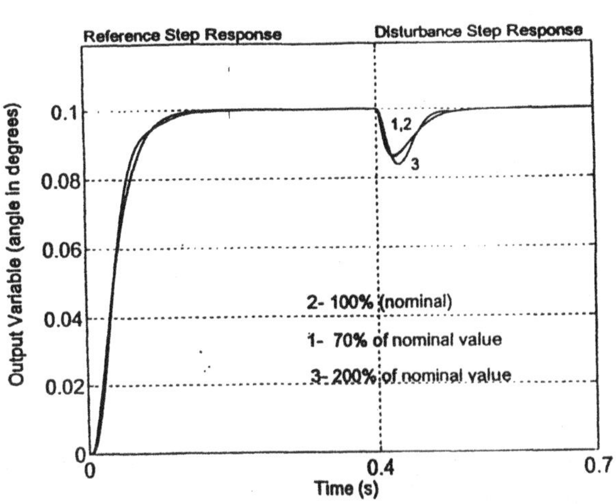

Fig.8. Output variable step responses for various uncertainty parameter values.

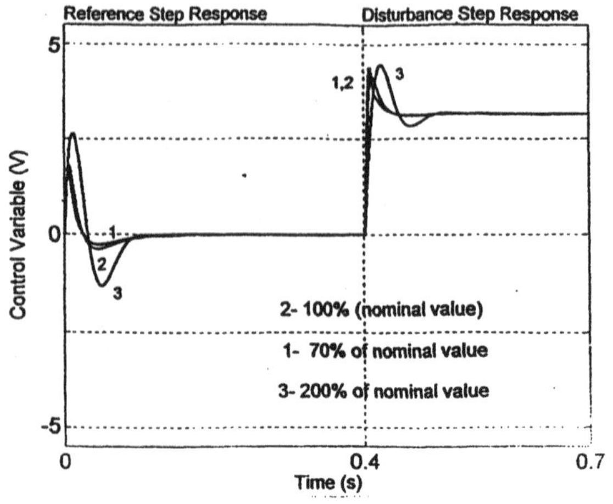

Fig.9. Control Variable step responses for various uncertain parameter values.

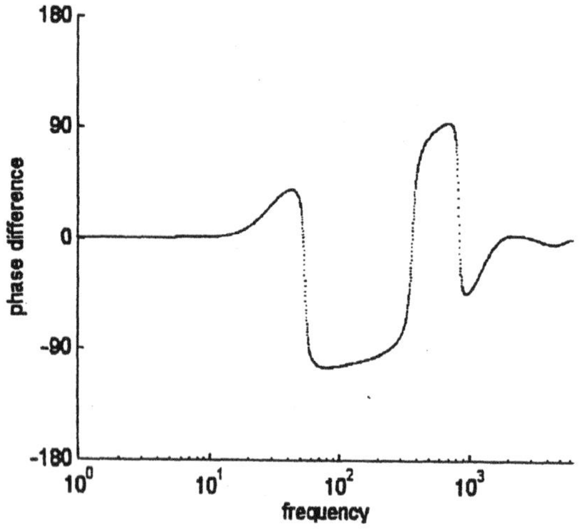

Fig.10.A typical plot of the phase difference of characteristic function value set for system that is Robustly Region-Stable for given uncertain parameter interval from 70% to 200% of nominal value.

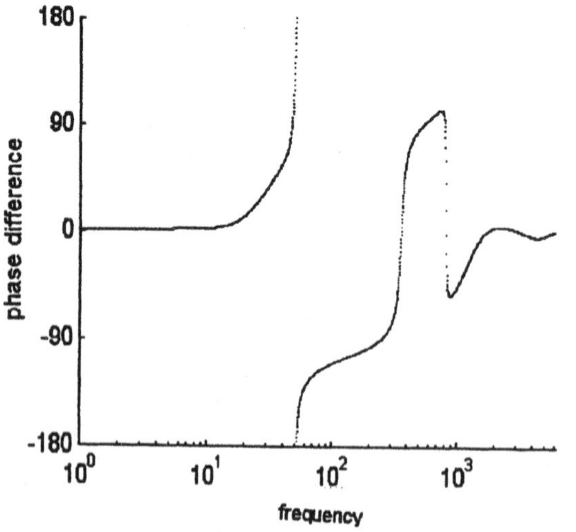

Fig.11.Typical plot for system that is not Robustly Region-Stable. The uncertain parameter interval is from 70% to 205% of nominal value.

algorithm has been designed which guarantees the following properties: **1)** Settling time about 0.2 s (it corresponds to the system bandwidth about 31 rad/s, - a typical requirement for position servosystems). 2)The reference tracking for step and ramp types of reference signals. **3)** The aperiodic responses with zero overshoot for all admissible values of uncertain parameter. **4)** The control system regards the time delay caused by the control variable computation time. **5)** The disturbance rejection for step type disturbance signals.

The control algorithm has been designed and implemented in a computer with sample period T=1ms. The resulting polynomials of the controller are: P=[1.000, -1.748, 1.136, -0.422, 0.0341], Q =[14.71, -30.98, 18.51, -2.233], R =[0.0109]. The γ-region was chosen according to Fig.6. The number of discrete closed loop poles is n=7. The vector b=1.03•[1,1,1,3,9.2,1] has been selected in a short iteration process. It guarantees the shape of root locus which lies in the prescribed region γ. See Fig.7. The system is γ–stable for J from *70%* to 200% of nominal value and stable from 27% to 1010% of nominal value. Corresponding time responses of the control system can be seen in Fig. 8 and Fig. 9. The results of application of Lemma 1 is shown in Fig.10 and Fig.11. Fig.10 proves that the designed control system is robustly region-stable for the given interval. Fig.11 brings the case when the tested uncertain parameter interval is longer than interval for which the controller was designed. The proposed method of robust design and analysis has been successfully tested also for other processes. For example: The robust benchmark problem dealing with two mass spring system, The position servosystem design with flexible joints.

6. REFERENCES

Ackermann, J. (1993). *Robust Control - Systems with Uncertain Physical Parameters.* Springer-Verlag London Limited.

Budinský, M. and Š. Kozák (1994). Robust Control Design Using the Algebraic Polynomial Approach. *Preprints of the 1st IFAC Workshop on New Trends in Design of Control Systems,* pp.167-172. Smolenice, Slovak Republic.

Budinský, M. and L. Jetto (1995). Some Issues on Deadbeat Control and Related Set of CAD procedures. *IFAC Conference on System Structure and Control.* Nantes, France.

Budinský, M. and Š. Kozák (1996). Robust Algebraic Digital Controllers for Systems with Uncertain Physical Parameters. *Proceeding of the second International conference on Process Control Řip 1996,* p.p. 28-34. University of Pardubice, Czech Republic.

Kučera, V. (1979). *Discrete Linear Control: The Polynomial Equation Approach.* Chichester, J. Wiley & son.

Kučera V., (1980).A deadbeat Servoproblem. *Int. J. Control,* **Vol. 32,** pp. 107-113.

SOFTWARE PACKAGE FOR POLYNOMIAL MATRIX OPERATIONS IN CONTROL

Michael Šebek

Institute of Information Theory and Automation
182 08 Prague, Czech Republic
(msebek@utia.cas.cz)

Abstract: A package of MATLAB macros for polynomial matrix operations and equations as well as control design procedures based on them has been developed that is called Polynomial Toolbox. The package covers basic operations with polynomial matrices, special forms of polynomial matrices as well as several alternative routines to solve various linear polynomial matrix equations and polynomial matrix factorization problems. In addition, procedures for solving both standard and modern control design problems for linear multi-input multi-output systems based on polynomial matrix operations and equations are supported. The current contents of the Toolbox is reported together with a definition of the polynomial matrix format.

Keywords: Linear systems; Polynomial methods; Polynomial matrices; Software package; Numerical algorithms.

1. INTRODUCTION

Polynomial techniques are now recognized as one of two main theoretical approaches to linear control theory. They have also been successfully used to solve various problems of signal filtering. It is characteristic for the approach that design procedures are reduced to the solution of linear and quadratic equations in polynomials or polynomial matrices. Hence it is relatively simple and easy to learn for students. On the other hand, some work is still to be done in related numerical algorithms for the matrix polynomial algorithms. To solve each particular equation, there usually existed several routines of different nature that are not yet tested, compared and evaluated. For some special equations, computational methods are missing. Hence, to create a toolbox of unified and well tested numerical algorithms seemed to be the next logical step to do in the field. The resulting program package is called Polynomial Toolbox. It is available to help the students as well as practicing control engineers and filter designers to profit from the advantages of the polynomial methods.

2. POLYNOMIAL TOOLBOX ON WWW

At the time of the conference, the first public version of Polynomial Toolbox will be released. It will be made available as a freeware. The readers are welcome to visit us[1], to register and download the Toolbox along with its hypertex manual. The preliminary version of the Polynomial Toolbox home page is shown on Fig 1.

3. PROJECT

The Toolbox development was supported by a joint[2] project[3] of the Department of Applied Mathematics at Twente and the Institute of Information Theory and Automation, Prague, Czech Republic. In this project, algorithms were developed for solving different kind of control problems based on a polynomial matrix approach. MAT-

[1] Please try `http://www.math.utwente.nl/~twhuib/pobox/` or contact the paper author via e-mail.

[2] The Polynomial Toolbox team includes Huibert Kwakernaak, Michael Šebek, Didier Henrion, Ferda Kraffer, Sonja Pejchová and Rens Strijbos

[3] Algorithms for CAE based on modern methods in control, EC COPERNICUS project CP 93:2424, 1994-1996.

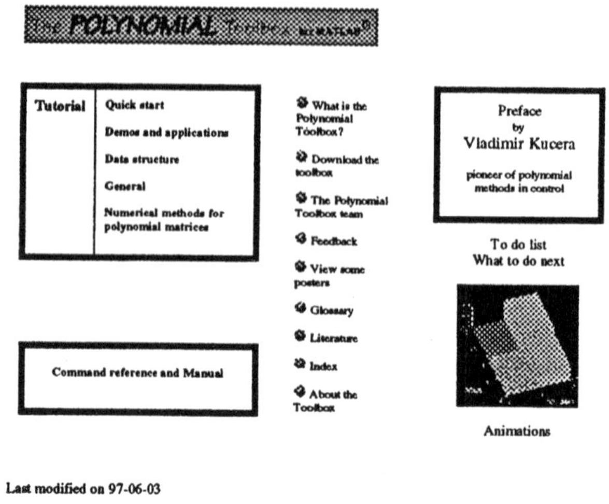

Figure 1: Polynomial Toolbox Web Site

LAB was chosen as a programming environment for developing and testing accurate and reliable algorithms in this research area. Special arrangements have to be made however, to allow comfortable handling of polynomial matrices. This is the subject of the next paragraph. The algorithms together with general supporting routines make a coherent MATLAB-toolbox. Incomplete and preliminary versions of the Toolbox were described in the reports [1],[2] and [3]. A description of some specialized algorithms in this area may be found in [4],[5],[6],[7] and [8]. In this paper, the current contents of the Toolbox is reported.

4. POLYNOMIAL MATRIX FORMAT

The representation of linear control systems may be given in different formats. In the time domain we use the state space description

$$\dot{x} = Ax + Bu, \qquad (1)$$

$$y = Cx + Du, \qquad (2)$$

with A, B, C and D constant matrices of appropriate dimensions, reflecting the number of state variables x, controls u and output y. In the frequency domain we use the polynomial matrix fraction descriptions

$$Y(s) = N(s)D^{-1}(s)U(s), \qquad (3)$$

or

$$Y(s) = P(s)^{-1}Q(s)U(s). \qquad (4)$$

where s is the Laplace variable and $U(s)$ and $Y(s)$ the Laplace transforms of u and y. The relation between the two descriptions is given by

$$N(s)D^{-1}(s) = P^{-1}(s)Q(s) = C(sI - A)^{-1}B + D \qquad (5)$$

The system matrices A, B, C and D are constant matrices which can be handled directly in MATLAB. The polynomial matrices N, D, P and Q however are polynomial matrices. Because MATLAB only implements two-dimensional matrices, there is no ready-to-use data structure available to store polynomial matrices. Different ways are open to store a polynomial matrix in a constant two-dimensional matrix. A first setup may be to store the coefficients of the polynomial matrix in a constant matrix and the degree of the polynomial matrix in a constant, both with names resembling the name of the polynomial matrix, i.e. the coefficients of the polynomial matrix $N(s)$ of degree n

$$N(s) = N_0 s^0 + N_1 s^1 + \cdots + N_n s^n \qquad (6)$$

are stored in a constant matrix N

$$N = [\ N_0 \mid N_1 \mid \cdots \mid N_n\ ] \qquad (7)$$

and the degree in the variable

$$\deg N = n \qquad (8)$$

Another way to represent a polynomial matrix in MATLAB is to store both the coefficients and the degree in a single constant matrix. The advantage is clear. All the information about the coefficients and the degree of the polynomial matrix is stored in one single data element. The disadvantage is of course one has to invent a way to differentiate between a constant matrix and a polynomial matrix both stored in the same way. A similar problem is encountered by the designers of the μ-analysis toolbox. They have chosen the permanent variable Inf as a tag to differentiate between different type of MATLAB data elements. We follow a similar approach by choosing the permanent variable NaN as a mark. As a result the polynomial matrix $N(s)$ may be stored in the single MATLAB element N given by

$$N = \left[\begin{array}{c|c|c|c|c} N_0 & N_1 & \cdots & N_n & \begin{matrix} n \\ 0 \\ \vdots \\ \vdots \\ 0 \end{matrix} \\ \hline 0 \cdots 0 & 0 \cdots 0 & 0 \cdots 0 & 0 \cdots 0 & NaN \end{array}\right] \qquad (9)$$

The coefficient matrix N of equation (7) is extended with an extra column and row. The degree of the polynomial matrix is placed in the first position and the mark NaN in the last position of the additional column. Note we need an extra row to handle polynomial matrices consisting of a single row only. All unused elements in the extra row and column are filled with zeros and do not play any role at this time. They may be used to discriminate between constant and polynomial matrices.

To comply with its mathematical definition ($\deg 0 - \infty$), a zero polynomial matrix is defined by $n = -\mathtt{Inf}$ and $N_0 = 0$.

Summing up, we distinguish two type of matrices: constant matrices, stored in a regular MATLAB two-dimensional data structure; and polynomial matrices, stored in an extended data structure defined by equation (9).

5. UTILITIES

The very fundamental macros to handle polynomial matrices involve basic identification (**pinfo**), various degrees and coefficients (**pdegco**), packing (**ppack**) and unpacking (**punpck**), creating and modifying (**pget, pput, pcoljoin, prowjoin**) and displaying (**pdisp, pshow, pdp**).

As and example, the polynomial matrix 4 × 4 polynomial matrix $B(s)$ of degree 3

$$B = \begin{bmatrix} 2+2s+2s^2 & 1+s & 1+s+5s^2 & 3+4s^2+3s^3 \\ 0 & 5+4s & 1+4s^2 & 1 \\ 0 & 0 & 3 & 4+s \\ 0 & 0 & 0 & 5 \end{bmatrix}$$

can be displayed as follows:

```
>> % Its internal description looks quite awful
>> B

B =

  Columns 1 through 12

    2     1     1     3     2     1     1     0     2     0     5     4
    0     5     1     1     0     4     0     0     0     0     4     0
    0     0     3     4     0     0     0     1     0     0     0     0
    0     0     0     5     0     0     0     0     0     0     0     0
    0     0     0     0     0     0     0     0     0     0     0     0

  Columns 13 through 17

    0     0     0     3     3
    0     0     0     0     0
    0     0     0     0     0
    0     0     0     0     0
    0     0     0     0   NaN

>> % but one can display it by its matrix coefficients
>> pshow(B)
            (4 x 4) polynomial matrix of degree 3.

    Coefficient matrix at power  0
    2     1     1     3
    0     5     1     1
    0     0     3     4
    0     0     0     5

    Coefficient matrix at power  1
    2     1     1     0
    0     4     0     0
    0     0     0     1
    0     0     0     0
%


%
    Coefficient matrix at power  2
    2     0     5     4
    0     0     4     0
    0     0     0     0
    0     0     0     0

    Coefficient matrix at power  3
    0     0     0     3
    0     0     0     0
    0     0     0     0
    0     0     0     0
```

```
>>
```

In addition, one can get its 2-D plot (Fig 1) or

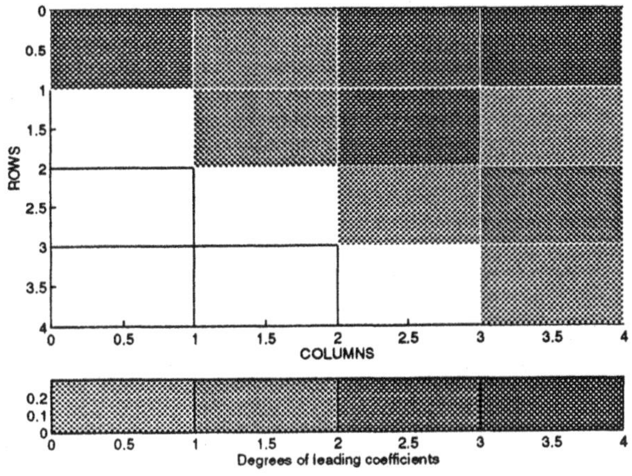

Figure 2: 2-D plot of the matrix $B(s)$.

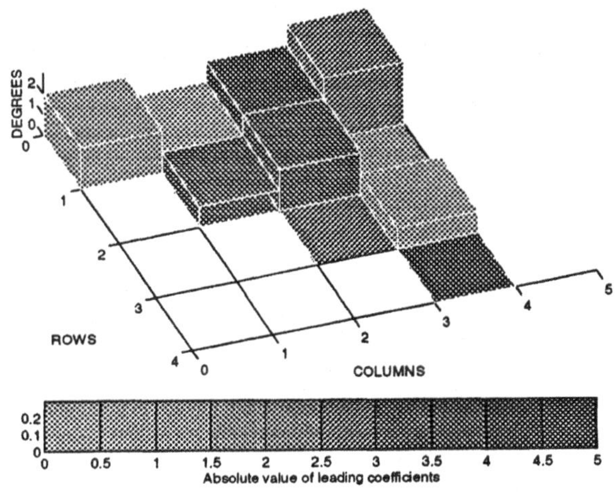

Figure 3: 3-D plot of the matrix $B(s)$.

6. BASIC OPERATIONS AND FORMS

All basic mathematical operations with polynomial matrices are supported such as addition (**padd**), subtraction (**psub**), multiplication (**pmul**), division (**prdiv**) and scaling (**pscl**). A matrix can be transformed to various particular forms including (row, column, diagonal) reduced (**pred**), triangular (**stairs**), diagonal (**pdiagonal**), Hermite (**hermite**) and Smith (**smith**).

7. MODELING FUNCTIONS

Left and right polynomial matrix fraction to state space conversion (**lmf2ss, rmf2ss**), state space to left and right polynomial matrix fraction conversion (**ss2lmf, ss2rmf**).

8. EQUATIONS

Various macros apply to solve typical linear equations with polynomial matrices such as $AX + BY = C$, $XA + YB = C$, $AX + YB = C$, $AX = B$, $XA = B$, $A^*X + X^*A = 2B$, etc.

9. FACTORIZATION

Spectral factorization of various types will also be supported such as $XX^* = B$ (the positive definite case), $XJX^* = B$ (J-spectral factorization) and $XY = B$ (the non-symmetric case).

10. SPECIAL FUNCTIONS

Special polynomial matrix functions supported include null space (**pnull**), determinant (**pdet**) and roots (**proots**), adjoint matrix (**adj**), matrix evaluation (**pval**), common divisors and multiples (**gld, grd, llm, lrm**).

11. NUMERICAL ALGORITHMS

Typical numerical algorithms employed in Polynomial Toolbox include elementary operations, indeterminate coefficients, interpolation and state-space-based procedures. In many routines, several algorithms are offered that can be chosen via string input arguments.

12. CONTROL DESIGN

Control design methods to be included in the next step will include the very best of polynomial theory of linear control. Typical representatives are: the Youla-Kučera parametrisation, pole placement, deadbeat control, H_2 optimal design and H_∞ design via polynomial techniques.

13. APPLICATION

The Toolbox is currently being used by the paper author in its course of linear control theory. The following short MATLAB session illustrates how it works.

Here a continuous system is sampled with a sample interval of 0.25 [s] and a dead-beat compensator, represented by the right matrix fraction description $Y(z)X^{-1}(z)$, is designed using macros of the Polynomial Toolbox. For clearness the macros which are part of the Polynomial Toolbox are written in capital letters.

```
% Dead-beat compensator design
% Define the continuous time system

A=[0 1  0; 0 0  0; 0 0 -2];
B=[0 0; 1 0; 0 1];
C=[1 0 1; 0 0 1];
D=[0 0; 0 0];

% Make discretization via
% Control System Toolbox

[a,b,c,d]=c2dm(A,B,C,D,0.25,'zoh');

% Calculate row-reduced left-prime
% representation of discrete time
% system and display result

[N,D]=SS2LMF(a,b,c,d);
PDISP(N),PDISP(D)

Columns 1 through 2

          0                  0.197
 0.0312 + 0.0312s      -0.274 + 0.197s


Columns 1 through 2

          0               -0.607 + s
  1 - 2s + s^2              -0.155

% Define the closed-loop characteristic
% polynomial matrix

C=[0 0  0 0  1 0  0 0  0 0
   0 0  0 0  0 0  0 0  0 1 ];
C = PPCK(C,4);

% Solve the closed-loop characteristic
% matrix equation D(z)X(z)+N(z)Y(z)=C

[X,Y]=AXBYC(D,N,C);

% Make state space representation
% of the compensator

[f,g,k,h]=RMF2SS(Y,X);

% construct the closed loop system

[as,bs,cs,ds]=series(a,b,c,d,f,g,k,h);
[ac,bc,cc,dc]=cloop(as,bs,cs,ds,-1);
```

```
% Plot impulse response matrix
dimpulse(ac,bc,cc,dc);
```

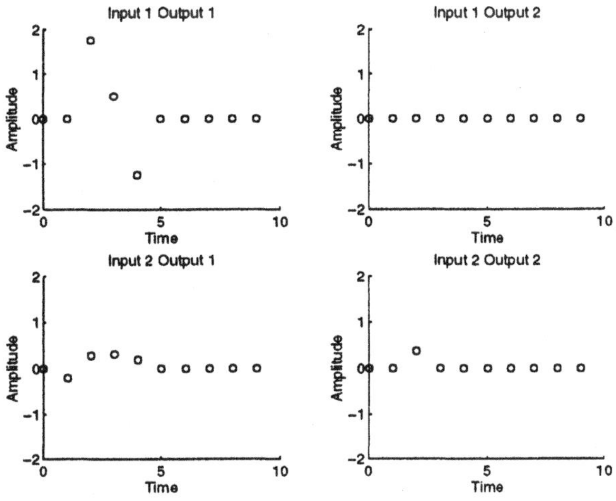

References

[1] Kwakernaak, H. (1990). MATLAB macros for polynomial H_∞ control system optimization. Memorandum 881. Department of Applied Mathematics, University of Twente.

[2] Kwakernaak, H. (1993). State space algorithms for polynomial matrix computations. Memorandum 1168. Department of Applied Mathematics, University of Twente.

[3] Šebek, M. (1990). An algorithm for spectral factorization of polynomial matrices with any signature. Memorandum 912. Department of Applied Mathematics, University of Twente.

[4] Kraffer, F., S. Pejchová and M. Šebek (1994). Polynomial operations: Numerical performance in Matrix Diophantine equations. *Kybernetika* vol. 30, no. 6, pp 745-753, Prague.

[5] Pejchová, S. (1993). Software Package for Polynomial Operations I. Research Report no. 1765. Institute of Information Theory and Automation, Prague.

[6] Šebek, M, S. Pejchová and F. Kraffer (1994). Testing algorithms for linear polynomial equations. Proceedings of the 33rd IEEE Conference on Decision and Control. Orlando, Florida.

[7] Šebek, M., F. Kraus, S. Pejchová and H. Kwakernaak (1995). Numerical Methods for Zeros and Determinant of Polynomial Matrix. Proceedings of 34th IEEE Conference on Decision and Control, New Orleans.

[8] Strijbos, R.C.W. (1996) Calculation of Right Matrix Fraction Descriptions; an Algorithm. Proceedings 4th IEEE Mediterranean Symposium, Crete, Greece.

NEW ASYMPTOTIC STABILITY CRITERIONS FOR NONLINEAR SYSTEMS IN CRITICAL CASES

Liu Xiangdong Fang Bo Huang Wenhu

*Department of Astronautical Engineering and Mechanics,
Harbin Institute of Technology, PRC*

Abstract: In this paper, asymptiotic stability of a nonlinear system in critical cases (NSICC) is studied. Two asymptotic stability criterions for NSICC with either a pair of pure imaginary eigenvalues or a simple zero eigenvalue are obtained by means of Lyapunov stability analysis and center manifold reduction technology . The two criterions proposed here mainly have two advantages: (1)the computation concerns only the second order derivatives instead of tedious manipulation of matrix; (2)the representation of the criterions reveals the relationship between the asymptotic stability of a NSICC and the modes of its linearized system.

Keywords: nonlinear systems, stability analysis, asymtotic stability, lyapunov methods, stability criterion , modes.

1. INTRODUCTION

Recently, nonlinear systems in critical cases (NSICC) have been paid significant attention in a large amount of literatures (e.g.Mees and Chua, 1979; Allwright and Mees, 1979; Abed and Fu, 1986, 1987; Behtash and Sastry, 1988; Fu and Abed 1993a, 1993b).The great upsurge of incentive for the study of NSICC is mainly due to two factors : rapid development of dynamical system theory, bifurcation analysis and chaotic dynamics, and the urgent requirement of engineering practice.

Early results on stability in critical cases are obtained by Lyapunov(1966), Malkin(1958), Laselle (1961), and Krasovskii(1963). Roughly speaking, Two stability analysis methods extensively employed are the well-known lyapunov stability analysis and the Project Method (Iooss and Joseph 1980). In the light of construction of a lyapunov function , Mees and Chua(1979) obtained a sufficient condition of stability for a two dimensional system with a pair of pure imaginary eigenvalues, Fu and Abed (1993a) emplicitly construct families of lyapunov functions for assessing the asymptotic stability of NSICC, By appealing to the Project Method, Abed and Fu (1987) derived the bifurcation formulae in their study of NSICC with a simple zero eigenvalue. Although the two methods have been successfully applied in stability analysis of NSICC, they must be examined with some limitations as follows : (1) The familiar shortcoming of Lyapunov stability analysis is that up to now there is still no general approach to construct a lyapunov function for a nonlinear system, (2) the results obtained by the Project Method must carry out tedious derivation and complicated manipulation.

The purpose of this paper is twofold. First we want to obtain the sufficient conditions of asymptotic stability for a NSICC with state variable equation in a more general form, second and more important , we want to find the relationship between the asymptotic stability of a NSICC and the modes of its linearization. More specifically only two critical cases are considered in this paper. In case(I), the linearization is characterized by the occurrence of a simple zero eigenvalue, with all remaining eigenvalues having strictly negative real parts, in case(II) , the linearization is assumed to possess a complex conjugate pair of simple, pure imaginary eigenvalues, with the remaining eigenvalues lying in the open left half complex plane. In the light of Lyapunov stability analysis and center manifold reduction method, this paper gives two asymptotic stability criterions , with a brief , representation and simple computation concerning the second or third order derivatives, for a NSICC with either a pair of pure imaginary eigenvalues or a simple zero eigenvalue, and successfully establishes the bridge between the asymptotic stability of NSICC with the modes of its linearized system.

This paper is organized as follows, section II provides some definitions and the center manifold theorem that will be used later. Main results of this paper are given in section III and section IV , asmptotical stability criterion for a NSICC whose linearization has a simple zero eigenvalue is contained in section III, section IV contains an analogous result for a NSICC in case II . Conclusion is given in section V.

2. MATHEMATICAL PRELIMINARY

Consider the following nonlinear autonomous system

$$\dot{x} = f(x), \qquad x \in R^n \qquad (1)$$

Where $f(x)$ is sufficiently smooth in x , the origin 0 is one equilibrium of (1) , i.e., $f(0) = 0$, $\varphi_t(x_0) = x(t, x_0)$is the flow generated by (1), the linearized system at 0 is given by

$$\dot{\xi} = A\xi, \qquad \xi \in R^n \qquad (2.2)$$

where $A = Df(0)$. Before giving the next result, we need a couple of definitions:

Definition 2.1 a set $s \subset R^n$ is said to be a local invariant set if for all $x_0 \in s$, there exist $T>0$, such that $\varphi_t(x_0) \in s$, for $|t|<T$, if T can be chosen to be ∞, then s is said to be an invariant set.

Definition 2.2 the local stable and unstable manifold of the origin 0, $w^s_{loc}(0), w^u_{loc}(0)$ are defined as follows

$$w^s_{loc}(0) = \{x \in U | \varphi_t(x) \to 0 \quad as \ t \to \infty, \quad and \ \varphi_t(x) \in U \quad for \ all \ t \geq 0\}$$

$$w^u_{loc}(0) = \{x \in U | \varphi_t(x) \to 0 \quad as \ t \to -\infty, \quad and \ \varphi_t(x) \in U \quad for \ all \ t \leq 0\}$$

where $U \subset R^n$ is a neighborhood of the origin 0.

Definition 2.3 the origin 0 is called a hyperbolic equilibrium point , if A has no eigenvalues with zero real part; otherwise, the origin 0 is a nonhyperbolic equilibrium point. In the case of nonhyperbolic equilibrium point, we partition the spectrum of A as:

$$\sigma(A) = \sigma^s \cup \sigma^u \cup \sigma^c$$

where

$$\mathrm{Re}\lambda \begin{cases} < 0 & if \ \lambda \in \sigma^s \\ = 0 & if \ \lambda \in \sigma^c \\ > 0 & if \ \lambda \in \sigma^u \end{cases}$$

let the (generalized) eigenspace of σ^s, σ^c and σ^u be E^s, E^c and E^u respectively, namely, $R^n = E^s \oplus E^u \oplus E^c$. With these definitions , we now state.

Theorem 2.1 (Center Manifold Theorem) Consider the nonlinear system given by (1), let f be a C^r vector field on R^n vanishing at the origin. Then there exist C^r stable and C^r unstable invariant manifolds w^s_{loc} and w^u_{loc} tangent to E^s and E^u at 0 and a C^{r-1} center manifold w^c_{loc} tangent to E^c at 0. The manifolds w^s_{loc} , w^u_{loc} and w^c_{loc} are all invariant for the flow of f. The stable and unstable manifolds are unique, but w^c_{loc} need not be.

For simplicity, and because it is actually the most interesting case, we assume that the unstable manifold is empty:

$$\begin{aligned} \dot{u} &= Au + W(u,v) \\ \dot{v} &= Bv + V(u,v) \end{aligned} \qquad (u,v) \in R^n \times R^m \qquad (3)$$

Where $A \in R^{n \times n}$ and $B \in R^{m \times m}$ have eigenvalues with, respectively, zero real parts and negative real parts. $W(u, v)$ and $V(u, v)$ vanish, alone with their first partial derivatives, at the origin($u = 0$, $v = 0$). Since the center manifold is tangent to E^c (the $v = 0$) we can represent it as a local graph

$$w^c_{loc} = \{(u,v) | v = h(u)\}; Dh(0) = 0; h(0) = 0 \qquad (4)$$

where h: $U \to R^m$ is defined on some neighborhood $U \subset R^n$ of the origin. Taking (4) into (3) we can obtain the reduced system of (3)

$$\dot{u} = Au + W(u,v) \qquad (5)$$

In the vicinity of the origin, since the flow on the stable manifold of (3) has the same characteristic as its linearized contracting flow, the local characteristic of (3) near the origin is then determined by its follow on the center manifold. The following theorem connects the stability of systems (3) to its reduced system (5)

Theorem 2.2 (Guckenheimer and Holmes 1983) if the origin ($u = 0$) of (5) is locally stable (resp., unstable, asymptotically stable), then the origin ($u = 0$, $v = 0$) of (3) is also locally stable (resp., unstable, asymptotically stable).

To solve for $h(u)$, we use the fact that $v = h(u)$ is invariant under the flow of (3), we can obtain

$$Dh(u)[Au + W(u,h(u))] = Bh(u) + V(u,h(u)) \qquad (6a)$$

or

$$N(h(u)) = Dh(u)[Au + W(u,h(u))] - Bh(u) - V(u,h(u)) = 0 \qquad (6b)$$

That is, $h(u)$ must satisfy the partial differential equation (6a) or (6b) with boundary conditions $h(0) =$

0; $Dh(0) = 0$. the partial differential equation (6a) can be approximated arbitrarily closely as a Talor series at $u = 0$

Theorem 2.3 (Guckenheimer and Holmes 1983) if a function $\varphi(u)$, with $\varphi(0) = 0$, $D\varphi(0) = 0$, can be found such that $N(\varphi(u)) = 0(p)$ for some $p>1$ as $| u | \to 0$, then it follows that

$$h(u) = \varphi(u) + 0(p) \qquad as \quad |u| \to 0$$

therefore, Theorem 2.3 allows us to approximate $h(u)$ by a polynomial in u to any desired accuracy.

By means of these results stated above, we are now ready to study the asymptotical stability of NSICC. Since the linear part of the reduced system (5) has all its eigenvalues on jw axis, we need to study the higher order term of the vector field. The following two sections contain the main results of this paper, let us introduce the asymptotic stability criterion for NSICC with a simple zero eigenvalue at first.

3. ASYMPTOTIC STABILITY CRITERION FOR NSICC IN CASE (I)

Firstly, let us consider the local signal difiniteness of a scalar function given by

$$p(x) = p_2 x^2 + p_3 x^3 + p_4 x^4 + \cdots\cdots, x \in R \qquad (7)$$

where $p(0) = 0$, $p(x)$ is sufficiently smooth.

Corollary 3.1 (1) if $p_2>0$, then there must exist a neighborhood κ of the origin ($x=0$) , for any $x \in \kappa$, $p(x)>0$; (2) if $p_2<0$, then there must exist a neighborhood κ' of the origin ($x = 0$), for any $x \in \kappa'$, $p(x)<0$.

Consider a scalar nonlinear system given by

$$\dot{x} = f(x), \qquad x \in R \qquad (8)$$

Where $f(x) = \delta x^2 + \varepsilon x^3 + 0(3), f(0) = 0, f(x)$ is sufficiently smooth in x. 0(3) stands for higher terms whose order is larger than 3.

Theorem 3.1: Consider the system(8) the stability of the equilibrium point ($x = 0$) is given by the following statement:

(1) if $\delta \neq 0$, the equilibrium point of (8) is locally unstable;

(2) if $\delta = 0$, and if $\varepsilon > 0$, then the equilibrium point of (8) is locally unstable; but if $\varepsilon < 0$, the equilibrium point of (8) is locally asymptotically stable.

(proof): (1) Without loss of generality, suppose $\delta > 0$, we can select a lyapunov function, $v(x) = \ln(x+1)$, it is obvious that $v(x)$ is positively definite for $|x| < 1$, and according to corollary (3.1), there must exist a $\varepsilon_0 > 0$, for $x < \varepsilon_0, f(x) > 0$, $\varepsilon_1 = \min(1, \varepsilon_0)$, the time derivative of $v(x)$, $\dot{v}(x) = \dfrac{f(x)}{1+x} > 0$, for $|x| < \varepsilon_1$; if $\delta < 0$, then a Lyapunov function for (8) can be chosen as $v(x) = \ln(1-x)$, there must exist a $\varepsilon'_0 > 0$, for $x < \varepsilon'_0, f(x) < 0, \varepsilon'_1 = \min(1, \varepsilon'_0)$, then $v(x)>0$, $\dot{v}(x) = \dfrac{f(x)}{x-1} > 0$, for $|x| < \varepsilon'_1$. Thus, the equilibrium point of (8) is locally unstable for $\delta \neq 0$.

(2) the proof is direct from selecting Lyapunov function $v(x) = x^2$.

this completes the proof.

A NSICC in case (I) can be characterized by the following state variable equation:

$$\begin{cases} \dot{x} = f(x, y) \\ \dot{y} = \Lambda y + g(x, y) \end{cases} \qquad (x, y) \in R \times R^{n-1} \qquad (9)$$

where

$$\Lambda = \begin{bmatrix} -\lambda_1 & & & 0 \\ & -\lambda_2 & & \\ & & \ddots & \\ 0 & & & -\lambda_{n-1} \end{bmatrix},$$

$$g(x,y) = \begin{bmatrix} g_1(x, y_1, y_2, \cdots, y_{n-1}) \\ g_2(x, y_1, y_2, \cdots, y_{n-1}) \\ \vdots \\ g_{n-1}(x, y_1, y_2, \cdots, y_{n-1}) \end{bmatrix}$$

$\mathrm{Re}\{\lambda_i\} \neq 0$, $\mathrm{Im}\{\lambda_i\}=0$, $i=1,2,\cdots,n-1$, $f(x, y)$ and $g(x, y)$ are sufficiently smooth , $f(x, y)$ and $g(x, y)$ vanish, alone with their first order partial derivatives , at the origin ($x = 0$, $y = 0$) , According to the Center Manifold Theorem, there must exist a center manifold

$$y = h(x) = \begin{bmatrix} h_1(x) \\ h_2(x) \\ \vdots \\ h_{n-1}(x) \end{bmatrix} \qquad (10)$$

$h(x)$ must satisfy the boundary conditions $h(0)$ 0 , $Dh(0) = 0$.

Then the asymptotical stability criterion for a NSICC in case (I) is given by the following theorem.

Theorem. 3.2 Consider a nonlinear system characterized by (9) , and let

$$a = f_{11} \qquad (11)$$

$$\Delta_1 = \frac{1}{3!} f_{111} + \frac{1}{2} \sum_{i=1}^{n-1} \frac{1}{\lambda_i} f_{1i} g^i_{11} \qquad (12)$$

where

$$f_{11} = \left. \frac{\partial^2 f}{\partial x^2} \right|_{x=0}, f_{111} = \left. \frac{\partial^3 f}{\partial x^3} \right|_{x=0},$$

$$f_{1i} = \left. \frac{\partial^2 f}{\partial x \partial y_i} \right|_{\substack{x=0 \\ y=0}}, g^i_{11} = \left. \frac{\partial^2 g_i}{\partial x^2} \right|_{\substack{x=0 \\ y=0}}$$

(1) if $a \neq 0$, then the equilibrium point of (9) is locally unstable;

(2) if $a = 0$, and if $\Delta_1 > 0$, the equilibrium point of (9) is locally unstable;

but if $\Delta_1 < 0$, the equilibrium point of (9) is locally asymptotically stable.

(proof): According to the fact that the center manifold is invariant under the flow generated by (9), $y = h(x)$ satisfy the following partial differential equation

$$Dh(x) f(x, h(x)) = \Lambda h(x) + g(x, h(x)) \qquad (13)$$

We can obtain from (13) that

$$h_{11}^i = \frac{1}{\lambda_i} g_{11}^i \quad (14a)$$

$$h_{111}^i = \frac{1}{\lambda_i} g_{111}^i - 3 f_{11} h_{11}^i + \frac{3}{\lambda_i} \sum_{j=1}^{n-1} g_{1j}^i \cdot h_{11}^j \quad (14b)$$

Thus, substitution (14a), (14b) and (10) into (9), we can obtain the reduced system of (9) as follows:

$$F(x) = \frac{a}{2} x^2 + \Delta_1 x^3 + 0(3) \quad (15)$$

It follows at once from **theorem 2.2** and **corollary 3.1.**

4. ASYMPTOTIC STABILITY CRITERION FOR NSICC IN CASE (II)

Consider a NSICC in case (II), which is characterized by the following nonlinear system

$$\begin{aligned} \dot{x} &= Wx + \tilde{f}(x,y) \\ \dot{y} &= \Lambda y + \tilde{g}(x,y) \end{aligned} \quad (x,y) \in R^2 \times R^{n-2} \quad (16)$$

where

$$W = \begin{pmatrix} 0 & -w \\ w & 0 \end{pmatrix}, \tilde{f}(x,y) = \begin{pmatrix} \tilde{f}_1(x_1, x_2, y_1, \cdots, y_{n-2}) \\ \tilde{f}_2(x_1, x_2, y_1, \cdots, y_{n-2}) \end{pmatrix}$$

$$\Lambda = \begin{bmatrix} -\lambda_1 & & & 0 \\ & -\lambda_2 & & \\ & & \ddots & \\ 0 & & & -\lambda_{n-2} \end{bmatrix}$$

$$\tilde{g}(x,y) = \begin{bmatrix} \tilde{g}_1(x_1, x_2, y_1, \cdots, y_{n-2}) \\ \vdots \\ \tilde{g}_{n-2}(x_1, x_2, y_1, \cdots, y_{n-2}) \end{bmatrix}$$

$\mathrm{Re}\{\lambda_j\} \neq 0, I_m\{\lambda_j\} = 0, j = 1,2,\cdots,n-2, \tilde{f}(x,y)$ and $\tilde{g}(x,y)$ are sufficiently smooth. According to theorem 2.1, there exist a center manifold

$$y = h(x) = \begin{pmatrix} h_1(x_1, x_2) \\ \vdots \\ h_{n-2}(x_1, x_2) \end{pmatrix} \quad (17)$$

$h(x)$ must satisfy the partial differential equation

$$Dh(x)[Wx + \tilde{f}(x, h(x))] = \Lambda h(x) + \tilde{g}(x, h(x)) \quad (18)$$

with boundary conditions $h(0) = 0$, $Dh(0) = 0$.

Then, the asymptotic stability criterion for NSICC in Case (II) is characterized by the following result.

Theorem 4.1: Consider a NSICC in case (II) descripted by (4.1), let

$$16\Delta_2 = \Delta_n + \sum_{j=1}^{n-2} l_j^T q_j r_j \quad (19)$$

where

$$\Delta_n = \tilde{f}_{111}^1 + \tilde{f}_{122}^1 + \tilde{f}_{112}^2 + \tilde{f}_{222}^2 + \frac{1}{w}\{\tilde{f}_{12}^1(\tilde{f}_{11}^1 + \tilde{f}_{22}^1) - \tilde{f}_{12}^2(\tilde{f}_{11}^2 + \tilde{f}_{22}^2) - \tilde{f}_{11}^1 \tilde{f}_{11}^2 + \tilde{f}_{22}^1 \tilde{f}_{22}^2\}$$

$$l_j = \begin{bmatrix} \tilde{g}_{11}^j \\ \tilde{g}_{12}^j \\ \tilde{g}_{22}^j \end{bmatrix}, \quad r_j = \begin{bmatrix} \tilde{f}_{1j}^1 \\ \frac{1}{2}(\tilde{f}_{2j}^1 + \tilde{f}_{1j}^2) \\ \tilde{f}_{2j}^2 \end{bmatrix}$$

$$q_j = \frac{1}{\lambda_j(\lambda_j^2 + 4w^2)} \begin{bmatrix} 3\lambda_j^2 + 8w^2 & -4w\lambda_j & \lambda_j^2 + 8w^2 \\ 4w\lambda_j & 4\lambda_j^2 & -4w\lambda_j \\ \lambda_j^2 + 8w^2 & 4w\lambda_j & 3\lambda_j^2 + 8w^2 \end{bmatrix}$$

$$\tilde{f}_{lmn}^p = \left.\frac{\partial^3 \tilde{f}_p}{\partial x_l \partial x_m \partial x_n}\right|_{\substack{x=0\\y=0}} \qquad \tilde{f}_{lm}^p = \left.\frac{\partial^2 \tilde{f}_p}{\partial x_l \partial x_m}\right|_{\substack{x=0\\y=0}}$$

$$\tilde{g}_{lm}^j = \left.\frac{\partial^2 \tilde{g}_j}{\partial x_l \partial x_m}\right|_{\substack{x=0\\y=0}} \qquad \tilde{f}_{lj}^p = \left.\frac{\partial^2 \tilde{f}_p}{\partial x_l \partial y_j}\right|_{\substack{x=0\\y=0}}$$

$$l,m,n,p = 1,2 \qquad j = 1,2,\cdots,n-2$$

if $\Delta_2 \neq 0$, it follows that:

(1) the equilibrium point of (16) is locally asymptotically stable, if $\Delta_2 < 0$;

(2) the equilibrium point of (16) is locally unstable, if $\Delta_2 > 0$.

In order to proof **theorem(4.1)**, we need some other results at first. Consider the following two-dimensional autonomous nonlinear system

$$\begin{cases} \dot{z}_1 = -wz_2 + \bar{f}_1(z_1, z_2) \\ \dot{z}_2 = wz_1 + \bar{f}_2(z_1, z_2) \end{cases} \quad (z_1, z_2) \in R \times R \quad (20)$$

where, $w > 0$, $\bar{f}_1(0,0) = \bar{f}_2(0,0) = 0, D\bar{f}_1(0,0) = D\bar{f}_2(0,0) = 0$, let

$$16\sigma = \frac{1}{w}\{\bar{f}_{12}^1(\bar{f}_{11}^1 + \bar{f}_{22}^1) - \bar{f}_{12}^2(\bar{f}_{11}^2 + \bar{f}_{22}^2) - \bar{f}_{11}^1 \bar{f}_{11}^2 + \bar{f}_{22}^1 \bar{f}_{22}^2\} + \bar{f}_{111}^1 + \bar{f}_{122}^1 + \bar{f}_{112}^2 + \bar{f}_{222}^2 \quad (21)$$

Theorem 4.2 suppose $\sigma \neq 0$, then the following statements about the stability of the equilibrium point of (20) is true:

(1) the equilibrium point of (20) is locally asymptotically stable, if $\sigma < 0$;

(2) the equilibrium point of (20) is locally unstable, if $\sigma > 0$.

(proof) According the Lemma 1 in Mees and Chua (1979), the proof is obvious. Next, let us proof the theorem 4.1

(proof):

since $h(x)$ must satisfy the partial differential equation

$$Dh(x)(wx + \tilde{f}(x, h(x))) = \Lambda h(x) + \tilde{g}(x, h(x)) \quad (22)$$

after some lengthy manipulations, we can obtain

$$h_{11}^i = \frac{2}{\lambda_i(\lambda_i^2 + 4w^2)}[(\frac{\lambda_i^2}{2} + w^2)\tilde{g}_{11}^i - w\lambda_i \tilde{g}_{12}^i + w^2 \tilde{g}_{22}^i] \quad (23a)$$

$$h_{12}^i = \frac{1}{\lambda_i(\lambda_i^2 + 4w^2)}[w\lambda_i \tilde{g}_{11}^i + \lambda_i^2 \tilde{g}_{12}^i - w\lambda_i \tilde{g}_{22}^i] \quad (23b)$$

$$h_{22}^i = \frac{2}{\lambda_i(\lambda_i^2 + 4w^2)}[w^2 \tilde{g}_{11}^i + w\lambda_i \tilde{g}_{12}^i - (\frac{\lambda_i^2}{2} + w^2)\tilde{g}_{22}^i] \quad (23c)$$

Taking (23a) ~ (23c) into (16), we can obtain the reduced system of (16)

$$\begin{cases} \dot{x}_1 = -wx_2 + \hat{f}_1(x_1, x_2) \\ \dot{x}_2 = wx_1 + \hat{f}_2(x_1, x_2) \end{cases} \quad (24)$$

We can derive Δ_2 from (17) and (24), by **Theorem**

2.2 and **Theorem 4.2** , we can obtain the result. This complete the proof.

5. CONCLUSION

Two new asymptotic stability criterions for NSICC whose linearization possesses either a pair of pure imaginary eigenvalues or a simple zero eigenvalue, are presented in this paper. compared with the other results, these criterions mainly have two advantages:

(1) computation concerns only the second or third derivatives of vector fields;

(2) the simple representation of the two criterions deeply reveals the close relation ship between the asymptotic stability of NSICC and the modes of its linearization.

REFERENCES

Allwright, D.J., and A.I. Mees. (1979). Lyapunov method applied to the Hopf bifurcation, *J.London Math. Soc.* , **2**, 293-299.

Abed, E.H. and J.H. Fu. (1986). Local feedback stabilization and bifurcation control, I.Hopf bifurcation. *system & Control letters*, 7, 11-17.

Abed, E.H., and J.H. Fu. (1987). Local feedback stabilization and bifurcation control, II. Stationary bifurcation, *system & control letters*, **8**, 467-473.

Behtash, S., and S.S. Sastry. (1988). Stabilization of nonlinear systems with uncontrollable linearization, *IEEE Trans. Aut. Control*, **33**, 585-590.

Fu, J.H., and E.H. Abed. (1993a). Families of Lyapunov functions for nonlinear systems in critical cases, *IEEE Trans, Aut, Control*, **38**, 3-16.

Fu, J.H., and E.H. Abed, (1993b). Linear Feedback Stabilization of Nonlinear systems with an uncontrollable critical Mode, *Automatica,* **29**, 999-1010.

Guckenheimer, J., and P. Holmes (1983). *Nonlinear Oscillations, Dymamical systems and Bifurcations of Vector Fields*. Springer-Verlag, New York.

Iooss, G., and D.D. Joseph. (1980). *Elementary Stability and Bifurcation Theory*. Springer-Verlag, New York.

Krasoviskii, N.N. (1963). *Stability of Motion: Applications of Lyapunov's Second Method to differential systems and equations with delay.* Stanford Univ. Press, Stanford.

LaSalle, J.P., and S.Lefschetz (1961). *Stability by Liapunov's Direct Method with Applications* Academic Press, New York

Lyapunov, A. M. (1966). *Stability of Motion* Academic Press, New York

Malkin, I.G.(1958). Some basic Theorem of the stability of motion in the critical case, *AMS Translations* , **38**.

Mees , A.I., and L.O. Chua. (1979). The Hopf bifurcation theorem and its applications to nonlinear oscillations in circuits and systems , *IEEE Trans. Circuits syst.* **26**, 235-254.

TERMINALLY STABLE SLIDING MODE CONTROL FOR SECOND-ORDER SYSTEMS

Andrzej Bartoszewicz

Institute of Automatic Control, Technical University of Łódź
18/22 Stefanowskiego St., 90-924 Łódź, Poland
email: andpbart@lodz1.p.lodz.pl

Abstract: In this paper a nonlinear and time-varying switching line is introduced and a novel variable structure control algorithm for second-order dynamic system subject to unknown disturbance is designed. The algorithm eliminates reaching phase and it guarantees monotonic error convergence to zero in finite time, without violating constraints on the state of the system. A new technique to attenuate chattering in the system is presented. The technique employs simple, first-order dynamical devices, i.e. an integrator or a low-pass filter, to compensate in part for unknown and changing disturbance. As a result of this compensation, the magnitude of the discontinuous control term is essentially reduced without any deterioration of the system performance.

Keywords: Variable structure systems, sliding mode control, switching surfaces.

1. INTRODUCTION

In recent years much of the research in the area of control systems theory focused on the design of a discontinuous feedback which switches the structure of the system according to the evolution of its state vector. This technique, usually called sliding mode control, provides an effective and robust means of controlling nonlinear plants (DeCarlo *et al.*, 1988; Hung *et al.*, 1993). The main advantage of this technique is that once the system state reaches a sliding surface, the system dynamics remain insensitive to a class of parameter variations and disturbances.

However, robust tracking is guaranteed only after the system state reaches the sliding surface i.e. robustness is not guaranteed during the reaching phase. Provided a conventional time-invariant sliding plane is considered, the advantage of the sliding mode control, namely the desired dynamic behaviour of the system, is not obtained for some time since the beginning of motion. Furthermore, usually for the given initial conditions there is a trade-off between the short reaching phase and the fast system response in the sliding phase. In order to overcome these problems the idea of the time-varying switching planes was introduced in (Choi *et al.*, 1994) and further discussed in (Bartoszewicz, 1995). The control algorithm proposed by Bartoszewicz (1995) eliminates the reaching phase and guarantees fast error convergence rate for second order uncertain systems with arbitrary initial conditions. However, similarly to the conventional algorithm with the fixed switching plane, both control algorithms proposed by Choi, *et al.* (1994) and Bartoszewicz (1995) can only assure asymptotic error convergence to zero.

In order to enhance convergence properties of dynamical systems, the concept of terminal attractors was introduced in (Zak, 1988) and further developed in (Venkataraman and Gulati, 1993). It was shown that application of a particular class of time-invariant (fixed) sliding lines to the control of second-order systems results in finite time error convergence. The lines which belong to this class are called terminal sliders and their advantageous properties will be further studied in this paper.

On the other hand, the favourable performance of the sliding mode control systems is usually achieved at the price of chattering, i.e. high-speed switching of the control signal. The chattering is undesirable not only because it may excite unmodelled high-frequency dynamics of the plant, but also because it will result in unnecessary wear and tear on the actuator components. In order to alleviate the problem of chattering continuous approximations of the discontinuous control laws are usually proposed (Rayan and Corless, 1984; Slotine and Li, 1991). In this case, the total invariance properties associated with the ideal sliding mode are lost, the system performance (at least to some extent) deteriorates and neither finite-time nor asymptotic error convergence to zero can be ensured.

In this paper a second-order nonlinear system with constraints on its state variables is considered. A continuously time-varying nonlinear switching line adaptable to initial conditions of the system is introduced and the existence of a sliding mode on the line is proved. The line is designed in such a way that the system error converges fast, and reaches zero in finite time. At time $t=t_0$ the line passes through the representative point specified by the initial conditions of the system in the error state space. Afterwards the line moves smoothly, in finite time, to reach its predetermined shape and location. By this means it is possible to avoid reaching phase and to force the representative point of the system to stay always on the line. As a consequence our control is robust with respect to external disturbance and parameter uncertainties from the very beginning of the motion, and it ensures finite time error convergence of the considered system. After presenting our basic control scheme, we propose a new technique to attenuate chattering without compromising the system performance. The technique is based on the observation that when the system is in the sliding mode, information about the disturbance can be extracted from the equivalent value of the discontinuous control. Therefore, we introduce an auxiliary continuous control signal which partly compensates for the disturbance. The signal is generated by a single integrator or a low-pass filter according to the equivalent value of the discontinuous control term. Consequently, the discontinuous term can be essentially reduced without any deterioration of the system performance. In fact the magnitude of the discontinuous control term is determined only by the uncertainty of the disturbance at the initial time $t=t_0$.

The remainder of the paper is organised as follows. The problem considered in the paper is formulated in Section 2. Section 3 presents details of the switching line design procedure and demonstrates the performance of the system restricted to the line. Control laws appropriate to attenuate the undesirable chattering in the system and to maintain its representative point on the line are proposed in Section 4. Finally, Section 5 comprises conclusions of the paper.

2. PROBLEM STATEMENT

Let us consider a single input single output, time-varying and nonlinear, second-order system

$$\begin{aligned} \dot{x}_1 &= x_2 \\ \dot{x}_2 &= f(x_1, x_2, t) + bu + d(t) \end{aligned} \tag{1}$$

subject to the state constraint

$$|x_1| \le X_1 \tag{2}$$

$$|x_2| \le X_2 \tag{3}$$

where X_1 and X_2 are a'priori known positive constants. In equation (1) x_1 and x_2 are the state variables of the considered system, t denotes time, f is a known nonlinear function of time and the system state, $b\neq 0$ is a known constant, d(t) is an unknown disturbance and u is the control input. The system is supposed to track the demand trajectory $x_1(t)= x_d(t)$ which is determined in such a way that

$$|x_d(t)| \le X_{1d} \tag{4}$$

$$|\dot{x}_d(t)| \le X_{2d} \tag{5}$$

where X_{1d} and X_{2d} are known positive constants and $X_{1d} < X_1$, $X_{2d} < X_2$. Let us denote

$$\beta_1 = X_1 - X_{1d} \tag{6}$$

and

$$\beta_2 = X_2 - X_{2d} \tag{7}$$

Furthermore, define the trajectory tracking error e(t) $= e_1(t) = x_1(t) - x_d(t)$ and its derivative de(t)/dt $= e_2(t) = dx_1(t)/dt - dx_d(t)/dt$. In this paper we assume that at the initial time $t=t_0$, the tracking error is $-\beta_1 \le e(t_0) = e_0 \le \beta_1$ and its derivative $e_2(t_0) = 0$. Notice that the assumption $e_2(t_0) = 0$ holds perfectly for the most typical regulation problem of the system which is initially in a steady state and its desired output changes instantaneously between two constant values. An example of such a problem is regulation of link positions in point-to-point robot arm control.

The purpose of this paper is to design a sliding mode control strategy for the system (1) which causes as little chattering in the system as possible and:
a) Makes the system insensitive to the disturbance

d(t) for any time $t \geq t_0$, i.e. from the very beginning of the system motion.

b) Drives the system error to zero monotonically and in a finite time. In other words, we require terminal stability without oscillations, overshooting or undershooting.

c) Satisfies constraints (2) and (3) for any initial tracking error, and any admissible demand trajectory (that is for any trajectory for which relations (4) and (5) hold).

3. SWITCHING LINE DESIGN

Let us introduce the following nonlinear and time-varying switching line

$$s(e_1, e_2, t) = 0 \tag{8}$$

where

$$s(e_1, e_2, t) = e_2(t) + sgn[e_1(t)]\alpha(t)[e_1(t)]^{2/3}$$
$$\alpha(t) = \begin{cases} a(t - t_0) & \text{for} \quad t_0 \leq t \leq t_f \\ a(t_f - t_0) & \text{for} \quad t > t_f \end{cases} \tag{9}$$

and a is a positive constant. If the representative point of the system (1) is restricted to lie on the line (8) then evolution of the system tracking error is governed by the following equation

$$\frac{de}{dt} = -sgn[e(t)]\alpha(t)[e(t)]^{2/3} \tag{10}$$

with the initial condition $e(t_0) = e_0$. Solving this equation and assuming $t_f \geq t_0 + \sqrt{6a^{-1}|e_0^{1/3}|}$, we get

$$e(t) = \begin{cases} \left[e_0^{1/3} - \frac{1}{6} a\, sgn(e_0)(t - t_0)^2\right]^3 & \text{for} \quad t < t_0 + \sqrt{6a^{-1}|e_0^{1/3}|} \\ 0 & \text{for} \quad t \geq t_0 + \sqrt{6a^{-1}|e_0^{1/3}|} \end{cases} \tag{11}$$

and

$$\dot{e}(t) = \begin{cases} -sgn(e_0)a(t - t_0)\left[e_0^{1/3} - \frac{1}{6} a\, sgn(e_0)(t - t_0)^2\right]^2 & \text{for} \quad t < t_0 + \sqrt{6a^{-1}|e_0^{1/3}|} \\ 0 & \text{for} \quad t \geq t_0 + \sqrt{6a^{-1}|e_0^{1/3}|} \end{cases} \tag{12}$$

Relations (11) and (12) clearly show that the system error monotonically converges to zero in finite time $t = t_0 + \sqrt{6a^{-1}|e_0^{1/3}|}$. Since $|e_0| \leq \beta_1$, taking into account relations (11) and (12), we conclude that for any $t \geq t_0$, $|e(t)| \leq \beta_1$, and consequently constraint (2) is always satisfied. On the other hand, constraint (3) is satisfied if the following inequality holds

$$|\dot{e}(t)| \leq \beta_2 \tag{13}$$

Notice that relation (13) is sufficient but not a necessary condition to satisfy constraint (3). Calculating maximum of the absolute value of the system error derivative we get

$$MAX[|\dot{e}(t)|] = \left|\dot{e}\left(t = t_0 + \sqrt{1.2 \cdot a^{-1}|e_0^{1/3}|}\right)\right| = \frac{16}{25} e_0^{2/3} \sqrt{1.2 \cdot a |e_0^{1/3}|} \tag{14}$$

and therefore, to satisfy relations (13) and (3) we choose

$$a = \frac{3125(\beta_2)^2}{1536|e_0|^{5/3}} \tag{15}$$

This choice of the constant *a* not only guarantees the shortest error convergence time, but also minimises the integral of the absolute value of the tracking error over the whole period of the control action. This can be easily verified as follows

$$J = \int_{t_0}^{\infty} |e(\tau)| d\tau = \int_{t_0}^{t_0 + \sqrt{6a^{-1}|e_0^{1/3}|}} \left[e_0^{1/3} - \frac{1}{6} a\, sgn(e_0)(t - t_0)^2\right]^3 d\tau \tag{16}$$

$$J = \frac{16}{35} e_0^{2/3} \sqrt{6a^{-1}|e_0|} \tag{17}$$

Equation (17) shows that increasing a results in decreasing criterion J. Therefore, putting a given by equation (15) we get the smallest integral of the absolute value of the system tracking error achievable without violating constraints (13) and (3). If the initial value of the tracking error is known precisely then this value is substituted for e_0 in equation (15). Otherwise, i.e. when this value is not known exactly, then in order to satisfy constraint (3) even in the case of the most adverse initial conditions we substitute β_1 for e_0. By this means we get

$$a = \frac{3125(\beta_2)^2}{1536(\beta_1)^{5/3}} \tag{18}$$

and in order to avoid excessive controller gains we choose

$$t_f = t_0 + \sqrt{6a^{-1}(\beta_1)^{1/3}} = t_0 + \frac{96\beta_1}{25\sqrt{5}\beta_2} = \\ = t_0 + 3.84\sqrt{0.2}\,\frac{\beta_1}{\beta_2} \tag{19}$$

This choice of t_f is motivated by the fact that increasing this value does not change anything in the system dynamics (since for $t \geq t_0 + \sqrt{6a^{-1}\left|e_0^{1/3}\right|}$, e(t)=0) but results in unnecessary increase of the controller gains.

4. CONTROL LAW

In order to achieve the favourable performance discussed in the previous section, the representative point of the system (1) must be restricted to the switching line (8). This is possible if the following control law is applied

$$u = u_0 + u_1 \tag{20}$$

When the tracking error $e_1(t)\neq 0$ is not equal to zero, u_0 is defined as follows

$$u_0 = b^{-1}\left[\dot x_{2d}(t) - f(x_1,x_2,t)\right] - b^{-1} a\, sgn(e_1)\cdot \begin{cases} e_1^{2/3} + \frac{2}{3}(t-t_0)\frac{e_2}{e_1^{1/3}} & \text{for } t_0 \leq t < t_f \\ \frac{2}{3}(t_f - t_0)\frac{e_2}{e_1^{1/3}} & \text{for } t \geq t_f \end{cases} \tag{21}$$

Otherwise, i.e. when the tracking error $e_1(t)=0$, then for any $t\geq 0$

$$u_0 = b^{-1}\dot x_{2d}(t) - b^{-1} f(x_1,x_2,t) \tag{22}$$

Furthermore

$$u_1 = -b^{-1} D\, sgn(s) \tag{23}$$

where D is such a constant that

$$|d(t)| < D \tag{24}$$

for any $t\geq t_0$. Notice that the control signal (21) cannot be applied for an arbitrary error state $[e_1(t), e_2(t)]^T$ because it might be unbounded when $e_1(t)$ approaches zero. We will show later that this is never the case in our control scheme. In fact, since $e_1(t)$ and $e_2(t)$ are not independent in the scheme, the signal u_0 approaches a finite value $u_0 \to b^{-1}\left[\dot x_{2d}(t) - f(x_1,x_2,t)\right]$ as $e_1(t) \to 0$ tends to zero. In order to prove stability of the sliding mode on the line (8) in the system (1) controlled according to (20)-(23), we check if the following condition is satisfied

$$s \neq 0 \;\Rightarrow\; s\dot s < 0 \tag{25}$$

Substituting (1), (8), (20), (21) and (23) into (25), for any $e_1(t)\neq 0$ we get

$$s\dot s = s\left[\dot e_2 + sgn(e_1)\dot\alpha e_1^{2/3} + \frac{2}{3} sgn(e_1)\alpha\frac{e_2}{e_1^{1/3}}\right] \tag{26}$$

$$s\dot s = \begin{cases} s\left\{\dot e_2 + a\, sgn(e_1)\left[e_1^{2/3} + \frac{2}{3}(t-t_0)\frac{e_2}{e_1^{1/3}}\right]\right\} & \text{for } t_0 \leq t \leq t_f \\ s\left[\dot e_2 + a\, sgn(e_1)\frac{2}{3}(t_f - t_0)\frac{e_2}{e_1^{1/3}}\right] & \text{for } t > t_f \end{cases} \tag{27}$$

$$s\dot s = s\left[d(t) - D\, sgn(s)\right] \leq 0 \tag{28}$$

which by definition (24) is negative for any $s\neq 0$. This implies that, if e(t) is different from zero, then with the proposed control the representative point (RP) of the system (1) always approaches the line (8). In particular, if at some instant of time, the RP belongs to the line, then it cannot leave the line as long as e(t) is different from zero. Suppose that the RP leaves the line at $t=\tau_1$. This implies that $e(\tau_1)=0$ and $de(\tau_1)/dt$ is different from zero, and so for τ_2 immediately after τ_1, $e(\tau_2)$ becomes different from zero, and the RP returns to the line. Thus, we conclude that the sliding mode on the line (8) is stable (actually exists) and the system is insensitive for any $t\in[t_0,\infty)$. Consequently, in the proposed scheme, tracking error and its derivative are not independent and the control signal (21) can be expressed as

$$u_o = \frac{1}{b}\left[\dot x_{2d}(t) - f(x_1,x_2,t)\right] - \frac{a\, sgn(e_1)}{b}\cdot \begin{cases} e_1^{2/3} - \frac{2}{3} a\, sgn(e_1)(t-t_0)^2\left[e_1(t)\right]^{1/3} & \text{for } t_0 \leq t < t_f \\ 0 & \text{for } t \geq t_f \end{cases} \tag{29}$$

It can be easily seen from (29) that u_0 is always bounded and $u_0 \to b^{-1}[\dot{x}_{2d}(t) - f(x_1, x_2, t)]$ as $e_1(t) \to 0$ tends to zero.

Unfortunately, in many practical applications the bound D on the disturbance d(t) may be difficult to determine. Therefore, the discontinuous control u_1 given by (23) may be designed to conservatively causing, as a result, excessive chattering. In the sequel we propose two techniques to attenuate the undesirable chattering in the system.

4.1. Integral compensation

In this subsection we do not require the disturbance d(t) to be bounded by a known constant at any time $t \geq t_0$. The only assumptions we make are:

i) the initial disturbance $d(t_0)$ is bounded

$$d_1 < d(t_0) < d_2 \tag{30}$$

ii) the disturbance does not change infinitely fast, i.e. there exists a possibly very big constant δ such that at any time $t \geq t_0$

$$|\dot{d}(t)| < \delta \tag{31}$$

We introduce the following notation

$$d_0 = 0.5(d_1 + d_2) \qquad k = 0.5(d_2 - d_1) \tag{32}$$

Note that if the initial disturbance is known precisely then $d_0 = d(t_0)$ and *k* is an arbitrarily small constant. Now we replace control (20) with the following switched feedback strategy

$$u = u_0 + u_d + u_c \tag{33}$$

where u_0, the so-called equivalent control in the absence of disturbance is defined by (21) and (22), u_d is a discontinuous term

$$u_d = -kb^{-1}\, sgn(s) \tag{34}$$

and u_c is a continuous compensation term

$$\dot{u}_c = T^{-1} u_d \tag{35}$$

with an initial condition $u_c(t_0) = -d_0/b$ and $T>0$.

Definition 1: The equivalent value $u_{deq}(t)$ of the discontinuous term $u_d(t)$ is defined as follows: if the system is not in sliding mode $u_{deq}(t) = u_d(t) = \pm b^{-1} \cdot k$; otherwise, i.e. when the system is in sliding mode, $u_{deq}(t)$ is a continuous control signal which drives the system along the discontinuity surface in the error state space. In other words $u_{deq}(t)$ is a continuous average of $u_d(t)$.

It can be verified that the condition $s\dot{s} < 0$ for the existence of the sliding mode is satisfied by the system (1) controlled according to (21), (22), (33), (34) and (35) at the initial time $t=t_0$. To this end, for any $e_1(t) \neq 0$, we take

$$s\frac{ds}{dt} = s\dot{e}_2 + \frac{2}{3} s \cdot sgn(e_1)\alpha \frac{e_2}{e_1^{1/3}} + = \begin{cases} s \cdot sgn(e_1) a e_1^{2/3} & \text{for} \quad t_0 \leq t < t_f \\ 0 & \text{for} \quad t \geq t_f \end{cases} \tag{36}$$

Then substituting (1) and (21) into (36) we get

$$s\dot{s} = s[b(u_d + u_c) + d(t)] = = s[-k \cdot sgn(s) + bu_c + d(t)] \tag{37}$$

Since at $t=t_0$, $u_c(t_0) = -d_0/b$ then

$$s(t_0)\dot{s}(t_0) = -k|s(t_0)| + s(t_0)[-d_0 + d(t_0)] < 0 \tag{38}$$

which is by definition (32) negative. Consequently, the system is in the sliding mode at $t=t_0$ and

$$|u_{deq}(t_0)| < k \cdot |b|^{-1} \tag{39}$$

On the other hand, when the system is in the sliding mode

$$b[u_{deq}(t) + u_c(t)] + d(t) = 0 \tag{40}$$

Differentiating equation (40) we obtain

$$b\dot{u}_{deq}(t) + b\dot{u}_c(t) = -\dot{d}(t) \tag{41}$$

then substituting relation (35) and replacing $u_d(t)$ with $u_{deq}(t)$ we get

$$T\dot{u}_{deq}(t) + u_{deq}(t) = -b^{-1} T\dot{d}(t) \tag{42}$$

Since this equation describes a typical first order dynamics, taking into account (31), we conclude that $u_{deq}(t)$ evolves in such a way that for any $t \geq t_0$

$$|u_{deq}(t)| \leq |u_{deq}(t_0)| < |b|^{-1} k \quad \text{or} \quad |u_{deq}(t)| < |b|^{-1} T\delta \tag{43}$$

On the other hand, it follows directly from Definition 1 that the system is in the sliding mode if

$$|u_{deq}(t)| < k \cdot |b|^{-1} \tag{44}$$

Therefore, taking into account (43), we conclude that the system will always remain in the sliding mode if

$$T \leq k\delta^{-1} \tag{45}$$

By this means we proposed a new control strategy which attenuates chattering and guarantees existence of the sliding mode on the switching line (8). In this strategy, the discontinuous control term does not depend on the magnitude of disturbance or its change rate. The only factor which determines the magnitude of the discontinuous control term is the uncertainty of the disturbance at the initial time $t=t_0$.

4.2. Low-pass filter compensation

In this subsection we again consider system (1) and we assume that inequalities (24), (30) and (31) hold. However, on the contrary to the control law (20)-(23), now the upper bound D on the disturbance d(t) will not affect the magnitude of the discontinuous control term. Similarly, as in the previous subsection, we propose the control law (21), (33) and (34). However, now we introduce a modified continuous compensation term

$$u_c + p\dot{u}_c = Ku_d \tag{46}$$

with an initial condition $u_c(t_0) = - d_0/b$, $p>0$ and $K>0$. Now we recall relations (36)-(38) to point out that the considered system with compensation term (46) is in the sliding mode at $t=t_0$. Therefore, we have $|u_{deq}(t_0)|<k/|b|$. On the other hand, again when the system is in the sliding mode relations (40) and (41) hold. Substituting these relations into (46) and replacing $u_d(t)$ with $u_{deq}(t)$ we obtain

$$p\dot{u}_{deq}(t) + (K+1)u_{deq}(t) = -b^{-1}\left[p\dot{d}(t) + d(t)\right] \tag{47}$$

Since equation (47) represents first order dynamics, $u_{deq}(t)$ evolves in such a way that for any $t \geq t_0$

$$\left|u_{deq}(t)\right| \leq \left|u_{deq}(t_0)\right| < k|b|^{-1} \quad \text{or}$$

$$\left|u_{deq}(t)\right| < max\left|\frac{d(t)+p\dot{d}(t)}{(K+1)b}\right| < \frac{D+p\delta}{(K+1)|b|} \tag{48}$$

Therefore, the system will always remain in the sliding mode if the following condition is satisfied

$$K \geq k^{-1}(D+p\delta) - 1 \tag{49}$$

This relation is consistent with the assumption $K>0$, since D is always equal to or grater than k. In this strategy, again the discontinuous control term does not depend on the magnitude of disturbance or its change rate, but only on the uncertainty of the disturbance at the initial time $t=t_0$. Finally, let us point out that (49) is sufficient but not a necessary condition for the existence of the sliding mode in the system.

5. CONCLUSIONS

This paper presents a new variable structure control strategy for second order uncertain system subject to constraints on the system state. A nonlinear, time-varying switching line is introduced. Initially, the line passes through the representative point of the system in the error state space, and then moves smoothly to its predetermined shape. Application of the line eliminates reaching phase, ensures finite time convergence of the system error and guarantees robustness from the very beginning of the controlled motion. The strategy proposed in this paper employs simple continuous tools to partly compensate for the unknown disturbance. Consequently, the discontinuous control term is essentially reduced. This term does not depend on the magnitude of the disturbance or its change rate but only on the uncertainty of the disturbance at the initial time $t=t_0$.

ACKNOWLEDGEMENTS

This work was supported under KBN grant 8 T11A 016 12.

REFERENCES

Bartoszewicz, A. (1995). A comment on 'A time-varying sliding surface for fast and robust tracking control of second-order uncertain systems'. *Automatica*, **31**, 1893-1895.

Choi S. B., D. W. Park and S. Jayasuriya (1994). A time-varying sliding surface for fast and robust tracking control of second-order uncertain systems. *Automatica*, **30**, 899-904.

DeCarlo R.S., S. Żak and G. Mathews (1988). Variable structure control of nonlinear multivariable systems: a tutorial. *Proceedings of IEEE*, **76**, 212-232.

Hung J.Y., W. Gao and J.C. Hung (1993). Variable structure control: a survey. *IEEE Trans. on Ind. Electron.*, **40**, 2-22.

Ryan E. P. and M. Corless (1984). Ultimate boundedness and asymptotic stability of a class of uncertain dynamical systems via continuous and discontinuous control. *IMA J. of Math. Control and Information*, **1**, 223-242.

Slotine J. J. and W. Li (1991). *Applied Nonlinear Control*. Prentice-Hall International Editions.

Venkataraman S. and S. Gulati (1993). Control of nonlinear systems using terminal sliding modes. *Trans. of the ASME*, **115**, 554-560.

Zak M. (1988). Terminal attractors for content addressable memory in neural networks. *Physics Letters*, **133**, 218-222.

MODELLING AND PARAMETER IDENTIFICATION OF DISCONTINUOUS WIENER SYSTEMS

Jozef Vörös

Slovak Technical University
Faculty of Electrical Engineering and Information Technology
Department of Automation and Control
Ilkovicova 3, 812 19 Bratislava, Slovakia

Abstract. The paper deals with the decomposition of block-oriented nonlinear dynamic systems having discontinuous asymmetric nonlinearities and with their identification using the Wiener model. Multiple application of a simple decomposition technique provides a special form of Wiener model with the minimum number of parameters. Putting some parameters of the proposed model to zero, models for the block-oriented nonlinear dynamic systems with pure preload nonlinearities, dead zone or two-segment piecewise-linear asymmetric nonlinearities are obtained. To demonstrate the feasibility of the identification method, more illustrative examples are included.

Keywords: Nonlinear system; identification; discontinuous nonlinearities, Wiener model.

1. INTRODUCTION

In the previous papers a simple decomposition technique using the so-called key term separation principle was applied to the block-oriented nonlinear dynamic systems where the nonlinear static blocks were approximated by proper polynomials or characterized by special nonlinearities (Vörös, 1985, 1991a, 1991b). It provided the nonlinear models of Hammerstein and Wiener type, respectively, in special forms being linear in all the model parameters and supplemented with equations for internal variables being linear-in-parameters, too.

This paper deals with modelling and identification of discontinuous nonlinear dynamic systems which can be represented by the Wiener model with the nonlinear static block containing the discontinuous function shown in Fig. 1. Although such nonlinear systems exist in practice, there are no results to be published on their modelling and identification. Possibly the dead zone causes the major problem.

A simple decomposition of nonlinear system description based on multiple application of the key term separation principle is presented that provides a special form of discontinuous Wiener model with the minimum number of parameters. The derived nonlinear model is linear in all the nonlinear and linear block parameters. The parameter estimation algorithm is a bootstrap one, with internal variables' estimation.

An advantage of the proposed form of Wiener model is that the parameters' estimation can be carried out with only one record of measured input/output data despite the dead zone of nonlinear characteristics. As the model parameters are identical with those of linear and nonlinear blocks, respectively, the model can be used in algorithms for direct adaptive control.

Putting some parameters of the proposed model to zero, models for the block-oriented nonlinear dynamic systems with pure preload nonlinearities, dead zones or two-segment piecewise-linear asymmetric nonlinearities are obtained. To demonstrate the feasibility of the identification method, more illustrative examples are included.

2. KEY TERM SEPARATION PRINCIPLE

Let *f*, *g* and *h* be one-to-one mappings defined on nonempty sets U, X, and Y as follows:

$$f : U \rightarrow X \,, \qquad (2.1)$$
$$g : X \rightarrow Y \,, \qquad (2.2)$$
$$h = g \circ f : U \rightarrow Y \,. \qquad (2.3)$$

Assume that the composite mapping *h* characterizes a system where U is the set of inputs, Y is the set of outputs, and for any output $y(t) \in Y$ there exists an input $u(t) \in U$ and an internal variable $x(t) \in X$ such that the equations

$$y(t) = g[x(t)] = g[f[u(t)]] = h[u(t)] \qquad (2.4)$$

are valid. The outputs can be written as the corresponding function of inputs (using *g o f*) or internal variables (using *g*). The mapping *h*, because of its composition, is generally more complex then those of *f* and *g*. This complicates the identification and control of the system described by *h* = *g o f*. The proposed decomposition described below can lead to useful simplifications.

The basic idea in the proposed technique is the so-called *half-substitution*. Let *g* be an analytic nonlinear mapping which can be rewritten into the following additive form

$$y(t) = g[x(t)] = x(t) + G[x(t)] \,, \qquad (2.5)$$

which consists of the separated variable x(t), plus the remainder of the original mapping assigned as G(.). According to (2.4) we can write

$$x(t) = f[u(t)] \qquad (2.6)$$

and if we substitute (2.6) for x(t) in (2.5) the original form of composite mapping *h* occurs without simplification.

However, if (2.6) is substituted only for the first term in (2.5) (for the separated variable) a new form of the same mapping appears:

$$y(t) = f[u(t)] + G[x(t)] \,, \qquad (2.7)$$

where the output depends on both the input u(t) and the internal variable x(t). Hence the original composite mapping (2.4) is replaced by (2.6) and (2.7). Note that the inner mapping appears both explicitly and implicitly in the outer one.

The above described simple decomposition of composite mapping and the subsequent half-substitution for the separated variable (or key term) constitute the so-called key term separation principle. Note, that the separated key term may be more complex than a simple variable.

3. DISCONTINUOUS NONLINEARITY DECOMPOSITION

For the decomposition of the discontinuous nonlinear characteristics depicted in Fig. 1, the key term separation principle can be applied to obtain such a description, that simplifies the estimation of parameters involved.

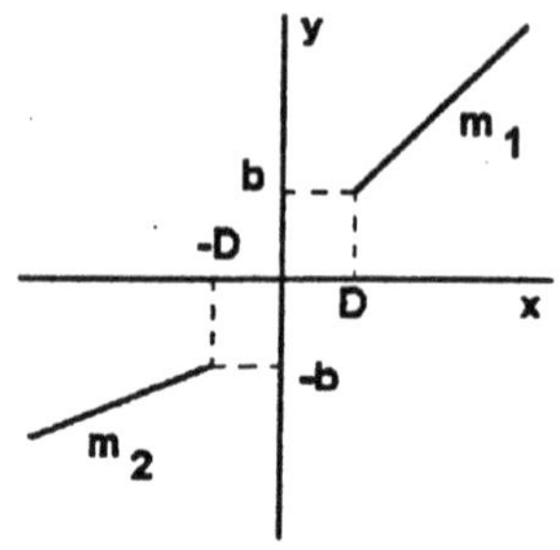

Fig. 1

The output of assumed nonlinearity y(t) according to Fig. 1 depends on the sign and magnitude of input x(t) and can be written as

$$y(t) = \begin{cases} \underline{y}(t) \,, & \text{if } |x(t)| > D \,, \\ 0 \,, & \text{if } |x(t)| < D \,, \end{cases} \qquad (3.1)$$

$$\underline{y}(t) = K(t)\{x(t) - D\,\mathrm{sgn}[x(t)]\} + b\,\mathrm{sgn}[x(t)], \qquad (3.2)$$

$$K(t) = m_1 + (m_2 - m_1)\, h(t) \qquad (3.3)$$

assuming that $|m_1| < \infty$, $|m_2| < \infty$ are the corresponding segment slopes, $0 \leq D < \infty$ is the dead zone and $|b| < \infty$ is the preload constant. The switching function h(t), or more precisely h[x(t)], is defined as follows:

$$h(t) = h[x(t)] = \begin{cases} 0 \,, & \text{if } x(t) > 0 \,, \\ 1 \,, & \text{if } x(t) < 0 \,. \end{cases} \qquad (3.4)$$

and it switches between two linear segments.

However, a similar kind of switching appears for the output y(t) in (3.1) because of the dead zone D and we can write

$$y(t) = h[D - |x(t)|]\, \underline{y}(t) \qquad (3.5)$$

instead of (3.1). Assuming the following equivalence for (3.4)

$$h[x(t)] = 0.5\{\, 1 - \mathrm{sgn}[x(t)] \,\} \qquad (3.6)$$

(for $x(t) \neq 0$) is applicable for that of (3.5) analogously, we can write (3.5) as:

$$y(t)=g[x(t),\underline{y}(t)]=0.5[1-sgn(D-|x(t)|)]\underline{y}(t). \qquad (3.7)$$

As g() in (3.7) is a composite mapping we can apply the key term separation principle. First, we separate the internal variable $\underline{y}(t)$ as the key term rewriting (3.7) as follows:

$$y(t)=\underline{y}(t)-0.5[1+sgn(D-|x(t)|)]\underline{y}(t). \qquad (3.8)$$

Then we substitute (3.2) for the separated term only (half-substitution), i.e.:

$$y(t)=K(t)\{x(t)-Dsgn[x(t)]\}+bsgn[x(t)] - 0.5\ [\ 1 + sgn(D - |x(t)|)\ ]\ \underline{y}(t). \qquad (3.9)$$

To avoid multiplication of parameters in (3.9) we can again apply the key term separation principle. Now we separate K(t)x(t) in (3.2) as the key term and then assigning

$$\xi(t)=K(t)x(t)=m_1x(t)+(m_2-m_1)h(t)x(t), \qquad (3.10)$$

we substitute (3.10) for the separated (first) term into (3.2)

$$\underline{y}(t)=m_1x(t)+(m_2-m_1)h(t)x(t)- -Dsgn[x(t)]\xi(t)/x(t) + b\ sgn[x(t)]. \qquad (3.11)$$

Finally, we can rewrite (3.9) as follows:

$$y(t)= m_1\ x(t) + (m_2 - m_1)\ h(t)\ x(t) - - D\ sgn[x(t)]\ \xi(t)/x(t) + b\ sgn[x(t)] - - 0.5\ [1 + sgn(D - |x(t)|)]\ \underline{y}(t). \qquad (3.12)$$

After two successive applications of the key term separation principle we have replaced the original composite mapping, describing the discontinuous characteristics of Fig. 1, with the equation (3.12) and those of (3.10) and (3.11) defining the imposed internal variables $\underline{y}(t)$ and $\xi(t)$, respectively. These equations are linear in all the parameters characterizing the discontinuous nonlinearity, except D appearing both linearly and nonlinearly. Hence (3.12), supplemented with (3.10) and (3.11) can be used as an appropriate model for the identification of the given nonlinearity.

It is worthy to note, that this description can be used for the nonlinearities representing the special forms of general discontinuous nonlinearity of Fig. 1, i.e. pure preload nonlinearities, dead zone or two-segment piecewise-linear asymmetric nonlinearities.

4. DISCONTINUOUS WIENER MODEL

The Wiener model is given by a linear dynamic system followed by a static nonlinearity block and is shown in Fig. 2. The difference equation model of its linear block can be given as

$$B(q^{-1})x(t) = A(q^{-1})u(t), \qquad (4.1)$$

where u(t) and x(t) are the inputs and outputs, respectively, $A(q^{-1})$ and $B(q^{-1})$ are scalar polynomials in the unit delay operator q^{-1}

$$A(q^{-1}) = a_0 + a_1q^{-1} + \ldots + a_mq^{-m}, \qquad (4.2)$$
$$B(q^{-1}) = 1 + b_1q^{-1} + \ldots + b_nq^{-n}. \qquad (4.3)$$

The output of linear block is identical with the input of nonlinear one. However, a direct substitution of x(t) from (4.1) into (3.12) would result in a very complex expression. Therefore, the key term separation principle can be applied again to simplify the model equation.

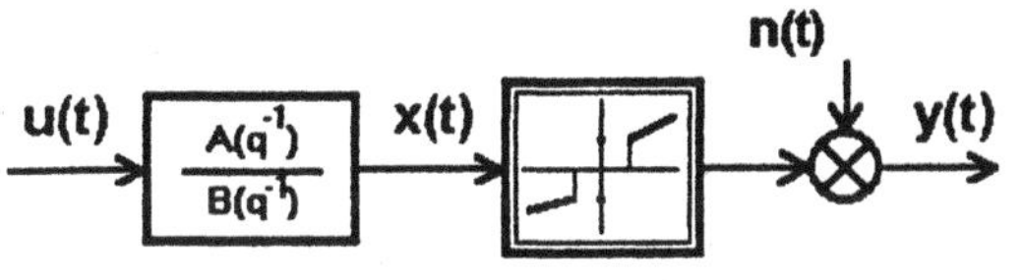

Fig. 2

Assuming that $m_1 = 1$ (this is always possible in the given system description) we separate the variable x(t) as the key term of the nonlinear mapping (3.12). Then after the substitution of x(t) from (4.1) into (3.12) for the key term only, the system output corrupted with additive noise n(t) is given in the form

$$y(t)=A(q^{-1})u(t)+[1-B(q^{-1})]x(t)+mh(t)x(t) - -Dsgn[x(t)]\xi(t)/x(t)+bsgn[x(t)] - - 0.5\{1+sgn[D-|x(t)|]\}\underline{y}(t)+n(t), \qquad (4.4)$$

where $m = m_2 - 1$. The equation (4.4) and those of (4.1), (3.10) and (3.11) defining the imposed internal variables x(t), $\xi(t)$, and $\underline{y}(t)$ represent a special form of Wiener model with discontinuous nonlinearity as the result of multiple application of the key term separation principle.

The Wiener model with discontinuous nonlinearity given by (4.4) can be put into a concise form

$$y_c(t) = \Phi^T(t)\ .\ \Theta_o + n(t), \qquad (4.5)$$

where the data vector is defined as

$$\Phi^T(t)=[u(t),u(t-1),\ldots, u(t-m), -x(t-1), -x(t-2), \ldots,-x(t-n),h(t)x(t), -sgn[x(t)]\xi(t)/x(t),sgn(x(t))], \qquad (4.6)$$

the vector of parameters is

$$\Theta_o^T=[a_0, a_1, \ldots, a_m, b_1, \ldots, b_n, m, D, b], \qquad (4.7)$$

and the output with a kind of dead zone influence correction is

$$y_c(t)=y(t)+0.5\{1+sgn[D-|x(t)|]\}\underline{y}(t). \quad (4.8)$$

5. IDENTIFICATION ALGORITHM

An identification algorithm using the derived special form of Wiener model was developed enabling the simultaneous estimation of all the model parameters using only one input/output record. As the internal variables $x(t)$, $\xi(t)$, and $\underline{y}(t)$ are unmeasurable and must be estimated, an iterative estimation process is considered, as in the case of Hammerstein model (Vörös, 1994).

The proposed iterative algorithm is based on the use of the preceding estimates of model parameters for the estimation of unmeasurable internal variables. Assigning as

$${}^{s}x(t) = {}^{s}B(q^{-1})\, u(t) + [1 - {}^{s}A(q^{-1})]\, {}^{s}x(t), \quad (5.1)$$

$${}^{s}\xi(t) = {}^{s}x(t) + {}^{s}m\, h(t)\, {}^{s}x(t)\, , \quad (5.2)$$

$${}^{s}\underline{y}(t) = {}^{s}x(t) + {}^{s}m\, h(t)\, (t) - {}^{s}D\, sgn[{}^{s}x(t)]\, {}^{s}\xi(t)/{}^{s}x(t) + {}^{s}b\, sgn[{}^{s}x(t)], \quad (5.3)$$

$${}^{s}y_c(t) = y(t) + 0.5\, \{1 + sgn[{}^{s}D - |{}^{s}x(t)|]\}\, {}^{s}\underline{y}(t)\, , \quad (5.4)$$

the estimates of internal variables and the corrected output in the s-th step and defining the vector of estimated data

$${}^{s}\Phi^{T}(t) = [u(t),u(t-1),\ldots,u(t-m),-{}^{s}x(t-1), -{}^{s}x(t-2),\ldots,-{}^{s}x(t-n),\ h(t)\,{}^{s}x(t), -sgn[{}^{s}x(t)]\,{}^{s}\xi(t)/{}^{s}x(t),\ sgn[{}^{s}x(t)]], \quad (5.5)$$

and the vector of parameter estimates

$${}^{s}\Theta^{T} = [\,{}^{s}a_0, {}^{s}a_1,\ldots, {}^{s}a_m, {}^{s}b_1,\ldots, {}^{s}b_n, {}^{s}m, {}^{s}D, {}^{s}b]\, , \quad (5.6)$$

we obtain from (4.5) the error to be minimized

$$e(t) = {}^{s}y_c(t) - {}^{s}\Phi^{T}(t).{}^{s+1}\Theta\ . \quad (5.7)$$

The various steps in the iterative procedure may be now stated as follows:

i) Minimizing a proper criterion based on (5.7) the estimates of both linear and nonlinear block parameters ${}^{s+1}\Theta$ are yielded using the corrected outputs ${}^{s}y_c(t)$ and ${}^{s}\Phi^{T}(t)$ with the s-th estimates of internal variables.

ii) Using (5.1)-(5.4) the estimates of internal variables ${}^{s+1}x(t)$, ${}^{s+1}\xi(t)$, ${}^{s+1}\underline{y}(t)$, and corrected outputs ${}^{s+1}y_c(t)$ are evaluated by means of the recent estimates of model parameters.

iii) If the estimation criterion is met, the procedure ends, else it continues repeating the steps i) and ii).

Convergence of this estimation algorithm using the Wiener model cannot be generally guaranteed (Ljung, 1987). Nevertheless, the testing of proposed algorithm has shown good results. Naturally, this procedure can be used for the previously mentioned special cases of discontinuous nonlinearities, but with one exception. Namely the case, when $m_1 = m_2$ and both D and b are nonzero, is unsolvable with the proposed form of model.

6. EXAMPLE

The proposed method for the identification of nonlinear dynamic systems with discontinuous nonlinearities using Wiener model was implemented and tested by means of MATLAB packages. Several systems were simulated and the estimations of all the model parameters (those of linear and nonlinear blocks) were carried out on the basis of input and output records as well as estimated internal variables.

To illustrate the feasibility of the identification method using the described technique, the following examples show the results of parameter estimation process for some types of discontinuous nonlinearities and the linear dynamic system given by the equation

$$[1-0.2q^{-1}+0.35q^{-2}]x(t)=[q^{-1}+0.5q^{-2}]u(t).$$

In the first case the nonlinear block of Wiener model is given by the characteristics shown in Fig. 3, i.e. all the parameters of discontinuous nonlinearity are nonzero. The process of parameter estimation is included in the Table 1. The true values of system parameters are written in the last row. The parameters' estimates are satisfactory after 10 iterations.

If we put $b = 0$ in the special Wiener model we obtain a two-segment piecewise-linear asymmetric nonlinearity with dead zone depicted in Fig. 4. In this case the process of parameter estimation is included in the Table 2 and the estimates are satisfactory after only 9 iterations.

In the case of the preload nonlinearity without dead zone with $D = 0$ and $b \neq 0$ shown in Fig. 5, even fewer iterations are required, see the results in Table 3.

These nonlinear dynamic systems with discontinuous nonlinearities were simulated on PC AT computer for t = 1,2, ... ,400 using a random input with |u(t)| 2. The noise n(t) was generated as Gaussian white noise with zero mean. In all the cases the least squares method has been employed for the simultaneous estimation of both linear and nonlinear block parameters with zero initial values (they are not included in the tables).

7. CONCLUSIONS

A new approach to modelling and identification of nonlinear dynamic systems having discontinuous asymmetric nonlinearities using the Wiener model has been presented. Multiple application of a simple decomposition technique based on the key term separation principle has provided a special form of Wiener model with the minimum number of parameters and enabling the identification to be carried out with only one record of measured input/output data.

Putting some parameters of nonlinear block to zero we can obtain models for the systems with pure preload nonlinearity, dead zone, two-segment piecewise-linear asymmetric nonlinearities, or their combinations.

The included examples of identification process will demonstrate the feasibility of proposed technique for different types of discontinuous nonlinear characteristics.

REFERENCES

Kung,M.C. and B.F. Womack (1984a). Discrete time adaptive control of linear dynamic systems with a two-segment piecewise-linear asymmetric nonlinearity. IEEE Trans. Automatic Control, Vol. AC-29, No.2, pp. 170-172.

Kung,M.C. and B.F. Womack (1984b). Discrete time adaptive control of linear systems with preload nonlinearity. Automatica, Vol. 20, No.4, pp. 477-479.

Ljung,L. (1987). System Identification : Theory for the User. PRENTICE-HALL, New Jersey.

Vörös,J. (1985). Nonlinear system identification with internal variable estimation. Proceedings IFAC Symp. Identification and Sys. Parameter Estimation, York, pp. 439-443.

Vörös, J. (1991a). Adaptive control of block-oriented systems with preload nonlinearities. Preprints 1st IFAC Symp. Design Methods in Control Systems, Zurich, pp. 606-611.

Vörös, J. (1991b). Identification and control of block-oriented systems with special nonlinearities. Preprints 9th IFAC Symp. Identification and System Parameter Estimation, Budapest, Vol. 1, pp. 622-627.

Vörös,J. (1994). Identification of discontinuous block-oriented nonlinear dynamic systems. Preprints IFAC Workshop on New Trends in Design of Control Systems, Smolenice, pp. 189-194.

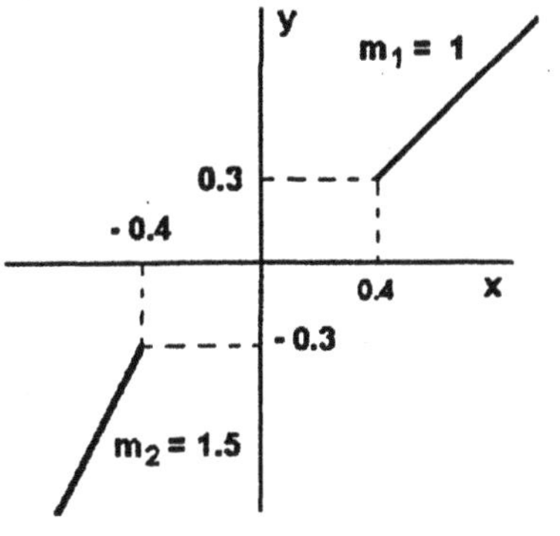

Fig. 3

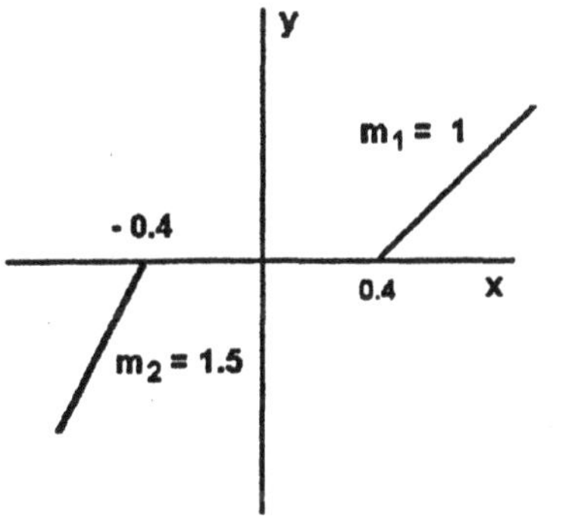

Fig. 4

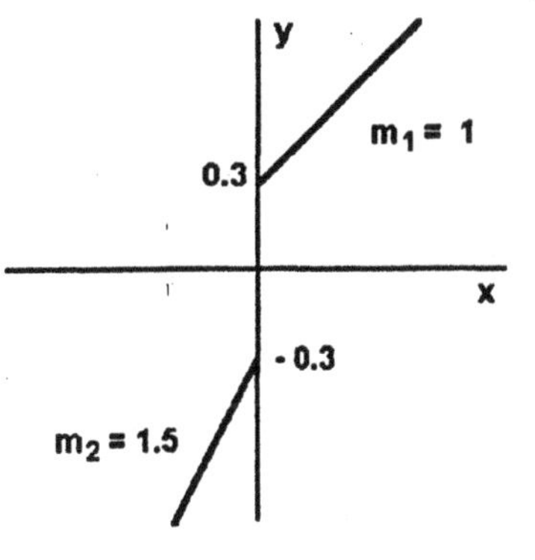

Fig. 5

Table 1

It.	a_1	a_2	b_1	b_2	m	D	b
1	1.1149	0.5461	-0.2580	0.4299	0.2640	0.2197	-
2	1.1189	0.5916	-0.1713	0.3438	0.3299	0.3952	0.1670
3	1.1226	0.5607	-0.1930	0.3491	0.4047	0.4918	0.2767
4	1.0942	0.5409	-0.2029	0.3483	0.4508	0.4917	0.3082
5	1.0653	0.5222	-0.2054	0.3509	0.4782	0.4715	0.3182
6	1.0360	0.5114	-0.2074	0.3535	0.4925	0.4438	0.3124
7	1.0224	0.5015	-0.2099	0.3531	0.4973	0.4203	0.3007
8	1.0142	0.5002	-0.2061	0.3529	0.4981	0.4163	0.3057
9	1.0045	0.4978	-0.2062	0.3535	0.5005	0.4091	0.3059
10	1.0014	0.4988	-0.2002	0.3528	0.5019	0.4022	0.3031
t	1.0000	0.5000	-0.2000	0.3500	0.5000	0.4000	0.3000

Table 2

It.	a_1	a_2	b_1	b_2	m	D
1	0,8915	0,4497	-0,2527	0,4360	0,4152	0,2815
2	0,9188	0,5037	-0,1681	0,3547	0,4676	0,3178
3	0,9477	0,4958	-0,1930	0,3638	0,5050	0,3464
4	0,9602	0,4980	-0,1924	0,3576	0,5078	0,3599
5	0,9696	0,4974	-0,1962	0,3568	0,5086	0,3693
6	0,9759	0,4976	-0,1978	0,3556	0,5075	0,3756
7	0,9804	0,4976	-0,1990	0,3550	0,5064	0,3802
8	0,9836	0,4977	-0,1999	0,3546	0,5056	0,3834
9	0,9859	0,4978	-0,2004	0,3543	0,5049	0,3858
t	1,0000	0,5000	-0,2000	0,3500	0,5000	0,4000

Table 3

It.	a_1	a_2	b_1	b_2	m	b
1	1,0022	0,4812	-0,1453	0,2498	0,3299	0,2939
2	1,0093	0,3640	-0,3874	0,4906	0,5242	0,2584
3	1,1189	0,7645	-0,0064	0,2533	0,4740	0,2128
4	1,0293	0,4078	-0,3023	0,3785	0,4522	0,2517
5	1,0264	0,5909	-0,1299	0,3316	0,4897	0,2762
6	1,0066	0,4709	-0,2279	0,3492	0,4875	0,2932
7	1,0025	0,5100	-0,1937	0,3553	0,5018	0,2976
8	0,9986	0,4997	-0,2011	0,3471	0,4999	0,3021
t	1,0000	0,5000	-0,2000	0,3500	0,5000	0,3000

PREDCON: A PACKAGE FOR MULTIOBJECTIVE CONTROLLER DESIGN [1]

A. S. Rykov * I. I. Vinogradova * and A. G. Kuznetsov **

* *Faculty of Economics & Information Engineering,*
Moscow Institute for Steel & Alloys, Leninsky prosp. 4, 117936 Moscow, Russia
alexandr@rykov.misa.ac.ru, irina@iv.misa.ac.ru
** *Department of Engineering Science, Oxford University,*
Parks Road, Oxford OX1 3PJ, United Kingdom
alexander.kuznetsov@eng.ox.ac.uk

Abstract. The package PREDCON offers facilities for dialogue-based control system design. It is developed in the form of a MATLAB toolbox and is based on optimisation methods proposed by the authors. The package is applicable for tuning parameters of various types of controller. It is particularly suitable for research and teaching control theory.

Keywords. Computer-aided control system design, multiobjective optimisation, dialogue methods.

1. INTRODUCTION

There has been a renewed interest in multiobjective design of controllers using parameter-search techniques. Several multiobjective computer-aided control system design (CACSD) packages are available, these include ANDECS (Grübel *et al.*, 1993), DELIGHT (Nye and Tits, 1986), CRITERIA (Bada, 1987), QDES (Boyd *et al.*, 1989) and MODCONS (Whidborne *et al.*, 1994). The motivation for creating PREDCON is based on the authors' impression that the existing packages do not exploit many opportunities provided by methods of multiobjective assessment.

The package PREDCON has been written in MATLAB for implementing the multiobjective optimisation approach to the controller design. The structure of the toolbox is somewhat similar to MODCONS (Whidborne *et al.*, 1994) and is shown in Fig. 1.

There are three main components in PREDCON:

- graphical user interface (GUI) (see Section 2);
- objective function calculator (see Section 3);
- search methods (see Section 4).

All components are interconnected by information flows which provide exchange of quantitative and qualitative data. Let us describe each of these parts.

2. USER INTERFACE

The GUI is a very important component of the design package. It provides the interaction between search method and the designer, and enables the user to observe the progress of the design process and to alter it.

[1] The authors acknowledge the support of the Commission of the European Communities (COPERNICUS grant CP941174)

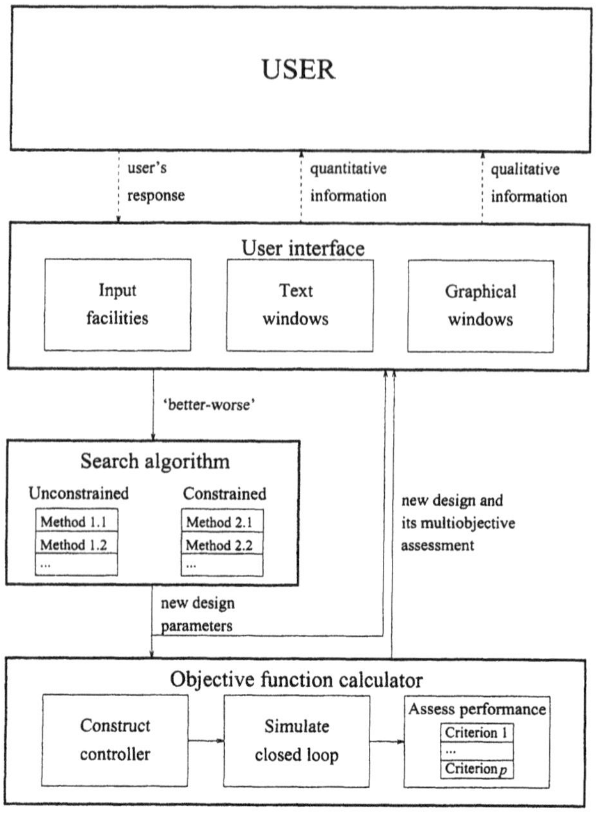

Fig. 1. Structure of the package.

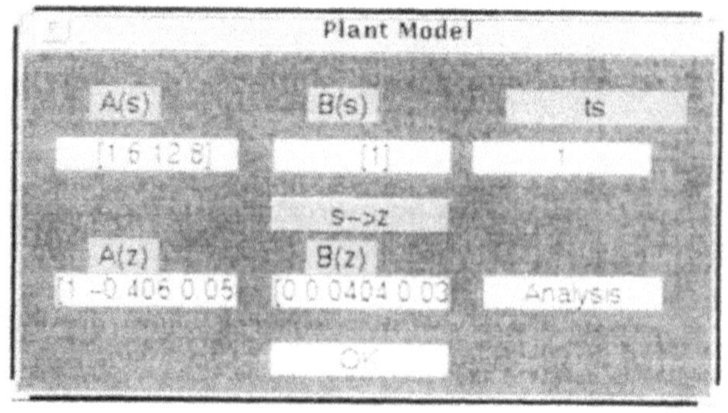

Fig. 2. The Plant model window

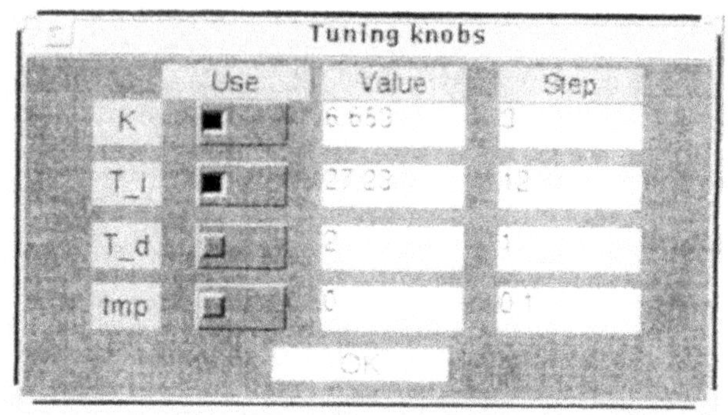

Fig. 3. The Tuning knobs window

The GUI consists of several windows in which data relevant to different aspects of tuning can be viewed and altered. The interaction procedure begins with input facilities. There are several windows in which information about the plant model, tuning knobs, optimisation methods etc. is input (Figs. 2 and 3). These windows contain fields (boxes) in which data should be entered. All polynomials are entered in the polynomial vector form, i.e. as vectors whose elements are the polynomial coefficients. The coefficients of discrete time polynomials are entered in order of decreasing powers of z (e.g. enter `[1 2 2]` for $1+2z^{-1}+2z^{-2}$) and those of continuous time polynomials are entered in order of decreasing powers of s (e.g. enter `[1 0 3]` for s^2+3). MATLAB expressions may also be entered in any entry box; they are evaluated automatically. Data are assigned to the parameters corresponding to the considered entry box only when Ctrl-Enter is pressed or when the mouse is clicked outside the entry box.

The user can input data through these windows directly or load previously saved files. The Load and Save facilities use conventions accepted in the operating system under which PREDCON is being run.

There are additional facilities providing the user with model analysis and automatic conversion of the continuous time model to its discrete time equivalent. The user can select the parameters to be tuned and the performance criteria to be taken into account.

After all necessary data have been prepared, the dialogue design procedure can begin. The user can see information about the controllers obtained during the design process both in graphical and text windows. The windows with graphical information provide the user with plots of the step response of the system for the set point of 100% and of the control signal sequence (Fig. 4). The window containing text information shows the values of the controller parameters being tuned and of the performance criteria (Fig. 5).

The interaction between the designer and the search algorithm consists in the following. The designer deals with two kinds of choice offered by the program:

- dividing the closed-loop responses presented in the graphical windows (using the values of performance criteria) into the groups, and
- choosing the better controller of the two given in the graphical windows using both graphical and text information.

The GUI of PREDCON hides the peculiarities of the optimisation method employed from the user.

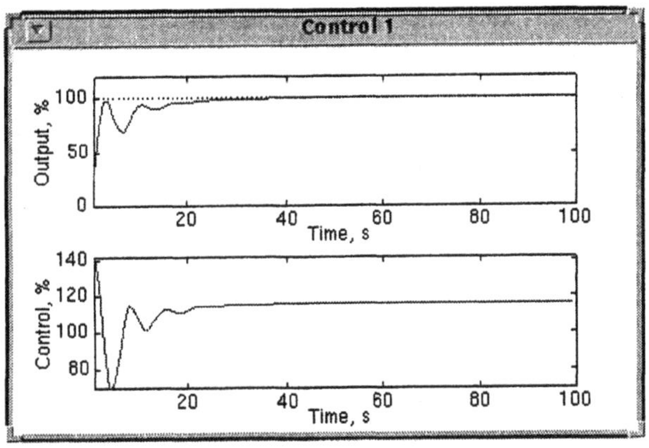

Fig. 4. The window with graphical information.

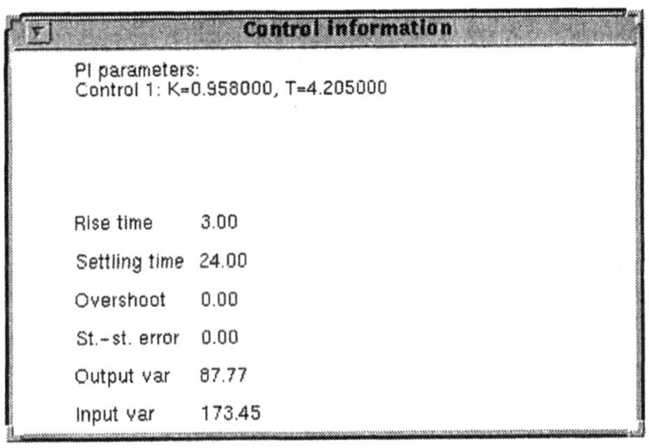

Fig. 5. The window with text information.

3. OBJECTIVE FUNCTION CALCULATOR

The second component is the objective function calculator. It is invoked by the search algorithm with the current design parameters and returns the values of the objective functions f_i, $i = \overline{1,p}$, which represent the following performance criteria: rise time t_r, settling time t_s, overshoot magnitude M_o, steady-state error e_{ss}, output and input variances σ_y^2 and σ_u^2. The objective function calculator performs the following actions. It

- constructs the controller from the design parameters $\boldsymbol{x}$;
- computes the closed-loop step response of the controlled plant model;
- evaluates the performance criteria values $f_i(\boldsymbol{x})$, $i = \overline{1,p}$.

The objective function calculator program can also be written by the user.

4. SEARCH METHOD

The search method is the core of the design package. After the designer has defined the plant model, the structure of the controller, n tuning knobs and the initial values of parameters, a search method is used to find the best (in the user's opinion) controller design.

The package can use various search methods of unconstrained as well as constrained optimisation. At the moment, the dialogue procedure of mirror reflections, the methods of deformed configurations and the Pareto-point method are included. The choice of the method depends on the dimensionality of the problem (the number of design parameters). There is also an option to determine designer's preferences by means of the Pareto-point method, and to perform an automatic search afterwards.

The first approach is based on the dialogue search method is described in (Rykov *et al.*, 1997). This approach is used if the dimensionality of the problem is low (less than 7 ± 2, depending on the number of performance criteria (Miller, 1956)). At the initial stage, the algorithm generates $n+1$ variants of the controller design near the initial point and calculates the closed-loop step responses of the plant model. Also, the values of the performance criteria for all the controllers are calculated. This information is supplied to the designer. The program offers him to divide the controllers into three groups of 'bad', 'moderate' and 'good' ones. Then the interactive search process begins. The $n+1$ configuration vertices corresponding to the variants of design are being mapped in the set of the feasible controller parameters. The program asks the user to choose the better controller of the two exhibited in the graphical windows. This choice is important for improving performance of the controller. The new iteration of the method begins after the algorithm detects that no improvement is possible at this stage. Then again $n+1$ responses with auxiliary information are offered to the designer for dividing them into the groups and the same procedure is repeated. The program stops when the designer is satisfied with the results.

The second approach is described in (Rykov *et al.*, 1996), and it can be used for problems of higher dimensionality and for extracting the designers preferences to be used later. At the initial stage the program generates two Pareto-optimal solutions (designs). The closed-loop step responses and the performance criteria values for each of the trial designs are offered to the user for comparison. The program asks the user to choose the better controller of the two exhibited in the graphical windows. Then the search process goes on according to a version of the Gauss-Seidel method (see (Rykov *et al.*, 1996)). At this stage of the search, the program asks the user to compare pairs of the trial designs. The search goes on until the designer is satisfied with the results. An option exists to extract the designer's preferences during this dialogue procedure, to store the preference vector and to use it later for an automatic design imitating the choice of that particular user for a similar problem.

The third approach is used for the automatic design, and is based on the preference (weight) vector obtained by the Pareto-point algorithm. The quality function is constructed, which has the form of a weighted sum of the performance criteria. This quality index is used as an objective function of the single-objective optimisation problem which is solved by procedures from the class of the deformed configuration techniques.

5. AN EXAMPLE: TUNING OF PI CONTROLLERS

Although the design of PI controllers is not the main purpose of this software package, this problem is used as an illustrative example of its application. Let $\boldsymbol{x} = (K, T_i)^T$ be the vector of independent variables, where K is the controller gain and T_i is the integral time. The obvious constraints $X = \{\boldsymbol{x} | K > 0,\ T_i > 0\}$ are added. The design quality is expressed by a multiobjective performance index (vector) whose elements are some well-known performance criteria such as rise time t_r, settling time t_s, overshoot magnitude M_o, steady-state error, output and input variances σ_y^2 and σ_u^2.

The classical approach for tuning of PI controllers has been introduced by Ziegler and Nichols (see any textbook on automatic control, e.g. (Franklin *et al.*, 1986)). However, this approach for tuning PI does not use natural multiobjective features of the formulated problem.

The results obtained by the method of Ziegler-Nichols and by the methods suggested by the authors and implemented in the PREDCON package were compared. We quote here some of the simulation results.

Let us consider a plant with the transfer function

$$G_1(s) = \frac{1}{(s+2)^3}.$$

The values of PI parameters obtained by the Ziegler-Nichols rules were taken as the starting point for the mirror reflection method. The simulation results are presented in Fig. 6. The dotted lines correspond to the responses obtained by the Ziegler-Nichols method, the solid lines — by the mirror reflection algorithm.

The significant improvement was achieved for the rise and settling time: $t_r = 2$ s, $t_s = 62$ s for the mirror reflection method against $t_r > 100$ s, $t_s = 91$ s for the Ziegler-Nichols rules. The small overshoot ($M_o = 4.22\%$) resulted after applying the mirror reflection technique does not deteriorate the performance. The input and output variances are approximately the same in the

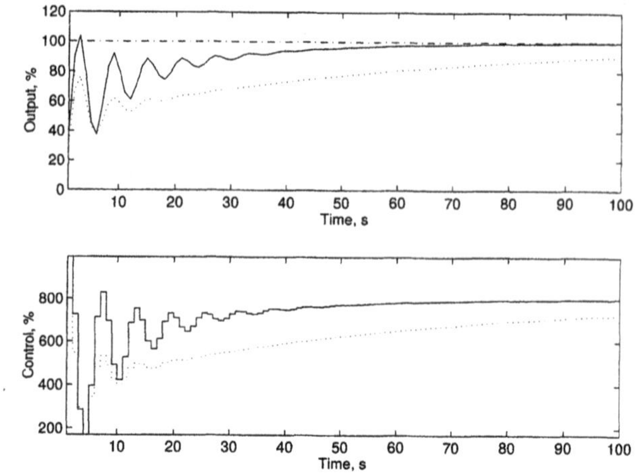

Fig. 6. Test case G_1. Comparison of Ziegler-Nichols (dotted lines) and mirror reflection (solid lines) methods.

both cases. Thus it can be seen that the mirror reflection method does not make design worse in comparison with the classical approach and, in most of the cases, improves it.

To demonstrate the Pareto-point method in action, let us consider a plant with the transfer function

$$G_2(s) = \frac{1}{s^2+2s+1} = \frac{1}{(s+1)^2}.$$

The plant has the natural frequency $\omega_n = 1$ and the damping ratio $\zeta = 1$ (double lag). Let the design quality be expressed by a reduced multiobjective performance index $\boldsymbol{z} = (-t_r, -t_s, M_o)^T$.

The designer was asked to tune the PI controller such that the closed-loop system had the fastest response to the unit step input. Thus, the design objective was

$$t_r \rightarrow \min, \quad t_s \rightarrow \min, \quad \text{ignore } M_o.$$

The tuning was performed in the dialogue mode by the Pareto-point method with simultaneous extracting of the designer's preferences. The resulting response is presented in the upper plot in Fig. 7. The obtained designer's preference (weight) vector was

$$\boldsymbol{\alpha}_d = (0.3627, 0.6127, 0.0246)^T.$$

Then this vector was used to construct a single-objective problem

$$F(\boldsymbol{x}) = \boldsymbol{\alpha}_d{}^T \boldsymbol{z} \rightarrow \max$$

for automatic tuning of the controller for the plant

$$G_3(s) = \frac{1}{s^2+s+1},$$

with $\omega_n = 1$, $\zeta = 0.5$. The results are presented in the lower plot in Fig. 7 (dashed line). The solid line corresponds to the response with $\boldsymbol{\alpha}^* = (0.5, 0.5, 0)^T$ representing the 'ideal' weight vector. It can be seen that the responses almost coincide.

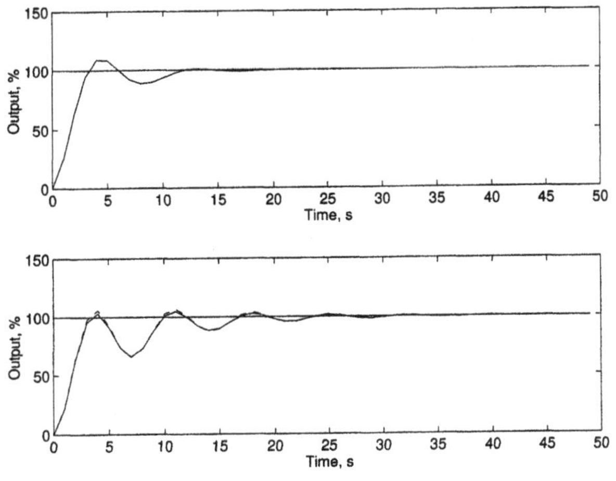

Fig. 7. Closed-loop responses for the plant G_2 (*top*) and for G_3 (*bottom*) tuned with α^* (*solid*) and α_d (*dashed*) with the objective $t_r, t_s \rightarrow \min$.

6. CONCLUSIONS

PREDCON is a package for selecting the tuning parameters of various types of controller. The package combines the power of the selected optimisation methods with the convenience of the GUI developed for the MATLAB environment. The key features of the PREDCON package are:

Interactivity The package is based on the new dialogue methods for solving problems of multiobjective design. Since these procedures are interactive, the designer can observe the process of controller tuning step by step, and the GUI enables him to make decisions about the tuning process.

Generality An advantage of the dialogue design techniques included in the PREDCON package is their generality. The multiobjective approach can be applied to design of controllers of different types.

PREDCON can be useful for teaching and research as a convenient tool for tuning controllers of any type. Experiments with tuning PI controllers (Rykov *et al.*, 1996; Rykov *et al.*, 1997) have been successful and give promising results.

7. REFERENCES

Bada, A. T. (1987). CRITERIA: A package for computer-aided control system design. *Computer-aided design* **19**, 466–474.

Boyd, S., N. Khraishi, C. Barratt, S. Norman, V. Balakrishnan and D. Meyer (1989). *QDES.* Information Systems Laboratory, Electrical Engineering Dept., Stanford University.

Franklin, R. F., J. D. Powell and A. Emami-Naeini (1986). *Feedback control of dynamic systems.* Addison-Wesley.

Grübel, G., H.-D. Joos, M. Otter and R. Finsterwalder (1993). The ANDECS design environment for control engineering. In: *Proceedings of the 12th IFAC World Congress.* Vol. 6. Sydney, Australia. pp. 447–454.

Miller, G. A. (1956). The magical number seven, plus or minus two: Some limits on our capacity for processing information. *Psychological Review* **63**, 81–97.

Nye, W. T. and A. L. Tits (1986). An application-oriented, optimization-based methodology for interactive design of engineering systems. *International Journal of Control* **43**(6), 1693–1721.

Rykov, A. S., I. I. Vinogradova and A. G. Kuznetsov (1996). Algorithms and software for multiobjective controller design. In: *Proceedings of the COPERNICUS spring school on adaptive and predictive control.* Oxford, UK. pp. 69–78.

Rykov, A. S., I. I. Vinogradova and A. G. Kuznetsov (1997). Dialogue method of mirror reflections for multiobjective tuning of controllers. In: *Proceedings of the 7th IFAC Symposium on Computer-Aided Control Systems Design CACSD'97.* Ghent, Belgium. pp. 151–156.

Whidborne, J. F., D.-W. Gu and I. Postlethwaite (1994). MODCONS — A MATLAB toolbox for multiobjective control system design. Technical Report 94-26. Department of Engineering, Leicester University.

Controllability Analysis and Control Structure Selection

Sebastian Engell

Lehrstuhl für Anlagensteuerungstechnik, Fachbereich Chemietechnik
Universität Dortmund, D-44221 Dortmund, Germany

Abstract: For the control of processing units or of overall plants, the selection of a good control structure is at least as important as the controller design. In this paper, a overview of quantitative measures of controllability which can be used for the decision on the control structure is presented. These criteria are all based on the proporties of the open-loop system and therefore only provide qualitative insight. For the final selection, the closed-loop system must be considered, i.e. a controller must be computed.

Keywords: Controllability, control structure, performance bounds, process control

1. INTRODUCTION

The classical meaning of controllability is that a system can be steered from any initial state to the origin in any prescribed period of time by some suitable control input. For linear time-invariant systems, this turned out to be equivalent to the possibility to assign arbitrary eigenvalues of the closed-loop system by constant state feedback. As some eigenvalues may be assigned freely in certain systems while others are fixed under state feedback, by a slight abuse of notation, the term "controllable eigenvalues" or "controllable modes" was introduced in order to allow a more detailed description of systems where not all eigenvalues can be assigned arbitrarily. The associated potential eigenvectors, the controllable subspace of the state space, can be characterized in "geometrical" terms (Wonham, 1979). If a linear system is controllable and also observable, i.e. the state vector can be reconstructed from the knowledge of the outputs over some period of time, the closed-loop system can be given arbitrary dynamics by dynamic output feedback.

This concept of controllability and the dual concept of observability have been immensely fruitful in the analysis both of linear and of nonlinear systems in the last 35 years. However, for the design of a control system, this notion of controllability is not very useful. The fact that it is possible, in principle, to assign all eigenvalues of the closed-loop system arbitrarily, is neither necessary nor sufficient for good control of a technical system. On one hand, only some of the eigenvalues - together with some of the zeros of the system which do not enter into the classical controllability analysis unless they happen to coincide exactly with one of the eigenvalues - determine the dynamic response and only these eigenvalues must be modified by feedback control. This corresponds to the classical observation, that the transient response is determined by the frequency response in some frequency range (around the crossover frequency) whereas its shape for very high and for very low frequencies hardly matters. On the other hand, to shift the eigenvalues to the desired locations may require control moves which cannot be realized or will drive the system beyond the range where the linear model is approximately valid. Also, the amplification of sensor noise caused by trying to achieve the desired fast closed-loop poles may lead to large unproductive control moves.

Hence there is a need to characterize the controllability of a plant quantitatively rather than just by a yes/no criterion.

Control Structure Selection

In a processing plant or in a module (unit) of such a plant, there usually exists a large number of variables which influence the operation of the plant as well as a number of variables which are measured at different intervals with different time-delays and different accuracy. E.g. in a fluidized catalytic cracker, a key unit in refineries which is analysed in great detail in a series of papers by Shinnar *et al.* (Arbel *et al.* 1995a,b, 1996,1997), there are at least 8 variables which can be manipulated, and about 10 product properties which should be controlled as well as at least 10 variables which have to satisfy certain constraints.

Control structure selection consists of two steps:

(a) choice of sets of manipulated and regulated variables used in continuous or quasi-continuous automatic feedback control

(b) choice of the controller structure (single-loop controllers, single loop controllers with compensation, multivariable controllers, linear or nonlinear controllers, etc.).

The choice of the control structure has a much stronger influence on the resulting performance than the design of the control algorithms. Clearly, the selection of a control structure should be based on some quantitative notion of controllability. The best set of manipulated and controlled variables is the one for which the plant is best controllable. Naively speaking, a plant has good controllability if a controller which is simple and easy to design achieves good performance. Control engineers, and control scientists in particular like plants which are difficult to control, because only for such plants, they can demonstrate the value of control theory. On the other hand, it is more important economically to use the knowledge on dynamics and control to help to avoid the design of plants which are difficult to control than to try to control an ill-designed plant (cf. Arbel *et al.*, 1997).

One more point should be added here: the relevant aspect from the point of view of the company which runs the process is not the quality of feedback control but the performance of the plant, i.e. the amount and the quality of the products in relation to the production cost. The performance is largely determined by the infrequent higher level actions taken either by the operators based on their experience or computed by a model-based optimizer, e.g. the choice of the operating point, the decision when to switch between different modes of operation etc.. Feedback control in the traditional sense, i.e. continuous or quasi-continuous control often is only a means to achieve good *operability*, i.e. to enable the operators or the higher level control algorithm to establish the desired operating regime smoothly and to maintain it with a minimum of deviations from the desired setpoint, despite poorly known static and dynamic relationships between inputs and outputs, changing flows into the unit, aging of the unit etc..

In the sequel, we adopt the view that the quantitative controllability of a plant for a certain choice of controlled outputs and manipulated inputs is the same as the possible quality of regulatory and servo feedback control for this particular control structure under real conditions:

- actuator range and speed limitations and bounds on internal variables
- measurement delays and limited measurement accuracy and frequency
- limited knowledge of the behaviour of the plant.

Of course, the performance can only be measured if assumptions on the disturbances and on the expected setpoint variations are made. Even then, performance often is difficult to quantify. The simplest case is to consider regulatory control against stochastic, independent disturbances. Then one may take a weighted sum of the residual variances of the regulated variables as a measure of the quality of control. Still the weighting factors must be chosen which is equivalent to scaling of the regulated variables. Stochastic disturbances are an idealized model of the small fluctuations which occur in any plant. However, there usually are also setpoint changes and/or characteristic, nonstochastic disturbances which often affect a number of outputs simultaneously. If there are constraints, one has to decide how constraint violations are considered in the performance criterion (hard or soft constraints).

2. CONTROLLABILITY I: STEADY STATE ANALYSIS

2.1 Description of the plant and of the objectives

We assume, as it is usually true in the control of large processing units, that there are three types of inputs to the plant (cf. Arbel *et al.*, 1996, Trierweiler, 1997):

- $\mathbf{u}_c$: p_c inputs which are used for automatic feedback control
- $\mathbf{u}_m$: p_m inputs which are available for manual or infrequent automatic adjustment
- $\mathbf{u}_d$: p_d inputs which vary but cannot be controlled (disturbances, setpoint changes).

Similarly, among the variables which characterize the operation of the process, we have

- $\mathbf{y}_c$: variables which are regulated to their setpoints by automatic feedback control
- $\mathbf{y}_b$: variables which have to be kept with in certain bounds, $\mathbf{y}_b \in \mathbf{Y}_b$

- y_r: variables which should be kept at certain values to assure optimal operation of the process (relevant variables).

Some or all elements of $\mathbf{y}_c$ may be elements of $\mathbf{y}_r$.

This plant description is shown graphically in Fig. 1.

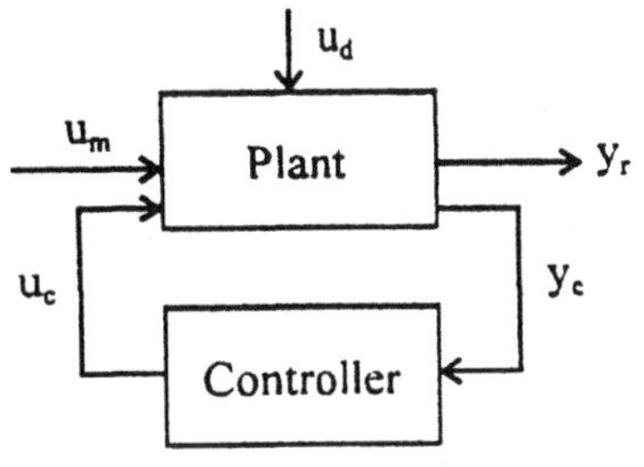

Figure 1: Plant under partial automatic control

For $\mathbf{u}_c$ and $\mathbf{u}_m$, $\mathbf{U}_c$, $\mathbf{U}_m$ indicate the availabe range of the respective inputs and $\mathbf{U}_d$ is the expected range of the disturbances. We measure the attainable steady state performance of a control configuration (which depends on the choice of the variables $\mathbf{u}_c$ and $\mathbf{y}_c$ alone) by the resulting maximal steady state errors over the assumed range of disturbances.

The steady state behaviour of the plant is given as

$$\mathbf{f}(\mathbf{x}^S, \mathbf{u}_c^S, \mathbf{u}_m^S, \mathbf{u}_d^S) = 0 \quad (1)$$

where $\dim(\mathbf{f}) = \dim(\mathbf{x}^S)$, and the outputs are related to the internal variables $\mathbf{x}$ and the inputs by

$$\mathbf{y}_c = \mathbf{g}_c(\mathbf{x}, \mathbf{u}_c, \mathbf{u}_m, \mathbf{u}_d) \quad (2a)$$

$$\mathbf{y}_b = \mathbf{g}_b(\mathbf{x}, \mathbf{u}_c, \mathbf{u}_m, \mathbf{u}_d) \quad (2b)$$

$$\mathbf{y}_r = \mathbf{g}_r(\mathbf{x}, \mathbf{u}_c, \mathbf{u}_m, \mathbf{u}_d). \quad (2c)$$

(1), (2a) may have multiple solutions ($\mathbf{x}_i^S$, $\mathbf{u}_{ci}^S$) for fixed $\mathbf{y}_c$, $\mathbf{u}_m$, $\mathbf{u}_d$. This situation is called an input multiplicity. Also, it may turn out that some components of $\mathbf{x}$ can assume arbitrary values. In this case, the control structure must be modified or extended because this corresponds to uncontrolled inventories in the plant.

Let $\|\mathbf{y}_r\|_W$ be a suitable weighted norm on the space of relevant outputs, e. g.

$$\|\mathbf{y}_r\|_W = \left(\mathbf{y}_r^T \mathbf{W} \mathbf{y}_r\right)^{1/2}$$

or

$$\|\mathbf{y}_r\|_W = \max_i\left(w_i y_{r_i}\right)$$

where the diagonal weighting matrix $\mathbf{W}$ is chosen such that the importance of the deviations of the components of $\mathbf{y}_r$ from their desired values is properly reflected.

2.2 Steady state accuracy with ideal measurements

We define the attainable steady state regulation error of a control configuration (ASSRE) as

$$\text{ASSRE} = \max_{\mathbf{u}_d^S} \min_{SS} \left\|\mathbf{y}_r^S - \mathbf{y}_r^D\right\|_W \quad (3)$$

under the following constraints:

$$\mathbf{f}\left(\mathbf{x}^S, \mathbf{u}_c^S, \mathbf{u}_m^S, \mathbf{u}_d^S\right) = 0 \quad (4)$$

$$\mathbf{u}_m^S = \mathbf{u}_m^D \quad (5)$$

$$\mathbf{u}_d^S \in \mathbf{U}_d \quad (6)$$

$$\mathbf{y}_b^S \in \mathbf{Y}_b \quad (7)$$

and

$$\mathbf{y}_c^S = \mathbf{y}_c^D \quad (8a)$$

if this can be achieved for some $\mathbf{u}_c^S \in \mathbf{U}_c$, otherwise

$$\mathbf{u}_c^S = \arg\min_{\mathbf{U}_c} \left\|\mathbf{y}_c^S - \mathbf{y}_c^D\right\|, \quad (8b)$$

i. e. if the control inputs reach their limits, it is assumed that the imputs are chosen such that a norm of the regulation error is minimized.

$\mathbf{y}_c^D$, $\mathbf{u}_m^D$ are the values of $\mathbf{y}_c$, $\mathbf{u}_m$ at the desired operation point.

Similarly, we define the attainable steady state error with manual correction (ASSREM) as

$$\text{ASSREM} = \max_{\mathbf{u}_d^S} \min_{\mathbf{u}_m^S} \min_{SS} \left\|\mathbf{y}_r^S - \mathbf{y}_r^D\right\|_W \quad (9)$$

under the constraints (4), (6), (7), (8a,b) and

$$\mathbf{u}_m^S \in \mathbf{U}_m. \quad (10)$$

2.3 Sensor accuracy

If sensor accuracy is a problem, it can be incorporated in the above analysis. As a variable cannot be controlled more tightly than the measurement precision, the real values of the controlled variables $\mathbf{y}_c$ must be assumed to deviate from their desired values $\mathbf{y}_c^D$ by the measurement accuracy. Measurement errors usually are relatively small, so the resulting error in $\mathbf{u}_c$ may be computed from the incremental gain matrix between $\mathbf{u}_c$ and $\mathbf{y}_c$

which results from the linearization of (1), (2a) for the values $\mathbf{u}_c^D$, $\mathbf{u}_m^S$, $\mathbf{u}_d^S$:

$$\Delta \mathbf{y}_c^S \approx \mathbf{G}_c\left(\mathbf{u}_c^D, \mathbf{u}_m^S, \mathbf{u}_d^S\right) \cdot \Delta \mathbf{u}_c^S .$$

So in the computation of $\mathbf{y}_r^S - \mathbf{y}_r^D$,

$$\mathbf{u}_c^S = \mathbf{u}_c^D - \mathbf{G}_c^{-1} \cdot \mathbf{e}_c \qquad (11)$$

must be used where $\mathbf{e}_c$ is a vector of measurement errors.

If $\mathbf{G}_c$ is nearly singular, small measurement errors my cause large errors in $\mathbf{u}_c^S$ which is not critical for the control of $\mathbf{y}_c$ because $\mathbf{y}_c^S$ is insensitive to these errors, but $\mathbf{y}_r$ may be strongly affected.

2.4 Plant directionality, coupling and scaling

The standard measure of directionality of a multivariable operator is its condition number γ. Let

$$\bar{\sigma} = \sigma_1 \geq \sigma_2 \geq \ldots \geq \sigma_{p_c} = \underline{\sigma}$$

be the singular values of $\mathbf{G}_c$. Then the steady state condition number of the plant, γ_c, is given by

$$\gamma_c = \frac{\bar{\sigma}}{\underline{\sigma}} . \qquad (12)$$

A large condition number indicates that the plant gain varies strongly with the direction of the inputs, the gain matrix then is called ill-conditioned. It is popular belief, that plants with an ill-conditioned gain matrix are hard to control and that such configurations should not be chosen. However, it was demonstrated in (Gjoesaeter and Foss, 1997) that this is not necessarily true as long as only the control of $\mathbf{y}_c$ by $\mathbf{u}_c$ is concerned (see also the discussion of the dynamic aspects below). However, a large condition number may lead to problems with measurement noise as mentioned above.

The condition number of the plant depends on the scaling of the inputs and of the outputs. A change of the units of the inputs and outputs may reduce the condition number. Thus one should distinguish between ill-conditioned and ill-scaled plants (Trierweiler, 1997). The directionality of the plant can be measured by the condition number of the optimally scaled plant:

$$\gamma^* = \min \gamma\,(\mathbf{L}_c\, \mathbf{G}_c\, \mathbf{R}_c) \qquad (13)$$

where $\mathbf{L}_c$, $\mathbf{R}_c$ are diagonal real matrices.

The optimal choice of $\mathbf{L}_c$, $\mathbf{R}_c$ gives a best-case characterization of the directionality of the plant. However, in reality, the physical scalings introduced by the ranges of actuators and sensors are of primary importance. Their ranges often cannot be chosen freely, at best there are distinct steps in which they can be adjusted, at worst, no devices with the desired ranges exist (or not at reasonable costs). So after the optimal scaling is determined mathematically, it must be checked how the real ranges of actuators and sensors can be chosen, and the singular values of the gain matrix must be computed after a normalization of the inputs and outputs according to these ranges.

The standard measure of the couplings in a plant which is frequently used in the process industry is the relative gain array (RGA, Bristol (1966)).

The relative gain array of the process is defined as

$$\Lambda = \mathbf{G}_c \circ \left(\mathbf{G}_c\right)^{-T} \qquad (14)$$

where $\circ$ denotes element-by-element multiplication and $\mathbf{M}^{-T}$ denotes the transpose of the inverse of $\mathbf{M}$, thus

$$\lambda_{ij} = g_{ij} \cdot \hat{g}_{ji} . \qquad (15)$$

λ_{ij} is the ratio of the open-loop gain in channel $j \rightarrow i$ to the gain in this channel when all other loops are perfectly controlled.

The elements in each column and in each row of the RGA sum up to 1. If one RGA value per column is close to 1 and the remaining ones are small, the interaction in the plant is low and a decentralized control structure in which the inputs and outputs corresponding to the single large entry in each column are paired will work satisfactorily at least for small closed-loop bandwidths. Conversely, a decentralized control structure where one of the corresponding RGA values is negative will suffer from the problem that the controller settings which achieve a stable closed-loop system for decentralized controllers with integral parts will give an unstable system if loops are opened (Chiu and Arkun, 1990). The property that all control loops can be detuned by reducing the gain from the design value to zero without causing instability is called decentralized integral controllability. If the gain matrix is diagonal or triangular, the RGA is the identity matrix.

Although the RGA is widely used, the loop pairing proposed by the RGA-analysis is not always the best one. One problem is that the RGA may change considerably after one loop is closed so that other pairings are more favourable with this loop closed then when feedback is not taken into account (see e.g. (Häggblom, 1997). Even for for 2x2-problems, not even the criterion that pairings should be avoided where the RGA value is negative is always correct. Arbel *et al.* (1996) show that for the fluidized catalytic cracker, the opposite pairing is preferable to the one proposed by the RGA because it gives faster and physically more sensible control and the disadvantage of loss of stability when one loop fails does not matter too much because of the presence of skilled operators and the slow reaction of the loop which goes unstable.

The RGA is a scaling-independent measure of coupling and of plant directionality. It is related to the minimized condition number γ^* by (Nett and Manousiouthakis, 1987)

$$\|RGA\| \stackrel{\Delta}{=} \max\left\{\max_j \sum_i |\lambda_{ij}|, \max_i \sum_j |\lambda_{ij}|\right\}$$
$$\gamma^*(\mathbf{G}(j\omega)) + \frac{1}{\gamma^*(\mathbf{G}(j\omega))} \geq 2\|RGA\|, \tag{16}$$

where λ_{ij} is the element ij of RGA. For 2 x 2 systems (16) reduces to an identity. Thus if the RGA has large elements, γ^* will be large and the gain matrix is ill-conditioned. The RGA thus provides not only a measure of the couplings in the plant but also of the directionality of the process.

2.5 Plant gain variation

It is clear that if the plant gain varies considerably over the range of operation, more complex controllers are necessary. In particular, configurations where the minimal singular value of the gain matrix in certain parts of the operating region becomes very small should be avoided because this practically means a loss of controllability in this part of the operating region which is likely to cause actuator saturation and necessitates special care by the operators.

3. CONTROLLABILITY II: DYNAMIC CONSIDERATIONS

3.1 Unstable dynamics

Clearly, the sets of controlled and manipulated variables must be chosen such that the unstable or marginally stable plant dynamics can be stabilized. In particular, all variables which are not self-regulating, in particular inventories, must be controlled. The control structure selection problem is often simplified by first closing these basic loops and then analyzing the remaining quality control problem (Seborg *et al.*, 1989).

3.2 RHP zeros, delays, unstable poles

The attainable performance of linear control systems has been investigated since 1985 by many authors. In the sequel, we state important qualitative results without referring to weighting functions for the performance which again depend on the desired performance. These can be derived from function theoretical results and factorizations of the plant dynamics (Engell, 1987a, 1988), see also e.g. (Boyd and Desoer, 1985, Freudenberg and Looze, 1988, Skogestad and Hovd, 1997, Chiu, 1997).

Linear multivariable control: The bandwidth where the largest singular value of the closed-loop system is small cannot be made larger than

- .5 times the smallest rhp zero of the plant
- the reciprocal value of the smallest delay between all inputs and outputs.

If these rules are violated, the control loop will be extremely sensitive to modelling errors and have a very oscillatory response to some disturbances.

Individual linear control loops: The bandwidth of each loop is restricted to .5 times the smallest rhp zero of the transfer function resp. the reciprocal value of the pure delay in the channel.

These bounds for the SISO-loops do not apply if a full multivariable controller is used, because the zeros of a multivariable system are not related to the zeros of the individual transfer functions unless the system is diagonal. Practically, however, even if a multivariable controller is used. these bounds become relevant as soon as saturation occurs in some channel or during startup and shutdown.

Unstable poles conversely imply a certain minimal bandwidth (2-5 times larger than the modulus of the poles) of the controlled system. Plants with unstable poles and significant (i.e. not much larger than the desired bandwidth of the closed-loop system) rhp zeros or significant time delays are extremely problematic. Special care also is required for unstable plants with actuator rate limitations.

3.3 Variation of the plant dynamics

If a fixed linear controller is used, the relative model uncertainty and the relative variation of the dynamics over the range of operation should not exceed 50% over the frequency range where the loop gain is high in order to provide a sufficient margin to instability due to changes of the sign of the gain. In the crossover frequency range of the loop, variations of the plant dynamics are amplified in the closed-loop and hence should be small to avoid large variations of the behaviour of the controlled system. The bandwidth must not exceed the frequency range over which a reliable dynamic model is available resp. the plant dynamics do not vary much.

3.5 Attainable speed of response due to actuator limitations

Besides model uncertainty, actuator limitations are the main source of bandwidth limitations for linear control. In a SISO control loop with symmetric input constraints $|u_c| \le M_c$, the attainable maximal output move after time T can be computed as

$$|\Delta y_c|_{max} = M_c \cdot \int_0^T |g_c(\tau)| d\tau . \qquad (17)$$

From this, the minimal rise time and thus the bandwidth of the closed-loop system can be determined if an assumption on the size of the disturbance for which the control loop should behave linearly is made. (It is not reasonable to assume that the input should never saturate for all possible disturbances or setpoint variations because then the actuators are never fully used).

In the multivariable case with differently effective inputs, the estimation of the appropriate bandwidth is difficult in general. One may resort to a simulation of a time-optimal controller to estimate the attainable rise times.

3.6 Directionality and coupling

By inserting the frequency response matrix $G_c(j\omega)$ instead of the gain matrix, the condition number and the RGA can be generalized to dynamic quantities $\gamma(\omega)$ and $\Lambda(\omega)$. Using the dynamic condition number and the dynamic RGA, the qualitative analysis of the controllability of a particular control configuration can be refined considerably. It has been shown (cf. Hovd and Skogestad, 1992) and it corresponds to the standard wisdom in controller design, that the condition number and the RGA are much more important at the gain crossover frequency ω_g of the open-loop system than at ω=0.

If the condition number is high at ω=0 but then decreases to small values, dynamic control with a sufficiently large bandwidth can be possible without encountering problems with directionality and sensitivity (Gjoesaeter and Foss, 1997). This does not, however, mean, that the points discussed in the section on steady state controllability become invalid. If the plant is ill-conditioned at low frequencies, there is the possibility of adverse effects of sensor noise on the control of the relevant quantities even if the control of the primarily controlled variables y_c is satisfactory. Also, the fact that the controlled plant becomes unstable if control loops are opened when the corresponding RGA-values are negative, remains true.

In order to provide a more detailed characterization of the difficulty of the controller design problem, new measures of controllability were introduced by Trierweiler (1997). A summary and case studies can be found in Trierweiler and Engell, (1997a, b).

Definition 1: The *Robust Performance Number* of a multivariable plant with transfer matrix $\mathbf{G}(s)$ is defined as

$$\mathrm{RPN} \stackrel{\Delta}{=} \Gamma_{sup}(\mathbf{G}, \mathbf{T}, \omega) = \sup_{\omega \in R} \{\Gamma(\mathbf{G}, \mathbf{T}, \omega)\} \qquad (18)$$

$$\Gamma(G,T,\omega) \stackrel{\Delta}{=} \sqrt{\bar{\sigma}([I - T(j\omega)]T(j\omega))\left(\gamma^*(G(j\omega)) + \frac{1}{\gamma^*(G(j\omega))}\right)} \qquad (19)$$

where $\gamma^*(\mathbf{G}(j\omega))$ is the *minimized* condition number of $\mathbf{G}(j\omega)$ and $\sigma([\mathbf{I}-\mathbf{T}]\,\mathbf{T})$ is the maximal singular value of the transfer function $([\mathbf{I}-\mathbf{T}]\,\mathbf{T})$. $\mathbf{T}$ is the (attainable) desired output complementary sensitivity function which is determined for the nominal model $\mathbf{G}$.

The RP-number is determined by two factors:

1. $\bar{\sigma}([\mathbf{I}-\mathbf{T}]\,\mathbf{T})$. This term acts as a weighting function. It is well-known that the system uncertainties close to the crossover frequency are more important for robust stability and robust performance than the uncertainties at low and high frequencies (Engell, 1988). For example, a system can have large uncertainty at low frequencies, but nevertheless present no stability and performance problems. This fact is taken into account by the weighting function $\sigma([\mathbf{I}-\mathbf{T}]\,\mathbf{T})$, which has its peak value in the crossover frequency range. The choice of $\mathbf{T}$ depends on the desired closed loop bandwidth, sensor noise, input constraints, and in particular on the nonminimum-phase part of $\mathbf{G}$, i.e., rhp-zeros, rhp-poles, and pure time delays.

2. $\gamma^*(G)+1/\gamma^*(G)$. The origin of this term is the analysis of robust performance (RP) of inverse-based controllers (Stein and Doyle, 1991). The value of μ for RP is a function of $\sqrt{\gamma(\mathbf{G}) + 1/\gamma(\mathbf{G})}$ multiplied by weighting functions that represent the performance and uncertainty terms (see, e.g., Skogestad and Postlethwaite, 1996)). In the definition of the RPN, $\sigma([\mathbf{I}-\mathbf{T}]\,\mathbf{T})$ is used as the weighting function, which depends on the desired attainable closed-loop transfer function $\mathbf{T}$.

The RP-number is a measure of how potentially difficult it is for a given system to achieve the desired performance robustly with a linear controller. The easiest way to design a controller is to use the inverse of the plant model. An inverse-based controller will have potentially good performance robustness only when the RP-number is small. As inverse-based controllers are simple and effective, it can be concluded that a good candidate control structure is one with a small RP-number. Examples showed that the design method described in Engell and Müller (1993) where an ideal inverting controller is approximated by a low order structured controller works well if RPN is small.

In the definition of RPN, an optimal, frequency depending scaling of the system was assumed. The RPN measures the difficulty of the controller design for the chosen plant configuration in the best case.

If the closed-loop system is nearly decoupled due to the use of an inverse-based controller, the scalings can be integrated easily into the controller and thus also frequency dependent scalings could be implemented, at least approximately. However, usually one is interested in simple controllers which will not achieve almost exact decoupling and dynamic scaling is not feasible. As discussed above, the real implementation of the controller will be done with physical scalings which correspond to the ranges of the sensors and the actuators used. Then the important quantity for the controller design is the performance number with constant scalings as defined below:

Definition 2: Let $\gamma(j\omega) \overset{\Delta}{=} \gamma(\mathbf{L}_s G(j\omega) \mathbf{R}_s)$ where $\mathbf{L}_s$, $\mathbf{R}_s$ are fixed scaling matrices.

Then

$$\Gamma_{LR}(G) \overset{\Delta}{=} \sqrt{\bar{\sigma}\left(\mathbf{L}_s[\mathbf{I} - \mathbf{T}(j\omega)]\mathbf{T}(j\omega)\mathbf{L}_s^{-1}\right)\left(\gamma(j\omega) + \frac{1}{\gamma(j\omega)}\right)}, \quad (20)$$

$$\text{RPN}_{LR} \overset{\Delta}{=} \sup_{\omega \in R} \{ \Gamma_{LR}(\mathbf{G}) \}. \quad (21)$$

As mentioned frequently , the scaling of the transfer matrix is important for the the controllability of a system. Good scaling also facilitates the controller design significantly. As the RPN provides a frequency dependent indicator of controllability, the determination of suitable scalings should be done for the frequency where the RPN has its peak value (Trierweiler 1997).

A detailed discussion of the application of the robust performance numbers together with an analysis of the effect of actuator ranges and rhp zeros to a CSTR benchmark control problem can be found in (Trierweiler and Engell, 1997b).

4. DETAILED ASSESMENT OF THE ATTAINABLE PERFORMANCE

All criteria discussed so far only give qualitative indications whether a control structure is likely to give good performance or not. The most promising candidate structures must then be compared by designing and simulating or testing a controller for this structure. The performance of course depends on the design and on the designer. It is desirable to have a more objective criterion to compare the control structures without reference to a particular design method, controller order etc.. Such a comparison is possible under mild assumptions on the problem setting and with reasonable computation times using convex optimization (Boyd and Barrat, 1991, Webers, 1997). The key idea is to use the Youla parametrization of all closed-loop systems which is affine in a free transfer matrix and to optimize the free parameter for convex contraints (such as maximal values of inputs and outputs for certain test signals and bounds on certain frequency responses to ensure robustness) and convex performance criteria (e. g. the mean squared regulation error).

In (Webers and Engell, 1997), the CSTR benchmark was treated using this method and it was shown that a decentralized controller can achieve almost the same performance as the optimal full one. Generally, the analysis of the effect of the choice of the *controller* structure is difficult unless all indicators (RGA, RPN) show that the problem is quite simple. The design method proposed by (Engell and Müller, 1993) allows the designer to test several structures of the controller with little effort, but the final assessment of the structure can only be made by simulation of the closed-loop system.

5. CONCLUDING REMARKS

The steady state considerations, i.e. the issues of actuator ranges, measurement accuracies, plant gain, and characterization of the desired operating point by means of the controlled variables usually are more important than detailed dynamic considerations. Among the criteria which refer to the plant dynamics, the issue of delays (in particular in the measurements) and rhp zeros is most important. If the dynamics contain a number of lags (large phase shift), they should be considered as having a pure delay of adequate size, because even though a compensation of a high order plant is possible theoretically whereas the compensation of a pure delay is not, practically it is not feasible in most cases because it implies high gain high bandwidth feedback with all the associated problems, as sensitivity to sensor noise and model uncertainty, and violation of actuator limitations.

Only after these basic criteria are investigated, a more detailed analysis of the dynamics in the interesting frequency range is worthwhile. This requires some a priori knowledge about the attainable resp. the desired bandwidth which may come from the specifications of the process or be derived from the basic performance limitations.

The controller design for a multivariable plant is facilitated considerably if the bandwidths of the loops are different because then a sequential design of a decentralized controller often gives good results. This however is not reflected in any of the criteria mentioned.

The selection of a control structure using open-loop information only is not recommended in general because feedback modifies the dynamics qualitatively. Indicators of plant controllability only provide qualitative information which can be used to reduce the number of alternatives which then must be explored in more detail, either by performing a design study or by optimization.

REFERENCES

Arbel, A., Z. Huang, I.H. Rinard, R. Shinnar and A.V. Sapre (1995a). Dynamics and Control of Fluidized Catalytic Crackers. 1. Modeling of the Current Generation of FCCs. Ind. Eng. Chem. Res. 34, 1228-1243.

Arbel, A., I.H. Rinard, R. Shinnar and A.V. Sapre (1995b). Dynamics and Control of Fluidized Catalytic Crackers. 2.Multiple Steady States and Instabilities. Ind. Eng. Chem. Res. 34, 3014-3026.

Arbel, A., I.H. Rinard and R. Shinnar (1996). Dynamics and Control of Fluidized Catalytic Crackers. 3. Designing the Control System: Choice of Manipulated and Measured Variables for Partial Control. Ind. Eng. Chem. Res. 35, 2215-2233.

Arbel, A., I.H. Rinard and R. Shinnar (1997). Dynamics and Control of Fluidized Catalytic Crackers. 4. The Impact of Design on Partial Control. Ind. Eng. Chem. Res. 36, 747-759.

Boyd, S. and C.H. Barrat (1991). Linear Controller Design: Limits of Performance. Prentice Hall.

Boyd, S. and C.A. Desoer (1985). Subharmonic functions and performance bounds in linear multivariable feedback systems. IMA J. Math. Contr. Info, 2, 153-170.

Bristol, E.H. (1966). On a New Measure of Interactions for Multivariable Process Control. IEEE Tr. on AC 11, 133-134.

Chen, J. (1997). Sensitivity Integrals and Transformation Techniques: A New Perspective. IEEE Tr. on AC-42, 1037-1042.

Chiu, M.S. and Y. Arkun (1990). Decentralized Control Structure Selection Based on Integrity Considerations. Ind. Eng. Chem. Res. 269-373.

Engell, S. (1987): On the limits of the performance of linear control systems. Preprints 10th IFAC World Congress, München, Vol. 9, 104-109.

Engell, S. (1988). Optimale lineare Regelung, Springer, Berlin.

Engell, S. and R. Müller (1993). Multivariable Controller Design by Frequency Response Approximation. Proc. 2nd European Control Conference, 1715-1720.

Freudenberg, J.S. and D.P. Looze (1988). Frequency Domain Properties of Scalar and Multivariable Feedback Systems. Springer, New York.

Gjoesaeter, O.B. and B.A. Foss (1997). On the Use of Diagonal Control Versus Decoupling for Ill-conditioned Processes. Automatica 33, 427-432.

Häggblom, K.E. (1997). Control Structure Selection via Relative Gain Analysis of Partially Controlled Systems. Proc. 1997 European Control Conference, WE-M-H7.

Havre, K. and S. Skogestad (1997). Effect of RHP Zeros and Poles on Performance in Multivariable Systems. Proc. 1997 ECC, TU-A-H1.

Hovd, M. and S. Skogestad (1992). Simple Frequency-Dependent Tools for Control System Analysis, Structure Selection, and Design. Automatica 28, 989-966.

Nett, C.N. and V. Manousiouthakis (1987). Eucledian Condition and Block Relative Gain. IEEE Trans. on Autom. Control, AC-32(5), 405-407.

Seborg, D.E., T.F. Edgar and D.A. Mellichamp (1989). Process Dynamics and Control. John Wiley.

Skogestad, S. and Postlethwaite, I, (1996). Multivariable Feedback Control - Analysis and Design. John Wiley.

Stein, G. and Doyle, J.C (1991). Beyond Singular Values and Loop Shapes, AIAA Journal of Guidance and Control, 14(1), 5-16.

Trierweiler, J. (1997). A Systematic Approach to Control Structure Design. Dissertation, Fachbereich Chemietechnik, Universität Dortmund.

Trierweiler, J.O. and S. Engell (1997a). The Robust Performance Number: A New Tool For Control Structure Design. Proc. PSE'97/ESCAPE-7, Computers & Chem. Eng. 21, Suppl., 409-414.

Trierweiler, J.O. and S. Engell (1997b). Controllability Analysis via the Robust Performance Number for a CSTR with Complex Reaction Scheme. Proc. ECC '97, TU-A-H3.

Webers, K. (1997). Bestimmung der erreichbaren Regelgüte durch konvexe Optimierung. Dissertation, Fachbereich Chemietechnik, Universität Dortmund.

Webers, K. and S. Engell (1996): Comparison of Multivariable and Decentralized Control for a CSTR by Convex Optimization. Proc. IFAC World Congress '96, San Francisco, Vol. M, 391-396.

Wonham, W.M. (1979). Linear Multivariable Control: a Geometric Approach. Springer, Berlin.

OUTPUT FEEDBACK CONTROL DESIGN FOR NONLINEAR SYSTEMS

C.H. MOOG
IRCyN, UMR CNRS 6597
1 rue de la Noë, BP 92101
44321 Nantes Cedex 3, France
E-mail: Claude.Moog@lan.ec-nantes.fr

X. XIA
7th Research Division
Beijing University of
Aeronautics and Astronautics
Beijing 100083, China

Abstract. We investigate such control problems as disturbance decoupling when static or dynamic output feedback solutions are sought.

Keywords. Nonlinear systems, output feedback, disturbance decoupling

1. INTRODUCTION

A large domain of modern control theory is devoted to state feedback schemes. Whenever the full state is not available for measurement, then either observers have to be designed, or measurement feedback is directly sought for solving the control problem under interest. In this paper, we are consider this latter issue and are concerned with the problem of disturbance decoupling and the noninteracting control problem for nonlinear systems.

Disturbance decoupling

The goal in this paper is to develop checkable conditions for the solvability of static output feedback disturbance decoupling problem of general SISO nonlinear systems with a single measurement.

The approach of the paper is independent from the *geometric* approach in (Andiarti and Moog, 1995) and may be used for other output feedback control problems which got a small amount of contributions and which constitute a challenge for future research. The paper is organized as follows. Section 2 is devoted to the static output feedback disturbance decoupling problem. The geometric role of conditioned invariance is highlightened in Section 3. Examples illustrate the solutions in Section 4.

2. DISTURBANCE DECOUPLING BY STATIC OUTPUT FEEDBACKS

Consider Σ

$$\begin{aligned} \dot{x} &= f(x,u,q) \\ y &= h(x) \end{aligned} \tag{1}$$

where $x \in \mathbb{R}^n$ is the state, $u \in \mathbb{R}$ is the input, $y \in \mathbb{R}$ is the controlled output, $q \in \mathbb{R}$ is the disturbance; f and h are meromorphic functions of their arguments. Consider also the measured output of the system (1)

$$z = h_m(x).$$

where $z \in \mathbb{R}$ and h_m is a meromorphic function.

Let $\mathcal{K}$ denote the field of meromorphic functions of x, u, q, and a finite number of derivatives of u and q. And define

$$E = span_{\mathcal{K}}\{\mathrm{d}\xi \mid \xi \in \mathcal{K}\}.$$

Recall that the relative degree r of y, with respect to input u, is defined by

$$r = \min\{k \in \mathbb{N} \mid \mathrm{d}y^{(k)} \notin span_{\mathcal{K}}\{\mathrm{d}x, \mathrm{d}q, \ldots, \mathrm{d}q^{(k-1)}\}\}.$$

A system Σ is said to be disturbance decoupled if the input u influences the output y, whereas the disturbance q does not. More precisely, the system is said to be disturbance decoupled (Nijmeijer and Van der Schaft, 1990) if the relative degree r of y

is finite, and for any $k \in I\!N$,

$$\frac{\partial y^{(k)}}{\partial q} = 0. \qquad (2)$$

This is equivalent to

$$span_{\mathcal{K}}\{dy, d\dot{y}, \ldots, dy^{(k)}, \ldots,\} \subset \mathcal{X} + \mathcal{U}, \quad (3)$$

in which

$$\mathcal{X} = span_{\mathcal{K}}\{dx\}$$

$$\mathcal{U} = span_{\mathcal{K}}\{du, d\dot{u}, \ldots, du^{(k)}, \ldots\}.$$

Proposition 2.1 Assume the relative degree r of y is finite, then (2) is equivalent to (3), which is, in turn, equivalent to

$$dy^{(r+k)} \in span_{\mathcal{K}}\{dx, du, \ldots, du^{(k)}\}, \qquad (4)$$

for any $k \in I\!N$.

Define

$$\Omega = \begin{array}{l}\{\omega \in \mathcal{X} \mid \forall k \in I\!N : \omega^{(k)} \in \\ span_{\mathcal{K}}\{dx, dy^{(r)}, \ldots, dy^{(r+k-1)}\}\}.\end{array} \qquad (5)$$

For affine nonlinear systems, Ω is shown in (Huijberts et al., 1995) to be the annihilator of some controllability distribution contained in $\ker dh$. What is important now is the fact that Ω is finitely computable by the following algorithm.

$$\Omega^0 = span_{\mathcal{K}}\{dx\},$$

$$\Omega^{k+1} = \{\omega \in \Omega^k \mid \dot{\omega} \in \Omega^k + span_{\mathcal{K}}\{dy^{(r)}\}\} \ (k \in I\!N).$$

Proposition 2.2 There exists an integer $k^* \in I\!N$ such that

(i) $\forall k \geq k^* + 1, \Omega^k = \Omega^{k^*}$;

(ii) $\Omega = \Omega^{k^*}$.

Suppose

$$u = \alpha(x, v) \qquad (6)$$

is a regular static state feedback, i.e. $\dfrac{\partial \alpha}{\partial v} \neq 0$. Let us investigate how Ω changes under the action of this feedback.

Denote $\bar{\mathcal{K}}$ the field of meromorphic functions of x, v, q and finite number of derivatives of v and q, and

$$\bar{\mathcal{E}} = span_{\bar{\mathcal{K}}}\{d\bar{\xi} \mid \bar{\xi} \in \bar{\mathcal{K}}\}.$$

Note that (6) defines in a natural way a *field isomorphism* ϕ from $\mathcal{K}$ to $\bar{\mathcal{K}}$. Also from (6), we have

$$\left\{\begin{array}{l} du = \dfrac{\partial \alpha}{\partial x}dx + \dfrac{\partial \alpha}{\partial v}dv \\ \vdots \\ du^{(k)} = *dx + *dq + \cdots + *dq^{(k-1)} \\ \quad + * dv + \cdots + *dv^{(k-1)} + \dfrac{\partial \alpha}{\partial v}dv^{(k)} \\ \vdots \end{array}\right. \qquad (7)$$

where $*$'s are meromorphic functions in $\bar{\mathcal{K}}$. Since $\frac{\partial \alpha}{\partial v} \neq 0$, (7) defines an isomorphism Φ from $\mathcal{E}$ to $\bar{\mathcal{E}}$, which is *compatible* with the field isomorphism ϕ, i.e., for any $\omega_1, \omega_2 \in \mathcal{E}$, and any $\theta_1, \theta_2 \in \mathcal{K}$, Φ has the property:

$$\Phi(\theta_1\omega_1 + \theta_2\omega_2) = \phi(\theta_1)\Phi(\omega_1) + \phi(\theta_2)\Phi(\omega_2).$$

Immediately, we have

Proposition 2.3 Let $\bar{\Omega}$ be the subspace defined as in (5) for the composite system of (1) and the feedback (6), then

$$\bar{\Omega} = \Phi(\Omega).$$

Now returning to disturbance decoupling. From (4) and the definition of the relative degree r of y, we can write,

$$dy^{(r)} = \omega_0 + \xi \cdot du, \qquad (8)$$

in which $\omega_0 \in \mathcal{X}$, $\xi \in \mathcal{K}$. And $\xi \neq 0$.

Lemma 2.4 Assume the relative degree r of y is finite, then the following statements are equivalent:

(i) The system Σ is disturbance decoupled;

(ii) $\forall k \in I\!N$, $\omega_0^{(k)} \in span_{\mathcal{K}}\{dx, du, \ldots, du^{(k-1)}\}$;

(iii) $dy^{(r)} \in \Omega + span_{\mathcal{K}}\{du\}$.

The main result of this section my now be stated.

Theorem 2.5 Suppose the output y has a finite relative degree r. Then the system (1) is disturbance decouplable by a static output feedback if and only if

(i) $dy^{(r)} \in \Omega + span_{\mathcal{K}}\{dz, du\}$;

and, when $dz \notin \Omega$,

(ii)

$$d\omega \wedge \omega = 0.$$

where $\omega \in span_{\mathcal{K}}\{dz, du\}$, is such that $dy^{(r)} - \omega \in \Omega$.

3. GEOMETRIC INTERPRETATION

In this section, we generalize some of the previous ideas to study a relevant differential geometric concept of (h, f, g) invariance. As mentioned earlier, it is still an open problem to find checkable conditions for (h, f, g) invariance in the most general case. Our study will bring some light to the solvability of this problem, and will present a differential geometric interpretation for some of the previous results.

To begin with, we need the following simple lemma.

Lemma 3.1 Suppose $a = (a_1, \ldots, a_\lambda)$, $b = (b_1, \ldots, b_\rho)$, $c = (c_1, \ldots, c_\tau)$ are such that (a, b, c) is a set of interdeterminates. And assume that ω_i, $i = 1, \ldots, q$ are q differential one forms satisfying

$$\omega_i \in \text{span}_{K(a,b,c)}\{da_1, \ldots, da_\lambda\}. \tag{9}$$

Then there locally exist q functions $\phi_i(a, c)$, $i = 1, \ldots, q$, such that

$$\text{span}_{K(a,b,c)}\{\omega_1, \ldots, \omega_q\} =$$

$$\text{span}_{K(a,b,c)}\{\sum_{j=1}^{\lambda} \frac{\partial \phi_i}{\partial a_j} da_j, i = 1, \ldots, q\}$$

iff

$$d\omega_i \wedge \omega_1 \wedge \cdots \wedge \omega_q \wedge dc_1 \wedge \cdots \wedge dc_\tau = 0.$$

For simplicity, we consider only control affine systems in the following.

$$\dot{x} = f(x) + g(x)u, \tag{10}$$
$$y = h(x) \tag{11}$$

where $x \in I\!R^n$, $u \in I\!R^m$ and $y \in I\!R^p$, and everything is assumed to be smooth.

Given an involutive distribution Δ. The following regularity conditions are assumed to hold through the discussion: Δ, $G = \text{span}\{g_1, \ldots, g_m\}$, $\Delta + G$, $\ker dh$ and $\Delta + \ker dh$ are of constant dimensional.

Having the above reqularity conditions, we assume, without loss of generality, that

$$\Delta + \ker dh = \ker \text{span}\{dh_1, \ldots, dh_\tau\}. \tag{12}$$

The next lemma gives a criterion for (h, f, g) invariance for Δ satisfying

$$\dim(\Delta + G) - \dim \Delta = m. \tag{13}$$

Lemma 3.2 Δ is (h, f, g) invariant iff

1. Δ is (f, g) invariant;
2. Δ is (h, f) invariant;
3. For any m functions T_i, $i = 1, \ldots, m$, satisfying $dT_i \in \Delta^\perp$ and
$$\text{rank}[L_g T_1, \ldots, L_g T_m]^T = m.$$
Let $\omega_i \in \text{span}\{dh_{\tau+1}, \ldots, dh_p, du_1, \ldots, du_m\}$ be the differential one form such that $\dot{T}_i - \omega_i \in \Delta^\perp$, then
$$d\omega_i \wedge \omega_1 \wedge \cdots \wedge \omega_m \wedge dh_1 \wedge \cdots \wedge dh_\tau = 0.$$
for $i = 1, \ldots, m$.

Without the assumption (13), we have only the following nonconstructive criterion for (h, f, g) invariance whose proof can be similarly worked out.

Lemma 3.3 Δ is (h, f, g) invariant iff

1. Δ is (f, g) invariant;
2. Δ is (h, f) invariant;
3. Denote $\sigma = \dim(\Delta + G) - \dim \Delta$. Consider σ functions T_i, $i = 1, \ldots, \sigma$, satisfying $dT_i \in \Delta^\perp$ and
$$\text{rank}[L_g T_1, \ldots, L_g T_\sigma]^T = \sigma.$$
Let $\omega_i \in \text{span}\{dh_{\tau+1}, \ldots, dh_p, du_1, \ldots, du_m\}$ be the differential one form such that $\dot{T}_i - \omega_i \in \Delta^\perp$, for $i = 1, \ldots, \sigma$. Then there exist $m - \sigma$ differential one forms $\omega_i \in \text{span}\{dh_{\tau+1}, \ldots, dh_p, du_1, \ldots, du_m\}$, $i = \sigma + 1, \ldots, m$, such that
$$d\omega_i \wedge \omega_1 \wedge \cdots \wedge \omega_m \wedge dh_1 \wedge \cdots \wedge dh_\tau = 0.$$
for $i = 1, \ldots, m$.

Having Lemma 3.2, Theorem 2.5 can be given the following geometric interpretation.

Theorem 3.4 Suppose the system (1) is affine, and its disturbance decoupling problem is solvable. Then it is disturbance decouplable by a static output feedback if and only if the maximal controllability distribution R^* contained in $\ker dh$ is (h_m, f, g) invariant. Note that R^* is isomorphic to $= \Omega^\perp$.

4. EXAMPLES

Examples are worked out to show the application and limitation of the above results.

Example 4.1

$$\dot{x}_1 = x_2^2 + x_2(x_3 + u)$$

$$\begin{aligned}\dot{x}_2 &= (\cos x_2)(x_3 + u)\\ \dot{x}_3 &= q\\ y &= x_1\\ z &= x_3\end{aligned}$$

One computes $\Omega = span\{\mathrm{d}x_1, \mathrm{d}x_2\}$ and $\mathrm{d}\dot{y} = (2x_2+x_3+u)\mathrm{d}x_2+x_2\mathrm{d}(x_3+u) \in \Omega+span\{\mathrm{d}z, \mathrm{d}u\}$. The form ω in Theorem 2.5 equals $\omega = x_2\mathrm{d}(x_3+u)$ which is integrable. Consequently, setting $v = (x_3+u)$, one concludes that the static output feedback $u = -z+v$ solves the disturbance decoupling problem.

When a static output feedback does not exist, then one can look for a dynamic output feedback solving the disturbance decoupling. This done the following.

Example 4.2 Consider the system

$$\begin{aligned}\dot{x}_1 &= x_2 + \sin x_1\\ \dot{x}_2 &= x_1x_2x_6 + x_3 + x_5 + u\\ \dot{x}_3 &= x_1 + x_2 + x_3 + x_4 + x_5 + x_6\\ &\quad +(1 + x_5)u\\ \dot{x}_4 &= -x_3 + \cos x_5 + x_6 \sin(x_1x_2)\\ \dot{x}_5 &= q\\ \dot{x}_6 &= x_1\\ y &= x_1\\ z &= x_5.\end{aligned}$$

It is easily computed that

$$\Omega = span\{\mathrm{d}x_1, \mathrm{d}x_2, \mathrm{d}x_6\},$$

$$\Omega + \dot{\Omega} + \ddot{\Omega} = span\{\mathrm{d}x_1, \mathrm{d}x_2, \mathrm{d}x_6, \mathrm{d}(x_3 + x_5 + u), \mathrm{d}(x_4 + x_5u + q + \dot{u})\},$$

$$span\{\mathrm{d}z, \mathrm{d}\dot{z}, \mathrm{d}\ddot{z}, \mathrm{d}u, \mathrm{d}\dot{u}, \mathrm{d}\ddot{u}\} = span\{\mathrm{d}x_5, \mathrm{d}q, \mathrm{d}\dot{q}, \mathrm{d}u, \mathrm{d}\dot{u}, \mathrm{d}\ddot{u}\}.$$

Note that

$$(\Omega + \dot{\Omega} + \ddot{\Omega}) \cap span\{\mathrm{d}z, \mathrm{d}\dot{z}, \mathrm{d}\ddot{z}, \mathrm{d}u, \mathrm{d}\dot{u}, \mathrm{d}\ddot{u}\} = \emptyset,$$

thus, ω in (iii) is uniquely defined by

$$\omega = (1 - \sin z + \dot{u})\mathrm{d}z + (1 + \dot{z})\mathrm{d}u + u\mathrm{d}\dot{z} + z\mathrm{d}\dot{u} + \mathrm{d}\ddot{z} + \mathrm{d}\ddot{u}.$$

The reader may check that the following dynamic output feedback is a decoupling feedback:

$$\begin{aligned}u &= -z + \eta_1,\\ \dot{\eta}_1 &= \eta_2 - \eta_1 z + z^2,\\ \dot{\eta}_2 &= \eta_1 - \cos z + v.\end{aligned}$$

5. CONCLUSION

In this paper a new NSC for *static output feedback disturbance decoupling* was obtained.

Comparing with geometric approach taken in (Isidori et al., 1981) or in (Andiarti and Moog, 1995), this paper provides an algebraic approach in quite an independent manner. Note that Necessary conditions as well as Sufficient conditions are established in (Moog and Xia, 1997). As similar approach can be used to tackle noninteracting control.

Acknowledgments:This work has been performed while X. Xia was with Laboratoire d'Automatique de Nantes with the financial support of Région "Pays de la Loire."

REFERENCES

Andiarti, A. and C.H. Moog (1995). "Output feedback disturbance decoupling in nonlinear systems," *IEEE Trans. Automat. Contr.*, vol. 41, pp. 1683-1689, 1996.

Battilotti, S. (1994). *Noninteracting control with stability for nonlinear systems*, Berlin: Springer Verlag.

Battilotti, S. (1997). *Proc. European Contr. Conf.*, Brussels.

Huijberts, H. J. C., L. Colpier and P. Moreau (1995). "Nonlinear input-output decoupling by static output feedback," *Proc. European Control Conf.*, Roma.

Isidori, A. (1995). *Nonlinear control systems* (Third Edition), Berlin: Springer Verlag, 1995.

Isidori, A., A.J. Krener, C. Gori-Giorgi and S. Monaco (1981). Nonlinear decoupling via feedback: a differential geometric approach, *IEEE Trans. Automat. Control*, AC-26, pp. 331-345.

Moog,C.H. and X. Xia (1997). *Proc. European Contr. Conf.*, Brussels.

Nijmeijer, H. and A.J. van der Schaft (1982)."Controlled invariance by static output feedback for nonlinear systems," *Systems & Control Letters*, Vol. 2, pp. 39-47.

Nijmeijer, H., and A.J. van der Schaft (1990). *Nonlinear dynamical control systems*, New York: Springer Verlag.

Glumineau, A., C.H. Moog and F. Plestan (1996). "New algebro-geometric conditions for the linearization by input-output injection," *IEEE Trans. Automat. Contr.*, vol. 41, pp. 598-603.

Plestan, F., (1995). *Linéarisation par injection d'entrée-sortie généralisée et synthèse d'observateurs*, PhD. Thesis Dissertation, Ecole Centrale de Nantes, Université de Nantes.

MODEL MATCHING WITH STABILITY FOR PERIODIC DISCRETE-TIME SYSTEMS

Patrizio Colaneri* and Vladimir Kučera**

* *Dipartimento di Elettronica e Informazione del Politecnico di Milano*
Piazza L. da Vinci 32, 20133 Milano (Italy), e-mail: colaneri@elet.polimi.it

** *Institute of Information Theory and Automation*
P.O. Box 18, 182 08 Praha (Czech Republic), e-mail:kucera@utia.cas.cz

Abstract: In this paper the model matching problem with stability is studied for discrete-time periodic systems. It is shown that, under some zeros matching conditions and stabilizability/detectability assumptions, it is possible to assign a target input-output periodic system by means of a periodic controller. Most remarkably, this allows to find conditions under which a periodic system can be converted into a closed-loop time invariant system via periodic control.

Keywords: Linear systems, periodic systems, model matching, periodic control.

1. INTRODUCTION

A major tool in the study of periodic systems consists in the use of suitable time-invariant representations. The most popular, among them, is the so-called *lifted reformulation* which dates back to an early work of Krank [6] and was later refined by Meyer and Burrus in [4]. More recently, an equivalent time-invariant reformulation, henceforth referred to as *cyclic reformulation*, was introduced in [10] and [2]. The two shift invariant representations are obviously in a one-to-one correspondence and one can use indifferently any of them depending on the nature of the problem at hand. Of course a time-invariant system must satisfies a structure constraint in order to be considered as a time-invariant representation of a periodic system. This is important if one wants to exploit the theory of shif invariant systems for control purposes. To this aim a complete answer to the realization problem of periodic system has been recently given in [5].

In this paper we only make use of the cyclic reformulation with the aim of solving the exact model matching problem with stability to discrete-time periodic systems. The same problem in the time-invariant case, without the closed-loop stability requirement, was originally solved in [9], see also [7] for an updated view. More recently, the problem of model matching via open loop precompensator was shown to be equivalent to a modified disturbance rejection problem, see [13], and the requirement of stability in a geometric framework was considered in [12].

The present paper develops further the conclusions reached in the previous work [14] where the state feedback model matching problem without the stability requirement was addressed. Here the state-feedback and output feedback model matching problems for periodic systems are both investigated and the stability requirement is included. Precisely, the main problem addressed herein is the existence of a periodic controller such that the closed-loop system is stable and equals a given input-output map, which can be expressed by means of the transfer function of the cyclic reformulation of the given periodic system to be matched. Necessary and sufficient conditions are given in terms of usual zeros matching conditions and stabilizability/detectability assumptions. Most interesting is the particular case in which one wants the closed-loop system to be internally stable and time-invariant. It turns out that this is always possible if the underlying periodic system is minimum phase.

The paper results take advantages of the various theoretical achievements of periodc systems and control theory, in particular for what specifically concern the definition and characterizations of zeros, see [1] and [3], and the structural properties [11].

2. THE CYCLIC REFORMULATION

The system under consideration is a T-periodic discrete-time system described by the state-space equations

$$x(t+1) = A(t)x(t) + B(t)u(t) \quad (1)$$

$$y(t) = C(t)x(t) + D(t)u(t) \quad (2)$$

where $t \in \mathcal{Z}$, the input $u(t)$ is m-dimensional and the output $y(t)$ is p-dimensional. Matrices $A(\cdot)$, $B(\cdot)$, $C(\cdot)$ and $D(\cdot)$ are periodic matrices of period $T \in \mathcal{Z}^+$.

In the cyclic reformulation the dimensions of the state, input and output variables are augmented. Precisely, given a signal $v(t) \in \mathcal{R}^q, t \in \mathcal{Z}$ and an initial time instant τ, define an augmented signal $\bar{v}(t) \in \mathcal{R}^{qT}, t \in \mathcal{Z}$ in the following way

$$\bar{v}(t) = \begin{bmatrix} \bar{v}_1(t)' & \bar{v}_2(t)' & \cdots & \bar{v}_T(t)' \end{bmatrix}'$$

where

$$\bar{v}_i(t) = v(t), \quad t = kT + \tau + i - 1, \quad i = 1, 2, \cdots, T$$

In this way any subvector of $\bar{v}(t)$ equals $v(t)$ once a period. The *cyclic reformulation* at time τ is given by

$$\bar{x}(t+1) = \bar{A}\bar{x}(t) + \bar{B}\bar{u}(t)$$
$$\bar{y}(t) = \bar{C}\bar{x}(t) + \bar{D}\bar{u}(t)$$

where

$$\bar{A} = \begin{bmatrix} 0 & \cdots & 0 & A(\tau+T-1) \\ A(\tau) & \cdots & 0 & 0 \\ 0 & \ddots & 0 & 0 \\ 0 & \cdots & A(\tau+T-2) & 0 \end{bmatrix}$$

$$\bar{B} = \begin{bmatrix} 0 & \cdots & 0 & B(\tau+T-1) \\ B(\tau) & \cdots & 0 & 0 \\ 0 & \ddots & 0 & 0 \\ 0 & \cdots & B(\tau+T-2) & 0 \end{bmatrix}$$

$$\bar{C} = \begin{bmatrix} C(\tau) & 0 & \cdots & 0 \\ 0 & C(\tau+1) & \cdots & 0 \\ 0 & 0 & \ddots & 0 \\ 0 & 0 & \cdots & C(\tau+T-1) \end{bmatrix}$$

$$\bar{D} = \begin{bmatrix} D(\tau) & 0 & \cdots & 0 \\ 0 & D(\tau+1) & \cdots & 0 \\ 0 & 0 & \ddots & 0 \\ 0 & 0 & \cdots & D(\tau+T-1) \end{bmatrix}$$

The transfer function from $\bar{u}$ to $\bar{y}$ will be denoted by $\bar{W}(z)$ and will be referred to as the cyclic transfer function of system (1),(2) at time τ. This transfer function changes with τ in a very simple way. Precisely, letting $\bar{W}_{\tau+1}(z)$ and $\bar{W}_\tau(z)$ be the transfer functions of the cyclic reformulations of the periodic system at $\tau+1$ and τ, respectively, it follows, see e.g. [11]:

$$\bar{W}_{\tau+1}(z) = \Delta_p' \bar{W}_\tau(z) \Delta_m \quad (3)$$

where

$$\Delta_i = \begin{bmatrix} 0 & I_i \\ I_{i(T-1)} & 0 \end{bmatrix}$$

Remark 0.1 *It is apparent from (3) that the zero structure (fiinite and infinite) of $\bar{W}(z)$ does not change with τ. Hence one can define the cyclic zero structure of a periodic system as the zero structure of any associated cyclic reformulation without any further specifications.*

Remark 0.2 *The Markov parameters $\bar{D}$ and $\bar{C}\bar{A}^{kT+i}\bar{B}$, $i = 0, 1, \cdots, T-1$, $k \geq 0$ of the transfer function of a cyclic reformulation have a very particular structure. Indeed, it turns out that*

$$\bar{W}(z) = \begin{bmatrix} f_{11}(z^T) & \cdots & f_{1T}(z^T)z^{T-1} \\ \cdots & \vdots & \cdots \\ f_{T1}(z^T)z^{1-T} & \cdots & f_{TT}(z^T) \end{bmatrix} \quad (4)$$

where $f_{ij}(z^T)$ are $p \times m$ rational functions in z^T.

Remark 0.3 *The cyclic reformulations can be associated with a time invariant system as well, provided a period T is given. Let us denote by $W(z) \in \mathcal{C}^{p \times m}$ the transfer function of the original time invariant system. The transfer function $\bar{W}(z) \in \mathcal{C}^{pT \times mT}$ of the cyclic reformulation does not depend on τ and is given by (4) with*

$$f_{ij}(z^T) = \frac{z^{i-j}}{T} \sum_{k=0}^{T-1} W(z\phi^k)\phi^{i-j}, \quad \phi = e^{j2\pi/T}$$

In particular it can be verified that if λ is a zero of $W(z)$ then λ^T is a zero of $\hat{W}(z)$ and all the T-th roots of λ^T are zeros of $\bar{W}(z)$. Similar relations hold for the poles.

3. MODEL MATCHING WITH STABILITY

In the following we will consider the model matching problem in two situations: in the first one the state is assumed available for feedback whereas in the second only the output variable is measured. In order to render the entire exposition easier, we limit the problem to the case when the periodic system is left invertible. This means that from now on we assume that $m \leq p$ and that the cyclic tranfer function $\bar{W}(z)$ of the system is full column rank.

Now consider first a state feedback control law acting on system (1) of the form

$$u(t) = K(t)x(t) + G(t)v(t) \qquad (5)$$

where $K(t)$ and $G(t)$ are T-periodic design matrices. To simplify the solution of the main problem below, we make the assumption that the matrix $G(t)$ is nonsingular for each t.
Feedback (5) can be also rewritten in terms of the cyclic reformulation. Recalling the definition of the signals in such a reformulation, it is apparent that (5) is equivalent to

$$\bar{u}(t) = \bar{K}\bar{x}(t) + \bar{G}\bar{v}(t)$$

where $\bar{K} \in \mathcal{R}^{mT\times nT}$ and $\bar{G} \in \mathcal{R}^{mT\times mT}$ are block diagonal matrices with entries $K(i) \in \mathcal{R}^{m\times n}$ and $G(i) \in \mathcal{R}^{m\times m}$, $i = 0, 1, \cdots, T-1$ respectively. The periodic feedback control system can be redrawn in terms of the cyclic reformulation. To this purpose, let $\bar{T}(z) = (zI - \bar{A})^{-1}\bar{B}$ denote the transfer function from the input $\bar{u}$ to the state $\bar{x}$. Moreover, let us denote by $\bar{W}_m(z) \in \mathcal{C}^{pT\times mT}$ the transfer function (from $\bar{v}$ to $\bar{y}$) of the cyclic reformulation associated with a given closed-loop (periodic) system (with input v and output y). Notice that $\bar{W}_m(z)$ can also derive from a $p \times m$ transfer function $W_m(z)$. In this case, the aim is to work out a periodic feedback such that the closed-loop system is time-invariant. In any case, it turns out that

$$\bar{W}_m(z) = \bar{W}(z)(I - \bar{K}\bar{T}(z))^{-1}\bar{G}$$

Theorem 0.1 *Let a T-periodic system be given with the left invertible full column rank $\bar{W}(z) \in \mathcal{C}^{pT\times mT}$ as input-output cyclic transfer function and assume that the periodic pair $(A(\cdot), B(\cdot))$ is stabilizable. There exist a periodic feedback matrix $K(\cdot)$ and a periodic matrix $G(\cdot)$, the latter non singular for all t, such that the closed loop system is internally stable and its cyclic transfer function is $\bar{W}_m(z)$ if and only if*

(i) The two transfer functions

$$\left[\ \bar{W}(z) \quad \bar{W}_m(z)\ \right]$$

and

$$\bar{W}_m(z)$$

have the same infinite zero structure

(ii) The two transfer functions

$$\left[\ \bar{W}(z) \quad \bar{W}_m(z)\ \right]$$

and

$$\bar{W}(z)$$

have the same infinite zero structure

(iii) The two transfer functions

$$\left[\ \bar{W}(z) \quad \bar{W}_m(z)\ \right]$$

and

$$\bar{W}_m(z)$$

have the same finite zero structure

(iv) The two transfer functions

$$\left[\ \bar{W}(z) \quad \bar{W}_m(z)\ \right]$$

and

$$\bar{W}(z)$$

have the same finite unstable zero structure.

Sketch of the proof. As for sufficiency, notice that conditions (1) and (ii) imply the existence of a proper $m \times m$ rational function $\bar{R}(z)$ such that $\bar{R}(z)^{-1}$ is still proper and $\bar{W}_m(z) = \bar{W}(z)\bar{R}(z)$. For the time-invariant case see e.g. [7], [8]. Moreover, $\bar{W}_m(z)$ and $\bar{W}(z)$ being the cyclic reformulations of periodic systems, the rational matrix $\bar{R}(z)$ can be considered as the cyclic reformulation of a periodic system too. As such, $\bar{R}^{-1}(\infty)$ is block diagonal. Moreover, let $\bar{T}(z) = N(z)A(z)^{-1}$ be a right coprime factorization of the strictly proper function $\bar{T}(z)$. Condition (iii) ensures that $\bar{R}(z)^{-1}A(z)$ is indeed a polynomial matrix. Moreover $\bar{R}^{-1}(z) - \bar{R}^{-1}(\infty)$, as well as $\bar{T}(z)$, have both the structure discussed in Remark 0.2. Due to this fact the equation

$$\bar{R}^{-1}(z) - \bar{R}^{-1}(\infty) = \bar{H}\bar{T}(z) \qquad (6)$$

admits a block-diagonal constant solution matrix $\bar{H} \in \mathcal{R}^{mT\times nT}$. Then, let

$$\bar{G} = \bar{R}(\infty), \quad \bar{K} = -\bar{G}\bar{H}$$

Matrix $\bar{G}$ is block diagonal so that $\bar{K}$ is block diagonal as well. The block entries of the matrices $\bar{K}$ and $\bar{G}$ yield the solution of the problem. We only need to show that $\bar{A} + \bar{B}\bar{K}$ is stable. Indeed this would correspond to stability of $A(\cdot) + B(\cdot)K(\cdot)$. The stability actually follows form condition (iv), which is in fact equivalent to the stability of the polynomial matrix $\bar{R}^{-1}(z)A(z)$.
The necessity of conditions (i), (ii) and (iii) immediately follows from the necessity of the same conditions for the time-invariant case, see Theorem 6.4 in [7], whereas necessity of (iv) is obvious.

Remark 0.4 *As already indicated the interest of Theorem 0.1 mainly consists in the case where one* *wants to match a closed-loop system which is actually time-invariant, with a certain transfer function $W_m(z) \in \mathcal{C}^{p \times m}$. In this case, $\bar{W}_m(z)$ is nothing but the transfer function of the cyclic reformulation of $W_m(z)$ (recall Remark 0.2 and the relation between the zeros of $W_m(z)$ and those of $\bar{W}_m(z)$).*

Now, let us turn to the problem of output feedback model matching. An easy way to build a stabilizing compensator which yields the desidered closed loop input-output operator calls for introduction of a stable periodic state observer. Actually, if the additional assumption of periodic system detectability is made, there exists a periodic matrix $L(\cdot)$ such that the observer dynamic matrix $A(\cdot) + L(\cdot)C(\cdot)$ is stable. Now, consider the observer

$$\begin{aligned}\hat{x}(t+1) = A(t)\hat{x}(t) + B(t)u(t) + \\ + L(t)(C(t)\hat{x}(t) - y(t) + D(t)u(t))\end{aligned} \tag{7}$$

The state error $e(t) = \hat{x}(t) - x(t)$ satisfies the unforced equation

$$e(t+1) = (A(t) + L(t)C(t))e(t)$$

which is stable. Now, let $K(\cdot)$ be constructed as in Theorem 0.1 and consider the control law

$$u(t) = K(t)\hat{x}(t) + G(t)v(t) \tag{8}$$

Equations (7), (8) are those of the output feedback compensator. The closed-loop input-output map coincides with the one given by the state feedback $u(t) = K(t)x(t) + G(t)v(t)$ since the state-error variables are completely unreachable. This map is the target one.

Theorem 0.2 *Let a T-periodic system be given with left invertible and full column rank $\bar{W}(z) \in \mathcal{C}^{pT \times mT}$ as input-output cyclic transfer function. Assume that the periodic pair $A(\cdot), B(\cdot))$ is stabilizable and the periodic pair $A(\cdot), C(\cdot))$ is detectable. There exists a output periodic controller such that the closed-loop system is internally stable and its cyclic transfer function input-output is $\bar{W}_m(z)$ if and only if conditions (i)-(iv) of Theorem 0.1 are fulfilled.*

4. EXAMPLES

In this section two illustrative examples are presented for the state-feedback case. In the first, the underlying periodic system is minimum phase (there are no finite unstable zeros) , so that it is possible to work out a periodic state-feeedback control law which match a internally stable time-invariant system. The second example, on the contrary, illustrates the impossibility of achieving a stable closed-loop time invariant target function, when the system at hand possesses non minimum phase zeros.

example 1

Consider system (1),(2) with $T = 2$, and

$$A(0)\ b = \begin{bmatrix} 1 & 0 \\ 1 & 1 \end{bmatrix},\quad A(1) = \begin{bmatrix} 1 & 0 \\ 0 & 1 \end{bmatrix},$$
$$B(0) = \begin{bmatrix} 1 \\ 0 \end{bmatrix},\quad B(1) = \begin{bmatrix} 0 \\ 1 \end{bmatrix}$$

$$C(0) = \begin{bmatrix} 1 & 0 \end{bmatrix},\quad C(1) = \begin{bmatrix} 0 & 1 \end{bmatrix},$$
$$D(0) = 0,\quad D(1) = 0$$

Then $\bar{T}(z)$ and $\bar{W}(z)$ are as follows:

$$\bar{T}(z) = \begin{bmatrix} z^2-1 & 0 \\ 1 & z^3-z \\ z^3-z & 0 \\ z & z^2-1 \end{bmatrix} \frac{1}{(z^2-1)^2}$$

$$\bar{W}(z) = \begin{bmatrix} z^2-1 & 0 \\ z & z^2-1 \end{bmatrix} \frac{1}{(z^2-1)^2}$$

Suppose that we want the closed loop system to be time invariant with

$$W_m(z) = \frac{1}{z^2}$$

as target transfer function. Hence, its cyclic version is

$$\bar{W}_m(z) = \frac{1}{z^2}\begin{bmatrix} 1 & 0 \\ 0 & 1 \end{bmatrix}$$

Then,

$$\bar{R}^{-1}(z) = \begin{bmatrix} z^2(z^2-1) & 0 \\ z^3 & z^2(z^2-1) \end{bmatrix} \frac{1}{(z^2-1)^2},$$
$$\bar{G} = \bar{R}(\infty) = \begin{bmatrix} 1 & 0 \\ 0 & 1 \end{bmatrix}$$

Conditions (i)-(iv) are fulfilled. Equation (6) is satisfied by

$$\bar{H} = \begin{bmatrix} 1 & 0 & 0 & 0 \\ 0 & 0 & 1 & 1 \end{bmatrix}$$

so that

$$\bar{K} = -\bar{G}\bar{H} = \begin{bmatrix} -1 & 0 & 0 & 0 \\ 0 & 0 & -1 & -1 \end{bmatrix}$$

The periodic control law is then

$$u(t) = K(t)x(t) + G(t)v(t)$$

where

$$G(0) = 1,\quad G(1) = 1$$
$$K(0) = \begin{bmatrix} -1 & 0 \end{bmatrix},\quad K(1) = \begin{bmatrix} -1 & -1 \end{bmatrix}$$

Notice that as a consequence of condition (iv), the closed-loop system has two characterustic multipliers equal to zero, i.e. is stable.

Example 2

Consider system (1),(2) with $T = 2$, and

$$A(0) = \begin{bmatrix} 3 & 0 \\ 0 & 2 \end{bmatrix}, \quad A(1) = \begin{bmatrix} 0 & 1 \\ 3 & 0 \end{bmatrix},$$
$$B(0) = \begin{bmatrix} 0 \\ 1 \end{bmatrix}, \quad B(1) = \begin{bmatrix} 1 \\ 0 \end{bmatrix}$$

$$C(0) = \begin{bmatrix} 1 & -1 \end{bmatrix},$$
$$C(1) = \begin{bmatrix} 1/3 & 2 \end{bmatrix}, \quad D(0) = 0, \quad D(1) = 0$$

Then $\bar{T}(z)$ and $\bar{W}(z)$ are as follows:

$$\bar{T}(z) = \begin{bmatrix} z^2 & z^3 \\ 9 & 9z \\ 3z & 3z^2 \\ z^3 & 18 \end{bmatrix} \frac{1}{(z^4 - 18)}$$

$$\bar{W}(z) = \begin{bmatrix} z^2 - 9 & z^3 - 9z \\ 2z^3 + z & z^2 + 36 \end{bmatrix} \frac{1}{(z^4 - 18)}$$

Suppose we want to match a time-invariant target system and let us choose

$$W_m(z) = \frac{1}{z}$$

Then

$$\bar{W}_m(z) = \frac{1}{z^2} \begin{bmatrix} 0 & z \\ z & 0 \end{bmatrix}$$

$$\bar{R}^{-1}(z) - \bar{R}^{-1}(\infty) = \frac{1}{z^4 - 18} \begin{bmatrix} z^2 + 36 & z^3 + 36z \\ z^3 - 9z & -9z^2 + 180 \end{bmatrix}$$

Conditions (i)-(iii) are fulfilled. Hence equation (6) has a block diagonal solution matrix

$$\bar{H} = \begin{bmatrix} 1 & 4 & 0 & 0 \\ 0 & 0 & -3 & 1 \end{bmatrix}$$

Therefore

$$\bar{G} = \bar{R}(\infty) = \begin{bmatrix} 0.5 & 0 \\ 0 & 1 \end{bmatrix}$$

$$\bar{K} = -\bar{G}\bar{H} = \begin{bmatrix} -0.5 & -2 & 0 & 0 \\ 0 & 0 & 3 & -1 \end{bmatrix}$$

The periodic control law is then

$$u(t) = K(t)x(t) + G(t)v(t)$$

where

$$G(0) = 0.5, \quad G(1) = 1$$
$$K(0) = \begin{bmatrix} -0.5 & -2 \end{bmatrix}, \quad K(1) = \begin{bmatrix} 3 & -1 \end{bmatrix}$$

Notice however that in this case the resulting closed-loop system characteristic multipliers are 0 and 9 and therefore the system is internally unstable. This is due to the fact that condition (iv) is not satisfied. It is also simple to show that in this case it is impossible to match a time-invariant system without relaxing the closed-loop stability requirement (iv). This conclusion follows from direct investigation of condition (iv) by taking in mind the structure of the cyclic reformulation of a time-invariant system (recall Remarks 0.2 and 0.3). The reason for this negative conclusion is the fact the the periodic system is nonminimum phase.

CONCLUDING REMARKS

The exact model matching problem with stability for periodic systems has been solved in this paper. The main point addressed here was to discuss the conditions under which a periodic system matches a time-invariant one while asymptotic stability is ensured.

REFERENCES

[1] Bolzern P., P.Colaneri and R.Scattolini (1986). Zeros of discrete-time linear periodic systems. *IEEE Trans. Automatic Control*, **Vol. AC-31**, pp. 1057-1058.

[2] Flamm D.S. A new shift-invariant representation for periodic system (1991). *Systems and Control Letters*, **Vol.17**, pp. 9-14.

[3] Grasselli O.M. and S.Longhi. Zeros and poles of linear periodic multivariable systems. *IEEE Trans. Circuits, Systems and Signal Processing*, **Vol. CSSP-7**, pp. 361-380.

[4] Meyer R.A. and C.S.Burrus (1976). Design and implementation of multirate digital filters. *IEEE Trans. Acoustics, Speech and Signal Processing*, **Vol. ASSP-1**, pp. 53-58.

[5] Colaneri P. and S.Longhi (1995). The realization problem for discrete-time periodic systems. *Automatica*, **Vol. 31**, pp. 775-779.

[6] Krank G.M.(1957). Input-output analysis of multirate feedback systems. *IRE Trans. Automatic Contr.*, **Vol. AC-32**, pp. 21-28.

[7] Kucera V. (1991). *Analysis and Design of Discrete Linear Control Systems*, Prentice Hall, London.

[8] Hautus M.L.J. and M.Heymann (1978). Linear feedback, an algebraic approach", *SIAM J. Control and Optimization*, **Vol. 16**, pp. 83-105.

[9] Wang S.H. and C.A.Desoer (1972). The exact model matching of linear multivariable systems. *IEEE Trans. Automatic Control*, **Vol. AC-17**, pp. 347-349.

[10] Park B. and E.I.Verriest (1989). Canonical forms of discrete linear periodically time-varying systems and a control application. *Proc. 28th Conference on Decision and Control*, Tampa (USA), pp. 1220-1225.

[11] Bittanti S. and P.Colaneri (1986). Analysis of discrete-time periodic systems. In: *Control and Dynamic Systems: Digital Control and Signal Processing - Systems and Techniques*, C.T.Leondes Ed., Academic Press, pp. 313-339.

[12] Malabre M. and J.C.M.Garcia (1993). The modified disturbance rejection problem with stability: a structural approoach", Proc. of the *European Control Conference*, pp.1119-1124.

[13] Emre E. and M.J.Hautus (1980). A Polynomial characterization of $(A; B)$-invariant and reachable subspaces. *SIAM J. Control and Optimization*, **Vol. 18**, pp. 420-436.

[14] Colaneri P. and V.Kučera. The model matching problem for discrete-time periodic systems. *IEEE Trans. Automatic Control*, to appear.

ACKNOWLEDGMENTS

The paper was partially supported by the project "Periodic control and filtering techniques of dynamical systems" of the Italian National Research Council (CNR), by MURST and by the Grant Agency of the Czech Republic under contract 102/97/086.

ROBUST PID CONTROL: APPLICATION TO A CONTINUOUS-STIRRED TANK REACTOR

Roman Prokop, Petr Dostál, Zdenka Prokopová

Department of Automatic Control, Faculty of Technology Zlín,
Nám. TGM 275, 762 72 Zlín, Czech Republic
tel., fax: ++420 67 7211521
E-mail: prokop@zlin.vutbr.cz

Abstract: A tuning method for a class of PID-like controllers and a Matlab-Simulink toolbox were developed for robust control SISO continuous-time systems. Controllers are obtained via solutions of diophantine equations in the ring of proper and stable rational functions. Uncertainty and robustness are studied through the infinity norm H_∞. A scalar parameter $m > 0$ was proposed as a tuning knob for controller design with respect to robustness and uncertainty evaluation. An application to a nonlinear chemical reactor is simulated and studied.

Keywords: Robustness, uncertainty, PID controllers, norms, algebraic approaches.

1. INTRODUCTION

Robust controllers and plant uncertainty became a useful and popular discipline in control theory during the last decade. The necessity of robust control was naturally developed by the situation when the nominal plant (used in control design) differs from the real (perturbed) one. A suitable tool for parameter uncertainty is the infinity norm H_∞. Hence, a polynomial description of transfer functions had to be replaced by another one. A convenient description adopted from Vidyasagar (1995), Kučera (1993), Doyle et al (1992) is a factorization approach where transfer functions are expressed as a ratio of two Hurwitz stable and proper rational functions (R_S). Then conditions of robust stability can be easily formulated in algebraic parlance and all controllers are obtained via linear diophantine equations in an appropriate ring.

A class of PID-like controllers is obtained by this way for first and second order SISO systems. The methodology is proposed and more precisely analyzed in Prokop and Dostál (1995), Prokop and Mészáros (1996), Prokop and Corriou (1997). The algebraic approach gives nontraditional PID structures proposed in e.g. Aström et al. (1991), Morari and Zafiriou, (1989). Moreover, for SISO controllers a scalar parameter $m > 0$ can be defined through the appropriate ring. This parameter is utilized as a tuning knob for a class of proposed SISO continuous-time controllers and it strongly influences the dynamics of the feedback system as well as the sensitivity of proposed controllers.

2. SYSTEM DESCRIPTION AND UNCERTAINTY

The plant uncertainty can be expressed conveniently in terms of the fractional description. It means that linear transfer functions are no more represented as a ratio of two polynomial but two elements of an another ring. For the purposes of robust control in the sense of this contribution any transfer function $H(s)$ of a (continuous-time) linear system can be expressed as a ratio of two elements:

$$H(s)=\frac{b(s)}{a(s)}=\frac{B(s)}{A(s)}$$
$$A(s)=\frac{a(s)}{m(s)};\ B(s)=\frac{b(s)}{m(s)} \tag{1}$$
$$m(s)=(s+m)^n; \quad n=\max\{\deg a;\deg b\}$$

where a, b are polynomials. Elements A, B,... constitute a subring $R_m(s)$ of the ring of all Hurwitz stable and proper rational functions $R_s(s)$. A class of PID like controllers is generated by first and second order systems and all controlled plants will be approximated by such systems. Systems with time delays will be approximated by the following fractional representation in $R_m(s)$:

$$H(s)=\frac{b_0e^{-\theta s}}{a_2s^2+a_1s+a_0}\approx\frac{b_0(1-\theta s+\frac{\theta^2}{2})}{a_2s^2+a_1s+a_0}=$$
$$=\frac{\dfrac{b_2s^2+b_1s+b_0}{(s+m)^2}}{\dfrac{a_2s^2+a_1s+a_0}{(s+m)^2}}=\frac{B(s)}{A(s)} \tag{2}$$

Systems of the first order are approximated in similar way and systems without time delays are simply obtained by putting $\theta=0$. For $m=0$ subring $R_m(s)$ expands to the traditional ring $R_s(s)$ (see Vidyasagar, 1985). An infinity norm (convenient for uncertainty) in both rings is defined by

$$\|H\|=\sup_{\mathrm{Re}\,s\geq 0}|H(s)|=\sup_{\omega\in E}|H(j\omega)|$$
$$\left\|H_1\ H_2\right\|=\left\|\begin{matrix}H_1\\H_2\end{matrix}\right\|=\sup_{\mathrm{Re}\,s\geq 0}\{|H_1(s)|^2+|H_2(s)|^2\}^{\frac{1}{2}} \tag{3}$$

This norm is the radius of the smallest circle containing the Nyquist plot of the transfer function. Almost all mathematical models differs from physical systems. Let $H(s)=\dfrac{B(s)}{A(s)}$ be a nominal plant and consider a family of perturbed systems $H'(s)=\dfrac{B'(s)}{A'(s)}$ where

$$\|A-A'\|\leq\varepsilon_1\quad\|B-B'\|\leq\varepsilon_2$$
$$\text{or}\quad\|A-A'\quad B-B'\|\leq\varepsilon \tag{4}$$

3. A CONTINUOUS-STIRRED TANK REACTOR MODEL

Continuous stirred tank reactors (CSTRs) are units frequently used in chemical industry. From the system engineering point of view they belong to a class of nonlinear lumped parameter systems and their mathematical models are described by nonlinear differential equations. It is well known that traditional control methods are often unsatisfactory for a CSTR control. Some control problems are due to the nonlinearity and high sensitivity of the output variables to the input changes. Moreover, the dynamic characteristics of a CSTR may exhibit non-minimum phase behavior, the time delay and sometimes also a changing sign of the gain in various operating points. The basic properties of the reactor depend always on the reaction type taking place inside.

Here, a perfectly mixed CSTR with a cooling jacket is considered, in which an exothermic consecutive reaction $A\xrightarrow{k_1}B\xrightarrow{k_2}C$ takes place. The product is the component B. The reactant volume as well as the reactant volumetric flow rate are assumed to be constant. The reactant and coolant densities as well as specific heat capacities are also constant. A perfectly mixed cooling jacket is assumed with constant coolant volume. The dynamics of the metal jacket wall is neglected since the time constant of the wall is very small in comparison with both reactant and coolant time constants.

Under the above assumptions, the CSTR model can be described by four differential equations of the form

$$\frac{dc_A}{dt}=-\left(\frac{Q_r}{V_r}+k_1\right)c_A+\frac{Q_r}{V_r}c_{Af} \tag{5}$$
$$\frac{dc_B}{dt}=-\left(\frac{Q_r}{V_r}+k_2\right)c_B+k_1c_A+\frac{Q_r}{V_r}c_{Bf} \tag{6}$$
$$\frac{dT_r}{dt}=\frac{h_r}{(\rho c_p)_r}+\frac{Q_r}{V_r}(T_{rf}-T_r)+\frac{A_hU}{V_r(\rho c_p)_r}(T_c-T_r) \tag{7}$$
$$\frac{dT_c}{dt}=\frac{Q_c}{V_c}(T_{cf}-T_c)+\frac{A_hU}{V_c(\rho c_p)_c}(T_r-T_c) \tag{8}$$

with initial conditions $c_A(0)=c_A^s$, $c_B(0)=c_B^s$, $T_r(0)=T_r^s$ and $T_c(0)=T_c^s$.

In Eqs. (5)-(8) c_A and c_B are component A and B concentrations, T_r and T_c are reactant and coolant temperatures, respectively, V are volumes, Q are volumetric flow rates, ρ are densities, c_p are specific heat capacities, A_h is the area for heat transfer and U is the overall heat transfer coefficient. The subscripts are r for the reactant and c for coolant. The feed concentration c_{Af}, and feed temperatures T_{rf}, T_{cf} are assumed to be known and constant, the feed concentration c_{Bf} is zero. The specific reaction rates are given by the Arrhenius equation

$$k_j = k_{0j} \exp\left(\frac{-E_j}{RT_r}\right), \qquad j = 1, 2 \tag{9}$$

where E_j are activation energies and R is the gas constant. The heat of reaction is expressed as

$$h_r = h_1 k_1 c_A + h_2 k_2 c_B \tag{10}$$

where h are reaction enthalpies.

For the next calculation and control simulation the component B concentration takes place of the controlled output and the coolant volumetric flow rate is used as the control input, both as deviations from their steady-state values so that

$$y(t) = c_B(t) - c_B^s \tag{11}$$

$$u(t) = Q_c(t) - Q_c^s \tag{12}$$

All parameter values, feed values and initial conditions of the model are shown in Table 1.

Table 1 Parameter values, feed values and initial conditions

$V = 1.2$ m^3	$\rho_r = 985$ kg m^{-3}
$V_c = 0.64$ m^3	$\rho_c = 998$ kg m^{-3}
$A_h = 5.5$ m^2	$c_{pr} = 4.05$ kJ kg^{-1}
$Q_r = 0.08$ m^{3}min^{-1}	$c_{pc} = 4.18$ kJ kg^{-1}
$U = 43.5$ kJ m^{-2}K^{-1}min^{-1}	
$k_{01} = 5.616e16$ min^{-1}	$E_2/R = 15290$ K
$k_{02} = 1.128e18$ min^{-1}	$h_1 = 4.8e4$ kJ kmol^{-1}
$E_1/R = 13477$ K	$h_2 = 2.2e4$ kJ kmol^{-1}
$c_{Af}^s = 2.85$ kmol m^{-3}	$T_{rf}^s = 323$ K
$Q_c^s = 0.03$ m^{3}min^{-1}	$T_{cf}^s = 293$ K
$c_A^s = 0.165$ kmol m^{-3}	$T_r^s = 350.2$ K
$c_B^s = 0.944$ kmol m^{-3}	$T_c^s = 330.5$ K

The dependence of the output variable upon the control input in the steady-state is shown in Fig. 1. Two operating points were chosen for estimation, approximation and control of the CSTR model. Both are depicted in Fig. 1. and they represent steady-state values of $c_B^s = 0.9435$ and $c_B^s = 1.3413$, respectively.

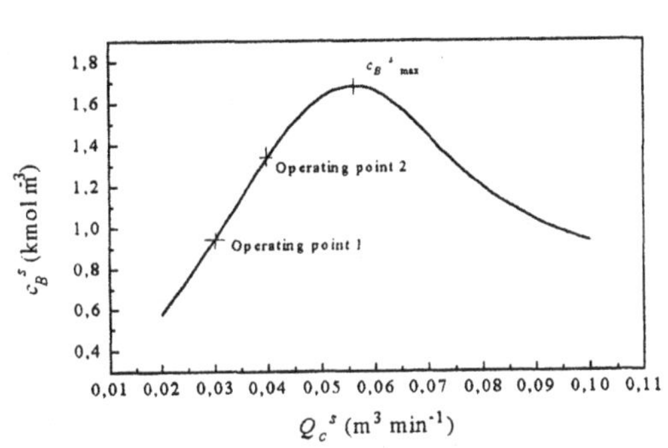

Fig. 1. Steady-state input-output behavior of CSTR.

For the operating point 1, the output y time responses to various step changes of input u are shown in Fig. 2. The nonlinearity of the controlled system is clearly documented.

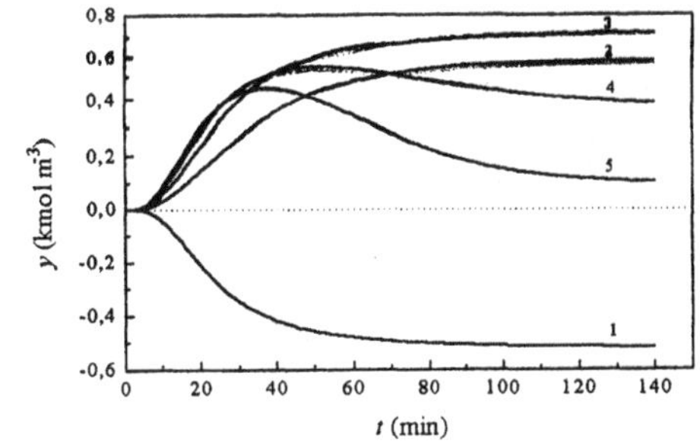

Fig. 2. The output y time responses to input u step changes: u = 1) -0.015, 2) 0.015, 3) 0.030, 4) 0.045, 5) 0.060 [m^{3}min^{-1}] .

4. CONTROL DESIGN

A control system depicted in Fig. 3. is considered for design of a class of PID - like controllers.

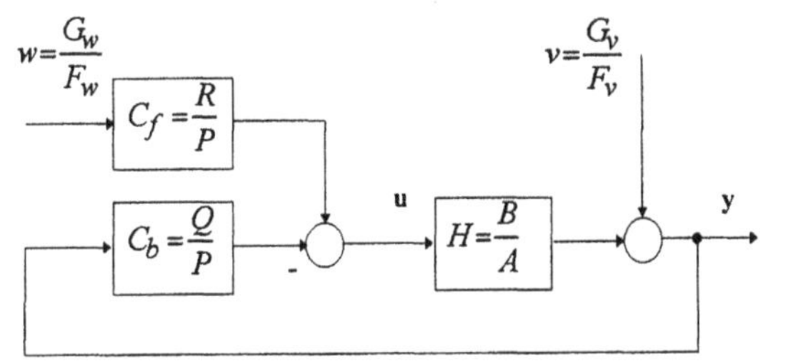

Fig. 3. General control system.

Note that for $Q = R$ a traditional feedback regulator operating on the tracking error is obtained. Transfer functions $w = \frac{G_w}{F_w}$ and $v = \frac{G_v}{F_v}$ represent the reference and disturbance signal, respectively. The control design in $R_m(s)$ is proposed in Prokop and Corriou (1997), Prokop and Mészáros (1996), Prokop and Dostál (1995). For the given nominal plant $H(s)$ in the form of (1) it consists of the following steps:

Step 1: All stabilizing (feedback) controllers are given by all solutions of the equation:

$$A\,P + B\,Q = 1 \tag{13}$$

in the parameter form expressed by the fraction:

$$\frac{Q}{P} = \frac{Q_0 - AT}{P_0 + BT} \tag{14}$$

where $P_0, Q_0 \in R_m(s)$ are particular solutions and T is arbitrary in $R_m(s)$.

Step 2: Feedback controllers (if necessary) are given in a similar way by all solutions of

$$F_w S + BR = 1 \qquad (15)$$

Step 3: Compute $P \in R_m(s)$ such that

F_w divides P for asymptotic tracking problems
F_v divides P for disturbance rejection problems
(16)

Step 4: For perturbed plants (4) choose such P, Q in (13), (14) which fulfill the conditions

$$\varepsilon_1 \|P_0 + BT\| + \varepsilon_2 \|Q_0 - AT\| \leq 1 \qquad (17.a)$$

$$\text{or} \quad \varepsilon \left\| \begin{matrix} P_0 + AT \\ Q_0 - AT \end{matrix} \right\| \leq 1 \qquad (17.b)$$

Steps 1 - 3 represents the first and simplest case of robust design. All solutions of diophantine equations in $R_m(s)$ are expressed as functions of the parameter $m > 0$. The value of this parameter strongly influences dynamics as well as robustness of the proposed control system. For a deeper insight into robustness the notion of the sensitivity function:

$$\in = \frac{y}{v} = \frac{1}{1 + HC_b} = A(P_0 + BT) \qquad (18)$$

can be utilized as the second possibility. Similarly, for SISO systems $\in$ is also a function of a scalar parameter $m > 0$ and it can be minimized by a simple scalar optimization method. In this way the „most robust" controller for a given nominal plant is obtained. The third case covers situations where perturbed plants are known and the norms in (4) can be evaluated. Then inequalities (17) yield generally nonlinear inequalities for the scalar parameter $m > 0$ which can be solved numerically.

5. ROBUST AND SENSITIVITY TUNING

A MATLAB-like toolbox with simulation support in SIMULINK was developed for nominal plants:

$$F_1(s) = \frac{b_0}{s + a_0}$$

$$F_2(s) = \frac{b_0 e^{-\theta s}}{s + a_0} \approx \frac{b_0(1 - \theta s)}{s + a_0}$$

$$F_3(s) = \frac{b_0}{s^2 + a_1 s + a_0} \qquad (19)$$

$$F_4(s) = \frac{b_0 e^{-\theta s}}{s^2 + a_1 s + a_0} \approx \frac{b_0\left(1 - \theta s + \frac{\theta^2}{2} s^2\right)}{s^2 + a_1 s + a_0}$$

which yield a class of generalized PID-like control structures. The program system enables design and simulation of a wide spectrum of robust control problems.

The simulation of the perturbed plant with the regulator computed for the nominal transfer function is carried out in standard Simulink environment. Simulation parameters as simulation horizon, reference, load disturbance and disturbances, input constraints, ... can also be defined by the user. All simulation variables can be stored and transferred out of the Matlab workspace and ITAE, IAE or IE can be calculated as a tool of comparison and quality evaluation of the control behavior.

The proposed methodology is illustrated by the example with the first order system and a time delay. Suppose a nominal plant in $R_m(s)$ given by

$$H(s) = \frac{\beta e^{-\theta s}}{s + a_0} \approx \frac{b_1 s + b_0}{s + a_0} = \frac{B(s)}{A(s)} == \frac{\frac{b_1 s + b_0}{s + m}}{\frac{s + a_0}{s + m}}, m \geq 0 \qquad (20)$$

The control goal is to design in a robust way:

a) an asymptotic tracking feedback controller (Q=R) for a stepwise reference signal
b) the same as in *a)* but with feedback and feedforward parts (Fig. 3.)

The denominator of the reference generator in $R_m(s)$ in the form $F_w = \frac{s}{s+m}$.

a) All solutions of AP+BQ = 1 in the parameter form are:

$$P = p_0 + \frac{b_1 s + b_0}{s + m} T; \quad Q = q_0 - \frac{s + a}{s + m} T; \qquad (21)$$

where $p_0 = \frac{b_0 - b_1 m}{b_0 - a_0 b_1}$; $q_0 = \frac{m - a_0}{b_0 - a_0 b_1}$ and T is arbitrary in $R_m(s)$. The simplest but not robust solution is given for T= 0. Robust controllers are given for P divisible by F_w which is obtained for $T = -\frac{m}{b_0} \cdot \frac{b_0 - b_1 m}{b_0 - a_0 b_1}$. Then the feedback (only) controller is given by

$$P_1 = \frac{s}{s + m} \frac{b_0 - m b_1}{b_0} p_0;$$

$$Q_1 = \frac{(q_0 + \frac{m}{b_0} p_0)s + m(q_0 + \frac{a_0 p_0}{b_0})}{s + m} \qquad (22)$$

and the control law is generated by the differential equation:

$$u = \frac{1}{\tilde{p}_1}[\tilde{q}_1(w-y) + \tilde{q}_0 \int (w-y)dt] \qquad (23)$$

where $\tilde{p}_1 = \frac{b_0 - mb_1}{b_0} p_0; \quad \tilde{q}_1 = q_0 + \frac{m}{b_0} p_0;$

$\tilde{q}_0 = (q_0 + \frac{a_0}{b_0} p_0)m$.

Naturally, equation (23) represents a PI regulator.

b) All feedforward parts are given by solutions of $FS + BR = 1$:

$$R = \frac{m}{b_0} - \frac{s}{s+m} Z \qquad (24)$$

where Z is again arbitrary in $R_m(s)$. For the simplest Z=0 we get the control law:

$$u = \frac{1}{\tilde{p}_1}[\frac{m}{b_0} w + \frac{m^2}{b_0} \int w dt - \tilde{q}_1 \int y dt - \tilde{q}_0 y] \quad (25)$$

which represents a generalized PI controller proposed e.g. by Aström et al. (1991) in the form:

$$u = k(\beta w - y) \quad + \quad \frac{k}{T_i} \int (w-y) d\tau \qquad (26)$$

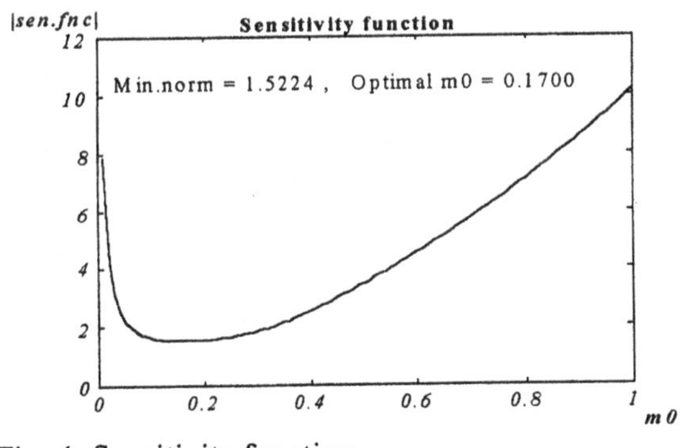

Fig. 4. Sensitivity function.

Fig. 4. illustrates the sensitivity function for $H(s) = \frac{1.25 e^{-5s}}{2s+1}$ and PI regulator as a function of $m > 0$ with the optimal value $m_0 = 0.17$. It is clear that all controller parameters are functions of $m > 0$. A deeper insight into conditions (17) can be found in Prokop and Corriou (1997), Prokop and Mészáros (1996), Prokop, Bobál and Prokopová (1997).

6. MODEL APPROXIMATION AND SIMULATION RESULTS

Dynamics of the CSTR described in section 3 in both operating points is approximated by the second order system (with and without time delay) in the form:

$$H(s) = \frac{20e^{-\theta s}}{400s^2 + 40s + 1} \approx \frac{B(s)}{A(s)}$$
$$\frac{B(s)}{A(s)} = \frac{\frac{b_2 s^2 + b_1 s + b_0}{(s+m)^2}}{\frac{400s^2 + 40s + 1}{(s+m)^2}}, m \geq 0 \qquad (27)$$

where the time delay is either $\theta = 0$ or $\theta = 5$. Coefficients b_i are computed according to (2). The approximation was obtained by a four-parameter method adopted from Aström and Hägglund (1995) applied to step responses in Fig. 2. Because the CSTR model is non-linear, the approximation is rather rough and it depends on a given operating point. In spite of this situation, it is desirable to find a PID-like controller which is „sufficiently acceptable" for both operation points. More precisely, the control goal is to design (in a robust way) a FB and FBFW controllers according to above mentioned methodology for nominal plants (27) and choose such values of $m > 0$ which give control responses without overshooting. From the technological point of view, aperiodic responses after step changes of the set point are inherent and natural conditions for chemical reactors control.

The „most robust controller" was obtained by scalar minimization of the sensitivity function (18) for the nominal plant (27) without time delay ($\theta = 0$). Both FB and FBFW control schemes gave the optimal value $m_0 = 0.05$. Derivation of controller parameters does not bring any problems but for the second order the resulting expressions are a bit more complex than the ones in (21-26).

Fig. 5. to 8. show the results of *Simulation 1* to *Simulation 4* of proposed controllers, respectively. The following list describes the conditions of control:

<u>*Simulation 1:*</u> Operating point 1, FB controller, without time delay ($\theta = 0$), set points $w = +0.5; +0.1$.
<u>*Simulation 2:*</u> Operating point 1, FBFW controller, with time delay ($\theta = 5$), set points $w = +0.5; +0.1$.
<u>*Simulation 3:*</u> Operating point 2, FB controller, without time delay ($\theta = 0$), set points $w = \pm 0.32$.
<u>*Simulation 4:*</u> Operating point 2, FBFW controller, with time delay ($\theta = 5$), set points $w = \pm 0.32$.
Simulation results document that there does not exist a feedback PID controller which enables the aperiodic control responses under given circumstances. Fig. 5. and 7. show control responses in both operating points with the optimal FB PID controller obtained by sensitivity minimization. The

situation was improved by involving a feedforward part of the control law (see Fig 6. and 8.). The resulting control law is of a generalized PID-type in the sense of (26) and Aström et al. (1991).

7. CONCLUSIONS

A design method based on the fractional representation was developed for SISO continuous-time systems generally with time delay. Control laws for second order systems yield a class of PID - like controllers. The proposed methodology enables to tune and influence the robustness and control behavior by a single scalar positive parameter. The design methodology was successfully applied to a nonlinear chemical reactor.

REFERENCES

Aström, K.J., Hang, C.C., Persson, P., Ho, W.K. (1991): Towards Intelligent PID Control. *Automatica*, **28**, pp.1-9.

Aström, K.J., Hägglund, T. (1995): *PID Controllers: Theory, Design and Tuning*, 343 p. Instrument Society of America.

Doyle J.C., Francis B.A., Tannenbaum A.R.(1992): *Feedback Control Theory*. Maxwell McMillan.

Kučera, V. (1993): Diophantine Equations in Control - A Survey, *Automatica*, **29**, No.6, pp.1361-75.

Kučera, V. (1995): Algebraic Methods in Control - Theory and Applications. Prepr. *IFIP/IFAC/IFORS Conf. System Modelling and Optimization*, Prague.

Morari, M., Zafirou, E. (1989): *Robust Process Control*. Prentice Hall, New Jersey.

Prokop, R., Dostál, P.(1995): Design of Simple Robust Controllers. Prepr. *10th Int. Conference.on Process Control '95*, Tatranské Matliare, pp.103-108.

Prokop, R., Prokopová Z. (1996): Design of Robust Control of SISO Continuous-time Systems, Prepr. *Conf. RIP-PC '96*, Horní Bečva.

Prokop, R., Corriou, J.P. (1997): Design and Analysis of Simple Robust Controllers, *Int. Journal of Control*, Vol. **66**, No. 6, pp. 905-921.

Prokop, R., Bobál V. and Prokopová Z. (1997): Design of Robust Controllers for Time - delay Systems, Prepr. *Conf. CIS '97, IFAC-IFIP-IMACS*, Belfort.

Venkatashankar, V., Chidambaram, M. (1994): Design of P and PI Controllers for Unstable Firs-order Plus time Delay Systems, *Int. J.of Control*, Vol. **60**, No.1, p.137-144.

Vidyasagar, M. (1985): *Control System Synthesis: A Factorization Approach*. MIT Press, Cambridge.

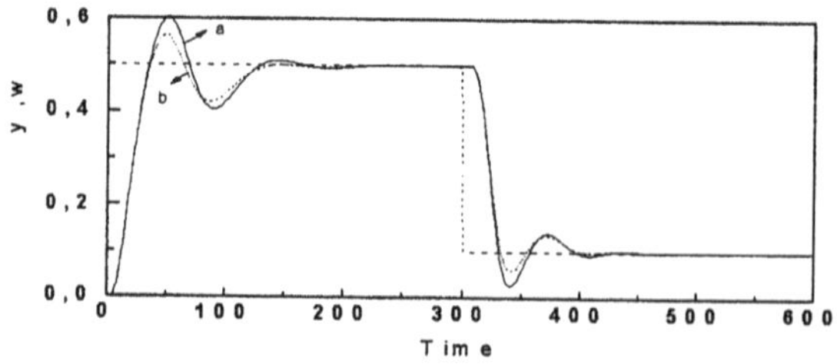

Fig. 5. Simulation 1 - output and reference

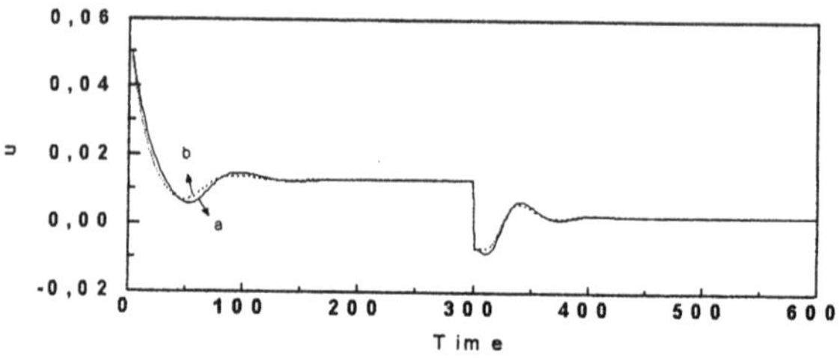

Fig. 5u. Simulation 1 - input

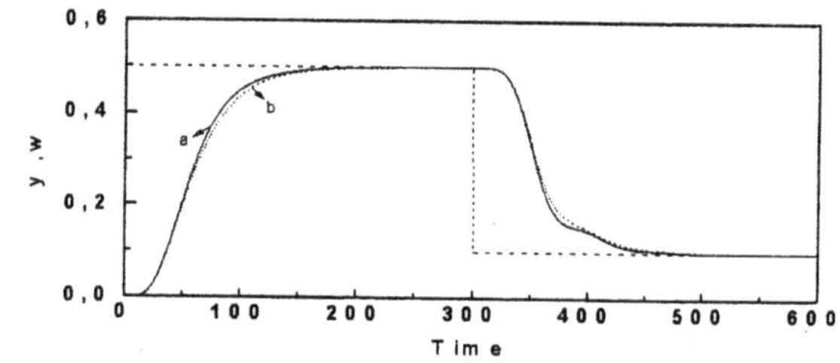

Fig. 6. Simulation 2 - output and reference

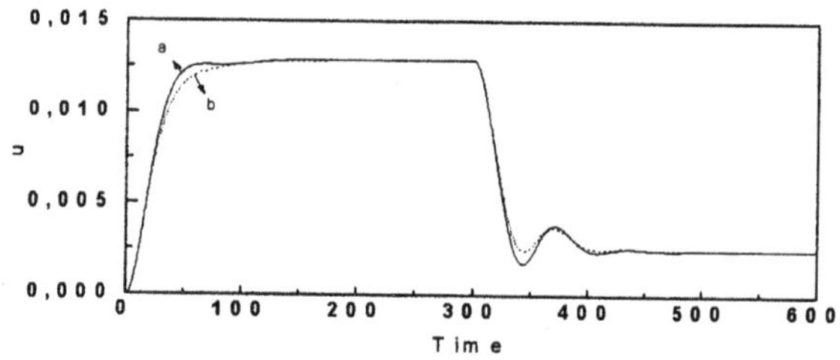

Fig. 6u. Simulation 2 - input

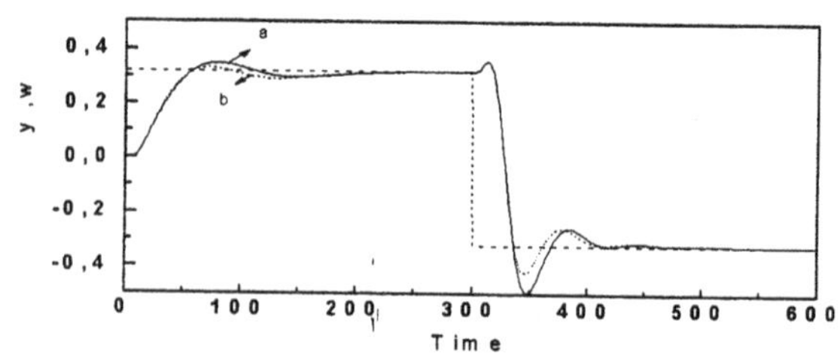

Fig. 7. Simulation 3 - output and reference

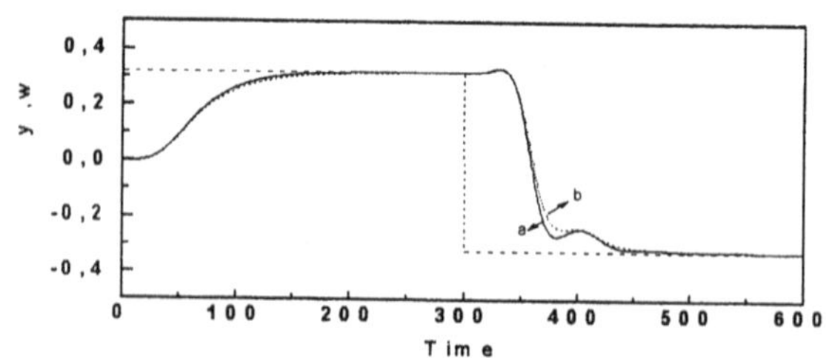

Fig. 8. Simulation 4 - output and reference

MODEL REFERENCE LINEARIZATION BASED ON H_∞ CONTROL

Bertran, E. and G. Montoro

Department of Signal Theory and Communications.
Universitat Politècnica de Catalunya.
C/ Jordi Girona, 1-3.
08034-Barcelona (Spain).
email: tscgml@eupbl.upc.es

Abstract: The use of H_∞ optimization methods in model reference structures is a new approach to system linearization. This paper presents a design methodology to linearize a system in this way. The resulting system is based on output error feedback (linear), and it has properties of robustness. The only information required about the nonlinearity is a bound of its gain (induced norm). Thus, it is a useful method for solving the tracking problem in uncertain systems (unmodelled dynamics). Theoretical results are applied to the linearization of a radiofrequency bandpass power amplifier.

Keywords: H-infinity control, Linearization, Model reference control, Nonlinear control systems, Saturation control.

1. INTRODUCTION

Nonlinearities in a system involve distortion and harmonics generation, not always removable with the use of linear filters. If the nonlinear effects occupy a frequential band outside of the desired signal one, then they can be neglected with a conventional filter. However, the problem is difficult to solve when signal and distortion spectra are in the same frequential band. In this case more advanced control techniques are required.

Some advanced linearization methods are difficult to use except when a detailed model of the nonlinear system is available. Moreover, there is the added difficulty of formulating the state-space model in physical variables in order to avoid nonlinear observers in further hardware developments. Variable structure methods are useful in front of unmodelled dynamics, but they present chattering in the control signal (Bertran and Montoro, 1995). Apart from aspects of implementation, chattering is not allowable in some applications due to its high frequency content (as is the case of communication designs).

Adaptive systems are a widely used solution for reducing nonlinear effects. Discrete-time solutions are a powerful way if the system bandwidth is reduced enough with regard to the controller speed (Nyquist condition). So, indirect adaptive systems suffer severe restrictions in implementation in high frequency systems. Model reference adaptive systems based on Lyapunov or Hyperstability designs can be implemented in continuous time, and the results are very good from the time domain point of view. However, the spectral analysis of the control signal reveals intermodulations and an increase of high frequency harmonics, very dangerous in applications such as vibration control or radiofrequency control systems.

Nonlinear systems have recently been controlled minimizing the L_2 induced norm. The L_2 gain is similar to H_∞ norm for linear systems. But this approach has some drawbacks (Ball, *et al.*, 1993), (Schaft, 1992), such as the nonlinearity of the controller (which hinders the continuous-time implementation) or the necessity of a detailed formulation of the nonlinear system.

In this paper, starting from a model reference structure, a continuous-time robust control strategy is studied and applied to the linearization problem. The structure obtained combines model reference structure and robustness, and no detailed modelling of the nonlinearities is necessary. The resulting controller is linear, and the only required design information about the nonlinearities is a bound related to the energy (2-norm) gain of the nonlinearities. The design method guarantees the robustness of the overall system. Finally, the method is applied to the linearization of a radiofrequency amplifier.

2. PRELIMINARIES

The classic way of linearization introduced by Zames (Zames, 1963) was based on a constant feedback in the nonlinear closed-loop, where the nonlinearity is static and sectorially bounded and the linear part of the system presents a second order bandpass response. The closed-loop relation is given by a functional as:

$$y = M(x)\ , \quad where\ M = M_o + \Delta M \qquad (1)$$

where y is the system output, and x its input. Zames found the best constant feedback in order to reduce the difference between the closed loop system M and the nominal (desired) system M_o. In this way he introduced the concept of *"system gain"* to establish a stability criterion, with a formulation and results similar to those obtained by using small-gain theory.

Since the work of Zames, the design of controllers capable of achieving insensibility to system uncertainties (both with linear and nonlinear unmodelled dynamics) has been an active area of research. Usually the uncertain system is characterized by a multiplicative model, as:

$$P = P_o + P_o\Delta_p = P_o(1 + \Delta_p) \qquad (2)$$

where P_o represents the transfer function of the linear nominal part and Δ_p is the uncertainty. So, the control objective is twofold: first, a tracking condition, and second, the closed-loop stability must be assured with regard to the uncertainties. Due to the internal uncertainties, an excessive control signal could unstabilize the overall system. This compromise between two control objectives is known as the *mixed sensitivity problem* (Doyle, *et al.*, 1992), and it has a well-known solution for linear uncertainties. However, for nonlinear uncertainties it is not directly applicable, as is the case of our linearization objective.

3. PROPOSED SOLUTION

The design of MRAC is systematic for linear systems with parametric (and bounded) uncertainty. However, if the plant has nonlinearities or structural uncertainty the systematic designs (Lyapunov or hyperstability based) do not assure zero tracking error and, in certain conditions, stability is not assured.

Model reference control systems are appropriate for solving tracking control objectives (Yaesh and Shaked, 1991). So, our proposal is to design a model reference system with robust performance in the presence of nonlinearities. Figure 1 represents the proposed linearizer structure. The linearization is done if the output of the nonlinear system follows the output of a linear reference model. For this proposal, infinity norm optimization theory is used in order to obtain the best linear feedback of the output error. The design methodology is detailed below.

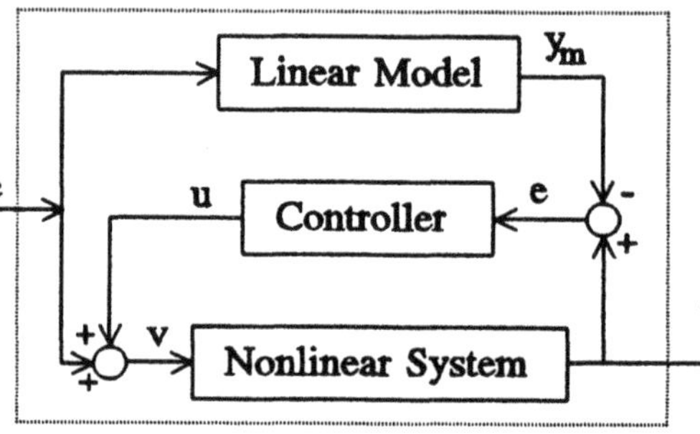

Fig. 1. Model reference structure.

Let y_m and y_p be the outputs of the reference model and the nonlinear system to be linearized, respectively. Their expressions are:

$$y_p = P(u+u_c) = P_o(1+\Delta_p)(u+u_c)$$
$$y_m = P_o(u_c) \qquad (3)$$

where u_c is the reference input and u is the control signal to be designed. The linear reference model is characterized by the linear functional P_o, so it is operable as a transfer function. To characterize the nonlinear system a multiplicative model is assumed, where Δ_p is a nonlinear functional. The error system is formulated as follows:

$$e = y_p - y_m = P_o(u + \Delta_p(u+u_c)) = P_o(u+n)$$
$$being:\ n = \Delta_p(u+u_c) \qquad (4)$$

where n represents an internal disturbance due to the nonlinear effects, which depends on $u+u_c$. Its output energy is assumed to be bounded as follows:

$$\|n\|_2 \leq \alpha \|v\|_2 \ \rightarrow\ \|n\|_2 \leq \alpha\ \|u+u_c\|_2 \leq \alpha\ \|u\|_2 + \alpha\ \|u_c\|_2 \qquad (5)$$

so, α is a bound of the gain between the energies of n and the total nonlinear system input $u+u_c$.

The control objective can be established as the minimization of the effect of the nonlinearity on a cost signal z related to the error and the control signal. The vector z is defined as:

$$z=\begin{bmatrix} z_1 \\ z_2 \end{bmatrix}=\begin{bmatrix} e \\ \beta_1 u \end{bmatrix} \tag{6}$$

where z_1 is the error signal e to be minimized, equation (4), and z_2 is a weighting factor of the control signal. By adjusting the parameter β_1 we can modify the importance of u in order to obtain a set of solutions with different control signal norms. Moreover, due to the dependency of the nonlinear effects on u, the choice of β_1 makes it possible to determine the degree of nonlinear effects and, consequently, a stability degree is established. This will be proved below.

The control signal u is obtained from the controller K, whose input is a noisy measurement e_y of the error e. Hence, it is given by:

$$\begin{aligned} e_y &= e+\beta_2 d = P_o(u+n)+\beta_2 d \\ u &= K e_y \end{aligned} \tag{7}$$

where β_2 is a weight of the measurement disturbance (d). If the weight β_2 were not considered, the obtained solution could have an infinite bandwidth, with the consequent problems of high frequency noise (Montoro, 1996).

The linear nominal system (P_0), strictly proper, is formulated in the state-space in order to obtain, as usual, the matrix A and the vectors b and c:

$$P_o=P_o(s)=c(sI-A)^{-1}b:=\left[\begin{array}{c|c} A & b \\ \hline c & 0 \end{array}\right] \tag{8}$$

Next, the problem is formulated in a similar way to the "standard" open-loop problem in H_∞ norm optimization (Doyle, *et al.*, 1989). Using equations (6) and (7), the open-loop transfer matrix (G) is obtained:

$$\begin{bmatrix} z_1 \\ z_2 \\ e_y \end{bmatrix}=\begin{bmatrix} P_0 & 0 & P_0 \\ 0 & 0 & \beta_1 \\ P_0 & \beta_2 & P_0 \end{bmatrix}\begin{bmatrix} n \\ d \\ u \end{bmatrix} \rightarrow \begin{bmatrix} z \\ e_y \end{bmatrix}=G\begin{bmatrix} w \\ u \end{bmatrix} \tag{9}$$

$$\textit{being:}\quad z=\begin{bmatrix} z_1 \\ z_2 \end{bmatrix} \quad \textit{and} \quad w=\begin{bmatrix} n \\ d \end{bmatrix}$$

The vector w includes the nonlinearity effects and the measurement noise. In our formulation is considered a noisy measurement of the error, e_y.

The open-loop *generalized system* can be formulated in state-space as:

$$G=G(s):=\left[\begin{array}{c|cc} A & B_1 & B_2 \\ \hline C_1 & 0 & D_{12} \\ C_2 & D_{21} & 0 \end{array}\right];\ B_1=[b\ \ 0];\ B_2=[b]; \tag{10}$$

$$C_1=\begin{bmatrix} c \\ 0 \end{bmatrix};\ C_2=[c];\ D_{12}=\begin{bmatrix} 0 \\ \beta_1 \end{bmatrix};\ D_{21}=[0\ \ \beta_2];$$

The equations obtained in (7) and (9) make it possible to formulate the closed loop transfer matrix as:

$$\begin{bmatrix} z_1 \\ z_2 \end{bmatrix}=\begin{bmatrix} \dfrac{P_o}{1-P_oK} & \dfrac{P_oK}{1-P_oK}\beta_2 \\ \dfrac{P_oK}{1-P_oK}\beta_1 & \dfrac{K}{1-P_oK}\beta_1\beta_2 \end{bmatrix}\begin{bmatrix} n \\ d \end{bmatrix} \rightarrow z=Tw \tag{11}$$

The controller K that minimizes the H_∞ norm of the closed-loop transfer matrix T is obtained by using the standard methodology (γ-iteration) (Doyle, *et al.*, 1989), based on the solution of two Riccati algebraic equations. So, the least γ which accomplishes $\|T\|_\infty<\gamma$ is obtained. In our case, the Hamiltonian matrices associated with the Riccati equations are:

$$X_\infty=Ric\left(\begin{bmatrix} A & \dfrac{B_1B_1^T}{\gamma^2}-\dfrac{B_2B_2^T}{\beta_1^2} \\ -C_1^TC_1 & -A^T \end{bmatrix}\right)$$

$$Y_\infty=Ric\left(\begin{bmatrix} A^T & \dfrac{C_1^TC_1}{\gamma^2}-\dfrac{C_2^TC_2}{\beta_2^2} \\ -B_1B_1^T & -A \end{bmatrix}\right) \tag{12}$$

Moreover, if there is no measurement noise, $d=0$. It follows, from equation (11):

$$\|z_1\|_2=\|e\|_2\le\gamma_e\|n\|_2 \qquad \textit{where:}\ \gamma_e=\left\|\frac{P_o}{1-P_oK}\right\|_\infty$$

$$\frac{\|z_2\|_2}{\beta_1}=\|u\|_2\le\gamma_u\|n\|_2 \qquad \textit{where:}\ \gamma_u=\left\|\frac{P_oK}{1-P_oK}\right\|_\infty \tag{13}$$

Consequently, from equations (5) and (13), the following relations are obtained:

$$\|n\|_2\le\frac{\alpha}{1-\gamma_u\alpha}\|u_c\|_2;\qquad \|u\|_2\le\frac{\gamma_u\alpha}{1-\gamma_u\alpha}\|u_c\|_2;$$

$$\|e\|_2\le\frac{\gamma_e\alpha}{1-\gamma_u\alpha}\|u_c\|_2; \tag{14}$$

from which is settled the search for the best controller, according to one method (Montoro, 1996):

i) Initially, choose a fixed β_2 value and a candidate one for β_1.

ii) By γ-iteration, find the controller that minimizes the H_∞ of the closed-loop transfer matrix T.

iii) Obtain the closed-loop transfer matrix, equation (11). Thus, the bounds γ_e and γ_u can be computed, for instance by measuring the peak of the frequency response.

iv) There is a compromise between tracking and stability. The condition $\gamma_u\alpha<1$ is necessary to achieve stability, equation (14), and a reduced γ_e is necessary to have reduced error. If the stability condition is accomplished, then β_1 can be decreased, in order to decrease γ_e, and go to *ii)* to compute another solution.

v) Finally, it could be necessary to adjust the desired controller bandwidth. If β_2 is increased, then the controller bandwidth tends to decrease. However, changing β_2 involves another controller which may change the γ_u and γ_e bounds. Hence, there is a compromise between tracking, stability and controller bandwidth.

4. PRACTICAL APPLICATION

This application focuses on the problem of the linearization of a radiofrequency amplifier with bandpass frequency response. The amplifier can be considered as a transistorized stage where the signal amplification is performed, followed by a frequency limitation stage. Thus, a bandpass power amplifier can be modelled as a static nonlinearity in cascade with a bandpass filter. This is shown in Figure 2, where the static nonlinear block $\phi(\cdot)$ is connected to a linear system $H(s)$.

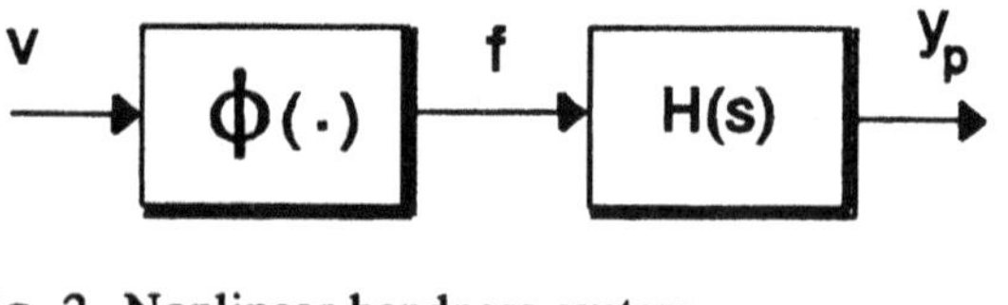

Fig. 2. Nonlinear bandpass system.

The transistorized power stage is characterized by considerable amplification in front of small signal inputs, which is reduced when the input signal level increases (it has a compression effect). From experimental tests, we consider the model of the static nonlinearity shown in Figure 3. The curve (1) is the nonlinearity, and (2) is a straight line that represents the linear behaviour, only true for small signals.

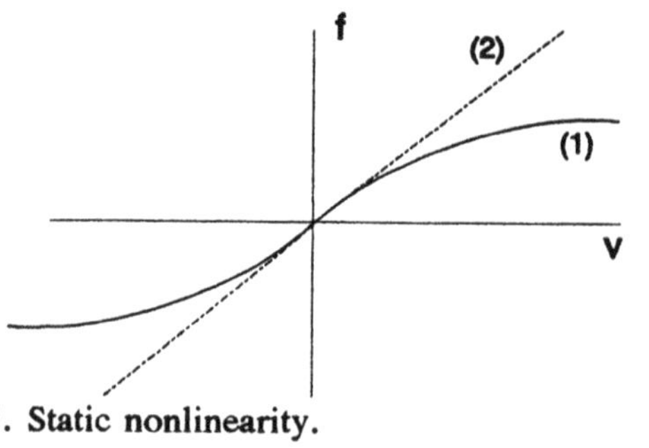

Fig. 3. Static nonlinearity.

We can approximate the nonlinearity curve (1) by the following hyperbolic tangent function, expanded in a Taylor series:

$$f = 10\tanh(v) \approx 10\left(v - \frac{v^3}{3} + \frac{2v^5}{15} - \ldots\right) \qquad (15)$$

The linear approximation around the origin is an amplification by *10*. The linear block is characterized by the following normalized (and coefficient rounded) bandpass function:

$$H(s) = \frac{0.2s}{s^2+0.2s+1} \qquad (16)$$

Let us now state the problem in order to structure it as detailed above to apply the proposed linearization method. Linear and nonlinear blocks are:

$$\textit{linear block:}\ \ y_p = \frac{0.2s}{s^2+0.2s+1} f \qquad (17)$$

$$\textit{nonlinear block:}\ \ f = 10v + 10n$$

So, the system to be linearized is:

$$y_p = \frac{2s}{s^2+0.2s+1}(v+n) \qquad (18)$$

$$n = \tanh(v) - v$$

The reference model is chosen to coincide with the linear part (nominal system, P_0) of the global nonlinear system:

$$y_m = \frac{2s}{s^2+0.2s+1} u_c \qquad (19)$$

It is straightforward to deduce, by choosing $v=u+u_c$, the error system formulation:

$$e = y_p - y_m = \frac{2s}{s^2+0.2s+1}(v+n-u_c) = \frac{2s}{s^2+0.2s+1}(u+n) \qquad (20)$$

The next step is to find a bound of the relation between the norms of the signals v and n. The graphical representation of n as a function of v is shown in Figure 4, and the following inequality between norms is revealed:

$$|n| \le |v| \;\rightarrow\; \|n\|_2 \le \|v\|_2 \;\rightarrow\; \alpha = 1 \qquad (21)$$

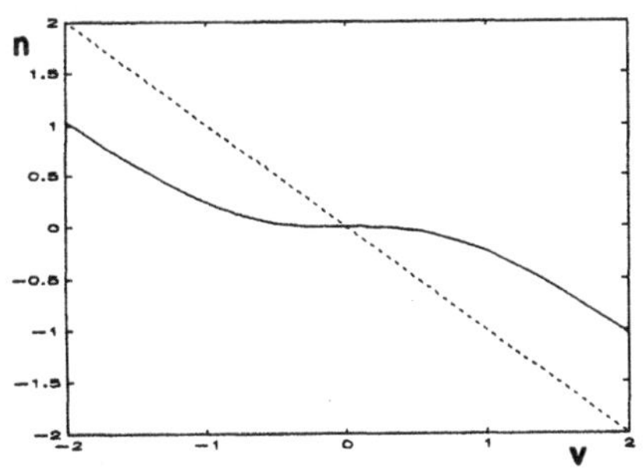

Fig. 4. Nonlinear perturbation characteristic, sectorially bounded.

Now, the problem is established as proposed above. The generalized system, whose infinity norm will be minimized, is:

$$A=\begin{bmatrix}-0.2 & -1\\ 1 & 0\end{bmatrix};\; B_1=\begin{bmatrix}1 & 0\\ 0 & 0\end{bmatrix};\; B_2=\begin{bmatrix}1\\ 0\end{bmatrix};\; C_1=\begin{bmatrix}2 & 0\\ 0 & 0\end{bmatrix};$$

$$C_2=[2\;\;0];\; D_{12}=\begin{bmatrix}0\\ \beta_1\end{bmatrix};\; D_{21}=[0\;\;\beta_2];$$

(22)

where β_1 and β_2 are selected according to a compromise between tracking, stability and controller bandwidth.

5. SIMULATION RESULTS

Numerical results, for a fixed $\beta_2=4$, are shown in the Table 1. By choosing a value of $\beta_1=1$ there is sufficient guarantee of $\gamma_u\alpha<1$, to ensure closed-loop stability. This value of β_1 corresponds to a suficiently small value of γ_e. The corresponding controller is:

$$K(s)=\frac{-2.837s}{s^2+4.967s+1} \qquad (23)$$

β_1	γ_e	γ_u
0.1	1.4	0.86
1	1.49	0.85
3	2.35	0.76
10	5.21	0.48

Table 1. Several solutions.

The overall system, including the system to be linearized, the reference model, and the controller, is shown in Figure 5. Graphical simulation results of the error signal $e(t)$ taking as command signal $u_c(t)$ a sinusoid normalized to $w=1$ and unity amplitude are shown in Figure 6.

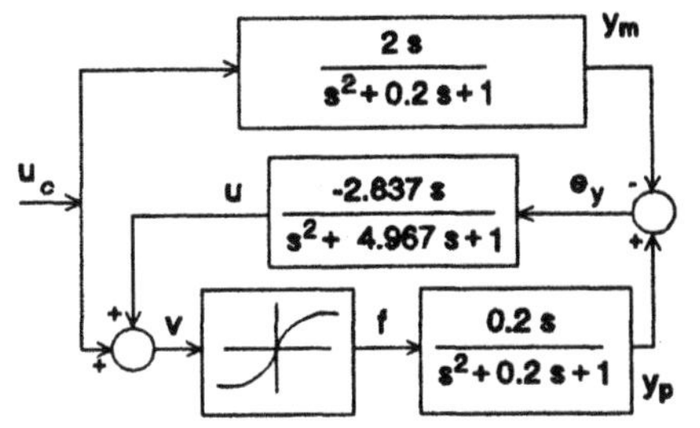

Fig. 5. Global system.

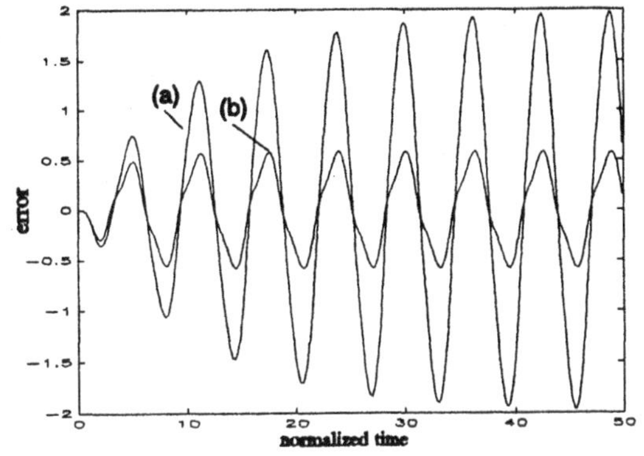

Fig. 6. Error signal for sinusoid excitation. (a) without controller, (b) with controller.

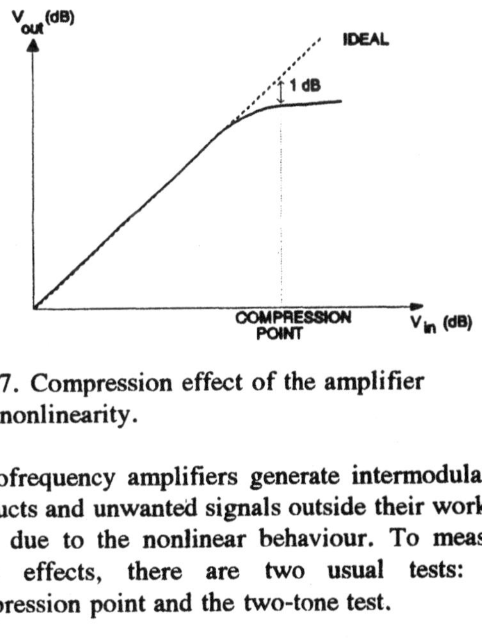

Fig. 7. Compression effect of the amplifier nonlinearity.

Radiofrequency amplifiers generate intermodulation products and unwanted signals outside their working band due to the nonlinear behaviour. To measure these effects, there are two usual tests: the compression point and the two-tone test.

The compression point test has been performed by increasing the input signal amplitude until the difference in amplitude between the fundamental harmonic in the output and the output amplitude of an ideal linear amplifier with the same amplification is 1 dB. It is a measure of the linear margin of the amplifier (Figure 7).

With the unlinearized system, the 1 dB compression point is obtained when the input signal (u_c) has an amplitude of *0.71* volts, while with the linearized system the input signal can be increased to *1.3* volts. This implies an improvement of *5.25 dB* in the compression point. The two-tone test is used to generate intermodulation terms. In this case the excitation has been (normalized frequencies):

$$u_c(t)=A\sin(\omega_1 t)+A\sin(\omega_2 t)$$
$$\omega_1=0.95; \quad \omega_2=1.05; \tag{24}$$

The amplifier output and the spectra of y_p are shown in Figure 8 and Figure 9 respectively. The amplifier fed by the optimum controller presents an attenuation of *4.55 dB* and *11.89 dB* in the intermodulation terms closest to the signal band.

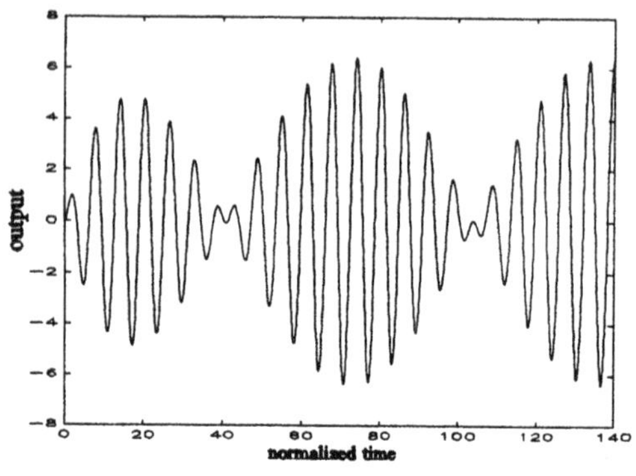

Fig. 8. Two tone response (y_p) of the linearized amplifier.

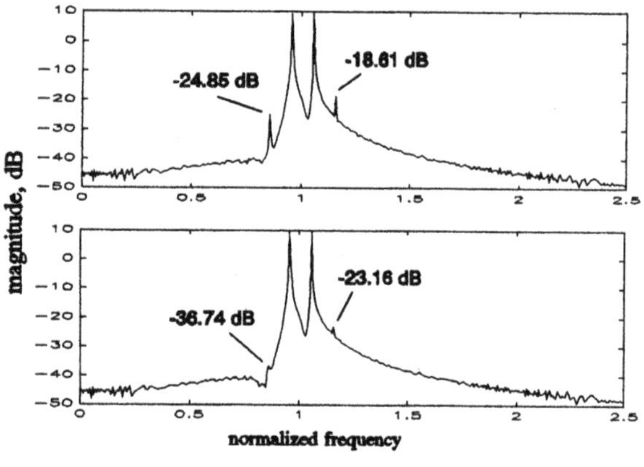

Fig. 9. Intermodulation test results, without (above) and with (below) the linearizer system.

6. CONCLUSIONS

The presented linearization solution is based on a model reference structure with linear output error feedback. This involves a suggestive simplification of the control system implementation. In the linearization field, this is a new alternative to other approaches, which presents advantages in system implementation and in the required design information of the nonlinear phenomena.

Closed-loop stability is considered in the design methodology. The obtained controller ensures stability and a bounded norm of the error between the model and the system to be linearized.

The theoretical results have been experimented on a communications amplifier, which yielded an improvement in the linearity margin (the compression point is increased). Furthermore, the intermodulation terms are reduced.

ACKNOWLEDGEMENTS

This work was supported by the spanish PRONTIC/CICYT under Grant TIC-95-1022-CO5-01.

REFERENCES

Ball, J.A., J.W. Heldon and M.L. Walker (1993). H_∞ Control for Nonlinear Systems with Output Feedback. *IEEE Transactions on Automatic Control*, **AC-38**, 546-559.

Bertran, E. and G. Montoro (1995). Adaptive VS-MRAC for Disturbance Cancellation. *IEE Electronics Letters*, **31**, 142-144.

Doyle, J.C., K. Glover, P.P. Khargonekar and B.A. Francis (1989). State-Space Solutions to Standard H_2 and H_∞ Control Problems. *IEEE Transactions on Automatic Control*, **AC-34**, 831-847.

Doyle, J.C., B.A. Francis and A.R. Tannenbaum (1992). *Feedback Control Theory*, Maxwell Macmillan International Editions.

Montoro, G. (1996). *Contribution to the Study and Development of Control Techniques Applied to System Linearization*, Ph.D. Thesis.

Schaft, A.J. (1992). L_2-Gain Analysis of Nonlinear Systems and Nonlinear State Feedback H_∞ Control. *IEEE Transactions on Automatic Control*, **AC-37**, 770-784.

Yaesh, I. and U. Shaked (1991). Two-Degree-of-Freedom H_∞ Optimization of Multivariable Feedback Systems. *IEEE Transactions on Automatic Control*, **AC-36**, 1272-1276.

Zames, G. (1963). *Functional Analysis Applied to Nonlinear Feedback Systems*, IEEE Transactions on Circuit Theory, 392-404.

ROBUST CONTROL OF VARIABLE FLOW PROCESSES

A.J. Niemi [*] **L. Tian** [**] **R. Ylinen** [***] **K. Zenger** [*]

[*] *Helsinki University of Technology, Control Engineering Laboratory, Otakaari 5 A, FIN-02150 Espoo, Finland*
[**] *Valmet Automation, P.O.Box 237, FIN-33101 Tampere, Finland*
[***] *University of Oulu, Systems Engineering Laboratory, P.O.Box 444, FIN-90571 Oulu, Finland*

Abstract: Models of continuous flow processes with variable parameters are studied and methods developed for their robust control. Such processes have proven invariant under variable flow and volume, if they are presented as functions of composite variables which have been introduced recently. Their finite impulse response (FIR) models are obtained e.g. by physical testing or identification, and expressed in terms of discrete, equidistant values of the appropriate variable. In the present study, such a model of a time-variable process with a long time delay is considered a high-order model; this is subsequently reduced to a low-order state-space model by means of the optimal Hankel-norm approximation. The robust feedback control is developed in terms of the same variable on the basis of the low-order model. The application of the control algorithm to the simulated process proves the approach superior to such conventional and robust control methods which have the regular time as their independent variable under unsteady conditions.

Keywords: Finite impulse response, continuous flow process, time-variable systems, Hankel-norm approximation, H_∞ control, prediction error method.

1. INTRODUCTION

Transfer of components of feed material through a continuous flow system is characterized by time-variable dynamics, if the total flow or amount of material in the system is not constant. Because this is often the case in industrial plants which are submitted to flow variations, the control of processes by time-invariant controllers cannot be considered satisfactory.

A method of managing the complexities caused by variable process parameters was introduced by Niemi (1977, 1988). He showed that by introducing a new integrated variable ("z") computed continuously on the basis of measurements of the flow or flow and volume, and by expressing the model as a function of this variable, the explicit flow, volume and time, being incorporated in z, were completely removed from the model. The time variable residence time distribution or impulse response was thereby replaced by a residence distribution which was a constant parameter function of z.

For modelling of flow processes in discrete variables, the residence distribution is discretized and truncated to a finite impulse response (FIR). This can be estimated under variable flow and volume by means of parametric methods (Tian *et al.* 1992, Niemi *et al.* 1993) or tracer tests (Niemi *et al.* 1996). A further step is the construction of controllers which operate as functions of z (Niemi 1991, Zenger *et al.* 1996).

The present paper deals with methods of modelling and robust control of time variable processes on the basis of the variable z, while the varying flow is assumed to be available by measurements. The FIR of flow system is obtained experimentally by means of chemical tracer and considered the initial result of modelling. This is then compressed, to reach a low-order description for controller design, by means of the optimal Hankel-norm approximation (Glover, 1984).

The compressed model is used for robust controller design. The control design objective includes both the robust performance requirement and the robust stability requirement (Doyle 1989).

As a difference to robust control using the regular time variable, the controller using the variable z either samples the main input and output variables at constant increments of z, i.e. at variable intervals of time, or produces equi-spaced sequences in z by interpolation. The approach has been used earlier to development of robust controls for variable parameter plants which were first identified by means of simulated and superimposed, stochastically generated data (Tian *et al.* 1994).

2. MODELS OF CONTINUOUS FLOW PLANTS

Concentration c_i of a constituent at the inlet of a continuous flow vessel is related to that c_o at the outlet by the unit impulse response g of the vessel. Their functional relationship is expediently described by an integral model:

$$c_o(t) = \int_{-\infty}^{t} g(t,\tau)c_i(\tau)d\tau \qquad (1)$$

Ii steady operational conditions the flow Q and volume V are constant and the model degenerates to the corresponding invariant function $g(t,\tau) = g(t-\tau)$. Analytical models of this type are available in the literature (Levenspiel 1962). They are often expressed as functions of a dimensionless time variable, the ratio of the running time to the mean residence time, which further is the ratio of the constant volume V and flow Q. The impulse response is equal to the residence time distribution (RTD) in terms of such a variable and only one distribution is needed to describe the dynamics at any point of operation within a larger range of flow and volume.

The flow and volume change continuously in most industrial systems affecting the residence of material within them, and their impulse responses are accordingly variable parameter functions $g(t,\tau)$. Since the conventional identification and control methods do not immediately apply to such processes, more practical models have been developed recently, starting from studies of the basic phenomena.

In order to develop the input-output model of a property like concentration of the fluid, the axial movement of the medium is studied. Assuming the flow and volume are known by standard measurements, an expression is first obtained for the average location $x_{av}(t,\tau)$ of a fluid element which has entered at time τ. If $V(t) = A(t)L$, where L is the length of the vessel, then

$$z(t,\tau) \triangleq x_{av}(t,\tau)/L = \frac{1}{V(t)}\int_{\tau}^{t} Q_i(\mu)d\mu \quad (2)$$

The flow pattern is then assumed invariable which is understood to imply that the distribution of the relative axial velocity and therefore that of the relative axial location $p(x(t,\tau)/x_{av}(t,\tau))$ of the fractions of such an element is invariable, irrespective of the values and changes of flow and volume. Through a change of variable another invariable distribution $p(z) = p(x_{av}(t,\tau)/L)$ at a fixed cross-section and particularly at the outlet end of the vessel where $x(t,\tau) = L$ is obtained. This is the impulse response of the concentration process or the residence distribution as a function of z. - For a detailed derivation, see Niemi, (1988).

The dimensionless variable $z = z(t,0)$ runs concurrently with the time t, but at a different pace. It has been shown that the substitution of z (2) into most analytical residence time distributions eliminates the explicit flow $Q(t)$, volume $V(t)$ and time t from them and brings the model (1) to a constant parameter form

$$c_o(z) = \int_{-\infty}^{z} p(z-\zeta)c_i(\zeta)d\zeta \qquad (3)$$

where $z = z(t,0)$ and $\zeta = z(\tau,0)$. This has been proven e.g. in cases of plug flow, Gaussian velocity distribution, laminar flow etc., and it can be used in many non-analytical flow patterns.

Because prior knowledge of the flow pattern is usually limited or almost nonexistent, it is difficult to parametrize an arbitrary residence distribution p, and black-box modelling methods have to be used to model the process. Therefore, the discretized form

$$c_o(z) = \sum_{k=1}^{d} p(kT_z)c_i(z-kT_z)T_z \qquad (4)$$

of the convolution (3) is written after a change of the direction of integration. Because p is a positive pulse function which approaches zero at high values of the argument, it is truncated to a finite impulse response (FIR) at a suitable value dT_z. The parameter vector $[p(T_z), \ldots, p(dT_z)]^T$ is invariant and can be estimated using experimental records as shown in Niemi *et al.* (1993) or by simulation or tracer tests.

The process modelled for control studied consisted of 10 m long rubber tube which was provided with a continuous flow of pure, ion-exchanged water. An impulselike input of chemical tracer was effected by fast injection of 1 ml of saline solution. The process response was measured through the electrolytic conductivity of its outlet flow.

Several such tests were made, at different flow rates including sinusoidally variable flow, and at such chemical contents that the input/output relationship was essentially linear. After normalization for a unit amount of tracer, the responses as functions of time differed (Fig. 1), but as they were expressed in terms of z (where the volume had always the same, constant value V) (Fig. 2), they converged reasonably close to a single function which resembles closely the response of the corresponding, theoretical laminar flow vessel. One representative response was chosen for use in the simulation studies (Fig.3).

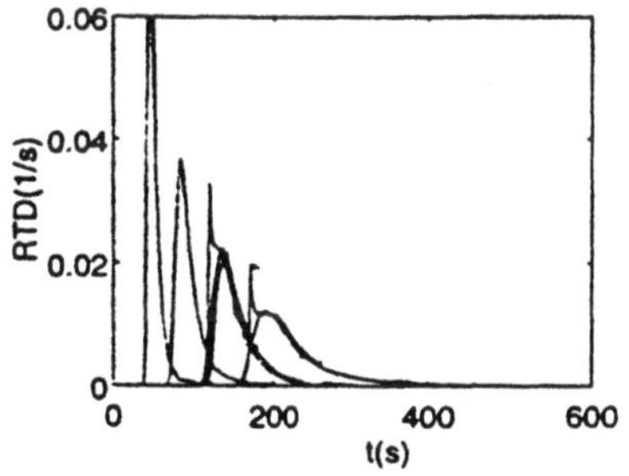

Fig. 1. RTD in time domain.

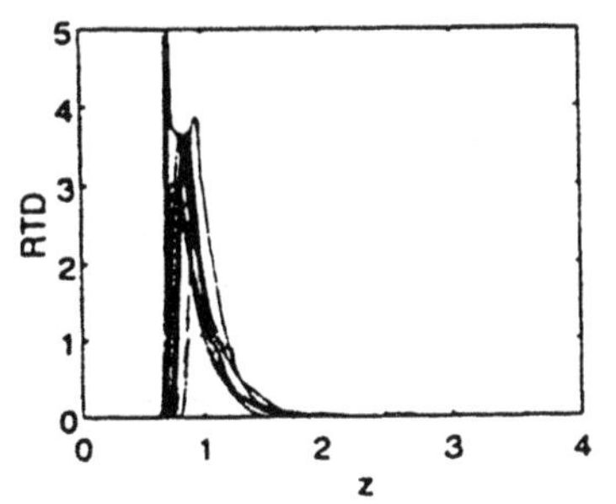

Fig. 2. RTD in z-domain.

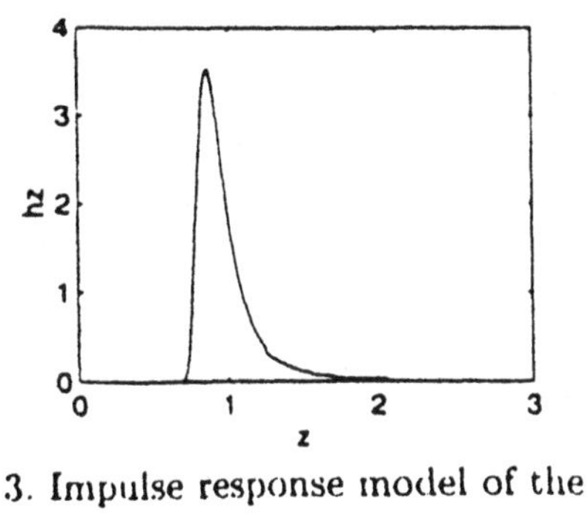

Fig. 3. Impulse response model of the process

In order to obtain a FIR model for control design, the impulse response is truncated at $z = 2.16$ and discretized using sampling interval $T_z = 0.144$ (see Fig. 3). This leads to a model of order 15, which is too high for controller design and must be reduced.

3. MODEL REDUCTION FOR CONTROLLER DESIGN

The model (4) has a state-space realization

$$x(z + T_z) = \tilde{A}x(z) + \tilde{B}u(z)$$
$$y(z) = \tilde{C}x(z) + \tilde{D}u(z) \quad (5)$$

$$\tilde{A} = \begin{bmatrix} 0 & 1 & \ldots & 0 \\ \ldots & \ldots & \ldots & \ldots \\ 0 & 0 & \ldots & 1 \\ 0 & 0 & \ldots & 0 \end{bmatrix}, \quad \tilde{B} = \begin{bmatrix} 0 \\ \ldots \\ 0 \\ 1 \end{bmatrix}, \quad (6)$$

$$\tilde{C} = [p(dT_z)T_z, \ \ldots, \ p(T_z)T_z], \qquad \tilde{D} = 0.$$

This discrete model is converted into a state-space realization of the continuous transfer function $G(s) = C(sI - A)^{-1}B + D$ e.g. via a bilinear (Tustin) transformation.

The Hankel singular values of the continuous transfer function $G(s)$, $\sigma_1 \geq \sigma_2 \geq \ldots \geq \sigma_n \geq 0$, can be computed from the controllability gramian and the observability gramian (Glover 1984). The Hankel-norm of $G(s)$ is equivalent to the maximal Hankel singular value

$$\| G(s) \|_H = \sigma_1(G(s)) \quad (7)$$

By using the algorithm developed by Glover (1984), the optimal Hankel-norm approximation $\hat{G}(s) = \hat{D} + \hat{C}(sI - \hat{A})^{-1}\hat{B}$ of order k minimizing $\| G(s) - \hat{G}(s) \|_H$ can be calculated.

If the approximation error is weighted by an appropriate function $W(s)$), the approximation problem becomes to the minimization of $\| (G(s) - \hat{G}(s))W(s) \|_H$. In particular, a proper $W(s)$ with zeros and poles in the right half domain is efficient in approximation. The weighted transfer function can be decomposed to stable and unstable parts (Safonov *et al.* 1987)

$$GW = [GW]_- + [GW]_+ \quad (8)$$

The approximation $\hat{G}_W$ of $[GW]_-$ is computed by the Glover algorithm, and the final approximation $\hat{G}$ is obtained by taking another stable and unstable decomposition

$$\hat{G} = [\hat{G}_W W^{-1}]_- \quad (9)$$

For flow processes it is important to have a correct gain (usually equal to one). Therefore, it is preferable to add a large weighting to the low frequency region. In this case $W(s)$ was chosen as:

$$W(s) = \frac{(s - 21.83)^2}{(s - 10.92)^2} \qquad (10)$$

The Hankel singular values of the stable part of $G(s)W(s)$ are plotted in Fig. 4. The order 6 gives a good approximation.

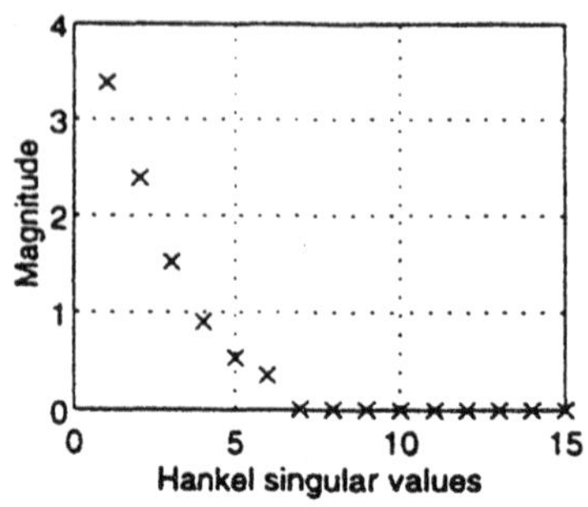

Fig. 4. Hankel singular values of $[G(s)W(s)]_-$.

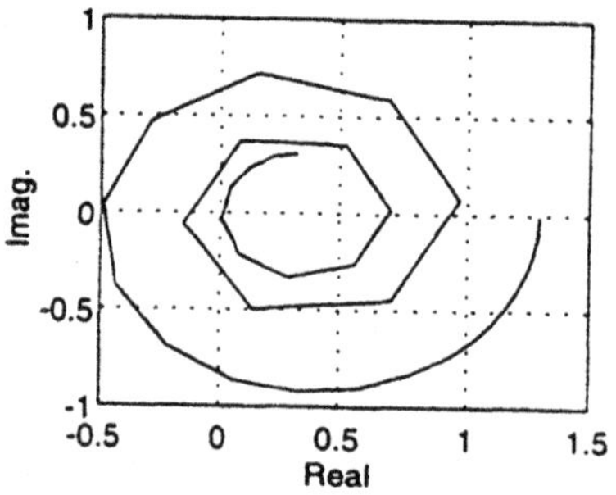

Fig. 5. Nyquist plot of $\hat{G}$.

Thus the following state-space representation of the flow process in z resulted:

$$\hat{G}: \begin{cases} \dfrac{dx(z)}{dz} = \hat{A}x(z) + \hat{B}c_i(z) \\ c_o(z) = \hat{C}x(z) + \hat{D}c_i(z) \end{cases} \qquad (11)$$

$$\hat{A} = \begin{bmatrix} -32.3 & -46.4 & 36.0 & -38.4 & -12.7 & -18.5 \\ 0 & -15.0 & 29.9 & -25.8 & -8.0 & -12.4 \\ 0 & -4.3 & -10.3 & 21.7 & 6.9 & 10.6 \\ 0 & 0 & 0 & -11.8 & 4.9 & -12.4 \\ 0 & 0 & 0 & -11.6 & -1.1 & -2.9 \\ 0 & 0 & 0 & 0 & 0 & -3.8 \end{bmatrix}$$

$$\hat{B} = \begin{bmatrix} -4.5 & -3.0 & 2.6 & -3.7 & -0.3 & -4.3 \end{bmatrix}^T$$

$$\hat{C} = \begin{bmatrix} 3.8 & 3.2 & -2.4 & 2.6 & 0.9 & 1.2 \end{bmatrix} \qquad (12)$$

$$\hat{D} = 0.6$$

Eqs. (11) and (13) represent a stable, non-minimum phase plant. Fig.5 shows its frequency response.

4. INVARIANT CONTROL OF TIME VARYING PROCESSES

The z-invariant model G ($\hat{}$ dropped for brevity) of the flow process can be used for control system synthesis. The control loop is here described as that in Fig. 6. The relative modelling error

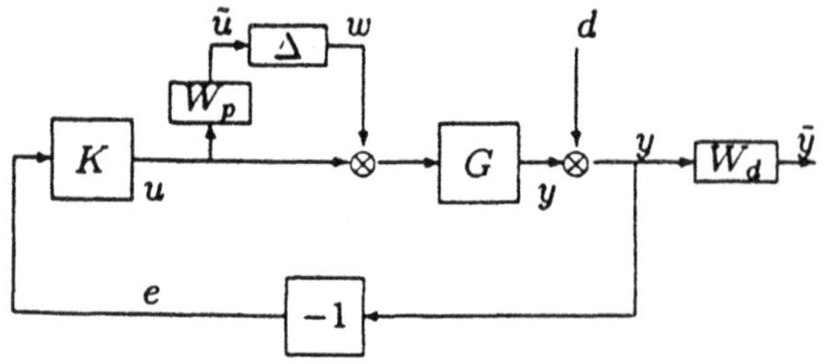

Fig. 6. Feedback control system.

is represented by a multiplicative perturbation characterized by a stable bounded perturbation Δ with a frequency-domain weighting W_p.

$$\Delta_i(i\omega) = \Delta(i\omega)W_p(i\omega), \quad \| \Delta \|_\infty < 1/\gamma \qquad (13)$$

The transfer function matrix without the modelling error from the perturbance w and the disturbance d to the control u and the output y is

$$\begin{bmatrix} u \\ y \end{bmatrix} = \begin{bmatrix} P_{11} & | & P_{12} \\ -- & + & -- \\ P_{21} & | & P_{22} \end{bmatrix} \begin{bmatrix} w \\ d \end{bmatrix} \qquad (14)$$

$$\begin{aligned} P_{11} &= -K(1 + GK)^{-1}G \\ P_{12} &= -K(1 + GK)^{-1} \\ P_{21} &= (1 + GK)^{-1}G \\ P_{22} &= (1 + GK)^{-1} \end{aligned} \qquad (15)$$

Here P_{22} and P_{11} are the sensitivity and the complementary sensitivity, respectively. The modelling error adds the feedback

$$w = \Delta W_p u \qquad (16)$$

to the model. Then the robust stability requires that the transfer function P_{11} satisfy

$$\| W_p P_{11} \|_\infty \leq \gamma \qquad (17)$$

The feedback changes the transfer function from d to y to the form

$$\hat{P}_{22} = P_{22} + P_{21}\Delta W_p(1 - P_{11}\Delta W_p)^{-1}P_{12} \qquad (18)$$

The robust performance problem is to keep this small which by a suitable choice of the weighting function W_d (Fig. 6) can be written as

$$\| W_d\hat{P}_{22} \|_\infty \leq \gamma \qquad (19)$$

Balas *et al.* (1991) have shown that if $\| \Delta \|_\infty < 1/\gamma$ and

$$\sup_\omega \mu(P(i\omega)) \leq \gamma \qquad (20)$$

the closed loop system will be internally stable, and the required robustness conditions (17) and

(19) are always satisfied. Here μ is the structured singular value defined by Doyle (1982). H_∞ controllers K satisfying the condition (20) are designed by K-D iteration (Doyle *et al.* (1989), Balas *et al.* 1991).

The relative modelling error is usually larger on higher frequencies. Furthermore, the disturbance attenuation at low frequencies is important for good control performance. Therefore, for the example plant (11-13) the following weighting functions were chosen:

$$W_p(s) = \frac{500(s+4)}{(s+4000)}, \ \gamma = 1 \qquad (21)$$

$$W_d(s) = \frac{0.5(s+0.6)}{s+0.0006} \qquad (22)$$

With these weighting functions a 10th order H_∞ controller K satisfies the stability and performance requirements. The maximal singular value of P and the structured singular value μ achieve at low frequencies the value $0.975 < 1 = \gamma$ (Fig. 7). The sensitivity and complementary sensitivity are displayed in Fig. 8.

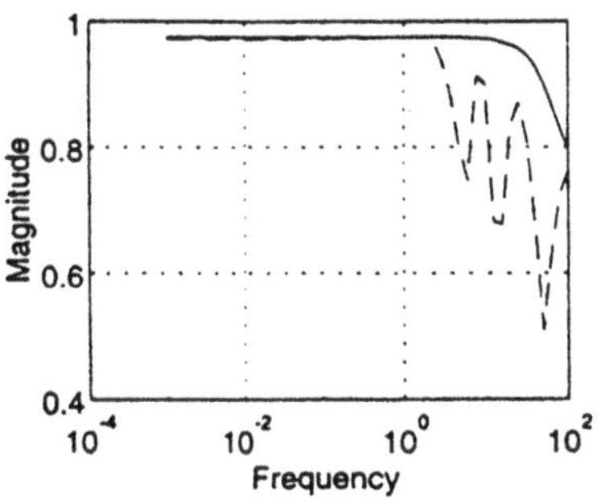

Fig. 7. Singular value $\bar{\sigma}(P(i\omega))$ (solid curve) and structured singular value $\mu(P(i\omega))$ (dashed curve).

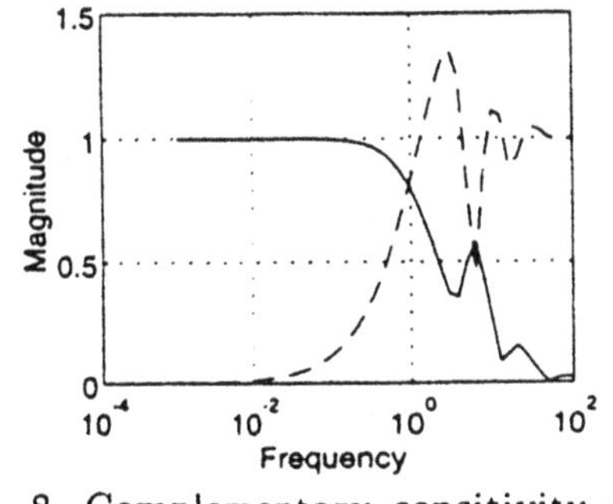

Fig. 8. Complementary sensitivity (solid curve) and sensitivity (dashed curve).

The invariant H_∞ controller designed is supposed to run in variable z, and has the continuous state-space realization:

$$K_z: \begin{cases} \dfrac{dx_k(z)}{dz} = A_k x_k(z) + B_k e(z) \\ c_i(z) = C_k x_k(z) + D_k e(z) \end{cases} \qquad (23)$$

In order to have a constant parameter controller, the closed-loop signals are preferably sampled at constant intervals of z which implies variable intervals of time. If the flow and volume are measured continuously and t_0 stands for the last sampling instant, the next, real or calculated sampling instant t is determined by the constant increment T_z, so that

$$T_z = \frac{1}{V(t)} \int_{t_0}^{t} Q_i(\mu)d\mu \qquad (24)$$

This includes also the case of variable volume. The discrete version $\tilde{K}_z$ of the controller can be calculated via the bilinear transformation, and then implemented in a digital computer. At every instant determined by (24), the controller samples data and issues control commands.

5. CONTROL UNDER VARIABLE CONDITIONS

The application of the designed H_∞ controller to flow processes is illustrated by means of simulations under variable conditions and inputs. The performance of the H_∞ controller is compared to the performance of a conventional PID-controller

$$u(z) = K_p(e(z) + \frac{1}{T_i}\int_0^z e(\zeta)d\zeta + T_d\frac{de(z)}{dz})(25)$$

both in constant flow (z proportional to t) and in variable flow cases. Here K_p, T_i and T_d are the standard tuning parameters of the PID controller. In all cases the PID-controller was tuned using several different tuning methods and the one resulting in the best possible control was chosen.

The process is assumed to operate under conditions of constant volume $V = 0.7854$ liter and variable flow $Q_i(t) = Q_o(t) = 0.5(1 + 0.75\sin(0.7t - 100))$ liter/min.

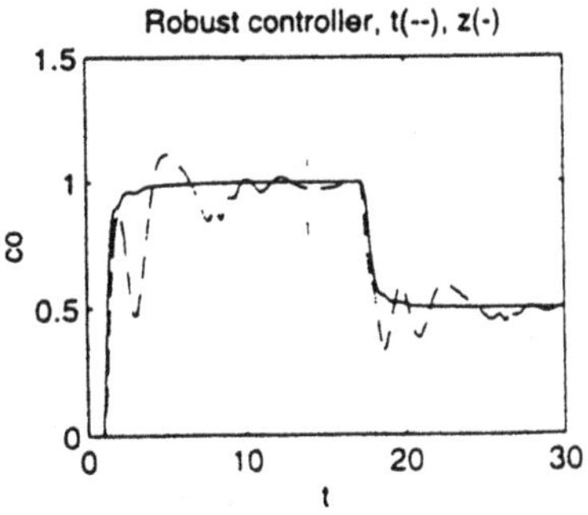

Fig. 9. Step responses under variable flow through the plant, and control by time-invariant (dashed) and z-invariant robust controllers (solid).

The robustness of the H_∞ controller is illustrated by the step responses in Fig.9. By using $\tilde{K}_z$

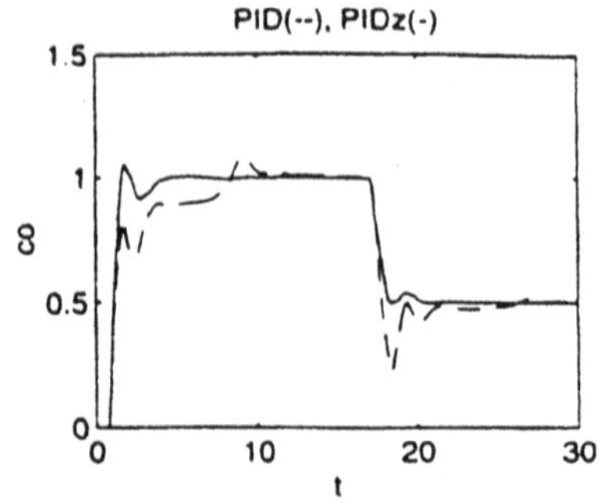

Fig. 10. Step responses under variable flow through the plant and control by time-invariant (dashed) and z-invariant PID controllers (solid).

instead of the time invariant controller tuned for the case of constant flow $Q_i(t) = Q_o(t) = 0.5$ liter/min the closed-loop response is much better and almost ideal. However, in spite of the 75 % changes of the flow, also the constant parameter controller still maintains the stability.

The results obtained by means of the z-invariant controller (25) are clearly better than those obtained with the time-invariant controller. However, the z-invariant robust controller is performing even better (Fig.9). On the other hand, in time-invariant control of the time-variable plant, the PID-controller seems to perform better than the H_∞ controller. It remains to be checked, if this is related e.g. to the large amplitude ($\pm 75\%$) of the flow disturbance.

6. CONCLUSIONS

Measured, numerically expressed responses to momentary inputs like tracer injections can be considered time-invariant process models of high order. these can be reduced to low order models by means of the optimal Hankel-norm approximation. The robust controller is then designed on the basis of the reduced state space model.

If the plant flow rate changes, the process model can be brought to a linear, constant parameter form by introduction of the appropriate new, integrated variable. The robust control can then be designed in terms of this variable, in the same manner as described above for the case of the time-invariant plant. If the same variable is used as the independent variable of the controller designed, this produces under flow and volume variations an equal result as the time-based control produces under constant flow and volume only. The procedure presented has therefore a large potential in control of industrial continuous flow processes which operate under variation of flow, volume or contained mass measured by standard instruments, even under variable flow patterns .

REFERENCES

Balas,J., J.C. Doyle, K. Glover, A.K. Packard, and R.S. Smith (1991). *μ-Analysis and Synthesis Toolbox (μ-TOOLS)*. The MathWorks Inc., MUSYN Inc.

Doyle,J.C. (1982)."Analysis of feedback systems with structured uncertainties." *Proc. IEE-D*, pp.242-250.

Doyle,J.C., K. Glover, P.P. Khargonekar, and B.A. Francis (1989). "State-space solution to standard H_2 and H_∞ control problem." *IEEE Trans. on Automat. Contr.*, vol. AC-34, no.8, pp.831-847, 1989.

Glover,K. (1984). "All optimal Hankel-norm approximations of linear multivariable systems and their L^∞-error bounds." *Int. J. Control.* **39**, pp.1115-1193.

Levenspiel,O. (1962). *Chemical Reaction Engineering*. Wiley, New York.

Niemi,A.J. (1977). "Residence time distributions of variable flow processes." *Int. J. of Applied Radiation and Isotopes,* **28**, pp.855-860.

Niemi,A.J. (1988). "Variable parameter model of the continuous flow-vessel." *Mathl Comput. Modelling,* **11**, pp.32-37.

Niemi,A.J. (1991). "Process control under variable flow and volume." *Proc. IFAC Int. Symp. on Intelligent Tuning and Adaptive Control (ITAC91)*, Pergamon Press, Oxford, pp.55-61.

Niemi,A.J., L. Tian, P. Jutila, and R. Ylinen (1993). "Experimental identification of variable parameter flow processes." Preprints of *IFAC 1993 World Congress,* Sydney, **9**, pp.455-460.

Niemi,A.J., K. Zenger, J. Thereska, and J. Griffith Martinez (1996) "Residence time distributions of processes under variable flow and volume." *12th Int. Congr. of Chem. and Process Engg CHISA '96*, Praha, 10p.

Safonov,M.G., E.A. Jonckheere, M. Verma, and D.J.N. Limebeer (1987). "Synthesis of positive real multivariable feedback systems." *Int. J. Control,* no.3, pp.817-842.

Tian,L., A.J. Niemi, and R. Ylinen (1992)."Recursive process identification under variable flow and volume." *Dynamics and Control of Chemical Reactors, Distillation Columns and Batch Processes, (DYCORD+'92)*, Pergamon Press, Oxford, pp.339-344.

Tian,L., A.J. Niemi, and R. Ylinen (1994)."Robust control under variable flow and volume." *Proceedings of the Third IEEE.Conf. on Control Applications*, Glasgow, pp. 781-786.

Zenger, K., A.J. Niemi, and M. Martikainen (1996)."Analysis and pilot scale testing of a class of process control algorithms with variable parameters." *Preprints of the 13th IFAC World Congress*, San Francisco, USA.

ROBUSTNESS AND REACHABLE BANDWIDTH VERSUS CONTROL AND IDENTIFICATION PERFORMANCES

L. Keviczky and **Cs. Bányász**

Computer and Automation Institute, Hungarian Academy of Sciences
H-1111 Budapest, Kende u 13-17, HUNGARY
Phone: +361-1665-435; Fax: +361-1667-503
e-mail: h10kev@ella.hu, cs.banyasz@sztaki.hu

Abstract: This paper demonstrates the necessity of iterative controller refinement methods in control engineering practice. A *generic optimal controller* structure and regulator design method, where the identification and control errors are identical, are used in the investigation. It is shown that this iterative scheme maximizes a robustness measure, too. A new uncertainty relationship is introduced limiting the product of modeling accuracy and robustness measures.

Keywords: combined identification and control, iterative controller refinement, robust control

1. INTRODUCTION

The need to design high performance control systems has not lost the importance inspite of the thousands of methods and algorithms published in the past decades. The huge number of papers indicates that no unique or best method was found. The solution depends on the model, criterion, uncertainties, disturbances, constraints, etc. (sometimes even on the designer's taste).

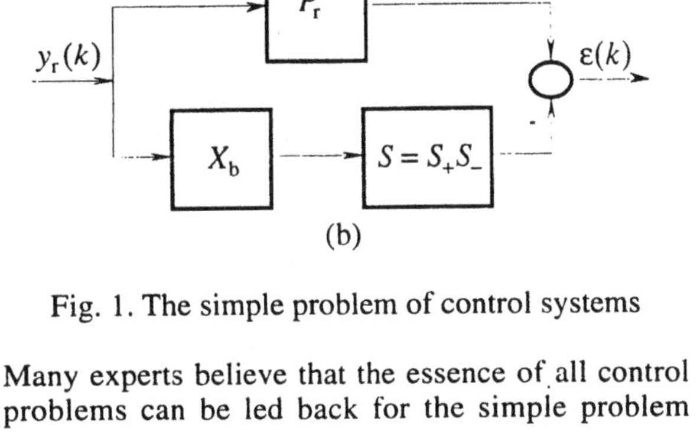

Fig. 1. The simple problem of control systems

Many experts believe that the essence of all control problems can be led back for the simple problem shown in Fig. 1a, i.e., how to choose a serial compensator transfer function X_a to a stable S ensuring a unity dynamic transfer (the influence of the so-called reference model P_r as a design goal for the overall transfer function has secondary importance now). The trivial solution $X_a = S^{-1}$ is not always applicable because Y is not invertible (inverse unstable: *IU*). This is mostly the case if the control is discrete time based on higher order sampled linear dynamic processes and/or with a delay time not integer multiple of the sampling period. In a general case the system $S = S_+S_-$ is factorable for an inverse stable (*IS*) S_+ and an *IU* S_- components, where $S_-(z)$ is a polynomial having all its zeros outside the unit disc and $S_+(z)$ is the remaining part (to get a unique factorization it is reasonable to ensure that $S_-(z)$ is monic). Because S_- can not be eliminated by simple cancellation mechanism it is sometimes called invariant system component. A heuristic but widely applied solution is to choose $X_a^h = S_+^{-1}$, when only the *IS* part is canceled, however, the invariant *IU* factor is untouched. Several controller design schemes are based on this method, in spite of there is no optimality connected. Unfortunately the remaining invariant factor can sometimes cause not tolerable dynamics, so its effect must be attenuated. This can be done, if an appropriate criterion is used for this purpose. Figure 1b shows a reduced form of the well-known model matching paradigm (see Kucera, 1981 or Wang and Chen, 1988), where X_b should be selected by minimizing a cost function as

$$X_b = \arg\min_{X_b}\{J_2\}$$
$$J_2 = \|P_r - X_b S\|_2 = \|P_r - X_b S_+ S_-\|_2 \quad (1)$$

Note that the above minimization in H_2-norm space is equivalent to the minimum mean square error (*MSE*) problem (the classical Wiener paradigm of optimal stochastic systems)

$$X_b = \arg\min_{X_b}\left\{E\left\{\varepsilon^2(k)\right\}\right\} = \arg\min_{X_b}\left\{E\left\{\left[(P_r - X S_+ S_-) y_r(k)\right]^2\right\}\right\} \quad (2)$$

if the external excitation $y_r(k)$ is a white noise sequence. Here E{...} stands for expectation. Be

$$P_r = \frac{B_r}{A_r} \quad (3)$$

a stable proper reference model with coprime B_r and A_r. it can easily be proved (Keviczky and Bányász, 1997) that the minimum of J_2

$$J_2 = \left\|\frac{X_1}{S_-(z)}\right\|_2 \quad (4)$$

can be reached by the optimal selection

$$X_b = \frac{X_2}{A_r S_-(z^{-1}) S_+(z)} = X_b(P_r, S) \quad (5)$$

where X_1 and X_2 are computed as the solutions of the Diophantine equation

$$X_1 A_r + X_2 S_-(z) = S_-(z^{-1}) B_r \quad (6)$$

Note that the order of this equation is generally much lower than the one applied in the general model matching paradigm.

Let us repeat the above optimization for the scheme of Fig. 1a, i.e., for the cost function

$$J_2 = \left\|\frac{B_r}{A_r}(1 - X_b S_+ S_-)\right\|_2 = \left\|\frac{B_r}{A_r} - \frac{B_r}{A_r} X_b S_+ S_-\right\|_2 \quad (7)$$

It is easy to check that the following optimal compensator is obtained

$$X_a = \frac{X_2}{B_r S_-(z^{-1}) S_+(z)} = \frac{X_2}{A_r S_-(z^{-1}) S_+(z)} \frac{A_r}{B_r} = X_b \frac{1}{P_r} = \frac{G_r}{S_+(z)} = X_a(P_r, S) \quad (8)$$

Because $X_b = P_r X_a$ it is obvious that Fig. 1a and Fig. 1b are equivalent if the serial compensator is optimized by minimizing the above J_2 in H_2-norm space. This equivalence, of course, holds in *MSE* sense, too, when a white noise y_r excitation is applied. It is easy to check that this equivalence holds for all excitation y_r having rational spectra, however, in general the optimal compensator will depend on the reference signal, too: $X_a = X_a(P_r, S, y_r)$ and $X_b = X_b(P_r, S, y_r)$. If y_r is derived from a white noise ξ by the stable rational filter

$$y_r(k) = \frac{B_y}{A_y}\xi(k) \quad (9)$$

then the Diophantine equation to be solved is

$$X_1 A_r A_y + X_2 S_-(z) = S_-(z^{-1}) B_r B_y \quad (10)$$

and

$$X_b = \frac{X_2}{A_r A_y S_-(z^{-1}) S_+(z)} = \frac{G_r}{P_r A_y S_+(z)} = X_b(P_r, S, y_r) \quad (11)$$

So the above introduction showed that in all cases it is possible to find the solution of the model matching paradigm without the loss of degree of freedom in a general form of

$$P_r(1 - G_r S_-) \quad (12)$$

where

$$G_r = \frac{X_2}{B_r S_-(z^{-1})} = G_r(P_r, S_-) \quad (13)$$

or $G_r(P_r, S_-, y_r)$ computed from (11) if y_r is not white noise.

It is also easy to check that the trivial $X_b = P_r / S = X_b(P_r, S) = P_r X_a(S)$ solution is obtained for *IS* processes. This solution does not depend on the y_r excitation and consequently $G_r = 1$ and $G_w = 1$ (see G_w later). Remember that these G_r and G_w filters are used generally in heuristic approaches and they can give quite nice results, however, please note that the approximation of the true optimality strongly depends on the applied reference signal, the output disturbance and the invariant process factors.

2. THE GENERIC CONTROLLER STRUCTURE

There are many variants of the two degree of freedom feedback control systems published since the classical works as, e.g., Horowitz (1963). Several of them were connected to pole placement design and partly to self-tuning and other type of adaptive regulators (see, e.g., Åström and Wittenmark, 1984). Some iterative controller refinement techniques are also based on this basic principle. In our recent works we have been discussing a *generic optimal controller* structure and regulator design method (Keviczky and Bányász, 1994; Keviczky, 1995; Keviczky and Bányász, 1995b), where the controller polynomials can be obtained without the solution of a Diophantine equation and in the iterative scheme suggested for this structure the identification and control errors are identical (see later).

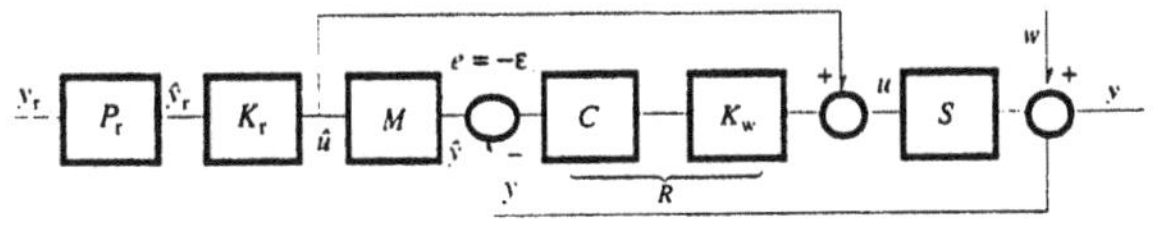

Fig. 2. *Generic scheme* for all cases

The *generic scheme* can be summarized for all cases (e.g., for inverse stable or unstable plants) according to Fig. 2. Here y_r, u, y and w are the reference, process input, output and disturbance signals in the closed-loop, where discrete-time representations are considered for computer controlled systems, i.e., the argument k of variables will mean the integer value discrete time (integer multiple of the sampling period) and z^{-1} means the backward shift operator $\left(z^{-1}y(k) = y(k-1)\right)$. S, M, P_r and R are the process, its model, a desired tracking (servo) reference model and the regulator transfer functions, respectively. The ideal overall transfer functions of the closed-loop can be formulated as

$$y^o(k) = P_r y_r(k) + (1 - P_w)w(k) = y_t^o(k) + y_d^o(k) \quad (14)$$

So the desired tracking (servo) transfer function y_t^o could be provided by the P_r and the desired regulating (or disturbance rejection) behavior y_d^o by the P_w reference models. (If P_w is the best LS predictor of w, then the obtained regulator is the minimum variance (MV) regulator.) Selecting the regulator as

$$R = CK_w = \frac{P_w}{1 - P_w G_w M_-} G_w M_+^{-1} \quad (15)$$

where

$$C = \frac{P_w}{1 - P_w G_w M_-}, \qquad K_w = G_w M_+^{-1} \quad (16)$$

and using the serial compensator

$$K_r = G_r M_+^{-1} \quad (17)$$

provide the

$$\begin{aligned} y(k) &= P_r G_r S_- y_r(k) + (1 - P_w G_w S_-)w(k) = \\ &= y_t(k) + y_d(k) \end{aligned} \quad (18)$$

reachable *"best"* overall transfer functions in case of the theoretically ideal exact model matching when $M = S$. Anyway the above method gives the unique explicit computation of the minimal order controller polynomials formulated in the general infinite number solutions of the optimal closed-loop model matching paradigm of Kucera (1981) , Wang and Chen (1988). Here

$$S = S_+ S_- \quad \text{and} \quad M = M_+ M_- \quad (19)$$

are assumed stable, S_+ and M_+ mean the IS, S_- and M_- the IU factors, respectively. It can be well seen that the ideal tracking $y_t^o(k) = P_r y_r(k)$ and regulating $y_d^o(k) = (1 - P_w)w(k)$ behaviors cannot be reached exactly for IU processes because of the invariant factor S_-. The influence of S_- can be attenuated and for given excitations can be optimized by the extra serial compensators $G_r(P_r, S_-, y_r)$ and $G_w(P_w, S_-, w)$. (In the last compensator the role of y_r should be substituted by w.) Express the deviation between the ideal $y^o(k) = y_t^o(k) + y_d^o(k)$ and the reachable *"best"* closed-loop as

$$\begin{aligned} \Delta y &= y^o(k) - y(k) = \\ &= P_r(1 - G_r S_-)y_r(k) - P_w(1 - G_w S_-)w(k) \end{aligned} \quad (20)$$

and it is easy to see that the optimality of Δy in H_2-norm space is provided by computing G_r and G_w according to (13) as it was shown in the Introduction.

If the process has a z^{-d} time delay it is reasonable to consider the factorization of S in the form of

$$\begin{aligned} S &= S_+ S_- = (S_+)(\overline{S}_- z^{-d}) = \frac{B}{A} z^{-d} = \\ &= \frac{B_+ B_-}{A} z^{-d} = \left(\frac{B_+}{A}\right)(B_- z^{-d}) \end{aligned} \quad (21)$$

where $S_- = \overline{S}_- z^{-d}$ is the non-invertible part where the discrete time delay z^{-d} is also a factor whose inverse z^d is not realizable. Using this form the overall transfer functions of (18) will be slightly modified as

$$\begin{aligned} y(k) &= P_r G_r \overline{S}_- z^{-d} y_r(k) + \\ &+ (1 - P_w G_w \overline{S}_- z^{-d})w(k) = y_t(k) + y_d(k) \end{aligned} \quad (22)$$

which clearly shows the influence of the uncompensable process time delay besides the influence of the invariant factor $\overline{S}_-$.

The final optimal controller is already very general because it covers the most critical processes where the design is not trivial. In the practice we should rather use the model M instead of the system S in the formulas, so the design procedure should apply

$$
\begin{aligned}
M &= M_+ M_- = (M_+)\left(\overline{M}_-\, z^{-d_m}\right) = \frac{\hat{B}}{\hat{A}} z^{-d_m} = \\
&= \frac{\hat{B}_+ \hat{B}_- z^{-d_m}}{\hat{A}} = \left(\frac{\hat{B}_+}{\hat{A}}\right)\left(\hat{B}_- z^{-d_m}\right)
\end{aligned}
\tag{23}
$$

instead of (21). Here $M_+, \overline{M}_-$ and z^{-d_m} are the inverse stable, inverse unstable factors and the delay time of the model, respectively.

Thus the first visible advantage of the presented new *generic* controller scheme is that it gives a clear evolving structure how to obtain the optimal controller in a step by step reasonable way. First one can reach a heuristic suboptimal solution by selecting $G_r = 1$ and $G_w = 1$. This solution does not need the solution of any Diophantine equation, but the regulator (15), (16) and compensator (17) transfer functions can be calculated directly. It has no optimality, besides the very important basic feature, that the desired poles and zeros can be placed by P_r and P_w. The invariant factors appear in the overall transfers unattenuated. If the optimality is really required then G_r and G_w can be optimized in the knowledge of the expected external excitations y_r and/or w. These optimal compensators need the solution of the special low order Diophantine equation presented in the Introduction both for y_r and/or for w.

3. ITERATIVE CONTROLLER REFINEMENT

Figure 3 using the *generic controller structure* shows an obvious way how to perform the identification step in a combined identification and control scheme (Keviczky, 1995), i.e., we should use a regular identification algorithm between the auxiliary variable $\hat{u}$ and the measured controlled variable y. Note that $\hat{u}$ and y must be obtained from the closed-loop operated by the *generic* optimal controller structure with the optimal regulator R and compensator K_r.

The model based version of the *generic scheme* shown in Fig. 3. is the long searched ideal one for the combined identification and control problem, because in this case *the control and modeling errors are identical*. It is interesting to show that the common modeling and control error is

$$
\begin{aligned}
\varepsilon &= y - \hat{y} = \frac{P_r G M_-}{1+RS}\ell y_r + \frac{1}{1+RS} w = \\
&= \frac{P_r G}{1+RS}\ell_+ y_r + \frac{1}{1+RS} w = -e
\end{aligned}
\tag{24}
$$

where

$$
\ell = \frac{\Delta}{M} = \frac{S-M}{M} \quad \text{and} \quad \ell_+ = \frac{\Delta}{M_+} \tag{25}
$$

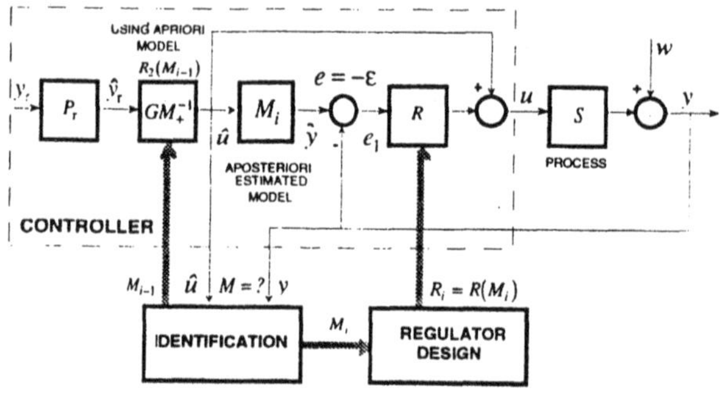

Fig. 3. Model based version of the *generic controller scheme* for iterative controller refinement

Because $\hat{u}$ depends on the model M, only an iterative (Keviczky, 1995) control refinement procedure can be performed.

4. *GENERIC SCHEME* ERROR PROPERTIES AND REGULATOR SENSITIVITIES

Figure 4 represents the uncertainty model corresponding to error equation (24). Straightforward calculations give the following equivalent form of the common error ε

$$
\varepsilon = \underbrace{\frac{G_r M_- \ell}{1+RS} P_r y_r}_{\varepsilon_r} - \underbrace{\frac{G_w M_- \ell}{1+RS} P_w w}_{\varepsilon_w} + \underbrace{\left(1 - P_w G_w M_-\right) w}_{\varepsilon_o} \tag{26}
$$

Here ε_r and ε_w are the contributions of y_r and w to the error ε, furthermore ε_o is the remaining uncompensable error (in case of ideal process model matching: $\ell = 0$).

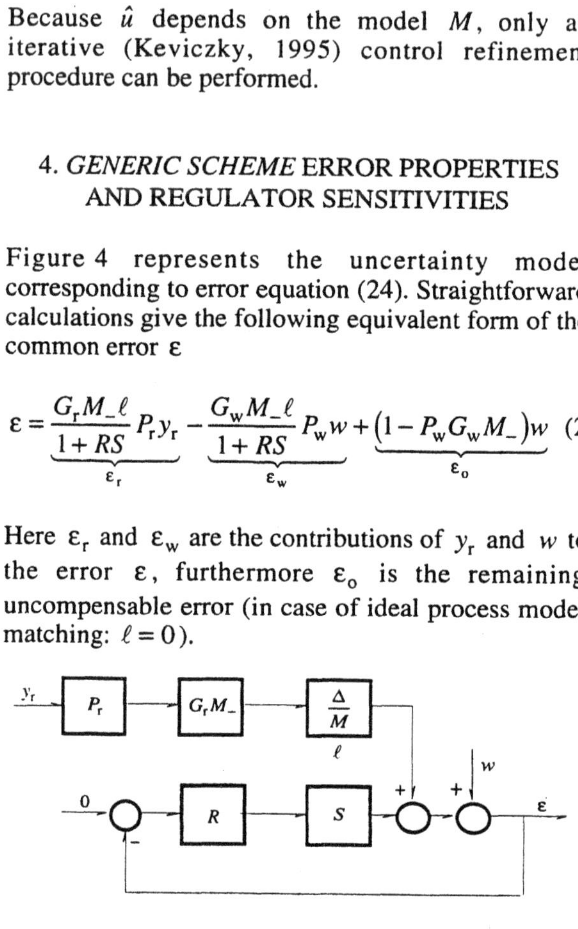

Fig. 4. Uncertainty model of the *generic scheme*

Because generally P_w is also not the ideal, it is an estimate only of the true (or optimal) P_w^o predictor, the above formula can be further rewritten as

$$
\begin{aligned}
\varepsilon &= \underbrace{\frac{G_r M_- \ell}{1+RS} P_r y_r}_{\varepsilon_r} - \underbrace{\frac{G_w M_- \ell}{1+RS} P_w w}_{\varepsilon_w} + \\
&+ \underbrace{\ell_w G_w M_- P_w w}_{\varepsilon_{\Delta P_w}} + \underbrace{\left(1 - P_w^o G_w M_-\right) w}_{\varepsilon_o^o}
\end{aligned}
\tag{27}
$$

Here $\varepsilon_{\Delta P_w}$ occurred due to the not ideal predictor P_w and ε_o^o is the theoretical residual error (when both $\ell = 0$ and $\ell_w = 0$), where

$$\ell_w = \frac{P_w^o - P_w}{P_w} = \frac{\Delta_w}{P_w} \tag{28}$$

$$\varepsilon_r = \frac{G_r M_- \ell}{1+RS} P_r y_r = \frac{M_- \ell}{1+RS} \hat{y}_r \tag{29}$$

$$\varepsilon_w = \frac{G_w M_- \ell}{1+RS} P_w w = \frac{M_- \ell}{1+RS} \hat{w} \tag{30}$$

$$\varepsilon_o^o = \left(1 - P_w^o G_w M_-\right) w \tag{31}$$

are introduced, furthermore

$$\hat{y}_r = P_r G_r y_r \quad \text{and} \quad \hat{w} = P_w G_w w \tag{32}$$

are filtered predicted variables based on the P_r and P_w reference models. Finally

$$\begin{aligned}\varepsilon &= \frac{M_- \ell}{1+RS}(\hat{y}_r - \hat{w}) + M_- \ell_w \hat{w} + \varepsilon_o^o = \\ &= \frac{M_- \ell}{1+RS}(\hat{y}_r - \hat{w}) + \varepsilon_{\Delta P_w} + \varepsilon_o^o\end{aligned} \tag{33}$$

is obtained, where

$$\varepsilon_{\Delta P_w} = G_w M_- \ell_w P_w w = M_- \ell_w \hat{w} \tag{34}$$

The detailed version of the uncertainty model corresponding to (33) is shown in Fig. 5.

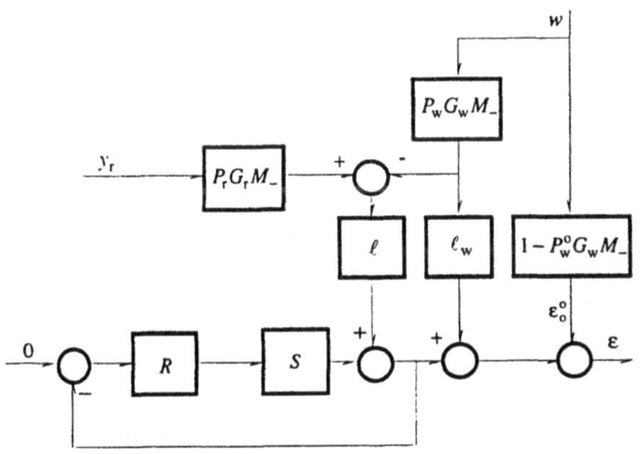

Fig. 5. The detailed uncertainty model of the *generic scheme*

One can recall that if the additive output noise is driven by

$$w(k) = \frac{C}{D}\varsigma(k) \tag{35}$$

where ς is white noise, then the well-known d-step ahead LS predictor can be obtained in the form of

$$w(k+d) = \frac{L}{D} w(k) + F\varsigma(k) = P_w^o\, w(k) + F\varsigma(k) \tag{36}$$

Here

$$D = LC + z^{-d} F \tag{37}$$

and the true (optimal) predictor is

$$P_w^o = \frac{L}{D} \tag{38}$$

Investigate the sensitivity of the optimal regulator when the process is *IS*: so $G_r = G_w = 1$, $M = M_+$ and $M_- = 1$ Using (15) it is easy to obtain that

$$\ell_R^o = \frac{\Delta R}{R_o} = \frac{R_o - R}{R_o} = -\ell \quad \text{for } \forall\omega \tag{39}$$

This means that any identification method minimizing $\|\ell\|_\infty$ will provide the most robust regulator

$$\left\|\ell_R^o\right\|_\infty = \left\|\frac{\Delta R}{R_o}\right\|_\infty = \left\|\frac{\Delta}{M}\right\|_\infty = \|\ell\|_\infty \tag{40}$$

If the uncertainty in R is related to R instead of R_o as above the following formula is obtained

$$\ell_R = \frac{\Delta R}{R} = -\frac{\ell}{1+\ell} = -\frac{\Delta}{S} \quad \text{for } \forall\omega \tag{41}$$

The last equation has much less importance, because S is not known, so any norm of $\|\Delta/S\|$ can not be minimized easily, in spite of $\|\Delta R/R\|_\infty = \|\Delta/S\|_\infty$. The relationship between the two regulator relative uncertainties is

$$\ell_R = -\frac{\ell}{1+\ell} = -\ell + \left.\frac{\ell^2}{1+\ell}\right|_{\ell \ll 1} \approx -\ell = \ell_R^o \tag{42}$$

These equations calculate only the uncertainty coming from the process model. To calculate the influence of the uncertainty in P_w the whole first order difference of ΔR should be formed as

$$\begin{aligned}\Delta R &\cong \frac{\partial R}{\partial M}\Delta M + \frac{\partial R}{\Delta P_w}\Delta P_w = -RM^{-1}\Delta M + \\ &+ R\frac{P_w^{-1}}{1-P_w}\Delta P_w\end{aligned} \tag{43}$$

Thus the relative uncertainty can be obtained by

$$\ell_R = \frac{\Delta R}{R} \cong -\frac{\Delta M}{M} + \frac{1}{1-P_w}\frac{\Delta P_w}{P_w} = -\ell + \frac{1}{1-P_w}\ell_w \tag{44}$$

5. PRODUCT INEQUALITIES FOR PERFORMANCE AND ROBUSTNESS

Let us denote the well-known robustness measure: the distance between the point $(-1+0j)$ and the Nyquist curve of the open-loop frequency characteristics $Y = RS$ by ρ

$$\rho = \rho(\omega, R) = |1 + RS| \qquad (45)$$

If $|Y(\omega = \infty)| = 0$, then $0 \le \rho \le 1$. Denote the magnitude of the disturbance rejection rate $|e(j\omega)|/|w(j\omega)|$, which is a possible measure of control performance by δ

$$\delta = \delta(\omega, R) = \frac{|e(j\omega)|}{|w(j\omega)|} = \left|\frac{1}{1 + RS}\right| \qquad (46)$$

Observe that δ is the well-known classical sensitivity function. It is obvious that $\delta\rho = 1$ for all frequencies. One can easily get the same relationship denoting their minimal and maximal values with respect to ω by

$$\left.\begin{aligned} \rho_m(R) &= \min_\omega\{\rho(\omega, R)\} = \min_\omega(\rho) \\ \delta_M(R) &= \max_\omega\{\delta(\omega, R)\} = \max_\omega(\delta) \end{aligned}\right\} \Rightarrow \delta_M\, \rho_m = 1 \qquad (47)$$

Denote the worst (maximum) values with respect to R by

$$\left.\begin{aligned} \hat{\rho}_m &= \max_R\left\{\min_\omega[\rho(\omega, R)]\right\} = \max_R\left\{\min_\omega(\rho)\right\} \\ \breve{\delta}_M &= \min_R\left\{\max_\omega[\delta(\omega, R)]\right\} = \min_R\left\{\max_\omega(\delta)\right\} \end{aligned}\right\} \Rightarrow \qquad (48)$$

$$\Rightarrow \hat{\rho}_m\, \breve{\delta}_M = 1$$

Summarizing the above three basic products we got

$$\delta\,\rho = 1 \quad ; \quad \delta_M\, \rho_m = 1 \quad ; \quad \breve{\delta}_M\, \hat{\rho}_m = 1 \qquad (49)$$

where the following simple calculations prove (47) and (48)

$$\rho_m(R) = \min_\omega|1 + RS| = \frac{1}{\max_\omega\left|\frac{1}{1 + RS}\right|} = \frac{1}{\left\|\frac{1}{1 + RS}\right\|_\infty} = \frac{1}{\delta_M(R)} \qquad (50)$$

$$\breve{\delta}_M = \min_R\left\{\max_\omega\left|\frac{1}{1 + RS}\right|\right\} = \frac{1}{\max_R\left\{\max_\omega\left|\frac{1}{1 + RS}\right|\right\}} = \frac{1}{\max_R\left\{\min_\omega|1 + RS|\right\}} = \frac{1}{\hat{\rho}_m} \qquad (51)$$

Using (46), (47) and (48) one can easily obtain the following basic inequalities, too

$$\breve{\delta}_M \le \delta_M(R) \quad ; \quad \rho_m(R) \le \hat{\rho}_m$$

$$\begin{aligned} \frac{1}{\hat{\rho}_m} &= \breve{\delta}_M \le \delta_M(R) = \frac{1}{\rho_m(R)} \\ \frac{1}{\delta_M(R)} &= \rho_m(R) \le \hat{\rho}_m = \frac{1}{\breve{\delta}_M} \end{aligned} \qquad (52)$$

The above results are not very surprising, they simply follow from certain basic classical properties of the closed-loop sensitivity function. However, we can get very interesting special properties of the *generic scheme*, because here the modeling and control errors are the same, i.e. $\varepsilon = -e$, so the above inequalities also limit the modulus of the modeling error. Because the norm two $\|\varepsilon\|_2$ of the modeling error is more generally applied than its modulus (norm one) let us introduce the relative accuracy

$$\sigma = \frac{\|\varepsilon\|_2}{\|w\|_2} = \frac{\|e\|_2}{\|w\|_2} \quad ; \quad \|w\|_2 \ne 0 \qquad (53)$$

as another measure for the common modeling and control performance. This measure is bounded by the well known relationship

$$\sigma = \frac{\|\varepsilon\|_2}{\|w\|_2} = \frac{\|e\|_2}{\|w\|_2} \le \left\|\frac{1}{1 + RS}\right\|_\infty = \delta_M(R) \qquad (54)$$

It is also possible to define similar min/max relationships for σ as we did it for δ. So obvious definitions could be $\sigma_M(R) = \max_\ell\{\sigma(\ell, R)\}$ and $\breve{\sigma}_M = \min_R\left\{\max_\ell[\sigma(\ell, R)]\right\}$. By these notations one can easily get

$$\breve{\sigma}_M \le \sigma_M(R) \le \delta_M(R) = \frac{1}{\rho_m(R)} \qquad (55)$$

6. ROBUSTNESS MEASURES USING THE GENERIC OPTIMAL CONTROLLER SCHEME

In this section the properties of robustness measures for the *generic scheme* is discussed for inverse stable processes, when similarly to the process $M_+ = M$, $M_- = 1$ and $G = 1$. (The results are very similar for inverse unstable processes, however, the necessary formulas are much more complicated.)

Straightforward calculations give that

$$\rho_m(R) = \min_\omega|1 + RS| = \frac{1}{\left\|\frac{1 - P_w}{1 + P_w\,\ell}\right\|_\infty} \qquad (56)$$

and the upper limits are easily obtained as

$$\frac{\rho_m(R)}{\hat{\rho}_m^{o}} \le \|1 + P_w\,\ell\|_\infty \le 1 + \|P_w\|_\infty\,\|\ell\|_\infty \qquad (57)$$

where

$$\hat{\rho}_{\mathrm{m}}^{\mathrm{o}} = \hat{\rho}_{\mathrm{m}}(\ell = 0) = \frac{1}{\left\|1 - P_{\mathrm{w}}\right\|_{\infty}} = \min_{\omega}\left|\frac{1}{1 - P_{\mathrm{w}}}\right| \quad (58)$$

so $\hat{\rho}_{\mathrm{m}}^{\mathrm{o}}$ is the theoretically best optimal (maximal) robustness measure obtained if $\|\ell\|_{\infty} \to 0$ (i.e., for exact model matching case) during the iterative controller refinement.

Because it is easy to construct lower limits, too, the following inequalities hold during the iteration

$$1 - \|P_{\mathrm{w}}\|_{\infty}\|\ell_i\|_{\infty} \le \min_{\omega}|1 + P_{\mathrm{w}}\,\ell_i| \le \frac{\rho_{\mathrm{m}}(R_i)}{\hat{\rho}_{\mathrm{m}}^{\mathrm{o}}} \le \quad (59)$$
$$\le \|1 + P_{\mathrm{w}}\,\ell_i\|_{\infty} \le 1 + \|P_{\mathrm{w}}\|_{\infty}\|\ell_i\|_{\infty}$$

which means that

$$\rho_{\mathrm{m}}(R_i) \to \hat{\rho}_{\mathrm{m}}^{\mathrm{o}} = \max_{R \in \mathcal{R}}\left\{\min_{\omega}[\rho(\omega, R)]\right\} \quad (60)$$

providing convergence if $\|\ell_i\|_{\infty} \to 0$.

7. CONCLUSIONS

This paper demonstrates the necessity of iterative controller refinement methods in control engineering practice. A *generic optimal controller* structure and regulator design method, where the identification and control errors are identical, are used in the investigation. The generic scheme can be easily optimized in H_2-norm space if the regulator and two special serial compensators are selected in a special way. This optimization depends on the applied reference signal and the outpout disturbance. It is shown that this iterative scheme maximizes a robustness measure, too. A new uncertainty relationship is introduced limiting the product of modeling accuracy and robustness measures. This new inequality family opens a possible future new way to quantitatively analyze inverse relationship between modeling accuracy, control performance and robustness instead of the formerly known general qualitative dependence.

REFERENCES

Åström, K.J. and B. Wittenmark (1984). *Computer Controlled Systems-Theory and Design*. Prentice Hall, Englewood Cliffs, N.J.

Gevers, M. (1991). Connecting identification and robust control: a new challenge, *9th IFAC/IFORS Symp. SYSID'91*, Budapest, Hungary, 1-10.

Horowitz, I.M. (1963). *Synthesis of Feedback Systems*. Academic Press, N.Y. and London.

Keviczky, L. and Cs. Bányász (1994). A new structure to design optimal control systems, *IFAC Workshop on New Trends in Design of Control Systems*, Smolenice, SK, 102-105.

Keviczky, L. (1995). Combined identification and control: another way, (Invited plenary paper.) *5th IFAC Symp. ACASP'95*, Budapest, Hungary, 13-30.

Keviczky, L. and Cs. Bányász (1995a). A new optimal regulator design method. *European Control Conference*, Rome, Italy, 3359-3364.

Keviczky, L. and Cs. Bányász (1995b). A new pole-placement regulator design method. *34th IEEE Conf. Decision and Control*, New Orleans, USA, 3408-3413.

Keviczky, L. and Cs. Bányász (1997). A generic optimal controller scheme. *Int. J. Systems Sci.*, . (to be published).

Kucera, V. (1981). Exact model matching, polynomial equation approach. *Int. J. Systems Sci.*, **12,** 12, 1477-1484.

Wang. S. and B. Chen (1988). Optimal model matching control for time-delay systems. *Int. J. Control*, **47**, 3, 883-894.

This work was supported in part under US ARO Grant No. *DAAH04-96-1-0068* and the Hungarian NSF (OTKA).

ONE APPROACH TO CONTINUOUS-TIME PREDICTIVE CONTROL

Boris Rohaľ - Ilkiv *,[1]

* *Department of Automatic Control and Measurement*
Faculty of Mechanical Engineering, Slovak University of Technology
Nám. Slobody č. 17, 812 31 Bratislava, Slovak Republic

Abstract: In this paper an alternative approach to model - based predictive control of a class of continuous - time linear systems is formulated. The approach is based on a spline approximation of given continuous - time cost function and spline approximation of convolution type system description with structured and unstructured parts. Proposed predictive controller produces the control input signal in the form of a polynomial spline of given order. The various physical constraints on the signal can be easily set.

Keywords: Adaptive control, Identification, Spline approximation, Filtration, Predictive control.

1. INTRODUCTION

The majority of the model - based predictive control methods rests upon a discrete - time approach. Only a few works deal with the methods in a continuous - time framework, see (Demircioğlu and Gawthrop, 1991),(Demircioğlu and Clarke, 1991). The aim of this contribution is to design a *continuous - time* predictive controller using spline functions for modelling, filtering and control. The idea of *spline - based* self - tuning controllers was first formulated in (Kárny *et al.*, 1990) and further elaborated in (Rohaľ-Ilkiv and Zelinka, 1996), (Zelinka *et al.*, 1996).

The starting point is a spline approximation of given continuous - time, convolution integrals representation of the system with a *structured* and a *unstructured* part. The approximation leads to a simple and easily identified model using filtered data. Based on a type of the approximation, several *non-recursive* spline filters are constructed. The control signal is considered to be collected from *polynomial segments* of given order which are constrained to form polynomial spline function. The predictive control algorithm consider a continuous - time cost function. After spline approximation the function is transformed to matrix quadratic form. To minimize the form the quadratic programming is necessary.

[1] Supported by ECC Commission of the European Communities (COPERNICUS grant CP941174)

2. MATHEMATICAL BACKGROUND

Throughout the paper the spline functions are frequently used, therefore in the section some necessary notions will be reminded.

2.1 *Polynomial splines*

Let

$$(\lambda_0 =)0 < \lambda_1 < \ldots < \lambda_q < T_x(= \lambda_{q+1}) \quad (1)$$

be a subdivision of the closed (finite time) interval $[0, T_x]$ by q distinct (time) points. The points

$\lambda_i, i = 1, 2, \ldots, q$ are the (interior) knots of the spline function which is defined as follows:

The spline function $s(t)$, of the order r (degree $r-1$) and the defect d is a function which

(i) in each open interval $[\lambda_i, \lambda_{i+1}], i = 0, 1, \ldots, q$ is a polynomial of degree $\leq r-1$

(ii) has continuous derivatives up to the order $r-d-1$ in the open interval $(0, T_x)$

For each fixed set $\lambda = (\lambda_1, \lambda_2, \ldots, \lambda_q)$ of knots of the form (1), the class of such splines is a linear space of functions with dimension $z = (\alpha_1 + \alpha_2 + \ldots + \alpha_q) + r$, where α_i is a multiplicity of the knots λ_i. Let's denote the linear space of spline functions as $P_{r,\lambda,\alpha}, \alpha = (\alpha_1, \alpha_2, \ldots, \alpha_q)$. If all interior knots λ_i, are simple, i.e. $\alpha_i = 1, i = 1, 2, \ldots, q$, then $P_{r,\lambda,\alpha} = C^{r-2}[(0, T_x)]$. Any spline $s(t)$ can be expressed in the terms of a space basis:

$$s(t) = \sum_{i=1}^{z} c_i M_i(t) = c^T m(t) \qquad (2)$$

$$c^T = [c_1, \ldots, c_z], \; m^T(t) = [M_1(t), \ldots, M_z(t)]$$

where $M_i(t)$ are the base functions of the spline space and c_i are coefficients of the spline function $s(t)$. In this paper as the base functions $M_i(t)$ we shall use B-splines (see e.g. (Boor, 1978) for definition and basic concepts). For B-spline basis the expansion (2) is called *B-representation* of the spline function $s(t)$.

2.2 *Conversion from B to pp-representation*

By its definition polynomial spline can be written as:

$$s(t) = p_i(t) = \sum_{j=1}^{r} p_{ij}(t - \lambda_i)^{j-1} \qquad (3)$$

$$t \in (\lambda_i, \lambda_{i+1}), \; i = 0, 1, \ldots, q$$

where $p_i(t)$ are polynomial pieces which represent the spline $s(t)$ on interval $[0, T_x]$. Relation (3) is called piecewise polynomial representation (pp-representation) of spline $s(t)$. It is evident that pp-representation of spline $s(t)$ is completely determined by the $(q+1).r$-dimensional vector p of the polynomial coefficients:

$$p^T = [p_{01}, p_{02}, \ldots, p_{0r}, \ldots, p_{q1}, p_{q2}, \ldots, p_{qr}] \; (4)$$

Given the B-representation (2) of a spline $s(t)$, it is relatively an easy matter to compute the vector p of its pp-representation:

$$p = P \; c \qquad (5)$$

the rows of matrix P can be simply found by differentiation of the vector m in knots t_i:

$$P = [\frac{1}{(j-1)!} D^{j-1} m^T(t_i)]_{j=1,2,3,\ldots,r}^{i=0,1,2,\ldots,q}$$

and $D^{j-1}m^T(t_j) = [D^{j-1}M_1(t), \ldots, D^{j-1}M_z(t)]$, $D^i f(t)$ - means $i-th$ derivative of $f(t)$. Given the vector p, the conversion from pp-representation to B-representation is more difficult because of left inversion of matrix P. But if it is known by some means (and it is our case) that approximated function lies in $P_{r,\lambda,\alpha}$ for a certain r, λ, α, then the left inversion of matrix P (matrix P^L) can be determined uniquely and

$$c = P^L \; p \qquad (6)$$

2.3 *Spline interpolation*

Given a set of k distinct data points $(t_i, y_i), i = 1, 2, \ldots, k, t_i \in [0, T_x]$ and a prescribed set λ of q interior knots (where q, r, d are assumed to be chosen such that the dimension z of resulting spline space $P_{r,\lambda,\alpha}$ equals k), the spline interpolation problem is to determine the coefficients vector c in the B-representation (2) of a spline interpolant $s_y(t)$ such that $s_y(t_i) = y_i, i = 1, 2, \ldots, k$. In matrix form the equations can be expressed as M c = y, where $i, j-th$ entry of the matrix M is the value $M_j(t_i)$ of the $j-th$ B-spline at t_i and vector y holds $y_i, i = 1, 2, \ldots, k$. The necessary and sufficient conditions for the existence and uniqueness of $s_y(t)$ is provided by Schoenberg-Whitney theorem. Under this conditions the system of linear equation M c = y is invertible and

$$c = M^{-1} y \qquad (7)$$

2.4 *Smoothing spline*

Now let the y_i are approximate values $y_i = f(t_i) + \epsilon_i$ of some supposed smooth function $f(t)$ measured at data points $t_1, t_2, \ldots, t_k$, $t_i \in [0, T_x]$. Assume that $f(t)$ is smooth in the sense that $f(t) \in W_2^m[0, T_x]$ and ϵ_i are zero mean uncorrelated random variables with a same variance. The task is to recover $f(t)$ from these data by constructing the function $s(t) = s_\beta(t)$ which for given parameter $\beta > 0$ (to be chosen somehow), minimizes the following criterion

$$k^{-1} \sum_{j=1}^{k} (y_j - s(t_j))^2 + \beta \int_0^{T_x} (D^m s(t))^2 \, dt \quad (8)$$

over all functions $s(t)$ with m derivatives. Minimization of (8) established a sort of compromise between our desire to stay close to the given data ($\beta \to 0$) and our desire to obtain a smooth function ($\beta \to \infty$). The minimizing function, $s_\beta(t)$, is the smoothing spline and parameter β is called the smoothing parameter. The solution $s_\beta(t)$ is a spline of order $r = 2m$, with simple interior knots and *natural* end conditions $D^j s_\beta(0) = D^j s_\beta(T_x) = 0$, for $j = m, \ldots, r-2$. The coefficient vector c of the smoothing spline $s_\beta(t)$ can be found in the form

$$c = (M^T M + k\,\beta S)^{-1} M^T y \tag{9}$$

The $i, j - th$ entry of matrix S is defined by integral:

$$S_{ij} = \int_0^{T_x} D^m M_i(t) D^m M_j(t) dt$$
$$i, j = 1, 2, \ldots, k$$

2.5 *Least squares spline*

In the case of high dimensional data sets (t_i, y_i) corrupted by noise $\epsilon_i, i = 1, 2, \ldots, k, k > z$, the least square approximation is very useful. The task is to find a least square spline $s_\lambda(t)$ which minimizes quadratic criterion:

$$k^{-1} \sum_{i=1}^{k} (y_i - s_\lambda(t_i))^2 \tag{10}$$

with respect to chosen set $\lambda = (\lambda_1, \lambda_2, \ldots, \lambda_q)$. If we define the matrix M as $M = [M_j(t_i)]_{i=1,2,\ldots,k}^{j=1,2,\ldots,z}$ and if M has rank $r + q$, which we will assume to be the case, then

$$c = [M^T M]^{-1} M^T y \tag{11}$$

The set λ is actually a smoothing parameter which must be suitably selected. For details see (Boor, 1978), (Schumaker, 1981).

Notation. All the introduced quantities, vectors and matrices necessary for an approximation of some function $f()$ will be indexed by the superscript f.

3. APPROXIMATION OF SYSTEMS USING SPLINE FUNCTIONS

3.1 *Controlled systems description*

Consider that the task is to design a controller for some *true* linear, time-invariant continuous SISO system described by

$$Y(s) = G(s)\, e^{-sT_d}\, U(s) + H(s)\, E(s) \tag{12}$$

where $U(s)$ and $Y(s)$ are the Laplace transforms of the control $u(t)$ and output $y(t)$ respectively, $G(s)$ and $H(s)$ are proper rational transfer functions, T_d is the system time delay. $H(s)$ is assumed stable and strictly proper to guarantee a finite variance of the stochastic part of $y(t)$. The term $E(s)$ represents the transform of a disturbance signal $e(t)$, which is considered to be constituted by a white stochastic process with finite intensity and uncorrelated with the input $u(t)$.

The problem is to create a model of the system (12) based on discrete-time observations of the input and output signals. The literature on this problem is very extensive and usually deals with filtering schemes necessary for handling the signals time derivatives, see (Unbenhauen and Rao, 1987), (Sagara and Zhao, 1991). The crucial point is an acceptable choice of a model set for the noise filter $H(s)$. As shown in (Wahlberg, 1988) noise filters with relative degree grater then zero lead to need differentiate the measured data. The problem can be avoided be using noise models of relative degree zero or by introducing prefilters. This paper discusses the problem from slightly different point of view

- we apply the apparatus of spline functions for the signals approximation and for calculation of their derivatives
- we restrict ourself only to a class of continuous-time control input signals which belongs to a space of polynomial functions of given order; this is a key theoretical and technical assumption - *Assumption* A_1

Consider the system (12) in following, theoretically equivalent form

$$H(s)^{-1} Y(s) = H(s)^{-1} G(s) e^{-sT_d}\, U(s) + E(s) \tag{13}$$

Using the Euclidean algorithm the rational functions $H(s)^{-1}$, $H(s)^{-1}G(s)$ can be expressed as unique sums of their **strictly proper** parts $H_{sp}(s)$, $HG_{sp}(s)$ and their **quotients** $k(s)$, $l(s)$

$$H(s)^{-1} = H_{sp}(s) + k(s)$$
$$H(s)^{-1} G(s) = HG_{sp}(s) + l(s) \tag{14}$$

$k(s) = k_0 + k_1 s + \ldots + k_{nk} s^{nk}$
$l(s) = l_0 + l_1 s + \ldots + l_{nl} s^{nl}$

substituting (14) to (13) we have

$$k(s)Y(s) = -H_{sp}(s)Y(s) + l(s)e^{-sT_d}U(s) + HG_{sp}(s)e^{-sT_d}U(s) + E(s) \tag{15}$$

Letting D be the differentiation operator and applying the convolution theorem to (15) we arrive to the next continuous-time form of the system description

$$k(D)y(t) = -\int_0^t a(t-\tau)y(\tau)d\tau + l(D)u(t-T_d) + \\ + \int_0^t b(t-\tau)u(\tau-T_d)d\tau + e(t) \qquad (16)$$

where

$$a(\tau) = \mathcal{L}^{-1}\,[H_{sp}(s)] \\ b(\tau) = \mathcal{L}^{-1}\,[HG_{sp}(s)] \qquad (17)$$

$\mathcal{L}^{-1}$ [ ] - symbol of the inverse Laplace's transform.

From modelling point of view this form of the system representation consists of *structured* and *unstructured* parts. Note, that depending on the noise transfer function various special cases of (15), known under special names, can be examined e.g. continuous - time ARMAX, ARX, FIR structures.

3.2 *Output signal and kernels approximation*

In this subsection we will discuss how to apply the general convolution description (16) for the system modelling based on discrete time output data and continuous time input data (*Assumption* A_1). There are some advantages in using (16), any (stable) system can be modelled without the need for accurate information on the *true* plant order and on the noise filter structure. Unavoidable apriory information refers mainly to:

- degrees nk, nl of the polynomials $k(s)$, $l(s)$ (relative orders of the transfer functions $H(s)^{-1}$, $H(s)^{-1}G(s)$)
- time supports of kernels $a(\tau)$, $b(\tau)$ (impulse responses belonging to $H_{sp}(s)$, $HG_{sp}(s)$); they are assumed to be of finite lengths T_a, T_b - *Assumption* A_2

Due to *Assumption* A_1 the required derivatives of the input signal $u(t)$ in (16) are immediately available in an explicits form - they are calculated from actual polynomial pieces of the input signal. On the other side the derivatives of the output signal $y(t)$ are not directly available, they must be determined numerically. In order not to amplify any high frequency noise present in the discrete time observed data, we will use spline functions for approximation of the underlying signal trend on a chosen time interval. Derivatives of the spline approximant are then the estimated derivatives of the signal $y(t)$.

In subsequent we will specify a *parametric* class of models where we want to find a model of the system. Let's approximate the kernel $a(t-\tau)$ and the signal $y(\tau)$ on time interval $[t-T_a, t]$ and the kernel $b(t-\tau)$ on time interval $[t-T_b, t]$ by suitably chosen spline functions:

$$s_a(t-\tau) = c_a^T m_a(t-\tau), \quad s_y(\tau) = c_y^T m_y(\tau) \\ s_b(t-\tau) = c_b^T m_b(t-\tau) \qquad (18)$$

and write the control input signal in (16) as a spline function

$$s_u(\tau - T_d) = c_u^T m_u(\tau - T_d) \qquad (19)$$

Substitute for the signals and kernels in (16) the spline approximants (18), (19) and integrate over the vector of B-splines. Select (without loss of generality) the leading coefficient of the polynomial $k(D)$ unity, then we can obtain following *parametric* form of the system description:

$$D^{nk}s_y(t) = -k^T ds_y(t) - c_a^T Q_a(t)c_y + \\ + l^T ds_u(t-T_d) + c_b^T Q_b(t)c_u + o + \epsilon(t) \quad (20)$$

where:

- $k = [k_0, \dots, k_{nk-1}]^T$, $l = [l_0, \dots, l_{nl}]^T$ are vectors of unknown parameters relating to the structured part of the model
- c_a, c_b are vectors of spline coefficients of the kernels $a(\tau)$, $b(\tau)$, i.e. the vector of unknown parameters relating to the unstructured part of the model
- c_u, c_y are vectors of spline coefficients of the input and output signals $u(\tau - T_d)$, $y(\tau)$ relating to the time intervals $[t-T_b, t]$, $[t-T_a, t]$
- $ds_y(t) = [s_y(t), Ds_y(t), \dots, D^{nk-1}s_y(t)]^T$ $ds_u(t) = [s_u(t), Ds_u(t), \dots, D^{nl}s_u(t)]^T$ are vectors of the *spline* derivatives of the output and input signals at the time instant t
- $Q_a(t)$, $Q_b(t)$ are $(za.zy)$, $(zb.zu)$ dimensional matrices consist of convolution integrals of the splines basis functions

$$Q_a(t) = \int_{t-T_a}^{t} m_a(t-\tau)m_y^T(\tau)d\tau \qquad (21)$$

$$Q_b(t) = \int_{t-T_b}^{t} m_b(t-\tau)m_u^T(\tau - T_d)d\tau$$

- o is an model offset term
- $\epsilon(t)$ is noise term of the model modified by the approximation errors and including also unmodelled dynamics.

The kernels $a(\tau)$, $b(\tau)$ and signal $y(\tau)$ can be approximated with uniformly or nonuniformly located knots. The knots of the spline control signal are assumed to be placed into the control instants t_k ($t_k = k.T$, $k = 0, 1, \ldots,$ T is control period) in which the coefficients of new polynomial segment of the control signal are calculated by the controller. Let's write the model (20) for the time instants t_k. The matrices $Q_a(t_k)$, $Q_b(t_k)$ depend only on the knot sequences $\lambda_a, \lambda_y, \lambda_b, \lambda_u$ used for approximation in the intervals $[t_k - T_a, t_k]$, $[t_k - T_b, t_k]$. For increasing time instants $t_k, k = 0, 1, \ldots,$ the relative position of all these knots does not change with time (the knot sequences are shifting in time always about the value T), so the matrices $Q_a(t_k)$, $Q_b(t_k)$ do not depend on the actual time t_k, $Q_a(t_k) = Q_a$, $Q_b(t_k) = Q_b$. They can be calculated in advance and integrations in (21)can be performed only on intervals $[0, T_a]$, $[0, T_b]$.

4. FILTRATION AND IDENTIFICATION

Let's arrange the output signal values observed during the time interval $[t_k - T_a, t_k]$ into a vector

$$y_k^T = [y(t_{k-a}), y(t_{k-a+1}), \ldots, y(t_k)] \qquad (22)$$

where $a + 1$ is a number of the sampled values, ($t_{k-a} \equiv t_k - T_a$). Depending on the type of approximation, the vector of spline coefficients c_y can be then calculated using equality (7),(9) or (11). Given the vector c_y the derivatives of the output signal necessary for identification immediately follows from next relation

$$\begin{aligned} D^j s_y(t_k) &= \sum_{i=1}^{zy} c_{yi} D^j M_{yi}(t_k) \\ &= c_y^T D^j m_y(t_k) \end{aligned} \qquad (23)$$

$$\begin{aligned} c_y^T &= [c_{y1}, \ldots, c_{zy}] \\ D^j m_y^T(t_k) &= [D^j M_{y1}(t_k), \ldots, D^j M_{zy}(t_k)] \\ j &= 1, \ldots, nk - 1 \end{aligned}$$

The control signal $s_u(t)$ is generated by a controller as a piecewise polynomial function of given order and defect with knots at time instants t_k, $k = 0, 1, \ldots$. In each interval $[t_k - T_b, t_k]$ the signal $s_u(t)$ can be represented by a vector p_k of coefficients of all its polynomial segments

$$p_k^T = [p^T(k-b), \ldots, p^T(k-1), p^T(k)] \quad (24)$$

where vector $p^T(i) = [p_{i1}, p_{i2}, \ldots, p_{ir}]$, $i = k, k - 1, \ldots, k - b$ contains coefficients of $i - th$ polynomial segment which was applied for the control between the time instants $[t_{i-1}, t_i]$ ($b + 1$ is the number of polynomial segments used in $[t_k - T_b, t_k]$). The coefficients of the last actual polynomial segment directly define the derivations $D^j s_u(t_k)$, $j = 0, 1, \ldots, nl$.

Now let's introduce the concept of *spline filtering* and define so-called *filtration matrices.* Note that the formulas (7), (9),(11) used for approximation of the system output based on observations (22) can be write formally as

$$c_y = \mathcal{M}\, y_k \qquad (25)$$

where

- for interpolation $\mathcal{M} = M^{-1}$
- for smoothing $\mathcal{M} = (M^T M +$
  $+ (a+1)\beta S)^{-1} M^T$
- for l. s. approximation $\mathcal{M} = (M^T M)^{-1} M^T$

Denote the product of matrix Q_a with matrix $\mathcal{M}$ as a *filtering* matrix F_a,$F_a = Q_a\, \mathcal{M}$ and the product of matrix Q_b with P^L as a *filtering* matrix F_b, $F_b = Q_b\, P^L$. Then the operation of the spline filtering simply means calculation of vectors of *filtered* outputs and inputs:

$$y^f(k) = F_a y_k \qquad p^f(k) = F_b p_k \qquad (26)$$

at each time instant $t_k, k = 1, 2, \ldots$. If we make use of the filtered quantities we can establish the model (20) to more simple and more useful for identification form

$$\begin{aligned} D^{nk} s_y(t) = &-k^T ds_y(t) - c_a^T y^f(k) + \\ &+ l^T ds_u(t - T_d) + c_b^T p^f(k) + o + \epsilon(k) \end{aligned} \qquad (27)$$

It's easy to see that the problem of estimating the model (27) from experimental data - *filtered inputs, outputs* and *derivatives* - simplifies to a standard linear regression least squares estimation problem. Shifting the interval with time t_k its possible to organize on-line identification of the system. The techniques (Kulhavý, 1987) of restricted forgetting was used for parameters estimation and tracking.

Note that selection of orders, numbers and positions of knots of the above spline functions is an important but not easy problem. In principle, the positions of the knots are free parameters occurring in a nonlinear way. For detailed discussion and some guidelines see (Rohaľ-Ilkiv and Zelinka, 1996), (Zelinka *et al.*, 1996).

5. PARAMETER FILTERING

Having identified the model (27) the task is how to use it for formulation of prediction equations. The main problem is how to deal with the signals derivatives. For this purpose let's utilize the relation (23) for calculation of spline derivatives and substitute it to (27), after minor rearrangement we obtain

$$(c_a^T Q_a + dm_y^T)c_y = (c_b^T Q_b(t) + dm_u^T)c_u + + o + \epsilon(k) \quad (28)$$

where

$$dm_y = \sum_{i=0}^{nk} k_i \; D^i m_y(t_k), \qquad k_{nk} = 1$$

$$dm_u = \sum_{i=0}^{nl} l_i \; D^i m_u(t_k)$$

Notice that B-functions and their derivatives don't alter with shifting time, consequently the vectors $D^i m_x(t_k), x = u, y,$ are constant and it's possible to determine them in advance. Using (6) and (25) we can instal the *original* observed data to the model

$$(c_a^T F_a + dm_y^T \mathcal{M}) \; y_k = (c_b^T F_b + dm_u^T P^L) \; p_k + + o + \epsilon(t) \quad (29)$$

and are ready to introduce so-called *filtered model parameters* by following relations

$$c_a^f = F_a^T \; c_a \qquad c_b^f = F_b^T \; c_b \quad (30)$$
$$dm_y^f = \mathcal{M}^T \; dm_y \qquad dm_u^f = (P^L)^T \; dm_u$$

Then it is possible to convert (29) into next simple form

$$cs_a^T \; y_k \; = cs_b^T \; p_k + o + \epsilon(k) \quad (31)$$

where

$$cs_a = (c_a^f + dm_y^f) \qquad cs_b = (c_b^f + dm_u^f)$$

Normalize the equation (31) by the last value of the vector cs_a (which stands next to the observed value $y(t_k)$) and write it in standard regression structure:

$$y(t_k) \; = \theta^T \; \varphi(t_k) + \epsilon_n(k) \quad (32)$$

where
θ - is a known vector of the normalized filtered parameters
$\varphi(t_k)$ - is a known data vector
$\epsilon_n(k)$ - is a normalized value of $\epsilon(k)$

Given a control (or output) horizon the prediction equations can be easily derived from (32) via simple algebraic manipulation (Clarke *et al.*, 1987), (Clarke and Mohtadi, 1989).

6. PREDICTIVE CONTROL SYNTHESIS

The predictive control scheme considered in this article is based on minimization of a continuous - time quadratic cost function written in the form:

$$J(u) = \frac{1}{T_h} \int_{t_k}^{t_k+T_h} [(y(t) - y_r(t))^2 w_y(t) + + (u(t) - u_r(t))^2 w_u(t)]dt \quad (33)$$

subject to

$$u(t) \in U_{ad}(t_k)$$

where:
$y_r(t)$,$u_r(t)$ are reference signals; $w_y(t)$, $w_u(t)$ are weighting functions; T_h is a control horizon and $U_{ad}(t_k)$ is a space of allowed control signals (expressing the user's demands on the control including constraints).

The task is to find the continuous - time control signal $u(t) \in U_{ad}(t_k)$ such that the predicted output signal is as close as possible to the reference signal and the cost function reaches its minimal value. In general this task is not trivial one. To make the optimization problem more simpler we shall take following approximation of the cost function. Let's consider the input and output signal in (33) (together with reference values) in the form of spline function. The output signals $y(t)$, $y_r(t)$ will be represented with the help of relation (2) by vectors c_h, c_h^r of their coefficients and by a vector $m_h(t)$ of their base functions. The projected spline control signal $s_u(t)$ will be calculated in its pp-representation (4). Then after substitution to (33) we find that:

$$J(p_h) = (c_y - c_y^r)^T Q_1 (c_y - c_y^r) + +(p_h - p_h^r)^T Q_u (p_h - p_h^r), \quad p_h \in C_{ad}(T_h) \quad (34)$$

Vector p_h contains the coefficients of all polynomial segments which form the projected control input signal $s_u(t)$ on $[t_k, t_k + T_h]$:

$$p_h^T = [p^T(t_k+1), \ldots, p^T(t_k+q)] \quad (35)$$

q is the number of this polynomial segments. Vector p_h^r organizes reference values. The set $U_{ad}(t_k)$ is expressed by a set $C_{ad}(T_h)$ of allowed values for vectors p_h. The *penalty* matrices Q_1, Q_u are given by

$$Q_1 = \frac{1}{T_h} \int_0^{T_h} m_h(t) m_h^T(t) w_y(t) dt \quad (36)$$

$$Q_u = \begin{bmatrix} X(1) & & & \\ & X(2) & & \\ & & \ddots & \\ & & & X(q) \end{bmatrix}$$

and $X(i)$, $i = 1, 2, \ldots, q$ by

$$X(i) = \frac{1}{T_h} \int_0^{T_u} n(\tau) n^T(\tau) \; w_u(t_k +$$

$$+\tau + (i-1)T_u)d\tau \quad (37)$$

with
$n^T(\tau) = [1, \tau, \tau^2, \ldots, \tau^{r-1}]$, T_u is a distance between the knots of the projected control input signal. Applying (7) to the first member of (34) we obtain the resulting discrete form of the continuous loss function (33):

$$J(p_h) = (y_h - y_h^r)^T Q_y (y_h - y_h^r) +$$
$$+(p_h - p_h^r)^T Q_u (p_h - p_h^r), \quad p_u \in C_{ad}(T_h) \quad (38)$$

and for the *penalty* matrix Q_y holds:

$$Q_y = [M^{-1}]^T Q_1 M^{-1} \quad (39)$$

The vector y_h includes the *sampled* values of the continuous-time output signal $y(t)$ which must be predicted over the interval $[t_k, t_k + T_h]$. Our design of the predictive controller can be outlined as follows:

- predict the future plant's outputs (vector y_h) over the interval $[t_k, t_k + T_h]$ based on the filtered process parameters
- assuming that the future setpoint (vector y_h^r) is known, choose a *continuous time* spline input signal (vector p_h) which minimizes (38) for given $C_{ad}(T_h)$
- use the first segment of vector p_h, i.e. $p(t_k+1)$ as a current input and repeat this procedure at the next time instant (receding horizon strategy).

Based on the model (32) and assuming zero future noise we can write the prediction equations for y_h in next well known form

$$\hat{y}_h = Gp_h + f \quad (40)$$

where banded lower-triangular matrix G is fully determined by the model parameters θ and vector f depends on data which are known at time t_k. With this equation the loss function (38) can be written:

$$J(p_h) = p_h^T \nabla J(0) + \frac{1}{2} p_h^T H p_h$$
$$p_h \in C_{ad}(T_h) \quad (41)$$

where

$$\nabla J(0) = 2[G^T Q_y (f - y_h^r) - Q_u p_h^r]$$
$$H = 2[G^T Q_y G + Q_u]$$

Because the loss function (41) is quadratic, the quadratic programming (QP) technique is well suited for solving the minimization problem. In our simulations we utilized the efficient QP technique of (Powell, 1985), where the requirement $p_h \in C_{ad}(T_h)$ was transformed to a system of linear inequalities $A.p_h \geq v$. The matrix A and the vector v were fully determined by the types of constraints set on the control input signal represented by p_h. Two classes of the constraints are simultaneously used:

- constraints which are inevitable to formulate the vector p_h so that $s_u(t) \in P_{r,\lambda,\alpha}$ for chosen r, λ and α (control signal is spline function of given order)
- constraints due to physical limitations on the actuator or process

This briefly described QP technique can easily cover other types of constraints which are interesting for practice.

Remarkable *tuning knob* in the control synthesis is the distance $T_u = k.T$, $k = 1, 2, \ldots,$ between the knots of the projected control signal in the interval $[t_k, t_k + T_h]$. Proper choice of T_u can reduce the computational burden of QP (due to smaller dimension of vector p_h) and enables us to set $w_u(t) = 0$ even for the control of non-minimum phase systems.

7. ILLUSTRATIVE EXAMPLES

Following two examples illustrate robustness of the controller against noise and unmodelled dynamics. The system is simulated using transfer function

$$G(s) = \frac{1 - 0.2s}{1 + 0.9s + 0.45s^2 + 0.1s^3}$$

and the output signal is corrupted by noise with transfer functions

- Experiment A:

$$H(s) = \frac{1}{(1 + 0.1s)^2}$$

- Experiment B:

$$H(s) = \frac{1 + 0.1s}{1 + 0.9s + 0.45s^2 + 0.1s^3}$$

The system step responses are depicted on Fig.1. A, B . The l.s. approximation was chosen for spline filtering of the system output signal measured with sampling period $T = 10$ (a number of numerical integration steps in the system simulation).

Main design parameters for the model and controller:

$r_a = 3$, $r_b = 3$, $r_y = 3$, $r_u = 3$ - parabolic splines
number of derivatives - $nk = 1$, $nl = 0$
control period - $T = 10$, kernel supports - $T_a = 8T$, $T_b = 16T$

control horizon - $T_h = 12T \quad T_u = 4$,

Simple inspection of the parameter values nk, nl shows that the structure of models used during simulations is slightly different from the structure of simulated systems in both experiments.

It is seen from figure Fig.1. A, B that the controller exhibits satisfactory closed-loop behaviour.

8. CONCLUDING REMARKS

In the paper a predictive control algorithm with spline control signal is suggested. The control signal is of the continuous time nature and belongs to the space of polynomial spline functions of defined order. The introduced spline filtration make it possible to perform identification on law order models and reduces the effects of the noise and the unmodelled dynamics.

9. ACKNOWLEDGEMENT

This work has been supported by grant CP 941174 under COPERNICUS scheme.

10. REFERENCES

Boor, C. De (1978). *Practical Guide to Splines.* Springer-Verlag. New York.

Clarke, D.W. and C. Mohtadi (1989). Properties of generalized predictive control. *Automatica* **25**, 859–875.

Clarke, D.W., C. Mohtadi and P. Tuffs (1987). Generalized predictive control:part i and ii. *Automatica* **23**, 137–160.

Demircioğlu, H. and D.W. Clarke (1991). Cgpc with guaranteed stability properties. *IEE Proceedings-D* **139**, 371–380.

Demircioğlu, H. and P.J. Gawthrop (1991). Continuous-time generalized predictive controÌ. *Automatica* **27**, 55–74.

Kárny, M., I. Nagy, J. Bohm and A. Halousková (1990). Design of spline-based self-tuners. *Kybernetika* **26**, 17–30.

Kulhavý, R. (1987). Restricted exponential forgetting in real-time identification. *Automatica* **23**, 589–600.

Powell, M.J.D (1985). On the quadratic programming algorithm of goldfarb and idnani. *Math. Programming Study* **25**, 46–61.

Rohal̆-Ilkiv, B. and P. Zelinka (1996). Spline based predictive control: The introduction. In: *Spring school on: Adaptive and Predictive Control* (A. Kuznetsov, Ed.). pp. 53–58. Oxford university. Oxford.

Sagara, S. and Y.Z. Zhao (1991). Filtering approaches to identification of continuous systems. In: *9th IFAC/IFORS Symposium on Identification and System Parameter Estimation.* Vol. II. pp. 1331–1339. Pergamon press. Budapest.

Schumaker, L. (1981). *Spline Functions. Basic Theory.* John Wiley. New York.

Unbenhauen, H. and G.P. Rao (1987). *Identification of Continuous Systems.* North-Holland. Amsterdam.

Wahlberg, B. (1988). *On the identification of continuous time dynamical systems.* Linkoping university. Linkoping.

Zelinka, P., B. Rohal̆-Ilkiv, B. Benčurik and V. Holička (1996). Spline based predictive control: Practical experiments. In: *Spring school on: Adaptive and Predictive Control* (A. Kuznetsov, Ed.). pp. 59–67. Oxford university. Oxford.

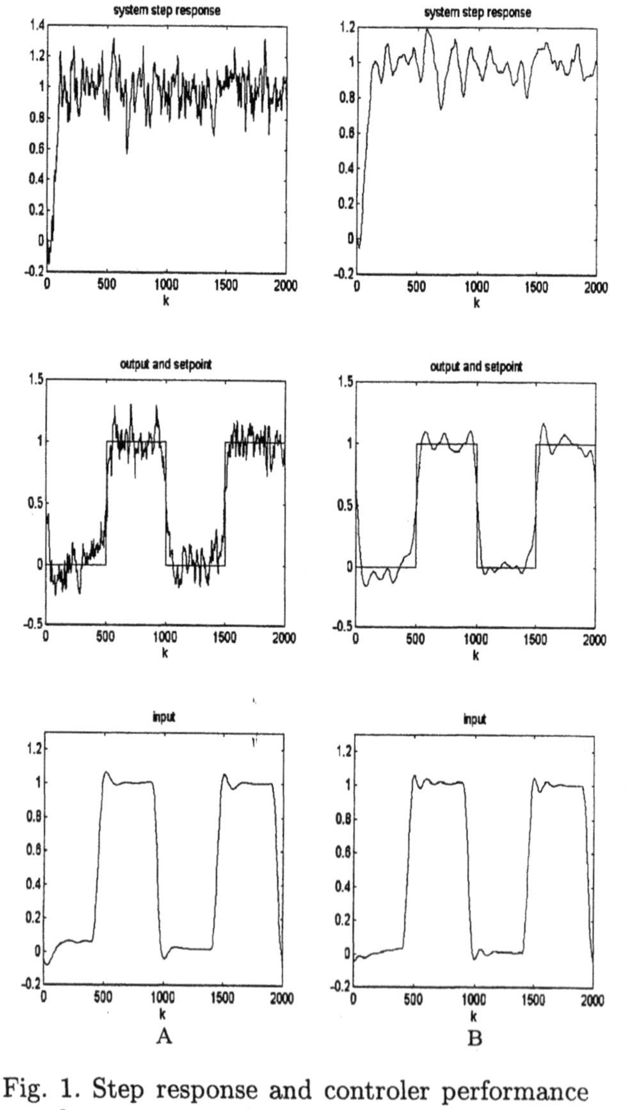

Fig. 1. Step response and controler performance for experiment A, B

NONLINEAR CONTROL SYNTHESIS

Antonín Víteček - Miluše Vítečková

*Technical University of Ostrava, 17. listopadu St., CZ-708 33 Ostrava - Poruba,
Czech Republic, Phone: +420 69 699 3485, Fax: +420 69 691 6490,
E-mail: antonin.vitecek@vsb.cz, miluse.viteckova@vsb.cz*

Abstract: The aggregation method in nonlinear control synthesis for the standard nonlinear dynamic subsystems is considered. The proposed approach is relatively simple and effective but nevertheless it does not demand a deep knowledge of the optimal control theory. The merit of the described nonlinear control synthesis is the design of nonadaptive, adaptive or combined controls which ensure feedback linearization of closed-loop control systems. An application is shown in an example.

Keywords: Nonlinear systems, multivariable control systems, control system synthesis, control algorithms.

1. INTRODUCTION

The problem of nonlinear control is still of interest even though there are excellent books dedicated to this subject (Canudas de Wit et al.,1996; Isidori, 1995; Lin, 1994; Nijmeijer and Van der Schaft, 1990, Spong and Vidyasagar, 1989).

This paper deals with nonlinear control synthesis which is based on the aggregation method. Its advantages are relative simplicity, the possibility of use for a wide class of the standard nonlinear dynamic subsystems and feasibility for the design of feedback linearizing nonadaptive, adaptive or combined control. The proposed approach is described in more detail in the publications (Víteček, 1991; Víteček *et al.*,1996).

2. STANDARD DYNAMIC SUBSYSTEMS

Consider controlled nonlinear dynamic subsystems with mathematical models in component form

$$\left.\begin{array}{ll}
\dot{x}_1 = x_2, & x_1(0) = x_{10}, \\
\vdots & \vdots \\
\dot{x}_{r_1} = f_{r_1}(\mathbf{x}) + \sum_{j=1}^{m} g_{r_1 j}(\mathbf{x}) u_j, & x_{r_1}(0) = x_{r_1 0}, \\
\dot{x}_{r_1+1} = x_{r_1+2}, & x_{r_1+1}(0) = x_{r_1+1,0}, \\
\vdots & \vdots \\
\dot{x}_{r_2} = f_{r_2}(\mathbf{x}) + \sum_{j=1}^{m} g_{r_2 j}(\mathbf{x}) u_j, & x_{r_2}(0) = x_{r_2 0}, \\
\dot{x}_{r_2+1} = x_{r_2+2}, & x_{r_2+1}(0) = x_{r_2+1,0}, \\
\vdots & \vdots \\
\dot{x}_n = f_n(\mathbf{x}) + \sum_{j=1}^{m} g_{nj}(\mathbf{x}) u_j, & x_n(0) = x_{n0}, \\
r_j = \sum_{i=1}^{j} n_i, \quad j = 1,2,\ldots,m, \quad r_m = n &
\end{array}\right\} \tag{1a}$$

or in vector form

$$\dot{\mathbf{x}} = \mathbf{f}(\mathbf{x}) + \sum_{j=1}^{m} \mathbf{g}_j(\mathbf{x}) u_j, \ \mathbf{x}(0) = \mathbf{x}_0, \tag{1b}$$

or in vector-matrix form

$$\dot{\mathbf{x}} = \mathbf{f}(\mathbf{x}) + \mathbf{G}(\mathbf{x})\mathbf{u}, \qquad \mathbf{x}(0) = \mathbf{x}_0, \qquad (1c)$$

that is

$$\mathbf{x} = [x_1, x_2, \ldots, x_n]^T, \ \dim \mathbf{x} = n,$$
$$\mathbf{u} = [u_1, u_2, \ldots, u_m]^T, \ \dim \mathbf{u} = m,$$
$$\mathbf{f} = \left[x_2, \ldots, f_{r_1}, x_{r_1+2}, \ldots, f_{r_2}, x_{r_2+2}, \ldots, f_n\right]^T, \ \dim \mathbf{f} = n,$$

$$\mathbf{G} = \begin{bmatrix} 0 & 0 & \ldots & 0 \\ \ldots & \ldots & \ldots & \ldots \\ g_{r_1 1} & g_{r_1 2} & \cdots & g_{r_1 m} \\ 0 & 0 & \ldots & 0 \\ \ldots & \ldots & \ldots & \ldots \\ g_{r_2 1} & g_{r_2 2} & \cdots & g_{r_2 m} \\ 0 & 0 & \ldots & 0 \\ \ldots & \ldots & \ldots & \ldots \\ g_{n1} & g_{n2} & \cdots & g_{nm} \end{bmatrix} = [\mathbf{g}_1, \mathbf{g}_2, \ldots, \mathbf{g}_m],$$

$$\dim \mathbf{G} = (n, m),$$

$$\mathbf{g}_j = \left[0, \ldots, g_{r_1 j}, 0, \ldots, g_{r_2 j}, 0, \ldots, g_{nj}\right]^T, \ \dim \mathbf{g}_j = m,$$

where $\mathbf{x}$ is the vector of the state variables (state), $\mathbf{u}$ - the vector of the control variables (control), $\mathbf{f}$ - the smooth vector function, $\mathbf{g}_j$ - the smooth vector function, n - the order of a nonlinear dynamic subsystem, n_i - the partial order, m - the number of control variables, T - the transpose symbol, dim - the dimension of a vector or the order of a matrix.

The dynamic subsystems with mathematical models (1) are of standard types, when they satisfy the condition (Víteček, 1991; Víteček *et al.*,1996)

$$\operatorname{rank} \mathbf{G} = m \ . \qquad (2)$$

The standard types of the mathematical models are in fact the models in the controller canonical forms (Kaczorek, 1983).

Nonstandard mathematical models can often be transformed to the standard models by a suitable choice of the state variables or an appropriate transformation (Isidori, 1995; Lin, 1994; Nijmeijer and Van der Schaft, 1990, Spong and Vidyasagar, 1989; Víteček, 1991; Víteček *et al.*,1996).

Example 1

The mathematical model of the controlled dynamic subsystem is not standard (Víteček *et al.*,1996)

$$\left.\begin{aligned} \dot{q}_1 &= q_2, & q_1(0) &= q_{10}, \\ \dot{q}_2 &= f_2'(q_1, q_2, q_3, q_5)\ , & q_2(0) &= q_{20}, \\ \dot{q}_3 &= q_4, & q_3(0) &= q_{30}, \\ \dot{q}_4 &= f_4'(\mathbf{q}) + g_{41}'(\mathbf{q})u_1 + g_{42}'(\mathbf{q})u_2, & q_4(0) &= q_{40}, \\ \dot{q}_5 &= q_6, & q_5(0) &= q_{50}, \\ \dot{q}_6 &= f_6'(\mathbf{q}) + g_{61}'(\mathbf{q})u_1 + g_{62}'(\mathbf{q})u_2, & q_6(0) &= q_{60} \end{aligned}\right\} \qquad (P1.1)$$

so and it is necessary to transform the model to the standard type.

Solution:

Write the mathematical model (P1.1) in the form

$$\dot{\mathbf{q}} = \mathbf{f}'(\mathbf{q}) + \mathbf{G}'(\mathbf{q})\mathbf{u}, \quad \mathbf{q}(0) = \mathbf{q}_0, \qquad (P1.2)$$

where

$$\mathbf{q} = [q_1, q_2, \ldots, q_6]^T,$$
$$\mathbf{f}' = [q_2, f_2', q_4, f_4', q_6, f_6']^T,$$
$$\mathbf{u} = [u_1, u_2]^T,$$
$$\mathbf{G}' = \begin{bmatrix} 0 & 0 \\ 0 & 0 \\ 0 & 0 \\ g_{41}' & g_{42}' \\ 0 & 0 \\ g_{61}' & g_{62}' \end{bmatrix}^T = [\mathbf{g}_1', \mathbf{g}_2'].$$

After the substitution of the original state vector $\mathbf{q}$ with the new state vector $\mathbf{x}$ in accordance with the relations (Víteček *et al.*,1996)

$$\left.\begin{aligned} x_1 &= q_1, \\ x_2 &= \dot{x}_1 = \dot{q}_1 = q_2, \\ x_3 &= \dot{x}_2 = \dot{q}_2, \\ x_4 &= \dot{x}_3 = \ddot{q}_2, \\ x_5 &= q_5, \\ x_6 &= \dot{x}_5 = \dot{q}_5 \end{aligned}\right\} \qquad (P1.3)$$

it can be obtained

$$\left.\begin{aligned} \dot{x}_1 &= x_2, \\ \dot{x}_2 &= x_3, \\ \dot{x}_3 &= x_4, \\ \dot{x}_4 &= f_4(\mathbf{x}) + g_{41}(\mathbf{x})u_1 + g_{42}(\mathbf{x})u_2, \\ \dot{x}_5 &= x_6, \\ \dot{x}_6 &= f_6(\mathbf{x}) + g_{61}(\mathbf{x})u_1 + g_{62}(\mathbf{x})u_2, \end{aligned}\right\} \quad \text{(P1.4a)}$$

$$\left.\begin{aligned} x_1(0) &= x_{10} = q_{10}, \\ x_2(0) &= x_{20} = q_{20}, \\ x_3(0) &= x_{30} = f_2'\big|_0, \\ x_4(0) &= x_{40} = \left[\frac{\partial f_2'}{\partial \mathbf{q}}\right]^T \mathbf{f}'\Bigg|_0, \\ x_5(0) &= x_{50} = q_{50}, \\ x_6(0) &= x_{60} = q_{60}, \end{aligned}\right\} \quad \text{(P1.4b)}$$

where

$$f_4(\mathbf{x}) = \left\{[\mathbf{f}']^T \frac{\partial^2 f_2'}{\partial \mathbf{q}^2} + \left[\frac{\partial f_2'}{\partial \mathbf{q}}\right]^T \frac{\partial \mathbf{f}'}{\partial \mathbf{q}}\right\}\mathbf{f}'\Bigg|_{\mathbf{x}},$$

$$g_{41}(\mathbf{x}) = \left[\frac{\partial f_2'}{\partial \mathbf{q}}\right]^T \frac{\partial \mathbf{f}'}{\partial \mathbf{q}} \mathbf{g}_1'\Bigg|_{\mathbf{x}},$$

$$g_{42}(\mathbf{x}) = \left[\frac{\partial f_2'}{\partial \mathbf{q}}\right]^T \frac{\partial \mathbf{f}'}{\partial \mathbf{q}} \mathbf{g}_2'\Bigg|_{\mathbf{x}},$$

$$f_6(\mathbf{x}) = f_6'\big|_{\mathbf{x}},$$

$$g_{61}(\mathbf{x}) = g_{61}'\big|_{\mathbf{x}},$$

$$g_{62}(\mathbf{x}) = g_{62}'\big|_{\mathbf{x}},$$

where $|_0$ or $|_x$ means that all variables have the initial values for t = 0 or that all the components of the vector $\mathbf{q}$ must be substituted with the components of the new vector $\mathbf{x}$ in accordance with the relations (P1.3).

The transformed mathematical model (P1.4) is standard already, where $n_1 = 4$, $n_2 = 2$, $n = 6$ and $m = 2$.

3. CONTROL SYNTHESIS PROBLEM AND ITS SOLUTION

Suppose that the control synthesis problem is to design the feedback control

$$\mathbf{u}^* = \mathbf{u}(\mathbf{x}, \mathbf{x}^w) \quad (3)$$

which ensures that the nonlinear dynamic subsystem (1) is tracking the desired state trajectory $\{\mathbf{x}^w(t)\}$ at the minimal value of the object quadratic functional

$$J = \int_0^\infty (\mathbf{e}^T\mathbf{Q}\mathbf{e} + \dot{\mathbf{e}}^T\mathbf{Q}_0\dot{\mathbf{e}})dt, \quad \mathbf{e} = \mathbf{x}^w - \mathbf{x}, \quad (4a)$$

$$\lim_{t\to\infty} \mathbf{e}(t) = \lim_{t\to\infty} \dot{\mathbf{e}}(t) = \mathbf{0}, \quad (4b)$$

where $\mathbf{x}^w$ is the vector of the desired state variables (dim $\mathbf{x}^w$ = n), $\mathbf{e}$ - the error vector (dim $\mathbf{e}$ = n), $\mathbf{Q}$ and $\mathbf{Q}_0$ - the weighting matrices which are positive-definite or positive-semidefinite of order n.

The formulated control synthesis problem can be solved easily, when the aggregate error vector is used (Víteček, 1991)

$$\mathbf{s} = -\mathbf{D}\mathbf{e}, \quad (5)$$

where $\mathbf{D}$ is the nonnegative aggregation matrix of the order (m,n) which satisfies the condition

$$\text{rank}\,(\mathbf{DG}) = m\,. \quad (6)$$

Then the functional (4) can be substituted which the considerably simplier one

$$J = \int_0^\infty (\mathbf{s}^T\mathbf{s} + \dot{\mathbf{s}}^T\mathbf{T}^2\dot{\mathbf{s}})dt, \quad (7a)$$

$$\lim_{t\to\infty} \mathbf{s}(t) = \lim_{t\to\infty} \dot{\mathbf{s}}(t) = \mathbf{0}, \quad (7b)$$

where $\mathbf{T}$ is the diagonal matrix of the positive time constants T_j of order m

$$\mathbf{T} = \text{diag}\,[T_1, T_2, \ldots, T_m]^T. \quad (8)$$

From the comparison of the functionals (4) and (7) the equalities can be obtained

$$\mathbf{Q} = \mathbf{D}^T\mathbf{D}, \quad \mathbf{Q}_0 = \mathbf{D}^T\mathbf{T}^2\mathbf{D}. \quad (9)$$

On the basis of the relations (9) it is obvious, that for the given matrices $\mathbf{D}$ and $\mathbf{T}$ the corresponding matrices $\mathbf{Q}$ and $\mathbf{Q}_0$ always exist, and therefore both functionals (4) and (7) are equivalent. For given matrices $\mathbf{Q}$ and $\mathbf{Q}_0$ both functionals will be equivalent on condition that the decomposition of matrices (9) exist.

The simple form of the quadratic functional (7) enables its direct minimization. Integrating the functional (7) by per partes method the result

$$J = \int_0^\infty \left(s^T s + \dot{s}^T T^2 \dot{s}\right) dt = \int_0^\infty (T\dot{s} + s)^T (T\dot{s} + s) dt + s_0^T T s_0 \quad (10)$$

can be obtained. Therefore the value

$$J^* = s_0^T T s_0 = e_0^T D^T T D e_0 \quad (11)$$

is constant (and depends only on initial conditions), it is obvious that the object quadratic functional (7) [consequently the functional (4) as well] reaches its global minimum (11) on the aggregate trajectory {s(t)} given by solution of the aggregate differential equation

$$\mathbf{T\dot{s}} + \mathbf{s} = \mathbf{0}, \qquad \mathbf{s}(0) = \mathbf{s}_0 \quad (12)$$

which in accordance with (5) has form for origin variables

$$\mathbf{TD\dot{e}} + \mathbf{De} = \mathbf{0}, \ \mathbf{e}(0) = \mathbf{e}_0 . \quad (13)$$

The differential equation (13) describes dynamics of the optimal closed-loop control system. It can seen, that the closed-loop control system is linear and its properties are defined by the matrices **D** and **T**. It is very important and simultaneously advantageous, because the choice of the weighting matrices **Q** and $\mathbf{Q}_0$ in the quadratic functional (4) is very demanding without knowledge of the direct influence on the dynamic behaviour of the closed-loop control system (13).

If the differential equation of the closed-loop control system (13) is written in the common form

$$\mathbf{\dot{e}} = \mathbf{Ae}, \qquad \mathbf{e}(0) = \mathbf{e}_0 \quad (14)$$

and if the elements of the aggregate matrix **D** are chosen in accordance with the relations (j = 1, 2,...,m; i = 1, 2,..., n)

$$\left.\begin{aligned} & d_{ji} = 0 \ \text{for} \ i \le r_{j-1} \quad \text{or} \quad i > r_j, \\ & d_{ji} \ne 0 \ \text{for} \ r_{j-1} < i \le r_j, \\ & d_{jr_j} = 1, \end{aligned}\right\} \quad (15)$$

that it can be shown that the closed-loop control system (14) has a characteristic polynomial

$$N(s) = \det(s\mathbf{I} - \mathbf{A}) = \prod_{j=1}^{m} N_j(s), \quad (16a)$$

$$N_j(s) = \left(d_{j,r_{j-1}+1} + d_{j,r_{j-1}+2}s + \cdots + d_{j,r_j-1}s^{n_j-2} + s^{n_j-1}\right)\left(\frac{1}{T_j} + s\right), \quad (16b)$$

where s is the complex variable, N_j - characteristic polynomial of the j-th autonomous control subsystem of order n_j, **I** - the unit matrix of order n, det - the determinant of a matrix.

It is obvious that in this case the optimal closed-loop control system (14) [or (13)] is created by m autonomous linear control subsystems of which the desired dynamic properties can be ensured by the suitable choice of the coefficients d_{ji} of their characteristic polynomial (16b).

On the basis of (1c), (6) and (13) for $\mathbf{u} = \mathbf{u}^*$ the optimal feedback control

$$\mathbf{u}^* = [\mathbf{DG}(\mathbf{x})]^{-1}\left\{\mathbf{T}^{-1}\mathbf{De} + \mathbf{D}\left[\dot{\mathbf{x}}^w - \mathbf{f}(\mathbf{x})\right]\right\} \quad (17)$$

can be obtained.

If disturbances influence the work of the nonlinear dynamic subsystem (1) then its mathematical model can have the form

$$\dot{\mathbf{x}} = \mathbf{f}(\mathbf{x},\mathbf{v}) + \mathbf{G}(\mathbf{x},\mathbf{v})\mathbf{u}, \ \mathbf{x}(0) = \mathbf{x}_0, \quad (18)$$

where **v** is the disturbance vector.

From the assumption of the full measurability of the disturbance vector **v**, in accordance with the relation (17) can be written

$$\mathbf{u}^* = [\mathbf{DG}(\mathbf{x},\mathbf{v})]^{-1}\left\{\mathbf{T}^{-1}\mathbf{De} + \mathbf{D}\left[\dot{\mathbf{x}}^w - \mathbf{f}(\mathbf{x},\mathbf{v})\right]\right\}. \quad (19)$$

Often if the mathematical model of the controlled dynamic subsystem is not known or the disturbances **v** can not be measured, then it is possible to use the following approach (Víteček, 1991).

Arrange the differential equation of the optimal closed-loop control system (13) and because $\dot{\mathbf{e}} = \dot{\mathbf{e}}(\mathbf{u}^*)$, therefore

$$\mathbf{m}(\mathbf{u}^*) = \mathbf{0}, \quad (20a)$$

can be written, where

$$\mathbf{m}(\mathbf{u}^*) = \mathbf{D}\dot{\mathbf{e}}(\mathbf{u}^*) + \mathbf{T}^{-1}\mathbf{De}. \quad (20b)$$

The equation (20a) can be solved by the fixed point iteration method

$$\begin{aligned} & \mathbf{u}_{k+1} = \mathbf{u}_k - \mathbf{Km}(\mathbf{u}_k), \\ & k = 0, 1, 2, \ldots, \end{aligned} \quad (21)$$

where **K** is the suitably chosen square matrix of order m, which ensures sufficiently fast convergence

$$\lim_{k\to\infty} \mathbf{u}_k = \mathbf{u}^* . \qquad (22)$$

The difference equation (21) can be substituted with the differential equation

$$\dot{\mathbf{u}}(t) = \Theta \mathbf{m}[\mathbf{u}(t)],\ \mathbf{u}(0) = \mathbf{u}_0 , \qquad (23)$$

where Θ is the suitably chosen square matrix of order m, $\mathbf{u}_0$ - the initial control.

Integrating the differential equation (23) the following control can be obtained

$$\mathbf{u}^{\times}(t) = \Theta \int_0^t \mathbf{m}[\mathbf{u}(\tau)]d\tau + \mathbf{u}_0 . \qquad (24)$$

If the used matrix Θ is chosen suitably (it can often be diagonal), this means that convergence

$$\lim_{t\to\infty} \mathbf{u}^{\times}(t) = \mathbf{u}^* \qquad (25)$$

must be sufficiently fast, then the control (24) may be treated as suboptimal and therefore it is designed by $\times$.

The initial control $\mathbf{u}_0$ can be computed for t = 0 on the basis of the relation (17) or (19). It may be determined very roughly, for example from the equilibrium state, etc because the control (24) has a integration behaviour character.

It is interesting that in the relation (24) the left side of the mathematical model of the optimal closed-loop control system [see (20) and (13)] is an integrand. After the substitution (20b) in (24) and the following modification

$$\mathbf{u}^{\times}(t) = \Theta\left[\mathbf{D}(\mathbf{e}-\mathbf{e}_0) + \mathbf{T}^{-1}\mathbf{D}\int_0^t \mathbf{e}d\tau\right] + \mathbf{u}_0 \qquad (26)$$

can be obtained. Because the mathematical model of the controlled dynamic subsystem (1) or (18) does not appear in the relation (26) and the initial control $\mathbf{u}_0$ can be roughly estimated, the control (26) can be considered as an adaptive control and the previous control (17) or (19) as a nonadaptive control. Therefore the square matrix Θ can be termed an adaptation matrix.

Sometimes the combined control may be useful

$$\mathbf{u}^c = \alpha\mathbf{u}^* + (1-\alpha)\mathbf{u}^{\times} ,\ \alpha \in \langle 0,1\rangle , \qquad (27)$$

where the coefficient α can be a constant or a time function. For example, in the beginning when a mathematical model of a controlled dynamic subsystem is known, α can be close to one and later when unpredictable behaviour changes appear it may approach zero.

From the relation (27) it is obvious, that the nonadaptive control component $\alpha\mathbf{u}^*$ uses known information about the properties of the controlled subsystem and the adaptive control component $(1 - \alpha)\mathbf{u}^{\times}$ carries out the suitable corrections which are given from the inaccuracy of the mathematical model and the unknown disturbances.

Example 2

For the nonlinear dynamic subsystem from example 1 it is necessary to design control on the basis of the aggregation method.

Solution

Write the standard mathematical model (P1.4) in the form (1c), where

$$\left.\begin{aligned} &\mathbf{x} = [x_1, x_2, \ldots, x_6]^T,\ \mathbf{u} = [u_1, u_2]^T, \\ &\mathbf{f} = [x_2, x_3, x_4, f_4, x_6, f_6]^T, \\ &\mathbf{G} = \begin{bmatrix} 0 & 0 \\ 0 & 0 \\ 0 & 0 \\ g_{41} & g_{42} \\ 0 & 0 \\ g_{61} & g_{62} \end{bmatrix} = [\mathbf{g}_1, \mathbf{g}_2]. \end{aligned}\right\} \qquad (P2.1)$$

Suppose that

$$\text{rank}\,\mathbf{G} = 2 . \qquad (P2.2)$$

A nonadaptive control is given by formula (17), where [see (15) for j = 1, 2; i = 1, 2,..., 6; $n_1 = 4$, $n_2 = 2$, n = 6, m = 2]

$$\mathbf{D} = \begin{bmatrix} d_{11} & d_{12} & d_{13} & 1 & 0 & 0 \\ 0 & 0 & 0 & 0 & d_{25} & 1 \end{bmatrix},\ \dim \mathbf{D} = (2,6), \qquad (P2.3)$$

$$\mathbf{T} = \begin{bmatrix} T_1 & 0 \\ 0 & T_2 \end{bmatrix},\ \dim \mathbf{T} = (2,2). \qquad (P2.4)$$

A characteristic polynomial of the closed-loop control system is given by relations (16), (P2.3) and (P2.4)

$$N(s) = N_1(s) N_2(s) , \qquad (P2.5)$$

where

$$N_1(s) = \left(d_{11} + d_{12}s + d_{13}s^2 + s^3\right)\left(\frac{1}{T_1} + s\right),$$

$$N_2(s) = \left(d_{25} + s\right)\left(\frac{1}{T_2} + s\right).$$

In accordance with the assignment it can be supposed for example the critically damped responses and therefore

$$\left.\begin{aligned} &N_1(s) = \left(\frac{1}{T_1} + s\right)^3\left(\frac{1}{T_1} + s\right) \Rightarrow \\ &\Rightarrow d_{11} = \frac{1}{T_1^3},\ d_{12} = \frac{3}{T_1^2},\ d_{13} = \frac{3}{T_1}, \\ &N_2(s) = \left(\frac{1}{T_2} + s\right)\left(\frac{1}{T_2} + s\right) \Rightarrow d_{25} = \frac{1}{T_2}. \end{aligned}\right\} \quad \text{(P2.6)}$$

From the assumption it can be written that the desired trajectory is at disposal in the component form

$$\{x_1^w(t)\},\ \{\dot{x}_1^w(t)\} = \{x_2^w(t)\},\ \{\ddot{x}_1^w(t)\} = \{\dot{x}_2^w(t)\},$$
$$\{x_5^w(t)\},\ \{\dot{x}_5^w(t)\} = \{x_6^w(t)\},\ \{\ddot{x}_5^w(t)\} = \{\dot{x}_6^w(t)\},$$

namely

$$\left.\begin{aligned} \mathbf{x}^w &= \left[x_1^w, x_2^w, x_3^w, 0, x_5^w, x_6^w\right]^T, \\ \dot{\mathbf{x}}^w &= \left[\dot{x}_1^w, \dot{x}_2^w, 0, 0, \dot{x}_5^w, \dot{x}_6^w\right]^T, \end{aligned}\right\} \quad \text{(P2.7)}$$

then the differential equation of the optimal closed-loop control system will be

$$\left.\begin{aligned} &\dot{e}_1 = e_2, && e_1(0) = e_{10}, \\ &\dot{e}_2 = e_3, && e_2(0) = e_{20}, \\ &\dot{x}_3 = x_4, && x_3(0) = x_{30}, \\ &\dot{x}_4 = \frac{1}{T_1^4}e_1 + \frac{4}{T_1^3}e_2 + \frac{6}{T_1^2}e_3 - \frac{4}{T_1}x_4, && x_4(0) = x_{40}, \\ &\dot{e}_5 = e_6, && e_5(0) = e_{50}, \\ &\dot{e}_6 = -\frac{1}{T_2^2}e_5 - \frac{2}{T_2}e_6, && e_6(0) = e_{60}. \end{aligned}\right\} \quad \text{(P2.8)}$$

The adaptive control can be obtained on the basis of the relations (26), (P2.3), (P2.4) and (P2.6), where

$$\mathbf{u}_0 = \left[\mathbf{DG}(\mathbf{x}_0)\right]^{-1}\left[\mathbf{T}^{-1}\mathbf{De}_0 - \mathbf{Df}(\mathbf{x}_0)\right],$$

$$\Theta = \begin{bmatrix} \theta_1 & 0 \\ 0 & \theta_2 \end{bmatrix}.$$

The diagonal elements θ_j of the adaptation matrix Θ can be determined by simulation (Víteček, 1991). The combined control is given by the linear combination of the nonadaptive control and the adaptive one, that is by formula (27).

4. CONCLUSIONS

This paper is dedicated to the original method of control synthesis of the standard nonlinear dynamic subsystems. The described method allows proposals of the nonadaptive, adaptive and combined control algorithms. The nonadaptive algorithm can be used for subsystems with precisely known mathematical models, where disturbances can be fully measured. The adaptive algorithms also allow their use for subsystems with minimum information regarding their dynamics and disturbances. The combined control algorithms use advantages of both types. All described control algorithms were verified by digital simulation with good results.

The research work was performed from the financial support of grant № 101/96/0829 of GA ČR.

5. REFERENCE

Canudas de Wit, C. and Siciliano, B.- Bastin, G. (1996). *Theory of Robot Control*. Springer - Verlag London Limited, Berlin - Heidelberg - New York

Isidori, A. (1995). *Nonlinear Control Systems*. The third edition. Springer - Verlag London Limited, Berlin - Heidelberg

Kaczorek, T. (1983). *Teoria wielowymiarowych układów dynamicznych*. Wydawnictwa Naukowo-Techniczne, Warszawa

Lin, C. F. (1994). *Advanced Control Systems Design*. PRT Prentice Hall, Englewood Cliffs - New Jersey

Nijmeijer, H. and Van der Schaft, A. J. (1990). *Nonlinear Dynamical Control Systems*. Springer - Verlag, Berlin - Heidelberg

Spong, M. W. and Vidyasagar, M. (1989). *Robot Dynamic and Control*. John Wiley & Sons, New York - Chichester

Víteček, A. (1991). *Optimal Path Control Synthesis by the State Variable Aggregation Method*. Habilitation Thesis, Technical University of Ostrava (in Czech), Ostrava

Víteček, A. *et al.* (1996). *New Approaches to Control of Complex Nonlinear Dynamic Subsystems*. Research Report of Grand Project GAČR № 101/95/1085, Technical University of Ostrava (in Czech), Ostrava

50 YEARS OF ROOT LOCUS : SOME NEW THOUGHTS

Annraoi de Paor

Department of Electronic and Electrical Engineering
University College Dublin
Belfield, Dublin 4, Ireland.

Abstract: Famous named curves are generated as Root Loci and used to produce optimum and pseudo-optimum stability designs for realistic processes and controllers. A problem in non-minimum-phase control is considered via a *principle of topological certainty* and an unusual root locus involving complex breakpoints is illustrated. A problem in dielectric theory, initially recast into an imaginary-parameter root locus problem, is solved via real parameter theory. Principles of *continuation, translation of axes* and *scaling* are invoked throughout. It is hoped to illustrate the versatility and power of root-locus inspired thinking.

Keywords: Root Locus, Stability, Control, Co-ordinate transformations, Non-minimum phase systems.

1. INTRODUCTION

The Root Locus Method is a set of theorems and techniques for visualising how the roots of the real-coefficient polynomial

$$P(s) = N(s) + kM(s) \quad (1),$$

where N(s) and M(s) are monic of degrees n and m, respectively, with $n \geq m$, move about the s plane as k varies over some defined range. The method was hinted at by Walter Evans in 1948, developed by him in 1950, and in 1954 given a definitive exposition. The status of the Root Locus Method was summarised by O. I. Elgerd (1967): "The Root Locus Method, introduced by Evans in 1948, won early popular acclaim and has since become our most important synthesis technique for single-input-single-output linear servomechanisms. "

The Evans approach is purely geometrical, based on the polar form of

$$P(s) = 0 \quad (2),$$

which leads to two relations

$$Arg(N(s)) - Arg(M(s)) = Arg(-k) = l\pi \quad (3)$$

$$|N(s)| / |M(s)| = |k| \quad (4)$$

In (3), l is an integer, odd for k>0 and zero or even for k<0. The positive parameter locus (k>0) is much more common than that for negative parameter, but the latter, via the *principle of continuation*, can be very valuable in design studies.

Analytical study of the Root Locus Method is well developed in the former Soviet sphere of influence (e.g. Teodorchik and Bendrikov (1963)), but largely neglected in Western Europe and the USA. The Evans method obscures the fact that the locus equation is often simply computed, and only for n+m>6 requires factorization of an equation more complicated than a quadratic (Harley and Power, 1973). Only two Western textbooks with which the author is familiar attempt to introduce geometrical and graphical approaches on an equal footing (Lehnigk, 1966; Power and Simpson, 1978).

The analytical approach is based on the Cartesian form of (2). By letting

$$s = \sigma + j\omega \qquad (5),$$

expanding (1) into real and imaginary parts (noting the restriction here to real k), equating both separately to zero in (2) and eliminating k, giving

$$\mathrm{Re}\,N\,\mathrm{Im}\,M - \mathrm{Im}\,N\,\mathrm{Re}\,M = 0 \qquad (6)$$

Equation (6) always contains ω as a factor, so that the real axis ($\omega = 0$) is part of the locus -- with different segments belonging to k positive and k negative. The remaining portion (7), describes branches off or at isolated crossings of the real axis.

$$\mathrm{Re}\,N(\mathrm{Im}\,M / \omega) - (\mathrm{Im}\,N / \omega)\,\mathrm{Re}\,M = 0 \qquad (7)$$

This paper explores the equations and curves yielded by (7) for various constellations of "poles" (roots of N(s)) and "zeros" (roots of M(s)). Configurations which seem at first to have little or no physical significance give rise to famous named curves (Archibald, 1957) which have simple constructions (Rektorys, 1964). These may, by invoking

- *continuation*
  i.e. if $k=k_0+\delta k$ then
  $P(s)=(N(s)+k_0M(s))+\delta kM(s))$
- *translation of axes*
- *scaling of axes*

yield controller designs for realistic processes which possess the "optimum stability property" that the ***rightmost eigenvalue is as deep in the left half plane as possible***. (This is a feature which the author has used with success over many years in many studies, e.g. Power (1971a,b) and de Paor(1985).)

2. WINDMILLS ON THE MIND: THE TRISECTRIX

In 1980 de Paor invented a novel aerodynamic theory of the 'optimum actuator disc' and later extended it to a design tool for real wind turbines (de Paor, 1982). Equation (8) had to be analysed:

$$P(s) = (s-1)^3 + k(s - 1/3) \qquad (8).$$

Continuation is invoked by letting

$$k = -3 + \delta\,\mathrm{k} \qquad (9);$$

axes are then *translated* by

$$s = \hat{s} + 3 \qquad (10);$$

and finally the plot is scaled by

$$\hat{s} = 3T\tilde{s} \qquad (11);$$

This leaves the effective problem of exploring

$$P(s) = (s + 1/T)^2 s + k(s + 8/(9T)) \qquad (12).$$

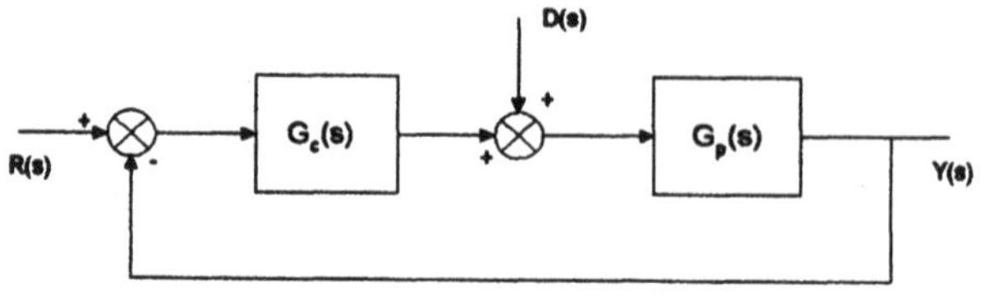

Fig. 1. Basic Feedback System

This is the characteristic polynomial of the feedback system shown in Fig. 1, with

$$G_p(s) = k_1 \Big/ \left(s + \frac{1}{T}\right)^2$$
$$G_c(s) = k_2\left(1 + \frac{8}{9Ts}\right) \qquad (13).$$
$$k = k_1 k_2$$

The problem is now PI control of a double lag process. The root locus of (12) is shown on Fig. 2.

The closed loop system is asymptotically stable for all $0<k<\infty$. There is a triple root at $s=-2/(3T)$ and choosing it as the design point gives *optimum stability* for

$$k_2 = 1/\left(3T^2 k_1\right) \qquad (14)$$

The analytical equation for the root locus of (12) is

$$\omega^2 = 4(\sigma + 1/T)^2 / ((\sigma + 5/(9T))9T) \qquad (15).$$

This is the Trisectrix of Colin MacLaurin, invented to facilitate trisection of an angle (Archibald, 1957).

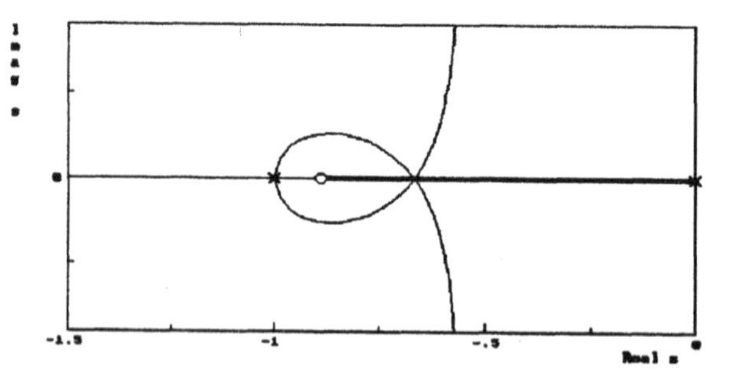

Fig. 2. Root Locus of the Trisectrix System.

3. PID CONTROL OF A DOUBLE INTEGRATOR PROCESS: THE LIMAÇON.

The process

$$G_p(s) = k_1 / s^2 \quad (16)$$

is considered under PID control:

$$G_c(s) = k_2 (s+a)^2 / s \quad (17).$$

The double integrator test case for controller design runs right through the book by Åström and Wittenmark (1990). The characteristic polynomial is

$$P(s) = s^3 + k(s+a)^2 \quad (18).$$

whose positive parameter locus is plotted in full line on Fig. 3. The system is asymptotically stable for

$$k > a/2 \quad (19).$$

given k>0, a>0 as necessary conditions. The complex roots meet as far to the left as possible in a double root at s = -3a for

$$k = 27a/4 \quad (20).$$

Equation (20) is a *pseudo-optimum stability design.* The analytical equation for the root locus of (18) is the Limaçon of Etienne Pascal (Boyer, 1991);

$$(\sigma^2 + \omega^2 + 2a\sigma)^2 = a^2(\sigma^2 + \omega^2) \quad (21).$$

This only requires repeated factorization of a quadratic in ω^2 to plot ω versus σ. The dotted negative parameter locus is the minor lobe.

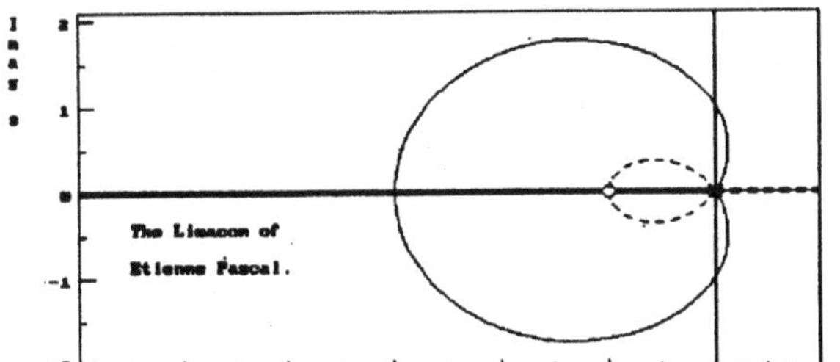

Fig. 3. PID control of a double integrator process.

4. FILTERED PID CONTROL OF A TWO-LAG PROCESS: STROPHOID AND CIRCLE

To begin, the root locus of the polynomial

$$P(s) = s^4 + k(s+a)^2 \quad (22).$$

is considered. The analytical equation of this is

$$(\sigma + a)^2 + \omega^2 = a^2 \quad \text{(circle)}$$
$$\omega^2 = \sigma^2(\sigma + a)/(a - \sigma) \quad \text{(strophoid)}$$
(23).

The circle belongs to k<0 and the strophoid to k>0. The leftmost breakpoint at s = -2a corresponds to k = $-16a^2$. Applying *continuation* with $k_0 = -16a^2$, which gives a real root at s = $2a(1+\sqrt{2})$, *translating* the origin to that real root, *scaling* to bring the leftmost poles, now at s = $-2a(2+\sqrt{2})$, to the value s = -b, and finally reverting to the notation of (1) gives

$$P(s) = (s+b)^2\left(s+\frac{2\sqrt{2}b}{2+\sqrt{2}}\right)s + k\left(s+\frac{(3+2\sqrt{2})b}{2(2+\sqrt{2})}\right)^2$$
(24).

The positive-parameter locus of (24), Fig. 4, applies to filtered PID control of a two-lag process:

$$G_p(s) = \frac{k_1}{(s+b)^2}$$
$$G_c(s) = \frac{k_2\left(s+\frac{(3+2\sqrt{2})b}{2(2+\sqrt{2})}\right)^2}{s\left(s+\frac{2\sqrt{2}b}{2+\sqrt{2}}\right)} \quad (25).$$

The quadruple eigenvalue has the *optimum stability property*, for which

$$k_2 = \frac{4b^2}{k_1(2+\sqrt{2})^2} \quad (26).$$

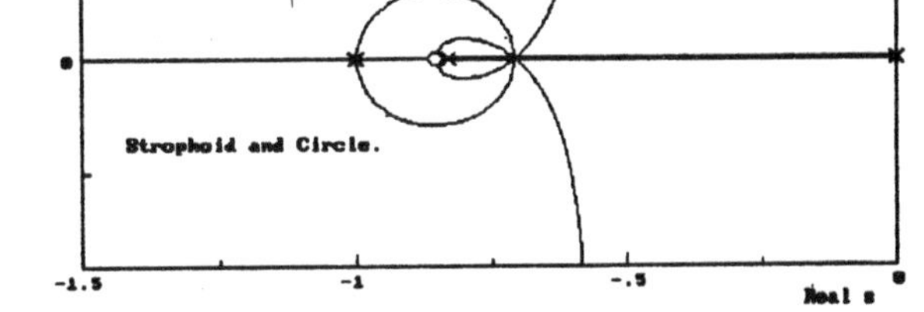

Fig. 4. Filtered PID control of a two-lag process

5. OPTIMUM STABILITY PI CONTROL OF A NON-MINIMUM-PHASE SYSTEM

A process and controller

$$G_p(s) = k_1(-s+b)/(s+a)^2$$
$$G_c(s) = k_2(s+c)/s \qquad (27).$$
$$(a,b,c>0)$$

are now considered, giving

$$P(s) = s(s+a)^2 - k(s-b)(s+c) \qquad (28),$$

so that there results a negative parameter root locus. The existence of an optimum stability design is tackled by exploring all possible topologies for c>0. The author invokes here the *principle of topological certainty*, which follows from an approximation theorem for root locus topology (Power, 1969). This asserts that all topologies which do not violate Evans' rules are certain to occur for some parameter values. If c is chosen so that for some value of k there is a triple root, then those values of k and c give an optimum stability design.

To find c and k, it is imagined that P(s) is a perfect cube, i.e.

$$P(s) = s^3 + (2a-k)s^2 + \left(a^2 - k(c-b)\right)s + kbc$$
$$= s^3 + 3ds^2 + 3d^2s + d^3 \qquad (29)$$

Equation (29) yields the following equations for $k = k_1k_2$ and c:

$$k^3 - (9b+6a)k^2 + (12a^2 + 35ab + 27b^2)k$$
$$-(8a^3 + 9a^2b) = 0 \qquad (30)$$
$$c = (2a-k)^3/(27kb)$$

For a, b>0, the pattern of signs in the first member of equation (30) establishes that it has only one real positive root for k.

As a simple example, taking a=b=1, there follows k = 0.2377968, and c = 0.8523119. The corresponding root locus plot is shown on Fig. 5.

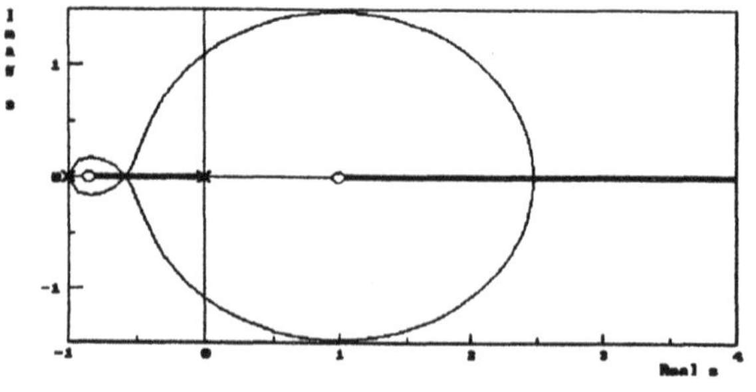

Fig. 5. Optimum stability PI control of a non-minimum phase system.

6. PID CONTROL OF A THREE-LAG PROCESS WITH INTEGRATOR: A SINGULAR BREAKPOINT BEHAVIOUR

The characteristic polynomial

$$P(s) = s^2(s+a)^3 + k(s+b)^2 \qquad (31),$$

arises from

$$G_p(s) = k_1/(s(s+a)^3)$$
$$G_c(s) = k_2(s+b)^2/s \qquad (32),$$

The breakpoint equation,

$$N(s)(dM/ds) = M(s)(dN/ds) \qquad (33),$$

simplifies to

$$s^2 + (5b/3)(s+2a/5) = 0 \qquad (34).$$

The positive b-parameter root locus for this is the circle centred on s = -2a/5, with radius 2a/5. The breakpoint at the origin simply reflects the fact that P(s) has a double root there for the trivial case k=0, whereas the other real breakpoint at s = -4a/5 corresponds to b = 0.96a. Root locus analysis (or application of the Routh-Hurwitz test) shows that this value of b gives an unstable system for all k>0. Now a very singular behaviour emerges: it turns out that every other point on the circle, with the sole exception of that corresponding to

$$b = 0.36a \qquad (35),$$

belongs to k *complex*. The condition b = 0.36a itself gives complex breakpoints through which the locus is passing for

$$k = k_1k_2 = 0.8a^3 \qquad (36).$$

The corresponding root locus is shown on Fig. 6. This is an *optimum stability design*. The breakpoint behaviour seems to merit special analysis. By using (34) repeatedly any polynomial in *s* of degree ≥2 can be reduced to one of first degree. Hence, in the equation P(s)= 0, one arrives at an expression of the form

$$k = (cs+d)/(es+f) \qquad (37)$$

Any point s = g + jh on the circle described by (34) is now selected and the problem is posed: under what condition will it give k real ? Substituting into (37), the required condition is obtained as

$$cf = ed \quad (38).$$

Now, c, f, e and d are all functions of a and b and working through the algebra yields

$$b^3 - (\frac{28}{25})ab^2 + (\frac{1031}{625})a^2b - (\frac{236}{625})a^3 = 0$$

which factorises to

$$(b - 0.36a)(b - 0.96a)(b - a) = 0 \quad (39).$$

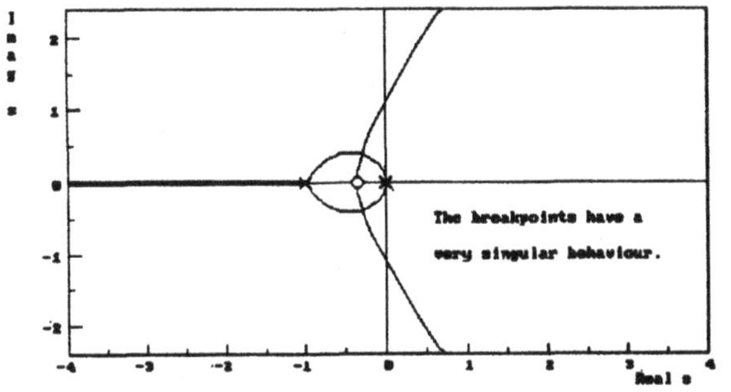

Fig. 6. Optimum stability PID control of a three-lag process with integrator.

7. EIGENVALUE LOCALISATION: GERSHGORIN CIRCLES AND CASSINI OVALS

In state space analysis of systems in vector-matrix form

$$dx / dt = Ax + Bu$$
$$y = Cx \quad (40),$$

it is sometimes convenient to get a preliminary idea of the locations of the eigenvalues of **A** by invoking the theorem of Gershgorin (Horn and Johnson, 1985). If R_i is the sum of the moduli of the off-diagonal elements from the ith row (or column) of **A**, the eigenvalues lie on or within the union of the circles

$$|s - a_{ii}| = R_i, \quad (i = 1, \ldots, n) \quad (41).$$

The circle (41) is the off-real-axis portion of the positive-parameter root locus of

$$P(s) = (s - a_{ii} - R_i)^2 + k(s - a_{ii}) \quad (42).$$

A. Brauer (Horn and Johnson, 1985), showed that the eigenvalues of **A** lie on or within the union of ½n (n-1) Cassini ovals

$$|s - a_{ii}||s - a_{jj}| = R_i R_j$$
$$(i = 1, \ldots, n-1 \quad (43).$$
$$j = i + k, \ k = 1, \ldots, n - i)$$

The main problem in applying Brauer's theorem, is the sheer tedium of drawing ovals. However, root locus theory provides a way forward. The typical oval described by (43) is the *negative-parameter* root locus of the polynomial

$$P(s) = (s^2 - (a_{ii} - a_{jj})s + a_{ii}a_{jj} + R_iR_j)^2$$
$$+ k(s - a_{ii})(s - a_{jj}) \quad (44).$$

Cassini ovals for $a_{ii} = 1$, $a_{jj} = -1$ and the three values of R_iR_j, 2, 1 and 0.3 are shown on Fig. 7. The intermediate value $R_iR_j = 1$ gives the Lemniscate of Bernoulli.

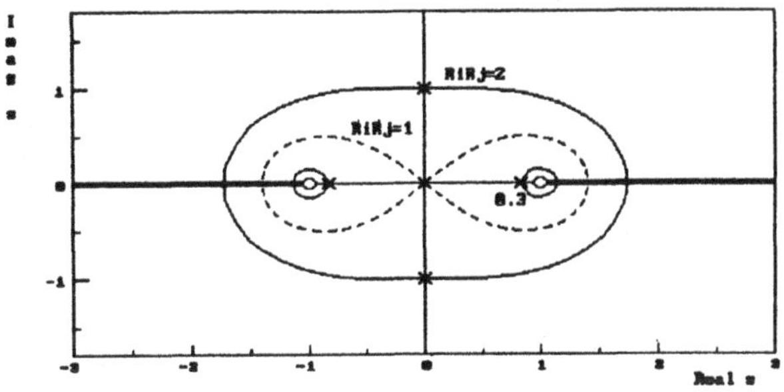

Fig. 7. Cassini's Ovals and the Lemniscate of Bernoulli.

8. A PROBLEM IN DIELECTRIC THEORY

Here is a problem in complex permittivity, posed to the author by Professor B.K.P. Scaife of Trinity College, Dublin. In developing a theory of complex permittivity in liquid dielectrics, Professor Scaife came up with the equation:

$$(s - 1)(2s + 1) / s = A / (1 + j\omega T) \quad (45).$$

The problem is whether, for $0<\omega<\infty$, the frequency response of s lies inside or outside a semicircle, for this has a great bearing on physical acceptability. The model is reasonable if the locus lies outside a semicircle and unreasonable otherwise. It is not difficult to rearrange (45) as

$$P(s) = ((s - 1)(s + \frac{1}{2}) - (\frac{A}{2})s) + j\omega T(s - 1)(s + \frac{1}{2})$$
$$= N(s) + j\omega TM(s) = 0 \quad (46)$$

By studying the imaginary parameter root locus,

$$ArgM(s) - ArgM(s) = -\pi / 2 \quad (47),$$

one finds that the model *is* reasonable. Imaginary parameter Root Locus is not available in any package with which the author is familiar. To study the problem via real parameter theory, (46) is recast as

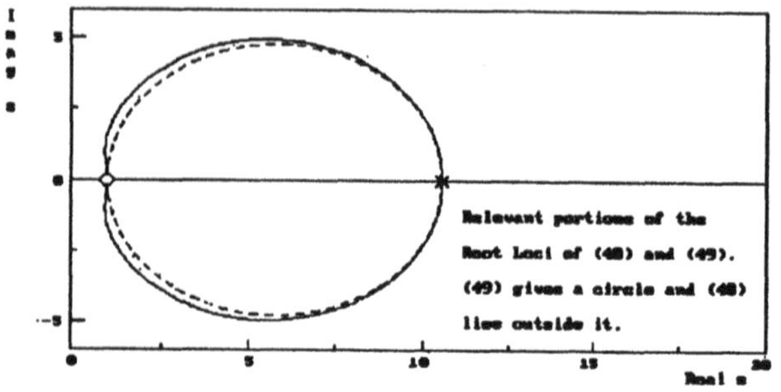

Fig. 8. A problem in Dielectric Theory.

$$N(s) = -j\omega TM(s),$$

which leads to

$$N^2(s) + \omega^2 T^2 M^2(s) = 0 \qquad (48).$$

The right half plane portion of the root locus of (48), drawn for $k = \omega^2 T^2$, is shown for A = 20 by the full curve of Fig. 8. Only the lower half has physical interest. The broken line shows the circle generated from the relevant poles and zeros in (46):

$$P'(s) = \left(s - \frac{A+1}{4} - \frac{1}{4}\sqrt{(A+1)^2 + 8}\right)^2 + k(s-1)^2 \qquad (49).$$

The full curve does indeed lie outside the circle.

9. CONCLUSION

Some unusual perspectives on the root locus method have been given by using famous named curves, along with *continuation, translation of axes* and *scaling*, to deduce optimum and pseudo-optimum stability designs for various processes. Topological differentiation has been illustrated in the problem of PID control of a three-lag-plus-integrator process. The author hopes that this may shed light on complex breakpoints in general in real-parameter loci. It has been shown that a non-real-parameter root locus problem may be converted to a real-parameter one and it is hoped that this feature also may hold the kernel of future developments. Two problems treated arose from outside the domain of control theory, and may help to illustrate the versatility of the Root Locus outside the domain in which it was originally proposed.

REFERENCES

Åström, K. J. and Wittenmark, B. (1990). *Computer-Controlled Systems: Theory and Applications* Addison Wesley, Reading, Mass.

Archibald, R.C. (=RCA).(1957) . Curves, special . *Encyclopaedia Britannica*, **6**, 887-899.

Boyer, C.B.(1991). *A History of Mathematics.* Wiley, New York. (revised by U.C. Merzbach).

de Paor, A.M. (1980) Teoiric úr don teascghníomhróir agus don mhuileann gaoithe (in Irish= A new theory for the actuator disc and the windmill)*Technology Ireland*, **11**, 35-38.

de Paor, A. M. (1982) Aerodynamic design of optimum wind turbines. *Applied Energy*, **12**, 221-228.

de Paor, A.M. (1985). A modified Smith predictor and controller for unstable processes with time delay . *Int. J. Control*, **41**, 1025-1036.

Elgerd, O.I. (1967). *Control Systems Theory.* McGraw-Hill, New York.

Evans, W.R. (1948). Graphical analysis of control systems. *Trans. AIEE*, **67**, 547-551.

Evans, W.R. (1950). Control system synthesis by the root locus method. *Trans. AIEE*, **69**, 66-69.

Evans, W.R. (1954). *Control System Dynamics.* McGraw-Hill, New York.

Harley, M. and H.M. Power, (1973). Analytical equations and geometrical constructions for some common root loci, *Int. J. Elec. Eng. Education*, **11**, 45-59.

Horn, R.A. and C.R. Johnson. *Matrix Analysis.* Cambridge University Press, Cambridge,UK.

Lehnigk, S.H. (1966). *Stability Theorems for Linear Motions.* Prentice Hall, Englewood Cliffs, New Jersey.

Power, H.M. (1969). Approximation theory for establishing root locus topology. *Electronics Letters*, **5**, 97-99.

Power, H.M. (1971a). Analytical root locus design study of servomechanisms with inertial damping. *Int. J. Elec. Eng. Education*, **9**, 411-424.

Power, H.M. (1971b). Analytical design of servomechanisms with spring-coupled-inertial dampers. *Int. J. Control*, **14**, 497-511.

Power, H.M. and R.J. Simpson (1978). *Introduction to Dynamics and Control.* McGraw-Hill, Maidenhead, UK.

Rektorys, K. (Ed.) (1964). *Survey of Applicable Mathematics.* Iliffe, London.

Teodorchik, K.F. and F.A. Bendrikov (1963). The method of plotting root paths of linear systems and for qualitative determination of the path type. In: *Theory of Continuous Linear Control Systems*, 7-10. Butterworth, London.

NEW TRENDS IN CONTROL ENGINEERING EDUCATION

Christian Schmid

Control Engineering Laboratory, Faculty of Electrical Engineering
Ruhr-University Bochum, D-44780 Bochum
Phone: ++49 234 7004093, Fax: ++49 234 7094101
E-mail: cs@esr.ruhr-uni-bochum.de

Abstract: Conventional lectures and laboratory courses show limitations in representing networks of complex dynamic subjects. Examples, methods and tools cannot be visualised or tested by the student sufficiently in this environment. First of all, this contribution discusses different ways in using the World Wide Web for control engineering education. Then in the main part about a software pilot project is reported. Both theoretical reflection and visual feedback, also together with practical experience are important aspects in control engineering instruction, which is supported by the Virtual Control Lab (VCLab) system developed during this project. The Web based hypertext technique is combined with tools for computer-aided control system design. The VCLab approach integrates plugins and Java applets which use the powerful computational engine of MATLAB/SIMULINK to do extensive simulations and real interactive animation of control engineering experiments driven by local or Web wide information.

Keywords: Control education, Educational aids, Laboratory education, Software, Teaching

1. INTRODUCTION

The merging of hypertext and wide area network functionality in the World Wide Web (WWW) is the main reason for exploiting Web technology for educational purposes to implement open and distributed hypermedia. The ability to integrate multi-media elements has significantly enhanced the ability to train and educate electronically. Whether the materials are a stand-alone tutorial or a full-fledged online workshop, the Web provides significant new functionality in transmitting information to the student and providing forums for exchange. The Web provides an effective mechanism for integrating design tools into a single interface and is an ideal tool for information which itself is rapidly changing. The Web is revolutionising some areas of study through increased opportunities for learning and alternative formats for information. The courseware based on the Web technology is not intended to replace other teaching/learning methods but rather to complement them.

Just browsing through the Web is already a pedagogical experience in itself. This kind of accidental learning is learning that happens at an unexpected moment about an unexpected subject. But there are structured ways in which the Web can be used for education and which are more adequate for the purpose of control engineering education. Using the advantage of the hypertext capabilities on a closed set of educational material, the „hyper courses“ can enable courseware designers to create educational material that has most of the characteristics of conventional courseware running on a local computer but with the advantage of being easily accessible from remote locations. In an open set of materials the pedagogical use of the Web is to exploit the information that is accessible on the Web and which must be managed to be useful for learners. E.g., the WWW Virtual Library is a distributed catalogue of Internet resources accessible on the Web, which helps navigating in this ocean of information. As useful as this open set may be for researchers or post-graduate students for their exploration of a subject domain, these resources are not suitable for more basic educational purposes. For the purpose of control engineering education a closed set of material consisting of explanatory text, examples, experiments and links to more expansive material available on the Web is more adequate.

One characteristic of the WWW that can appear as a limitation is the fact that the HTTP protocol is stateless, that is, there is no direct relationship of any kind between two consecutive requests to the same server, even if the queries come from the same user. However, an intelligent Computer Assisted Learning (CAL) system should take into account the background of each learner to tailor its behaviour to the learner's capabilities and past history, and to provide appropriate remediation to learners who experience difficulties with some concepts (Sens, 1992). To achieve this, the educational software has to keep track of the users' states and actions, which seems to contradict the statelessness of the Web. This limitation can be overcome by adding an assistant component based on an expert system, which can be installed at the server site.

This paper reports about a project (Müller, 1997; Schmid and Müller, 1997) which is in the current stage primarily not intended in developing a full CAL system but rather in exploiting the Web technology for educational purposes in control engineering. Therefore, Web technology will be used exclusively for which components are available as cheap commercial products, as shareware or freeware. In order to illustrate examples or experiments and increase distinctness and motivation, the possibility of user-controlled animation has to be considered. This is only practicable having a widespread system for Computer-Aided Control System Design (CACSD) tied together with the Web components to a hypertext-based learning system. The main question in this context is, how to putty together Web components with software for CACSD with a minimum of expense in costs and programming. Due to the data standards used in the Web technology, such a system is platform independent and thus an attractive candidate for distance learning where students, teachers and possibly equipment may be in geographically different locations.

2. CONTROL ENGINEERING EDUCATIONAL ASPECTS

An important part of control engineering education is the combination of theoretical knowledge with practical experience. The first is conventionally taught during lectures and exercises and based on lecture notes and/or textbooks or based on tutorials distributed by the Web - as recently shown by Pagel, *et al.* (1997). The practical experience is obtained during separate laboratory courses which are resource intensive and thus it can take time and money to have innovative control experiments. Control engineering instruction should combine theory and practice in each lesson. The students must be able to learn modelling of systems in order to develop controllers. The design process together with simulations and observing the dynamics of a physical implementation gives the student a valuable insight.

Design and visual feedback is an important aspect of control engineering instruction. For this reason, students have to be in a laboratory to gain laboratory experience. Today, communication techniques become available such that students can be anywhere. This kind of telepresence may be based on telerobotic systems, virtual reality and simulation systems. By Bohus, *et al.* (1996) a telerobotic approach is used to run real control experiments over the Internet. A special purpose remote lab user interface has been developed to enable the students to conduct experiments. This remote lab approach includes a control software development tool, a video and audio interface together with a collaboration tool. This is necessary for giving the remote user the feeling of being in the laboratory. An excellent example of a remote lab with sound feedback is available at the University of Tennessee (Henry, 1996).

The remote lab approach is only one possibility for running control engineering experiments. Its elements may be embedded as hypermedia elements into a Web based system. An other approach is to surpass the laboratory in demonstrating control applications and to help the student to discover the application of theories learned in the classroom. This may be realised by the simulation of real world systems and by animation of experiments. Such a lab has been developed by Kozura, *et al.* (1996) and extended to include adaptive control (Kaufman, *et al.*, 1997) by using MATLAB/SIMULINK, which is running only locally on the computer, but is an excellent way in contributing to the student's intuitive understanding of the theory behind the experiment. Similar teachware approaches for small industrial problems can be found in (Mansour and Schaufelberger, 1989) or (Schaufelberger, 1990). A supplementary virtual lab during the conventional lab course and within additional distance education in the form of a course offered across the Web, will fully engage the students in the learning process through an interactive, dynamic environment involving the student, on-line materials and an instructor. The combination of these approaches in using a widespread software tool or CACSD system together with hypertext classroom materials and other hypermedia Web components leads to the conception of an integrated lab. The Virtual Control Lab (VCLab) described in the following chapters shows these features.

There are different requirements that have to be fulfilled and some guidelines to be observed for an educational system of this kind which resides distributed on the Web. These requirements refer to interaction, simulation, virtual experiments and visualisation. Examples are given here:

1. All materials (text, graphics, videos, sounds, CACSD programs, VR objects) are in hypertext form and must be stored on Web servers to be accessible by clients via Web browsers.
2. The local Web browser is the only user interface in this environment. Each functionality is embedded in this Web tool.
3. In order to support calculation and simulation, the local installation of a CACSD system (MATLAB/SIMULINK) is necessary.
4. Hypertext material is organised like a textbook with small chapters which makes navigation easier. Essential keywords or parts are either linked together or they are linked to documents outside of this textbook; maybe to other textbooks or to related documents in the Web.
5. Text materials have to show all information including mathematics.
6. For illustrations using diagrams, static graphic elements should be avoided. The diagrams should be directly generated by the CACSD system. E.g., for a section dealing with the influence of a damping factor in a system, let the student actively explore this influence by integrating a Bode plot and a slider-gain element directly into the text.
7. For complicated interrelations, the animation by simulation is the right tool. The visualisation of objective effects of experiments is an important factor. The laboratory plant has to be modelled describing both the dynamics and e.g. the mechanical construction. The latter is for showing the plant on the screen and acting with this plant in a virtual reality (VR) scene.

3. TECHNIQUES USED FOR THE VCLAB

The general idea is to provide a common platform to this courseware system. For this reason the Web and a Web browser (Netscape Navigator 3.0) are chosen. This makes the system independent of a specific platform. The communication on the Web is organised in a client/server structure. The servers provide information and data and the clients are used to consume and to display this information (Fig. 1). HTTP transports any file type with the same protocol over the same line. The browser already provides general mechanisms to gather and send information through the Internet. The Web supports all infrastructure. For the VCLab three further important design details and requirements are given:

1. Calculation and simulation is performed locally by local software. Here MATLAB/SIMULINK is used, which must be installed on each client. Only the courseware will be send through the network. E.g., this includes also MATLAB M- or MEX-files. In order to install this additional functionality, MATLAB has to be linked to the Web browser.
2. The Web browser is the only user interface. All results are represented with this Web tool. Thus the courseware Web pages have to be designed in such a way that they can contain all necessary information for displaying results and for having interactive elements.
3. Authoring of the courseware must be as simple as possible. A compact programming technique on a high degree of abstraction is necessary.

The basic supporting language for the courseware is the Hyper Text Markup Language (HTML). On the today's standard level 3.2 the Web browser is capable of including Java applets (= Java programs for a restricted set of tasks) into its window. This enables the author to create and use interactive and animated elements in the courseware. Since such applets are programmed in Java they are portable to all Web browsers which support Java. Powerful supplemental browser components are the dynamically loaded modules, called plugins. These plugins are used for two different reasons. First, they extend the browser's capabilities to view different file formats (animation, images, sounds, etc.). The more important part, however, is that they can be implemented in a native language, e.g. C++. This makes it possible to write applications that are capable of using all computer resources and communication methods. In the following sections some technical details are il-

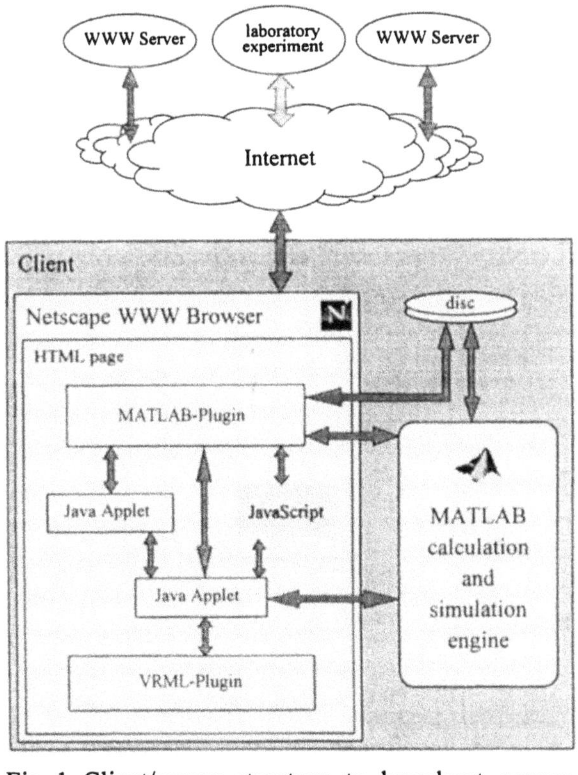

Fig. 1. Client/server structure to broadcast courseware, client configuration and interconnection between CACSD tool and Web elements.

lustrated using small examples.

4. SELECTED REALISATION ELEMENTS

The simple example for generating a Bode plot of a system in state-space representation in Fig. 2 illustrates the use of the MATLAB plugin for executing MATLAB routines over the Internet. This plugin may have an in-line window where the MATLAB command window output is displayed. The text in the plugin windows shows that „system1" is executed. This has been already performed while the Web page is being loaded. At the corresponding place the HTML code in Fig. 3 contains a hyperlink to the MATLAB M-file „system1.m" on the server. This file is automatically transferred and sent to the MATLAB engine for execution. In MATLAB plugins any M-files free of parameters can be used for the hyperlink. In order to use this mechanism also for a set of MATLAB routines, e.g. for a toolbox, and also for data and MEX-files, the new class of MM-files has been introduced. MM-files are combined M/MEX library/data files, which are created using the ZIP method without compression.

In order to show control engineering problems in the courseware one needs to display mathematical expressions, but the Web is lacking of this support so far. The „HotEqn" Java applet has been developed. This powerful applet uses the familiar LaTeX notation to encode mathematics as used in scientific word processing. Font sizes are variable, thus the applet can be used to display in-line as well as stand-alone expressions. The applet is capable of doing inter-applet communications as well. The variables in the expressions may be variable names of variables from the MATLAB workspace. This allows to move variables and their mathematical representations as well as their numerical values between different expressions. On the left-hand side in Fig. 2, the Eqns. (1) to (6) show mathematical equations. The systems matrices are displayed in Eqns. (1) to (4). It is possible to assign values to matrices in two different ways. They can be loaded by the MATLAB plugin or they can be put into the MATLAB workspace by HotEqn applets. Eqns. (5) and (6) show a symbolic form of an LTI system. The variables do not have explicit numerical values yet. Since the matrices A,B,C,D already exist in the workspace, one can click on both equations to see the representation change and to appear with their numerical values.

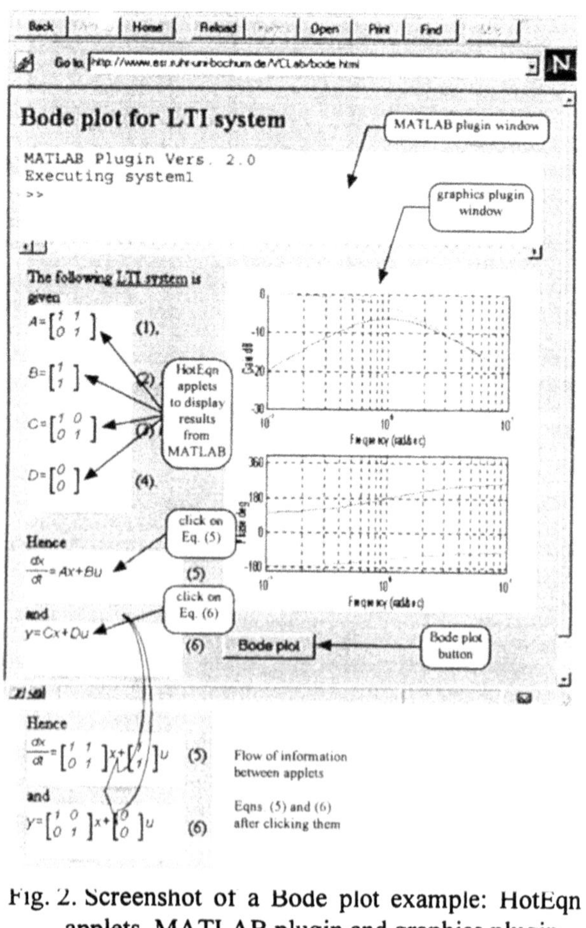

Fig. 2. Screenshot of a Bode plot example: HotEqn applets, MATLAB plugin and graphics plugin.

A further feature is the embedding of MATLAB statements into the HTML code. This integration is performed by using the JavaScript extension of HTML. In the example from Fig. 2, the button „Bode plot" (see Fig. 3) starts „onClick" three different methods. The first one does a Bode plot in MATLAB. The second command transfers the Bode plot graphics to the clipboard. The third command finally calls a method from the graphics plugin which in turn maps the contents of the clipboard to the browser window.

Fig. 3 shows the HTML code for the Bode plot example. The code consists of three main elements: the embed tag for the MATLAB plugin, the HotEqn applet tags containing different parameters to create and store matrices and finally the embed tag to get the result of the calculations of the button „Bode plot" back from MATLAB. All applets and plugins are represented by an unique name and are configured with certain parameters. This example demonstrates the compactness of the VCLab approach. Simple textbook Web pages for courseware materials can now be generated easily by using these few basic Web elements.

In order to support the simulation and animation aspect in the courseware and to construct a laboratory with plants in virtual reality, a supplementary set of new Web elements has been developed. This is demonstrated by the second example, Fig. 4, which shows the animated simulation of a level control system with three tanks. The task is to explore the dynamical behaviour of this plant by simulation studies under varying conditions. The browser is used both for providing the textual explanation of the problem and for controlling the simulation. The mechanical construction and the environment has been modelled by using the Virtual Reality Modelling

```
<html><head><title>Bode Plot</title></head>
<body>
<h2>Bode plot for LTI system</h2>
<embed src="system1.m" name="matlab" width=500 height=100>
The following <a href="LTI_system.html">LTI system</a> is given: <br>
                              :
Hence <br>
<applet code=HotEqn.class height=50 width=150
        name="eqn5"           align=middle MAYSCRIPT>
<param  name="LeftSide"       value="\frac{dx}{dt}">
<param  name="RightSide"      value="Ax + Bu">
<param  name="prev0"          value="eqn1">
<param  name="prev1"          value="eqn2">
</applet> (5)<br>
                              :
<embed type="image/emf" width=300 height=330 name="Bode_plot">
<form>
<input type="button" value="Bode plot"
        onClick="document.matlab.engEvalString('bode(A,B,C,D)');
                 document.matlab.engEvalString('print -dmeta');
                 document.Bode_plot.showClipBrd();">
</form>
</body></html>
```

MATLAB plugin

HotEqn applet for Eqn. (5)

graphics plugin

Bode plot button

Fig. 3. HTML code of the Bode plot example from Fig. 2.

Language (VRML). This VR scene is visualised by the VRML plugin that comes with the Netscape browser. The plugin enables the user to watch the plant during the simulation from any viewpoint or angle and it even generates the plant's noise as well. The plugin provides a control interface to move objects within the scene or to handle user actions during walking through the scene or e.g. pressing buttons or opening/closing valves. The scene consists of a couple of different objects: there is the lab environment with desks, chairs und the plant itself with the three interconnected tanks, two pumps and the valves.

Fig. 4. Screenshot of an animated simulation example of a SIMULINK model.

The plant's dynamics are described by an ordinary SIMULINK model, as shown in Fig. 5. For the animation purpose a new output block has been developed. This browser block can be added to the SIMULINK diagram and is used for sending simulation signals to the browser and in turn controls the simulation by receiving information from the browser. On the browser side this block is interfaced to the Web page by a Java applet. This animation applet is configured by various parameters which specify the number of inputs (outputs from the browser block) and by the transformations which apply to the inputs for moving the objects of the VR scene. It also defines callback actions which apply to the user's interaction with the VRML model. In this example the user can start and stop the simulation by pressing the „Power button" in the scene or open and

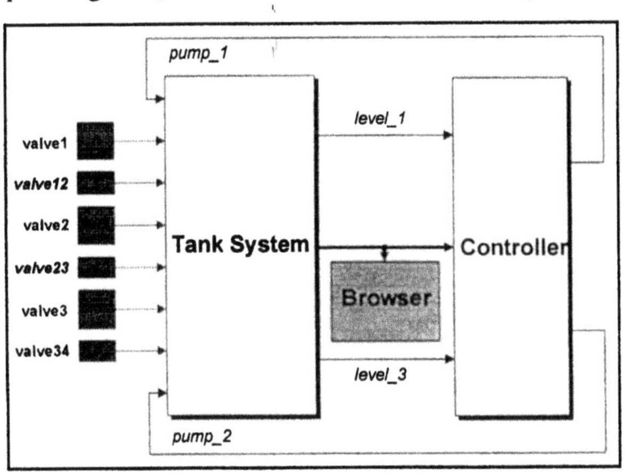

Fig. 5. SIMULINK diagram overview of the tank plant including a browser block for animation.

close the valves by mouse actions. The Web page shown in Fig. 4 contains a hidden hyperlink to a MM-file, which contains all files and data for the plant simulation. This link is managed by a hidden MATLAB plugin.

An additional option, not shown in this examples, is the augmenting of VR scenes by MATLAB generated graphics elements. With this feature it is possible to convert any MATLAB graphics to a VR object and post it to a scene displayed in the browser's window. One can take the mouse and walk through or fly over and visit, e.g. a 3D Bode plot.

The authoring of the VCLab courseware is still a manually process since all previous authoring software for Web pages supports only general tags for applets and plugins but does not allow one to do further programming. The VCLab approach provides the author with a variety of functional interfaces, which make it possible to concentrate on the development of the control issues, but not on the programming in Java or C++.

The implementation of VCLab relies on the condition that the browser can be tied up with MATLAB. Right now, the system runs with the Netscape Navigator 3.0 on Windows95 and on WindowsNT 3.x platforms. Due to the current rapid change in the Web browser technology it might be possible to use a standardised programming interface for all platforms in the near future.

5. CONCLUSIONS

This contribution demonstrates, that it is possible to overcome the static character of Web pages and to make them attractive for the control engineering education. The advantage of the VCLab approach is the use of existing MATLAB/SIMULINK models. E.g., only adding of one block to a SIMULINK diagram extends this tool to perform simulations in the Web environment.

For control engineering students it is easy to use and there is no need to learn MATLAB/SIMULINK for using courseware based on the VCLab approach. The user can focus on the learning and understanding of problems. The usage of a locally installed MATLAB/SIMULINK shows fast simulations without the need to have high bandwidth connections to a remote Web server. This is especially important for animation like the tank example. The hypertext approach enables to link different control topics. Since MATLAB stores all systems and variables users just need to read or display an HTML document and click on buttons or equations which enables them to check on certain conditions or properties of the given systems. To evaluate a new control approach or to understand the control of a machine one only needs to follow a hyperlink.

Further developments beyond this first stage of the VCLab project will be the extension to the remote lab approach using LiveVideo. An other important task will be the development of specific authoring tools for generating comprehensive courseware. They should support the automatic generation of VCLab enabled HTML code. This is generally not an easy job as shown by Zekl, 1995.

For those who want to evaluate VCLab:
`http://www.esr.ruhr-uni-bochum.de/VCLab/`

REFERENCES

Bohus, C., L.A. Crowl, B. Aktan and M.H. Shor (1996). Running control cngineering experiments over the Internet. Prepr. IFAC World Congress, San Francisco USA, paper 4c-01 5.

Henry, J. (1996). Controls laboratory teaching via the World Wide Web. ASEE Annual Meeting, Washington, paper 3513.

Kaufman, H., J. Kozura, K. Thayer and S. Ozcelik (1997): Animated emulations for illustrating adaptive control principles. Prepr. American Control Conference, Albuquerque.

Kozura, J., H. Kaufman, J. Reber, R. Shamayev, B.W. Bequette and D. O'Loughlin (1996). Computer animated control laboratory emulation modules. Prepr. IFAC World Congress, San Francisco USA, paper 4c-02 6.

Mansour, M. and W. Schaufelberger (1989). Software and laboratory experiments using computers in control education. IEEE Control Systems Magazine 9, No. 3, p. 19-28.

Müller, S. (1997). Untersuchungen zum Einsatz des World Wide Web in der regelungstechnischen Ausbildung. Diploma Thesis ESR-9615, Ruhr-University Bochum. Also available on http://www.esr.ruhr-uni-bochum.de/VCLab/software/Mueller/Diplomarbeit.htm

Pagel, J., Y. Sun, D. Tilbury, L. Oms, M. Suri and W. Messner (1997). Control tutorials for Matlab on the World Wide Web. Prepr. American Control Conference, Albuquerque.

Schaufelberger, W. (1990). Design and implementation of software for control education. IEEE Trans. on Education, vol. 33, No. 3, p. 291-297.

Schmid, C. and S. Müller (1997). A contribution to control engineering education on the Web. Prepr. IFAC Workshop ACE'97, Istanbul.

Sens, U. (1992). Rechnerunterstützte Wissensvermittlung für Automatisierungsingenieure beim Erlernen der methodischen Grundlagen und der Handhabung von rechnerunterstützten Werkzeugsystemen. PhD Dissertation, University of Stuttgart.

Zekl, A. (1995) Rechnerunterstützung für Autoren bei der Erstellung von Lernprogrammen für „Intelligente" Tutorielle Systeme. PhD Dissertation, University of Stuttgart.

NEW APPROACH TO OPTIMAL CONTROL THEORY

Pavel Žampa, Jiří Mošna and Pavel Prautsch

University of West Bohemia, Faculty of Applied Sciences,
Department of Cybernetics
Univerzitní 22, 306 14 Plzeň, Czech Republic,
Tel.: +420-19-2742252, Fax.: +420-19-279050
e-mail: zampa@kky.zcu.cz, mosna@kky.zcu.cz,
prautsch@kky.zcu.cz

Abstract: A view on the synthesis of optimal control systems is presented here. It stems from a newly outlined system theory and leads to a general optimizing recursion. Some interesting and fundamental results for general type of loss function and their connection to the usual solution are shown.

Keywords: optimal control, system theory, stochastic systems, dynamic programming, probability density function

1. INTRODUCTION

The theory of discrete-time control systems has been developed from two directions. First is the area of engineering applications and second is the one of sequential statistical decision theory. Despite their quite general approach (Striebel, 1975), there may be found some restrictions in the formulation. The loss function is selected as additive and the control is *a priori* assumed to depend on the system state, provided that the state can be measured (Shreve and Bertsekas, 1978).

New knowledge of the system theory (Žampa, 1996) has a wide application in exact formulation and statement of particular problems of automatic control. This fact enables to consistently derive numerous presently existing principles and methods as a consequence of assumed causal characteristics of the system. Consistent use of the system theory enables to cope with many paradoxes an contradictions of present theory and often yields new and more correct solutions.

Problem of the synthesis of an an optimal discrete-time control system can be formulated and solved by the newly outlined system theory.

It is assumed that the control system consists of two parts that are in mutual interaction via information connections. The structure of the control system with given causal probabilities of the controlled part enables to define a class of all generally stochastic controllers that are capable of feedback interaction with the controlled part.

By consistent use of results of the system theory the well known Bellman optimality principle is derived as a consequence of causality principle. Solution for an arbitrary loss function leads to a generalized optimizing recursion.

Standard solution of the problem of stochastic optimal control in the form of Bellman equation is then obtained from the general recursion under an suitable specification of the general formulation.

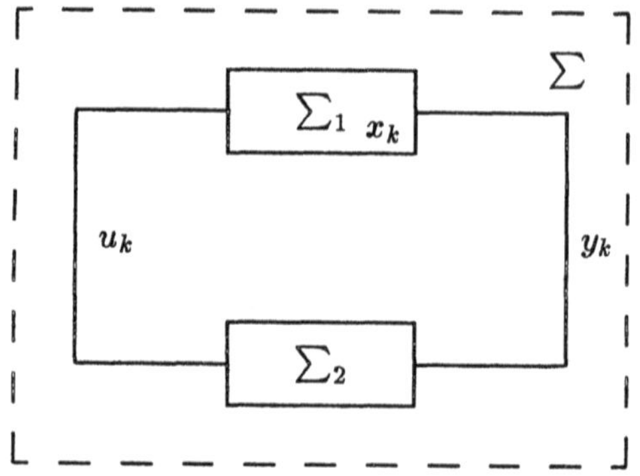

Σ ... control system
Σ_1 ... generalized controlled component
Σ_2 ... controlling component
u_k ... control variables at time t_k
U_k ... set of all control variables
x_k ... inner nonmeasurable variables
y_k ... output (observed) variables (of the Σ_1 component)

Fig. 1. Stochastic feedback control system

2. CONTROL SYSTEM AND ITS BEHAVIOUR

Control theory deals with mathematical models of real systems, that are called abstract systems. The control system is composed of a generalized controlled component, that comprises all surroundings relevant to the given problem, and a controlling component (Fig. 1). Both are mutually connected only by information relations and are not influenced by the environment of Σ.

Stochastic control systems are assumed here. The controlled component Σ_1 is considered to be completely defined if all its causal probability density functions (pdfs)

$$p_k(x_k, y_k : u_0^{k-1}, x_0^{k-1}, y_0^{k-1}) \qquad (1)$$

are known. Function p_k is a pdf of a vector (x_k, y_k) provided that sequence $(u_0^{k-1}, x_0^{k-1}, y_0^{k-1})$ is given, where $x_0^k = (x_0, x_1, \ldots, x_k)$ is a sequence of vectors for $k = 0, 1, 2, \ldots, F$. The control horizon F is assumed to be finite. Thus the general form of equation (1) includes a state approach too.

Causal pdfs of the controlled part determine a class of all admissible stochastic components Σ_2, which are generally given by causal pdfs

$$f_k(u_k : y_0^k, u_0^{k-1}), \quad k = 0, 1, 2, \ldots, F. \qquad (2)$$

They describe a causal mechanism by which the controlling component Σ_2 assigns a generally random value of control u_k defined on a set of control values U_k to every information segment of observed variables (y_0^k, u_0^{k-1}).

Function f_k will be called the control law at the time instant t_k. Control law for a sequence $t_k, t_{k+1}, \ldots, t_j$, $k \leq j$ is indicated by $f_k^j = (f_k, f_{k+1}, \ldots, f_j)$. The set of all admissible strategies at time t_k is denoted by Φ_k and the set of all admissible strategies Φ_k^j on the interval $\langle t_k, t_j \rangle$ on the sequence $t_k, t_{k+1}, \ldots, t_j$, $k \leq j$ is obviously

$$\Phi_k^j = \Phi_k \times \Phi_{k+1} \times \ldots \times \Phi_j. \qquad (3)$$

Causal pdfs of the controlled and controlling components completely define behaviour of the control system. Since the controlled component Σ_1 is given, expected behaviour of the control system depends on $f_0^F \in \Phi_0^F$. Dependance of a particular characteristic of the control systems on the chosen control law segment f_k^j will be indicated explicitly by '; f_k^j' as an argument of the given characteristic.

Choosing a general loss function $L(u_0^F, x_0^F, y_0^F) \mapsto R$, its expected value for any selection of $f_0^F \in \Phi_0^F$ is given by

$$J\left[f_0^F\right] = E\left\{L(u_0^F, x_0^F, y_0^F); f_0^F\right\}, \qquad (4)$$

where the symbol $E\{\cdot\}$ stands for the mean value operator. Aim of the control system synthesis is to find a component Σ_2 with a control law $f_0^{*F} \in \Phi_0^F$ for the controlled component Σ_1 that would minimize the criterion $J\left[f_0^F\right]$ of the control system quality on the set of all admissible strategies.

Causal pdfs are parametric density functions parametrized by their complete and imminent cause. Thus it must hold

$$p(x_k, y_k | u_0^{k-1}, x_0^{k-1}, y_0^{k-1}; f_0^F) = p_k(x_k, y_k : u_0^{k-1}, x_0^{k-1}, y_0^{k-1}) \qquad (5)$$

$$p(u_k | x_0^{k-1}, y_0^k, u_0^{k-1}; f_0^F) = f_k(u_k : y_0^k, u_0^{k-1}) \qquad (6)$$

for all alternative strategies $f_0^F \in \Phi_0^F$ and all $k = 0, 1, 2, \ldots, F$. That implies that the pdf of the consequence, conditioned by the knowledge of the whole system history, is equal to the causal probability function, no matter what control law has been applied.

Two fundamental probabilistic relations can be obtained from the causality properties of the control system. They refer to the density of an initial segment of the system trajectory (u_0^k, x_0^k, y_0^k) and to the probability of the initial segment of the nonmeasured variable x_0^k conditioned by observed variables (u_0^k, y_0^k).

Lemma 1. Pdf of the initial segment of the system trajectory at time t_k is independent on future control law f_{k+1}^F:

$$p(u_0^k, x_0^k, y_0^k; f_0^F) = p(u_0^k, x_0^k, y_0^k; f_0^k). \qquad (7)$$

Proof. See the Appendix.

Lemma 2. Aposteriori pdf of the initial segment of nonmeasured variable x_0^k at time t_k conditioned by the observed segment (u_0^k, y_0^k) does not depend on the control law f_0^F and it is also independent on the value of control u_k:

$$p(x_0^k | u_0^k, y_0^k; f_0^F) = p(x_0^k | u_0^{k-1}, y_0^k). \quad (8)$$

Proof. See the Appendix.

It follows from (5) and Lemma 2 that the pdf of the observed variable y_k at time t_k, conditioned by the data segment observed up to the time t_k is independent on the control law too

$$p(y_k | u_0^{k-1}, y_0^{k-1}; f_0^F) = p(y_k | u_0^{k-1}, y_0^{k-1}). \ (9)$$

Behaviour of the control system can be investigated from a view of an outer observer. Information segment (y_0^k, u_0^{k-1}) is known at time t_k and control u_k is to be generated. Future development till the time instant $t_j, j = k+1, \ldots, F$ is of interest.

Lemma 3. Complete description of the expected development of observed variables is given by pdfs:

$$p(y_{k+1}^j, u_k^j | y_0^k, u_0^{k-1}; f_0^F) = p(y_{k+1}^j, u_k^j | y_0^k, u_0^{k-1}; f_k^j), \quad (10)$$

$$p(y_{k+1}^{j+1}, u_k^j | y_0^k, u_0^{k-1}; f_0^F) = p(y_{k+1}^{j+1}, u_k^j | y_0^k, u_0^{k-1}; f_k^j). \quad (11)$$

Proof. See the Appendix.

3. GENERAL OPTIMIZING RECURSION

In order to transform the synthesis of the optimal control system to a recurrent succession of simpler optimization tasks, general optimizing recursion will be derived as a consequence of causal characteristics of the set of all competitive control systems. Term of a final segment of the control law is essential for further derivation.

At the time instant t_k the control system has applied the control law f_0^{k-1}. It is searched for the best future control $f_k^F \in \Phi_k^F$ that minimizes the expected value of the loss function for observed data.

Future behaviour of the control system is given by a conditional pdf

$$p(x_0^F, y_{k+1}^F, u_k^F | y_0^k, u_0^{k-1}; f_0^F) = p(x_0^F | y_0^F, u_0^F; f_0^F) \times p(y_{k+1}^F, u_k^F | y_0^k, u_0^{k-1}; f_0^F). \quad (12)$$

Substituting equations (8) and (10) to (12) the behaviour pdf has a form

$$p(x_0^F, y_{k+1}^F, u_k^F | y_0^k, u_0^{k-1}; f_0^F) = p(x_0^F | y_0^F, u_0^F) \times p(y_{k+1}^F, u_k^F | y_0^k, u_0^{k-1}; f_k^F). \quad (13)$$

Apparently the expected value of the loss function is independent on the chosen law f_0^{k-1}, *ie* on the past way of control. The optimality criterion J_k of the final segment of the control law at time t_k can be written as

$$J_k \left[y_0^k, u_0^{k-1}; f_k^F\right] = E\left\{L(u_0^F, x_0^F, y_0^F) | y_0^k, u_0^{k-1}; f_k^F\right\}. \quad (14)$$

Final segment of law $f_k^F(k)$ that minimizes the criterion J_k with respect to $f_k^F \in \Phi_k^F$ is an optimal final segment of the control law at time t_k. Corresponding value of the criterion J_k will be called the generalized Bellman function W_k

$$W_k\left[y_0^k, u_0^{k-1}\right] = J_k\left[y_0^k, u_0^{k-1}; f_k^F(k)\right], \quad (15)$$

for $k = 0, 1, \ldots, F$. Optimality principle follows from the assumed causality properties of the control system.

Theorem 1. (Optimality principle) Every optimal final segment $f_j^F(j)$ of a control law at time t_j is equal to a final segment $f_j^F(k)$ of the optimal control segment $f_k^F(k)$; $\{k \leq j\}$ in the sense that they lead to the same minimal value of the optimality criterion

$$J_k\left[y_0^k, u_0^{k-1}; f_k^{j-1}(k), f_j^F(J)\right] = J_k\left[y_0^k, u_0^{k-1}; f_k^{j-1}(k), f_j^F(j)\right]. \quad (16)$$

Proof. See the Appendix.

If $f_0^F(0)$ minimizes the value of $J_0\left[y_0; f_0^F\right]$, then it also minimizes the criterion $J[f_0] = \int J_0\left[y_0; f_0^F\right] p(y_0) dy_0$ and the optimal final segment $f_0^F(0)$ is the searched optimal control law f_0^F.

General method of dynamic programming directly follows from the optimality principle. Suppose, we have found an optimal final segment $f_{k+1}^F(k+1)$ at time t_{k+1}. If $f_k^* \in \Phi_k$ satisfies

$$J_k\left[y_0^k, u_0^{k-1}; f_k^*, f_{k+1}^F(k+1)\right] \leq J_k\left[y_0^k, u_0^{k-1}; f_k, f_{k+1}^F(k+1)\right], \quad (17)$$

for all $f_k \in \Phi_k$, then the law $\left(f_k^*, f_{k+1}^F(k+1)\right)$ is (according to the optimality principle) equivalent to the optimal final segment of the control law $f_k^F(k)$.

Optimality criterion at time t_k can be written in a form

$$J_k\left[y_0^k, u_0^{k-1}; f_k, f_{k+1}^F(k+1)\right]$$
$$= \int J_{k+1}\left[y_0^{k+1}, u_0^k; f_{k+1}^F(k+1)\right]$$
$$\times p(y_{k+1}, u_k | y_0^k, u_0^{k-1}; f_k) \mathrm{d}(y_{k+1}, u_k). \quad (18)$$

Using equation (15) an important relation is got

$$J_k\left[y_0^k, u_0^{k-1}; f_k, f_{k+1}^F(k+1)\right]$$
$$= \int E\left\{W_{k+1}\left[y_0^{k+1}, u_0^k\right] | y_0^k, u_0^k\right\}$$
$$\times f_k(u_k : y_0^k, u_0^{k-1}) \mathrm{d}u_k, \quad (19)$$

where

$$E\left\{W_{k+1}\left[y_0^{k+1}, u_0^k\right] | y_0^k, u_0^k\right\}$$
$$= \int W_{k+1}\left[y_0^{k+1}, u_0^k\right]$$
$$\times p(y_{k+1} | y_0^k, u_0^k) \mathrm{d}y_{k+1} \quad (20)$$

is independent on f_k. For the sake of conveniency $\Gamma_k\left[y_0^k, u_0^k\right]$ is denoted as

$$\Gamma_k\left[y_0^k, u_0^k\right] = E\left\{W_{k+1}\left[y_0^{k+1}, u_0^k\right] | y_0^k, u_0^k\right\}. \quad (21)$$

Substituting (21) to (19) the searched law f_k^* is found as:

$$f_k^*(u_k : y_0^k, u_0^{k-1})$$
$$= \operatorname*{argmin}_{f_k \in \Phi_k} E\left\{\Gamma_k\left[y_0^k, u_0^k\right] | y_0^k, u_0^{k-1}; f_k\right\} \quad (22)$$

and the value of the optimality criterion is then equal to the generalized Bellman function

$$W_k\left[y_0^k, u_0^{k-1}\right]$$
$$= \min_{f_k \in \Phi_k} E\left\{\Gamma_k\left[y_0^k, u_0^k\right] | y_0^k, u_0^{k-1}; f_k\right\}. \quad (23)$$

When the backward recursion, which starts at the boundary condition

$$W_{F+1}\left[y_0^F, u_0^F\right] = E\left\{L(x_0^F, y_0^F, u_0^F) | y_0^F, u_0^F\right\}, \quad (24)$$

arrives to the time $k = 0$ the optimal final segment $f_0^F(0)$ is obtained, as well as the value of optimality criterion J:

$$J\left[f_0^F(0)\right] = E\left\{W_0\left[y_0\right]\right\}. \quad (25)$$

Control law obtained by (22), (23) with the boundary condition (24) guarantees the best possible future continuation of the control at any time t_k, no matter what the past control was, so it is robust in this sense.

Optimal law with this attribute will be called the Bellman optimal law. Any law found by the optimizing recursion is optimal. On the other hand it can be shown that generally there may exist a solution of the control system synthesis problem even in such cases, where this recursion has no solution.

Presented optimizing recursion enables to solve the optimization problems even if the set of competitive strategies $\Phi_k'^F$ is a subset of the set of all permissible control strategies Φ_k^F. The usual case, when the optimal control law is deterministic is shown in the next theorem.

Theorem 2. Let the set of all strategies Φ_k at time t_k be equal to the set of all pdfs. Then the optimal law may be chosen deterministic and the control u_k^* is given by a function

$$u_k^* = G_k(y_0^k, u_0^{k-1}). \quad (26)$$

Hence the optimizing recursion has a form:

$$G_k(y_0^k, u_0^{k-1})$$
$$= \operatorname*{argmin}_{u_k \in U_k} E\left\{W_{k+1}\left[y_0^{k+1}, u_0^k\right] | y_0^k, u_0^k\right\}, \quad (27)$$
$$W_k\left[y_0^k, u_0^{k-1}\right]$$
$$= \min_{u_k \in U_k} E\left\{W_{k+1}\left[y_0^{k+1}, u_0^k\right] | y_0^k, u_0^k\right\}, \quad (28)$$

with boundary condition (24). Equation (25) holds for the value of optimality criterion.

Proof. See the Appendix.

Recurrent equation (28) with the boundary condition (24) can be called the generalized Bellman recursion. It was derived under very general assumptions about the loss function and control law. As one of the results, an interesting fact follows. In the case that the system state is measurable, it is a folklore in automatic control that the optimal control law depends just on the measured state at time t_k. However, this statement stays true only under a specific assumption about structure of the loss function.

4. ADDITIVE LOSS FUNCTION

We show briefly that the usual form of dynamic programming recursion can be derived from the general optimizing recursion by a special selection of the loss function L in an additive form

$$L(u_0^F, x_0^F, y_0^F) = \sum_{i=0}^{F} l_i(u_i, x_i, y_i). \quad (29)$$

Such arrangement leads to significant simplification of the calculation.

Optimality criterion J_k of final segment of the control law defined by (14) yields a form

$$J_k\left[y_0^k, u_0^{k-1}; f_k^F\right] = D_k\left[y_0^k, u_0^{k-1}; f_k^F\right]$$
$$+ I_k\left[y_0^k, u_0^{k-1}; f_k^F\right], \quad (30)$$

where I_k is the final part of the optimality criterion given by

$$I_k\left[y_0^k, u_0^{k-1}; f_k^F\right] = E\left\{\sum_{i=k}^{F} l_i(\cdot)|y_0^k, u_0^{k-1}; f_k^F\right\}, \quad (31)$$

and D_k represents its complement

$$D_k\left[y_0^k, u_0^{k-1}; f_k^F\right] = E\left\{\sum_{i=0}^{k-1} l_i(u_0^i, x_0^{i+1}, y_0^{i+1})|y_0^k, u_0^{k-1}; f_k^F\right\} (32)$$

Variables $(u_0^{k-1}, x_0^k, y_0^k)$ are necessary to compute (32). Pdf $p(x_0^k|y_0^k, u_0^{k-1}; f_k)$ is, according to Lemma 2., independent on the control law. Hence, function D_k can be written for all f_k^F

$$D_k\left[y_0^k, u_0^{k-1}; f_k^F\right] = D_k\left[y_0^k, u_0^{k-1}\right]. \quad (33)$$

Bellman function V_k is defined by substituting (31) and (32) to the generalized Bellman function (23) and employing the independence property of D_k. It is equal to the final value of the optimality criterion I_k for all $k = 0, \ldots, F$

$$V_k\left[y_0^k, u_0^{k-1}\right] = \min_{f_k^F \in \Phi_k^F} \left\{I_k\left[y_0^k, u_0^{k-1}; f_k^F\right]\right\}. \quad (34)$$

Substituting to general Bellman recursion (28) with boundary condition (24) it follows

$$V_k(y_0^k, u_0^{k-1}) = \min_{u_k \in U_k} E\left\{V_{k+1}\left[y_0^{k+1}, u_0^k\right] + l_k(y_0^{k+1}, u_0^k)|y_0^k, u_0^k\right\} \quad (35)$$

for $k = F, F-1, \ldots, 0$, with boundary condition

$$V_{F+1}(y_0^F, u_0^F) = 0. \quad (36)$$

Optimal control law

$$u_k^* = G_k(y_0^k, u_0^{k-1}) \quad (37)$$

is given by an equation

$$G_k(y_0^k, u_0^{k-1}) = \operatorname*{argmin}_{u_k \in U_k} V_k(y_0^k, u_0^{k-1}). \quad (38)$$

The optimality criterion for the optimal control law gains a value

$$J\left[f_0^{*F}(0)\right] = E\left\{V_0\left[y_0\right]\right\}. \quad (39)$$

Recurrent solving of the Bellman equation under its boundary condition enables to obtain the best possible control law of the control system from all admissible control strategies. Optimal law is deterministic.

5. CONCLUSION

Formulation of the control system synthesis based on a new system theory is presented here and leads to a general solution of the optimization problem. It provides systematic guidelines for understanding the nontrivial cases of optimal control design.

Applying the system theory on the class of competitive control systems the principle of optimality was derived as a consequence of the causality properties of the control system. Solution of the synthesis problem leads to a general optimizing recursion that does not have the usual form of Bellman equation. Obtained result guarantees every final segment of the optimal law to be optimal. That implies that the control at time t_k is optimal with no respect to the former control of the system.

The relation comes to usual Bellman equation under special conditions on structure of the loss function. It is also worth to mention that if the loss function has the property of additivity, it does not change the evaluation of the past in the future. That is basically the reason, why an optimal controller, having complete information about the system state available, need to exploit the immediate value of the controlled component state only.

6. ACKNOWLEDGMENTS

The research was partially supported by the Grant Agency of The Czech Academy of Sciences under grant GA AV A2147701/1997 and the Grant Agency of Czech Republic under grant GA ČR 102/97/0466.

7. REFERENCES

Shreve, S. and D. Bertsekas (1978). Dynamic programming in borel spaces. In: *Dynamic Programming and its Applications* (M.L. Puterman, Ed.). pp. 115–130. Academic Press. New York.

Striebel, C. (1975). *Optimal Control of Discrete Time Stochastic Systems.* Vol. 110 of *Lecture Notes Econom. Math. Systems.* Springer-Verlag. Berlin.

Žampa, P. (1996). On a new system theory and its new paradigms. In: *Cybernetics and Systems' 96.* Vol. 1. Austrian Society for Cybernetic Studies. Vienna. pp. 3–7.

8. APPENDIX

Proof of Lemma 1. Probability density of the initial segment of the system trajectory (u_0^k, x_0^k, y_0^k) can be written

$$p(u_0^k, x_0^k, y_0^k; f_0^F) = \prod_{i=0}^{k} p(u_i | u_0^{i-1}, x_0^i, y_0^i; f_0^F) \times p(x_i, y_i | u_0^{i-1}, x_0^{i-1}, y_0^{i-1}; f_0^F). \tag{40}$$

Substituting the conditional probabilities by causal (see equations (5) and (6)) a relation

$$p(u_0^k, x_0^k, y_0^k; f_0^F) = \prod_{i=0}^{k} f_i(u_i : y_0^i, u_0^{i-1}) \times \prod_{i=0}^{k} p_i(x_i, y_i : u_0^{i-1}, x_0^{i-1}, y_0^{i-1}). \tag{41}$$

is obtaineda nd the statement of Lemma 1 is now obvious. □

Proof of Lemma 2. For conditional probability of the internal variable x_0^k holds

$$p(x_0^k | u_0^k, y_0^k; f_0^F) = p(u_0^k, x_0^k, y_0^k; f_0^F) / p(u_0^k, y_0^k; f_0^F). \tag{42}$$

Where, since f_i is not a function of x_0^k, it can be written

$$p(u_0^k, y_0^k; f_0^F) = \prod_{i=0}^{k} f_i(u_i : y_0^i, u_0^{i-1}) \times \int \prod_{i=0}^{k} p_i(x_i, y_i : u_0^{i-1}, x_0^{i-1}, y_0^{i-1}) \mathrm{d}x_0^k. \tag{43}$$

Substituting (43) to (42) yields an equation

$$p(x_0^k | u_0^k, y_0^k; f_0^F) = \frac{\prod_{i=0}^{k} p_i(x_i, y_i : u_0^{i-1}, x_0^{i-1}, y_0^{i-1})}{\int \prod_{i=0}^{k} p_i(x_i, y_i : u_0^{i-1}, x_0^{i-1}, y_0^{i-1}) \mathrm{d}x_0^k} \tag{44}$$

which leads to Lemma 2. □

Proof of Lemma 3. It holds

$$p(y_{k+1}^j, u_k^j | y_0^k, u_0^{k-1}; f_0^F) = [\prod_{i=k+1}^{j} p(u_i | y_0^i, u_0^{i-1}; f_0^F) \times p(y_i | y_0^{i-1}, u_0^{i-1}; f_0^F)] p(u_k | y_0^k, u_0^{k-1}; f_0^F). \tag{45}$$

Substituting (6) and (9) to (45) yields an equation

$$p(y_{k+1}^j, u_k^j | y_0^k, u_0^{k-1}; f_0^F) = [\prod_{i=k+1}^{j} f_i(u_i : y_0^i, u_0^{i-1}) \times p(y_i | y_0^{i-1}, u_0^{i-1})] f_k(u_k : y_0^k, u_0^{k-1}) \tag{46}$$

that leads to equation (10). Since it also holds

$$p(y_{k+1}^{j+1}, u_k^j | y_0^k, u_0^{k-1}; f_0^F) = p(y_{j+1} | y_0^j, u_0^j; f_0^F) \times p(y_{k+1}^j, u_k^j | y_0^k, u_0^{k-1}; f_0^F), \tag{47}$$

using relations (9) and (10) the relation (11) can be written. □

Proof of Theorem 1. Applying property (13) of the final segment of the control law to the criterion of the final segment yields

$$J_k [y_0^k, u_0^{k-1}; f_k^F] = \int J_{F+1} [y_0^F, u_0^F] \times p(y_{k+1}^F, u_k^F | y_0^k, u_0^{k-1}; f_k^F) \mathrm{d}(y_0^k, u_0^{k-1}), \tag{48}$$

where

$$J_{F+1} [y_0^F, u_0^F] = \int L(u_0^F, x_0^F, y_0^F) \times p(x_0^F | y_0^F, u_0^F) \mathrm{d}x_0^F. \tag{49}$$

The value of $J_{F+1} [y_0^F, u_0^F]$ is independent on the control law.

Applying equations (10) and (11) for $0 \le k \le j \le F$ results to a relation, which can be symbolically written as

$$J_k [y_0^k, u_0^{k-1}; f_k^F] = E\left\{ J_j \left[y_0^j, u_0^{j-1}; f_j^F\right] | y_0^k, u_0^{k-1}; f_k^{j-1}) \right\}. \tag{50}$$

Now suppose that an optimal final segment $f_j^F(j)$ of the control law at time t_j is known. Thus it holds for all $f_j^F \in \Phi_j^F$ independently on f_0^{j-1}

$$J_j \left[y_0^j, u_0^{j-1}; f_j^F(j)\right] \le J_j \left[y_0^j, u_0^{j-1}; f_j^F\right]. \tag{51}$$

Inequality (51) remains when multiplied by a positive conditioned pdf $p(y_{k+1}^j, u_k^{j-1} | y_0^k, u_0^{k-1}; f_k^{j-1})$ and integrated:

$$J_k \left[y_0^k, u_0^{k-1}; f_k^{j-1}, f_j^F(j)\right] \le J_k \left[y_0^k, u_0^{k-1}; f_k^{j-1}, f_j^F\right]. \tag{52}$$

It follows from (52) that a law $f_k^{j-1}(j-1) \in \Phi_k^{j-1}$ found for a known segment $f_j^F(j)$, which for all $f_k^{j-1} \in \Phi_k^{j-1}$ satisfies a condition

$$J_k \left[y_0^k, u_0^{k-1}; f_k^{j-1}(j-1), f_k^F(j)\right] \le J_k \left[y_0^k, u_0^{k-1}; f_k^{j-1}, f_k^F(j)\right], \tag{53}$$

the segment of the control law

$$f_k^F(k) = (f_k^{j-1}(j-1), f_j^F(j)) \tag{54}$$

is an optimal final segment of the control law at time t_k for all $0 \le k \le j \le F$. □

Proof of Theorem 2. is omitted because of its similarity to the proof of Theorem 1. □

CONTROL SYSTEM DESIGN STUDY FOR A MICROMACHINED ACCELEROMETER

Michael Kraft, Christopher P. Lewis, Thomas G. Hesketh

Coventry University, School of Engineering
Priory St., Coventry, CV1 5FB, UK
Tel.: +44 1203 838842, Fax: +44 1203 838949
email: cey143@coventry.ac.uk

Abstract: In this paper an analogue, closed loop accelerometer based upon a bulk-micromachined, capacitive sensing element is described. The sensing element is embedded in a closed loop control system using electrostatic forces as a reset mechanism. A mathematical model is derived which allows to employ conventional linear control theory. The validity of the model is discussed. Particular attention is paid to the controller design; here a PID controller is used. The simulated results are compared to measurements on the actual hardware prototype.

Keywords: Accelerometers, Microsystems, Control Theory, Closed Loop Sensor, Modelling

1. INTRODUCTION

The market for micromachined silicon accelerometers is continuously increasing, the majority of which are relatively simple open loop devices with a variety of mechanical structures. In this paper a 'servo' or closed loop transducer is described. A bulk-micromachined sensing element is depicted in fig. 1. It typically consists of two fixed outer electrodes and a movable centre plate or seismic mass which acts as a common centre electrode of two capacitors electrically connected in series. Any acceleration causes the seismic mass to move from its central position, this results in a differential imbalance in capacitance which is proportional to the acceleration for small deflections of the seismic mass.

Such an open loop accelerometer has limited performance in terms of bandwidth, linearity and dynamic range. Additionally, the device suffers from inherent nonlinear effects due to the damping, (Zhang *et. al.*, 1992, Marco *et. al.*, 1993), and the electrostatic forces which are introduced by the signal pick-off; these effects increase in magnitude with the deflection of the seismic mass. One method of reducing these nonlinearities and improving the performance is to use the sensing element in a closed loop control system in which a reset force is applied to the seismic mass to keep it at the central position between the electrodes. This force has to counterbalance the accelerating force; the magnitude of which provides an accurate measure for the acceleration. In micromachined sensing elements the dimensions are as such that electrostatic

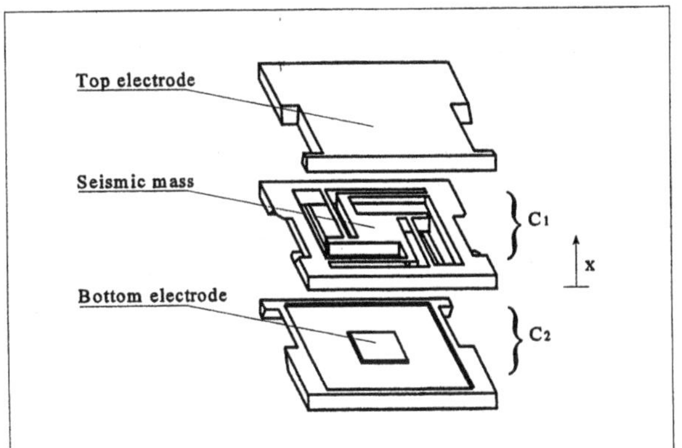

Fig. 1: A typical bulk-micromachined, capacitive sensing element.

forces can be used to generate the restoring force. In the design approach chosen in this work the seismic mass or centre plate had no floating charge being effectively grounded; with this arrangement it is difficult to generate a linear feedback signal because of the dependency of the electrostatic force on the square of the voltage and the inverse square of the gap between the electrodes. The most common solution to this problem is to superimpose two electrostatic forces on the seismic mass which are in balance (van Kampen *et. al.*, 1994, Zimmermann *et. al.*, 1995, Boxenhorn and Greiff, 1990, Warren, 1991). However, this leads only to a linear feedback relationship if very small displacements of the seismic mass are assumed. A further complication is that the micromachined, capacitive sensing element had only three connections (top and bottom electrodes and seismic mass), care had to be taken that the feedback signals did not interact with the excitation signals required for signal pick-off. Generally two arrangements are possible: separation in either the time or the frequency domains, (Burstein and Kaiser, 1996). Usually, for an analogue, closed loop accelerometer the latter approach is employed; this relies on the use of a high frequency carrier applied to the outer plates of a capacitive half bridge, the centre plate of which is used for the signal pick-off, (Kroemer *et. al.*, 1995, Kuehnel and Sherman, 1994). The magnitude of the subsequent signal is proportional to the imbalance in capacitance, which is, in turn, proportional to the deflection of the seismic mass from its rest position for small motion. This signal is then subjected to a phase-sensitive demodulator and a low-pass filter. The output signal of the filter provides a measure of the imbalance in capacitance.

It is obvious that the design of a closed loop analogue accelerometer is by no means a trivial task, each of the problems outlined above has to be carefully addressed. In this paper the particular problems associated with the development of an analogue servo accelerometer are considered. For brevity only the small signal conditions are discussed in detail here, however particular problems arising under large signal operation are identified.

2. MATHEMATICAL MODELLING

In order to provide suitable feedback force, F_{el}, a bias voltage, V_B, an excitation voltage, V_1, and the feedback signal, V_F were applied to the electrodes as in eq. (1).

$$\begin{aligned} V_{TP} &= V_1 - V_B + V_F \\ V_{BP} &= -V_1 + V_B + V_F \end{aligned} \qquad ...(1)$$

Here, the resultant voltages applied to the top and bottom plates are V_{TP} and V_{BP} respectively. This arrangement produces an electrostatic force on the seismic mass as in eq.(2).

$$F_{el} = \frac{1}{2}\varepsilon_0 A \left(\frac{\left(\frac{V_{1,pp}}{2}\sin\omega t - V_B + V_F\right)^2}{(d-x)^2} - \frac{\left(-\frac{V_{1,pp}}{2}\sin\omega t + V_B + V_F\right)^2}{(d+x)^2} \right) \qquad ...(2)$$

where 2d is the nominal gap between the outer electrodes, x is the deflection of the seismic mass from the central position, A is the area of the plates and ε_0 is the absolute permittivity of free space.

Making the assumption that the seismic mass deflections are small, it can be shown that the feedback force is proportional to the applied voltage, V_F, as in eq.(3), (van Kampen *et. al.*, 1994).

$$F_{el} \underset{x \to 0}{\overset{\lim}{=}} 2\varepsilon_0 A \frac{V_B}{d^2} V_F \qquad ...(3)$$

In the simplest case the feedback voltage, V_F, is proportional to the output voltage of the low-pass filter and, in turn, the position of the seismic mass.

The bias voltage is chosen according to the desired dynamic range of the accelerometer; to support a force produced by an acceleration 'a' the minimum bias voltage required is given by eq.(4).

$$V_B = \sqrt{\frac{mgd_0^2}{\varepsilon_0 A}} \qquad ...(4)$$

m is the mass of the centre plate. Eq. (4) simplifies to $V_B/\sqrt{a}$= 8.7 V/$\sqrt{}$m/s for the accelerometer under development.

It was then possible to plot the net electrostatic force acting on the seismic mass as a function of the deflection for different bias voltages as in fig. 2. The solid line shows the accurate solution of eq. (2), the dashed line the linearized solution according to eq. (3). The pick-off gain used was 1.6 V/µm (this value was obtained from measurement of the actual hardware), a proportional gain factor of 6 was assumed.

It is obvious that the electrostatic force has a useful linear range for deflections of approximately ±0.5 µm. If the seismic mass is deflected further, the

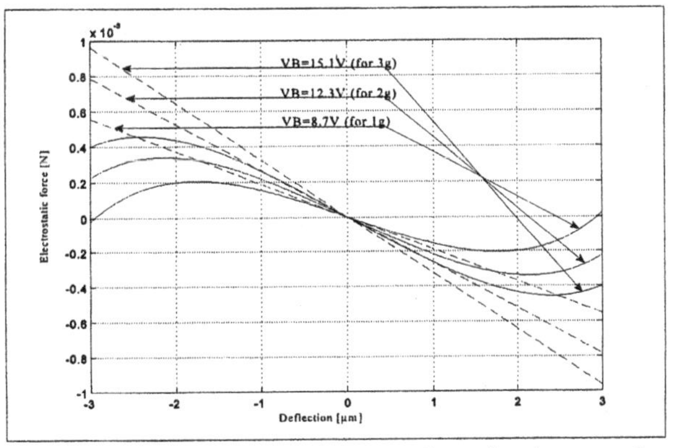

Fig. 2: Net electrostatic force on the seismic mass for different bias voltages and proportional feedback. The solid line shows the exact solution according to eq. (2), the dashed line the approximated solution for small deflections according to eq. (3).

feedback relationship becomes nonlinear, ultimately the feedback changes polarity and the seismic mass is attracted to one of the outer electrodes. This is a fundamental problem and a serious drawback of an analogue, closed loop system. In a real system such a situation could arise if either of the two drive amplifiers supplying the voltage to the top and bottom electrodes saturate due to an acceleration magnitude larger than the dynamic range of the accelerometer; this could occur as a result of a shock condition, (Kraft, 1997).

If small deflections of the seismic mass are assumed the mathematical model for the sensing element can be assumed to be a second order system, (Lewis *et. al.*, 1996), since the nonlinear damping effects due to viscous flow or 'squeeze film' damping can be neglected. As a first approximation the signal pick-off can be assumed to be a simple gain factor. A second order Butterworth low-pass filter was used with a corner frequency of 600 Hz. A PID controller was introduced for compensation, the output signal of which constituted the output signal of the accelerometer. Fig.3 shows the mathematical model of the analogue, closed loop accelerometer where m, b and k are sensing element parameters (b: damping coefficient, k: spring constant); KP, KI and KD are the controller settings.

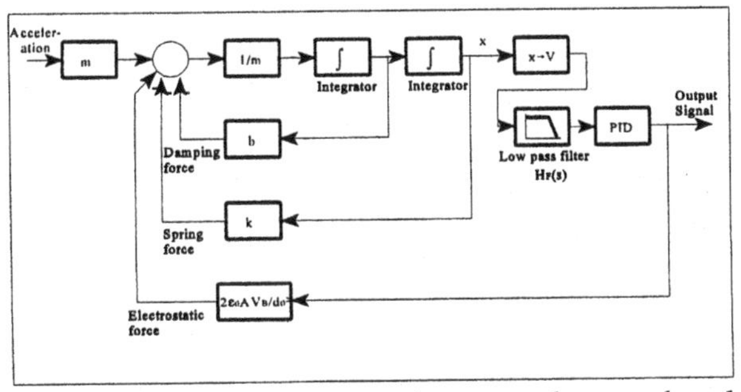

Fig. 3: Small signal model for an analogue, closed loop accelerometer.

It was now possible to determine the open loop and closed loop transfer functions, the feedback gain being abbreviated by $K_F=2\varepsilon_0 A V_B/d_0^2$:

Open Loop TF:

$$F_{OL}(s) = K_{po} m H_F(s) \frac{K_D s^2 + K_P s + K_I}{m s^3 + b s^2 + k s} \quad ...(5a)$$

where HF(s) is the filter transfer function.

Closed loop TF:

$$F_{CL}(s) = K_{po} m (K_D s^2 + K_P s + K_I)/(m a_2 s^5 + (m a_1 + b a_2) s^4 + (m + b a_1 + k a_2) s^3 + (b + k a_1 + K_F K_{po} K_D) s^2 + (k + K_F K_{po} K_P) s + K_F K_{po} K_I) \quad ...(5b)$$

where a1, a2 are filter parameters.

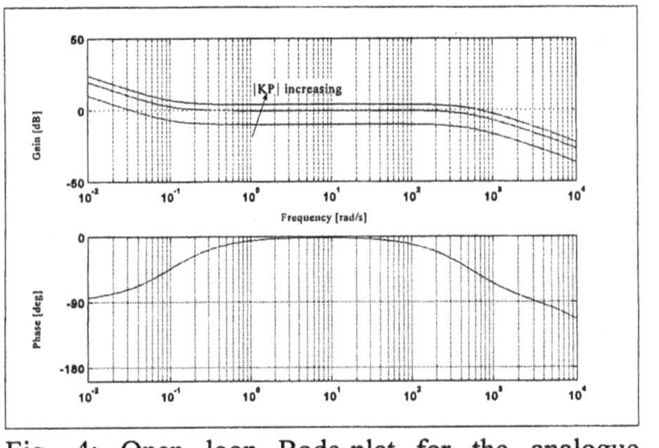

Fig. 4: Open loop Bode-plot for the analogue accelerometer for small deflections and for different settings for the proportional and integral gain.

From inspection of the closed loop transfer function it can be seen that the smaller the parameters of the sensing element (spring constant k and damping factor b) the more significant the influence of the PID parameters become, hence for closed loop operation a micromachined sensing element with small spring and damping constants is desirable.

An accelerometer having a dynamic range of 2 g, (which necessitated a bias voltage, VB = 12.3 V according to eq. (4)) has a Bode-plot as in fig. 4; this shows the simulated results with proportional gain constant KP values of 2, 6 and 10. The integral gain constant was set to KP/10 for each run and zero differential gain was assumed. Inspection of fig. 4 reveals that the -3 dB cut-off frequency is about 500 rad/s (80 Hz). Additionally, it can be seen that the phase margin is sufficient large for all three simulation runs, consequently the closed loop system is predicted to be stable.

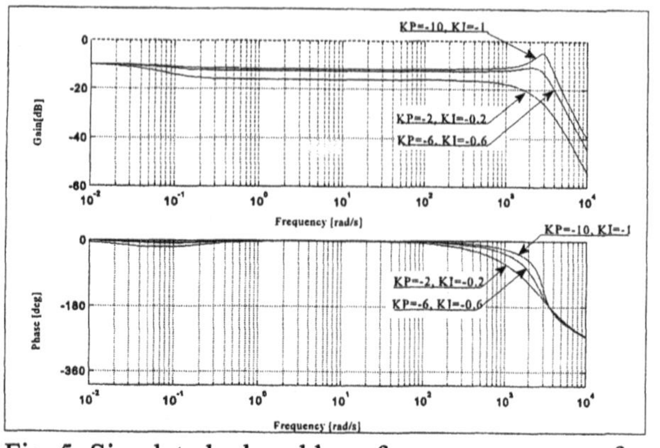

Fig. 5: Simulated, closed loop frequency response for the analogue accelerometer for small deflections and for different settings of the proportional and integral gain.

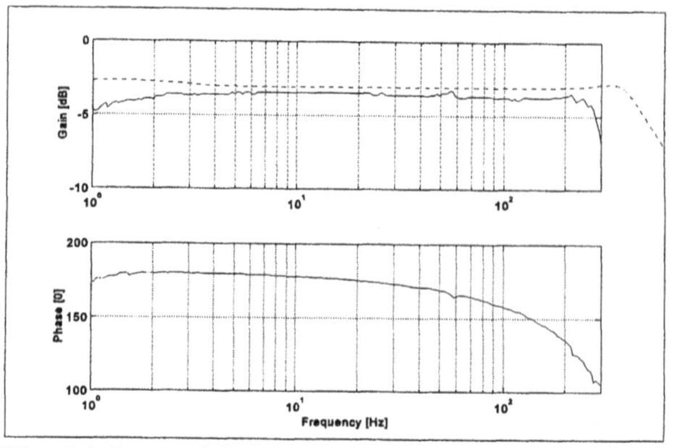

Fig. 6: Measured frequency response of the analogue, closed loop accelerometer. The dashed line shows the simulated response for $K_P = 6$ and $K_I = 0.6$.

For verification the corresponding simulated, closed loop frequency responses are shown in fig. 5. It can be seen that the bandwidth of the accelerometer was increased considerably. As expected the higher the proportional gain setting, the higher the bandwidth and the flatter the frequency response. A proportional gain of 2 resulted in unsatisfactory performance, this was too low to counteract the integral gain required for accurate steady state performance. For a proportional gain of 6 the bandwidth was increased to 465 Hz and for gain of 10 the bandwidth was 357 Hz, this was due to a resonant peak. Obviously, derivative gain was not required here.

3. MEASUREMENT RESULTS

The micromachined accelerometer was tested dynamically by mounting it, together with a reference, Sundstrand Q-flex accelerometer on a vibration table. The reference accelerometer used had a flat frequency response up to 300 Hz. The measured, closed loop response is shown in fig.6 together with the simulated response for $K_P = 6$ and $K_I = 0.6$, the entire frequency range from 0 to 300 Hz lies within the 3 dB window, consequently the cut-off frequency is increased from 80 Hz in the open loop condition to above 300 Hz. Generally, the bandwidth of the device was satisfactory for the requirements specified in this research project.

The measured frequency response agreed quite well with the theoretically predicted performance. With the test method used it is difficult to quote exact figures for the degree of agreement since the measured frequency response is only valid up to the bandwidth of the reference accelerometer. A further drawback this test method is the use of a vibration table without closed loop control of acceleration; the acceleration magnitude does not stay constant over the frequency range tested but increases with the square of the frequency. Consequently, it is difficult to generate sufficient acceleration at low frequencies and at higher frequencies care has to be taken that the transducer does not saturate. The frequency response analyser used (Voltech TF 2000) related the signal from the accelerometer under test to the reference transducer to get a measured frequency response directly referring to acceleration..

An obvious discrepancy between the simulated and measured results is that in the simulation a slight decrease of gain is predicted at low frequencies due to the integral action of the controller where as the measured frequency response shows an increase of gain. Since the acceleration magnitudes are very small at low frequencies, the increase in gain at low frequencies is believed to originate from the way the measurement was taken. A mechanical resonance of the mounting of the sensing element and the reference accelerometer on the vibration table can also cause mid-range measurement errors.

Further experiments with different PID gain settings were carried out, these did not result in any improvement in performance. Another discrepancy between the hardware implementation and the mathematical model was that if the gain settings of the PI controller were increased considerably, the system started to exhibit an unstable oscillation; such behaviour is believed to originate from effects which are not considered in the mathematical model, e.g. cross-axis sensitivity, non-ideal parallel electrodes etc.

4. CONCLUSIONS

In this paper a mathematical model for an accelerometer based upon a micromachined, capacitive sensing element is developed. In order to apply conventional control theory the assumption that the seismic mass stays close at the central position between the electrodes has to be made. There was good agreement between measured and theoretical results. A closed loop control system, as expected, showed improved performance, the bandwidth was increased from 80 Hz to above 300 Hz. Under certain conditions the polarity of the feedback signal can change, this drawback can only be overcome by applying a different control strategy, e.g. use of oversampling conversion or sigma-delta modulation. Such a device not only results in an improved stability but also yields a direct digital output signal, (Lewis and Kraft, 1996, Smith *et. al.*, 1994, de Coulon *et. al.*, 1993, Henrion *et. al.*, 1990, Lu *et. al.,* 1995). However, the design of the control system cannot be approached any longer with conventional control theory as these systems have an inherently nonlinear, discrete data structure.

REFERENCES

Boxenhorn, B. and Greiff, P. Monolithic silicon accelerometer. *Sensors and Actuators*, **A21-A23**, pp. 273-277, 1990.

Burstein, A. and Kaiser, W. J. Mixed analog-digital highly sensitive sensor interface circuit for low-cost microsensors. *Sensors and Actuators*, **A52**, pp. 193-197, 1996.

de Coulon, Y., Smith, T., Hermann, J., Chevroulet, M. and Rudolf, F. Design and test of a precision servoaccelerometer with digital output. *7th Int. Conf. Solid-State Sensors and Actuators (Transducer '93), Yokohama*, pp. 832 - 835, 1993.

Henrion, W., Disanza, L., Ip, M., Terry S. and Jerman, H. Wide dynamic range direct digital accelerometer. *IEEE Solid State Sensor and Actuator Workshop,* pp. 153-157, Hilton Head Island, 1990.

Kraft, M. Closed loop digital accelerometer employing oversampling conversion. *Coventry University, Ph.D. dissertation*, 1997.

Kroemer, O., Fromhein, O., Gemmeke, H., Kühner, T., Mohr, J. and Strohrmann, M. High-precision readout circuits for LIGA acceleration sensors. *Sensors and Actuators*, **A46-A47**, pp. 196-200, 1995.

Kuehnel, W. and Sherman, S. A surface micromachined silicon accelerometer with on-chip detection circuitry. *Sensors and Actuators*, **A45**, pp. 7-16, 1994.

Lewis, C. P., Kraft, M. and Hesketh, T. G. Mathematical Model for a Micromachined Accelerometer. *Trans. Inst. of Meas. and Control.*, **Vol. 18, No.2**, pp. 92-98, 1996.

Lewis, C.P. and Kraft, M. Simulation of a Micromachined Digital Accelerometer in SIMULINK and PSPICE. *UKACC Int. Conf. on Control,*, **Vol 1**, pp. 205 - 209, Exeter UK, 1996.

Lu, C., Lemkin, M. and Boser, B. A monolithic surface micromachined accelerometer with digital output. *IEEE J. Solid-State Circuits*, **Vol. 30, No. 12**, pp. 1367-1373, 1995.

Marco, S., Samitier, J., Ruiz, O., Herms, A. and Morante, J. Analysis of electrostatic damped piezoresistive silicon accelerometer. *Sensors and Actuators*, **A 37-38**, pp. 317-322, 1993.

Smith, T., Nys, O., Chevroulet, M., de Coulon, Y. and Degrauwe, M. Electro-mechanical sigma-delta converter for acceleration measurements. *IEEE International Solid-State Circuits Conference,* pp. 160-161, 1994.

van Kampen, R. P., Vellekoop, M., Sarro, P. and Wolffenbuttel, R. F. Application of electrostatic feedback to critical damping of an integrated silicon accelerometer. *Sensors and Actuators*, **A 43**, pp. 100-106, 1994.

Warren, K. Electrostatically force-balanced silicon accelerometer. *J. of the Inst. of Nav.*, **V. 38, No. 1**, pp. 91-99, 1991.

Zhang, L., Cho, D., Shiraishi, H. and Trimmer, W. Squeeze-film damping in micromechanical systems. *ASME, Micromechanical Systems,* **DSC-Vol. 40**, pp. 149-160, 1992.

Zimmermann, L., Ebersohl, J., Le Hung, F., Berry, J. P., Baillieu, F., Rey, P., Diem, B., Renard, S., Caillat, P. Airbag application: a microsystem including a silicon capacitive accelerometer, CMOS switched capacitor electronics and true self-test capability. *Sensors and Actuators*, **A 46-47**, pp. 190-195, 1995.

TUNING OF DIGITAL PID CONTROLLERS BASED ON ZIEGLER - NICHOLS METHOD

Vladimír Bobál*, Jiří Macháček, Roman Prokop***

**Technical University of Brno, Faculty of Technology in Zlín,
Department of Automatic Control, Nám. T. G. Masaryka 275,
762 72 Zlín, Czech Republic,
e-mail: bobal@zlin.vutbr.cz*

***University of Pardubice, Faculty of Chemical Technology,
Department of Process Control and Computer Techniques, Nám. Čs. legií 565,
532 10 Pardubice, Czech Republic,
e-mail: machacek@hlb.upce.cz*

Abstract: A unified algorithm for auto-tuning of digital PID controllers for n-order process model have been derived in this paper. Tuning procerdure uses the identification of ARX process model parameters using recursive least squares method (RLSM) with adaptive directional forgetting. The parameter estimates serve for computing of the ultimate (critical) proportional gain and period of oscillations on the basis of modified Ziegler-Nichols method for digital control loops. The relations for setting procedures in case of the 2nd and the 3rd order model are derived in the form of analytical formulas. In case of the higher order ARX models the MATLAB Symbolic Math Toolbox was used. These algorithms are suitable for automatic tuning of digital PID controllers or of the adaptive control of technological processes. The digital PID controller with weghting factor and noise filtering constant in the derivative component has been applied for temperature control of a laboratory fermentor. There has been developed the MATLAB-Toolbox ATC PID for designing, testing, simulating and real time of auto-tuning digital PID controllers.

Keywords: PID controllers, digital control, stability limits, recursive least squares, self-tuning control, computer aided design.

1. INTRODUCTION

PID controllers are still widely used for control of the most part industrial processes. The reason for it is in their robustness, simplicity and good experience of operators with their tuning. The popular Ziegler - Nichols (1942) (Z-N) method is based on knowledge of the ultimate proportional gain and period of oscillations. The current classical methods for determination of ultimate parameters consist in manual setting of the proportional controller gain until process is on the stability boundary. The faster method is rele fedback auto - tuning (Åström and Hägglund, 1984), which uses rele instead of the P controller. The process input is then a periodic square wave and the output approaches more or less sinusoidal shape according to filterring ability of process. The ultimate gain K_{pu} is then calculated as a ratio of the first harmonic of input signal amplitude and output signal amplitude.

The discrete ARX model which is obtained from

recursive identification is used in this papers to appoint of ultimate parameters. The identification is realized by recursive least square method with imroved numerical stability by means of LD decomposition and with directional forgetting (Kulhavý, 1987). The parameter estimates and sampling periode T_0 are used for the computation of the ultimate gain K_{pu} and the ultimate period of oscillations T_0. The ultimate parameters are calculated from condition that the poles of closed loop of the prorocess and P controller lie on the unit circle of the z-plane, see Bobál and Vašek (1987), Macháček and Drábek (1990), Bobál (1994a) and Bobál (1995). From ultimate parameters are calculated the PID controller parameters. The digital PID control may be then adaptive. Identification procedure can be switched off when the parameter estimates reach their true values. It is in principle a self-tuning approach with switching off of the recursive identification after the parameter estimates are converged.

The Z-N heuristic tuning formula for PID controllers gives various closed loop responses according to the type of controlled process. The good characteristic of process dynamics is normalized process gain κ defined as product of the process steady-state gain $G(0)$ and the ultimate gain K_{pu} (Åström et al., 1992)

$$\kappa = G(0)K_{pu} \tag{1}$$

The recommendation for refinements of the Z-N formula is given in Hang, Aström and Ho (1991). The parameters of PID controller (proportional gain K_p, integral T_i and derivative T_d time constants) are changed according to normalized process gain. For reduce of great overshoot of the set-point response was proposed method of set-point weighting. The parameter β which weights of set-point in the proportional term is also changed with parameter κ.

2. DETERMINING OF ULTIMATE PARAMETERS

The discrete process models describe dynamic behaviour including the sampling and holding operations and then ultimate parameters change with the change of sampling period T_0. These parameters are further designed as $K_{pu}(T_0)$ and $T_u(T_0)$.

The relations for setting of discrete PID controllers will be now derived. Let the process be described by the single input single output (SISO) system model in the form of the discrete transfer function

$$G_p(z) = \frac{Y(z)}{U(z)} = \frac{z^{-d}B(z^{-1})}{A(z^{-1})} \tag{2}$$

where $U(z)$ and $Y(z)$ are the z-transforms of the controller output and the process output, respectively d denotes the time delay expressed as an integer multiple of the sampling period T_0 and A, B are n-degree polynomials defined by

$$A(z^{-1}) = 1 + \sum_{i=1}^{n} a_i z^{-i}; \quad B(z^{-1}) = \sum_{i=1}^{n} b_i z^{-i} \tag{3}$$

Let the process (2) is controlled by the proportional controller

$$G_C(z) = \frac{U(z)}{E(z)} = K_p \tag{4}$$

where $E(z) = W(z) - Y(z)$ is z-transform of the control error, $W(z)$ is z-transform of the reference signal. When solving the equations (2) and (4) we get the transfer function of the closed control loop

$$G_w(z) = \frac{Y(z)}{W(z)} = \frac{z^{-d}K_pB(z^{-1})}{A(z^{-1}) + z^{-d}K_pB(z^{-1})} \tag{5}$$

The denominator of the transfer function (5) is the characteristic polynomial

$$D(z) = z^{n+d}[A(z^{-1}) + z^{-d}K_pB(z^{-1})] \tag{6}$$

whose poles determine the behaviour of the closed control loop. The poles of the characteristic polynomial (6) lie on the unit circle and inside unit circle when the condition $K_p = K_{pu}(T_0)$ is fulfilled. Two possibilities of the pole placement of the characteristic polynomial on the unit circle may occur so that the closed control loop is on the stability boundary:

- The characteristic polynomial (6) includes a pair of complex conjugate poles $z_{1,2} = \alpha \pm j\beta$ for which holds $\alpha^2 + \beta^2 = 1$. Then the characteristic polynomial (6) includes a quadratic three-term $C_1(z) = z^2 - 2\alpha z + 1$.
- The characteristic polynomial (6) includes one or more real poles $z_j = \alpha$ (imaginary part $\beta = 0$) where $\alpha = -1$. The positive pole $\alpha = 1$ will not put the control loop into oscillation. Then characteristic polynomial (6) includes j-terms $C_j(z) = (z+1)^j$, $j = 1,2,\ldots$.

The characteristic polynomial must be divisible by one of these terms and it leads in the first case to the solution of the polynomial equation

$$z^{n+d}\left[A(z^{-1})+z^{-d}K_{pu}(T_0)B(z^{-1})\right]= \\ =(z^2-2\alpha z+1)z^{n+d-2}E(z^{-1}) \tag{7}$$

where

$$E(z^{-1})=1+\sum_{i=1}^{n+d-2}e_i z^{-i} \tag{8}$$

and unknown parameters are $K_{pu}(T_0)$, α and coefficients e_i of the polynomial $E(z)$. The equation (7) can be solved e. g. by comparison of terms with the same powers of variable z.

In the second case it leads to solution of the polynomial equation

$$z^{n+d}\left[A(z^{-1})+z^{-d}K_{pu}(T_0)B(z^{-1})\right]=(z+1)^j z^{n+d-j}F(z^{-1}) \tag{9}$$

where

$$F(z^{-1})=1+\sum_{i=1}^{n+d-j}f_i z^{-i} \tag{10}$$

and unknown parameters are $K_{pu}(T_0)$ and coefficients f_i of the polynomial $F(z)$.

The solution of equation (9) can be made more simply. The pole $\alpha=-1$ forms the oscillating component $(-1)^k$ which corresponds to a continuous function $\cos\frac{\pi}{T_0}t$ with ultimate period $T_u(T_0)=2T_0$. The variable z is defined in z-transformation as

$$z=e^{jwT_0}=\cos\omega T_0+j\sin\omega T_0 \tag{11}$$

The equation (11) gives for $\omega_u=\frac{2\pi}{T_u(T_0)}=\frac{\pi}{T_0}$ the solution $z=-1$. The ultimate gain for the open-loop system is then given by

$$K_{pu}(T_0)=-\frac{1}{G_P(-1)} \tag{12}$$

The term $C_1(z)=z^2-2\alpha z+1$ has its equivalent in the denominator $z^2-2\cos\omega\, T_0 z+1$ of z-transformation of the harmonic function $\cos\omega t$ or $\sin\omega t$. The real component α can be then expressed for $\omega=\omega_u$ as

$$\alpha=\cos\omega_u T_0=\cos\frac{2\pi T_0}{T_u(T_0)} \tag{13}$$

The equation for computing of the ultimate period can be easily derived from equation (13)

$$T_u(T_0)=\frac{2\pi T_0}{\cos^{-1}\alpha} \tag{14}$$

The determining of ultimate parameters also consists in the following procedure:

- Ultimate gain $K_{pu}(T_0)$ - the right solution must be selected from equations (7) or (12) according to the conditions that $K_{pu}(T_0)\rangle 0$ and the all other poles of the characteristic polynomial (6) are stable.
- Ultimate period $T_u(T_0)$ - for complex conjugate pole is $T_u(T_0)$ computed according to equation (14), for real poles $\alpha=-1$ $T_u(T_0)=2T_0$.

2.1 Ultimate gain for the second order model

The relations for calculation of the ultimate parameters for the second order process model have been proposed by Bobál (1995). Let the process be described by the second order model (n = 2, d = 0 in equations (2) and (3)). The characteristic polynomial is in the form

$$D(z)=z^2+(a_1+b_1K_P)z+(a_2+b_2K_P) \tag{15}$$

In the first case equation (7) for the second order model is given in the form

$$z^2\{1+a_1z^{-1}+a_2z^{-2}+K_{pu}(T_0)[b_1z^{-1}+b_2z^{-2}]\}= \\ = z^2-2\alpha z+1 \tag{16}$$

By comparison of same powers of z then two equations are obtained in the form

$$a_1+K_{pu}(T_0)b_1=-2\alpha;\quad a_2+K_{pu}(T_0)b_1=1 \tag{17}$$

The equations (17) give relations for computation the ultimate gain

$$K_{Pu}(T_0)=\frac{1-a_2}{b_2} \tag{18}$$

and for computation the real part of complex conjugate poles

$$\alpha=-\frac{a_1+K_{Pu}(T_0)b_1}{2} \tag{19}$$

The ultimate period $T_u\ (T_0)$ is computed according to relation (14).

In the second case equation (9) for the second order model is given in the form

$$z^2\{1+a_1z^{-1}+a_2z^{-2}+K_{pu}(T_0)[b_1z^{-1}+b_2z^{-2}]\}= \\ = (z+1)(z+f_1)$$

(20)

The polynomial equation (20) gives another relation for computing the critical gain

$$K_{Pu}(T_0)=\frac{a_1-a_2-1}{b_2-b_1} \tag{21}$$

The ultimate gain $K_{Pu}\ (T_0)$ is then computed either from equation (18) or (21) according to fulfilment of the condition

$$[a_1+b_1K_{Pu}(T_0)]^2-4[a_2+b_2K_{Pu}(T_0)] \leq 0 \tag{22}$$

It is evident that when fulfilling condition (22), the characteristic polynomial (15) includes a pair of complex conjugates poles and for the computation $K_{Pu}\ (T_0)$ relation (19) is therefore used. If the condition (22) is not fulfilled, the characteristic polynomial includes one or more real poles $\alpha = -1$ and $K_{Pu}\ (T_0)$ is computed according to relation (21).

The relation (21) is possible derived from equation (12), too.

2.2 Ultimate gain for the third order model

The relations for calculation ultimate parameters for the third process model (n = 3, d = 0 in equations (2) and (3)) can be derived by some analytical way. Characteristic polynomial is in the form

$$z^3\{1+a_1z^{-1}+a_2z^{-2}+a_3z^{-3}+K_{pu}(T_0)[b_1z^{-1}+ \\ +b_2z^{-2}+b_3z^{-3}]\}=(z^2-2\alpha z+1)z(1+e_1z^{-1}) \tag{23}$$

By comparison of same powers z can be by easy way derived three equations with three unknown parameters $K_{pu}(T_0)$, α and e_1:

$$\begin{aligned} a_1+b_1K_{pu}(T_0)&=-2\alpha+e_1 \\ a_2+b_2K_{pu}(T_0)&=1-2\alpha e_1 \\ a_3+b_3K_{pu}(T_0)&=e_1 \end{aligned} \tag{24}$$

The equations (24) give solution

$$K_{pu\,1,2}(T_0)=\frac{-r_1\pm\sqrt{d}}{2r_2} \tag{25}$$

where

$$\begin{aligned} r_0&=a_3(a_3-a_1)+a_2-1 \\ r_1&=b_3[(2a_3-a_1)+b_2-a_3b_1 \\ r_2&=b_3(b_3-b_1) \\ d&=r_1^2-4r_0r_2 \end{aligned} \tag{26}$$

Right solution is given for

$$K_{pu}(T_0)\rangle 0; \quad |e_1|\langle 1 \tag{27}$$

If conditions (27) are not fulfiled neither for $K_{pu1}(T_0)$ nor $K_{pu2}(T_0)$ the ultimate gain must be computed either equation (9) or equation (12). Then ultimate gain is given

$$K_{pu}(T_0)=\frac{a_1-a_2-a_3-1}{b_2-b_3-b_1} \tag{28}$$

Bobál (1994a) used for the calculation of the poles of the third order process model the Cardan formulas. Calculation of the ultimate gain for the fourth and higher order process model is necessary realize by numerical way, e.g. using MATLAB. It is suitable use for deriving of the relations Symbolic Math Toolbox.

3. DIGITAL PID CONTROLLER

Let the continuous PID controller be described by equation

$$u(t)=K_p\left[\beta w(t)-y(t)+\frac{1}{T_I}\int_0^{\tau} e(t)dt-T_d\frac{y_f(t)}{dt}\right] \tag{29}$$

where β is the weithing factor <0, 1> and y_f is output signals filtered by system of 1st order with transfer function

$$\frac{Y_f(s)}{Y(s)}=\frac{1}{1+s\frac{T_d}{N}} \tag{30}$$

The noise filtering constant is choose in the range of 3 - 15.

To get a digital version of the PID controller, it should discretize the integral and derivative component at the equation (29). For discretizing the integral component, usually forward rectangular method (FRM), backward rectangular method (BRM) or trapezoidal method (TRM) are used. The derivative component is mostly replaced by the 1st order difference (two-point difference). For practical use the incremental control algorithms which compute the actual value of the controller output $u(k)$ from the previous value $u(k-1)$ and from compensation increment are suitable. By discretizing TRM has control algorithm form

$$u(k) = u_{pi}(k) + u_d(k) \tag{31}$$

where

$$u_{pi}(k) = u_{pi}(k-1) + \beta\, w(k)K_p - y(k)K_p + e(k)\frac{K_p T_0}{2T_i} + $$
$$+ e(k-1)\frac{K_p T_0}{2T_i} - \beta\, w(k-1)K_p + y(k-1)K_p \tag{32}$$

and

$$u_d(k) = \frac{T_d}{T_d + T_0 N} u_d(k-1) - K_p \frac{T_d N}{T_d + T_0 N} y(k) +$$
$$+ K_p \frac{T_d N}{T_d + T_0 N} y(k-1) \tag{33}$$

The Ziegler-Nichols formula gives for controller (31) - (33) these relations:

$$K_p = 0.6 K_{pu}(T_0); \quad T_i = 0.5 T_u(T_0); \quad T_d = 0.125 T_u(T_0) \tag{34}$$

4. TEMPERATURE CONTROL OF LABORATORY FERMENTOR

The digitale PID controller (31) - (34) has been applied for temperature control of a experimental fermentor for the enzymatic hydrolysis of chromium containing leather wastes in laboratory conditions. The results of the control process are shown in Figs. 1 - 4. The temperature step response with static constraint of the controller output (electrical resistive heating) is shown in Fig. 1. This agorithm has been used for static constraint of the controller output:

$u(k) = u_R(k)$ for $u_{min} \le u_R(k) \le 0.5\, u_{max}$
$u(k) = u_{min}$ for $u_R(k) < u_{min}$
$u(k) = 0.5\, u_{max}$ for $u_R(k) > 0.5\, u_{max}$

where $u(k)$ is real controller output, $u_R(k)$ is generated controller output and u_{min}, u_{max} are permissible constraints of the controller.

The temperature step response with dynamic constraint is shown in Fig. 2. Dynamic constraint has been realized by filtration of the controller output

$$u_p(k) = u(k) - [u(k) - u_C]e^{-ck} \tag{35}$$

where u_C is central value of the controller output and $c > 0$ is speed constant. The parameters of the filter was chosen $u_C = 0.1$, $c = 0.1$. In both cases sampling period $T_0 = 800$ s, $N = 10$, $\beta = 0.1$ and the initial vector of parameter estimates was $\hat{\Theta}^T(0) = [0.1, 0.1, 0.1, 0.2]$.

The course of the parameter estimates $\hat{\Theta}$ and of the PID controller parameters are presented in the Fig. 3 and 4.

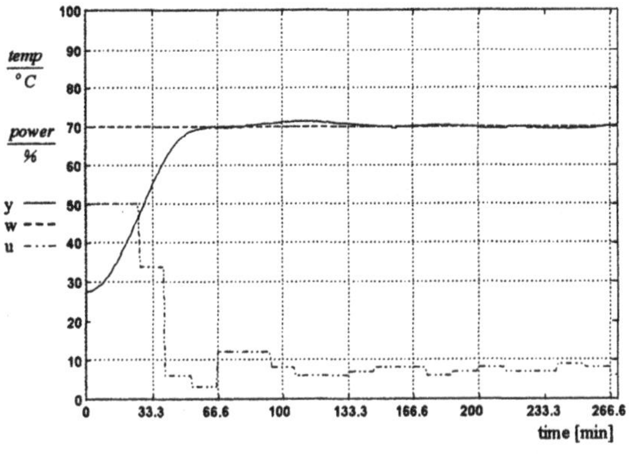

Fig. 1 Temperature control with static constraint of controller output

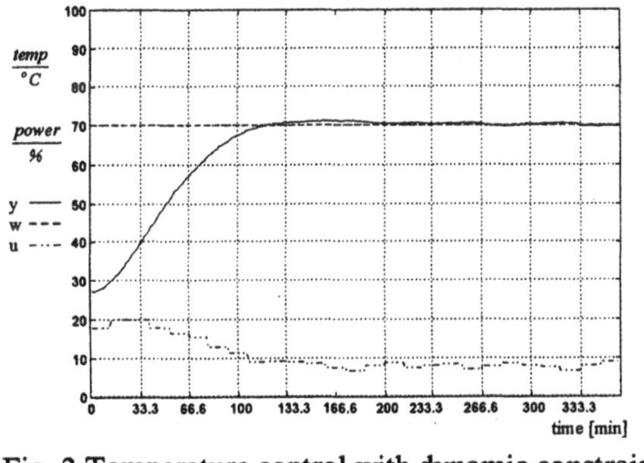

Fig. 2 Temperature control with dynamic constraint of controller output

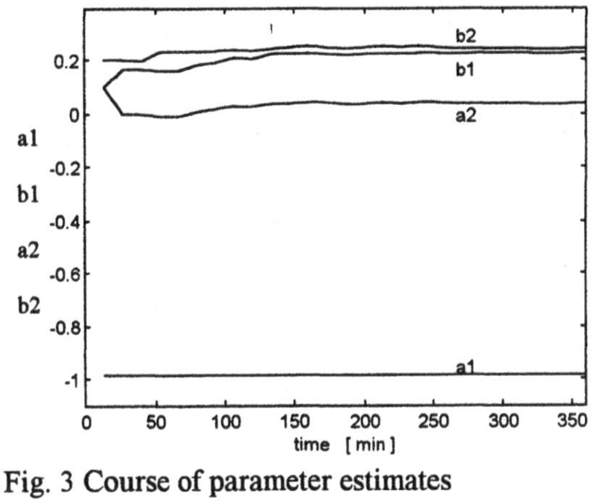

Fig. 3 Course of parameter estimates

5. CONCLUSIONS

For designing , testing, simulating and real time control has been developed the MATLAB - Toolbox ATC PID. This MATLAB - Toolbox is written in m-files using the functions of the Control System

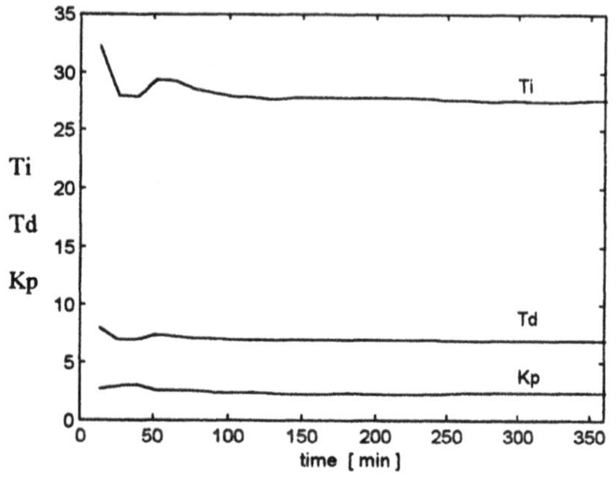

Fig. 4 Course of PID controller parameters

Toolbox and Symbolic Math Toolbox in environment of the MATLAB 4.2 for Widows. This Toolbox includes tools which facilitate the simulation of controlled process, recursive identification, controller PID design and real time control.

The designed auto-tuning PID algorithms (Bobál, 1994b) which are included in the MATLAB - Toolbox ATC PID are simple, sufficiently robust and suitable for control of a large class of controlled systems (proportional with minimum and nonminimum phase, integral and with time delay, too). For this reason, this Toolbox can be preferably used for supporting the project of the auto-tuning industrial PID controllers. The particular designed auto-tuning PID controllers are programmed as the C++ functions for the industrial applications, too. The PID controller (Takahashi et al., 1971) with auto-tuning procedure for the second order model has been successfully tested in pilot plant for the control of tube wall-thickness in the plastic extrusion (Bobál, 1990) and Z - N PID controller for the control of pilot distillation column (Macháček et al., 1995). The PID controller derived for the third order model has been applied by Pivoňka (1993, 1995) for the control of the 200 MW power plant.

ACKNOWLEDGEMENTS

This work was supported by the Grant Agency of the Czech Republic under grant No. 102/96/0927.

REFERENCES

Åström, K.J. and T. Hägglund (1984). Automatic tuning of simple regulators with specifications on phase and amplitude margins. *Automatica*, **20**, 645-651.

Åström, K. J., C. C. Hang, P. Pearson and W. K. Ho (1992). Towards intelligent PID control. *Automatica*, **28**, 1-9.

Bobál, V. (1990). Algorithms of self-tuning PID controllers (in Czech). *Hability thesis*, Technical University of Brno, Faculty of Technology in Zlín, Czechoslovakia.

Bobál, V. (1994a). Robust self-tuning PID controller. In: *Proc. 1st IFAC Workshop on New Trends in Design of Control Systems,* Smolenice, Slovak Republic, 312-317.

Bobál, V. (1994b). Auto-tuning of digital PID controllers using recursive identification. *Internal Report ESR 9404*, Ruhr-University Bochum, Control Engineering Laboratory, Germany.

Bobál, V. (1995). Self-tuning Ziegler-Nichols PID controller. *International Journal of Adaptive Control and Signal Processing*, **9**, 213-226.

Bobál, V. and V.Vašek (1987). Self-tuning PID controller (in Czech). *Automatizace*, **30**, 204-207.

Hang, C. C., K. J. Åström and W. K. Ho (1991). Refinements of the Ziegler-Nichols tuning formula. *IEE Procedings-D*, **138**, 111-118.

Kulhavý, R. (1987). Restricted exponential forgetting in real time identification. *Automatica*, **23**, 586-600.

Macháček, J. and O. Drábek (1990). Calculation of critical gain and critical frequency for the PSD-controller design by Ziegler-Nichols method (in Czech). *Automatizace*, **33**, 142-144.

Macháček , J. , O. Drábek and S. Krejčí (1995). Adaptive discrete PID controller. In: *Proc. Scientific Papers of the University of Pardubice*, Czech Republic, Series A, **1**, 221-229.

Pivoňka, P. (1993). Modelling and control of power plant with self-tuning controller. In: *Proc. Int. Workshop on Applied Automatic Control WAAC'93*, Czech Technical University Prague, 19-23.

Pivoňka, P. (1995). Modelling, adaptive, neural and fuzzy control of coal power plants. In: *Proc. 2nd IFAC Symposium on Power Plants and Power Systems*, Cancún, México.

Takahashi, Y., C. Chan and D. Auslander (1971). Parametereinstellung bei linearen DDC-Algorithmen. *Regelungstech. Prozessdatenver.*, **19**, 237-284.

Ziegler, J. G. and N. B. Nichols (1942). Optimum settings for automatic controllers. *Trans. ASME*, **64**, 759-768.

ON THE CONTROL OF A BUILDING MODEL SUBJECT TO SEISMIC EXCITATION

Laura Menini,[1] **Monica Pasca,**[2] **Antonio Tornambè**[3] and **Luca Zaccarian**[1]

[1] *Dipartimento di Informatica, Sistemi e Produzione, Università di Roma Tor Vergata, via della Ricerca Scientifica, 00133 ROMA, Italy*

[2] *Dipartimento di Ingegneria Strutturale e Geotecnica, Università di Roma "La Sapienza", via Eudossiana 18, 00184 ROMA, Italy*

[3] *Dipartimento di Informatica e Automazione, Università di Roma Tre, via della Vasca Navale 79, 00146 ROMA, Italy*

Abstract. The control problem of a building model, subject to seismic excitation, is considered in this paper. First, a feed-forward compensator is designed to compensate the ground acceleration; secondly, an LQG optimal feedback controller is added to the control scheme to improve the control performances. In order to reduce the complexity of the resulting controller, the computations design procedure is carried out with respect to an approximate model of the building, obtained by eliminating some of the state variables from a balanced state space realization of the original model. Simulation runs confirm the effectiveness of the proposed control structures.

Key Words. Model reduction, Feedforward compensation, Active compensation, LQG control

1. INTRODUCTION

A remarkable increase has occurred in the last few years regarding the active control of civil structures (Housner *et al.*, 1996). Generally, active control strategies have been proposed under the assumption that all the state variables (i.e., the structural displacements and velocities) are available for feedback, as well as additional acceleration measurements. Due to the presence of ground movements, absolute measurements of story displacements and velocities are difficult to be actually obtained. As a matter of fact, during seismic activity, the foundation of the building moves rigidly with the ground, and the reference frame, which is referred to the foundation of the building, is not inertial at all. The assumption that direct measurement of all the story displacements and velocities are available for feedback is, therefore, unacceptable. On the contrary, accelerometers suitably located on the building, may actually provide absolute acceleration measurements.

In this paper, a building model, representing a three-story, single-bay structure, which is equipped with an active mass driver (AMD) system for control purposes, is considered. The linear time-invariant model of the considered structure has been taken from Dyke et al. (Dyke *et al.*, 1996b), who validated such a model with experimental results in the frequency range $0 \div 100\ Hz$. Although a limited number of degrees of freedom has been considered in the model, it is constituted of 28 linear differential equations of the first order. Such a model is endowed with two inputs: the control input, which is the input of the hydraulic motor actuating the active mass, and the disturbance input, which is the ground acceleration. The problem considered in this paper is the attenuation of the effects due to the ground acceleration by controlling the active mass properly.

First, a feed-forward linear time-invariant controller is used to partially compensate the ground acceleration; secondly, an optimal LQG feedback control law is used in order to improve the performances obtained by means of the feed-forward action.

In order to allow an actual implementation of the proposed controller, physical constraints on the actuator are taken into account. By considering that the main seismic excitation generally lasts less than 10 seconds, the complexity of the controller has been taken as low as possible so to allow its actual implementation.

It is well known that, in structural seismic protection, control objectives should be the reduction of both the interstorey relative displacements, in order to limit stresses and therefore damage, and floor maximum accelerations, for comfort purposes. These aims are pursued by considering a direct compensation of the ground acceleration on a scalar output that should be able to represent, as much as possible, the behaviour of the whole system: the interstorey drifts are analysed as candidate outputs and, in particular, due to minimum phase properties, the relative displacement between the third and the second floor is chosen, as it can also be a good measure of the first vibration mode of the structure. Furthermore, the actuator acceleration is the most limiting factor; thus, the scalar output is chosen as a linear combination of the relative displacement between the third and the second floor, and the absolute acceleration of the actuator mass.

Quite good results are achieved in terms of reduction of the interstorey drifts and of the story accelerations, but simulations under the given excitations, namely the N-S record of the 1940 El Centro earthquake and the N-S record of the 1969 Hachinohe one, suitably scaled on the model, have shown that the peak values of the acceleration of the actuator mass would be unacceptable.

This open-loop controller is taken as a starting point (pre-filter) for the realization of an optimal control law based on the classical LQG technique, in which the absolute acceleration of the third floor, the mass acceleration and the controller input have been minimised. This further control strategy allows the performances of the overall control system to be improved, and, in particular, to reduce the acceleration of the actuator mass.

2. DYNAMIC MODEL

2.1. *Evaluation model*

The structure considered in this paper is an actively controlled three-story single-bay model building, based on the test structure implemented at the Structural Dynamic and Control/Earthquake Engineering Laboratory at the University of Notre Dame, IN (USA), constituted by an experimental scaled model building tested also at the National Center for Earthquake Engineering Research at SUNY-Buffalo, NY (USA).

The laboratory structure is a 1:4 scale model, tested by Chung et al. in state feedback experiments (Chung *et al.*, 1989), and by Dyke *et al.*, with a tendon/pulley control system (Dyke *et al.*, 1994) and an active mass driver (Dyke *et al.*, 1996a). The model is a 158 cm high steel frame, constituted by three floors, whose weight is approximately 73 kg each. The frequencies of the first three modes of the structure are 1.16 Hz, 3.54 Hz and 5.71 Hz, respectively. An active mass driver (AMD), which consists of a moving steel mass driven by a single hydraulic actuator, whose weight is 5.2 kg (approximately 1.7% of the total mass of the structure), is placed on the third floor. Sensors are placed on the system in order to measure the position and the absolute acceleration of the AMD, the absolute accelerations of the three floors and the ground acceleration. The two inputs of the system are assumed to be a disturbance input $\ddot{x}_g(t)$, consisting of an unidirectional earthquake excitation, and the control input $u(t)$ of the hydraulic actuator. In order to allow the actual implementation, the proposed controller must satisfy the following constraints:

$$|u(t)| \leq 3\,V, \tag{2.1a}$$
$$|x_m(t)| \leq 9\,cm, \tag{2.1b}$$
$$|\ddot{x}_{am}(t)| \leq 6\,g, \tag{2.1c}$$

where constraint (2.1a) is related to the working range of the actuator, constraint (2.1b) is related to the maximum displacement of the active mass driver piston and constraint (2.1c) is related to its maximum acceleration.

As reported in (Spencer *et al.*, 1997), a 28 state space linear model, which takes into account also the interactions between the structure and the actuator in the system dynamics description, has been evaluated and validated by laboratory tests (Dyke *et al.*1996b). The AMD controller has been included in the model with assigned physical characteristics. The linear time-invariant state space representation of the model is given by:

$$\dot{\mathbf{x}}(t) = \mathbf{A}\,\mathbf{x}(t) + \mathbf{B}\,u(t) + \mathbf{E}\,\ddot{x}_g(t)\,, \tag{2.2a}$$
$$\mathbf{y}(t) = \mathbf{C}_y\,\mathbf{x}(t) + \mathbf{D}_y\,u(t) + \mathbf{F}_y\,\ddot{x}_g(t) + \mathbf{v}(t), \tag{2.2b}$$
$$\mathbf{z}(t) = \mathbf{C}_z\,\mathbf{x}(t) + \mathbf{D}_z\,u(t) + \mathbf{F}_z\,\ddot{x}_g(t)\,, \tag{2.2c}$$

where $\mathbf{x}(t) \in \mathbb{R}^{28}$ is the state vector having dimension 28, $\ddot{x}_g(t)$ is the scalar ground acceleration, $u(t)$ is the scalar control input, $\mathbf{y}(t) = [x_m(t), \ddot{x}_{a1}(t), \ddot{x}_{a2}(t), \ddot{x}_{a3}(t), \ddot{x}_{am}(t), \ddot{x}_g(t)]^T$ is the vector of the responses that can be directly measured, and $\mathbf{z}(t) = [x_1(t), x_2(t), x_3(t), x_m(t), \dot{x}_1(t), \dot{x}_2(t), \dot{x}_3(t), \dot{x}_m(t), \ddot{x}_{a1}(t), \ddot{x}_{a2}(t), \ddot{x}_{a3}(t), \ddot{x}_{am}(t)]^T$ is the vector of the variables to be taken under control; $\mathbf{v}(t)$ is the vector of the measurement noise, and $\mathbf{A}$, $\mathbf{B}$, $\mathbf{E}$, $\mathbf{C}_y$, $\mathbf{D}_y$, $\mathbf{F}_y$, $\mathbf{C}_z$, $\mathbf{D}_z$ and $\mathbf{F}_z$ are real matrices of appropriate dimensions, whose explicit expressions are omitted for the sake of brevity. In particular, $x_i(t)$ and $\dot{x}_i(t)$ are the displacement relative to the ground and the velocity of the i-th floor, respectively, $i = 1, 2, 3$; $x_m(t)$ is the displacement of the AMD relative to the third floor; $\ddot{x}_{ai}(t)$ is the absolute acceleration of the i-th floor, $i = 1, 2, 3$ and $\ddot{x}_{am}(t)$ is the absolute acceleration of the AMD mass. Define the interstorey drifts $d_1(t) := x_1(t)$, $d_2(t) := x_2(t) - x_1(t)$ and $d_3(t) := x_3(t) - x_2(t)$; $d_1(t)$ is the relative position of the first floor with respect to the ground, and $d_i(t)$ is the relative position of the i-th floor with respect to the $(i-1)$-th one, $i = 2, 3$.

2.2. *Reduced order model*

In order to reduce the system dimensions, a similarity transformation, represented by matrix $\mathbf{T}$, is determined so that system (2.2), expressed in the new coordinates $\hat{\mathbf{x}}(t) = \mathbf{T}\,\mathbf{x}(t)$, is in balanced form (*i.e.*, its observability and controllability Gramian matrices are equal and diagonal):

$$\dot{\hat{\mathbf{x}}}(t) = \hat{\mathbf{A}}\,\hat{\mathbf{x}}(t) + \hat{\mathbf{B}}\,u(t) + \hat{\mathbf{E}}\,\ddot{x}_g(t)\,, \tag{2.3a}$$
$$\mathbf{y}(t) = \hat{\mathbf{C}}_y\,\hat{\mathbf{x}}(t) + \mathbf{D}_y\,u(t) + \mathbf{F}_y\,\ddot{x}_g(t) + \mathbf{v}(t), \tag{2.3b}$$
$$\mathbf{z}(t) = \hat{\mathbf{C}}_z\,\hat{\mathbf{x}}(t) + \mathbf{D}_z\,u(t) + \mathbf{F}_z\,\ddot{x}_g(t)\,, \tag{2.3c}$$

where $\hat{\mathbf{A}} := \mathbf{T}\mathbf{A}\mathbf{T}^{-1}$, $\hat{\mathbf{B}} := \mathbf{T}\mathbf{B}$, $\hat{\mathbf{E}} := \mathbf{T}\mathbf{E}$, $\hat{\mathbf{C}}_y := \mathbf{C}_y\mathbf{T}^{-1}$, $\hat{\mathbf{C}}_z := \mathbf{C}_z\mathbf{T}^{-1}$.

Given the Gramian matrix of the balanced system (2.3), small elements in the main diagonal indicate the state variables that can be removed from the system to reduce its order properly. By removing all the state variables having the corresponding value in vector $\mathbf{g}$ (which contains the elements of the main diagonal of the Gramian matrix) smaller than 10, the following reduced-order model of order 6 is obtained:

$$\dot{\mathbf{x}}_r(t) = \mathbf{A}_r\,\mathbf{x}_r(t) + \mathbf{B}_r\,u(t) + \mathbf{E}_r\,\ddot{x}_g(t)\,, \qquad (2.4a)$$

$$\mathbf{y}_r(t) = \mathbf{C}_{yr}\mathbf{x}_r(t) + \mathbf{D}_y u(t) + \mathbf{F}_y \ddot{x}_g(t) + \mathbf{v}(t)\,, \qquad (2.4b)$$

$$\mathbf{z}_r(t) = \mathbf{C}_{zr}\,\mathbf{x}_r(t) + \mathbf{D}_z\,u(t) + \mathbf{F}_z\,\ddot{x}_g(t)\,, \qquad (2.4c)$$

where the explicit expressions of matrices $\mathbf{A}_r \in \mathrm{I\!R}^{6\times 6}$, $\mathbf{C}_{yr} \in \mathrm{I\!R}^{6\times 6}$, $\mathbf{C}_{zr} \in \mathrm{I\!R}^{12\times 6}$, and vectors $\mathbf{B}_r \in \mathrm{I\!R}^{6\times 1}$, $\mathbf{E}_r \in \mathrm{I\!R}^{6\times 1}$, $\mathbf{D}_y \in \mathrm{I\!R}^{6\times 1}$, $\mathbf{F}_y \in \mathrm{I\!R}^{6\times 1}$, $\mathbf{D}_z \in \mathrm{I\!R}^{12\times 1}$ and $\mathbf{F}_z \in \mathrm{I\!R}^{12\times 1}$, are omitted for the sake of brevity.

The chosen model dimensions are obtained by a trade-off between a sufficiently low order and a sufficiently high accuracy of the reduced-order model (2.4); in the following, a compensator will be designed on the basis of such a model, while its performances will be tested with reference to the whole evaluation model (2.2).

Quantities related to the reduced-order model are identified in the rest of the paper by subscript r.

3. CONTROL DESIGN

3.1. *Feed-forward controller*

As a first step, a scalar output $\varphi_r(t)$ of the reduced-order system (2.4) is identified so to be representative, as much as possible, of the effects of the ground acceleration on the behaviour of all the output variables to be taken under control. By considering that the effects of the ground acceleration can be seen mainly on the interstorey drifts $d_{ri}(t)$, $i = 1, 2, 3$, that the mass acceleration $\ddot{x}_{am}(t)$ is a critical parameter in the controller design, and that both transfer functions from $u(s)$ to $d_{r3}(s)$ and $\ddot{x}_{am}(s)$, respectively, are minimum phase, whereas both transfer functions from $u(s)$ to $d_{r1}(s)$ and $d_{r2}(s)$, respectively, are non-minimum phase, the output $\varphi_r(t)$ to be rendered as insensitive as possible with respect to the external excitation $\ddot{x}_g(t)$, has been chosen as follows:

$$\varphi_r(t) := d_{r3}(t) + 0.02\,\ddot{x}_{am}(t)\,, \qquad (3.1)$$

where the coefficient value 0.02 has been chosen as a trade-off between the reduction of the interstorey drift $d_{r3}(t)$ and the reduction of the absolute acceleration of the mass $\ddot{x}_{am}(t)$.

The relationship between the output $\varphi_r(s)$ and the control and disturbance inputs $u(s)$ and $\ddot{x}_g(s)$, respectively, is given by

$$\varphi_r(s) = W_1(s)\,\ddot{x}_g(s) + W_2(s)\,u(s)\,, \qquad (3.2)$$

where $W_1(s)$ and $W_2(s)$ are computed on the basis of the reduced-order model (2.4); in particular, the output $\varphi_r(s)$ has been chosen so that the transfer function $W_2(s)$ is bi-proper and

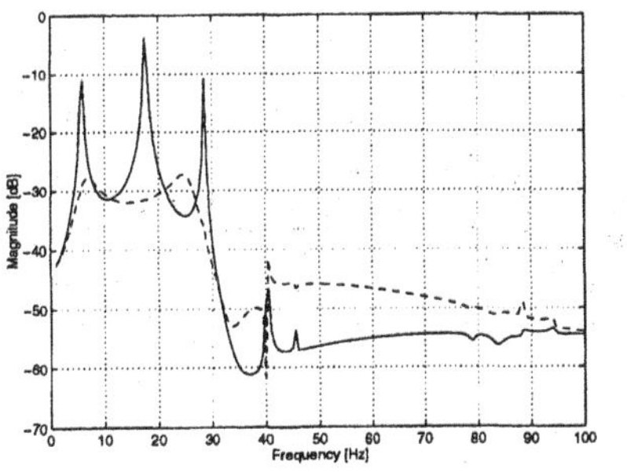

Fig. 1. Magnitude of the transfer functions from $\ddot{x}_g(s)$ to $\varphi(s)$ in the uncontrolled case (continuous lines) and in the case in which the feed-forward control law (3.3) is applied to system (2.2) (dashed lines).

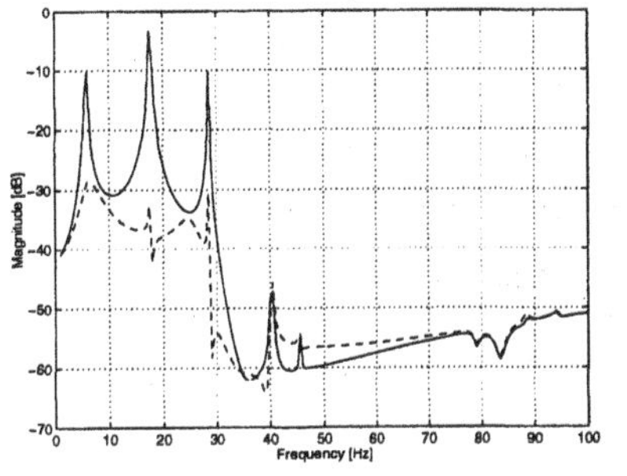

Fig. 2. Magnitude of the transfer functions from $\ddot{x}_g(s)$ to $d_3(s)$ in the uncontrolled case (continuous line) and in the case in which the feed-forward control law (3.3) is applied to system (2.2) (dashed line).

minimum-phase, while $W_1(s)$ results to be bi-proper too. Since the input $\ddot{x}_g(s)$ is measured, it is easy to see from equation (3.2) that the following control law, which uses the knowledge of $\ddot{x}_g(s)$,

$$u(s) = G_F(s)\,\ddot{x}_g(s) + \tilde{u}(s)\,, \qquad (3.3)$$

where $\tilde{u}(s)$ is the new control input and $G_F(s) := -W_1(s)/W_2^{-1}(s)$ is a bi-proper asymptotically stable transfer function, whence, the realisability of the feed-forward compensator is granted. Moreover, due to the asymptotic stability of the evaluation model (2.2), the open-loop control scheme in Figure 4 may actually be implemented. Under the action of the controller $G_F(s)$, the disturbance input $\ddot{x}_g(s)$ is totally compensated in the scalar output $\varphi_r(s)$ with respect to the reduced-order model; as a matter of fact, equation (3.2) under control law (3.3) becomes:

$$\varphi(s) = W_2(s)\,\tilde{u}(s)\,. \qquad (3.4)$$

It is important to stress that the behaviour represented by (3.4) is not actually achieved on the whole evaluation model (2.2), due to the drastic approximations that have been introduced in deriving the reduced-order model (2.4). As a matter of fact, the ground acceleration $\ddot{x}_g(t)$ still affects the output $\varphi(t)$.

In particular, Figure 1 shows the Bode diagrams of the magnitudes of the transfer functions from $\ddot{x}_g(s)$ to $\varphi(s)$, computed on the basis of the whole evaluation model (2.2), in the uncontrolled case

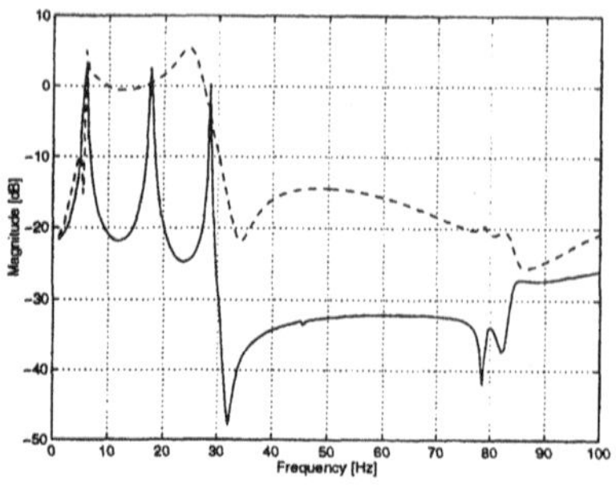

Fig. 3. Magnitude of the transfer functions from $\ddot{x}_g(s)$ to $\ddot{x}_{am}(s)$ in the uncontrolled case (continuous line) and in the case in which the feed-forward control law (3.3) is applied to system (2.2) (dashed line).

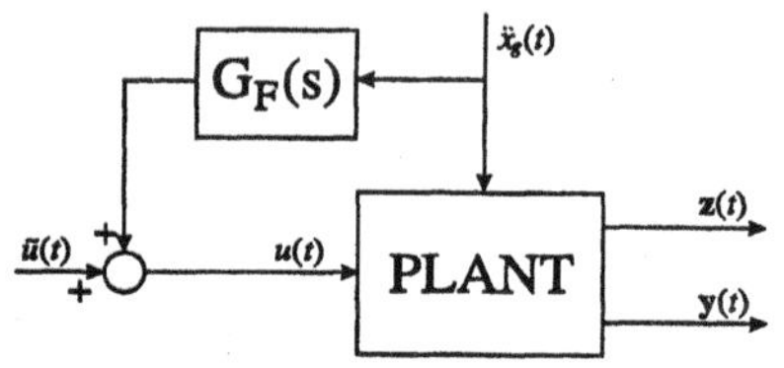

Fig. 4. Block diagram of system (2.2) under the action of the feed-forward control law (3.3).

(continuous lines), and in the case in which the feed-forward control law (3.3) is applied to the evaluation model (2.2) (dashed lines). Figures 2 and 3 show the magnitude of the transfer functions from the ground acceleration $\ddot{x}_g(t)$ to the third drift $d_3(t)$ and to the mass absolute acceleration $\ddot{x}_{am}(t)$. A general reduction of the magnitudes can be observed at low frequencies; similar behaviour can be found on the other transfer functions but are here omitted for the sake of brevity.

The overall control system has been simulated using the 1940 El Centro N-S record and the 1969 Hachinohe N-S record as excitation signals, whose durations are $T_E = 10\,sec$ and $T_H = 7.2\,sec$, respectively. The performances of the control action (3.3) on system (2.2) are tested on the basis of the above simulations by means of the indices reported in the following. The first three indices are related to the interstorey drifts, normalised with respect to d_{i0}^E and d_{i0}^H, which represent the maximum values of the i-th interstorey drift, $i = 1, 2, 3$, of the uncontrolled structure, under the El Centro and the Hachinohe excitations, respectively:

$$I_1 = \max_t \left\{ \frac{|d_1^E(t)|}{d_{10}^E}, \frac{|d_1^H(t)|}{d_{10}^H} \right\}, \quad (3.5a)$$

$$I_2 = \max_t \left\{ \frac{|d_2^E(t)|}{d_{20}^E}, \frac{|d_2^H(t)|}{d_{20}^H} \right\}, \quad (3.5b)$$

$$I_3 = \max_t \left\{ \frac{|d_3^E(t)|}{d_{30}^E}, \frac{|d_3^H(t)|}{d_{30}^H} \right\}, \quad (3.5c)$$

where $d_i^E(t)$, $d_i^H(t)$, $i = 1, 2, 3$, are the interstorey drifts of the controlled structure, under the El Centro and the Hachinohe signals, respectively.

The following indices I_4, I_5 and I_6 are related to the absolute acceleration of the floors, normalised with respect to the maximum story accelerations,

Index	I_1	I_2	I_3	I_4
Value	0.785	0.769	0.181	0.704
Index	I_5	I_6	I_7	I_8
Value	0.788	0.696	20.53	0.024

Table 1 Performance indices obtained using the control structure in Figure 4.

$\ddot{x}_{ai0}^E$ and $\ddot{x}_{ai0}^H$, in the uncontrolled case:

$$I_4 = \max_t \left\{ \frac{|\ddot{x}_{a1}^E(t)|}{\ddot{x}_{a10}^E}, \frac{|\ddot{x}_{a1}^H(t)|}{\ddot{x}_{a10}^H} \right\}, \quad (3.6a)$$

$$I_5 = \max_t \left\{ \frac{|\ddot{x}_{a2}^E(t)|}{\ddot{x}_{a20}^E}, \frac{|\ddot{x}_{a2}^H(t)|}{\ddot{x}_{a20}^H} \right\}, \quad (3.6b)$$

$$I_6 = \max_t \left\{ \frac{|\ddot{x}_{a3}^E(t)|}{\ddot{x}_{a30}^E}, \frac{|\ddot{x}_{a3}^H(t)|}{\ddot{x}_{a30}^H} \right\}, \quad (3.6c)$$

where $\ddot{x}_{ai}^E(t)$, $\ddot{x}_{ai}^H(t)$, $i = 1, 2, 3$, are the absolute accelerations of the three floors of the controlled structure, under the El Centro and the Hachinohe excitations, respectively.

Furthermore, the following indices I_7 and I_8 are related to the evaluation of the active mass displacement and absolute acceleration, normalised with respect to the maximum values of the third floor absolute position and acceleration in the uncontrolled case, denoted by x_{30}^E and $\ddot{x}_{a30}^E$ under the El Centro excitation, and by x_{30}^H and $\ddot{x}_{a30}^H$ under the Hachinohe excitation:

$$I_7 = \max_t \left\{ \frac{|x_m^E(t)|}{x_{30}^E}, \frac{|x_m^H(t)|}{x_{30}^H} \right\}, \quad (3.7a)$$

$$I_8 = \max_t \left\{ \frac{|\ddot{x}_{am}^E(t)|}{\ddot{x}_{a30}^E}, \frac{|\ddot{x}_{am}^H(t)|}{\ddot{x}_{a30}^H} \right\}, \quad (3.7b)$$

where $x_m^E(t)$, $x_m^H(t)$, and $\ddot{x}_{am}^E(t)$, $\ddot{x}_{am}^H(t)$, are the displacements and the absolute accelerations of the active mass, under the El Centro and the Hachinohe excitations, respectively.

The values of the resulting indices are listed in Table 1; such indices show that the interstorey drifts and the floor accelerations are satisfactorily damped; from the value of index I_3, it can be seen that the maximum value of the drift $d_3(t)$ is highly reduced by the control action. The third interstorey drift time response obtained from a simulation under the El Centro historical earthquake record is depicted in Figure 5, which shows a strong reduction of this quantity with respect to the uncontrolled case; similar results have been obtained using the Hachinohe historical earthquake record.

The controller, which is a 6-state dynamic system, gives satisfactory performances concerning the structure parameters. As for the control constraints, the maximum displacement of the AMD, $\max_t\{|x_m(t)|\} = 3.128\,cm$, achieved under the Hachinohe earthquake, satisfies the controller constraint defined in equation (2.1b), as well as the maximum value of the input sig-

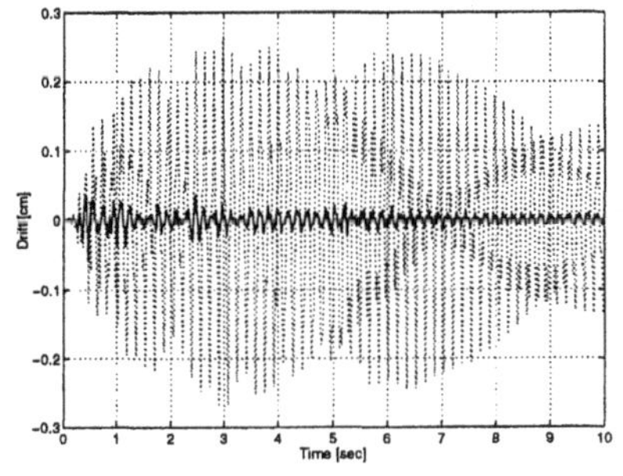

Fig. 5. Time histories of the third interstorey drift $d_3(t)$ in the uncontrolled case (dotted line) and in the case in which feed-forward control law (3.3) is applied to system (2.2) (solid line) under the El Centro earthquake N-S record.

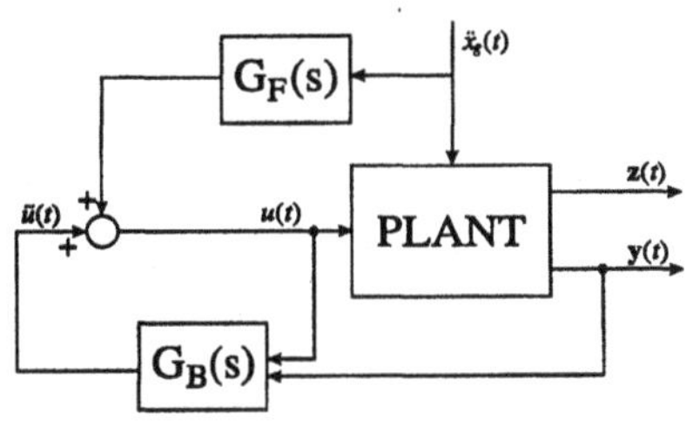

Fig. 6. Block diagram of system (2.2) under the combined action of the feed-forward control law (3.3) and the feedback control law (3.9).

nal, $\max_t\{|u(t)|\} = 1.024\,V$, which does not exceed constraint (2.1a), whereas, unfortunately, the constraint on the mass acceleration (2.1c) is not satisfied, as the following peak values of the acceleration, $\max_t\{|\ddot{x}_{am,E}(t)|\} = 12.04\,g$ and $\max_t\{|\ddot{x}_{am,H}(t)|\} = 8.12\,g$, are obtained, under the El Centro and the Hachinohe input signals, respectively. Whence, the control structure given in Figure 4 should be modified to satisfy such a control constraint.

3.2. *Feedback control law*

In order to reduce the acceleration of the AMD (which exceeds constraint (2.1c) in the control structure employed in the previous subsection), a feedback LQG optimal controller $G_B(s)$ (designed on the basis of an extended system constituted by the plant (2.4) and the feed-forward controller (3.3)) is added to the control scheme of Figure 4, obtaining the overall scheme reported in Figure 6.

The output $\mathbf{y}(t)$ of the plant is augmented with the signal $u(t)$, thus obtaining the new output $\tilde{\mathbf{y}}(t) = [\mathbf{y}^T(t)\; u(t)]^T$ of the system (2.4), (3.3).

Note that the input of the controller $G_B(s)$, used in the control scheme shown in Figure 6, consists of the measured output $\mathbf{y}(t)$ of the plant (2.4) and of the input signal $u(t)$ of the plant itself. Such a connection allows to reduce the magnitude of signal $u(t)$, which plays a dominant role in the determination of the mass acceleration. Assuming $\ddot{x}_g(t)$ to be a stationary white noise with covariance $S_{\ddot{x}_g\ddot{x}_g}$, the performance index of the controller design is chosen as the infinite horizon per-

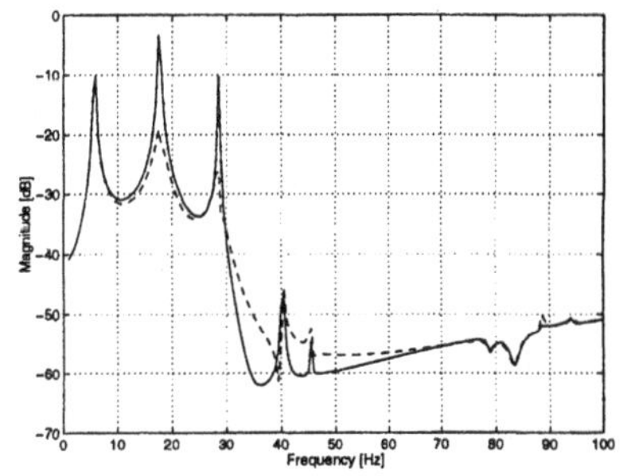

Fig. 7. Magnitude of the transfer functions from $\ddot{x}_g(s)$ to $d_3(s)$ in the uncontrolled case (continuous line) and in the case in which the control law (3.9) is applied to system (2.2) (dashed line).

formance index

$$J := \lim_{T\to\infty} \frac{1}{T}\mathrm{E}\left\{\int_0^T \tilde{\mathbf{y}}^T(t)\mathbf{Q}\tilde{\mathbf{y}}(t) + r\tilde{u}^2(t)\,dt\right\}, \quad (3.8)$$

where $\mathbf{Q} = \mathrm{diag}\{0, 0, 0, 1, 10, 0, 20\}$ and $r = 0.1$ are the weights related to the output and to the input of the overall system (2.4), (3.3), respectively. Now, assuming that $S_{\ddot{x}_g\ddot{x}_g} = 25$ and that the measurement noise $\mathbf{v}(t)$ is a Gaussian white noise having the 6-dimensional identity matrix as covariance matrix, the control law that minimises the index J given in (3.8) can be determined as a static feedback control law from the state $\hat{\mathbf{x}}_a(t)$ of a Kalman filter, which provides the optimal estimate of the actual state $\mathbf{x}_a(t)$ of system (2.4), (3.3) under the action of the inputs $\ddot{x}_g(t)$ and $\mathbf{v}(t)$. The state $\hat{\mathbf{x}}_a(t)$ of the resulting controller, which is actually the state of the Kalman filter, has then dimension 12 (namely the same dimension of the state $\mathbf{x}_a(t)$ of system (2.4), (3.3)).

Combining together the feed-forward and the feedback control actions, a complete feedback compensator, having order 18, is obtained. For practical purposes, a reduction of the state dimension of the compensator to $n_c = 12$ is introduced. Such a reduction is carried out by means of the same procedure adopted in Subsection 3.1 removing from the state of the compensator transformed in a balanced form the last six components, thus reducing its dimensions to $n_c = 12$:

$$\dot{\mathbf{x}}_c(t) = \mathbf{A}_c\mathbf{x}_c(t) + \mathbf{B}_c\mathbf{y}(t), \quad (3.9a)$$

$$u(t) = \mathbf{C}_c\mathbf{x}_c(t) + \mathbf{D}_c\mathbf{y}(t), \quad (3.9b)$$

The explicit expressions of the matrices $\mathbf{A}_c \in \mathbb{R}^{12\times12}$, $\mathbf{B}_c \in \mathbb{R}^{12\times6}$, $\mathbf{C}_c \in \mathbb{R}^{1\times12}$ and $\mathbf{D}_c \in \mathbb{R}^{1\times6}$, are omitted for the sake of brevity. Since all the eigenvalues of matrix $\mathbf{A}_c$ have negative real parts, the overall controller is asymptotically stable, as well as the resulting closed-loop system.

Similarly to the previous subsection, the effects of the controller (3.9) on the outputs of interest have been studied by computing the Bode diagrams of the related transfer functions. Figures 7 and 8 show the magnitude of the transfer functions from the ground acceleration $\ddot{x}_g(t)$ to the third interstorey drift $d_3(t)$ and the mass absolute acceleration $\ddot{x}_{am}(t)$. Comparing the Bode diagrams in Figures 7-8, with those in Figures 2-3

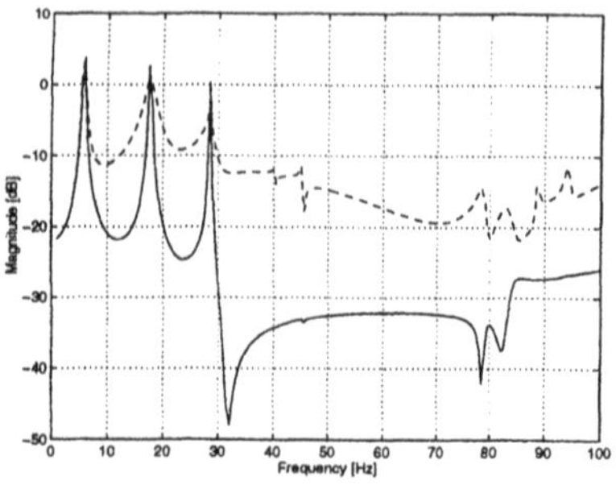

Fig. 8. Magnitude of the transfer functions from $\ddot{x}_g(s)$ $\ddot{x}_{am}(s)$ in the uncontrolled case (continuous line) and in the case in which control law (3.9) is applied to system (2.2) (dashed line).

Index	I_1	I_2	I_3	I_4
Value	0.762	0.708	1.244	0.683
Index	I_5	I_6	I_7	I_8
Value	0.720	0.679	8.008	0.013

Table 2 Performance indices obtained using the control structure in Figure 6.

(which represent the same transfer functions when the control action is performed by the only feed-forward compensator), it can be seen that the feedback control action deteriorates the behaviour of the Bode diagrams of the transfer function from $\ddot{x}_g(t)$ to $d_3(t)$; however, it has been noticed that the maximum values of the other transfer functions (which are not reported here) are mainly unchanged, whence the decrease of performance of the control action is not expected to be significant on all performance indices. On the other hand, the feedback control action strongly reduces the high values of the transfer functions from $\ddot{x}_g(t)$ to $\ddot{x}_{am}(t)$ and $x_m(t)$ (the latter is not reported for the sake of brevity), which are related to the constrained variables of the system; whence, highly decreased values for the indices I_7 and I_8 are expected to be found using this new controller.

The performances of the controller on the whole evaluation model (2.2) have been tested by means of simulations using both the earthquakes described in Subsection 3.1. The resulting performance indices, which are reported in Table 2, show a better performance with respect to the 6-order feed-forward controller designed in Subsection 3.1, in almost all cases. In particular, while the peak value of the third interstorey drift is lightly increased, the AMD respects all the constraints (2.1).

The two plots depicted in Figure 9 show the time responses of the third floor absolute acceleration responses $\ddot{x}_{a3}(t)$ under the Hachinohe excitation, when control laws (3.3) and (3.9), respectively, are applied to system (2.2). It can be observed from this figure, that the decrease of the peak values performed by control law (3.9) with respect to control law (3.3), is upheld by a general improvement of the system behaviour over the total earthquake interval. Similar results have been obtained for the El Centro earthquake input signal, but are not represented here for the sake of brevity.

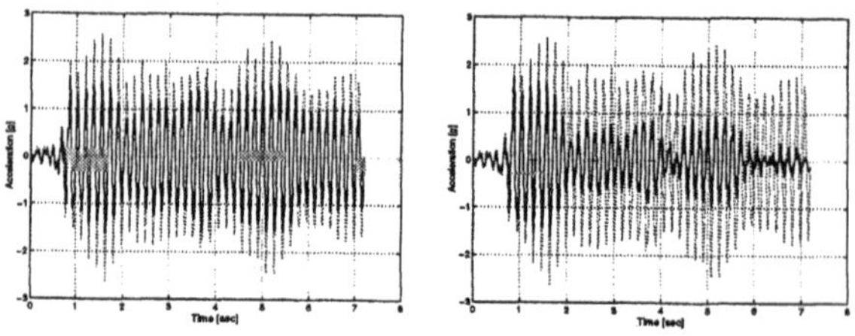

Fig. 9. Time histories of the third floor acceleration $\ddot{x}_{a3}(t)$ under the Hachinohe earthquake N-S record in the uncontrolled case (dotted lines) and in the cases in which (a) the feed-forward control law (3.3) or (b) control law (3.9) is applied to system (2.2).

4. CONCLUSIONS

An active control action on a building model under seismic excitation is performed by an active mass driver placed on the last floor of a three story building. A dynamic controller is proposed, as the combination of a feed-forward compensation of the ground acceleration with a feedback LQG optimal control law, based on an extended system constituted by the actual plant and the feed-forward controller. With the aim of maintaining the order of the compensator lower than 13, a reduction procedure is employed, for both the model of the structure and the resulting controller. Simulation runs show a satisfactory behaviour of the controlled system.

REFERENCES

Spencer B.F. Jr., Dyke S.J. and Deoskar H.S. (1997). Benchmark problems in structural control. Part I: Active Mass Driver system, to appear in a special issue of *Earthquake eng. struct. dyn.*.

Chung L.L., Lin R.C., Soong T.T. and Reinhorn A.M. (1989). Experiments on active control for MDOF seismic structures, *J. Engrg. Mech.* **115**, 1609-1627.

Dyke S.J., Spencer B.F. Jr., Quast P., Sain M.K., Kaspari D.C. Jr. and Soong T.T. (1994). Experimental verification of acceleration feedback control strategies for an active tendon system, *National Center for Earthquake Engineering Research Technical Report NCEER-94-0024*.

Dyke S.J., Spencer B.F. Jr., Quast P., Kaspari D.C. Jr. and Sain M.K. (1996a). Implementation of an active mass driver using acceleration feedback control, *Microcomputers in Civil Engrg.* **11**, 305-323.

Dyke S.J., Spencer B.F. Jr., Quast P., Sain M.K., Kaspari D.C. Jr. and Soong T.T. (1996b). Acceleration feedback control of MDOF structures, *J. Engrg. Mech.* **122**, 907-917.

Housner G.W., Soong T.T. and Masri S.F. (1996). Second generation of active structural control in civil engineering, *Microcomputers in Civil Engrg.* **11**, 289-296.

DYNAMIC BEHAVIOUR OF FRÖHLICH'S BRAIN-WAVE MODEL

G. J. Hyland*, C.P. Lewis and A. Ucar****

* *The University of Warwick, Department of Physics, Coventry, CV4 7AL, UK*

** *Coventry University, School of Engineering, Nonlinear Systems Design Group, Q Block, Coventry, CV1 5FB, UK*
Fax: 44 1203 838949; e-mail: cex199@coventry.ac.uk

Abstract: Motivated by the empirical finding that the *EEG* corresponding to varying degrees of consciousness exhibit deterministic chaos to a corresponding degree, Fröhlich's Brain-Wave model is analysed for the first time in its full generality (3 degrees of freedom). Some new results are presented concerning the response of the system to external stimuli - both *ac* and *dc*. In contrast to an earlier claim that a restricted version of the model (containing only 2 degrees of freedom) exhibits a chaos *only* when the external stimulus contains both *ac* **and** *dc* components, we find that *ac* component **alone** is sufficient. Readmitting the 3rd degree of freedom can result in much more complex bifurcation scenarii , as is readily seen from consideration and comparison of the control space basins.

Keywords: Chaos; Fröhlich; Brain-waves; Control Space Basin

1. INTRODUCTION

Fröhlich's model (Fröhlich 1977) for the generation of brain-waves is based on self-sustaining collective chemical oscillations between the excited states and ground-states of a system of enzymes which are assumed to be spatially distributed throughout the Greater Membrane of the brain in crudely ordered fashion, possibly *via* their attachment to microtubules. The energy required to activate the enzymes is considered to derive from their interaction with surrounding substrate molecules, in the course of which some of the latter are destroyed. If in their excited state the enzymes possess an electric dipole moment which is absent in the ground-state then associated with these chemical oscillations will be an oscillatory electric polarisation field; it is this field which is identified with the brain-waves.

Despite the undeniable novelty of Fröhlich's Brain-Wave model and the variety of scenarii afforded by the underlying non-linear dynamics, all analyses to date have been conducted within an approximation which essentially neglects the dynamics of that part of the system constituted by the non-activated enzymes. This approximation, which reduces the number of degrees of freedom from three to two, removes the possibility of chaotic behaviour in the case where the system is isolated from external stimuli. It is known, however, from time series analysis of the electroencephalogram (EEG), that the brain *does* exhibits chaotic behaviour (Gallez and Babloyantz 1990); this imposes a severe constraint on any admissible model of brain-wave generation.

In this paper the dynamic behaviour of Fröhlich's model is investigated and the possibility of chaotic behaviour discussed.

2. BRAIN-WAVE MODEL

Fröhlich's model is based upon a system of substrate molecules (of number S) and enzymes, N of which are in the highly activated state and Z of which are not excited.

The rate of increase in the number of activated enzymes is proportional to (*i*) their concentration, N, (*ii*) the concentration Z of the remaining unexcited enzymes and (*iii*) the concentration, S, of the substrates. Including the decrease in N due to spontaneous transitions back to the ground state (characterised by a rate constant β), the following equation for first state variable, N, is obtained:

$$\dot{N} = \alpha NZS - \beta N, \tag{1}$$

where α is a constant parameter. On the other hand, for every enzyme excited (to the highly polar state) one substrate molecule is chemically destroyed; the supply is assumed to be maintained, however, by an influx of new substrates. If the rate of attraction of new substrates is γ, then the second state equation is

$$\dot{S} = -\alpha NZS + \gamma S. \tag{2}$$

Finally, in his initial publication (Fröhlich 1977) Fröhlich presented a third equation

$$\dot{Z} = -\alpha NZS + \beta N + \lambda(A - Z), \tag{3}$$

where λ is a constant parameter, and the final term on the RHS, $+\lambda(A\text{-}Z)$, is considered to arise from selective long-range interactions, the origin of which does not concern us here.

The equilibrium points of the steady-state solution of the Brain-Wave model are obtained by equating (1), (2) and (3) to zero, which yields

$$N_0 = \frac{\gamma}{\alpha Z_0},\ S_0 = \frac{\beta}{\alpha Z_0},\ Z_0 = A \tag{4}$$

It may be noted that if equation (3) is neglected (which is equivalent to assuming that the concentration, Z, of unexcited enzymes remains constant = A), the problem simplifies to a Lotka-Volterra predator (N) - prey (S) system, such as is often used to model biological phenomena such as biological clocks and time-dependent neural networks. This assumption will be made at the beginning of Section 3, but will be relaxed later.

To examine the stability of the equilibrium point (N_0, S_0, Z_0), small changes n, s, z are considered about this point. Substituting the new variables $N=N_0+n$, $S = S_0+s$ and $Z = Z_0+z$ into (1), (2) and (3) yields

$$\dot{n} = \gamma s + \alpha Ans + \frac{\gamma\beta}{\alpha A^2}z + \frac{\beta}{A}nz + \frac{\gamma}{A}zs + \alpha nsz \tag{5}$$

$$\dot{s} = -\beta n - \frac{\gamma\beta}{\alpha A^2}z - \frac{\beta}{A}nz - \alpha Ans - \frac{\gamma}{A}zs - \alpha nsz \tag{6}$$

$$\dot{z} = -\lambda z - \dot{n}, \tag{7}$$

where the constants are all positive.

Neglecting any terms *non*-linear in n, s, and z, we obtain

$$\ddot{n} + \gamma\beta n = 0, \tag{8}$$

which is a simple harmonic oscillator of natural frequency $\omega_0 = (\beta\gamma)^{1/2}$.

Associated with the oscillatory electric polarisation is an electric current which, being subject to resistance, gives a negative contribution to dn/dt (see below). On the other hand, a positive non-linear contribution arises from interaction between the excited (dipolar) enzyme states which favours a non-zero average polarization. Including these effects, equation (5) generalises to

$$\dot{n} = \gamma s + \alpha Ans + \frac{\gamma\beta}{\alpha A^2}z + \frac{\beta}{A}nz + \frac{\gamma}{A}zs + \alpha nsz + (c^2 e^{-\Gamma^2 n^2} - d^2)n \tag{9}$$

where c, d, Γ are constant parameters. Provided $c^2>d^2>0$, the non-linear system represented by equations (6), (7), and (9) exhibits limit cycling behaviour as shown in Fig 1, for $c^2 = 5$, $d^2 = 1$, $\Gamma^2=0.01$, $\alpha A = 0.1$, $\beta = \gamma = 5$, and $\lambda = 60$.

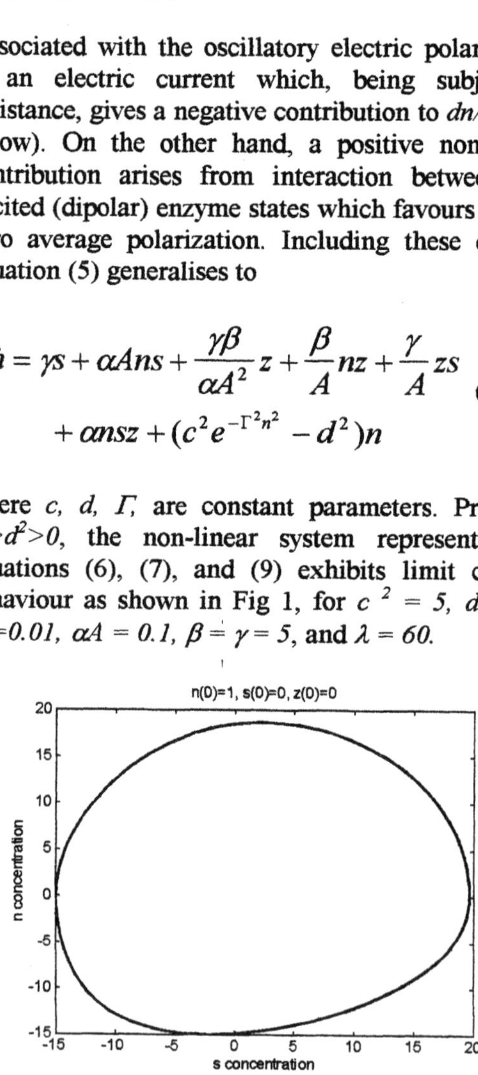

Fig. 1: The n-s phase portrait of the unforced system.

The dependence of the maximum value of the fluctuating z variable on the value of the parameter λ is shown in Fig 2.

It will be seen that z is significantly larger than zero only for $1 < \lambda < 20$; for $\lambda > 500$, z is effectively zero, and we retrieve the two component (n,s) system which has already been extensively investigated by Kaiser (1992).

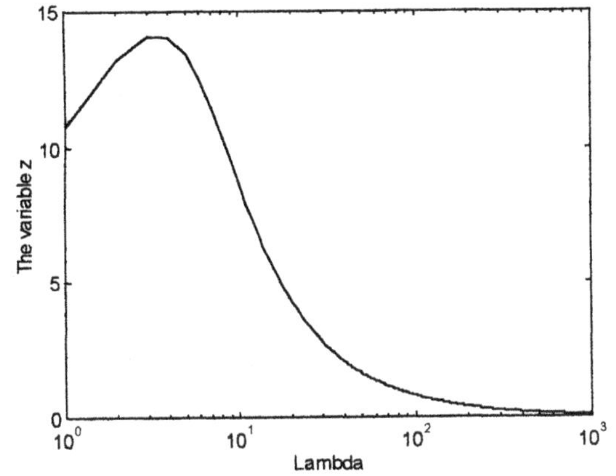

Fig. 2: The concentration z versus λ.

Since the set of non-linear equations (6), (7) and (9) contain *three* variables (n, s, z), the possibility of *chaotic* solutions exists, as is indeed necessary if Fröhlich's model is to afford a realistic description of brain-waves. To be *quantitatively* correct, however, his model must be generalised to include at least *two extra* degrees of freedom, since the minimum embedding dimension is known to be five (Babloyantz & Destexhe, 1986). Despite this, it is still of interest to analyse equations (6), (7) and (9) for chaotic solutions, and this is currently in progress.

3. EFFECTS OF EXTERNAL STIMULUS

Of particular interest is the response of the system to an external electric field stimulus, $F(t)$, which interacts directly (*via RHS* of (9)) with the dynamic polarisation associated with the activated enzyme state. $F(t)$ includes both the external applied field, $F_{ext}(t)$, and the thermal fluctuating component arising within the system itself, $F_T(t)$, the mean square magnitude of which may, of course, be much larger than that of F_{ext} (Fröhlich and Hyland, 1995).

To date, only when z is *neglected* has the response of the system been studied. Within this restricted (n,s) model, is has been claimed (Kaiser, 1992) that *chaos* arises *only* when the external field contains a *dc* component, F_0 ; our analysis, however, shows that this is *not* the case.

Thus for $F_{ext}(\equiv F_1 \cos\omega t) = 146 \cos 5t$, and the same parameter values as used above, the time response of the n and s concentrations are shown in Fig 3, and the associated phase portrait and Poincaré section in Figs 4 and 5, respectively ; these clearly indicate the presence of deterministic chaos , the existence of which is confirmed by the calculation of a *positive* Lyapunov exponent for the concentration n.

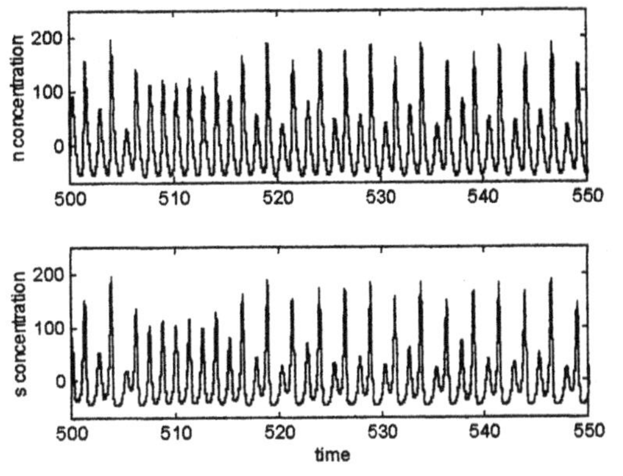

Fig. 3: The time response of enzyme concentrations, n and s, when n is driven with an external stimulus $F_{ext} = 146 \cos 5t$.

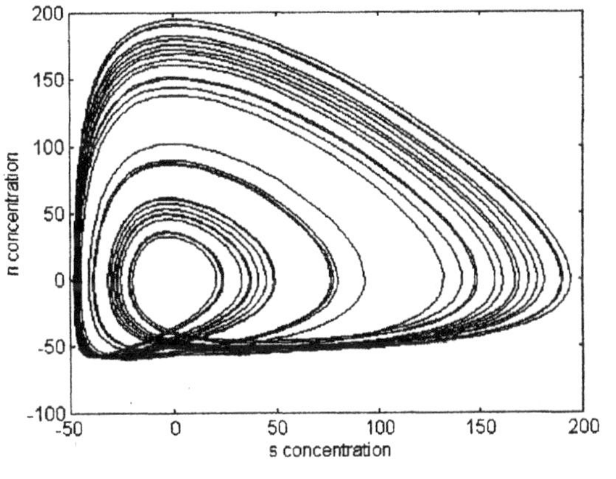

Fig. 4: The phase portrait of n-s system ($Z = A$) when driven with an external stimulus $F_{ext} = 146 \cos 5t$ applied to the variable n.

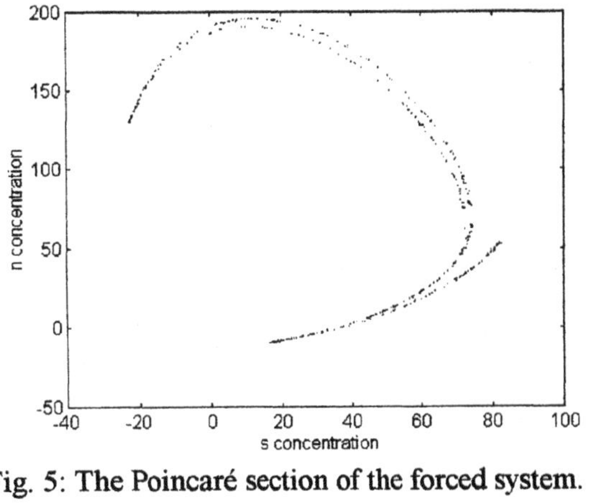

Fig. 5: The Poincaré section of the forced system.

It is, of course, possible - as shown in Fig 6 - to preserve limit cycle and quasi-periodic behaviour, for appropriate choices of amplitude of the external stimulus for the (fixed) frequency, $\omega = 5$ *rad/s*.

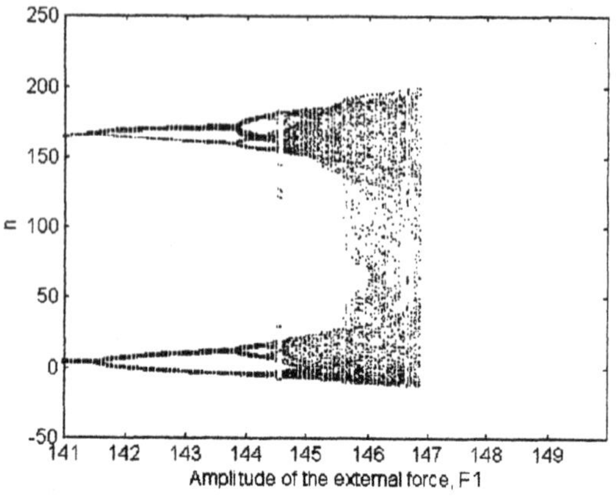

Fig. 6: The bifurcation diagram of the restricted version of the model in which z and F_o are neglected .

Fig 6 indicates the presence of a small chaotic region between $F_1 = 144.5$ and $F_1 = 146.9$, beyond which the system becomes unstable. *Thus, contrary to Kaiser (1992), we find the possibility of deterministic chaos* ***without*** *the necessity of a dc component, F_o, in the external stimulus.*

In view of this difference, it is necessary to re-examine the case $F_o \neq 0$.

For $F_o=100$, chaos indeed exists (see Fig 7), but does not set in until $F_1 = 315$; evidently, the effect of the *dc* component, F_o, is actually to *stabilize* the system *against* chaos!

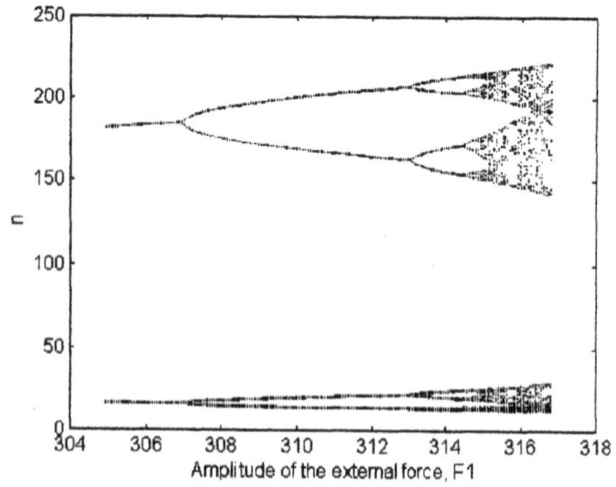

Fig. 7: The bifurcation diagram of the restricted version of the model in which z is neglected, but $F_o \neq 0$.

It is now of interest to investigate the effect on the bifurcation scenario of *re*-admitting the variable z, both for $F_o \equiv 0$ and $F_o \neq 0$.

(i) $F_o \equiv 0$: The bifurcation diagram shown in Fig 8 indicates that for $z \neq 0$ ($\lambda = 200$), chaos still sets in near the *same* value of F_1 as when $z \equiv 0$ (see Fig 6), but persists for a rather smaller range of F_1, beyond which the system again becomes unstable.

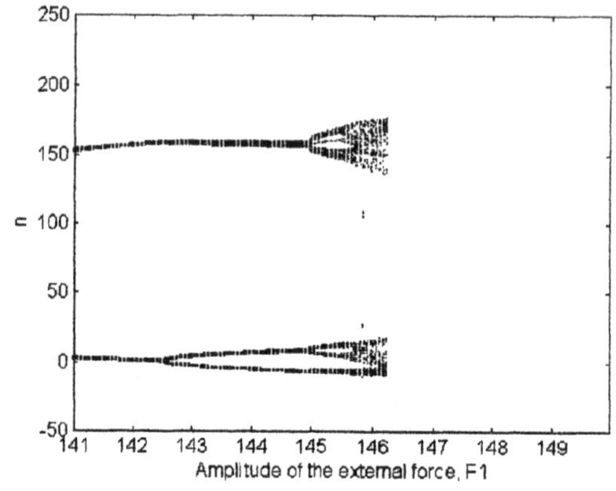

Fig. 8: The bifurcation diagram of the forced system ($F_o \equiv 0$) when z is re-admitted ($\lambda = 200$).

(ii) $F_o \neq 0$: For $F_o = 100$, the effect of $z \neq 0$ ($\lambda=200$) is to slightly extend the chaotic region shown in Fig7, corresponding to $z \equiv 0$. Similarly, as with $z \equiv 0$, the effect of $F_o \neq 0$ is to delay the onset of chaos, from

$F_1 \approx 146$ to 314.

It is to be emphasised that the above discussion is relevant only to the particular value of λ (=200) chosen. The response of the system for much smaller values of λ - in particular in the vicinity of the peak shown in Fig 2 - is much more complex, and will be reported on in a future paper.

4. STABILITY (CONTROL SPACE BASIN)

An alternative approach to the investigation of the dynamic behaviour of nonlinear systems is the control space basin (CSB); this has been applied to the study of the stability of marine structures (Thompson *et al.*, 1990), transducers and chaotic systems (Lewis *et al.*, 1997a). This approach illustrates the *long term* behaviour of a forced system (Lewis *et al.*, 1997b), whose stability and possible bifurcation scenarii are depicted in an input amplitude - frequency plane.

A typical CSB diagram is shown in Fig 9 for the simplified *n-s* model derived from equations (6) and (9) by taking $z \equiv 0$, whilst those for the generalised model with $\lambda = 5$ and $\lambda = 200$ are shown in Figs 10 and 11, respectively. All three diagrams indicate bifurcation phenomena leading to chaos, although a great similarity between Figs 9 and 11 will be noted - *i.e* for such a large value of $\lambda (=200)$, the third order system closely approximates to the second order one. The situation in the case of $\lambda = 5$ (Fig 10) is much more complex, especially at higher frequencies. In particular, the value of the resonance frequency is reduced to approximately 3 *rad/s,* whilst the bifurcation scenario leading to chaos is quite different.

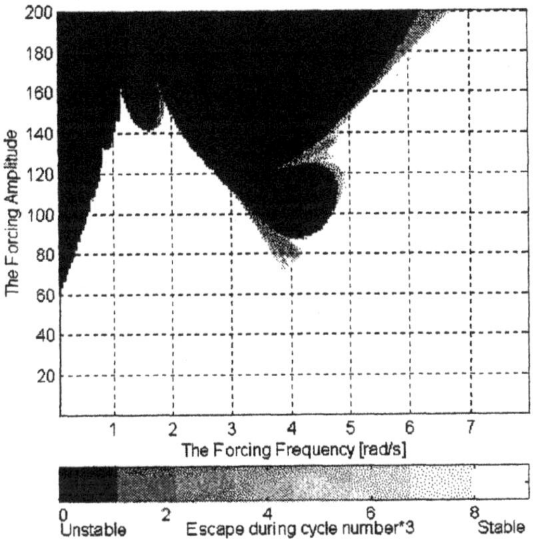

Fig. 9: The CSB diagram of the simplified (*n-s*) model.

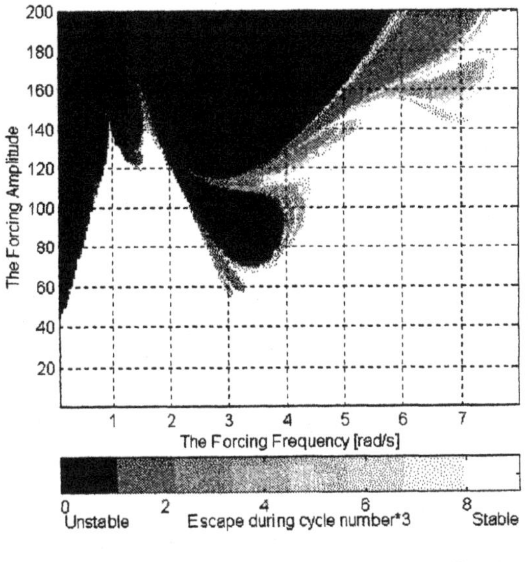

Fig. 10: The CSB diagram of the generalised model with $\lambda = 5$.

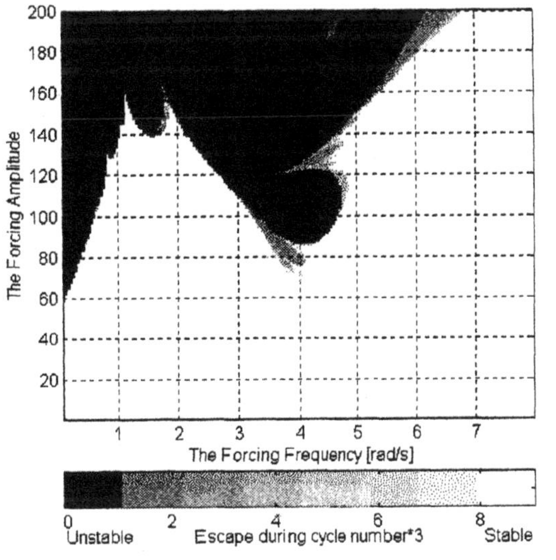

Fig. 11: The CSB diagram of the generalised model with $\lambda = 200$.

5. CONCLUSIONS

Fröhlich's model for Brain Waves (EEG) in its most general form has been analysed for the first time, and, under appropriate conditions, is found to exhibit deterministic chaos.

In particular, the response of the model to weak external, time-periodic stimuli has been investigated, thus extending existing analyses (Kaiser 1992) based on the restricted (*n,s*) model.

In the case of the generalised model, the bifurcation scenario is found to be somewhat different-particularly for small $\lambda \leq 10$ - as can most easily be seen by comparing the respective CSB diagrams.

The study of chaos in the brain is of practical importance, and methods of controlling such behaviour based on a control strategy (Ott, *et al.,* 1990) normally used in industrial systems have been suggested (Schiff, *et al.,* 1994). In this regard, the modelling and simulation of Fröhlich's Brain-Wave model described in this paper should facilitate the study of new control system strategies.

REFERENCES

Babloyantz, A., and A. Destexhe (1986). Low-dimensional chaos in an instance of epilepsy *Proc. Natl. Acad Sci.* **83**, pp. 3513-3517.

Fröhlich, H., (1977). Possibilities of long and short range electric interactions of biological systems *MIT Neuroscience Res. Prog. Bull.* **15**, pp 67-72.

Gallez, D., and A. Babloyantz (1991). Predictability of human EEG: a dynamical approach *Biol. Cybern* Vol. **64**, pp. 381-391, 1991; see also '*Chaos in Brain Function*', Basar E. (Ed), (1990). Springer-Verlag, Berlin and Heidelberg,.

Fröhlich, F., and G.J. Hyland (1995). *Fröhlich Coherence at the Mind-Brain Interface,'* in Scale in Conscious Experience: Is the Brain Too Important To Be Left to Specialists to Study?; King, J., and Pribram, K.H., (Eds), Lawrence Erlbaum Ass., Mahwah New Jersey; Hove UK.

Thompson, J.M.T., R.C.T. Rainey and M. S. Soliman (1990). Ship stability criteria based on chaotic transients from incursive fractals. *Phil. Trans. R. Soc. Lond. A.*, **332**, pp. 149-167.

Lewis, C.P., A. Ucar and S.R. Bishop (1997a). Dynamic behaviour of complex nonlinear systems., presented In MBD-MST'97 *1st Int. Symposium on Multi-Body Dynamic, Monitoring & Simulation Techniques'* Bradford, UK.

Lewis, C.P., A. Ucar and S.R. Bishop (1997b). Stability of Nonlinear Systems and the Effect of Time Delay Control. submitted to *Trans. of Institute of Measurement and Control.*

Kaiser, F. (1992). Biophysical models related to Fröhlich excitations', *Nanobiology,* **1**, pp. 149-161.

Ott, E., C. Grebogi and J.A. York (1990). Controlling Chaos. *Phys. Rev. Lett.*, **64** (11), pp. 1196-1199.

Schiff S.J., K Jerger, D.H. Duong, T. Chang, M. L. Spano and W.L. Ditto (1994). Controlling chaos in the brain. *Nature* **370** pp. 615-620.

ROBUST CONTROL OF A FLUIDIZED BED COMBUSTOR FOR SEWAGE SLUDGE

Andreas Wolff and Jan Lunze

Technical University Hamburg-Harburg
Institute of Control Engineering
D-21071 Hamburg
{wolff, lunze}@tu-harburg.d400.de

Abstract. A robust controller of a fluidized bed combustor is developed. The most important difficulty of this control task results from severe uncertainties of the dynamic input/output behaviour of the system. Therefore, methods for describing incompletely known systems and for analysing the robustness of the closed loop will be used. The results include the development of a control structure for fluidized bed combustors, the proof of the reachable control performance and the practical test of the controller.

Keywords. Robust Control, identification, industrial control, multivariable control, robust stability, robust performance

1. INTRODUCTION

Sewage sludge is incinerated in order to generate electrical energy and to reduce the volume of the sludge. Sewage sludge combustors can only be used industrially, if the gaseous emissions are guaranteed to be kept within the strict legal limits. Nowadays sewage sludge combustion plants with a bubbling fluidized bed are controlled by hand. In this paper it will be described how this control task can be solved by an automatic controller.

The control of sewage sludge combustion has not been investigated until now. The control methods used for coal fired fluidized bed combustors give only a few hints for the solution of this control task, because this kind of combustor differs considerably in the construction and therefore in the dynamic behaviour from sewage sludge combustors. For coal fired fluidized bed combustors good set point behaviour is desired, whereas for sewage sludge combustors good disturbance rejection is desired to compensate the disturbance caused by a varying sludge composition.

The severe difficulty in solving the control problem is the lack of knowledge about the dynamic behaviour of the process. The modelling of the chemical reactions in the bed area and the free board is still a subject of research (Werther, 1994). Therefore, to solve the control problem, methods are applied, which need only an approximate model of the combustor and take the model uncertainties into account. The plant is descibed by a very simple model and upper bounds of the model uncertainties which are identified by experiments. The uncertainties are taken into account during the controller design. Therefore the desired performance of the closed–loop system can also be guaranteed at the real plant, as long as the model errors are below the model uncertainty bounds.

The methods for robustness analysis can be found in (Lunze, 1988). They are based on the description of the model uncertainties by upper bounds which estimates the possible deviation between the input/output behaviour of the nominal model and the real plant. In the literature, this is called an unstructured description of the model uncertainties. This method is quite suitable for modelling this plant, because the uncertainties can not be associated directly to parameter uncertainties as demanded by methods described in (Ackermann, 1994).

This paper describes the control problem, the control structure and the design of the robust controller. The results are illustrated by experimental tests of the controller.

2. CONTROL TASK

Figure 1 shows the stationary fluidized bed combustor (FBC) operated by the Process Engineering Department of the Technical University of Hamburg-Harburg. The important process engineering features of the combustor are described in (Ogada, 1995). The unit combusts wet sewage sludge (with a water content of about 80 %) and propane in a bubbling fluidized bed.

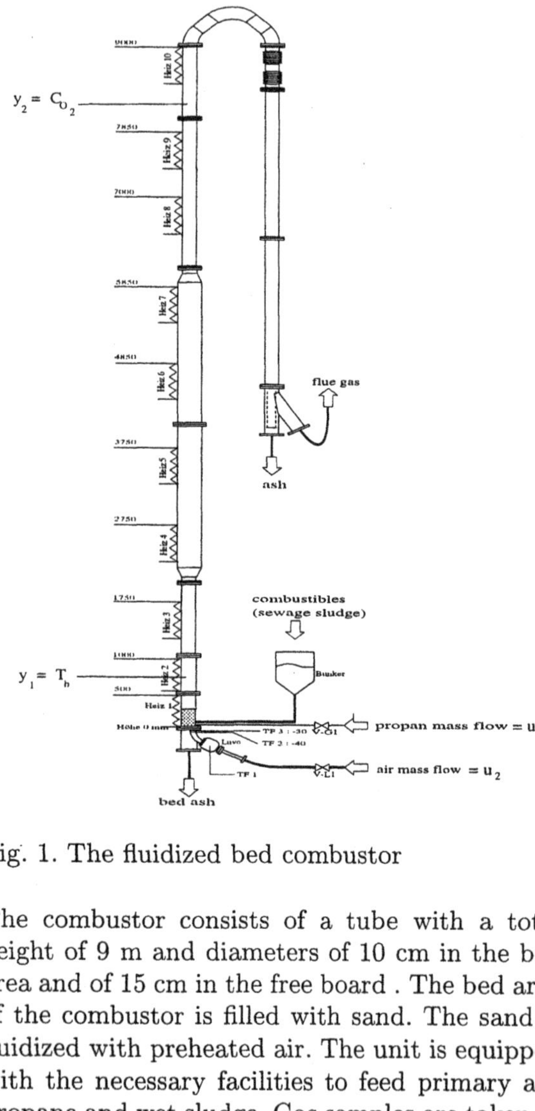

Fig. 1. The fluidized bed combustor

The combustor consists of a tube with a total height of 9 m and diameters of 10 cm in the bed area and of 15 cm in the free board . The bed area of the combustor is filled with sand. The sand is fluidized with preheated air. The unit is equipped with the necessary facilities to feed primary air, propane and wet sludge. Gas samples are taken at the top of the riser of the FBC, where the oxygen concentration C_{O_2} and the concentrations of CO_2, CO, NO_2 and SO_2 are measured. The combustor is heated by electrical heatings in the combustor walls to compensate the heat loss, which occurs due to the extreme height/diameter ratio of the combustor. Because of these heatings the combustor has a dynamical behaviour similar to industrially used combustors, which have a lower heat loss and, therefore , do not need heating zones.

The controller should influence the combustion process in a way such that the emissions are kept within the legal limits. The most important disturbances are the varying sludge composition and the time dependent water content of the sludge. It is prescribed by law that the average concentration of NO or CO over a day should be less than $200\frac{mg}{m^3}$ or $50\frac{mg}{m^3}$, respectively. The average value of CO over a half hour should be less than $100\frac{mg}{m^3}$ and 90% of the measured CO-peaks within 24 hours should be less than $150\frac{mg}{m^3}$ (Hansmann, 1991).

The control task is difficult for several reasons:

- The dynamic behaviour of the system is highly unknown. The current process engineering research results give only a few hints about the chemical reactions taking place in the combustor and about the hydrodynamics.
- Only rough on-line-measurements of the most important quantities are available. Therefore, the measurement performance limits the performance of the controller.
- The combustion process shows considerable fluctuations. The slow actuators can not compensate these fluctuations.

The practical constraints have two substantial consequences for solving the control task. First it is impossible to identify a sufficiently accurate model by means of the well known identification methods. Hence, model uncertainties have to be taken into account. Second, the structural properties of the plant have to be taken into account during the controller design.

3. MODEL STRUCTURE

The structure of the model is shown in Figure 2. It is determined by decomposing the furnace according to its physical structure taking the theoretical and experimental work of (Ogada, 1995) into consideration. The model consists of several subsystems which describe the heat balance of the bed area, the heat balance of the free board, the overall oxygen balance, the overall NO and CO balance, and the gas velocity.

Each subsystem has inputs and outputs. The inputs of the heat balance are the total power $P_{el,bed}$ of the electrical heating system of the bed area, the air mass flow $\dot{m}_l$ and its temperature T_{air}, the sludge mass flow $\dot{m}_s$ and the propane mass flow $\dot{m}_p$. The output is the resulting bed temperature T_{bed}. $\dot{m}_l$, $\dot{m}_p$, $\dot{m}_s$ are also inputs of the oxygen balance. The resulting output is the O_2 concentration C_{O_2}. The subsystem described by the heat balance of the free board has the input T_{bed} coming from the heat balance of the bed area. The other inputs are the total electrical power $P_{el,free}$ of the free board and the sludge mass flow $\dot{m}_s$. The resulting output is the average temperature of the free board T_{free}. C_{O_2}, T_{free}, $\dot{m}_s$ and gas velocity u are inputs of the NO and CO balance.

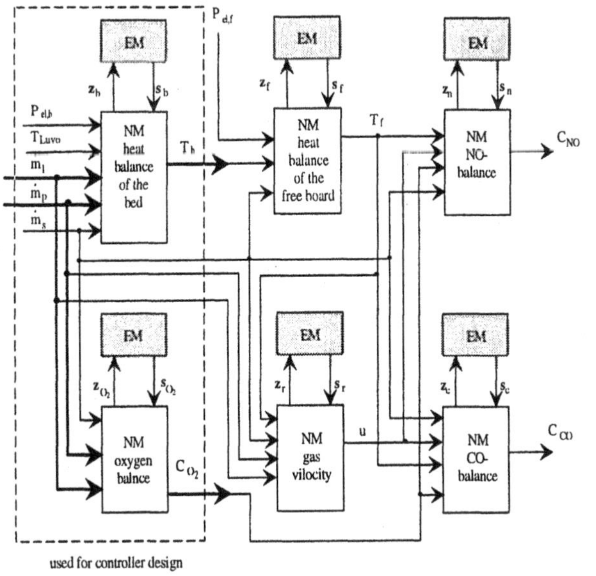

Fig. 2. Model of the fluidized bed combustor

The resulting outputs are the NO-concentration C_{NO} and the CO-concentration C_{CO}.

4. CONTROL STRUCTURE

The control aims described in Section 2 refer to the outputs C_{NO} and C_{CO}. The model shows that on the one hand the quantities depend very strongly on the input signals. On the other hand they depend on the oxygen concentration C_{O_2}, the bed temperature T_b, the freeboard temperature T_f and the gas velocity. The block diagram shown in Figure 2 shows that the coupling between the mentioned quantities are one-sided. The model consists of a serial–parallel connection of the subsystems.

These properties have a structural nature, i.e. they do not depend on the particular parameter values. This suggests to solve the control task by means of a cascaded controller, where the inner loop controls the bed temperature and the oxygen concentration and the outer loop prescribes the set points of the inner controller in a way that the gaseous emissons are kept within the legal limits despite of varying sludge composition. The control structure is shown in Figure 3. The choice of the suitable input signals will be described in Section 6.1.

The design of the entire controller will be divided into two steps. First, the controller will be designed in order to achieve good command response and second the outer controller will be designed to achieve good disturbance rejection, so that the entire controller fulfills the desired performance specification. The inner controller has to fulfil the following performance specifications:

(1) Stability
(2) Asymptotic tracking for stepwise set point changes and disturbances
(3) Dynamic specifications:
   - Settling time of the oxygen concentration below 5 minutes
   - Settling time of the bed temperature below 30 minutes
   - The two closed–loop systems should be decoupled.

In both design steps the model uncertainties bring about the major difficulties of the controller design. The performance specifications have to be met despite the model uncertainties. How the uncertainties are handled and the controller design is carried out, will be shown for the inner closed–loop system. Therefore, only the left part of the model will be further considered.

5. MODELLING

Because of considerable model uncertainties, each subsystem is described by a nominal model and an error model as shown in Figure 2. The models are connected via the signal $\mathbf{s}$ and $\mathbf{z}$. The models have the following form

$$\mathbf{y}(s) = \hat{\mathbf{G}}_{yu}(s)\mathbf{u}(s) + \mathbf{f}(s). \qquad (1)$$

The transfer function matrix $\hat{\mathbf{G}}_{yu}(s)$ describes the approximate I/O behaviour, in which the error $\mathbf{f}(s)$ appears. One way to describe the error is

$$\mathbf{f}(s) = \delta\mathbf{G}(s)\mathbf{u}(s) \qquad (2)$$

in which $\delta\mathbf{G}(s)$ is not known. It is assumed, that an upper bound $\mathbf{V}(s)$ of $\delta\mathbf{G}(s)$ is known so that

$$|\delta\mathbf{G}(s)| \leq \mathbf{V}(s) \qquad (3)$$

holds. In this relation $\mathbf{V}(s)$ is an element–by–element bound of $\delta\mathbf{G}(s)$. Equations (1) – (3) describe a set $\mathcal{G}$ of dynamic systems, to which the real system

$$\mathbf{y} = \mathbf{G}\mathbf{u} \qquad (4)$$

belongs:

$$\mathbf{G} \in \mathcal{G} = \{\hat{\mathbf{G}}_{yu} + \delta\mathbf{G} \;:\; |\delta\mathbf{G}(s)| \leq \mathbf{V}(s)\}. \qquad (5)$$

This kind of model will be used for the entire system as well as for all subsystems in which $\mathbf{u}$ und $\mathbf{y}$ are different signals for each subsystem.

For identifiction of the subsystems in Figure 2 N experiments are carried out. Each experiment yields a transfer function

$$g^i_{kl}(s) = \frac{k^i_{s,kl}}{T^i_{kl}s + 1} e^{-sT^i_{T,kl}} \quad (i = 1, 2, \ldots, N). \qquad (6)$$

The index i marks the number of the experiment, the indices k and l mark the elements (k, l) of the matrix $\hat{\mathbf{G}}_{yu}(s)$. $k^i_{s,kl}$ marks the static gain, T^i_{kl} the time constant and $T^i_{T,kl}$ the time delay. All determined transfer functions form a set

$$\mathbf{g}_{kl} = \left\{ g^1_{kl}(s), g^2_{kl}(s), \ldots, g^N_{kl}(s) \right\} \tag{7}$$

of transfer functions, for which a model (5) can be set up:

$$\mathbf{g}_{kl} \supseteq \tilde{\mathbf{g}}_{kl} = \left\{ \hat{g}_{kl} + \delta g_{kl}, |\delta g_{kl}| \leq v_{kl} \right\}. \tag{8}$$

The nominal model

$$\hat{g}_{kl}(s) = \frac{\hat{k}_{kl}}{\hat{T}_{kl}s + 1} e^{-p\hat{T}_{T,kl}} \tag{9}$$

has the parameters

$$\hat{k}_{s,kl} = \frac{1}{2}\left\{ \max_i(k^i_{s,kl}) + \min_i(k^i_{s,kl}) \right\} \tag{10}$$

$$\hat{T}_{kl} = \frac{1}{2}\left\{ \max_i(T^i_{kl}) + \min_i(T^i_{kl}) \right\} \tag{11}$$

$$\hat{T}_{T,kl} = \frac{1}{2}\left\{ \max_i(T^i_{T,kl}) + \min_i(T^i_{T,kl}) \right\}. \tag{12}$$

The upper error bound is given by

$$v_{kl}(s) = \max_i |g^i_{kl}(s) - \hat{g}_{kl}(s)|. \tag{13}$$

such that

$$|g^i_{kl}(s) - \hat{g}_{kl}(s)| \leq v_{kl}(s) \tag{14}$$

holds. Using Eqns. (9) – (13) one gets the following nominal model for the heat balance and oxygen balance

$$\begin{pmatrix} T_b \\ C_{O_2} \end{pmatrix} = \underbrace{\begin{pmatrix} \hat{g}_{11} & \hat{g}_{12} & \hat{g}_{13} & \hat{g}_{14} & \hat{g}_{15} \\ 0 & \hat{g}_{22} & \hat{g}_{23} & \hat{g}_{24} & 0 \end{pmatrix}}_{\hat{\mathbf{G}}_{yu}} \begin{pmatrix} P_{el,b} \\ \dot{m}_p \\ \dot{m}_s \\ \dot{m}_l \\ T_{Luvo} \end{pmatrix} + \mathbf{f}. \tag{15}$$

Because the error $\mathbf{f}$ depends on five input signals, the upper bound of the error $\mathbf{V}(s)$ is a $(2,5)$–matrix. The bound can not be described analytically, but it is given numerically for selected frequency points $s = j\omega$.

6. CONTROLLER DESIGN

6.1 *Selection of the Controller Inputs*

Suitable inputs have to be chosen for the design of the oxygen and bed temperature controller. This is done by examining the normalized static transition matrix (Korn, 1991)

$$\mathbf{K}_{ns} = \frac{1}{k^2_{s,max}} \begin{pmatrix} [\underline{k}^2_{s,11}, \overline{k}^2_{s,11}] & \cdots & [\underline{k}^2_{s,1m}, \overline{k}^2_{s,1m}] \\ \vdots & \ddots & \vdots \\ [\underline{k}^2_{s,r1}, \overline{k}^2_{s,r1}] & \cdots & [\underline{k}^2_{s,rm}, \overline{k}^2_{s,rm}] \end{pmatrix} \tag{16}$$

of the system (15), in which

$$k^2_{s,max} = \max_{i,j} (k^2_{s,ij}) \tag{17}$$

$$k_{s,ij} = \lim_{s \to 0} g_{ij}(s) \tag{18}$$

hold. Since the system is incompletely known, $\mathbf{K}_{ns}$ is an interval matrix. For $k_{s,ij}$, the intervals $[\underline{k}_{s,ij}, \overline{k}_{s,ij}]$ are given, which take the model uncertainties into consideration. As a suitable input for a specified output j the input i with the highest normalized static reinforcement is chosen.

Therefore, the propane mass flow $\dot{m}_p$ is chosen to control T_b and $\dot{m}_l$ to control O_2. The resulting control structure is shown in Figure 3. Hence,

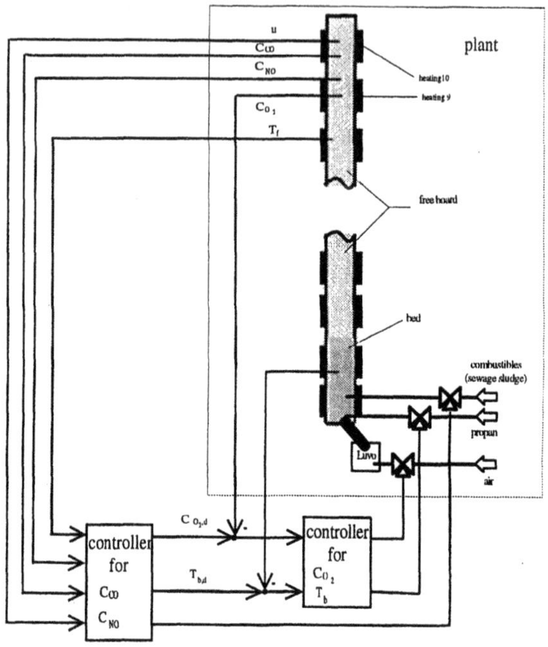

Fig. 3. Control struktur

the nominal model of the plant is diminished to

$$\begin{pmatrix} T_b \\ C_{O_2} \end{pmatrix} = \underbrace{\begin{pmatrix} \hat{g}_{12} & \hat{g}_{14} \\ \hat{g}_{22} & \hat{g}_{24} \end{pmatrix}}_{\hat{\mathbf{G}}_B} \begin{pmatrix} \dot{m}_p \\ \dot{m}_l \end{pmatrix} + \mathbf{f}. \tag{19}$$

6.2 *Controller Structure and Parameter*

Now the controller

$$\begin{pmatrix} \dot{m}_p \\ \dot{m}_l \end{pmatrix} = \mathbf{K}_c(s)\mathbf{K}(s) \begin{pmatrix} T_{b,d} - T_b \\ C_{O_2,d} - C_{O_2} \end{pmatrix} \tag{20}$$

has to be designed. The reduced normalized static transition matrix $\mathbf{K}_{ns,B}$ indicates considerable cross-coupling, if one compares the diagonal entries with the other elements. In order to reach a good controller performance, the decoupling matrix

$$\mathbf{K}_c = \begin{pmatrix} f & -\dfrac{\hat{k}_{s,14}}{\hat{k}_{s,12}} f \\ -\dfrac{\hat{k}_{s,22}}{\hat{k}_{s,24}} f & f \end{pmatrix} \tag{21}$$

with

$$f = \frac{1}{1 - \frac{\hat{k}_{s,14}\hat{k}_{s,22}}{\hat{k}_{s,12}\hat{k}_{s,24}}}. \quad (22)$$

was chosen according to (Föllinger, 1992) and afterwards the bed temperature and oxygen controller

$$\mathbf{K}(s) = diag\left(k_{p,ii}\left(1 + \frac{1}{T_{I,ii}s}\right)\right) \quad (23)$$

was designed.

6.3 *Robustness Analysis*

The controller (20) was designed with the help of the nominal model in such a way that the desired performance was reached. Due to the great model uncertainties, it has to be examined if the controller satisfies the performance specifications for all other systems $\mathbf{G} \in \mathcal{G}$.

The robustness analysis is carried out with the method described in (Lunze, 1988), which is based on a element–by–element estimation of the model uncertainties. This method includes the stability analysis as well as the analysis of the I/O-performance of the closed–loop system as follows. With the system (1) and the controller

$$\mathbf{u} = -\mathbf{K}(\mathbf{y} - \mathbf{w}) \quad (24)$$

one gets the following relation of the closed–loop system

$$\mathbf{y} = (\mathbf{I} + \hat{\mathbf{G}}\mathbf{K})^{-1}\hat{\mathbf{G}}\mathbf{K}\mathbf{w} + (\mathbf{I} + \hat{\mathbf{G}}\mathbf{K})^{-1}\mathbf{f} \quad (25)$$
$$\mathbf{u} = (\mathbf{I} + \mathbf{K}\hat{\mathbf{G}})^{-1}\mathbf{K}\mathbf{w} - (\mathbf{I} + \mathbf{K}\hat{\mathbf{G}})^{-1}\mathbf{K}\mathbf{f}, \quad (26)$$

and in short

$$\mathbf{y} = \mathbf{G}_{yw}\mathbf{w} + \mathbf{G}_{yf}\mathbf{f} \quad (27)$$
$$\mathbf{u} = \mathbf{G}_{uw}\mathbf{w} + \mathbf{G}_{uf}\mathbf{f} \quad (28)$$

For $\mathbf{f} \equiv \mathbf{0}$ the approximate model of the closed–loop system

$$\hat{\mathbf{y}} = \mathbf{G}_{yw}\mathbf{w} \quad (29)$$

is obtained, which equals the controlled nominal system. Therefore, it is stable and fulfills the performance specifications.

An estimate how far the original output $\mathbf{y}$ may deviate from the approximate output $\hat{\mathbf{y}}$ follows from the Eqns. (1), (3) and (27), (28). Eqns. (27) and (29) yield

$$|\mathbf{y} - \hat{\mathbf{y}}| \le |\mathbf{G}_{yf}||\mathbf{f}| \quad (30)$$

and together with (1) and (3)

$$|\mathbf{f}| \le \mathbf{V}|\mathbf{u}|. \quad (31)$$

Using (28) the following can be proven (Lunze, 1988):

Lemma 1. If the approximate model (29) and the error model (3) are stable, then the following assertions hold:

- The closed–loop system is robustly stable, if

  $$\lambda_{max}(|\mathbf{G}_{uf}(j\omega)|\,\mathbf{V}(j\omega)) < 1 \quad \text{for all} \quad \omega \quad (32)$$

  is valid.
- Under the condition (32) the closed–loop system is described by

  $$\mathbf{y} = (\mathbf{G}_{yw} + \delta\mathbf{G}_{yw})\mathbf{w} \quad (33)$$

  with

  $$|\delta\mathbf{G}_{yw}| \le |\mathbf{G}_{yf}|\,\mathbf{V}\,(\mathbf{I} - |\mathbf{G}_{uf}|\,\mathbf{V})^{-1}\;|\mathbf{G}_{uw}|.$$

The designed controller fulfills the assumptions of this lemma, because the nominal system is stabil. As shown in Figure 4, the condition (32) is fulfilled for $k^* = 1$ and hence, the closed–loop is system is stable.

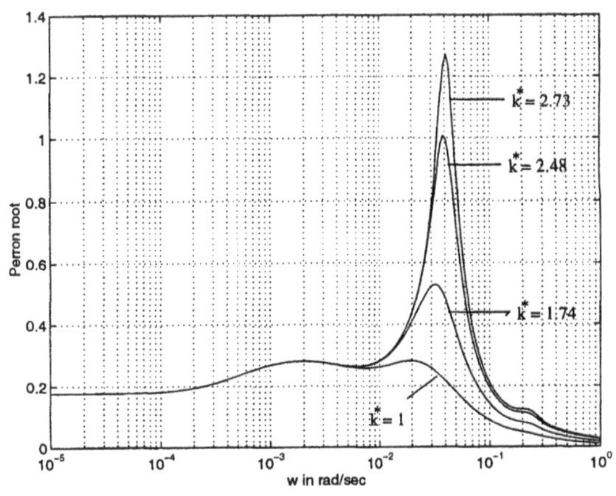

Fig. 4. Examination of the robust stability of the improved controller

Taking Eqns. (33) and (1) into account, the bounds of the magnitude shown in Figure 5 are obtained.

For the used error bounds the frequency response is described by this bounds according to Lemma 1. The upper bound shows that the cross–coupling and the coupling $w_2 \mapsto y_1$ is low, the plant output $y_1 = T_b$ shows no peak and $y_2 = C_{o_2}$ only a small peak. The settling times are for the oxygen controller approximatelly 5 minutes and for the bed temperature controller approximatelly 15 minutes.

The model and the controller were found by searching for a model with sufficient accuracy with the help of the robustness analysis. First the described method starts with a very coarse model to test the first controller. The robustness analysis shows that

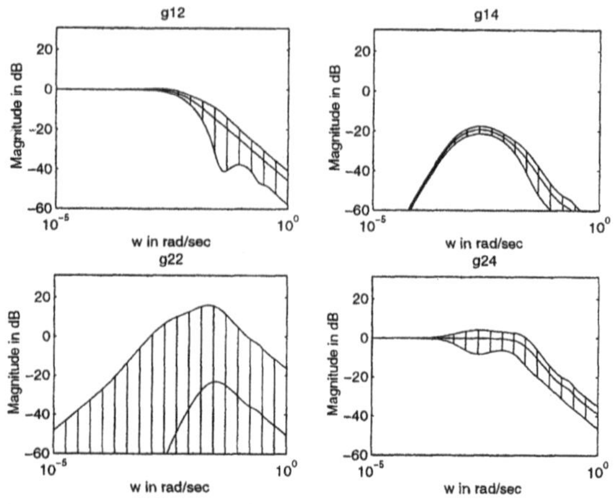

Fig. 5. Command step response of the improved controller

the model has to be improved especially at low frequencies. At those frequencies the model was reidentified and the controller was redesigned as described above. For more details see (Wolff, 1996).

7. EXPERIMENTAL VERIFICATION

The experimental results are shown in Figure 6, namely the reaction of the combustor due to a change in the oxygen concentration. NO- and CO-

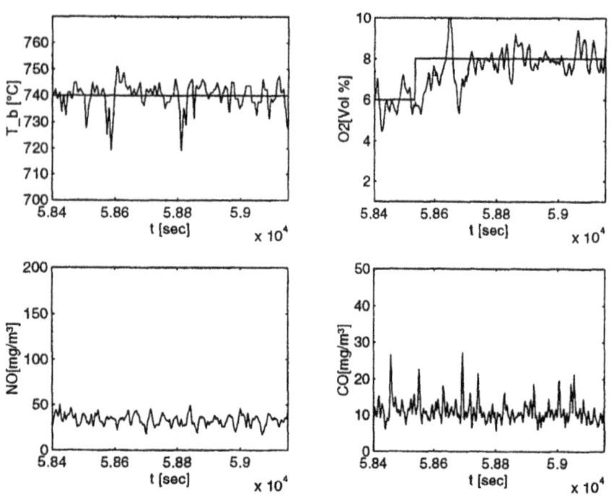

Fig. 6. Behaviour of the closed–loop system with the improved controller due to change in the oxygen command signal

concentrations are quite below the legal limits of $200\frac{\mathrm{mg}}{\mathrm{m}^3}$ NO and 50 $\frac{\mathrm{mg}}{\mathrm{m}^3}$ CO .
The minimal reachable settling time is determined by the time delay. Up to $k^* = 2.48$ the controller reinforcement can be increased without loosing the robust stability (Figure 4). For achieving an even shorter settling time, the time delay of the oxygen measurement system has to be reduced, for example, by using an optical measurement system. The time constant of the actuators used for industrial combustors are in the order of several minutes and this gives also a lower bound of the settling time. In order to react faster on disturbances of O_2– and CO-concentrations, faster measurement systems are necessary. One possibility is to use a faster CO–measurement system which detects CO–peak directly above the bed area and then to inject oxygen in the free board as used in refuse incineration plants.

8. CONCLUSION

A controller for a sewage sludge combustor with a bubbling bed was designed and tested at a pilot-plant. Because of the complexity of this combustor no sufficiently accurate model of the plant could be derived. Therefore, with the help of the robustness analysis a model with sufficient accuracy was searched for. First, the described method starts with very coarse model to test the first controller. The robustness analysis shows that the model has to be improved especially at low frequencies. The designed controller reaches good experimental results.

Acknowledgement

The experiment have been carried out at the pilot-plant of the department of process engineering (Prof. Dr. Werther) at the Technical University of Hamburg-Harburg. The project was supported by the Deutsche Forschungsgemeinschaft within the Sonderforschungbereich 238 "In-situ measuring techniques and dynamic modelling of multiphase flow systems ".

9. REFERENCES

Ackermann, J. u.a. (1994). *Robuste Regelung.* Springer–Verlag. Berlin.

Föllinger, O. (1992). *Regelungstechnik.* Hüthig–Verlag. Heidelberg.

Hansmann, K. (1991). Bundes-Immissionschutzgesetz und ergänzende Vorschriften.

Korn, U.; Jumar, U. (1991). *PI-Mehrgrößenregler.* R. Oldenbourg Verlag. München.

Lunze, J. (1988). *Robust Multivariable Feedback Control.* Prentice–Hall. London.

Ogada, T. (1995). Emissions and Combustion Characteristics of Wet Sewage Sludge in a Bubbling Fluidized Bed. PhD thesis. TU Hamburg–Harburg.

Werther, J.; Ogada T.; Phillippek Ch. (1994). Combustion of municipal sewage sludge in the fluidized bed. In: *Proc. 2nd Int. Interfluid Sympow Fluidized Bed Combustion.* Nagoya. pp. 9–36.

Wolff, A.; Lunze, J. (1996). Robuste Regelung einer Wirbelschicht für Klärschlamm. **at**-*Automatisierungstechnik* **11**, 522–532.

PARAMETER ESTIMATION OF US MACROECONOMIC ALTERNATIVE MODELS

Jan Štecha, Kateřina Satoriová , Osvald Vašíček

*Jan Štecha, Department of Automatic Control,
Faculty of Electrical Engineering, Czech Technical University of Prague,
Charles Square 13, 121 35 Prague 2
E-mail: stecha@control.felk.cvut.cz*

*Kateřina Satoriová, Department of Economics, Management and Academic Sciences,
Faculty of Electrical Engineering, Czech Technical University of Prague,
Zikova 2, 160 00 Prague 6
E-mail: satoriov@lab.felk.cvut.cz*

*Osvald Vašíček, Department of applied Mathematics,
Faculty of Economics and Administration, Masaryk University,
Zelný trh 2/3, 65790 Brno
E-mail: osvald@merkur.econ.muni.cz*

Abstract: Simple model of US macroeconomy is considered in this paper. Extended Kalman filter is used for state and parameter estimation of this simple macroeconomic model. Smoothing procedure is also used to obtain estimates conditioned by whole set of data. Alternative models of parameter development are used and their probability is evaluated based on the given data. Some interesting results are obtained.

Keywords: Macroeconomic models, Kalman filter, Smoothing, Alternative models, Bayesian approach.

1. INTRODUCTION

It is great challenge to model macroeconomic phenomena. Having good macroeconomic model it is possible to explain some phenomena and at the same time make forecasts for the future development of some important time series.

Simple macroeconomic model of American economy is described in (Pindyck, Rubinfeld, 1990). Our approach uses the structure of this model. Alternative models of parameter development is the contribution of the paper.

Extended Kalman filter with smoothing is used for estimation of state and parameter development. At the same time probabilities of alternative models are evaluated based on given data.

The article is organized in the following way. In the next section simple macroeconomic model is described. Extended Kalman filter and smoothing procedure are explained next. Then follows the probability of alternative models computation. Simulation results of state and parameter estimation and probability of alternative models are presented in the final section.

2. USA MACROECONOMIC MODEL

Simple macroeconomic model of American economy is described in (Pindyck, Rubinfeld, 1990). Such model considers three-sector economy and so gross national product $Y(t)$ equals sum of consumption of households $C(t)$ investment expenditures $I(t)$ and government expenditures $G(t)$.

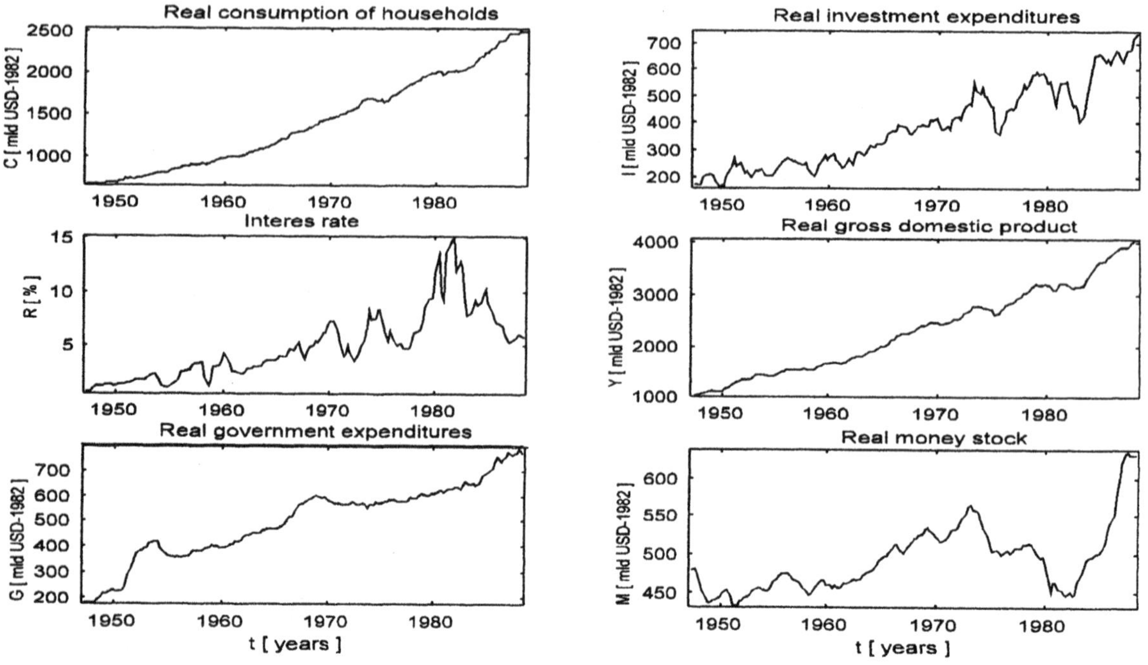

Figure 1: Data of US economy

So

$$Y(t) = C(t) + I(t) + G(t).$$

Time unit is the quarter of the year. Equation for consumption function

$$C(t) = c_0 + c_1 Y(t) + c_2 C(t-1)$$

expresses that consumption $C(t)$ is proportional to gross domestic product $Y(t)$ and consumption in previous quarter of the year $C(t)$.

Equation for investment function

$$I(t) = i_0 + i_1\left[Y(t-1) - Y(t-2)\right] + i_2 Y(t) + \\ + i_3 R(t-4)$$

expresses that investment expenditures $I(t)$ is proportional to gross domestic product $Y(t)$ and first difference of gross domestic products in previous quarter of the year $[Y(t\text{-}1)\text{-}Y(t\text{-}2)]$ and is also proportional to interest rate a year ago $R(t-4)$.

Equation for interest rate

$$R(t) = r_0 + r_1 Y(t) + r_2\left[Y(t) - Y(t-1)\right] + \\ + r_3\left[M(t) - M(t-1)\right] + r_4\left[R(t-1) + R(t-2)\right]$$

expresses that interest rate $R(t)$ is proportional to gross domestic product $Y(t)$ created at the same quarter of the year and also to the first difference of domestic products in the current and previous quarter of the year $[Y(t)\text{-}Y(t\text{-}1)]$, further to the first difference of money supply in the current and previous quarter of the year $\left[M(t) - M(t-1)\right]$ and finally to the sum of interest rates in two previous quarters of the year $\left[R(t-1) + R(t-2)\right]$.

Input variables of the model are real government expenditures $G(t)$ and real money stocks $M(t)$. Output variables of the model are real gross domestic product $Y(t)$, real consumption of households $C(t)$, real investment expenditures $I(t)$ and interest rate $R(t)$.

Unknown parameters c_0, c_1, c_2, i_0, i_1, i_2, i_3, r_0, r_1, r_2, r_3, r_4 in previous equations are subject of identification. Parameters with index $_0$ are considered to be constant. The rest of parameters can be time dependent.

For estimation of parameters of macroeconomic model time series of data of US economy was taken beginning from the first quarter of the year 1947 to the first quarter of the year 1988. The time series of macroeconomic data was taken from (Pindyck, Rubinfeld, 1990). The time development of the time series is shown in Fig. 1.

3. EXTENDED KALMAN FILTER AND SMOOTHING

The problem of the state estimation is possible to solve for non-linear system. The state equations of this system are:

$$\mathbf{x}(t+1) = \mathbf{f}(\mathbf{x}(t), \mathbf{u}(t), t) + \mathbf{v}(t)$$
$$\mathbf{y}(t) = \mathbf{g}(\mathbf{x}(t), \mathbf{u}(t), t) + \mathbf{e}(t),$$

where the non-linear function $\mathbf{f}$ and $\mathbf{g}$ have finite first derivation with respect to the state $\mathbf{x}(t)$. The stochastic factors $\mathbf{v}(t)$ and $\mathbf{e}(t)$ are stationary and Gauss processes with zero means and with covariance matrices $\mathbf{R}_v$ and $\mathbf{R}_e$. Also initial state $\mathbf{x}(t_0)$ is the stochastic quantity with the mean μ_{x0} and covariance $\mathbf{R}_{x0}$.

Extended Kalman filter is used for state estimation of nonlinear system. The estimation has two steps: filtration and prediction.

State filtration equals to:

$$\hat{\mathbf{x}}(t|t) = \hat{\mathbf{x}}(t|t-1) + \mathbf{R}_{xx}\mathbf{C}^T(\mathbf{C}\mathbf{R}_{xx}\mathbf{C} + \mathbf{R}_e)^{-1} \left[\ \mathbf{y}(t) - \mathbf{g}(\hat{\mathbf{x}}(t|t-1), \mathbf{u}(t), t)\ \right]$$

$$\mathbf{R}_{xx}(t|t) = \mathbf{R}_{xx} - \mathbf{R}_{xx}\mathbf{C}^T(\mathbf{C}\mathbf{R}_{xx}\mathbf{C}^T + \mathbf{R}_e)^{-1}\mathbf{C}\mathbf{R}_{xx}$$

where $\hat{\mathbf{x}}(t|\tau)$ is the mean of the state in time t based on data till time τ, $\mathbf{R}_{xx} = \mathbf{R}_{xx}(t|t-1)$ is the covariance matrix of data in time t based on data till time $(t-1)$ and $\mathbf{C} = \mathbf{C}(t)$ is a matrix obtained by the linearization of output equations in the point $\hat{\mathbf{x}}(t|t-1)$:

$$\mathbf{C}(t) = \frac{\partial\, \mathbf{g}(\mathbf{x}(t), \mathbf{u}(t), t)}{\partial\, \mathbf{x}(t)}$$

State prediction equals to:

$$\hat{\mathbf{x}}(t+1|t) = \mathbf{f}(\hat{\mathbf{x}}(t|t), \mathbf{u}(t), t)$$

$$\mathbf{R}_{xx}(t+1|t) = \mathbf{A}\mathbf{R}_{xx}(t|t)\mathbf{A}^T + \mathbf{R}_v$$

where $\mathbf{A} = \mathbf{A}(t)$ is obtained by the linearization of state equations in the point $\hat{\mathbf{x}}(t|t)$:

$$\mathbf{A}(t) = \frac{\partial\, \mathbf{f}(\mathbf{x}(t), \mathbf{u}(t), t)}{\partial\, \mathbf{x}(t)}$$

The iterative algorithm is started with the initial condition

$$\hat{\mathbf{x}}(t_0|t_0-1) = \mu_{x0}$$

$$\mathbf{R}_{xx}(t_0|t_0-1) = \mathbf{R}_{x0}$$

Smoothing is state estimation based on the whole set of data. Smoothing is realized by forward and backward filter run. Forward run is realized by usual extended Kalman filter. Backward filter uses predictive and filtered estimates. The backward filter for smoothing is

$$\hat{\mathbf{x}}(t|T_f) = \hat{\mathbf{x}}(t|t) + \mathbf{F}(t)\left[\hat{\mathbf{x}}(t+1|T_f) - \hat{\mathbf{x}}(t+1|t)\right]$$

$$\mathbf{R}_{xx}(t|T_f) = \mathbf{R}_{xx}(t|t) - \mathbf{F}(t)\left[\mathbf{R}_{xx}(t+1|t) - \mathbf{R}_{xx}(t+1|T_f)\right]\mathbf{F}^T(t)$$

where $\hat{\mathbf{x}}(t|T_f)$, $\mathbf{R}_{xx}(t|T_f)$ are the smoothed values of the mean and covariance matrix and the gain of the backward filter equals to:

$$\mathbf{F}(t) = \mathbf{R}_{xx}(t|t)\mathbf{A}^T\left[\mathbf{R}_{xx}(t+1|t)\right]^{-1}$$

Initial conditions for the backward filter run $\hat{\mathbf{x}}(T_f|T_f)$, $\mathbf{R}_{xx}(T_f|T_f)$ are equal to the final conditions of the forward filter. Note that for backward filter run no data are used; all the information necessary is contained in the statistics of the predictive and filtered estimates.

4. ALTERNATIVE MODELS PROBABILITIES

In some cases more then one model structure is considered. Let's have several hypotheses $H^{(i)}$, i=1,2,... about model structures. We assign the probability to each hypothesis.

Suppose that in the time t of the algorithm we know the probability of the hypothesis $H^{(i)}$ based on data D_{t-1} (data till time $(t-1)$):

$$p(H^{(i)}(t)|D_{t-1})$$

The probability of the hypotheses updating is calculated by the following way:

$$p(H^{(i)}(t)|D_t) = \frac{1}{\alpha} p(\mathbf{y}(t)|H^{(i)}(t), D_{t-1})\, p(H^{(i)}(t)|D_{t-1})$$

where

$$p(\mathbf{y}(t)|H^{(i)}(t), D_{t-1}) = \frac{1}{\sqrt{2\pi\, \det(\mathbf{R}_{yy}^{(i)})}} e^{-\frac{1}{2}(\mathbf{y}(t)-\hat{\mathbf{y}}^{(i)})^T(\mathbf{R}_{yy}^{(i)})^{-1}(\mathbf{y}(t)-\hat{\mathbf{y}}^{(i)})}$$

$$\hat{\mathbf{y}}^{(i)} = \hat{\mathbf{y}}^{(i)}(t|t-1) = \mathbf{C}^{(i)}\hat{\mathbf{x}}^{(i)}(t|t-1)$$

$$\mathbf{R}_{yy}^{(i)} = \mathbf{R}_{yy}^{(i)}(t|t-1) = \mathbf{C}^{(i)}\mathbf{R}_{xx}^{(i)}(t|t-1)(\mathbf{C}^{(i)})^T$$

This is the filtration step in hypotheses probability updating. In prediction step we want to find probability in the time $t+1$ based on the same set of data. This step is usual based on model of hypotheses development. Because such model is usually not known we can use the following formula:

$$p(H^{(i)}(t+1)|D_t) = \lambda\, p(H^{(i)}(t)|D_t) + (1-\lambda)\, p(H^{(i)})$$

where $p(H^{(i)})$ is a prior probability of the hypotheses (with no data).

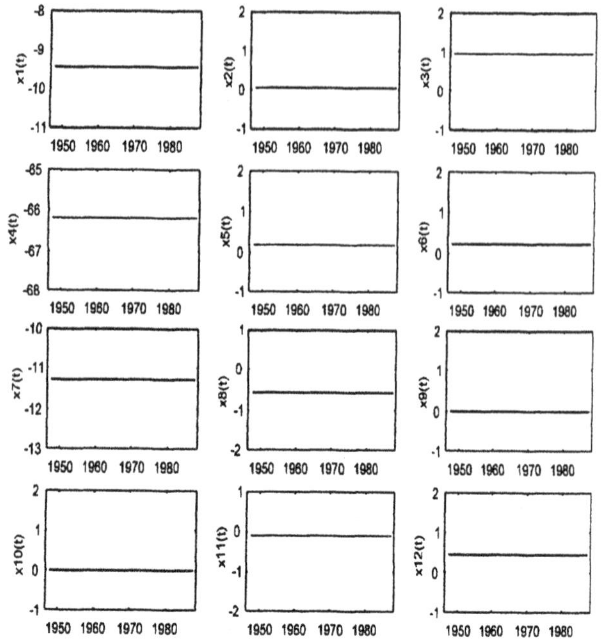

Figure 2: Time development of the parameters

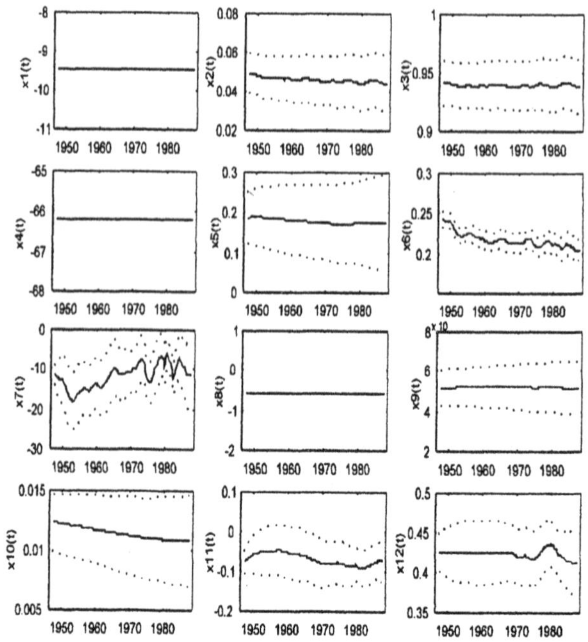

Figure 4: Time development of the parameters

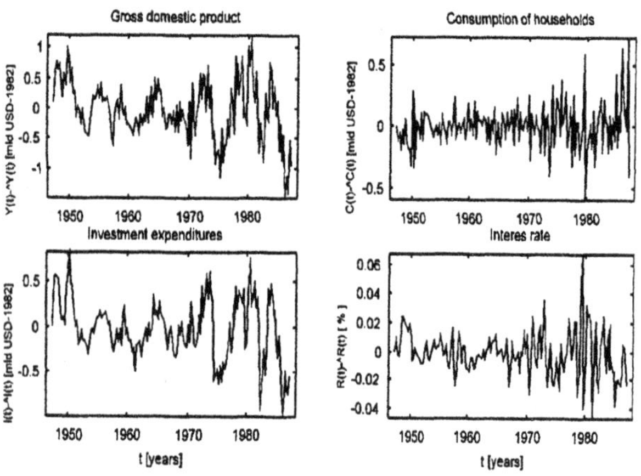

Figure 3: Outputs errors of constant parameters

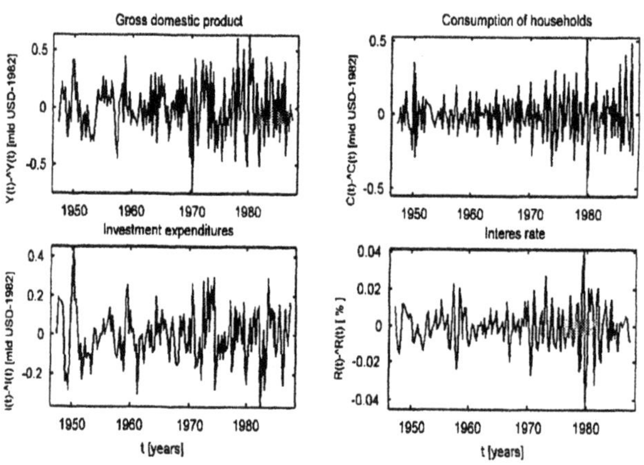

Figure 5: Outputs errors of variable parameters

5. REALIZATION OF THE MODEL

The equation of macroeconomic models must be transformed to the form of state equation of the nonlinear discrete time stochastic system

$$\mathbf{x}(t+1) = \mathbf{f}(\mathbf{x}(t), \mathbf{u}(t), t) + \mathbf{v}(t),$$
$$\mathbf{y}(t) = \mathbf{g}(\mathbf{x}(t), \mathbf{u}(t), t) + \mathbf{e}(t),$$

where output vector $\mathbf{y}(t) = [Y(t)\ C(t)\ I(t)\ R(t)]^{\mathrm{T}}$ and input vector $\mathbf{u}(t) = [G(t)\ M(t)]^{\mathrm{T}}$.

The states of the system are formed from the unknown parameters $c_i(t)$, $i_i(t)$, $r_i(t)$ and also from the shifted outputs. Formal state equation of parameter development is a random walk equation $x_i(t+1) = x_i(t) + v_i(t)$ and the noise $v_i(t)$ is introduced to obtain time dependent parameter estimates.

6. SIMULATION RESULTS

In (Pindyck, Rubinfeld, 1990) are given estimated parameters of the macroeconomic model obtained by minimization of sum of square errors between model and reality.

The constant parameters estimates are

$c_0 = -9{,}454$	$i_0 = -66{,}195$	$r_0 = -0{,}5561$
$c_1 = 0{,}0541$	$i_1 = 0{,}1684$	$r_1 = 0{,}00051$
$c_2 = 0{,}9260$	$i_2 = 0{,}2181$	$r_2 = 0{,}0135$
	$i_3 = -11{,}256$	$r_3 = -0{,}0853$
		$r_4 = 0{,}4259.$

The same estimates are obtained by Kalman filter considering zero noise variances in state equations of parameter development. The time development of the parameters and of the outputs errors is shown in Fig. 2 and Fig. 3.

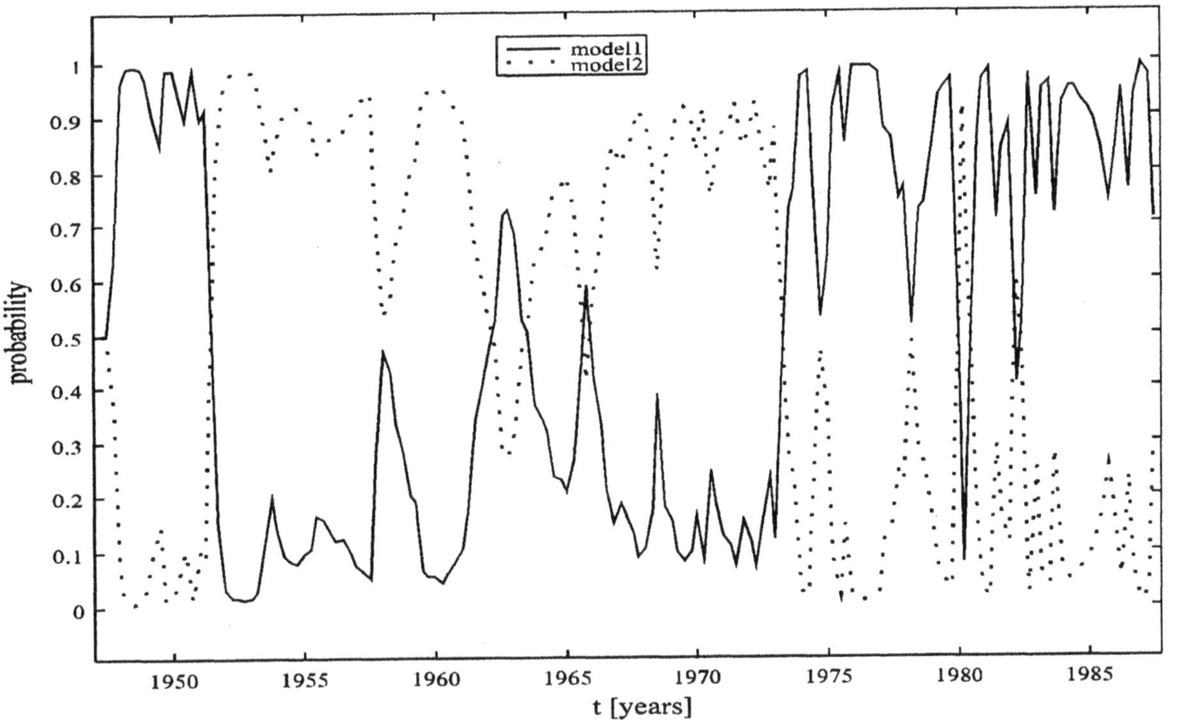

Figure 6: Alternative models probabilities

Better estimates (with lover sum of mean square errors) are obtained with time dependent parameters - see Fig. 4 and Fig. 5.

To compare both approaches - constant and time dependent parameters - alternative models approach was used. One model is considered with zero noise covariance in state equation of parameter development and the second model is considered with nonzero noise covariance which enable time dependent parameter estimates. At the same time the probability of these two alternative models were computed. The results are presented in Fig. 6. Solid line is the probability of the first model with time varying parameters, while dotted line is the probability of the model 2 with constant parameters. Prior probabilities $p(H^{(1)}) = p(H^{(2)}) = 0{,}5$ were used to have no model preferences.

Model 2 with constant parameters has higher probability in time period from the year 1952 to 1974, it is the time period when US economy developed with no great disturbances. The only exception is the year 1963 - the year of president Kennedy assassination.

Model 1 with time varying parameters has higher probability in time period 1945 to 1952 - the time period after the second world war and during Korean war. Model 1 has higher probability also in time period 1975 to 1988 - the period of two recessions in 1975 and 1982.

Based on the previous results the change (jumps) of parameters is enabled only in the time of the biggest changes in economy - in the year 1950, 1975 and 1982. The time development of parameters of the macroeconomics model and of the outputs errors is shown in Fig. 7 and Fig. 8.

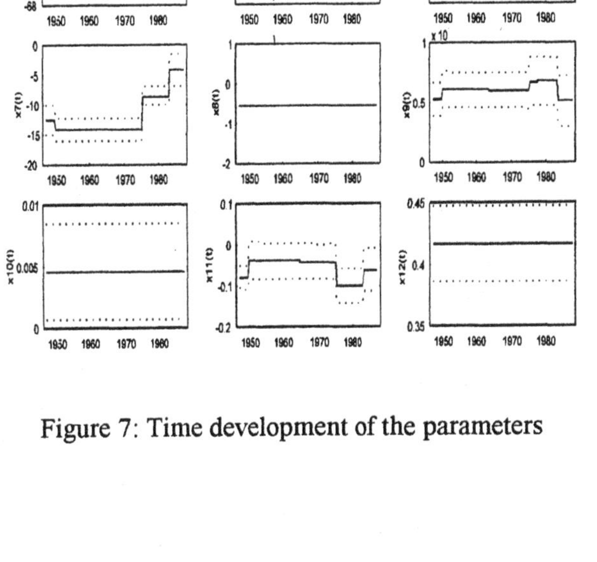

Figure 7: Time development of the parameters

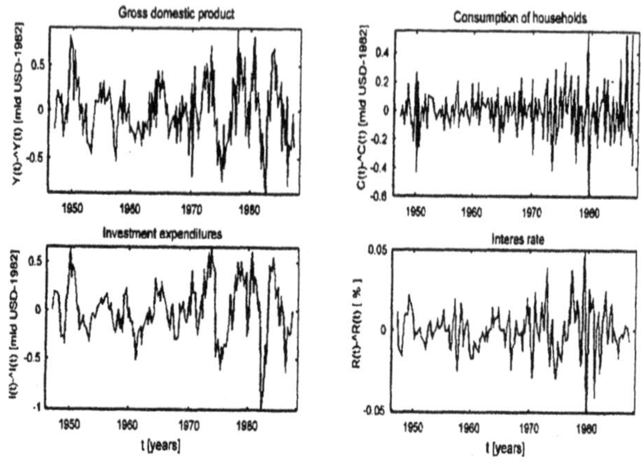

Figure 8: Outputs errors

7. CONCLUSION

Kalman filtering is used for state and parameter estimation of simple macroeconomic model based on real data. Interesting results were obtained using alternative models. The data show the shocks in the economy.

Our effort is now aimed on models of Czech economy where only short data set is available and the economy is still in the transition state. So the problem of modeling and estimation is more difficult in such case.

REFERENCES

Havlena, V. and J. Štecha (1993). Fault Detection Based on Alternative Models of Parameter Development. Proceedings of the Int. *Workshop on Applied Automatic Control*, WAAC'93, pp. 44-47, Prague.

Lewis, F. L. (1986). *Optimal Estimation*. J. Wiley, New York.

Havlena, V., J. Štecha and F.J. Kraus (1996). Decision and Approximation Based Algorithms for Identification with Alternative Models. Proceedings of *IFAC'96 International Conference* in San Francisco.

Peterka, V. (1981). Bayesian Approach to System Identification. In P. Eykhoff (Ed.), *Trends and Progress in System Identification*. Pergamon Press, Oxford.

Pindyck, R.S. and D.L. Rubinfeld (1990). *Econometric Models and Economic Forecasts*. McGRAW - HILL, Inc., New York.

Satoriová, K. (1996). *Utilization of Alternative Models for Identification of Systems in Economy*. MSc thesis, Czech Technical University of Prague, FEL, Department of Automatic Control.

Štecha, J., V. Havlena and O. Vašíček (1993). Smoothing with Alternative Models of Parameter Development. Proceedings of the *12th IASTED Int. Conference "Modeling, Identification and Control"*, Innsbruck, Austria, pp. 179-182.

Štecha, J. and O. Vašíček (1993). *Odhad měnicích se parametrů ekonomicko-matematických modelů bayesovskými postupy*. (Parameter Estimation of the Economy Models by Bayessian Approaches). Acta Oeconomica Pragensia, pp. 55-70.

Štecha, J. and V. Havlena (1994). Parameter Tracking with Alternative Noise Models. Preprints of the *European IEEE CMP'94 Workshop on Computer - Intensive Methods in Control and Signal Processing*, Prague, September, pp. 171-176.

Štecha, J. and V. Havlena (1995). Smoothing in Simultaneous State and Parameters Estimation. Proceedings of *Third European Control Conference*, ROMA, Vol. 4, pp. 2165-2170.

MULTIVARIABLE PREDICTIVE CONTROL

Peter Zelinka *,[1]

* *Department of Automatic Control and Measurement*
Faculty of Mechanical Engineering, Slovak University of Technology
Nám. Slobody . 17, 812 31 Bratislava, Slovak Republic

Abstract: This paper presents some real-time laboratory experiments with a multivariable predictive control law based on the Generalized Predictive Control (GPC) and Constrained Receding Horizon Predictive Control (CRHPC) to demonstrate their performance under various conditions. The proposed controllers rests upon a set of MISO models of multivariable system. The control signal is generated subject to amplitude and increment constraints.

Keywords: Generalized predictive control,Multivariable systems,ARMA models,Criterion function,Constrains

1. INTRODUCTION

The motivation of this contribution was to demonstrate the applicability of the multivariable predictive control approach introduced in (Chow, 1996) to various multivariable systems and to gain our experiences (B. Rohál-Ilkiv, 1994) in this type of control before experiment on real plant. The real-time experiments were performed on laboratory scale models constructed in a control laboratory at the department. Two systems were selected from a group of scale models in the lab:

- *water power plant* - this system is characterized by a considerable time delay, stochastic uncertainties and nonlinearities
- *aerothermal system* - idea of this system was taken from multizone reheating furnace .

The paper is organized as follows: section 2 describes model of multivariable plant and derivation of k-step-ahead predictor. The design of controller is given in section 3. Section 4 deals with some implementation aspects. Some results from real laboratory experiment are presented in section 5. Section 6 concludes the paper.

2. PLANT DESCRIPTION

A suitably parametrized model, utilizable for a derivation of k-step-ahead predictor is fundamental for design of predictive control. It is possible to use several types of predictors, which forms depend on formulation: singlestep, multistep, step-recursive formulation, minimum variance ... (G. Favier, 1990). For the purposes of this paper, we will assume a multivariable CARIMA (Controlled-Auto-Regressive-Integrated-Moving-Average) model in incremental form :

$$\mathbf{A}(q^{-1})\Delta\mathbf{y}(t) = \mathbf{B}(q^{-1})\Delta\mathbf{u}(t-1) + \mathbf{T}(q^{-1})\xi(t) \qquad (1)$$

where $\mathbf{A}(q^{-1}), \mathbf{B}(q^{-1}) and \mathbf{T}(q^{-1})$ are polynomial matrices in the backward shift operator,

$$\begin{aligned} \mathbf{y}(t) &= [y_1(t),\ y_2(t)\ \ ...]^T \\ \mathbf{u}(t) &= [u_1(t),\ u_2(t)\ \ ...]^T \\ \xi(t) &= [\xi_1(t),\ \xi_2(t)\ \ ...]^T \end{aligned} \qquad (2)$$

[1] Supported by ECC Commission of the European Communities (COPERNICUS grant CP941174)

are the output vector, input vector and white noise vector respectively. q^{-1} is the backward-shift operator and $\Delta = 1 - q^{-1}$ the differencing operator. If the plant has a non-zero dead-time the leading elements of polynomials in $\mathbf{B}(q^{-1})$ are zero. It is assumed that model is observable and reachable.

For simplicity in the sequel we consider square system pxp:

$$\begin{aligned} \mathbf{y}(t) &= [y_1(t),\ y_2(t)\ \ldots\ y_p(t)]^T \\ \mathbf{u}(t) &= [u_1(t),\ u_2(t)\ \ldots\ u_p(t)]^T \\ \xi(t) &= [\xi_1(t),\ \xi_2(t)\ \ldots\ \xi_p(t)]^T \end{aligned} \tag{3}$$

and diagonal structure of polynomial matrices $\mathbf{A}(q^{-1}) = diagA_i(q^{-1}), \mathbf{T}(q^{-1}) = diagT_i(q^{-1}) \forall i \in [1,p]$. The lead-in the diagonal structure of polynomial matrices, modify considered MIMO model (1)to an interrelated set of p MISO models. Their interrelations are respected by including the other present outputs into separate models. Formally, for $i = 1...p$, the model of i-th output can be written in the form :

$$\begin{aligned} A_i(q^{-1})\Delta y_i(t) = &\mathbf{B}_i(q^{-1})\Delta\mathbf{u}(t-1) + \\ &T_i(q^{-1})\xi_i(t) \end{aligned} \tag{4}$$

where $\mathbf{B}_i(q^{-1})$ is i-th row of polynomial matrix $\mathbf{B}(q^{-1})$:

$$\mathbf{B_i} = [B_i1, B_i2...B_ip] \tag{5}$$

For brevity, the argument (q^{-1}) is dropped from now. Considered model is sufficiently general for adequate representation most of processes. The deterministic parts of MIMO description and the set of MISO models are equivalent, it means, that exist unambiguous relationship between deterministic parts of model (1) and (4)(In fact transformation MIMO to MISO model depends on noise properties (M. Kárny, 1992). Furthermore, a number of estimated parameters of model (4) is smaller than that of a model with full matrices **A** and **T**. It makes MISO models more suitable in adaptive application, when on-line estimators are used. A proposed structure of considered model simplifies the specification of terminal constraints and implementation of predictor and controller.

As is typical for GPC type of controller, polynomials of T matrix are used for optimal disturbance rejection and robust design (T.-W. Yoon, 1995),therefore identification of T is assumed not necessary.

k-step-ahead predictor

Output predictions based on set of MISO CARIMA models is possible carry out in a way similar to the SISO case (D. W. Clarke, 1987); (Soeterboek, 1994).

In order to derive equation of predictor, we rewrite the process model (4)to the form:

$$\begin{aligned} y_i(t+k) = &\frac{\mathbf{B_i}}{A_i\Delta}\Delta\mathbf{u}(t+k-1) + \\ &\frac{T_i}{A_i\Delta}\xi_i(t+k) \end{aligned} \tag{6}$$

The $y_i(t+k)$ is prediction of i-th output in time $t+k$. Let us divide the member $\frac{T_i}{A_i\Delta}\xi_i(t+k)$ to two parts: a part known at time t (this part is possible to express from measured data) and future signal:

$$\frac{T_i}{A_i\Delta} = E_i{}^k + q^{-k}\frac{F_i{}^k}{A_i\Delta} \tag{7}$$

and to replace in equation (6):

$$\begin{aligned} y_i(t+k) = &\frac{\mathbf{B_i}}{A_i\Delta}\Delta\mathbf{u}(t+k-1) + \\ &E_i{}^k\xi_i(t+k) + \frac{F_i{}^k}{A\Delta}\xi_i(t) \end{aligned} \tag{8}$$

Because the term $E_i{}^k\xi_i(t+k)$ involves future (unpredictable) noise $(\xi_i(t+k),..,\xi_i(t+1))$ only and because $\xi_i(t)$ is assumed to be discrete white noise with zero mean, the minimum variance k-step ahead predictor can simply be obtained by eliminating this term from (8) :

$$\begin{aligned} \hat{y}_i(t+k) = &\frac{\mathbf{B}_i}{A_i\Delta}\Delta\mathbf{u}(t+k-1) + \\ &\frac{F_i^k}{A_i\Delta}\xi_i(t) \end{aligned} \tag{9}$$

where the symbol $\hat{}$ in (9) indicates that $\hat{y}(t+k)$ is an estimate of $y(t+k)$.

It is evident, that the member $\frac{F_i^k}{A_i\Delta}\xi_i(t)$ express effect of last disturbances and unmodelled dynamics on predicated output.

We can try to express it based on known measured data y_i and $\Delta\mathbf{u}$ from (8) :

$$\frac{F_i^k}{A_i\Delta}\xi_i(t) = \frac{F_i^k}{T_i}y_i(t) - \frac{F_i^k\mathbf{B}_i}{A_i\Delta T}\Delta\mathbf{u}(t-1) \tag{10}$$

From (7) follows

$$-\frac{F_i^k}{A_i\Delta} = q^k[E_i^k - \frac{T_i}{A_i\Delta}] \tag{11}$$

Substituting (11) in (10) results in

$$\begin{aligned} \frac{F_i^k}{A_i\Delta}\xi_i(t) = &\frac{F_i^k}{T_i}y_i(t) + \frac{E_i^k\mathbf{B}_i}{T_i}\Delta\mathbf{u}(t+k-1) - \\ &\frac{\mathbf{B}_i}{A_i\Delta}\Delta\mathbf{u}(t+k-1) \end{aligned} \tag{12}$$

Further, utilizing (12) for substitution $\frac{F_i^k}{A_i\Delta}\xi_i(t)$ in term (9) yields:

$$\hat{y}_i(t+k) = \frac{E_i^k\mathbf{B}_i}{T_i}\Delta\mathbf{u}(t+k-1) + \frac{F_i^k}{T_i}y_i(t) \qquad (13)$$

The influence of disturbances is express by measured data in this equation.

Notice, that term $\frac{E_i^k\mathbf{B}_i}{T_i}\Delta\mathbf{u}(t+k-1)$ determining the transfer of manipulated signal $\Delta\mathbf{u}$ to output signal y_i contains 'old' (determined) and 'new' values of inputs (,they will be designed by control synthesis). In order to split up the k-step ahead predictor (13) in parts that is known at time t and future signals , we introduce next Diophantine equation:

$$\frac{E_i^k\mathbf{B}_i}{T_i} = \mathbf{G}_i^k + q^{-k}\frac{\mathbf{H}_i^k}{T_i} \qquad (14)$$

where

$$\mathbf{G}_i^k = [g_{i1}^0 + g_{i1}^1 q^{-1} + \ldots + g_{i1}^{k-1}q^{-(k-1)}, \ldots \\ g_{ip}^0 + g_{ip}^1 q^{-1} + \ldots + g_{ip}^{k-1}q^{-(k-1)}]$$

In case $T_i = 1$ the g_{ij}^k is k-th co-ordinate of the step response of the subsystem $\frac{B_{ij}}{A_i}$.

Using (14) in equation (13) results in:

$$\hat{y}_i(t+k) = \mathbf{G}_i^k\Delta\mathbf{u}(t+k-1) + \frac{\mathbf{H}_i^k}{T_i}\Delta\mathbf{u}(t-1) + \frac{F_i^k}{T_i}y_i(t) \qquad (15)$$

where $\mathbf{G}_i^k\Delta\mathbf{u}$ and the last two terms are *forced* and *free* responses of the process respectively. Note that the forced response is caused only by controls yet to be applied and free response rely on the known past values of $\Delta\mathbf{u}$ and presented and past values of y_i:

$$f_i(t+k) = \frac{\mathbf{H}_i^k}{T_i}\Delta\mathbf{u}(t-1) + \frac{F_i^k}{T_i}y_i(t) \qquad (16)$$

And finally, resulting equation of k-step ahead predictors of y_i becomes:

$$\hat{y}_i(t+k) = \mathbf{G}_i^k\Delta\mathbf{u}(t+k-1) + f_i(t+k) \qquad (17)$$

Matrix notation

Collecting the k-step ahead predictors for $k = 1, .., N+m$ in a matrix notation yields:

$$\begin{bmatrix}\hat{\mathbf{Y}}\\ \underline{\hat{\mathbf{Y}}}\end{bmatrix} = \begin{bmatrix}\mathbf{G}\\ \underline{\mathbf{G}}\end{bmatrix}\Delta\tilde{\mathbf{u}} + \begin{bmatrix}\mathbf{F}\\ \underline{\mathbf{F}}\end{bmatrix} \qquad (18)$$

where

$$\begin{aligned}
\hat{\mathbf{Y}} &= [\hat{y}_1(t+1), \ldots \hat{y}_1(t+N), \ldots \\ &\quad [\hat{y}_p(t+1), \ldots \hat{y}_p(t+N)\ ]^T \\
\underline{\hat{\mathbf{Y}}} &= [\hat{y}_1(t+N+1), \ldots \hat{y}_1(t+N+m), \ldots \\ &\quad \hat{y}_p(t+N+1), \ldots \hat{y}_p(t+N+m)\ ]^T \\
\Delta\tilde{\mathbf{u}} &= [\Delta u_1(t), \ldots \Delta u_p(t), \ldots \\ &\quad \Delta u_1(t+N_u), \ldots \Delta u_p(t+N_u)\ ]^T
\end{aligned}$$

$$\mathbf{G} = \begin{bmatrix} g_{11}^0 & \ldots & g_{1p}^0 & \ldots & 0 \\ g_{11}^1 & \ldots & g_{1p}^1 & \ldots & 0 \\ & \ldots & & \ldots & \\ g_{p1}^{N-1} & \ldots & g_{pp}^{N-1} & \ldots & g_{pp}^{N-N_u} \end{bmatrix}$$

$$\underline{\mathbf{G}} = \begin{bmatrix} g_{11}^{N} & \ldots & g_{1p}^{N} & \ldots & g_{1p}^{N-Nu+1} \\ g_{11}^{N+1} & \ldots & g_{1p}^{N+1} & \ldots & g_{1p}^{N-N_u+2} \\ & \ldots & & \ldots & \\ g_{p1}^{N+m-1} & \ldots & g_{pp}^{N+m-1} & \ldots & g_{pp}^{N-N_u+m} \end{bmatrix}$$

$$\begin{aligned}
\mathbf{F} &= [f_1(t+1), \ldots f_1(t+N), \ldots \\ &\quad [f_p(t+1), \ldots f_p(t+N)\ ]^T \\
\underline{\mathbf{F}} &= [f_1(t+N+1), \ldots f_1(t+N+m), \ldots \\ &\quad f_p(t+N+1), \ldots f_p(t+N+m)\ ]^T
\end{aligned}$$

f_i is possible to calculate from (16).
Note that in processes with time delay d, an input $\mathbf{u}(\mathbf{k})$ has effect on $\mathbf{y}(t+d+1), \mathbf{y}(t+d+2), \ldots$ only. It means that if the dead-time is presence the first d rows of $\mathbf{G}$ is null.

3. CONTROL LAW

We first write a control law for GPC strategy. Then, we introduce additional features to this control law in order to obtain the CRHPC controller.

Consider the following GPC cost function:

$$J_1(t) = \sum_{i=1}^{p}\Big(\sum_{k=N_1}^{N_2}\mu_i(k)e_i^2(t+k) + \sum_{k=1}^{N_u}\lambda_i(k)[\Delta\tilde{u}_i(t+k-1)]^2\Big), \qquad (19)$$

where $N_1 \geq 1$ and $N_2 \geq N_1$ are *lower,upper costing horizon* respectively and $N_u \leq N_2$ is *control horizon.* $\mu_i(k)$ and $\lambda_i(k)$ are weights on the predicted tracking error and control moves. $\Delta\tilde{u}_i$ is postulated control signal . The tracking error e_i is given by:

$$e_i(t+k) = \hat{y}_i(t+k) - w_i(t+k) \qquad (20)$$

where $w_i(t+k)$ is setpoint of i-th output in time $(t+k)$.

From the definitions of vectors above and expectation of the cost function (19) can be written:

$$J_1(t) = (\mathbf{G}\Delta\tilde{\mathbf{u}} + \mathbf{F} - \mathbf{W})^T\mathbf{M}(\mathbf{G}\Delta\tilde{\mathbf{u}} + \mathbf{F} - \mathbf{W}) + \Delta\tilde{\mathbf{u}}^T\mathbf{\Lambda}\Delta\tilde{\mathbf{u}} \qquad (21)$$

or in the simple quadratic form:

$$J_1(t) = \Delta\tilde{\mathbf{u}}^T\mathbf{Q}\Delta\tilde{\mathbf{u}} + \Delta\tilde{\mathbf{u}}^T\mathbf{L} + K \qquad (22)$$

where

$$\begin{aligned} \mathbf{Q} &= \mathbf{G}^T\mathbf{M}\mathbf{G} + \mathbf{\Lambda} \\ \mathbf{L} &= 2\mathbf{G}^T\mathbf{M}(\mathbf{F} - \mathbf{W}) \\ K &= (\mathbf{F} - \mathbf{W})^T\mathbf{M}(\mathbf{F} - \mathbf{W}) \end{aligned} \qquad (23)$$

$$\begin{aligned} \mathbf{M} = diag\ & [\mu_1(t+1), \quad ...\ \mu_1(t+N), \quad ... \\ & [\mu_p(t+1), \quad ...\ \mu_p(t+N) \quad]^T \\ \mathbf{\Lambda} = diag\ & [\lambda_1(t+1), \quad ...\ \lambda_p(t+1), \quad ... \\ & [\lambda_1(t+N_u), \quad ...\ \lambda_p(t+N_u) \quad]^T \end{aligned}$$

The vector of setpoints has a similar structure as vector $\mathbf{Y}$:

$$\begin{aligned} \mathbf{W} = & [w_1(t+1), \quad ...\ w_1(t+N), \quad ... \\ & [w_p(t+1), \quad ...\ w_p(t+N) \quad]^T \end{aligned} \qquad (24)$$

The minimalization of the cost function we can perform by quadratic programming or analytically setting $\frac{\delta J1(t)}{\delta\Delta\tilde{\mathbf{u}}}$ to zero:

$$\frac{\delta J_1(t)}{\delta\Delta\tilde{\mathbf{u}}} = 2\mathbf{G}^T\mathbf{M}\mathbf{G}\Delta\tilde{\mathbf{u}} + 2\Delta\tilde{\mathbf{u}}^T\mathbf{G}^T\mathbf{M}(\mathbf{F} - \mathbf{W}) = 0 \qquad (25)$$

from where optimal GPC control sequence

$$\begin{aligned} \Delta\tilde{\mathbf{u}} &= -\mathbf{Q}^{-1}\mathbf{L}(t) = \\ &= (\mathbf{G}^T\mathbf{M}\mathbf{G} + \mathbf{\Lambda})^{-1}\mathbf{G}^T\mathbf{M}(\mathbf{W} - \mathbf{F}) \end{aligned} \qquad (26)$$

CRHPC cost function can be obtained from (19) with conditions:

- $N_1 = 1$
- $N_2 = N$
- $N_u = N - d + 1$, where d is time delay
- The Minimalization of the cost function is subjected to equality constraints on controls and tracking error:

The control objective may thus be written as follows :

$$minJ(t) = \sum_{i=1}^{p}(\sum_{k=1}^{N}\mu_i(k){e_i}^2(t+k) + \lambda_i(k)\sum_{k=1}^{N_u}(\Delta\tilde{u}_i(t+k-1))^2) \qquad (27)$$

$$\begin{aligned} & sub\,ject\ to \\ & \quad \Delta\tilde{u}_i(t + N_u + k) = 0,\ \forall k \geq 0, \forall i \\ & \quad e_i(t+N+k) = 0, \qquad \forall k \in [1,m], \forall i \end{aligned} \qquad (28)$$

N marks *costing horizon* and m defines the number of *end-point* equality constraints on the tracking error ($0 \leq m \leq N_u$; $m = max\ \delta(A_i) + 1), i \in [1,p]$.

By the use of the matrix notation (18) and Lagrangian multipliers $\mathbf{\Upsilon}$we can form augmented cost:

$$J_a(t) = (\mathbf{G}\Delta\tilde{\mathbf{u}} + \mathbf{F} - \mathbf{W})^T\mathbf{M}(\mathbf{G}\Delta\tilde{\mathbf{u}} + \mathbf{F} - \mathbf{W}) + \Delta\tilde{\mathbf{u}}^T\mathbf{\Lambda}\Delta\tilde{\mathbf{u}} + \mathbf{\Upsilon}(\underline{\mathbf{G}}\Delta\tilde{\mathbf{u}} + \underline{\mathbf{F}} - \underline{\mathbf{W}}) \qquad (29)$$

The vector $\underline{\mathbf{W}}$ correspond with $\underline{\hat{\mathbf{Y}}}$:

$$\begin{aligned} \underline{\mathbf{W}} = & [w_1(t+N+1), \ ...\ w_1(t+N+m), \ ... \\ & w_p(t+N+1), \ ...\ w_p(t+N+m) \ ]^T \end{aligned}$$

The minimum of cost function has to satisfy conditions:

$$\frac{\delta J_a}{\delta\Delta\tilde{\mathbf{u}}} = 2(\mathbf{G}\mathbf{\Delta}\tilde{\mathbf{u}} + \mathbf{F} - \mathbf{W})^T\mathbf{M}\mathbf{G} + 2\mathbf{\Lambda}\Delta\tilde{\mathbf{u}} + \mathbf{\Upsilon}\underline{\mathbf{G}}^T = 0 \qquad (30)$$

$$\frac{\delta J_a}{\delta\mathbf{\Upsilon}} = \underline{\mathbf{G}}\Delta\tilde{\mathbf{u}} + \underline{\mathbf{F}} - \underline{\mathbf{W}} = 0 \qquad (31)$$

The solution of equations (30),(31) is optimal sequence $\Delta\tilde{\mathbf{u}}$

$$\Delta\tilde{\mathbf{u}} = \tilde{\mathbf{G}}(\mathbf{I} - \mathbf{G}^T(\underline{\mathbf{G}}\tilde{\mathbf{G}}\underline{\mathbf{G}}^T)^{-1}\underline{\mathbf{G}}\tilde{\mathbf{G}})\mathbf{G}^T\mathbf{M}(\mathbf{W} - \mathbf{F}) + \tilde{\mathbf{G}}\underline{\mathbf{G}}^T(\underline{\mathbf{G}}\tilde{\mathbf{G}}\underline{\mathbf{G}}^T)^{-1}(\underline{\mathbf{W}} - \underline{\mathbf{F}}) \qquad (32)$$

where

$$\tilde{\mathbf{G}} = (\mathbf{G}^T\mathbf{M}\mathbf{G} + \mathbf{\Lambda})^{-1} \qquad (33)$$

The solution can be obtain also via quadratic programing minimizing cost function in simple quadratic form :

$$J_1(t) = \Delta\tilde{\mathbf{u}}^T\mathbf{Q}\Delta\tilde{\mathbf{u}} + \Delta\tilde{\mathbf{u}}^T\mathbf{L} + K \qquad (34)$$

where

$$\begin{aligned} \mathbf{Q} &= \mathbf{G}^T\mathbf{M}\mathbf{G} + \mathbf{\Lambda} \\ \mathbf{L} &= 2\mathbf{G}^T\mathbf{M}(\mathbf{F} - \mathbf{W} + \mathbf{\Upsilon}\underline{\mathbf{G}}) \\ \mathbf{\Upsilon} &= (\underline{\mathbf{G}}\tilde{\mathbf{G}}\underline{\mathbf{G}}^T)^{-1}((\underline{\mathbf{Y}} - \underline{\mathbf{F}}) - \\ & \quad (\underline{\mathbf{G}}\tilde{\mathbf{G}}\mathbf{G}^T)^{-1}\mathbf{M}(\mathbf{Y} - \mathbf{F})) \end{aligned}$$

K notes cost constant term. It has no influence on determination of optimal control sequence $\Delta\tilde{\mathbf{u}}$.

4. INEQUALITY CONSTRAINTS AND SET-POINT CONDITIONING

The important aspect of predictive control - time domain formulation allows to include an inequality constraints on process variables (which are linearly dependent on the controls) in the control law. These inequality usually relate to hardware limitations, safety conditions ,or even define performance goal, minimize fuel consumption ...

Consider a convex compatible set $\mathcal{U}_i \subset R^{pxN_u}$ of inequality constraints and a nonempty set $\mathcal{U}_e \subset R^{pxN_u}$ of equality constraints (28) in the form :

$$\underline{\mathbf{G}}\Delta\tilde{\mathbf{u}} + \underline{\mathbf{F}} - \underline{\mathbf{W}} = 0 \tag{35}$$

In constrained GPC or CRHPC the objective is still to minimize the cost(22),(34), subject to satisfaction of the inequality constraints $\mathcal{U}_i$. The use of same form of search based optimalization - typically QP is required to perform the cost minimalization.

In case of CRHPC can arise difficulties caused by incompatibility between the equality and inequality constraints $\mathcal{U}_i \cap \mathcal{U}_e = 0$. It means, that required setpoint sequence can't be achieved with the control moves $\Delta\tilde{\mathbf{u}} \subset \mathcal{U}_i$. One of the several solution of this problem is *setpoint conditioning* (Scokaert, 1994). The key idea of this solution is to implement feasible set points sequence $\underline{\mathbf{W}}_{\mathbf{f}}$ which is the closest to $\underline{\mathbf{W}}$ and its values are such that the end-point equality constraints are compatible with inequality constraints. Some form of optimalization needs to be used whereby the quantity $|\underline{\mathbf{W}}_{\mathbf{f}} - \underline{\mathbf{W}}|$ is minimized subject to inequality constraints $\mathcal{U}_i$.

In (Chow, 1996) this problem is solved via linear programing. A alternative solution can be obtained minimizing the second norm $||\underline{\mathbf{W}}_{\mathbf{f}} - \underline{\mathbf{W}}||$ subject to $\mathcal{U}_i$. This task leads on problem of a finding a $\Delta\tilde{\mathbf{u}}_f \subset \mathcal{U}_i$, which minimize:

$$\begin{aligned} J_f &= (\underline{\mathbf{G}}\Delta\tilde{\mathbf{u}}_f + \underline{\mathbf{F}} - \underline{\mathbf{W}})^T(\underline{\mathbf{G}}\Delta\tilde{\mathbf{u}}_f + \underline{\mathbf{F}} - \underline{\mathbf{W}}) \\ &su\ bject\ to \\ &\mathcal{U}_i \end{aligned} \tag{36}$$

The feasible set points sequence is calculated from:

$$\underline{\mathbf{W}}_f = \underline{\mathbf{G}}\Delta\tilde{\mathbf{u}}_f + \underline{\mathbf{F}} \tag{37}$$

Then $\underline{\mathbf{W}}_f$ replace $\underline{\mathbf{W}}$ in $\mathcal{U}_e$ (35). The solution of this problem coincide with stability of the system. More about stability guaranties see in (Chow, 1996) .

5. PRACTICAL EXPERIMENTS

The next examples will demonstrate the basic features of the presented controller on the real-time laboratory experiments. In all experiments on-line identification of set of MISO models by conventional last-squares method with directional forgetting (Kulhavý, 1987) was used. According to recending horizon strategy only the first part of vector $\Delta\tilde{\mathbf{u}}$ was used to compute control signal $\mathbf{u}(t)$.

5.1 *Experiments on the water-power plant.*

The model is multivariable system with time delay due to transport of liquid in a hose. The system consists of two tanks, see Fig. 1. upper tank and lower tank, which are joined by a screw shaped flexible hose. The slope of hose is constant. The diameter of hose is sufficiently big to ensure a free flow of water. The water is taken from the upper tank on the water wheel which drives the electric generator and generate output signal y_1. A quantity of water is controlled by opening the valve (a position servo - input variable u_1) and by changing the level of water in the upper tank. From the lower tank the water is given back by a pump and run-back piping. The pump is controlled by signal u_2. The level of water in the upper tank is measured by capacity indicator (signal y_2). All variables are analog.

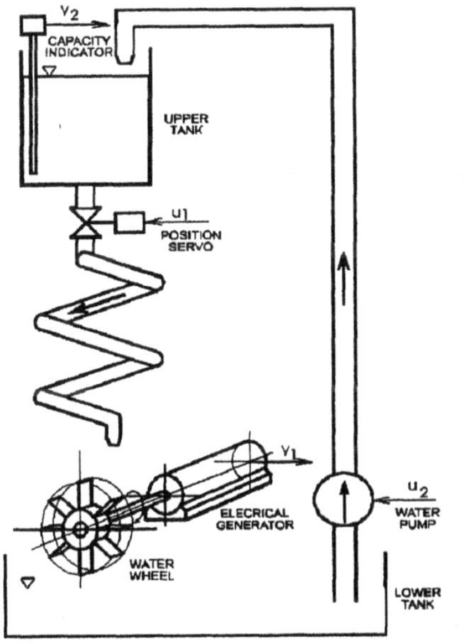

Fig. 1. Water power plant

Characteristics of system are known. The step responses, Fig. 3. shows, that the plant is combination a first order system with considerable time delay and astatic system. The estimate of raise time of the electric generator is 1.5 second and time delay varies from 6 to 8.5 seconds (depending on the level of the liquid in the upper tank). A presence of noise is an effect of a mechanical commutation of electrical generator.

A nonlinearity of the system is mainly due to nonlinear characteristic of the valve. In addition a friction of the position servomechanism cause a backlash.

The experiments with plant were performed with square wave setpoints of height of water and voltage of generator. The aim of the experiment was to test behavior of the multivariable predictive controller and its ability to *cover* the time delay of the controlled process by the *b-parameters* of the model.

A satisfactory results ,see Fig. 4., were obtained for GPC with control period $T = 2s$ and following design parameters:

$p = 2, N_1 = 1, N_2 = 15, N_u = 10$

$\lambda_i = 10, \mu_i = 1, \forall i \in [1, p]$

orders of the regression model:
$\delta B_{ij} = 12, \quad \delta A_{ij} = 6, \forall i, \forall j \in [1, p]$

The projected control signal was constrained in its amplitudes and increments.

The influence of intense penalization of controls $\lambda_i = 50, \mu_i = 1, \forall i \in [1,p]$ is shown on Fig. 5. Similar effect was obtained with $\lambda_i = 10, \mu_i = 1, \forall i \in [1,p]$ and hard restrictions of increments of controls $-0.2 \leq \Delta u_i(k) \leq 0.2, \forall i \in [1,p], \forall k \in [1, N_u]$, see Fig. 6.

5.2 *Experiments on the aerothermal system .*

This laboratory scale model, Fig. 2., is representative of a multivariable thermal process.

The physical basis of the proposed model is the flow of air through a *long* tube with a *small* constant cross section of a semicircle form. The flowing air is heated by two electrical heaters which are distributed in the length of the tube. The process output y_1, y_2 is a hot air temperature, which is measured by bead thermistors. The process inputs consist of two manipulated heating inputs u_1, u_2(electrical heaters) . This process is nonlinear, because the heating dynamics is different from the cooling dynamics.

A step responses of the system are on Fig. 7. It is clear, that cross influence of u_1 to y_2 is greater then u_2 to y_1 caused by air flow. Following design parameters were used in experiment with CRHPC controller, with control period $T = 3s$:

$p = 2, N_1 = 1, N_2 = 5, N_u = 6$

$\lambda_i = 1, \mu_i = 1, \forall i \in [1, p]$

orders of the regression model:
$\delta B_{ij} = 2, \quad \delta A_{ij} = 3, \forall i, \forall j \in [1, p]$

Results are shown on Fig.8.

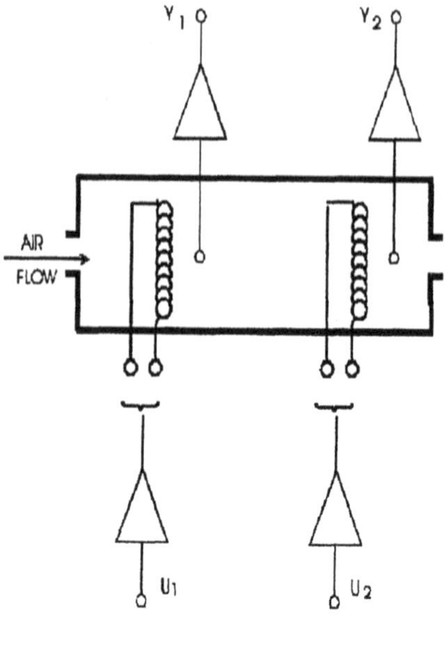

Fig. 2. Aerothermal plant

6. CONCLUSION REMARKS

All experiments were performed in the system MATLAB on PC 586/100. The presented controllers are composed of C and Fortran - coded MEX and M files, for dynamic running and for the setting of the controller.

The experiments confirm good stability properties of tested GPC and CRHPC controllers.

7. ACKNOWLEDGEMENTS

The author acknowledges the current financial support of the Commission of the European Communities (COPERNICUS grant CP941174) for the work on constrained predictive control.

8. REFERENCES

B. Rohál-Ilkiv, P. Zelinka, R. Richter (1994). Multivariable adaptive temperature ,control for a reheating furnace. In: *Colloquium on "Improvements in furnace control: current developments and new technologies"*. The Institution on Electrical Engineer. Savoy Place ,London.

Chow, C.-M. (1996). *Predictive control with constraints.* D.Phil thesis, Department of Engineering Science. Oxford University,Oxford.

D. W. Clarke, C. Mohtadi, P. S. Tuffs (1987). Generalized predictive control. Part I: The basic algorithm and Part II: Extensions and interpretations. *Automatica* **23(2)**, 137–160.

G. Favier, D. Dubois (1990). A review of k-step ahead predictors. *Automatica* **26(1)**, 75–84.

Kulhavý, R. (1987). Restricted exponential forgetting in real time identification. *Automatica* **23**, 598 - 600.

M. Kárny, A. Halousková, J. Bohm (1992). Mimo - a set of siso ?. multivariete system adaptively controlled as a set of single-input single-output models. In: *IFAC Workshop MICC '92*. Prague.

Scokaert, P. O. M. (1994). *Constrained predictive control.* D.Phil thesis, Department of Engineering Science. Oxford University,Oxford.

Soeterboek, R. (1994). *Predictive control. A unified approach.* D.Phil thesis, Delft University. Delft, The Netherlands.

T.-W. Yoon, D.W. Clarke (1995). Observer design in receding-horizon predictive control. *Int. J. Control* **61(1)**, 171–191.

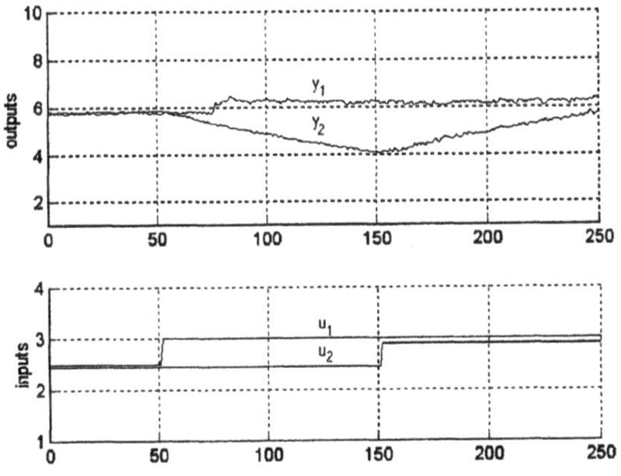

Fig. 3. The step response of water power plant

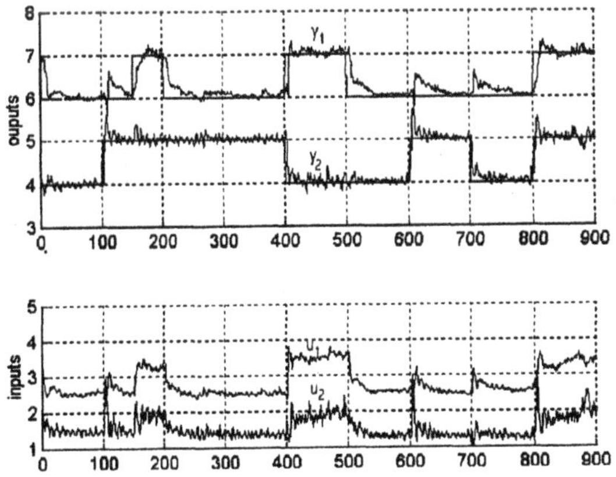

Fig. 4. The behaviors of GPC

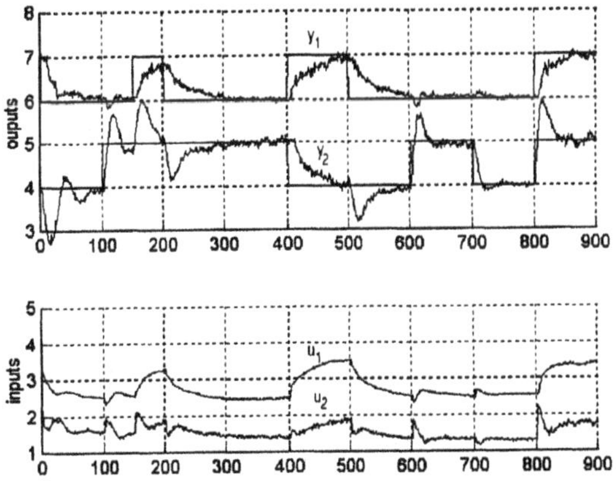

Fig. 5. The behaviors of GPC - the intense penalization of controls

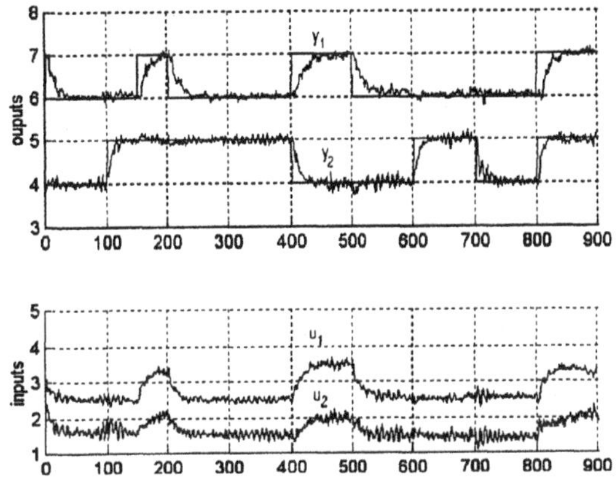

Fig. 6. The behaviors of GPC - the hard restrictions of increments of controls

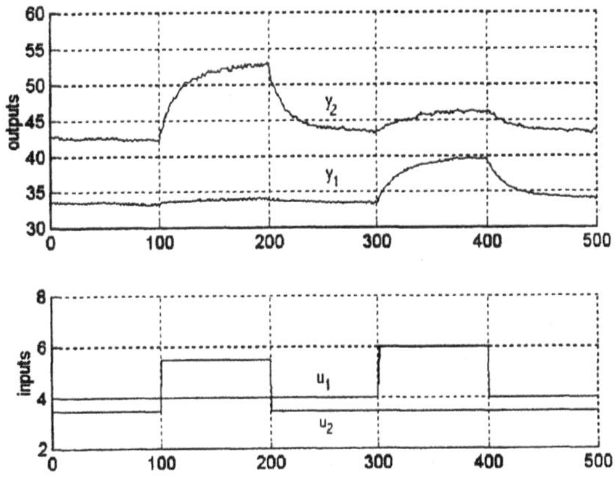

Fig. 7. The step response of aerothermal plant

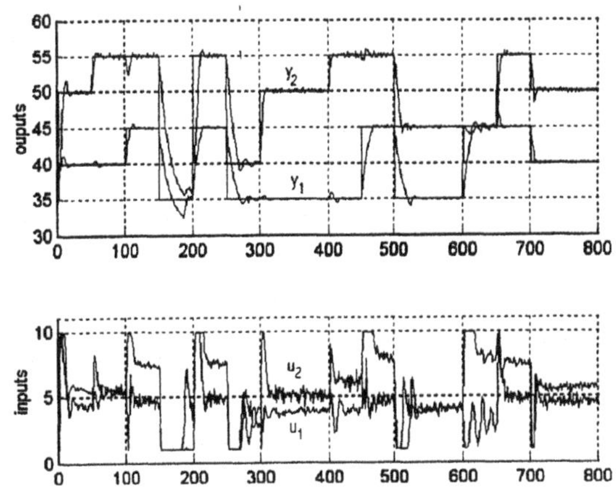

Fig. 8. The behaviors of CRHPC

THE ANALYSIS AND SYNTHESIS OF LQ REGULATOR BASED ON ORTHONORMAL SERIES REPRESENTATION MODEL

Jozef Hejdiš, Štefan Kozák

Department of Automatic Control, Faculty of Electrical Engineering and Information Technology, Slovak University of Technology, Ilkovičova 3, 812 19 Bratislava, Slovak Republic. E-mail: kozak@kasr.elf.stuba.sk

Abstract: In this paper the problem of modeling the Laguerre ladder network in the discrete-time domain is addressed. New algorithm of computing the discrete state-space model of the Laguerre ladder network, which reduces requirements on computer memory and provides a simple applicability to computer systems, is proposed. A receding horizon LQ regulator based on Laguerre function model is also presented.

Keywords: unstructured adaptive control, orthonormal series representation, orthonormal function model, Laguerre ladder network, discrete state-space model, LQ control.

1. INTRODUCTION

During the past decades, several adaptive control systems have been developed. Most of the systems are model-based. In particular, in the input-output case, ARMA or FIR models are widely used. Recently, the search for robust adaptive control systems, which require a minimal *a priori* information has led to the development of unstructured adaptive control. Using this approach, the ARMA or the FIR model is replaced by an orthonormal series representation of the plant dynamics. The main advantage of this method is that any stable plant can be modeled without knowledge of the structure, for example without assumption about the true plant order or time delay. Laguerre functions are usually used to model plant dynamics because of their simplicity and their similarity to transient signals, however, other orthonormal functions may be used, such as Laning-Battin functions. Compared with the ARMA or FIR model, using of the model based on orthonormal series representation has several advantages: a good approximation of a system with a delay; a tolerance to unmodeled dynamics and a reduced sensitivity of the estimated parameters; convenient filter network realization; good low frequency match between the estimated model and the true plant model; etc. One of the main problems when currently identfying plants using the orthonormal series representation is to express the model in the discrete-time domain. In this paper,a memory saving algorithms for computing the discrete state-space model are presented. One of the regulators, parameters of which can be computed using the linear state-space model, is a receding horizon LQ regulator. Using this regulator several iterations of the Riccati difference equation have to be made to evaluate the control action. Fast and memory saving algorithm of the adaptive LQ regulator, which is based on the Laguerre filter network model, is proposed at the end of the paper.

2. MODELLING BY MEANS OF ORTHONORMAL FUNCTIONS

2.1 Laguerre functions

Laguerre functions form an orthonormal set in $L_2\langle 0,\infty)$. In the continuous-time domain they may be expressed as

$$l_i(t) = \sqrt{2p}\,\frac{\exp(pt)}{(i-1)!}\frac{d^{i-1}}{dt^{i-1}}[t^{i-1}\exp(-2pt)],$$

$$i = 1, \ldots, N$$

where i is the order of the function and p is the time scale. Corresponding Laplace transform for this set is

$$L_i(s) = \sqrt{2p}\,\frac{(s-p)^{i-1}}{(s+p)^i}, \quad i = 1,\dots,N \tag{1}$$

From (1) it is evident that Laguerre functions can be generated using a simple ladder network of RC filters, which are called Laguerre filters Hudzovic (1989). According to the Riesz-Fisher theorem, for a given real continuous function q(t) in $L_2\langle 0,\infty)$, there exist a real number ε and an integer N such that

$$\int_0^{\infty} |q(t) - S(t)|^2 dt < \varepsilon \tag{2}$$

where

$$S(t) = \sum_{i=1}^{N} c_i l_i(t) \tag{3}$$

The coefficients c_i are called the Laguerre spectra gains and for deterministic signals they can be calculated by

$$c_i = \int_0^{\infty} h(t) l_i(t) dt \tag{4}$$

where h(t) is the impulse response of the plant. In practice the Laguerre spectra gains c_i are estimated using the least-squares parameter estimation.

2.2 Laning-Battin functions

Laning-Battin functions can be described in continuous time domain by

$$w_i(t) = \sum_{j=1}^{i} q_{ij} \exp(-ipt) \qquad i = 1,\dots,N$$

where

$$q_{ij} = \sqrt{2ip\frac{2j}{i+j}} \prod_{\substack{k=1\\k\neq j}}^{i} \frac{j+k}{j-k}$$

Laplace transform of this functions is

$$W_i(s) = \sqrt{2ip}\,\frac{1}{s+ip} \prod_{k=1}^{i-1} \frac{s-kp}{s+kp} \tag{5}$$

2.3 Generalized orthonormal functions

To show the similarity of Laguerre- and Laning-Battin function we express Laplace transform of Laguerre functions in the form

$$L_i(s) = \sqrt{2p}\,\frac{1}{s+p} \prod_{k=1}^{i-1} \frac{s-p}{s+p} \tag{6}$$

Comparing expressions (5) a (6) a new set of generalized orthonormal functions can be found in (Hudzovic, 1989). Their convenient representation in the frequency domain is given by

$$Z_i(s) = \sqrt{2[1+(i-1)m]p}\,\frac{1}{s+p} \prod_{k=1}^{i-1} \frac{s-[1+(k-1)m]p}{s+[1+(k-1)m]p} \tag{7}$$

where $m \in \langle 0,\infty)$. Generalized orthonormal functions can be described in the continuous time domain

$$z_i(t) = \exp(-pt) \sum_{k=1}^{i} q_{ik} \exp(-mkpt)$$

where

$$q_{ik} = \sqrt{2[1+(i-1)m]p}\,\frac{1}{m^{i-1}}\,\frac{\sum_{j=0}^{i-2}[2+(k+j)m]}{\sum_{\substack{j=0,\\j\neq i}}^{i-1}(k-j)}$$

Relations similar to (2), (3) and (4) can be proved for Laning-Battin functions and generalized orthonormal functions. Laning-Battin functions and generalized orthonormal functions can be generated in the same way as Laguerre functions using RC filters (Hudzovic 1989).

3. COMPUTATION OF THE ORTHONORMAL FILTER NETWORK MODEL

For the parameter estimaione it is necessary to express the orthonormal filter network in the discrete-time domain. From the practical viewpoint, it is convenient to describe the ladder network model in the state-space. Denoting the state vector $x^T(t) = \begin{pmatrix} x_1(t) & x_2(t) & \dots & x_n(t) \end{pmatrix}$, we can express the discrete state-space model of the filter network in the form

$$x(k+1) = Ex(k) + gu(k)$$
$$z(k) = Fx(k)$$

where E is the lower triangular $n \times n$ matrix, g is the n-dimensional vector, F is the diagonal $n \times n$ matrix, u(k) is the system input (control action) and z(k) is the n−dimensional vector of generalized orthonormal function values. We aimed at developing a simple analytical method of computing elements of the matrices E, F and the vector g which would be applicable in the digital control.

To derive a the algorithm, we shall convert the Laplace transform representation (7) into the state-space form

$$\dot{x}(t) = Ax(t) + bu(t)$$
$$z(t) = Fx(t)$$

where A is the lower triangular $n \times n$ matrix in the form

$$A = \begin{pmatrix} -p & 0 & 0 & \cdots & 0 \\ -2p & -(1+m)p & 0 & \cdots & 0 \\ -2p & -2(1+m)p & -(1+2m)p & \cdots & 0 \\ \vdots & \vdots & \vdots & \ddots & \vdots \\ -2p & -2(1+m)p & -2(1+2m)p & \cdots & -(1+(n-1)m)p \end{pmatrix}$$

$$= \begin{pmatrix} a_1 & 0 & 0 & \cdots & 0 \\ 2a_1 & a_2 & 0 & \cdots & 0 \\ 2a_1 & 2a_2 & a_3 & \cdots & 0 \\ \vdots & \vdots & \vdots & \ddots & \vdots \\ 2a_1 & 2a_2 & 2a_3 & \cdots & a_n \end{pmatrix} \quad (8)$$

b is the n-dimensional vector in the form

$$b = \begin{pmatrix} 1 & 1 & 1 & \dots & 1 \end{pmatrix}^T \quad (9)$$

and F is a diagonal n x n matrix in the form

$$F = \begin{pmatrix} \sqrt{2p} & 0 & \cdots & 0 \\ 0 & \sqrt{2(1+m)p} & \cdots & 0 \\ 0 & 0 & \cdots & 0 \\ \vdots & \vdots & \ddots & \vdots \\ 0 & 0 & \cdots & \sqrt{2(1+(n-1)m)p} \end{pmatrix} \quad (10)$$

Using the matrix A, the vector b, the matrix E and the vector g provides

$$E = \exp(AT), \quad g = (E - I)A^{-1}b \quad (11)$$

where T is the sampling period and exp(AT) is the exponential matrix function

$$\exp(AT) = \sum_{i=0}^{\infty} \frac{(AT)^i}{i!} \quad (12)$$

Because of a limited precision of the computer arithmetic only the first k elements of the exponential series (12) can be used to compute the matrix E. The matrix E is a lower triangular one in the form

$$E = \begin{pmatrix} e_{11} & 0 & 0 & \cdots & 0 \\ e_{21} & e_{22} & 0 & \cdots & 0 \\ e_{31} & e_{32} & e_{33} & \cdots & 0 \\ \vdots & \vdots & \vdots & \ddots & \vdots \\ e_{n1} & e_{n2} & e_{n3} & \cdots & e_{nn} \end{pmatrix}$$

Let us introduce the matrix $S^i = (AT)^i$ which is the i-th power of the matrix (AT). The matrix S^i is updated in the following way $S^{i+1} = S^i(AT)$, whereby $S^0 = I$.

Then, the exponential series (12) can be written as

$$E = \sum_{i=0}^{\infty} \frac{(AT)^i}{i!} = \sum_{i=0}^{\infty} \frac{S^i}{i!} \quad (13)$$

It is easy to prove that the matrix S^i is a lower triangular matrix in the form

$$S^i = \begin{pmatrix} s^i_{11} & 0 & 0 & \cdots & 0 \\ s^i_{21} & s^i_{22} & 0 & \cdots & 0 \\ s^i_{31} & s^i_{32} & s^i_{33} & \cdots & 0 \\ \vdots & \vdots & \vdots & \ddots & \vdots \\ s^i_{n1} & s^i_{n2} & s^i_{n3} & \cdots & s^i_{nn} \end{pmatrix}$$

The elements of the matrices S, which are not equal to zero, can be computed using a recursive formulas

$$s_{ij}(k+1) = -(1+(j-1)m)pT\left(s^k_{ij} + 2\sum_{l=j+1}^{i} s_{il}(k)\right) \quad (14)$$

$$i = n, \dots, 1$$
$$j = 1, \dots i$$
$$k = 1, \dots, \infty$$
$$S^1 = A$$

To compute the elements of the matrix E we introduce a n x n matrix H

$$h_{ij}(k+1) = h_{ij}(k) + \frac{s_{ij}(k+1)}{k!} \quad i = 1, \dots, n \quad (15)$$

$$j = 1, \dots, i$$
$$k = 1, \dots, \infty$$
$$H^1 = A$$

The relation between the elements of the matrix H and the matrix E can be expressed as

$$e_{ij} = \lim_{k \to \infty} h_{ij}(k)$$

Thus, the recurrent relations have been derived to compute the elements of the matrix E. As stated above, in practical realization it is necessary to carry out only a finite number of iterations using equations (14) and (15). The number of iterative loops needed depends on the time-scale size. An appropriate condition to terminate the iterative loop can be

$$\left|s_{ij}(k)\right| < \varepsilon, \quad i = 1, \dots, n; \quad j = 1, \dots i$$

where ε is an appropriately chosen small real number.

Now, we derive the algorithm for computing the vector g using expression (11). Multiplying the matrices, it can be proved that the inverse matrix A^{-1} corresponding to the matrix A is the lower triangular n x n matrix in the form

$$A^{-1} = \begin{pmatrix} -1 & 0 & \dots & 0 \\ \frac{2}{1+m} & -\frac{1}{1+m} & \dots & 0 \\ -\frac{2}{1+2m} & \frac{2}{1+2m} & \dots & 0 \\ \vdots & \vdots & \ddots & \vdots \\ \frac{2.(-1)^n}{1+(n-1)m} & \frac{2.(-1)^{n+1}}{1+(n-1)m} & \dots & \frac{2.(-1)^{2n}}{1+(n-1)m} \end{pmatrix} \frac{1}{p} \quad (16)$$

Using (9), (11) and (16) the following expression can be proved easily

$$g_i = (-1)^i\left(e_{ii} - 1\right)/(1+(i-1)m) + \\ + \sum_{j=1}^{i-1} (-1)^j e_{ij}/(1+(j-1)m)/p \quad i = 1,\dots,n \quad (17)$$

Thus, all relations necessary for computing the elements of the matrix E and the vector g of the discrete state-space model of generalized orthonormal filter network have been derived. The algorithm consists of the Eqs. (14), (15) and (17). Similar algorithms can be derived for Laguerre and Laning-Battin filter network, which are the special cases of generalized orthonormal filter network. The way of deriving model of the Laguerre filter network is described bellow.

3.1 Laguerre filter network modeling.

An algorithm, which is similar to the one described above, can be derived for computing matrices of the model of the Laguerre filter network. If we put m=0 in expressions (8) and (10), the continuous time state-space model of the Laguerre filter network can be described as

$$\dot{l}(t) = Al(t) + bu(t)$$

where A is the lower triangular n x n matrix of the form

$$A = \begin{pmatrix} -p & 0 & 0 & \dots & 0 \\ -2p & -p & 0 & \dots & 0 \\ -2p & -2p & -p & \dots & 0 \\ \vdots & \vdots & \vdots & \ddots & \vdots \\ -2p & -2p & -2p & \dots & -p \end{pmatrix} =$$

$$= \begin{pmatrix} a_n & 0 & 0 & \dots & 0 \\ a_{n-1} & a_n & 0 & \dots & 0 \\ a_{n-2} & a_{n-1} & a_n & \dots & 0 \\ \vdots & \vdots & \vdots & \ddots & \vdots \\ a_1 & a_2 & a_3 & \dots & a_n \end{pmatrix}$$

and b is the n-dimensional vector of the form

$$b = \begin{pmatrix} 1 & 1 & 1 & \dots & 1 \end{pmatrix}^T \sqrt{2p} \quad (18)$$

Laguerre discrete time state-space model can be written as

$$l(k+1) = El(k) + gu(k)$$

where g is the n−dimensional vector and E is the lower triangular n x n matrix. It is easy to prove, that matrix E has form

$$E = \begin{pmatrix} e_n & 0 & 0 & \dots & 0 \\ e_{n-1} & e_n & 0 & \dots & 0 \\ e_{n-2} & e_{n-1} & e_n & \dots & 0 \\ \vdots & \vdots & \vdots & \ddots & \vdots \\ e_1 & e_2 & e_3 & \dots & e_n \end{pmatrix} \quad (19)$$

Elements of matrix E can be computed using expression (13), in which matrix S^i can be written as

$$S^i = \begin{pmatrix} s_n & 0 & 0 & \dots & 0 \\ s_{n-1} & s_n & 0 & \dots & 0 \\ s_{n-2} & s_{n-1} & s_n & \dots & 0 \\ \vdots & \vdots & \vdots & \ddots & \vdots \\ s_1 & s_2 & s_3 & \dots & s_n \end{pmatrix}$$

The last rows of the matrices E and S^i respectively contain all the different values of their non-zero elements. Thus, to reduce the final algorithm requirements on the computer memory it is useful to save only the elements of the last row. To be able to compute them, it is necessary to transform recursive formulas (14) and (15). For Laguerre filter network they can be expressed as

$$s_i(k+1) = -pT\left(s_i(k) + 2\sum_{j=i+1}^{n} s_j(k)\right) \quad i = 1,\dots n \quad (20)$$
$$k = 1,\dots,\infty$$
$$s_i^1 = a_i$$

$$h_i(k+1) = h_i(k) + s_i(k)/k! \quad i = 1,\dots,n \quad (21)$$
$$k = 1,\dots,\infty$$
$$h_i^1 = a_i$$

Corresponding to (16) the inverse matrix A^{-1} can be written as

$$A^{-1} = \begin{pmatrix} -1 & 0 & 0 & \dots & 0 \\ 2 & -1 & 0 & \dots & 0 \\ -2 & 2 & -1 & \dots & 0 \\ \vdots & \vdots & \vdots & \ddots & \vdots \\ (-1)^n 2 & (-1)^{n+1} 2 & (-1)^{n+2} 2 & \dots & -1 \end{pmatrix} \frac{1}{p} \quad (22)$$

Using equations (11), (19) and (22) the following expression can be proved easily

$$g(i) = \left((-1)\left({}^i e_n - 1\right) + \sum_{j=n+1-i}^{n-1} (-1)^{i+j-1} e_j\right)\sqrt{\frac{2}{p}} \quad (23)$$
$$i = 1,\dots,n$$

Thus, all the relations necessary to compute the elements of the matrix E and vector g of Laguerre filter network model have been derived. The algorithm is given by Eqs. (20), (21) and (23). Using the elements e_i and g_i of the matrix E and the vector g the elements of the state vector of the Laguerre ladder network can be computed as follows

$$x_i(k+1) = g_i u + \sum_{j=1}^{i} x_j(k) e_{n-i+j},$$
$$i = 1, \ldots, n, \; k = 0, \ldots, \infty$$

4. LQ CONTROL USING THE LAGUERRE FILTER MODEL

LQ control is a large field studied in many papers and books and is well understood. Here, only an implementation of the receding horizon LQ regulator based on Laguerre function model is provided. The Laguerre function model has been chosen because of its simplicity and small demands on computer memory, but similar relations can be derived for the generalized orthonormal function model. The model of the plant to be controlled is asssumed in the form:

$$x(k+1) = Ex(k) + gu(k)$$
$$y(k) = cx(k)$$

where E is a lower triangular nxn matrix given by (19), g is the n-dimensional vector, elements of which are given by (23), x is the n-dimensional state vector, y is the output of the plant, u is the control action and c is the vector of the Laguerre spectra gains.

To ensure the offset-free zero-stady state it is necessary to extend the original model by introducing the integration term as the new state variable

$$x_{n+1}(k+1) = x_{n+1}(k) + u_r(k) - Cx(k)$$

The extended state-space model is

$$\begin{pmatrix} x(k+1) \\ x_{n+1}(k+1) \end{pmatrix} = \begin{pmatrix} E & 0 \\ -C & 1 \end{pmatrix} \begin{pmatrix} x(k \\ x_{n+1}(k) \end{pmatrix} +$$
$$+ \begin{pmatrix} g \\ 0 \end{pmatrix} u(k) + \begin{pmatrix} 0 \\ 1 \end{pmatrix} u_r(k)$$

$$y_c(k) = C_c x_c(k), \text{ where } C_c = \begin{pmatrix} C^T \\ 1 \end{pmatrix}$$

Denoting the new composite vector and matrices as

$$: x_c = \begin{bmatrix} x & x_{n+1} \end{bmatrix}^T$$

$$E_c = \begin{pmatrix} E & 0 \\ -c & 1 \end{pmatrix} \qquad L_c(k+1) = \begin{pmatrix} L(k+1) \\ L_{n+1}(k+1) \end{pmatrix}$$

The quadratic performance criterion to be minimized by the regulator is given by

$$J = qy_c^2(k+N) + \sum_{j=0}^{N-1} \left(qy_c^2(k+j) + ru^2(k+j) \right) =$$
$$= x_c^T(k+N) Q x_c(k+N) +$$
$$+ \sum_{j=0}^{N-1} \left(x_c^T(j+k) Q x_c(j+k) + ru^2(j+k) \right) \qquad (24)$$

The receding horizon LQ regulator, which minimizes the cost function (24), is given as

$$L_c(k+1) = g^T P_c(k+1) E_c$$
$$d(k+1) = g_c^T P_c(k+1) g_c + r$$
$$P_c(k) = E_c^T P_c(k+1) E_c + Q -$$
$$- L_c^T(k+1) L_c(k+1) / d(k+1)$$
$$\text{for } k = N-1, \ldots, 0$$

The control law is

$$u(k) = -1/d(0) L_c(0) x_c(k)$$

where $P_c(N) = Q(N)$

To save the computer memory and to make the evaluation of the control action faster only the lower triangular part of the symmetric matrices $P_c(k)$ is to be computed and stored in a $n \times (n-1)$-dimensional vector. Using matrix multiplication it can be shown easily that the recursive relations for computing values of the scalar d(k) and the elements of matrices $P_c(k)$ and vectors $L_c(k)$ are given by:

$$d(k+1) = r + \sum_{i=1}^{n} \sum_{j=1}^{n} g_i g_j p_{\max(i,j)\min(i,j)}(k+1)$$

$$l_i(k+1) = \sum_{m=1}^{n} g_m \Big(\sum_{k=i}^{n} e_{n+i-k} p_{\max(k,l)\min(k,l)}(k+1)$$
$$- P_{n+1,m}(k+1) c_{ci} \Big)$$

$$l_{n+1}(k+1) = \sum_{m=1}^{n} g_{cm} p_{n+1,m}(k+1)$$

$$p_{ij}(k) = (p_{n+1,n+1}(k+1) + q) c_i c_j +$$
$$\sum_{k=i}^{n} \sum_{l=j}^{n} e_{n+i-k} e_{n+j-l} p_{\max(k,l)\min(k,l)}(k+1) -$$
$$+ - c_j \sum_{k=i}^{n} e_{n+i-k} p_{n+1,k}(k+1) - c_i \sum_{k=j}^{n} e_{n+j-k} p_{n+1,k}(k) -$$
$$- l_i(k+1) l_j(k+1) / d(k+1) \quad i = 1, \ldots, n$$

$$p_{n+1,j}(k) = \sum_{l=j}^{m} e_{n+j-l} p_{n+1,l}(k+1) + p_{n+1,n+1}(k+1) c_j + c_j -$$
$$- \frac{l_{n+1}(k+1) l_j(k+1)}{d(k+1)}$$

$$p_{n+1,n+1}(k) = p_{n+1,n+1}(k) + 1 - \frac{l_{n+1}^2(k+1)}{d(k+1)}$$

for k=N-1,...,0

Following example demonstrates the applicability of the adaptive receding horizon LQ regulator using on algorithm described above.

5. CASE STUDY

Consider the plant

$$G(s) = \frac{s^2 + s + 1}{s^3 + 3s^2 + 2s + 2}$$

The self-tuning receding horizon LQ regulator is used to control the plant. Fig.1 shows the time response of the closed-loop system with weighting coefficients $q_n = q = 5, r = 50$ and control horizon $N = 10$. The controlled plant is modelled by a 4th-order Laguerre filter model with time scale $p = 1$ and the sample time $T = 0.2$. The LDFIL modification of the recursive least-squares with forgetting factor $\varphi = 0.995$ is used to identify the Laguerre spectra gains c. Figure 2 shows the controller output.

6. CONCLUSION

The algorithms for computing the discrete state-space model of the generalized orthonormal filter network and receding horizon LQ-regulator have been presented. To obtain the required precision of the mathematical model of the plant, in most practical applications it is necessary to choose the higher order of the Laguerre model which increases requirements on computer memory. That is why we tried to reduce them while creating the algorithms. The derived algorithms are memory saving and providing simple applicability even to small computer systems.

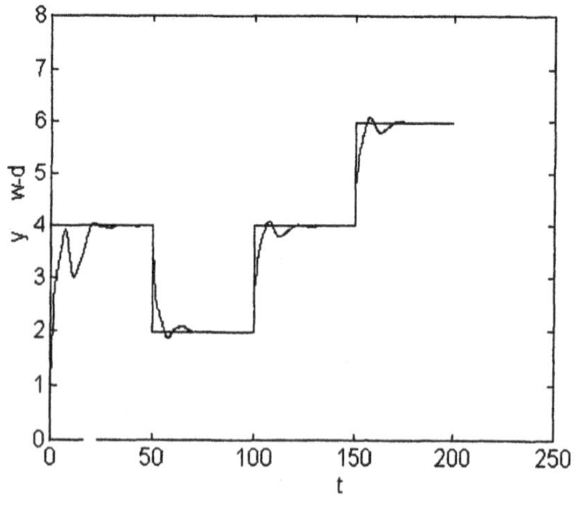

Figure 1 Set point and output for the case sudy using self-tuning LQR extended controller

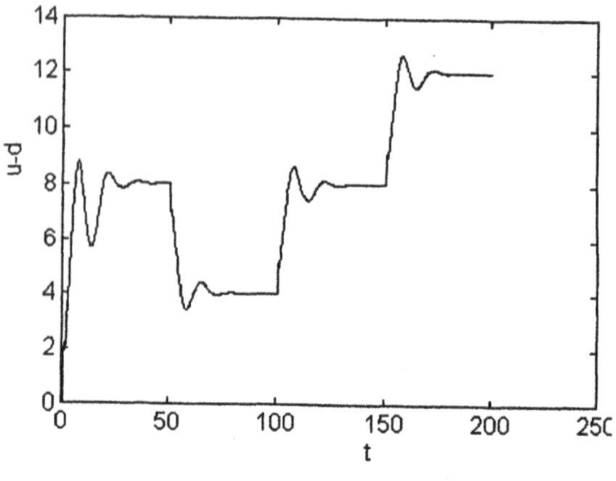

Figure 2 Controller output for set point change of Fig.1

7. REFERENCES

Zervos,C.C.and G.A.Dumont (1988). Deterministic adaptive control based on Laguerre series representation. *Int. J. Control*, **48**, 2333-2359.

Dumont, G.A., A. Elnaggar and A. Elshafei (1993). Adaptive predictive control of system with varying time delay, *International Journal of Adative Control and Signal Processing*, 7, 91--100.

Dumont, G.A., and Y.Fu (1993). Nonlinear adaptive control via Laguerre expansion of Voltera kernels. *International Journal of Adaptive Control and Signal Processing*, 7, 367 - 382.

Kozák, Š., and A. Belková, (199 3). Selftuning controllers with unmodelled dynamics of the processes. *Automatizace*, 7, 193-197.

Kozák, Š., (1994) . Nonlinear selftuning controller based upon Laguerre series representation.In *Proc.1st IFAC Workshop on New Trends in Design of Control Systems Smolenice*. Slovak Republic, 432-438.

Hudzovič, P. (1989) .Orthogonal exponential functions. *Internal project*, FEI STU.

Åström, K.J. , and B.Wittenmark (1984). *Computer Controlled Systems*. Prentice-Hall.

TWO DYNAMICAL CLASSES OF PI-CONTROLLERS FOR THE 1ST ORDER LOOPS

M. Huba, P. Kul'ha, P. Bisták and Z. Skachová

Slovak Technical University, Department of Automation and Control, Faculty of Electrical Engineering and Information Technology , SK-812 19 Bratislava.
Email: huba@elf.stuba.sk

Abstract: This paper presents modular design of P, I and PI controllers dedicated for systems with dominant 1st order dynamics in the control loop. It is shown that under presence of control signal constraints the traditionally used "linear" PI-controller guarantees a desired control quality only for a special distribution of the dynamical terms in the control loop - for the dominant 1st order dynamics concentrated in the feedback path (sensor or filter dynamics). For the more frequent situation with the dominant 1st order plant placed in the feedforward path, a new integrating "anti-windup" controller is derived. It is shown that the controller derived represents two different dynamical classes of controllers characteristic with none or one possible train of saturated control signal pulses in their response to a step change of the reference signal.

Keywords: Saturating Control, Antiwindup PI-Control, Disturbance Observer

1. INTRODUCTION

The PI controllers are known as the basic and most frequently used instruments in the industrial automation and control. Their history begins in the time after the World War I. According to their simplicity one could expect that all basic problems relevant for their utilisation have been already clarified. However, in fact, the opposite is true. Up to date, at practically each conference oriented to the control systems design, new papers devoted to an "optimal" design or tuning of PI controllers can be found. So, already this single moment is enough to indicate a till now hidden aspects necessary for an reliable controller design. This paper tries to identify as one of the key factors **the distribution of the dynamical terms in the control loop** and under consideration of the control signal saturation it also proposes new ways to really generally acceptable solutions.

The paper starts with analysing control problem denoted usually by heading **Antiwindup PI-Controller** in the last years. The treatment is divided into modules - the first one introduces and interprets the bang-bang and the pole-assignment control of a linear 1st order plant as two limit features which naturally complement each other in the **minimum time pole assignment P-controller**. Next it is shown that the serious problem of the false and overfluous integration is not only due to the control signal constraints - the reasons are hidden already in the usually used linear controller algorithm. This is shown to lead to the optimal behaviour only in control loops with dominant dynamics concentrated in the feedback. For both basic loop configuration the pole assignment controller are developed.

2. BANG-BANG AND P-CONTROL OF A 1ST ORDER PLANT

Consider a 1-st order plant with the output y and the control signal u

$$\frac{dy}{dt} = K_s u - ay \qquad (1)$$

whereby $T_p=1/a$ represents the plant time constant. In the case of digital control with the sampling period T the outputs y(k), k=0,1,... measured at the sampling instants t=kT can be described as

---

The work has been partially supported by Slovak Scientific Grant Agency, Grant No. 95/5195/198.

$$y(k+1) = Dy(k) + Ku(k)$$

$$D = e^{aT}; \quad K = K_s \frac{1-D}{a} \tag{2}$$

An **admissible control signal** is usually defined by

$$U_1 \leq u \leq U_2 \tag{3}$$

In order to maintain a desired constant output signal y=w, in its neighbourhood it must be possible to change the direction of the output signal changes in (1) by the admissible control signal (3). So, the **admissible reference signal value** has to be restrict e.g. by

$$\frac{K_s U_1}{a} < w < \frac{K_s U_2}{a}; \quad \frac{K_s}{a} > 0 \tag{4}$$

In controlling unstable systems (a<0) it is yet necessary to introduce notion of the **admissible initial states y_0**. It means that the sign of the output derivative has to be changeable by the admissible control also in the initial states, when

$$(K_s U_1 - a y_0)(K_s U_2 - a y_0) < 0$$

This, according to the convexity of the admissible control (3) and the linearity of the plant feedback (1) guarantees that the sign of the output derivation can be changed in all points of a trajectory between an admissible initial and the desired state.

Since the rate of control signal changes in (1) is proportional to the control signal values, the fastest possible transients towards a new desired output value correspond to the limit control signal values. The well known relay (bang-bang, or on-off) controller can be described e.g. by the control algorithm

$$u(k) = \begin{cases} U_1; & K_s e(k) < 0 \\ U_2; & K_s e(k) \geq 0 \end{cases} \tag{5}$$

whereby

$$e(k) = w(k) - y(k) \tag{6}$$

is the control error.

However, under the relay control, oscillations occur in the neighbourhood of the desired state. These may be suppressed by the more general pole assignment control.

One moment of the specification of the pole assignment control means that, under assumption of a piecewise constant reference signal w=const, the control error (6) is desired to decrease λ-times in each control step, i.e.

$$e(k+1) = \lambda e(k) \tag{7a}$$

Usually, the quotient λ is denoted as the closed loop pole. Expressed in the terms of the closed loop transfer function, the input/output pole assignment behaviour is described by

$$F_w(z) = \frac{1-\lambda}{z-\lambda} \tag{7b}$$

In order to achieve a monotonous control error decrease, the choice $\lambda \in \langle 0,1)$ has to be done. The demand (7) together with (2) leads to a P-controller algorithm with an additional bias signal u_R necessary to maintain the output at the reference value in the steady state:

$$u(k) = K_R(w - y(k)) + \frac{a}{K_S} w = K_R e(k) + u_R \tag{8}$$

$$u_R = wa/K_S$$

$$K_R = (D-\lambda)/K \quad , \quad a \neq 0$$
$$K_R = (1-\lambda)/K_S T \quad , \quad a = 0 \tag{9}$$

It is important to note that, with an admissible control signal (3), the regular (linear) decrease (7) can only be achieved in the proportional band - on a limited set of initial states surrounding an admissible set point value w. At more distant points the control value (8) exceeds the given limits (3) and, therefore, the decrease (7) cannon be achieved. The limitation of (8) can be expressed as

$$u(k) = sat\{K_R e(k) + u_R\} \tag{10}$$

whereby the function sat {} (saturation) is defined as

$$sat\{u\} = \begin{cases} U_2, & u > U_2 \\ u, & U_1 \leq u \leq U_2 \\ U_1, & u < U_1 \end{cases} \tag{11}$$

This second important aspect of the pole assignment P-controller with subsequent limiter is that in the region of saturation (outside of the proportional band) it gives the fastest rate of changes towards the set point - it acts similarly as the relay minimum time controller.

In other words, for an admissible initial state y_0, for admissible reference value w and $\lambda \in \langle 0,1)$ the output signal of the P-controller (8) can be limited to an arbitrary interval (3) containing u_R in its interior **without any parasitic effects like the overshoot, oscillations**, etc. The limits imposed only change the maximal possible rate of output variations and so more or less prolong the transient processes. The control sequence may consist of a train of saturated pulses followed by a non-saturated pulse sequence corresponding to (7) which finally takes the output to the desired value. In an ideal case corresponding to λ=0 (deadbeat response), the non-saturated pulse sequence can be reduced to a single step.

However, it is known that the minimum time controllers are very sensitive to the neglected time delays, quantisation ,noise, etc. Therefore, it is frequently required to decrease the controller gain and to slow down the transient responses by choosing an appropriate value $\lambda \in \langle 0,1)$. The desired closed loop pole λ is as a quotient characterising the **maximal allowed control error decrease per one control step.** The pole assignment P-controller (9) followed by the limiter (10) may be interpreted as the **minimum time pole assignment controller with the control error decrease limited by** $e(k+1)/e(k) \leq \lambda$,i.e. the minimum time controller giving the fastest possible transients towards the desired output value outside of the proportional band and guaranteeing the prescribed control error decrease per one sampling period (7a) within the proportional band.

From the controller equation (8) and from the constrained proportional band it follows that by limiting the control error decrease per one sampling period one simultaneously limits the possible control signal changes per one sampling period.

3. COMPENSATION OF DISTURBANCES

Usually, the possible plant disturbances are approximated by an equivalent input disturbance, since it can be simply compensated by an opposite signal added to the output of the controller. Similarly, an output disturbance may be compensated by a corresponding modification of the set point. Here, for the sake of brevity, only the compensation based on the input disturbance reconstruction and compensation will be considered.

An input disturbance changes the admissible reference signal set, as well as the admissible initial condition set, when e.g. only reference values and disturbances satisfying condition

$$u_R = wa/K_s - v \in (U_1, U_2) \qquad (12)$$

may be considered. The basic problem of real control loops is that the disturbances are mostly not measurable and not known in advance. Therefore, different observers for the reconstruction of the equivalent disturbances are used, denoted in connection with the disturbance compensation usually as **controllers with I-action**.

4. CLASSICAL PI-CONTROLLER

To explain problems of the classical PI-controller the transient responses corresponding to the control law

$$u_I(k) = u_I(k-1) + K_I e(k) \qquad (13)$$

$$u(k) = u(k-1) + K_R\left[e(k) - e(k-1)\right] + K_I e(k) \qquad (14)$$

will be analysed at first. For the sake of simplicity, let us start with considering an example of the single integrator as the plant. The only role of the integral action is to compensate a possible input disturbance v. If the proportional gain corresponding to the parameter λ is given by (9), the integral gain K_I corresponding to the double real closed loop pole of the loop with I-action is

$$K_I = K_R(1 - \lambda^{1/2})/(1 + \lambda^{1/2}) \qquad (15)$$

To cover all characteristic situations, it will be considered a unite step of reference signal in combination with 3 different disturbances

a) $v > 0$; b) $v = 0$; c) $v < 0$

Note, that at the beginning of the transient (when e>0) the output of integral part rises in positive direction in spite to the fact that for the compensation of a positive disturbance v>0, a contrary acting (negative) compensation signal is needed. This false integration stops only after exceeding the reference value. So, **already in the linear case** the step reaction curve of loops with permanently working parallel I-action is characterised by a high overshoot and a long-term balanced process (Fig.1a).

An additional overshoot occurs also for v = 0 when there is no integration needed at all and the I-action is overfluous.

Transients without overshoots are possible for v < 0 only in the case (c). Choosing an appropriate K_I the transients could be optimised. However, the optimisation procedure has to be repeated for each new disturbance value and to take into account the actual initial conditions. This, of course, has no practical sense, since the disturbance is supposed to be unknown!

Since the limitation of the control signal prolongs the duration of the initial phase of transients with the false integration, the undesired overshoot will increase. It has led to interpretation, that the false integration - **the windup effect** - is caused by the control signal saturation (Fig. 1b). As a result of this interpretation, many different anti-windup controller modifications, see (e.g., Peng, *et al.*, 1996; Kothare, *et al.*, 1994) have been developed with anti-windup additional loops activated in the case of the control signal limitation. It is interesting to note that although these methods usually declare their "mission statement" "to suppress the effect of the control signal saturation" and to recover the linear behavior, they often yield better performance as the original linear controller (compare curves b in Fig.3 with those in Fig.1a) and so they actually do not obey their mission!

From this simple example it is clear that an optimisation of the PI-controller with the parallel I-action has **no sense.** But, this is actually **the reason** why a lot of papers on optimal setting of PI-controller can be found in literature!

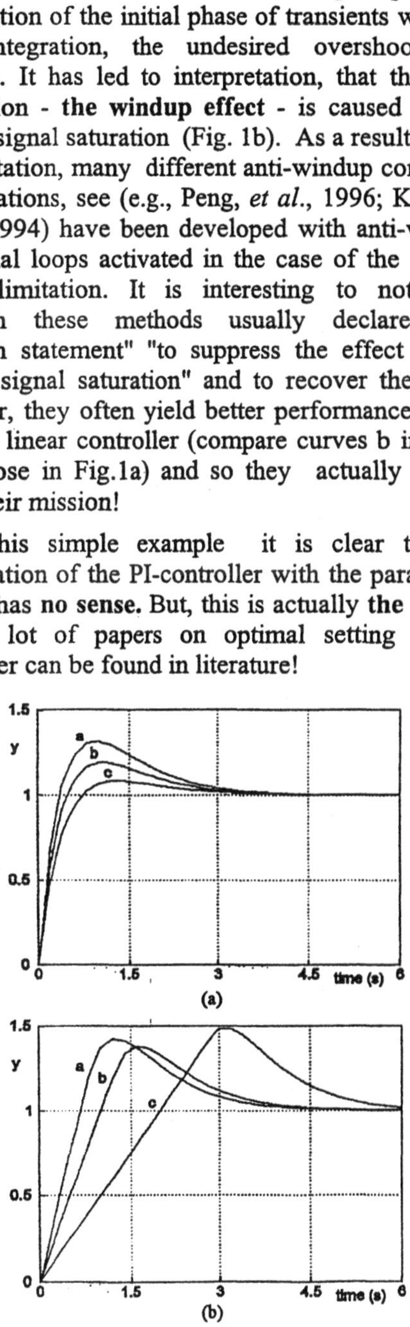

Fig. 1. Classical PI-controller (13-15): step response without (a) and with control signal saturation (b). a=0; T=0.04; $\lambda = 0.5$; $|u| \leq 0.2$.

5. ANTI-WINDUP PI_1-CONTROLLER

As it was introduced in the section above, one of the reasons to use the controllers with I-action is to compensate a possible disturbance causing under P-control a steady state control error. From the plant equation written for the actual input u_a (Fig.2)

$$y(k) = Dy(k-1) + Ku_a(k-1) \qquad (16)$$

it is easy to calculate the estimate of $\hat{u}_a(k-1)$ and then

$$v(k-1) = u_a(k-1) - u(k-1) = \\ = [y(k) - Dy(k-1)]/K - u(k-1) \qquad (17)$$

In order to compensate possible noise influence, it is recommended to filter the value of $\hat{v}$ according to

$$v_f(k) = av_f(k-1) + (1-a)v(k-1) \qquad (18)$$

The role of the filter pole α is similar to that of the closed loop pole in the P-controller design - by slowing down the reconstruction process in order to compensate noise influence the disturbance compensation works also slower.

In order to be able to compare the developed controller with the other already known ones it is interesting to derive the controller equations in the proportional band, when the saturation is not being active. Here, the nonlinear controller can be replaced by a two degree of freedom controller or by a linear PI-controller and an additional input filter. The role of input filters in suppressing the windup phenomenon is well known and the industrially produced controllers give an equivalent possibility of modified control algorithms, e.g. in the form (Takahashi *et al.*, 1971) with error acting on I only

$$u(k) = u(k-1) - K_R[y(k) - y(k-1)] + K_I e(k) \qquad (19)$$

(In standard textbooks this problem is mostly not presented, or explained very vaguely!) The developed anti-windup controller can also be expressed as a parallel connection of two PI-controllers having two different inputs (the control error and the plant output), as well as different parameters

$$U(z) = \left[K_R a + K_R(1-a)\frac{z}{z-1}\right]E(z) - \\ -\left[\frac{1-a}{K}D + \frac{a(1-a)}{K_s}\frac{z}{z-1}\right]Y(z) \qquad (20)$$

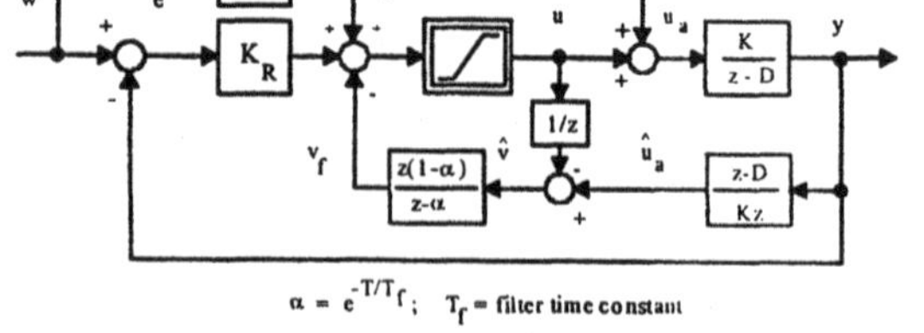

Fig. 2. Anti-Windup PI_1-controller: P-controller with the exact input disturbance reconstruction and compensation + additional filtration.

This form of the controller equations shows the relations with the weighted input controllers treated recently by several authors.

If starting with any linear form, it is problematic to get the modified nonlinear structure fully respecting the control signal saturation. Therefore, it is more appropriate to start with the nonlinear control structures giving a clear role of the control signal saturation, to extend it by the disturbance reconstruction and compensation and to prove the stability, e.g. by using the Popov or circle criterion.

6. SYSTEMS WITH DOMINANT DYNAMICS IN THE FEEDBACK PATH (I_0- AND PI_0-CONTROLLER)

Till now it has been shown that, in controlling the 1st order plant, the P-controller can be equipped by additional loops for disturbance reconstruction and compensation. Only within the proportional band, it can be transformed into an equivalent PI-controller with input filter, or into a parallel combination of two PI-controllers with control error and the plant output as inputs.

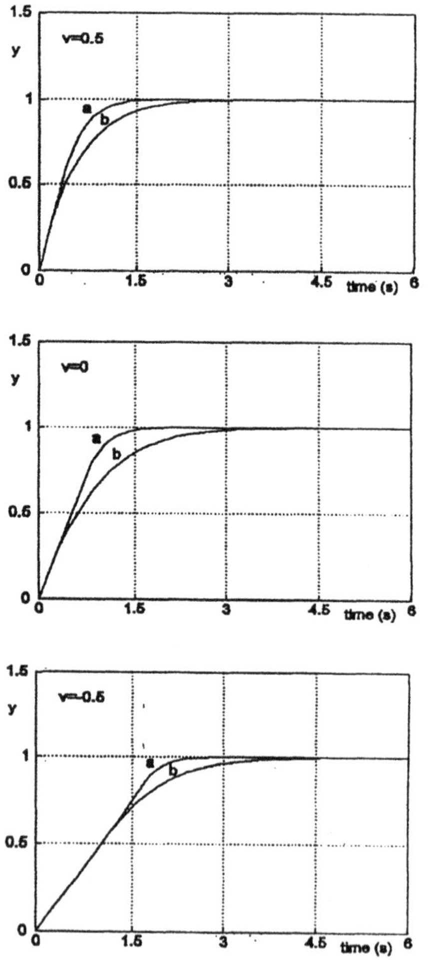

Fig. 3. Transient responses: Anti-Windup PI_1-controller (a) and linear PI-controller with saturation $|u| \leq 0.2$ and the anti-windup compensator according to Glattfelder and Schaufelberger (1986) (b).

Next it will be shown, that the structure of a single 'linear" PI-controllers can represent also an optimal solution. However, the dominant 1st order dynamics will now be placed in the **feedback** path.

6.1. Memoryless system and I_0 - controller.

Let us consider memoryless system described by the equation

$$y = K(u + v) = Ku_a \tag{21}$$

whereby v represents a non measurable and piecewise constant deterministic disturbance, which may also represent discrepancy between the supposed and the actual loop parameters.

In the case of negligible disturbances the output can be simply set to the reference value y=w through the feedforward control

$$u(k) = \frac{1}{K}w(k) \tag{22}$$

Under influence of a constant input disturbance v, the admissible reference signal set corresponding to the admissible control $u \in \langle U_1, U_2 \rangle$ is given as

$$w \in \left((U_1 + v)K; (U_2 + v)K \right) \tag{23}$$

The input disturbance can be reconstructed as the difference between the identified input $u_a(k-1)$ and the controller output u(k-1)

$$v(k-1) = \frac{1}{K}y(k-1) - u(k-1) \tag{24}$$

The control signal can now be computed by means of

$$u(k) = \frac{1}{K}w(k) - v(k-1) =$$
$$= u(k-1) + K_I e(k); \quad K_I = \frac{1}{K} \tag{25}$$

whereby for $w=w(k)=const$

$$e(k) = w - y(k-1) \tag{26}$$

is the control error.

Now, a straightforward solution would be to introduce a filter into the disturbance reconstruction. However, in order to compensate also the possible system's imperfection (fluctuations of the gain K), another solution will be found by desiring a systematic control error decrease defined by

$$e(k+1) = \lambda e(k) \quad ; \quad \lambda \in \langle 0, 1) \tag{27a}$$

In this case, the closed loop transfer function becomes

$$F_w(z) = \frac{(1-\lambda)z}{z-\lambda} \tag{27b}$$

and from the above equations it follows

$$K_I = (1-\lambda)/K \tag{28}$$

The algorithm (25) corresponds to the I-controller. In an ideal case (K_I=1/K) the process reaction curve has a stepwise character (deadbeat response) and so, for an admissible reference signal, **the saturation does not occur** - the control signal immediately acquires a new admissible steady state value! The saturation neither occurs for the controller setting corresponding to (27-8). So, **this dynamical class of controllers is naturally good treatable within the scope of the linear control theory!**

6.2 Compensation of one time constant: PI_0 - controller.

Each control loop naturally involves a lot of different time delays. Often it is e.g. necessary to consider a feedback time delay introduced by sensor or filter dynamics, when the controller receives values of a delayed signal $y_m(k)$ instead of the real output values y(k-1). The influence of the time delays increases for relatively short sampling periods, when the I-controller (25) leads to overshoot and oscillations. These can be suppressed in two ways: **(a)** through an appropriate choice of the closed loop pole λ (decrease of the controller gain), or **(b)** through the reconstruction of the actual output y(k-1). In the case of the 1st order (stable) dynamics described by

$$y_m(k) = Dy_m(k-1) + (1-D)y(k-1)$$
$$D = e^{-T/T_s} \tag{29}$$

with the sensor time constant T_s , the reconstructed output is given by

$$y(k-1) = \frac{y_m(k) - Dy_m(k-1)}{1-D} =$$
$$= y_m(k) + \frac{D}{1-D}[y_m(k) - y_m(k-1)] \tag{30}$$

The corresponding structure (Fig.4a) represents compensation by inverse transfer function and it can also be modified into the equivalent loop with PI-controller and input filter (Fig4b).

It is to note that, for admissible reference signals and correctly tuned controller $\left(K_I = (1-\lambda)/K;\ \lambda \in \langle 0,1) \right)$, the control signal does not attack the saturation limits. Therefore, the saturation nonlinearity can be omitted and the resulting controller replaced by the classical linear PI-controller. Also here it should be completed by an input filter or implemented by the modified algorithm (19) with error acting on I-only and K_R=$K_I D/(1-D)$.

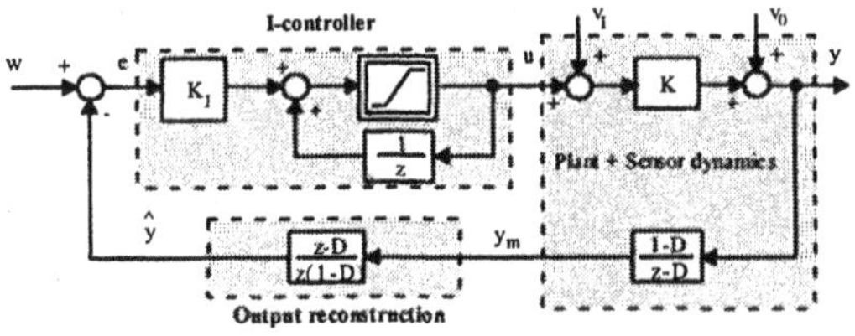

Fig. 4. PI_0-controller: I-controller with compensation of the feedback time constant.

6.3 Optimal and suboptimal dynamics of PI-controllers.

In the above control loop the sensor dynamics must be stable. In the deadbeat case, the control, as well as the output signal y change stepwise after a reference signal step. So, the measured output y_m converges exponentially to the new admissible reference value w without attacking the saturation. Therefore, the measured error is exponentially stable and the same controller can be used not only for controlling the output y, but also for controlling y_m. It means that **the linear PI_0-controller is also acceptable for structures with the 1st order plant dynamics in the feedforward path**. Of course, this configuration is no more the optimal one - the loop behaviour can be improved by different measures (e.g. changing experimentally the controller setting). However, these modifications do not lead to a generally valid optimum! Indeed, there exist infinitely many papers, where such an "optimum" has been found, but always only for a special combination of saturation limits, plant parameters, sampling periods, initial states, etc. Here, in "optimisation" of not well defined task the roots of inflation of the "optimal" controller settings may be found.

In order to distinguish the two different dynamical classes of the PI-controllers, the indices "0" and "1" have been introduced denoting the number of possible saturated pulse trains in the optimal control signal sequence. (Under circumstances the same denotation is being used also for other purposes in the literature!)

7. GENERALISATION

At the first place, it has to be mentioned that - after specifying the desired input/output behaviour by (7b) and choosing the observer pole a - the same controller can also be derived using the anti-windup controller modification heuristically proposed by Aström and Wittenmark (1984). However, the advantage of the newly developed strategy is that it can be easily extended for controlling higher order plants, while the Aström's method remains effective only in controlling 1st order plants in the feedforward path.

Preserving the above described control signal dynamics, a PID_0-controller could be derived from I_0-controller by compensating two stable time constants in the feedback. The algorithm can be developed from the difference equation of the delay, or by applying the standard synthesis procedure described e.g. by Aström and Wittenmark (1984) for the desired input/output behaviour (27b), desiring the deadbeat observer and not allowing a plant zeros cancellation.

For a possibly unstable 1st order plant in the feedforward path and a stable time constant in the feedback a PID_1-controller can be developed by the modified procedure by Aström and Wittenmark (1984): It is desired to specify the input/output behaviour (7b), to choose the observer pole α and not allow a plant zeros cancellation.

The dynamical group of PID_0-controllers will be characteristic by an exponential control signal response to a step change of the set point. For l=0 the control sequence will have a stepwise character.

The PID_1-controllers may have one saturated train of pulses in their control signal response.

Besides of these two dynamical groups a new group of controller arises that could be denoted as PID_2-controller: They are characteristic by the **two possible saturated trains of pulses** in their control signal response. This group corresponds to the control of the 2nd order plants placed in the feedforward path. However, the corresponding controllers are nonlinear and they cannot be derived by a simple modification of the linear methods described in standard textbooks.

8. CONCLUSIONS

It was shown that an optimal controller design depends on the distribution of the dynamical terms in the control loop. Two different types of PI-controllers have been identified for the two possible 1st order loops and the role of input filters, and or modified controller equation have been explained.

Similarly, for the control loops with 2nd order dominant dynamics, 3 different types of PID controllers may be developed.

Generalisation of the above results points out not only on the necessary changes in the control design strategies, but also on need of appropriate outputs given by the identification phase (when e.g. a measured process reaction curve does not yield sufficient information about distribution of the dynamic terms in the loop).

The above described missing aspect of the controller design could have been neglected for such a long time, since the "linear" PI-controller replacing the optimal one could under certain circumstances give "acceptable" results.

Of course, the existence of not acceptable solutions has been the driving force in development of different new modifications of optimal controller tuning and different anti-windup structures. The existence of two dynamical classes of PI-controllers and 3 dynamical classes of PID controllers explains also the roots of "inflation" of different "optimal" tuning procedures.

REFERENCES

Aström, K.J and B. Wittenmark (1984). *Computer Controlled Systems. Theory and Design*. Prentice Hall, Englewood Cliffs, N.J..

Glattfelder, A.H. and W. Schaufelberger (1986). Start-up performance of different proportional-integral-anti-wind-up regulators. *Int.J.Control*, **44**, No.2, 493-505.

Kothare, M.V., P.J. Campo, M. Morari and C.V. Nett (1994). A Unified Framework for the Study of Anti-windup Designs. *Automatica*, **30**, 1869-1883.

Peng, Y., D. Vrancic and R. Hanus (1996). Anti-Windup, Bumpless, and Conditioned Transfer Techniques for PID Controllers. *IEEE Control Systems*, **14**, 48-57.

Takahashi, Y., C.S. Chan and D.M. Auslander (1971). Parametereinstellung bei linearen DDC-Algorithmen. *rt19*, No.6, 237-244.

THEORETICAL ANALYSIS OF ZIEGLER-NICHOLS CONCLUSIONS

Katarina Žáková and Mikuláš Huba

Dept. of Automation and Control, Faculty of Electrical Engineering and Information Technology, Slovak University of Technology, Ilkovičova 3, 812 19 Bratislava, Slovak Republic.
E-mail: katarina@elf.stuba.sk, huba@elf.stuba.sk

Abstract: The paper analyses characteristics of the closed loop system with the P-, PI-, and PID-controllers and different models approximating the plant time lags. It extends some early analytical results derived by Oldenbourg and Sartorius (1951) in such a manner that they can be considered as a generalisation of the other well known classical method experimentally derived by Ziegler and Nichols (1942) and modified for discrete time systems by Takahashi at al. (1971).

Keywords: PID controller, continuous control, digital control, stability

1. INTRODUCTION

The method of Ziegler and Nichols (1942) belongs to the most popular methods used for settings of controllers. It enables simple controller setting also during the course of technological process. It was presented in two various forms. In the first one, the "ultimate sensitivity method" it is necessary to bring the system to the stability boundary by raising the sensitivity (gain) of a P controller and to measure the critical controller gain $K_{Cu} = K_u$ and the period of oscillation P_u. However, sometimes it is impossible to take the system to the stability boundary or such measurement is not allowed because of the safety and the technological reasons. In this case, a couple of parameters R and L used for setting of P, PI or PID controllers can be taken from the process reaction curve excited by the unit change of the manipulated variable (the "transient response method") (Fig.1). Symbol R is used for the slope of a tangent drawn to the point of inflection of the unit step response. For a non unit step response it has to be normalized by the actual change Δu of the manipulated variable. L is called the "total time lag" and defined as the time at which this tangent cuts the time axis.

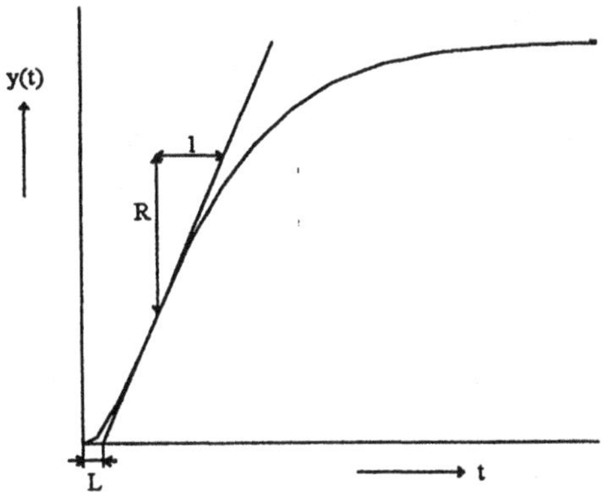

Fig. 1. Typical process reaction curve

Ziegler and Nichols have designed controllers to guarantee the "quarter amplitude ratio" (or a quarter amplitude damping). For basic types of linear continuous controllers:

* The work has been partially supported by the Slovak Scientific Grant Agency, Grant No. 1/4220/97.

P-controller $C(s) = K_C$

PI-controller $C(s) = K_C\left(1+\frac{1}{T_i s}\right)$

PID-controller $C(s) = K_C\left(1+\frac{1}{T_i s}+T_D s\right)$

they recommended following sets of parameters:

"Transient response" method

	K_C	$1/T_i$	T_D
P	$1/RL$	-	-
PI	$0.9/RL$	$0.3/RL$	-
PID	$1.2/RL$	$0.5/RL$	$0.5L$

"Ultimate sensitivity" method

	K_C	$1/T_i$	T_D
P	$0.5K_u$	-	-
PI	$0.45K_u$	$1.2/P_u$	-
PID	$0.6K_u$	$2/P_u$	$P_u/8$

For a typical application, relations between both types of measurement were defined by equations

$$L=\frac{P_u}{4} \tag{1}$$

$$R=\frac{8}{P_u K_u} \tag{2}$$

The method has been extended also for the setting of discrete controllers. Takahashi, Chan and Auslander (1971) came out from the algorithm of the continuous PID controller

$$u(t) = K_C\left[e(t)+\frac{1}{T_i}\int e(t)\,dt + T_D\frac{d}{dt}e(t)\right]$$

whereby they used a trapezoidal approximation for the integral and the derivation was substituted by a difference. In such a way they deduced similar relations as Ziegler and Nichols for the continuous case. In fact, only formulas for the "transient response" method were changed. We can see that, in comparison to their continuous counterparts, there is the sampling period T which enters into the equations in form of a contribution to the identified delay.

"Transient response" method - discrete case

	K_C	$1/T_i$	T_D
P	$\frac{1}{R(L+T)}$	–	–
PI	$\frac{0.9}{R\left(L+\frac{T}{2}\right)}$	$\frac{0.3}{L+\frac{T}{2}}$	–
PID	$\frac{1.2}{R\left(L+\frac{T}{2}\right)}$	$\frac{0.5}{L+\frac{T}{2}}$	$0.5\left(L+\frac{T}{2}\right)$

Following their experimental results, the authors didn't recommend to use the "transient response" method for $\frac{L}{T}\rightarrow 0$. Alike, the method of "ultimate sensitivity" is not recommended for the ratio $\frac{L}{T}\approx\frac{1}{4}$. Both methods give sufficient results for $\frac{L}{T}\geq 0.5$.

2. INTEGRATOR SYSTEM IN THE CONTROL STRUCTURE WITH PARASITIC DELAYS - CONTINUOUS CASE

For the sake of simplicity, Ziegler and Nichols have not analyzed a type of the time lags involved in the control loop. In order to show the possible consequences, the behavior of an integrator system combined with two basic parasitic delays: the dead time and the time constant delay will be analyzed at first. In difference to the "quarter amplitude damping" introduced by Ziegler and Nichols, the controller setting corresponding to the double real dominant pole and guaranteeing the fastest possible transients without overshoot will be used.

2.1 *Integral+dead time system*

Following the open loop transfer function

$$F(s) = K_C\,\frac{K_P\,e^{-T_d s}}{s} \tag{3}$$

the optimal gain of the controller can be determined very easily by means of a derivation (5) of the characteristic equation (4) according to the complex variable s

$$K_C K_P\,e^{-T_d s}+s=0 \tag{4}$$

$$\begin{aligned}&\tfrac{d}{ds}\left(K_C K_P\,e^{-T_d s}+s\right)=\\&=-K_C K_P\,T_d\,e^{-T_d s}+1=0\end{aligned} \tag{5}$$

Equations (4), (5) lead to the final formula for the controller gain

$$K_{Copt}=\frac{1}{eK_P T_d} \tag{6}$$

for which the dominant real pole is

$$s_{opt}=-1/T_d \tag{7}$$

To derive the critical closed loop gain we can use a substitution $s=j\omega$ in (4). In this way there arise a set of two equations

$$\begin{aligned}K_C K_P\cos(T_d\omega)&=0\\K_C K_P\sin(T_d\omega)&=\omega\end{aligned} \tag{8}$$

which finally get the critical controller gain

$$K_{Ccr} = \frac{\pi}{2K_P T_d} \tag{9}$$

The corresponding period of oscillation is

$$P_u = \frac{2\pi}{\omega} = 4T_d \tag{10}$$

2.2 *Integral+time constant system*

In the case of I_1T_1-system

$$F(s) = K_C \frac{K_P}{s(T_1 s + 1)} \tag{11}$$

the characteristic equation is

$$T_1 s^2 + s + K_C K_P = 0 \tag{12}$$

The double real closed loop pole corresponds to

$$K_{Copt} = \frac{1}{4K_P T_1} \tag{13}$$

whereby

$$s_{opt} = -\frac{1}{2T_1} \tag{14}$$

The critical controller gain for this system is going to the infinity, since all coefficients of the characteristic polynomial are positive and according to Routh-Shur test of stability it means that the system is stable for all $K < \infty$:

$$K_{Ccr} = \infty \tag{15}$$

2.3 *Adequacy of Ziegler and Nichols method*

Although Ziegler and Nichols didn't distinguished the type of parasitic time delays in the control structure, from some accepted rules it is possible to reconstruct the type of time lags they have used.

The dependence (1) between the period of oscillations P_u and the total time lag L is typical for the integral+dead time system. In the case of approximation of the process by integral+time constant system with the process reaction curve very similar to the above one, the closed loop doesn't oscillate how matter high values of controller gain will be set - the relation (2) doesn't hold and the "ultimate sensitivity" method cannot be used.

On the other hand, an active compensation of the time lags (Žáková, 1991) by the pre-act (derivative) control effect is typical for systems based on models with the integral and the dominant time constant. For an effective compensation of the dead time quite different controller structures derived by Smith (1958) have to be used.

The comparison of the optimal controller gains (6) and (13) with that one given by the "transient response method" shows that the approximations by the integral+dead time system or integral+time constant system yield the gains of the P-controller which are 2.71 or 4 times smaller than the values recommended by Ziegler and Nichols. It is caused by the different performance criterions: The double real pole corresponds to the transients without overshoot, while Ziegler and Nichols tried to set the gain as high as possible in order to reduce the offset and influences of the possible load changes. However, in spite to this differences, for practical reasons it is not so important to distinguish between both types of the time lags.

On the other hand, quite different situation arises by using the ultimate sensitivity method. According to (15) the method is not applicable for approximations based on integral+time constant models.

It can be seen from these few examples, that the acceptation of only one type of the dominant time lag would lead to contradictory results. Therefore, for the approximation of a sufficiently broad class of dynamical systems the method has to work on with models containing both types of the dominant time lags, which can be done e.g. by using the plant model

$$F(s) = \frac{Ke^{-T_d s}}{s(T_1 s + 1)} \tag{16}$$

In Huba and Žáková (1992), new relations for the optimal and the critical controller gain were introduced depending on the ratio $\frac{T_1}{T_d}$. Evaluation of the experimental results of the both forms of the method by Ziegler and Nichols enables to determine all designed parameters R, L, P_u and K_u. So it is possible to estimate the actual ratio between the both types of the parasitic lags and, after this identification, to design the optimal value of the controller gain.

2.4 *Setting of the Integral Action*

Up to here, only the design of the P-controller have been considered. The parasitic lags were compensated by decreasing the optimal controller gain. Due to this, an increase of the time delays causes an increase of the possible permanent control error. It can be eliminated by means of the integral action. The basic control loop (integrator + delays + P-controller) can be approximated by the second order dynamics:

$$F(s) = \frac{s_{opt}^2}{(s - s_{opt})^2} \tag{17}$$

Then, the characteristic equation of the basic loop extended by the loop with the I-controller ($1/T_i = K_i$) will be

$$s (s - s_{opt})^2 + K_i \, s_{opt}^2 = 0 \tag{18}$$

For this new structure, the optimal value of the integral gain K_i may be again determined to give the double real closed loop pole corresponding to the condition $\frac{dK_i}{ds} = 0$. Then

$$K_{i\_opt} = -\frac{4}{27} s_{opt} \tag{19}$$

or, particularly for the approximation with the dead time $L = T_d$:

$$K_{i\_opt} = 0.15/T_d \tag{20}$$

and for the time constant $L = T_1$:

$$K_{i\_opt} = 0.07/T_1 \tag{21}$$

The mixed case $L = T_d + T_1$ introduced in Huba and Žáková (1992) does not give simple analytical formulas.

Similarly as above the computed values of K_i are again approximately 2-times ($L = T_d$) or 4-times ($L = T_1$) smaller than the values determined by Ziegler and Nichols. As far as the setting of the proportional gain of the PI-controller is considered, Ziegler and Nichols proposed to decrease its value in comparison to the "pure" P-controller. It was caused by an increased inclination of the loop to oscillations in the presence of the I part of the controller.

2.5 *Setting of the Derivative Action*

The use of the derivative action is fully approved only for the compensation of a time constant. It was shown in Huba (1992) that for a time constant T_1 placed in the feedback loop it is appropriate to choose $T_D = T_1$, while for a time constant T_1 placed in the feedforward loop a better output behavior can be mostly reached by decreasing T_D to about $T_1/2$, i.e. to the value recommended by Ziegler and Nichols. For the compensation of the dominant dead time the derivative action doesn't bring essential advantages and here the control structures introduced by Smith (1958) have to be used. According to the stability problems, it is recommended to use digital controllers for this purpose where it is possible to reduce parasitic phenomenons by choosing appropriate sampling period.

3. DISCRETE TIME CASE

3.1 *Integral+dead time system*

For the sake of simplicity, only the control structure (3) with the dead time equal to an integer multiply of the sampling period $T_d = kT$ will be considered. Although these results does not completely characterize the systems behavior for non integer multiply, from the practical point of view it is mostly sufficient to use the derived results for an arbitrary delay.

The corresponding transfer function of the system (3) in z-domain is

$$G(z) = \frac{K_C K_P}{z^{k+1} - z^k} \tag{22}$$

The optimal controller gain is again computed with the aim to get a double real closed loop pole. It can be written in the form

$$K_{Copt} = \frac{1}{K_P T} \frac{k^k}{(k+1)^{k+1}} = \frac{1}{K_P} \frac{1}{T_d + T} \left[\frac{T_d}{T_d + T}\right]^{\frac{T_d}{T}} \tag{23}$$

whereby

$$z_{opt} = \frac{k}{k+1} = \frac{T_d}{T_d + T} \tag{24}$$

In determining the critical controller gain one can use a substitution $z = e^{j\omega}$ ($|z| = 1$). The resulting complex equation can be solved for the real and the imaginary part separately. Then

$$K_{Ccr} = \frac{2}{K_P T} \cos \frac{k\pi}{2k+1} = \frac{2}{K_P T} \cos \frac{T_d \pi}{2T_d + T} \tag{25}$$

and period of oscillation

$$P_u = 2(2k+1)T = 2(2T_d + T) \tag{26}$$

3.2 *Integral+time constant system*

The discrete transfer function of the I_1T_1-system (11) is

$$G(z) = K_C K_P \frac{b_1 z + b_0}{(z-1)(z-D)} \tag{27}$$

whereby

$$D = e^{-\frac{T}{T1}} \tag{28}$$

$$b_1 = T + T_1 D - T_1 \qquad (29)$$
$$b_0 = T_1 - TD - T_1 D$$

The optimal controller gain is

$$K_{Copt} = \frac{1}{K_P T} \frac{1 - D}{(1 + \sqrt{\tau})^2} \qquad (30)$$

whereby $\tau = \dfrac{T_1 (1 - D)}{T}$. This value can be reached for the pole

$$z_{opt} = \frac{D - \tau + (1 - D)\sqrt{\tau}}{1 - \tau} \qquad (31)$$

Similarly to the previous case, the critical controller gain is

$$K_{Ccr} = \begin{cases} \dfrac{1}{K_P T} \dfrac{2(1 + D)}{1 + D - 2\tau} & T_1 \leq 0.268T \\ \dfrac{1}{K_P T} \dfrac{1 - D}{\tau - D} & T_1 \geq 0.268T \end{cases}$$

and the corresponding period of oscillation is

$$T_0 = \begin{cases} 2T & T_1 \leq 0.268T \\ \dfrac{2\pi T}{\arccos\left[1 - \dfrac{(1 - D)^2}{2(\tau - D)}\right]} & T_1 \geq 0.268T \end{cases}$$

3.3 *Integral + generalized time delay*

As it was already mentioned above, it is recommended to consider both types of parasitic time lags in the control design. Such a system was described by (16) and its discrete counterpart is given by the following discrete function

$$G(z) = K_C K_P \frac{b_1 z + b_0}{z^k (z - 1)(z - D)} \qquad (32)$$

where D, $b_1$and b_0 are given by (28) and (29).

Relation for the optimal controller gain can be deduced from the characteristic equation of the system

$$\begin{aligned} &K(T_1 D + T - T_1) z + \\ &+K(T_1 - T_1 D - TD) + \\ &+ z^k (z - 1)(z - D) = 0 \end{aligned} \qquad (33)$$

and its derivation

$$\begin{aligned} &K(T_1 D - T_1 + T) + (2 + k)\, z^{k+1} - \\ &- (k + 1)(D + 1) z^k + kDz^{k-1} = 0 \end{aligned} \qquad (34)$$

Unfortunately, it is not possible to express the optimal controller gain by a simple analytical formula. However, for the known model parameters, one can easily find the controller gain by means of computer. As it was already mentioned above, these parameters, especially the ratio $\frac{T_1}{T_d}$, may be identified by combination of the both methods proposed by Ziegler and Nichols (see Huba and Žáková;1992).

After denoting the total time lag as

$$L = T_1 + T_d \qquad (35)$$

and introducing the substitution

$$\varepsilon = T_1 / T_d \qquad (36)$$

it can be expressed

$$T_1 = \frac{\varepsilon L}{\varepsilon + 1} \qquad (37)$$

$$k = \frac{T_d}{T} = \frac{T_1}{\varepsilon T} = \frac{L}{T(\varepsilon + 1)} \qquad (38)$$

$$D = e^{-\frac{T}{T_1}} = e^{-\frac{T(\varepsilon + 1)}{\varepsilon L}} \qquad (39)$$

Using these parameters in (33) and (34) enables to calculate for a chosen $\varepsilon = T_1/T_d$ the graphical dependence $K_{Copt} = f(\varepsilon, L, T)$ as shown on Fig.2.

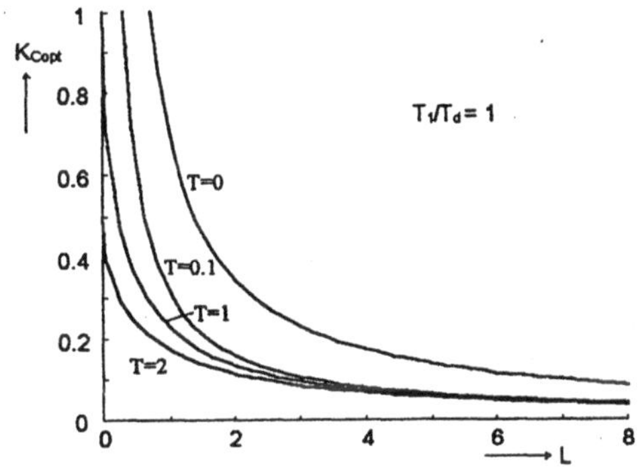

Fig. 2. Optimal P-controller gain .

Here, the global lag L is formed by the same amount of the transport delay and the time constant delay, respectively ($T_d = T_1 = L/2$). The dependence of the optimal gain is presented for various sampling periods.

The critical controller gain will be again determined by substituting $z = e^{j\omega}$ into the characteristic equation. From a set of equations for the real, as well as for the imaginary parts, the desired parameter can be received.

$$\begin{aligned} &\cos(k + 2)\omega - (1 + D)\cos(k + 1)\omega + \\ &+ D\cos k\omega + K(T_1 D - T_1 + T)\cos\omega + \\ &+ K(T_1 - T_1 D - TD) = 0 \end{aligned}$$

$$\begin{aligned} &\sin(k + 2)\omega - (1 + D)\sin(k + 1)\omega + \\ &+ D\sin k\omega + K(T_1 D - T_1 + T)\sin\omega = 0 \end{aligned}$$

The critical controller gain can be expressed as a function of the ratio $\varepsilon = T_1/T_d$, of the total lag L

and of the sampling period T. An example of the graphical dependence is presented in Fig.3.

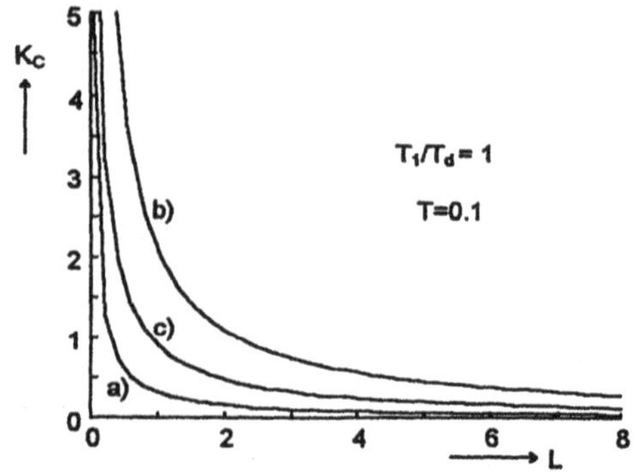

Fig. 3. Setting of P-controller a) optimal gain corresponding to the double real pole b) critical closed loop gain Kcr c) P-controller gain according to Takahashi et al.

For comparison, there is also presented the optimal controller gain corresponding to the double real pole of the loop (32) together with the gain modified for the discrete time systems by Takahashi et al (1971). It can be seen that the results correspond to the conclusions which have been done for the continuous case.

3.4 *PI controller*

The setting of the integral action can be determined in similar manner as in the continuous case. The dynamics of the loop with the P-controller can be approximated by the second order system

$$G(z) = \frac{(1 - z_{opt})^2}{(z - z_{opt})^2} \quad (40)$$

Considering an additional loop with the I-controller leads to the characteristic equation

$$(z-1)(z - z_{opt})^2 + K_I (1 - z_{opt})^2 = 0 \quad (41)$$

The double real pole corresponds to the condition $\frac{dK_I}{dz} = 0$ when

$$K_{I\_opt} = \frac{4}{27}(1 - z_{opt}) \quad (42)$$

For $L = T_d$

$$K_{I\_opt} = \frac{4}{27}\frac{T}{T_d + T}$$

and for $L = T_1$

$$K_{I\_opt} = \frac{4}{27}\frac{(1-D)\sqrt{T}}{\sqrt{T} + \sqrt{T_1(1-D)}} \quad (43)$$

An example of the optimal gain K_I normalized by the sampling period for the mixed time delay $L = T_d + T_1$ is shown in Fig.4.

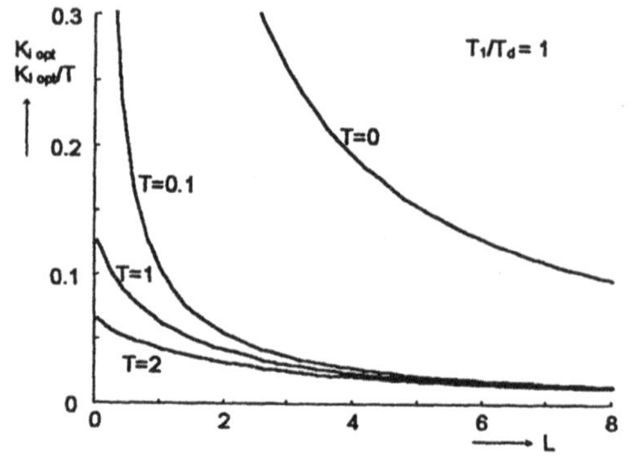

Fig. 4. Optimal integral action for PI-controller.

4. CONCLUSION

The control structure with integrator and two types of parasitic delays was analyzed and compared with the results of the method by Ziegler and Nichols and its modification for discrete systems. The analysis have been done for the continuous as well as for the discrete time case and it should cover all interesting situations occurring in controlling systems with monotonous process reaction curves by means of PID controllers. It has shown that by an approximation using the integrator in combination with a single type of the time delay it is not possible to cover all possible characteristic situations. The applicability of the method can be extended by introducing a mixed time delay. However, the number of experiments has to be increased and the results can not be expressed by simple analytical formulas.

5. REFERENCES

Huba, M. (1992). *Teória automatického riadenia III.* AS STU Bratislava.

Huba, M. and K. Žáková (1992). Possibilities of extended use of Ziegler and Nichols method. *Kybernetika a informatika*, **5**.

Oldenbourg, R.C. and H. Sartorius (1951). *Dynamik selbsttätiger Regelungen.* 2.Auflage, R.Oldenbourg-Verlag, München.

Smith,O.J.M (1958). Feedback control systems. McGraw-Hill.

Takahashi, J.G, C.S. Chan and D.M. Auslander (1971). Parametereinstellung bei linearen DDC-Algorithmen. *Regelungstechnik*, **19**, 237-244.

Ziegler, J.G. and N.B.Nichols (1942). Optimum settings for automatic controllers. *Trans. ASME*, 759-768.

Žáková, K. (1991). *Teoretická analýza experimentálnych výsledkov Zieglera a Nicholsa.* Diploma thesis.

MODELING, CONTROL AND DESIGN OF DISTRIBUTED PARAMETER SYSTEMS AS LUMPED INPUT AND DISTRIBUTED OUTPUT SYSTEMS WITH DEMONSTRATION IN MATLAB

G.Hulkó, M. Antoniová, C. Belavý, A. Heugerová, J. Belanský, J. Szuda, P. Végh

Department of Automatic Control and Measurement, Faculty of Mechanical Engineering, Slovak Technical University, Nám. Slobody 17, 812 31 Bratislava, Slovak Republic.

Abstract: The paper brings an overview of methods of modeling, control and design of distributed parameter systems with lumped input and distributed output. Selected results will be demonstrated using high-performance simulation programming tools offered by environment of MATLAB.

Keywords: Distributed parameter system, modelling, control, design, demonstration

1. INTRODUCTION

In technical but also other branches Distributed Parameter Systems (DPS) are mostly present as Lumped Input and Distributed Output Systems (LDS). Dynamics of such systems is described by lumped and distributed models. Control problems for LDS systems will be solved using open loop and closed loop structures as well as internal model control principle. Robust and adaptive control aspects will be also outlined. Design of spatial distribution and construction of DPS are also included.

Some results will be demonstrated under MATLAB compatible simulation system selected from the monograph of Hulkó G. et al.: Modeling, Control and Design of Distributed Parameter Systems as Lumped Input and Distributed Output Systems with Demonstration in MATLAB.

2. MODELING OF LDS

Linear LDS, as shown in fig. 1, is considered to be distributed on the interval ‹0,L› and its time-space variables can be depicted in a right-handed axonometric system.

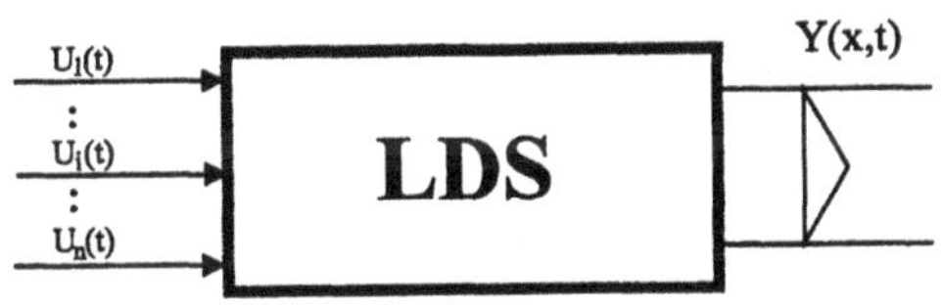

Fig. 1. Lumped input and distributed output system

Assuming both, zero initial and boundary conditions, influence of an input signal, $U_i(t)$, onto LDS is shown in fig. 2.

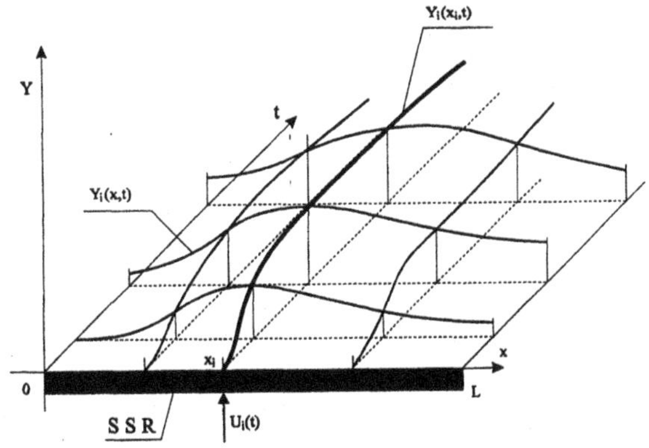

Fig. 2. Response of LDS to input $U_i(t)$

The system output is represented by the distributed variable $Y_i(x,t)$. The input-output relation can be described by convolution

$$Y_i(x,t) = \mathcal{G}_i(x,t) \otimes U_i(t) \tag{1}$$

where $\mathcal{G}_i(x,t)$ stands for impulse function of the LDS and $\otimes$ designates the operation of convolution.

For a selected $Y_i\,(x_i,t)$, at point x_i, it holds

$$Y_i(x_i,t) = \mathcal{G}_i(x_i,t) \otimes U_i(t) \tag{2}$$

where $\mathcal{G}_i\,(x_i,t)$ is the partial distributed impulse function of LDS at point x_i. When applying all lumped input variables, the following models can be obtained

$$\{Y_i(x,t) = \mathcal{G}_i(x,t) \otimes U_i(t)\}_{i=1,n} \tag{3}$$

$$\{Y_i(x_i,t) = \mathcal{G}_i(x_i,t) \otimes U_i(t)\}_{i=1,n} \tag{4}$$

while the overall distributed output is given as

$$Y(x,t) = \sum_{i=1}^{n} Y_i(x,t) = \sum_{i=1}^{n} \mathcal{G}_i(x,t) \otimes U_i(t) \tag{5}$$

In s-region, the following model equations hold

$$\{Y_i(x,s) = S_i(x,s) U_i(s)\}_i \tag{6}$$

$$\{Y_i(x_i,s) = S_i(x_i,s) U_i(s)\}_{i=1,n} \tag{7}$$

$$Y(x,s) = \sum_{i=1}^{n} Y_i(x,s) = \sum_{i=1}^{n} S_i(x,s)\, U_i(s) \tag{8}$$

Through discretization, assuming unit sampling period, we get

$$\{Y_i(x,k) = \mathcal{G}H_i(x,k) \oplus U_i(k)\}_{i=1,n} \tag{9}$$

$$\{Y_i(x_i,k) = \mathcal{G}H_i(x_i,k) \oplus U_i(k)\}_{i=1,n} \tag{10}$$

$$Y(x,k) = \sum_{i=1}^{n} Y_i(x,k) = \sum_{i=1}^{n} \mathcal{G}H_i(x,k) \oplus U_i(k) \tag{11}$$

where $\mathcal{G}H_i(x,k)$ is the distributed, discrete impulse function of LDS, involving a zero-order holds $\{H_i\}$ i = 1,n: HLDS. $\oplus$ designates the operation of discrete convolution.

In the z- region, further can be defined

$$\{Y_i(x,z) = SH_i(x,z)\, U_i(z)\}_{i=1,n} \tag{12}$$

$$\{Y_i(x_i,z) = SH_i(x_i,z)\, U_i(z)\}_{i=1,n} \tag{13}$$

$$Y(x,z) = \sum_{i=1}^{n} Y_i(x,z) = \sum_{i=1}^{n} SH_i(x,z)\, U_i(z) \tag{14}$$

Eq. (11) is frequently used in numerical simulations when distributed, discrete impulse responses are obtained through differences of shifted discrete distributed transient responses $\{\mathcal{H}H_i(x,k)\}_{i=1,n}$,

$$\{\mathcal{G}H_i(x,k) = \mathcal{H}H_i(x,k) - \mathcal{H}H_i(x,k-1)\}_{i=1,n} \tag{15}$$

For control synthesis purposes, let us introduce reduced distributed transient responses as

$$\{\mathcal{H}HR_i(x,\infty) = \mathcal{H}H(x,\infty)/\mathcal{H}H(x_i,\infty)\}_{i=1,n} \tag{16}$$

In steady-state, it holds then

$$Y(x,\infty) = \sum_{i=1}^{n} Y_i(x,\infty) = \sum_{i=1}^{n} Y_i(x_i,\infty)\, \mathcal{H}HR_i(x,\infty) \tag{17}$$

When constructing models in analytical form, the basic structure of LDS, given in fig. 3, can be considered where, besides the Distributed Input and Distributed Output System (DDS), blocks of actuating members $\{SA_i\}_{i=1,n}$ and generators $\{GU_i\}_{i=1,n}$ of distributed input signals: $\{U_i(\xi,p)\}_i$ are involved, while $U(\xi,p) = \sum_{i=1}^{n} U_i(\xi,p)$.

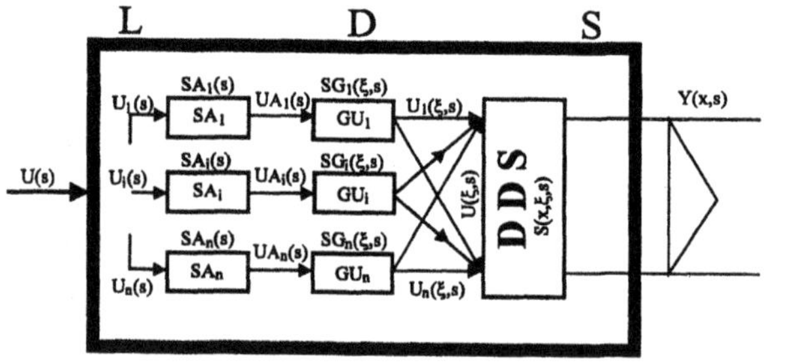

Fig. 3. Transfer blocks of LDS

The transfer functions from fig. 3 yield in s-region

$$\left\{ S_i(x,s) = \int_0^L S(x,\xi,s) SG_i(\xi,s) SA_i(s)\, d\xi \right\}_{i=1,n} \tag{18}$$

The transfer functions of the distributed input generators often take the form

$$\{SG_i(x,s) = SG_i(s) T_i(x)\}_{i=1,n} \tag{19}$$

where $\{T_i(x)\}_i$ mean shapers of the distributed input in subregions of the interval ‹0,L›.

The above equations confirm that lumped and distributed transfer functions can be used in analytical modeling of LDS.

For determination of both $\{\mathcal{H}H_i(x,k)\}_i$, and $\{\mathcal{G}H_i(x,k)\}_i$, in a complex region of definition, the Finite Element Method (FEM) can be used.

In description of LDS, the convention used in numerical solution of Partial Differential Equations (PDE) and in FEM has been fully adopted. The values of distributed variables are calculated in selected points of the region of definition of LDS and the other values are approximated using an interpolation technique. Since the desired accuracy is ensured by an appropriate choice of the method of approximation - there is no distinction between the approximated, Y(x,k), and the approximating, Y_a (x,k), variables. Further, only the distributed output, Y(x,k), will be mentioned.

Experimental identification of the relations between the lumped inputs and distributed outputs in selected points of the definition region is carried out using experimental identification methods, well established for lumped systems. Functions between the selected points are then interpolated in dependence on the desired accuracy, [ ].

To the end of this section, it can be concluded that for LDS all the methods of modelling and identification valid for the Lumped Parameter Systems(LPS) and DPS, can be utilized.

3. CONTROL OF DPS

Let the goal of control is to maintain the steady-state control error to be minimal, i.e.

$$\min\|W(x,\infty)-Y(x,\infty)\| = \|W(x,\infty)-\breve{Y}(x,\infty)\| \qquad (20)$$

where W(x,∞) is the control variable in steady-state, $\breve{Y}(x,\infty)$ is its optimal approximation and ||.|| is the quadratic norm.

Let us solve the above problem first in distributed open loop structure as it is shown in fig. 4.

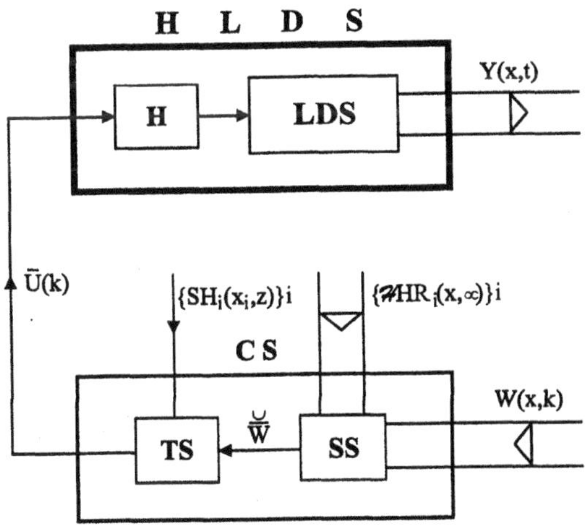

Fig. 4. Distributed parameter open loop control scheme

Let a non-zero, step-wise control signal, W(x,k) = W(x, ∞) is applied to the steady-state, zero initial condition system. In the control scheme, block SS solves the following approximation problem

$$\min\left\|W(x,\infty)-\sum_{i=1}^{n} W_i\,\mathcal{H}HR_i(x,\infty)\right\| = \qquad (21)$$
$$= \left\|W(x,\infty)-\sum_{i=1}^{n} \breve{W}_i\,\mathcal{H}HR_i(x,\infty)\right\|$$

The elements of the optimal solution vector, $\breve{\mathbf{W}} = \{\breve{W}_i\}_i$, in Control System (CS) influence in block TS SISO control loops $\{SH_i(x_i,z), R_i(z)\}_i$ as it is shown in fig. 5.

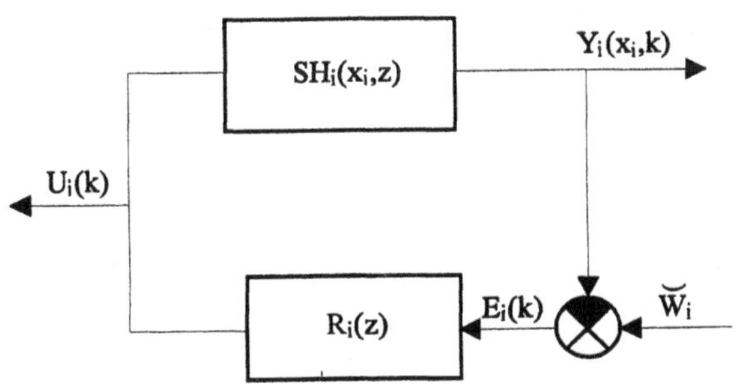

Fig. 5. The ith SISO control loop in block TS: $\{SH_i(x_i,z), R_i(z)\}_{i=1,n}$

The manipulated variables, $U_i(k)$ are inputs of HLDS. These control loops are set to have zero steady-state control errors. Then in steady-state $\{\breve{W}_i = \breve{Y}_i(x_i,\infty)\}_i$ holds. Eq. (17) yields

$$\left\| W(x,\infty) - \sum_{i=1}^{n} \breve{W}_i \mathcal{H}HR_i(x,\infty) \right\| =$$

$$\left\| W(x,\infty) - \sum_{i=1}^{n} \breve{Y}_i(x_i,\infty) \mathcal{H}HR_i(x,\infty) \right\| = \qquad (22)$$

$$= \left\| W(x,\infty) - \breve{Y}(x,\infty) \right\|$$

It means that, provided the solution of the approximation problem (21) does exist and is unique, $\breve{Y}(x,\infty)$ is also unique and ensures the best approximation of $W(x,\infty)$ in steady-state.

Following this philosophy of control synthesis, distributed parameter feedback control loops can be designed, also with robust or adaptive properties (see Fig. 6, 7, 8, 9).

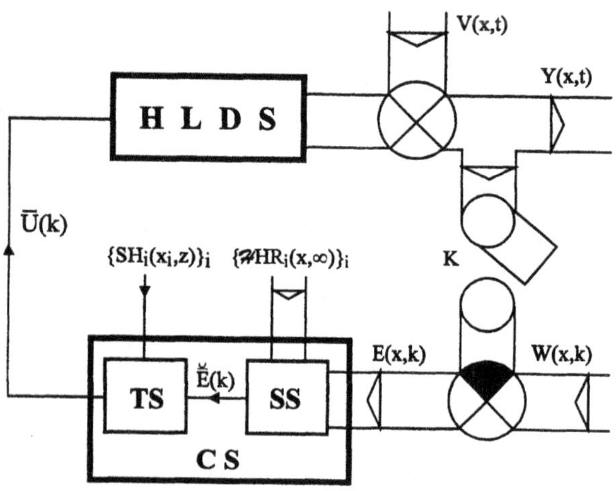

Fig.6. Distributed parameter feedback control scheme

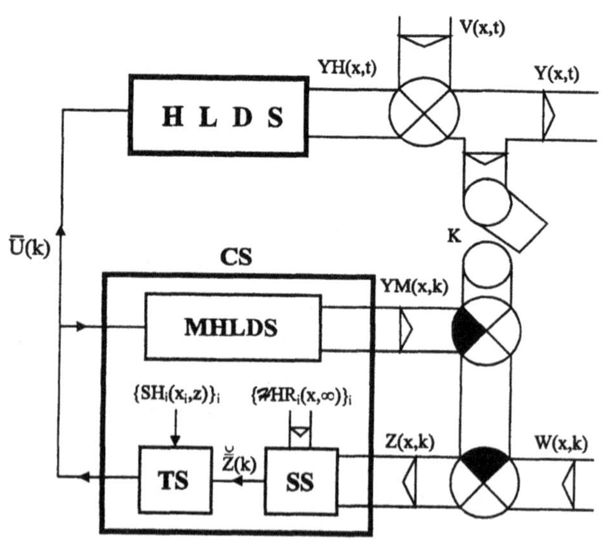

Fig. 7. Distributed parameter feedback control system with internal model

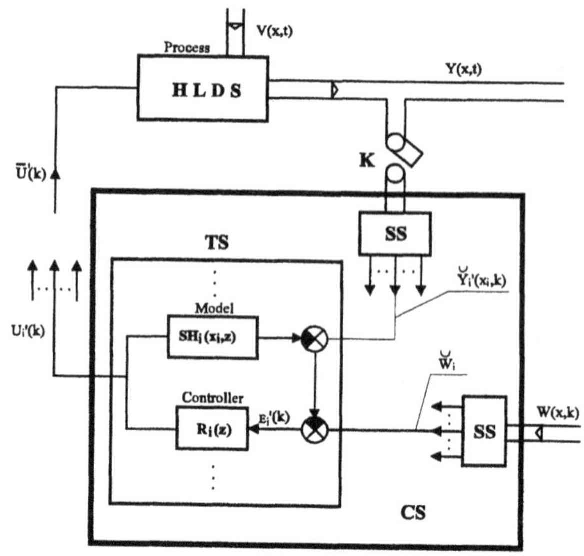

Fig. 8. Distributed parameter feedback robust system of control

In the case of adaptive control, the controlled system dynamics is modeled by a predictor with lumped input and distributed output. The control synthesis within the PD, SS and TS blocks is carried out using the estimated lumped and distributed dynamic characteristics.

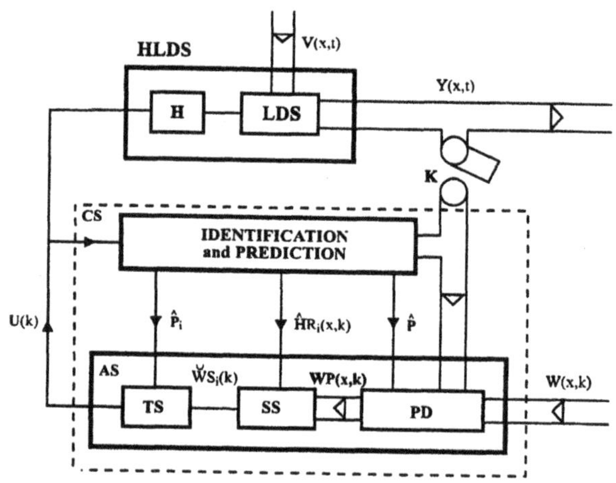

Fig. 9. Distributed parameter adaptive system of control

4. DESIGN OF DPS

Industrial objects with distributed parameters are often put into the role of distributed parameter systems of control. The optimization problem of such control system is decomposed into time and space region - just in the same way as it has been shown in control synthesis.

In time region, for control system optimization, techniques of lumped system optimization can be used as the time dependent control synthesis is carried out in SISO loops with lumped parameters.

In spatial dependence, one of the frequent tasks is the optimal sensors location problem. The minimal number of sensors can be found through minimization of the following approximation problem

$$\min\left\|Y(x,\infty)-Ya(x,\infty)\right\| \tag{23}$$

with the aim to determine the minimal number of unevenly located nodal points of the spline functions for the distributed output quantity approximation.

Another class of problems is connected with the spatial organization of the controlled DPS. This includes for example the selection of appropriate number and optimal location of the distributed input generators. This can be solved by searching the global minimum of the approximation problem

$$\min\left\|W(x,\infty)-\sum_{i=1}^{n}\breve{W}_i\,\mathcal{H}HR_i(x,\infty)\right\| \tag{24}$$

where the aim is to find the minimal number and optimal location of distributed input generators such that functional (24) is minimal.

5. DEMONSTRATION

LDS of parabolic type on interval $\langle 0,1\rangle$.

The transfer functions of the actuators and distributed input quantities generators , fig 3, are

$$\left\{SA_i(s)=\frac{1}{TA_i s+1}\right\}_i, \tag{25}$$

$$TA_1=10, TA_2=15, TA_3=10, TA_4=12$$

$$\left\{SG_i(s)=\frac{1}{TG_i s+1}\right\}_i, \tag{26}$$

$$TG_1=15, TG_2=25, TG_3=30, TG_4=20$$

The behavior of shaper element $\{T_i(x)\}_i$ is depicted in fig. 10.

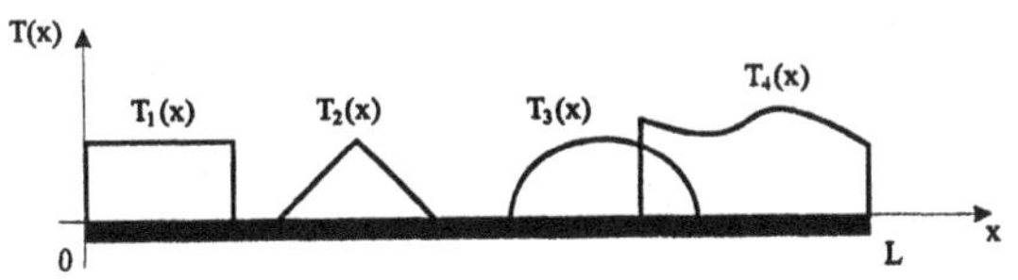

Fig. 10. Shaper elements of distributed input variables on interval <0,L>

The coefficients of transfer function S(x, ξ, s) take the values: a = 0.011, b = 0.0707

$$S(x,\xi,s)=\frac{1}{L(s+b)}+\frac{2}{L}\sum_{n=1}^{\infty}\frac{\cos\frac{n\pi x}{L}\cos\frac{n\pi\xi}{L}}{s+b+\frac{n^2\pi^2a^2}{L^2}} \tag{27}$$

The dynamic behavior of HLDS was simulated by using the CONTROL and SPLINE MATLAB toolboxes. The input-output relations of HLDS were modeled through discrete convolution (11).

The OPTIMIZ toolbox has been used to solution of approximation problem of the block SS while control loops of block TS were carried out using the SIMULINK toolbox function.

The control process, simulated according to the scheme in fig. 7, is shown in figs. 11 - 13.

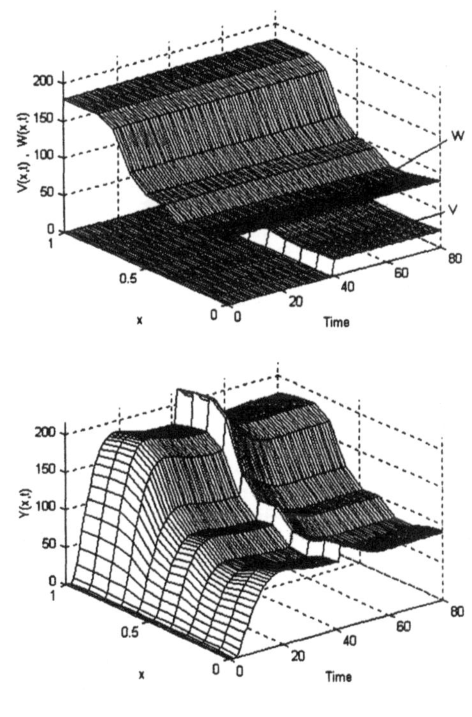

Fig. 11. Control process of LDS of parabolic type. W(x,k), Y(x,t), V(x,t) - control, controlled and disturbance quantities

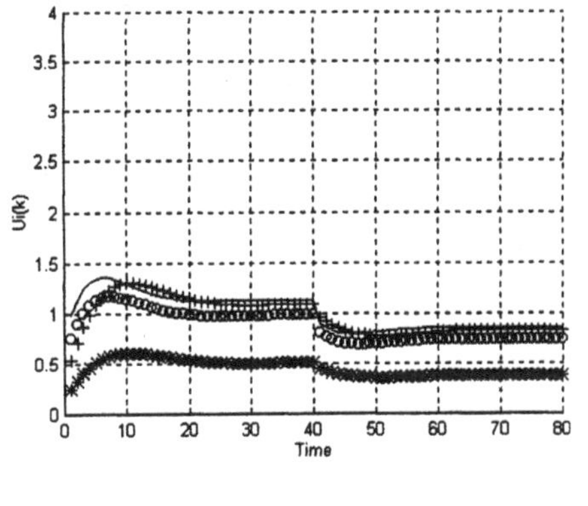

Fig. 12. Actuating quantities
* - U1, _ - U2, + - U3, °° - U4

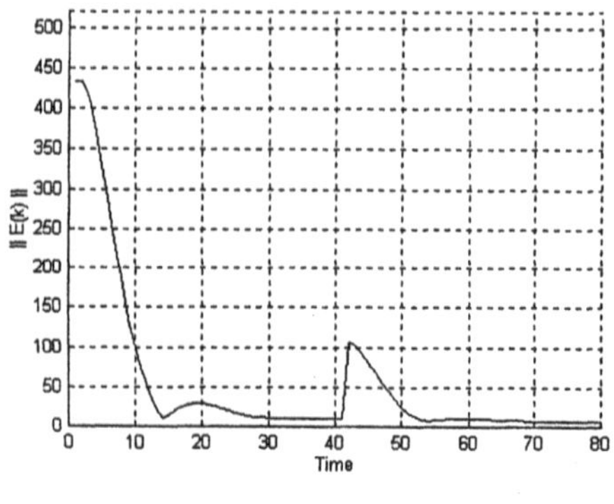

Fig. 13. Quadratic norm of distributed deviation

LDS on subregion Ω of E_2

If DDS block of LDS structure has a complicated from its dynamics is modeled on the basis of Finite Elements Method (FEM) - MATLAB PDE tool. On fig.14, 15 we have a metal slab and a beam of complicated forms as DDS blocks. The control processes of LDS of parabolic and hyperbolic type are on fig. 16 - 21.

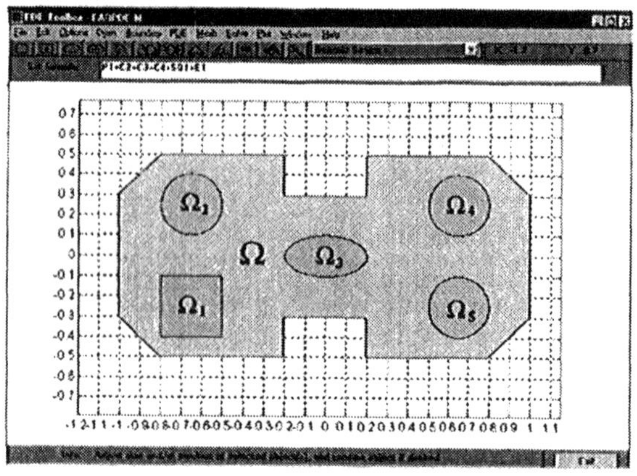

Fig. 14. The metal slab form Ω with supports of the distributed input quantities Ω1 - Ω5 in MATLAB PDE tool

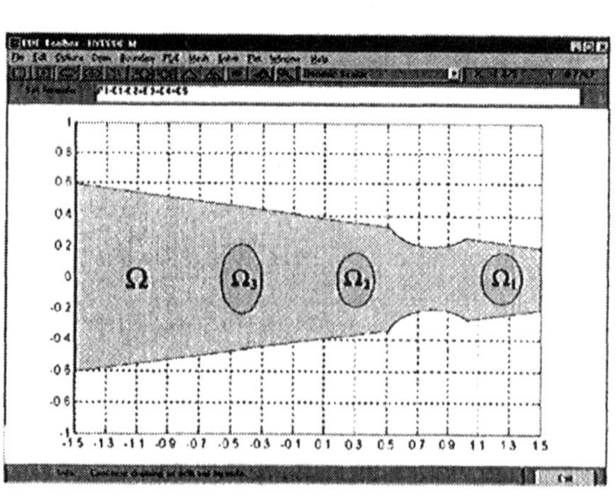

Fig. 15. The smart material beam of given form Ω with supports of the distributed input quantities Ω1 - Ω3 in MATLAB PDE tool

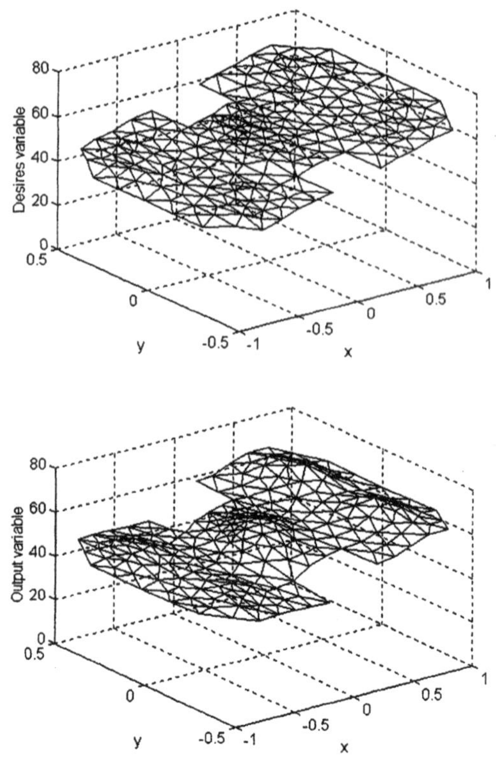

Fig. 16. Heating process of metal slab in steady state $W(x,\infty)/Y(x,\infty)$ desires/controlled quantities

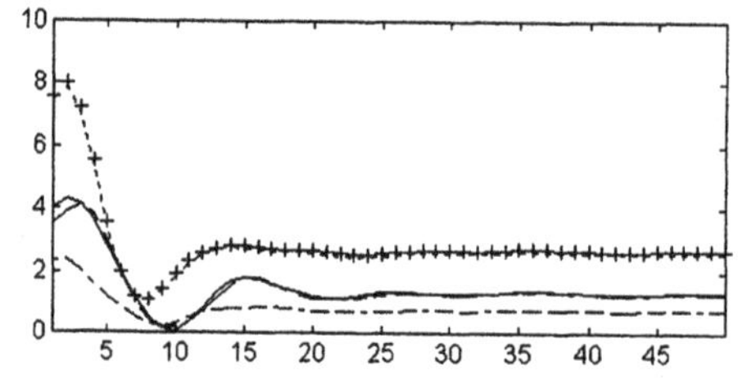

Fig. 17. Heating process input quantities
—- U1, --- - U2, -.- - U3, :: - -U4, ++ - U5

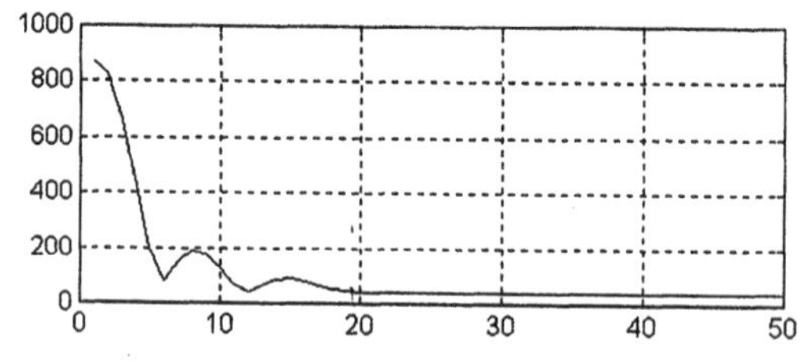

Fig. 18. Quadratic norm of distributed deviation

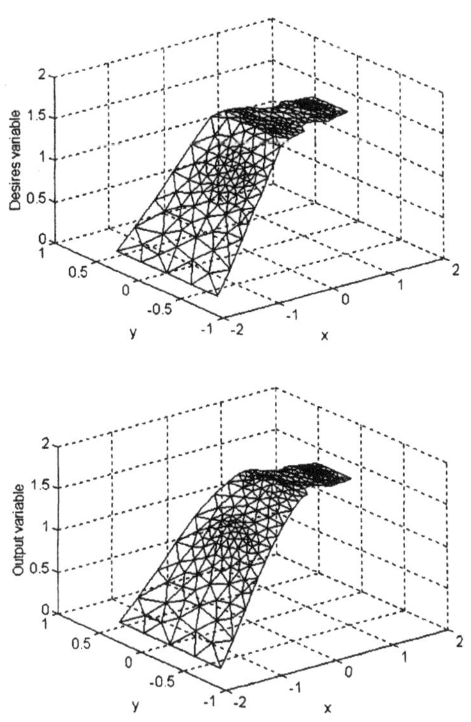

Fig. 19. Deviation process of beam in steady state
W(x,∞)/Y(x,∞) desires/controlled quantities

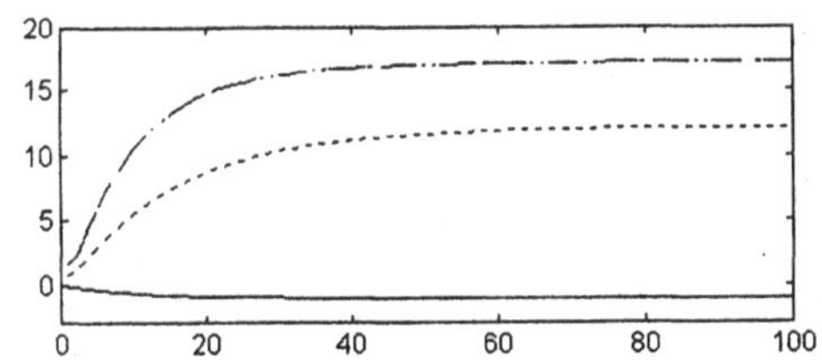

Fig. 20. Deviation process input quantities
—- U1, --- - U2, -.- - U3

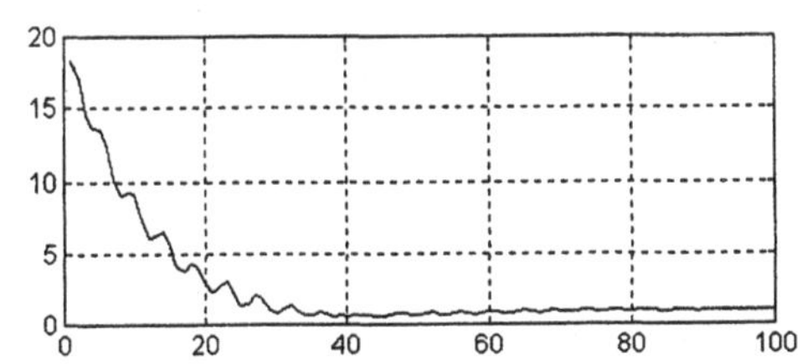

Fig. 21. Quadratic norm of distributed deviation

LDS physical model

The process of adaptive control is demonstrated on distributed parameter continual thermal process physical model, figs. 22 - 29.

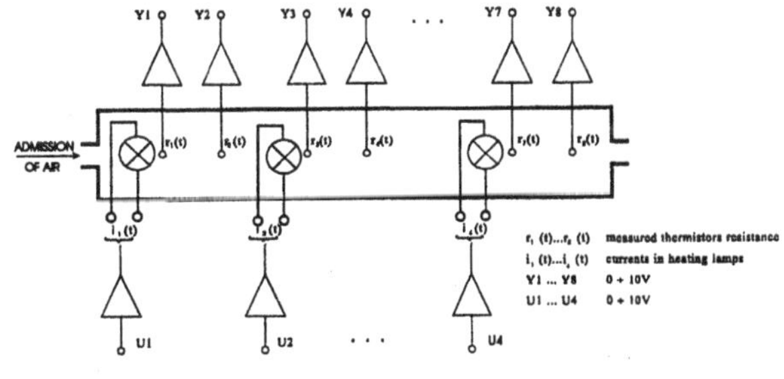

Fig. 22. Scheme of physical model of distributed parameter continual thermal process.

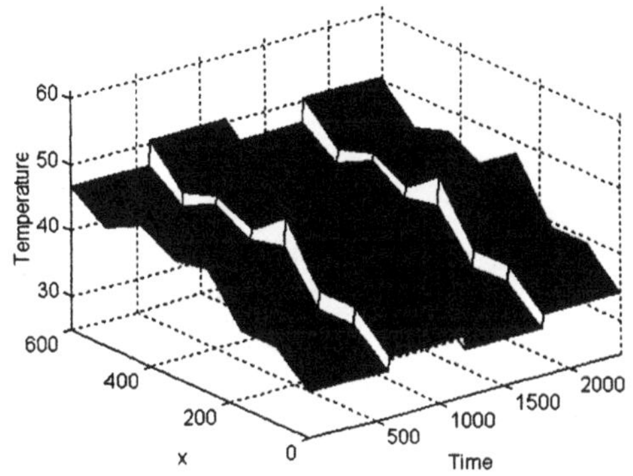

Fig. 23. Control quantity for adaptive control of physical model

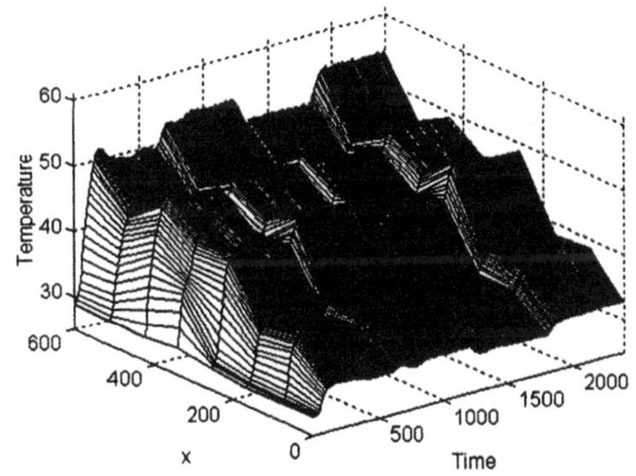

Fig. 24. Controlled quantity by LQG control of physical model

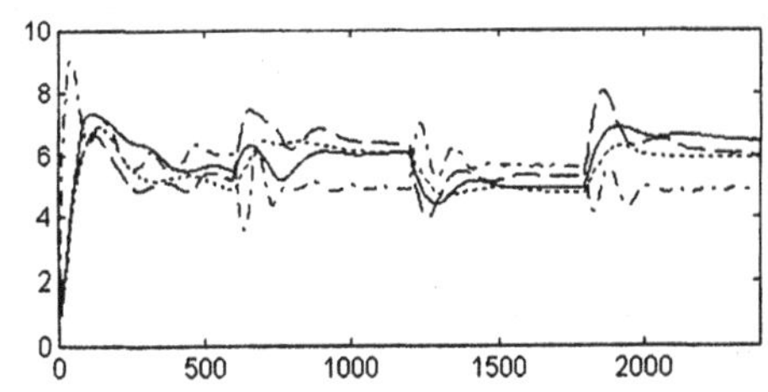

Fig. 25. Actuating quantities at LQG adaptive control of physical model
—- U1, -- - U2, -.- - U3, ...U4

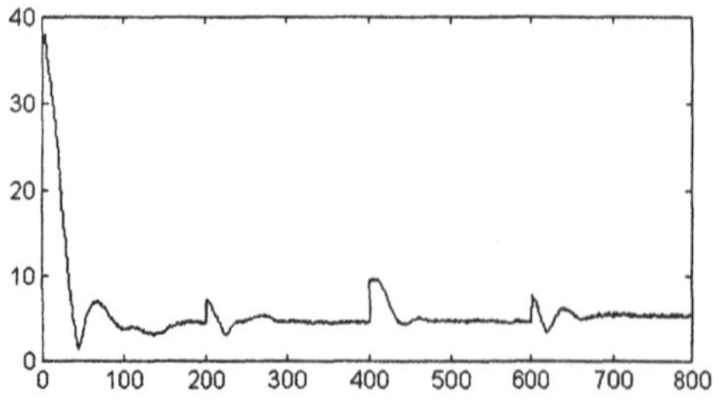

Fig. 26. Quadratic norm of distributed deviation of adaptive control of physical model

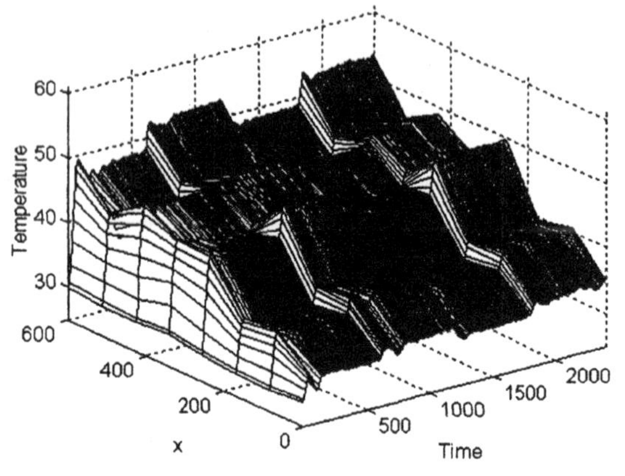

Fig. 27. Controlled quantities at GPC of physical model

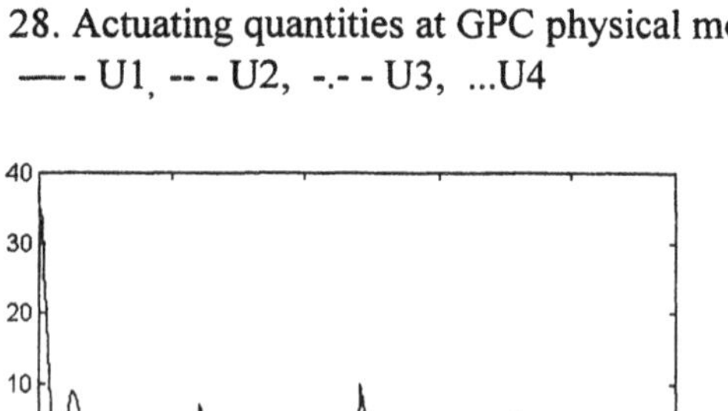

Fig. 28. Actuating quantities at GPC physical model —- U1, --- U2, -.- - U3, ...U4

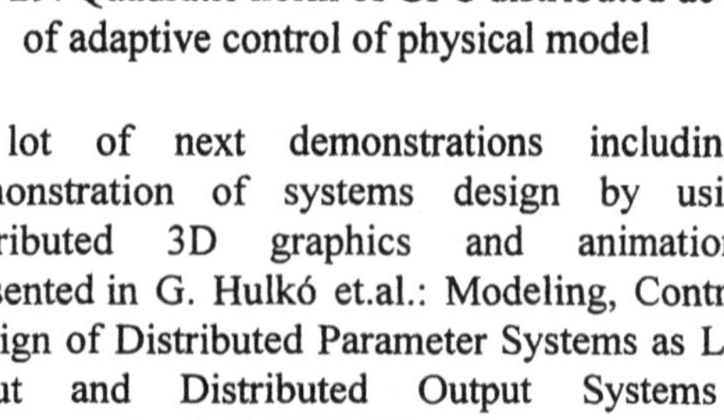

Fig. 29. Quadratic norm of GPC distributed deviation of adaptive control of physical model

A lot of next demonstrations including the demonstration of systems design by using of distributed 3D graphics and animations is presented in G. Hulkó et.al.: Modeling, Control and Design of Distributed Parameter Systems as Lumped Input and Distributed Output Systems with Demonstration in MATLAB.

ACKNOWLEDGEMENT

This work has been carried out under the financial support of the VEGA project Development of Automatic Control Methods (grant 95/5195/198) and the EC granted COPERNICUS project „Adaptive and Predictive Control with Physical Constraints" (grant CP 941174)

REFERENCES

Butkovskij, A.G. (1965). *Optimal control of distributed parameter systems*. Nauka, Moskva.

Butkovskij, A.G. (1982). *Green's functions and transfer functions handbook*. Ellis Horwood Limited Publishers Chichester.

Lions, J.L. (1971). *Optimal control of systems governed by partial differential equations*. Springer - Verlag.

Hulkó, G. (1987). *Distributed Parameter Systems Control by means of Multi - Input and Multi - Distributed - Output Systems I., II.* Preprints of IMACS / IFAC Symposium "Distributed Parameter Systems'87, Hiroshima.

Hulkó, G (1987). *Control of Distributed Parameter Systems by means of Multi-Input and Multi-Distributed - Output Systems*. Preprints of 10-th World Congress of IFAC, Munich.

Hulkó, G. (1989): *Control of Lumped Input and Distributed Output Systems*. Preprints of 5-th IFAC/IMACS/IFIP Symposium on Control of Distributed Parameter Systems, Perpignan.

Hulkó, G. *et al.* (1990). *Computer Aided Design of Distributed Parameter Systems of Control*. 11-th World Congress of IFAC, Tallinn,.

Hulkó, G. *et al.* (1995). *Identification and Modeling of Lumped Input and Distributed Output Systems of the Accuracy Requested*. Preprints of European Control Conference ECC 95, Roma.

Hulkó, G. *et al.* (1997). *Modeling, Control and Design of Distributed Parameter Systems as Lumped Input and Distributed Output Systems with Demonstration in MATLAB*. Publishing House of STU, Bratislava.

OBSERVATION ALGORITHMS FOR AN UNDER-ACTUATED CLOSED KINEMATIC CHAIN.

L. Menini*, A. Tornambè, L. Zaccarian***

**Dip. Informatica, Sistemi e Produzione, Università di Roma Tor Vergata*
Via della Ricerca Scientifica, 00133 Roma, Italy.
E-mail menini@disp.utovrm.it, zack@disp.utovrm.it
***Dip. Meccanica e Automatica, Università di Roma Tre*
via C. Segre 60, 00146, Roma, Italy.
Tel. + 39 6 7259 4405, Fax + 39 6 2020519
E-mail tornambe@tovvx1.ccd.utovrm.it

Abstract. A closed under-actuated kinematic chain is considered, which represents a simple model for the study of in-plane oscillations of a suspended cable. The aim of this paper is to develop a model-based asymptotic observer of the state variables of the chain. Simulation runs show the effectiveness of the proposed observation scheme.

Key Words. Observers, Mechanisms, Lyapunov methods

1. INTRODUCTION

The purpose of this paper is to develop a state observer for the under-actuated 3-link closed kinematic chain shown in Figure 1, under the assumption that only partial measurements of the state variables are available. The interest in such a mechanical system arises from practical applications.

Modelling and control of under-actuated chains is of great importance in industrial applications. For instance, relevant interest can be found in modelling and control of suspended cables. Because of the light intrinsic damping of these structures, their oscillations are a problem of practical relevance (see, e.g., (Fujino *et al.*, 1993; Fujino and Susumpow, 1994; Triantafyllou, 1987)). Due to the distributed nature of such systems, approximations are needed in the modelling process; if the attention is restricted to in-plane oscillations, the behaviour of the suspended cable can be approximated by the one of a planar chain, constituted by a finite number of rigid links connected by rotational joints. The kinematic chain studied in this paper can be thought of as a simple model of this kind.

Modelling and control of closed kinematic chains is of great interest in robotic applications; as a matter of fact, industrial manipulators containing closed kinematic chains can be designed for high precision tasks, and, moreover, several operating conditions, such as the case of multiple robots interacting with the same object, can be represented by closed kinematic chains.

A non-linear dynamic model for the system under consideration has been developed in (Menini *et al.*, 1996), where it was also shown that simple PD control laws succeed in asymptotically stabilising the system about equilibrium configurations. However, such control laws yield transient oscillations of remarkable amplitude, due to the presence of the proportional term, which is needed especially in view of possible parameter uncertainties. Hence, the need for more sophisticated control algorithms is evident. One of the main difficulties of such a task consists in the fact that measuring the state variables of the chain is impossible in practice, since they do not even correspond to any physical variable. To overcome such difficulties, in this paper the problem of estimating the state variables of the chain is dealt with: a non-linear full order observer is proposed, which asymptotically achieves state reconstruction for the considered model of the chain.

2. MODEL OF THE CLOSED KINEMATIC CHAIN

The mechanical system under consideration, represented in Figure 1, is a chain composed of three equal rigid links having length L and mass M (labelled by 1, 2 and 3) and of a cart modelled as a lumped dimensionless mass m. The mechanism lays on a vertical plane and is subject to the gravitational field (the modulus of the gravity acceleration is denoted by g). The first link is connected to the base body (inertial frame) by the rotational joint 1; the second link is connected to the first one by the rotational joint 2; the third link is connected to the second one by the rotational joint 5; the cart, modelled as the fusion of a rotational and a prismatic joint, is connected to the third link by the rotational joint 3 and to the base body by the prismatic joint 4. An inertial orthonormal reference frame XY is defined in the plane of motion. The cart is constrained to move along the X axis, which means that the axis of the prismatic joint 4 coincides with the X axis of the inertial reference frame. The only control input is the external force u applied to the cart along the X direction. Both viscous damping and friction at the rotational joints 1, 2, 3, and 5, and at the prismatic joint 4, have been assumed to be negligible.

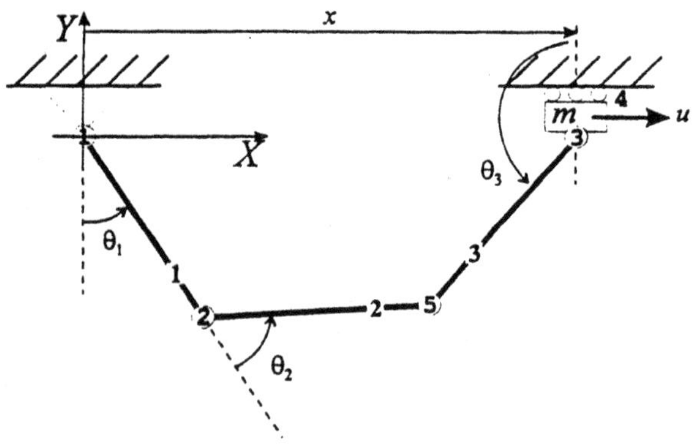

Fig. 1. The closed kinematic chain.

The first joint of the chain coincides with the origin of the inertial reference frame XY, whose Y axis is parallel to the gravitational field and has opposite direction. The angular positions of joints 1, 2 and 3 (defined as shown in Figure 1) are denoted by θ_1, θ_2 and θ_3, respectively, and the linear position of joint 4 is denoted by x.

Such a mechanism has two degrees of freedom, and a suitable vector of generalised coordinates describing the configuration of the system is $q = [x\,,\ \theta_1]^T$. It is well known that the dynamic model of the described mechanical system can be recast in the following equations of motion:

$$B(q)\,\ddot{q} + C(q\,,\ \dot{q})\,\dot{q} + h(q) = \begin{bmatrix} u \\ 0 \end{bmatrix}, \qquad (2.1)$$

the expressions of $B(q)$, $C(q\,,\ \dot{q})$ and $h(q)$ can be found in (Menini *et al.*1996). It is stressed that the model derived in (Menini *et al.*1996) is valid in a convenient region β of the working space of the mechanism. It is easy to see that, by varying the intensity of the input u, assumed constant, one obtains a set of operating points: let Γ be the intersection between such a set of operating points and β.

3. A NON-LINEAR OBSERVER FOR THE CHAIN

The aim of this section is the design of an asymptotic observer of the state $q = [x,\ \theta_1]^T$, $\dot{q} = [\dot{x},\ \dot{\theta}_1]^T$ of the closed kinematic chain in Figure 1, having as inputs the measured position $x(t)$ and velocity $\dot{x}(t)$ of the cart, and the force $u(t)$ applied to the chain, and as outputs the estimates $\hat{q}(t)$, $\dot{\hat{q}}(t)$ of the state variables.

The observer here proposed is a dynamical system constituted by a "copy" of the model of the chain, and a suitable linear output injection, namely

$$B(\hat{q})\,\ddot{\hat{q}} + C(\hat{q}\,,\ \dot{\hat{q}})\,\dot{\hat{q}} + h(\hat{q}) = \begin{bmatrix} u \\ 0 \end{bmatrix} + \begin{bmatrix} K_x\,(x-\hat{x}) + K_v\,(\dot{x}-\dot{\hat{x}}) \\ 0 \end{bmatrix}, \qquad (3.1)$$

being $K_x,\ K_v \in \mathbf{R}$, $K_x,\ K_v \geq 0$. It is stressed that, if the initial conditions $q(0)$, $\dot{q}(0)$ of the chain are known, then, by setting $\hat{q}(0) = q(0)$, $\dot{\hat{q}}(0) = \dot{q}(0)$, one obtains $\hat{q}(t) = q(t)$, $\dot{\hat{q}}(t) = \dot{q}(t)$, $\forall t \geq 0$.

3.1. *Asymptotic convergence of the error dynamics*

In order to prove the effectiveness of the proposed observer (3.1), some preliminaries are needed.

In the following, if u_0 is a constant value of the input u, denote by $q_0 = [x_0\ \theta_{1,0}]^T$ the equilibrium configuration of the chain, corresponding to u_0, and denote by $\tilde{q}(t) = [\tilde{x}(t)\ \tilde{\theta}_1(t)]^T$ the difference $\tilde{q}(t) := q(t) - q_0$, $\forall t \geq 0$. Moreover, define the estimation error at time t, as:

$$e(t) := \begin{bmatrix} e_x(t) \\ e_\theta(t) \end{bmatrix} := q(t) - \hat{q}(t).$$

By rewriting (3.1) in terms of $e(t)$ and by subtracting such an equation from (2.1), the following equation is obtained:

$$\begin{aligned} &B(q)\,\ddot{q} - B(q-e)\,(\ddot{q}-\ddot{e}) + \\ &C(q\,,\ \dot{q})\,\dot{q} - C(q-e\,,\ \dot{q}-\dot{e})\,(\dot{q}-\dot{e}) + \\ &h(q) - h(q-e) + \begin{bmatrix} K_x\,e_x + K_v\,\dot{e}_x \\ 0 \end{bmatrix} = 0\,, \end{aligned} \qquad (3.2)$$

which, together with (2.1), describes the system constituted by the chain and the observer. It is easy to see that the point $[q^T\ \dot{q}^T\ e^T\ \dot{e}^T]^T = [q_0^T\ 0\ 0\ 0\ 0\ 0\ 0]^T \in \mathbf{R}^8$ is an equilibrium point of such a system; for the subsequent derivations, it is useful to consider the linear system obtained by linearising equations (2.1) and (3.2) about $q = q_0$,

$\dot{q} = 0$, $e = 0$, $\dot{e} = 0$:

$$B_0\, \ddot{\tilde{q}} + H_0\, \tilde{q} = \begin{bmatrix} \tilde{u} \\ 0 \end{bmatrix}, \tag{3.3a}$$

$$B_0\, \ddot{e} + H_0\, e + \begin{bmatrix} K_x\, e_x + K_v\, \dot{e}_x \\ 0 \end{bmatrix} = 0, \tag{3.3b}$$

where $\tilde{u} := u - u_0$, $B_0 := B(q_0)$, and $H_0 := \left.\dfrac{\partial H(q)}{\partial q}\right|_{q=q_0}$. By inspection, it is immediate to conclude that the linearised system (3.3) is actually constituted by two decoupled subsystems, (3.3a) and (3.3b), having as state variables $[\tilde{q}^T\ \dot{q}^T]^T$ and $[e^T\ \dot{e}^T]^T$, respectively; only subsystem (3.3b) is of interest here. Asymptotic stability of subsystem (3.3b) is stated by the following lemma.

Lemma 1 If $K_x \geq 0$ and $K_v > 0$, then the linear subsystem (3.3b) is asymptotically stable.

Proof. The following positive definite function $W_L(\cdot, \cdot) : \mathbf{R}^2 \times \mathbf{R}^2 \to \mathbf{R}$:

$$W_L(e, \dot{e}) = \frac{1}{2}\dot{e}^T B_0 \dot{e} + \frac{1}{2} e^T \left(H_0 + \begin{bmatrix} K_x & 0 \\ 0 & 0 \end{bmatrix} \right) e,$$

can be taken as a candidate Lyapunov function for system (3.3b); as a matter of fact, for each operating point, B_0 is a positive definite matrix (it is the generalised inertia matrix of the mechanical system under consideration in the configuration q_0) and, for each operating point in Γ, H_0 is a positive definite matrix (this can be seen by computing its two principal minors).

Recalling that $B_0 = B_0^T$ and $H_0 = H_0^T$, it is easy to compute the total time derivative $\dot{W}_L(\cdot, \cdot)$ of $W_L(\cdot, \cdot)$ along the motion of system (3.3b):

$$\dot{W}_L(e, \dot{e}) = -\dot{e}^T \begin{bmatrix} K_v & 0 \\ 0 & 0 \end{bmatrix} \dot{e}. \tag{3.4}$$

Since $\dot{W}_L(\cdot, \cdot)$, given by (3.4), is negative semidefinite, by Theorem 26.1 of (Hahn, 1967), the largest invariant subset G_W contained in the set $\Omega_W := \{(e, \dot{e}) : \dot{W}_L(e, \dot{e}) = 0\}$ is attractive. In the following it is proved that G_W coincides with the origin of $\mathbf{R}^4$ for all $K_x \geq 0$, provided that $K_v > 0$, thus guaranteeing the asymptotic stability of subsystem (3.3b) for such K_x, K_v.

If $K_v > 0$, then it is easy to verify that $\Omega_W = \{(e, \dot{e}) : \dot{e}_x = 0\}$; hence, in Ω_W (and, therefore, in G_W) e_x is constant and $\ddot{e}_x = 0$. Equation (3.3b) can then be rewritten as follows along Ω_W:

$$B_0 \begin{bmatrix} 0 \\ \ddot{e}_\theta \end{bmatrix} + H_0 \begin{bmatrix} e_x \\ e_\theta \end{bmatrix} = \begin{bmatrix} -K_x\, e_x \\ 0 \end{bmatrix}; \tag{3.5}$$

in addition, since $K_x \geq 0$ and H_0 is positive definite, the matrix $\overline{H}_0$ defined as

$$\overline{H}_0 := H_0 + \begin{bmatrix} K_x & 0 \\ 0 & 0 \end{bmatrix},$$

is non-singular. Then, from (3.5), one has:

$$\begin{bmatrix} e_x \\ e_\theta \end{bmatrix} = -\overline{H}_0^{-1} B_0 \begin{bmatrix} 0 \\ \ddot{e}_\theta \end{bmatrix},$$

which allows the following relationships to be written:

$$e_x = g_x\, \ddot{e}_\theta, \tag{3.6a}$$

$$e_\theta = g_\theta\, \ddot{e}_\theta, \tag{3.6b}$$

where $[g_x\ g_\theta]^T$ is the second column of matrix $-\overline{H}_0^{-1} B_0$. From the expression of g_x as function of the operating point, it is easy to see that $g_x \neq 0$ for all $K_x \geq 0$ and for all operating points in Γ. Since e_x is constant in Ω_W (and, therefore, in G_W) condition (3.6a), with $g_x \neq 0$, implies that also $\ddot{e}_\theta$ is constant in G_W; hence, from (3.6b), it follows that e_θ is constant in G_W. In addition, the fact that e_θ is constant implies $\ddot{e}_\theta = 0$, which, from (3.6), implies that $e_x = 0$ and $e_\theta = 0$ in G_W. Finally, $e_\theta = 0$ in G_W implies that $\dot{e}_\theta = 0$ in G_W; hence, for each operating point in Γ, G_W coincides with the origin of $\mathbf{R}^4$ and the asymptotic stability of subsystem (3.3b) follows, for all $K_v > 0$, $K_x \geq 0$. □

Now, consider subsystem (3.3b) written in first order normal form

$$\begin{bmatrix} \dot{e} \\ \ddot{e} \end{bmatrix} = A_O \begin{bmatrix} e \\ \dot{e} \end{bmatrix},$$

where $A_O \in \mathbf{R}^{4\times 4}$ is defined as

$$A_O := \begin{bmatrix} 0 & I_2 \\ A_{O,2,1} & A_{O,2,2} \end{bmatrix},$$

with $A_{O,2,1}$, $A_{O,2,2} \in \mathbf{R}^{2\times 2}$ given by

$$A_{O,2,1} := -B_0^{-1} \left(H_0 + \begin{bmatrix} K_x & 0 \\ 0 & 0 \end{bmatrix} \right),$$

$$A_{O,2,2} := -B_0^{-1} \begin{bmatrix} K_v & 0 \\ 0 & 0 \end{bmatrix}.$$

Because of the asymptotic stability of subsystem (3.3b), it is well known that the following Lyapunov matrix equation admits a unique positive definite solution P:

$$A_O^T P + P A_O + I_4 = 0. \tag{3.7}$$

For convenience, for each real number $\alpha > 0$, define the following (closed and bounded by construction) set $\mathcal{P}_\alpha$:

$$\mathcal{P}_\alpha := \{z \in \mathbf{R}^4 : z^T P z \leq \alpha\}.$$

Now, the following result, concerning the convergence of the estimates $\hat{q}(t)$, $\dot{\hat{q}}(t)$ towards the state variables $q(t)$, $\dot{q}(t)$ of the chain, under suitable assumptions about the motion of the chain and about the initial conditions of the observer, can be given.

It is convenient to consider equation (3.2) as a time-varying system having $[e^T(t)\ \dot{e}^T(t)]^T$ as state vector: the dependence on time enters equation (3.2) through the values of $q(t)$, $\dot{q}(t)$ and $\ddot{q}(t)$, which satisfy equation (2.1).

Proposition 1 Assume that $u(t) = u_0, \forall t \geq 0$. Setting $K_x \geq 0$, and $K_v > 0$, there exists a positive real number d such that, if $q(t)$ is the solution of (2.1) from an initial condition $q(0)$, $\dot{q}(0)$, such that $\|q(0) - q_{1,0}\| < d$ and $\|\dot{q}(0)\| < d$, then $(e, \dot{e}) = (0, 0)$ is a uniformly asymptotically stable equilibrium point of (3.2).

Moreover, there exists a positive real number c such that $\mathcal{P}_c$ is an estimate of the region of attraction of such a point.

Proof. For convenience, define the linear function $F_\ell(\cdot, \cdot) : \mathbf{R}^2 \times \mathbf{R}^2 \to \mathbf{R}^2$, as follows:

$$F_\ell(e, \dot{e}) := -B_0^{-1}\left(H_0\, e + \begin{bmatrix} K_x\, e_x + K_v\, \dot{e}_x \\ 0 \end{bmatrix}\right),$$

so that the linear system (3.3b) can be recast as follows:

$$\ddot{e} = F_\ell(e, \dot{e}).$$

Moreover, if $\hat{\beta}_0 := \{z \in \mathbf{R}^2 : z + q_0 \in \beta\}$, define the function $F(\cdot, \cdot, \cdot, \cdot) : (\mathcal{I} \subset \mathbf{R}^2) \times \mathbf{R}^2 \times \hat{\beta}_0 \times \mathbf{R}^2 \to \mathbf{R}^2$, as follows:

$$\begin{aligned} F(e, \dot{e}, \tilde{q}, \dot{q}) :=& \\ &-(B(q-e))^{-1}\Big(B(q)\,\ddot{q} - B(q-e)\,\ddot{q} + \\ &C(q\,,\,\dot{q})\,\dot{q} - C(q-e\,,\,\dot{q}-\dot{e})\,(\dot{q}-\dot{e}) + \\ &h(q) - h(q-e) + \begin{bmatrix} K_x\, e_x + K_v\, \dot{e}_x \\ 0 \end{bmatrix}\Big), \end{aligned}$$

so that system (3.2) can be recast as:

$$\ddot{e} = F(e, \dot{e}, \tilde{q}, \dot{q}).$$

Function $F(\cdot, \cdot, \cdot, \cdot)$ is well defined, provided that the arguments of matrix $B(\cdot)$ in equations (2.1) and (3.2) belong to the set $\hat{\beta}$, projection of the working space β on the (x, θ_1) plane.

In order to guarantee that $q(t)$ belongs to $\hat{\beta}$, $\forall t \geq 0$, it is sufficient to require

$$d < d_0, \tag{3.8}$$

where d_0 is a suitable positive real number, whose existence is implied by the stability of the time-invariant system (2.1) (having as state variables $q(t)$, $\dot{q}(t)$) with $u(t) = u_0$, for all $t \geq 0$.

In order to guarantee that $q(t) - e(t)$ belongs to $\hat{\beta}$, it is required, in addition, that

$$\|e(t)\| < \overline{\delta}, \forall t \geq 0, \tag{3.9}$$

where $\overline{\delta}$ is a suitable real positive number that depends on d; hence, the set $\mathcal{I}$ can be defined as a ball in $\mathbf{R}^2$ having radius $\overline{\delta}$ and centre in the origin: $\mathcal{I} := \{z \in \mathbf{R}^2 : \|z\| < \overline{\delta}\}$.

Finally, consider the function $F_{n\ell}(\cdot, \cdot, \cdot, \cdot) : (\mathcal{I} \subset \mathbf{R}^2) \times \mathbf{R}^2 \times \hat{\beta}_0 \times \mathbf{R}^2 \to \mathbf{R}^2$, defined by:

$$F_{n\ell}(e, \dot{e}, \tilde{q}, \dot{q}) := F(e, \dot{e}, \tilde{q}, \dot{q}) - F_\ell(e, \dot{e}),$$

and notice that $F_{n\ell}(\cdot, \cdot, \cdot, \cdot)$ is an analytic function in its domain of definition.

The following property of function $F_{n\ell}(\cdot, \cdot, \cdot, \cdot)$ will be useful: for each (arbitrarily small) real positive number ν, there exist two real positive numbers δ_ν and d_ν, satisfying $0 < \delta_\nu < \overline{\delta}$ and $0 < d_\nu < d$, respectively, such that, if c_ν is the largest real number such that $\mathcal{P}_{c_\nu}$ is contained in the ball

$$\left\{\begin{bmatrix} e \\ \dot{e} \end{bmatrix} \in \mathbf{R}^4 : \quad \left\|\begin{bmatrix} e \\ \dot{e} \end{bmatrix}\right\| < \delta_\nu\right\},$$

then

$$\|F_{n\ell}\| \leq \nu \left\|\begin{bmatrix} e \\ \dot{e} \end{bmatrix}\right\|, \quad \forall \begin{bmatrix} e \\ \dot{e} \end{bmatrix} \in \mathcal{P}_{c_\nu}, \tag{3.10}$$

provided that $\|q(0) - q_{1,0}\| < d_\nu$ and $\|\dot{q}(0)\| < d_\nu$. This follows from a well known result of real analysis, by taking into account that $F_{n\ell}(0, 0, \tilde{q}, \dot{q}) = 0$, for all $\tilde{q} \in \hat{\beta}_0$, $\dot{q} \in \mathbf{R}^2$, if the numbers δ_ν and d_ν are chosen so that

$$\left\|\begin{bmatrix} \dfrac{\partial F_{n\ell}}{\partial e} & \dfrac{\partial F_{n\ell}}{\partial \dot{e}} \end{bmatrix}\right\| < \nu, \tag{3.11}$$

for all $t \geq 0$, for all $q(0)$, $\dot{q}(0)$ such that $\|q(0) - q_{1,0}\| < d_\nu$, $\|\dot{q}(0)\| < d_\nu$, for all e, $\dot{e}$ such that $\left\|\begin{bmatrix} e \\ \dot{e} \end{bmatrix}\right\| < \delta_\nu$. This choice can be actually done for any real positive number ν, by virtue of the stability of system (2.1), and by the fact that the entries of the Jacobian matrix in (3.11) are continuous functions and that such a matrix is equal to the null matrix, when evaluated at $e = 0$, $\dot{e} = 0$, $\tilde{q}$, $\dot{q}$. It is stressed that, in (3.11), the symbol $\|\cdot\|$ denotes the matrix norm induced by the Euclidean norm of vectors in $\mathbb{R}^2$ and $\mathbb{R}^4$.

Now, consider the following positive definite decrescent function (see (Hahn1967) for the definition of "decrescent"):

$$W(e,\,\dot e) = [e^T\,\dot e^T]\,P \begin{bmatrix} e \\ \dot e \end{bmatrix},$$

where P is the solution of the Lyapunov equation (3.7). Function $W(\cdot,\,\cdot)$ can be taken as a candidate Lyapunov function for system (3.2). Computing its total time-derivative $\dot W(\cdot,\,\cdot)$ along the motion of the time-varying system (3.2), one has

$$\begin{aligned} \dot W(e,\,\dot e) &= 2\,[e^T\,\dot e^T]\,P \begin{bmatrix} \dot e \\ \ddot e \end{bmatrix} \\ &= 2\,[e^T\,\dot e^T]\,P \begin{bmatrix} \dot e \\ F(e,\,\dot e,\,\tilde q,\,\dot q) \end{bmatrix} \\ &= 2\,[e^T\,\dot e^T]\,P \begin{bmatrix} \dot e \\ F_\ell(e,\,\dot e) \end{bmatrix} + \\ &\quad\; 2\,[e^T\,\dot e^T]\,P \begin{bmatrix} 0 \\ F_{n\,\ell}(e,\,\dot e,\,\tilde q,\,\dot q) \end{bmatrix}. \end{aligned}$$

By the definition of P in equation (3.7), it is easy to verify that

$$2\,[e^T\,\dot e^T]\,P \begin{bmatrix} \dot e \\ F_\ell(e,\,\dot e) \end{bmatrix} = -\left\| \begin{bmatrix} e \\ \dot e \end{bmatrix} \right\|^2,$$

while, recalling the properties of the Euclidean norm and of symmetric matrices, and relationship (3.10) it is easy to show that

$$\left| 2\,[e^T\,\dot e^T]\,P \begin{bmatrix} 0 \\ F_{n\,\ell} \end{bmatrix} \right| < 2\,\lambda_{\max}(P)\,\nu \left\| \begin{bmatrix} e \\ \dot e \end{bmatrix} \right\|^2,$$

where $\lambda_{\max}(P)$ indicates the maximum eigenvalue of matrix P.

Hence, assuming $\|\tilde q(0)\| < d_\nu$ and $\|\dot q(0)\| < d_\nu$, it follows that, $\forall \quad [e \quad \dot e]^T \in \mathcal{P}_{c_\nu}$:

$$\dot W(e,\,\dot e) \le -\,(1 - 2\,\lambda_{\max}(P)\,\nu) \left\| \begin{bmatrix} e \\ \dot e \end{bmatrix} \right\|^2.$$

The proof of the proposition is easily completed by setting $\nu < \dfrac{1}{2\,\lambda_{\max}(P)}$, $d = d_\nu$, $c = c_\nu$, and by Theorem 42.4 of (Hahn1967). □

Remark 1 In Proposition 1, for the sake of simplicity, the input function $u(t)$ has been assumed to be constant. However, the proof can be easily generalised to the case of a non-constant input function, satisfying suitable assumptions.

3.2. *Simulations*

In order to test the efficiency of the proposed observation algorithm (3.1), several simulations starting with different initial conditions for the chain and for the observer, with suitable input functions $u(\cdot)$, and with different choices for the parameters K_x and K_v of the observer, have been run. The case study reported here is relative to the time interval $T = [0,\,10\,s]$, to the constant input function $u(t) = 5.66\,N$, $\forall t \in T$ (corresponding to the equilibrium configuration $q_0 = [2\,m\;\pi/6\,rad]^T$), to the initial conditions $q(0) = [1\,m\;0]^T$, $\dot q(0) = [0\;0]^T$, $\hat q(0) = q_0$, $\dot{\hat q}(0) = 0$, and to the following choice for the parameters of the observer: $K_x = 4.5\,N/m$, $K_v = 3\,N\,s/m$.

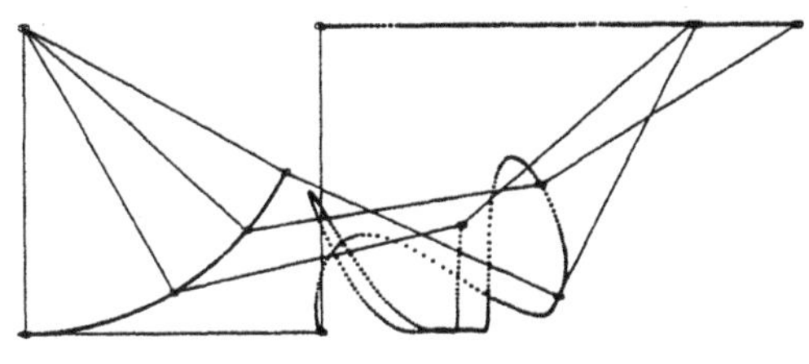

Fig. 2. The motion of the chain.

In Figure 2, four configurations assumed by the chain during the considered motion are depicted, together with the trajectories followed by the joints of the chain, restricted to the first few seconds of simulation.

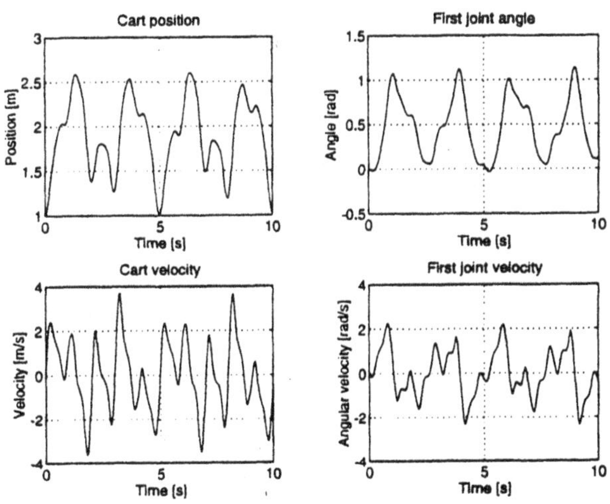

Fig. 3. Time behaviour of the state variables of the chain.

In Figure 3, the time responses of the four state variables of the chain, i.e., the cart position $x(t)$, the first joint angle $\theta_1(t)$, the cart velocity $\dot x(t)$, and the first joint velocity $\dot\theta_1(t)$, are reported, whereas the time behaviour of the four components of the estimation error $[e^T(t)\;\dot e^T(t)]^T$, i.e., the cart position error $e_x(t)$, the first joint angle error $e_\theta(t)$, the cart velocity error $\dot e_x(t)$, and the first joint velocity error $\dot e_\theta(t)$, are shown in Figure 4.

From the aforementioned plots, it is possible to

see that the observation errors are all negligible after about 4 seconds of simulated behaviour of the chain, and that this result has been obtained in correspondence to an "almost-periodic" free-motion of the chain, about its equilibrium position, of significant amplitude, with respect to the admissible motions for the proposed model.

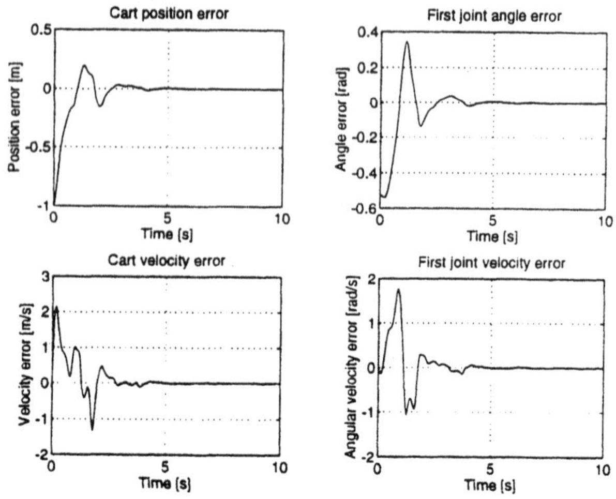

Fig. 4. Time behaviour of the observation errors.

4. CONCLUSIONS

With reference to the mechanical system in Figure 1, a non-linear observer has been proposed and has been proven to asymptotically attain state reconstruction. Simulation runs confirm the effectiveness of the observer in estimating the state vector.

Future work has to be done to apply the results here obtained for the design of suitable control schemes, in order to achieve better performances than those obtained by asymptotically stabilising the system by means of simple PD control laws.

REFERENCES

Fujino, Y. and T. Susumpow (1994). "An experimental study on active control of in-plane cable vibration by axial support motion". *Earthquake Eng. Struct. Dyn.* **23**, 1283–1297.

Fujino, Y., P. Warnitchai and B.M. Pacheco (1993). "Active stiffness control of cable vibration". *J. of Applied Mechanics* **60**, 984–953.

Hahn, W. (1967). *Stability of Motion.* Springer-Verlag.

Menini, L., A. Tornambè and L. Zaccarian (1996). Modelling and control of an under-actuated closed kinematic chain. In: *Proceedings of the 4th IEEE Symposium on New Directions in Control & Automation.* Maleme, Chania. pp. 156–161.

Triantafyllou, M. S. (1987). "Dynamics of cable and chains". *Shock and Vibration Digest* **19**, 3–5.

HARD REAL-TIME SCHEDULING FOR AGE CONSTRAINT COMPUTATIONS

W. Albrecht, D. Zöbel

*Institute for Informatics, University of Koblenz, Rheinau 1,
56075 Koblenz, Germany*

Abstract: In large scale monitoring and control systems typically many computations are requested in parallel. Explicit hard real-time scheduling has to construct a schedule off-line by interleaving all required computations in a way which meets the application's real-time requirements. Tasks are responsible to carry out the different computations. This paper will focus on the special case where real-time requirements impose so-called age constraints on tasks. Age constraint tasks are usually transformed into periodic tasks before schedule construction. A new scheduling scheme will be presented that avoids such unnecessary transformations. Its quality will be proven in comparison to the periodic schedulers.

Keywords: Computer controlled systems, Constraints, Deadlines, Monitored control system, Real-time systems, Response times, Scheduling Algorithms

1. INTRODUCTION

Computations carried out for monitoring and control have to keep up with the evolving state of the physical world. Such computations are subject to hard real-time requirements if a timing failure may lead to some intolerable catastrophic behavior. Thus a guarantee is needed that no timing failure will occur at runtime. As in large scale systems there may be many computations requested in parallel, one may timemultiplex the usage of a scarce computing resource, that is usually the processor. As computations will be carried out by associated tasks, there is the problem of interleaving the tasks' execution in such a way that the applications hard real-time requirements will still be met. It is assumed that preemptiblity (task's interruption by others) will be allowed. In order to solve the problem a so called explicit static schedule will be constructed. Such a schedule represents all reservations for the individual tasks on the processor and is constructed off-line. If successful, the schedule will be executed cyclic at run-time.

Depending on the application there may be different kinds of real-time requirements, imposing different kinds of timing constraints on the placement of the associated tasks in the schedule. This paper has the focus on hard real-time static scheduling for tasks where a so-called age constraint is imposed on each task. An age constraint forces a task τ_i to be repeatedly executed. This has to be done in such a manner that the distance between two consecutive executions of the same task τ_i is bound by the task's age constraint A_i. To be exact, A_i bounds the maximum distance allowed between the beginning of a task's execution and the end of the task's consecutive execution. It will follow some examples showing what kind of hard real-time requirements in monitoring and control applications will impose age constraints on the associated tasks.

The first example is about how to guarantee responsiveness in control. Looking at figure 1 assume a task τ_i to be responsible for computing some control instruction if a stimulus E_i occurs at some unforeseeable time. The instructions will be carried out as soon as the task's next starting execution, called τ_i^{k+1}, has finished. Thus an age constraint placement of τ_i will bound the response time – regardless of when E_i occurs.

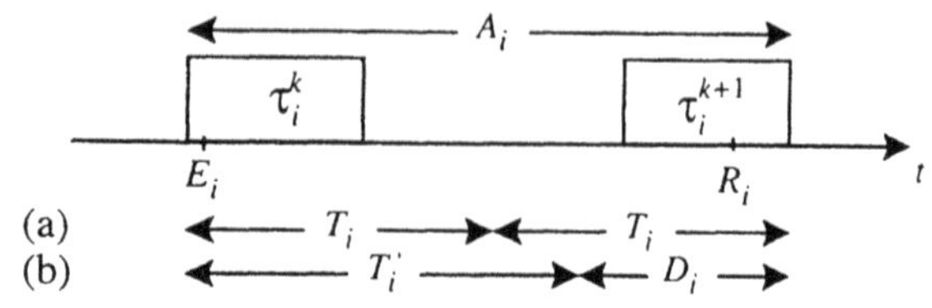

Fig. 1. Requirements imposing age constraints and period realizations (a) and (b)

As another example, assume task τ_i is responsible for refreshing a data value (may be sensor readings or derived information) representing state information about the physical world. Assume the time the data value will be requested is unforeseeable and the data value(s) are up-to-date when used as input by τ_i starting execution. Looking again at figure 1, the freshness of the computed data value – requested for example at R_i – depends on the distance from the beginning of the last already finished execution τ_i^k of τ_i. Thus the maximum 'age' (with respect to the real world), a state information is allowed to have, can be bound by an age constraint A_i.

The "real-time database" – a central concept of the distributed MARS-System (Kopetz 1986) – imposes age constraints in a similar way. Note that for distributed systems the age constraint scheduling of the shared communication line is necessary.

As a last example (Nakazato 1993) examines how age constraint tasks may keep up monitoring "significant changes" in the environment. In short: the more likely the real world value will change over time in a significant matter, the shorter must be the associated task's age constraint.

Note that none of the examples requires periodic placement of tasks, but only a constraint on the maximum distance between consecutive jobs.

As in (Albrecht 1996) we have shown that preemptive[1] scheduling of a set of age constraint tasks is NP-hard, one can not suspect an efficient optimal solution, but only heuristic ones of varying quality. The problem of hard real-time scheduling subject to age constraint tasks has been examined in (Mok 1983, Kopetz 1986, Audsley *et al.* 1992, Song and Liu 1992, Nakazato 1993). The first two doing explicit static scheduling, the others focusing on implicit dynamic scheduling, only assigning priorities to tasks before runtime. The scheme presented in the seminal work (Mok 1983) for scheduling age constraint tasks, i.e. transforming the tasks into periodic ones, is adopted by all latter ones. Getting rid of the NP-hard scheduling problem in an ad-hoc manner their focus was in different kind of synchronization making scheduling still NP-hard. In (Audsley *et al.* 1992) the proposal was made to use a task's deadline shorter than its period in order to satisfy age constraints. Although this was done with respect to implicit dynamic scheduling (deadline monotonic scheduling) it could be used for explicit static scheduling as well.

It will be shown later that transforming a task's age constraint into a periodicity constraint forces an unnessecary increase of utilization in most cases. That is way our scheduling scheme will avoid transforming task's age constraint into a corresponding periodic one in order to construct the schedule. The only other scheduling scheme avoiding period tasks we know about is in (Han 1992). There the focus is on a more loosely defined distance constraint, i.e. bounding the maximum distance between the finishing times of consecutive executions of a task.

The rest of this paper is organized as follows: Section 2 describes the underlaying model. Common solutions using the periodic tasks model will be revisited in section 3. In section 4 will present a new scheduling algorithm and discuss its quality in comparison to the other approaches. Concluding remarks are offered in section 5.

2. MODEL

An age constraint task $\tau_i = (C_i, A_i)$ is defined by its required maximum execution time C_i and its age constraint A_i. We force $C_i \leq A_i/2$ to hold. All Parameters in the model will be integers. Tasks are assumed to be preemptible. Simplifying assume there are no precedence or exclusion relations between different tasks. The repeated executions of a task τ_i is defined by its sequence $\tau_i^1, \tau_i^2, \tau_i^3, \ldots$ of jobs. A schedule for a set of age constraint tasks $\Theta = \{\tau_1, \ldots, \tau_n\}$ can be modeled as a function $s : \mathbb{N} \rightarrow \{0, 1, \ldots, n\}$, where $\mathbb{N}$ corresponds to the sequence of time slices to be divided among the tasks. A free timeslice will be marked by zero. For sake of brevity we will not discuss the cyclic representation of a schedule in this paper. With respect to a schedule s for each job τ_i^k let us mark b_i^k the beginning of the first and e_i^k the end of the last timeslice reserved for this job. An age constraint A_i of a task τ_i is *satisfied* with respect to schedule s if for any two consecutive jobs τ_i^{k-1}, τ_i^k ($k \in \mathbb{N}$): $e_i^k - b_i^{k-1} \leq A_i$ is true and the intra task precedence relationship $e_i^{k-1} \leq b_i^k$ ($k > 1$) is met. For initialization b_i^0 is defined to be zero. A schedule s is *feasible* for a

[1] A corresponding reduction presented in (Mok 1983) assumes tasks may not be preempted.

set of age constraint tasks Θ if all tasks $\tau_i \in \Theta$ satisfy their age constraint with respect to s. The minimum load imposed by an age constraint task τ_i can be as small as $C_i/(A_i - C_i)$ (if each job is executed without preemption and the distance between the beginning of two consecutive jobs is $A_i - C_i$). $\delta_i = A_i - C_i$ will be called the density of τ_i.

In the classic periodic process model a task $\tau_i = (C_i, T_i)$ is defined by its period T_i instead of an age constraint. A periodicity constraint T_i of a task τ_i is ***satisfied*** in a schedule s if for each job τ_i^k ($k \in \mathbb{N}$) there are C_i time slices reserved between its ready time $r_i^k = (k-1)T_i$ and its deadline $d_i^k = kT_i$. A schedule s is ***feasible*** for a set of periodic tasks Θ if all tasks $\tau_i \in \Theta$ satisfy there periodicity constraint with respect to s. There may be given an additional optional parameter for a task's deadline D_i. It overwrites the default assumption for the deadlines of the task's jobs according to $d_i^k = (k-1)T_i + D_i$. Note that one has to differenciate between the relative deadline of a task and the absolute deadlines of their jobs. If there is a periodic task $\tau_{i'} = (C_{i'}, T_{i'})$ such that any schedule satisfying its periodic constraint $T_{i'}$ will also satisfy the age constraint A_i of some task $\tau_i = (C_i, A_i)$ then it is said that τ_i can be *realized* by $\tau_{i'}$.

The next section will revisit the approaches relying on the periodic task model.

3. REALIZING AGE CONSTRAINTS BY PERIODIC TASKS

The two hard real-time schedulers proposed for realizing age constraint tasks using periodic ones are earliest deadline first (EDF) (Liu and Layland 73) and deadline monotonic scheduling (DMS) (Audsley 1990). They are both based on the concept of ready time and priority: At any instant of time out of all ready jobs the one with highest priority is selected for execution. While EDF assigns the priority with respect to indiviudal jobs according to their absolute deadlines, DMS assigns priority to tasks with respect to the their relative deadlines; all jobs of one task having the same so called fixed priority. DMS equals rate monotonic scheduling(Liu and Layland 73) if the tasks' deadlines are equal to the corresponding periods $D_i = T_i$. Feasibility can be tested easily (on base of the utilization imposed by the individual task set) for EDF and DMS if $D_i = T_i$ (Liu and Layland 73). An efficient necessary and sufficient condition for a set of periodic tasks with deadlines smaller than periods is given for DMS (for example (Audsley 1990)) but not for EDF [2]. This is because in standard fixed priority scheduling one only has to test, whether the first job of each task can be placed in the schedule before its deadline under critical phasing (Liu and Layland 73) – the whole schedule's length has not to be considered as for EDF. Critical phasing means that at one point in time t (assume $t = 0$) there is a job of each task getting ready just at t altogether. Now the suggested solutions will be examined in more detail in order to compare them with our new scheduler later on. For sake of briefity the possible use of offsets (Audsley 1991) will not be discussed.

In (Mok 1983) it is proposed that an age constraint task $\tau_i = (C_i, A_i)$ can be realized by a corresponding periodic one $\tau_{i'} = (C_i, T_{i'})$ setting its period $T_{i'} = A_i/2$ and then using EDF for schedule construction. This method reflects the fact that in one period a job τ_i^k may be scheduled just at the beginning of that period, while in the consecutive period there may be other jobs causing τ_i^k to finish not earlier than at the end of its period (see Fig. 1(a)). Instead of the optimum possible load of $C_i/(A_i - C_i)$ there will be as much as $2C_i/A_i$ load imposed by τ_i, thus making it more likely that there is not enough room for lower priority tasks. Of course the (worst-case) placement of jobs at the end of a period may not really ever happen. Assigning an additional deadline for a task can guarantee such a restriction in the task's placement. Then, if there is a deadline D_i smaller than the period for a periodic task τ_i, the period can be as small as $T_i' = A_i - D_i$ in order to still realize the age constraint A_i (see Fig. 1(b)). The problem now is to assign deadlines as small as possible, such that the periodicity constraint (getting harder because of the shorter deadlines) can still be satisfied in some schedule. This clearly depends on the scheduler used. In (Audsley *et al.* 1992) the proposal was made to use DMS in order to realize age constraint tasks.

Now it will be shown how to use scheduling analysis in order to look out for a task's deadline that can be met. The scheme will be called "age constraint deadline monotonic scheduling, according to tasks' density δ" (in short $\mathcal{D}(\delta)$). The scheme starts by first assigning priorities to the age constraint tasks with respect to their density (the smaller the density the higher the priority, ties being broken arbitrarily). Suppose there is the set $\Theta = \{\tau_1, \ldots, \tau_n\}$ of age constraint tasks $\tau_i = (C_i, A_i)$, $(1 \leq i \leq n)$ labelled with respect to their priority, τ_1 being the one with highest priority (i.e. smallest density). Suppose all tasks start in phase at $t = 0$, that is $r_i^1 = 0$, $1 \leq i \leq n$. Relying on the critical instant theorem (Liu and Layland 73) e_i^1 (i. e. the finishing time of τ_i's first job) will be the maximum possible time a job of task τ_i may ever be delayed. Thus one may assign

[2] (Leung and Merrill 1980) shows the NP-completeness of a related problem.

$D_i = e_i^1$ and know that this deadline will always be met. The period will be set to $T_i = A_i - D_i$, thus the derived periodic task will realize the age constraint. In order to compute the finishing time e_i^1 of task τ_i $(1 \le i \le n)$ the periods of the higher priority tasks had to be known. Because of this the scheme starts with $\tau_1, \tau_2, \ldots$.

The highest priority task τ_1 will not be disturbed by any other, thus τ_1's first job will finish at $e_1^1 = C_1$, its deadline D_1 and period T_1 are set according to the achievable deadline $D_1 = e_1^1$ and $T_1 = A_1 - D_1 = A_1 - C_1$. Now the scheme can succeed with $\tau_2, \tau_3, \ldots$ in the same manner.

The scheme will stop successfully if all tasks have been assigned a deadline and period, then finally construct a feasible schedule using DMS for the derived set of periodic tasks. The scheme will stop with failure if $e_i^1 > A_i/2$ for some task $\tau_i \in \Theta$; no feasible schedule will be constructed.

4. FIXED PRIORITY SCHEDULING FOR AGE REQUIREMENTS

Our static schedule construction scheme called "age constraint fixed priority scheduling, according to tasks' density δ" (in short $\mathcal{A}(\delta)$) works as follows. First the scheme assign priorities to each age constraint task with respect to its density, just as done for $\mathcal{D}(\delta)$, leading to the same priority sorted task set $\Theta = \{\tau_1, \ldots, \tau_n\}$ of age constraint tasks $\tau_i = (C_i, A_i)$, $(1 \le i \le n)$. Constructing the schedule proceeds from the task with highest priority to the lower ones by reversing time slices for their sequences of jobs $\tau_i^1, \tau_i^2, \ldots$, $(1 \le i \le n)$. The difference between ours and standard fixed priority scheduling is in the reservation scheme. For each job τ_i^k actually to be integrated into the schedule, its placement will not be adjusted to some ready time but according to its deadline d_i^k. This deadline will be set depending on the job's predecessor's start time b_i^{k-1} and the age constraint A_i, that is $d_i^k = b_i^{k-1} + A_i$. Because $b_i^0 = 0$ by definition, it follows $d_i^1 = A_i$. Having got d_i^k, the scheme then tries to reserve free time slices for τ_i^k going back in time from d_i^k until either C_i free time slices could have been reserved or until a timeslice is reached which has been reserved for the job's predecessor τ_i^{k-1}. In the latter case the scheme stops with no feasible schedule being constructed. Otherwise the scheme goes on until all tasks are integrated into the schedule.

An implementation of $\mathcal{A}(\delta)$ doesn't have to place all jobs of all higher priority tasks before a lower priority job but only those that might compete for the time slices, that the low priority job wants to have. The implemented scheduler will in fact generate a longer and longer schedule until a cycle is found or a failure occures.

Note that orienting the reservation scheme according to the deadline directly there is no more need for the concept of ready time. That is: no more need to guess when the time is right to begin to compete for the processor in order to be finished until deadline. The scheduler tries to place the consecutive jobs as far as possible into the future, thus trying to minimize the imposed load, in this way making it more amenable that lower priority tasks can satisfy there age requirement.

It follows an example of schedule construction comparing $\mathcal{D}(\delta)$ (a) and $\mathcal{A}(\delta)$ (b) for tasks τ_1, τ_2 with $A_1 = 4$, $A_2 = 6$ and $C_1 = C_2 = 1$. There

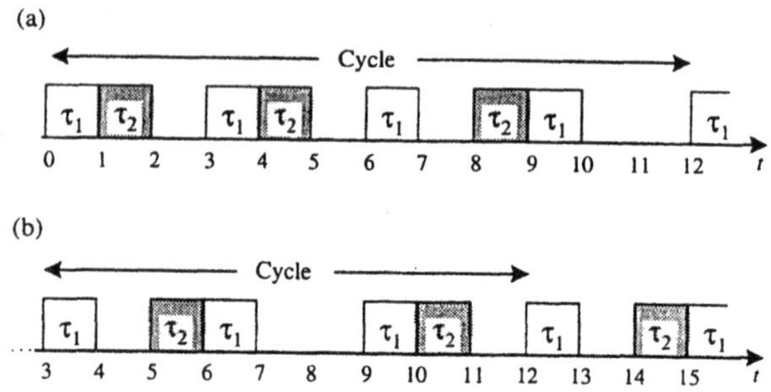

Fig. 2. Schedules using $\mathcal{D}(\delta)$ (a) or $\mathcal{A}(\delta)$ (b).

is a differences in cycle length 9 vs. 12 and load 5/9 vs. 7/12 for both schedulers. The difference in load of about 3 % in this small example[3] may cause another task τ_3 not to be feasibly scheduled by $\mathcal{D}(\delta)$ but by $\mathcal{A}(\delta)$; for example $C_3 = 4$, $A_3 = 16$ will do that. Now it will be shown that $\mathcal{A}(\delta)$ always performs as least as good or better than $\mathcal{D}(\delta)$.

Theorem 1. Given a set $\Theta = \{\tau_1, \ldots, \tau_n\}$ of age constraint tasks $\tau_i = (C_i, A_i)$, $1 \le i \le n$. If Θ can be feasibly scheduled using $\mathcal{D}(\delta)$, then it can be feasibly scheduled using $\mathcal{A}(\delta)$.

Proof. Let Θ_w (Ω_w) be the subset of Θ (Ω) consisting of the w age constraint tasks (substituting periodic tasks) with smallest density δ_i. They will have highest priority in $\mathcal{D}(\delta)$ and in $\mathcal{A}(\delta)$ as well. Let $s_w^{\mathcal{A}(\delta)}$ denote the schedule constructed according to $\mathcal{A}(\delta)$ for Θ_w. In order to prove the theorem by induction on w we establish the induction assumptions:

(1) If Θ_w can feasibly be scheduled using $\mathcal{D}(\delta)$ then it can feasibly be scheduled using $\mathcal{A}(\delta)$.
(2) The number of reserved time slices in $s_w^{\mathcal{A}(\delta)}$ will not exceed the number of requested time slices under critical instant using $\mathcal{D}(\delta)$, as given by $\sum_{\tau_{i'} \in \Omega_w} \lceil \frac{\Delta t}{T_{i'}} \rceil C_{i'}$, with respect to any interval length $\Delta t \in \mathbb{N}$.

[3] Note that no offset for one of the tasks can improve the result acheaved by $\mathcal{D}(\delta)$.

For the task set Θ_1, consisting of a single task τ_x with smallest density, both schedulers will not differ in placing their jobs at equidistant intervals of length $A_x - C_x$. Thus (1) and (2) are trivially satisfied. Now suppose both assumptions are true for Θ_w, we show they will hold for Θ_{w+1}. We need to show that (1) and (2) will hold for the only one additional task of lowest density, say $\tau_y = (C_y, A_y)$, being in Θ_{w+1} but not in Θ_w.

Following assumption (1), τ_y can be realized by a periodic task $\tau_{y'}$ using $\mathcal{D}(\delta)$. There must be C_y free time slices in $[0, A_y/2)$ at critical instant $t=0$. It follows $t_D \leq A_y/2$, with t_D marking the point in time just after the last one of the C_y first free time slices following Ω_w's critical instant. Following assumption (2), there are at least C_i free time slices in $s_w^{\mathcal{A}(\delta)}$ for any interval of length $t_D \leq A_y/2$. The reservation of time slices according to $\mathcal{A}(\delta)$ for a job τ_y^{k+1} will be no problem if its predecessor τ_y^k did not take up an interval longer than t_D for its reservations. An induction on job-index k can show that following

$$e_y^k \leq b_y^k + t_D \leq b_y^k + A_y - t_D = d_y^{k+1} - t_D \leq b_y^{k+1}$$

(see Fig. 3, upper part) there will be no overlap between any two consecutive jobs of τ_y. As $\mathcal{A}(\delta)$ fails iff an intra task precedence relation can not be meet, induction hypothesis (1) follows immediately.

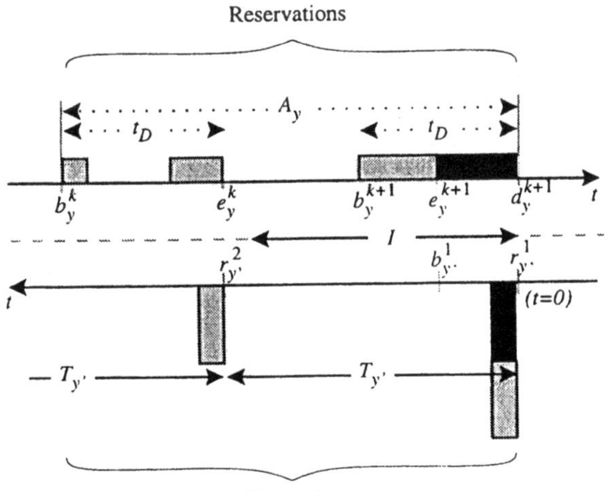

Fig. 3. Reservations done by $\mathcal{A}(\delta)$ and the mirror image of $\mathcal{D}(\delta)$'s critical instant.

In order to prove induction hypothesis (2) first observe that – because of induction assumption (2) – the repetition rate of $\mathcal{D}(\delta)$-Requests is equal to or larger than that of $\mathcal{A}(\delta)$-Reservations for all tasks in Ω_w. This is valid for Ω_{w+1} as well, because there is a repetition rate of $(b_y^{k+1} - b_y^k \geq A_y - t_D)$, for any two consecutive jobs τ_y^k, τ_y^{k+1} $(k \in \mathbb{N})$ in $s_w^{\mathcal{A}(\delta)}$ and in comparison there is $T_{y'} = A_y - t_D$, for the period of the substituting task $\tau_{y'}$ using $\mathcal{D}(\delta)$.

Although the repetition rate is larger for each individual task using $\mathcal{A}(\delta)$, we still have to observe the relative placements of the new task τ_y ($\tau_{y'}$) for both schedulers. This is because the ***reservations*** for two consecutive jobs of τ_y may in fact lay closer together, than the *requests* for two consecutive jobs of $\tau_{y'}$. It will be shown that the overall reservations of $\mathcal{A}(\delta)$ will stay below the overall requests of $\mathcal{D}(\delta)$ anyway.

Figure 3 shows a possible distribution of two jobs' execution using $\mathcal{A}(\delta)$. Assume it is $e_y^{k+1} < d_y^{k+1}$ because of the higher priority task(s). An interval I of arbitrary length ($Lgh(I)$) will have a maximum of reservations for τ_y if it ends at e_y^{k+1}. Note, there are more $\mathcal{A}(\delta)$-Reservations than $\mathcal{D}(\delta)$-Requests in I with respect to τ_y and $\tau_{y'}$. But, accounting for all tasks, we will get the maximum number of reserved time slices by $\mathcal{A}(\delta)$ for interval length $Lgh(I)$ if we move I up to d_y^{k+1}, because no time slice in $[e_y^{k+1}, d_y^{k+1})$ is free. Now look at the mirror image of the critical instant of the corresponding periodic tasks starting at $t=d_y^{k+1}$ (Fig. 3, lower part). Applying assumption (2), there must be a corresponding delay for $\tau_{y'}^1$ at critical instant using $\mathcal{D}(\delta)$; the number of requests of higher priority tasks at critical instant is equal to or exceeds $d_y^{k+1} - e_y^{k+1}$. Because of this and because the repetition rate of $\mathcal{D}(\delta)$-Requests is equal to or larger than that for $\mathcal{A}(\delta)$-Reservations, we conclude that there are at least as many requests with respect to $\mathcal{D}(\delta)$ than there are reservations with respect to $\mathcal{A}(\delta)$ within I. This is induction hypothesis (2). □

Besides this analytical proof we have done performance evaluation comparing $\mathcal{A}(\delta)$ with EDF for some randomly generated tasks sets in (Wisser 1996). For EDF tasks' periods are set to $A_i/2$, because we do not know of any efficient way to find good combinations of period and deadline, as it was possible for the DMS based scheme $\mathcal{D}(\delta)$. Performance evaluation shows that there is a very good chance to find a feasible schedule using $\mathcal{A}(\delta)$ even if the load imposed by the corresponding periodic tasks using EDF passes the 100% utilization bound, reaching about 120% and more; utilization being measured according to $\sum_{i:\tau_i \in \Theta} 2C_i/A_i$.

5. CONCLUDING REMARKS

While age requirements can be handled by periodic tasks as done in all known systematic schemes, it has been shown that this is a waste of the computational resource, thus leading to situations where not all timing requirements can be met due to heavy load which is unnessecary imposed into the schedule.

The scheduler $\mathcal{A}(\delta)$ has been proposed that does not construct a schedule on the basis of periodic tasks but directly on age constraints. This imposes less restrictions for the placement of jobs into the schedule. $\mathcal{A}(\delta)$'s quality has been shown by analytically comparing it with an adapted version of DMS, called $\mathcal{D}(\delta)$, and by performance evaluation in comparison to EDF (in (Wisser 1996)).

Future work will extend our performance evaluation comparing $\mathcal{A}(\delta)$ with $\mathcal{D}(\delta)$. Resulting cycle length and load of the schedules constructed by the different schedulers will be in focus of performance evaluation as well. After the theory of age constraint scheduling has been explored, we will have to focus on practical applications. For example the granularity of time slices has to be adjusted according to individual applications, especially if one will do scheduling for a shared communication medium instead for the processor. There is ongoing work developing a computer-aided tool that will guide the software developer in using ours (and others) scheduling schemes.

We think that our new scheduler will at least for some monitoring and control applications support the timemultiplexed use of a scarce computing resource; sometimes it will make the timemultiplexed use possible at all. This is because of its better capabilities in finding a feasible schedule if hard real-time requirements are given in form of age constraints.

6. REFERENCES

Albrecht, W. (1996). Echtzeitplanung für Alters- oder Reaktionszeitanforderungen. Technical report. Institut für Informatik, Universität Koblenz-Landau.

Audsley, N. (1990). Deadline monotonic scheduling. Technical report. Department of Computer Science, University of York.

Audsley, N. (1991). Optimal priority assignment and feasibility of static priority tasks with arbitrary start times. Technical report. Department of Computer Science, University of York.

Audsley, N., A. Burns, M. Richardson, K. Tindell and A. Wellings (1992). Absolute and relative temporal constraints in hard real-time databases. In: *Proc. of IEEE Euromicro Workshop on Real Time Systems.*

Han, Ching-Chih (1992). Scheduling real-time computations with temporal distance and separation constraints and with extended deadlines. PhD thesis. University of Illinois at Urbana-Champaign.

Kopetz, H. (1986). Design of real-time systems. In: *Computer Systems for Process Control* (R. Güth, Ed.). Plenum Press. New York.

Leung, Joseph Y-T. and M. L. Merrill (1980). A note on preemptive scheduling of periodic real-time tasks. *Information Processing Letters* **11**(3), 115–118.

Liu, C. L. and James W. Layland (73). Scheduling algorithms for multiprogramming in a hard-real-time environment. *Journal of the ACM* **20**(1), 46–61.

Mok, A. K. (1983). Fundamental design problems of distributed systems for the hard-real-time environment. PhD thesis. Massachusetts Institute of Technology.

Nakazato, Hidenori (1993). Issues on synchronizing and scheduling tasks in real-time database systems. PhD thesis. University of Illinois at Urbana-Champaign.

Song, X. and J. W. S. Liu (1992). Maintaining temporal consistency: Pessimistic vs. optimistic concurrency control. In: *Proc. of the IEEE Symposium on Computer-Aided Control System Design.* Napa, California.

Wisser, R. (1996). Verplanung von Prozessen mit Altersanforderungen. Studienarbeit, Institut für Informatik, Universität Koblenz-Landau.

DYNAMIC CONTROL SCHEMES FOR MANUFACTURING SYSTEMS

Francesco Martinelli, Salvatore Nicosia, Paolo Valigi
Dipartimento di Informatica, Sistemi e Produzione - Università di Roma "Tor Vergata"
via della Ricerca Scientifica, 00133 Roma, Italy - Tel. +39 6 72594405 Fax. +39 6 2020519
e-mail: {valigi,nicosia}@disp.utovrm.it

Abstract. In this paper, dynamic control schemes for manufacturing systems are studied. A novel state reconstruction algorithm is used as a basic tool for developing the control schemes, algorithm that allows to exactly compute the system performance under perturbed values of some system parameters. The control schemes are used to counteract the effects of unknown parameters characterizing the system, and/or to reject disturbances acting on it. In this paper, the specific case of a multi part-type manufacturing system is considered, and the proposed controllers are used to dynamically allocate buffer capacities, in order to maximize system throughput, despite uncertainties on the system. Simulation results are presented to illustrate the effectiveness of the approach, and to compare two different variants of a basic controller. In addition, a concurrent implementation of the algorithm is also presented.

Key Words. Discrete event systems, manufacturing, Kanban systems, dynamic control, state reconstruction.

1. INTRODUCTION

One of the most studied problems in control and optimization of manufacturing systems is that of buffer capacity allocation (see, e.g., (Glasserman and Yao, 1996) and references therein). In particular, the buffer allocation problem has been studied both in the static and in the dynamic case.
The problem of optimal allocation of buffer capacities is relevant also in the framework of Kanban systems, since several type of these systems can be described by means of finite capacity queueing networks, with the buffer capacity being the number of Kanbans assigned to nodes and/or different part-types (see, e.g., (Savsar, 1996; Di Mascolo *et al.*, 1991; Bard and Golany, 1991; Berkley, 1991; Mitra and Mitrani, 1991) and references therein).
The problem of optimal allocation of buffer storage in production lines was at the origin of the first perturbation analysis (PA) algorithm (Ho *et al.*, 1979). Since then, perturbation analysis techniques have been further applied to the buffer capacity allocation problem. Among others, in (Liberatore *et al.*, 1995*a*;*b*;1997) modified finite PA rules have been used to solve such a problem for general queueing systems, with application to Kanban manufacturing systems; in (Panayiotou and Cassandras, 1996) an exact algorithm for the analysis of single class queueing networks, under buffer capacity augmentation, has been proposed, in (Cassandras and Panayiotou, 1996) a scheme for concurrent sample path analysis similar to the reconstruction algorithm considered here has been studied.
In this paper, the *state reconstruction* algorithm proposed in (Martinelli *et al.*, 1997*a*;*b*) is used as a fundamental tool to develop dynamic buffer allocation policies.
The state reconstruction algorithm is based on the assumption that relevant information, such as the system initial state and the system "input sequence", can be extracted from the observation of the system over a finite horizon, and properly stored. Based on the knowledge of these data, and assuming an accurate queueing network model of the system is available, the state sequence and the performance indexes can exactly reconstructed, regardless of the magnitude of the parameter perturbations. The algorithm is inherently parallel, and can be seen as a concurrent approach to the solution of dynamic discrete optimization problems.
The reconstructed performance indexes then, can be used to implement dynamic control schemes, such as the ones proposed in this paper. In particular, two variants of a basic control scheme are presented. The first scheme, a global control scheme, is based on the reconstruction of the state for all the admissible system parameters, while in the second scheme, a local control scheme, only a reduced number of parameters, close to the nominal one, are considered.
The scheme is mainly intended for on-line control of real systems, and therefore the reduction in computational effort achieved by this approach, as compared to brute force simulation, is not the major issue. Nevertheless, also from this perspective the scheme is useful, since, with respect to brute force simulation, the proposed algorithm capitalizes on data obtained from the observation of the real system.
The organization of the paper is as follows. The state reconstruction problem is considered in Section 2, Section 3 presents the proposed dynamic control scheme, Section 4 presents the application to a simple manufacturing system, Section 5 illustrates a concurrent implementation of the algorithm. Some conclusions are drawn in Section 6.

2. STATE RECONSTRUCTION PROBLEM

In this paper it will be considered the quite general class of manufacturing systems that can be modeled by means of queueing networks with: (a) general service time at each node, (b) general routing policy at each node, (c) general scheduling policy at each node, (d) finite buffer capacity, (e) multi-class, (f) non-preemptive service at each node, (g) infinite arrival rate sources, (h) infinite capacity sinks.
Each node of a network comprises a server and several input queues or buffers; each server has an input queue for each class it can provide service to.
The algorithm for state reconstruction used in this paper is based on the dynamic equations of a queueing network, written following the GSMP formalism. A detailed description of the algorithm can be found in (Martinelli *et al.*, 1997*b*), while an application to dynamic optimal allocation of buffer capacity can be found in (Martinelli *et al.*, 1997*a*).
Let x be the *network extended state*, containing all queue

lengths, customers under service, blocked servers, and the service completion times of all the customers currently under service, and let $\mathcal{X}$ denote the extended state space of the queueing network.

A key element for the execution of the marking algorithm is the "input sequence", containing information on the stochastic nature of the system. Assume the system behavior has been observed for an infinite time-period, and let the infinite sequence $u_D(\cdot)$ be the realization of the random sequences determining customer service duration for all the nodes. Similarly, let the infinite sequences $u_R(\cdot)$ and $u_S(\cdot)$ be the sequences of realizations of the random routing and scheduling decisions. Then, the *input sequence* $u(\cdot)$ is the set of all the (infinite) sequences $u_D(\cdot)$, $u_R(\cdot)$ and $u_S(\cdot)$.

The behavior of a queueing network depends also on a number of *system parameters* θ. In this paper, the system parameter vector will coincide with the capacities of (some of) the buffers in the system. Let Θ be the parameter space, which is assumed to be finite.

Let the *extended state sequence* $x(\cdot)$ be the sequence of values taken by the extended state as the system evolves. Then, for a given input sequence $u(\cdot)$, for a given initial extended state x_0, and for a given nominal vector of system parameters θ, such a sequence $x(\cdot)$ can be formally computed as:

$$x(\cdot) = \Phi(x_0, u(\cdot), \theta), \tag{1}$$

where function Φ can be formally defined via the composition of the GSMP-like dynamic equations describing the evolution of a queueing network, and will be referred to as the *input to extended state map*. Of course, such a map is of no real interest, being based on infinite sequences.

To derive a physically implementable algorithm, a modification of the map (1) has been introduced: the *marking algorithm* (Martinelli *et al.*, 1997*b*). Let (with an abuse of notation) $u_D(\cdot)$ be a finite sequence of service time, $u_S(\cdot)$ be a finite sequence of scheduling decisions, and $u_R(\cdot)$ be a finite sequence of routing decisions. Moreover, let k be a vector, whose entries, properly ordered, are the number of elements of these three finite sequences. Finally, let $u_k(\cdot)$, be the *finite input sequence*, i.e., the set comprising the three finite sequences $u_D(\cdot)$, $u_S(\cdot)$, and $u_R(\cdot)$.

The marking algorithm allows to compute a (perturbed) finite extended state sequence $\hat{x}(\cdot)$, based on the knowledge of a finite input sequence $u_k(\cdot)$, of an (observed) initial extended state x_0, and of a (perturbed) vector of network parameter $\hat{\theta}$ (Martinelli *et al.*, 1997*b*). The algorithm is strongly based on the map (1), and uses a marking mechanism to detect servers for which state reconstruction cannot be performed any longer, because of lack of data, due to the finite length of the input sequence. In the following, the marking algorithm will be simply denoted by the mapping:

$$\hat{x}(\cdot) = \hat{\Phi}(x_0, u_k(\cdot), \hat{\theta}). \tag{2}$$

For queueing networks, the values that the state may assume are not independent from network parameters. Here, the entries of vector θ correspond to the maximum values of some of the components of the queueing state. To formalize such a relationship between network parameters and admissible values of the extended state, let $\mathcal{X}(\theta)$ denote the *admissible extended state space under parameter* θ. A given extended state x is referred to as an *admissible extended state under parameter* θ if $x \in \mathcal{X}(\theta)$.

The marking algorithm exactly reconstructs the longest extended state sequence that can be computed from the given observed input sequence and initial extended state, provided that the initial extended state is admissible under the desired parameter, $x_0 \in \mathcal{X}(\hat{\theta})$, and that some hypotheses are fulfilled by the probability distribution of the service durations of all the servers (this is proved in (Martinelli *et al.*, 1997*b*), where a detailed description of the reconstruction algorithm is presented). If the initial extended state is not an admissible state under the system parameter for which state reconstruction is required, one has to somehow modify such an initial state, in order to allow state reconstruction. In the following, if the given initial extended state is not admissible under the perturbed system parameter $\hat{\theta}$ under consideration, it will be made admissible by a suitable "projection" on the extended state space under parameter $\hat{\theta}$, i.e., by setting the queue lengths that are larger than the capacity determined by the perturbed system parameter, to a value equal to these perturbed capacities.

3. DYNAMIC CONTROL OF MANUFACTURING SYSTEMS

In this paper, the state reconstruction algorithm has been applied to the problem of dynamic allocation of buffer capacities in manufacturing systems. The problem is relevant *per se*, and also because, in most of the Kanban controlled manufacturing systems, the number of Kanban assigned to each node can be considered as the buffer capacity of a suitable queueing network model of the manufacturing system (see (Di Mascolo *et al.*, 1991; Berkley, 1991)).

The proposed control scheme is aimed at dynamically varying the buffer capacity of some queues in the system, in reaction to unknown changes in the production scenario, in order to optimize system performance.

The performance measure considered in this paper is the *long term system throughput* $T(H)$, computed over a large number of service customers:

$$T(H) := \frac{H}{\tau(H)}, \tag{3}$$

where $\tau(H)$ is the time at which the H-th serviced customer departs from the system.

It is assumed that a subset $\mathcal{Q}_C$ of the set of queues, the set of *controlled queues*, is such that their buffer capacity can be modified, while satisfying the following constraint, which defines the parameter space Θ:

$$\sum_{(i,\alpha)\in\mathcal{Q}_C} \theta(i,\alpha) = B_{tot}, \tag{4}$$

where $\theta(i,\alpha)$ is the buffer capacity of queue (i,α), and B_{tot} is the total buffer capacity available to the controlled queues. Constraint (4), defining the set Θ, is quite common is real manufacturing systems.

The basic idea of the control scheme is to perform periodic *control actions* on the system, each control action consisting in a possible modification of the buffer capacities of the system, and the *control period* (or *observation period*) corresponding to the time needed by the system to service a given constant number M of customers. Each control action is the result of a *control decision*, that can be taken on the basis of observation of the real system.

Suppose $m - 1$ control actions have been already implemented, $m \geq 0$, and assume the value of the network parameter θ, after the $(m-1)$-th control action (i.e., during the m-th control period), is $\theta = \theta_m$.

During the m-th observation period, the input sequence $u_k(\cdot)$, the initial and final extended states x_0 and x_f, respectively, are recorded (the dependence on the control period index m has been omitted to keep notation easy). Then, by means of the marking algorithm (2), at the end of such an m-th control period, the extended state sequence is reconstructed, for all $\theta \in \Theta_m$, with the subset $\Theta_m \subseteq \Theta$ defined as

$$\Theta_m := \{\theta \in \Theta : x_f \in \mathcal{X}(\theta)\}, \tag{5}$$

where Θ_m is the subset of all the buffer allocations for which the observed final extended state x_f is admissible. A modified initial extended state $\bar{x}_0$ must be selected with the projection rule defined in the previous section, to make the extended state reconstruction possible for all $\theta \in \Theta_m$. For each $\theta \in \Theta_m$, the marking algorithm (2), with initial extended state $\bar{x}_0$ and input sequence $u_k(\cdot)$, produces a different extended state sequence, with a different length, i.e., a different number of terms.

For each $\theta \in \Theta_m$, once extended state reconstruction has been completed, the *one-period perturbed throughput* can be computed as:

$$T_m(M_\theta) := \frac{M_\theta}{\tau_m(M_\theta) - \tau_{m-1}(M)}, \tag{6}$$

where M_θ is the number of customers serviced by the system in the path reconstructed with the parameter θ, $\theta \in \Theta_m$, and $\tau_{\bar{m}}(\bar{M})$ denotes the time at which the $\bar{M}$ customer of the $\bar{m}$-th control period departs from the system ($\tau_0(\cdot) \equiv 0$).

In addition, at the end of the m-th control period, the *observed one-period throughput* $T_m^O(M)$, defined as:

$$T_m^O(M) = \frac{M}{\tau_m(M) - \tau_{m-1}(M)}, \tag{7}$$

is computed as well.

The availability of the observed one-period throughput, and of the values of the perturbed one-step throughput for several perturbed network parameters, makes it easy to take control decisions, depending on the chosen performance criteria.

The control problem considered in this paper, is that of long-term throughput optimization, that is, the problem if that of finding the buffer allocation solving the constrained optimization problem:

$$\max_{\theta} T(H), \tag{8}$$

$$\text{subject to} \quad \sum_{(i,\alpha)\in\mathcal{Q}_C} \theta(i,\alpha) = B_{tot}. \tag{9}$$

In this case, at the end of a generic (e.g., the m-th) control period, the control decision can be taken by selecting the perturbed network parameter yielding the maximum value of the one-period perturbed throughput:

$$\theta^* = \arg\max_{\theta\in\Theta_m} T_m(\theta), \tag{10}$$

$$\theta_{m+1} = \theta^*. \tag{11}$$

Notice that, since the network parameter θ_m during the m-th control period is surely such that $\theta_m \in \Theta_m$ (by the definition of Θ_m), in (11) it is assumed that the throughput value $T_m(\theta_m)$ corresponding to the nominal network parameter θ_m, is the observed one-step throughput, computed according to equation (7).

4. AN APPLICATION: A MULTI-PART TYPE SYSTEM

The proposed control scheme has been applied to the manufacturing system depicted in figure 1, and simulation experiments have been carried out, to asses its effectiveness. It is a simple multi part-type manufacturing system, with the service duration of all the nodes independent of part type, nodes 3, 4, and 5 working only one part-type each. The routing probability from node 2 to nodes 3, 4, and 5 is a model of the mix of part type arriving at the system, and of a deterministic routing policy, based on part-type: each type to a different node. Each node has only one queue.

The set of controlled queues is assumed $\mathcal{Q}_C = \{2, 3, 4, 5, 6\}$, and also buffer capacity allocations that are largely different from the nominal one have been considered, since the state reconstruction algorithm is very accurate. In the following, both a *local control scheme* and a *global control scheme* are illustrated. In the former scheme, only a small number of buffer allocations, close to the nominal one (i.e., close to the parameter value θ_m), are considered at each control period, while in the latter all the possible control actions $\theta \in \Theta_m$ are compared at each control decision. In particular, here the local control scheme analyzes, at each control period, all the network parameters that can be obtained from the current one, by adding of subtracting 1 to any entry of θ, while satisfying the constraint (4).

In general, the number of possible control actions at each control decision, i.e., the number of different throughput values to be compared, has to be traded off with the computational complexity of the overall control scheme. Hence, local control schemes may turn to be more of particular interest. As a matter of facts, the simulation results reported in the following section, clearly show that achieved performance, in the case of local schemes, are similar to the ones of the global control scheme in the long range, and slightly worst over short time intervals. On the contrary, local control schemes have a considerably lower computation cost, hence they seem to be preferable in real time applications.

In the simulation experiments, all the service times are exponentially distributed, with mean service time equal to 0.5, 1, 3, 2, 1, and 1, for nodes from 1 to 6, respectively. The sum B_{tot} of the buffer capacity of the controlled queues is chosen equal to 8 (without taking into account customers under service), and the initial buffer allocation is equal to 1 for nodes 2 and 6, and equal to 2 for nodes 3, 4, and 5. The routing policy from node 2 to nodes 3, 4, and 5 is random. It is assumed that the routing probabilities are unknown, and

such that the initial buffer allocation is not optimal. Hence, those unknown probabilities can be considered as a disturbance acting on the system, that has to be rejected by the control scheme.
The control period M has been chosen as $M = 500$, while the long term throughput has been computed over $H = 30000$ customers, corresponding to 60 control actions.
As a preliminary analysis, the *uncontrolled system*, i.e., the given system without any control scheme active, and with constant routing probabilities equal to 0.05, 0.05, and 0.9, has been simulated. In particular, the uncontrolled system has been simulated under all the possible buffer allocations, in order to determine the optimal one, i.e., the one yielding the maximum throughput under constant buffer allocation. It turns out that the optimal buffer allocation is given by $\theta_{opt} = (1\,0\,0\,4\,3)$, for nodes 2 through 6, respectively. The long term throughput for the uncontrolled system, with optimal buffer capacity allocation, has been used as a comparison term (see figure 2, dotted line).
The long term throughput of the *globally controlled system*, i.e., of the system under the effect of the proposed global control scheme, is reported in figure 2 (dashed line), as a function of customers serviced by the system, and compared with the performance of the uncontrolled system, with fixed buffer capacity equal to 1 for queues 1 and 6, and equal to 2 for queues 3, 4 and 5 (solid line in fig. 2), and of the uncontrolled system with fixed buffer capacity, equal to the optimal one (dotted line in fig. 2). The diagram clearly indicates the strong improvement on system performance achieved by the proposed control scheme: as a matter of fact, the performances of the controlled system are almost equal to the optimal one.
From the point of view of the chosen control actions, it turns out that the global control scheme chooses only 8 different buffer allocations, out of a total of 495, and in particular only the following allocations are considered:

$$\begin{aligned}
\theta_1 &= (1\,0\,0\,3\,4), & \theta_2 &= \theta_{opt} = (1\,0\,0\,4\,3),\\
\theta_3 &= (1\,0\,0\,5\,2), & \theta_4 &= (1\,0\,1\,3\,3),\\
\theta_5 &= (1\,1\,0\,3\,3), & \theta_6 &= (1\,1\,0\,4\,2),\\
\theta_7 &= (1\,2\,2\,2\,1), & \theta_8 &= (2\,0\,0\,3\,2).
\end{aligned}$$

Figure 3 illustrates the percentage number of times each buffer allocation is chosen by the global control scheme, among all the 8 allocations considered by the scheme at least once. The chart indicates that almost 80% of the control actions coincide with the buffer allocation θ_2, which is the optimal one for the system without control. Hence, the system is almost always operated under the optimal buffer allocation, although a) such an allocation is not known a priori to the system, and b) the initial buffer allocation is quite different from the optimal one.
The system has been tested also under the effect of the local control scheme. Figure 4 compares the behavior of the long term throughput for the *locally controlled system* – i.e., for the system controlled by means of the local scheme (dashed line in Figure 4) – for the uncontrolled system, with buffer allocation equal to 1 for queues 1 and 6, and equal to 2 for queues 3, 4 and 5 (solid line in fig. 4), and for the uncontrolled system, with optimal buffer allocation (dotted line in fig. 4). Also in the case of the locally controlled system, the performance are almost equal to the one achieved with the optimal buffer allocation. Figure 5 reports the difference between the long term throughput for the globally controlled system and for the locally controlled system: the performance in the latter case are slightly better on the short term, and became identical on the long term.

5. A CONCURRENT IMPLEMENTATION

The basic idea of the control scheme considered in this paper is to use the reconstruction algorithm (2) to simultaneously estimate system performance for the set of parameter vectors Θ_m, based on data gathered from the observation of the actual system behavior. The process of simultaneous estimation is inherently parallel: as a matter of fact, it accounts to executing several times the algorithm (2), with the same observed input sequence, the same initial extended state, and different parameter vectors. Once the input sequence has been determined, the state reconstruction for all the parameters in Θ_m can be carried out independently, and in a concurrent manner. No message passing is required between the processors reconstructing the state for different parameters, once the initial extended state and the observed input sequence has been propagated to all the processors.
The parallel implementation of the reconstruction algorithm considered in this paper is based on PVM (Parallel Virtual Machine) (Geist *et al.*, 1993), a package that allows to use several single CPU computers, connected together via a LAN, as a single parallel machine. The chosen implementation is of the Master\Slave type. The master machine is responsible for coordinating the behavior of the overall system, gathers data from the field, determines the set Θ_m of parameters for which state reconstruction is required, distributes initialization data to all the other machines for state reconstruction, and eventually determines the optimal control action, taking into account the results of the concurrent computation carried out by the slave machines. In the results here presented, the master machine is also responsible for running the simulation emulating the real system.
The concurrent state reconstruction algorithm, at the m-th control decision, and assuming p processors are available (the master and $p-1$ slaves), can be described in terms of the following algorithm:

Concurrent State Reconstruction Algorithm (C-SRA)

Step 1. The master gathers data from the real system, computes the set Θ_m of control actions, and partitions the set Θ_m into p subsets $\Theta_{m,1}, \Theta_{m,2}, \ldots, \Theta_{m,p}$.

Step 2. The master initializes the computation in each slave machine, by passing the observed input sequence, the initial extended state, and a subset $\Theta_{m,i}$ of the desired parameters.

Step 3. Each slave machine (and the master as well) carries out the assigned computations, and returns the estimated performance indexes to the master machine.

Step 4. The master machine, upon completion of all the computations on the slave machines, determines the next control action. □

One of the major problem in parallel computation is workload assignment. Here, workload assignment is performed by simply partitioning the parameter set Θ_m into p subsets having the same cardinality. The purpose of this paper is not the investigation of optimal assignment policies, and therefore such a simple rule has been used. Nevertheless, more sophisticated rules, as well as the possibility of using a state reconstruction scheme also for workload assignment, could be further explored.

Besides the results reported in the previous section, aimed at testing the effectiveness of the proposed control scheme, additional simulation experiments have been carried out, to study the benefit arising from the concurrent implementation of the algorithm. The speed-up factor (i.e., the ratio between the time required to service H parts in the case of a single processor machine, and in the case of a parallel machine with several processors) has been considered as a performance index. The results are reported in the following figures 6 and 7. Configurations with one (dotted line), two (dashed lines), and three (solid lines) processors (unix single processor machines) have been considered.

The figures indicate that execution times are linear in the cardinality of the parameter space Θ_m, i.e., in the number of extended state sequences that have to be reconstructed, for all the number of processors.

The speed-up factor increases with the cardinality of Θ_m: this is mainly due to the fact that a larger cardinality corresponds to an increase in the computation effort required from each slave machine, while the amount of information exchanged between the master and each slave is almost constant. Additional results, not reported here for brevity, indicate that the speed-up increases also when the design parameter M is increased: also in this case, the computation effort required from each slave machine is increased, while the amount of exchanged information remains the same.

The results clearly indicate the considerable improvement that a parallel implementation of the algorithm may provide: in view of the inherent parallelism of the state reconstruction scheme, a speed-up close to the maximum theoretical one is achieved.

6. CONCLUSIONS

Dynamic control of manufacturing systems has been studied, by means of a novel state reconstruction algorithm, based on data observed from the real system, and on a model of the system itself. The control scheme is able to react to disturbances acting on the system, such as unknown routing probabilities at some nodes.

Simulation results have been presented to illustrate the effectiveness of the proposed scheme, and to compare alternative variants of it. In addition, a concurrent implementation of the proposed algorithm has been also presented.

Remain several open problems, such as the determination of the optimal initial state for the reconstruction algorithm, advanced marking algorithms, in order to make a better use of the input sequence, the definition of rules to select design parameters, such as the control period.

Acknowledgment

This work has been supported by MURST 40 % and 60 % funds. We would like to thank Mr. Alessandro Loretucci for having developed the simulator and run the simulation experiments.

REFERENCES

Bard, J.F and B. Golany (1991). Determining the number of Kanbans in a multiproduct, multistage production system. *Int. J. Prod. Res.* **29**(5), 881– 895.

Berkley, B.J. (1991). Tandem queues and Kanban controlled lines. *Int. J. Prod. Res.* **29**(10), 2057– 2081.

Cassandras, C.G. and C.G. Panayiotou (1996). Concurrent sample path analysis of discrete event systems. pp. 3332–3337.

Di Mascolo, M., Y. Frein, Y. Dallery and R. David (1991). A unified modelling of Kanban systems using Petri nets. *The Int. Journal of Flexible Manufacturing Systems* **3**, 275–307.

Geist, A., A. Beguelin, J. Dongarra, W. Jiang, R. Manchek and V. Sunderam (1993). PVM 3.0 user's guide and reference manual. Technical report. (available via anonymnous ftp).

Glasserman, P. and D. D. Yao (1996). Structured buffer-allocation problems. *Discrete Event Dynamic Systems: Theory and Applications* **6**, 9–41.

Ho, Y.C., M.A. Eyler and T. T. Chien (1979). A gradient technique for general buffer storage design in a production line. *Int. Journ. of Prod. Res.* **17**(6), 557– 580.

Liberatore, G., S. Nicosia and P. Valigi (1995*a*). Dynamic allocation of kanbans in a manufacturing system using perturbation analysis. In: *Proc. of the 1995 INRIA/IEEE Conference on Emerging Technologies and Factory Automation.* Vol. 3. Paris, France. pp. 595–600.

Liberatore, G., S. Nicosia and P. Valigi (1995*b*). Path contruction techniques for dynamic control of kanbans systems. In: *Proc. of the 34th Conference on Decision and Control.* New Orleans, LO. pp. 2610–2611.

Liberatore, G., S. Nicosia and P. Valigi (1997). Dynamic allocation of buffer capacity in discrete event systems. *Journal of Intelligent Automation and Soft Computing.*

Martinelli, F., S. Nicosia and P. Valigi (1997*a*). Dynamic control of manufacturing systems based on a novel state reconstruction algorithm. In: *1997 IEEE Int. Conference on Robotics and Automation.* Albuquerque, NM.

Martinelli, F., S. Nicosia and P. Valigi (1997*b*). State reconstruction techniques for discrete event dynamic systems: Theory and application. Technical Report R.01-97. Dipartimento di Informatica, Sistemi e Produzione, Università di Roma "Tor Vergata". Roma.

Mitra, D. and I. Mitrani (1991). Analysis of Kanban discipline for cell coordination in production lines - II: Stochastic demands. *Operations Research* **39**(5), 807– 822.

Panayiotou, C. G. and C.G. Cassandras (1996). Optimization of kanban-based production systems. In: *proc. of the WODES 96, International Workshop on Discrete Event Systems.* Edinburgh (Scothland): IEE Computing Control Division.

Savsar, M. (1996). Effects of kanban withdrawal policies and other factors on the performance of JIT systems – a simulation study. *Int. J. of Production Research* **34**(10), 2879–2899.

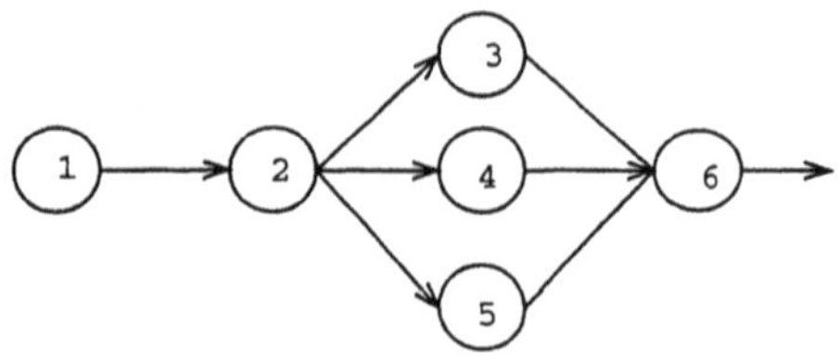

Fig. 1. The manufacturing system

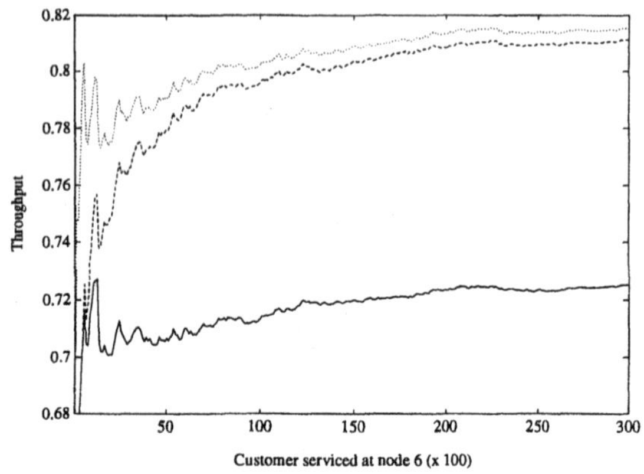

Fig. 2. Global control scheme: throughput behavior (dotted: optimal; dashed: controlled; solid: uncontrolled)

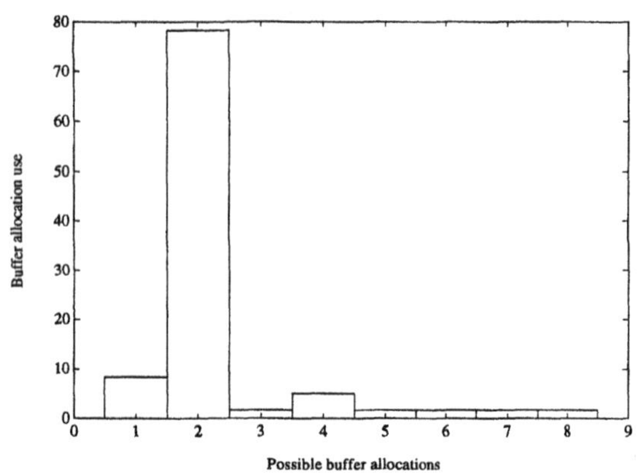

Fig. 3. Selected buffer allocations

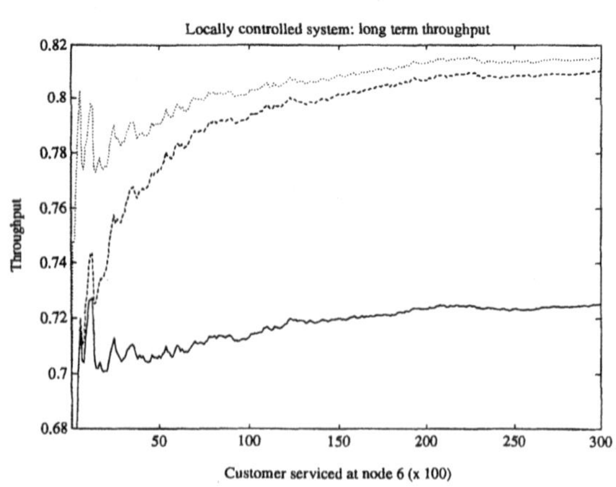

Fig. 4. Local control scheme: throughput behavior (dotted: optimal; dashed: controlled; solid: uncontrolled)

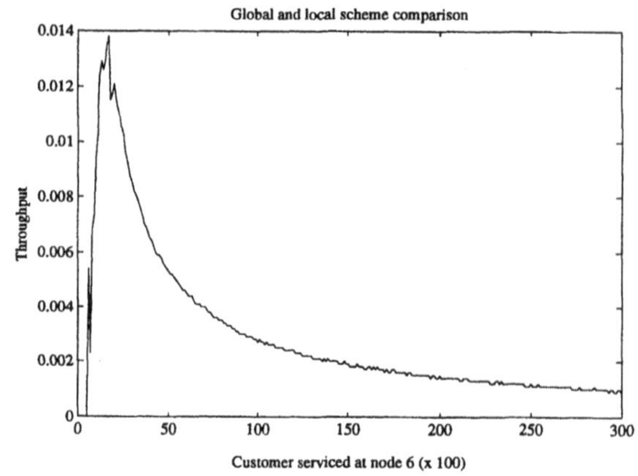

Fig. 5. Local control scheme: throughput comparison

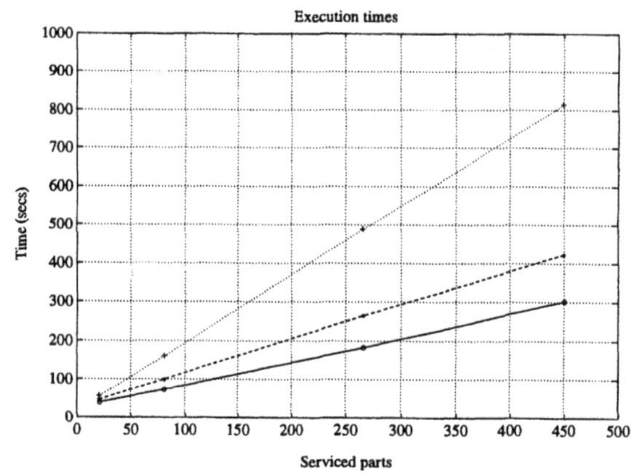

Fig. 6. Concurrent implementation: execution time (secs)

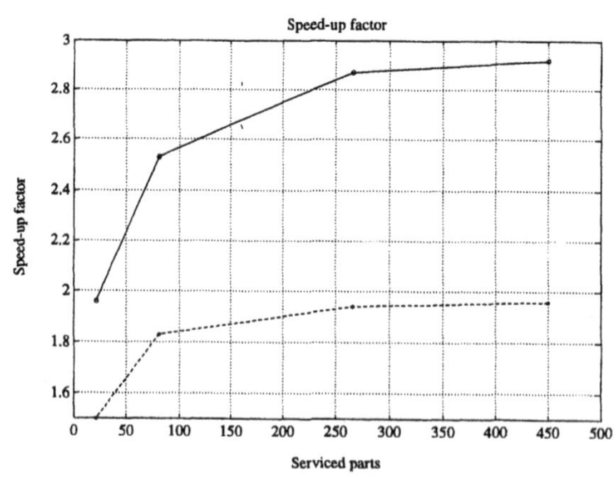

Fig. 7. Concurrent implementation: speed-up factor

MECHATRONICAL SYSTEMS OF ADAPTIVE ROBOT HAND

Václav Kalaš, Eduard Gers

Faculty of Electrical Engineering and Information Technology STU
Department of Automation and Control
Ilkovicova 3, 812 19 Bratislava, Slovakia
E-mail: edo@kar.elf.stuba.sk

Abstract: The contribution demonstrates the robot gripper, whose compressive force is automatically controlled to inevitable minimal value, also namely in dynamic modes, when besides gravity force also a rank of other forces and torques affect to the manipulated object.

Keywords: compressive force controller, variable force controller, sensoric systems of slip, tangential forces sensors, six-component force-torque sensors

1. INTRODUCTION

The recent period in robotics signalizes the remarkable improvement also in the area of the robot hands improving. The new kinematic structures, control and sensor systems are being developed, adaptivity and artificial intelligence are being applied. This fact could be presented *e.g.* by the grippers established on the basis of *SMA (Shape Memory Alloys)*. Encouraging results in the field of *artificial muscles* were achieved, grippers with sensibility close to the human hand, such as *The belgrade USC Robot hand,* resp. *Salisbury Hand* can be also mentioned, where the biomechanical principles, artificial reflexes control, teaching principles, neural nets and many other are applied.

This contribution is oriented to computer synthesis of the robot gripper with the ability of automatic jaws compressive force control, also in dynamic modes, when the manipulated object is affected besides the gravity force also by acceleration force, the environment resistance force, centrifugal force, Coriolis force, rotative torques trying to turn the object in the gripper, and others. Compressive force of the jaws is automatically controlled to inevitable minimal value toavoid object damaging or hurting, but to ensure the manipulated object not to shift, even not to drop out from the gripper.

The motion systems of the gripper are synthesized in cascade structures with the current, basic compressive force, variable compressive force control circuits as well as the speed and position control circuits.

The gripper is also able to identify the static friction coefficient by itself and is able to respect the alterations of this value in dynamic modes. It can also ensure the invariance of the system to the manipulated object mass within the certain bounds.

The gripper contains the force sensoric system, relative acceleration, speed and position sensoric system as well as the six-component force-torque sensors to obtain the information about tangential forces.

The implementation of such type of grippers can be found for example in the following areas: robotized manipulation in glass industry, microelectronic technologies, robotized laboratory operations, robotized fruits collection and sorting, robotized vaccination and manipulation with broilers, human robotics in extremities operations and robotized reconvalescent acts, and others.

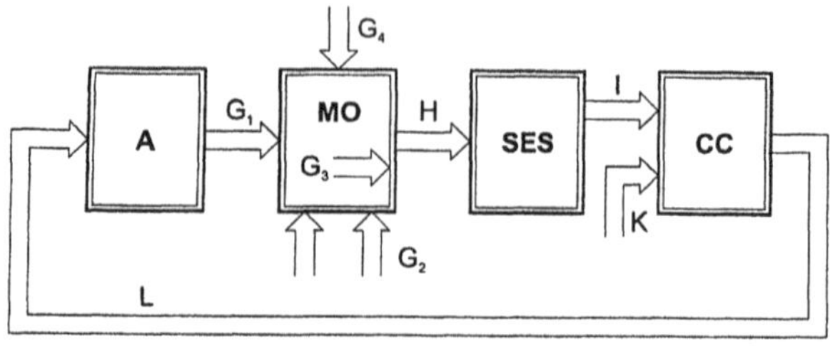

Fig. 1. Basic structure of gripper control systems.

2. BASIC STRUCTURE OF GRIPPER CONTROL SYSTEMS

Let us consider the manipulated object **MO** with the mass **m** to be grasped in the robot hand. An aggregate of actuating signals $\mathbf{G_1}$ generated by the actuator **A** affects to the object, like grasp compressive forces **F**, generated velocity of the gripper motion, its position, orientation etc. - Fig. 1. Further, also gravity force $\mathbf{F_g}$, disturbance forces $\mathbf{F_a}$ and torques $\mathbf{T_k}$ from the acceleration, i.e. an aggregate of signals $\mathbf{G_2}$ affects to the object, then the forces from Coriolis acceleration $\mathbf{F_{Col}}$, the resistance of environment $\mathbf{F_p}$, centrifugal forces $\mathbf{F_{cen}}$. Also rotary torques and the forces from technological operations can affect. In addition to mentioned influences, the internal forces and torques from the activity of manipulated object can cause slip, especially when manipulating biological objects.

Manipulated object **MO** is further influenced by the signals $\mathbf{G_4}$, that consist of the gravity field, magnetic field, vibrations, and also climatic parameters.

Real behaviour of the object is given by the aggregate of signals **H**, while some of them are sensed and processed by the sensoric system **SES**. On the basis of certain control strategy the control computer **CC** processes signals **I** and the reference states signals **K** of the manipulated object **MO**. By means of the signals **L** and actuators **A** the signals $\mathbf{G_1}$ are generated.

3. BASIC STRUCTURE OF GRIPPER KINEMATIC SYSTEMS

Let the manipulated object MO to be grasped in the robot gripper and affected by the mentioned forces and torques according to Figure 2.

In Figure 2, **1** means electric servomotor with the incremental position sensor **2**, with the gear **3-4**, cog pinion **5**, cog racks **6** that are kinematic component parts of the jaws **7**. The gripper disposes of compressive slabs **8**, on which from the side of the object MO the sensoric systems **9** are mounted. In the real conditions adynamic friction $\mu(v_s)$ can occure between the object MO and the active surface of sensors **9**, where v_s represents contingent relative slip velocity between the object MO and sensors **9**. In order to achieve soft grasp, the gripper disposes of critically dumped flexible elements **10**.

For the gripper activity, the following torque relationship is established, based on the motor features, as it can be seen from Figure 2:

$$(T_m - T_{fr0})i\eta_1\eta_2 - 2FR_p = J_r\frac{d\omega_m}{dt}, \qquad (1)$$

where T_m represents the internal motor rotary torque, T_{fr0} - friction torque of the motor, i - transmission ratio of the gear with transmission efficiencies η_1, η_2. R_p is the contact radius of the pinion **5**, F - compressive force of the gripper jaws, J_r - moment of inertia reduced to the motor shaft and ω_m represents the motor angular velocity.

For D.C. motor **1** with the torque-voltage constant C_u and the armature current I from (1) results

$$I = \frac{2FR_p + T_{fr0}i\eta_1\eta_2 + J_r\frac{d\omega_m}{dt}}{i\eta_1\eta_2 C_u} \qquad (2)$$

For the most unfavourable time progress of the force $F(t)$ and velocity $\omega_m(t)$ according to this relationship appropriate parameters of the motor **1** can be established.

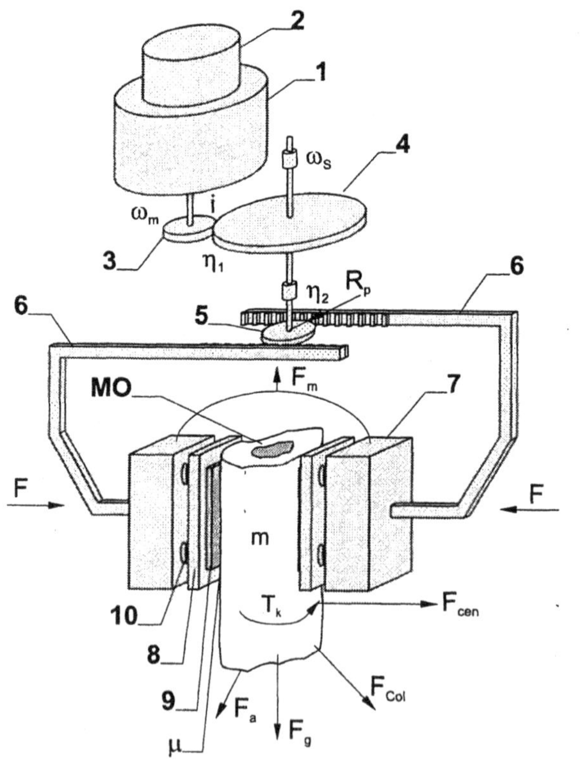

Fig. 2. Simplified kinematic structure of the robot hand with linear motion of jaws.

4. MOTION EQUATIONS FOR THE SYSTEM

Considering general motion of the robot hand and regarding all forces and torques mentioned above, it is very difficult to express the relationship for the compressive force F. The force control circuits have to automatically manage the effects of all kinds of forces and torques. For better understanding, let us reduce only to the case of vertical motion of the gripper.

Following the previous denotations for a symetric object, the compressive force F should satisfy the condition

$$F > \frac{F_g + ma + F_e(v)}{2\mu(v_s)}, \tag{3}$$

where $a = \frac{dv}{dt}$ means the acceleration of the gripper and object MO together, when the relative slip velocity $v_s=0$. The force $F_e(v)$ is the resistance of environment that affects to the object MO with the velocity v.

Let us consider the moving complex of the gripper with the mass m_{ch} to be affected by the motoric force $F_m(t)$. Moving upwards, since there is not any relative object movement in the gripper, it holds:

$$(m_{ch} + m)\, a = F_m(t) \tag{4}$$

The forces trying to shift the object in the gripper, are $F_g + ma + F_e(v)$. Against these forces the friction forces $2\mu F_0 + 2\mu v F_0$ affect to the object, where $2\mu F_0$ is the boundary friction force needful to ideally compensate the gravity force F_g and $2\mu v F_0$ is the reserve force, i.e. over-compensation of the gravity force F_g. Boundary, i.e. not-slipping acceleration of the system, ideally compensating the gravity force F_g, is $a_{cr}=(2\mu v F_0 - F_e(v))/m$.

Moving upwards, following equation for the relative object movement in the gripper jaws is established:

$$\begin{aligned} &\Delta F_m(t) + F_g + F_e(v - v_s) - \\ &-\left[2\mu(v_s)F_0 + 2\mu(v_s)vF_0 + 2\mu(v_s)\widetilde{F}\right] = \\ &= ma_s = m\frac{dv_s}{dt} = m\frac{d^2 s_s}{dt^2} \end{aligned} \tag{5}$$

while $\Delta F_m(t) = m(a - a_{MP})$, where a_{MP} is the object MO acceleration in the stationary reference frame. $2\mu(v_s)\widetilde{F}$ representing the variable compressive force will be considered later.

Moving the gripper downwards, perfectly compensating the gravity force effect, the critical acceleration a_{cr}, described by the following relationship, is considerably higher:

$$a_{cr} = \frac{4\mu F_0 + 2\mu v F_0 - F_e(v)}{m} \tag{6}$$

The relative object motion is described as follows:

$$\begin{aligned} &\Delta F_m(t) + F_e(v - v_s) - \\ &-\left[F_g + 2\mu(v_s)F_0 + 2\mu(v_s)vF_0 + 2\mu(v_s)\widetilde{F}\right] = \\ &= ma_s = m\frac{dv_s}{dt} = m\frac{d^2 s_s}{dt^2} \end{aligned} \tag{7}$$

5. THE GRIPPER COMPRESSIVE FORCE GENERATOR

Requesting automatic control of the compressive force to inevitable minimal value in dynamic modes, the compressive force generator has to be designed to reach extremely high dynamic parameters. To satisfy this condition, a torque servosystem was designed for the force $2F=40N$ and $m=2kg$ with the motor ALSTHOM RS120 whose parameters are: $P_n=30W$, $M_n=0.095Nm$, $n_n=3000$ r.p.m., $U_n=22V$, $I_n=1.53A$, $M_{fr0}=0.01Nm$, $C_u=0.042Nm/A$.

The actuator consists of the "H" scheme with PWM with the switching frequency of transistors 20kHz. The mass of the gripper jaws is $m_g=0.42kg$. The moment of inertia reduced to the motor shaft is $J_m=0.4948\,.10^{-5}\ kgm^2$. The transmission ratio i of the gear with efficiencies $\eta_1=0.92$, $\eta_2=0.9$ is close to its optimal value, $i=4$. The pinion radius is $R_p=5.10^{-3}m$. Other parameters are: $R_m=2.65W$, $K_I=1.78V/A$, $T_f=0.786ms$, $K_f=1$, $K_{ac}=2.5V/V$, $T_{ac}=0.1\mu s$, $T_e=0.415ms$.

The discrete time-optimal armature current controller with constrained control signal was designed using algebraic theory (fourth order discrete system). The global spring constant of the flexible elements in the gripper is $K_p=20.10^3N/m$ and the time constants are $T_a=1.434.10^{-4}s$, $T_b=2.395.10^{-4}s$.

To reduce the disturbance influence (especially the changes of efficiencies η_1, η_2), the system involves in cascade structure the superior force control loop - the analog PID controller whose parameters was experimentally tuned up . Parameters of this control loop are: $K_{FS}=0.15V/N$, $T_{FS}=T_r=0.01s$. Then, for comparison, also the discrete force controller was designed by the algebraic theory and is of sixth order. Control signal is constrained.

The jaws positioning structure is not mentioned in this paper.

Global block diagram of the force generator using classic current and force control can be seen in Figure 3. The transfer function of the flexible element is considered in the case of contact between the object MO and gripper jaws. While the object and jaws are not in contact, this transfer function becomes a double integrator. So the force generator represents a variable structure system here (Kalaš, *et al.*, 1994).

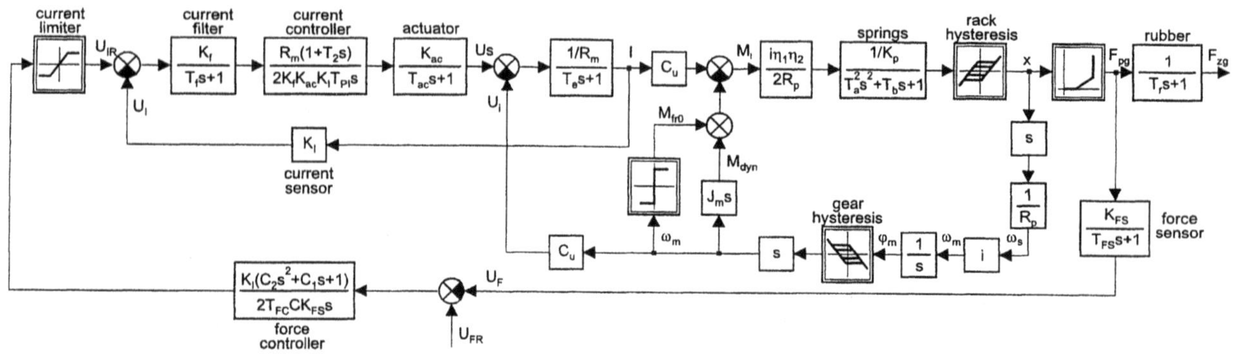

Fig. 3. The force generator block diagram.

Figure 4 shows the response of the force generator with classic and algebraic force controller to reference signal U_{FR} step change. It can be seen that the algebraic system is nearly two times faster than the classic system, and in both cases it can be said that the system reaches very good dynamics.

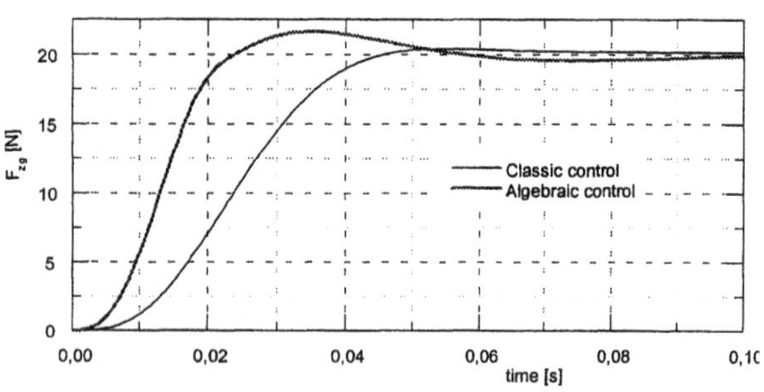

Fig. 4. Response of the gripper compressive force generator at the contact between jaws and the manipulated object.

6. AUTOMATIC CONTROL OF THE GRIPPER COMPRESSIVE FORCE WITH LITTLE SLIP OF THE MANIPULATED OBJECT

One of the ways to control the compressive force automatically and maintain it to its minimal value, is to apply the relative acceleration a_s (AS), velocity v_s (VS), or position s_s sensoric system (PS) that senses the relative movement of the object - Figure 5.

In the structure shown in Figure 5, there is a correcting block CB that ensures the required global control process quality, and nonlinear block NB with dead band for the movement upwards and downwards that results from the upper relationships. The compressive force generator FG shown in Figure 5 generates the constant component of the force $2\mu F_0(1+\nu)$ that consists of the gravity force F_g compensation and the gravity force partial safety over-compensation $2\mu\nu F_0$. Besides these forces the generator produces also the variable compressive force $2\tilde{F}$ that figures in equations (5), (7).

Let us consider the force generator FG working in a little neighborhood of the operating point to be a linear system and let $\mu(v_s)=\mu=const.$ Following the denotations in Figure 5, the relationship for the variable force control loop is established as follows:

$$\frac{2\mu\tilde{F}(s)}{\Delta F_m^*(s)} = \tag{8}$$

$$= \frac{K_s(1+T_s s)M_1(s)K_F N_1(s)\mu}{s^i m M_2(s)N_2(s) + K_s(1+T_s s)M_1(s)K_F N_1(s)\mu}$$

From (8), it results that when the relative velocity sensoric system is applied (when i=1) and considering non-persisting elements, i.e. $M_1(s)=M_2(s)=N_1(s)=N_2(s)=1$, then

$$\frac{2\mu\tilde{F}(s)}{\Delta F_m^*(s)} = \frac{K_s K_F \mu}{ms + K_s K_F \mu} \tag{9}$$

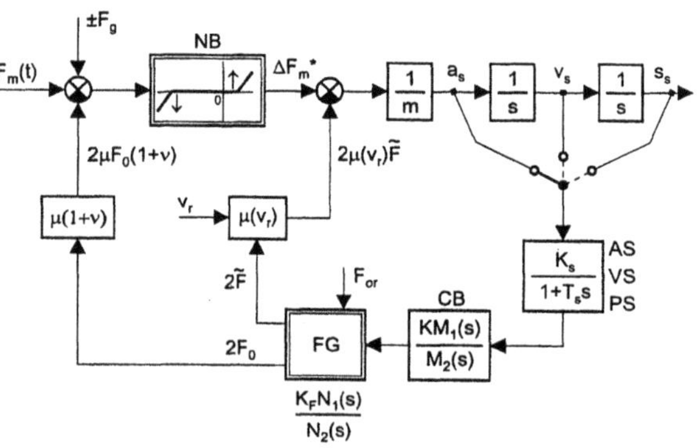

Fig. 5. Structure of the system with the minimal compressive force automatic control on the basis of slip variables.

Evidently, for high gain of the control loop, the system can be invariant to the changes of the object mass m.

Further, from (8) results that applying the position sensoric system, without any additional arrangements, the control loop becomes structural unstable. Using the relationship (8), appropriate correcting block CB can be designed, substituting the force generator by the second order system (from the response in Figure 4).

Figure 6 shows the progress of each state variable. The transfer function of the correcting block CB is $1+Ts$, where $T=0.1 \cdot 10^{-3}s$ (PD type). The global gain of the control loop is $K=8000$, object mass is $m=2kg$, $T_s \to 0$ and $\mu(v_s)=\mu = const=2$. The motoric force impulse to the system $F_m(t)$ has sinusoidal shape upward and downward as well (Fig. 6), the amplitude is 60N, and the duration is 1s with the 0.5s pause. The reserve force $2\mu v F_0$ was chosen as 10% of ideal compensating force $2\mu F_0$. Figure 6 also shows the progress of compensating variable force $2\mu\tilde{F}$ that can be notably smaller at downward movement. The force generator FG was substituted by the second order system here.

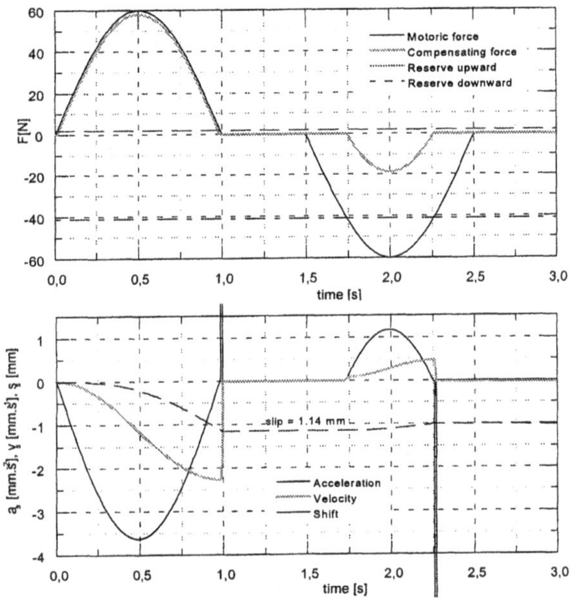

Fig. 6. The progress of input, output and selected state variables with little slip of the manipulated object

The lower part of Figure 6 shows the object slip acceleration, velocity and position progress. It is evident that the object MO has shifted about 1mm.

Since it is technically difficult to sense the slip acceleration a_s, miniature double-axes IRC sensoric position systems can be used, with the differentiation of the signal. More about sensoric systems that can be applied in this conception, is discussed by Havlík (1988), Hlisnikovský (1987a), Ryasuke, *et al.*, (1981) and in the prospects of Canon, USA.

7. GRIPPER COMPRESSIVE FORCE AUTOMATIC CONTROL WITH TANGENTIAL FORCES SENSING

One of disadvantages of the system that uses a slip of the manipulated object is that it may worsen the robot positioning in some technologies. It is caused by little slip that must rise to activate the force generator.

To eliminate this effect, the sensoric system with tangential forces (Hlisnikovský, 1987b) sensor can be applied, that senses the forces F_t between the gripper and the manipulated object MO (labelled **9** in Figure 2). In this case, the force generator initiates at the change of tangential forces, without any need to slip of the object MO. So the dynamics of such system can be considerably higher.

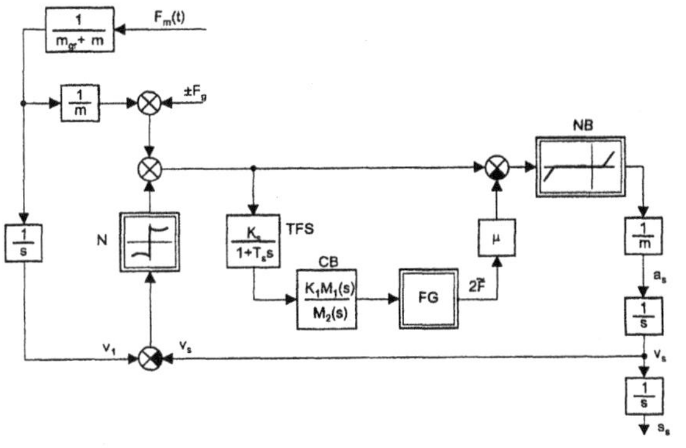

Fig. 7. Principal structure of the gripper with tangential forces sensoric system.

The principal structure of the control circuit for this case is shown in Figure 7, where TFS means six-component force-torque sensor, that can be used to sense both tangential forces and compressive force F. Nonlinear block is the same as in Figure 5. In this structure also a contingent slip caused by the friction of the environment resistance (represented by the nonlinearity N) is considered, but this can be ignored when the dynamics of the generator FG is sufficient.

Figure 8 shows the behaviour of such system using the tangential forces sensor, with the correcting block CB of PID type (P=0.0625, I=0.0045s, D=0.015s). All of the other variables are the same as in the previous case. The force generator was considered in the full structure shown in Fig. 3, but using discrete algebraic current and force control.

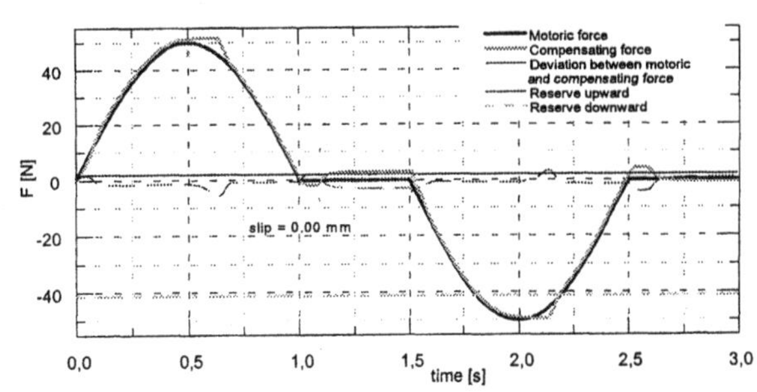

Fig. 8. The variables response in the adaptive system of the gripper with tangential forces sensors.

In this case, the object MO have not shifted in the robot gripper at all, and the compressive force F was as small as possible.

Using this philosophy, the system is invariant to the changes of the object mass m in wide range. When the gain of the variable force $\tilde{F}$ control loop is sufficient, the system is also invariant to the changes of the friction coefficient.

To automatically identify the friction coefficient μ at nought velocity, combination of both system can be used. While gradually increasing the compressive force, when the object MO stops slipping, the value $2\mu F_0$ is remembered and following the relationship

$$\mu = \frac{F}{2F_0}, \tag{11}$$

the value of the friction coefficient μ can be calculated. It is also possible to arrange a nonlinear block $\mu=\mu(v_s)$ to the force generator circuit.

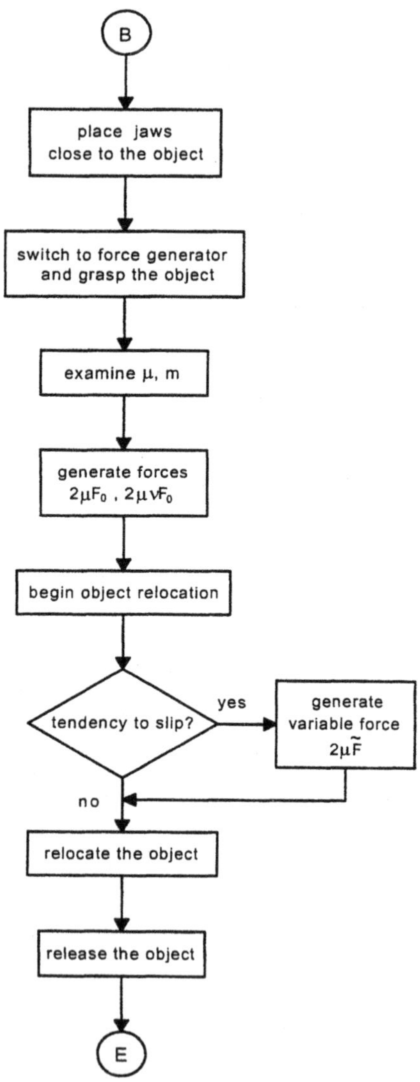

Fig. 9. Possible algorithm of the gripper operation.

When necessary, the information of the object mass could be automatically acquired from:

$$m = \frac{2\mu F_0}{g} \tag{12}$$

The global gripper operation is expressed by the program flow chart in Figure 9.

8. CONCLUSION

The contribution has demonstrated some basic questions concerned with the robot hand that is able to work with minimal compressive force affecting to the manipulated object, also in dynamic modes, when a number of miscellaneous forces and torques affects. Two kinds of conceptions have been offered. One with some little slip required and the second one that does not need any slip.

The quality criterion of each conception can be represented by the time integral of global compressive force $F(t)$ affecting to the manipulated object. In these conceptions the value of this integral can be multiple times lower than in conventional types of robot hands.

REFERENCES

Havlík, Š. (1988). *Snímač sklzu predmetov*, ČSSR invention N° 256190 from 15. November, (Slovak).

Hlisnikovský, I. (1987a). Metódy a spôsoby snímania sklzu predmetu v chápadle, In: *Adaptívne robotické systémy*, p.39, DT ČSVTS, Banská Bystrica (Slovak).

Hlisnikovský, I. (1987b). *Snímač tangenciálnych síl chápadla robota*, kandidátska dizertačná práca, Faculty of Mechanical Enginering SUT, Bratislava (Slovak).

Kalaš, V., J. Polakovičová (1994). Inteligentné chápadlo robota, In: *Kybernetika a umelá inteligencia*, pp.75-84, TU Košice (Slovak).

Ryasuke - Masuda (1981). Global Sensoric System for the Direct Robot Control and its Construction Approach, *Proc. of the 11th Int. Symp. on Industrial Robots*, pp. 159 - 166, Tokyo.

The prospects of Canon USA. *Laser Rotary Encoder*, New York Office, New York.

DYNAMICS, SERVOCONTROL AND STABILITY OF THE LOCAL STRUCTURE OF PUMA ROBOTS

T. Boiadjiev, G. Boiadjiev

Central Laboratory of Mechatronics and Instrumentation - BASC
Institute of Mechanics - BASC

Abstract: Problems connected with the control of the local structure of manipulating robots type RRR are discussed in the paper. Mathematical model based on the graph theory and the orthogonality principle is used. The equations obtained for the motion of the analyzed object are in the base of the control synthesis. The method for servocontrol with standard corrections is used and the stability of the synthesized control of the motion of the manipulating robot is proved. Computer simulation of the control is done and the presented results are analyzed.

Keywords: Robot, dynamics, computer testing, control, stability.

1. INTRODUCTION

Recently the problems connected with the so called contact tasks are paid a rising attention what is the next step to the practical application improvement of the manipulating robots. Their successful solving gives a new possibilities when complex processes have to be worked out especially for the tasks where the interaction between the manipulating system and its environment is taken in account. The starting point for such an investigation includes the following conditions: the system has to be positioned and oriented precisely before the beginning of the contact task. So, the questions about position and orientation precision at the end of the transitional process are very important as well, especially the ones connected with the interaction between the regional motion and the local one. The precise orientation made by the local structure leads to deviation of the precise position already made by the regional one. Then, (Akulenko, *et al.*, 1990) an evaluation of this deviation can be done. It is also proved (Akulenko, *et al.*, 1990) the local structure considered later has the best orientation possibilities what is made by criterion using the Riman surfaces and calculation of areas corresponding to orientation motion of different strucrures investigated there. But, before the investigation of that problems, the question of precise orientation have to be discussed, what is the subject of this work.

2. DYNAMIC MODEL

The equations of motion of the considered local structure which consists of three rotational joints with perpendicular axes R⊥R⊥R, are obtained by the dynamic model based on the graph theory and the orthogonality principle (Boiadjiev, 1991; Boiadjiev and Lilov, 1993; Boiadjiev, *et al.*, 1994). They are written in the form

$$\mathbf{A}\ddot{\mathbf{q}} = \mathbf{B} \qquad (1)$$

where the elements of the matrices $\mathbf{A}, \mathbf{B}$ have the following form (J_{sk}^{p} denote the inertia tensor elements of the body $\mathbf{p}$ written in its local coordinate system):

$$a_{11} = \frac{1}{4}m_1h^2 + 2m_2h^2 + J_{33}^3\cos^2 q_2 + J_{11}^2\sin^2 q_2 +$$
$$+J_{22}^3\sin^2 q_2\sin^2 q_3 + J_{11}^3\sin^2 q_2\cos^2 q_3 + J_{33}^1$$

$$a_{12} = \frac{1}{2}\left(J_{22}^3 - J_{11}^3\right)\sin q_2 \sin 2q_3$$
$$a_{13} = J_{33}^3 \cos q_2$$
$$a_{22} = J_{11}^3 \sin^2 q_3 + J_{22}^3 \cos^2 q_3 + J_{33}^2$$
$$a_{23} = 0$$
$$a_{33} = J_{33}^3$$

$$b_1 = -\left(\frac{1}{2}m_1 + m_2 + m_3\right)gh\cos q_1 - J_{11}^2 \dot{q}_1\dot{q}_2 \sin 2q_2 + \\ +\left(J_{11}^3 - J_{22}^3\right)\left(\dot{q}_1\dot{q}_3 \sin^2 q_2 \sin 2q_3 + \right. \\ \left. +\dot{q}_2\dot{q}_3 \sin q_2 \cos 2q_3\right) + J_{33}^3\left(\dot{q}_1\dot{q}_2 \sin 2q_2 + \right. \\ \left. +\dot{q}_2\dot{q}_3 \sin q_2\right) + \frac{1}{2}\left(J_{11}^3 - J_{22}^3\right)\dot{q}_2^2 \cos q_2 \sin 2q_3 - \\ -\left(J_{11}^3 + J_{22}^3\right)\dot{q}_1\dot{q}_2 \sin 2q_2 \sin^2 q_3$$

$$b_2 = \frac{1}{2}\left(J_{11}^2 - J_{33}^3\right)\dot{q}_1^2 \sin 2q_2 - J_{33}^3 \dot{q}_1\dot{q}_3 \sin q_2 + \\ +\left(J_{11}^3 - J_{22}^3\right)\left(\dot{q}_1\dot{q}_3 \sin q_2 \cos 2q_3 - \dot{q}_2\dot{q}_3 \sin 2q_3\right) + \\ +\frac{1}{2}J_{22}^3 \dot{q}_1^2 \sin^2 q_3 \sin 2q_2 + \frac{1}{2}J_{11}^3 \cos^2 q_3 \sin 2q_2$$

$$b_3 = J_{11}^3\left(\frac{1}{2}\dot{q}_2^2 \sin 2q_3 - \dot{q}_1\dot{q}_2 \sin q_2 \cos 2q_3\right) + \\ +J_{22}^3\left(2\dot{q}_1\dot{q}_2 \sin q_2 \sin 2q_3 + \dot{q}_2^2 \cos 2q_3\right) + \\ +\frac{1}{2}\left(J_{22}^3 - J_{11}^3\right)\dot{q}_1^2 \sin^2 q_2 \sin 2q_3 + J_{33}^3 \dot{q}_1\dot{q}_2 \sin q_2$$

3. SYNTHESIS OF THE CONTROLLED LAW

The dynamic control is effected on the basis of motion equations of the manipulating robots (Boiadjiev and Manolov, 1992; Boiadjiev, *et al.*, 1994) :

$$\mathbf{A}(\mathbf{q}(t))\ddot{\mathbf{q}}(t) + \mathbf{b}(\mathbf{q}(t), \dot{\mathbf{q}}(t)) = \mathbf{u}(t) \tag{2}$$

It is known that every orientation of the local structure can be presented by only-one-way corresponding values of the general coordinates whose in our case are the angles of the structure's kinematic joints. So, to obtain the required orientation is the same as to finish the motion of the structure starting from zero initial state of the joint angles up to some corresponding to this orientation angles for an interval of time [0,T]. Also, the additional requirement towards the transitional process is to have a critical-aperiodical character.
The synthesis of the controlled signal $\mathbf{u}(t)$ is executed in accordance with the method for servocontrol with standard corrections (Boiadjiev and Manolov, 1992). The controlled signal consists of three components :

$$\mathbf{u}(t) = \mathbf{u}_{fb}(t) + \mathbf{u}_d(t) + \Delta\mathbf{u}(t) \tag{3}$$

whose forming is executed as follows :

1). The component, formed from the feedback connections on the joint position, speeds and accelerations - $\mathbf{u}_{fb}(t)$

$$\mathbf{u}_{fb}(t) = \left(\hat{\mathbf{A}} - \mathbf{I}\right)\ddot{\mathbf{q}}(t) + \hat{\mathbf{b}} - \mathbf{K}_2\dot{\mathbf{q}}(t) - \mathbf{K}_1\mathbf{q}(t) \tag{4}$$

where $\hat{\mathbf{A}}$ and $\hat{\mathbf{b}}$ are valuations of the matrix $\mathbf{A}(\mathbf{q}(t))$ and the vector $\mathbf{b}(\mathbf{q}(t), \dot{\mathbf{q}}(t))$ from (2); $\mathbf{K}_1$ and $\mathbf{K}_2$ are diagonal matrices of the feedback connections of the joint positions and speeds.
After linearization of the motion equations of the local structure, the valuations have the following form :

$$\hat{a}_{11} = 2m_2h^2 + \frac{1}{4}m_1h^2 + J_{33}^1 + J_{33}^3$$
$$\hat{a}_{12} = 0$$
$$\hat{a}_{13} = J_3^{33}$$
$$\hat{a}_{22} = J_{33}^2 + J_{22}^3$$
$$\hat{a}_{23} = 0$$
$$\hat{a}_{33} = J_{33}^3$$

$$\hat{b}_1 = -\left(\frac{1}{2}m_1 + m_2 + m_3\right)gh$$
$$\hat{b}_2 = 0$$
$$\hat{b}_3 = J_{22}^3\dot{q}_2^2$$

The choice of the values of the diagonal matrices elements of the feedback connections are conformable with the requirement that the transitional process has critical-aperiodical character. Such values could be for example $k_{ii}^1 = 1,\ k_{ii}^2 = 2,\ i = 1,2,3$.

2). Component, which introduces information in the system for the desired motion $\mathbf{u}_d(t)$.

$$\mathbf{u}_d(t) = \ddot{\mathbf{q}}_d(t) + \mathbf{K}_2\dot{\mathbf{q}}_d(t) + \mathbf{K}_1\mathbf{q}_d(t) \tag{5}$$

In this particular case when that component is formed by the valuations of the joint positions, speeds and accelerations, the local structure should achieve to the end of the transitional process. They can be as follows :

$$q_{d1}(T) = 1\,rad;\quad q_{d2}(T) = 0.7\,rad;\quad q_{d3}(T) = 0.5\,rad$$
$$\dot{q}_{di}(T) = 0\,rad/s,\quad \ddot{q}_{di}(T) = 0\,rad/s^2,\quad i = 1,2,3.$$

If (4) and (5) are replaced in (2), then :

$$\Delta\ddot{\mathbf{q}}(t)+\mathbf{K}_2\Delta\dot{\mathbf{q}}(t)+\mathbf{K}_1\Delta\mathbf{q}(t)=\Delta\mathbf{u}(t)-\Delta\mathbf{f}(t) \quad (6)$$

where $\Delta\mathbf{f}$ marks the sum $\left[(\mathbf{A}-\widehat{\mathbf{A}})\ddot{\mathbf{q}}(t)+(\mathbf{b}-\widehat{\mathbf{b}})\right]$, and $\Delta\mathbf{q}(t)=\mathbf{q}(t)-\mathbf{q}_d(t)$. If the valuations are precise, i.e. $\widehat{\mathbf{A}}=\mathbf{A}$ and $\widehat{\mathbf{b}}=\mathbf{b}$, a necessity of a third component in the controlled signal should not be needed : in that case $\Delta\mathbf{f}(t)\equiv 0$, and equation (6) of the closed system should be :

$$\Delta\ddot{\mathbf{q}}(t)+\mathbf{K}_2\Delta\dot{\mathbf{q}}(t)+\mathbf{K}_1\Delta\mathbf{q}(t)=\mathbf{0} \quad (7)$$

where $\mathbf{0}\in R^n$ and the initial conditions $\mathbf{q}(0)=\mathbf{q}_d(0)$ and $\dot{\mathbf{q}}(0)=\dot{\mathbf{q}}_d(0)$ assure working out the desired motion.

To assure such precise valuations $\widehat{\mathbf{A}}$ and $\widehat{\mathbf{b}}$ there is need a big number of calculations, which is very hard to realize in the practice in real time.

Therefore the existence of $\Delta\mathbf{f}(\mathbf{q}(t),\dot{\mathbf{q}}(t),\ddot{q}(t))\neq\mathbf{0}$ in the right side of equation (6) is guiding to the appearance of different by valuation in the time deviations from the desired dynamics of the closed system, i.e. $\Delta\mathbf{q}(t)\neq\mathbf{0}, \Delta\dot{\mathbf{q}}\neq\mathbf{0}$, and $\Delta\ddot{\mathbf{q}}(t)\neq\mathbf{0}$.

3). Component, which realizes evaluation and makes corrections of the inexact and/or neglected members in the dynamic model $\Delta\mathbf{u}(t)$.

For the forming of the last component $\Delta\mathbf{u}(t)$ two steps must be realized :

a) generation of standard trajectories for the joint positions, speeds and acceleration $\mathbf{q}_e(t)$, $\dot{\mathbf{q}}_e(t)$ and $\ddot{\mathbf{q}}_e(t)$.

The generation of those standard trajectories is executed in accordance with the problem set up for solving - the desired final state of the system and the character of the transitional process as well as the choice of the elements of the constant matrices of the feedback connections of the joint positions and speeds.

b) obtaintion of a valuation, which characterizes the deviation of the system from the desired motion for each interval of discreting of the time $\mathbf{V}(t_i)$ and numerical integrating of the differential equattion

$$\ddot{\mathbf{z}}(t)+\mathbf{K}_2\dot{\mathbf{z}}(t)+\mathbf{K}_1\mathbf{z}(t)=\mathbf{K}\mathbf{V}(t) \quad (8)$$

where :

$$\mathbf{V}(t_i)=\mathbf{V}_1(t_i)+\mathbf{V}_2(t_i)+\mathbf{V}_3(t_i) \quad (9)$$

and

$$\begin{aligned}\mathbf{q}(t_{i-1})-\mathbf{q}_e(t_{i-1})&=\mathbf{V}_1(t_i)\\ \dot{\mathbf{q}}(t_{i-1})-\dot{\mathbf{q}}_e(t_{i-1})&=\mathbf{V}_2(t_i)\\ \ddot{\mathbf{q}}(t_{i-1})-\ddot{\mathbf{q}}_e(t_{i-1})&=\mathbf{V}_3(t_i)\end{aligned} \quad (10)$$

The choice of the elements of the diagonal matrix $\mathbf{K}$, connected with the size extent of the standard corrections, depends of the concrete problem executed by the manipulating robot.

The last component $\Delta\mathbf{u}(t)$ as a result of the additional feedback connection of the joint position, speeds and acceleration, is formed by the so formulated after integration of equation (6) $\mathbf{z}(t_i),\dot{\mathbf{z}}(t_i)$ and $\ddot{\mathbf{z}}(t)$ as :

$$\Delta\mathbf{u}(t_i)=\ddot{\mathbf{z}}(t_i)+\mathbf{K}_2\dot{\mathbf{z}}(t_i)+\mathbf{K}_1\mathbf{z}(t_i). \quad (11)$$

When $\Delta\mathbf{u}(t)$ is replaced with the expression (11) in equation (6), the motion of the system is described with

$$\ddot{\varepsilon}(t)+\mathbf{K}_2\dot{\varepsilon}(t)+\mathbf{K}_1\varepsilon(t)=-\Delta\mathbf{f}(t), \quad (12)$$

where $\varepsilon(t)=\Delta\mathbf{q}(t)-\mathbf{z}(t)$. In the last equation the sum $[\ddot{\mathbf{z}}(t)+\mathbf{K}_2\dot{\mathbf{z}}(t)+\mathbf{K}_1\mathbf{z}(t)]$ compensates the influence of the inexacts $\Delta\mathbf{f}$ with the help of corrections, introduced by $\mathbf{z}(t),\dot{\mathbf{z}}(t)$ and $\ddot{\mathbf{z}}(t)$ toward $\Delta\mathbf{q}(t),\Delta\dot{\mathbf{q}}(t)$ and $\Delta\ddot{\mathbf{q}}(t)$.

4.STABILITY OF THE CLOSED SYSTEM

The stability of the considered local structure is proved by the Direct method of Liapunov (Antchev and Lilov, 1981). The equations of the closed system can be written as (Manolov and Boiadjiev, 1993)

$$\ddot{\mathbf{q}}=\mathbf{P}(\mathbf{q})^{-1}\mathbf{H}(\mathbf{q},\dot{\mathbf{q}})=\hat{\mathbf{G}}(\mathbf{q},\dot{\mathbf{q}}), \quad (13)$$

where

$$\mathbf{H}(\mathbf{q},\dot{\mathbf{q}})=\left[\hat{\mathbf{b}}-\mathbf{b}(\mathbf{q},\dot{\mathbf{q}})\right]+(\mathbf{K}-\mathbf{K}_2)\dot{\mathbf{q}}+(\mathbf{K}-\mathbf{K}_1)\mathbf{q}+\mathbf{D}$$

and

$$\mathbf{D}=[d_1,d_2,d_3]^T,\quad d_i=d_i(q_{id},\dot{q}_{id},\ddot{q}_{id},q_{ie},\dot{q}_{ie}),\quad i=1,2,3.$$

Using the parameters of the local structure of PUMA 560 (Armstrong, *et al.*, 1986), the matrix $\mathbf{P}$ has the following form:

$$\mathbf{P}(\mathbf{q})=\widetilde{\mathbf{A}}(\mathbf{q})=\mathbf{A}(\mathbf{q})-\hat{\mathbf{A}}=\begin{bmatrix}\widetilde{a}_{11} & \widetilde{a}_{12} & \widetilde{a}_{13}\\ \widetilde{a}_{12} & \widetilde{a}_{22} & 0\\ \widetilde{a}_{13} & 0 & 0\end{bmatrix} \quad (14)$$

For the eigenvalues of the matrix

$$[\mathbf{P(q)} - \lambda.\mathbf{I}] = \begin{bmatrix} \widetilde{a}_{11} - \lambda & \widetilde{a}_{12} & \widetilde{a}_{13} \\ \widetilde{a}_{12} & \widetilde{a}_{22} - \lambda & 0 \\ \widetilde{a}_{13} & 0 & -\lambda \end{bmatrix}$$

we have:

$$\det[\mathbf{P(q)} - \lambda.\mathbf{I}] = \lambda^3 - (\widetilde{a}_{11} + \widetilde{a}_{22})\lambda^2 + $$
$$+(\widetilde{a}_{11}\widetilde{a}_{22} - \widetilde{a}_{13}^2 - \widetilde{a}_{12}^2)\lambda + \widetilde{a}_{13}^2\widetilde{a}_{22} = 0 \qquad (15)$$

If any of eigenvalues are equal to zero then we would have

$$\widetilde{a}_{13}^2\widetilde{a}_{22} = 0 \qquad (16)$$

In our case for the PUMA parameters:

$$\left[J_{33}^3(1 - \cos q_2)\right]^2\left[(J_{22}^3 - J_{11}^3)\sin^2 q_3 - J_{33}^2\right] = 0.$$

If there the following translation is made for the interval where the second coordinate of the structure is changed

$$q_2 = [-110°;+110°] \cong [10°;210°]$$

which is correct towards the size of the interval itself and only changes the initial value of this coordinate, then (16) is different from zero, i.e. $\lambda_i \neq 0$ and the inverse matrix of $\mathbf{P}$ exists. Next,

$$\det \mathbf{P}(q) = -\widetilde{a}_{13}^2\widetilde{a}_{22} = J_{33}^2(J_{33}^3)^2(1 - \cos q_2)^2 > 0$$

and

$$\mathbf{P}^{-1}(q) = -\frac{1}{\widetilde{a}_{13}^2\widetilde{a}_{22}}\begin{bmatrix} 0 & 0 & -\widetilde{a}_{13}\widetilde{a}_{22} \\ 0 & -\widetilde{a}_{13}^2 & \widetilde{a}_{12}\widetilde{a}_{13} \\ -\widetilde{a}_{13}\widetilde{a}_{22} & \widetilde{a}_{12}\widetilde{a}_{13} & \widetilde{a}_{11}\widetilde{a}_{22} - \widetilde{a}_{12}^2 \end{bmatrix}$$

The last vector of (13) has the form:

$$\hat{\mathbf{G}}(\mathbf{q},\dot{\mathbf{q}}) = \begin{bmatrix} -\dfrac{h_3}{\widetilde{a}_{13}} \\ \dfrac{h_2}{\widetilde{a}_{22}} \\ \dfrac{h_1\widetilde{a}_{13} - h_3\widetilde{a}_{11}}{\widetilde{a}_{13}^2} \end{bmatrix}, \qquad (17)$$

where h_1, h_2, h_3 are the elements of the vector $\mathbf{H(q,\dot{q})}$.

The equation (13) is written in the normal form as

$$[\dot{x}_1, \dot{x}_2, \dot{x}_3, \dot{x}_4, \dot{x}_5, \dot{x}_6]^T =$$
$$= \left[x_2, -\frac{h_3}{\widetilde{a}_{13}}, x_4, \frac{h_2}{\widetilde{a}_{22}}, x_6, \frac{h_1\widetilde{a}_{13} - h_3\widetilde{a}_{11}}{\widetilde{a}_{13}^2}\right]^T \qquad (18)$$

The function of Liapunov is chosen to be a positively defined one which for the concrete case here is as follows:

$$V = \frac{1}{2A}x_6^2\left(x_1^2 + x_2^2 + x_3^2 + x_4^2 + x_5^2\right) + \frac{1}{C}x_6^2, \qquad (19)$$

where A, C are positive constants which meaning should be cleared later.
The derivative of this function is

$$V' = \frac{1}{A}x_6\dot{x}_6\left(x_1^2 + x_2^2 + x_3^2 + x_4^2 + x_5^2\right) + \frac{2}{C}x_6\dot{x}_6 +$$
$$+\frac{1}{A}x_6^2\left(x_1\dot{x}_1 + x_2\dot{x}_2 + x_3\dot{x}_3 + x_4\dot{x}_4 + x_5\dot{x}_5\right)$$

If the following expression is denoted by

$$D = \left(x_1^2 + x_2^2 + x_3^2 + x_4^2 + x_5^2\right)$$

then the derivative becomes

$$V' = \frac{1}{A}x_6\dot{x}_6 D + \frac{1}{A}x_6^2 D' + \frac{2}{C}x_6\dot{x}_6. \qquad (20)$$

For the parameters of the considered structure (Armstrong,, *et al.*, 1986) :

$$\frac{h_1\widetilde{a}_{13} - h_3\widetilde{a}_{11}}{\widetilde{a}_{13}^2} < 0. \qquad (21)$$

In the meaning of (18) the derivative of the Liapunov's function is:

$$V' = \frac{1}{A}x_6\frac{h_1\widetilde{a}_{13} - h_3\widetilde{a}_{11}}{\widetilde{a}_{13}^2}\left(x_1^2 + x_2^2 + x_3^2 + x_4^2 + x_5^2\right) +$$
$$+\frac{1}{A}x_6^2\left(x_1x_2 - x_2\frac{h_3}{\widetilde{a}_{13}} + x_3x_4 + x_4\frac{h_2}{\widetilde{a}_{22}} + x_5x_6\right) +$$
$$+\frac{2}{C}x_6\frac{h_1\widetilde{a}_{13} - h_3\widetilde{a}_{11}}{\widetilde{a}_{13}^2} \qquad (22)$$

Taking in account that (20) is continuous and limited finction it is possible to choose the constant A, so that

$$\left|DA^{-1}\right| \leq \varepsilon_1, \ \left|A^{-1/2}D'\right| \leq \varepsilon_2, \ \left|x_6^2A^{-1/2}\right| \leq \varepsilon_3.$$

If we take $\varepsilon = \max(\varepsilon_i)$, $i = 1,2,3$, then

$$V' = 2x_6\dot{x}_6\left(D+\frac{1}{C}\right)+x_6^2 D' \leq x_6\dot{x}_6\left(\varepsilon+\frac{1}{C}\right)+\varepsilon^2 \leq 0 \ ,$$

because the following inequality

$$\varepsilon^2 \leq -x_6\dot{x}_6\left(\varepsilon+\frac{1}{C}\right)$$

holds when there is taken the constants (they exist what follows from (20) and (21))

$$-\delta_1 < x_6\dot{x}_6 < -\delta_2 < 0, \quad |x_6\dot{x}_6| \leq \delta ,$$

and the last constant C is got to satisfy the inequality

$$\frac{\varepsilon(\varepsilon-\delta)}{\delta} \leq \frac{1}{C}$$

Then, according with the Liapunov's theorem about stability (Antchev and Lilov, 1981), the closed system is stable.

5. COMPUTER SUMULATION

As parameters of considered local structure, needed for the execution of the computer simulation of the synthesized control law, there are used the data for the industrial robot PUMA 560 (Armstrong, *et al.*, 1986). The desired final state of the system as well as the evaluations of the matrices containing in the equation (2) are already discussed in control synthesis section. The values of the joint coordinates of the structure are presented on fig. 1.

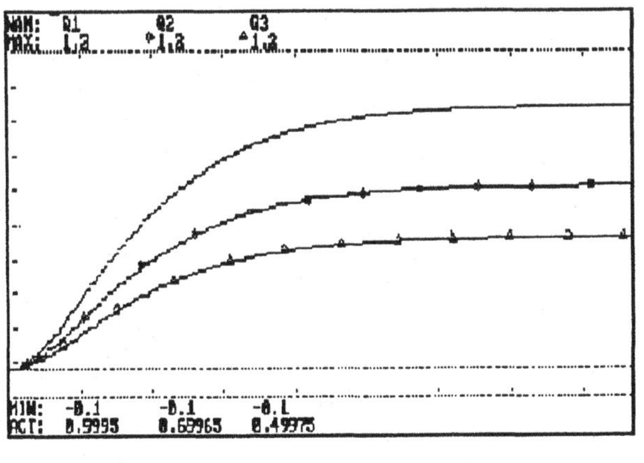

Fig. 1. Joint coordinates of the structure.

Here the following notations are made:

the curve with no mark - the grafics of q_1 - Q1

the curve with * - the grafics of q_2 - Q2

the curve with Δ - the grafics of q_3 - Q3

The last row of all the figures means the values at the end of the transitional process.

On the fig. 2. the standard corrections for the first joint of the structure are shown.

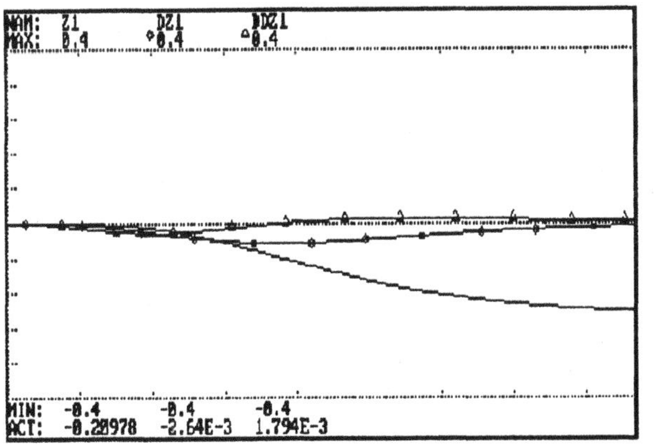

Fig. 2. Standard corrections for the first joint.

Here the following notations are made:

the curve with no mark - the grafics of standard corrections for joint position - Z1

the curve with * - the grafics of standard corrections for joint speed - DZ1

the curve with Δ - the grafics of standard corrections for joint acceleration - DDZ1

On the fig. 3. the joint positions are shown where the third component of the control signal is excluded for the first joint only.

On the fig. 4. the corresponding error for the first joint is shown in this case.

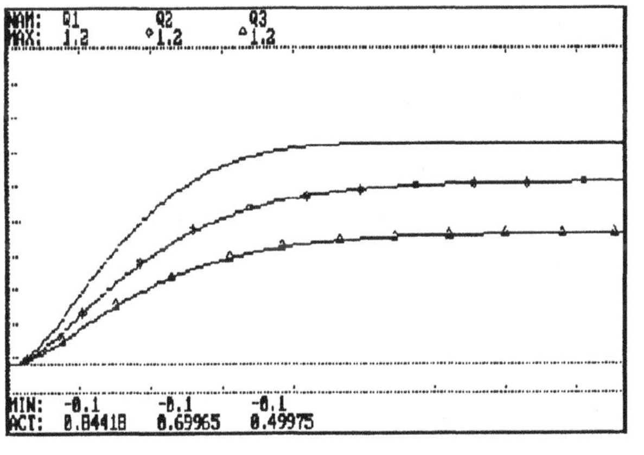

Fig. 3. The joint positions without the third component of the control signal for the first joint only.

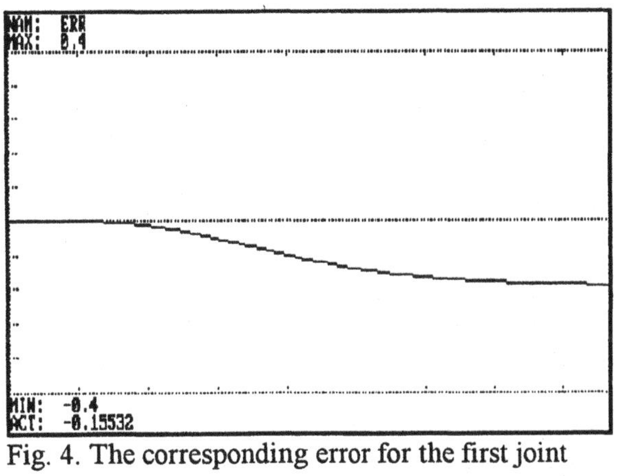

Fig. 4. The corresponding error for the first joint

The notations of this fig. 3 and fig. 4 are the same as the ones for the fig. 1.

6. CONCLUSIONS

The dynamic equations describing the local structure motion for manipulating robot type PUMA are formulated in the presented paper. The requirements towards the motion which have to be worked out by the structure are formulated as well. The control law is synthesized on the base of the dynamic model and the requirements mentioned above. The stability of the closed system is proved. A computer simulation is executed and the presented results ilustrate that the closed system implements precisely the desired trajectory of the motion.

REFERENCES

Akulenko, L., G. Boiadjiev, L. Lilov (1990). Wrist Kinematical Structures and Dynamics of the Orientation Motion of An Electromechanical Manipulation Robot. *Journal of Theoretical and Applied Mechanics*, Year XXI, **No 3**, pp. 24-36, Sofia (in Russian).

Antchev, A., L. Lilov (1981). *Stability of the motion,* chapter 3, pp. 52-63. Nauka i izkustwo, Sofia (in Bulgarian).

Armstrong, B., O. Khatib, J. Burdick (1986). The Explicit Dynamic Model and Inertial Parameters of Puma 560 Arm. In: *Proc. of the IEEE Int. Conf. on Robotics and Automation*, **Vol. I**, pp. 510-518.

Boiadjiev, G (1991). Modelling of electromechanical system. Ph. D. Thesis, Sofia (in Bulgarian).

Boiadjiev, G., L. Lilov (1993). Dynamics of multicomponent system based on the orthogonality principle. *Journal of Theoretical and Appllied Mechanics,* **Vol 1**, pp. 11-26.

Boiadjiev, T., G. Bojadjiev, O. Manolov, N. Shiwarov (1994). Dynamics, Servocontrol and Stability of the Regional Structure of PUMA Robots. In: *Proc. European Robotics and Intelligent Systems Conference EURISCON'94,* **Vol 2**, pp. 935-950, Malaga, Spain.

Boiadjiev, T., O. Manolov (1992). Servocontrol with Standard Corrections. In: *Proc. Second Int. Conf. ICARCV'92,* p.p. RO-4.7.1. - RO-4.7.6., Singapore.

Manolov, O., T. Boiadjiev (1993). Method for adaptive servo-control. Stability of closed system. In: *Proc. 12th IASTED Int. Conf. Modelling, Identification and Control,* pp. 213-214, Insbruck, Austria.

OBSERVABILITY AND MEASUREMENT PLACEMENT

El Hassan MAZZOUR*, Nicolas HERAUD**

** Laboratoire d'Automatique et de Procédés - ISMRA, 6 Bd Maréchal Juin - 14050 Caen Cedex*
Tél. : 02 31 45 27 11 Fax : 02 31 45 27 60 e-mail : mazzour@greyc.ismra.fr

*** CMCS - Université de Corse - CNRS URA 2053, BP 52 - Quartier Grossetti - 20250 Corte*
Tél. : 04 95 45 00 49 Fax 04 95 61 05 51 e-mail : heraud@univ-corse.fr

Abstract : The control of a process requires a minimum of representative information on its operating state. This information can be obtained by both direct sensors or indirect using the mathematical models. It frequently occurs that the number of direct measures is not sufficient to allow unmeasured variables to be deduced and additional sensors have to be installed to render the required variables observable. This paper presents a new technique to place sensors in systems described simultaneously by linear and bilinear equations. It takes into account placement constraints and a complete observability is obtained. Moreover it is pointed out that, from an observability point of view, a bilinear system may be studied by considering a succession of linear systems.

Keywords: sensor positioning, observability, redundancy equations, linear systems, bilinear systems.

1. INTRODUCTION

In all the studies, where physical phenomenon occur, the problem of state estimation is analysed. Indeed, the knowledge of whereat and reliable information is primordial because it puts in good conditions the results of every taking of decision. This problem consists in approaching the true sizes, with estimated sizes which verify the model equations and are as near the measures as possible. But this estimation can only be made if the considered proceeding is observable.

Many works have already been published about the analysis of the static system observability. The first works were about the study of linear systems and are probably due to [VAC 69], their, [STA 81] and [CRO 83] have largely helped to develop this analysis. The algorithms of observability which have been proposed are generally based on the graph theory [MAH 76] or on a classification of the variables, from a matrix of projection [CRO 89]. After the analysis phase, which puts into evidence the redundancies but also the eventual weaknesses of instrumentation system, it is proposed to study the kind of making observable a system, so that the cost breaded by the addition of sensors, is as weak as possible. To reach this result, one acts, in the on hand, on the localisation of the sensors and, in the other hand, on their number. At that time, only a few works on the conception of an instrumentation system which respect the constraints, such as these stated before, have been done. Works concerning the observability and the placing of optimal sensor on electrical networks have been proposed by [CLE 90]. This best puts into evidence two possible analysis : the numerical methods and the topological ones. [MON 86] has worked in a great part in the numerical domain area. [CHE 89] has directed this study by basing itself on the topology of networks. In the robotics domain, [CAM 90] shows that the judicious localisation of sensors allows to minimise the number of measures to take and to obtain the necessary information for the realisation of a work. We must also emerge the works of [MAD 92], who propose a solution based on the analysis of the incidence matrix of the graph associated to the processes, and the works of [IMM 94], who formalises the problem into terms of linear programmation, that is to say, which optimisation variables can be whole, permits to take into account the occurrence and the position of the measures in the optimisation. The more elaborated works are probably these ones presented by [ALI 93] and [LUO 94]. Based on the analysis of the cycles of the graph associated to the processes, these studies take at the same time into account the variables observability and the sensors reliability, which measure them, to propose an architecture of « optimal » instrumentation according to some criterion. [HER 91] and [MAQ 87] have formalised the problem under the constraints of costs in the linear case, by testing all the possible instrumentation configurations, which make observable the system and have computed the cost of each of them to find this one which corresponds to the minimal cost.

We propose another step, which emerges directly from the optimum solution. This is realised by inserting in the observability constraint, the constraints of cost and technology. This study concerns the linear and bilinear systems into static state. This models are often used when one establishes metrical results or energising results. We must notify that the considered study is only based on the structure of equations, which directs the proceeding.

2. PROBLEM FORMULATION

To illustrate the problems involved in the observability of a linear system and the problem of locating sensors, we consider the simple network shown on Figure 1 (for example a material transport process) where the measured vector (for example volumic flow rates) includes the following variables : $Xm = (X_1)$.

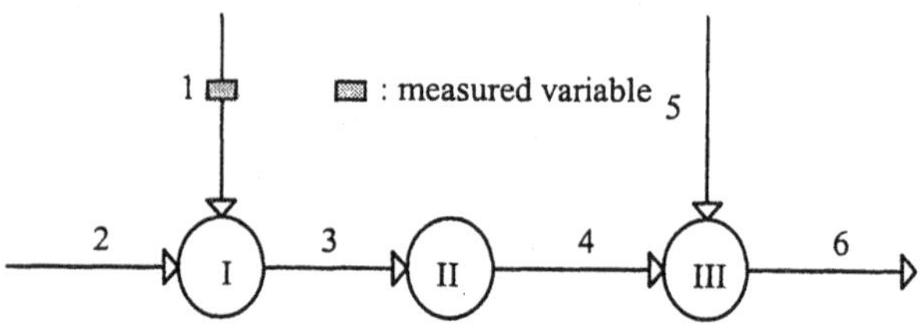

Figure 1. a simple process

This flowsheet can be described by a set of linear algebraic equations :

$$X_1^* + X_2^* - X_3^* = 0 \qquad (1)$$

$$X_3^* - X_4^* = 0 \qquad (2)$$

$$X_4^* + X_5^* - X_6^* = 0 \qquad (3)$$

where X_i^* (i = 1 to 6) is the true values of the variable X_i.

This description may be used for all processes described by the so-called incidence matrix M :

$$M X^* = 0$$

where $X^* = (X_1^* \ X_2^* \ X_3^* \ X_4^* \ X_5^* \ X_6^*)^T$.

Let us now re-examine the previous example. If we wish to make the system globally observable the most obvious solution is to measure the pair of variables (X_4,X_6). This solution is not unique and, here, we can list the six different solutions : (X_2,X_5); (X_2,X_6); (X_3,X_5); (X_3,X_6); (X_4,X_5); (X_5,X_6). The solution is not unique due to the fact that the greatest regular part of a matrix is not unique. We are therefore confronted with a choice when extracting this part which leads to use the constraints. These latter can be fixed by economical (cost of the measure) or technological considerations (topology of the plant). Two different problems therefore arise.

- Is it possible to make a system globally observable by adding new sensors knowing that we already have measure from different points in the system and that some variables cannot be measured (technological or economical problems) ?
- If we fix a cost function by weighting each measure, what set of measures will make the system fully observable at a minimum cost ?

3. INTUITIVE ANALYSIS

Now we associated to each variable, proportional weight to the corresponding sensor installation cost. The goal is therefore to determine the variable which needed to measure in order to respect the constraints according to the cost, with minimising the global installation cost. The associated costs at each measure are presented on table 1.

no-measured variables	**sensor weights**
2	10
3	15
4	20
5	*
6	25

Table 1. Measure costs

Note that (*) indicates that the installation cost of the sensor in the 5 is infinity, one can interprets this as no-measured variable for technological reasons.

The possible combinations of the variables to measure in order to ensure the minimal observability of the variables by eliminating the variable X_5 of the measure, thus that their corresponding costs are given on table 2.

Combinations	measure costs
X2,X6	35
X3,X6	40
X4,X6	45

Table 2. Associated costs corresponding to the combinations of measures.

From the above table, the observability with least cost is obtained with the measure of the variables X_2 and X_6. It is important to remark, from this example, that the minimization of the cost leads to the installation of two sensors which measure X_2 and X_6 (cost equals to 35).

4. DECOMPOSITION ACCORDING TO THE OBSERVABILITY WITH MINIMUM COST

Let a system described by its architecture, in other words by the ties between the different material supports. These connections express the of matter transfer, of energy or of information between the material supports associated to the activities and are associated to the system description variables. Among these variables which characterize the system, there are those which are measurable easily, those which are measurable with difficulty and those which are not measurable. They are blinded by equations plus or minus complex.

These equations can be ranged in the matrix M = (m_{ij}) with $1 \leq i \leq n$ and $1 \leq j \leq v$ with assuming that this system is described by n equations and v variables. In the case of an array of mater balance-sheet, n represents the number of knots and v the number of array lines.

For a system which is already partly instrumented we will sub-divide the vector of X variables into two distinct sub-vectors [RAG 91] :

$$X = (X_I \ X_{II})^T \qquad (4.a)$$

with X_I : vector of the measured variables,
X_{II} : vector of the unmeasured variables.

We are now confronted with the following problem : is the knowledge of the measure X_I sufficient in order to render the remaining variables of the system observable ? Otherwise, is it possible to obtain this observability by positioning other sensors ?

The incidence matrix of the network can then also be sub-divided according to the partition (4.a) :

$$M = \begin{bmatrix} M_1 & M_2 \end{bmatrix} \qquad (4.b)$$

where M_1 is a matrix of dimension (n.m) and M_2 is a matrix of dimension (n.(v-m)).

One can class the columns of the matrix M_2 in order that the measures which there are associated were arranged in decreasing order of cost. The inaccessibly points to the measure will be considered have been infinitely costs. Afterwards, we must write the regular large part of the matrix M_2, on the identity matrix. This transformation is realised by means of the pivot algorithm, with limitation of the pivot choice to the columns of M_2. Thus, we obtain the following decomposition :

$$M_2 = \begin{bmatrix} M_{21} & M_{22} \\ M_{23} & M_{24} \end{bmatrix} \qquad (5)$$

with

: dimension of M_{21} = r.r

: dimension of M_{22} = r.(v-m-r)

: dimension of M_{23} = (n-r).r

: dimension of M_{24} = (n-r).(v-m-r)

where M_{21} represents the largest regular part of M with rank less or equal to n.

And by separating M_1 according to the decomposition in rows of M_2, M becomes :

$$M = \begin{bmatrix} M_{11} & M_{21} & M_{22} \\ M_{12} & M_{23} & M_{24} \end{bmatrix} \qquad (6)$$

with

: dimension of M_{11} = r.m

: dimension of M_{12} = (n-r).m

Pre-multiplying matrix M by the regular matrix R

$$R = \begin{bmatrix} M_{21}^{-1} & 0 \\ -M_{23}M_{21}^{-1} & I_{n-r} \end{bmatrix}$$

gives the following equivalent form :

$$RM = \begin{bmatrix} M_{21}^{-1}M_{11} & I_r & M_{21}^{-1}M_{22} \\ M_{12} - M_{23}M_{21}^{-1}M_{11} & 0 & 0 \end{bmatrix} \qquad (7)$$

because $M_{24} - M_{23}M_{21}^{-1}M_{22} = 0$ ([FAY 87])

Leads to the final form :

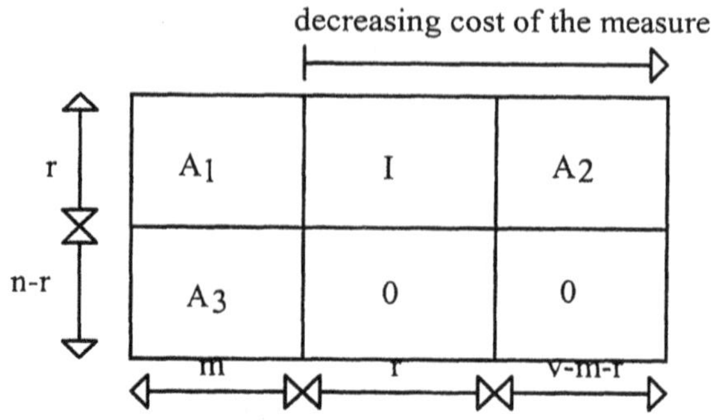

Figure 2. particular canonical form of the incidence matrix

This matrix can be read in the following way. The second row gives the redundancy equations between the variables already measured, notice that all the measured variables are not necessarily present in these equations. Finally the non-null rows of block A_2 prevent any further deductions. Thus the system is globally observable if block A_2 does not exist.

Transform these variables to become observables, needs to add at least (v-m-r) measure points. They correspond to weak cost emplacements because the variables are ranged in decreasing order of their measure cost. In other world, to obtain the globally observability of the industrial process with minimal cost, we must add on the installation these (v-m-r) supplementary sensors.

Remark

In the case of future instrumentation, or when it is convenience to propose the modification of existing sensor position according to the fixed goal, the sub matrix $M = M_2$.

4.1. Algorithm of minimization of the measure cost in case of linear systems

In this part we are interested in the linear systems described by a (8) form equation :

$$M X^* = 0 \tag{8}$$

where X^* is the vector of the real values of the X magnitude, M is the matrix combined with the system.

The algorithm must be, from the array topology, the measure point positions and the supplementary measure costs, given the observability of variables with loss cost and the equations which permit their estimation.

Remember that the present study is based on the structure of the process equations, it is the structural observability. In the following figure, the treatment which permits the study of the instrumentation architecture is summarized:

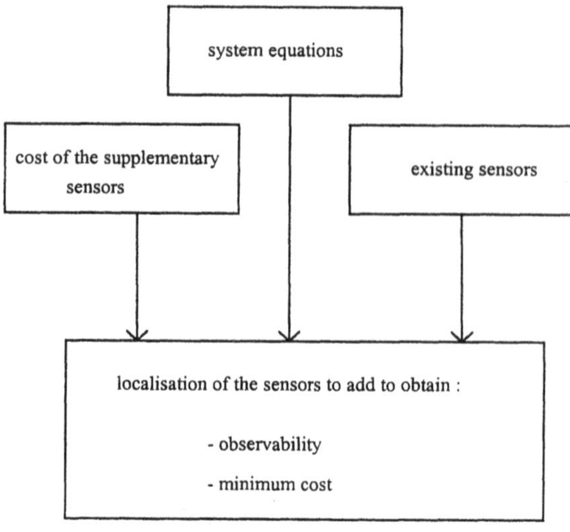

Figure 3. Instrumentation architecture

The proposed algorithm is analysed into 6 stages :

Stage 1 : The matrix M is arranged in order, on one hand according to the measured variables and on the other hand according to the non measured variables.

Stage 2 : A weight is associated with each non measured variable. This weight takes the measure cost and the technological constraints into account (in this last case the cost is regarded as being infinite).

Stage 3 : The columns of the M sub-matrix corresponding to the non measured variables are arranged in decreasing order according to the measure weight.

Stage 4 : In the column one looks for a pivot corresponding to the highest measure weight. If any pivot cannot be found in this column, we go to the following one.

Stage 5 : The non-zero terms of the pivot column are eliminated by the addition or subtraction of rows. The stage 4 operation is repeated until the last column.

Stage 6 : When the operations of the pivot are ended, the columns containing a pivot are permuted so that an identity matrix can come to light. Then a particular canonical form of the M matrix can be obtained. The variables corresponding to the columns that have a pivot will be deductible. Those corresponding to the other columns will have to be calculated. Then the instrumentation configuration at a minimal cost will be deduced.

Comment

In the applications considered throughout this study, the M matrix particular structure allows an easy implementation of this splitting up. Indeed this one being made up of elements equal to 0, 1 or -1, simple rows additions or subtractions allow the extraction of an identity block without any mounding off mistake. However our study can be applied to any type of matrix provided that those rows are not linear combination one for the other.

4.2. Application

We take the network of the first figure as an application. We note down in the matrix of M incidence that modelise this network, the columns numbers represent various variables. The indexes associated with a possible measured variable represents its cost. When the index is a star (*) the corresponding variable cannot be measured because of technological reasons (infinite cost). The weights associated with the sensors that we wish to add on the process are those from the first table.

The matrix M associated with the network of the first figure is written :

M =

Variable	1	2	3	4	5	6
Equation						
I	1	1	-1	0	0	0
II	0	0	1	-1	0	0
III	0	0	0	1	1	-1

Let's apply to this matrix the method of minimization of the cost of measure previously developed. The stages 1, 2 and 3 give M :

M =

Variable	1	5_*	6_{25}	4_{20}	3_{15}	2_{10}
Equation						
I	1	0	0	0	-1	1
II	0	0	0	-1	1	0
III	0	1	-1	1	0	0

The stage 4 and 5 give M as the following form :

M =

Variable	1	5_*	6_{25}	4_{20}	3_{15}	2_{10}
Equation						
I'	-1	0	0	0	1	-1
II'	-1	0	0	1	0	-1
III'	1	1	-1	0	0	1

and the stage 6 allow to obtain the particular canonical form of the M matrix :

M =

Variable	1	5_*	4_{20}	3_{15}	6_{25}	2_{10}
Equation						
III'	1	1	0	0	-1	1
II'	-1	0	1	0	0	-1
I'	-1	0	0	1	0	-1

So we obtain the observability of the process at a minimal cost measuring the variables X_6 and X_2. Thus the cost associated with these measures is worth (25 + 10) = 35.

Let's notice that the algorithm gives the same variables to measure as in the paragraph 3 and therefore it gives the same observability cost.

5. CASE OF BILINEAR SYSTEMS

Let's hold up as an example the case of systems in which we have two types of measures like a volumic flow and a volumic mass or a volumic flux and a temperature ... This very common case takes on particular importance as for as industry is concerned.

Generally the bilinear systems are described by equations like :

$$M\,X1^* = 0 \tag{9}$$

$$M\,(X1^* .* X2^*) = 0 \tag{10}$$

Where X1* and X2* are respectively the vectors of the magnitudes X1 and X2 real values and have the same dimensions. The product .* corresponds to the term to term product of two vectors.

Let's consider a developing stage process of which the structure and the equipment have been partly defined. The variables linked with this installation can be classified according to their accessibility to the measure. Thus several categories of variables could have been worked out : the variables measurable in X1 and X2, the variables only measurable with X1, those only measurable with X2 and the variables being non measurable.

Some economic or technical reasons can be at the origin of this partition. The problem can be set as follows : what is the minimal number of sensors and what is their location in order to get an overall observability of the installation at an optimal cost ?

Let's consider an elementary system (the M matrix is then reduced to a row). If we call n1 the number of the non measured variables in X1 and n2 the number of the non measured variables in X2 and knowing that the number of variables is equal to v (n1 and n2 $\leq$ v), the conditions of observability of an elementary bilinear system can be listed one by one [RAG 91] or be represented by a two level-tree (or shaft) [HER 91], and they can be written as follows :

$$a)\ n1 \leq 2 \text{ et } n2 \leq 1 \qquad (11)$$

$$b)\ n1 + n2 \leq 2 \qquad (12)$$

For a non elementary system (that is to say that the M matrix has got at least two rows), the study of the bilinear systems observability will be done according to a hierarchic diagram. At a first stage we complete the list of the observable variables in X1 (OBSX1 list) from putting the matrix M in a particular canonical form beside the X1 magnitude. Then at a second stage we do the same beside the X2 magnitude of which we take each equation (row) and we check for n2 = 0 and n2 = 1 if the relationship (13) proves to be true. If it does so, we complete the list of the observable variables in X2 (OBSX2 list) and the list of the observable variables in X1.

$$n1' + n2 \leq 2 \qquad (13)$$

where n2 is the number of non measured variables by equation in X2 into M2, n1' is the number of non observable variables for this equation in X1 (not belonging to the OBSX1 list) and n1 is the number of non measured variables by equation (row) in X1 into M1.

The algorithm to treat the observability of bilinear systems is hierarchical. It uses the particular canonical form of the matrix M defined above. During this study, we have to detect the equations for those the number of variables no-measured respect the observability constraints of bilinear system.

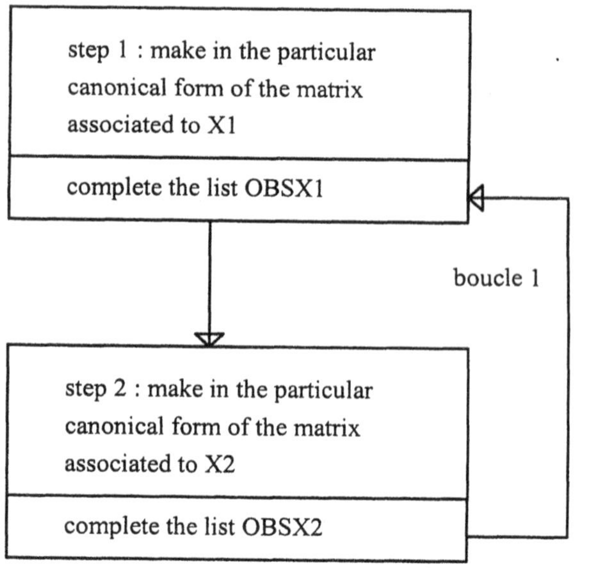

Figure 4. Treatment of the bilinear observability

In the first, we test the stability of the lists which contain the observability variables in X1 and X2. When the contain of these lists is not modified, we stop the treatment.

5.1. Algorithm of minimization of the cost of measure in the case of bilinear systems

The algorithm of minimization of the instrumentation cost of the bilinear systems is then written as :

Stage 1 : Looking for the particular canonical forms M1 and M2 that correspond to X1 and X2 magnitudes in the same way as in the linear system.
Stage 2 : Making in M2 the measure of all the variables corresponding to the columns that do not contain a pivot so that the maximum number of non measured variables par equation in X2 is n2 = 0 (redundancy equation) or n2 = 1 (deduction equation).
Stage 3 : Making in M1 the measure of all the variables of which the columns do not contain any pivot and for which we have more than two missing measures par equation. In this case the maximum number of measured variables in X1 is either n1 = 2, n1 = 1 (deduction equation) or n1 = 0 (redundancy equation).
Stage 4 : Launching the algorithm calculating OBSX1 and OBSX2. If OBSX1 and OBSX2 are complete, that is to say that all the variables are observable, we stop. Then we get the observability at an optimal cost by measuring the variables that need to be measured described in the second and third stages. Otherwise we carry out the following stage.
Stage 5 : We measure the least expensive variable in M1 or in M2. Various cases appear :

- if OBSX1 and OBSX2 are complete, we stop,
- if this variable measure does not modify the lists OBSX1 and OBSX2 we cancel its measure.

- if OBSX1 and OBSX2 are not complete we do the stage 5 again until we obtain in M1 only one variable of which the column does not contain any pivot.

Stage 6 : When only one variable of which the column does not contain any pivot remains in M1, two cases can appear :

- if the variable to be measured (minimum cost) is in M1, we measure it (OBSX1 and OBSX2 will be complete).

- if the variable(s) to be measured are in M2, we measure these variables as long as OBSX1 and OBSX2 are not complete but at the same time as soon as their measure costs exceed the one of the M1 variable of which the column does not contain any pivot, we cancel all these measures to measure M1, OBSX1 and OBSX2 then will be complete.

Comment

If at the same time two variables, one in M1 and the other in M2 have the same measure cost, we preferably work out the one in M1 and this especially when only one variable without pivot is missing in M1. Otherwise we lest the measure of each of them to take the one that give more observable variables.

5.2. Application

Let's consider a schematised process in a network form Figure 5 made up of 3 nodes and 7 rows.

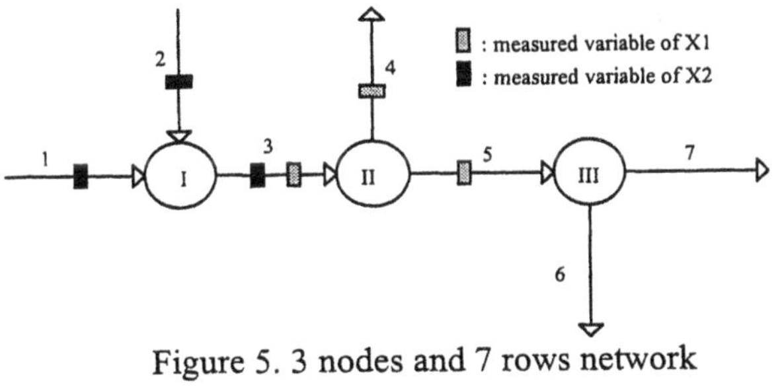

Figure 5. 3 nodes and 7 rows network

An instrumentation architecture of this process is given by the third table for the X1 magnitude and by the table 4 for the X2 magnitude.

no-measured variables on **X1**	Sensor weights
1	15
2	20
6	5
7	25

Table 3. Costs associated with the measures in X1

no-measured variables on **X2**	Sensor weights
4	10
5	30
6	20
7	*

Table 4. Costs associated with the measures in X2

From the tables 3 and 4, we will study the configurations which permit to obtain the observability of the array with the least cost. In the first we observe the different combinations which permit us to respect the observability constraints.

The following table 5 summarize these combinations.

combinations	measures of the variables of X1	measures of the variables of X2	measures costs
1	6	4 et 6 or 5 et 6	35 or 55
2	7	" "	55 or 75
3	1 and 6	" "	50 or 70
4	1 and 7	" "	70 or 90
5	2 and 6	" "	55 or 75
6	2 and 7	" "	75 or 95
7	6 and 7	4, 5 and 6	90
8	1 and 6	" "	80
9	2 and 6	" "	85
10	1 and 7	" "	100
11	2 and 7	" "	105

Table 5. Measure combinations costs

Afterwards, we take the combination with the minimum cost. The combination (1), which has for function cost equal to 35, gives the observability with minimum cost by doing the measure of the variable 6 at X1 and the variables 4 and 6 at X2.

Remark

In the fact that the variable 7 in X2 is no-measurable limits considerably the number observability combinations to treat.

Let's apply the algorithm of the cost minimization to this bilinear system :

The matrix of M incidence corresponding to the X2 magnitude is written :

	Variable	1	2	3	7*	5_{30}	6_{20}	4_{10}
	Equation							
	I	1	1	-1	0	0	0	0
M =	II	0	0	1	0	-1	0	-1
	III	0	0	0	-1	1	-1	0

The first stage applied to the M matrix gives its particular canonical form M2 :

	Variable	1	2	3	7*	5_{30}	6_{20}	4_{10}
	Equation							
	III'	0	0	-1	1	0	1	1
M2 =	II'	0	0	-1	0	1	0	1
	I'	1	1	-1	0	0	0	0

The matrix of M incidence corresponding to the X1 magnitude is written :

	Variable	3	4	5	7_{25}	2_{20}	1_{15}	6_{5}
	Equation							
	I	-1	0	0	0	1	1	0
M =	II	1	-1	-1	0	0	0	0
	III	0	0	1	-1	0	0	-1

M in its particular canonical form is written :

	Variable	3	4	5	7_{25}	2_{20}	1_{15}	6_{5}
	Equation							
	III'	0	0	-1	1	0	0	1
M1 =	I'	-1	0	0	0	1	1	0
	II'	1	-1	-1	0	0	0	0

The second stage gives the following result : in order that the number n2 of the variables not being measured by the equation in X2 is smaller than 1 or equal to 1, in M2, we measure the variables 6 and 4. Their measure costs are worth (20 + 10) = 30, therefore M2 becomes MII.

	Variable	1	2	3	4_{10}	6_{20}	7*	5_{30}
	Equation							
	III'	0	0	-1	1	1	1	0
MII =	II'	0	0	-1	1	0	0	1
	I'	1	1	-1	0	0	0	0

The third stage does not generate any alteration. Indeed the number n1 of the non measured variables by equation in X1 is smaller than two or equal to two, so we do not make a measure on the X1 in magnitude in M1. The 4th stage allow the calculations of the OBSX1 and OBSX2 lists (to simplify the writing we only make a note in those lists of the observable variables index) and we obtain : OBSX1 = (3, 4, 5).

In order to calculate OBSX2, in MII we repeat the equations for which all is calculated in X2 (equation I') and we check if the number n1' of the non observable variables corresponding in X1 is smaller than 2 or equal to 2. If it is so, we complete the OBSX1 and OBSX2 lists with these variables. In our case, this is expressed by :

- the analysis of the MII matrix equation I' that gives :

OBSX1 = (3, 4, 5, 1, 2), OBSX2 = (1, 2, 3).

In MII we spot the equations for which a variable is non measured in X2 (equations II' and III') and we check if the number n1' of non observable variables in X1 for these equations is smaller than 1 or equal to 1. If it is so, we complete the OBSX1 and OBSX2 lists.

- Then the analysis of the equation III' does not allow to complete OBSX1 = (3, 4, 5, 1, 2) and OBSX2 = (1, 2, 3) because the number of non observable variables in X1 for this equation is higher than 1.

-The analysis of the equation II' leads to the result : OBSX1 = (3, 4, 5, 1, 2) and OBSX2 = (1, 2, 3, 4, 5). Let's notice that OBSX1 and OBSX2 are not complete. Then the stage 5 is undertaken.

We are going to measure the least expensive variable in M1 or in MII, in our case we are going to measure the variable 6 in M1, its cost of measure is 5. In its particular canonical form M1 then becomes MI :

MI =

Variable	3	4	5	6_5	7_{25}	2_{20}	1_{15}
Equation							
III'	0	0	-1	1	1	0	0
I'	-1	0	0	0	0	1	1
II'	1	-1	-1	0	0	0	0

from MI and MII we complete OBSX1 and OBSX2 again. We obtain : OBSX1 = (3, 4, 5, 6, 7). The analysis of the MII equations is made up of :

- the analysis of the I' equation that allows to complete OBSX1 = (3, 4, 5, 6, 7, 1, 2) and OBSX2 = (1, 2, 3).

- the analysis of the III' equation that gives : OBSX1 = (3, 4, 5, 6, 7, 1, 2) and OBSX2 = (1, 2, 3, 4, 6, 7).

- the analysis of the II' equation that gives the final result : OBSX1 = (3, 4, 5, 6, 7, 1, 2) et OBSX2 = (1, 2, 3, 4, 6, 7, 5).

Let's note that OBSX1 and OBSX2 are complete that is to say that all the variables are observable in X1 and X2 at minimal cost. This minimal observability cost is worth (20 +10 +5) = 35. It is made up of : 20 for the measure cost of the variable 6 in X2, 10 for the measure cost of the variable 4 in X2 and 5 for the measure cost of the variable 6 in X1.

Note that the proposed algorithm, gives the same observability cost such that obtained by testing all the possible observability combinations (table 5).

6. CONCLUSION

This paper deals with the observability of a process described by a set of linear-bilinear algebraic equations with partial measurements. From this analysis, we found a classification of the variables and the equations of the process, for the unobservable subset we define the location of new sensors to increase the observability index. The proposed strategy has been found to be a very efficient way of establishing the number and location of additional sensors. Two situations can be dealt with : the first one is concerned with an already partially instrumented process which is not yet observable, the second one concerns processes with no instruments at all.

REFERENCES

[VAC 69] V. VACLAVEK, « Studies on system engineering III. Optimal choice of the balance measurements in complicated chemical engineering systems, » Chemical Engineering Science, vol. 24, 1969, pp. 947-955.

[STA 81] G.M. STANLEY, R.S.H. MAH, « Observability and redundancy in process data estimation, » Chemical Engineering Science, vol. 36, 1981, pp. 259-272.

[CRO 83] C.M. CROWE, Y.A. GARCIA CAMPOS, A. HRYMAK, « Reconciliation of process flow rates by matrix projection - part I. The linear case, » AICHE *Journal*, vol. 29, n° 6, 1983, pp. 881-888.

[MAH 76] R.S.H. MAH, G.M. STANLEY, D.M. DOWNING, « Reconciliation and rectification of process flow and inventory data, » Industrial Engineering Chemistry Process Design Development, vol. 15, n° 1, 1976, pp. 175-183.

[CRO 89] C.M. CROWE, « Observability and redundancy of process data for steady state reconciliation, » Chemical Engineering Science, vol. 12, 1989, pp. 2909-2917.

[CLE 90] K.A. CLEMENTS, "Observability methods and optimal meter placement," *Electronical Power & Energy Systems*, vol. 12, n° 2, 1990, pp. 88-93.

[MON 86] A. MONTICELLI, F. WU, "Observability analysis for orthogonal transformation based state estimation," *IEEE Transactions on Power Systems*, vol. PWRS-1, n°1, 1986, pp. 201-208.

[CHE 89] R.L. CHEN, "A fast integer algorithm for observability analysis using network topology," *16th PICA Conf. Proc.*, May 1989, pp. 41-48.

[CAM 90] A. CAMERON, H. DURANT-WHYTE, "A bayesian approach to optimal sensor placement," *The International Journal of Robotics Research*, vol. 9, n°5, October 1990, pp. 70-88.

[MAD 92] F. MADRON, V. VEVERKA, « Optimal selection of measuring points in complex plant by linear models, » AICHE *Journal*, vol. 38, n° 2, 1992, pp. 227-236.

[IMM 94] P.J. IMMONEN, « Classification and placement of measurements through LP/MIP algorithms, » in Proc. of IFAC/IMACS Symposium on Fault Detection, Supervision and Safety for Technical Processes Safeprocess'94, Espoo, Finland, June 13-15, 1994.

[ALI 93] Y.ALI, S.NARASIMHAN, « Sensor network design for maximizing reliability of linear processes, » AICHE *Journal*, vol. 39, n° 5, 1993, pp. 820-828.

[LUO 94] M. LUONG, D. MAQUIN, C.T. HUYNH, J. RAGOT, « Observability, redundancy, reliability and integrated design of measurement system, » in Proc. of 2nd IFAC Symposium on Intelligent Components and Instruments for Control Applications, SICICA'94, Budapest, Hungary, June 8-10, 1994.

[HER 91] N. HERAUD, "Validation et observabilité des systèmes multilinéaires," Thèse de doctorat de l'Institut National Polytechnique de Lorraine, 1991.

[MAQ 87] D. MAQUIN, "Observabilité, diagnostic et validation de données des procédés industriels," Thèse de doctorat de l'Université de Nancy I, 1987.

[RAG 91] J. RAGOT, D. MAQUIN, G. BLOCH, W. GOMOLKA, "Observability and variables classification in bilinear processes," *Journal A*, vol. 31, n° 4, 1991, pp. 17-24.

[FAY 87] J. FAYOLLE, "validation de données et diagnostic des procédés industriels," thèse de doctorat de l'université de Nancy I, 1987.

EVALUATION OF PRODUCTIVITY AND ENERGY EFFECTIVENESS AND PARAMETRIC OPTIMISATION OF ELECTRIC DRIVE SYSTEMS USED IN POSITIONING MECHANISMS

Svetoslav Yordanov[*] and Galin Daskalov[**]

[*] *Technical University, Electric Drives Department, Sofia, Bulgaria*
[**] *Technical University - branch Plovdiv, Control Systems Department, Plovdiv, Bulgaria*

Abstract: In the present paper the results from authors investigations, aiming at parametric optimisation of the discussed class of systems, are summarised in a unified approach for investigation and design. Proposed are analytical expressions and an algorithm for determining productivity and the defined by the authors efficiency in terms of mechanism parameters and law of motion taking into account the laws of distribution of displacements and loads. Shown are the results from investigations of mechanisms, typical for the discussed class systems, obtained by means of a developed software package of applied programs.

Keywords: Electric drive systems, cyclic duty, efficiency, productivity.

1. INTRODUCTION

Electric drive systems, used in cyclic duty mechanisms, carry out similar operations and meet the same technological requirements such as accuracy, productivity and energy effectiveness. This enables the deriving of common mathematical relationships for characteristic quantities as a foundation for developing a unified methodology for design and optimisation of the systems discussed. The investigations in the present paper are devoted to the development of this unified approach.

The considerable interest towards cyclic duty mechanisms since they have been in existence is due to their wide application and great industrial significance, as well as to the still existing and not utilised opportunities for a considerable improvement of their technical, economical and operational characteristics.

Concerning productivity, depending on operational conditions, it is expedient to optimise the electric drives of reversible mechanisms in respect with:

1) Maximum productivity;
2) Minimum cost price per unit productivity;
3) Minimum cost price at given productivity.

Energy processes (we have in mind the processes of converting, accumulating and dissipating of electric, mechanical, magnetic and thermal energy) in electric drive systems are attended by thermal, mechanical and electric loading of some of their components, which causes their ageing and in some cases destruction. The interest towards energy processes of electric drive systems in cyclic duty mechanisms is determined mainly by:

1) The trend towards energy consumption minimisation;
2) The need for developing efficient methods for thermal selection of their motors, operating at intensely varying random load;

3) Insufficient reliability of some widely used drives for heavy duty operation.

The problems concerning productivity increase and minimisation of operational energy consumption of electric drives (for the systems discussed) are solved mainly in the course of design by selecting:

1) The type of electric drive system;

2) The type of the law of motion (the relationship of speed in terms of time) - with or without a low positioning speed (for precise positioning);

3) The type of braking mode;

4) The value of the rated speed, maximum acceleration, the dynamic and static torques;

5) The types of starting and braking processes control (stepwise or not, with a constant driving or dynamic torque, without a device for reducing the positioning motion time, etc.).

The solutions to the above-mentioned problems influence the price and other important features of electric drive systems. Therefore these solutions should be accepted by applying well-grounded engineering and economic criteria, for example minimum expenses per unit productivity. In cases when the prices of electric drive system does not depend on the selection of its variant or parameter, it is expedient to apply engineering criteria, for example minimum energy consumption or minimum energy consumption per unit productivity.

In all cases, in order to make a decision, it is necessary to use exact quantitative relationships for productivity and energy consumption during operation in terms of the mechanism parameters and the law of motion with the laws of distribution of displacements and loads taken into consideration. The large number of varied parameters , the lack of analytical expressions for speed and torque in terms of time, as well as the problems connected with the displacement statistics consideration, make it difficult to find such relationships (Safiuddin, 1990). For the time being this problem has not been solved in a satisfactory way. The problem of minimising the operational energy consumption of electric drives used in cyclic duty mechanisms up to now has been put and solved partially by selecting a drive system, law of motion and breaking mode. With the exception of some special cases, the tasks and problems related to power optimisation of this class of electrical drives has not been formulated in literature.

The present material proposes a technique for investigating, design and set-up of electric drive systems, used in cyclic duty mechanisms. It represents a unified approach for parametric optimisation. In order to achieve this aim the following fundamental and interrelated problems should be solved:

1) Deriving of analytical expressions about the relationship of cyclic duty mechanisms productivity in terms of the type and parameters of the law of motion and electric drive system particularities;

2) Determining the relationship of cyclic duty mechanisms efficiency in terms of the electric drive system parameters and the type of the law of motion parameters;

3) Developing methods for evaluation of productivity and energy effectiveness, taking into account the mode of operation, as well as software packages for their implementation in practice in the course of investigation and design.

2. PRODUCTIVITY

It is convenient to express the productivity Pr of cyclic duty mechanisms as the reciprocal value of the time of cycle T_c:

$$Pr = 1 / T_c. \tag{1}$$

The time of cycle T_c is the sum of the time of motion of mechanism T_m and the total duration of pauses within one cycle- the time of pauses T_p . In general the time of pauses is conditioned by technological considerations and does not depend on the time of motion.

The time of motion T_m is determined mainly by the value of displacement and the type and parameters of the law of motion. Their selection is determined by:

1) The optimum criterion adopted- maximum productivity, minimum thermal loading of motor, minimum operational consumption of energy and others;

2) The restrictions imposed at the same time or in various combinations to: the error during braking, maximum speed, maximum acceleration, maximum value of the first derivative of acceleration, maximum power, etc.;

3) The type and structure of electric drive system, determining the type of law of motion.

The significant differences in productivity of cyclic duty mechanisms for different laws of motion are most of all connected with the braking process character and more precisely- with the presence or absence of motion with positioning speed for accurate positioning. For some of these systems the braking process is monotonous as a rule, for other the final braking is preceded by motion at speed, lower than the rated, and the rest utilise both variants of braking. In some of the systems controlled the first derivative of acceleration is restricted. It has been proven, that this restriction has a comparatively low influence on mechanism productivity (Yordanov, *et al.*, 1980). For this reason authors propose that transient processes of accelerating and decelerating should be assumed of uniform acceleration. Such are the laws of motion in the three-, four-, five-, and six-sector speed diagrams, shown in Fig. 1. The choice of the diagrams to represent this investigation does not confine the application of the results obtained because the method proposed is applicable to all other laws of motion. Moreover the speed diagrams proposed are typical of cyclic duty mechanisms and are characterised by visualness of their mathematical description.

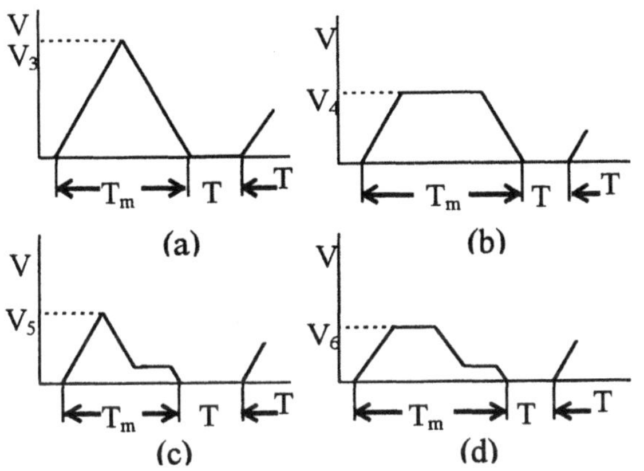

Fig. 1. Types of speed diagrams:
(a) Three-sector diagram;
(b) Four-sector diagram;
(c) Five-sector diagram;
(d) Six-sector diagram.

Let the values of displacement of mechanism and its accelerations during speeding up and slowing down are known. Under these conditions the time of motion T_m will be the smallest if speed V changes according the three-section diagram, that includes the sections of speeding up, slowing down and a pause (Fig. 1-(a)). In this case the time of motion T_{m3} is the sum of the acceleration and deceleration times. It is convenient to express the time of pauses T_p by means of T_{m3} and the relative continuance of operation with motion according the three-section diagram $\varepsilon_3 = T_{m3} / T_{c3}$, e.g.

$$T_p = T_{m3} \frac{1-\varepsilon_3}{\varepsilon_3}. \qquad (2)$$

With the three-section speed diagram the highest (for the specific displacement and accelerations) speed V_3 and productivity Pr_3 are achieved. From (1) and (2) it is obtained

$$Pr_3 = \left[T_{c3} \left(1 + \frac{1-\varepsilon_3}{\varepsilon_3} \right) \right]^{-1}, \qquad (3)$$

where T_{c3} is the time of cycle, corresponding to T_{m3}.

In spite of its providing the highest productivity, the motion according the three-section diagram is not economically advantageous, because the maximum speed V_3 is utilised momentarily- theoretically zero time, while in accordance with it the power, mass and thus the price of electric drive system are increased. The four-section speed diagram (Fig. 1-(b)), which in fact utilises the maximum speed V_4, is economically more advantageous. Moreover, the decisive condition for utilisation of the three- or four-section speed diagram is to meet the requirement of precise braking (if such a requirement exists). Otherwise, before it finally stops, the mechanism should be switched to a low positioning speed motion, selected according the requirement of precise braking. In this case the motion within one cycle follows the five- or six-section speed diagram (Fig. 1-(c) and (d)).

Expressions similar to (3), corresponding to various variants of the law of motion and taking into account parameters of mechanism and electric drive system, can be derived. When comparing the obtained relationships it is expedient to represent productivity Pr and rated speed V_r in relative units by the factor of utilisation of maximum productivity

$$f = Pr / Pr_3$$

and the factor of utilisation of maximum speed

$$\beta = V_r / V_3 .$$

It can be seen from the so defined coefficients f and β ,that the law of motion according the three-section speed diagram is assumed basical when estimating the enumerated parameters influence on electric drive system.

Considering the fact, that values of displacements and static loads of cyclic duty mechanisms are independent random variables, and the static load remains constant for each displacement, it can be written for productivity

$$f = \left[\frac{\sum_{l=1}^{k} q_l \sum_{i=1}^{n} p_i . T_{mli}}{\sum_{l=1}^{k} q_l \sum_{i=1}^{n} p_i . T_{m3li}} + \frac{1-\varepsilon_3}{\varepsilon_3} \right]^{-1}. \qquad (4)$$

Here p_i and q_l designate the probabilities of

occurrence of ith displacement and l[th] static load, and T_{mi} and T_{m3li} - the current values of times T_m and T_{m3}.

The application of (4) for calculation of productivity makes it possible when comparing different variants to take into account particularities of the real operational mode of mechanism, characterised mainly by the random character of displacement values and static loads. It should be noticed, that the calculation of even one value by (4) requires great attention and performance of a rather large and branched procedure, which is unacceptably labour-consuming for engineering practice. This can be done by use of programmes, which implement the generalised algorithm, proposed in the following part.

The present paper shows the results of investigation of typical systems from the discussed class, obtained by the developed software package of applied programmes. In Fig. 2 are shown the relationships $f(\beta)$ of the relative productivity f in terms of the factor of utilisation of maximum speed β with $\varepsilon_3 = 1$ for motion along four-section speed diagram (curves 1,3,5) and motion along six-section speed diagram (curves 2,4,6).

The graphical relationships shown are obtained as follows:

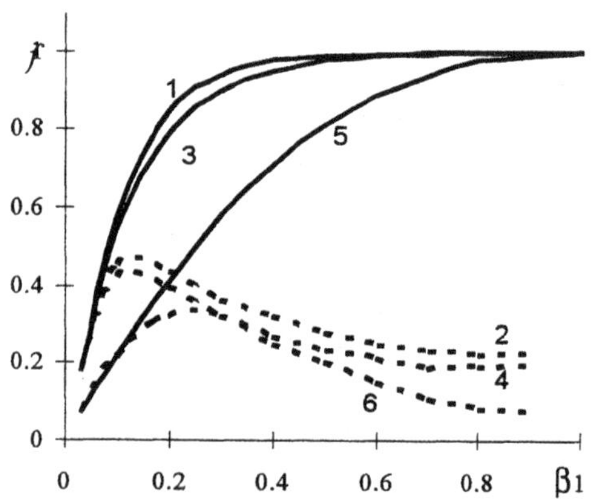

Fig 2. Relationships of productivity f in terms of speed β for different laws of motion and modes of operation.

a) curves 1 and 2- the real laws of displacement and static loads distributions are taken into account;

b) curves 3 and 4- a simplifying assumption is made, that the static load is determined constantly equal to the rated (M_{sr});

c) curves 5 and 6- simplifying assumptions are made, that static load and displacement values are constant and equal to M_{sr} and the maximum displacement respectively.

The following conclusions can be drawn from the results obtained:

1) The procedure of making a decision on the selection of type and parameters of electric drive system is correct under the concrete circumstances only if reliable data about the laws of displacement and load distribution is utilised. Information about displacement statistics is difficult to obtain, but its neglecting is inadmissible since it leads to large errors in the relationships obtained and makes investigation pointless. For most mechanisms the load statistics is well known and its taking into account is not difficult;

2) At low values of β the utilisation of smoothly controlled drives, providing precise braking without low positioning speed motion, does not in fact lead to higher productivity, that is technically and economically not advantageous;

3) The relationship of productivity in terms of the rated speed has a flat maximum, that makes it possible to select the rated speed considerably lower than optimum by implementing an economical or power criterion of optimum;

4) Utilisation of average-valued β is efficient only for electric drive systems, providing precise braking without a positioning speed motion and with the presence of a requirement toward high productivity.

3. ENERGY EFFICIENCY

3.1. Definition of power characteristics.

When applying engineering and economic criteria for evaluation and selection of variants, parameters it is necessary to calculate the energy component of the total annual costs. In this respect the expected annual energy consumption for each of the compared variants should be determined:

$$E_a = \int_0^{t_a} P_m(t)dt,$$

where $P_m(t)$ is the current value of power exchanged with mains. In this case the annual energy consumption E_a is the most informative quantity. It is, however, difficult to utilise E_a because it strongly depends on the operational parameter of total operation time t_a within the year, that varies in a wide range.

The universal parameter efficiency is extensively applied for evaluation of energy quantities. The respective average efficiency η_a for a sufficiently long time interval t is defined by the identity

$$\eta_a = \frac{W_u}{W_m} = \frac{W_u}{\int_0^t P_m dt}, \qquad (5)$$

where W_u and W_m are the useful work and the energy consumed from mains, respectively, and P_m is the power exchanged with mains.

In elevating mechanisms the defining of the term "useful work" W_u offers considerable difficulties when determining η_a. The attempts to define the useful work W_u as a sum of the products of the weights of loads displaced and their lifting and descending heights, the sum of the products of static torques and angular paths of motor or as a constant, do not yield good results.

If W_u is defined as the sum of the absolute values of the products of static torques M_s and angular displacements φ, it will give information on the actual work performed by motor. In horizontal-displacement mechanisms the products of static torques and angular displacements are positive and the ratio

$$\eta_a = \frac{W_u}{W_m} = \frac{\sum |M_s . \varphi|}{\int_0^t P_m dt} \qquad (6)$$

fully corresponds to the physical meaning of the term efficiency.

In elevating mechanisms it is possible to descend a load in generator mode operation of the electric drive, i.e. under certain conditions the energy consumed W_m can be less than the so defined useful work W_u. In this case η_a determined by (6) will be greater than unity. This is an irritating physical absurd on the part of a quantity termed "efficiency". For this reason the authors propose that η_a, determined by (6), be termed "coefficient of energy effectiveness " (K_e).

The above-mentioned example suggests the necessity of introducing a new evaluation characteristic. It could be the relative consumption of energy (per unit useful work), whose practical application is facilitated by the fact of being the reciprocal value of η_a in (6). The relative losses of energy (per unit productivity) can also be utilised for energy evaluation of electric drive systems. The bringing into use of these characteristics is hindered only by the established tradition to use the term "efficiency" for such purposes even when it offers problems.

As a result of the considerations carried out the authors find it expedient to introduce a new definition of the term "efficiency" η of electric drive systems used in cyclic duty mechanisms. It is defined as the integral over the sum of the power P_u, delivered to shaft and exchanged with mains in braking mode, to integral over the sum of the power P_c, consumed by shaft and consumed from mains in driving mode, during the same interval of time t

$$\eta_a = \frac{W_u}{W_c} = \frac{\int_0^t P_u dt}{\int_0^t P_c dt} \qquad (7)$$

Here W_u and W_c designate the energies corresponding to P_u and P_c. On considering the random character of displacement values and static loads, it is obtained finally

$$\eta = \frac{\sum_{l=1}^{k} q_l \sum_{i=1}^{n} p_i W_{uli}}{\sum_{l=1}^{k} q_l \sum_{i=1}^{n} p_i W_{cli}}. \qquad (8)$$

Here W_{uli} and W_{cli} designate useful work and the energy, consumed from mains, respectively, determined for the ith displacement and the lth static load.

3.2. Efficiency determination.

The determination of efficiency by (7) and (8) supposes:

1) Use of statistical data about:

a) the frequencies $q_1^*, \ldots, q_e^*, \ldots, q_k^*$ for static loads occurrence $M_{s1}, \ldots, M_{sl}, \ldots, M_{sk}$. The laws of transported loads distribution are exhaustively studied and standardised. Moreover in elevating mechanisms a probability $0.5q_e$ for motion with static load $M_s > 0$ (at lifting) and probability $0.5q_e$ for motion with static load $M_s < 0$ (at descending) will correspond to the probability q_e of transporting a load of mass m_e;

b) the law of distribution of displacements

$s_1,\ldots,s_i,\ldots,s_n$, performed by mechanism under specific operational conditions. Data about displacements distribution law can be obtained :

- by computations for mechanisms with finite number of points of equal probability;
- by measuring with specialised automated measuring device;
- by utilising already collected statistical data about the displacements distribution for mechanisms, operating under similar conditions.

2) Use of mathematical descriptions of relationships of speed $\omega_{li}(t)$ and load $M_{sli}(t)$ in terms of time within each ith displacement for each lth value of M_s in compliance with the type and parameters of the law of motion.

3) Representation of the power consumed from mains $P_{c1li}(t)$ in terms of time by the parameters of the law of motion and electric drive system. This requires the use of a mathematical description of the relationship of the power of losses in terms of angular velocity ω and the torque M_r on motor shaft, specific for every type of electric drive system. Losses components in practically all electric drive systems are investigated in detail. For every electric drive system the power P_{c1} , consumed from mains, can be represented as a polynomial, that includes coefficients, powers and products of angular velocity ω and torque M_r on motor shaft. For example the power, consumed from mains by cyclic duty mechanisms, driven by induction motors, is determined by the expression:

$$P_{cl}=a_1.M^2\pm a_2.M\pm a_3.M.\omega+a_4.\omega^{1.3}+a_5, \quad (9)$$

where M is the electromagnetic torque of motor, $a_1 - a_5$ are coefficients, determined by the specific values of the parameters of motor, regulator, mechanism, etc.

For parametric optimisation of electric drive systems, used in cyclic duty mechanisms, it is necessary to build the relationship of η in terms of several parameters, some of which are constrained (for example: maximum acceleration, minimum productivity, etc.). Such a problem can be solved in the following sequence:

1) Specified are:

a) electric drive system data- the type of motor, the type of motor speed control, operational modes of motor at acceleration, constant motion and deceleration, the way transient processes pass during acceleration and deceleration;

b) the type of mechanism- for horizontal or vertical displacement;

c) the conditions of operation- by means of statistics (for example the histogram) of displacements and statistics (for example the histogram) of loads;

d) the rated values of motor, mechanism and control device.

2) The optimum value of a given parameter x is found by the criterion $\eta=\eta_{max}$ by assigning the parameter several values and performing for each of them the calculations in 3), 4) and 5).

3) Calculated are for each lth value of loads M_{sl} the average (for all n displacements) values of:

time of motion of mechanism $T_l=\sum_{i=1}^{n}p_i.T_i$

useful energy $W_{ul}=\sum_{i=1}^{n}p_i.W_{ui}$

consumed energy $W_{cl}=\sum_{i=1}^{n}p_i.W_{ci}$,

where the index i designates quantities, corresponding to the ith displacement.

4) From T_e, W_{ul} and W_{cl} , calculated for all k values of loads, one obtains the respective values of:

time of motion $T=\sum_{l=1}^{k}q_l.T_l$

useful energy $W_u=\sum_{l=1}^{k}q_l.W_{ul}$

consumed energy $W_c=\sum_{l=1}^{k}q_l.W_{cl}$.

Here the index l designates quantities, corresponding to the lth static load.

5) From the results, obtained in 4), efficiency η is calculated by (8).

6) On finishing the given values of the optimising parameter the relationship $\eta=F_1(x)$ is built.

7) From the obtained in 4) results the relative (in respect to the maximum for the discussed case) productivity f is calculated by (4) and the relationship $f=F_2(x)$ is built.

8) From $\eta=F_1(x)$ and $f=F_2(x)$ the relationship $\eta=F_3(f)$ is built.

The large volume and complexity of computational work, necessary to implement the algorithm proposed, predetermine the only acceptable way of its realisation - by creating a programme for computation by personal computer. In this respect the authors have developed programmes for parametric optimisation of cyclic duty mechanisms.

In Fig. 3 are shown the relationships of efficiency η in terms of the maximum speed factor of utilisation β of horizontal displacement mechanisms for the cases of motion without (curves 1, 3, 5) and with (2,4,6) low positioning speed. The graphical relationships drawn are obtained for three types of control induction motors: frequency control with recuperative braking- curves 1 and 2; parametric with dynamic braking curves 3 and 4; parametric with braking by plugging- curves 5 and 6.

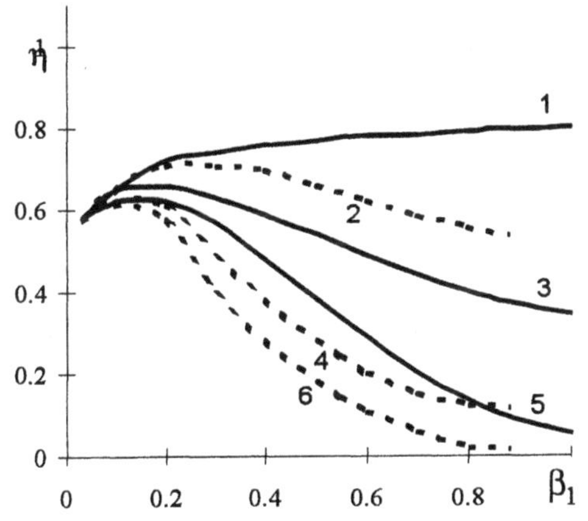

Fig. 3. Relationships of efficiency η in terms of speed β for different systems of electric drives and laws of motion.

In Fig. 4 are shown the relationships of efficiency η of the factor of utilisation of maximum productivity f for relative continuance of operation $\varepsilon_r = 1$. The numbers of curves are the same as the respective in Fig. 3.

In Fig. 5 are shown the relationships (1, 3, 5) of $\eta(\beta)$ from Fig. 3 for motion without positioning speed of horizontal displacement mechanisms and the respective relationships (2,4,6), obtained for vertical displacement mechanisms.

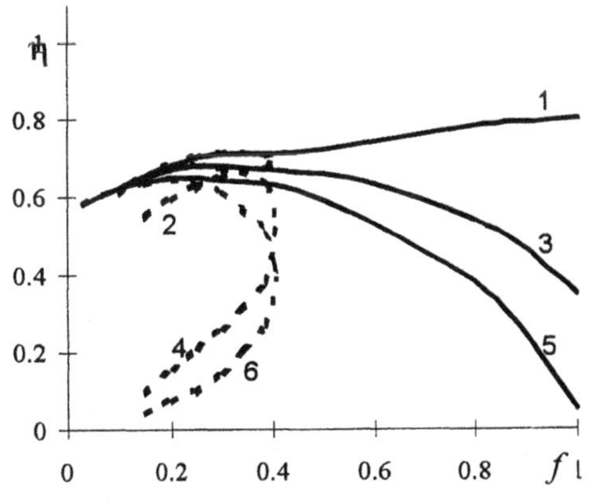

Fig. 4. Relationships of efficiency η in terms of productivity f for different systems of electric drives and laws of motion.

The influence of mode of operation on efficiency- the relationships $\eta(\beta)$, of one of the highest-quality electric drive systems- frequency converter-induction motor, was investigated. The results from investigation of horizontal displacement mechanisms are shown in Fig. 6, where: curve 3

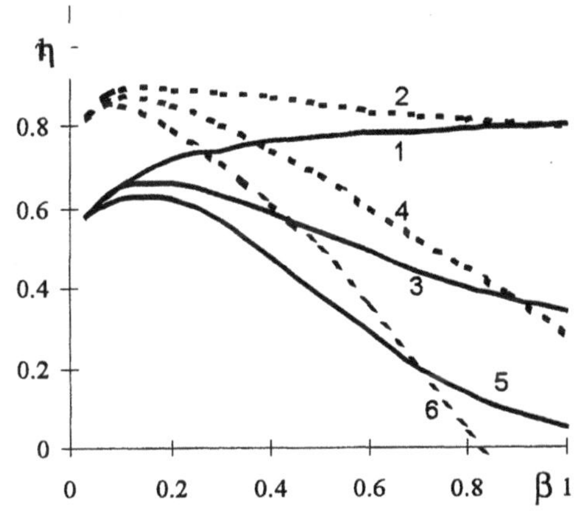

Fig. 5. Relationships of η in terms of β for different types of mechanisms and electric drive systems.

is obtained with taking into account the real laws of displacement and static loads; curve 2- under the simplifying assumption, that static load is constant and equal to the rated M_{sr} ; curve 3- under the simplifying assumptions, that static load and value of displacement are constant and equal to M_{sr} and the maximum displacement respectively.

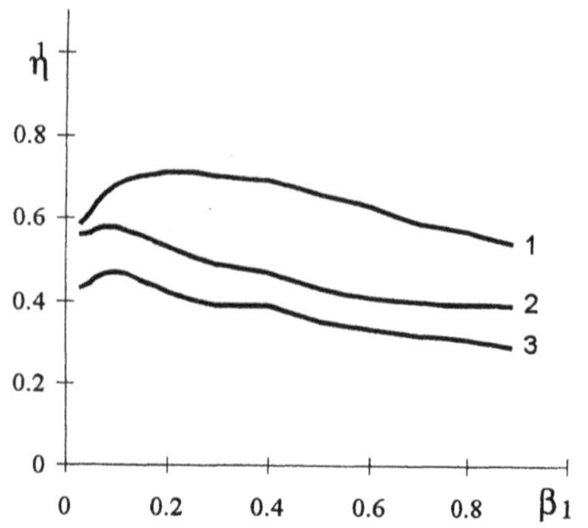

Fig. 6. Relationships of η in terms of β of system 'frequency converter- induction motor' for different modes of operation.

Analysis of obtained results:

1) The average efficiency of electric drive systems, used in cyclic duty mechanisms, has a maximum in the space of parameters of motor and mechanism;

2) The width of the function $\eta = f(\beta)$ maximum enables to take into account more restrictions and criteria without reducing efficiency practically. This makes it possible to optimise the system by the maximum efficiency criterion for a given productivity;

3) The laws of displacements and loads distributions have a strong influence on the relationship $\eta(\beta)$. For this reason parametric optimisation of electric drive systems, used in cyclic duty mechanisms, is correct only when the above mentioned laws are taken into account;

4) The difference in efficiency values is not a sufficient condition to prefer one system to another. In order to make a final, technically and economically grounded decision, it is necessary to determine the electrical energy annual consumption of the compared variants of electric drives.

4. CONCLUSIONS

1. Proposed is an approach for determining productivity and efficiency of electric drive systems, used in cyclic duty mechanisms, when parameters of the real operational mode are taken into account.

2. The application of the analytical relationships obtained and the method for their calculation during design and optimisation of the systems discussed makes it possible to open the objectively existing possibilities for increasing their technical, economical and operational characteristics.

3. A serious difficulty when solving the discussed optimisation tasks is the necessity to use statistical information about the laws of displacements distribution. It was shown, that the intuitive approach, that replaces this information with "expected" laws of distribution, leads to inadmissible errors and renders the investigation pointless. The explicit necessity of such information requires to unify efforts of different groups of researchers in order to exchange the data obtained as quickly and exhaustively as possible, and after they are accumulated, to search the representing laws of distribution.

REFERENCES

Safiuddin, M. (1990). Evaluation of Losses in DC Machines - A Statistical Approach. *IEEE Transactions On Industry Applications.* **2,** pp. 252-258.

Yordanov S., R. Rainov and D. Kraichev (1980). *Selection of Optimum Parameters of Reversible Electric Drives,* pp. 31-40. Tehnika, Sofia.

Continued from outside back cover

Title/Year of publication	Editor(s)	ISBN
1997 (continued)		
Algorithms and Architectures for Real-Time Control (W)	Ruano & Fleming	0 08 042930 0
Computer Aided Control Systems Design (S)	Boullart, Loccufier & Mattsson	0 08 042383 3
Control of Industrial Systems (C)	Grujić *et al*	0 08 042907 6
Advanced Control of Chemical Processes (S)	Shah & Arkun	0 08 043038 4
Singular Solutions and Perturbations in Control Systems (W)	Gurman, Miller & Dmitriev	0 08 042932 7
1998		
Intelligent Components and Instruments for Control Applications (S)	Foulloy	0 08 042608 5
Transportation Systems (S)	Papageorgiou & Pouliezos	0 08 042931 9
Robust Control Design (S)	Banyasz	0 08 042606 9
Intelligent Manufacturing Systems (W)	Jongwon Kim	0 08 043025 2
Distributed Computer Control Systems (W)	MacLeod & Kwon	0 08 042933 5
Manoeuvring and Control of Marine Craft (C)	Vukic & Roberts	0 08 042934 3
Mathematical and Control Applications in Agriculture and Horticulture (W)	Munack & Tantau	0 08 043037 6
System Identification (SYSID'97) (S)	Sawaragi & Sagara	0 08 042592 5
Advances in Control Education (S)	Bir *et al*	0 08 042591 7
Automation in the Steel Industry: Current Practice and Future Developments (W)	Kwon & Won	0 08 043029 5
Control of Power Plants and Power Systems (S)	Zhou Xiaoxin	0 08 042609 3
Fault Detection, Supervision and Safety for Technical Processes (S)	Patton & Chen	0 08 042381 7
Management and Control of Production and Logistics (C)	Binder, Hirsch & Aguilera	0 08 043036 8
Robot Control (S)	Guglielmi	0 08 043026 0
New Trends in Design of Control Systems (W)	Kozak & Huba	0 08 042935 1
Real Time Programming (W)	Maranzana	0 08 043045 7
Automated Systems Based on Human Skill (S)	Cernetic & Brandt	0 08 043024 4
Artificial Intelligence in Real Time Control (S)	Rauch	0 08 042927 0
System Structure and Control (C)	Ionescu & Popescu	0 08 043023 6
Advances in Automotive Control (W)	Rizzoni & Utkin	0 08 043226 3
Intelligent Components for Vehicles (W)	Ollero	0 08 043232 8
Intelligent Autonomous Vehicles (S)	Salichs & Halme	0 08 042929 7
Algorithms and Architectures for Real-Time Control (W)	Nocetti	0 08 043235 2
Artificial Intelligence in Agriculture (W)	Kozai	0 08 043030 9
Supplemental Ways for Improving International Stability (C)	Dumitrache, Kile & Kopacek	0 08 043231 X
Intelligent Assembly and Disassembly (W)	Kopacek & Noe	0 08 043042 2
Computer Applications in Biotechnology (C)	Shioya & Yoshida	0 08 043051 1
Control Applications in Post-Harvest and Processing Technology (W)	Farkas	0 08 043046 5
On-Line Fault Detection and Supervision in the Chemical Process Industries (W)	Dhurjati, Kramer & Cauvin	0 08 043233 6
Dynamics and Control of Process Systems (S)	Georgakis *et al*	0 08 043230 1
Control Applications and Ergonomics in Agriculture (W)	Sigrimis	0 08 043229 8
1999		
Nonsmooth and Discontinuous Problems of Control Optimization and Applications (W)	Batukhtin & Kirillova	0 08 043237 9
Real Time Programming (W)	Zhang & Halang	0 08 043243 3
Information Control Problems in Manufacturing (S)	Morel & Vernadat	0 08 042928 9
Computation in Economics, Finance and Engineering: Economic Systems (S)	Greenblatt & Holly	0 08 043048 1
Nonlinear Control Systems Design (S)	Huijberts *et al*	0 08 043049 X
Linear Time Delay Systems (W)	Dion	0 08 043047 3
System Structure and Control (S)	Lafay	0 08 043035 X
Large Scale Systems: Theory and Applications (S)	Groumpos & Koussoulas	0 08 043034 1
Automatic Control in Aerospace (S)	Lee	0 08 043041 4
Adaptive Control and Signal Processing (W)	Grimble & Johnson	008 043238 7
Automation in Mining, Mineral and Metal Processing (S)	Heidepriem	0 08 043031 7
Low Cost Automation (S)	Chen & Chai	0 08 043027 9
Distributed Computer Control Systems (W)	DePaoli & MacLeod	0 08 043242 5
Analysis, Design and Evaluation of Man-Machine Systems (S)	Nishida & Inoue	0 08 043032 5
Control in Natural Disasters (W)	Sano	0 08 043240 9
Motion Control (W)	Georges & Sename	0 08 043044 9
TELEMATICS - A New Dimension of Automation (W)	Brandt, Genser & Kopacek	0 08 043228 X
Artificial Intelligence in Real Time Control (S)	Pao & LeClair	0 08 043227 1
Space Robotics (W)	Rondeau	0 08 043050 3
Control Applications in Marine Systems (W)	Kijima & Fossen	0 08 043033 3
Intelligent Manufacturing Systems (W)	Pereira & Kopacek	0 08 0432395
14th Triennial World Congress (TWC)		

Customers wishing to obtain details of all available IFAC volumes, should contact their nearest Elsevier Science office.

INVERSE FORCE-TORQUE ANALYSIS OF OPEN SERIAL KINEMATIC STRUCTURES

K. Dobrovodský

Institute of Control Theory and Robotics, Slovak Academy of Sciences
Dúbravská cesta 9, 842 37 Bratislava, Slovakia
Tel: +421 7 373271, Fax: +421 7 376045, E-mail: utrrdobr@savba.sk

Abstract: Force-torque analysis of open serial kinematic structures is considered in this work. Vector representation of force and torque balance equalities embody the common starting point of both the direct and the inverse task of the static analysis. The direct task is to establish prismatic joint forces and revolute joint torques, caused by an external endpoint force, torque, and the gravity of links, in the given endpoint position and orientation. The inverse kinematics is involved in this problem. Redundant structures admit additional restrictions in the form of the cost function depending on joint variables. The inverse task is to establish the external force and torque, having measured forces in prismatic and torques in revolute joints in the given position of joints. The aim of the paper is to present the solution of the direct task and the way how to construct equations for the inverse force-torque task.

Keywords: robotics, kinematics, force-torque balance, algebraic approach

1. INTRODUCTION

The direct task of the force-torque analysis is to establish forces along axes of prismatic joints and torques about axes of revolute joints of the frictionless mechanism under the assumption that the force and the torque is applied to the endpoint of an open serial kinematic structure. If the weight of the links is neglected, the transpose of the Jacobian matrix is frequently used to express the relation between the joint force-torque n-tuple and the endpoint force and torque (Asada and Slotine, 1986). It will be shown how the force-torque balance equalities provide the two problem formulations depending on what is apriori given and what are the unknowns.

The direct force-torque problem involves the inverse kinematic problem. In the case of a given position of the end effector of a nonredundant 6 DOF manipulator one has a finite (up to 16) number of positions of the links. For the given external force and/or torque applied to the endpoint of the structure, forces and torques between links for each set of link positions are unambiguous. Thus there are at most 16 solutions. Moreover, redundant kinematic structures admit optimization of these forces and/or torques by some kind of the cost function (Dobrovodský, 1996; Lenarčič, 1994).

In the case of a given (measured) forces and torques in joints between the links, the external force and/or torque is the unknown, while the values of the joint variables are known. Thus the inverse force-torque problem involves the solution of the direct kinematic problem and leads to a set of linear equations with a matrix depending on the values of joint variables.

2. LINK TO LINK TRANSFORMATIONS

Suppose that one has a fixed coordinate system

(reference) and a moving coordinate system, and let both coordinate systems be identical at the instant. Denote $\mathbf{i}(\delta), \mathbf{j}(\delta), \mathbf{k}(\delta)$ elementary translations and $\mathbf{I}(\delta), \mathbf{J}(\delta), \mathbf{K}(\delta)$ elementary rotations in respect to rectangular coordinate axes $\mathbf{i}, \mathbf{j}, \mathbf{k}$, respectively. Parametric operators

$$\mathbf{i}(\delta), \quad \mathbf{j}(\delta), \quad \mathbf{k}(\delta) \tag{1}$$

$$\mathbf{I}(\delta), \quad \mathbf{J}(\delta), \quad \mathbf{K}(\delta) \tag{2}$$

may be represented by the well known 4x4 homogeneous transformation matrices:

$$\mathbf{i}(\delta) = \begin{pmatrix} 1 & 0 & 0 & \delta \\ 0 & 1 & 0 & 0 \\ 0 & 0 & 1 & 0 \\ 0 & 0 & 0 & 1 \end{pmatrix} \quad \mathbf{j}(\delta) = \begin{pmatrix} 1 & 0 & 0 & 0 \\ 0 & 1 & 0 & \delta \\ 0 & 0 & 1 & 0 \\ 0 & 0 & 0 & 1 \end{pmatrix}$$

$$\mathbf{k}(\delta) = \begin{pmatrix} 1 & 0 & 0 & 0 \\ 0 & 1 & 0 & 0 \\ 0 & 0 & 1 & \delta \\ 0 & 0 & 0 & 1 \end{pmatrix}$$

$$\mathbf{I}(\delta) = \begin{pmatrix} 1 & 0 & 0 & 0 \\ 0 & \cos(\delta) & -\sin(\delta) & 0 \\ 0 & \sin(\delta) & \cos(\delta) & 0 \\ 0 & 0 & 0 & 1 \end{pmatrix}$$

$$\mathbf{J}(\delta) = \begin{pmatrix} \cos(\delta) & 0 & \sin(\delta) & 0 \\ 0 & 1 & 0 & 0 \\ -\sin(\delta) & 0 & \cos(\delta) & 0 \\ 0 & 0 & 0 & 1 \end{pmatrix}$$

$$\mathbf{K}(\delta) = \begin{pmatrix} \cos(\delta) & -\sin(\delta) & 0 & 0 \\ \sin(\delta) & \cos(\delta) & 0 & 0 \\ 0 & 0 & 1 & 0 \\ 0 & 0 & 0 & 1 \end{pmatrix}$$

Let $\delta = \lambda_m$ are constants (link parameters) and $\delta = \vartheta_m$ are joint variables, $m = 1, 2 \ldots$ The following abbreviations are now introduced:

constant translations:

$$\mathbf{a}_m = \mathbf{i}(\lambda_m), \quad \mathbf{b}_m = \mathbf{j}(\lambda_m), \quad \mathbf{c}_m = \mathbf{k}(\lambda_m) \tag{3}$$

constant rotations:

$$\mathbf{A}_m = \mathbf{I}(\lambda_m), \ \mathbf{B}_m = \mathbf{J}(\lambda_m), \ \mathbf{C}_m = \mathbf{K}(\lambda_m) \tag{4}$$

variable translations:

$$\mathbf{x}_m = \mathbf{i}(\vartheta_m), \quad \mathbf{y}_m = \mathbf{j}(\vartheta_m), \quad \mathbf{z}_m = \mathbf{k}(\vartheta_m) \tag{5}$$

variable rotations:

$$\mathbf{X}_m = \mathbf{I}(\vartheta_m), \ \mathbf{Y}_m = \mathbf{J}(\vartheta_m), \ \mathbf{Z}_m = \mathbf{K}(\vartheta_m) \tag{6}$$

A symbolic sequence **cZcXazZcXcZc** defines the kinematic structure of a serial kinematic chain with seven links connected by six joints: $\mathbf{Z}, \mathbf{X}, \mathbf{z}, \mathbf{Z}, \mathbf{X}, \mathbf{Z}$. An implicit indexing of link parameters $\boldsymbol{\lambda} = (\lambda_1, \lambda_2, \ldots)$ and joint variables $\boldsymbol{\vartheta} = (\vartheta_1, \vartheta_2, \ldots)$ assumed from left to right uniquely defines the following parametric 4x4 homogeneous transformation between the endpoint and the base of the kinematic structure:

$$\mathbf{T}(7,0) = \mathbf{c}_1\mathbf{Z}_1\mathbf{c}_2\mathbf{X}_2\mathbf{a}_3\mathbf{z}_3\mathbf{Z}_4\mathbf{c}_4\mathbf{X}_5\mathbf{c}_5\mathbf{Z}_6\mathbf{c}_6 \tag{7}$$

Link to link relative transformations are as follows:

$\mathbf{T}(1,0)$	$=$	$\mathbf{c}_1\mathbf{Z}_1$	1st link to the base
$\mathbf{T}(2,1)$	$=$	$\mathbf{c}_2\mathbf{X}_2$	2nd link to the 1st link
$\mathbf{T}(3,2)$	$=$	$\mathbf{a}_3\mathbf{z}_3$	3rd link to the 2nd link
$\mathbf{T}(4,3)$	$=$	$\mathbf{Z}_4$	4th link to the 3rd link
$\mathbf{T}(5,4)$	$=$	$\mathbf{c}_4\mathbf{X}_5$	5th link to the 4th link
$\mathbf{T}(6,5)$	$=$	$\mathbf{c}_5\mathbf{Z}_6$	6th link to the 5th link
$\mathbf{T}(7,6)$	$=$	$\mathbf{c}_6$	endpoint to the 6th link

3. FORCE-TORQUE BALANCE EQUALITIES

The free rigid body of an individual link incorporated in an open kinematic chain will first be considered, (Asada and Slotine, 1986). Let $\mathbf{f}_m$ is the linear force exerted by the link $m-1$ and acts upon the link m. The force applied to link m by link $m+1$ is then given by $-\mathbf{f}_{m+1}$ since $\mathbf{f}_{m+1}$ is the force exerted by the link m and acts upon the link $m+1$. The gravity force acting at the centroid of the link m is denoted by $\mathbf{g}_m$ and is proportional to the scalar mass and the acceleration vector of gravity. The balance of linear forces acting at the link m is then given by

$$\mathbf{f}_m - \mathbf{f}_{m+1} + \mathbf{g}_m = \mathbf{0} \qquad m = 1, \ldots, n \tag{8}$$

Note that "link 0" is the base and "link $n+1$" is the external environment. External force acting on the last link n is $-\mathbf{f}_{n+1}$. Next, the balance of torques about selected points of links will be derived. Torques $\mathbf{t}_m$ will be denoted the same way as forces. The linear forces also cause torques about the origins of elements. The balance of torques with respect to the origin of the first coordinate system moving together with the following link is given by

$$\mathbf{t}_m - \mathbf{t}_{m+1} + \mathbf{h}_m \times \mathbf{g}_m - \mathbf{p}_{m+1} \times \mathbf{f}_{m+1} = \mathbf{0} \qquad m = 1, \ldots, n \tag{9}$$

where $\mathbf{p}_m$ is the position vector of the origin of the coordinate system of m-th element in respect to the origin of the coordinate system of $(m-1)$-th element and $\mathbf{h}_m$ is the position vector of the center of gravity of m-th element in respect to the origin of the coordinate system of m-th element. Note that all vectors are defined "point to point", thus no choice was made for coordinates of these vectors.

Equalities (8), (9) create the common starting point for several problem formulations. In these equalities the concrete kinematic structure is entering through the symbolic sequence like **cZcXazZcXcZc**.

4. FORCES AND TORQUES IN JOINTS

In the first force-torque problem, one is given the force $-\mathbf{f}_{n+1}$ and the torque $-\mathbf{t}_{n+1}$ acting upon the last element of the structure, the gravity forces $\mathbf{g}_m$, and the position vectors $\mathbf{h}_m$ of centers of gravity of links in respect to origins of coordinate frames attached to the links. The unknowns are the internal forces $\mathbf{f}_m$ and the torques $\mathbf{t}_m$. The vector form of the solution is

$$\mathbf{f}_m = \mathbf{f}_{n+1} - \sum_{l=m}^{n} \mathbf{g}_l \tag{10}$$

$$\mathbf{t}_m = \mathbf{t}_{n+1} - \sum_{l=m}^{n} \mathbf{h}_l \times \mathbf{g}_l + \sum_{l=m}^{n} \mathbf{p}_{l+1} \times \mathbf{f}_{l+1} \tag{11}$$

where $m = n, n-1, \ldots, 1$. Note that the existence of the solution does not depend on the values of joint variables. The coordinates of gravity forces $\mathbf{g}_l$ are given in respect to the basis $\mathbf{i}_0, \mathbf{j}_0, \mathbf{k}_0$, the coordinates of positions $\mathbf{h}_l$ in respect to the basis $\mathbf{i}_l, \mathbf{j}_l, \mathbf{k}_l$, respectively. The orthonormal bases should be expressed by the basis $\mathbf{i}_0, \mathbf{j}_0, \mathbf{k}_0$. Finally, coordinates of forces $\mathbf{f}_m$ and torques $\mathbf{t}_m$ should be transformed into coordinates in the basis $\mathbf{i}_m, \mathbf{j}_m, \mathbf{k}_m$ of the corresponding frame. The unknown joint force or torque is given by the single coordinate related with the joint axis in accordance with the kinematic structure specification string. For example, if $\mathbf{\Omega} = $ **ZcXazZcXcZc**, then the tuple of coordinates of forces in prismatic and torques in revolute joints is

$$\mathbf{t}_{1,z},\ \mathbf{t}_{2,x},\ \mathbf{f}_{3,z},\ \mathbf{t}_{4,z},\ \mathbf{t}_{5,x},\ \mathbf{t}_{6,z} \tag{12}$$

where the first index is the link number and the second index indicates the coordinate axis. In the case of nonredundant kinematic structure, the inverse kinematics will be solved first, and, having expressed the relations between orthonormal bases, one can evaluate coordinates of forces and torques according to expressions (10), (11) in the basis $\mathbf{i}_0, \mathbf{j}_0, \mathbf{k}_0$. Finally, transformation of the coordinates into the appropriate relative coordinates in the basis $\mathbf{i}_m, \mathbf{j}_m, \mathbf{k}_m$, respectively, has to be done. This procedure is straightforward, except the inverse kinematics, where up to 16 solutions may occur.

In the case of redundant kinematic structure, inverse kinematics cannot be solved first and separately. In this case, the cost function with values depending on joint variables can be defined. Since not only forces along and torques about the joint axes may be critical, the recommended cost function is the sum of square magnitudes of vector forces and torques (not the components of forces along and torques about joint axes only).

5. ENDPOINT FORCE AND TORQUE

In the inverse force-torque problem, (measured) forces along and/or torques about joint axes of redundant or nonredundant kinematic structure in a given joint position are given. The endpoint force $-\mathbf{f}_{n+1}$ and the torque $-\mathbf{t}_{n+1}$ acting upon the last element of the structure from the equalities (8), (9) are now to be obtained. Since single relative coordinates are given (measured), these equations cannot be solved in a compact vector form. First, the vectors in equalities (8), (9) will be expressed by the coordinates in respect to the base:

$$\mathbf{F}_0(m) - \mathbf{F}_0(m+1) + \mathbf{G}_0(m) = \mathbf{0} \tag{13}$$

$$\mathbf{T}_0(m) - \mathbf{T}_0(m+1) + \mathbf{H}_0(m) \times \mathbf{G}_0(m) - \\ -\mathbf{P}_0(m+1) \times \mathbf{F}_0(m+1) = \mathbf{0} \tag{14}$$

In the second step, the coordinates $\mathbf{F}_0, \mathbf{T}_0, \mathbf{H}_0, \mathbf{P}_0$ will be expressed by relative coordinates in respect to the appropriate system of coordinates with the origin on joint axes:

$$\mathbf{R}(m,0)\,\mathbf{F}_m(m) - \\ -\mathbf{R}(m+1,0)\,\mathbf{F}_{m+1}(m+1) + \mathbf{G}_0(m) = \mathbf{0} \tag{15}$$

$$\mathbf{R}(m,0)\,\mathbf{T}_m(m) - \mathbf{R}(m+1,0)\,\mathbf{T}_{m+1}(m+1) + \\ +\mathbf{R}(m,0)\,\mathbf{H}_m(m) \times \mathbf{G}_0(m) - \\ -\mathbf{R}(m+1,0)\,[\mathbf{P}_{m+1}(m+1) \times \mathbf{F}_{m+1}(m+1)] = \mathbf{0} \tag{16}$$

where $\mathbf{R}(m,0)$ are the 3x3 submatrices of link to link transformation matrices $\mathbf{T}(m,0)$ created from the first 3 rows and the first 3 columns (orientation matrices). In these equalities n relative coordinates, like for example (12), along joint axes are known and there are $6n$ scalar equations with $6(n+1)-n$ unknown relative coordinates of forces $\mathbf{f}_m$ and torques $\mathbf{t}_m$, $m = 1, \ldots, n+1$. The matrix of the resulting system of linear equations, and thus even the existence of the solution, depends on the known values of joint variables. In the case of a nonredundant full position–orientation structure ($n = 6$), the number of equations is equal to the number of the unknowns. If the kinematic structure is redundant, the system of scalar equations becomes to be overconstrained and the measured forces along and/or torques about joint axes should be considered to be erroneous. Pseudo inverse of the matrix of the system of scalar equations may be used to find a pseudo solution.

6. CONCLUSION

The static properties of a general open serial kinematic structure with the desired position and orientation and with the force and torque applied to the endpoint of the structure were studied. Force-torque balance equalities were expressed by vectors, having strictly distinguished between vectors and coordinates of vectors. Finally, the case of the measured forces and torques with the goal to specify the force and the torque acting upon the endpoint of the kinematic structure was discussed. Further steps led to the derivation of scalar equations for coordinates of vectors in the given position of joints leading to the final set of linear equations of the inverse force-torque static problem for any open serial kinematic structure.

REFERENCES

Asada, H. and J.-J.E. Slotine (1986). Robot Analysis and Control. John Wiley & Sons, New York, Chichester, Brisbane, Toronto, Singapore.

Dobrovodský, K. (1996). Quaternion Based Force-Torque Analysis of Open Serial Kinematic Structures. In: *Recent Advances in Robot Kinematics* (J. Lenarčič and V. Parenti-Castelli (Ed.)), pp. 235–244. Kluwer Academic Publishers, Dordrecht / Boston / London.

Lenarčič, J. (1994). Minimum Joint Torque Configurations of Planar Multiple-Link Manipulator. In: *Advances in Robot Kinematics and Computational Geometry* (J. Lenarčič and B. Ravani (Ed.)), pp. 421–426. Kluwer Academic Publishers, Dordrecht / Boston / London.

ALGORITHM FOR TASK-RESOURCE SELECTING AND MODELLING OF THE PRODUCTION LINE

I. Budinská , B. Frankovič

Institute of Control Theory and Robotics
Slovak Academy of Sciences
84237 Bratislava, Dúbravská c. 9 , Slovak Republic
e mail : utrrfran @ savba. sk

ABSTRACT: This paper deals with task-resource selecting and control of production line. For the solution of this problem, an algorithm based on the production cost and the resource characteristics was developed. There is shown the simulation way for the verification of designed algorithm in this paper.

KEYWORDS: Modelling, optimisation, simulation, scheduling, production line

1. INTRODUCTION

The main effort in the production system is to keep the best configuration and production in certain time periods. It is possible to obtain only if the system works reliable, its capacity is utilised by an optimal way so that the operation cost of production will be minimal and the quality of the products will be at the high level. From this point of view the objects and the approaches of the solution could be formulated as follows:

The object is the resource selection producing acceptable (or better optimal) performance at acceptable (or minimal) cost.

The approaches of solution:

- Model of production line which considers:
    - capital cost
    - production rate or capacity
    - cycle time (the time between order and delivery)
    - work in process
    - on-time performance
- Optimisation on the base of decision process utilisation to find an iteration algorithm for minimisation of the production cost.

In section 2 is described the formulation of task resource and of the control problem. In section 3 the formulated problem simulation with the obtained results is compared and in the conclusion the further way of the research is indicated.

2. ALGORITHM FOR TASK RESOURCE SELECTING

The problem of control of the flexible manufacturing lines may be devided into two parts (Frankovič et. all, 1997):

- For the problem of scheduling to develop the algorithm for task - resource selecting, working on the base of required technological process and criterion of the process performance.

- Determination of optimal operation sequences of production systems using the decision making process.

The decision algorithm for the selection of the designed production line configurations alternatives is based on the given technological task, available resources with their main characteristics, production requirement and some economical aspect which are to the disposition for the designer.

Supposed that (Gerschwin, 1996):

- a resource is any part of the production system
- an activity is a pair of tasks associated with a resource (it is supposed, that only one activity can appear at a resource at any time)
- an interruption is an undesirable activity

Then on the base of this suppositions, the activity state is defined as follows:

$\Gamma_{ij}(t)$ = **1**; when resource i is occupied by activity j
=**0**; otherwise

where t = 0, 1, 2, . . . T
for $i = 1, 2, \ldots n$
and $j = 1, 2, \ldots m$

This is a binary variable, where **1** denotes that resource i is occupied by activity j at time t and **0** otherwise.

The production rate is possible to express by the activity j frequency

$$u_{ij} = \frac{1}{T} N_{ij}(T) \qquad (1)$$

where $N_{ij}(T)$ is the total number of times that resource i is occupied by activity j in time T.
Denoted by τ_{ij} the average duration of activity j at resource i is possible to express the capacity limitation by relation

$$1 \geq \sum_{i=1}^{n} \sum_{j=1}^{m} \tau_{ij} u_{ij} = E \sum_{i=1}^{n} \sum_{j=1}^{m} \Gamma_{ij}(t) \qquad (2)$$

for all resources i,
where $E\Gamma_{ij}$ is the fraction of time that resource i is busy doing operations on task j.
It is necessary to express the cost of production.

The definition can be introduced:
$g(x)$ is the cost of surplus and backlog,
$\alpha(t)$ is the resource state (free or occupied from the aspect of its capacity, failured or repaired from the aspect of its reliability)
The Markov dynamics for α is

$$u(t) \in \Omega(\alpha(t)) \qquad (3)$$

where α depends on the characteristics of production demand , production cost and capacity of production resource.

The constraint to the dynamics of the system is

$$\Delta x = u(t) - d(t) \qquad (4)$$

where: $d(t)$ denotes the demand production rate
$x(t)$ denotes the real production rate; surplus/backlog production
$t = 0, 1, \ldots$ T

See Fig.1.

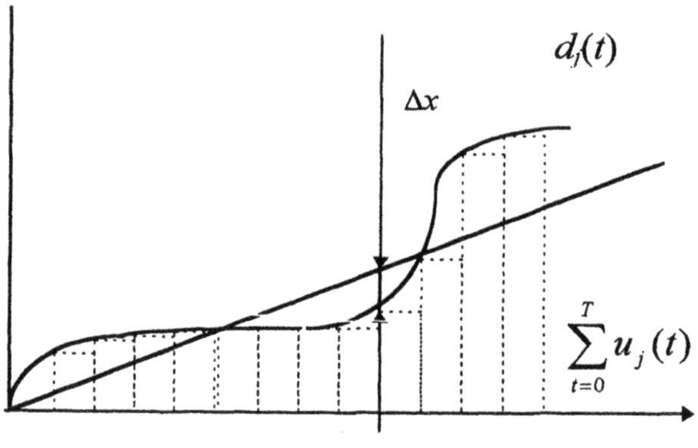

Fig.1. Cumulative production and demand

Then the criterion considered the cost of production may be expressed by the relation

$$Q(x, \alpha, t) = \min E\left\{\sum_{t=0}^{T} G(X(t))\right\} \qquad (5)$$

where $G(X)$ is the fraction function to function $g(x)$ and
$X(T)$ is the fraction function to function $x(t)$.

The goal is for large time T to find (u, α) to satisfy (5).

3. SIMULATION AS A TOOL OF SCHEDULING OF FMS

The production line can be viewed as a discrete event dynamic system (DEDS).

For simulation of production line can be used different types of models. The dynamics of the system can be simulated by using for i.e. state - charts (Fogel, 1996), Petri Nets, Neural Networks and stochastic models based on generalised semi-Markov processes (GSMP).
The simulation program developed on the basis of the stochastic model of DEDS was developed for producing sample path of the stochastic systems.
Consider a sample path of X with a finite length, $\{X_n;\ 0\leq\ n<L\}$, and define a sample performance measure

$$\eta_L = \frac{1}{L}\sum_{n=0}^{L-1} f(X_n) \qquad (6)$$

Since X is irreducible, aperiodic and finite, it is ergodic (Cinlar, 1975).
Thus (Cao et all, 1996)

$$\eta = \lim_{L\to\infty} \eta_L \qquad (7)$$

That is, η can be approximated by η_L provided large enough.

The modelling of timed DEDS on the base of GSMP is described in (Glynn, 1989). GSMP is particular type of stochastic process, which captures the essential dynamic structure of a DEDS.
A GSMP is characterised by the following elements:
- set of states, denoted S; $s\in S$
- set of events, denoted E; $e\in E$
- set of trigger events in state s, denoted E(s)
- family of (conditional) probability distribution functions of residual life of events, denoted $F(t;.,e)$
- family of (conditional) probability distribution functions of transition to a new state, denoted $p(s';.,e)$
State of the system is described by the state vector $\boldsymbol{s}$

Let the system is asynchronous automatic assembly system which is described in detail in (Kamath and Sanders, 1991). The system consists of K pallets, N workstations, L operators. The service time of workstations, the time between two jams of workstations and service time of operators are random. Let all these random variables are independent of the past. The Markov property can be expressed

$X_{n+1} = f_n(X_n\ ,\ \eta_{n+1})$
$X_0 = x_0$
where $\{\ \eta_n:\ \ n\geq 1\}$ is sequence of independent identically distributed random variables and independent with x_0.
In the AAAS the next physical state of the system depends on history only via present physical state and event e from the state of events.

The system can be described as a Markov chain (Valent et all, 1994).
The technological process can be described by procedure of the state transitions in the manufacturing cell. Transport devices move assembly components into the manufacturing cell (workstations) and final products out from the manufacturing cell. In case of failure, they are transport components for new manipulation and new producing. Manipulation robots are divided assembly components between three producing machines and an assembly robot and also between two checking and testing devices.

After design an optimal configuration of manufacturing line the simple path of the system could be simulated. That means, the state (the vector $\boldsymbol{s}$) of the production line in chosen time T could be obtained. This state must be evaluated from the point of queues in front of the stations. The optimal for the effectiveness is then, if the queues are about equal and not very long. In the case of a long queue a new decision task for design optimal production line, new decision task for scheduling must be started. Considering the scheduling, the best results could be obtained in case the long queues appear only in front of the some few workstations. The task is to choose one of three possibilities of treatment of this appearance. There we could add other devices from the resources, i.e. the original device could be substituted for with an other one which characteristic is better or finally we could us other workstation in the production line, which is idle in any time in the process of production. On doing the new decision the simulation will be done again.

3.1. Simulation results

Table 1 Selected resources (working time is in minutes; expense of operation is in $ for op./hour)

		TASK1		TASK2		TASK3		TASK4	
		working time for operation	expense of operatio	working time for operation	expense of operation	working time for operation	expense of operatio	workin time for operation	expense of operatio
Machining	MI	30	30						
	M2			35	38			25	20
	M3					40	45		
Manipulation	Ma1	10	8						
	Ma2			12	10				
	Ma3					15	12		
	Ma4							5	13
Transporting	Tr1	15	10			10	10		
	Tr2			12	12			10	12
Assembly	As	45	50					40	50
Checking and testing	Ts1	10	15					10	15
	Ts2	12	15	12	15	12	15		

There is given the technological process with two transport devices (Tr1, Tr2), three manipulation robots (Ma1, Ma2, Ma3), three machines (M1, M2, M3), one assembly machine (As) and two check and testing devices (Ts1, Ts2). We consider the configuration introduced in Tab.1., which is obtained on the base of equation (5).

The simulation gives us the good view of the manufacturing system in time T. It is the determined time of the simulation. The performance computation is carrying out in case of non-AAAS (asynchronous automatic assembly system). The physical state of the production system, that means numbers of customers in front of the workstations, is shown in Fig.2.

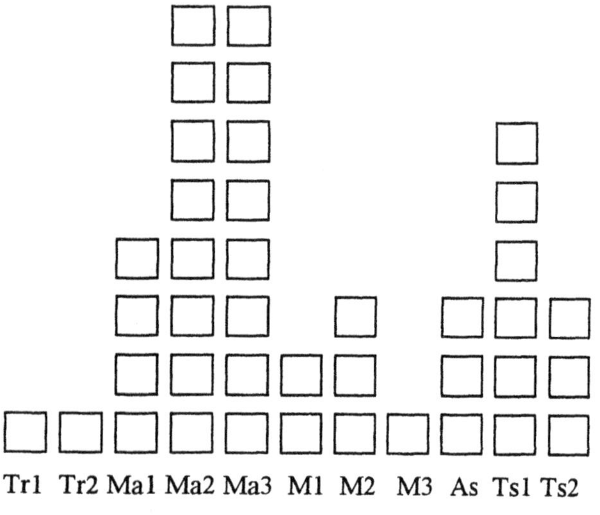

Fig. 2 The physical state vector (it means the queues in front of the workstations) of the production line in time T = 72,14 minutes and number of customers = 50.

The undesirable large queues appear in front of servers Ma2 and Ma3. That assumes the new decision task for these two devices, and the new selection has to change α (t) for each of this resources.

4. CONCLUSION

In this paper is demonstrated a possible approach to the solution of optimal task-resource selecting for the production line. The described algorithm considers with the production cost and the resource characteristics as the criterion for the design of optimal configuration of production line. For the verification is used a simulation program developed on the basis of the stochastic model of DEDS. The results obtained by simulation for a fictious production line shows where are the problematic points in the performance of production line requiring the changes in the task-resource selecting.

5. REFERENCES

Cao, X. R., Xue-Ming, Y., Qiu L.: (1996) A Single Sample Path-Based Performance Sensitivity Formula for Markov Chains, *Proc. 13th Triennial World Congress*, San Francisco, USA, , pp. 371

Cinlar, E.:(1975) *Introduction to Stochastic Processes*. Prentice-Hall, New Jersey,

Fogel, J.: (1996) Concurrent systems: Specification and verification of some real-time . Control properties. In *Proc. of Int. Conference modern Techniques of Statistical Data Analysis*, Košice pp. 177

Frankovič, B., Labátová, S., Budinská I.:(1997) Design of configuration and control of production systems using the decision making. In *Proc. of MCPL'97*, Campinas, Brasil (in press)

Gerschwin, S.B.:(1996) Design and operation of manufacturing systems - a control based synthesis. In *Proc. of CIMAT 96*, Conference on computer integrated manufacturing. Grenoble (plenary lection)

Glynn, P.W.: (1989)A GSMP formalism for discrete event system. In *Proc. IEEE*, Vol.77, 1989 pp.14

Kamath, M., J. Sanders, J. L.:(1991) Modeling Operator/Workstation Interference in Asynchronous Automatic Assembly Systems, In *Discrete Event Dynamic Systems: Theory and Applications 1*, pp.93

Valent, M., Lovász, Z., Frankovič, B.:(1994) Long-run behaviour of the asynchronous assembly automatic system, In *Proc.of New Trends in Design of Control Systems*, Smolenice, pp.374

H_∞-OPTIMAL POSITION CONTROL OF A HYDRAULIC DRIVE

V. Kneppová - Š. Kozák

Department of Automatic Control Systems, Faculty of Electrical Engineering and Information Technology, Slovak Technical University, Ilkovièova 3, 812 19 Bratislava, Slovakia. E-mail:kozak@kasr.elf.stuba.sk.

Abstract. In this paper an application of the H_∞-optimal position control of a hydraulic drive is presented. The treatment of model uncertainties in the field of robust control is demonstrated and an analysis for the two most important kinds of uncertainties is presented. Main goal is to portray the relation between the weighting functions and the performance and robustness of the controlled system. On the basis of this detailed analysis, a set of simple tuning rules for the appropriate selection of the weighting functions is derived. A software package for the analysis of the plant and the selection of the weighting functions has been developed and successfully applied to a hydraulic drive.

Keywords: weighting functions, sensitivity functions, H_∞-optimal control, discrete-time single-input/single-output systems, robust control, hydraulic drive

1. INTRODUCTION

The challenging task in the design phase of H_∞-control is the selection of weighting functions. More so for practical applications, since the performance and robustness of the controller are both strongly related to the adopted weighting functions, it is impossible to shape the unweighted frequency response of the closed-loop transfer function. In (Levine and Reichert, 1990) some qualitative rules for the selection of weighting functions, to match the sensitivity and complementary sensitivity functions, have been presented. In (Jang, *et al.*, 1994) an experimental solution for H_∞-weighting function selection, by exploiting matrix experiments, is given. The selection for a two-block mixed-sensitivity problem is presented in (Postlethwaite, *et al.*, 1990), Postlethwaite, Tsai and Gu, studying as how a constrained optimization can be posed so that improper and unstable continuous weighting functions can be used.

In this paper a frequency domain approach to the design of a discrete time H_∞-optimal servo-compensator is used (Kiffmeier and Unbehauen, 1993). The performance and robustness of the system are optimized according to the mixed-sensitivity H_∞-criterion presented in (Kwakernaak,1985).

2. SENSITIVITY AND WEIGHTING FUNCTIONS

The analysis of sensitivity functions gives quantitative information about how sensitive the nominal model is to uncertainties of the plant parameters. We define the set of admissible plant models $P_\Sigma(z)$ by

$$P_\Sigma(z) = \{ P_i(z) \, , \, i = 1,\ldots,n \}, \tag{1}$$

where $z = e^{j\omega T_s}$, and T_s being the sampling time. Because of the robustness of an H_∞-control design we should ensure that a controller $C(z)$ stabilizes all plants in the set $P_\Sigma(z)$.

2.1 Sensitivity Function S and Weighting Function W_e

Let us consider the standard configuration shown in Fig. 1. The problem is to regulate the output y of the nominal single-input/single-output plant $P_n(z)$ to follow some given reference signal w and to reject the disturbance d by designing a controller $C(z)$, z_e is a weighted system error. As ever, internal stability is the paramount condition. This aforesaid problem can be formulated as "minimizing" the amplitude of the transfer function from d to y over all frequencies. For the case $w = 0$ this transfer function is given by the sensitivity function

$$S(z) = (1 + P_n(z)C(z))^{-1}. \qquad (2)$$

By choosing an appropriate weighting function $W_e(z)$ the frequency response of $S(z)$ and the performance of the controlled system can be optimized, if the controller is designed such that the condition

$$|W_e(e^{j\omega T_s})S(e^{j\omega T_s})| < 1 \qquad \forall\omega \qquad (3)$$

holds. The H_∞-optimization technique allows to find such controllers.

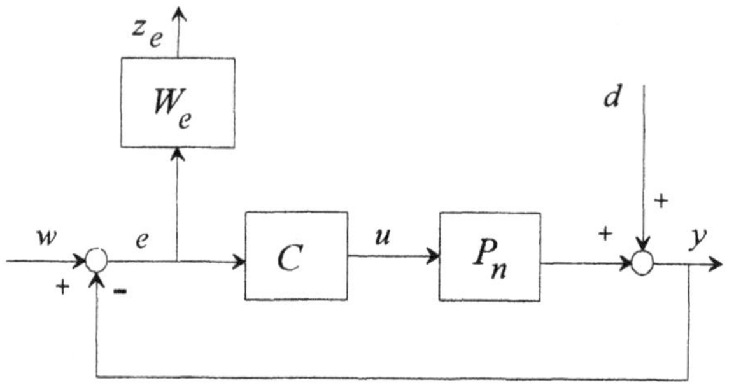

Figure 1 Standard configuration for sensitivity function S and weighting function We

2.2 Complementary Sensitivity Function T and Weighting Function Wy

Now consider the configuration shown in Fig. 2, where z_y is the weighted output of $P_n(z)$. The set of plant models $P_\Sigma(z)$ is defined in this case as

$$P_\Sigma(z) = (1 + \Delta_m(z))P_n(z), \qquad (4)$$

where $\Delta_m(z)$ is the multiplicative uncertainty. We define the complementary sensitivity function T of this configuration by

$$T(z) = P_n(z)C(z)(1 + P_n(z)C(z))^{-1}. \qquad (5)$$

Our goal is to achieve a stable closed-loop system for all possible uncertainties described by $\Delta_m(z)$.

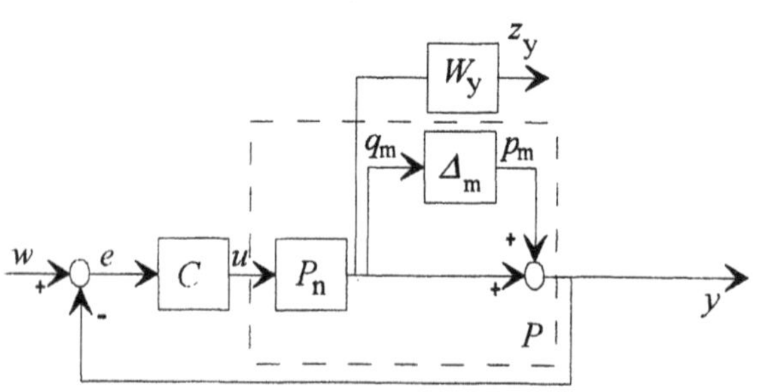

Figure 2 Standard configuration of multiplicative uncertainty and weighting function Wy

As the multiplicative uncertainty is not exactly known, hence we define the weighting function $W_y(z)$ which is the upper bound of the multiplicative uncertainty $\Delta_m(z)$. Then the controller has to be designed such that the condition

$$|\Delta_m(e^{j\omega T_s})| < |W_y(e^{j\omega T_s})| \qquad \forall\omega \qquad (6)$$

holds. A sufficient condition for the robust stability of the closed-loop system is given by

$$|W_y(e^{j\omega T_s})T(e^{j\omega T_s})| < 1 \qquad \forall\omega. \qquad (7)$$

2.3 Control Sensitivity Function R and Weighting Function W_u

Let us consider the configuration shown in Fig. 3, where z_u is the weighted manipulated variable u. The set of admissible plant models $P_\Sigma(z)$ is defined as

$$P_\Sigma(z) = P_n(z) + \Delta_a(z), \qquad (8)$$

where $\Delta_a(z)$ is the additive uncertainty. The control sensitivity function R of this configuration is given by

$$R(z) = C(z)(1 + P_n(z)C(z))^{-1}. \qquad (9)$$

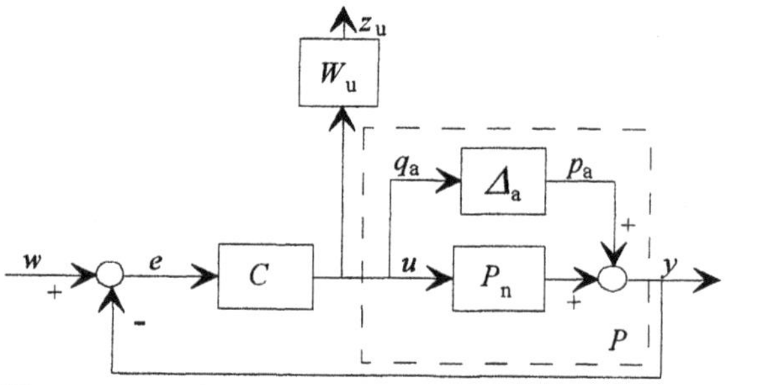

Figure 3 Standard configuration of additive uncertainty and weighting function Wu

The closed-loop system has to be stable for all possible additive uncertainties $\Delta_a(z)$. We define an upper bound of $\Delta_a(z)$ as

$$|\Delta_a(e^{j\omega T_s})| < |W_u(e^{j\omega T_s})| \qquad \forall\omega. \qquad (10)$$

A sufficient condition for the stability of the closed-loop system is given by

$$|W_u(e^{j\omega T_s})R(e^{j\omega T_s})| < 1 \qquad \forall\omega. \qquad (11)$$

2.4 Tuning Rules for Choosing the Parameters of the Weighting Functions

In Table 1 values of the parameters of the weighting functions W_e, W_u and W_y are proposed by the authors of this contribution. In many applications these parameters have proved to provide good results for most practical cases. The weighting functions are defined by the parameters $k{*}0$, $k{*}\infty$ and $\omega{*}$, where $k{*}0$ ($k{*}\infty$) is the amplitude for $\omega \to 0$ ($\omega \to \infty$) and $\omega{*}$ is the bandwidth of the weighting function, ω_S is the crossover frequency of the nominal system $P_n(z)$. The recommended value for the parameter k_{e0} is $1/e_\infty$, where e_∞ is the maximum steady-state control error of the closed-loop system. If we choose $e_\infty = 0$, then the amplitude of sensitivity function $S(z)$ will be zero at low frequencies. By choosing $ke0$ in this way, good tracking properties are achieved $(e = (1+C(z)P(z))^{-1}w << 1)$ *for* $\omega \rightarrow 0$.

A large bandwidth ω_e always involves a high control effort. For high frequencies the aspect of robustness becomes more important, because the model uncertainty usually increases with the frequency. These considerations lead to a trade-off between robustness and sensitivity in feedback control.

$W_e=$ $[ke0, ke\infty, \omega_e]$	$1/e\infty$	$<0.1;0.5>$	$2\omega_S$
$W_u=$ $[k_{u0}, k_{u\infty}, \omega_u]$	$<0.1; 0.5>$	∞	$5\omega_e$
$W_y=$ $[k_{y0}, k_{y\infty}, \omega_y]$	$<ke\infty, 1)$	$<10; 50>$	ω_e

Table 1 Parameters of the weighting functions We, Wu and Wy

There is also a trade-off between good tracking and robustness, i.e. $T(e^{j\omega T_s}) + S(e^{j\omega T_s}) = 1$. The larger the bandwidth ω_y of the complementary sensitivity function $T(z)$, the smaller the robustness of the system and hence the better are the tracking properties.

The best results have been obtained for bandwidths in the range $\omega_S \leq \omega_e \leq \omega_y \leq 2\omega_S$. The larger the bandwidth ω_y, the bigger is the overshoot of the controlled variable y. It is desirable to make the weighting function $Wu(z)$ *for* $\omega \rightarrow \infty$ as large as possible, because then the admissible additive uncertainty is large in the high frequency range. The larger the bandwidth ω_u, the less significant is the robustness of the system. If the bandwidth ω_u is chosen from the interval $3\omega_e \leq \omega_u \leq 5\omega_e$, then the control variable does not have high frequency oscillations.

The given parameters shown in Table 1, had been tested in practice with good results. It is worth mentioning that we have to keep the order of the weighting functions as low as possible, because the order of the optimal controller depends linearly on the order of the weighting functions. In almost all practical cases the order of the weighting functions is 1.

3. H∞- CONTROLLER DESIGN

The system will be optimized for the mixed sensitivity H_∞-criterion according to the (Kwakernaak,1985). Let us consider the classical servocompensator configuration designed by the H_∞-optimization algorithm (Kiffmeier and Unbehauen, 1993). The design procedure for single-input/single-output plants leads to the solution of a coupled system of diophantine and spectral factorization equations. The result of the computation is an H_∞-optimal controller. For more details see (Kiffmeier and Unbehauen, 1993) and (Kiffmeier and Unbehauen*, 1993).

4. DESIGN PROCEDURE FOR SISO PLANTS

The following steps for the design procedure are proposed:

1. Based on system identification, determine a nominal model $Pn(z)$ and the set of the plant models $P\Sigma(z)$. Then compute the multiplicative and additive uncertainties from:

$$\Delta mi(z) = Pi(z)/Pn(z)^{-1} \tag{12}$$

$$\Delta ai(z) = Pi(z) - Pn(z) \quad i = 1,..,n, \tag{13}$$

   where $P_i(z)$ are from the set $P_\Sigma(z)$.

2. Fit the initial values of the parameters for the weighting functions $We(z)$, $Wu(z)$ and $Wy(z)$ according to the desired system performance (3), the robustness conditions (7), (11) and Table 1.
3. Calculate the H_∞-controller.
4. Evaluate the system performance and robustness.
5. If they are not desirable return to step 2.

5. APPLICATION OF THE DESIGN METHODOLOGY

The structure of this laboratory system is presented in the Figure 4. It contains the spring-mass-damping system which position is controlled with the help of the hydraulic cylinder with the servo valve. The detailed describing of this system is given in (Du, 1989).

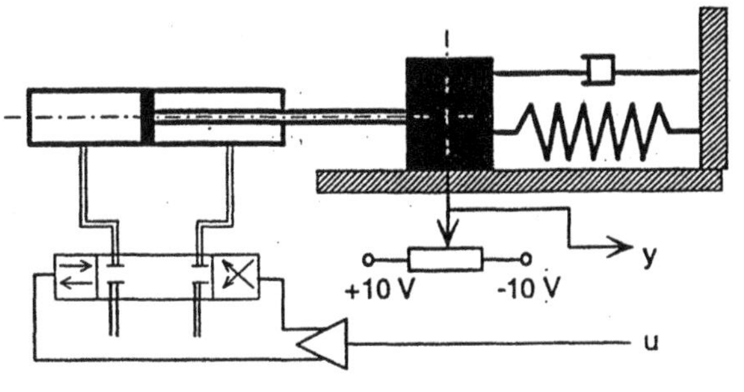

Figure 4 Simplified structure of the laboratory system of the position hydraulic control

This laboratory system has many non linearity's which cause the following effects:

- The damp works in the negative shift direction. In

the positive direction is the only small damping.

- The shift power of the hydraulic cylinder depends on the direction because the different effective areas on the sides of the hydraulic cylinder. On one side is the effective area smaller then on the other.
- The hydraulic elements for the setting of the cylinder contains the non linear lines and the capacity of the leads is variable. That effects on the frequency of the cycles which can be very high. It causes the high control effort .
- The non controlled system can be posed to the maximal left position by the discontinuous current in the servo valve.
- The spring is pretight. Therefore there are the Coulomb's forces in the damp, spring and charging.

On the basis of these non linearity's the describing of a dynamic behavior of the position control is possible only approximately by a linear transfer function. In (Du, 1989) is given the following plant model of this system

$$P_S = \frac{K_s \omega_0}{T_0\, s\left(s^2 + 2d\omega_0 s + \omega_0^2\right)}. \qquad (14)$$

The parameters of this transfer function depend on the load mass M and the viscose friction coefficient K_v in the damp. The following parameters are derived from the experimental frequency measure by the unloaded case:

$$K_S/T_0 = 170\ sec^{-2},\ \omega_0 = 100\ sec^{-1}\ and\ d = 0.25. \qquad (15)$$

The manipulated variable u and controlled variable y are normalized as following

$$y = \pm 1 \cong \pm 100\ mm, \qquad (16a)$$

$$u = \pm 1 \cong \pm 10\ V. \qquad (16b)$$

The typical operating point with the load

$$M = M_n = 10\ kg,\ K_v = K_{vn} = 750 \frac{N}{m/sec}, \qquad (17)$$

which is taken as the nominal case. Then the parameters are valued

$$K_{Sn}/T_{0n} = 120\ sec^{-2},\ \omega_0 = 70.7\ sec^{-1},\ d = 0.503. \qquad (18)$$

On the basis of the physical considerations the following dependencies of the real parameters on the nominal values are given:

$$\omega_0 = \sqrt{\frac{M_n}{M}}\,\omega_{0n},\ M = (0.5...2)M_n, \qquad (19a)$$

$$d = \frac{K_v}{K_{vn}}\sqrt{\frac{M_n}{M}}\,d_n,\ K_v = (0.3...4)K_{vn}, \qquad (19b)$$

$$\frac{K_S}{T_0} = f\sqrt{\frac{M_n}{M}}\frac{K_{Sn}}{T_{0n}},\ f = (0.66, 1.33). \qquad (19c)$$

The following system performance are desired:

- Guaranteed stability for all possible parameters according to the Eq. (19) (20a)
- Normalized manipulated variable u(t) has to be smaller then 2 for the step changes of the reference signal w(t) = 1.0, t ≥ 0. (20b)
- No steady control error for the step changes in the reference signal and disturbance in the input or output of the plant. (20c)
- As large as possible bandwidth of the control for the good tracking properties. (20d)

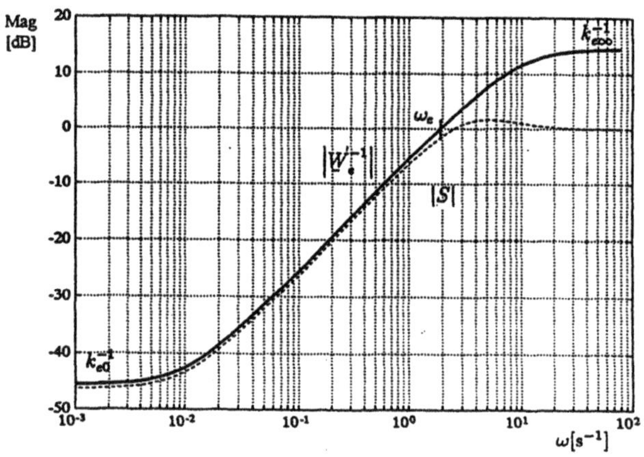

Figure 5 Sensitivity function S and the inverse weighting function W_e

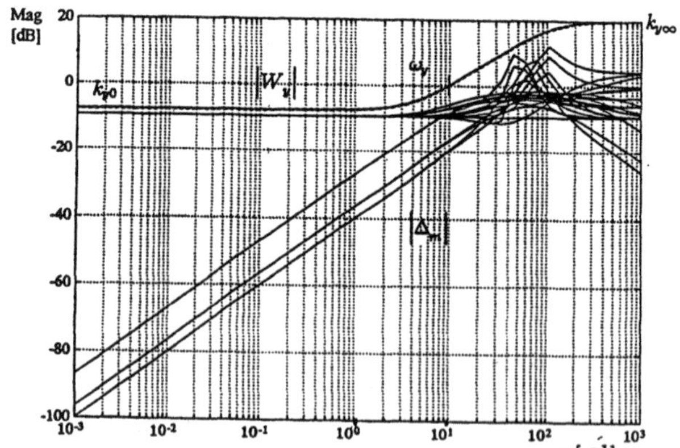

Figure 6 Multiplicative model uncertainties Δ_m and the weighting function W_y

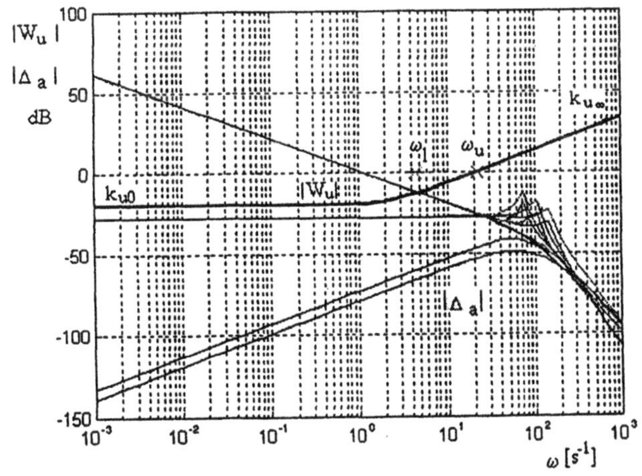

Figure 7 Additive model uncertainties Δ_a and the weighting function W_u

The selection of the weighting functions can be seen in the Figures 5, 6 and 7 and in Table 2. In order to achieve the good tracking properties the parameter k_{e0} has to be as small as possible. This is chosen according to the rules of the Table 1. On the other hand the bandwidth ω_e has to be as large as possible because of the good tracking properties. The robustness can be injured by the large bandwidth ω_e. The following discrete-time transfer function of the hydraulic drive is given by

$$P_S(z) = (0.016z^{-1} + 0.039z^{-2} + 0.0093z^{-3})/(1 - 0.495z^{-1} - 0.138z^{-2} - 0.368z^{-3}) \quad (21)$$

The plant was discretized using a sampling period of 0.04 sec.

	Servocompensator
Data for controller design	$W_e = [\infty, 0.31, 4.43, 1]$ $W_u = [0.1, \infty, 22.18, 1]$ $W_y = [0.31, 10, 4.43, 1]$
Controller degree	5
H_∞ -norm	1.27
Sampling time	40 ms

Table 2 Data of the H_∞ -optimal controller design by the hydraulic position control

There is the real time control with the servocompensator in two extreme operating points in the Figure 8:

OP1: minimal loading mass, minimal damping

$$(M = 5\text{kg}\ , K_v = 250 \frac{N}{m/\sec}) \text{ and}$$

OP2: maximal loading mass, maximal damping

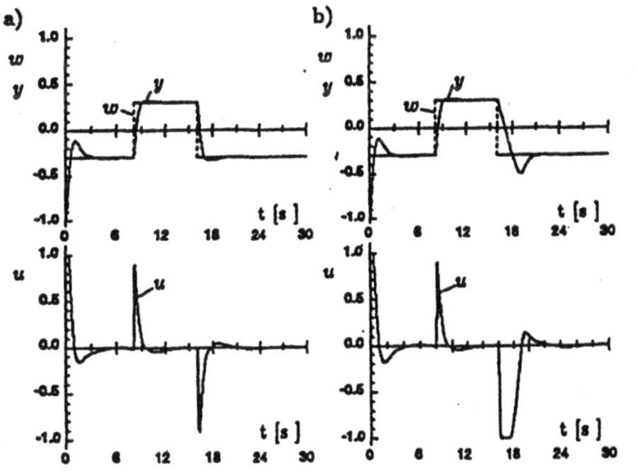

Figure 8 Real time control with the servocompensator by the hydraulic position control
a) minimal load (OP1)
b) maximal load (OP2)

$$(M = 20\ \text{kg}\ , K_v = 3000 \frac{N}{m/\sec}).$$

The nominal case is very similar to the case in the operating point OP1. The reference signal is changed between w = -0.3 (≅-30mm) and w = 0.3 (≅30mm).

6. CONCLUSIONS

The motivation for this work was to demonstrate how the parameters of the weighting functions can be chosen in H_∞- optimal robust control and how they influence the system performance and robustness. A software package has been developed for this purpose. The design of the H_∞- optimal controller is based on (Kiffmeier and Unbehauen, 1993) and (Kiffmeier and Unbehauen*, 1993). Furthermore, an H_∞- optimal servocompensator has been designed for the hydraulic position control, using the approach presented in (Kiffmeier and Unbehauen, 1993).

REFERENCES

Jang C.D., H.S. Ju, S.W. Liu (1994). Experimental Design of H∞-Weighting Functions for Flight Control Systems, *Proc. American Control Conference '94*, USA, 2516-2520

Kiffmeier U. and H. Unbehauen (1993). Design of discrete time H∞-optimal servo-compensators, *Proc. American Control Conference '93*, San Francisco, USA, 1062 - 1066

Kiffmeier U. and H. Unbehauen* (1993). Design of discrete-time H∞-optimal servo-compensators for MIMO systems, *Proc. Second European Control Conference*, Groningen, 2052-2057

Du P. (1989). Die Realisierung eines neuen adaptiven Reglers für hydraulische Vorschubantriebe mittels Mikrorechner, Dissertation, Ruhr-University Bochum, *Fortschritt- Berichte VDI*,

Reihe **8**, No. **195**, VDI Düsseldorf

Kwakernaak H. (1985). Minimax frequency domain performance and robustness optimization of linear feedback systems, *IEEE Trans. on Automatic Control*, Vol. **30**, pp. 994 - 1004

Levine W.S., R.T. Reichert (1990). An introduction to H∞ Control System Design, *Proc. of the 29th Conference on Decision and Control*, 2966-2974

Postlethwaite I., M.C. Tsai, D.W. Gu (1990). Weighting Function Selection in H∞ Design, *Proc. IFAC*, Tallinn, Estonia, Vol. **5**, 104-109

NUMERIC SOLUTION A COEFFICIENT IDENTIFICATION OF SYSTEM OF DIFFERENTIAL EQUATIONS USING OPTIMAL CONTROL ALGORITHM

Anna Jadlovská, Ing.[1]
Anna Hapáková,PaedDr.[2]

[1] *The Department of Cybernetics and Artificial Intelligence
Faculty of Electrical Engineering and Informatics,
Technical University of Košice, Slovakia*
[2] *The Department of Informatics, Mathematics and Physics,
Faculty of Production Technology
Technical University of Košice with residence* for
Prešov

Abstract: The contribution deals with the created algorithm for numeric solution of problems of optimal control with terminal criterion. Optimal control theorem and philosophy of gradient method is used for computation of increment of an objective function. It is showed, that the presented algorithm may be applied for seeking coefficients of system of differential equations, if this problem of coefficient identification is transformed into the task of optimal control with terminal criterion.

Keywords: system differential equations, optimal control, criterion function, identification algorithm, required accuracy.

1. INTRODUCTION

The paper addresses creation of algorithm for numerical solution problems of optimal control pattering

$$x'(t) = f(x(t), u(t), t), \quad 0 \le t \le t_1 \qquad (1)$$

$$x(0) = x_0 \qquad (2)$$

$$u(t) \in U \qquad (3)$$

$$J = F(x_1(t_1), \dots x_n(t_1)) \qquad (4)$$

where $x(t)=(x_1(t),..., x_n(t))$ is state variable vector and $x(0)$ is initial condition vector, t is time and t_1 is given time.
$u(t) = (u_1(t),..., u_r(t))$ is control vector of system and U is closed set of r - dimension space.
$f = (f_1,f_2,...f_n)$ is given vector of right side functions of differential equations and J is an objective function containing terminal variables.
The task is finding such a control vector u(t) of the system, that fulfills (1) and optimalizes (4).

Presented algorithm may be applied for solution a problem of coefficient identification of system of differential equations with initial conditions, if solved problem is transformed to the task pattering (1) to (4). Presented algorithm for computation of increment of an objective function uses optimal control theorem and philosophy of gradient method.

2. THE BASIS OF THE METHOD OF DEFINED PROBLEM SOLUTION

Unknown control vector u(t) may be approximated with assistance of polynomials $P_{ik}(t)$, (of Legendrov , Tschebischevov or generalized).

Let $P_{ik}(t)$ be given polynomials of variable t, degree k. Then control vector can be expressed in the form

$$u_i(t) = \sum_{k=1}^{n} a_{ik} P_{ik}(t), \qquad i=1,2,...,n \qquad (5)$$

where a_{ik} are unknown coefficients.

In the following we consider the case for control vector which contains only one component, so we can write

$$u(t) \approx u_1(t) = \sum_{k=1}^{n} a_k P_k(t)$$

where unknown coefficients a_k should be appointed such a way, that defined objective function shall (4) acquire minimum rating.

Furthermore, we consider the matrix M, whose elements are derivatives of the functions $f_1, f_2, \ldots, f_n$, that is

$$M = \begin{pmatrix} \frac{\partial f_1(x,u,t)}{\partial x_1}, & \cdots & , \frac{\partial f_1(x,u,t)}{\partial x_n} \\ \vdots & & \\ \frac{\partial f_n(x,u,t)}{\partial x_1}, & \cdots & , \frac{\partial f_n(x,u,t)}{\partial x_n} \end{pmatrix}$$

The elements of this matrix can be computed for given time t, if we know control vector u(t) and system status vector x(t) at that time.

According to the definition we can write an expression for adjunct system of differential equations:

$$\frac{dP(t)}{dt} = -M^t P(t) \qquad (6)$$

where unknown quantities are elements $p_1(t)$, $p_2(t)$, ...,$p_n(t)$ of vector P(t) and M^t is transposed matrix to matrix M. Concerning to the system (6) we can also add conditions defined at time $t = t_1$

$$P(t_1) = (p_1(t_1), p_2(t_1), \ldots, p_n(t_1))^t \qquad (7)$$

or

$$p_1(t_1) = \frac{\partial F(x_1(t_1), \ldots, x_n(t_1))}{\partial x_1}, \ldots,$$

$$p_n(t_1) = \frac{\partial F(x_1(t_1), \ldots, x_n(t_1))}{\partial x_n}$$

It is obvious, that $t = t_1$ is final time for the system (1), but initial time for the system of differential equations (6).

We know to compute x(t) by numeric integration (1), when u(t) is known, so we can solve the system of differential equations (6) with initial conditions (7) and gain values of vector elements P(t) for decreasing time from t_1 to 0.

Hamilton function is according to the definition expressed in the form

$$H(x(t), u(t), p(t), t) = \sum_{j=1}^{n} p_j(t) f_j(x(t), u(t), t) \qquad (8)$$

and its derivation according to the control vector is

$$\frac{\partial H}{\partial u_i} = \sum_{j=1}^{n} p_j(t) \frac{\partial f_j(x(t), u(t), t)}{\partial u_i} \qquad (9)$$

According to the optimal control theorem, the values of increment $\partial u_i(t)$ of control vector elements $u_i(t)$ make increment of objective function, δJ which is defined by the relation

$$\delta J = \int_0^{t_1} \frac{\partial H(x,u,p,t)}{\partial u_i} du_i(t) dt \qquad (10)$$

when we appoint increment to every parameter a_k then every element $u_i(t)$ has an increment pattering

$$\delta u_i(t) = \frac{\partial u_i(t)}{\partial a_k} \partial a_k \qquad (11)$$

After putting the relation (11) into (10), we will get an adequate value of increment of objective function and we will derive the relation (12) quite easily.

$$\frac{\partial J}{\partial a_k} = \int_0^{t_1} \sum_i \frac{\partial H(x,u,p,t)}{\partial u_i} \frac{\partial u_i(t)}{\partial a_k} dt \qquad (12)$$

The relation (12) enables numerical calculation of values of every derivation $\partial J / \partial a_k$ if we have already computed numeric values of state variables of vector x(t) and adjunct variables p(t), that are adequate to control vector u(t), (Kubík,et al.,1982;Mikleš and Hutla, 1986;Ralston, 1973; Legras,1980;Jadlovská and Sarnovský, 1996).

3. ALGORITHM OF THE METHOD

Defined problem of optimal control (1) to (4) can be solved by given algorithm, where k - iteration contains these steps k = 1,2,...:

K1: Solution a system of differential equations (1),(2), with selected initial control $u = u^k(t)$, aiming to appoint trajectory $x = x^k(t)$ at interval $<0,t_1>$

K2: Solution an adjunct system of differential equations (6),(7) at "inverted time" from $t = t_1$ to $t = 0$ where $u = u^k(t)$, $x = x^k(t)$ and determination of adjunct variables $p^k(t)$ at interval $<0,t_1>$

K3: Computation of value of objective function

K4: If absolute magnitude of value difference of objective functions (a new value and a value in the previous iteration) is less than required accuracy then end of computation, else continue to do K5.
K5: Storage of computed values of control vector $u^k(t)$ and value of objective function.
K6: Determination of increment of control and objective function; continue to do K1.

Solution a problem of optimal control requires effective algorithms (a good quality approximation, acceptable computing time) for computing of state vector element values x(t), adjunct vector p(t) as well as it is needed for computing of integrals for gradient definition. That is why at present we may use programming system MATLAB for solution concrete tasks.

4. IDENTIFICATION PROBLEM TRANSFORMED INTO THE TASK OF OPTIMAL CONTROL.

Let the system of differential equations be given in the form

$$x_1'(t) = u(t) \,.\, x_2(t)$$
$$x_2'(t) = x_1^2(t) - x_2^2(t) + u^2(t) \qquad (13)$$
$$x_1(0) = x(0) = 1, \qquad 0 \leq t \leq 2 \qquad (14)$$

The problem consists in finding such u(t), that solution $x_1(t)$ approximates function $x_1^*(t)$ the best way. The function $x_1^*(t)$ is defined by measured data. If is the reason for looking for minimum of objective function

$$J_1 = \int_0^2 (x_1(t) - x_1^*(t))^2 \, dt \qquad (15)$$

To transform mentioned task into the form (1) to (4), we will install such function $x_3(t)$:

$$x_3'(t) = (x_1(t) - x_1^*(t))^2 \qquad (16)$$
$$x_3(0) = 0, \qquad (17)$$
$$J_1 = x_3(2)$$

Then transformed task will include three differential equations (13), (16) with initial conditions (14), (17); and the task is to find such u(t) function, that maximizes objective function:

$$J = -J_1 = -x_3(2)$$

If the function u(t) is approximated by polynomial pattering

$$u(t) = a_0 + a_1 t + a_2 t^2 + a_3 t^3$$

then unknown quantities of our problem are coefficients a_0, a_1, a_2, a_3.
To use the algorithm of mentioned method, it is necessary to write also adjunct system of differential equations according to (6) and Hamilton function (8), (Alexejev, at al.,1991; Legras,1980;Jadlovská and Sarnovský,1996).

5. CONCLUSIONS

The paper deals with the method of solution a defined optimal control problem with terminal objective.
It is shown, that we may transform the coefficient identification problem of the system of differential equations with initial conditions to the defined optimal control problem.

Created algorithm of mentioned method is presented there. The algorithm does not use the method of maximum principle and therefore the method ranks among direct methods, that uses for solution a system of differential equations and for computation of gradient (which presupposes integral calculation) approximate numeric methods. Numeric experiments show satisfactory convergence of algorithm.

REFERENCES

Alexejev,V.M., V.M. Tichomirov and S.V. Fomin (1982). *Matematická teorie optimálních procesu*, ČSAV, Praha, pp.40-97.

Jadlovská,A.and J.Sarnovský (1996). *Riešenie úloh optimálneho riadenia*, AT&P Journal 6, ročník III. december 1996, MH Bratislava, pp.80-84.

Kubík,S. at al.(1982). *Teorie automatického řízení II.* SNTL, TKI Praha, pp.11-206.

Legras,J.: (1980). *Algorithmes et programmes d'optimisation non linéaire avec contraintes*, Masson, Paris, New York, pp. 50-151.

Mikleš,J. and V.Hutla (1986). *Teória automatického riadenia*, Alfa SNTL, Bratislava, Praha , pp. 174-394.

Ralston,A.(1973). *Základy numerické matematiky*, ČSAV, Praha, pp. 45-244.

A ROBUST NON-LINEAR CONTROL DESIGN FOR A TURBOGENERATOR

V. Vesely, J. Simovec
Department of Automatic Control Systems, Faculty of Electrical Engineering and Information Technology, Slovak University of Technology, Bratislava.
D. P. Papadopoulos
Department of Electrical Computer Engineering, Democritos University of Thrace, 67100 Xanthi, Greece.

Abstract: In this paper is presented a new approach to robust non-linear control design, which can guarantee a prescribed decay rate of exponential stability for known system uncertainties. The proposed approach doesn't employ matching conditions. The non-linear power system with deterministic uncertainties is chosen as demonstration example.

Keywords: robust, non-linear, exponential stability, semi-linear, power system.

1. INTRODUCTION

The control of the power systems consisting of interconnected networks of transmission lines linking generators and loads is an important problem.
Still present there are used the controllers based on linearised synchronous generator and turbine model.
But the very non-linear nature of the power system models and the resulting severe disturbances preclude the use of classical linear control design techniques.
Many investigators have tried to overcome the non-linearity by using various approaches. For example global robust adaptive control, Mudroncik, Vesely (1994); control design for the dynamic performance improvement Papadopoulos (1986); other approaches Kozak (1995);
Because of an important role of exponential stability in robust control, Dubrov, Zuber (1989), our main *goal* is to design controller for specified non-linear system, which can guarantee robust exponential stability of the system with prescribed properties.
The control design method here proposed is based on the semi-linear representation of the non-linear system model, definition of exponential generalised dynamic systems, polynomial formalism and direct Lyapunov method, Barbashin (1970), Krasovskij (1959) and Gaiduk (1995). As an example is used the model of the one machine - infinite bus power system with both input and system uncertainties for governor control design.
The paper is organised as follows. In Section 2 the problem formulation and some preliminary results are given. The main results for the design of a non-linear controller which guarantees the stability and prescribed decay rate of exponential stability of the closed-loop uncertain system is given in Section 3. In Section 4 is the method derived in Section 3 applied to model system. Simulation results of the resulting closed-loop system are presented in sections 5 and 6.

2. PROBLEM FORMULATION

Consider the following power system uncertain dynamic model

$$\dot{\mathbf{x}} = \boldsymbol{f}(\mathbf{x}) + \delta\boldsymbol{f}(\mathbf{x}) + (\boldsymbol{b}(\mathbf{x}) + \delta\boldsymbol{b}(\mathbf{x})).\mathbf{u} \qquad (1)$$

where $\mathbf{x} \in R^n$ is the state, $\mathbf{u} \in R^m$ is the control vector, $\boldsymbol{f}(\mathbf{x})$, $\delta\boldsymbol{f}(\mathbf{x})$: $R^n \to R^n$, $\boldsymbol{b}(\mathbf{x})$, $\delta\boldsymbol{b}(\mathbf{x})$: $R^n \to R^{n\times m}$ are continuous and uniformly bounded vector functions of class C^k, k>0 is sufficiently large, differentiable on the set $R_\rho \times R^m$ with respect to system variables **x**, **u**.

$$R_\rho = \{\mathbf{x} \in R^n : \|\mathbf{x}\| < \rho\}, \rho > 0$$

and $\boldsymbol{f}(0) = \delta\boldsymbol{f}(0) = 0$, $\boldsymbol{b}(0) + \delta\boldsymbol{b}(0) \neq 0$
$\boldsymbol{f}(\mathbf{x})$ and $\boldsymbol{b}(\mathbf{x})$ are supposed to be known. In the next developments we employ Lemma 1, Barbashin (1970).

Lemma 1

The following equality is true for a non-linear continuous and uniformly bounded vector functions of a class C^k $f(\mathbf{x})$

$$f(\mathbf{x}) = \mathbf{A}(\mathbf{x}).\mathbf{x} \qquad (2)$$

where $\mathbf{A}(\mathbf{x}) \in R^{n \times n}$ is a matrix with non-linear entries given as follows

$$a_{ij}(\mathbf{x}) = \int_0^1 f_{ij}(x_1, \ldots, \theta.x_j, 0, \ldots, 0)d\theta \qquad (3)$$

where $f_{ij}(\mathbf{x}) = \dfrac{\partial f_i}{\partial x_j}$, $i, j \in N = \{1,2,\ldots,n\}, \underline{x} \in R_p$

□

Owing to lemma 1 the uncertain dynamic system (1) can be rewritten in the form

$$\dot{\mathbf{x}} = (A(\mathbf{x}) + \delta\mathbf{A}(\mathbf{x})).\mathbf{x} + (\mathbf{b}(\mathbf{x}) + \delta\mathbf{b}(\mathbf{x})).\mathbf{u} \qquad (4)$$

The corresponding system without uncertainty, referred to as nominal model, is

$$\dot{\mathbf{x}} = \mathbf{A}(\mathbf{x}).\mathbf{x} + \mathbf{b}(\mathbf{x}).\mathbf{u} \qquad (5)$$

where $\mathbf{A}(\mathbf{x})$ and $\mathbf{b}(\mathbf{x})$ are supposed to be known. The unknown matrices $\delta\mathbf{A}(\mathbf{x})$ and $\delta\mathbf{b}(\mathbf{x})$ represent system uncertainties, respectively. Assume that the following assumption holds.

Assumption 1

There exist two symmetric positive definite constant matrices $\mathbf{Q} \in R^{n \times n}$, $\mathbf{R} \in R^{m \times m}$ which are supposed to be known and constants $\gamma_a \geq 0$ and $\gamma_b \geq 0$ such that

$$\begin{aligned} &\delta\mathbf{A}^T(\mathbf{x}).\delta\mathbf{A}(\mathbf{x}) \leq \gamma_a.\mathbf{Q} \\ &\delta\mathbf{b}^T(\mathbf{x}).\delta\mathbf{b}(\mathbf{x}) \leq \gamma_b.\mathbf{R} \quad , \mathbf{x} \in R_p \end{aligned} \qquad (6)$$

□

Assumption 1 provides the upper bounds on uncertainties. Let us note that the uncertainties can be non-linear and fast time-varying and no statistical information about them is assumed.

Let the control algorithm be

$$\mathbf{u} = -\mathbf{k}^T(\mathbf{x}).\mathbf{x} \qquad (7)$$

where $\mathbf{k}(\mathbf{x}) \in R^{n \times m}$. Then the closed-loop uncertain system is

$$\begin{aligned} \dot{\mathbf{x}} &= (\mathbf{A}(\mathbf{x}) - \mathbf{b}(\mathbf{x}).\mathbf{k}^T(\mathbf{x})).\mathbf{x} \\ &+ (\delta\mathbf{A}(\mathbf{x}) - \delta\mathbf{b}(\mathbf{x}).\mathbf{k}^T(\mathbf{x})).\mathbf{x} \end{aligned} \qquad (8)$$

The corresponding nominal closed loop system is

$$\dot{\mathbf{x}} = (\mathbf{A}(\mathbf{x}) - \mathbf{b}(\mathbf{x}).\mathbf{k}^T(\mathbf{x})).\mathbf{x} = \mathbf{A}_C(\mathbf{x}).\mathbf{x} \qquad (9)$$

Before the exact statement of the studied problem let us introduce the following definition, Krasovskij (1959).

Definition 1

The uncertain closed-loop system (8) is said to be robustly exponentially stable with decay rate $\alpha > 0$ if there exist a Lyapunov function $V(\mathbf{x}): R^n \to R^+$, $V(\mathbf{x}) \in C^1$, for the nominal closed-loop system (9) so that for all initial conditions $\mathbf{x}(t_0) = \mathbf{x}_0 \in R_p$ and for all admissible uncertainties given by (6) the following inequality holds along the solution of (8).

$$\frac{dV(\mathbf{x})}{dt} \leq -\alpha.V(\mathbf{x}) \qquad (10)$$

with conditions

$$\begin{aligned} &c_1\|\mathbf{x}\|^2 \leq V(\mathbf{x}) \leq c_2\|\mathbf{x}\|^2 \\ &\dot{V}(\mathbf{x}) \leq -c_3\|\mathbf{x}\|^2 \\ &grad_x V(\mathbf{x}) \leq c_4\|\mathbf{x}\| \end{aligned} \qquad (11)$$

where $(grad_x V(\mathbf{x}))^T = \left[\dfrac{\partial V}{\partial x_1}, \ldots, \dfrac{\partial V}{\partial x_n}\right]$,

$c_i > 0, i = 1,2,\ldots,n$ and $\|.\|$ denotes standard Euclidean norm.

□

Remark 1

A solution of (10) is

$$V(\mathbf{x}) \leq V(\mathbf{x}_0, t_0)e^{-\alpha.t} \qquad (12)$$

Using the inequalities (11)

$$\|\mathbf{x}\| \leq \sqrt{\frac{c_2}{c_1}}\|\mathbf{x}_0\|.e^{-\frac{\alpha}{2}t} \qquad (13)$$

for all $\mathbf{x} \in R_p$. Inequality (13) implies that if the closed-loop system (8) is exponentially stable with decay rate $\alpha > 0$, then $\|\mathbf{x}(t)\|$ exponentially converge with decay rate $\dfrac{\alpha}{2}$ for all $\mathbf{x} \in R_p$.

□

The following problem is studied in this paper. Find the control algorithm (7) so that the closed-loop system (8) with the bounded uncertainties (6) is robustly exponentially stable with a prescribed decay rate α.

3. ROBUST NON-LINEAR CONTROL SYSTEM DESIGN

In this section is used the analytical method with polynomial formalism for the design of the state feedback non-linear controller that guarantees the robust exponential stability of the uncertain closed-loop system (8) with bounded uncertainties (6). The non-linear controller design procedure is divided into two steps. First the $\mathbf{k}(\mathbf{x})$ is determined for the nominal model (9) and decay rate $\alpha_1 \geq \alpha$. The sufficient stability conditions for the closed-loop uncertain system (8) with uncertainties (6) guaranteeing the robust exponential stability with decay rate α are investigated in the second step. In this paper we assume that m=1, that is $\mathbf{k}(\mathbf{x}) \in R^{n \times 1}$.

Consider the nominal closed-loop system (9). The eigenvalues λ of the matrix $\mathbf{A}_C(\mathbf{x})$ are given as the roots of characteristic polynomial

$$\mathbf{A}_C(\lambda, \mathbf{x}) = \det(\lambda \mathbf{I} - \mathbf{A}_C(\mathbf{x})) = 0 \qquad (14)$$

Let the characteristic polynomial (14) be equal to the prescribed polynomial $A_d(\lambda, \mathbf{x})$

$$A_C(\lambda, \mathbf{x}) = A_d(\lambda, \mathbf{x}) \qquad (15)$$

where $A_d(\lambda, \mathbf{x}) = \prod_{i=1}^{n} (\lambda - \lambda_{di}(\mathbf{x}))$ and $\lambda_{di}(\mathbf{x}), i = 1,2,\ldots n$ are negative definite functions for all $\underline{x} \in R_\rho$. When all $\lambda_{di}(\mathbf{x}) \neq \lambda_{dj} \quad i,j = 1,2,\ldots n$ are chosen, the system (15) has solution to entries of $\mathbf{k}(\mathbf{x})$ if and only if, Gaiduk (1995)

$$\det\left[\mathbf{b}(\mathbf{x}), \mathbf{A}(\mathbf{x}).\mathbf{b}(\mathbf{x}), \ldots, \mathbf{A}(\mathbf{x})^{n-1}.\mathbf{b}(\mathbf{x})\right] \neq 0 \quad (16)$$

for all $\mathbf{x} \in R_\rho$. The sufficient stability conditions of (9) under conditions of (15) and (16) are given by the theorem of Gaiduk (1995). In the following it is assumed that the conditions of the above mentioned theorem are satisfied. From (15) it can be seen that by choosing functions $\lambda_{di}(\mathbf{x}), i = 1,2,\ldots,n$ it is possible to achieve quality and desired stability of the non-linear closed-loop uncertain system (8). In the next paragraphs our main results on exponential stability of the nominal as well as uncertain system are provided.

<u>Theorem 1</u>

The dynamic system (9) with $\mathbf{x} \in R_\rho$ is robustly exponentially stable with a decay rate $\alpha_1 \geq \alpha$ for some gain vector $\mathbf{k}(\mathbf{x})$ if the following sufficient conditions hold.

i. $\det\left[\mathbf{b}(\mathbf{x}), \mathbf{A}(\mathbf{x}).\mathbf{b}(\mathbf{x}), \ldots, \mathbf{A}(\mathbf{x})^{n-1}.\mathbf{b}(\mathbf{x})\right] \neq 0$ for $\mathbf{x} \in R_\rho$.

ii. There exist both negative definite functions $\lambda_{di}(\mathbf{x}), i = 1,2,\ldots,n$ for all $\mathbf{x} \in R_\rho$ and a matrix $\mathbf{H}_d(\mathbf{x}) \in R^{n \times n}$ with the characteristic polynomial

$$H_d(\lambda, \mathbf{x}) = \prod_{i=1}^{n} (\lambda - \lambda_{di}(\mathbf{x})) \qquad (17)$$

such that the matrix

$$\mathbf{H}_d^T(\mathbf{x}) + \mathbf{H}_d(\mathbf{x}) \qquad (18)$$

is stable, i.e. all real parts of eigenvalues of this matrix are in the left half plane.

iii. There exist a candidate Lyapunov function $V(\mathbf{x}): R^n \to R^+$ for nominal closed-loop system (9).

iv. The sufficient conditions given by Gaiduk's theorem (1995) are satisfied.

v. The entries of gain vector $\mathbf{k}(\mathbf{x})$ are determined from the following equation

$$H(\lambda, \mathbf{x}) \stackrel{!}{=} H_d(\lambda, \mathbf{x}) \qquad (19)$$

where $H(\lambda, \mathbf{x}) = \det(\lambda \mathbf{I} - \mathbf{H}(\mathbf{x})) = 0$

and

$$\begin{aligned} &\mathbf{H}(\mathbf{x}) = \mathbf{F}(\mathbf{x}) + \alpha_1 \mathbf{W}(\mathbf{x}) - \mathbf{B}(\mathbf{x}).\mathbf{k}^T(\mathbf{x}) \\ &\mathbf{F}(\mathbf{x}) = \mathbf{D}^T(\mathbf{x}).\mathbf{A}(\mathbf{x}) \in R^{n \times n} \\ &\mathbf{B}(\mathbf{x}) = \mathbf{D}^T(\mathbf{x}).\mathbf{b}(\mathbf{x}) \in R^{n \times 1} \\ &V(\mathbf{x}) = \mathbf{x}^T.\mathbf{W}(\mathbf{x}).\mathbf{x} \\ &grad_{\underline{x}} V(\mathbf{x}) = \mathbf{D}(\mathbf{x}).\mathbf{x} \end{aligned} \qquad (20)$$

□

The proof of Theorem 1 is omitted.

Let us now provide the sufficient conditions to guarantee the exponential stability of uncertain closed-loop system (8) with uncertainty bounds (6).

<u>Theorem 2.</u>

The uncertain dynamic system (8) with bounded uncertainties (6) and $\mathbf{x} \in R_\rho$ is exponentially stable with a decay rate α if the following conditions are satisfied

i. Condition of Theorem 1.

ii. There exist positive constants α_1, ε_1, ε_2, such that the matrix $\mathbf{M}(\mathbf{x})$ is negative semidefinite/definite.

$$\begin{aligned} \mathbf{M}(\mathbf{x}) = &-2(\alpha_1 - \alpha).\mathbf{W}(\mathbf{x}) + \varepsilon_1 \mathbf{Q} + \\ &+ \mathbf{D}^T(\mathbf{x}).\mathbf{D}(\mathbf{x}).\left(\frac{\gamma_a}{\varepsilon_1} + \frac{\gamma_b}{\varepsilon_2}\right) \\ &+ \varepsilon_2 \mathbf{k}(\mathbf{x}).\mathbf{R}.\mathbf{k}^T(\mathbf{x}) \end{aligned} \qquad (21)$$

□

The proof of Theorem 2 is omitted.

The design procedure of the non-linear controller is following:

Algorithm A:

a. Choose $\alpha_1 \geq \alpha$

b. Determine the characteristic polynomial

$$F_1(\lambda, \mathbf{x}) = \det(\lambda \mathbf{I} - \mathbf{F}_1(\mathbf{x}))$$
$$\mathbf{F}_1(\mathbf{x}) = \mathbf{F}(\underline{x}) + \alpha_1 \mathbf{W}(\mathbf{x})$$

c. The characteristic polynomial of matrix $\mathbf{H}(\mathbf{x})$ is in the form

$$\mathbf{H}(\lambda, \mathbf{x}) = \mathbf{F}_1(\lambda, \mathbf{x}) + \sum_{i=1}^{n} k_i(\mathbf{x}).v_i(\lambda, \mathbf{x}) \qquad (22)$$

where

$$v_i(\lambda, \mathbf{x}) = c_i.adj(\lambda \mathbf{I} - \mathbf{F}_1(\mathbf{x})).\mathbf{B}(\mathbf{x}) = \sum_{j=0}^{n-1} r_{ij}.\lambda^j$$

and $c_1 = [1,0,\ldots,0]$, $c_2 = [0,1,0,\ldots,0]$, ...

d. Determine the characteristic polynomial $H_d(\lambda, \mathbf{x})$.

e. Let the characteristic polynomial $H(\lambda, \mathbf{x})$ be equal to the prescribed polynomial $H_d(\lambda, \mathbf{x})$

$$H(\lambda, \mathbf{x}) = H_d(\lambda, \mathbf{x}) \text{ and}$$
$$H_d(\lambda, \mathbf{x}) - F_1(\lambda, \mathbf{x}) = \sum_{i=0}^{n-1} e_i(\mathbf{x}).\lambda^i$$

The following algebraic equation is obtained for unknown gain vector $\mathbf{k}(\mathbf{x})$

$$\mathbf{r}.\mathbf{k}(\mathbf{x}) = \mathbf{e}(\mathbf{x}) \qquad (23)$$

where $\mathbf{r} = \{\bar{r}_{i,j}\}_{n \times n}$,

$\mathbf{e}^T(\mathbf{x}) = [e_0(\mathbf{x}), e_1(\mathbf{x}), \ldots, e_{n-1}(\mathbf{x})]$,

and $\bar{r}_{i,j} = r_{i,j+1}$

f. Check the negative semidefinitness or definiteness of matrix $\mathbf{M}(\mathbf{x})$ (21). If there exist positive constants ε_1, ε_1, such that the matrix $\mathbf{M}(\mathbf{x})$ is negative semidefinite/definite then stop, else increase α_1 and go to step d. If there are some causes for not increasing α_1, the constants γ_a, γ_b could be decreased, then go to step a.

□

4. DESIGN OF THE NON-LINEAR CONTROLLER FOR THE TURBOGENERATOR.

The system under consideration will be one machine connected through a transmission line to an infinite bus.

The mathematical model of a stable machine power system is given by (24), Stroev et all (1995). Exciter and the other dynamics, which are not included in the power system model given below, will be included in the uncertainty of this power system.

$$\begin{aligned} \dot{x}_1 &= x_2 \\ \dot{x}_2 &= \frac{1}{T_j}(x_3 - Dx_2 - K_S(\sin(x_1 + \delta_0) - \sin\delta_0)) \\ \dot{x}_3 &= x_4 \\ \dot{x}_4 &= \frac{1}{a_1}\left(u - \frac{1}{\sigma}x_2 - x_3 - a_2 x_4\right) \end{aligned} \qquad (24)$$

Applying the Barbashin theorem to system (24) the model may be expressed in the form

$$\mathbf{A}(\mathbf{x}) = \begin{bmatrix} 0 & 1 & 0 & 0 \\ -\frac{P(\mathbf{x})}{T_j} & -\frac{D}{T_j} & \frac{1}{T_j} & 0 \\ 0 & 0 & 0 & 1 \\ 0 & -\frac{1}{\sigma.a_1} & -\frac{1}{a_1} & -\frac{a_2}{a_1} \end{bmatrix} \qquad (25)$$

$$\mathbf{b}^T(\mathbf{x}) = \begin{bmatrix} 0 & 0 & 0 & \frac{1}{a_1} \end{bmatrix}$$

where $P(\mathbf{x}) = -K_S \dfrac{\sin(\delta_0 + x_1) - \sin\delta_0}{x_1}$.

The system uncertainties are included in the matrices

$$\delta\mathbf{A}(\mathbf{x}) = \begin{bmatrix} 0 & 0 & 0 & 0 \\ \delta a_{21} & \delta a_{22} & \delta a_{23} & 0 \\ 0 & 0 & 0 & 0 \\ 0 & \delta a_{42} & \delta a_{43} & \delta a_{44} \end{bmatrix}$$

$$\delta\mathbf{b}^T(\mathbf{x}) = [0 \quad 0 \quad 0 \quad \delta b_4] \qquad (26)$$

For symbols notation see Stroev et all (1995)

Conditions of simulation

1. Simulations were carried out using the SIMNON simulation package with nominal parameters D=0.01307, K_S=1.029, T_j=0.01258, a_1=2.5, a_2=1, δ_0=0.76713,
Let σ=1000 and therefore does not affect the results of simulation.
Thus the σ parameter is not considered in other calculations.
2. Parameters K_S and D, which bring uncertainties into the system, can varied by ±20% from their nominal values.
3. Coefficients of prescribed characteristic polynomial $\mathbf{H}_d(\lambda, \mathbf{x}) = \lambda^4 + h_3\lambda^3 + h_2\lambda^2 + h_1\lambda + h_0$ are 20, 155, 510, 630, respectively. Parameter α is 3.
4. Initial state is $\mathbf{x}_0^T = [1 \quad 0 \quad 0 \quad 0]$.
5. Control input u in system (24) is not bounded.

6. System states and parameters are normalised and showed in unified units.
7. Figures of transient behaviour of the system show Euclidean norm of the state vector $\|\mathbf{x}\|$.

For checking control quality and theory proof there will be found bounding exponential function for norm of the state vector. It can be expressed in the form

$$env(\|\mathbf{x}\|) = c(\|\mathbf{x}_0\|).e^{-\beta.t} \tag{27}$$

where c is a constant dependent on the norm of the initial state and β is a decay rate. This is only a first approximation by the envelope of state vector norm, so it is possible to find some deviations from the exact bounding function.

System without control

For quality comparison of the designed regulators in Figure 1 are shown the transients of the system (24) without control. The dotted line is an approximation of the bounding exponential function, which determines self-exponential stability decay rate β_0 of the system. In this case β_0=0.5288

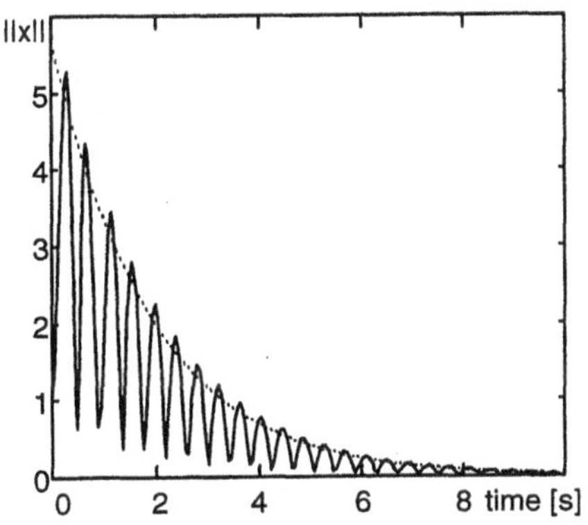

Fig. 1. Response of non-controlled system.

5. ROBUST CONTROLLER DESIGN - 1ST PHASE

In this phase the design of controller is accomplished, which satisfies exponential stability of the controlled system with prescribed decay rate α. The system here is taken without uncertainties, which thought will be considered in the 2nd phase of controller design.
Let the structure of controller be in the form

$$\mathbf{u} = -\mathbf{k}^T(\mathbf{x}).\mathbf{x}\text{, where } \mathbf{u} \in R^1 .$$

For the experiment requirement a Lyapunov function is needed, which is chosen in the form

$$\begin{aligned} V(\mathbf{x}) &= \mathbf{x}^T.\mathbf{W}(\mathbf{x}).\mathbf{x} \quad \mathbf{W}(\mathbf{x}) = \frac{1}{2}\mathbf{I}_4 \\ \dot{V}(\mathbf{x}) &= \mathbf{D}(\mathbf{x}).\dot{\mathbf{x}} \quad \mathbf{D}(\mathbf{x}) = \mathbf{I}_4 \end{aligned} \tag{28}$$

where $\mathbf{I}_4$ is the identity matrix of 4th order.

The Matrix $\mathbf{H}(\mathbf{x}) = \mathbf{D}^T\mathbf{A}(\mathbf{x}) - \mathbf{D}^T\mathbf{b}(\mathbf{x})\mathbf{k}^T(\mathbf{x}) + \alpha\mathbf{W}(\mathbf{x})$ of the system under study is after substitution

$\mathbf{H}(\mathbf{x}) =$

$$\begin{bmatrix} \frac{1}{2}\alpha & 1 & 0 & 0 \\ -\frac{P(\mathbf{x})}{T_j} & -\frac{D}{T_j}+\frac{1}{2}\alpha & \frac{1}{T_j} & 0 \\ 0 & 0 & \frac{1}{2}\alpha & 0 \\ -\frac{k_1(\mathbf{x})}{a_1} & -\frac{k_2(\mathbf{x})}{a_1} & -\frac{1+k_3(\mathbf{x})}{a_1} & -\frac{1+k_4(\mathbf{x})}{a_1}+\frac{1}{2}\alpha \end{bmatrix} \tag{29}$$

After comparing the characteristic polynomials of the matrices $\mathbf{H}(\mathbf{x})$ and $\mathbf{H}_d(\mathbf{x})$ one obtains a set of equations. They can be solved with respect to the entries of the vector $\mathbf{k}(\mathbf{x})$, which gives the equations for controller parameters $\mathbf{k}(\mathbf{x})$.

Results of simulation

Transients of the controlled system are shown in Figure 2. The system response with maximum considered uncertainties is in Figure 3.

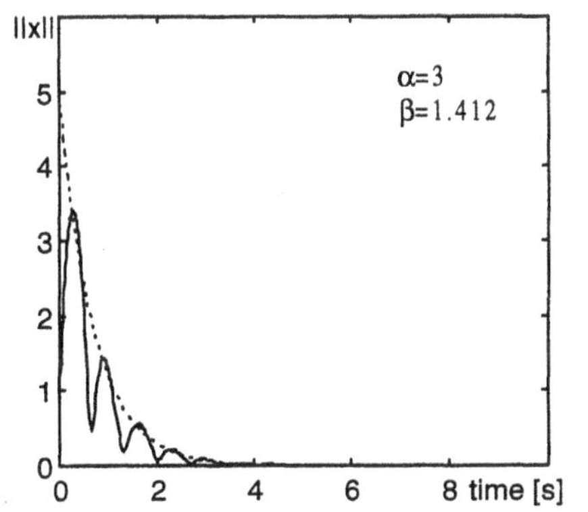

Fig. 2. Response of system with prescribed α.

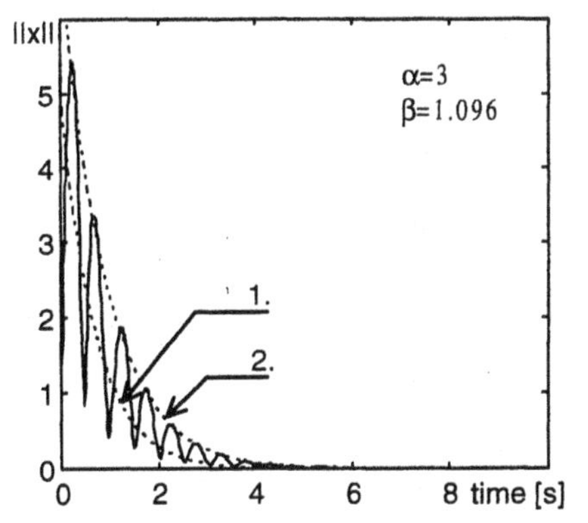

Fig. 3. The response of controlled system with uncertainties.

The control process quality of the uncertain system is worse as of the system without uncertainties. The two dotted lines in Figure 3 shows this comparison in the Graph. The first represents β of the system without uncertainties, whereas the second one represents β of the system with maximum considered uncertainties.

6. ROBUST CONTROLLER DESIGN - 2ND PHASE

There is continued in design of robustly exponential stable controller here. Additionally the system uncertainties, (26) must be considered.

For the prescribed system parameters we get the upper bound of uncertainties

$$\mathbf{Q} = \begin{bmatrix} 0.3414 & 0.4742 & 0 & 0 \\ 0.4742 & 0.6586 & 0 & 0 \\ 0 & 0 & 0 & 0 \\ 0 & 0 & 0 & 0 \end{bmatrix}, \ \gamma_a=383.7479$$

$$\mathbf{R=0},\ \gamma_b=0 \qquad (30)$$

As follows from algorithm A, for given α we have to choose $\alpha_1>\alpha$ so, that there exist positive ε_1, ε_2 such that the matrix $\mathbf{M(x)}$ from (21) is negative semidefinite.

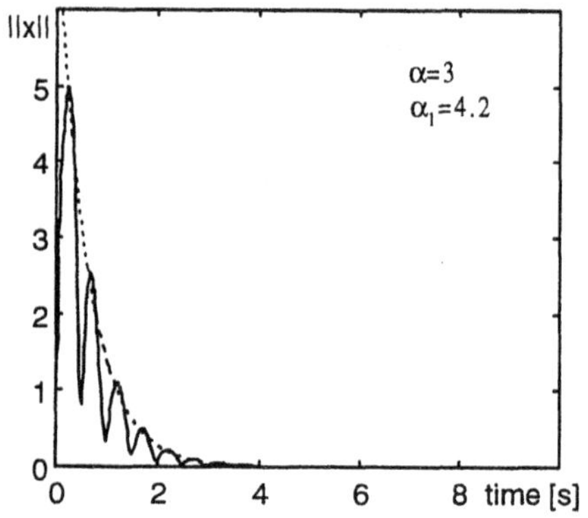

Fig. 4. System response with the robust controller.

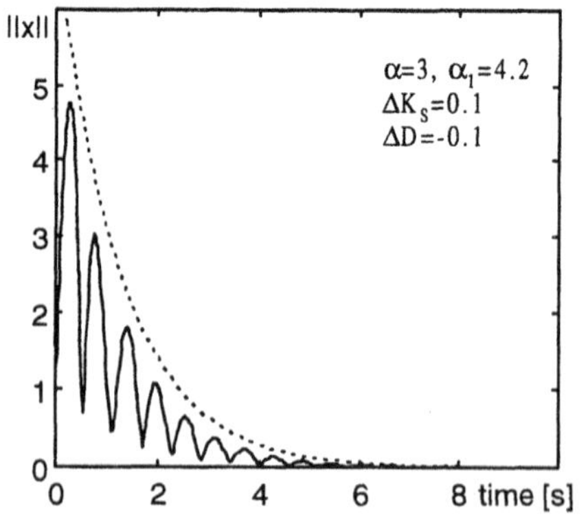

Fig. 5. System response for small uncertainties.

After several trials we get α_1=4.2 for α=3. The results of the simulation with the applied uncertainties are shown in Figure 4.

The design of the robust exponentially stable controller guarantees desired decay rate for the worst case of the parameter changes. This property can be seen in Figure 5 where are shown the results of the system simulations when the parameter changes are inside the bounds.

For purposes of comparison the dotted exponential functions in Figure 5 show bounding function of the system with maximum uncertainties.

7. CONCLUSION

In this paper is demonstrated the design procedure of robust exponentially stable controllers with prescribed decay rate for non-linear systems.

In the first phase the non-robust controller, the properties of which are verified on wide range of prescribed parameters, is designed. This can be seen in Figures 2, 3.

In the second phase the design of the robust controller is intended to find a control which guarantees prescribed properties with uncertainties being mentioned.

The main problem of this phase was to choose proper α_1, ε_1, ε_2 to guarantee negative semidefinitness of the matrix $\mathbf{M(x)}$ (21). We used the graphical method, which is based on the graph construction of dependence between prescribed α and β obtained from simulation. (Not showed in this paper.)

8. REFERENCES

[1] Barbashin E. A. (1970) *Lyapunov Functions.* Nauka, Moscow (in Russian)

[2] Dubrov A. M., Zuber I. E. (1989): Design of Exponentially Stable Control Systems of Nonlinear Processes. In: *Automation and Remote Control.* **No. 8**, 33-40.

[3] Gaiduk A. R. (1995): Analytical Non-linear Control Systems Design. In: *Proc. of 3rd EC Conference - Rome*, 1503-1505.

[4] Krasovskij N. N. (1959) *Some Tasks from Theory and Stability Motion.* Fizmatgiz Moscow.

[5] Kozak S. (1995). Simple and Robust PID Controller. In: *Applied Mathematics and Computation.*, **No. 70**, 2-8.

[6] Mudroncik D., Vesely V. (1994). Power System Stabilization using the Dominant Subsystem Principle. In: *IEE Proc. on Generation, Transmission and Distribution.* **Vol. 141 No. 3**, 243-247.

[7] Papadopoulos D. P. (1986): Excitations Control of Turbogenerators with Output Feedback. In: *Electrical Power and Energy Systems*, 176-180.

[8] Stroev V. A., Sharov Y. V., Asembeer V. S. (1995). Representation of Control Devices in the Calculation of Transients in Power Systems. In: *Stockholm Power Tech*, **June 18-22**, 392-397.

AN APPROACH TO FAULT DIAGNOSIS OF METALLURGICAL PROCESS

El Hassan MAZZOUR*, Nicolas HERAUD**

* *Laboratoire d'Automatique et de Procédés - ISMRA, 6 Bd Maréchal Juin - 14050 Caen Cedex*
Tél. : 02 31 45 27 11 Fax : 02 31 45 27 60 e-mail : mazzour@greyc.ismra.fr

** *CMCS - Université de Corse - CNRS URA 2053, BP 52 - Quartier Grossetti - 20250 Corte*
Tél. : 04 95 45 00 49 Fax 04 95 61 05 51 e-mail : heraud@univ-corse.fr

Abstract : Acquisition of coherent and representative data about the state of processes remains a major problem of process control. This paper proposes an approach which ensure coherence in measurements on an operational industrial process. The basic technique employed relies on balance equilibration of resource mass flow. An interesting particularity of the set up is the ease with which it can be adapted to check productivity or detect sensor failures. In the case of sensor failure, it suffices to examine the evaluations before and after balance equilibration. The algorithm was tested by means of computer simulation using data from an industrial metallurgical flow-sheet.

Keywords : data reconciliation, measurement error, diagnosis.

1. INTRODUCTION

Accurate and reliable material and energy accounting in physical and chemical processes is of great importance for process monitoring as well as for decision-making. Generally measured flows in processes are corrupted by random errors but also by systematic errors resulting from instrument failures and biases, the latter resulting from miscalibrations and drifts. Even small random errors may be unacceptable in the case of cumulative effect which can be quite significant. As a general result of these errors, the process data set will be inconsistent in regard to mass and energy conservation laws.

Sensor and fault detection have received considerable attention this last decade. If a process model is exactly known the fault detection problem may be solved by a filter generating a residual from calculated and measured outputs. This problem is known as FDI (Fault Detection and Isolation) and for details and comprehensive surveys we point out to the articles of Frank [FRANK 1990] and Iserman [ISERMAN 1984].

A wide variety of failure detection approaches have been developed ; these include simple limit-checking, pattern recognition physical redundancy of sensors, redundancy due to the relationships between variables (or functional redundancy), artificial intelligence techniques. The most important approach is based on the redundancy concept and principally uses analytical redundancy. This approach uses intensively the explicit models of the process ; the raw data collected through measurement devices are tested accordingly with the prior information embodied in the model. Let us however notice that this strategy supposes that the model must be exact. For this reason, it is a common practice to separate the models into an exact structural one (such as mass or energy balance equations) and the empirical models obtained by mathematical and statistical approach ; in the last case the sensitivity to modelling errors is the key problem in the application of analytical redundancy.

A traditional scheme for data filtering is to provide a reconciliation of all the measurement data ; the so-called balance equilibration procedure has been well investigated by many authors including Vaclavek [VACLAVEK 1976], Almasy [ALMASY 1984], Mah [MAH 1987]. A recent survey of these techniques can be found in Ragot [RAGOT 1990]. Data reconciliation techniques using redundancies in resource mass flow equations are now well known.

Reviews of data reconciliation applied to chemical processes have been published in [HOLLY 1989] and [HODOUIN 1989]. Algorithms of detection of gross errors in process constraints have been developed in [ROMAGNOLI 1980] and [CROWE 1989]. Nevertheless, their implementation on production sites poses a number of major problems. These often arise from :

- the overall complexity of processes owing to the large number of variables involved,
- the difficulty or impossibility of acquiring data at certain points of processes,
- the presence of biased data,
- the non linearity of process equations.

A number of techniques can be employed to overcome these difficulties. Among these :

- the use of hierarchical computation: applicable to processes involving many variables or with non linear equations. In the non linear case, a full range of the techniques of numerical computation are resorted to : uncoupling or linearisation of governing equations ; relaxation ; direct iteration and so on,
- the use of process decomposition based on the criterion of observability in cases where measurements cannot be obtained,
- the detection and the identifying of measurement errors with wide amplitude in systematically unbalanced processes.

In the ensuing development, we shall give a concrete example to illustrate most of these problems. In particular, we shall consider total or partial mass flow evaluations governed by trilinear equations, as well as evaluations involving unavailable data measurements.

The proposed example is about a section of acid uranium pulp filtration of the uranium processing plant at Jouac, France, owned by the French company: Total Compagnie Minière France (T.C.M.F.). On this section, we have measurements for volumic flow, density, and uranium contents in both the solid and liquid phases of flow. The results are then examined station by station and then in their chronological order.

2. METHOD FOR MULTILINEAR BALANCE EQUILIBRATION

The principle of data reconciliation consists of estimating the true values which verify the flow equations and are as close as possible to the measurements. This problem comes to minimizing a weighted least squares problem (2.3) with the respect of the contraints (2.4 to 2.8).

The following preliminary hypotheses are assumed :

a) the model is exact : the conservation principle of matter or energy holds ; the different units cannot accumulate matter : should accumulation occur, the overall internal stock is assumed to be constant,

b) measurement error distribution is gaussian with a zero mean and a known variance (see equ. 2.1).

Now, let us consider a network with n nodes and v streams. Four types of measurement are available from each stream. These include : the volume and mass flow rates as well as liquid and solid phase contents. The appropriate balance equations can be written from these measurements. In general, measurements will not be available for all variables. We can write the measurement equations :

$$X = \mathbf{H}\, X^{*} + \varepsilon \qquad (2.1)$$

The correct values, i.e. those that satisfy the material balance equations have to be estimated from these measurements. This can be done by maximizing the likelihood function of the measurement X ; in the case of one observation this function is given by :

$$V = (2^{\prime})^{\frac{-n}{2}} (\mathrm{Det}(\mathbf{V}))^{\frac{-n}{2}} \exp - \frac{1}{2} \| X - \mathbf{H}\, X^{*} \|^{2}_{\mathbf{V}^{-1}} \qquad (2.2)$$

where n represents the dimension of the vector X.

The best estimator $\hat{X}$ of the true value X^* is the one which maximizes the likelihood function V with the respect of the contraints 2.4 to 2.8. Since the matrix **V** is known and diagonal, the problem of estimation is reduced to find the minimum with respect to $\hat{X}$ of :

$$\phi = \frac{1}{2} \| \mathbf{H} \hat{X} - X \|^{2}_{\mathbf{V}^{-1}} \qquad (2.3)$$

subject to :

the volume flow rate constraints :

$$\mathbf{M_Q} \hat{Q} = 0 \qquad (2.4)$$

the mass flow rate constraints :

$$\mathbf{M_D} (\hat{Q} \;.\!*\; (\mathbf{S1}^T \hat{\rho})) = 0 \qquad (2.5)$$

the solid phase mass flow constraints :

$$(\mathbf{M_S} \hat{Q}) \;.\!*\; \{(\mathbf{S2}^T \rho_S) \;.\!*\; (\mathbf{S1}^T \hat{\rho} - \mathbf{S2}^T \rho_L) \;./\; (\mathbf{S2}^T(\rho_S - \rho_L))\} = 0 \qquad (2.6)$$

the corresponding liquid phase volume constraints :

$$(\mathbf{M_S} \hat{Q}) \;.\!*\; \{(\mathbf{S2}^T \rho_S - \mathbf{S1}^T \hat{\rho}) \;./\; (\mathbf{S2}^T (\rho_S - \rho_L))\} + \mathbf{M_L} \hat{Q} = 0 \qquad (2.7)$$

and finally, the constraints for the uranium mass flow :

$$(\mathbf{M_{TLTS}} \hat{Q}) \;.\!*\; \{(\mathbf{S2}^T \rho_S) \;.\!*\; (\mathbf{S1}^T \hat{\rho} - \mathbf{S2}^T \rho_L) \;./\; (\mathbf{S2}^T (\rho_S - \rho_L)) \;.\!*\; (\mathbf{S2}^T \hat{T}_S) + (\mathbf{S2}^T \rho_S - \mathbf{S1}^T \hat{\rho}) \;./\; (\mathbf{S2}^T (\rho_S - \rho_L)) \;.\!*\; (\mathbf{S4}^T \hat{T}_L)\} + (\mathbf{M_{TL}} \hat{Q}) \;.\!*\; (\mathbf{S4}^T \hat{T}_L) = 0 \qquad (2.8)$$

where the vector operations .* and ./ denote respectively componentwise multiplication and division of two vectors of the same dimension.

Equations (2.4) through (2.8) may be written as :

$$\mathbf{F}(\hat{X}) = 0 \qquad (2.9)$$

We have a non linear constrained least squares problem (see equ. 2.4 and 2.8). This can be solved by linearizing the constraint equations and using an iterative estimation procedure. At step i of the iteration, the linearized constraints may be put in the form :

$$A_{i-1}\hat{X}_i = b_{i-1} \qquad (2.10)$$

where on the one hand the matrix A_{i-1} is explicitly given by :

$$A_{i-1} = \begin{bmatrix} A11_{i-1} & 0 & 0 & 0 \\ A21_{i-1} & A22_{i-1} & 0 & 0 \\ A31_{i-1} & A32_{i-1} & A33_{i-1} & A34_{i-1} \\ A41_{i-1} & A42_{i-1} & 0 & 0 \\ A51_{i-1} & A52_{i-1} & 0 & 0 \end{bmatrix} \qquad 2.11)$$

where the $A_{kl_{i-1}}$ are given by :

$$A11_{i-1} = M_Q \qquad (2.12)$$

$$A21_{i-1} = M_D \otimes (S1^T \hat{\rho}_{i-1}) \qquad (2.13)$$

for the operation $\otimes$, a precise definition is given below :

Let us consider a matrix **A** and vector B of respective dimensions p.q and q.1, A_j is the jth row of A. The jth row of product $\mathbf{A} \otimes \mathrm{B}$ is defined by: $(\mathbf{A} \otimes \mathrm{B})_j = [A_j{}^T .* B]^T$. The result of $\mathbf{A} \otimes \mathrm{B}$ is a p.q matrix.

$$A31_{i-1} = M_{TL} \otimes (S4^T \hat{T}_{L_{i-1}}) + M_{TL} \otimes \{(S1^T \hat{\rho}_{i-1}) .* (S2^T (\rho_S .* \hat{T}_{S_{i-1}} - S3^T \hat{T}_{L_{i-1}}) ./ (\rho_S - \rho_L)) + S2^T ((\rho_S ./ (\rho_S - \rho_L)) .* (S3^T \hat{T}_{L_{i-1}} - \rho_L .* \hat{T}_{S_{i-1}}))\} \qquad (2.14)$$

$$A41_{i-1} = M_S \otimes \{(S2^T \rho_S - S1^T \hat{\rho}_{i-1}) ./ (S2^T(\rho_S - \rho_L))\} + M_L \qquad (2.15)$$

$$A51_{i-1} = -M_S \otimes \{((S1^T \hat{\rho}_{i-1}) .* (S2^T \rho_S) - S2^T (\rho_S .* \rho_L)) ./ (S2^T(\rho_S - \rho_L))\} \qquad (2.16)$$

$$A22_{i-1} = \{S1\,(M_D \otimes \hat{Q}_{i-1})^T\}^T \qquad (2.17)$$

$$A32_{i-1} = \{S1\,(M_{TLTS} \otimes \{\hat{Q}_{i-1} .* (S2^T \{(\rho_S .* \hat{T}_{S_{i-1}} - S3\,\hat{T}_{L_{i-1}}) ./ (\rho_S - \rho_L)\})\})^T\}^T \qquad (2.18)$$

$$A42_{i-1} = -\{S1\,(M_S \otimes \{\hat{Q}_{i-1} .* (S2^T \{\rho_S ./ (\rho_S .* (\rho_S - \rho_L))\})\})^T\}^T \qquad (2.19)$$

$$A52_{i-1} = M_S \otimes \{\hat{Q}_{i-1} .* (S2^T \{\rho_S ./ (\rho_S - \rho_L)\})\} \qquad (2.20)$$

$$A33_{i-1} = \{S4\,(M_{TLTS} \otimes \{\hat{Q}_{i-1} .* (S2^T \{(-S2 \{S1^T \hat{\rho}_{i-1} + \rho_S\}) ./ (\rho_S - \rho_L)\})\})^T\}^T \qquad (2.21)$$

$$A34_{i-1} = \{S2\,(M_{TLTS} \otimes \{\hat{Q}_{i-1} .* (S2^T \{\rho_S ./ (\rho_S - \rho_L) .* (S2 \{S1^T \hat{\rho}_{i-1} + \rho_L\})\})\})^T\}^T \qquad (2.22)$$

and the b_{i-1} on the other hand is defined by :

$$b_{i-1} = \begin{bmatrix} 0 \\ b1_{i-1} \\ b2_{i-1} \\ b3_{i-1} \\ b4_{i-1} \end{bmatrix} \qquad (2.23)$$

where the $b_{k_{i-1}}$ are also given by :

$$b1_{i-1} = M_D (\hat{Q}_{i-1} .* (S1^T \hat{\rho}_{i-1})) \qquad (2.24)$$

$$b2_{i-1} = M_{TL} \{\hat{Q}_{i-1} .* (S4^T \hat{T}_{L_{i-1}})\} + M_{TLTS}\{\hat{Q}_{i-1} .* (S2^T \{\rho_S ./ (\rho_S - \rho_L) .* (2S2 (S1^T \hat{\rho}_{i-1} - \rho_L)).* (S3\,\hat{T}_{L_{i-1}})\})\} \qquad (2.25)$$

$$b3_{i-1} = -M_S \{\hat{Q}_{i-1} .* (S1^T \hat{\rho}_{i-1}) .* (S2^T (\rho_S ./ \{\rho_S .* (\rho_S - \rho_L)\}))\} \qquad (2.26)$$

$$b4_{i-1} = M_S \{\hat{Q}_{i-1} .* (S1^T \hat{\rho}_{i-1}) .* (S2^T \{\rho_L ./ (\rho_S - \rho_L)\})\} \qquad (2.27)$$

Thus, at step i we have to minimize :

$$\phi = \frac{1}{2} \| H\hat{X}_i - X \|^2_{V^{-1}} \qquad (2.28)$$

with respect to $\hat{X}_i$, subject to :

$$A_{i-1}\hat{X}_{i-1} = b_{i-1} \qquad (2.29)$$

To solve the above problem, we can employ the penalization functions to ensure numerical stability. For this, we inject a weighted form of the constraints into the criterion to get :

$$\tilde{\phi} = \frac{1}{2} \| H\hat{X}_i - X \|^2_{V^{-1}} + k^2 \| A_{i-1}\hat{X}_i - b_{i-1} \|^2 \qquad (2.30)$$

where k is a sufficiently large scalar constant chosen to respect sufficiently the constraints. They are respected when k is infinite. In practice, we have a good approximation with k = 100.

The minimum of the new criterion $\tilde{\phi}$ is given by :

$$\frac{\partial\tilde{\phi}}{\partial\hat{X}_i} = H^T V^{-1} (H\hat{X}_i - X) + k^2 A_{i-1}^T (A_{i-1}\hat{X}_i - b_{i-1}) = 0 \qquad (2.31)$$

For a globally observable network, Darouach in [DAROUACH 1986] established the existence of the matrix :

$$G_{i-1} = (\mathbf{H}^T \mathbf{V}^{-1} \mathbf{H} + k^2 \mathbf{A}_{i-1}^T \mathbf{A}_{i-1})^{-1} \qquad (2.32)$$

We deduce from (2.30) and (2.31) :

$$\hat{X}_i = \mathbf{G}_{i-1} (\mathbf{H}^T \mathbf{V}^{-1} X + k^2 \mathbf{A}_{i-1}^T b_{i-1}) \quad (2.33)$$

The solution (2.33) is iterative. The estimate $\hat{X}$ of the true value is obtained when we have the convergence of the criterion. It can be proved that for k infinite, the solution (2.33) corresponds to the solution directly obtained by solving problem (2.3) under the constraints (2.4) to (2.8) with the Lagrange method.

The calculus of the estimate is valid only in the case of random errors. The presence of gross errors or systematic errors in the raw data totally invalidate the least-square procedure described above. Then, it is a necessity to detect, isolate and identify such errors in order to remove them from the measurements. Gross error detection may be achieved a priori before data reconciliation or a posteriori after data reconciliation. In the first case, detection is operated on the equation imbalance residuals and in the second case the residuals from the least-square procedure are tested. In both cases, detection involves the use of statistical tests based on the assumption that the random errors in the data are normally distributed ; tests on the generated residuals have been developped to detect outliers.

In the following we express the statistical properties of the estimate ; in particular, the variance of the estimated values is established in order to construct the normalized adjustments on which outlier analysis will be performed.

Statistical properties of $\hat{X}$

The linearized constraint (2.29) near the point X^* is written :

$$\mathbf{A}^* X^* = b^* \qquad (2.34)$$

where $\mathbf{A}^*$ and b^* are functions of X^*.

Using the above developments, we obtain :

$$\hat{X} = \mathbf{G}^* (\mathbf{H}^T \mathbf{V}^{-1} X + k^2 \mathbf{A}^{*T} b^*) \qquad (2.35)$$

where $\mathbf{G}^* = (\mathbf{H}^T \mathbf{V}^{-1} \mathbf{H} + k^2 \mathbf{A}^{*T} \mathbf{A}^*)^{-1}$ (2.36)

and we deduce:

$$\text{Exp}(\hat{X}) = \mathbf{G}^* (\mathbf{H}^T \mathbf{V}^{-1} \mathbf{H} + k^2 \mathbf{A}^{*T} \mathbf{A}^*) X^* = X^* \qquad (2.37)$$

Then the estimator $\hat{X}$ is unbiased.

Let us turn now to the calculation of the variance-covariance matrix of $\hat{X}$. By definition :

$$\hat{\mathbf{V}} = \text{Exp}\left[(\hat{X} - \text{Exp}(\hat{X}))(\hat{X} - \text{Exp}(\hat{X}))^T\right] \quad (2.38)$$

and we deduce :

$$\hat{\mathbf{V}} = \mathbf{G}^* \mathbf{H}^T \mathbf{V}^{-1} \mathbf{H} \mathbf{G}^{*T} \qquad (2.39)$$

We do not know the true value X^* but it is near the estimation $\hat{X}$. Thus we approach the true value X^* by the estimation $\hat{X}$ and we obtain :

$$\hat{\mathbf{V}} = \mathbf{G} \mathbf{H}^T \mathbf{V}^{-1} \mathbf{H} \mathbf{G}^T \qquad (2.40)$$

3. APPLICATION TO URANIUM PULP FILTRATION SECTION

3.1. Description of the section

The network in figure 1 represents one of the most complex part of the acid uranium pulp filtration process. It comprises 22 streams and 6 nodes. Each node represents a specific processing unit. The input stream to the network, stream 1, injects a two-phase acid pulp while the output stream represented by stream 12 delivers a liquid uranium solution, with stream 11 evacuating a two-phase waste product.

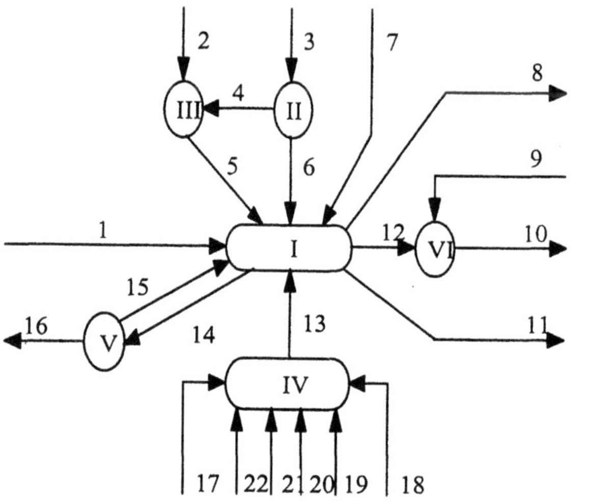

node I	:	acid pulp filter
node II	:	water input
node III	:	preparation of flocculating agent
node IV	:	preparation of acidified cleaning water
node V	:	recycling
node VI	:	tank

figure 1. network for pulp filtration section

3.2. Balance equilibration

The balance equilibration for the filtration section is summarized in table 1. The first column indicates the type of measurement and the stream. The second column registers the measurements where available, the third the estimated values while the fourth and the fifth give the corresponding standard deviations. The last column gives the adjustment ratio. For each measurement, it is computed according to the formula:

$$ADJ = (\mathbf{H}\hat{X} - X) ./ X \qquad (3.1)$$

STREAM	MEASURE-ADJUSTMENT MENT	ESTIMATE	STANDARD DEVIATION	ESTIMATED STANDARD DEVIATION	RATIO %
VOLUME FLOW RATE m^3/unit time					
1	588.000	636.538	47.000	38.393	8.255
2	0.074	0.074	0.010	0.010	-0.000
3		266.207		47.528	
4	33.200	33.054	7.000	6.997	-0.439
5		33.128		6.997	
6	240.000	233.153	48.000	47.051	-2.853
7	504.000	474.280	100.000	91.112	-5.897
8	228.700	225.841	18.000	17.881	-1.250
9	6.000	5.999	0.500	0.500	-0.012
10	993.000	926.959	80.000	66.141	-6.651
11		277.756		63.371	
12		920.960		66.143	
13	588.000	581.465	47.000	46.111	-1.111
14	504.000	617.190	200.000	114.666	22.458
15	84.000	83.182	17.000	16.959	-0.974
16		534.008		114.227	
17		90.975		67.071	
18	178.000	177.999	14.000	14.000	-0.001
19	0.021	0.021	0.010	0.010	-0.000
20	192.000	192.000	40.000	40.000	0.000
21	1.470	1.470	0.600	0.600	-0.000
22	120.000	120.000	24.000	24.000	0.000
DENSITY Kg/m^3					
1	1.406	1.371	0.060	0.055	-2.475
2	1.000	1.000	0.002	0.002	-0.000
3	1.000	1.000	0.002	0.001	0.000
4	1.000	1.000	0.002	0.001	0.000
5	1.000	1.000	0.002	0.001	-0.000
6	1.000	1.000	0.002	0.002	0.000
7	1.000	1.000	0.002	0.002	-0.000
8	1.030	1.030	0.040	0.040	0.000
11	1.727	1.722	0.070	0.070	-0.297
12	1.030	1.030	0.040	0.040	0.000
13		1.000		0.068	
14		1.019		0.019	
15	1.140	1.140	0.050	0.050	0.000
16	1.000	1.000	0.020	0.020	0.000
17	1.000	1.000	0.002	0.002	-0.000
18		1.000		0.224	
19	1.000	1.000	0.020	0.020	-0.000
20	1.000	1.000	0.020	0.020	-0.000
21	1.800	1.800	0.004	0.004	-0.000
22	1.000	1.000	0.040	0.040	0.000
LIQUID PHASE URANIUM CONTENT g/l					
1	4.260	4.397	0.200	0.186	3.222
8	2.000	1.757	0.400	0.378	-12.128
9	0.020	0.020	0.001	0.001	0.000
10	1.980	1.980	0.004	0.004	-0.005
11	0.037	0.037	0.001	0.000	-0.003
12		1.993		0.001	
13		0.002		0.000	
16	0.021	0.021	0.002	0.000	-0.068
17	0.002	0.002	0.000	0.000	0.000
18	0.001	0.001	0.000	0.000	0.001
20	0.002	0.002	0.000	0.000	0.001
22	0.002	0.002	0.000	0.000	0.000
SOLID PHASE URANIUM CONTENT ppm					

Table 1 continued

1	0.062	0.062	0.004	0.004	0.056
18	0.062	0.062	0.002	0.002	-0.014

table 1. results of balance equilibration

Starting with a set widely heterogenous measurements comprising :

- 16 volume flow rates,
- 17 stream densities,
- 10 liquid phase uranium content measurements,
- 2 solid phase uranium content measurements,

we were able to compute the estimates for :

- 22 volume flow rates,
- 20 stream densities,
- 12 liquid phase uranium content measurements,
- 2 solid phase uranium content measurements.

From these, we deduced the values for the different flow across the section using balance equilibration :

- 1960 m^3/unit time for the volume flow,
- 2197 T/unit time for the mass flow,
- 329 T/unit time for the solid flow,
- 1833 m^3/unit time for the liquid flow,
- 2259 Kg/unit time for the uranium flow.

These results were calculated using $\rho_S(1) = 2.6$, $\rho_L(1) = 1.07$, $\rho_S(18) = 2.6$ et $\rho_L(18) = 1$ as solid and liquid component densities of streams 1 and 18.

The adjustment ratios provide a preliminary basis for appraisal : measurements with larger adjustment ratios are those that deviate the most from their true values than those with smaller ratios. Thus the parameters Q(1), Q(7), Q(10), Q(14), ρ(1), TL(1) and TL(8) need much greater adjustments to reconcile them. To obtain a finer appraisal of these results requires a chronological analysis of the balance equilibration.

3.3. Chronology of balance equilibration

Now we want to detect if there are eventual perturbations on the measurements and their evolutions. For this, an auxiliary randomly distributed parameter is computed : the normalized adjustment. It is normally distributed with a zero mean and a unit variance. A given stream is suspect if its normalized adjustment lies outside [-2, 2].

The normalized adjustment for each measurement is computed according to the formula :

$$T_{CN} = (\mathbf{H}\hat{X} - X) ./ (\text{Diag}(\mathbf{V}) - \text{Diag}(\mathbf{H}\hat{\mathbf{V}})).^{**}(1/2) \quad (3.2)$$

where .**(1/2) corresponds to the square root of each component.

Figures 2, 3 and 4 present side by side the evolution of the volume flow rates and the associated normalized adjustments for the principal streams of the network, namely streams 1, 8 and 10. For the volume flow rate graphs, the continous line type represent the measurements while the estimated values are symbolized by the cross, just as on the graph for the normalized adjustments. The continous line types of the latter represent the critical threshold values.

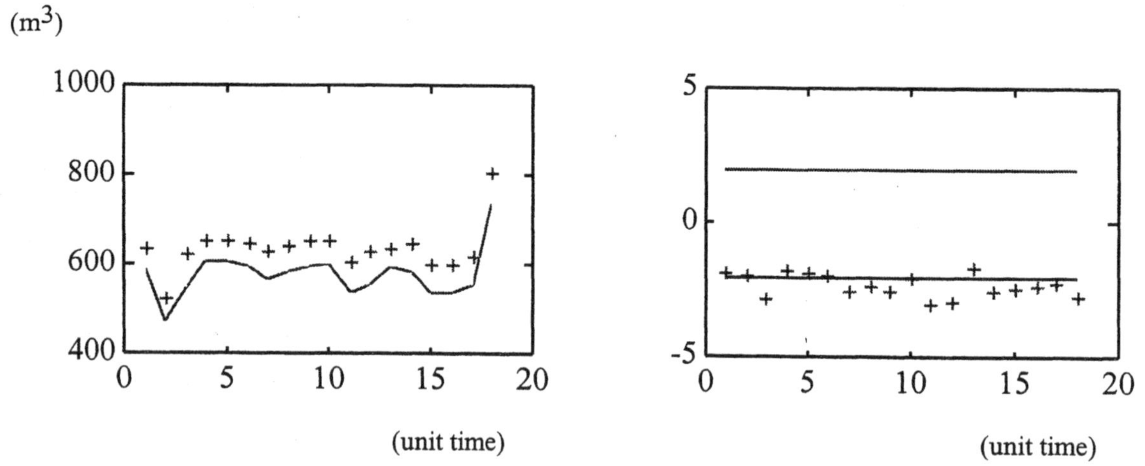

figure 2. stream 1 flow rate and normalized adjustment

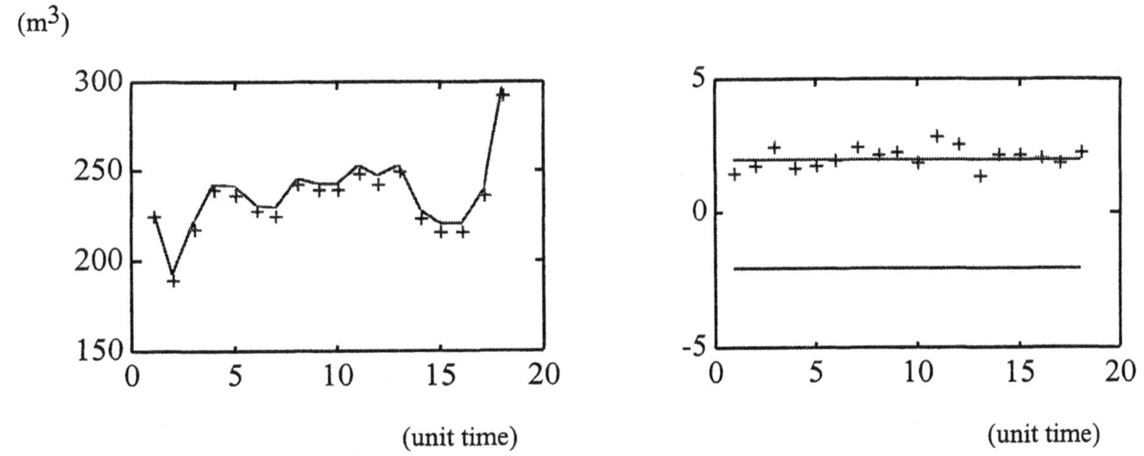

figure 3. stream 8 flow rate and normalized adjustment

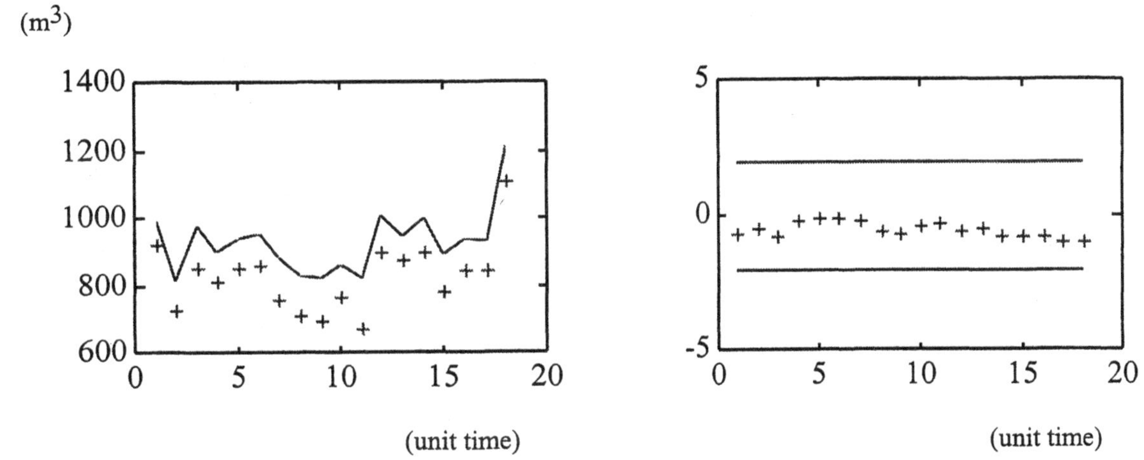

figure 4. stream 10 flow rate and normalized adjustment

We see from the comparison of normalized adjustment graphs that stream 1 has a much higher normalized adjustment. The associated sensor is therefore suspect. We also notice that the standard deviation of its distribution is near to unity. The mean is different from zero. However, as this tendancy persists througth 18 sucessive intensive measurements, we deduce that the flow-meter associated with this stream indicates a measurement about 50 (m^3 /unit time) in excess.

4. CONCLUSION

The presented methodology has been successfully applied to reconciliate raw data and to detect gross errors in nonlinearly constrained data. On different simulations it has been obtained a high percentage of gross errors detection. On the real flowsheet under consideration, we sought to ensure coherence to the different measurements and in the process we were able to detect a false measurement. We have seen that the analysis of a balance equilibration provides a preliminary appraisal on the validity of its measurements. The chronological analysis of the measurements greatly improves and sharpens our appraisal.

The evolution of the normalized adjustements which affords a straigthforward interpretation, not only facilitates the detection of incoherent measurements, but can lead, in some cases, to the determination of the nature of sensor failures.

Notation

n : number of nodes of the network

v : number of streams of the network

Q(i) : the volume flow rate vector of the stream i (m^3/unit time)

$\rho(i)$: the density measurement of the stream i (Kg/m^3)

$T_L(i)$: the liquid phase content measurement of the stream i (g/l)

$T_S(i)$: the solid phase content measurement of the stream i (ppm)

$\rho_S(i)$: the density of the stream i for the solid involved in the concentrate balance equation (Kg/m^3)

$\rho_L(i)$: the density of the stream i for liquid involved in the concentrate balance equation (Kg/m^3)

$\mathbf{M_Q}$: (n.v) incidence matrix associated with the volume flow rates
$\mathbf{M_D}$: (n_1 .v) incidence matrix associated with the mass flow rates
n_1 : number of independant relations mass flow rates
$\mathbf{M_{TL}}$: (n_2 .v) incidence matrix associated with the concentrate mass flow for the liquid phase streams only
n_2 : number of independant concentrate mass flow relations
$\mathbf{M_{TLTS}}$: (n_2 .v) incidence matrix associated with the concentrate mass flow for the two phased streams
$\mathbf{M_S}$: (n_3 .v) incidence matrix for solid phase flow rates
n_3 : number of independant relations solid phase mass flow
$\mathbf{M_L}$: (n_3 .v) incidence matrix for liquid phase flow rates
Q : (q.1) volume flow rate measurement vector
$\hat{Q}$: (v.1) volume flow rate estimation vector
ρ : (d.1) density measurement vector
$\hat{\rho}$: (d_c .1) density estimation vector
T_L : (tl.1) liquid phase content measurement vector
$\hat{T}_L$: (tl_c .1) solid phase content estimation vector
T_S : (ts.1) solid phase content measurement vector
$\hat{T}_S$: (ts_c .1) liquid phase content estimation vector
ρ_L : (ts.1) density vector for liquid involved in the concentrate balance equation
ρ_S : (ts.1) density vector for the solid involved in the concentrate balance equation
X : ((q + d + tl + ts).1) measurement vector
$\hat{X}$: ((v + d_c + tl_c + ts_c).1) estimation vector
ε : the measurement error vector
V : measurement variance diagonal matrix of size (q + d + tl + ts).(q + d + tl + ts)
$\hat{\mathbf{V}}$: variance diagonal estimation matrix of size (v + d_c + tl_c + ts_c).(v + d_c + tl_c + ts_c)
X^* : the true value of X
S1 : (d_c .v) volume flow rate selection matrix for the mass flow balance equation
S2 : (ts_c .v) volume flow rate selection matrix for the two-phase concentrate flow balance equation
S3 : (ts_c .tl_c) solid content selection matrix for the two-phase concentrate flow balance equation
S4 : (tl_c .v) volume flow rate selection matrix for the one-phase (liquid) concentrate flow balance equation
H : (q + d + tl + ts).(v + d_c + ts_c + tl_c) selection measurement matrix
Det(**V**) : the determinant of the matrix **V**
Diag(**V**) : the diagonal of the matrix **V**
F : constraint function
Exp : the mathematical expectation
ADJ : ((q + d + tl + ts).1) adjustment ratio for the measurement vector
T_{CN} : ((q + d + tl + ts).1) normalized adjustment for the measurement vector

REFERENCES

[ALMASY 1984]
G.A. ALMASY, R.S.H. MAH. Estimation of measurement error variances from process data. Industrial Engineering Chemistry Process Design Development, vol. 23, p. 779-784, 1984.

[CROWE 1989]
C.M. CROWE. Test of maximum power for detection of gross errors in process constraints. AIChE Journal, vol. 35, n° 5, p. 869-872, 1989.

[DAROUACH 1986]
M. DAROUACH. Observabilité et validation des données de systèmes de grande dimension. Application à l'équilibrage des bilans de mesures. Thèse de doctorat d'état de l'Université de Nancy I, 1986.

[FRANK 1990]
P.M. FRANK. Fault diagnosis in dynamic systems using analytical and knowledge-based redundancy - A survey and some new results. Automatica, vol. 26, n° 3, p. 459-474, 1990.

[HODOUIN 1989]
D. HODOUIN, F. FLAMENT. New developments in material balance calculations for mineral processing industry. Society of Mining Engineers annual meeting, paper 89-147, Las Vegas, february 27 - march 2, 1989.

[HOLLY 1989]
W. HOLLY, R. COOK, C.M. CROWE. Reconciliation of mass flow rate measurements in a chemical extraction plant. The Canadian Journal of Chemical Engineering, vol. 67, p. 595-601, 1989.

[ISERMAN 1984]
R. ISERMAN. Process fault detection based on modelling and estimation methods : a survey. Automatica, vol. 20, p. 387-404, 1984.

[MAH 1987]
R.S.H. MAH. Data screening. Foundations of computer aided process operations. G.V. Reklaitis and H.D. Springgs ed., p. 67, Elsevier, 1987.

[RAGOT 1990]
J. RAGOT, M. DAROUACH, D. MAQUIN, G. BLOCH. Validation de données et diagnostic. Traité des nouvelles technologies, série diagnostic et maintenance, Hermès, Paris, 1990.

[ROMAGNOLI 1980]
J.A. ROMAGNOLI, G. STEPHANOPOULOS. On the rectification of measurement errors for complex chemical plants, steady-state analysis. Chemical

Engineering Science, vol. 35, n° 5, p. 1067-1081, 1980.

[SERTH 1986]
R.W. SERTH, W.A. HEENAN. Gross error detection and data reconciliation in steam-metering systems. AIChE Journal, vol. 32, n° 5, p. 733-742, 1986.

[STANLEY 1977]
G.M. STANLEY, R.S.H. MAH. Estimation of flows and temperatures in process networks. AIChE Journal, vol. 23, n° 5, p. 642-650, 1977.

[VACLAVEK 1976]
V. VACLAVEK, M. LOUCKA. Selection of measurements necessary to achieve multicomponent mass balances in chemical plant. Chemical Engineering Science, vol. 31, n° 12, p. 1199-1205, 1976.

DYNAMIC BEHAVIOUR OF NONLINEAR MEASUREMENT SYSTEMS

C.P. Lewis*, A. Ucar*, S. R. Bishop**

** Coventry University, School of Engineering, Nonlinear Systems Design Group, Q block, Coventry, CV1 5FB, UK*
Tel: 44 1203 838842, Fax: 44 1203 838949
e-mail: cey159@coventry.ac.uk

*** University College London, Centre for Nonlinear Dynamics*
Civil Engineering Building, Gower Street, London, WCIE 6BT, UK

Abstract: A computer based technique has been developed to test the performance of complex nonlinear systems, using Matlab/Simulink™ and C/C++. In this approach a dynamic system, represented by its mathematical model, is stimulated by a sequence of sinusoidal signals and the system stability is observed by a stability diagram. Furthermore the technique is developed to test the degree of accuracy and sensitivity of transducers. Such devices are complicated nonlinear, closed-loop, systems in their own right. In this paper the long term stability of a typical pressure transducer model with hard and soft spring effects is examined. The range of use is determined by a shaded stability diagram. In addition performance of a micromachined, silicon, accelerometer is examined. Important properties of the accelerometer such as the degree of accuracy and sensitivity are considered

Keywords: Stability, Accuracy, Sensitivity, Transducers, Nonlinear systems

1. INTRODUCTION

In this paper a technique is introduced which may be used to observe the long term stability of dynamic systems. Such a system, represented by its mathematical model, is stimulated by a sequence of sinusoidal signals having a range of amplitudes and frequencies. The time domain motion of the system is observed over several excitation cycles for both fixed values of forcing amplitude and frequency. The results of the simulation are depicted in a stability, or control space, diagram. The degree of stability is depicted by the colour of the pixel (or monochrome shading). A commercially developed computer simulation package, Matlab/Simulink and C/C++ programme have been used which may readily be applied to a wide variety of highly nonlinear continuous and discrete dynamic systems.

As an example the stability of a typical pressure transducer (Pu, *et al.*, 1995) with a nonlinear spring force is examined and the safe regions of excitation variables defined. The effects of both hard and soft springs suspension are shown on the stability diagram. From the practical engineering point of view an approach has been developed to investigate not only the stability of closed loop analogue and digital transducers but also the accuracy and sensitivity of these complex devices. An analogue, micromachined, silicon, accelerometer has been chosen as a further example (Lewis, *et al.*, 1996).

2. DESCRIPTION OF THE TECHNIQUE

We consider a mathematical model of a nonlinear dynamic system described as

$$\dot{x} = f(x(t)), \qquad x(t_0) = x_0 \tag{1}$$

where $\dot{x}(t) = dx/dt$, $x(t) \in \Re^n$ is the state at time t, and $f: \Re^n \rightarrow \Re^n$ is called the nonlinear *vector field*.

The system behaviour, when subjected to a range of sinusoids of differing frequencies and amplitudes, give rise to the stability (control space) diagram (Thompson, 1990). In practice bounds are imposed on the critical state variables; when the bound is reached the degree of the stability is determined by the input cycles $k \in N$, which is depicted by an appropriate pixel colour or shading. This procedure is repeated over the range of input signal frequencies of interest. Then the amplitude of the signal is increased and a new frequency sweep takes place. It has been shown that when studying the overall morphology of the long term behaviour, the escaping set for $k=\infty$ is approximated with good scientific accuracy by $k \in [8,64]$ (Thompson, 1996). The system transient phase space diagrams are obtained for the forced system with a grid of velocity and displacement values to investigate system stability with in the face of disturbances (Stewart, *et al.*, 1995).

The system in equation (1) may represent any dynamic system which may be modelled in Matlab/Simulink. The control statement for selecting the excitation magnitude and frequency is written in a C/C++ programme.

The accuracy of the stability estimation is dependent on the modelling accuracy, the number of the excitation signal frequencies and the numerical step size of the integration process.

3. EXAMPLES

3.1 A typical pressure transducer

Pressure transducers are inherently nonlinear due to the effects of gravity, drag and pressure forces on the manometer dynamics. The nonlinearity in the system dynamics may be in the damping or spring force (Doebelin, 1990). Here a typical pressure transducer represented by a second order, single degree of freedom, system with mass-spring damping is considered (Pu, *et al.*, 1995).

The damping factor, α, has been chosen to be 0.01 and the motion of the system is described by

$$\begin{aligned} \dot{x}_1 &= x_2 \\ \dot{x}_2 &= -\alpha x_2 - k_1 x_1 - k_2 x_1^{\,3} + F\cos(\omega t) \end{aligned} \tag{2}$$

where x_1 and x_2 are the displacement and velocity respectively of the mass and F and ω are the amplitude and angular frequency of the forcing signal. In the system of (2) k_1 is the stiffness of the linear part and k_2 is the coefficient of the nonlinear part of the restoring force.

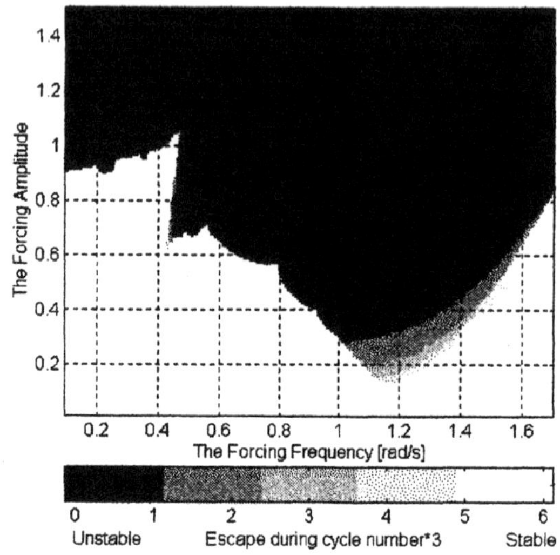

Fig. 1. The stability diagram of the system with hard spring in the amplitude and frequency *(F, ω)* domain.

The stability of the system of (2) was investigated with the technique described above in section **2**. The motion of the displacement is physically limited to $|x_1|=1.1$ and this is used as the stability limit. The system with the hard spring has been simulated and the stability diagram is depicted in Fig 1. The diagram was generated by numerical time integration from a grid of 200 by 200 forcing conditions using Runge-Kutta, numerical integration algorithm with a 0.05 step size.

In the simulation process the integration cycle was halted when the displacement exceeded the bound and the next input values were then taken. The stability diagram shows the degree of stability in coloured, or shaded, pixels in the input amplitude and frequency domain with zero initial conditions. The unshaded area represents the conditions under which stable behaviour may be expected. In the approach used the maximum number of the forcing cycles was *24*.

Fig 2 shows the effect of the soft spring for the system of (2). This diagram was obtained using the same system parameters as with the hard spring; the useful range is unshaded.

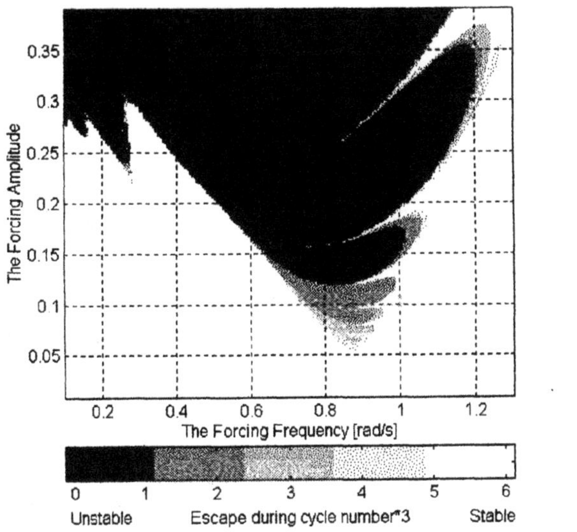

Fig. 2. The system stability diagram with soft spring in the amplitude and frequency *(F, ω)* domain.

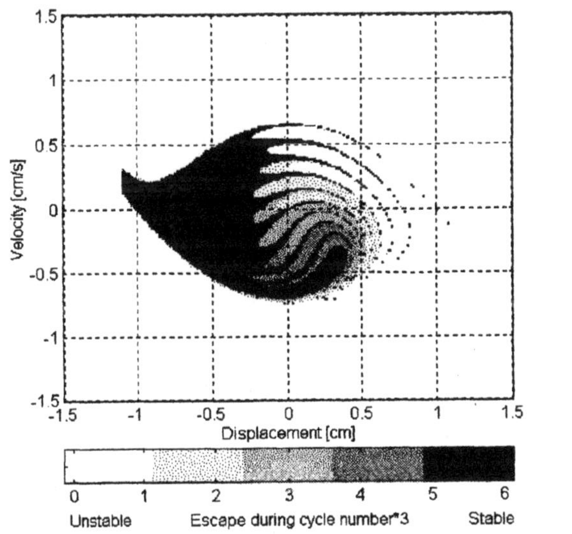

Fig. 3. The transient phase space basin in the initial conditions space $[x_1(t_0), x_2(t_0)]$ for $\omega = 0.9$ *rad/s* and $F = 0.08$.

The transient dynamic behaviour of the system was observed by a cell-to-cell mapping algorithm (Hsu, 1987). This method was used to determine the attractor basin phase portrait for a given grid of initial conditions. The degree of stability is depicted in the Poincare plane, for fixed amplitude and frequency of the forcing signal. This is achieved by observing the number of forcing cycles before the bounds of stability for velocity or displacement is reached. We refer to this diagram as a transient phase space basin diagram Fig 3. The bifurcation curve and fractal nature of the boundary in *(F, ω)* is clear. The system with hard springing has significantly larger useful range than that with the soft spring, since, at high frequencies the nonlinear effects are not significant due to the small displacements.

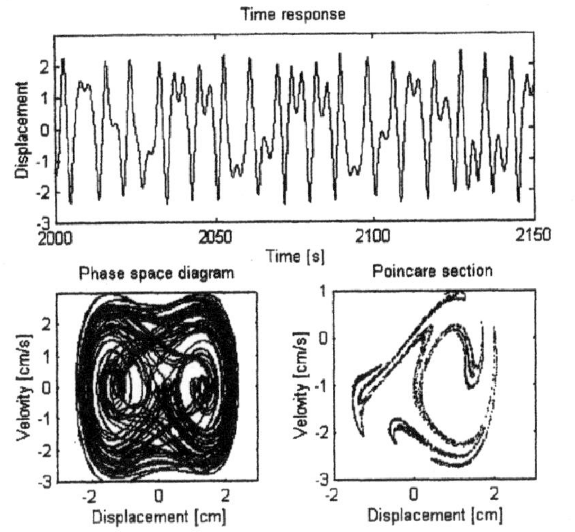

Fig. 4. A chaotic behaviour of the system of (2).

The behaviour of the system in the *(F, ω)* plane is obviously complex. After an initial transient response, the motion of the system of (2) settles onto an attractor, which can be of one of the forms: point, periodic, quasiperiodic or chaotic. Typical chaotic behaviour of the system is depicted in Fig 4. A time delay control structure designed to impose regular behaviour onto a one of chaotic motion is given in (Lewis, *et al.*, 1997b).

3.2. A micromachined accelerometer.

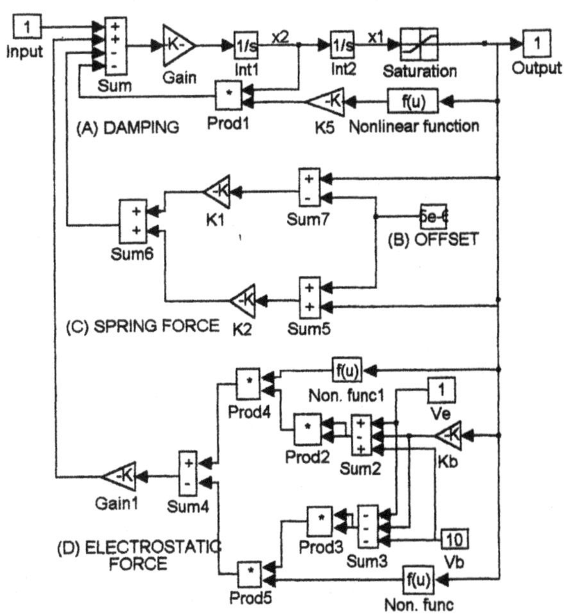

Fig. 5. A closed-loop, silicon, accelerometer.

The second example involves the development of a new approach since the properties of a closed-loop transducer is examined.

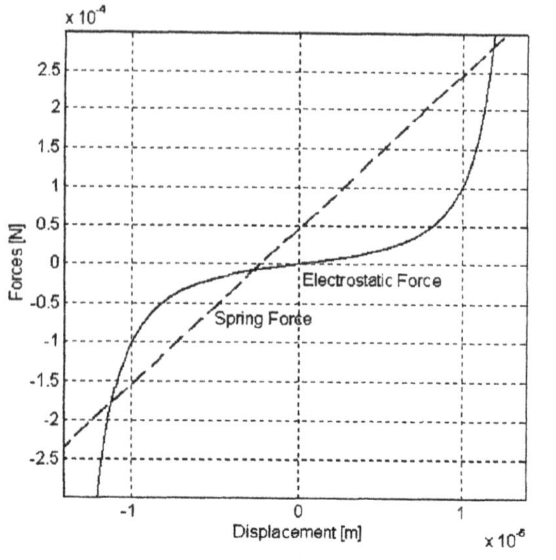

Fig. 6. Electrostatic and unequal spring forces on the seismic mass.

A typical micromachined accelerometer consists of a seismic mass suspended between two plates. Nonlinearities occur in the squeeze film damping, a function of velocity and displacement, and the methods of signal pick off and reset. A mathematical model for this device is given by Lewis *et al* (1996). The closed-loop block diagram of the system is depicted in Fig 5 in which the feedback signal, *(A)*, represents the effects of the nonlinear damping. Due to manufacturing tolerances the mass may not be at the mid position between the plates, this introduces an offset effect *(B)* on the motion. Here the system with worst offset value, $5\mu m$ is considered. This degree of offset would be unacceptable in practice.

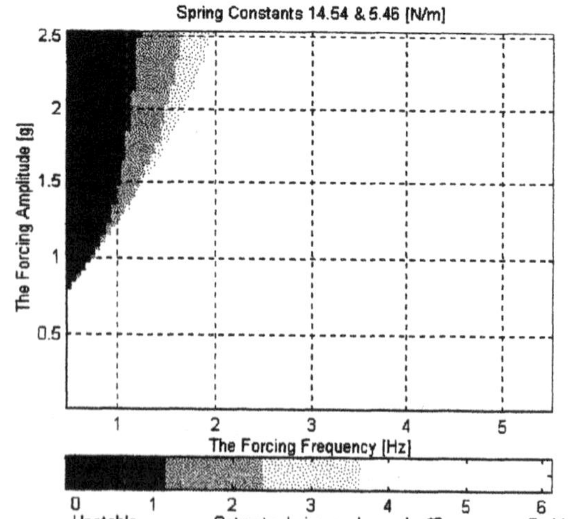

Fig. 7. Stability diagram of the accelerometer with offset.

The electrostatic force *(D)* and spring force *(C)* are depicted versus the displacement in Fig 6. When the electrostatic force exceeds the spring force the open loop behaviour is unstable. The stability of the closed-loop behaviour is considered using technique is given in section **2** for unequal spring constants ($K_1 \neq K_2$) in Fig 5 and is depicted in Fig 7.

The stability bound is defined such that when the deflection of the motion, x_1, from the central position reaches the saturation limit ($\pm 15.10^{-6} m$), the seismic mass locks up to one of the plate electrodes, which would be desisted in an initial navigation application. Using the above approach it may be shown that the stability of the system is decreased by the effect of the offset

4. ACCURACY AND SENSITIVITY

The degree of the accuracy and sensitivity of the sensors are of crucial importance and software has been developed to meet this need. The accuracy of such devices may be defined as the distortion of the output signal as a percentage of the input signal magnitude; this is an important consideration for measurement system.

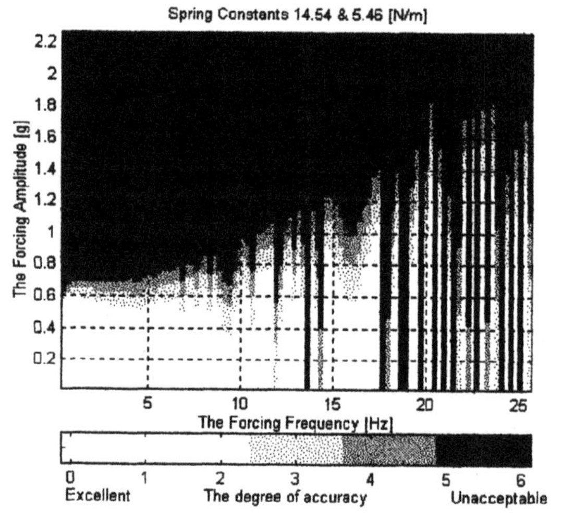

Fig. 8. The accuracy diagram of the accelerometer with offset.

The digital Fourier transform was used to obtain the harmonic power components of the output signal; the distortion being defined as 'the noise power as a percentage of the fundamental power'. The degree of distortion is again denoted by a coloured or shaded pixel in the amplitude/frequency plane.

Fig 8 show the accuracy diagram of the accelerometer for unequal spring constants. In this figure the darkest shaded area represents the region in which the output signal is heavily distorted *>5%*.

The system response is examined in term of accuracy and sensitivity under steady state conditions.

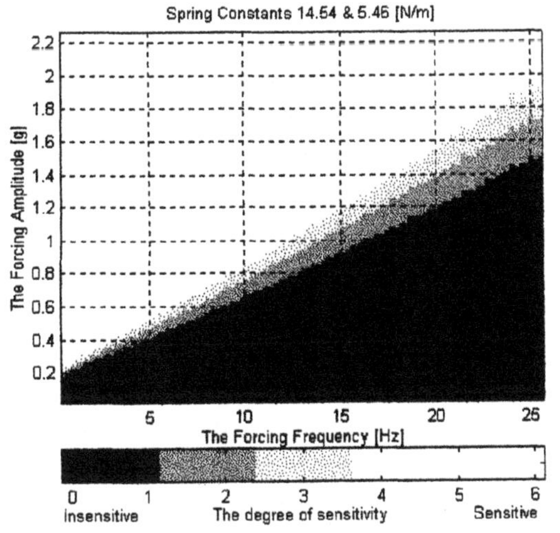

Fig. 9. Sensitivity diagram with offset.

Sensitivity is defined as 'the ratio of the fundamental of the output signal to the input signal amplitude' displayed in the amplitude/frequency plane.

Since, at high frequencies, the output signal is reduced by the low pass filtering of the dynamic system, it is necessary to obtain a measure of the sensitivity of the accelerometer as in Fig 9.

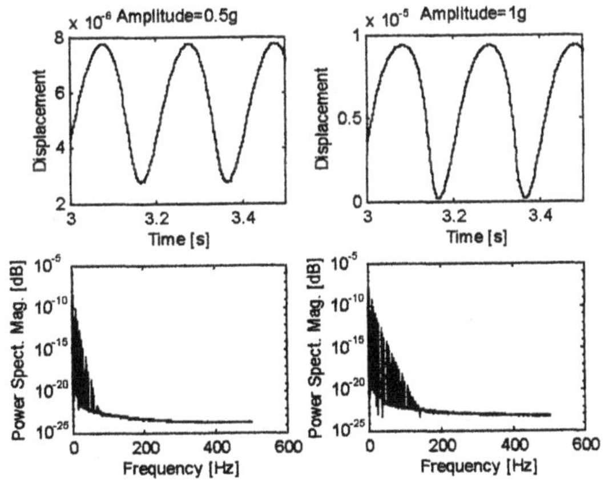

Fig. 10. The system output responses in the time and frequency domains for equal spring constants.

Fig 10 shows the corresponding time responses when a 5 *Hz* forcing signals of amplitudes *0.5* and *1 g* are used; also shown are the associated periodograms. The corresponding percentage distortion was *2.09* and *7.49* and the corresponding sensitivities *5.62* and *35.9* respectively. Fig 8 and 9 depict accuracy and sensitivity under dynamic condition. This information may be used in the design of a sensor.

5 ADVANTAGES OF THE APPROACH

The practical advantages of the techniques are:
1. pressure transducer stability diagrams are depicted which may be used to observe the useful range of operation;
2. since powerful software is used, any highly nonlinear system may be modelled and the stability examined;
3. the accuracy and sensitivity of a system, such as accelerometer, may be observed;
4. since the results are depicted visually, they are applicable for use in practical applications;
5. high order and discrete-data systems, which are notoriously difficult to analyse, can now be investigated.

6 SYSTEM COMPENSATION

The method outline above is not restricted, of course, to transducer of measurement applications. An aid to the choice of system parameters is to define a stability ratio (SR) which is the ratio of the are of stable (unshaded) to the are of the unstable region (shaded). The authors have used the safe basin and the concept of SR to investigate various types of compensation, for example, time delay control of chaotic systems which has particular application to modern electronic signal processing (Lewis *et al.*, 1997a).

7 CONCLUSIONS

The design and production of advanced sensors and data fusion has been identified as being of particular importance to the defence and aerospace industries. A new approach is made to determine the long term stability of measurement and other types of dynamic system. The use of a readily available simulation package enables this technique to be applied to a wide range of nonlinear systems (Lewis *et al.*, 1997b). The simulations were obtained using a simple PC with Pentium'90 processor, with the rapid increase in computing power the authors believe that such an approach is both viable and necessary. Obviously the basic technique could be applied directly using *C/C++* programming language with a huge reduction in simulation time, however the advantage of using a software, which is probably going to be used in study of nonlinear system dynamics in any case, would be lost.

This approach is being used for the study not only of analog accelrometers but newly developed digital accelerometers and pressure transducers. In addition it has been found to be effective in the study of the control of chaotic systems (Lewis *et al.*, 1997c).

REFERENCES

Doebelin E. O. (1990). *Measurement systems application and design*. McGraw-Hill.

Hsu, C. S. (1987). *Cel-to-Cell Mapping: A Method of Global Analysis for Nonlinear Systems, Springer-Verlag*, New York.

Lewis C. P., Kraft. M., and Hesketh, T. G. (1996) Mathematical model for a micromachined accelerometer. *Trans. Inst. Measurement & Control*. **Vol. 18**, 92-98.

Lewis C. P., Ucar A., and Bishop, S. R, (1997a). Stability of nonlinear systems and the effects of time delay control.' Submitted to *Trans. Inst. Measurement & Control*.

Lewis, C. P., Ucar, A., and Bishop, S. R. (1997b). Dynamic behaviour of complex non-linear systems, in *Multi-Body Dynamics: Monitoring and Simulation Techniques*, (ed. H. Rahnejat and R. Whalley), MEP, London, pp. 201-208.

Lewis C. P., Ucar A., and Bishop, S. R, (1997c). An approach to the investigation of robustness and stability of delay feedback control of nonlinear systems. exepted to be presented in *the 15th IMACS World Congress on scientific computation, modelling and applied mathematics*, Berlin.

Pu, X. Z., Zhu, M. W., and Fan, M. J. (1995). The study of dynamic characteristics in nonlinear systems. *IEEE Trans. Instrumentation and Measurement.*, **Vol. 44**, No. 3, pp. 652-656.

Stewart, H. B., Thompson, J. M. T., Ueda, Y., and Lansbury, A. N. (1995). Optimal escape from potential wells-patterns of regular and chaotic bifurcation, Physica D 85, 259-295.

Thompson, J. M. T., F. R. S., and Soliman, M. S. (1990). Fractal control boundaries of driven oscillators and their relevance to safe engineering design. *Proc. R. Soc. Lond.* ***A*** **428**, 1-13.

Thompson J. M. T. (1996). Global Dynamics of Driven Oscillators: Fractal Basin and Indeterminate Bifurcations. In *Nonlinear Mathematics and its Applications*, ed. P. J Aston, Cambridge University Press, Cambridge, pp. 1-47.

D.E.D.S. MODELLING AND KNOWLEDGE-BASED CONTROL[1]

František Čapkovič

Institute of Control Theory and Robotics, Slovak Academy of Sciences
Dúbravská cesta 9, 842 37 Bratislava, Slovak Republic
e-mail: utrrcapk@nic.savba.sk, http://www.savba.sk/~utrrcapk/capkhome.htm

Abstract. An alternative approach to modelling the discrete-event dynamic systems (DEDS) and to knowledge-based control of them is presented in this paper. Petri nets and oriented graphs are simultaneously used in order to express in analytical terms the new combined model of DEDS and to represent analytically knowledge about the control task specifications (like criteria, constraints, etc.) by means of such a new combine approach. Both the DEDS model and the knowledge base are simultaneously used in the procedure of the control system synthesis.

Key Words. Control system synthesis, discrete-event dynamic systems, knowledge representation, knowledge-based control, modelling, oriented graphs, Petri nets

1. INTRODUCTION

In order to synthetize the control of DEDS (discrete-event dynamic systems) two basic requirements cannot be usually avoided - a correct model of the system to be controlled and a domain oriented knowledge base (KB) expressing knowledge about the control task specifications. DEDS are systems discrete in nature, usually asynchronouos. Their state is changed owing to the occurrence of discrete events. There are two kinds of discrete events in DEDS - spontaneous events (they cannot be influenced from outside) and controllable ones (they can be influenced from outside and, consequently, they are chosen to be the DEDS control variables). Although DEDS (like manufacturing systems, transport systems, communication systems, etc.) are artificial systems as to their structure and the variability of possibilities of their development, there is not any uniform technique of their mathematical description in the control theory. In addition to this, the existence of the spontaneouos discrete events makes the DEDS modelling and control synthesis more difficult. The need[1] of the knowledge representation flows from the fact that the control task specifications (like criteria, constraints, different circumstances concerning the surrounding influences, etc.) are, as a rule, given not in analytical terms but e.g. verbally. Because the DEDS control synthesis consists in finding a suitable sequence of the controllable discrete events (at respecting the spontaneous ones, of course), the joint activity of both the system model and the KB makes possible to do it better. The potentional elementary control possibilities are generated by means of the DEDS model in any step of the control synthesis procedure. These elementary control possibilities are tested with respect to existence conditions. When there are several control possibilities satisfying the existence conditions the most suitable control possibility is chosen by means of the knowledge base. Summarized, this paper presents the combined approach (based on Petri nets and oriented graphs) to both the DEDS modelling and the knowledge representation. Such an approach contributes to simplifying the process of handling both information about the system to be controlled and knowledge about the control task specifications. The simultaneous utilizing of the DEDS model and the KB makes the procedure of the knowledge-based control synthesis more qualified. This paper represents the contribution to the previous author's works Čapkovič (1991-1997d) and respects the undestanding of the fuzzy Petri nets presented by Looney(1988).

2. THE COMBINED MODEL OF DEDS

Although there are many different approaches how to model DEDS the Petri nets (PN) are used very frequently. In order to describe the new approach

[1] The work was partially supported by the Slovak Grant Agency for Science (VEGA) in frame of the project No. 2/4036/97

combining PN and oriented graphs, that is proposed in this paper, consider PN to be (as to their structure) the directed bipartite graphs - i.e. the graphs with two kinds of nodes (positions and transitions) and two kinds of edges (i.e. the arcs oriented from positions to transitions and the arcs oriented from transitions to positions). On the other hand, the PN can be considered to be the ordinary oriented graphs - i.e. the graphs with one kind of nodes (positions) and one kind of edges (oriented arcs among positions). In this idea the PN transitions are understood to be fixed on the oriented arcs among positions. The combined structure where one part of the PN is described by means of the bipartite oriented graph and another part by means of the ordinary oriented graph, can be formally described as follows

$$\langle P, T, F, G, \Delta \rangle; \; P \cap T = \emptyset; \; F \cap G = \emptyset \quad (1)$$

where

$P = \{p_1, ..., p_n\}$ is a finite set of the PN positions with p_i , $i = 1, n$, being the elementary PN positions.

$T = \{t_1, ..., t_m\}$ is a finite set of the PN transitions with t_j , $j = 1, m$, being the elementary PN transitions.

$F \subseteq P \times T$ is a set of the oriented arcs emerging from the positions and entering the transitions. It can be expressed by means of the arcs incidence matrix $\mathbf{F} = \{f_{ij}\}$, $f_{ij} \in \{0, 1\}$, $i = 1, n$; $j = 1, m$. Its element f_{ij} represents the absence (when 0) or presence (when 1) of the arc oriented from the position p_i to its output transition t_j.

$G \subseteq T \times P$ is a set of the oriented arcs emerging from the transitions. The arcs incidence matrix $\mathbf{G} = \{g_{ij}\}$, $g_{ij} \in \{0, 1\}$, $i = 1, m$; $j = 1, n$ expresses the occurrence of the arc oriented from the transition t_i to its output position p_j.

$\Delta \subseteq P \times P$ is a set of the oriented arcs (implicitly including the PN transitions) among the positions. The arcs incidence matrix $\mathbf{\Delta} = \{\delta_{ij}\}$, $i = 1, n$; $j = 1, n$ is the functional matrix. Its element δ_{ij} expresses not only the occurrence of the arc oriented from the position p_i to another position p_j but also the transition function of the corresponding transition fixed on this arc (it will be explained below in the connection with modelling the spontaneous discrete events).

Using the analogy between the PN positions and the DEDS subprocesses as well as the analogy between the PN transitions and the DEDS discrete events (e.g. starting or ending the subprocesse, etc.) we can represent the DEDS model on three different ways.

2.1. *The PN-based model*

Such a model can be expressed in the following form

$$\mathbf{x}_{k+1} = \mathbf{x}_k + \mathbf{B}.\mathbf{u}_k \quad , \quad k = 0, N \quad (2)$$

$$\mathbf{B} = \mathbf{G}^T - \mathbf{F} \quad (3)$$

$$\mathbf{F}.\mathbf{u}_k \leq \mathbf{x}_k \quad (4)$$

where

k is the disrete step of the DEDS dynamics development.

$\mathbf{x}_k = (\sigma^k_{p_1}, ..., \sigma^k_{p_n})^T$; $k = 0, N$ is the n-dimensional state vector of the DEDS in the step k; $\sigma^k_{p_i}$, $i = 1, n$ is the state of the elementary subprocess p_i in the step k (in the PN analogy it is the state of the elementary position p_i) - its activity is expressed by 1 and its passivity by 0.

$\mathbf{u}_k = (\gamma^k_{t_1}, ..., \gamma^k_{t_m})^T$; $k = 0, N$ is the m-dimensional control vector (it represents the discrete events) of the DEDS in the step k; $\gamma^k_{t_j}$, $j = 1, m$ is the state of occurring the discrete event t_j in the step k. In the PN analogy it is the state of the elementary transition t_j(1 - enabled, 0 - disabled).

$\mathbf{B}$ is the $(n \times m)$-dimensional structural matrix of constant elements expressing the causal relations between subprocesses and discrete events. It is given by means of the $(n \times m)$-dimensional matrix $\mathbf{F}$ of constant elements expressing causal relations between the subprocesses and the discrete events and $(m \times n)$-dimensional matrix $\mathbf{G}$ of constant elements expressing the mutual causal relations between the discrete events and the subprocesses.

T symbolizes the matrix or vector transposition.

This model is most suitable in case when all of the discrete evets are controllable.

2.2. *The model based on oriented graphs*

This kind of the model is the following

$$\mathbf{x}_{k+1} = \mathbf{\Delta}.\mathbf{x}_k \quad , \quad k = 0, N \quad (5)$$

where

$\mathbf{\Delta} = \{\delta_{ij}\}$, δ_{ij} , $i = 1, n$; $j = 1, n$ is the $(n \times n)$-dimensional functional structural matrix expressing the causal relations between subprocesses that are influenced by *spontaneous* discrete events occurring in the system. The elements of this matrix represent the transition functions of the transitions representing the spontaneouos discrete events - e.g. $\delta_{ij} = \gamma_{p_j|p_i}$ is concerning the transition fixed on the arc oriented form p_i to p_j.

This model is most suitable in case when all of the discrete evets are spontaneous, i.e. uncontrollable, and consequently, they represent the system parameters.

2.3. *The combined model*

When there are simultaneously present both the spontaneous discrete events and the controllable ones in the DEDS to be modelled, the combined model is most suitable. In such a case the set of transitions T representing the discrete events consists of two subsets T_c (representing controllable discrete events) and T_u (representing uncontrollable spontaneous discrete events). It means that $T = T_c \cup T_u$, $T_c \cap T_u = \emptyset$. Consequently, the transition functions of the transitions representing the controllable discrete events create the components of the control vector $\mathbf{u}_k$ while the transition functions of the transitions representing the spontaneous discrete events create the elements of the functional matrix $\mathbf{\Delta}$ - i.e. the system parameters. Hence, the DEDS model can be expressed as follows

$$\mathbf{x}_{k+1} = \mathbf{A}.\mathbf{x}_k + \mathbf{B}.\mathbf{u}_k \quad , \quad k = 0, N \qquad (6)$$

$$\mathbf{A} = \mathbf{I}_n + \mathbf{\Delta} \qquad (7)$$

$$\mathbf{B} = \mathbf{G}^T - \mathbf{F} \qquad (8)$$

$$\mathbf{F}.\mathbf{u}_k \leq \mathbf{x}_k \qquad (9)$$

where

$\mathbf{A}$ is the $(n \times n)$-dimensional functional matrix - the system matrix consisting of the $(n \times n)$-dimensional identity matrix $\mathbf{I}_n$ and the functional matrix $\mathbf{\Delta}$ that is the $(n \times n)$-dimensional functional matrix expressing the causal relations between subprocesses that are influenced by ***spontaneous*** discrete events in the system.

$\mathbf{B}$ is the $(n \times m)$-dimensional structural matrix of constant elements expressing the causal relations between subprocesses and ***controllable*** discrete events. It is given by means of the $(n \times m)$-dimensional matrix $\mathbf{F}$ of constant elements expressing causal relations between the subprocesses and the controllable discrete events and $(m \times n)$-dimensional matrix $\mathbf{G}$ of constant elements expressing the mutual causal relations between the controllable discrete events and the subprocesses.

3. THE KNOWLEDGE REPRESENTATION BY MEANS OF THE COMBINED APPROACH

An analogy to the above introduced combined approach to DEDS modelling can be utilized also for the knowledge representation. However, the analogy between the set of statements (some pieces of knowledge) S_i, $i = 1, n$ and the set of the PN positions must be made as well as the analogy between the set of *IF-THEN* rules R_j, $j = 1, m$ and the set of the PN transitions. The mutual causal interconnections between the statements and the rules or among the statements are understood to be analogical to the mutual causal interconnections between the PN positions and transitions or among the PN positions. In addition to this the ordinary PN are replaced by means of logical PN (LPN) or/and fuzzy PN (FPN). Consequently, we have also three possibilities of the knowledge representation. In order to avoid any ambiguity with respect to previous symbolics used in the DEDS modelling, let us denote here the state vector to be $\mathbf{\Phi}_K$, control vector to be $\mathbf{\Omega}_K$, and incidence matrices to be $\mathbf{\Psi}$ and $\mathbf{\Gamma}$.

3.1. *The PN-based model*

Such a kind of the knowledge representation can be analytically described as follows

$$\mathbf{\Phi}_{K+1} = \mathbf{\Phi}_K \; \underline{or} \; \mathbf{B} \; \underline{and} \; \mathbf{\Omega}_K, \quad K = 0, N_1 \qquad (10)$$

$$\mathbf{B} = \mathbf{\Gamma}^T \; \underline{or} \; \mathbf{\Psi} \qquad (11)$$

$$\mathbf{\Psi} \; \underline{and} \; \mathbf{\Omega}_K \leq \mathbf{\Phi}_K \qquad (12)$$

where

$\mathbf{\Phi}_K = (\phi^K_{S_1}, ..., \phi^K_{S_n})^T$; $K = 0, N_1$ is the state vector of the KB truth propagation in the step K. $\phi^K_{S_i}$, $i = 1, n$ is the state of the truth of the elementary statement S_i in the step K. T symbolizes the vector or matrix transposition and N_1 is an integer representing formally the number of different state vectors occurring during the KB dynamics development.

$\mathbf{\Omega}_K = (\omega^K_{R_1}, ..., \omega^K_{R_m})^T$; $K = 0, N_1$ is the "control" vector of the KB expressing the rules enabling (or better the rules evaluability, i.e. the readiness of the rules to be evaluated) in the step K. $\omega^K_{R_j}$, $j = 1, m$ is the state of enabling the elementary rule R_j to be fired in the step K.

$\underline{and}$, $\underline{or}$ are, respectively, the operator of logical multiplying and additioning.

The inference mechanism of the statements truth propagation can be analytically expressed as follows

$$\overline{\mathbf{\Phi}}_K = \underline{neg} \; \mathbf{\Phi}_K = \mathbf{1}_n - \mathbf{\Phi}_K \qquad (13)$$

$$\mathbf{v}_K = \mathbf{\Psi}^T \underline{and} \; \overline{\mathbf{\Phi}}_K \qquad (14)$$

$$\mathbf{\Upsilon}_K = \underline{neg} \; \mathbf{v}_K = \mathbf{1}_m - \mathbf{v}_K = = \underline{neg}(\mathbf{\Psi}^T \; \underline{and}(\underline{neg} \; \mathbf{\Phi}_K)) \qquad (15)$$

where

neg is the operator of logical negation.

$\mathbf{v}_K$ is a m-dimensional auxiliary vector pointing out (by its nonzero elements) the rules that cannot be evaluated, because there is at least one false statement among its input statements. This declaration is qualified only in the analogy with the LPN. In the analogy with FPN any statement is always true with a fuzzy measure.

$\mathbf{\Upsilon}_K$ is a m-dimensional vector pointing out the rules that have all their input statements true and, consequently, they can be evaluated in the step K of the KB dynamics development. This vector is a base of the inference, because it contains information about the rules that can contribute to obtaining the new knowledge. These rules correspond to the nonzero elements of the vector $\mathbf{\Upsilon}_K$. This is also qualified only in the analogy with the LPN. In the analogy with the FPN any rule is always evaluable with a fuzzy measure - i.e. it always contributes to obtaining the new knowledge. If there is not any important obstacle in utilizing of complete knowledge yielding by $\mathbf{\Upsilon}_K$, this vector can be used as the KB "control" vector, i.e. $\mathbf{\Omega}_K = \mathbf{\Upsilon}_K$.

$\mathbf{1}_n$ is n-dimensional constant vector with all its elements equal to 1.

3.2. *The model based on oriented graphs*

The approach based on the oriented graphs yields the following form of the knowledge representation in analytical terms

$$\mathbf{\Phi}_{K+1} = \mathbf{\Delta}\ \underline{and}\ \mathbf{\Phi}_K \quad , \quad K = 0, N_1 \tag{16}$$

The inference mechanism is implicitly included owing to the fact that the rules (as the analogy with the PN transitions) are fixed on the oriented arcs among the statements. This model is suitable especially for expressing both-sided implications.

3.3. *The combined model*

Combining both of the previous approaches the following knowledge representation can be obtained

$$\mathbf{\Phi}_{K+1} = \mathbf{A}\ \underline{and}\ \mathbf{\Phi}_K\ \underline{or}\ \mathbf{B}\ \underline{and}\ \mathbf{\Omega}_K \tag{17}$$
$$K = 0, N_1$$
$$\mathbf{A} = \mathbf{I}_n\ \underline{or}\ \mathbf{\Delta} \tag{18}$$
$$\mathbf{B} = \mathbf{\Gamma}^T\ \underline{or}\ \mathbf{\Psi} \tag{19}$$
$$\mathbf{\Psi}.\mathbf{\Omega}_K \leq \mathbf{\Phi}_K \tag{20}$$

The inference mechanism is the same like in the PN-based model introduced above.

4. THE DEDS CONTROL SYNTHESIS

The problem of the DEDS control synthesis is that of finding the most suitable sequence of the control vectors $\mathbf{u}_k$, $k = 1, N$. Usually, there are several possibilities of the further development of the system dynamics in any step k. Hence, the control synthesis procedure has to be able to generate these possibilities. It is performed by means of the DEDS model. Consequently, the most suitable possibility has to be chosen by means of the KB. Because the KB operates in each step k of the DEDS model development, the step of the KB dynamics development was denoted as K in order to distinguish them. The KB has to consider the situation in any step k very thorougly from the aspect of the global development of the controlled object. The KB decision in any step k depends on both the actual situation existing in the system itself - i.e. on the present state of the controlled system $\mathbf{x}_k$ corresponding to the control vector in the previous step $\mathbf{u}_{k-1}$ - and the actual final state $\mathbf{\Phi}_K$ of knowledge inference in the step k. Several steps $K = 0, N_1$ of the KB dynamics development may be necessary in any step k of the system dynamics development. The KB decision directly influences the choice of the control vector $\mathbf{u}_k$. This fact can be formally expressed as follows

$$\mathbf{u}_k = \mathcal{F}(\mathbf{x}_k,\ \mathbf{u}_{k-1},\ \mathbf{\Phi}_K) \tag{21}$$

or more exactly

$$\mathbf{u}_k = \mathcal{F}(\mathbf{\Phi}_K(\mathbf{\Omega}_{K-1}(\mathbf{\Phi}_{K-1},\ \mathbf{u}_{k-1},\ \mathbf{x}_k))) \tag{22}$$

where

$\mathcal{F}(.)$ is a symbolical operator. It cannot be expressed in analytical terms. It depends on the actual case of the DEDS, on actual case of the control task specifications, etc.

The off-line approach to control synthesis is given on Fig. 1 and the on-line one on Fig. 2. In these

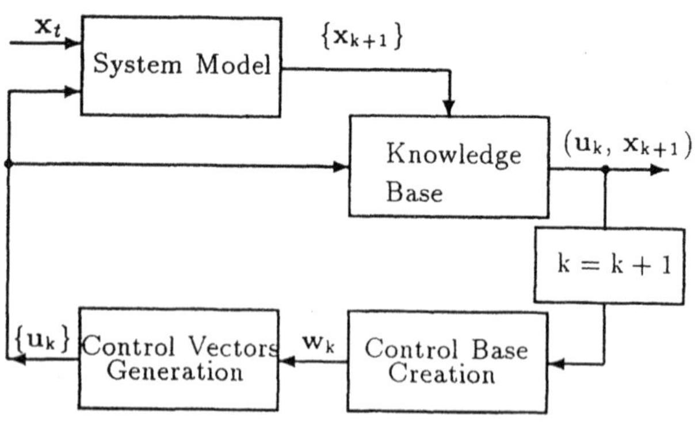

Fig. 1. The principial scheme of the control synthesis (off line approach)

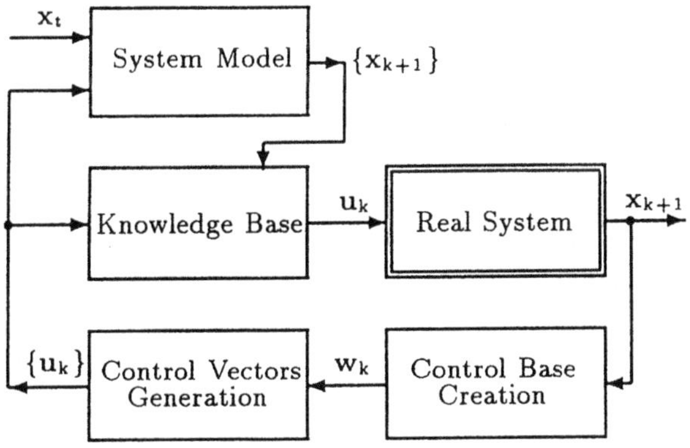

Fig. 2. The principial scheme of the control synthesis (on line approach)

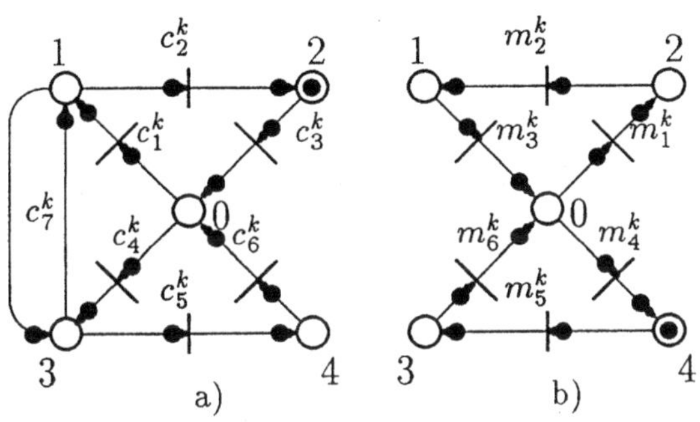

Fig. 3. The new representation of the maze. a) possible behaviour of the cat; b) possible behaviour of the mouse

figures the vector $\mathbf{x}_t$ symbolizes the desirable terminal state of the controlled object and the vector $\mathbf{w}_k$ symbolizes the control base vector (Čapkovič, 1994c) pointing out all of the enabled discrete events in the step k. The proposed combine approach to the DEDS modelling yields a simplier solution of the control synthesis problem. It will be illustrated on the following example.

5. THE ILLUSTRATIVE EXAMPLE

Consider the example defined by Wonham and Ramadge (1987). It is the same maze problem that was solved in (Čapkovič, 1994c) by means of the PN-based approach. To demonstrate the advantages of the approach proposed in this paper the combined model of the animals behaviour will be used. The KB expressing the control tasks specifications stays the same. The new combined model of the maze can be graphically expressed by means of Fig. 3. It can be seen that the uncontrollable door of the maze is replaced not by means of two PN transitions c_7^k, c_8^k like before - i.e. in (Čapkovič, 1994c), but by means of the edges of the oriented graph. Hence, the parameters of the cat model are expressed by means of the below

Table 1 The results of the control synthesis

Cat Behaviour	Mouse Behaviour
Step k = 0	
${}^c\mathbf{x}_0 = (0\,0\,1\,0\,0)^T$	${}^m\mathbf{x}_0 = (0\,0\,0\,0\,1)^T$
${}^c\overline{\mathbf{x}}_0 = (1\,1\,0\,1\,1)^T$	${}^m\overline{\mathbf{x}}_0 = (1\,1\,1\,1\,0)^T$
${}^c\mathbf{w}_0 = (0\,0\,1\,0\,0\,0)^T$	${}^m\mathbf{w}_0 = (0\,0\,0\,0\,1\,0)^T$
${}^c\mathbf{u}_0 = {}^c\mathbf{w}_0$	${}^m\mathbf{u}_0 = {}^m\mathbf{w}_0$
${}^c\mathbf{x}_1 = (1\,0\,0\,0\,0)^T$	${}^m\mathbf{x}_1 = (0\,0\,0\,1\,0)^T$
${}^c\mathbf{x}_1 \neq {}^m\mathbf{x}_1$	
Step k = 1	
${}^c\mathbf{w}_1 = (1\,0\,0\,1\,0\,0)^T$	${}^m\mathbf{w}_1 = (0\,0\,0\,0\,0\,1)^T$
the control possibilities are: $\{\underline{c_1}, \underline{c_4}, \underline{m_6}\}$	
all of them are possible	
${}^c\mathbf{u}_1^1 = (1\,0\,0\,0\,0\,0)^T$	${}^m\mathbf{u}_1 = {}^m\mathbf{w}_1$
${}^c\mathbf{x}_2^1 = (0\,1\,0\,0\,0)^T$	${}^m\mathbf{x}_2 = (1\,0\,0\,0\,0)^T$
${}^c\mathbf{u}_1^2 = (0\,0\,0\,1\,0\,0)^T$	= n o n e =
${}^c\mathbf{x}_2^2 = (0\,0\,0\,1\,0)^T$	= n o n e =
${}^c\mathbf{x}_2^1 \neq {}^m\mathbf{x}_2$; ${}^c\mathbf{x}_2^2 \neq {}^m\mathbf{x}_2$	
Step k = 2	
${}^c\mathbf{w}_2^1 = (0\,1\,0\,0\,0\,0)^T$	${}^m\mathbf{w}_2 = (1\,0\,0\,1\,0\,0)^T$
${}^c\mathbf{w}_2^2 = (0\,0\,0\,0\,1\,0)^T$	= n o n e =
the control possibilities are:	
$\{\underline{c_2}, \not{m}_1, \underline{m_4}\}$; c_2 has priority to m_1	
$\{\not{c}_5, \not{m}_1, \underline{m_4}\}$; m_4 has priority to c_5	
${}^c\mathbf{u}_2^1 = (0\,1\,0\,0\,0\,0)^T$	${}^m\mathbf{u}_2 = (0\,0\,0\,1\,0\,0)^T$
${}^c\mathbf{u}_2^2 = (0\,0\,0\,0\,0\,0)^T$	= n o n e =
${}^c\mathbf{x}_3^1 = (0\,0\,1\,1\,0)^T$	${}^m\mathbf{x}_3 = (0\,0\,0\,0\,1)^T$
${}^c\mathbf{x}_3^{11} = {}^c\mathbf{x}_0$	${}^m\mathbf{x}_3 = {}^m\mathbf{x}_0$
${}^c\mathbf{x}_3^2 = (0\,1\,0\,1\,0)^T$	= n o n e =

introduced matrices ${}^c\mathbf{F}$, ${}^c\mathbf{G}$, ${}^c\mathbf{A}$ and the parameters of the mouse model are expressed by means of the following matrices ${}^m\mathbf{F}$, ${}^m\mathbf{G}$, ${}^m\mathbf{A}$.

$$n = 5 \qquad m_c = 6$$

$$ {}^c\mathbf{F} = \begin{pmatrix} 100100 \\ 010000 \\ 001000 \\ 000010 \\ 000001 \end{pmatrix} \quad {}^c\mathbf{G} = \begin{pmatrix} 01000 \\ 00100 \\ 10000 \\ 00010 \\ 00001 \\ 10000 \end{pmatrix} $$

$$n = 5 \qquad m_m = 6$$

$$ {}^m\mathbf{F} = \begin{pmatrix} 100100 \\ 001000 \\ 010000 \\ 000001 \\ 000010 \end{pmatrix} \quad {}^m\mathbf{G}^T = \begin{pmatrix} 001001 \\ 010000 \\ 100000 \\ 000010 \\ 000100 \end{pmatrix} $$

$$ {}^c\mathbf{A} = \begin{pmatrix} 1 & 0 & 0 & 0 & 0 \\ 0 & 1 & 0 & 1 & 0 \\ 0 & 0 & 1 & 0 & 0 \\ 0 & 1 & 0 & 1 & 0 \\ 0 & 0 & 0 & 0 & 1 \end{pmatrix} \quad {}^m\mathbf{A} = \begin{pmatrix} 1 & 0 & 0 & 0 & 0 \\ 0 & 1 & 0 & 0 & 0 \\ 0 & 0 & 1 & 0 & 0 \\ 0 & 0 & 0 & 1 & 0 \\ 0 & 0 & 0 & 0 & 1 \end{pmatrix} $$

The results of the problem solving are introduced in a comprehensive form in Table 1. The control synthesis process is simplier, because the cat control vector has a smaller dimensionality. The solution of the control synthesis problem is the same like that introduced in (Čapkovič, 1994c), of course.

6. CONCLUSIONS

The new alternative approach to modelling of DEDS and to knowledge representation about the control task specifications was presented in this paper. It is based on combination of both the Petri net-based approach and the approach based on ordinary oriented graphs. The results achieved by means of the presented approach were compared with the results achieved in the previous author's approach based on Petri nets. It was performed on the same illustrative example.

7. REFERENCES

Čapkovič, F. (1991). Petri nets-based computer aided synthesis of control systems for discrete event dynamic systems. In: *Preprints of the 5-th IFAC Symposium on Computer Aided Design in Control Systems CADCS'91, Swansea, U.K., July 1991* (H.A. Barker, Ed.), pp. 409-414. Pergamon Press, Oxford.

Čapkovič, F. (1993). Modelling and justifying discrete production processes by Petri nets. *Computer Integrated Manufacturing Systems*, **6**, No 1, 21-35.

Čapkovič, F. (1994a). Computer-aided design of intelligent control systems for discrete event dynamic systems. In: *Proceedings of the IEEE/IFAC Joint Symposium on Computer-Aided Control System Design CACSD'94, Tucson, Arizona, USA, March 1994* (S.E. Mattsson, J.O. Gray, F.E. Cellier, Eds.), pp. 55-60. IEEE Service Center, Piscataway.

Čapkovič, F. (1994b). A rule-based representation of fuzzy knowlege for control purposes by means of Petri nets. In: *Proceedings of the 2nd European Congress on Intelligent Techniques and Soft Computing EUFIT'94, Aachen, Germany, September 1994* (H.J. Zimmermenn, Ed.), Vol. 3, pp. 1451-1455. ELITE, Aachen.

Čapkovič, F. (1994c). A Petri net-based approach to synthesis of control systems for discrete event dynamic systems. In: *Preprints of the 1st IFAC Workshop New Trends in Design of Control Systems - NTDCS'94, Smolenice, Slovak Republic, September 7-10, 1994*, pp. 398-403.

Čapkovič, F. (1995a). Fuzzy knowledge representation by means of the Petri net-based model with variable parameters. In: *Proceedings of the 3rd European Congress on Intelligent Techniques and Soft Computing EUFIT'95, Aachen, Germany, August 28-31, 1995*, (H.J. Zimmermenn, Ed.), Vol.III, pp. 1780-1784, ELITE, Aachen.

Čapkovič F. (1995b). Petri nets-based approach to intelligent control synthesis of FMS. In: *Proceedings of the 1995 INRIA/IEEE Symposium on Emerging Technologies and Factory Automation - ETFA'95, Paris, France, October 1995*, Vol. 1, pp. 323-331, IEEE Computer Society Press, Los Alamitos, CA, USA.

Čapkovič, F. (1995c). Using fuzzy logic for knowledge representation at control synthesis. *BUSEFAL*, **63**, 4-9.

Čapkovič, F. (1996a). A Petri net approach to analysis of complex discrete event systems and to synthesis of their control. In: *Proceedings of the 13th European Meeting "Cybernetics and Systems'96", Vienna, Austria, April 9-12, 1996*, (R. Trappl, Ed.), Vol. 1, pp. 83-88.

Čapkovič, F. (1996b). Knowledge-Based Control of DEDS. In: *Proceedings of the 13th IFAC World Congress 1996, San Francisco, USA, June 30-July 5, 1996*, Vol. J, paper J-3c-02.6, pp. 347-352, Compact Disc, Elsevier Science Ltd., Pergamon.

Čapkovič, F. (1997a). Petri nets and oriented graphs in fuzzy knowledge representation for DEDS control purposes. *BUSEFAL*, **69**, 21-30.

Čapkovič F. (1997b). Synthesis of knowledge-based control systems of flexible manufacturing systems. In: *Proceedings of the 2nd World Congress on Intelligent Manufacturing Processes & Systems - IMPS'97, Budapest, Hungary, June 10-13, 1997* (L. Monostori, Ed.), pp. 303-308, Springer Verlag, Budapest, Hungary.

Čapkovič F. (1997c). Synthesis of the flexible manufacturing systems control by means of simulation. In: *Proceedings of the 13th International Conference on Computer-Aided Production Engineering - CAPE'97, Warsaw, Poland, June 1997* (J.A. McGeough, S. Strzelczak, Eds.), pp. 359-366, Warsaw University of Technology, Poland.

Čapkovič F. (1997d). Fuzzy knowledge in DEDS control synthesis. In: *Proceedings of the 7th International Fuzzy Systems Association World Congress - IFSA'97, Prague, Czech Republic, June 25-29, 1997* (M. Mareš, R. Mesiar V. Novák, J. Ramík, A. Stupňanová, Eds.), pp. 550-554, Academia, Prague, Czech Republic.

Looney, C.G. (1988). Fuzzy Petri nets for rule-based decisionmaking. *IEEE Trans. Syst. Man Cybern.*, **SMC-18**, No 1, 178-183.

Wonham, W.M. and P.J.Ramadge (1987). On the supremal controllable sublanguage of a given language. *SIAM J. Control and Optimization*, **25**, No 3, 637-659.

A DESIGN METHOD OF THE CONFLICT-FREE PETRI NET MODELS FOR THE MANUFACTURING SYSTEMS CONTROL

B. Hrúz

Slovak University of Technology, Faculty of Electrical Engineering and Information Technology
Department of Automatic Control Systems
Ilkovičova 3, 812 19 Bratislava, Slovak Republic
Phone: +42 7 791698, e-mail: hruz@kasr.elf.stuba.sk

Abstract. A certain class of Petri nets interpreted for control, shortly CTRL - Petri nets, are introduced and defined in the paper. Three basic elementary CTRL - Petri net structures are described reflecting three main control situations with respect to conflict decision problem in the flexible manufacturing systems (FMS). Two main sources of the conflicts in the FMS are analyzed and specified. Then a method for removal of the conflicts in the FMS modelling CTRL - Petri net is presented. The method is illustrated on a practical example.

Keywords: flexible manufacturing systems, control, discrete event systems, Petri nets, conflict-free Petri net.

1. INTRODUCTION

Petri nets are powerful tools for the dynamic behaviour specification of the discrete event systems (Murata, 1989; Abel, 1990). It is mainly due to their ability to represent distributed, concurrent and parallel activities, states or events in the discrete event dynamic systems (DEDS).

The distribution or branching of the system activities brings about, of coarse, a decision problem. Petri nets in their essence are capable to enhance variety of possible dynamic system developments. Fig. 1 shows an elementary Petri net structure where a place has more than one outgoing arcs and where all transitions $t_1, t_2, \ldots, t_k$ are fireable by the marking $M(p) = 1$. After firing of one transition the other ones are not more firable. Such a Petri net is usable to represent a situation when more events can occur and one event eliminates the other. On the other hand the situation is not usually acceptable in the system control because it contains indeterminism to be somehow removed. The problem should be coped with very often in the flexible manufacturing systems (FMS).

M. C. Zhou et al. (1992) proposed a hybrid methodology for the Petri net modelling of the flexible manufacturing systems. A main goal of the work was to create Petri net models using specific building blocks so that the final model does not include indeterminism. It means the decision-free choice model is obtained. Zhou et al. did not solve directly the problem of the manufacturing system control, though their methodology supports creation of the Petri net models suitable for the manufacturing systems. Those models then constitute a base for the control design. In this paper another approach to the solution of the

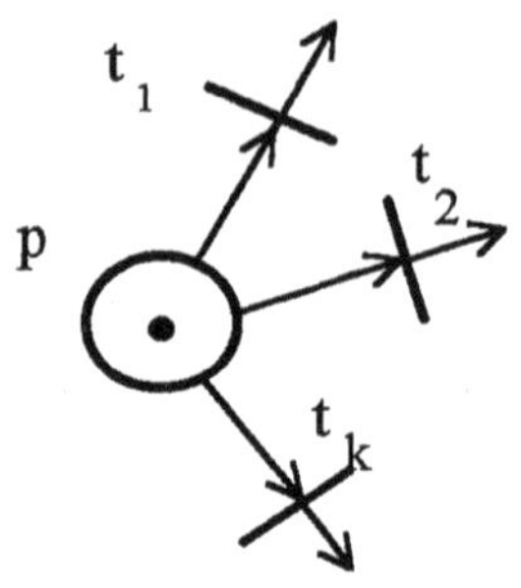

Fig. 1. Decision problem modelled by a Petri net.

conflict or decision problem in the context of the control is proposed. Direct aiming at the control is substantiated by the fact that the FMS control can be formulated usually directly without priorily having models of the controlled technolgy itself. It is because the function and evolution of the typical discrete manufacturing system is determined step by step by the control commands. In what follows basic Petri net objects are defined, then sources of conflicts in the FMS are analyzed and the corresponding Petri nets structures are described. Using those structures and having designed a Petri net model of control the conflicts can be removed using a systematic procedure.

2. CONFLICTS IN THE PETRI NETS INTERPRETED FOR CONTROL

As stated above the aim of this paper is a contribution to the FMS control design. To use Petri nets for that purpose requires supplements to the general Petri net theory specific for control. One main issue of them is definition of the so-called Petri nets interpreted for control. In what follows a necessary piece of knowledge is introduced which presents an approach based on the works of König and Quäck (1988), David and Alla (1992), and Hrúz (1994).

<u>Definition 1.</u>
The Petri net interpreted for control, shortly CTRL - Petri net is given by the 10-tuple

$$CP = (P,T,F,W,C,\mathbf{m}_0,LOG,COM,INH,IND) \qquad (1)$$

where
P is the set of places $P = \{p_1, p_2, \ldots, p_n\}$
T is the set of transitions $T = \{t_1, t_2, \ldots, t_m\}$
F is the binary relation $F \subseteq (P \times T) \cup (T \times P)$
W is the weight function $W: F \to \mathbf{I}^+$ (positive integers)
C is the capacity function $C: P \to \mathbf{I}^+ \cup \{\omega\}$, ω is symbol of infinity in $\mathbf{I}^+$, the capacities of all places can be aggregated into the vector $\mathbf{c} = (C(p_1), C(p_2), \ldots, C(p_n))^T$
$\mathbf{m}_0$ is the initial marking vector whose components are given by the function $M: P \to \mathbf{N}$ so that $\mathbf{m}_0 = (M(p_1), M(p_2), \ldots, M(p_n))^T$
LOG is the function which maps transitions to logical conditions (including empty condition)
COM is the function which maps places to control commands (including empty command)
INH is the set of inhibitors given by relation $INH \subseteq P \times T$
IND is the set of incidentors given by relation $IND \subseteq P \times T$

Some definitions necessary for the paper subject will be introduced.

<u>Definition 2.</u>
The incidence matrix of the given CTRL - Petri net is given by

$$A = (\mathbf{t}_1\ \mathbf{t}_2 \ldots \mathbf{t}_m) \text{ where } \mathbf{t}_i = \begin{pmatrix} t_{1i} \\ t_{2i} \\ \cdot \\ \cdot \\ \cdot \\ t_{ni} \end{pmatrix}, \qquad (2)$$

$$t_{ki} = \begin{cases} W(t_i, p_k) \text{ if } (t_i, p_k) \in F \\ -W(p_k, t_i) \text{ if } (p_k, t_i) \in F \\ 0 \text{ otherwise} \end{cases},$$

$t_i \in T, p_k \in P, i = 1,2,\ldots,m;$
$k = 1,2,\ldots,n$

<u>Definition 3.</u>
Let a CTRL - Petri net be given. The marking is the function

$$M: P \to \mathbf{N} \text{ (natural numbers)} \qquad (3)$$

The marking can be expressed by the vector $\mathbf{m} = (M(p_1), M(p_2), \ldots, M(p_n))^T$. The numbers $M(p_k)$ are graphically visualized as the tokens inserted in the places p_k.

Definition 4.
Let an actual marking M for a CTRL - Petri net be given. The marking flow rule is given by equation

$$\mathbf{m}' = \mathbf{m} + \mathbf{t}_i \tag{4}$$

iff

$$\begin{cases} (0 \le \mathbf{m} + \mathbf{t}_i \le \mathbf{c}) \wedge \\ (\text{all logical conditions mapped to } \mathbf{t}_i \text{ by } LOG \text{ are true}) \wedge \\ \wedge\begin{pmatrix} \text{pre - places of all inhibitors pointing to } \mathbf{t}_i \\ \text{are zero marked} \end{pmatrix} \wedge \\ \wedge\begin{pmatrix} \text{pre - places of all incidentors pointing to } \mathbf{t}_i \\ \text{are non - zero marked} \end{pmatrix} \end{cases}$$

Prior to definition of the conflict a notion of the reachability is to be considered. The reachable marking is the marking which can be achieved in a CTRL - Petri net by the repeated use of the marking flow rule starting with the initial marking. The set of all reachable markings will be denoted as usual by $R_N(\mathbf{m}_0)$.

Definition 5.
Let a CTRL -Petri net be given. Further let an actual marking $\mathbf{m}$ be given. Two or more transitions firable by $\mathbf{m}$ are said to be in conflict if firing of one transition causes that all other transitions lost fireability by the next marking, i.e. formulated for two transitions

$$\begin{aligned} & t, t' \in T \text{ are in conflict by } \mathbf{m} \in R_N(\mathbf{m}_0) \\ & \text{if} \\ & (\text{both } t, t' \text{ are firable by } \mathbf{m}) \wedge \\ & \wedge (t' \text{ is not fireable by } \mathbf{m}' = \mathbf{m} + \mathbf{t}) \end{aligned} \tag{5}$$

A standard way how to consider dynamics of the Petri net models is to assume that only one transition can fire in a time point. If parallel or simultaneous firing of more transitions would be allowed it is possible to describe the conflict as follows. If two or more transitions in a CTRL - Petri net cannot be fired simultaneously they present the CTRL - Petri net conflict (Starke, 1990).

3. BASIC PETRI NET STRUCTURES FOR MANUFACTURING CONTROL

A standard situation for a flexible manufacturing system is to peform a technological job or operation for which some necessary resources should be available like semiproducts, robots, manipulators, conveyors, processing machines, quality inspection centers etc. There are two kinds of resources, namely those processed in an operation like semiproducts and those performing the operation like machines. The following three figures shows the basic CTRL - Petri net structures corresponding to three basic situations with respect to conflicts in which the resources can be involved (Hrúz, 1995; 1996).

Fig. 2a shows a typical situation without any conflict. Places $p_{11}, p_{12}, \ldots, p_{1k}$ specify necessity of the semiproducts $s_1, s_2, \ldots, s_k$ for the job realization. If the places are occupied with tokens the required semiproducts are available. Tokens in the places $p_a, p_b, \ldots, p_z$ mean that the second kind of resources like processing machines, robots etc. are available for the job performance. The job or operation itself is represented by the place p_3. The command for the start of the operation in this CTRL - Petri net is S:=1. Place p_2 specifies required transfer of the semiproducts in the input of the processing machine - command I:=1, p_4 with the command O:=1 is connected

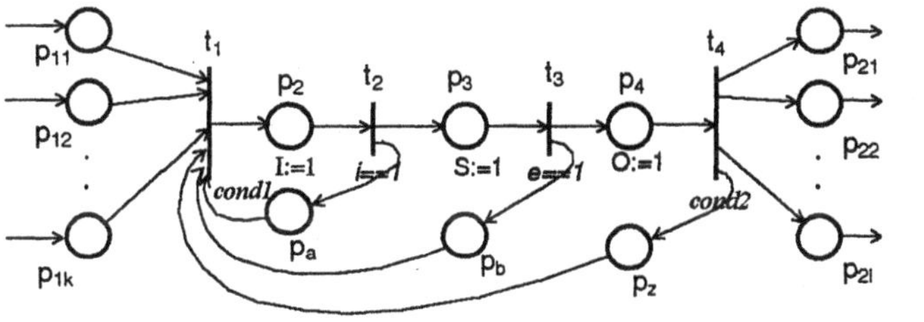

Fig. 2a. Resources and operations without conflict

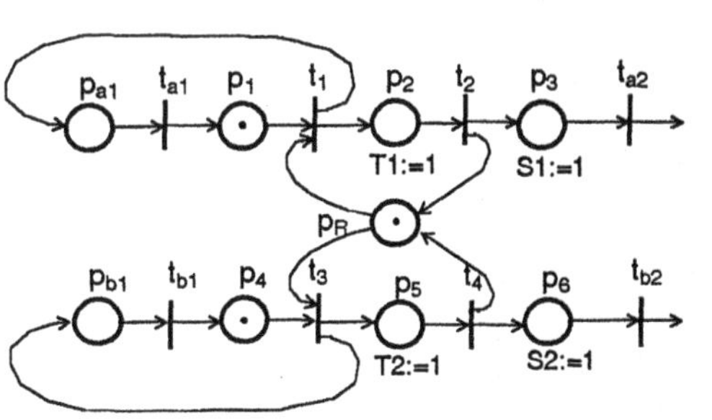

Fig. 2b. Conflict situation due to shared resources

with the transfer out of the manufacturing cell. Various events as semiproduct presence signals, start or end of the operations etc. can be associated with the transitions. More output places $p_{21}, p_{22}, \ldots, p_{2l}$ mean that the operation brings about more separate workpieces as the output of the operation. Let us follow a portion of the model dynamics. After firing of t_1 a token comes in the place p_2 and the command I:=1 is generated forcing the transfer of the semiproducts to the processing machine, start of conveyors etc. When process variable i has the value 1, transition t_2 fires and the token comes in the place p_3 and p_a. The command S is set to the value one which triggers the operation. Analogously after end of operation (process variable $e == 1$) the semiproducts obtained in the operation are transferred to the output and the machine is free. Outgoing arcs from the transitions t_2, t_3, t_4 specify that after corresponding events the resources are available again.

With respect to structure Petri net in Fig. 2a belongs to the the Petri net class called marked graphs. Marked graphs possess no conflicts.

The next typical situation in the FMS are modelled by the CTRL - Petri net depicted in Fig. 2b. One resource e.g. robot (available if the place p_R is occupied with a token) is used or shared for two operations (Zhou, 1991), namely a workpiece transfer to the first machine (place p_2, control command $T1 := 1$) and a workpiece transfer to he second machine (place p_5, control command $T2$:=1). Starts of the operations are initiated by means of the commands $S1$, $S2$. There can be more shared resources. For the brevity the elementary structure in Fig. 2b is limited only to one shared resource. Finally Fig. 2c shows the situation when job decision or job choice occurs in

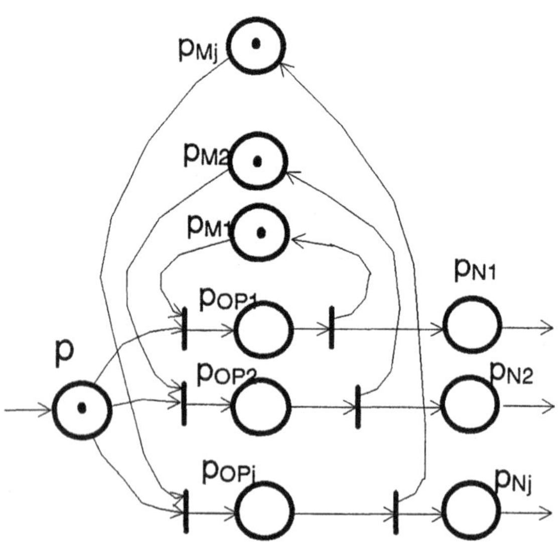

Fig. 2c. Conflict situation due to operation choice

the system. There place p specifies availability of a semiproduct for the next processing which can be done either by the operation $O1$ (place p_{OP1}) or by the operation $O2$ (place p_{OP2}) or etc. up to the operation Oj (p_{OPj}). As it is clear from Fig. 2c the machine $M1$ is required for the operation $O1$, the machine $M2$ for $O2$ etc.

4. CONFLICT-FREE CTRL - PETRI NET MODELS OF FMS

Assume a CTRL - Petri net is built up of the elementary structures as described in the preceding section. The conflicts which can be present in the CTRL - Petri net should be eliminated in order to reach through the control a deterministic required behaviour. The Zhou's et al. methodology can be applied also for the CTRL - Petri nets used for the control specification.

Combining the described elementary building blocks (Fig. 2a,b,c) can bring about aggregated and more complex Petri net structures containing conflicts.

It is assumed that a reversible, bounded and live (on level L4, see Murata, 1989) CTRL - Petri nets are designed specifying the desired control. Conflicts due to the shared resources with or without operation choice can be eliminated with help of inhibitors. The elimination method consists in the following steps.

Step 1.
The given CTRL - Petri net is inspected and all places with two or more outgoing arcs are found. Let the set of those places be denoted as P_O .Then the set of places with more than one incoming arcs is formed and denoted as P_I . The sets may not be disjunctive, i.e. it may be $P_O \cap P_I \neq \emptyset$. If the CTRL - Petri net model contains operation choice it is assumed that according to the elementary structure of Fig. 2c the resources required for the operations are represented in the net by the corresponding places.

Step 2.
A set of the outgoing arcs for each place $p_k \in P_O$ is

$$A_k = \{(p_k, t_{j_1}), (p_k, t_{j_2}), \ldots, (p_k, t_{j_k})\} \qquad (6)$$

where

$$(p_k, t_{j_1}) \in F, \quad (p_k, t_{j_2}) \in F, \ldots, \quad (p_k, t_{j_k}) \in F.$$

Step 3.
A set of the incoming arcs for each place $p_m \in P_I$ is

$$B_m = \{(t_{j_1}, p_m), (t_{j_2}, p_m), \ldots, (t_{j_m}, p_m)\} \qquad (7)$$

where

$$(t_{j_1}, p_m) \in F, \quad (t_{j_2}, p_m) \in F, \ldots, \quad (t_{j_m}, p_m) \in F .$$

Step 4.
The elements of the sets A_k , B_m wil be denoted as

$A_k = \{a_{k1}, a_{k2}, \ldots, a_{kj_k}\}$ and $B_m = \{b_{m1}, b_{m2}, \ldots, b_{mj_m}\}$, respectively.

All combinations of two elements (arcs) of the set A_k are taken into consideration. Let one combination be taken and denoted by $\{a_u, a_v\}$, $a_u = (p_k, t_u)$, $a_v = (p_k, t_v)$. The arcs a_u , a_v point to the transitions t_u , t_v respectively. Further the pre-places $\tilde{p}_u$ and $\tilde{p}_v$ of t_u and t_v other than p_k are considered. An inhibitor going out of the $\tilde{p}_u$ is created pointing to the post-transition of the pre-place $\tilde{p}_v$. Depending on the required system behaviour it can be done inversely. If there is more pre-places it is sufficient to create only one inhibitor for one combination of arcs.

This step is repeated for all combinations formed from the set A_k . The procedure is repeated for all sets A_k , k = 1,2,...,. If necessary it is repeated also for the sets B_m , m = 1,2,.....,.In the case of the shared resources it is sufficient to treat only the sets A_k .

It should be noted that by formal application of the described method redundant inhibitors can be generated because more inhibitors can eliminate the same conflict. After the addition of inhibitors the obtained Petri net should be analyzed on reversibility, boundedness and liveness.

5. EXAMPLE

Fig. 3 shows as an exmaple a robotic cell with one robot R, two processing machines M1 and M2, input (C1) and output (C2) belt conveyors. Workpieces gather at the stopping S where photosensor P1 detects a workpiece to be prepared for the processing in the cell. The machines perform different jobs with different operation times. The prepared workpiece is transferred to the free machine. If both machines are free the machine M1 has priority. After finishing a job the robot transfers the processed workpiece on the output conveyor C2 (assuming there is free room for it).

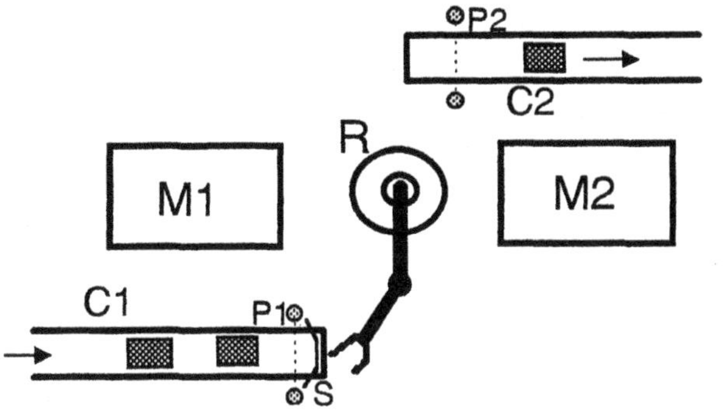

Fig. 3. A robotic manufacturing cell.

In this example the robot is a typical shared resource. The corresponding place is p_R in the CTRL - Petri net depicted in Fig. 4 specifying the robotic cell coordination control. Place p_B if occupied with a token specifies control situation in which the command is gnerated for the return of the robot to the basic position. Places p_{M1} , p_{M2}

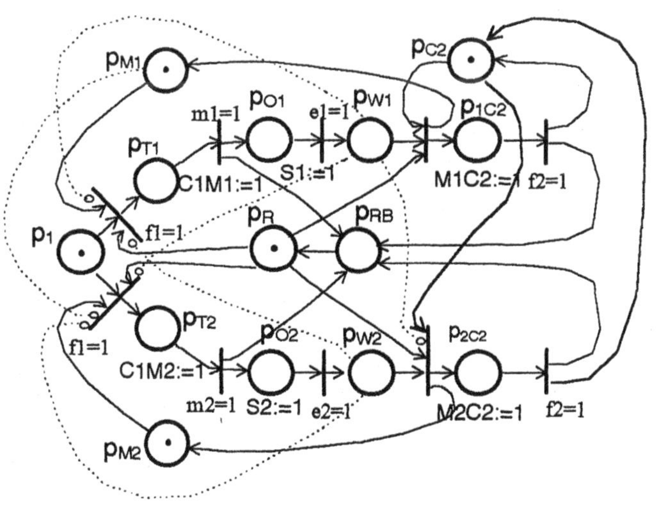

Fig. 4. CTRL-Petri net for the control of robotic cell

stand for availability of the machines M1 and M2. Analogously place p_{e2} for the conveyor C2. Condition $f1 = 1$ signals the presence of A workpiece at the input.

The control commands are associated with places. E. g. the command C1M1:=1 starts the transfer of the workpiece from the fixed position at the conveyor C1 into the machines M1, the command C1M2 into the machine M2.

6. CONCLUSION

A contribution to the design of Petri nets specifying control of discrete event systems was presented in the paper. As a next step in the resaerch would be reasonable systematically to analyze various Petri net control structures in order to apply the presented procedure and to develope additional ones so that the process of DEDS control design could be effective. Computer support tools based on those procedures would be very desirable.

7. REFERENCES

Abel, D. (1990). *Petri-Netze für Ingenieure.* Springer-Verlag, Berlin.

David, R., and H. Alla (1992). *Petri nets and Grafcet.* Prentice Hall, New York.

Hrúz, B. (1994). Discrete event systems modelling and real-time control. *Journal of Electrical Engineering,* **45**,363-370.

Hrúz, B. (1995). Control of discrete event dynamic systems in the context of flexible manufacturing systems and Petri nets. *Report no. 100 of the Control Engineering Laboratory, Helsinki University of Technology.*

Hrúz, B., A. Niemi, and T. Virtanen (1996). Composition of conflict-free Petri net models for control of flexible manufacturing systems. *Proc. of the 13th IFAC World Congress,San Francisco*, **vol. B**, 37-42.

König, R., and L. Quäck (1988). *Petri-Netze in der Steuerungs- und Digitaltechnik.* R. Oldenbourg Verlag, München, Wien.

Murata, T. (1989). Petri nets: properties, analysis and applications. *Proc. IEEE*, **77**, 541-580.

Zhou, M. C., F. DiCesare, and A. A. Desrochers (1992). A hybrid methodology for synthesis of Petri net models for manufacturing systems. *IEEE Trans. Robotics and Automation*, **8**, 350-361.

Zhou, M. C., and F. DiCesare (1991). Parallel and sequential mutual exclusions for Petri net modelling for manufacturing systems. *IEEE Trans. Robotics and Automation*,**7**, 515-527.

A CONCEPT OF AN ALGEBRAIC GENERALISATION OF PETRI NETS

Gabriel Juhás

Institute of Control Theory and Robotics, Slovak Academy of Sciences
Dúbravská 9, 842 37 Bratislava, Slovak Republic
Tel. +421-7-378 3216 *Fax +421-7-376045* *Email* utrrjuhi@savba.sk

Abstract: In the paper we introduce a concept of a simple useful extension of place/transition Petri nets based on a generalisation of the algebra used in the net dynamics. This extension enables to use not only nonnegative integers as states (markings) of places and mainly not exclusively the operation of addition in the net dynamics, but elements of an arbitrary commutative monoid $(H, \oplus)$, in which for every a, b from H the equation $x \oplus a = b$ has maximally one solution from H, to be states and the operation of the monoid in the net dynamics as well. By other words in presented extension the state set and the operation used in the net dynamics are parameters. Such extended definition preserves the advantages given by the state equation and covers standard place/transition nets, which are understood as its special subclass.

Keywords: Petri-nets, Discrete event systems, Concurrency, Algebraic approaches, Mathematical models

1. INTRODUCTION

Petri net models (Reisig, 1985; Murata, 1989) are a modelling tool for describtion, analysis, simulation as well as control of Discrete Event Systems that are characterised as being concurrent, asynchronous, distributed, parallel and non-deterministic.

Main advantages of place/transition Petri nets consist in the facts that they grant the well-arranged graphical expression of modelled systems as well as the powerful mathematical tool of modelled systems analysis in form of an algebraic equation called the state equation.

There are many extensions of place/transition nets - many types of high-level nets (Jensen and Rozenberg, 1991) that enable to use different types of tokens (e.g. coloured Petri nets (Jensen, 1992; Jensen, 1994), self-modifying nets (Valk, 1981) that can dynamically change their structure - to mention only a few of them. In the paper we introduce a concept of a simple useful extension of place/transition nets based on a generalisation of the algebra used in the net dynamics.

To follow (Reisig, 1985; Murata, 1989), place/transition net (P/T net) may be interpreted as a bipartite oriented graph, where one type of nodes represents system quantities (so called places) and the other one represents the events (so called transitions). They are interconnected by directed arcs labelled by positive integer weights. Only different types of nodes can be connected. In P/T nets nonnegative integers (numbers of tokens) are used to be states (called markings) of single places. Changes of markings of a P/T net are caused by firing of transitions. Firing of a transition subtracts from the marking of every place connected with the transition by an arc directed to the transition the number of tokens determined by the weight of the arc and adds to the marking of every place connected with the transition by an arc directed to the place the number of tokens given by the weight of the arc. The transition can fire if and only if the number of tokens in every place connected with the transition by an arc directed to the transition is not less than the weight of the arc, i.e. iff there is enough tokens for firing. Evidently, (partial) operation of subtraction of nonnegative integers can be derived from addition as well. Thus, in P/T nets the algebra of nonnegative integers with the operation of addition is crucial. It is also clear that for every nonnegative integers a, b the equation $x + a = b$ has nonnegative solution if and only if $b \geq a$ and then this solution is unique. Consequently, the algebra of nonnegative integers with addition is a commutative monoid in which for every nonnegative integers a, b the equation $x + a = b$ has maximally one nonnegative solution. So, in P/T nets definition the number of places, number of transitions, a flow relation (i.e. set of arcs), values of weights of arcs and an initial marking are parameters. The set from which markings and weights take the values (nonnegative integers) together with the operation (addition of nonnegative integers) used in the P/T net dynamics (i.e. the algebra used in the net) is fixed.

The extension presented in the paper is based on the fact that it enables to choose an arbitrary commutative monoid $(H, \oplus)$, in which for every a, b from H the equation $x \oplus a = b$ has maximally one solution from H, to be an algebra used in the net dynamics. Then the states and weights take values from the carrier of the monoid and in the net dynamics the operation of the monoid is used. The algebra used in the net dynamics and the state set are now parameters. Consequently, the definition of Algebraically Generalised Place/ Transition Nets is introduced in the paper, when a state of an algebraically generalised P/T net is represented by a function that maps the places on elements of an arbitrary commutative monoid with the mentioned property. A state equation is reformulated applying the monoid operation, the definition is further developed and examples of useful types of algebraically generalised P/T nets are presented in the paper.

2. ALGEBRAIC BACKGROUND

As usual, in the paper **Z** denotes integers, $\mathbf{Z}^+$ denotes positive integers and **N** denotes nonnegative integers (i.e. $\mathbf{N} = \mathbf{Z}^+ \cup \{0\}$). The notation $[A \to B]$ denotes for arbitrary non empty sets A and B the set of all functions from A to B, formally $[A \to B] = \{f; f: A \to B\}$ (we use $[A \to B]$ rather than B^A). If $A = \{a_1, .. a_k\}$ ($k \in \mathbf{Z}^+$), then for an arbitrary function $f \in [A \to B]$ symbol $\mathbf{f}$ denotes the vector of its function values, $\mathbf{f} = [f(a_1), . ., f(a_k)]' \in \mathrm{N}^k$ ($[]'$ denotes transpose of a vector []). For an arbitrary non empty set A we denote by A^* the set of all finite sequences (strings) of elements from A (including the empty string). As usual, we introduce the notion a *multi-set* over a non empty set A as a function b: $A \to \mathbf{N}$. In this way, each finite sequence $\sigma = \{u_i\}_{i=1}^{k} \in A^*$ is uniquely associated with a multi-set over A denoted by b_σ determining the multiplicity of each element of A in σ, formally $\forall\, a \in A$: $b_\sigma(a) = \#(I_a)$, where $I_a = \{i;\ i \in \{1\ ..\ k\} \wedge u_i = a\}$ and $\#(I_a)$ is cardinal number of the set I_a.

Let $\Gamma = (H, \oplus)$ be an arbitrary *commutative monoid* (i. e. Γ is a set H called *carrier* with a binary relation $\oplus$: $H \times H \to H$ such that: $\oplus$ is associative, i. e. $\forall\, x, y, z \in H$: $(x \oplus y) \oplus z = x \oplus (y \oplus z)$;$\exists\, \varepsilon \in H$: $\forall\, h \in H$: $h \oplus \varepsilon = \varepsilon \oplus h = h$ (ε is called *identity* or *neutral element* in Γ); $\oplus$ is commutative, i.e. $\forall\, x, y \in H$: $x \oplus y = y \oplus x$). Let A be an arbitrary set and let $k \in \mathbf{Z}^+$. We define:

- for an arbitrary sequence $\{h_i\}_{i=1}^{k}$ of elements of set H: $\sum_{i=1}^{k}{}^{\oplus} h_i = h_1 \oplus .. \oplus h_k$;
- *right scalar multiplication* $*_\oplus$: $H \times \mathbf{N} \to H$ of elements of H by nonnegative integers given recursively according to $n \in \mathbf{N}$ by equations: $h \in H$: $h *_\oplus 0 = \varepsilon \wedge h *_\oplus (n+1) = h *_\oplus n \oplus h$;

  Remark 1. This scalar multiplication has higher priority than operation $\oplus$.
- *right scalar multiplication* $\bullet_\oplus$: $H^k \times \mathbf{N}^k \to H$ of a k-dimensional vector of elements of the set H by a k-dimensional nonnegative integer vector as follows: $\forall\, \mathbf{V} = [v_1, . ., v_k]' \in H^k\ \forall\, \mathbf{X} = [x_1, .., x_k]' \in \mathbf{N}^k$: $\mathbf{V}' \bullet_\oplus \mathbf{X} = \sum_{i=1}^{k}{}^{\oplus} v_i *_\oplus x_i$;
- unique *multi-set extension* of a function $f \in [A \to H]$ to the linear function $\hat{f} \in [A_{MS} \to H]$ such that $\forall\, b \in A_{MS}$: $\hat{f}(b) = \sum_{a \in A}{}^{\oplus} f(a) *_\oplus b(a)$;
- binary operation $\oplus^{[A \to H]}$ on $[A \to H]$ and binary operation $\oplus^k$ on H^k such that $\forall\, f, f' \in [A \to H]$: $\forall\, a \in A$: $(f \oplus^{[A \to H]} f')(a) = f(a) \oplus f'(a)$, and $\forall\, \mathbf{h}, \mathbf{h}' \in H^k$: $\forall\, i \in \{1, .., k\}$: $(\mathbf{h} \oplus^k \mathbf{h}')_i = h_i \oplus h_i'$.

Consequently, $([A \to H], \oplus^{[A \to H]})$ is a commutative monoid. In addition, let *CMONOIDS* denote the class of all commutative monoids. Let for a function $G \in [A \to \mathit{CMONOIDS}]$ the function value $G(a)$ of an element $a \in A$ be denoted as $G(a) = \Gamma_a = (H_a, \oplus_a)$. The symbol ε_a denotes identity element in the monoid Γ_a. Then the set $[G] = \{g;\ g \in [A \to \bigcup_{a \in A} H_a] \wedge \forall\, a \in A$: $g(a) \in H_a\}$ together with binary operation $\oplus^{[G]}$ on $[G]$ such that $\forall\, g, g' \in [G]$: $\forall\, a \in A$: $(g \oplus^{[G]} g')(a) = g(a) \oplus_a g'(a)$ forms a commutative monoid. If A is a finite indexed set, then also set $[\mathbf{G}] = \{\mathbf{g};\ g \in [G]\}$ with operation $\oplus^{[G]}$ such that $\forall\, \mathbf{g}, \mathbf{g}' \in [\mathbf{G}]$: $\mathbf{g} \oplus^{[G]} \mathbf{g}' = \mathbf{g}''$, where $g'' = g \oplus^{[G]} g'$, forms a commutative monoid.

3. PETRI NETS - BASIC DEFINITIONS

In accordance with (Reisig, 1985) a net can be defined as follows:

Definition 2. A *net* is an ordered tuple $N = (P, T, F)$, where: $P = \{p_1, .. , p_m\}$ is a finite and non empty set of *places*; $T = \{t_1, .. , t_n\}$ is a finite and non empty set of *transitions*; $P \cap T = \emptyset$; $F \subseteq P \times T \cup T \times P$ is a *flow relation* (set of arcs).

Then place/transition net is a net together with the weight function labelling the elements of the flow relation by positive integers and with the initial marking.

Definition 3. A *place/transition net* is an ordered tuple $PN = (P, T, F, W, M_0)$, where: (P, T, F) is a net; W: $F \to \mathbf{Z}^+$ is a *weight function*; M_0: $P \to \mathbf{N}$ is an *initial marking*.

Let PN be a P/T net. Then the flow relation and the weight function determine (Lautenbach, 1987) an *input function* I: $T \to [P \to \mathbf{N}]$ and an *output function* O: $T \to [P \to \mathbf{N}]$ given as follows:
$\forall\, t \in T\, \forall\, p \in P$:
$I(t)(p) = W(p, t)$ if $[p, t] \in F$ else $I(t)(p) = 0 \wedge$
$O(t)(p) = W(t, p)$ if $[t, p] \in F$ else $O(t)(p) = 0$.
The input and output functions may be interpreted also as integer $m \times n$ dimensional integer matrices denoted by $\boldsymbol{I}$ and $\boldsymbol{O}$, which i-th columns are formed by vectors $\boldsymbol{I}(t_i)$ and $\boldsymbol{O}(t_i)$ for $i \in \{1,.., n\}$.

Remark 4. Naturally, input and output functions (matrices) of a P/T net PN may uniquely replace the flow relation and weight functions of PN in definition 3. Traditionally P/T nets have K-function determining the capacity of places. In the paper we omit it, because it is not crucial in the approach. For the concept of K-function in algebraically generalised P/T nets see e.g. (Juhás, 1996).

A function C: $T \to [P \to \mathbf{Z}]$ such that $\forall\, t \in T\, \forall\, p \in P$: $C(t)(p) = O(t)(p) - I(t)(p)$ (integer $m \times n$ dimensional matrix given by difference $\boldsymbol{C} = \boldsymbol{O} - \boldsymbol{I}$) is called *incidence function* (*incidence matrix*) of PN. A state of PN is expressed by a function M: $P \to \mathbf{N}$ called *marking* (evidently, M is a multi-set over P). Change of marking of a P/T net PN is caused by a transitions occurrence (firing of transitions). Transition $t \in T$ is *enabled to fire at a marking* M iff holds: $\forall\, p \in P$: $M(p) \geq I(t)(p)$. A *firing of an enabled* $t \in T$ *at a marking* M causes the change of the marking M to the marking M' such that $\forall\, p \in P$:
$M'(p) = M(p) + O(t)(p) - I(t)(p)$ or equivalently $M'(p) + I(t)(p) = M(p) + O(t)(p)$. We also write $M[t\rangle M'$.

Remark 5. The condition determining if a transition is enabled to fire may have such interpretation, that the change caused by the input function causally precedes the change caused by the output function. Thus, for any enabled $t \in T$ at a marking M we always have that for a marking M' such that $M[t\rangle M'$ holds: $\exists\, \boldsymbol{X} \in \mathbf{N}^m$: $\boldsymbol{X} + \boldsymbol{I}(t) = \boldsymbol{M} \wedge \boldsymbol{M}' = \boldsymbol{X} + \boldsymbol{O}(t)$.

A finite sequence $\sigma = \{u_i\}_{i=1}^{k} \in T^*$ is called *enabled occurrence sequence at marking* M_0 iff $\forall\, i \in \{1 .. k\}$: transition u_i is enabled at the marking M_{i-1}, where inductively $M_i = M_{i-1} + C(u_i)$ or equivalently $M_i + I(u_i) = M_{i-1} + O(u_i)$ (in accordance with the previous section we should write $+^{[P\to\mathbf{N}]}$ instead of + to denote addition of functions in the equation). A *firing of an enabled occurrence sequence* σ *at marking* M_0 leads to the marking M such that (in matrix form)

$$\boldsymbol{M} = \boldsymbol{M}_0 + \boldsymbol{C} \cdot \boldsymbol{b}_\sigma, \text{ i.e. } \boldsymbol{M} + \boldsymbol{I} \cdot \boldsymbol{b}_\sigma = \boldsymbol{M}_0 + \boldsymbol{O} \cdot \boldsymbol{b}_\sigma$$

or (in functional form)

$$M = M_0 + \hat{C}(b_\sigma), \text{ i.e. } M + \hat{I}(b_\sigma) = M_0 + \hat{O}(b_\sigma),$$

where b_σ is the multi-set over T determined by the sequence σ and $\hat{C}, \hat{I}$ and $\hat{O}$ are multi-set extensions of functions O, I and C (also in accordance with previous section we should write $+^m$ and $\bullet_{+^m}$ instead of + and $\cdot$, respectively). Previous equation is called *state equation* of P/T net PN. It is also written $M_0[\sigma\rangle M$ and vector $\boldsymbol{Y} = \boldsymbol{b}_\sigma$ is called a *firing vector*. So, a marking M is *reachable from marking* M_0 iff there exists a sequence $\sigma \in T^*$ enabled at M_0 such that holds $\boldsymbol{C} \cdot \boldsymbol{b}_\sigma = \boldsymbol{M} - \boldsymbol{M}_0$ or equivalently $\boldsymbol{M} + \boldsymbol{I} \cdot \boldsymbol{b}_\sigma = \boldsymbol{M}_0 + \boldsymbol{O} \cdot \boldsymbol{b}_\sigma$ (i.e. such that $M_0[\sigma\rangle M$). The set of all markings reachable from an initial marking M_0 is denoted as $[M_0\rangle$.

Manipulation with matrix equations arising from matrix representation of P/T net and its dynamics leads to the creation of powerful algebraic methods of structural analysis of P/T nets (necessary condition for reachability, P and T invariants etc.), for more details see e.g. (Desel *et al.*, 1996; Juhás and Kocian 1995; Kruckeberg and Jaxy, 1987; Lautenbach 1987; Memmi and Roucariol, 1980). Briefly, *P-invariant* of a P/T net PN is a multi-set X over P (vector $\boldsymbol{X} \in \mathbf{N}^m$) for which holds: $\boldsymbol{C}' \cdot \boldsymbol{X} = \boldsymbol{0}$, i.e. $\boldsymbol{O}' \cdot \boldsymbol{X} = \boldsymbol{I}' \cdot \boldsymbol{X}$ (where $\boldsymbol{0}$ is zero vector of length n). *T-invariant* of a P/T net PN is a multi-set Y over T (vector $\boldsymbol{Y} \in \mathbf{N}^n$) for which holds: $\boldsymbol{C} \cdot \boldsymbol{Y} = \boldsymbol{0}$, i.e. $\boldsymbol{O} \cdot \boldsymbol{Y} = \boldsymbol{I} \cdot \boldsymbol{Y}$ (where $\boldsymbol{0}$ is zero vector of length m).

The commutative monoid $(\mathbf{N}, +)$ used in P/T net dynamics and the form of P/T nets dynamics grant interesting properties. For any two markings for which a transition is enabled it causes the same change of marking. The commutativity of addition causes that if two sequences of transitions associated with the same multi-set are enabled from a marking, then the change of marking caused by firing of one is the same as the change caused by firing of the other. This enables to formulate the state equation in which the multi-set determined by the firing sequence is used.

4. ALGEBRAICALLY GENERALISED PLACE/TRANSITION PETRI NETS

In this section algebraically generalised P/T nets are introduced. The extension is based on the observation that many of useful properties of standard P/T nets are not given by using nonnegative integers and their addition exclusively, but by the fact that nonnegative integers with addition form a commutative monoid in which for every $a, b \in \mathbf{N}$ the equation $x + a = b$ has maximally one solution $x \in \mathbf{N}$.

Definition 6. An *algebraically generalised place/transition net* is an ordered tuple $GPN = (P, T, F, \Gamma, W, M_0)$, where: (P, T, F) is a net; $\Gamma = (H, \oplus)$ is an arbitrary commutative monoid in which holds $\forall x, y, z \in H$: $(y \oplus x = z \oplus x \Rightarrow y = z)$; $W: F \rightarrow H \setminus \{\varepsilon\}$ is a *weight function* (where ε is identity element in monoid Γ); $M_0: P \rightarrow H$ is an *initial state.*

The input and output function together with firing rule and state equation of algebraically generalised P/T nets are defined in strict analogy with standard P/T nets, when only operation $\oplus$ of the monoid Γ and elements of the set H are used instead of the operation of addition and nonnegative integers.

Let GPN be an algebraically generalised P/T net. Then the *input function* $I: T \rightarrow [P \rightarrow H]$ and *output function* $O: T \rightarrow [P \rightarrow H]$ are defined as follows:
$\forall t \in T \, \forall p \in P$: $[p, t] \in F \Rightarrow I(t)(p) = W(p, t) \wedge [p, t] \notin \mathrm{F} \Rightarrow I(t)(p) = \varepsilon$ and
$\forall t \in T \, \forall p \in P$: $[t, p] \in F \Rightarrow O(t)(p) = W(t, p) \wedge [t, p] \notin F \Rightarrow O(t)(p) = \varepsilon$
($\forall$ i $\in$ {1, . ., n} i-th columns of $m \times n$ dimensional matrices $\boldsymbol{I}$ and $\boldsymbol{O}$ are formed by vectors $\boldsymbol{I}(t_i)$ and $\boldsymbol{O}(t_i)$).

A state of GPN is expressed by a function $M: P \rightarrow H$. A transition $t \in T$ is *enabled to fire at a state M* of GPN iff holds: $\forall p \in P: \exists x \in H: x \oplus I(p)(t) = M(p)$. A *firing of an enabled $t \in T$ at a state M* causes the change of the state M of GPN to the state M' such that $\forall p \in P: M'(p) \oplus I(t)(p) = M(p) \oplus O(t)(p)$. As usual, we write $M[t\rangle M'$. A sequence $\sigma = \{u_i\}_{i=1}^{k} \in T^*$ is *enabled to fire at M_0* iff $\forall i \in \{1 .. k\}$: transition u_i is enabled at the state M_{i-1}, where inductively $M_i \oplus^{[P\rightarrow H]} I(u_i) = M_{i-1} \oplus^{[P\rightarrow H]} O(u_i)$. It is clear, that a *firing of an enabled sequence σ at state M_0* leads to the state M for which holds

$\boldsymbol{M} \oplus^m \boldsymbol{I} \bullet_{\oplus^m} \boldsymbol{b}_\sigma = \boldsymbol{M}_0 \oplus^m \boldsymbol{O} \bullet_{\oplus^m} \boldsymbol{b}_\sigma$ or equivalently

$M \oplus^{[A \rightarrow H]} \hat{I}(b_\sigma) = M_0 \oplus^{[A \rightarrow H]} \hat{O}(b_\sigma)$.

A state M is *reachable from state M_0* if and only if there exists a sequence $\sigma \in T^*$ enabled at M_0 such that $\boldsymbol{M} \oplus^m \boldsymbol{I} \bullet_{\oplus^m} \boldsymbol{b}_\sigma = \boldsymbol{M}_0 \oplus^m \boldsymbol{O} \bullet_{\oplus^m} \boldsymbol{b}_\sigma$. The set of all markings reachable from initial state M_0 in GPN is denoted as $[M_0\rangle$. A multi-set X over P (a vector $X \in \mathbf{N}^m$) is called *P-invariant* of GPN iff holds: $\boldsymbol{O}^t \bullet_{\oplus^n} X = \boldsymbol{I}^t \bullet_{\oplus^n} X$ and a multi-set Y over T (vector $Y \in \mathbf{N}^n$) is called *T-invariant* of GPN iff holds: $\boldsymbol{O} \bullet_{\oplus^m} Y = \boldsymbol{I} \bullet_{\oplus^m} Y$.

Remark 7. Evidently, the validity of the implication $\forall x, y, z \in H$: $(y \oplus x = z \oplus x \Rightarrow y = z)$ in monoid Γ is crucial for the possibility to define dynamics of the net in the presented form. For instance, it guarantees deterministic firing of transitions, i.e. it guarantees that a firing of an enabled $t \in T$ at an M causes the change to the unique marking M' given by the equation $M' \oplus^m I(t) = M \oplus^m O(t)$. It also guarantees that the use of the firing rule defined anallogically with standard P/T nets preserve the natural interpretation, that the change caused by the input function causally precedes the change caused by the output function. By other words, for any enabled $t \in T$ at a state M we always have that for a state M' such that $M[t\rangle M'$, i.e. for M' such that $\boldsymbol{M}' \oplus^m \boldsymbol{I}(t) = \boldsymbol{M} \oplus^m \boldsymbol{O}(t)$) holds: $\exists \boldsymbol{X} \in H^m$: $\boldsymbol{X} \oplus^m \boldsymbol{I}(t) = \boldsymbol{M} \wedge \boldsymbol{M}' = \boldsymbol{X} \oplus^m \boldsymbol{O}(t)$. Finally, as a consequence, it enables to use the state equation in presented form.

Consequently, standard P/T nets form a subclass of the class of algebraically generalised P/T nets such that $\Gamma = (\mathbf{N}, +)$.

We could also use an exention of Γ to a commutative group in order to introduce an incidence function as it is done in standard P/T nets definition, where group $(\mathbf{Z},+)$ is such extension of monoid $(\mathbf{N},+)$).

Algebraically generalised P/T nets may be also further specified by other requirements on properties of the monoid Γ . Thus, the presented definition enables to determine many useful properties of the nets that are given by the type of the used algebra Γ as it is shown on the following simple illustrative example.

Lemma 8. *Let GPN be an algebraically generalised P/T net such that Γ is a commutative group. Then $\forall M \in [P \rightarrow H]$: $\forall t \in T$: t is enabled at M.*

Proof: From $M(p), I(t)(p) \in H$ for every $M \in [P \rightarrow H]$: $t \in T, p \in P$, and $(H, \oplus)$ is a commutative group $\Rightarrow \forall M \in [P \rightarrow H] \, \forall t \in T \, \forall p \in P: \exists x \in H: x \oplus I(t)(p) = M(p)$, i.e. $\forall M \in [P \rightarrow H] \, \forall t \in T$: t is enabled at M □.

Corollary 9. *Let GPN be an algebraically generalised P/T net such that Γ is a commutative group. Then $\forall \sigma \in T^*$: σ is enabled at M_0.*

Corollary 10. *Let GPN be an algebraically generalised P/T net such that Γ is a commutative group. Then a state M is reachable from state M_0 (M*

$\in [M_0\rangle)$ *if and only if there exists such* $Y \in [T \to \mathrm{N}]$ *that* $\boldsymbol{M} \oplus^m \boldsymbol{I} \bullet_{\oplus^m} Y = \boldsymbol{M}_0 \oplus^m \boldsymbol{O} \bullet_{\oplus^m} Y$.

So, for an algebraically generalised P/T net in which Γ is a commutative group we have that in each state all transitions can concurrently occur. As a consequence, the collorary 10 says that if we choose a commutative group to be algebra used in the net dynamics, then the existence of the solution of the state equation $\boldsymbol{M} \oplus^m \boldsymbol{I} \bullet_{\oplus^m} Y = \boldsymbol{M}_0 \oplus^m \boldsymbol{O} \bullet_{\oplus^m} Y$ is necessary and sufficient condition for reachability of M from M_0.

In order to enable the use of different algebras for single places (or subsets of set P) the definition of algebraically generalised P/T nets may be further developed as follows.

Definition 11. A *high algebraically generalised P/T net* is an ordered tuple $HGPN = (P, T, F, G, W, M_0)$, where: (P, T, F) is a net; G: $P \to CMONOIDS$ such that $\forall\, p \in P$: $\forall\, x, y, z \in H_p$: $(y \oplus_p x = z \oplus_p x \Rightarrow y = z)$; W: $F \to \bigcup_{p \in P} H_p$ such that $\forall\, [p, t] \in F$: $W(p, t) \in (H_p \setminus \{\varepsilon_p\}) \wedge \forall\, [t, p] \in F$: $W(t, p) \in (H_p \setminus \{\varepsilon_p\})$ is a *weight function*; $M_0 \in [G]$ is an *initial state*.

Let $HGPN$ is a high algebraically generalised P/T net. Then its input and output functions (matrices) are defined analogically as for algebraically generalised P/T nets, only identity elements of related monoids are used in correspondent expressions. A state of $HGPN$ is expressed by a function $M \in [G]$. A transition $t \in T$ is *enabled to fire at a state M* iff holds: $\forall\, p \in P$: $\exists\, h \in H_p$: $h \oplus_p I(t)(p) = M(p)$. A *firing of* $t \in T$ *at a state M* causes the change to the state M' such that $M' \oplus^{[G]} I(t) = M \oplus^{[G]} O(t)$. A *firing of an enabled sequence* σ (defined analogically as for P/T nets) *at state* M_0 leads to the state M such that $\boldsymbol{M} \oplus^{[G]} \boldsymbol{I} \bullet_{\oplus^{[G]}} \boldsymbol{b}_\sigma = \boldsymbol{M}_0 \oplus^{[G]} \boldsymbol{O} \bullet_{\oplus^{[G]}} \boldsymbol{b}_\sigma$ or equivalently $M \oplus^{[G]} \hat{I}(b_\sigma) = M_0 \oplus^{[G]} \hat{O}(b_\sigma)$.

It is possible to define T-invariants for $HGPN$ but P-invariant can not be generally defined for $HGPN$ in strict analogy with standard P/T nets. It is only possible to define partial P-invariants for subsets of set of places P, for which a function G is constant. To solve this problem it is possible to define general P-invariants of $HGPN$ in similar way as they are defined for coloured Petri nets (see (Jensen, 1994)).

Evidently, algebraically generalised P/T nets are a special case of high algebraically generalised P/T nets, when function G is constant on P.

For better illustration, two useful types of algebraically generalised P/T nets are defined as follows.

Definition 12. An algebraically generalised P/T net GPN, where $\Gamma = (\mathbf{Z}_k = \{0, .., k-1\}, +_{\mathrm{mod}\,k})$ $(k \in \mathrm{Z}^+$, $+_{\mathrm{mod}\,k}$ is modulo k addition (modulo k addition of integers a and b is the $(a + b)/k$ residue) see e.g. (MacLane and Birkhoff, 1968)), is called *modulo k additive P/T net*.

It is suitable to use modulo k P/T net in applications, when modelled system components have k-states changed in a cycle - as is e.g. modelling of cyclic shift register with length k, modelling of overflow etc.

Example 13. On Fig. 1 there is a modulo 8 additive P/T Net (in which places model e.g. the 3-bit registers) with 3 places and 1 transition, input matrix $\boldsymbol{I} = [2, 5, 0]'$, output matrix $\boldsymbol{O} = [0, 0, 4]'$ and initial marking $\boldsymbol{M}_0 = [3, 2, 5]'$ before and after firing of transition t_1. Formally, we have:

$\boldsymbol{M}_1 +_{\mathrm{mod}\,8^3} \boldsymbol{I}(t_1) = \boldsymbol{M}_0 +_{\mathrm{mod}\,8^3} \boldsymbol{O}(t_1)$, i.e.

$\boldsymbol{M}_1 +_{\mathrm{mod}\,8^3} [2, 5, 0]' = [3, 2, 5]' +_{\mathrm{mod}\,8^3} [0, 0, 4]' \Rightarrow$

$\boldsymbol{M}_1 = [1, 5, 1]'$ or alternatively using incidence matrix given by modulo 8 difference of matrices $\boldsymbol{O}$ and $\boldsymbol{I}$:

$\boldsymbol{M}_1 = \boldsymbol{M}_0 +_{\mathrm{mod}\,8^3} \boldsymbol{C}(t_1) = [3, 2, 5]' +_{\mathrm{mod}\,8^3} [6, 3, 4]' = [1, 5, 1]'$.

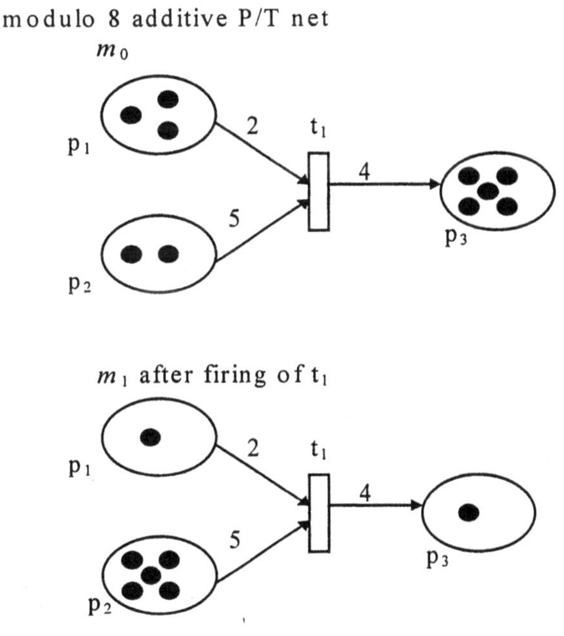

Fig. 1. Graphical expression of a modulo 8 additive Petri net.

Definition 14. An algebraically generalised P/T net GPN, where $\Gamma = (\mathbf{Z}^+, \cdot)$ ($\cdot$ denotes standard multiplication) is called *positive integer multiplicative P/T net*.

Remark 15. In accordance with previous definitions standard P/T nets may be also called *nonnegative integer additive P/T nets*. Generally, we can form the name of a P/T net type by adding of names of a carrier and an operation of a monoid before.

Example 16. On Fig. 2 there is a multiplicative P/T net with 1 place, 3 transitions, $\boldsymbol{I} = [1, 1, 4]$, $\boldsymbol{O} = [4, 2, 1]$ and $\boldsymbol{M}_0 = 1$ before and after firing of sequence of transitions $\sigma = \{t_2, t_1, t_1, t_3, t_3\}$. Formally, we have $\boldsymbol{M}_5 \cdot \boldsymbol{I} \bullet. \boldsymbol{b}_\sigma = \boldsymbol{M}_0 \cdot \boldsymbol{O} \bullet. \boldsymbol{b}_\sigma$, i.e.
$\boldsymbol{M}_5 \cdot [1, 1, 4] \bullet. [2, 1, 2]' = 1 \cdot [4, 2, 1] \bullet. [2, 1, 2]'$, i.e. $\boldsymbol{M}_5 \cdot 1^2 \cdot 1^1 \cdot 4^2 = 1 \cdot 4^2 \cdot 2^1 \cdot 1^2$, i.e. $\boldsymbol{M}_5 \cdot 16 = 32 \Rightarrow \boldsymbol{M}_5 = 2$ or alternatively using incidence matrix given by (element by element) division of matrix $\boldsymbol{O}$ by matrix $\boldsymbol{I}$: $\boldsymbol{M}_5 = \boldsymbol{M}_0 \cdot \boldsymbol{C} \bullet. \boldsymbol{b}_\sigma = 1 \cdot [4, 2, 1/4] \bullet. [2, 1, 2]' = 1 \cdot 4^2 \cdot 2^1 \cdot (1/4)^2 = 32/16 = 2$.

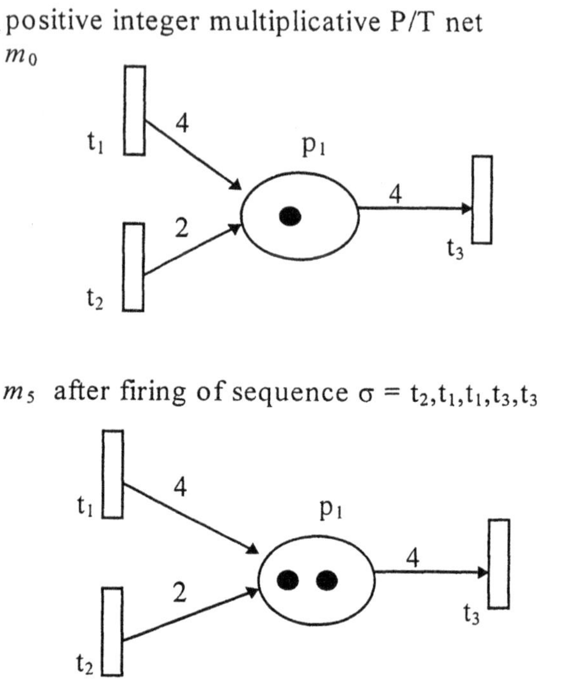

Fig. 2. Graphical expression of a positive integer multiplicative P/T net.

5. CONCLUSIONS

A concept of an extension of place/transition Petri nets based on a generalisation of the algebra used in the net dynamics is presented in the paper. The aim of the extension is to grant more general class of objects that preserves main advantages of standard P/T nets, which consist of graphical and linear algebraic description of a modelled system. A formal definition of Algebraically Generalised P/T Net is given in the paper. A non-traditional forms of P/T nets were obtained using this extension. Main subjects of the further research consist in finding of other algebras that could be eligible for the use in algebraically generalised P/T nets (e.g.monoid of polynoms (over a commutative ring) with multiplication, etc.) and in development of analytical methods in the general level as well as in the level of single types of algebraically generalised nets, such as mentioned modulo P/T nets and multiplicative P/T nets. The other direction of research presented in (Juhás, 1996a) may investigate the possibility of introduction of the state equation into theory of automata. The possible direction of the further research lies also in the generalisation of the algebra used in other types of Petri nets, as it is made for coloured Petri nets in (Juhás, 1996b).

REFERENCES

Desel, J., K. P. Neuendorf and M. D. Radola (1996). Proving nonreachability by modulo-invariants. *Theoretical Computer Science* **153** (1-2), 49-64.

Jensen, K. (1992). *Coloured Petri nets. Volume 1: Basic concepts*. Springer, Berlin.

Jensen, K. (1994). *Coloured Petri nets. Volume 2: Analysis methods*. Springer, Berlin.

Jensen, K. and G. Rozenberg (1991). *High level nets: theory and application*. Springer, Berlin.

Juhás, G. (1996a). An algebraic generalisation of Petri nets, In *Proc. 30th International Conference Modelling and Simulation of Systems, MOSIS'96* (J. Stefan, Ed), 69-74. MARQ, Ostrava, CZ.

Juhás, G. (1996b). On algebraic structures based extension of coloured Petri nets. In *Proc 2nd Conference Process Control'96*, 137-140, University of Pardubice, Pardubice, CZ.

Juhás, G. and M. Kocian (1995). Invariant relation behaviour in Petri nets, In: *Proc. of the 1995 European Simulation Multiconference, ESM '95* (M. Snorek, M. Sujansky, and A. Verbraeck, Eds.), 160-164. [SCS] Publication, Ghent.

Kruckeberg, F. and M. Jaxy (1987). Mathematical methods for calculating invariants in Petri nets, In: *Advances in Petri Nets 1987, Lecture Notes in Computer Science* **266** (G. Rozenberg, Ed.), 104-131. Springer, Berlin.

Lautenbach, K. Linear algebraic techniques for place/transitions nets (1986) In: *Proc. of an Advanced Course on Petri Nets 1986, Lecture Notes in Computer Science* **254** (W. Brauer, W. Reisig, G. Rozenberg, Eds.), 142-167. Springer, Berlin.

MacLane, S. and G. Birkhoff (1968). *Algebra.* The Macmillan Company, New York.

Memmi G. and G. Roucariol (1980). Linear algebra in net theory. In: *Net theory and applications, Lecture Notes in Computer Science* **84** (W. Brauer, ed.). Springer, Berlin.

Murata, T. (1989). Petri nets: properties, analysis and applications. *Proc. of the IEEE* 77 (4), 541-588.

Reisig, W. (1985) *Petri nets: an introduction.* Springer, Berlin.

Valk, R. (1981). Generalizations of Petri nets. In: *10th Symp. Mathematical Foundations of Computer Science, Lecture Notes in Computer Science* **118,** 140-155. Springer, Berlin.

A MODEL BASED APPROACH TO THE SUPERVISORY CONTROL OF FMS

Jaroslav Fogel

Institute of Control Theory and Robotics, Slovak Academy o Sciences, Dubravska 9, 842 37 Bratislava, Slovakia

Abstract. A model based approach to the supervisory control of FMS is presented in the paper. The proposed model comprises both the structural and the behavioural specification of the manufacturing system. The statecharts visual formalism is proposed for the structural specification, and temporal logic operators are proposed for the behavioural specification. The behavioural specification is translated into the statecharts syntax as the guarded constraints on the transition firing. The closed- loop control consists of monitoring module, controller module, synchronization block, and mentioned model of the system. The control decisions of the controller module are given according to the nominal path, and the actual state configuration of the model.

Keywords. Events, Statecharts, Temporal logic, Supervisory control, Concurrent systems, Manufacturing processes, Model.

1. INTRODUCTION

The manufacturing system is the productive research area of many researchers during last several years. Many effort of the research community is devoted to develop suitable model from the point of view of the control purposes. The supervisory control consists of several tasks, e.g., the monitoring of the system behaviour via sensory feedback, control evaluation in accordance with a supervisory control law, and control implementation via appropriate device programs on the local control level. The existing supervisory controllers are based on several techniques such as Petri nets (Zhou and DiCesare, 1993), controlled automata (Brandin, 1996), (Lauzon, *et al.*, 1996), and temporal logic (Lin and Ionescu, 1994).

The proposed supervisory controller described in this paper is based on the statecharts model of the manufacturing cell with the behaviours specified by the temporal logic operators. The goal of the control is to reach the desired state following the nominal path.

2. FMS MODELLING

The flexible manufacturing systems (FMS) belong to the set of discrete event dynamic systems (DEDS), the individual devices of which can be naturally modelled by finite state automata (Brandin, 1996), (Lauzon, *et al.*, 1996). In (Lauzon, *et al.*, 1996), the shuffle product of the individual resources' models is used to create the automaton plant model. In the paper (Fogel, 1997), the different approach based on the statecharts formalism (Harel,*et al.*, 1987) is presented. This model will be used in the next parts of the paper.

Shortly, according to the mentioned citation, the device of FMS is modelled as OR -state representing the state life cycle of it. The plant model is defined.

Definition :

The plant model is the parallel composition of models $M_1, M_2, ...M_n$, and is defined as AND-state composed from OR-substates $M_1, M_2, ...M_n$, in which

(1) $M_i, i = 1, ...n$ are models of the devices

(2) The state configuration x is the maximal orthogonal set of basic states, where maximal orthogonal set is a set of states that the system can be in simultaneously.
(3) The system configuration associated with the time instant t_{i+1} is a tuple $SC = (x, E_{ext}, E_{int}, TM, C)$, where x is a maximal state configurations, E_{ext} is a set of external events that have occurred by the environment in $< t_i, t_{i+1})$, E_{int} is a set of internal events occurring in the model in this time interval, TM is a set of the timeout events and their time of occurrence related to the current time interval, C is a set of current value of conditions

According to the semantics of statecharts described in (Harel and Naamad, 1995), the asynchronous time model assumes that system reacts whenever an external change occurs, allowing for several external changes to occur simultaneously, and allowing several steps to take place within a single point in time.

3. THE BEHAVIOURAL SPECIFICATION OF FMS

The behavioural specifications of FMS can be classified into (Brandin, 1996):

- logic-based specification such as safety (absence of deadlock, mutual exclusion, etc.), liveness (termination, total correctness, cycles), synchronization conditions, operations sequencing conditions, part routing conditions
- temporal production specification, e.g., production time
- utility optimality specifications, e.g., costs.

Temporal logic is appropriate for specification of the behaviour for the system where the operational or denotational semantics is too fine-grained to make useful statements about the overall behaviour of a complex system. In (Manna and Pnueli, 1993), temporal logic formalism has been proposed for specification and verification of reactive systems. Recently, it has been applied to the modelling and design of DEDS (Lin and Ionescu, 1994).
The syntax and semantics of the temporal logic operators used in this paper are given in (Fogel, 1997). Temporal logic is a modal logic, and each temporal formula is interpreted as true or false against a model of the world as in (Clarke, *et al.*, 1986). In the next part, such model based on the reachability tree will be defined.

The definition of the reachability tree for statecharts formalism is given in (Masiero,*et al.*, 1994) together with an algorithm for computing it. In the tree, each node represents a state configuration. Two configurations in the tree are connected by the arc if there is an event triggering the transition among them.

The notion of reachability of two configurations is also defined in (Masiero,*et al.*, 1994). The state configuration x_i is reachable from another state configuration x_j, if there is at least one sequence of enabled events (each event may cause a transition firing yielding the configuration change) $e_1, e_2, ...e_n$, and a set of intermediate state configurations $x_j, x_{j+1}, ...x_{j+n} = x_i$ such that $x_{j+k-1} \rightarrow^{e_k} x_{j+k}, \forall k = 1, 2, ...n$. The reachable state configuration must be the node of the reachability tree. It is easily to show that the reachability tree is a state machine defined as a triple $SM = (X, R, P)$, where

(1) X is a finite set of state configurations.
(2) R is a binary relation on X, defining the possible transitions between the configurations; the arc $(x_i, x_j) \in R$ indicates that there is a direct path between x_i and x_j. Obviously, if there is no such relationship in R, then there is no arc connecting two configurations. Every configuration must be connected to at least one other configuration.
(3) P assigns to each configuration the set of atomic propositions true in this configuration.

The path $\alpha(x_0)$ is an infinite sequence of configurations starting at $x_0; \alpha(x_0) = (x_0, x_1, x_2, ...)$ such that each adjacent tuple of configurations (x_i, x_{i+1}) is in R.

In (Fogel, 1997), the syntax and semantics of some behaviours expressed as temporal propositions, against a state machine SM are given. These propositions are concerning mainly the following behaviours: mutual exclusion, safety- precedence, synchronization conditions, operations sequencing, and temporal operation specification.

4. THE TRANSLATION SCHEME

The idea is from (Peron, 1994) in which the author showed that a transition structure completely describes a statecharts. When the basic dynamic properties of a system are given by using a specification language of higher level (for instance, by logical formulas), he proposed to use the transition structure for providing implementation of such specification, and finally, to shift it to the domain of statecharts. The transition structure is defined as an fourtuple
$TS = (T, \rightarrow, \triangleright, L)$,
where T is a set of transitions, $\rightarrow \subseteq T \times T$ is a sequence relation, $\triangleright \subseteq T \times T$ is a independence relation, and L is a set of transitions labels.

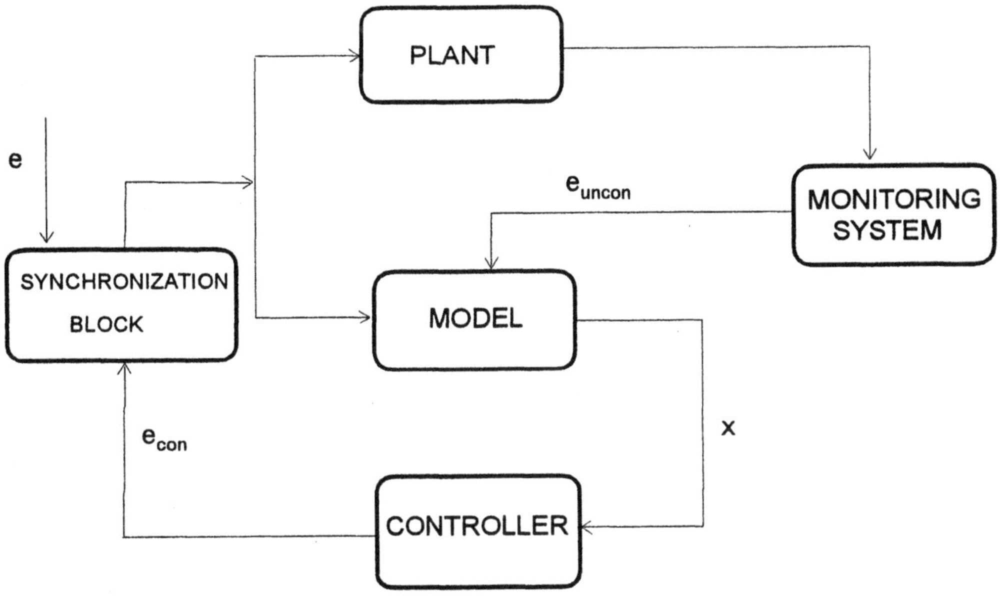

Fig. 1. Model- based supervisory control scheme

The rules for the translation of the temporal logic propositions to a semantically equivalent TS specification are given in (Fogel, 1997).

5. SUPERVISORY CONTROL OF FMS

Two hierarchical levels of FMS, the scheduling level and the control level are considered. The scheduling level determines the order in which components are loaded onto the system, and it runs the methods used to determine when the components are moved between workstations comprised in the system. The control level covers those decisions concerning the response to disturbances in the real- time operation of the system. The control action must be made in response to machine breakdown and tool wear or breakage.

The decisions coming from the scheduling level are controllable input events of the plant. All controllable events are assumed to be disabled by default. The machine breakdown or tool breakage are uncontrollable accidental events obtained by monitoring of the plant via sensory feedback, to which the system must react through the closed- loop supervisory control. Further uncontrollable events are events generated in the plant, e.g., the sensor information (for example if part is present in the machine or not) or the information about operation finishing, etc. Figure 1 shows the rough block diagram of the closed- loop control, where e_{con} are controllable events, e_{uncon} are uncontrollable events generated in the plant, and x is the state configuration of the model. The goal of the controller is to do such decisions that the plant will reach the desired state configuration. The nominal path leading from the initial configuration to the desired is determined from the reachability tree analysis. The result of this is a sequence of enabled events, and a set of intermediate state configurations fulfilling the specified behaviours.

The control law is constructed as a mapping from the set of state configurations in the set of events, determined by the nominal path. In the case of disturbances, the nominal path has to be reconstructed by the alternate path given by the model. An output event from the controller can enable the controllable input event of the plant through the transition which is triggered only if both events are synchronized (Lin and Ionescu, 1994). It is a special case of synchronization since two systems use the same event. In the opposite case, the input event of the plant is disabled. The algorithm for the closed- loop control is described in the following steps:

- Off- line part
  - (1) Construction of the composite model of the plant (Def. 1).
  - (2) Specification of the model properties by temporal logic operators.
  - (3) Translation of the properties into the guarded constraints of the transition.
  - (4) Construction of the reachability tree.
  - (5) Set the initial and final configurations.
  - (6) Find the nominal path in the tree, and set the *length_of_path*.
- On-line part
  - (1) Set the initial time interval $< t_i, t_{i+1})$, and the initial model configuration x_i^{model} for $i = 0$

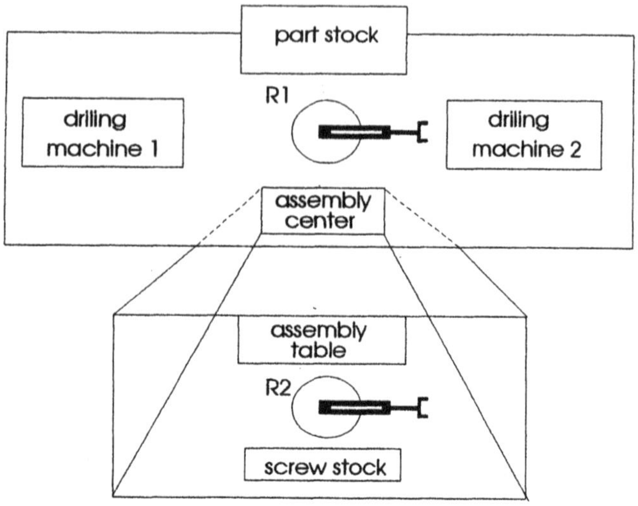

Fig. 2. Working cell description

(2) Set $k = 1$

(3) If $x_k^{path} = x_i^{model}$ then continue by next step else start procedure with x_i^{model} (actual model configuration) as a parameter to choose an alternate path from the tree, go to step 2

(4) Choose an event e given by the nominal path

(5) If e is a controllable event associated with the actual time interval, continue by next step else
wait for time interrupt, increment time interval index $i = i + 1$, and go at the beginning of this step else
if e is an uncontrollable event which has already occurred, go to step 8 else
if it has not occurred, wait till then it occurrs, increment time interval index about number of waiting time units $i = i + h$, and go to step 8 else
if it has not occurred during d-time units, then timeout signaling the faulty situation interrupts the waiting, increment time interval index $i = i + d$, go to step 3

(6) Synchronize the output event e of the controller with the event from the scheduling level

(7) If event e is enabled continue by next step else disable input event of the plant, start rescheduling procedure, and go to step 1

(8) Set $k = k + 1$

(9) If $k > length\_of\_path$ then continue by next step else go to step 3

(10) end

In the case of cyclic production the initial and the final configurations of the nominal path may coincide. In this case, the test for the algorithm finishing is needless.

6. IMPLEMENTATION ISSUE

The supervisory controller must have the capability to generate the control strategy in real time, based on feedback information on the occurrence of events in the workcell. On the host computer, where the control algorithm resides several tasks have to operate concurrently in the multitasking environment with the different priority. For example, the highest priority will be given to the task which handles the interrupts from real -time clock synchronizing timers of both levels, control and scheduling. The monitoring task which gains the information from the workcell will have the second highest priority. This information enters the model in the form of uncontrollable events. These events can be accidental (as breakdown signals) or responses to requests sent by supervisory controller (signals about operation finishing). The supervisor considers a change of state only when a response is received from the plant. The controller task which generates the control event in dependence on the actual model configuration and the nominal path, will have the third highest priority. It is supposed here that on the local level of control, the local controllers of resources are responsible for the execution of the supervisor requests.

7. A CASE STUDY

The example describes a working cell (Combacau and Courvoisier, 1990) constituted by a feeding robot R_1, two drilling machines M_1, M_2, an assembly table on which another robot R_2 works, a local stock of parts and, a local stock of screws. Figure 2 shows a view of these elements. The operating sequence to be executed is the following:

- a part of type 1 is drilled on M_1 then tapped on M_2
- a part of type 2 is drilled on M_1
- after two parts have been machined, they are assembled and screwed together on the assembly table by R_2
- after finishing an assembly operation, the assembly table rotates to the empty position

Fig 3 shows the parallel composition of the models, where state *WorkingCell* is AND decomposed into OR substates expressing resources' models. The transitions are labelled by events. The meaning of the events follows:

f : R_1 grasps a part and moves it to M_1
g : R_1 grasps a part1 from M_1 and moves it to M_2
k : R_1 grasps a part2 from M_1 and moves it to the assembly table
n : R_1 grasps a part1 from M_2 and moves it to the assembly table
m, u, y : start of moving operations of R_1 to the position *atIdlePos*

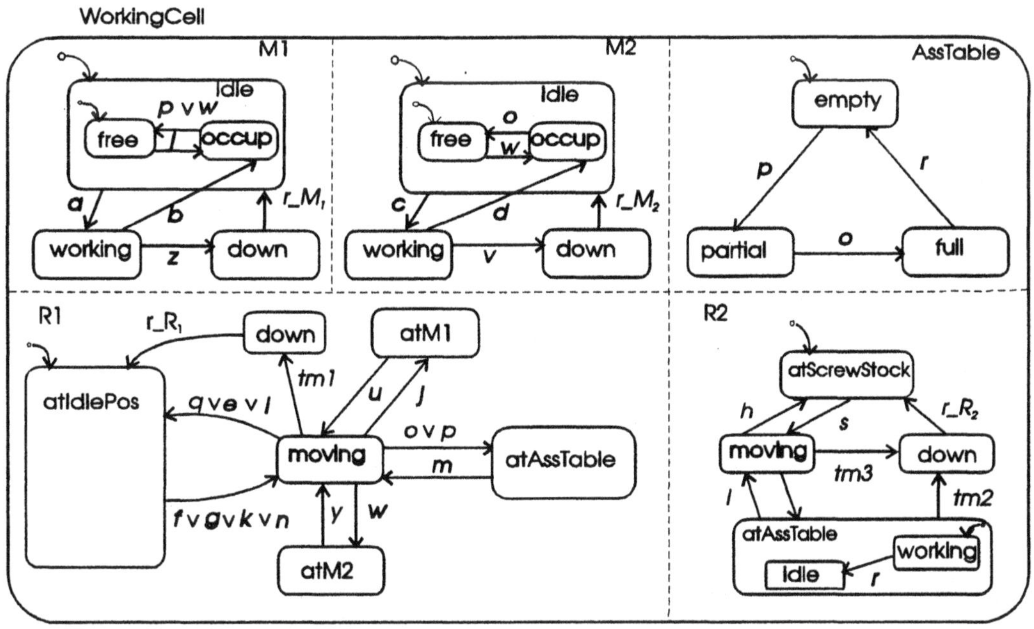

Fig. 3. Working cell in a statecharts representation

s : R_2 grasps a screw, moves it to the assembly table, and performs an assembly operation
l : R_2 moves to the position *atScrewStock*
r_M1 : start of repairing operation and return M_1 to service
r_M2 : start of repairing operation and return M_2 to service
r_R1 : start of repairing operation of robot R_1 and return it to service
r_R2 : start of repairing operation of robot R_2 and return it to service
a : sensor signal part is presented in M_1 starting the drilling operation
c : sensor signal part is presented in M_2 starting the tapping operation
z : sensor signal breakdown on M_1
v : sensor signal breakdown on M_2
j : finish of moving operation of part to the M_1
w : finish of moving operation of part1 to the M_2
b : finish of drilling operation
d : finish of tapping operation
o : finish of moving operation of part2 to the position *atAssTable*
p : finish of moving operation of part1 to the position *atAssTable*
r : finish of assembly operation, and rotating the assembly table to the *empty* position
h : finish of moving operation of R_2 to the position *atScrewStock*
q, e, i : finish of moving operation of R_1 to the position *atIdlePos*
$tm1 : \neg(a \wedge c \wedge (o \vee p))$
$tm2 : \neg r$
$tm3 : \neg h$

The behavioural specification of the working cell follows:

The mutual- exclusion propositions specify the assembly table as a critical region for both robots, it means that the moving operations op_k, op_s, and op_n, op_s started by the events k, s and n, s, respectively, can be never performed simultaneously.
These propositions are
$\Box(\neg op_k \vee \neg op_s)$
$\Box(\neg op_n \vee \neg op_s)$

Both the drill machines have a buffer of size one. Next two safety- precedence propositions explicitly specify that neither machine may start operating on a new workpiece before the workpiece currently on the machine has been transferred onwards.
$(entered(s1) \wedge \neg exit(s1))\ P\ f$
$(entered(s2) \wedge \neg exit(s2))\ P\ g$
where $s1 = M_1.Idle.free$ and $s2 = M_2.Idle.free$.
Next three safety- precedence propositions specify parts routing on the assembly table which has a buffer of size two. Only, if parts of both types are presented on the table they can be assembled.
$(entered(s1) \wedge \neg exit(s1))\ P\ k$
$(entered(s2) \wedge \neg exit(s2))\ P\ n$
$(entered(s3) \wedge \neg exit(s3))\ P\ s$
where $s1 = AssTable.partial$, $s2 = AssTable.empty$, $s3 = AssTable.full$.

Synchronization propositions are:
$\Box(a \wedge l)$, $\Box(c \wedge y)$, $\Box(f \wedge s)$
These propositions mean that operations started by events a, l or c, y or f, s can be performed in parallel.

The sequencing propositions specify the order of the operations in two successive steps. The part of type 1 must be worked on machine M_1, and then on machine M_2, and subsequently evacuated

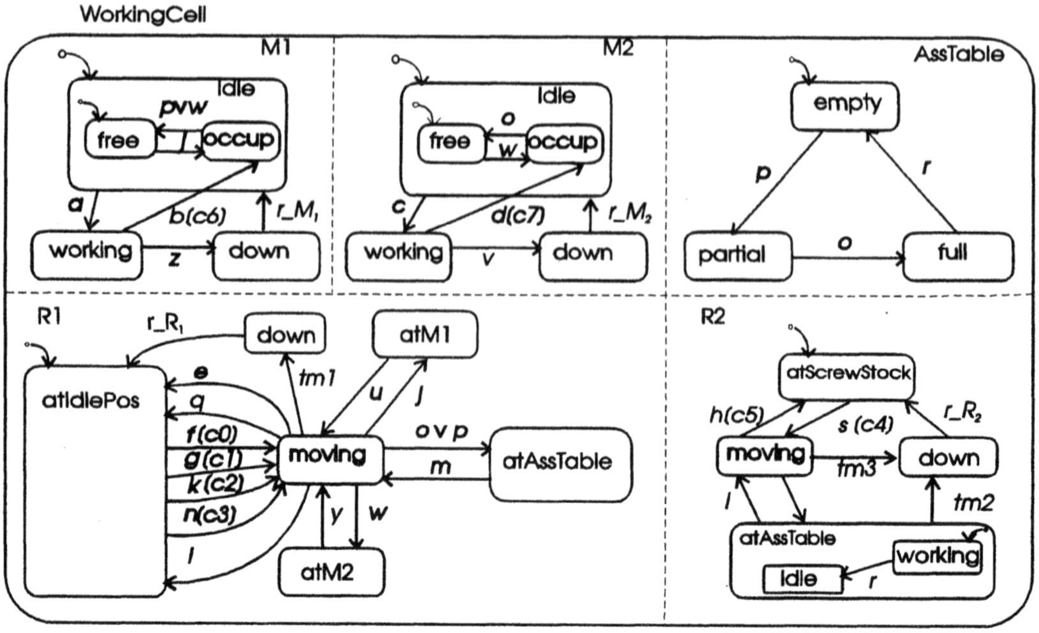

Fig. 4. Functional specification of the working cell in a statecharts representation

to the assembly table. The part of type 2 must be worked on machine M_1, and subsequently evacuated to the assembly table. The operations sequencing for the part of type 1 is specified by the propositions:
$\Box(j \rightarrow \bigcirc(a)),\ \Box(b \rightarrow \bigcirc(g)),$
$\Box(w \rightarrow \bigcirc(c)),\ \Box(d \rightarrow \bigcirc(n))$
Similarly, for the part of type 2 by the propositions:
$\Box(j \rightarrow \bigcirc(a)),\ \Box(b \rightarrow \bigcirc(k))$
Also, the following operations can be in sequence:
$\Box(i \rightarrow \bigcirc(f)),\ \Box(q \rightarrow \bigcirc(s)),$
$\Box(q \rightarrow \bigcirc(f)),\ \Box(h \rightarrow \bigcirc(k))$
These propositions enable the parallel performance of the operations of both robots and machines and speed up the assembly process.

The temporal propositions specify the upper bounds of the operations duration. The duration of the drilling operations on the machines M_1, M_2 is specified:
$\Box(a \rightarrow \Diamond_{t_1}(b)),\ \Box(c \rightarrow \Diamond_{t_2}(d))$
The duration of the moving operations of the robots R_1, R_2 is specified by propositions:
$\Box(f \rightarrow \Diamond_{t_3}(j)),\ \Box(g \rightarrow \Diamond_{t_4}(w)),$
$\Box(k \rightarrow \Diamond_{t_5}(o)),\ \Box(n \rightarrow \Diamond_{t_6}(p)),$
$\Box(u \rightarrow \Diamond_{t_7}(e)),\ \Box(y \rightarrow \Diamond_{t_8}(i)),$
$\Box(m \rightarrow \Diamond_{t_9}(q)),\ \Box(l \rightarrow \Diamond_{t_{10}}(h))$
The duration of the assembly operation of the robot R_2 is specified by
$\Box(s \rightarrow \Diamond_{t_{11}}(r))$

Figure 4 shows the model with the behavioural specification. The conditions in the square brackets which create the guarded constraints on the transition firing are defined by the following equations:
$c_0 = in(M_1.Idle.free)$
$c_1 = in(M_2.Idle.free)$
$c_2 = in(AssTable.partial \wedge \neg in(R_2.atAssTable)$
$c_3 = in(AssTable.empty)$
$c_4 = in(AssTable.full) \wedge in(R_1.atIdlePos)$
$c_5 = in(R_1.atIdlePos) \wedge in(M_1.Idle.occup)$
$c_6 = in(R_1.atIdlePos)$
$c_7 = in(R_1.atIdlePos)$
The timeout events are defined:
$tm1 : (timeout(f, t_3) \vee timeout(g, t_4) \vee timeout(k, t_5) \vee timeout(n, t_6))$
$tm2 : timeout(s, t_{11})$
$tm3 : timeout(l, t_{10})$

The nominal path of the example according to the previous specification is following
$x_0 \rightarrow^f x_1 \rightarrow^j x_2 \rightarrow^{a,u} x_3 \rightarrow^e x_4 \rightarrow^b x_5 \rightarrow^g x_6 \rightarrow^w x_7 \rightarrow^{c,y} x_8 \rightarrow^i x_9 \rightarrow^f x_{10} \rightarrow^j x_{11} \rightarrow^{a,u} x_{12} \rightarrow^e x_{13} \rightarrow^d x_{14} \rightarrow^n x_{15} \rightarrow^p x_{16} \rightarrow^m x_{17} \rightarrow^q x_{18} \rightarrow^b x_{19} \rightarrow^k x_{20} \rightarrow^o x_{21} \rightarrow^m x_{22} \rightarrow^q x_{23} \rightarrow^{s,f} x_{24} \rightarrow^j x_{25} \rightarrow^{a,u} x_{26} \rightarrow^e x_{27} \rightarrow^b x_{28} \rightarrow^g x_{29} \rightarrow^r x_{30} \rightarrow^l x_{31} \rightarrow^w x_{32} \rightarrow^{c,y} x_{33} \rightarrow^h x_8$

The control strategy is given by the nominal path for those configurations in which controllable events must occur. Table 1. shows these configurations. In the table, the following notation for the state configuration is used $x = (M_1, M_2, R_1, R_2, AssTable)$. The positions, in the state configuration which are not defined uniquely are marked by star. Last four events in the table have meaning start of repairing operations after the model reaches *down* state. In this case, the alternate path must be generated to reach the desired configuration. For instance, when the signal a does not occur during t_3 time units, timeout event triggers the transition of the robot R_1 to the *down* state. The alternate path in this case will be $x_{d\_R1} \rightarrow^{r\_R1} x_0 \rightarrow^f x_1 \ldots$

state configuration x	event
$x_0, x_9, x_{23} = (M_1.Idle.free, *, R_1.atIdlePos, *, *)$	f
$x_5, x_{28} = (M_1.Idle.occup, M_2.idle.free, R_1.atIdlePos, *, *)$	g
$x_{19} = (M_1.Idle.occup, *, R_1.atIdlePos, R_2.atScrewStock, AssTable.partial)$	k
$x_{14} = (*, M_2.Idle.occup, R_1.atIdlePos, R_2.atScrewStock, AssTable.empty)$	n
$x_2, x_{11}, x_{25} = (M_1.idle.occup, *, R_1.atM1, *, *)$	u
$x_7, x_{32} = (*, M_2.idle.occup, R_1.atM2, *, *)$	y
$x_{16}, x_{21} = (*, *, R_1.atAssTable, R_2.atScrewStock, *)$	m
$x_{23} = (*, *, R_1.atIdlePos, R_2.atScrewStock, AssTable.full)$	s
$x_{30} = (*, M_2.Idle.free, R_1.moving, R_2.atAssTable.Idle, AssTable.empty)$	l
$x_{d\_R1} = (*, *, R_1.down, *, *)$	r_R1
$x_{d\_R2} = (*, *, *, R_2.down, *)$	r_R2
$x_{d\_M1} = (M_1.down, *, *, *, *)$	r_M1
$x_{d\_M2} = (*, M_2.down, *, *, *)$	r_M2

Table 1.

8. CONCLUSIONS

In the future, much attention must be devoted mainly in the development of the algorithm for the automatic generation of the alternate path from the reachability tree for the case, if the cause of the fault is not explicitly determined in the model. Such situation probably will require the error diagnosis and the error recovery modules which extend proposed approach by knowledge engineering techniques.

The author is grateful to the Slovak Grant Agency for Science VEGA (grant No, 2/1185/97) for partial supporting of this work.

9. REFERENCES

Brandin, B. A. (1996). The Real-Time Supervisory Control of an Experimental Manufacturing Cell, *IEEE Transactions on Robotics and Automation,* Vol., 12, No.1., pp. 1- 13.

Clarke, E. M., E. A. Emerson, A. P. Sistla (1986). Automatic Verification of Finite- State Concurrent Systems Using Temporal Logic Specifications, *ACM Transaction on Progr. Languages and Systems,* Vol. 8, No. 2, pp. 244-263.

Combacau, M. M., Courvoisier (1990). A Hierarchical and Modular Structure for FMS Control and Monitoring, *1st AI Simulation and Planning in High Autonomy Systems,* Arizona.

Fogel, J. (1997). A Statecharts approach to the modelling of discrete nanufacturing systems, *7th Symposium on Computer Aided Control System Design,* Gent, Belgium, April, pp. 249- 255.

Harel, D. A. Pnueli, J. P. Schmidt, R. Sherman (1987). On the Formal Semantics of Statecharts, *Proc. 2nd IEEE Symp. Logic in Comp. Science,* New York, pp. 54- 64.

Harel, D. and A. Naamad (1995). The STATEMENT Semantics of Statecharts, *Technical report, i-Logix, Inc.,* October, pp. 1- 30.

Lauzon, S. C., A.K. Ma, J.K. Mills, B. Benhabib (1996). Application of Discrete- Event System Theory to Flexible Manufacturing, *IEEE Control Systems* February, pp. 41- 47.

Lin, J., D. Ionescu (1994). A Reachability Synthesis Procedure for Discrete Event Systems in a Temporal Logic Framework, *IEEE Transactions on Systems, Man, and Cybernetics,* Vol. 24, No. 9., pp. 1397- 1406.

Manna. Z., A. Pnueli (1993). A Temporal Proof Methodology for Reactive Systems, *Program Design Calculi* (Ed. Manfred Broy), Springer-Verlag, pp. 287- 323.

Masiero, P. C., J. C. Maldano, I. G. Boaventura (1994). A reachability tree for statecharts and analysis of some properties, *Information and Software Technology,* 36 (10), pp. 615- 624.

Peron, A. (1994). Statecharts, Transition Structures and Transformations, *Proc. TAPSOFT'95 Lecture Notes in Computer Science 789,* Springer, Berlin, pp. 806- 821.

Zhou, M. and F. DiCesare (1993). Petri Net Synthesis for Discrete Event Control of Manufacturing Systems, *Kluver Academic Publishers,* Boston.

PROGRAM SOLUTION OF SUPERVISORY CONTROL BASED ON PETRI NETS

Jana Flochová, Branislav Hrúz, Peter Jirsák

Automatic Control Systems Department
Faculty of Electrical Engineering and Information Technology
Slovak Technical University
Ilkovičova 3, 812 19 Bratislava, Slovak Republic
phone: +42-7-791698, e-mail: hruz@kasr.elf.stuba.sk

Abstract: This paper describes methods based on Petri nets analysis techniques for constructing a supervisory controller of discrete event dynamic systems. Two methods (method based on reachability tree analysis and on-line method based on matrix equations) have been designed. Programs in C++ have been written to solve supervisory control problem and real time control of a discrete event dynamic system.

Keywords: Discrete event dynamic systems, Supervisory control, Petri-nets, Reachability, Reachable states, Computer control system design, Computer testing, Programs, Program controllers

1. INTRODUCTION

The problem of supervisory control of discrete event dynamic system (DEDS) was presented by Ramadge and Wonham (1987) and since then has been studied extensively. Ramadge and Wonham, (1989), Ho (1990) described the dynamic behaviour of DEDS with a possibly infinite set of states and a finite set of events. The adopted level of abstraction assumes that events appear spontaneous, asynchronous and instantaneous and that always only one event occurs in some instant of time. For a given DEDS, it is of interest to synthesise a supervisor that prevents the occurrence of undesirable states of the DEDS and that guarantees that certain termination states are reached. Disabled events are certainly prevented from occurring and enabled events are not necessarily forced to occur. The mathematical framework of Ramadge and Wonham's approach is based on finite state machines and formal languages. A DEDS is regarded as a finite state machine formal language generator.

Balemi et al. (1993) studied the problem how to develop the methods of Ramadge and Wonham and how to add the new one in order to improve the application possibilities of the practical control design. In their opinion, the lack of automatic synthesis applications that base themselves directly on the Ramadge and Wonham framework is due to three classes of key problems:

1. Model interpretation. The logical plant model proposed in supervisory control theory assumes a plant that "generates" event spontaneously. This model is not appropriate for most real systems. Real systems usually react to commands as inputs with responses as outputs. One of the main points there is to embed the input-output system structure in the modelling of systems.
2. Computational complexity and state-space explosion.
3. Supervisor implementation. The theory provides a few indication how to implement the control software for a DEDS. The synthesis produces an abstract supervisor model, which is directly not

usable in its form for the control software of a real system.

Balemi presented an application of supervisory control theory to a semiconductor manufacturing piece of equipment and gave an input-output interpretation of supervisory control theory and performed a sample synthesis for an oxide growth technology where states of the order of 10^6 occurred.

Petri nets are graphical and mathematical modelling tool for describing and studying information processing systems that are characterised as being concurrent, asynchronous, distributed, parallel, nondeterministic, and/or stochastic. Petri nets are, mainly in their graphical form, a transparent and understandable tool for the supervisory control problem. The advantage of Petri nets over simpler models, such as state machines, is given by the fact that there often exist analysis techniques permitting an efficient solution of the control problems based on the net structure.

Many researchers have used Petri net as a tool for modelling, analysing and synthesising control law for a DEDS. Holloway and Krogh (1990) introduced cyclic controlled marked graphs as a special class of controlled Petri nets, Screenivas and Krogh (1992) used Petri nets with inhibitor arcs to model infinite state supervisor, Guia and DiCesare (1994) have done work on Petri net languages and their use in supervisory control theory, Yamalidou et al. (1996) describe a computationally efficient method for constructing a Petri net feedback controller based on place invariants - a structural Petri net property, Juhas define Algebraically Generalised Coloured Petri Nets (AGCPNs), the extension enables to use not only non-negatives integers (numbers of tokens) as places colours states, but elements of subsets of arbitrary commutative groups (spaces over ring of integers). Hrúz et al. (1996) developed method for removing conflicts from the Petri net models specifying control of the flexible manufacturing systems.

2. AUTOMATA AND REGULAR LANGUAGES BASED SUPERVISORY CONTROL

Ramadge and Wonham defined a controlled DEDS (a plant) as a 5-tuple

$$G = (Q, \sum(q), \delta, q_0, Q_m) \qquad (1)$$

where Q is the set of the states q_i, Σ is a finite alphabet used to label transitions (events) between states, $\sum(q) \subseteq \sum$ is a possible subset of the transitions for the state q, $\delta: \sum(q) \times Q \to Q$ is a partial transition function , $q_0 \in Q$ is the initial state and $Q_m \subseteq Q$ is a subset of states called marker states. The transition function δ can be extended to a partial function on $\sum^* \times Q \to Q$ by defining $\delta(\varepsilon, q)=q$, $q \in Q$ and $\delta(s\sigma, q) = \delta(\sigma, \delta(s,q))$ whenever $q' = \delta(s,q)$ and $\delta(\sigma, q')$ are both defined and $\sum^*$ is the set of all possible strings of events including also the empty string e. The language generated by G is

$$L(G) = \left\{ w: w \in \sum{}^* \; and \; \delta(w, q_0) \in Q \right\}. \qquad (2)$$

The language marked by G is

$$L_m(G) = \left\{ w: w \in L \; and \; \delta(w, q_0) \in Q_m \right\}. \qquad (3)$$

$L(G)$ is the set of all possible finite sequences of events that can "occur", while $L_m(G) \subset L(G)$ is a distinguished subset of the generated sequences that may be "marked" (e.g. completed tasks).

To model a supervisory control Ramadge and Wonham partition the set of events Σ into uncontrollable and controllable events $\sum = \sum_u \bigcup \sum_c$. The events in $\sum_c$ can be disabled at any time, while those in $\sum_u$ model events over which the controller has no influence, e.g., machine breakdown in a manufacturing system etc.

Let $\Gamma_c = \{0,1\}^{\Sigma_c}$ denote the set of all different binary assignments to the elements of Σ_c. Each assignment $\gamma_c \in \Gamma_c$, i.e., each function $\gamma_c : \sum_c \to \{0,1\}$ is a control pattern. An event (with label) $\sigma \in \sum_c$ is said to be enabled by γ_c when $\gamma_c(\sigma) = 1$, and is disabled otherwise. It is convenient to extend each γ_c to a pattern function $\gamma : \sum \to \{0,1\}$ by defining $\gamma(\sigma) = 1$ for each $\sigma \in \sum_u$. It means, each uncontrollable event is always enabled. We can extend Γ_c to Γ and define $\omega: L \to \Gamma$. A supervisor is a system that changes the control pattern dynamically. The supervisory control consists simply in switching the control pattern through a sequence of elements of Γ, in order to achieve a behaviour of DEDS given by a sublanguage $K \subseteq L$.

3. PETRI NETS

Petri nets being graphical and mathematical modelling tools for describing and studying information processing systems, may be interpreted as a bipartite oriented graph, where one type of nodes represents a system conditions (so called places) and other one represents the events (so called transitions). They are interconnected by oriented arcs. The number of tokens in a place marks the particular system state. Then an occurring of an event (firing of a transition) changes the states of places according to the weights of graph arcs - i.e. some tokens are added or removed from each place according to the weight and the direction of related arcs.

Formal definition of a Petri net: (Murata 1989, David and Alla 1994). A Petri net is a 5-tuple

$$PN = (P, T, F, W, M_0) \qquad (4)$$

where:

1. $P = \{p_1, p_2, ... p_m\}$ is a finite set of places,
2. $T = \{t_1, t_2, ... t_n\}$ is a finite set of transitions,
3. flow relation $F \subseteq (P \times T) \cup (T \times P)$ is a set of arcs,
4. $W: F \rightarrow \{1,2,3...\}$ is a weight function of arcs,
5. $M_0: P \rightarrow \{0,1,2,3...\}$ is the initial marking,

Methods of analysis for Petri nets may be classified into the following four groups: the coverability (reachability) tree method, the matrix-equation approach, reduction or decomposition techniques and languages.

Interpreted Petri nets. The expression 'interpreted Petri nets' can be applied to various interpretations according to the use wished to be made of it, e.g. interpretations to the description of software, hardware, logic controllers, formal languages etc. A class of Petri nets interpreted for the supervisory control design described by Flochová and Hrúz (1996) and denoted as C-Petri net is a 7-tuple

$$PNC = (P, T, F, W, M_0, CC, A) \qquad (5)$$

where CC is a set of dynamic logical control conditions and an element of CC should be assigned to each controlled transition of the Petri net PNC by the bijection $A: T \rightarrow CC$. The rules for token flow are similar to those of generalised or ordinary PN with the addition that a transition t_j is firable if control condition $cc_j \in CC$ is true.

4. SUPERVISORY CONTROL DESIGN

Our approach consists of the following steps:

1. The system to be controlled is modelled by a Petri net (consisting of one or more subnets) and the reachability graph for this net is designed
2. Analysis of the oriented paths in the reachability graph and determination of inadmissible nodes (markings) and not allowed edges (transitions)
3. Determination of values of the dynamic control conditions cc_j for all node-markings m_i and for all enabled transitions t_j in every node
4. Real time control of analysed discrete event dynamic system

The reachability tree design and analysis and supervisory control design are solved by programs written in Borland C++. The programs need processors PC 386 and higher with mathematical coprocessor. The user interface is in colour graphic VGA mode 640*48O. The design of the programs is in *Fig. 1-2*. The user interface of the programs is written in English and/or Slovak language. The modules enable to solve following problems:

1. The complete Petri net analysis of the system (plant) to be controlled, reachability tree construction, incidence matrix design and correction, matrix analysis
2. The supervisory control problem formulation
3. The supervisor design and testing

Two methods of supervisory control (direct based on reachability tree analysis and on-line computation algorithm) have been solved by programs. A part of program solutions are control of sample applications and on line help.

The main menu is fully keyboard and/or mouse driven and consists of the following items:

- Incidence matrix
    - design
    - display
    - print
    - modify and/or correct
- Reachability three
    - input of the initial marking
    - reachability tree construction
- Two methods of supervisory control (algorithm based on reachability tree analysis and on line algorithm)
- Control of guided vehicles in a manufacturing cell used by Holloway and Krogh 1990, Li and Wonham 1994, Yamalidou at al. 1996
- Random real events generators
- On line help
- Quit

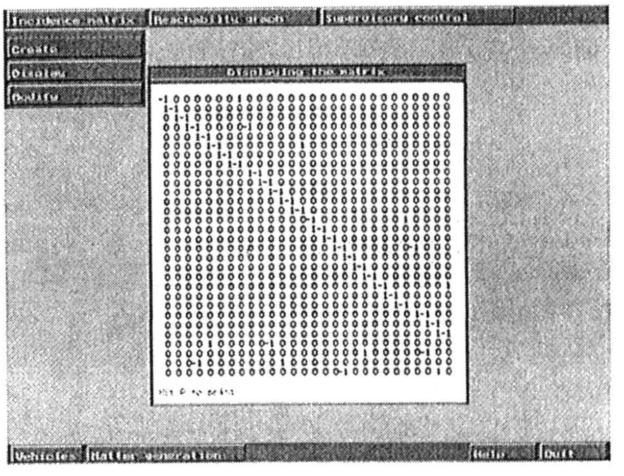

Fig 1. Interface of the supervisory control program.

4.1. The supervisory control method based on reachability tree analysis

The algorithm consists of the following steps:

Without possibility to return to home states or nodes.

Inadmissible nodes in the reachability graph are labelled as IA_nodes.

Repeat until new inadmissible nodes or new not allowed edges exist:

If the edge of the reachability three (RT) is uncontrollable and its successor is IA_node the preceding node is labelled as IA_node.

If the edge of the RT is controllable and its successor is IA_node the edge is not allowed and labelled as NA_edge.

If all the edges going out of a node are not allowed the node is labelled as IA_node.

The labels of edges going out of a node determine the values of the dynamic control conditions cc_j

With possibility to return to home states or nodes.

Begin in home state. Analyse from which admissible nodes it is possible to reach home nodes using only allowed edges. The admissible nodes from which it is not possible to reach home nodes are converted into inadmissible ones.

4.2. The on line supervisory control method

The on-line supervisory control algorithm is the modification of the algorithm in paragraph 4.1. for the case of unbearable state-space explosion. The program control a process without enumerating all reachable nodes-markings with help of the set of all known inadmissible markings $\boldsymbol{RI}(PN, M_0)=\boldsymbol{RI}$. It is possible to express the set $\boldsymbol{RI}$ as a list of states or analytically with mathematical equations. The user draws the Petri nets of the system to by controlled, gives the program following inputs data: incidence matrix $\boldsymbol{D}$, initial marking $\boldsymbol{M_0}$, the set of inadmissible markings $\boldsymbol{RI}$, the number and indexes of uncontrollable events. Program analyses the topical system state, extends the set RI, allows or forbids events with help of the fundamental-flow equation of the controlled system calculating all possible new markings and comparing them with the set $\boldsymbol{RI}$.

The control of following three discrete event dynamic systems was solved by the programs.

The cat and mouse example introduced by Ramadge and Wonham (1987) involves a maze of five rooms where cat and mouse can circulate (Fig 2).

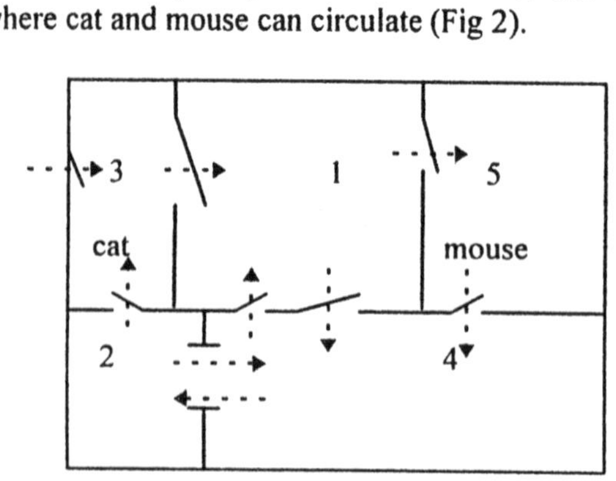

Fig.2 Example with a maze

The rooms are connected with doors through which the animals can pass. The problem is to control the doors so that the cat and mouse can never meet in the same room. Each animal should be able to return to its original room and should have maximum freedom of movement. The example specifies that all doors except of the door between room 2 and 4 are controllable. Both methods (direct method using RT construction and on line method) are suitable for control and were successfully used to solve the cat and mouse maze problem.

The on-line control of flexible manufacturing cell (Holloway and Krogh 1990) was solved with help of a set of algebraic equations, mathematical descriptions of forbidden states. The example illustrates the control of three workstations (two part receiving stations and one completed parts station) and five automated guided vehicles, that transport material between the stations and of four shaded forbidden zones in which vehicles trajectories cross the floor. A forbidden situation arises when two vehicles are present in a zone at the same time. In this system The set of reachable markings consists of 7 741 440 elements and the number of forbidden states overflows one million (1 345 563). The controller of this system is responsible for co-ordinating the departures of vehicles to prevent meeting of two vehicles in the same zone. This problem was solved with help of the adjusted feedfordward control algorithm. The forbidden states were specified as the set of four equations and. The values of the dynamic control conditions cc_j were set according to the equations checking.

A robotic cell shown in Fig.3.. Two robots R1 and R2 transfer two kinds of parts (A and B) in the cell.

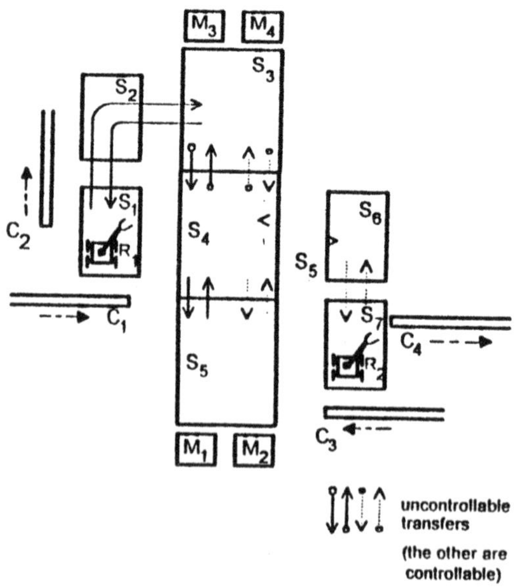

Fig.3 Example with a robotic cell.

Robot R1 picks up a part A from the input conveyor C1 and transports it either through the rooms S2 and S3 to the machine M3 or through S4 and S5 to the machine M1. Then after machining R1 transports the part in the same way back to S1 and puts it on the output conveyor C2. The robot R2 transports a part B from input conveyor C3 through S7, S6, S4, and S3 to the machine M4 and back onto C4 , or through S7, S6, S4 and S5 to the machine M2 and back onto S4. All transitions between rooms are controllable except those between the rooms S3 and S4. Both methods described in sections 4.1. and 4.2. were used to solve this problem.

CONCLUSIONS

This paper presents two program oriented methods for solving supervisory control problems. First method is based on reachability tree analysis and guaranties non-existence of forbidden states and return to home states. Second on-line method is based on matrix equations and it is used in control problems with extensive state explosion. Programs written in C++ can be used to solve the real time supervisory control of a discrete event dynamic system.

REFERENCES

Balemi,S., Hoffman,G.J., Gyugyi, P., Wong-Toi, H. and Franklin, G.F. (1993): Supervisory control of a rapid thermal multiprocessor. *IEEE Trans. on Automatic Control,* **38**, pages 1040-1059.

Čapkovič,F. (1994): A Petri net-based approach to synthesis of control systems for discrete event dynamic systems. *Preprints of the 1st IFAC Workshop on New trends in design of control systems*, Bratislava, September 7-10, 1994, pages 398-403.

David,R. and Alla,H. (1994): Petri nets for modelling of dynamic systems-a survey. *Automatica*, 30, pages 175-202.

Flochová, J., Hrúz, B.: Supervisory control for discrete event dynamic systems based on Petri nets. *Proceeding of International conference on Process control,* Horní Bečva, Jun 1996, Vol.2, pages 80-83

Giua,A., DiCesare,F.(1994): Blocking and controllability of Petri Nets in supervisory control. *IEEE Trans. on Automatic Control*, **39**, pages 818-823.

Ho, Y.C. (1990): Analyzing complexity and performance in a man-made world. An introduction to discrete event dynamic systems. *Preprints of the 11th IFAC World Congress*, Tallin, August 13-17,1990,Vol.1, pages 66-70.

Holloway L.E., Krogh, B.H. (1990): Synthesis of Feedback Control Logic for a Class of Controlled Petri Nets . *IEEE Trans. on Automatic Control*, **35**, pages 514-523.

Hrúz,B.,Niemi, A.J., Virtanen, T.: (1996): Composition of conflict-free Petri net models for control of flexible manufacturing systems. *Proceeding of the 13 th World Congress*, San Francisco, 1996.

Hrúz,B. (1994): Discrete event systems modelling and real-time control. Journal of electrical engineering, **45**, pages 363-370

Juhás,G.: On algebraic structures based extension of coloured Petri nets. *Proceeding of International conference On Process control,* Horní Bečva, Jun 1996,**2**, pages 137-140

Li, Y., Wonham, W.M. (1993): Control of Vector Discrete-Event Systems I - The base model. *IEEE Trans. on Automatic Control*, **38**, pages 1214-1227.

Ramadge,P.J., and Wonham,W.M. (1989): The control of discrete event systems. *Proceeding of the IEEE*, **77**, pages 81-98.

Ramadge,P.J., and Wonham,W.M.(1987): On the supremal controllable sublanguage of a given language, *SIAM J. Control and Optimisation*, **25**, pages 637-659.

Sreenivas, R.S., Krogh, B.H. (1992): On Petri Nets Models of Infinite State Supervisors. *IEEE Trans. on Automatic Control*, **37**, pages 274-277.

Yamalidou,K., Moody, J. ,Lemmon, M., Antsaklis, P.(1996): Feedback Control of Petri Nets Based on Place Invariants. *Automatica*, **32**, pages 15-28.

STATISTICAL APPROACH TO CREATION OF MACROSITUATIONS OF COMPLEX SYSTEMS

E.OCELIKOVA

Technical University of Kosice, Department of Cybernetics and AI, 041 20 Kosice, Slovakia
e-mail:ocelike@ccsun.tuke.sk

Abstract. The contribution deals with the problem of creation of macrosituations of complex systems by methods of cluster analysis. It presents method of analysis of probability distribution function for the creation of clusters of multidimensional data. It describes experience and results obtained during the classification of extensive sets of the situations which were represented by real pictures of the Earth surface.

Key Words. Complex systems, multicriterial classification, situational control, cluster analysis, probability distribution function

1. INTRODUCTION

The core of the control of complex systems is to cover an axtensive set of situations in the complex system by a number of control strategies. Situation of a complex system means the entire state of the complex system with respect to the external surroundings.

We assume that the state can be describe by a set of symptomps so that each situation is characterised by a m-dimensional scale vector. As the number of control strategies is in most of the case significantly lower than the number of possible situation in complex system, it is necessary to create classes of related situations, so called macrosituations number of which corresponds to the number of available strategies.

The main goal of the situation control of the complex systems is therefore to sort situations into classes with respect to their similarities and differences and to assign control strategies to these classes.

Methods of classification of the situations of complex system to the competent macrosituations may be various in accordance with a different classifying points of view like statistical, multicriterial, fuzzy, heuristic, etc.

In the following we will look at the cluster analysis which grants suitable means of the creation of macrosituation.

2. MACROSITUATIONS AND CLUSTER ANALYSIS

Cluster analysis is defined (Anderberg, 1973; Lukasova and Sarmanova, 1985; Hartigan, 1975) as a general logical approach formulated as a procedure that groups the objects into the classes - clusters. Clustering is based on the similarity and the unsimilarity of the objects.

One of the basic problems of cluster analysis is the conception of the mutual similarity between the objects and its quantitative expression. The commonest way to express this similarity is through the metrics.

Because the objects are characterized by m-dimensional vectors that are the points of m-dimensional euclidean space, it is possible to use euclidean distance as a measure of the similarity. The shorter distance is between objects in the space, the more similar objects are. Besides Euclidean metrics there are also another metrics suitable for judging of the mutual similarity between the objects, e.g. Sokal's metrics, Minkowski's metrics, sup-metrics, etc.

In case of the situational control the situations of the complex system are the objects of cluster analysis. Created clusters represent the classes of the situations, the macrosituations.

Let $\mathbf{x} = (x_1, x_2, \ldots, x_m)$ and $\mathbf{y} = (y_1, y_2, \ldots, y_m)$ be two situations of complex system. The coefficients of their similarity are evaluated by following formulas of metrics mentioned before:

- Euclidean metrics

$$d_E(x,y) = [\sum_{i=1}^{m}(x_i - y_i)^2]^{1/2} \quad (1)$$

- Sokal's metrics

$$d_S(x,y) = [1/m \sum_{i=1}^{m}(x_i - y_i)^2]^{1/2} \quad (2)$$

- Minkowski's metrics

$$d_M(x,y) = \sum_{i=1}^{m}|x_i - y_i| \quad (3)$$

- sup-metrics

$$d_\infty(x,y) = \max_i (|x_i - y_i|). \quad (4)$$

3. MODES ANALYSIS METHODS

From the mathematical statistics point of view the set of objects assigned for clustering represents a selection from a statistical set represented by a multidimensional random vector. It is logical to think of a statistical law of probability distribution in the defined set. Clustering methods based on an attribute estimation of distribution functions of random vectors are established on probabilistic apprehension of cluster creation. The clusters can be defined according to the existence and the position of probability distribution function modes. The mode is a certain value of random variable where the probability distribution function has a local maximum. The function determination for a discrete random variable is the same as for a continuous variable. The variational interval of values is divided to subintervals where the frequency of occurence of the random variable values is monitored. Followingly the corresponding histogram gives a concept of a course of the distribution function, the existence and the position of modes probability distribution function, fig.1.

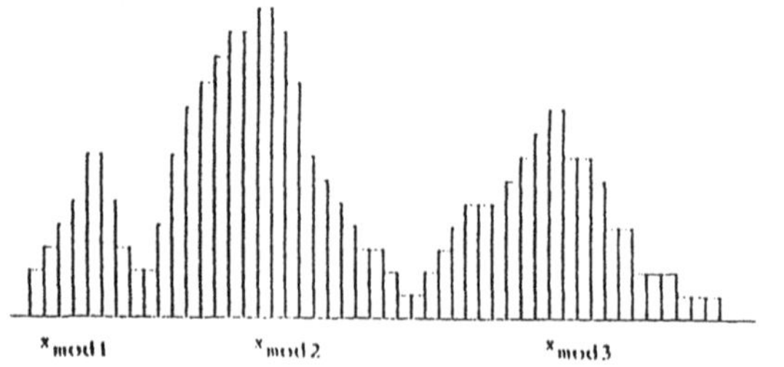

Fig.1 The histogram showing the number and the position of modes

The clusters are searched according to "dense areas", i.e. such areas where a relatively big number of points is located while surroundings is almost is without points. From a cluster analysis point of view the dense occurence of points is expected near the probability distribution function mode of a particular random vector. Consequently the mode location corresponds to the location of the typical cluster point.

In case multidimensional random vector the position of probability distribution function modes is estimated the same way as in one-dimensional random variable. A m-dimensional block of Euclidian m-dimensional space is the model for area of all possible realizations of m-dimensional random vector. That area have to be divided to m-dimensional subintervals. Their models are also m-dimensional blocks where the frequency of point occurance is monitored. The intervals with the highest frequency of occurance are separated from themselves by intervals with the relatively low occurance frequency. Those intervals enable to maken an estimation of modes location in a defined set.

One of the methods that seeks the dense areas in models space is a modified Kittler's method of seeking modes - ANMO method.

ANMO method requires giving one control parameter α - the surroundings of an examined point. At the beginning of clustering process there must be specified the number of points in α-surroundings of each point. I.e. for O_i point the α-surroundings is defined as the set of all points $O_j \in \mathbf{O}$, $d(O_i,O_j) < \alpha$, where $\mathbf{O}$ is the set of points that represent a certain set of objects and d is suitable coefficient of object unsimilarity. The size of α-surroundings affects the size of created clusters.

4. MULTIDIMENSIONAL DATA CLUSTERING

The process of clustering has been applied in creating macrosituations from the extensive sets of situations. Individual situations of the complex system were represented by real pictures of the Earth surface. The pictures were obtained from a distance exploration of the Earth by scaning in seven spectral zones. Hence the situations had multispectral character and they were represented by seven dimensional vectors. Obtained Earth pictures were separated into various samples. All pictures in one sample represented one object type, f.e. forest, water, soil, etc. Four sets were available: MIX1.dat, MIX2.dat, PODA.dat, VODA.dat with the whole number of situations 1797.

Considering the size of α-surroundings is the control parameter in ANMO method and it affects final distribution of situations into classes so it was required to find out the influence of this parameter on clustering results - the number of clusters and their quality.

Approximately 40 clustering experiments with different α-surroundings sizes have been done. The size was gradually lowered always by one. If for a lowered value the number of clusters significantly raised up the clustering was ended on the previous value of surroundings. It led to determination of a certain 'sensitive' limit from where the clusters started to split to small clusters and it upset the compactness of primary created clusters. If for a lowered α-surroundings size the number of clusters did not significantly raised up the clustering process continued by additional lowering the size until finding the "sensitive" limit.

5. CONCLUSION

As the description of ANMO method implies this method tries to include maximum of points in a cluster by gradual spreading in a set of points through α-surroundings that were mentioned above. During narroving the examined surroundings can happen that the points at the cluster edge are already not "reached" by more distant points. It means they are not located in α-surroundings of marginal points. Therefore those points become the bases of another clusters or stay isolated points. It is depends on an analyst to choose such α-surroundings that would enable to create an expected number of clusters while a quality of created clusters will be good.

6. REFERENCES

Anderberg, M.R.(1973). Cluster Analysis for Applications. Academic press, New York.

Hartigan, J.A.(1975). Clustering Algorithms. J.Wiley, New York-London-Sydney-Toronto.

Jardine, N. and R. Sibson (1968). The Construction of Hierarchic and Non-Hierarchic Classifications. *Computer J.*, No.11, pp.177-185.

Lukasova, A. and J. Sarmanova (1985). Metody shlukove analyzy. SNTL, Praha.

Ocelikova, E.(1992). Significance of Multicrite-rialClassification for the selection of a Suitable Control Strategy of Complex Systems. *Elektrotechnicky casopis* 44, No.12,pp.372-374.

Ocelikova,E.(1994). Classification of multispectral data. In: Proc. MOSIS '94, Zabreh na Morave, pp.207-211.

AN INTELLIGENT TRANSITION CONTROLLER BASED ON PHYSICAL NEURAL STATES

Jozef Mazák

Nuclear Power Plant Research Institute, Inc.
Division of Information Technologies
Okružna 5, 918 64 Trnava, Slovakia
E–mail: mazak@vuje.com

Abstract: In this paper we propose a complete scheme for a nonlinear transition controller design to steer the process state between different operating points. Main features of the controller are that it includes a nonlinear neural state observer and a nonlinear neural dynamic state feedback. The process disturbances are considered in the design as well. The controller is designed for a wide range of nonlinear process dynamics where the performance of linear controllers dramatically deteriorates. The state observer includes a nonlinear state-space simulation model of the process being estimated from measured input/output data. The state vector is partitioned into two parts which reflect our a priori knowledge about the process dynamics. The design algorithm is verified using a simulation model of a fluidized bed polymerization reactor.

Keywords: gray box modelling, state-space, nonlinear control, neural networks

1. INTRODUCTION

Let us assume a partly known nonlinear dynamic system described in discrete time by

$$x(k+1) = f[x(k), u(k), w(k)] \qquad \text{(1a)}$$
$$y(k) = h[x^0(k)] + v(k) \qquad \text{(1b)}$$

where $x \in \mathcal{R}^n, u \in \mathcal{R}^m, y \in \mathcal{R}^p$ are the state, control and output respectively, $w \in \mathcal{R}^r, v \in \mathcal{R}^p$ are the process disturbance and measurement noise respectively. The state map f is assumed to be a general analytical function and is assumed to be only partly known. The known part of f will represent our a priori knowledge about the process dynamics. The unknown part of the state map f will be then a subject of modelling. The output map h is assumed to be a known general analytical function. That means that we assume that we know how some of the process states relate to the outputs. With respect to the above discussion we assume that the state of the system (1) is partitioned into two components $x = (x^0, x^1)$. The first part of the state vector x^0 has an exact physical meaning and reflects our a priori knowledge about the process dynamics. The physical meaning of x^0 states is determined by relations of these states to outputs either through the function h or by known mutual relations of different state components contained in x^0 which are fixed by the known part of the state map f. The complementary part of the state vector x^1 contains the remaining components of the state vector kept for modelling purposes.

Fluidized bed polymerization reactors belong to one of the most important processes in petrochemical industry. These processes are unstable due to the exothermic polymerization reaction. The process dynamics show a nonlinear behavior caused by an exponential dependency of the reaction rate on the temperature (Choi and Ray, 1985). The functional dependence of different quantities in the reactor is quite complicated and exact mathematical models describing these relations are not available. What is supposed to be available are only measured input and output data. Based on principal relations which hold any time in the reactor, like heat and mass balances in different phases of the process, we developed a simulation model of this process which is then used for the design and for testing of our control algorithm.

Different nonlinear functions are in our design approximated by neural networks. The neural network is considered as a static nonlinear map

$$\mathcal{N} : z_0 \rightarrow z_1 \rightarrow z_2 \rightarrow \ldots \rightarrow z_{N_L}, \qquad z_i \in \mathcal{R}^{N_{l_i}}$$

where z_i represent outputs of the neural network layers. If we relate z_0 with the neural network inputs and z_{N_L} is then the output of the neural network then $\mathcal{N}$ is a neural network consisting of N_L layers of N_{l_i} processing (activation) functions, namely

$$s(u) = 1/(1 + e^{-u})$$

The mapping performed by different layers is a weighted and biased sum of the previous layer outputs

$$z_{i+1} = s(W_i z_i + B_i)$$

where W_i is a $N_{l_{i+1}} \times N_{l_i}$ matrix and B_i is a $N_{l_{i+1}} \times 1$ matrix of network weights and the function s is applied to a vector element wise. For $i = 0$ the corresponding layer is called the input layer and similarly for $i = N_L$ the output layer.

The transition controller design proposed in this paper is a three stage design including an output error identification, a state observer design and a state feedback design. The Section 2 tackles the problem of state estimation for a nonlinear process including a nonlinear output error identification and nonlinear filtering. In Section 3 we will describe the state feedback structure we propose. In Section 4 we will briefly introduce the process we control. Finally, in Section 5 we will present the simulation results.

2. STATE ESTIMATION

Let us assume that there is available a set of measured I/O data in a form of multivariate time series

$$D = \{u(k), y(k)\}, \quad k = 0, \ldots, N \qquad (2)$$

where u and y are inputs and outputs of the system (1) respectively, k is the discrete time and N is the number of measured data points. The problem is to find a mathematical model which will describe the process behaviour underlying the data (2).

Let us parameterize the mathematical model of the process as follows

$$\hat{x}(k+1) = \hat{f}[\hat{x}(k), u(k), \Theta_f] + \hat{g}[\hat{y}(k) - y(k), \Theta_g] \qquad (3a)$$

$$\hat{y}(k) = h[\hat{x}^1(k), u(k)] \qquad (3b)$$

where $\hat{f}$ is a nonlinear function which is a combination of a known analytical part and a black box neural network which weights are denoted by Θ_f. The function $\hat{g}$ stands for another neural network which weights are denoted by Θ_g. This model incorporates the effects of process disturbances which are modelled in an additive manner. The equation (3) approximates a nonlinear state estimator of the process and is similar to e.g. (Suykens, 1995). Here we estimate the Θ_f and Θ_g parameters separately as described in the following subsections.

2.1 *Output-error identification*

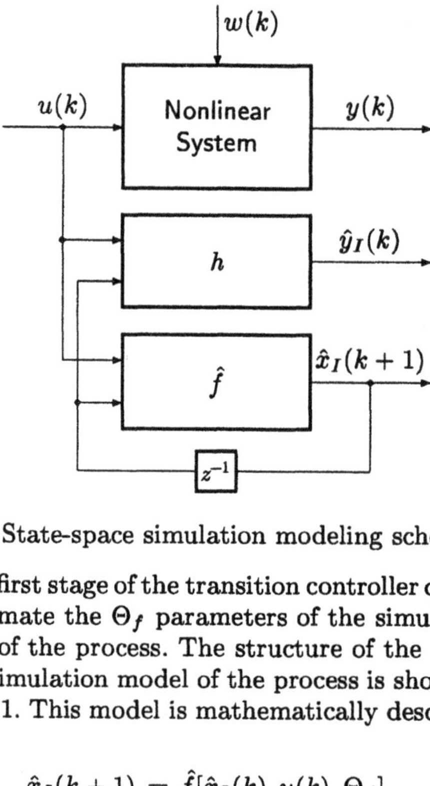

Fig. 1. State-space simulation modeling schema

At the first stage of the transition controller design we estimate the Θ_f parameters of the simulation model of the process. The structure of the state-space simulation model of the process is shown in Figure 1. This model is mathematically described by

$$\hat{x}_I(k+1) = \hat{f}[\hat{x}_I(k), u(k), \Theta_f] \qquad (4a)$$

$$\hat{y}_I(k) = h[\hat{x}_I^0(k), u(k)] \qquad (4b)$$

where the parameters Θ_f are optimized with respect to a suitable choice of the state vector dimension n and the neural network complexity by minimizing the following cost function

$$J(\Theta_f) = \frac{1}{2Np} \sum_{k=0}^{N} (\hat{y}_I(k, \Theta_f) - y(k))^T (\hat{y}_I(k, \Theta_f) - y(k)) \qquad (5)$$

The subscript I in equations (4) and (5) is used to denote some intermediate values. The above introduced cost function can be minimized in principle using any of the non-convex minimization techniques. Obtained results are usually validated on a separate data set and from these results we have to decide about the final values for both the state vector dimension and the network complexity.

2.2 *Observer estimation*

Once we have completed the identification procedure we can estimate the nonlinear observer filter gain $\hat{g}$ in the state estimator (3). This term corrects the a priori state estimation $\hat{x}_I(k)$, obtained by $\hat{f}$ only, for the effect of process disturbances at the time instance $k-1$ which are seen through the output errors $e(k) = \hat{y}(k) - y(k)$. The only free parameters at this stage to be optimized are Θ_g representing the complexity of the neural network which approximates $\hat{g}$. The cost function is now given by the equation (5) where $\hat{y}$ takes the place of $\hat{y}_I$. The Θ_f parameters are fixed during this minimization to already optimized values. The complexity of the $\hat{g}$ neural network has to be judged from the validation results. Figure 2 shows the final structure of the proposed state observer.

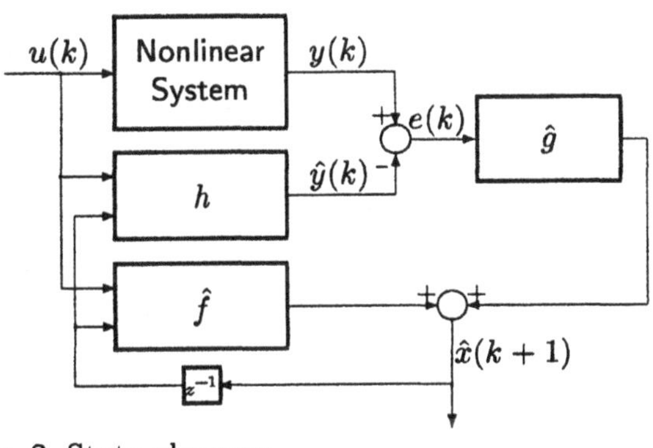

Fig. 2. State observer

3. TRANSITION CONTROLLER

By the transition controller we mean a controller which steers the process state from one operating point to another operating point. An operating point of the process is assumed to be expressed by specific values of the state vector components. Due to the process limitations and the validity of the estimated mathematical model of the process we must consider a set of constraints with respect to all state and control variables. To guarantee that an operating point of the process be an equilibrium point the following equation

$$x_e = \hat{f}[x_e, u_e, \Theta_f] \tag{6}$$

must be satisfied.

The controller is basically a nonlinear static state feedback, denoted by $\hat{k}$, which is extended with

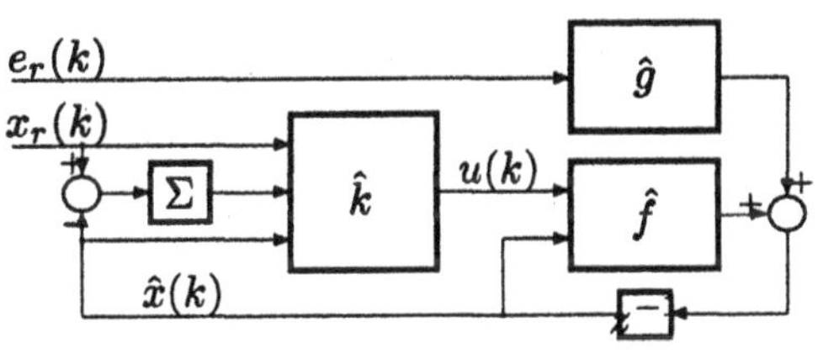

Fig. 3. Controller estimation diagram

a dynamic component in a form of a summation term $z(k)$. This term is used to remove the final steady-state errors in tracking of constant non-zero reference values. The structure of the state feedback is mathematically given by

$$z(k+1) = az(k) + x_r^0(k) - \hat{x}^0(k) \tag{7a}$$
$$u(k) = \hat{k}[x_r(k), z(k), \hat{x}(k), \theta_k] \tag{7b}$$

where $z(k)$ is the state of the controller explicitly defined as an integrator state for the choice of $a = 1$. The dimension of this vector is chosen with respect to controllability of the system. The reference signal x_r is typically specified a priori for the known part of the state vector components x^0. The problem of determination of the reference trajectories for the complementary state vector components x^1 is described in details in (Mazák, 1996) The controller estimation set-up is shown in Figure 3. As the state feedback is optimized off line it is not possible to compute the filter contribution to the state simulation. However, this effect can be simulated by defining an additional reference signal $e_r(k)$ with statistical properties of the innovation sequences we have obtained from the observer design step. The parameters $\hat{k}$ are tuned by minimizing an index defined as follows

$$J = \sum_{k=1}^{N_c} \Big((\hat{x}(k) - x_r(k))^T Q (\hat{x}(k) - x_r(k)) \Big) + \sum_{k=1}^{N_c} \Big(\Delta u(k)^T R \Delta u(k) \Big) \tag{8}$$

where $\Delta u(k) = u(k) - u(k-1)$. Note that we can not penalize in the cost function (8) the absolute values of neither state nor control variables because we would introduce final tracking errors.

The minimization of the cost function is performed numerically. As we optimize a cost function over a finite time horizon the stability of the closed loop is not guaranteed in general. By an introduction of the constant $a < 1$ in (7a) we explicitly define stability of the controller dynamics which helps us to optimize the controller such that the closed loop remains stable.

4. PROCESS SIMULATION MODEL

The mathematical modelling of the reactor is in detail described in (Mazák and Balseiro, 1996). In this process we have defined inputs as follows:

(1) T_{set} - the setpoint for the temperature of the emulsion
(2) qc_{set} - the catalyst input flow
(3) $C_x/C_{2_{set}}$ - ratio of co-monomer and monomer at the input of the reactor
(4) P_{set} - setpoint of the pressure at the top of the reactor

The outputs were defined as follows:

(1) Q_0 - the production rate
(2) T_e - the temperature of emulsion
(3) P_{C2} - partial pressure of the monomer
(4) C_x/C_2 - ratio of the co-monomer and monomer in the gascap

And finally the disturbances were defined as follows:

(1) w_{T_w} - water temperature disturbance
(2) w_{qc} - catalyst flow disturbance reflecting impurities in the catalyst and irregularities in its activity
(3) $w_{C_{flow}}$ - disturbance on a total input monomer and co-monomer flow reflecting impurities and temperature changes.

If we define as a known part of the state vector in the model (3) process variables Q_0, T_e, C_2, C_x than we can use our a priori knowledge of the relation of these states to the measured outputs which is given as follows

$$\begin{aligned} Q_0 &\stackrel{\text{def}}{=} \hat{y}_1 = \hat{x}_1 \\ T_e &\stackrel{\text{def}}{=} \hat{y}_2 = \hat{x}_2 \\ \eta_2 C_2 T_e \approx P_{C2} &\stackrel{\text{def}}{=} \hat{y}_3 = \hat{x}_3/\hat{x}_2 \Rightarrow \hat{x}_3 \equiv C_2 \\ P_{Cx}/P_{C2} = C_x/C_2 &\stackrel{\text{def}}{=} \hat{y}_4 = \hat{x}_4/\hat{x}_3 \Rightarrow \hat{x}_4 \equiv C_x \end{aligned}$$

where η_2 is a known constant.

5. RESULTS

We simulated the polymerization process for 601 process hours. As a test signal we used a low pass filtered uniformly distributed random signal filtered by a fourth order Butterworth filter with a cut-off frequency $0.2/T_s$, where $T_s = 5$min was the sampling time. The process dynamics ware modeled by a nonlinear grey-box state-space model of 6-th order. The number of inputs was 4 and the number of outputs was also 4. The state map was approximated by a neural network with one hidden layer. The output layer was composed of four linear nodes which belongs to the x^0 part of the state vector and the rest of output nodes was chosen nonlinear. The output map in (4) was fixed as follows

$$h[\hat{x}^0] = \begin{bmatrix} \hat{x}_1^0 \\ \hat{x}_2^0 \\ \hat{x}_3^0 \hat{x}_2^0 \\ \hat{x}_4^0/\hat{x}_3^0 \end{bmatrix} \tag{9}$$

and results from our a priori process knowledge. We estimated a set of models by varying the number of hidden nodes in the state map between 6 and 15. Next these models were validated on a validation data set, different from the estimation set. At the end we have chosen a model with the best performance measured as follows

$$\eta = \frac{||\hat{y} - y||}{||y||} \cdot 100\% \tag{10}$$

where $||.||$ is the usuall Euklidian vector norm. The best performance was found for a model with 12 nodes. The estimation and validation results are shown in Figure 4. For a comparison we estimated also a linear state-space model which showed a considerably worse performance. The performance of the nonlinear model was 0.12% and 0.13% on the estimation and validation data sets respectively, and 16.51% (17.03%) was the performance achieved by the linear model on the estimation (validation) data.

The filter gain in the observer was approximated by a neural network with one hidden layer. The neural network had 4 inputs standing for the output errors and 6 outputs standing for the state correction terms. The number of the hidden layer nodes was varied from 4 to 9 and as the final network was chosen a configuration with 4 nodes. The spectra of innovation sequences obtained by this filter are shown in Figure 4. As can be seen from this plot the filter tries to whitened the a priori state estimates obtained by the process model.

The controller was estimated in two steps. In the first instance we have chosen $e_r(k) = 0$. We have chosen three working points in correspondence with (6). As a reference signal we have defined a sequence of low pass filtered unit steps defining all possible transitions among these operating points. The time constants for the first order low pass filters used here were chosen independently for all state vector components partially using our a priori process knowledge as follows: $[20, 20, 40, 90, 40, 40]$min. During the optimization we varied the integration constant a in the folowing order $a = 0.9, 0.99, 0.998, 1$. The penalty matrices in (8) where chosen experimentally with respect to the state and control bounds as diagonal matrices $Q = \text{diag}(1, 1, 1, 1, 0, 0)$, $R = \text{diag}(0.1, 1, 0.5, 0.01)$. The results obtained at this stage are shown in Figures 5 and 6. The integrator states z are not shown. They are simply obtained by filtering the tracking errors with a first order filters with the time constant a.

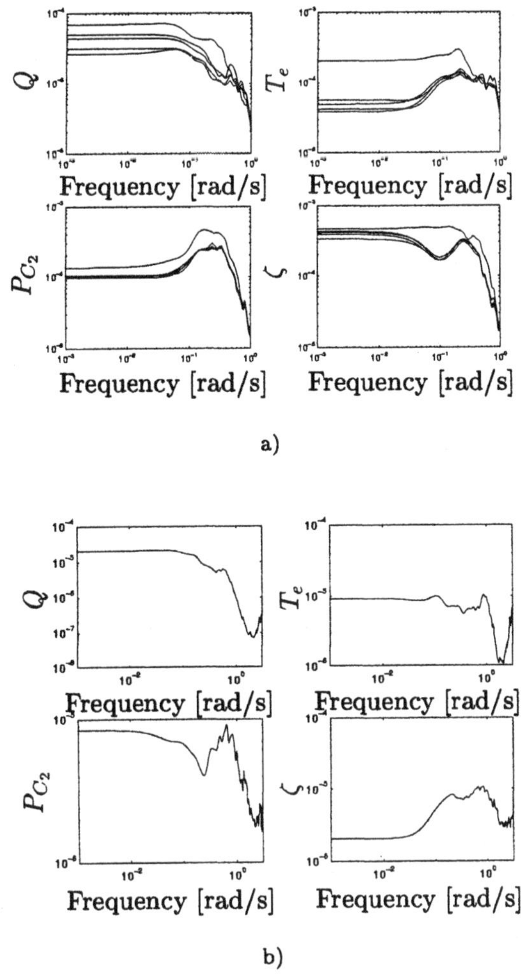

Fig. 4. a) Estimated model output error spectra for validation data; b) Observer validation results: Spectra plots of prediction errors for all four outputs.

The last step of the controller design is a re-estimation of the controller together with the filter gain by setting $e_r(k) \neq 0$. We simulated these sequences as a Gaussian white noise. For the simplicity we included into the controller neural network training set three examples of each transition for the different realization of the noise. The controller neural network was trained by further minimization of the cost function as the one used in the previous step. These results are similar to those already shown and due to the excessive data sets the Figure 7 shows only a small part of the resulting state trajectories.

Note, that the controller is designed for the estimated model of the process. Therefore the validation of the controller was done by applying it to the original process, in our case by a simulation in SIMULINK. The validation of the whole controller showed the correctness of the previously obtained results and are not shown here.

6. CONCLUSIONS

The ethylene polymerization reactor, as introduced in this paper, is a very difficult process to control. It is a stiff nonlinear process, which is characterized by a very broad range of dynamics. It is a multivariable process, with a high interaction between inputs and outputs and a complex dynamics containing both minimum and non-minimum phase behaviors. The dynamics of the process can be a priori only described in main lines, as well as the disturbance influences. The ranges for manipulated variables are very narrow. It was shown, that a complex nonlinear controller can be successfully designed if we take necessary precautions concerning the stability and constraints violations.

The partitioning of the state vector gives us a possibility for better analysis of estimated results and to take better decisions about the solution of different problems during such a complex controller design problem as the one proposed in this paper.

What remains a big problem for this kind of design are the excessive data sets used for training of different networks. Due to the non-convex nature of the training process it has to be repeated many times, until a satisfactory performance is obtained. In reality this is a very time consuming process. However, our approach seems to be a powerful approach to considerably improve the existing linear control techniques used for this process as we could already see from the estimation of a linear model for this process.

7. REFERENCES

Choi, K. Y. and W. H. Ray (1985). Fluidized bed reactors for ethylene and propylene polymerization. *Chemical Engineering Science* **40**(12), 2261–2279.

Mazák, J (1996). Transition Control Based on Grey, Neural States. PhD thesis. Eindhoven University of Technology. Eindhoven, The Netherlands.

Mazák, J. and J. L. F. Balseiro (1996). Modeling of a fluidized bed reactor for ethylene polymerization. Research project. Eindhoven University of Technology, Dept. of Electrical Engineering. Eindhoven, The Netherlands.

Suykens, J. (1995). Artificial neural networks for modelling and control of nonlinear systems. PhD thesis. Katolieke Universiteit Leuven. Leuven.

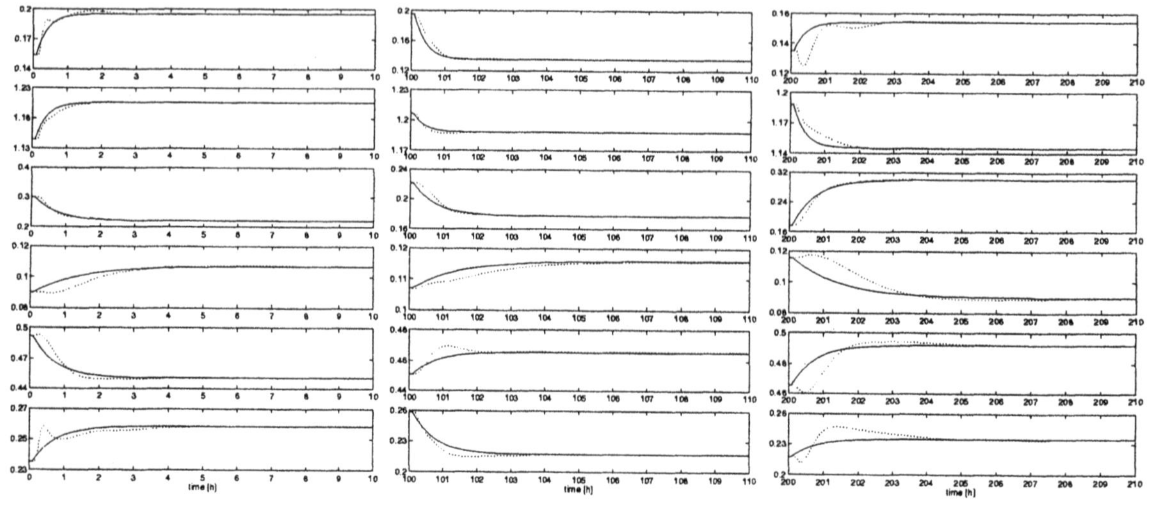

Fig. 5. Three columns of state transitions. Solid line – reference, Dotted line – state.

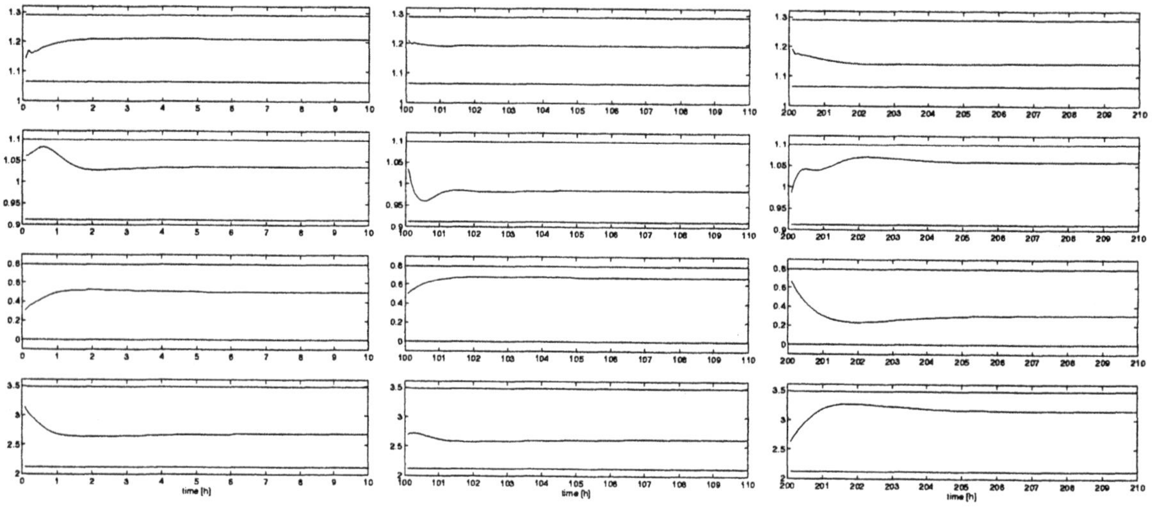

Fig. 6. Corresponding control inputs

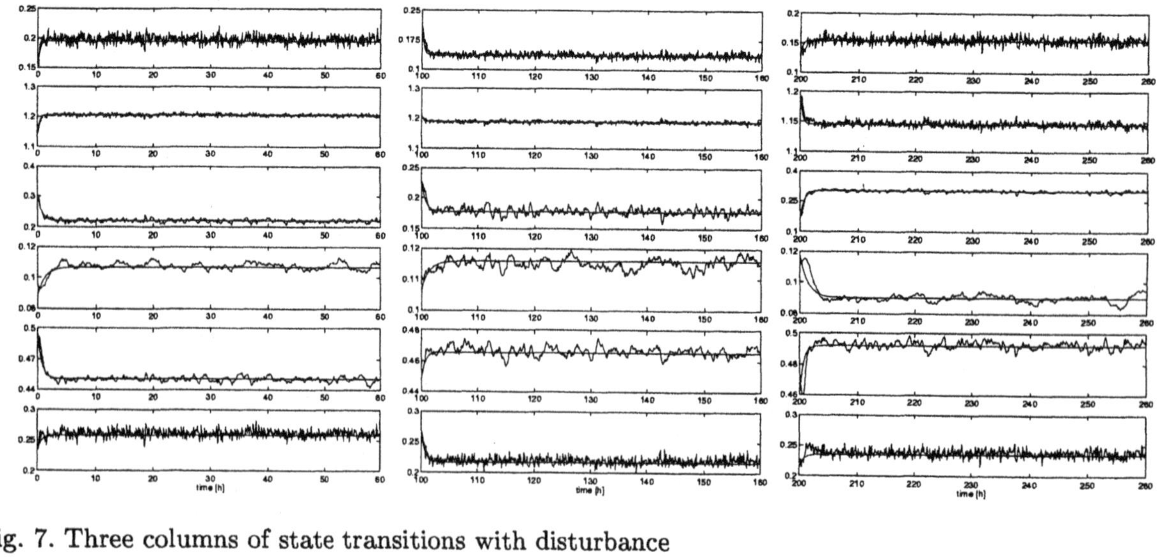

Fig. 7. Three columns of state transitions with disturbance

AN ADAPTIVE HYBRID FUZZY - NEURAL CONTROLLER

Michail Petrov **Todor Proychev** **Andon Topalov**

Control Systems Department, Technical University - Plovdiv
61, St.Petersburg blv., 4000 Plovdiv, Bulgaria, e-mail:mpetrov@tu-plovdiv.bg

Abstract: This paper presents investigations toward adaptive fuzzy control using neural networks for the control of nonlinear and time variable physical processes. By using a hybrid structure, the fuzzy logic controller can be composed of its traditional components and two neural networks. The first neural network works as an adaptive rule estimator for online acquisition and modification of the fuzzy rules as it was presented in a previous work. Here a second neural network is proposed as a further development of the hybrid structure which implements the defuzzification function and works as an adaptive neural defuzzifier. The standard one step gradient descent learning procedure is implemented and the output from a conventional feedback controller is used for tuning the parameters of the neural defuzzifier. Comparison with the traditional fuzzy technique is discussed. The results obtained by simulation show the better performance of the proposed hybrid control structure.

Keywords: Adaptive Fuzzy Systems, Neural Network, Fuzzy Hybrid Systems.

1. INTRODUCTION

Fuzzy logic and neural networks are two of the most promising research areas in the modern control systems theory. They offer very large capabilities concerning complex system modeling, prediction, control, and performance. Fuzzy logic provides a powerful tool to capture the uncertainties associated with human cognitive processes. Neural networks, on the other hand, offer great advantages in learning, adaptation, fault tolerance, parallelism and generalization. Recently great efforts have been made toward integrating these two techniques into a unified framework. Many applications have already been reported using a combination of these two techniques [Gupta, 1994], [Berenji, 1992], [Jang, 1992], [Lin, 1994], [Buckley, 1994].

Fuzzy logic controllers (FLC) have been applied successfully in many cases for industrial process control. It has been shown that FLC is a very effective control scheme in controlling these plants whose detailed and accurate mathematical descriptions are not available. There are various applications that confirm the capability of the FLC [Mamdani, 1974]. However, the FLC design still has several well-known difficulties: (1) The fuzzy control rules are experience oriented and the suitable membership functions have to be obtained by time-consuming trial and error procedure; (2) The characteristic of the control system can not be pre-specified; (3) There is no criterion to determine optimality or at least sub-optimality of the FLC.

These drawbacks could be partially eliminated with the recently introduced neuro-fuzzy networks which combine the advantages of neural networks and fuzzy systems. Thanks to the implementation of supervised and unsupervised learning methods, it becomes possible to optimize both the antecedent and consequent parts of a linguistic rule-based fuzzy system. Concerning the antecedent part, almost all of the learning methods applied to the fuzzy neural networks are limited only to the tuning of the membership functions of the fuzzy sets. They do not change the fuzzy rules. However, the process of selecting an adequate fuzzy relation matrix remains subjective and time consuming, generally completed by trial-and-error approach. It is therefore important to establish a mechanism for automatic adjusting of the fuzzy relation matrix in order to make the controller performance robust. In [Petrov and Topalov, 1994] a hybrid neural-fuzzy reasoning model is presented, taking advantage of the learning capability of the neural network.

This paper presents a further development of the proposed hybrid structure of the fuzzy controller where the defuzzification is performed by a neural network. Defuzzification plays a central role in the implementation of the fuzzy logic controller. Typical methods are the center of area method (COA) and the mean of maximal method (MOM). The problems that usually arise with these methods are rooted in situations where two or more distinct modes of solution exist and the combining of these distinct modes leads to an unsatisfactory conclusion. By implementing a neural network as an adaptive defuzzifier some of the inconveniences with the traditional methods could be avoided and the system performance could be improved.

The standard one step gradient descent learning procedure is then implemented and the output from a conventional feedback controller is used for tuning the parameters of the neural defuzzifier. Comparison with the traditional fuzzy technique is discussed. The results obtained by simulations show the better performance of the proposed hybrid control structure.

2. STRUCTURE OF THE ADAPTIVE FUZZY CONTROL SYSTEM

The configuration of the proposed adaptive hybrid fuzzy-neural controller (AHFNC) is shown in Fig. 1. The proposed fuzzy-neural controller incorporates two neural networks, one for rule base estimation and second for defuzzification implementation. Gomi and Kawato (1993) have proposed learning schemes using feedback error learning for a neural network model applied to an adaptive nonlinear feedback controller. In these learning schemes, a conventional feedback controller (CFC) is provided both as an ordinary feedback controller to guarantee global asymptotic stability in a compact space and as an inverse reference model of the response of the controlled object. This approach is implemented here in order to tune the neural defuzzifier.

The AHFNC implements the basic control function in the system, and the CFC implements a helping function. The first controller generates a control action u_{NN} received from its neural network part, and the second controller generates a computed control action u_{PD} . The final control signal is $u = u_{NN} + u_{PD}$. The control action u_{PD} slowly decreases during the time, therefore the main rule performs the signal u_{NN}. The system is organized as a SISO system with a set point $r(t)$ and an output $y(t)$.

2.1 The traditional part of the adaptive hybrid fuzzy-neural controller.

The structure of the AHFNC includes several blocks from the traditional fuzzy controller as it is shown in Fig. 1. They are: fuzzifier, fuzzy rule base and inference engine. The fuzzifier performs the function of fuzzification and converts input data from an observed input space into proper linguistic values of fuzzy sets through predefined input membership functions. The AHFNC works with input signals of the system error e and the change rate of error de as usually. The state error is defined as the output difference between set point $r(t)$ and plant output $y(t)$ at time t, or

$$e(t) = r(t) - y(t) \tag{1}$$

Hence the change rate of error de at time t will be:

$$de(t) = e(t) - e(t-1) \tag{2}$$

The rule base consists of a set of preliminarily chosen fuzzy logic rules in the form of 'if-then' to describe the control policy of expert knowledge. The conventional FLC observes inputs $e(t)$ and $de(t)$, then adjusts the output activation $u(t)$ by 'if-then' rules which are represented as:

$$R^n: \text{ if } e \text{ is } A_i \text{ and } de \text{ is } B_i \text{ then } u \text{ is } C_i \tag{3}$$

where e, de, u are linguistic variables. Their universes of discourse are [-E,E], [-dE,dE] and [-U,U].

The fuzzy inference engine should match the output of the fuzzifier with the fuzzy logic rules and perform fuzzy implication and approximate reasoning to decide a fuzzy control action. The fuzzy implication is realized by means of the max-min composition [Mamdani, 1974]:

$$\mu_i = \max\{\min[\mu_{Aj}, \mu_{Bk}]\} \tag{4}$$

where μ_{Aj} and μ_{Bk} specify the membership values received from the intersection and union operations upon fuzzy sets of the input signals.

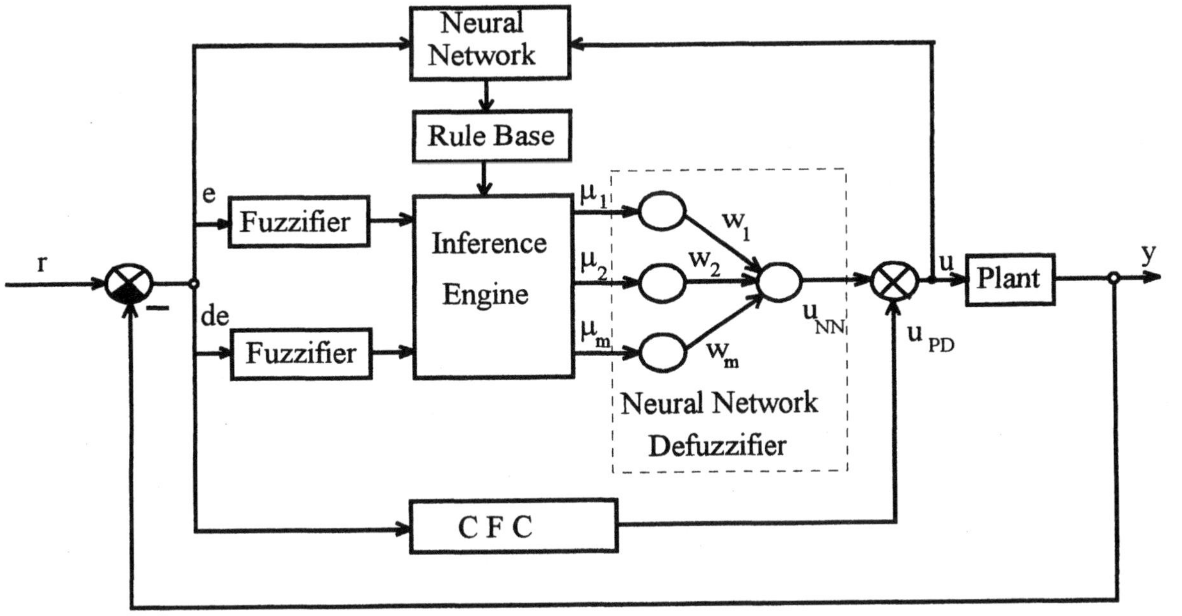

Fig. 1. The structure of the control system with the proposed hybrid fuzzy controller.

2.2 The neural network rule estimator.

This part of the AHFNC was investigated in our previous paper [Petrov, Topalov, 1994]. It contains two elements: neural network for product space clustering of input vector (*e, de, u*) and a part for statistical decision of a new rule base. A variety of neural networks for solving such a task were investigated. The most appropriate are the neural networks with unsupervised learning. For example: competitive learning, learning vector quantization, self-organizing maps etc. Their topology (Fig. 2) for such a case has two layers: input field *Fx* with three neurons for three fuzzy variables: $x_1= e$, $x_2= de$, $x_3=u$, and an output field *Fy* with neurons whose number depends on the number of fuzzy sets for each variable.

The algorithm used for adaptive rule estimation by neural network, called adaptive vector quantization (AVQ), was introduced by Kosko, [1992]. AVQ systems adaptively quantize pattern clusters in R^r. Stochastic competitive learning systems are neural AVQ systems. Neurons compete for the activation induced by randomly sampled patterns. The corresponding synaptic vectors adaptively quantize the pattern space R^r. The *s* synaptic vectors w_j define the *s* columns of the synaptic connection matrix **W. W** interconnects the *r* inputs or linear neurons in the input neuronal field *Fx* to the *s* competing nonlinear neurons in the output field *Fy* (Fig. 2).

The learning algorithm estimates the unknown probability density function *p(x)*, which describes the distribution of patterns in R^r. This distribution corresponds to the density of appearance of separate fuzzy set values of the output control variable ***u***.

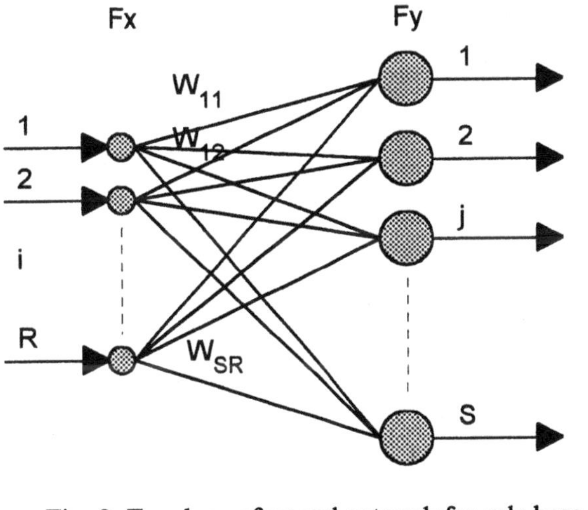

Fig. 2. Topology of neural network for rule base estimation

The AVQ system compares Euclidean distance of the current vector sample ***x(t)*** and single *s* columns (the *s* synaptic vectors $w_1(t), w_2(t), \ldots w_s(t)$) of the synaptic connection matrix **W**. If the *j*-th synaptic vector $w_j(t)$ is the closest to ***x(t)***, the *j*-th output neuron "wins" the competition for activation at the moment *t* and the synaptic vector changes. The "winner" corresponds to the fixed value of the fuzzy control variable in the fixed cell of the fuzzy rule matrix. The algorithm counts the number of these appearances and, after learning, computes the density function *p(x)*. From values of density function are calculated the new fuzzy set values of the output *u* in each fuzzy rule.

2.3 The neural network defuzzifier.

The proposed neural defuzzifier (Fig. 1) consists of an input layer and an output layer where the output layer contains a set of adjustable parameters to learn. The input layer receives the information from the fuzzy inference system with membership values μ_i for each fuzzy set of the output signal u and then calculated by the fuzzy implication (4). The number of the neurons in the input layer is equal to the number of the fuzzy sets of the output signal. The output layer contains only one neuron. Its input function is:

$$f = \sum_{i=1}^{m} w_i \mu_i \qquad (5)$$

where μ_i are the input signals, w_i are the synaptic weights which correspond to the fuzzy sets centers of the fuzzy singletons of the output signal u_{NN}.
The activation of the output neuron is given by formula:

$$a = \frac{\sum_{i=1}^{m} w_i \mu_i}{\sum_{i=1}^{m} \mu_i} \quad \text{or } a = \sum_{i=1}^{m} w_i \overline{\mu_i} = u_{NN} \qquad (6)$$

This way the final control action of the fuzzy controller u_{NN} is computed as the weight average of the outputs μ_i from the inference engine. Both control actions from the neural network defuzzifier u_{NN} and from conventional feedback controller u_{PD} are unified and form the output signal u.

$$u = u_{PD} + u_{NN} \qquad (7)$$

The learning algorithm for the neural defuzzifier is based on the minimization of an error measurement function which is defined as

$$E = \frac{\varepsilon^2}{2} \qquad (8)$$

where ε is calculated by equation

$$\varepsilon = u - u_{NN} = u_{PD} \qquad (9)$$

where u denotes desired control action and u_{NN} is calculated by the neural network.
This algorithm performs one step gradient descent learning procedure. Assuming that w_i is an adjustable parameter in output neuron (e.g. the center of a membership function or a fuzzy singleton), the general parameter learning rule used is [Lin, 1994]:

$$w_i(t+1) = w_i(t) + \eta\left(-\frac{\partial E}{\partial w_i}\right) \qquad (10)$$

where η is the learning rate, and

$$\frac{\partial E}{\partial w_i} = \frac{\partial E}{\partial a} \cdot \frac{\partial a}{\partial f} \cdot \frac{\partial f}{\partial w_i} \qquad (11)$$

The computation of the first derivative is:

$$\frac{\partial E}{\partial a} = -\varepsilon = -(u - u_{NN}) = -u_{PD} \qquad (12)$$

The computation of the second derivative is:

$$\frac{\partial a}{\partial f} = \frac{1}{\sum_{i=1}^{m} \mu_i} \qquad (13)$$

The computation of the third derivative is:

$$\frac{\partial f}{\partial w_i} = \frac{\partial\left(\sum_{i=1}^{m} w_i \mu_i\right)}{\partial w_i} = \mu_i \qquad (14)$$

The recurrent equation for adjustable parameter is:

$$w_i(t+1) = w_i(t) + \eta u_{PD} \overline{\mu_i} \qquad (15)$$

3. SIMULATION RESULTS AND DISCUSSION

Investigations have been carried out by simulations of an adaptive control system. A second order model of the plant was used:

$$W(s) = \frac{K}{(1+sT_1)(1+sT_2)} \qquad (16)$$

It was interpreted as a discrete time single-input, single-output model of the form:

$$a_0 y(t) + a_1 y(t-1) + a_2 y(t-2) = Ku(t) \qquad (17)$$

where u and y are the input and output of the plant. The typical results were received with initial values of the parameters of the plant K=2.5, T_1=10s, T_2=5s. It is possible to add a stochastic noise to the coefficients a_0, a_1, a_2 with an amplitude of 30% from the initial value and the fuzzy controller can still perform robustly. It is the same with the output of the system. The default initial parameters of the fuzzy membership functions of the input and output controller variables are shown in Fig. 3 and Fig. 4.

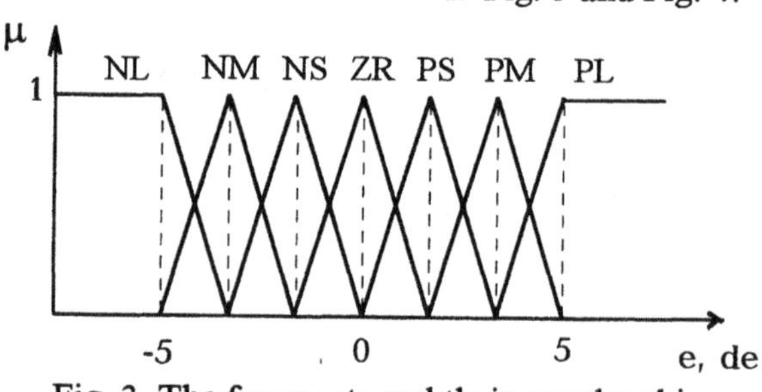

Fig. 3. The fuzzy sets and their membership functions for input variables.

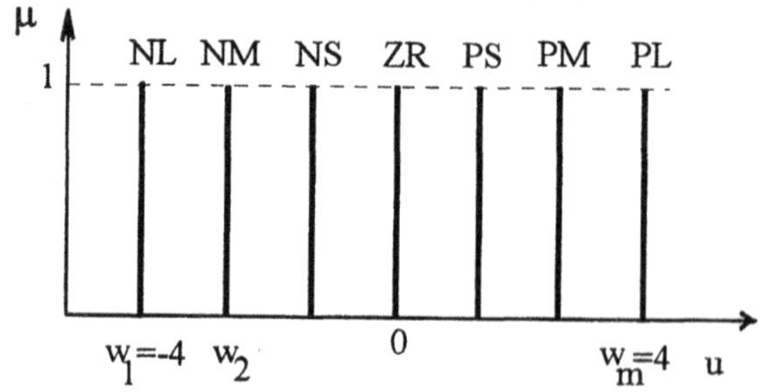

Fig. 4. The fuzzy sets and their membership functions for the output activation.

The number of fuzzy subsets for A_i and B_i from (3) is chosen to be seven (NL,NM,NS,ZE,PS,PM,PL) with triangular membership functions as it is shown in Fig. 3. The fuzzy subsets C_i are also seven with singleton functions as it is shown in Fig. 4. Hence there are 49 fuzzy rules, as it is shown in Fig. 5. The values of the fuzzy subsets C_i in the rule - base matrix (Fig. 5) can be changed adaptively by the neural network estimator and improve the system performance.
For convenience and simplicity the CFC is realized as a discrete PD controller with equation:

$$u_{PD}(t) = K_p e(t) + K_d de(t) \tag{18}$$

where the controller parameters which are used during simulation are $K_p = 1.5$ and $K_d = 0.3$. The output signal u_{PD} is used mainly in the learning algorithm (15). Its control function in (7) is very small and decreases during the time.

de \ e	**NL**	**NM**	**NS**	**ZR**	**PS**	**PM**	**PL**
NL	NL	NL	NM	NM	NS	NS	ZR
NM	NL	NM	NM	NS	NS	ZR	PS
NS	NM	NM	NS	NS	ZR	PS	PS
ZR	NM	NS	NS	ZR	PS	PS	PM
PS	NS	NS	ZR	PS	PS	PM	PM
PM	NS	ZR	PS	PS	PM	PM	PL
PL	ZR	PS	PS	PM	PM	PL	PL

Fig. 5. Rule base matrix with initial fuzzy logic rules

Both the neural network rule estimator and the adaptive neural defuzzifier work together in real-time mode with a discrete period T_s=0.05. The learning rate for the first one is η=0.03 and the learning duration is approximately 1000 discrete steps. After this learning period the rules in the rule base matrix are changed by the algorithm according to the system performance during the learning period. The neural defuzzifier learns in each discrete step with a stable learning rate η=0.01. The obtained results have shown that a learning epoch consisting of 1000 steps is enough to obtain satisfactory system performance. The initial synaptic matrix $\mathbf{W_0}$ for the neural net estimator is prepared preliminarily according to initial rule base matrix from Fig. 5. The initial synaptic weights or adjustable parameters for neural net defuzzifier are chosen according to Fig. 4. Simulations were provided in various cases with different plant parameters. Typical results are shown in Figs. 6, 7, 8, 9, 10. All of figures show the behavior of the set point r (curve 1), the system output y (curve 2) and the control action u (curve 3).
Fig. 6 shows the system performance when the controller has not been learned yet. This corresponds to the function of the conventional FLC and a steady state error is observed. The learning process in the AHFNC improves the situation (Fig. 7) and the steady state error is eliminated. The adaptive controller can succeed even if a stochastic noise is added to the certain plant parameter (a_0, a_1 or a_2) (Fig. 8). It works correctly after learning if an additive noise is added to the system output (Fig. 9). Fig. 10 shows the system performances when a plant parameter in the model (16) has been changed during the simulation.

4. CONCLUSIONS

The structure of an adaptive hybrid fuzzy - neural controller was introduced. By combining the techniques from a neural network rule base estimator proposed in the previous paper and a neural network defuzzifier described in this paper, an improved approach was outlined. Independent implementation of fuzzy controllers and neural networks for control systems is not effective enough. A promising approach is to combine them into hybrid structures in order to obtain a better performance.
The computer simulations carried out with a second order plant with changeable parameters satisfactorily verified the validity and the better performance of the proposed structure of the hybrid controller comparing with a conventional controller. The adaptive neural defuzzifier plays a very useful function in this structure. This is evident from the obtained results, where the transient responses of the system output have shown an improvement after several learning epochs with this neural network.

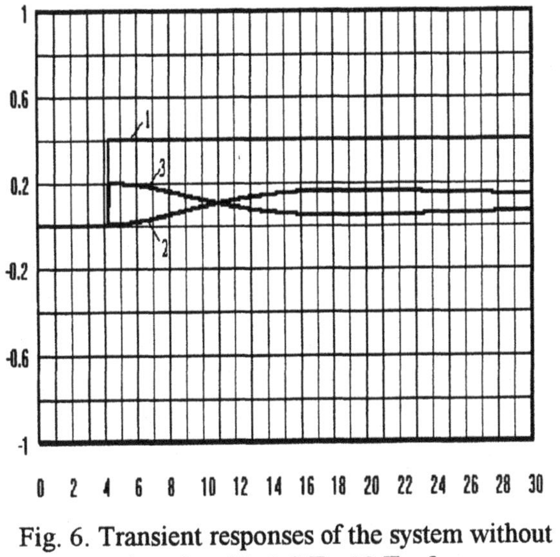

Fig. 6. Transient responses of the system without learning. K=2.5,T_1=10,T_2=5.

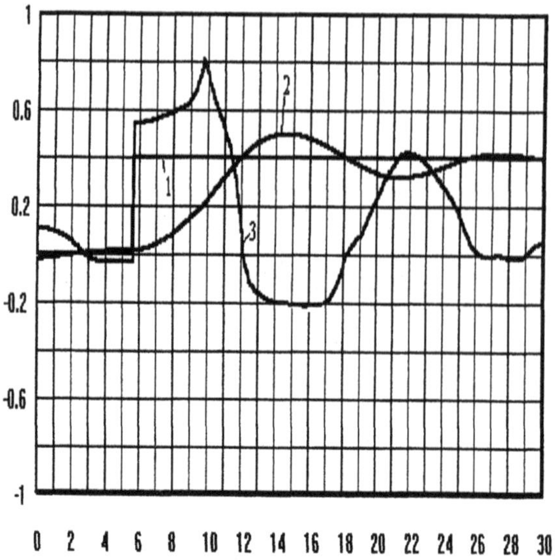

Fig. 7. Transient responses of the system after learning.

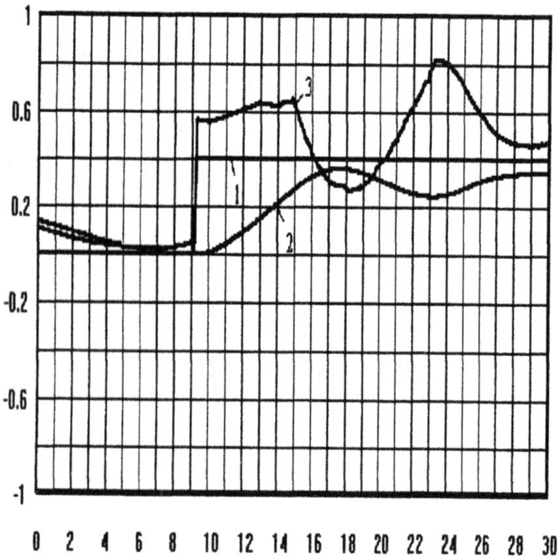

Fig. 8. Transient responses of the system with learning, when noise was added to a plant parameter.

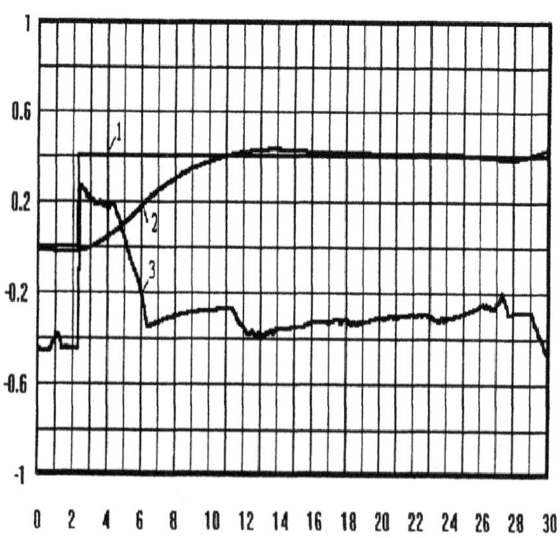

Fig. 9.Transient responses of the system with learning and additive noise.

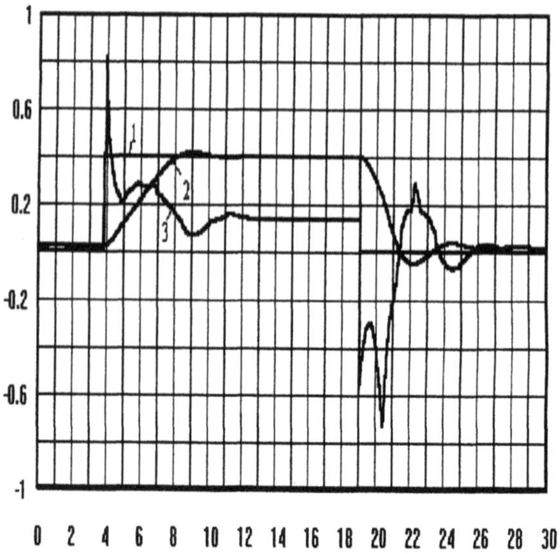

Fig. 10. Improved transient responses after 2 learning epochs with a changed plant parameter $K=2.5$, $T_1=10 \rightarrow T_1=1$, $T_2=5$.

5. REFERENCES

Berenji, H.R. (1992). Fuzzy and neural control. In: *An Introduction to Intelligent and Autonomous Control* (Passino and Antsaklis, (Ed.)), Kluwer Academic Publishers.

Berenji, H. (1992). Learning and Tuning Fuzzy Logic Controllers Through Reinforcements. *Proceedings IFAC AORTIC '92*, Delft, The Netherlands, June 1992.

Buckley, J.J., Y. Hayashi.(1994). Fuzzy Neural Networks, A Survey. *Fuzzy Sets and Systems*, **vol. 66**, pp.1-13.

Gomi, H. and M. Kawato (1993). Neural Network Control for a Closed–Loop System Using Feedback–Error–Learning. *Neural Networks*, **vol. 6**,.

Gupta M.M. and D.H. Rao (1994). On the Principles of Fuzzy Neural Networks. *Fuzzy Sets and Systems*, **vol. 61**, pp.1-18.

Jang, J.R. (1992). Self-Learning Fuzzy Controllers Based on Temporal Back Propagation. *IEEE Transactions on Neural Networks*, **Vol. 3**, No. 5, September 1992.

Kosko, B. (1992). *Neural Networks and Fuzzy Systems: A dynamical System Approach to Machine Intelligence*. Prentice - Hall, Englewood Clifs, NJ.

Lin, C.T. (1994). *Neural Fuzzy Control Systems with Structure and Parameter Learning*. World Scientific, NY.

Mamdani, E.H. (1974). Application of fuzzy algorithms for control of simple dynamic plant. *Proceedings IEEE*, **Vol. 121**, No. 12,.

Petrov, M.G., A.V.Topalov (1994). On Adaptive Fuzzy Control Using Neural Networks. *IFAC Workshop on New Trends in Design of Control System*, Smolenice, Slovak Republic, Sept. 1994.

NEURAL NETWORK BASED CONSTRAINED PREDICTIVE CONTROL USING STOCHASTIC APPROXIMATION ALGORITHM

A. Rusnák°, K. Najim*, A. Mészáros°, and M. Fikar°

°Slovak University of Technology, Faculty of Chemical Technology, Radlinského 9, 81237 Bratislava, Slovakia, Phone: (+42) 7 395 269, Fax: (+42) 7 393 198
**E.N.S.I.G.C. Chemin de la loge, 31078 Toulouse Cedex, France*
Phone: (+33) 62 25 23 69, Fax: (+33) 62 25 23 18

Abstract: This paper presents a constrained predictive control strategy using artificial neural networks (ANN). In this control scheme two ANN are used. The recurrent ANN is used as a multi-step ahead predictor. The control action is provided by the multilayer feedforward ANN. The weights of this ANN are estimated at each control step using a stochastic approximation (SA) algorithm by minimizing a quadratic control objective which is based on a series of the future predictions and future control actions, and by preventing violations of process constraints. Simulation results demonstrate the usefulness and the robustness of this predictive control algorithm.

Keywords: Constraints, Fermentation processes, Neural networks, Predictive control, Stochastic approximation

1. INTRODUCTION

There has been a growing awareness over the last decades that constrained predictive control can produce performance in real applications. Constrained predictive control techniques are useful to deal with nonminimum phase and time-varying (time-varying time delay, steady-state gain, etc.) systems , and plant model mismatch. They are of particular interest from the point of view of both broad applicability and implementation simplicity.

The use of ANNs in model predictive control, which has been recognized as an effective tool for handling some of difficult control problems, has recently attracted a great deal of attention, primarily because ANN appear to provide a convenient means for modelling complicated non-linear processes at low cost. Several studies on the use of ANN as the basis for model predictive controllers (with finite prediction horizons), have been published (Temeng *et al.* 1995; Rusnák *et al.* 1996).

To avoid the use of locally linearized models of the process to be controlled as in the classical GPC approach (Clarke *et al.* 1987), and complex optimization techniques, this paper presents a new simple solution of deriving constrained predictive control based on ANN. The design of this solution is based on the training of a recurrent and a feedforward neural networks. The recurrent and the feedforward neural networks are respectively used as a multi-step predictor and for calculating the control signal (neural controller). The multi-layer feedforward ANN is trained so as to achieve the control objective. The main idea presented in this study concerns the use of stochastic recursive approximation techniques as learning tool for the design of neural networks controllers to solve both unconstrained and constrained (minimization of a constrained quadratic cost function and preventing violations of process constraints) predictive control problems. The control approach described below is general and does not depend on the structure of the control objective and the constraints.

2. PREDICTIVE CONTROL

An interesting formulation of this control approach has been given by Clarke *et al.* (1987). This formulation is based on a CARIMA model, and on the minimization of the conditional expectation of a quadratic function measuring the control effort and the distance between the predicted system output and some predicted reference sequence over the receding horizon, i.e.

$$J = E\left\{ \sum_{j=N_1}^{N_2} \left(y^*(t+j) - y(t+j)\right)^2 + \right. \\ \left. + \sum_{j=1}^{N_2} \lambda(j)\left(\Delta u(t+j-1)\right)^2 / \Im_t \right\} \quad (1)$$

where $y(.)$, $\Delta u(.)$, and $y^*(.)$ are the controlled variable, future control increments, and set-point, respectively. E is the conditional expectation with respect to the σ-algebra $\Im_t$ generated by the available data up to time t. N_1, and N_2, are respectively, the minimum and the maximum prediction horizon. The weighting factors $\lambda(j)$ $(j=1, N_2)$ serve for penalization of the future control increments Δu. Usually $\lambda(j)$ is considered as a constant λ. In common with most constrained predictive control strategies, it is assumed that beyond the control horizon N_u, the future control actions are constant.

It is required that the set-point is tracked with zero offset. The original GPC is based on CARIMA model and provides offset-free control in the presence of nonstationary step-like disturbances as long as the closed-loop system is stable.

Since constrained predictive control techniques are based on linear models, their performance is mainly affected by modelling errors, which are principally due to uncertainties and nonlinearities. So it would be useful to develop a non-linear model for controller design and to derive a suitable control strategy.

The next section deals with the use of neural networks for prediction and control.

3. ARTIFICIAL NEURAL NETWORKS

The ability of ANN to represent non-linear relations leads to the idea of using networks in the design of control strategies (Hunt *et al.* 1992). For processes, which are difficult or perhaps impossible to model, the application of ANN seems to be very attractive, since a trained ANN is a non-linear and a non-parametric model (Barron 1993).

If system input/output relationship can be written as:

$$y = f(u) \quad (2)$$

where y, u is a plant outputs and inputs respectively, and $f(.)$ is an unknown, possibly non-linear function. Our aim is to design an ANN which generates some functions $g(.)$ which closely approximates the unknown function $f(.)$, by using input/output data from process.

It is possible to carry out the identification which can be described as follows:

$$\hat{y} = g(U, W) \quad (3)$$

where $\hat{y}$ is the ANN prediction of y from eqn.(1), U is the ANN input vector and W is the weights vector to be identified. Each layer in the ANN structure consists of an appropriate number of neurons and all neurons in a given layer are usually fully connected to the neurons in the next layer.

Every neuron consists of synaptic inputs connections and a single output. The neuron's output is calculated as follows:

$$out = f(inp) = f\left(\sum_{i=0}^{n-1} x_i w_i + w_n\right) \quad (4)$$

where x_i are the neuron's inputs, w_i are the weights, and w_n is an additional input called bias. The function $f(.)$ is the activation function of the neuron. The sigmoid function is commonly used as the activation function.

Consider the problem concerning the design of an algorithm which at time t predicts simultaneously the outcomes of the process $\{y(t)\}$ at time t, $t+1$, $t+2$, ... , $t+N_2$ where N_2 is the prediction horizon. Recurrent ANNs are more appropriate for application in model predictive control (Su and McAvoy 1992; Werbos 1990) than the feedforward ANNs. In this study a recurrent ANN is used as a multi-step predictor. Recurrent ANN include delayed information flow back to preceding layers. An example of the proposed structure of the ANN predictor is depicted in Fig. 1a.

The ANN inputs consist of the previous and current values of process inputs and outputs. The training process of this ANN predictor is carried out using a backpropagation through the time algorithm (Werbos 1990).

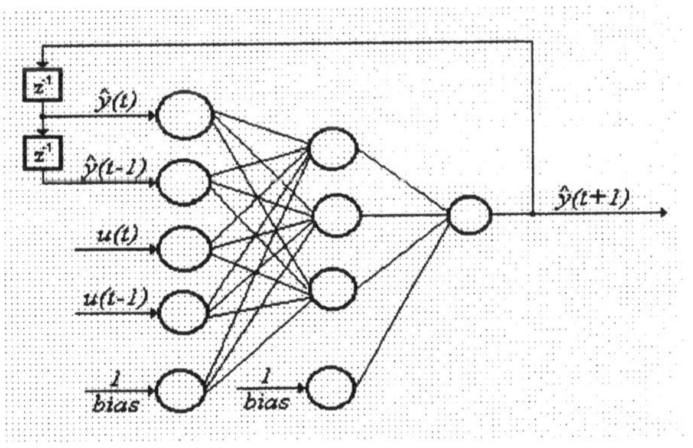

Fig. 1a. Structure of the ANN predictor.

The controller inputs consist of the plant predictions which are provided by the recurrent neural network (long-term predictor) and the desired value of the plant output. The outputs correspond to the present and future increments of the control signal. The weights of this ANN controller are updated directly using a stochastic approximation algorithm which minimizes the control objective (1) subject to eventually some constraints (any control objective can be considered in this control approach). These weights are considered as the controller parameters. Since control action is based on the prediction of the plant behaviour, offset can occur due to disturbances and model mismatch when the ANN is used as a dynamic model of the controlled process. Therefore the plant output is predicted at each sampling time as follows:

$$y(t) = g(U,W) + d(t) \tag{5}$$

where U are the ANN inputs, W is a vector of the weights to be optimized, and d is a disturbance. This disturbance (correction) d of the prediction is computed by the following equation:

$$d(t+i) = d(t) = y(t) - \hat{y}(t) \tag{6}$$

where y is the current value of the plant output and $\hat{y}$ is the prediction of y generated by the ANN predictor. The disturbance is assumed to be constant over the prediction horizon. The proposed structure of this ANN controller is depicted in Fig. 1b.

The general schematic diagram of the predictive controller is depicted in Fig. 2. At each sampling period the following signals are fed into the ANN predictor: last and present plant outputs, last values of control actions applied to the process and the calculated future control sequence from the last sampling period.

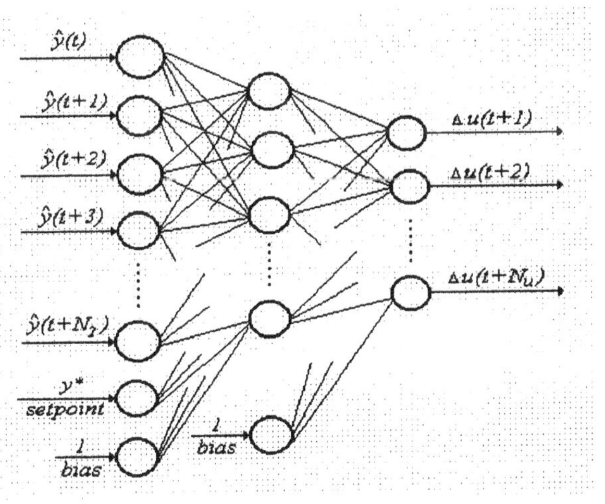

Fig. 1b. Structure of the ANN controller.

The predictor calculates predictions of the plant outputs over the relevant horizon. These are corrected with the calculated deviation between actual process output and predicted process output at time t. Next the predictions are fed together with the set-point value (or sequence of future set-point values for programmed set-point) into the ANN controller. This minimizes the criterion (1), constructs the future control increments, and closes the inner loop. This procedure is repeated until the future control increments converge.

The algorithm used for training this neural network controller is described in the following section.

4. STOCHASTIC APPROXIMATION

The synthesis of the neural network controller is formulated as the determination of their associated weights which minimizes the unconstrained (constrained) control objective function J, i.e.

$$W = \arg\min_{W} J(W) \tag{7}$$

where W is the weights vector, and solved using stochastic approximation techniques.

The stochastic approximation (SA) techniques belong to the class of random search techniques (Kushner and Clark 1978; Spall 1992; Ljung *et al.* 1992; Najim and Poznyak 1994).

The SA algorithm is based on an approximation of the gradient formed from (generally noisy) measurements of the objective function. There are several ways in which the gradient can be approximated (Spall 1992). In our study SA algorithm approximates the gradient using simultaneous perturbation technique.

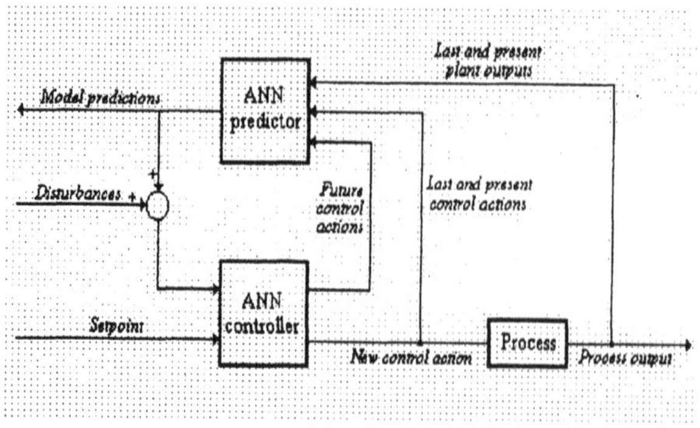

Fig. 4. The control loop structure.

The convergence analysis of stochastic approximation techniques can be found in several studies (Kushner and Clark 1978; Ljung *et al.* 1992).

A key feature of many practical control problems is the presence of constraints on both controlled and manipulated variables. Inequality constraints arise commonly in process control problems due to physical limitations of plant equipments. The constrained optimization (control) problem will be formulated and solved as follows:

Let us consider the following constrained optimization problem:

$$\min_{W} J_o(W) \qquad (8)$$

under the constraints

$$J_i(W) \le 0 \quad (i = 1, \dots, m) \qquad (9)$$

In this study, $J_o(W)$ is associated with the control objective defined by (1). The constraints $J_i(W) \le 0$, $(i = 1, \dots, m)$ are usually associated with process physical limitations of valves, reactor volume, etc.

Let us introduce the Lagrange function (Walk 1983-84; Ljung *et al.* 1992) defined by

$$L(W,\Phi) = J_o(W) + \sum_{j=1}^{m} \varphi_j J_j(W) \qquad (10)$$

where $\Phi = [\varphi_1, \dots, \varphi_m]^T$ is the Khun-Tucker vector.

To solve the optimization problem ((8)-(9)), an iterative algorithm based on stochastic approximation techniques has been proposed by Walk (1983-84). This algorithm maximizes simultaneously $L(W,\Phi)$ with respect to Φ and minimizes it with respect to W . It is presented in what follows:

Let the estimates (W_n, Φ_n) are available at time n where W_n are k-dimensional random vector, $\Phi_n = (\varphi_{ni})$ $i = 1, \dots, m$ are m-dimensional random vector with $\varphi_{ni} \ge 0$ $(n \in N)$ on a probability space (Ω, Ψ, P). Let a_n, c_n $(n \in N)$ be real positive sequences tending to zero and satisfying:

$$\sum a_n^2 c_n^{-2} \langle \infty, \quad \sum a_n = \infty, \quad \sum a_n c_n \langle \infty \qquad (11)$$

The observation noise (the contamination of function values) is modelled by a square integrable real random variables
V_{nl}^{i} $(i = 0, \dots, m;\ l = 1, \dots, k;\ n \in \mathrm{N})$.

The optimization algorithm is given by Walk (1983-84):

$$\Phi_{n+1} = \Phi_n + a_n \tilde{D}_\phi L(W_n, \Phi_n) \qquad (12)$$

where

$$(\tilde{D}_\phi L(W_n, \Phi_n))_i = \max\{J_i(W_n) - V_n^i, \ -a_n^{-1}\varphi_{ni}\}$$

$$(i = 1, \dots, m) \qquad (13)$$

and

$$W_{n+1} = W_n - a_n \tilde{D}_w L(W_n, \Phi_n) \qquad (14)$$

where

$$(\tilde{D}_w L(W_n, \Phi_n))_l = (2c_n)^{-1}[J_0(W_n + c_n e_l) - J_0(W_n - c_n e_l) - V_{nl}^0] + \sum_{j=1}^{m} \varphi_{nj} (2c_n)^{-1}[J_j(W_n + c_n e_l) - J_j(W_n - c_n e_l) - V_{nl}^j] \qquad (15)$$

(e_l k-dimensional unit vector with 1 as *l-th* coordinate, $l=1, \dots, k$).

With

$$\Psi_n = f(W_1, \Phi_1, V_1^i, \dots, V_{n-1}^i \ (i = 1, \dots, m), V_{1l}^i, \dots, V_{n-1,l}^i \ (i = 0, \dots, m;\ l = 1, \dots, k))$$

it is assumed:

$$\forall\, n,l,i \quad E(V_{nl}^i | \Psi_n) = 0, \quad E(V_{nl}^i | \Psi_n) = 0 \qquad (16)$$

In this algorithm, the components of the gradients of the Lagrange function toward the weights W and the Kuhn-Tucker parameters Φ are estimated by finite differences.

5. SIMULATION RESULTS

To demonstrate the performance and the feasibility of this approach, it was applied to control a non-linear SISO model of a continuous-flow, stirred biochemical reactor. This model describes the growth of *Saccharomyces cerevisiae* on glucose with continuous feeding. The complete description of used model can be found in Mészaros *et al.* (1995).

The experiments described here are intended to illustrate the use of the predictive control algorithm described previously when applied to control the dissolved oxygen concentration, $c_o(t)$, using the dilution rate, $D_g(t)$, as the manipulated variable.

The structure of the ANN predictor used was [*6, 5, 1*] - six neurons in the input layer with inputs [$y(t-1), y(t-2), y(t-3), u(t), u(t-1), u(t-2)$], five neurons in the hidden layer and one neuron in the output layer. The sampling period was set equal to *0.5h*. The training set contained *600* input/output pair samples. The structure of the ANN controller used was [*14, 8, 4*]. The inputs of the ANN controller consist of the future predictions of the process behaviour obtained using the recurrent ANN predictor. The output layer generates the control action such that the minimization of the control performance criterion is achieved.

The remain of the parameters were set as follows:

$N_1 = 1,\ N_2 = 13,\ \lambda = 0.1,\ a_n = 0.3,\ c_n = 0.01.$

Notice that a_n has to decrease in order to remove the influence of disturbances, but not so rapidly (Kushner and Clark 1978). In the noise free case, a_n can be either constant or decreasing sequence that converges to a constant value.

The first simulation experiment, the result of which is shown in Fig.3, does not consider any constraints. The lower graph of Fig. 3 represents the variation of the control signal, $u(t)$, derived from the predictive control calculation. Fig. 3 demonstrates the regulation and tracking performance of the predictive control algorithm described above.

In the second experiment, constraints on the variation of the control variable ($|\Delta D_g(t)| \leq 0.1$) has been considered. Fig. 4 shows the evolution of the dissolved oxygen concentration (c_o), the dilution rate (D_g) and its variation. The control signal has no energic variations which corresponds to industrial requirements.

In practice, it is impossible to obtain perfect measurement and uniform dissolved oxygen concentration.

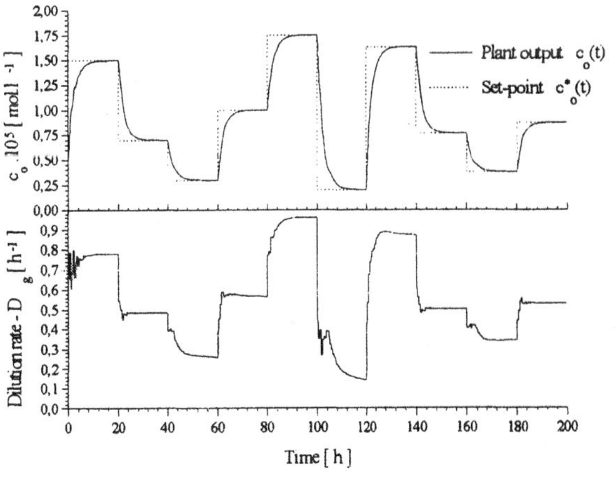

Fig. 3. Variation in the output, desired output and the manipulated variable for predictive control.

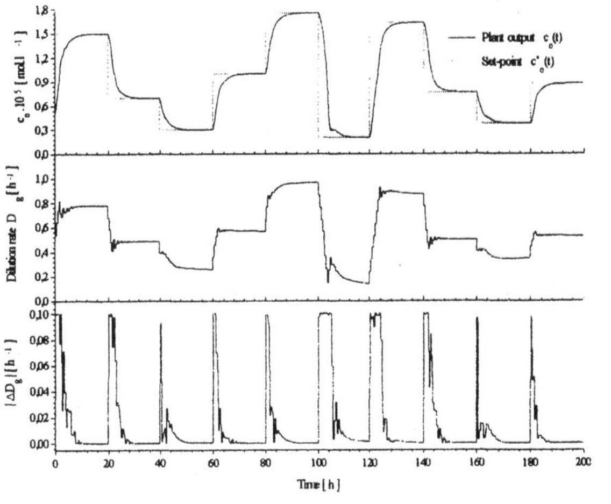

Fig. 4. Variation in the output, desired output and the manipulated variable for constrained predictive control with constraint on the manipiulated variable of the process ($|\Delta Dg(t)| \leq 0.1$).

Therefore a measurement noise, $\xi(t)$, has been introduced with zero mean and variance equal to *0.02*. Fig. 5 shows the time variations of the dissolved oxygen concentration and the dilution rate, respectively,. Obtained results show the ability of the predictive control algorithm to adapt itself to the variations of the desired output. Notice that the set-point changes lead to the variation of the bioprocess dynamics (steady-state gain change, etc.). These results also demonstrate that the presented control algorithm has excellent control (regulation and tracking) performance.

6. DISCUSSION AND CONCLUSIONS

This paper has been concerned with constrained predictive control based on neural networks. The process output prediction over the prediction horizon were provided by a recurrent network trained by the backpropagation through time algorithm.

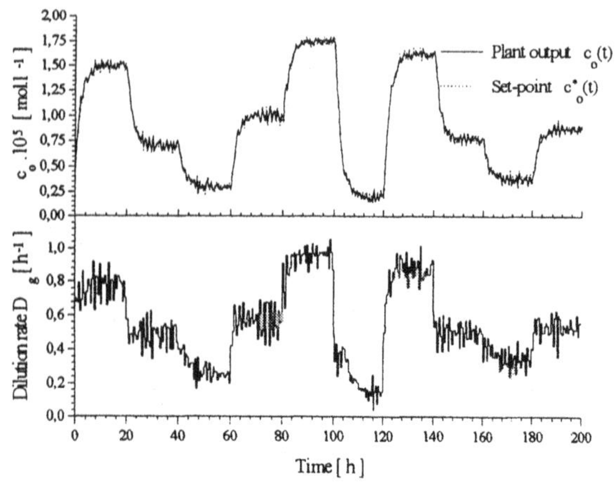

Fig. 5. Variation in the output, desired output and the manipulated variable for constrained predictive control in the case of noisy measurements $(y(t) = co(t) + \xi(t))$.

A multilayer feedforward network has been used as a neural controller and was trained using stochastic approximation algorithms for which convergence as well as convergence rate results are available. Both unconstrained and constrained predictive control strategies have been considered. Numerical experiments related to a fermenter illustrate the feasibility and the performance of this predictive control algorithm. Unlike other constrained predictive strategies, this control algorithm is easy to implement and few design parameters are associated to it. This control strategy is more flexible for use in adaptive systems design than those available in the literature, and does not need to include "special tricks" in a supervisory shell as it is the case for self-tuning regulators and model reference adaptive controllers. Finally, this predictive control algorithm can be easily extended to the multivariable case by increasing the size (number of neurons in the layers) of the neural networks used in this control design.

ACKNOWLEDGEMENT

The authors are pleased to acknowledge the financial support of the Scientific Grant Agency of Ministry of Slovak Republic and Slovak Academy of Sciences under grants N° 1/2149/95 and N° 95/5195/198

REFERENCES

Barron, A. (1993). Universal approximation bounds for superpositions of a sigmoidal function. *IEEE Trans. on Information Theory*, **39**, 930-940.

Clarke, D.W. Mohtadi, C. and Tuffs, P.S. (1987). Generalized predictive control - Part I. The basic algorithm, Part II. Extensions and interpretations. *Automatica*, **23**, 137-160.

Hunt, K. J. D. Sbarbaro, D. Zbikowski, R. and Gawthrop, P.J. (1992). Neural networks for control systems - a survey. *Automatica*, **28**, 1083 - 112.

Kushner, H. J. and Sanvicente, E. (1975). Stochastic approximation for constrained systems with observation noise on the system and constraints. *Automatica*, **11**, 375-380.

Kushner, H. J. and Clark, D.S. (1978). *Stochastic approximation methods for constrained and unconstrained systems.* Springer-Verlag, Berlin.

Ljung, L. Pflug, G. and Walk, H.(1992). *Stochastic approximation and optimization of random systems.* Springer-Verlag, Berlin.

Meszaros, A. Brdys', M. Tatjewski, P. and Lednicky, P. (1995). Multilayer adaptive control of continuous bioprocesses using optimising control technique. Case study: Baker's yeast culture. *Bioprocess Engineering*, **12**, 1-9.

Najim, K. and Poznyak, A.S. (1994). Neural networks synthesis based on stochastic approximation algorithm. *Int. J. of Systems Science*, **25**, 1219-1222.

Rusnák, A. Fikar, M. and Meszaros, A. 1996, (1996).Model predictive control using neural network plant model. *2nd Conference Rizeni Procesu RIP '96, Horni Becva, Czech Republic, June 3-6.*

Spall, J. C. (1992). Multivariate stochastic approximation using a simultaneous perturbation gradient approximation. *IEEE Trans. Auto. Control*, **37**, 332-341.

Su, H. T. and McAvoy, T.J. (1992). Long-term predictions of chemical processes using recurrent neural networks: a parallel training approach. *Ind. Eng. Chem. Res.* **31**, 1338-1352.

Temeng, K. O. Schnelle, P.D. and McAvoy, T.J. (1995). Model predictive control of an industrial packed bed reactor using neural networks. *J. Proc. Cont.* **5**, 19-27.

Walk, H. (1983-84). Stochastic iteration for a constrained optimization problem. *Commun. Statist.-Sequential Analysis*, **2**, 369-385.

Werbos, P. (1990). Backpropagation through time: what it does and how to do it. *Proceedings of the IEEE*, **78**, 1550-1560.

INTELLIGENT MOTION PLANNING AND CONTROL FOR A SIMULATED MOBILE ROBOT

Andon V. Topalov, Diana D. Tsankova, Michail G. Petrov and Todor Ph. Proychev

Control Systems Dept., Technical University Sofia - branch Plovdiv, 61, St. Petersburg blv., 4000, Plovdiv, Bulgaria, tel./fax.: +359-32-221104, e-mail: topalov@tu-plovdiv.bg

Abstract: A complete motion planning and control procedure for mobile robot is presented. A collision-free path is calculated using a neural-net motion planner. The output of the planner is then transformed into a time indexed data sequence which is fed into a tracking controller that takes into account the complete dynamics of the mobile base. A locomotion control structure based on the integration of a kinematic controller and an adaptive fuzzy-net torque controller is proposed. An evolutionary *feedback-error-learning* method for automatic elicitation of knowledge in the form of fuzzy if-then rules is developed. The results of the simulations show the effectiveness of the proposed approach.

Keywords: Mobile robots, path planning, neural networks, fuzzy logic, genetic algorithms.

1. INTRODUCTION

Recently, a lot of work on motion planning and nonlinear control design for wheeled mobile robots has been done. In contemporary mobile robots, moving with relatively high speed, good performance can not be obtained without taking into account the vehicle dynamics (Fierro and Lewis, 1995; Zulli *et al.*, 1995). In addition, in many practical cases, exact knowledge of the mobile robot dynamics is unattainable, e.g., friction is very difficult to model by conventional techniques (Kim *et al.*, 1996). To solve this problem robust-adaptive control methods combining conventional approaches with new learning approaches are required.

In this paper a feasible neural-net motion planning procedure is integrated with a non-linear tracking controller that takes into account the complete dynamics of the mobile base. An appropriate interface has been designed for linking the motion planning and the control design procedures. This interface transforms the purely geometric output of the motion planner into a time-indexed sequence of points, assigning a desired linear and angular velocity. While preserving the main idea of the back stepping control approach (Fierro and Lewis, 1995; Zulli *et al.*, 1995), a suitable control input for velocity following is given by an adaptive fuzzy-net controller (FNC). A genetic-based *feedback-error learning* approach is developed and then implemented for on-line tuning the membership functions and the other parameters of the FNC's rule base to achieve a desired input-output mapping.

2. DESCRIPTIONS AND MODELLING OF THE NON-HOLONOMIC MOBILE ROBOT

The robot constitutes of a rigid trolley equipped with two non-deformable driving wheels mounted on the same axis, and a front free wheel. The motion and orientation are achieved by independent actuators, e.g., DC motors providing the necessary torques to the driving wheels.

The robot in the world possesses three degrees of freedom in its positioning which are represented by a posture,

$$\mathbf{q} = \begin{bmatrix} x_c & y_c & \theta \end{bmatrix}^{\mathrm{T}}, \qquad (1)$$

where (x_c, y_c) are the coordinates of the reference point C and the heading direction θ is taken counterclockwise from the X – axis.

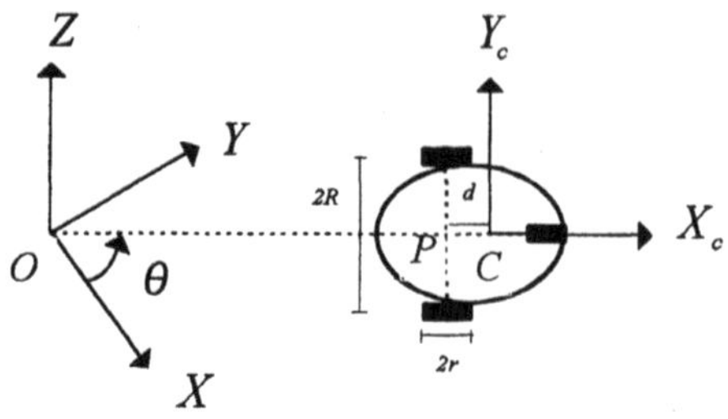

Fig. 1. Geometric model of a non-holonomic mobile platform.

The vehicle's motion is controlled by its linear velocity υ and its angular velocity ω, which are also function of time

$$\mathbf{v} = [\upsilon \quad \omega]^T = [v_1 \quad v_2]^T . \qquad (2)$$

The mobile robot's kinematics is defined by a Jacobian matrix $\mathbf{J}(\theta)$ that transforms velocities $\mathbf{v}$ expressed in mobile base coordinates to velocities $\dot{\mathbf{q}}$ expressed in Cartesian coordinates:

$$\begin{bmatrix} \dot{x}_c \\ \dot{y}_c \\ \dot{\theta} \end{bmatrix} = \dot{\mathbf{q}} = \mathbf{J}\mathbf{v} = \begin{bmatrix} \cos\theta & -d\sin\theta \\ \sin\theta & d\cos\theta \\ 0 & 1 \end{bmatrix} \begin{bmatrix} v_1 \\ v_2 \end{bmatrix}, \qquad (3)$$

where $|v_1| \le \upsilon_{max}$ and $|v_2| \le \omega_{max}$. υ_{max} and ω_{max} are the maximum linear and angular velocities of the mobile robot. System (3) is called the steering system of the vehicle.

In general the dynamical equations of the mobile robot system having an n-dimensional configuration space with generalized coordinates $(q_1, \ldots, q_n)$ and subject to m constraints can be described by (Fierro and Lewis, 1995):

$$\mathbf{M}(\mathbf{q})\ddot{\mathbf{q}} + \mathbf{V}_m(\mathbf{q},\dot{\mathbf{q}})\dot{\mathbf{q}} + \mathbf{F}(\dot{\mathbf{q}}) + \mathbf{G}(\mathbf{q}) + \boldsymbol{\tau}_d = \mathbf{B}(\mathbf{q})\boldsymbol{\tau} - \mathbf{A}^T(\mathbf{q})\boldsymbol{\lambda}, (4)$$

where $\mathbf{M}(\mathbf{q}) \in R^{n \times n}$ is a symmetric, positive definite inertia matrix, $\mathbf{V}_m(\mathbf{q},\dot{\mathbf{q}}) \in R^{n \times n}$ is the centripetal and coriolis matrix, $\mathbf{F}(\dot{\mathbf{q}}) \in R^n$ denotes the surface friction, $\mathbf{G}(\mathbf{q}) \in R^n$ is the gravitational vector, $\boldsymbol{\tau}_d$ - denotes bounded unknown disturbances including unstructured unmodelled dynamics, $\mathbf{B}(\mathbf{q}) \in R^{n \times r}$ is the input transformation matrix, $\boldsymbol{\tau} \in R^r$ is the input vector, $\mathbf{A}(\mathbf{q}) \in R^{m \times n}$ is the matrix associated with the constraints, and $\boldsymbol{\lambda} \in R^m$ is the vector of constraint forces.

3. THE COLLISION-FREE PATH PLANNING

The Collision-Free Path Planning can be stated as: Given an object with an initial position and a goal position, and a set of obstacles. The problem is to find a continuous path from the initial position to the goal position which avoids colliding with obstacles along the trajectory. The motion planning procedure proposed in the paper is based on the theoretical work of Meng and Picton (1992). The path for the object is represented by a set of N via points. The path finding algorithm is equivalent to optimizing a cost function, defined in terms of the total path length and the collision penalty, by moving the via points in the direction that minimizes the cost function. A two-layer log-sigmoid/log-sigmoid backpropagation neural network is used to produce the collision penalty. The environment area is divided into 2D grid cells and a binary value is assigned to each grid cell to indicate an obstacle presence: "0" means that the cell is fully unoccupied, and "1" - the cell is occupied. To take into account the robot's geometry the obstacles are modified by growing its size isotropically by the robot's radius plus a small tolerance.

The x, y coordinates of the cells' centres and the corresponding assigned binary values are used as learning patterns for the 2-input/1-output neural network. The output of the network represents the collision penalty for the current position (x, y) of the object. The collision penalty of a path is defined as the sum of individual collision penalties of all the via points. The energy function for collision is defined as :

$$E_C = \sum_{i=1}^{N} C_i , \qquad (5)$$

where C_i is the collision penalty for the ith via point. The energy function for the path length is defined as the sum of squares of all the segments' lengths connecting the via points:

$$E_L = \sum_{i=1}^{N} L_i^2 = \sum_{i=1}^{N} [(x_i - x_{i-1})^2 + (y_i - y_{i-1})^2], \qquad (6)$$

In accordance with Meng and Picton (1992) the total energy is

$$E = E_C + E_L . \qquad (7)$$

In this work the equation (7) was modified by multiplying both energy functions with positive weight coefficients k_C and k_L to express our preference for the collision criterion or for the path length criterion, respectively:

$$E = k_C E_C + k_L E_L \ . \tag{8}$$

The dynamical equation for a via point is chosen to make time derivative of the energy be negative along the trajectory, because the low energy implies less collision and a shorter path. The time derivative of E is

$$dE/dt = d(k_L E_L + k_C E_C)/dt =$$
$$= \sum_{i=1}^{N} (k_L \partial L_i^2 / \partial x_i + k_C \partial C_i / \partial x_i) dx_i / dt + \tag{9}$$
$$+ \sum_{i=1}^{N} (k_L \partial L_i^2 / \partial y_i + k_C \partial C_i / \partial y_i) dy_i / dt \ .$$

Let

$$dx_i / dt = -(k_L \partial L_i^2 / \partial x_i + k_C \partial C_i / \partial x_i) \ ,$$
$$dy_i / dt = -(k_L \partial L_i^2 / \partial y_i + k_C \partial C_i / \partial y_i) \ , \tag{10}$$

then

$$dE/dt = -\sum_{i=1}^{N} [(dx_i/dt)^2 + (dy_i/dt)^2] < 0 \ . \tag{11}$$

$dE/dt = 0$ if and only if $dx_i/dt = 0$ and $dy_i/dt = 0$. Therefore, all via points move along trajectories, decreasing the energy, and finally reach equilibrium positions. In (10)

$$\partial C_i / \partial x_i = f'(I2) \sum_{j=1}^{S} W2_j f'(I1_j) W1^x{}_j \ ,$$
$$\partial C_i / \partial y_i = f'(I2) \sum_{j=1}^{S} W2_j f'(I1_j) W1^y{}_j \ , \tag{12}$$

where C_i is the output of the network, $I2$ is the input of the output layer neuron, $I1_j$ is the input of the jth hidden layer neuron, S is the number of hidden layer neurons, $W1^x{}_j$ is the weight coefficient of the input "x" with respect to j^{th} hidden layer neuron, $W2_j$ is the weight coefficient of jth hidden layer neuron's output with respect to the output layer neuron. From (10) and (12), the dynamical equations for x_i and y_i are derived as:

$$dx_i / dt = -[k_L (2x_i - x_{i-1} - x_{i+1}) +$$
$$+ k_C f'(I2) \sum_{j=1}^{S} W2_j f'(I1_j) W1^x{}_j] \ ,$$
$$dy_i / dt = -[k_L (2y_i - y_{i-1} - y_{i+1}) +$$
$$+ k_C f'(I2) \sum_{j=1}^{S} W2_j f'(I1_j) W1^y{}_j] \ . \tag{13}$$

4. CONTROL DESIGN

The trajectory tracking problem for non-holonomic vehicles is posed as follows: Given a prescribed reference cart

$$\dot{x}_r = \upsilon_r \cos\theta_r, \ \dot{y}_r = \upsilon_r \sin\theta_r, \ \dot{\theta}_r = \omega_r,$$
$$\mathbf{q}_r = [x_r \quad y_r \quad \theta_r]^{\mathrm{T}}, \ \mathbf{v}_r = [\upsilon_r \quad \omega_r]^{\mathrm{T}} \ , \tag{14}$$

with $\upsilon_r > 0$ for all t, find a smooth velocity control $\mathbf{v}_c$ such that $\lim_{t\to\infty}(\mathbf{q}_r - \mathbf{q}) = 0$, and compute the torque input $\boldsymbol{\tau}(t)$ for (4) by some means such that $\mathbf{v} \to \mathbf{v}_c$ as $t \to \infty$.

The proposed structure for the mobile robot tracking control system is presented in Fig. 2. It is assumed that the solution to steering system tracking problem as found in Kanayama *et al.*, (1990) is available. This is denoted as $\mathbf{v}_c(t)$.

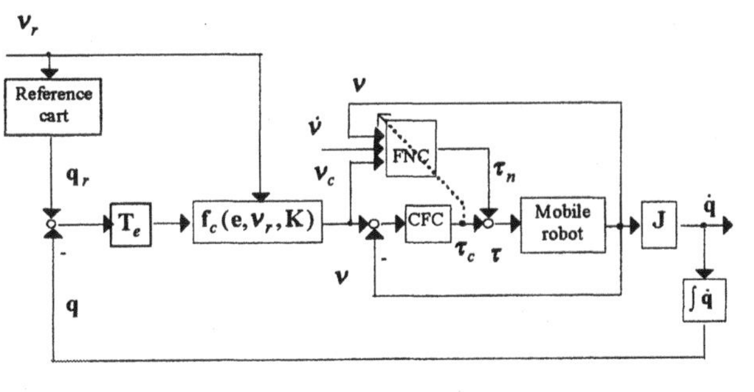

Fig. 2. Block diagram of fuzzy-net tracking control of mobile robot.

The tracking error posture is expressed in the basis of a frame linked to the mobile platform (Fierro and Lewis, 1995)

$$\mathbf{e} = \mathbf{T}_e(\mathbf{q}_r - \mathbf{q}) \ ,$$
$$\begin{bmatrix} e_1 \\ e_2 \\ e_3 \end{bmatrix} = \begin{bmatrix} \cos\theta & \sin\theta & 0 \\ -\sin\theta & \cos\theta & 0 \\ 0 & 0 & 1 \end{bmatrix} \begin{bmatrix} x_r - x \\ y_r - y \\ \theta_r - \theta \end{bmatrix} . \tag{15}$$

and the auxiliary velocity control input that achieves tracking for (3) is given by

$$\mathbf{v}_c = \begin{bmatrix} \upsilon_r \cos e_3 + k_1 e_1 \\ \omega_r + k_2 \upsilon_r e_2 + k_3 \upsilon_r \sin e_3 \end{bmatrix},$$
$$\mathbf{v}_c = \mathbf{f}_c(\mathbf{e}, \mathbf{v}_r, \mathbf{K}) \ . \tag{16}$$

Given the desired velocity $\mathbf{v}_c(t) \in R^{n-m}$, the auxiliary velocity tracking error can be defined as

$$\mathbf{e}_c = \mathbf{v}_c - \mathbf{v} \ . \tag{17}$$

In the proposed velocity control scheme, no preliminary knowledge of the cart dynamics is

assumed. The velocity control input vector $\mathbf{v}_c$ is transformed into desired left and right wheel velocities and compared to the current wheel velocities of the mobile robot. Each wheel velocity is under independent control. In accordance with the *feedback-error-learning* approach (Gomi and Kawato, 1993), a conventional PD feedback controller (CFC) is used both as an ordinary feedback controller to garantee global asymptotic stability during the learning period, and as a reference model for the responses of the controlled object. To obtain the desired response during tracking-error convergence movement by compensating for the nonlinear object dynamics, the fuzzy-neural network is trained to become a nonlinear regulator. The output of the CFC is fed to the FNC as the error signal used for learning.
There is two FNCs with identical structure for each robot's wheel velocity. The structure of the FNC is presented in Fig. 3. It has three inputs (x_1, x_2, and x_3) corresponding to the actual wheel velocity and acceleration, and the desired wheel velocity, respectively.

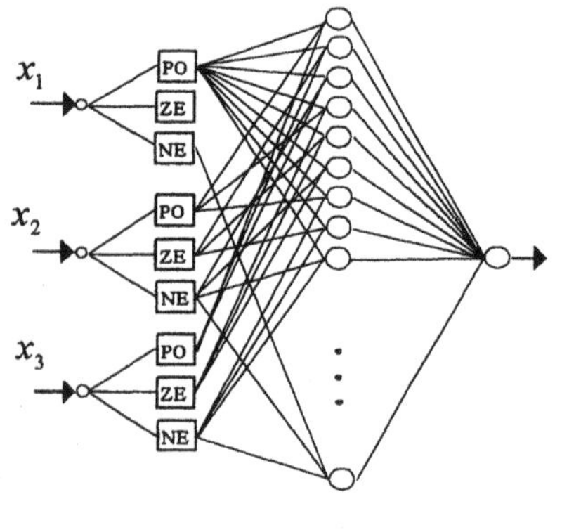

Fig. 3. Structure of the fuzzy-net model.

Each FNC input is associated with three fuzzy labels. FNC's rule base consists of fuzzy if-then rules of Takagi and Sugeno's type. The corresponding i,j,k th fuzzy rule can be expressed as

If x_1 *is* A_i *and* x_2 *is* B_j *and* x_3 *is* C_k *then* $f_{ijk} = d_{ijk}$,

where $i,j,k = 1,2,3$. The firing strength of i,j,k th rule is obtained as the T-norm of the membership values on the premise part (by using a multiplication operator):

$$w_{ijk} = \mu_{A_i}(x_1)\mu_{B_j}(x_2)\mu_{C_k}(x_3)\,. \qquad (18)$$

The membership functions of the linguistic labels A_i, B_j, and C_k are scaled Gaussian functions. The overall output is computed as the weighted average of each rule's output.

5. THE EVOLUTIONARY FEEDBACK-ERROR-LEARNING APPROACH

The classical optimization algorithms cannot be relied upon to find the optimal (global) solution in multi-parameter search spaces. Genetic Algorithms (GAs) are adaptive methods widely used in searching the optimal point in a highly nonlinear and complex space. They are parallel, global search techniques that emulate natural genetic operations (Michalewicz, 1992). Because they simultaneously evaluate many points in the parameter space, GAs are more likely to converge toward the global solution. They need not assume that the search space is differentiable or continuous.
Considering these advantages of GAs, a genetic-based *feedback-error learning* approach has been developed to provide on-line learning of the FNC. Each parameter of the controller is encoded by 8 bits. FNC's parameters are the widths σ, the means c of the membership functions and the outputs f of the fuzzy if-then rules. The candidate solutions of the FNC's parameters are represented by binary strings. Each string of the candidate solutions has the following form:

$$\begin{array}{l} |\sigma_{A_1},\sigma_{A_2},\sigma_{A_3},\sigma_{B_1},\sigma_{B_2},\sigma_{B_3},\sigma_{C_1},\sigma_{C_2},\sigma_{C_3}| \\ |c_{A_1},c_{A_2},c_{A_3},c_{B_1},c_{B_2},c_{B_3},c_{C_1},c_{C_2},c_{C_3}|f_{111},\ldots,f_{ijk},\ldots,f_{333}| \end{array} \qquad (19)$$

where σ_{A_i}, σ_{B_j}, σ_{C_k} represent the binary values of the widths, and c_{A_i}, c_{B_j}, and c_{C_k} - the binary values of the means of the membership functions, and f_{ijk} - the binary value of the output of the ijk^{th} fuzzy if-then rule. The real values of these parameters are limited to [-5, 5]. The population size is taken 50. The fitness function is defined as follows:

$$F^i(k) = 10/\{0.001 + \left[\tau^i(k-1) - \tau_n^i(k)\right]^2\}\,, \qquad (20)$$

where superscript i means left or right robot's wheel, $\tau_n(k)$ is the output of FNC at time instant k, $\tau(k-1)$ is the sum of the output signals of CFC and FNC (the total control input signal) at time instant $k-1$. The constant value, 10, is introduced to enlarge the fitness difference between any two offspring. This fitness function is time dependent and learning is continuing on-line. Initial population of candidate solutions is generated randomly. Evaluation is expected for the fitness value of each solution to rank the trial solutions. The candidate with higher value has the higher opportunity to reproduce new population by crossover and mutation. During the evolution, the elite preserving strategy is adopted. Candidates are chosen from both of the offspring cluster and the former solution cluster if their fitness values are higher than the other chromosomes. Thus a new group of candidate solutions is formed whose size remains the same as that before the reproduction. The reproduction process iterates until the convergence criteria is satisfied. A convergence criteria is adopted according to the condition that the maximum of the absolute value of the error within the interval of 15 previous time steps for the current time step k is below 0.002.

A three-point crossover technique was used here. The strong mixing effect caused by the crossover operator is helpful to overcome the local optima.

Mutation operation is another method to avoid the local optimum. Mutation helps ensure that no point in the search space has a zero probability of being explored. In this paper the mutation probability was set to 0.001.

At every time step the solution with the best fitness function is used for the current representation of the FNC parameters.

GAs have the property that they usually converge to the optimal point slower than back-propagation algorithm after reaching the near-optimal area due to the their randomness. So further fine adjustment of the FNC parameters (when the convergence criteria is satisfied and the reproduction process is stopped) can be obtained by using a dynamic back-propagation algorithm and the error defined in (20). In this case, as it has been proposed by Xu and Wang, (1995), only the outputs of the fuzzy if-then rules can be tuned additionally. Such hybridization schemes (Topalov *et al.*, 1996) are also useful in some cases when the computational power of the mobile robot's on-board computer is not sufficient to perform on-line evolutionary *feedback-error-learning*. Rough tuning of the FNC parameters could be made by using simulations with only fine on-line tuning implemented on the real robot.

6. SIMULATION RESULTS

To illustrate the performance of the proposed path planning and control schemes they were simulated in MATLAB environment and tested on several examples. In accordance with Fig. 1 the following parameters of the geometric model were chosen: r=0.025 m, R= 0.036 m, d=0.02 m. The mass of the mobile robot was set to be 0.5 kg. In order to handle non-smooth reference paths without causing the robot's slippage, the target velocities (υ_c, ω_c) were limited by constants $(\hat{\upsilon}_c, \hat{\omega}_c)$. The values of these constants were adopted to be $(\hat{\upsilon}_c, \hat{\omega}_c)$=(0.4 m/sec, 1.57 rad/sec). The posture controller gains were chosen so that the closed-loop system exhibited a critical damping behavior (Kanayama *et al.*, 1990): $k1$= 10/sec, $k2$= 64/m^2, and $k3$= 16/m. The robot's sampling time was set to T_s= 8 ms. The following parameters for the PD controller were used: $k_P = 1$ and $k_D = 0.01$. The linear reference velocity was set to 0.3 m/sec. The neural network from the motion planning algorithm has been trained with backpropagation with momentum and adaptive learning rate. The number of neurons in the hidden layer was S = 25. The workspace was divided into 20×20 grid cells, and neural network used the obtained 400 training samples. The sum squared error of the final learning epoch was 0.001. After training the neural network weights and biases were obtained and then used to find a path described by a set of $N = 25$ via points (without initial and target points). The workspace consisted of two obstacles. The initial position was (0,0), and the goal position was (2,2). The initial path, needed for solution of equations (13) was chosen as a straight line from the initial position to the goal position. The weight coefficients in (13) are chosen to be $k_L = 1$ and $k_C = 5$. Fig.4 shows the approximation performance of the neural network (dotted area), the piece-wise liner path obtained by the described above algorithm (dashed line with o-marks), and the mobile robot movement from the initial position to goal one along the already calculated path (bold line). Clearly, the mobile platform was able to track the reference trajectory.

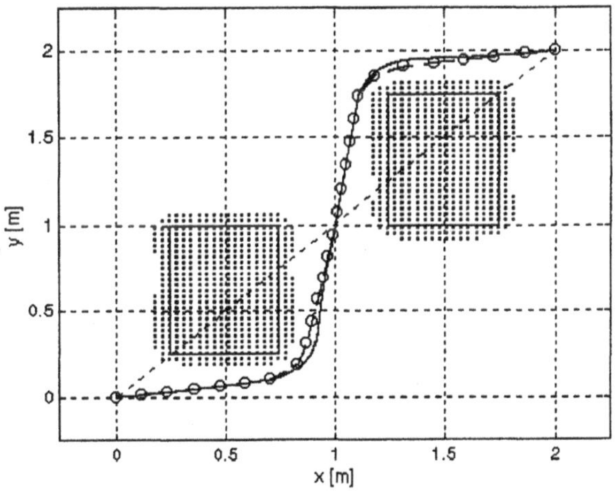

Fig. 4. Simulation of the mobile robot movement.

In order to demonstrate the adaptation capability of the control scheme to execute desired motions in the face of unforeseen parameter variations, an example when the robot pick up a load during the trajectory tracking task was carried out. In this example the mobile robot was required during the following a specified trajectory to pick up an unknown load after the first 4 sec (500 time steps). The pay-load is considered as an increasing of the robot's mass from 0.5 kg to 0.75 kg and a change of the moment of inertia from 0.00065 kgm^2 to 0.00126 kgm^2. The limited target velocities for the simulation are $(\hat{\upsilon}_c, \hat{\omega}_c)$ = (0.4 m/s, 0.8 rad/s). The performance delivered by the proposed control scheme under these conditions is depicted in Fig.5 where the actual trajectory is indicated in solid line while the reference trajectory indicated in dashed line. It can be seen that the tracking performance of the mobile robot with changed parameters after the third angular discontinuity is almost the same like in the beginning (after the first angular discontinuity), when the robot had its parameters unchanged.

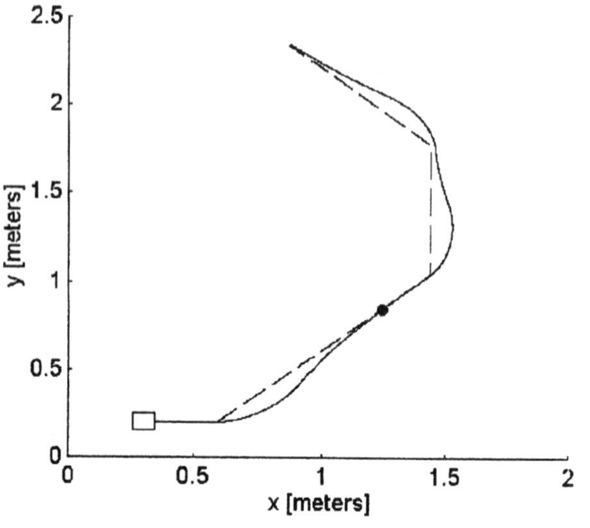

Fig. 5. Trajectory tracking with change of the robot's mass and moment of inertia. (The position where robot undergoes changes of its mass and moment of inertia is the black point).

7. CONCLUSIONS

A complete motion planning and control procedure for mobile robot is presented. A collision-free path is calculated using a neural-net motion planner. In general the discussed path planning method may be involved in the group of the off-line potential field methods (variational planning (Latombe, 1991)). The computation of the shortest path is its advantage. The neural network can automatically build up the collision penalty function (as an obstacles' potential function approximation) after training by using the information of the environment, no matter how many obstacles there are and what kinds of shape obstacles look like. This makes the algorithm effective in practice. Since the algorithm provides massive parallelism in computation, the path planning can be done in real-time by making good use of parallel hardware.

The output of the planner is then transformed into a time indexed data sequence which is fed into a tracking controller that takes into account the complete dynamics of the mobile base. A locomotion control structure based on the integration of a kinematic controller and an adaptive fuzzy-net torque controller is presented. It is associated with a new on-line genetic-based *feedback-error-learning* method for automatic elicitation of knowledge in the form of fuzzy if-then rules. The on-line learning of the FNC not only facilitates adaptivity of control to changes in the robot dynamics and/or external conditions, but also eliminates problems associated with unmodeled dynamics.

The simulation experiments carried out showed the effectiveness of the proposed path planning and control strategy. It will be further verified on a real mobile robot.

REFERENCES

Fierro, R and F.L. Lewis (1995). Control of a Nonholonomic Mobile Robot: Backstepping Kinematics into Dynamics. *Proc. of 34th Conf. on Decision and Control,* pp. 3805-3810.

Gomi, H. and M. Kawato (1993). Neural Network Control of a Closed-Loop System Using Feedback-Error-Learning. *Neural Networks,* Vol. 6, pp. 933-946.

Kanayama, Y., Y. Kimura, F. Miyazaki and T. Noguchi (1990). A Stable Tracking Control Method for an Autonomous Mobile Robot. *Proc. IEEE Int. Conf. on Robotics and Automation,* Vol. 1, pp. 384-389.

Kim, J.-H., H.-K. Chae, J.-Y. Jeon and S.-W. Lee (1996). Identification and Control of Systems with Friction Using Accelerated Evolutionary Programming, to appear in *IEEE Control Systems Magazine*.

Latombe, J.-C. (1991). *Robot Motion Planning.* Kluwer Academic Publishers, pp.316-317.

Meng, H. and P.D. Picton (1992). A Neural Network for Collision-Free Path Planning. In: *Artificial Neural Networks* (I. Aleksander and J. Taylor, Ed.), Vol. 2, pp. 591-594.

Michalewicz, Z. (1992). *Genetic Algorithms + Data Structures = Evolution Programs.* Springer-Verlag.

Topalov, A.V., K.-C. Kim, J.-H. Kim, and B.-K. Lee (1996). Fast Genetic On-Line Learning Algorithm for Neural Network and Its Application to Temperature Control. *Proc. IEEE Int. Conf. on Evolutionary Computation,* pp. 649-654.

Xu, J.-X. and X. Wang (1995). Fast Tuning of Fuzzy-Neural Network Structure by Genetic Algorithm. *Proc. IEEE Singapore Int. Conf. on Intelligent Control and Instrumentation,* pp. 87-91.

Zulli, R, R. Fierro, G. Conte and F.L. Lewis (1995). Motion Planning and Control for Non-Holonomic Mobile Robots. *Proc. IEEE Int. Symp. on Intelligent Control,* pp. 551-557.

FUZZY STATE CONTROL OF THE NONLINEAR BALANCE CONTROL SYSTEM

Nenad Muškinja, Boris Tovornik

University of Maribor, Faculty of Electrical Engineering and Computer Science, Smetanova 17, 62000 Maribor, Slovenia

Abstract: Using fuzzy control in state space and controlling the state of the system with a fuzzy controller opens new possibilites for use of the theory of fuzzy sets. In this paper, we represented the basic idea of a fuzzy state space control and compared it with the classic state space control through solving the problem of balance control.

While experimenting on the balance system, we are forced to use the modern theory of control. While developing the optimal state space control, the largest obstacle we face is to set up an acceptable linear mathematical model. This model will become useless in the system when the parameters are variable. Development of the fuzzy state space controller, on the other hand, does not require mathematical model of the proces, but fuzzy control rules must be obtained from human exerts.

Keywords: Fuzzy control, Nonlinear control systems, State-space models

1. INTRODUCTION

The idea of implementing fuzzy control in state space originated in the development of optimal control for the process of balance control.

1.1 Model description

The Balance control system, reel following a prescribed reference position along a bar mounted on a pivot is shown in figure 1 (Camo A/S, 1991).

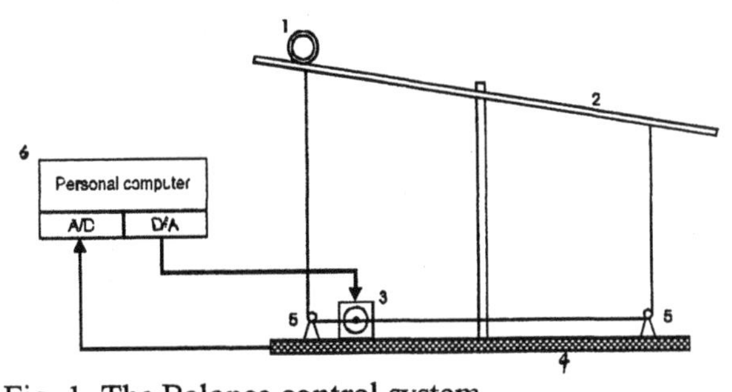

Fig. 1. The Balance control system

The significant parts of the system are: 1. Reel moving along the bar 2, 2. Bar mounted on a pivot, 3. DC motor, 4. Base, 5. Pulley, 6. Personal computer.

The problem is to find the optimal strategy for controlling the reference curve of the reel along the balanced bar. By developing the linear state space controller, we have to set linear mathematical model which is suitable for the given non-linear process of the balance control. (See Section 2.) The linear state space controller (see Section 3) works optimally only around the working point, for which the linear model of the the process was derived; this was also seen in the realization of the real process. We included the state observer because all states of the system are not directly measurable (R. Isserman, 1981). We were interested to see how the fuzzy control performs in the same situation. First, we developed the fuzzy PI controller for the position of the reel (D. Donlagic, 1994); however, it turned out that without the information of the angle of the bar, stable execution of the balance system cannot be insured. Consequently, we arrived at the idea that the fuzzy controller would control each of the state variables separately. Expertise of the mathematical model of the process is not necessary for the development of the fuzzy state controller (Section 4); we need only to know the behavior of the process. In other words, we must know how to set the reference variable according to the states of the process. Estimation of the nonmeasurable state variables was represented by the simple differentiation of the reel position, which insured stability for the execution of the closed loop system, even in the real time realization.

The following will illustrate the basic structure of the fuzzy state control. It is based on the assumption that the fuzzy state controller is a nonlinear controller which, without any additional dynamic parts, peforms nonlinear static characteristics. The analysis of the closed loop system is based on the knowledge of the system matrices of the linear part and of the nonlinear function of the fuzzy state controller. From this, we get the system of the non-linear differential equations, which is solved by the theory of the non-linear systems, like the method of harmonic balance.

1.2 Basic Structure of the Fuzzy State Controller

The basic idea is to set the fuzzy controller into the state space. Figure **2** shows the basic structure of the fuzzy state space controller.

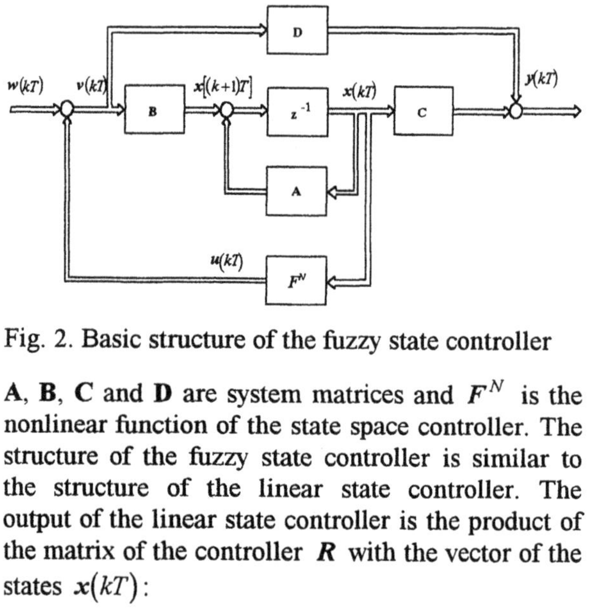

Fig. 2. Basic structure of the fuzzy state controller

A, **B**, **C** and **D** are system matrices and F^N is the nonlinear function of the state space controller. The structure of the fuzzy state controller is similar to the structure of the linear state controller. The output of the linear state controller is the product of the matrix of the controller $\boldsymbol{R}$ with the vector of the states $\boldsymbol{x}(kT)$:

$$\boldsymbol{u}(kT) = \boldsymbol{R}\boldsymbol{x}(kT), \tag{1}$$

the output of the fuzzy state controller is given with the non-linear function $\boldsymbol{F}^N$.

$$\boldsymbol{u}(kT) = \boldsymbol{F}^N\{\boldsymbol{x}(kT)\}. \tag{2}$$

DEFINITION 1:

We have an m-dimensional state vector $\boldsymbol{x}(kT) = [x_1(kT), x_2(kT), \ldots, x_m(kT)]$, whose elements are the input linguisitic variables from the space $U = [U_1, U_2, \ldots, U_m]$, and an n-dimensional output vector $\boldsymbol{u}(kT) = [u_1(kT), u_2(kT), \ldots, u_n(kT)]$, whose elements are the output linguistic variables from space $V = [V_1, V_2, \ldots, V_n]$.

For each input vector $\boldsymbol{x}(kT)$ there exists an output vector $\boldsymbol{u}(kT)$, such that:

$$\begin{aligned} B_i(kT) &= \text{infer}_i(\boldsymbol{x}(kT)) \quad i = 1 \ldots n \\ u_i(kT) &= \text{defuz}(B_i) \end{aligned} \tag{3}$$

Therefore:

$$\boldsymbol{u}(kT) = \begin{bmatrix} \text{defuz}(\text{infer}_1(\boldsymbol{x}(kT))), \\ \text{defuz}(\text{infer}_2(\boldsymbol{x}(kT))), \\ \vdots \\ \text{defuz}(\text{infer}_n(\boldsymbol{x}(kT))) \end{bmatrix}, \tag{4}$$

where $B_i(kT)$ represents the i-th output fuzzy set, **infer** equals the inference mechanism which forms the fuzzy set (based on the given rules and membership functions), and **defuz** equals the defuzziation function of the output fuzzy set.

Example 1:

We have two input linguisitic variables (m=2) and one output linguistic variable (n=1), therefore:

$$\boldsymbol{x}(kT) = [x_1(kT), x_2(kT)], \quad x_1(kT) \in U_1, \quad x_2(kT) \in U_2$$

and

$$\boldsymbol{u}(kT) = u(kT) = \text{defuz}(\text{infer}(\underline{x}(kT))), \quad u(kT) \in V$$

This means that F^N =**defuz·infer**.

In the example of the linear state controller, the matrix of the controller equals:

$$\boldsymbol{R} = [r_1, r_2]$$

And the output of the controller is a linear function of the system's states, which results in a flat surface in the $\mathbb{R}^3$ space (Figure **3**).

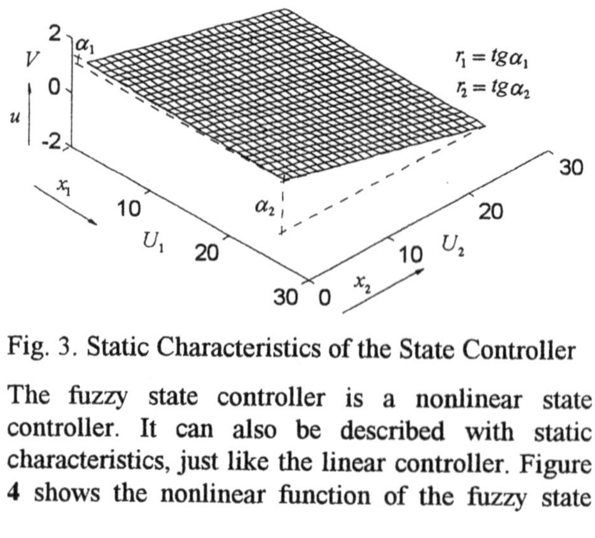

Fig. 3. Static Characteristics of the State Controller

The fuzzy state controller is a nonlinear state controller. It can also be described with static characteristics, just like the linear controller. Figure **4** shows the nonlinear function of the fuzzy state space controller's output. Digressions from the surface represent the digressions from the linear state controller.

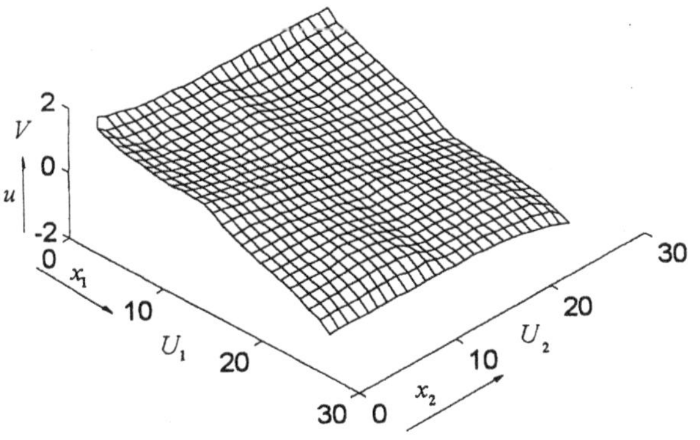

Fig. 4. Static Characteristics of the Fuzzy State Controller

We can conclude from Example 1 that the fuzzy state controller actually is a nonlinear state controller. The output of the fuzzy state controller, in general, is a nonlinear function of the state variables (equation (2) and (4)). The fuzzy state controller forms a static nonlinear surface in $\mathbb{R}^{m+1}$ space for each of the outputs $u_i(kT)$, without any additional dynamic parts.

The state equations of the closed loop system, from Figure 2, are:

$$\begin{aligned} \boldsymbol{x}((k+1)T) &= \boldsymbol{A}\boldsymbol{x}(kT) + \boldsymbol{B}\boldsymbol{w}(kT) + \boldsymbol{B}\boldsymbol{u}(kT) \\ \boldsymbol{y}(kT) &= \boldsymbol{C}\boldsymbol{x}(kT) + \boldsymbol{D}\boldsymbol{w}(kT) + \boldsymbol{D}\boldsymbol{u}(kT) \end{aligned} \tag{5}$$

In considering equation (2), we get:

$$\begin{aligned} \boldsymbol{x}((k+1)T) &= \boldsymbol{A}\boldsymbol{x}(kT) + \boldsymbol{B}\boldsymbol{w}(kT) + \boldsymbol{B}\boldsymbol{F}^N\{\boldsymbol{x}(kT)\} \\ \boldsymbol{y}(kT) &= \boldsymbol{C}\boldsymbol{x}(kT) + \boldsymbol{D}\boldsymbol{w}(kT) + \boldsymbol{D}\boldsymbol{F}^N\{\boldsymbol{x}(kT)\} \end{aligned} \tag{6}$$

The analysis of the closed loop system becomes a problem of solving the nonlinear differential equation. For the stability analysis, according to Lyapunov, we need an explicit form for the nonlinear surface $\boldsymbol{F}^N\{\boldsymbol{x}(kT)\}$ (H.Kendl and J.Ruger, 1993).

2. BALANCE CONTROL

Design of mathematical models are often regarded a difficult task. If you apply the physical approach you have to specify the differential equations with their parameters describing the dynamics of the system. You have to do the right assumptions and simplifications. Plant measurement recordings may be used for parameter estimation. Using plant recordings you may also find the process model through an identification procedure.

In order to express the system equations in state space form:

$$\dot{x} = \boldsymbol{A}x + \boldsymbol{B}u \tag{7}$$

we define the following state space vector:

$$\boldsymbol{x} = \begin{bmatrix} x_1 \\ x_2 \\ x_3 \\ x_4 \end{bmatrix} = \begin{bmatrix} x \\ \dot{x} \\ \alpha \\ \dot{\alpha} \end{bmatrix} \begin{matrix} \text{position of the reel} \\ \text{velocity of the reel} \\ \text{angle of the bar} \\ \text{angular velocity of the bar} \end{matrix} \tag{8}$$

We get the following differential equations:

$$\begin{aligned} \dot{x}_1 &= x_2 \\ \dot{x}_2 &= \frac{2}{3} g x_3 \\ \dot{x}_3 &= x_4 \\ \dot{x}_4 &= -\frac{1}{T} x_4 + bu \end{aligned} \tag{9}$$

or:

$$\dot{\boldsymbol{x}} = \begin{bmatrix} 0 & 1 & 0 & 0 \\ 0 & 0 & \frac{2}{3}g & 0 \\ 0 & 0 & 0 & 1 \\ 0 & 0 & 0 & -\frac{1}{T} \end{bmatrix} \boldsymbol{x} + \begin{bmatrix} 0 \\ 0 \\ 0 \\ b \end{bmatrix} u \tag{10}$$

where g is acceleration due to gravity, the time constant T and the control gain b are experimentally estimated to $T = 0.1\,\text{s}$ and b=4.4.

3. DEVELOPMENT OF THE OPTIMAL STATE CONTROLLER

We used the program package MATLAB 4.2 and CONTROL Toolbox (function dlqr) for the development of the optimal controller. We tested the linear state controller, developed in such a way, in real time with the BALANCE CONTROL model.

3.1 Control Strategy

The optimal strategy is based on the optimal control strategy. We derive the feedback gain matrix with the use of the pole placement method. First, we define the sample time T_S=0.055s and we initialize the system matrices $\boldsymbol{A}, \boldsymbol{B}, \boldsymbol{C}$ and $\boldsymbol{D}$ in the discrete space. Next, it is necessary to set the poles of the closed loop and, with the help of the **dlqr** function, calculate the feedback gain vector $\boldsymbol{R}$.

$$\boldsymbol{R} = [0.8954 \quad 1.7499 \quad 8.2528 \quad 0.8161],$$

Vector $\boldsymbol{R}$ is used as the optimal linear state space controller.

The block scheme of the close loop balance system is represented in the $\boldsymbol{S}$ space in Figure 5. The control of the system was executed from the relaxed state with the position $x_1 = 1$ into the reference state $x_1 = 0$. We kept in mind that the states x_2 and x_3 must be estimated because they are not directly measurable.

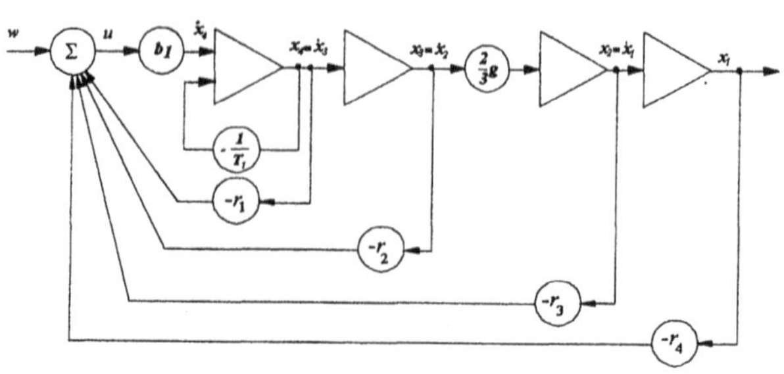

Fig. 5. The Control Scheme of the Balance System

From the control scheme in Figure **5**, it is obvious that we can calculate the nonmeasurable states of the system from the measured values x_1 and x_4. The speed of the reel x_2 equals the differentiation of the measured position x_1, which we get with the first difference.

$$x_2\left((k+1)T_S\right)=\left(x_1\left((k+1)T_S\right)-x_1\left(kT_S\right)\right)/T_S \quad (11)$$

The angle of the bar x_3 equals the integral of the measured angle speed x_4 and, according to the rectangle method of integration, equals: a)

$$x_3\left((k+1)T_S\right)=x_3\left(kT_S\right)+\left(x_4\left(kT_S\right)+\left(x_4\left((k+1)T_S\right)-x_4\left(kT_S\right)\right)/2\right)\cdot T_S \quad (12)$$

The time responses of the balanced reel from position 1 to position 0 are shown in Figure **6**.

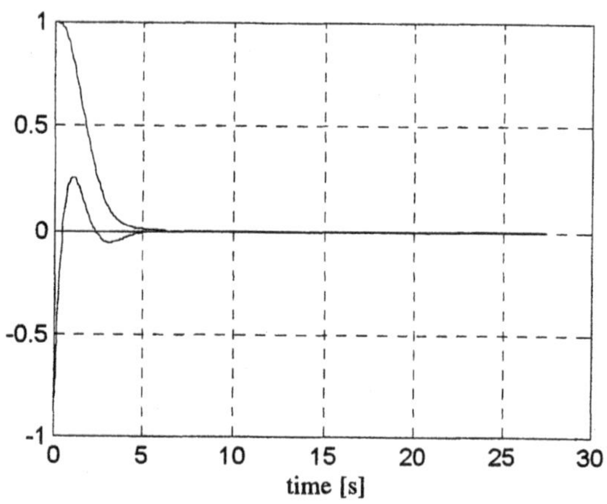

Fig. 6. Time response of the reel position and reference signal

3.2 Control in Real Time

Control in real time was exectued with the ADDA card PCL711, personal computer PCAT 486, and the programming tool Turbo C.

It is recognizable, from Figure 7, that the linear state controller used on the real balance system cannot stabilize the reel. The biggest problem presented is the correct estimation of the angle of the bar, because in real time we never have a totally relaxed system. Thus, the integration method for estimating the angle of the balanced bar does not produce the desired results.

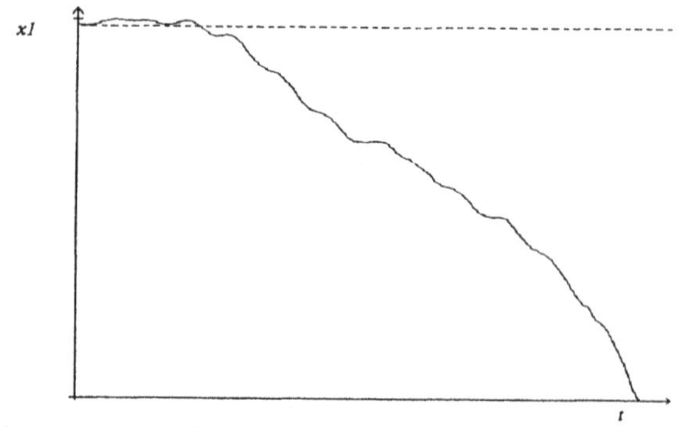

Fig. 7. Control of the Reel with the Linear State Controller

We can get the optimal solution by using the observer of the system's states. In Figure **8**, this observer is shown with the control of the reel on the reference curve.

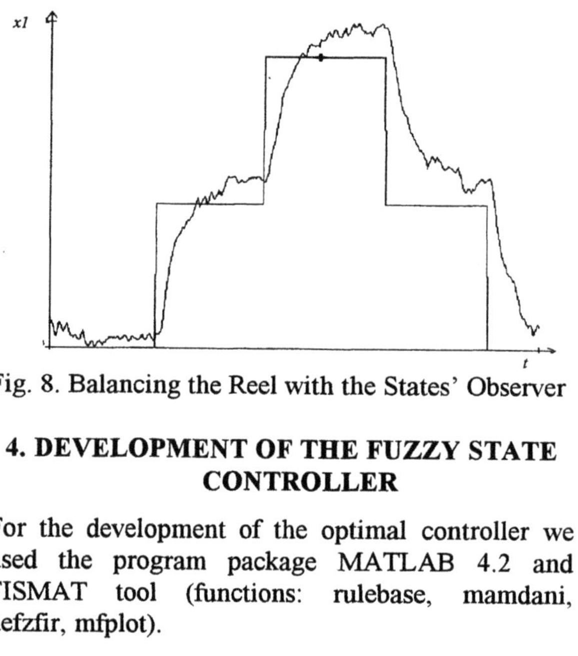

Fig. 8. Balancing the Reel with the States' Observer

4. DEVELOPMENT OF THE FUZZY STATE CONTROLLER

For the development of the optimal controller we used the program package MATLAB 4.2 and FISMAT tool (functions: rulebase, mamdani, defzfir, mfplot).

4.1 Control Strategy

The control strategy is based on a good knowledge of the system's behavior. We included three input linguistic variables (the reel position on the bar, the speed of the reel and the angle of the bar) and one output linguistic variable (the angle speed of the motor), which was used as the control signal for the system.

The form and the number of the membership functions are defined with the function **rulebase()**, which is also used for the creation of the control rules. We have three input linguistic variables: x_i, $i=1\ldots3$ and one output linguistic variable u. Their membership functions have the form of a triangle and are placed evenly throughout the whole defined space U_i and V, as shown in Figure **9**.

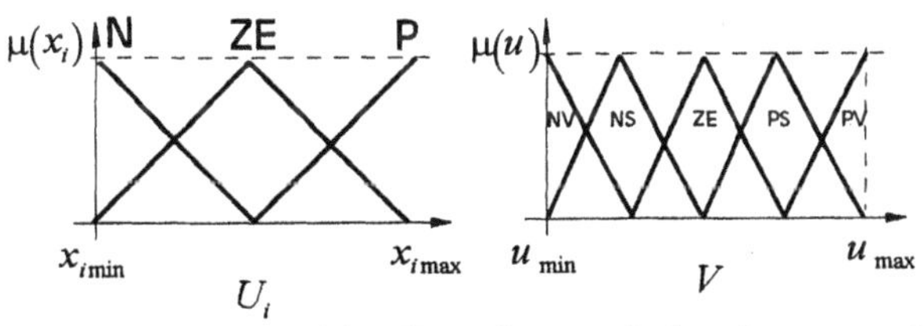

Fig. 9. Membership Functions of the Input and Output Linguistic Variables

We calculate the output fuzzy set B in each iteration with the **mamdani()** function:

We then get the control singal $u(kT_S)$ with the defuzzification of the output fuzzy set. For that, we use the **defzfir()** function with the weight method (COG):

$$u(kT_S) = \text{defzfir}(B,V,1)$$

where B equals the output fuzzy set, V equals the definition space, and 1=COG is the defuzzification method.

Figure **10** shows the creation of the rules with the **rulebase()** and the printing of the membership functions, using the **mfplot()** function.

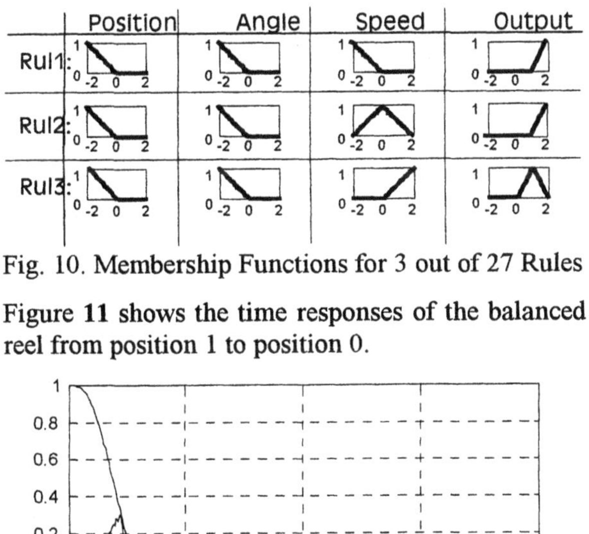

Fig. 10. Membership Functions for 3 out of 27 Rules

Figure **11** shows the time responses of the balanced reel from position 1 to position 0.

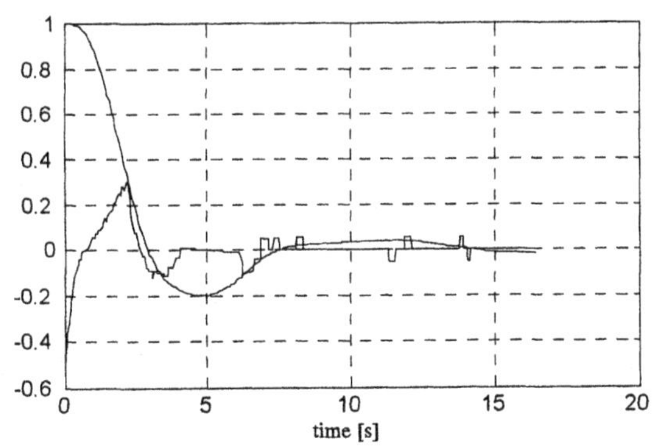

Fig. 11. Time response of the reel position and reference signal

4.2 Control in Real Time

Execution in the real time was performed with tan ADDA card PCL 711, fuzzy module OMRON FS-30AT, personal computer PC AT 486 and the programming tool Turbo C.

In section 3, we discovered that the integration method for the estimation of the angle of the bar is unreliable. A more accurate estimation of the angle of the bar is supplied by differentiating the reel position twice:

$$x_3((k+1)T_S) = \left(x_1(kT_S) - 2x_1((k-1)T_S) + x_1((k-2)T_S)\right)/T_S^2 \quad (13)$$

The integration method for the estimation of the angle of the bar is less reliable than the differentiation method. This can be proven by caluclating the measurement's margin of error. We know that when we are integrating, the absolute errors of each measurement are adding on compoundly; thus, while the time increases, the measurement's margin of error of indirect measurement of the angle of the bar increases simultaneously. The absolute error is also added on compoundly when we are differentiating. However, the measurement's margin of error for the indirect measurement of the angle of the bar is not increasing with time, but is dependent on the level of differentiation (number of subtractions and additions).

We used the same matrix of the rules (Table 1) as in the simulations. Figure **12** shows the results of the control of the reel following the reference curve. It turns out that the fuzzy state controller performs the same way as in the simulations. The result of our work was an optimal state controller with a very high level of robustness. The reel followed the reference curve quickly and the static error was minimal.

Table 1 Matrix of the Rules for the Fuzzy State Controller

		POSITION N			POSITION ZE			POSITION P		
		SPEED N	SPEED ZE	SPEED P	SPEED N	SPEED ZE	SPEED P	SPEED N	SPEED ZE	SPEED P
ANGLE	N	PV	PV	PS	PS	PS	ZE	ZE	NS	NS
ANGLE	ZE	PV	PS	PS	PS	ZE	NS	NS	NS	NV
ANGLE	P	PS	PS	ZE	ZE	NS	NS	NS	NV	NV

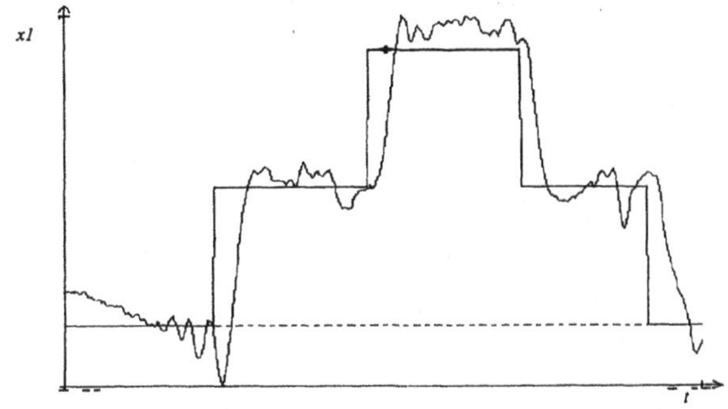

Fig. 12. Balancing of the Reel with the Fuzzy State Controller

The fuzzy state space controller for the balance control was based on three input linguistic variables which were the state variables of the system at the same time. Each of those input linguistic variables has three membership functions which are triangular in shape. The angle and the distribution of the memberhsip functions are symetrical according to the definition space of each linguistic variable. The execution of the fuzzy controller can be optimized by correctly forming the rules and the output nonlinear function of the fuzzy state regulator. Increasing the number of the membership functions does not drastically improve the quality of the control; however, it increases the number of rules, which then causes obscurity in the development of the controller.

5. CONCLUSION

In this work, we showed the basic ways of how to develop the fuzzy state space controller. We concluded that the controlling of the system's states with the fuzzy state controller is a good alternative for controlling the demanding nonlinear processes. The assumption that the fuzzy state controller is a nonlinear controller, being described by a nonlinear function of the state variables, makes possible the analysis of the closed loop system on the system of nonlinear differential equations.

The reel balance control on the balanced bar requires the use of the modern control techniques, due to the fact that the system is constantly on the stability boundary and not with a minimal phase. Realization of the optimal controller in state space requires knowledge of the process' mathematical model and the measurable (estimated) state variables. Execution of the fuzzy state controller is also limited to the measurable (estimates) state variables in the system, but does not require the mathematical model of the process. The simulations show that both of the controllers supply good results when the reference value of the position is changed; however, the fuzzy controller is much more robust when controlled in real time because no additional identification of the linear model is required.

The aim of the work was to show the usefulness of the fuzzy controllers in processes which are very complicated to control. However, we are still limited to the experiments while seeking the optimal solution for the fuzzy controller. The biggest advance of the fuzzy state controller is that we are controlling just those states of the stystem which have the biggest influence on the system's execution.

REFERENCES

CAMO A/S: *Balance Control Work Bench Version 2, User's Guide.*, Trondheim, Norway, 1991.

Đonlagic D.: "*Osnove snovanja mehkih (fuzzy) regulacij,*" Faculty of Electrical Engineering and Computer Science, Maribor, 1994.

Fölinger O.: "*Nichtlineare Regelungen I, Grundlagen und Harmoniche Balance,*" Oldenburg-Verlag, München, Wien, 1969.

Isserman, R.: "*Digital Control Systems,*" Springer-Verlag, Berlin, Heidelberg, New York, 1981.

Kendl H. and Rüger J.: "*Verfahren zum Entwurf und Stabilitätsnachweis von Regelungssystemen mit Fuzzy-Reglern,*" Oldenburg-Verlag, Automatisierungstechik vol. 41, pp. 138-144, 1993.

FUZZY LOGIC CONTROLLERS AND THEIR REPLACEMENT BY A NON-FUZZY APPROXIMATION ALGORITHM

Ivan Sekaj

Department of Automatic Control Systems, Faculty of Electrical Engineering and Information Technology, Slovak University of Technology, Ilkovicova 3, 812 19 Bratislava, Slovak republic

Abstract : As an alternative to the defuzzification a new computation method of fuzzy logic controller output is presented. In connection with this method an other procedure is proposed, which allowes to substitute also for fuzzification. It uses the same knowledge base in the form of a rule-table like the fuzzy logic controller, but without using the fuzzy logic. The proposed algorithm is simple and the results which where verified in simulation and in the real-time control as well are the same like with fuzzy logic controllers.

Key words : Fuzzy logic controller, PID controller, knowledge base, control surface, defuzzification, control surface approximation, fuzzy logic replacement

1. INTRODUCTION

Fuzzy logic and the theory of fuzzy sets is an important tool in the process control. One of their most important application are fuzzy logic controllers (FLC). FLC's are knowledge-based controllers which extend the conventional control with knowledge or experiences of an expert/operator. The fuzzy logic is the means for interpreting this existing process and control knowledge. FLC's are able to extend the basic process variable set (control error and its derivatives or integral, e.t.c.) of other measurable or computed variables, thus extending the conventional control strategies. Generaly the FLC is a nonlinear controller which can imitate the strong nonlinear behaviour of a human-operator, for example in the case of nonlinear systems control in several working points, in which the control using conventional controllers is problematic or unfeasible.

2. FUZZY LOGIC CONTROLLERS AND THEIR ANALOGY WITH PID CONTROLLERS

The principle of a conventional FLC is evident from Fig.1 (where IP is input processing, F is fuzzification, KB is the knowledge base, D is defuzzification and OP is output processing). After the measurement, all variables are calculated and normalized for the use in the fuzzy part.

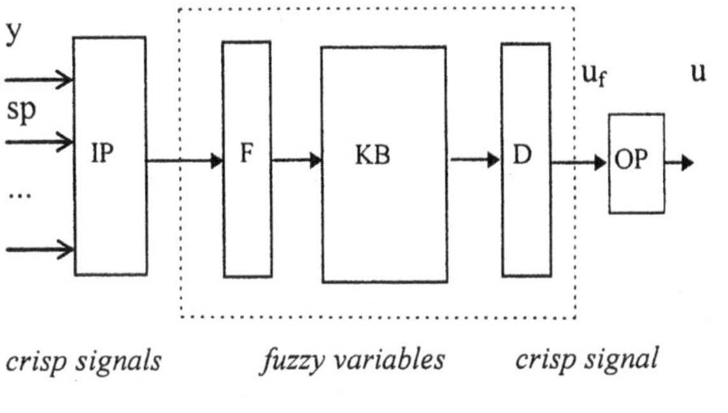

Fig. 1. Block scheme of the FLC

Usually following variables are used: e - control error, de - 1st derivative of e, d^2e - 2nd derivative of e, ie - integral of e. In special cases we can use also other variables or signals. For the fuzzification different types of membership functions are used. Without loos of generality let us consider symmetric- and triangle-type membership functions like in Fig.2 (where NB-negative big, NM-negative medium, NS-negative small, ZE-zero, PS-positive small, PM-positive medium, PB-positive big).

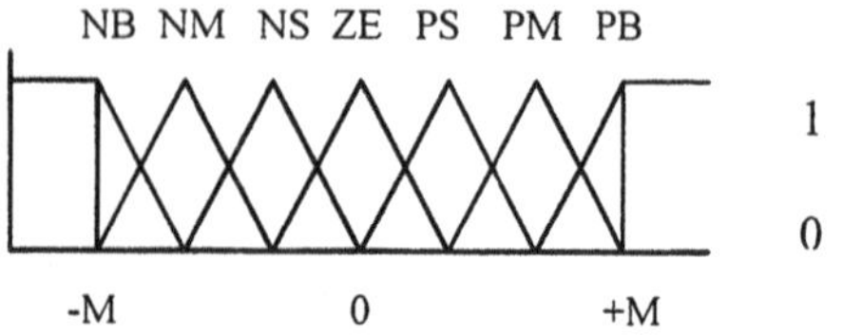

Fig. 2 . Symmetric- and triangle-type membership functions

The controller characteristics can be changed by dislocating and deforming of the membership functions (Driankov, et all., 1993; Steinbauer 1996) and also by changing of the scaling factor M, what will be used later. Basic influence on the performance has the knowledge base. In FLC's the knowledge representation of the process dynamics and of the control strategy , quantified in a limited number of points in the form of a n-dimensional matrix is a table. The interpolation mechanism between these points represents here the fuzzy logic. The dimension of the rule-table is determined by the number of input variables. The number of membership functions, which determines the size of the rule-table increases with the increase of the control strategy complexity and/or with the process nonlinearity degree. In the last procedure of the fuzzy part after the inference - in the defuzzification the most frequently used method is the Center of Gravity method or the simplier Singleton method. More details concerning the defuzzification methods can be found in (Driankov, et all., 1993). The control with FLC's is in fact the assignment of a crisp defuzzificated output value to the input vector. In other words, for the input vector an corresponding approximate output value is computed, which is represented by a point on the control surface (characteristic surface). This value represents directly or after integration (see Table 2) the controller output. The surface is a form of a knowledge base representation (for example see Fig.5 and Fig.6). Depending on which input signals have been used, and how the defuzzificated value u_f, has been processed the FLC presents an appropriate behaviour. Under some conditions , the behaviour of a FLC could be analogous or even equivalent to a PID controller. Now let us consider, that the input membership functions are in the form given in Fig.2, the knowledge base is a „quasilinear" table like Tab.1, and the Mamdani's inference and the Center of Gravity defuzzification method are used. Basic types of Fuzzy-PID controllers (FPID) are in Tab. 2, where $u(k)$ - is the FLC output in the step k, $u_f(k)$ - is the output from the fuzzy part after defuzzification. The main advantage of a FLC consists in that it makes possible to change the control strategy to achieve a specific closed loop behaviour. However, tuning of such a controller is a more complex problem, because the FLC has much more degrees of freedom than the PID controller. To obtain analogous characteristics with a PID controller consider „standard" membership functions (Fig.2) and a linear rule-base (Tab.1). The analogy with the PID parameters tuning (proportional, integral and derivative) for a FLC is the change of the input and output membership functions scaling factors M. The exact conditions for the equivalence of Fuzzy-PID and PID, to reach the equivalent time responses of the closed loop are given in (Lichtenberg, 1994).

de / e	**NB**	**NM**	**NS**	**ZE**	**PS**	**P**	**PB**
NB	*NB*	*NB*	*NB*	*NB*	*NM*	*NS*	*ZE*
NM	*NB*	*NB*	*NB*	*NM*	*NS*	*ZE*	*PS*
NS	*NB*	*NB*	*NM*	*NS*	*ZE*	*PS*	*PM*
ZE	*NB*	*NM*	*NS*	*ZE*	*PS*	*PM*	*PB*
PS	*NM*	*NS*	*ZE*	*PS*	*PM*	*PB*	*PB*
PM	*NS*	*ZE*	*PS*	*PM*	*PB*	*PB*	*PB*
PB	*ZE*	*PS*	*PM*	*PB*	*PB*	*PB*	*PB*

Tab.1 . Linear rule-table with bounded control output

Type	**Inputs**	**Tab. dim.**	**Output**
P	*e*	*1*	$u(k) = u_f(k)$
I	*e*	*1*	$u(k) = u(k-1) + u_f(k)$
PD-pos.	*e, de*	*2*	$u(k) = u_f(k)$
PI-incr.	*e, de*	*2*	$u(k) = u(k-1) + u_f(k)$
PI-pos.	*e, ie*	*2*	$u(k) = u_f(k)$
PID-pos.	*e, ie, de*	*3*	$u(k) = u_f(k)$
PID-incr.	e, de, d^2e	*3*	$u(k) = u(k-1) + u_f(k)$

pos. - positional algorithm
incr. - incremental algorithm

Tab.2. Basic types of Fuzzy-PID controllers

The aim of using fuzzy controllers is not to imitate PID's , but to provide new posibilities. With the application of available knowledge the linear rule-table changes to generally nonlinear one and the equivalent PID controller changes to a special controller which is optimised with connection to the process dynamics and the control aim. The greatest disadvantage consists perhaps in that there is still not avaiable a compact and general theory for the design and tuning of such systems . Some methods are described in (Zheng,1992; Koch, et all., 1994; Boverie, et all., 1991). An other disadvantage is the need for some apriori knowledge about the controlled process. These problems could be mostly overcome using adaptive systems (Wang, 1994; Driankov, 1993), Neuro-Fuzzy systems (Preuss, 1994; Hensel, et all., 1995) or other principles from the field of artifical intelligence.

3. ALTERNATIVE TO THE DEFUZZIFICATION

The most frequently used defuzzification method is the Centre of Gravity method or the Singleton method. Let us describe an other way of approximating the fuzzy output which is similar to the Sugeno-type fuzzy system. Consider a rule-table, which does not contain lingvistic variables values, which are defuzzificated to a crisp output value, but it contains directly a matrix of output values (numbers) $b_{j,i}$, $j=1,2,...,m,$ $i=1,2,...,n$ (see Tab.3). Following approximation of the output value is proposed. Without looss of generality, consider two input variables x and y with membership functions given in Fig. 2 and one output variable u_f.

	x						
	-M						+M
-M	-3	-3	-2	-2	-1	-1	0
	-3	-2	-2	-1	-1	0	1
	-2	-2	-1	-1	0	1	1
y	-2	-1	-1	0	1	1	2
	-1	0	1	1	1	1	2
	0	1	1	1	1	2	2
+M	0	1	1	1	1	2	3

Tab.3. Example of a numerical rule-base

Suppose, that the measured input vector $[x_m, y_m]$ activates the membership functions F_i, F_{i+1} and F_j, F_{j+1} with values $\mu_{x_i}, \mu_{x_{i+1}}$ and $\mu_{y_j}, \mu_{y_{j+1}}$ (Fig.3). The corresponding table-values are $b_{j,i}$, $b_{j,i+1}$, $b_{j+1,i+1}$, $b_{j+1,i}$. Let the membership functions be in the normalized form

$$\mu_{x_i} + \mu_{x_{i+1}} = 1$$
$$\mu_{y_j} + \mu_{y_{j+1}} = 1$$

Then let the output be

$$u_f = \mu_{x_i}\mu_{y_j}b_{j,i} + \mu_{x_{i+1}}\mu_{y_j}b_{j,i+1} + \\ +\mu_{x_{i+1}}\mu_{y_{j+1}}b_{j+1,i+1} + \mu_{x_i}\mu_{y_{j+1}}b_{j+1,i} \tag{1}$$

That means, that the output value is equal to the sum of partial products of activated memberhip function values with the corresponding table values.

An example of a projection of a control surface, defined by Tab.3, and computed by the singleton method is in Fig.5. Fig.6 depicts the surface computed using (1). In this case the surface is smoother. The solution of a 3-dimensional (n-dimensional) table is analogous.

4. DESIGN OF A CONTROL SURFACE APPROXIMATION WITHOUT USING FUZZY LOGIC

As it has been already mentioned a FLC is a system, generating in every sampling period an output value (or increment) which is equal to the value of the corresponding control surface point. This control surface is determined by fix points - values of the rule-table. Fuzzy logic is the approximation (interpolation) mechanism between these points (Wang, 1994; Preuss, et all., 1994). Consider, like in section 3, that the membership functions are of symmetrical and triangle type (but it is not a condition) and the rule-base is defined by a numerical table. The following method is an analogy to the one in the section 3. For illustration let us onesmore consider 2 inputs x and y (see Fig.3). Let the measured input vector $[x_m, y_m]$ determines the position of the point M and its corresponding value u_a on the control surface that we want to approximate. The absolute coordinates of the points B_1, B_2, B_3, B_4 and their table values $b_{j,i}$, $b_{j,i+1}$, $b_{j+1,i+1}$, $b_{j+1,i}$ are known. Let the relative coordinates (in relation to the relative origin B_1) are $B_1=[0,0]$, $B_2=[h,0]$, $B_3=[h,h]$, $B_4=[0,h]$, where h is the step of the table. The relative coordinates of the point M are $[x_r, y_r]$. According to section 3, the approximation of the u_a is (1). Without use of membership functions we can obtain an analoguous approximation

$$u_a = \{(h-x_r)(h-y_r)b_{j,i} + \\ +[h-(h-x_r)](h-y_r)b_{j,i+1} + \\ +[h-(h-x_r)][h-(h-y_r)]b_{j+1,i+1} + \\ +(h-x_r)[h-(h-y_r)]b_{j+1,i}\} / h$$

or

$$u_a = [(h-x_r)(h-y_r)b_{j,i} + x_r(h-y_r)b_{j,i+1} + \\ +x_r y_r b_{j+1,i+1} + (h-x_r)y_r b_{j+1,i}] / h \tag{2}$$

The approximation of the control surface defined by Tab.3 using this method gives the same results like (1) (Fig.6) and also the same results like the more computationally pretentious bilinear interpolation method (Matlab[TM]). The solution of the 3-dimensional (n-dimensional) table where we have

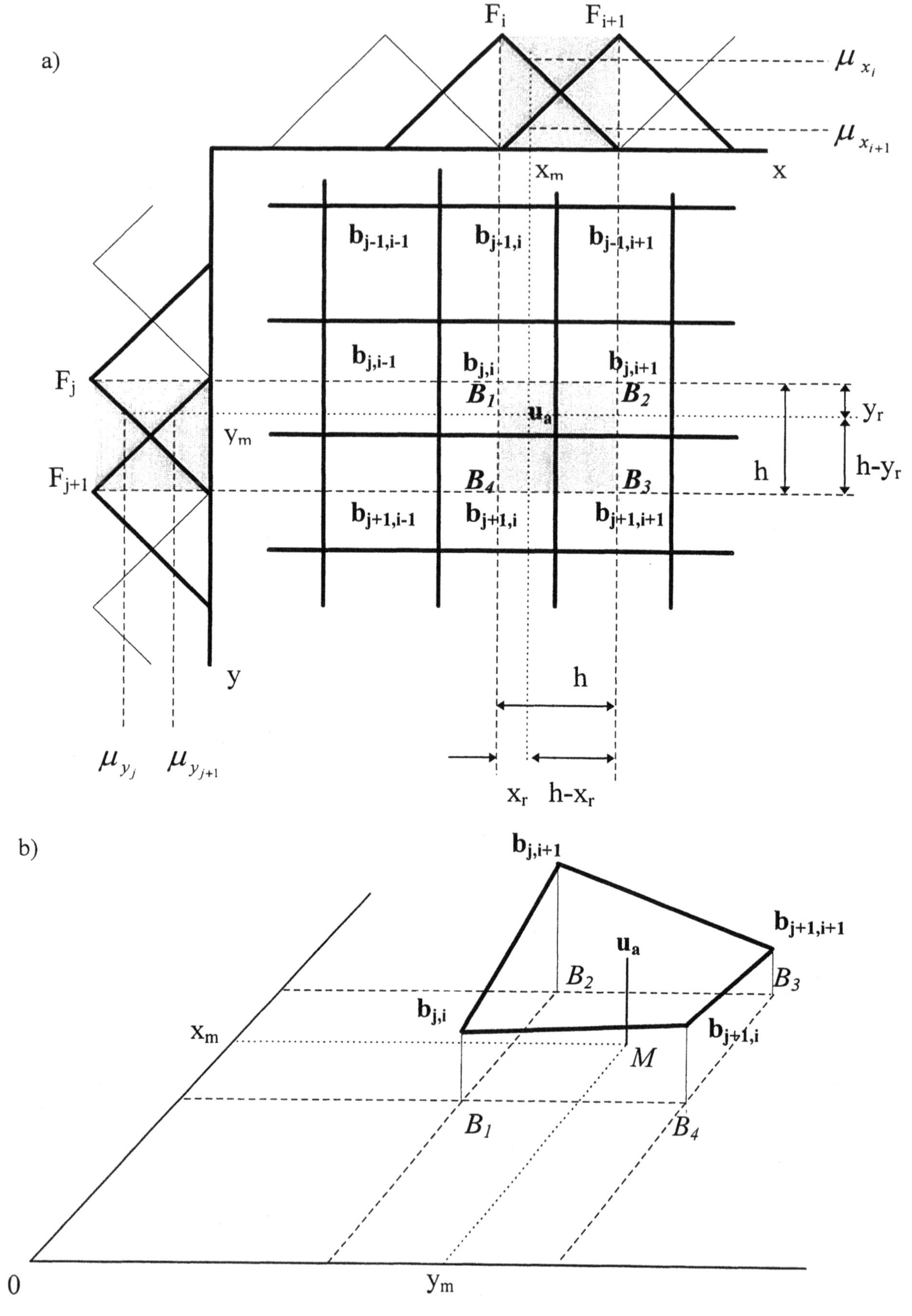

Fig.3. Control surface approximation

to approximate among 8 points (2^n-points) in the space is analogous. In the case of FLC's with nonsymmetrical or non triangular membership functions or in the case of time-variant (adaptive) membership functions we can modify (2) using appropriate parameters or functions. The control algorithm to be carried out in one controller sampling period is as follows:

1. measuring, normalisation and computation of input variables
2. allocation in the table
3. approximation of the control surface point u_a using (2)
4. controller output u processing (see Tab.2)

5. EXPERIMENTAL RESULTS

The proposed approaches has been verified in real-time experiments on a DC drive speed control. Time response to setpoint jumps with the FLC based on the algorithm (1) is shown in the Fig.7. The control surface has been designed with the aim to achieve a slow increase and a fast decrease of the speed without ovrshoot. Time response to the setpoint jump using the approximation algorithm based on the expression (2) is in Fig.8. This method can be used with advantage also for other types of applications. It was used for example in the adaptation mechanism of knowledge-based self-tuning PID controllers. Different types of this knowledge-based adaptation mechanisms could use different types of input signals. The e, de, d^2e can be employed to tune the controller prameters continuously during the transient mode. Other kind of information like overshoot, settling time, rise time, setting ratio (Pfeiffer, et all., 1994), damping e.t.c. can be used to tune the controller parameters after the end of every transient process. Fig.9 shows simulation results of the second mentioned principle in the case of a knowledge-based selftuning PI controller with a 2nd order system.

6. CONCLUSION

The method proposed in the section 3, which substitutes defuzzification, is simple and in FLC's practicaly applicable. The control surface approximation method, described in the section 4, allows without fuzzy logic to obtain the same results like with a FLC. Such a controller uses the same knowledge base and the corresponding interpretation algorithm is fast and simple. This method can be used whereever we want to use a knowledge base of this type.

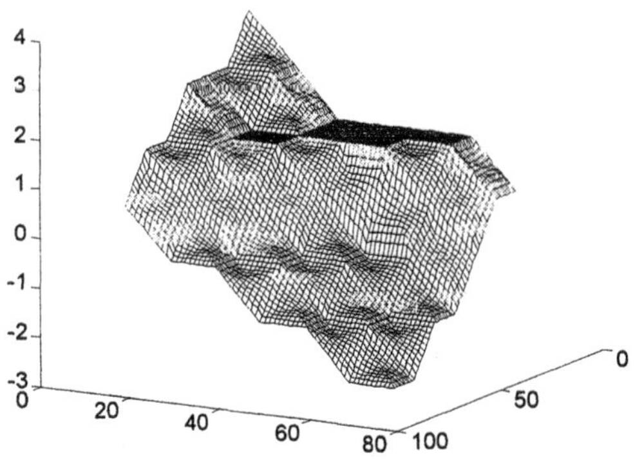

Fig.5. Control surface of the FLC computated by the singleton defuzzification method

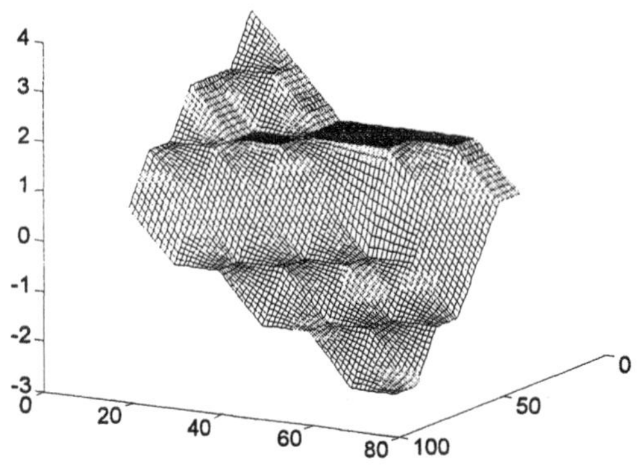

Fig. 6 . Control surface computated by the proposed methods

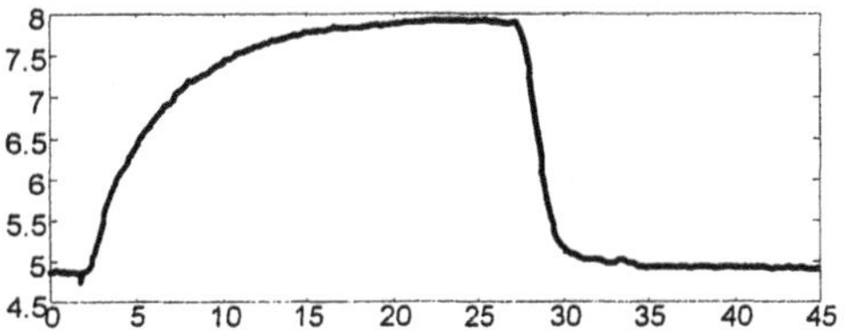

Fig.7. Time response of the DC drive under the proposed FLC

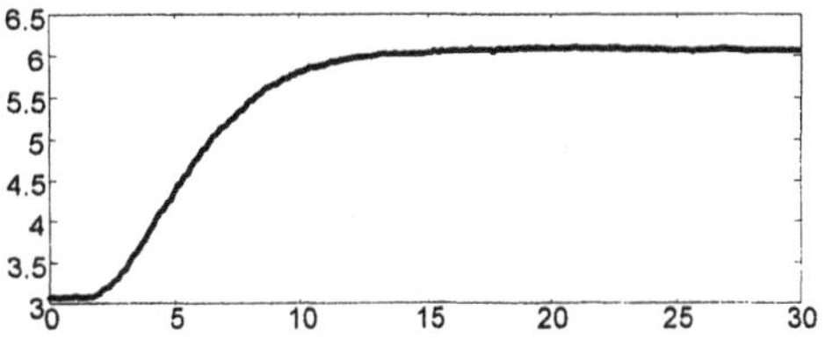

Fig.8 . Time response of a DC drive under the approximation algorithm controller

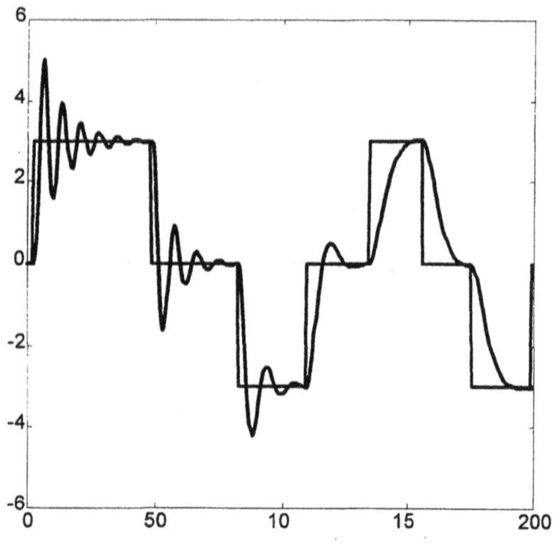

Fig. 9. Simulated set-point responcses of the knowledge-based selftuning PI controller

REFERENCES

Driankov, D., Hellendoorn H., Reinfrank M (1993). An Introduction to Fuzzy Control. Spinger-Verlag Berlin Heidelberg,

Steinbauer, P. (1996). Experimentování s fuzzy PI regulátory. *Automatizace 11/1996 (in czech),*

Lichtenberg, M. (1994). Generierung kennlinien-gleicher Fuzzy-Regler aus den Parametern konventionellen (PID-) Regler. *Automatisierungstechnik, 12/94,* Oldenburg Verlag.

Wang, Li-Xin (1994). Adaptive Fuzzy Systems and Control. PTR Prentice Hall.

Preuss, H.P., Tresp V. (1994). Neuro-Fuzzy. *Automatisierungstechnische Praxis 5/94,* Oldenburg Verlag.

Zheng, Li (1992). A Practical Guide to Tune of P and PI Like Controllers. IEEE .

Koch, M., Kuhn Th., Wernstedt J. (1994). Methods for Optimal Design of Fuzzy Controllers. TU Ilmenau, Germany.

Boverie, S., Demaya B., Titli A. (1991). Fuzzy Logic Control Compared with other Automatic Control Aproaches. *Proc. of the Conf. on Decision and Control,* Brighton, England.

Hensel, H., Holzmznn H., Pfeiffer B.M. (1995). Optimierung von Fuzzy-Control mit Hilfe Neuronaler Netze. *Automatisierungstechnische Praxis 11/95,* Oldenburg Verlag.

Galichet, S., Foulloy L. (1995). Fuzzy Controllers: Synthesis and Equivalences. *IEEE on Fuzzy Sys.,* **Vol.3, No.2.**

Pfeiffer, B.M., Isermann R. (1994). Selftuning of Classical Controllers with Fuzzy-logic. *Mathematics and Computers in Simulation 37,* Conference, Brussels.

Ng, K.Ch., Li Y. (1996). Reduced Rule-base and Direct Implementation of FLC systems. *13th Triennial World Congres,* San Francisco, USA.

DSP IMPLEMENTATION OF SPEED QUANTITY NEURAL NETWORK ESTIMATION OF CONTROLLED SOLID STATE DRIVE

Dobrucký B., Chyla P., Vanek R.*, Balažovič P.

Faculty of Electrical Engineering,
**Faculty of Management Science and Informatics,*
University of Žilina, Slovak Republic

Abstract: The paper deals with a identification of solid state drive using three-layer artificial neural network realized by fixed point digital signal processor (DSP) of the type of TMS320C25 - 100 ns. Back propagation learning process has been done for speed quantity of slid state drive, consisting of separately excited DC motor fed by four quadrant converter (chopper) unit. The input and output signals were obtained by sigma-delta A/D and D/A converters including both anti-aliasing and digital filter blocks. Comparisons of carried-out real estimated- (by DSP), simulated- (by PC) and measured samples of output signal are given in the paper.

Keywords: digital signal processor, neural network, control algorithm, estimation, solid state drive, A/D conversion, four quadrant converter

1. TASK FORMULATION

A classical solution of automatic control tasks in electric drives is determined by knowledge of mathematical model of regulated system either in transfer function form or in state-space equations (mainly at state-space control using observer techniques). Unlike this, the neural- and fuzzy logic control need not above mentioned models. In certain sense, the artificial neural network (ANN) can imitate human behavior, whereby the whole system is controlled by a man of such a way, that he determines a manipulated variable by watching the system response using experiences learned before. ANN represents artificial connections of numerous non-linear neurons whereby the processing of information is parallel one. Besides, the other features of ANN are the non-linear mapping, self-learning and self-organizing. By means of supervised training is able to learn a relationship between input and output patterns of signals in off-line process. The goal of this learning process is to minimize output error by its back propagation to the input layer of neurons. It's known, that realization of any transfer function need at least one hidden layer of the neurons, i.e. the whole ANN can be realized by three-layer network of the neurons. Self-learning properties of neural network make it possible to use the ANN for identification and control of various dynamic system, e.g. as solid state drive.

In this paper, the neural network was realized by digital signal processor (DSP) type of TMS320 C25/100 ns, in opposite of the papers of Bertoluzzo et al. (1994) and Veerasooriya et al. (1993). **To obtain output sample sets of the signal using experimental neural identification (by DSP) and computer simulation has been a main task.** The three-layer neural network has been implemented by means of above mentioned DSP using control program written in assembly language. The software implementation and verification has been done using a LSI prototype board from Loughborough Sound Imagine Ltd., U.K.

2. DESCRIPTION OF USED NEURAL NETWORK

Artificial neural networks are computing systems which are inspired by human brain behavior and its architecture. There exists many types of artificial neural networks but this papers deals with using multi-layer feedforward neural networks.

2.1 Feedforward neural network

Feedforward neural network is a specific neural network architecture where sets of neurons are arranged in layers. Activation signals of neurons in one layer are transmitted to the next layer through a set of links which either attenuate or amplify the signals based on respective weights. The single layer symbolic representation of the feedforward neural network is depicted at the following figure:

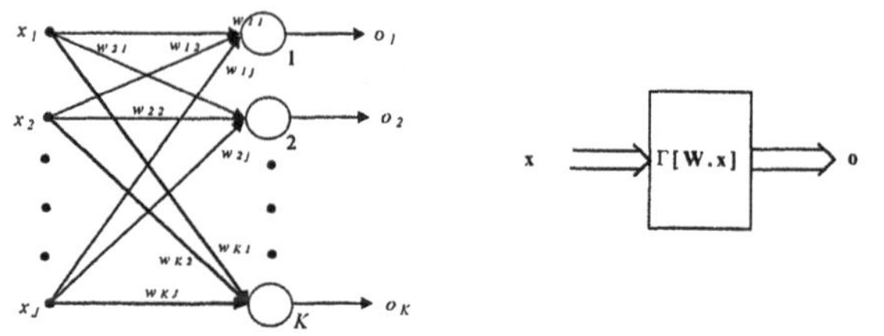

Fig. 1. A single layer of neural network and its block diagram.

Each neural network layer consists of numerous simple processing units termed as *neurons* or *nodes*. Every neuron model consists of a processing element with synaptic input connections and a single output. The neuron output signal is given by the following relationship :

$$o_k = f(\mathbf{w}'_k.\mathbf{x}) \quad \text{and} \quad net_k = (\mathbf{w}'_k.\mathbf{x}) \qquad (1)$$

where $\mathbf{w}'_k$ is weight vector and $\mathbf{x}$ is input vector of the neural network. The function $f(\mathbf{w}'_k.\mathbf{x})$ eq. (1) is often referred to as an activation function. Its domain is the set of activation values, *net*, of the neuron model. The variable *net* is defined as a scalar product of the weight and input vector eq. (1).

The neuron as a processing node performs the operation of summation of its weighted inputs and subsequently it performs the nonlinear operation *f(net)* through its activation function.

The proposed transfer function eq. (2) used in this papers is bipolar continuous :

$$f(net) = \frac{2}{1+\exp(-\lambda.net)} - 1 \qquad (2)$$

An elementary feedforward neural network architecture of K neurons receiving J inputs, as shown in Fig. 1, has its input and output vectors respectively :

$$\mathbf{x} = [x_1, x_2, .., x_n]'$$
$$\mathbf{o} = [o_1, o_2, .., o_m]'$$

Weight w_{kj} connects the *k'th* neuron with the *j'th* input. The transformation, performed by each of the k neurons in the network, is strongly nonlinear mapping expressed by the following matrix equation as :

$$\mathbf{o} = \Gamma[\mathbf{W}.\mathbf{x}] \qquad (3)$$

where $\mathbf{W}$ is weight matrix and Γ is nonlinear matrix operator.

The mapping of an input pattern into an output pattern as shown in eq. (3) is of the feedforward instantaneous type. This type of network, as shown in Fig. 1, can be connected in cascade to create a multi-layer network. In such network, the output of a layer is the input to the following layer.

2.2 Training of neural network

Neural networks have been found to be effective systems for learning functional mappings from examples. This is done by adjusting the weights and node thresholds of set of interconnected neurons according to a specific learning rule, especially steepest descent algorithm version.

In its fundamental form, error backpropagation is a specific implementation of the steepest descent algorithm given in (Zaruda,1992).

The objective is to find a set of weights so as to optimize predefined error function E based on a set of P training patterns.

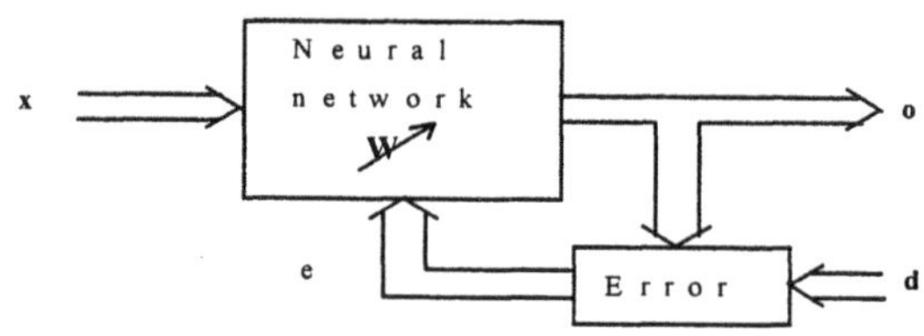

Fig. 2 Block diagram of basic training mode of supervised learning.

The mean square error E eq. (4) is defined as :

$$E_p = \frac{1}{2}\sum_{k=1}^{J}\left(d_{pk} - o_{pk}\right)^2 = \frac{1}{2}\left\|\mathbf{d}_p - \mathbf{o}_p\right\|^2 \qquad (4)$$

The updating formula eq. (5) for the weights is given by :

$$w'_{kj} = w_{kj} + \Delta w_{kj} \qquad (5)$$

It requires the weight adjustment by computing individual weight updating as follows :

$$\Delta w_{kj} = -\eta.\frac{\partial E}{\partial w_{kj}} \qquad (6)$$

It is obvious that the gradient component $\partial E/\partial w_{kj}$ depends only on the net_k of a single neuron therefore using the chain rule it is valid the following expression in (Zaruda, 1992) :

$$\frac{\partial E}{\partial w_{kj}} = \frac{\partial E}{\partial(net_k)} \cdot \frac{\partial(net_k)}{w_{kj}} \tag{7}$$

The second term of the product in eq. (7) can be written as :

$$\frac{\partial(net_k)}{\partial w_{kj}} = x_j \tag{8}$$

The weight adjustment formula eq. (6) can now be rewritten using the error signal δ_{ok}

$$\Delta w_{kj} = \eta . \delta_{ok} . x_j \tag{9}$$

To adapt the weights, the error signal term delta, introduced in (Zaruda , 1992), defined as :

$$\delta_{ok} = -\frac{\partial E}{\partial(net_k)} \tag{10}$$

needs to be computed for *k'th* neural net layer. The final formula for the weight adjustment can now be obtain from eq. (9) as :

$$\Delta w_{kj} = \eta . (d_k - o_k) . f'(net_k) . x_j \tag{11}$$

The constant η is known as the learning step. This value is an application specific and is selected arbitrarily based on experience. In this paper, the update formula eq. (11) is used for neural network training.

3. SOLID STATE DRIVE EXPERIMENTS

DC motors are always a good proving base for advanced control algorithms because the theory is extendable to other types of motors.

3.1 Estimated system theory description

Non-linear system will be separately-excited DC motor supplied by the PWM converter unit so that input variables are armature current and its PWM controlled voltage and considered system output is motor revolution. Produced non-linear manner is due to partly non-linear converter output voltage (u_a), partly by the multiplication two state variables, and by considering non-linear magnetizing curve ($\phi(i_b)$) into system model. Differential equations are given by the following expressions :

$$\begin{aligned} \frac{di_a}{dt} &= \frac{1}{L_a}\left(u_a - R_a i_a - c\phi(i_b)\omega\right) \\ \frac{d\omega}{dt} &= \frac{1}{J}\left(c\phi(i_b)i_a - m_p\right) \\ \frac{di_b}{dt} &= \frac{1}{L_b}\left(u_b - R_b i_b\right) \end{aligned} \tag{12}$$

This state space model is simulated on PC machine for obtaining required input/output patterns for the first time weight coefficients tuning but then used data patterns for off-line learning process is obtained from experimental laboratory measurement. This required training data is applied at back-propagation learning process, as shown in Fig. 3, in three layer network because it has been proved that only a one hidden layer is sufficient to approximate any continuos function.

The conventional wisdom is to start from a small number of hidden neurons and grow a network of appropriate size until a sufficiently small value is obtained for the error function *E*. A good starting point is to set the number of hidden neurons equals to the input dimension. Large number of hidden neurons is unnecessary burden on both on-line and off-line training and could cause overdetermination of the mapping.

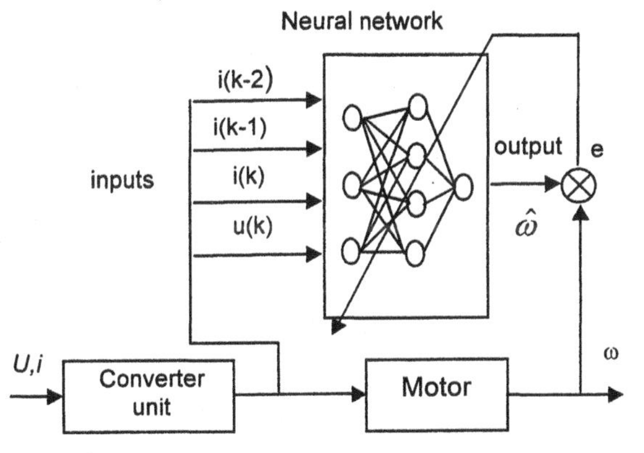

Fig. 3. Neural network configuration for training process.

The neural network's role is to identify the unknown mapping among current and voltage and speed of the motor, see Fig. 3. The neural estimator can emulate eq. (13) the motor dynamics as follows :

$$\hat{\omega} = N\big(i(k), i(k-1), i(k-2), u(k)\big) \tag{13}$$

where $\hat{\omega}$ is observed speed for the set of the inputs to the network.

The feedforward neural network is trained by presenting it with the set of input/output patterns obtained from the experimental laboratory measurement. During the generation of training data set, random sequence of terminal voltage is applied to converter unit. The resulting motor speed should be cover all intended regions of DC motor operation including starting, accelerating and breaking in both forward and reverse direction. The representative learning data set enables the neural estimator to generalize its training to new operating points not seen during training. The data set has to be normalized between 1 and -1.

The architecture and stopping criteria are proposed according to learning algorithm presented in section 2.2. The four inputs-, eight hidden- and one output neurons are chosen. After the error equals to 0.0015, the training process is stopped and the weight coefficients are taken from the neural network software simulation realized by PC IBM compatible machine. The program is written in high performance computing language.

The degree of learning and generalization of the trained feedforward neural network is investigated through real time test procedure as shown at figures in the next section. The number of training patterns and learning coefficient η are selected based on experience.

The weights achieved from simulation are used in neural speed estimator realized by DSP type of TMS320C20/100 ns. Behavior of that configuration is precisely described in the next section together with experimental results.

3.2 Laboratory measurement configuration

DC motor is separately-excited, name plate rating of armature voltage 110 V, armature current 19 A, and revolution 1400 min^{-1}. The load is an induction motor supplied by DC voltage, connected through coupling to the DC motor.

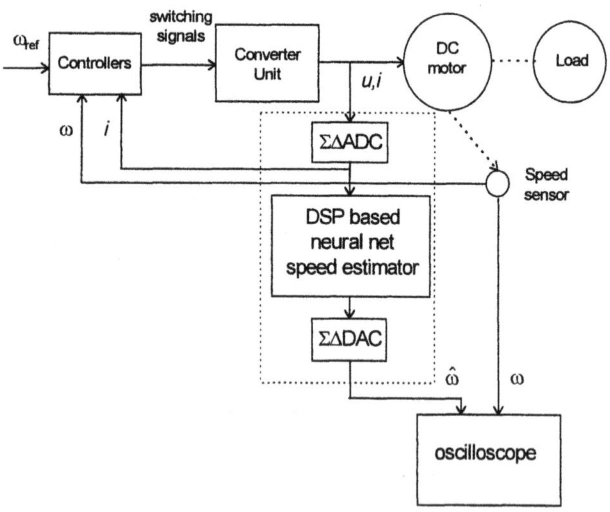

Fig. 4 Experimental laboratory scheme of solid state drive.

Variable terminal voltage of the dc motor is supplied by a PWM controlled dc/dc converter. The converter unit consists of four IGBTs arranged in a H bridge configuration as shown in Fig. 5. Parallel fast recovery diodes are added for commutation. Terminal voltage is regulated by pulse width modulation (PWM) of the IGBTs. The switching frequency is set up to 8 kHz.

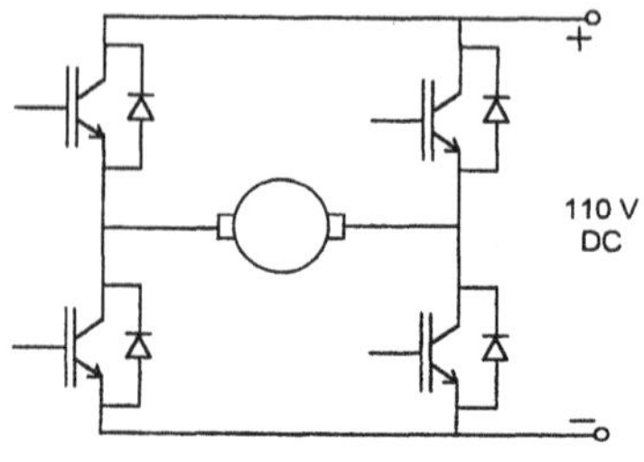

Fig. 5 H-bridge configuration of converter unit.

The drive has been controlled by ordinary analog regulators of angular speed and armature current. The first of them has been type of PI algorithm, the second one of hysteresis type. Neural-based estimator of angular speed is denotes dashed box in the Fig. 4. It has been realized by prototype board from LSI Loughborough (U.K.) with TMS320C25 fixed point 16-bit digital signal processor and sigma-delta A/D and D/A converters (Burr-Brown - 12 bits). Using of those type of converters it has not been necessary to use digital filtration of input signals, because of they involve anti-aliasing and digital filter. Assembly programming language was used for development of program for neural network based estimator described in Chap. 2. Both estimated and measured signals of angular speed of the motor have been observed at HP5601 digital storage oscilloscope and the screens was transferred to PC (MS Word).

3.3 Estimation results

The interesting estimation results are presented in this section.

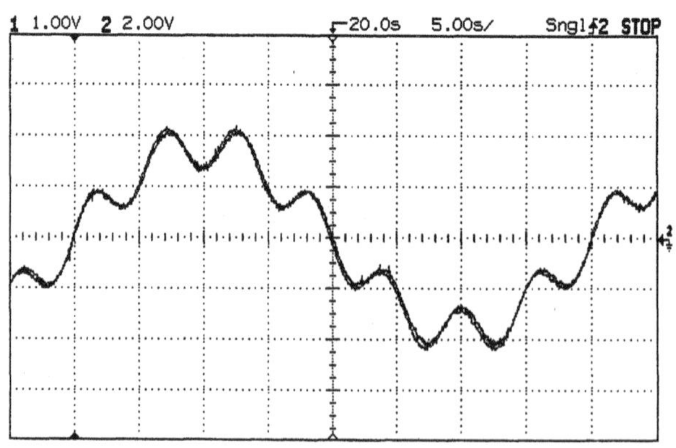

Fig. 6 No-load speed response of neural estimator and actual angular motor speed.

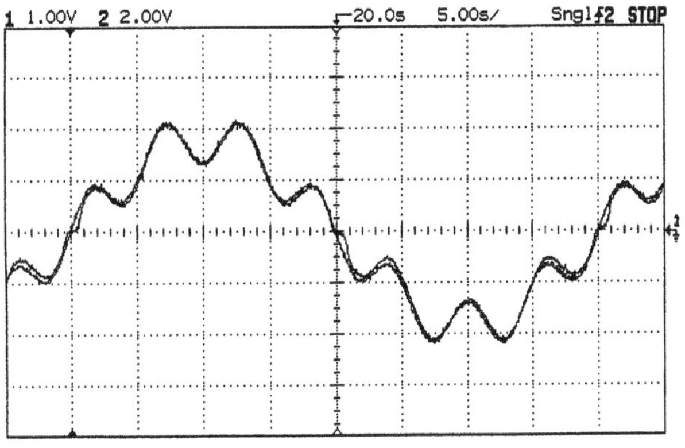

Fig. 7 Neural estimator speed response by applying load to the motor and motor speed.

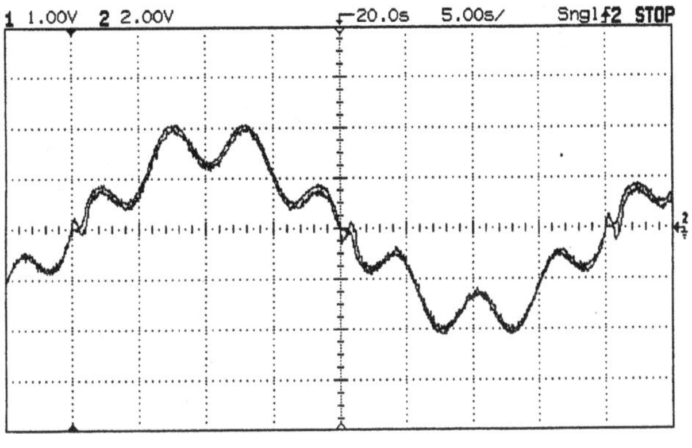

Fig. 8 On-line neural estimator weight updating at the beginning process of learning.

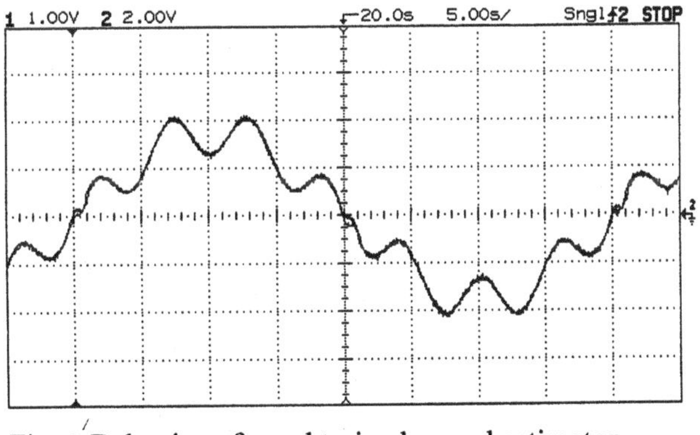

Fig. 9 Behavior of good trained neural estimator.

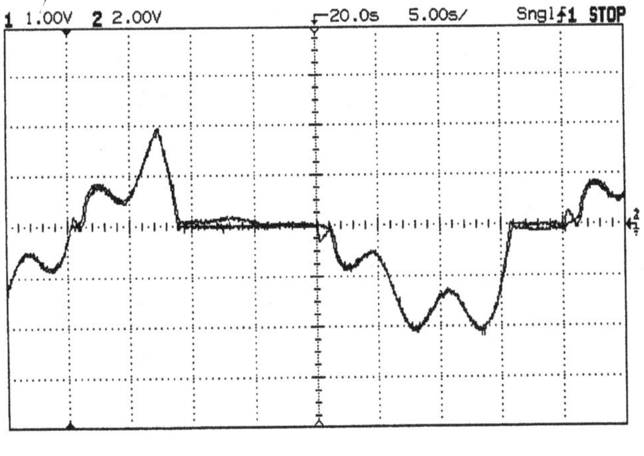

Fig.10 Neural estimator speed response at failure of supply voltage.

In Fig. 6 there is shown output angular speed of DC motor, carried-out from measurements by angular speed sensor and also angular speed estimated by DSP-based neural estimator, both at zero load torque of motor. The weight coefficients neural network have been determined by computer simulation of whole drive with neural-based estimator, at zero load. In this first case, the neural estimator observes actual DC motor speed very precisely because neural net is very well trained for that input/output patterns, it is obvious at Fig. 6.

In Fig. 7 are depicted angular speed time dependencies as in previous case, but at certain value of load torque. It can be seen from the results, that estimated value of angular speed has a small error at reverse changes of speed from positive to negative one. After short time the both time waveforms (estimated and measured) are very close to each other.

In both cases of those in Fig. 6 and Fig. 7 the neural network has been learned at off-line back-propagation process. In order to be the neural estimator invariabled and non-sensitive to any changes and external disturbances, it has to be continuously learning at on-line process. Some results of on-line learning experiments are shown in following pictures.

In Fig. 8 and Fig. 9 there are depicted estimated and measured speed dependencies after on-line learning process, which is continued during the whole time of run of the drive. Unambiguously better features of neural speed estimator are obvious from all presented experimental measurements by on-line learning process.

Well trained neural network based estimator, like it is in this case, can predict actual value of angular speed even at failure of supply voltage (see Fig. 10). The picture shows the behavior of neural estimator at failure of supply of DC drive. It's clear from the figures, that estimated angular speed is very close to the measured one. This dedicates to use the neural estimator instead of mechanical speed sensor. The estimated signal can be also used as input for control block of solid state drive in the Fig. 4.

4. CONCLUSIONS

A simulation and design of neural network for identification of electric drive has been presented in the paper. Three layer neural network with saturated linear transfer function of neurons has been designed and realized by fixed point digital signal processor TMS320C25/100 ns.. The neural network has been trained by back propagation algorithm at off-line learning process. As regulated system has been considered an electric drive with DC motor fed by four quadrant converter. The estimation of output signal was obtained by three different ways: simulation by PC, estimation by DSP neural network and finally, real time measurements of the DC drive. Identification process was made over input patterns of voltage and current and output signal of system angular speed..

The comparison of simulation experiments, results from DSP neural network estimation and as well as laboratory measurements has been done in the paper. As it can be seen that all performances are very similar and show close agreement of time waveforms. That means, that design of neural network and its properties are fulfilling the user requirements. Identification process has shown that DSP neural network estimation does not depend on internal structure of investigated system. Once learned neural network can be also used for inversion model control of the investigated system. Neural network estimator can be applied in any similar system of solid state drives (induction motor drive, synchronous motor drive), switched mode power supplies, uninterruptible power supplies and many others.

5. ACKNOWLEDGEMENT.

The paper has been supported by grant project of Ministry of Education of Slovak Republic No. 1/3234/96 „Implementation of Control Algorithms for Complex Systems of Control in Electroenergetics" and No. 1/4243/97 „Research of Properties of Active Filters with Special Focusing on Negative Influences of Power Electronic Converters, Acting in Traction Network, on Energetic Supply System".

The authors would like also to thank University of East London for lending of LSI prototype DSP card.

REFERENCES

Balažovič P. (1996). Using Neural Network Toolbox in MatLab Environment for Controlling of PES. *Diploma project.* University of Zilina, Faculty of Electrical Engineering

Bertoluzzo M., Buja G.S., Todesco F. (1994). Neural Network Adaptive Control of a DC Drive. In: *Proceedings of IECON'94 Conference.* Bologna, **Vol.** 2, pp. 1232-1237

Chyla P. (1995). Neural Network Implementation in Power Electronic System Control. *Project of PhD Thesis.* University of Zilina, Faculty of Electrical Engineering

Dobrucky B., Chaloupka F., Kucer P., Racek V. (1995). A neural Fuzzy Control with Stator Resistance Estimator for IM Drive with DSC. In: *Proceedings of EPE'95 Conference.* Sevilla, **Vol. 1**, pp. 1.789-1.794

Narendra K.S., Parthasarathy K. (1990). Identification and Control of Dynamical Systems Using Neural Networks. *IEEE Transactions on Neural Networks.* **Vol. 1**. No. 1, pp. 4-27

Skalicky J. (1995). Servodrive Control in Mechatronic Systems. In: *Proceedings of AE'95 Symposium.* Pilsen, pp. 73-76

Vanek R. (1994). Introduction to Neural Networks. *Textbook of University of Zilina*

Weerasooriya S., El-Sharkawi M. (1993). Laboratory Implementation of a Neural Network Trajectory Controller for a DC Motor. *IEEE Transaction on Energy Conversion.* **Vol. 8**, No. 1, pp. 107-113

Zaruda J.M. (1992). Introduction to Artificial Neural Systems. *West Publishing Company.* New York

INTELLIGENT CONTROL OF SINGLE INVERTED PENDULUM

P. Sinčák*, M. Kopčík*, N. Kopčo*, R. Jakša*

**Technical University of Košice, Slovakia, Faculty of Electrical Engineering and Informatics, Department of Cybernetics and Artificial Intelligence Laboratory of Artificial Intelligence, Letná 9, 040 01 Košice, Slovakia*
contact E-mail: sincak@ccsun.tuke.sk

Abstract: The paper deals with neuro–fuzzy tools for non–linear system control. Inverted pendulum was chosen as a typical non-linear model for control. The control task was defined as lifting inverted pendulum from up–side-down position into standing position by a cart movement and keep this position of the pendulum stable. Two different controllers were observed, ANFIS (Adaptive Neural Inference System) and NARA (Neural Approximate Reasoning Approach). Finally a simple neuro-controller based on feed-forward neural network with classical back-propagation adaptation was studied. Results show that ANFIS seems to be a powerful tool for this kind of task. In the paper an extensive report about experiments with pendulum control is presented.

Keywords: Neural Networks; On–line Control; Optimal Controller; Mathematical Model; Intelligent Control

1. INTRODUCTION

Neural networks and the fuzzy controllers represent two main parts of intelligent technologies. Main features of these technologies are knowledge processing and its utilization in control task. In general AI methods can be divided according to the knowledge representation used as it is shown in table 1.

There are many advantages in neural approach as well as in fuzzy controller design (Kosko, 1992; Vaščak, 1996). Neuro–fuzzy approach presents a big challenge to utilize the advantages of both approaches into one control tool. One of them is a fact that with neuro–fuzzy approach the fuzzy rule base design is avoided as it is created automatically according to training data.

Neuro–fuzzy approach of studied methods represents in fact a neural implementation of Takagi–Sugeno fuzzy controller. Inference principles are presented on figure 3. The main characteristic of the controller is a crisp part of the consequent rule.

Table 1. AI–methods according to knowledge representation

	Knowledge	
structural	Expert Systems	Fuzzy Controller
non–structural	–	Neural Network
	symbolic	non–symbolic

ANFIS represents a tool based on adaptive network[1]. NARA is based on semi–modular neural network consisting of more or less independent networks. Number of experiments have been done observing controller behavior in process of controlling simple inverted pendulum. The logical aspect of experiments was to simulate human action to control a simple inverted pendulum concerning mentioned task. Therefore a human control by

[1] not a neural network!

expert was done and obtained data were used to control inverted pendulum.

2. INVERTED PENDULUM MODEL

The pendulum model is shown on figure 1. The pivot of the cart allows the pendulum to be in up–side-down position (hanging downwards). Control task was defined to use a cart motion to lift and balance pendulum into stand–up position. This position should be kept stable by cart motion. The system has parameters as follows:

- M_v is a cart mass
- M_1 is a pendulum mass
- R_1 is a pendulum length
- K_v is factor of friction against moving cart
- K_1 is factor of friction against pendulum rotation

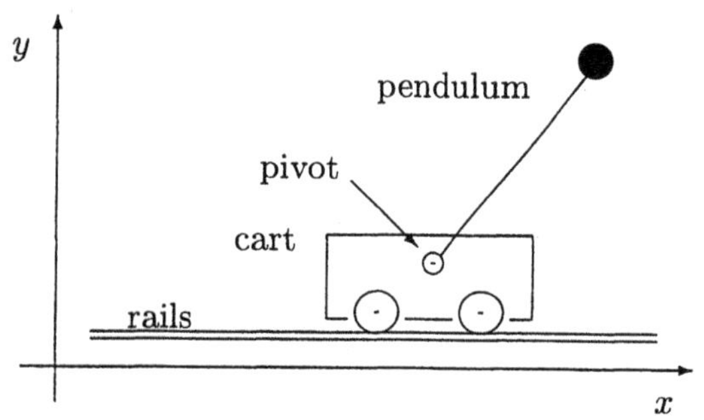

Fig. 1. Model of pendulum on a cart

These model parameters settings can be arbitrarily changed. Every change will cause some minor changes in controller design. Controller has the following inputs representing the controlled system state parameters:

- ϕ is pendulum angle between current and desired position
- x_1 is a current cart position
- x_v is desired cart position

2.1 *Equations of movement*

In derivation of dynamic model (Riedmiller, 1993; Kawaji *et al.*, 1992) a variant principles of mechanics were applied where the movement of primary elements of the system is described by Lagrange–Euler equation

$$\frac{d}{dt}\left(\frac{\partial L}{\partial \dot{q}_i}\right) - \frac{\partial L}{\partial q_i} = Q_i, \tag{1}$$

where L is Lagrange function defined as a difference of dynamic inertia E_k and potential energy E_p

$$L = E_k - E_p, \tag{2}$$

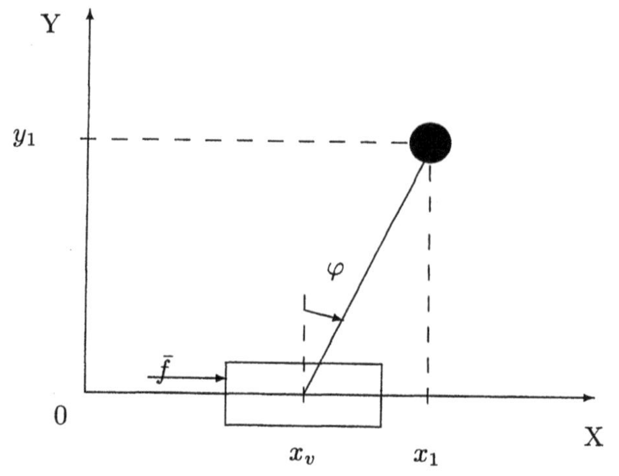

Fig. 2. Coordinates designation

where i is degree of freedom, q_i is i-th general coordinate and Q_i is i-th generalized force.
From figure 2 it is possible to write for the coordinates of significant points

$$\begin{aligned} x_1 &= x_v + R_1 \sin\varphi \\ y_1 &= R_1 \cos\varphi, \end{aligned} \tag{3}$$

where x_v is the position of the cart and φ is the angle of the pendulum turn.

For the overall dynamic inertia can be written

$$\begin{aligned} E_k &= E_k^{cart} + E_k^{pendulum} \\ &= \frac{1}{2}M_v\dot{x}_v^2 + \frac{1}{2}M_1 v_1^2, \end{aligned} \tag{4}$$

where $\dot{x}_v$ a v_1 represent speed of the cart and of the end-point of the pendulum. Following equations (3) can be written for v_1

$$v_1^2 = \dot{x}_1^2 + \dot{y}_1^2 = \dot{x}_v^2 + 2R_1\dot{x}_v\dot{\varphi}\cos\varphi + R_1^2\dot{\varphi}^2 \tag{5}$$

Then the dynamic inertia can be expressed as

$$\begin{aligned} E_k^{cart} &= \frac{1}{2}M_v\dot{x}_v^2 \\ E_k^{pend} &= \frac{1}{2}M_1(\dot{x}_v^2 + 2\dot{x}_v R_1\dot{\varphi}\cos\varphi + {R_1}^2\dot{\varphi}^2) \end{aligned} \tag{6}$$

and the potential energy can be expressed as

$$E_p = E_p^{cart} + E_p^{pendulum}, \tag{7}$$

where

$$\begin{aligned} E_p^{cart} &= 0 \\ E_p^{pendulum} &= M_1 g R_1 \cos\varphi, \end{aligned} \tag{8}$$

where g represents the gravity. Based on (2) and the above mentioned equations it is possible to solve a system of Lagrange equations,

$$\frac{d}{dt}\left(\frac{\partial L}{\partial \dot{x}_v}\right) - \frac{\partial L}{\partial x_v} = f \tag{9}$$

$$\frac{d}{dt}\left(\frac{\partial L}{\partial \dot{\varphi}}\right) - \frac{\partial L}{\partial \varphi} = 0, \tag{10}$$

where f is the outside force acting on the cart. The derivation is as follows.

$$\frac{\partial L}{\partial \dot{x}_v} = \dot{x}_v(M_v + M_1) + R_1 M_1 \dot{\varphi} \cos\varphi$$

$$\frac{d}{dt}\left(\frac{\partial L}{\partial \dot{x}_v}\right) = \ddot{x}_v(M_v + M_1) + \ddot{\varphi} R_1 M_1 \cos\varphi - $$
$$-\dot{\varphi}^2 R_1 M_1 \sin\varphi \tag{11}$$

$$\frac{\partial L}{\partial x_v} = 0$$

and similarly

$$\frac{\partial L}{\partial \dot{\varphi}} = R_1 M_1 \dot{x}_v \cos\varphi R_1^2 M_1 \dot{\varphi}$$

$$\frac{d}{dt}\left(\frac{\partial L}{\partial \dot{\varphi}}\right) = R_1 M_1 (\ddot{x}_v \cos\varphi - \dot{x}_v \dot{\varphi} \sin\varphi +$$
$$+ R_1 \ddot{\varphi}) \tag{12}$$

$$\frac{\partial L}{\partial \varphi} = R_1 M_1 \sin\varphi (g - \dot{x}_v \dot{\varphi}).$$

After substitution of equations (11) into (9) and substitution of (12) into (10) the equations of movement are obtained not considering the losses caused by the movement. If the friction of the air against the cart and certain breaking momentum of the bar in axis depending on speed of revolving is considered, the resulting equations of movement will be defined as follows

$$\ddot{x}_v(M_v + M_1) + \ddot{\varphi} R_1 M_1 \cos\varphi -$$
$$-\dot{\varphi}^2 R_1 M_1 \sin\varphi + \dot{x}_v K_v = f$$
$$\ddot{x}_v R_1 M_1 \cos\varphi + \ddot{\varphi} R_1^2 M_1 -$$
$$-R_1 M_1 g \sin\varphi + \dot{\varphi} K_1 = 0. \tag{13}$$

A system of two non-linear differential equations of second order (13) represents the sought mathematical model of given mechanical system.

2.2 *Representation in State Space*

After transformation of (13) and some further modifications including substitution of variables $x_1 = x_v, x_2 = \dot{x}_v, x_3 = \varphi, x_4 = \dot{\varphi}, det = M_v + M_1 \sin^2\varphi$ the following state description of the system is obtained

$$\begin{pmatrix} \dot{x}_1 \\ \dot{x}_2 \\ \dot{x}_3 \\ \dot{x}_4 \end{pmatrix} =$$

$$\begin{pmatrix} 0 & 1 & 0 & 0 \\ 0 & \frac{-K_v}{det} & 0 & \frac{x_4 R_1^2 M_1 \sin\varphi + K_1 \cos\varphi}{R_1 det} \\ 0 & 0 & 0 & 1 \\ 0 & \frac{K_v \cos\varphi}{R_1 det} & 0 & \frac{-x_4 R_1^2 M_1^2 \sin\varphi\cos\varphi - (M_v + M_1)K_1}{R_1^2 M_1 det} \end{pmatrix}$$

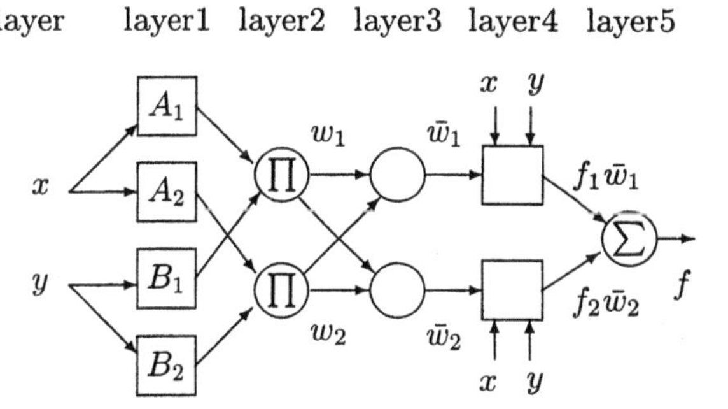

Fig. 3. Structure of Takagi–Sugeno fuzzy model in ANFIS

$$\begin{pmatrix} x_1 \\ x_2 \\ x_3 \\ x_4 \end{pmatrix} + \begin{pmatrix} 0 \\ \frac{-M_1 g \sin\varphi\cos\varphi}{det} \\ 0 \\ \frac{(M_v + M_1) g \sin\varphi}{R_1 det} \end{pmatrix} + \begin{pmatrix} 0 \\ \frac{1}{det} \\ 0 \\ \frac{-\cos\varphi}{R_1 det} \end{pmatrix} \tag{14}$$

3. ADAPTIVE NEURAL INFERENCE SYSTEM - ANFIS

ANFIS (Jang, 1993) is considered to be an **adaptive network** which is very similar to **neural networks** (Jang and Sun, 1995). Adaptive network has no synaptic weights but so called adaptive and non-adaptive nodes. It must be said that adaptive network can be easily transformed to neural networks architecture with classical feed-forward topology.

ANFIS is adaptive network which works like adaptive network simulator of Takagi–Sugeno's fuzzy controllers. This adaptive network has predefined adaptive network topology. The basic topology description is as follows:

(1) layer #1 – each node of this layer is adaptive and it's function is:

$$O_i^1(x) = \mu_{A_i}(x), \tag{15}$$

where x is input and $\mu_{A_i}(x)$ is output which represents membership function of input to level linguistic variable A_i. Usually bell membership function is used:

$$\mu_{A_i}(x) = \frac{1}{1 + [(\frac{x - c_i}{a_i})^2]^{b_i}} \tag{16}$$

where a_i, b_i, c_i are parameters which are determining membership function shape. These are premise parameters.

(2) layer #2 – nodes in this layer are not adaptive. They realize only simple multiplication Π of inputs. These outputs represent a rule strength w_i.

(3) layer #3 – nodes in this layer are not adaptive and they calculate a ratio of rule strengths as follows:

$$\bar{w}_i = \frac{w_i}{w_1 + w_2}, i = 1, 2. \tag{17}$$

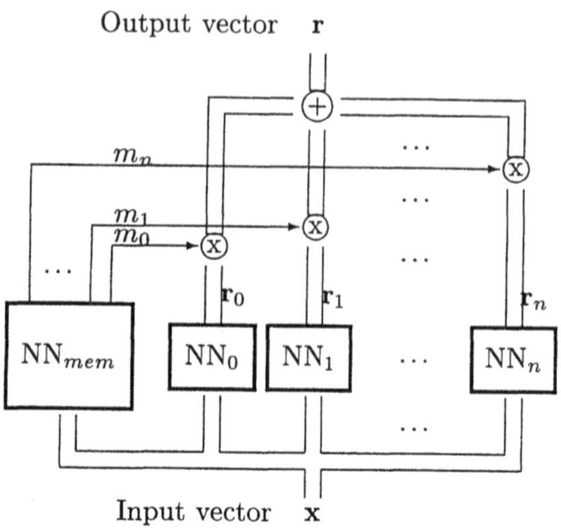

Fig. 4. Structure of NARA type of NNs

(4) layer #4 – nodes in this layer are adaptive and so the transfer function will be as follows:

$$fi(x) = \bar{w}_i f_i = \bar{w}_i(p_1 x + q_i y + r_i), \quad (18)$$

where p_i, q_i, r_i are consequent parameters.

(5) layer #5 – nodes are non-adaptive and just make addition of inputs as follows:

$$f(x) = \sum_i \bar{w}_i f_i = \frac{\sum_i w_i fi}{\sum_i w_i}. \quad (19)$$

As mentioned above there are premise parameters a_i, b_i, c_i and consequent parameters p_i, q_i, r_i.

This network is trained in sense of supervised learning. So there are training data and our goal is to train adaptive network to be able to approximate unknown function given by training data. Training data were gathered from manual control of pendulum. Our goal is to adapt and find the right value of above parameters for function approximation.

4. NARA – NEURAL ADAPTIVE REASONING APPROXIMATOR

NARA type of Neural network is a sub-modular NN built in accordance with IF-THEN structure of fuzzy inference rules (Han and Moraga, 1996). It is suitable for control of non-linear systems especially of systems that exhibit behavior which can be described as a set of distinct behavioral characteristics groups.

The basic structure of NARA neural network is shown on figure 4. This neural network consists of subnetworks of two kinds: NN_{mem} sub-network representing IF (premise) part of fuzzy inference rules and NN_s representing THEN (consequent) part of the s-th fuzzy inference rule. The NN_s subnetworks are mutually independent.

The goal of the system is to apply NNs on design and generation of member functions for fuzzy inference rules. The advantage of this type of NN is that it is possible to structurally analyze the NN and include some knowledge into it in order to improve the performance of the network.

Usually a Kohonen type of NN is used as the NN_{mem} subnetwork and the NN_s networks are implemented in form of feedforward networks using backpropagation llearning.

Two different approaches are possible in the process of design and learning of this kind of NN. There is an algorithm which describes how to build this NN automatically (not making use of some expert knowledge) and how to learn it (so-called Automatic design). It is also possible to design this kind of NN strictly by application of some a priory knowledge of the controlled system and the strategy needed to control it (so-called Expert design).

5. EXPERIMENTS AND RESULTS

The main goal of this work is to present a comparative study of different approaches in control of a non-linear system (in this case a simple pendulum) with emphasis on neuro-fuzzy control systems. The task of the controller is to wobble the pendulum hanging downwards, get it into stand-up position and stabilize it in the middle point of the rails.

Three neuro-fuzzy control systems were compared:

- **ANFIS**
- **NARA**
  - using Automatic design
  - using Expert design
- **Feedforward NN** using Bacpropagation learning algorithm

In order to evaluate the ability of networks to generalize, on all the networks data obtained by recording the control actions used for this task by a human trying to control the system using mouse which allowed to apply a different force on the cart in different moments. The data describing final phase of control (i.e. the phase of stabilizing the pendulum in the stand-up position) were generated using optimal controller based on quadratic criterion. The patterns were thoroughly prepared in order to cover as broad range of states, probable to occur in course of control, as possible.

5.1 *Description of experiments on ANFIS*

An ANFIS topology consisting of four input neurons was used and the number of member functions for the angle was enlarged in comparison to the other inputs. The topology of the net was

3-3-5-3 (No. of member functions belonging to position, No. of member functions belonging to speed of cart, No. of member functions belonging to angle of pendulum, No. of member functions belonging to angle speed of pendulum). The consequent parameters were initialized to "0" so the unlearned controller output was equal to zero. The course of learning is described on figure 5. Although the resulting learning error was quite

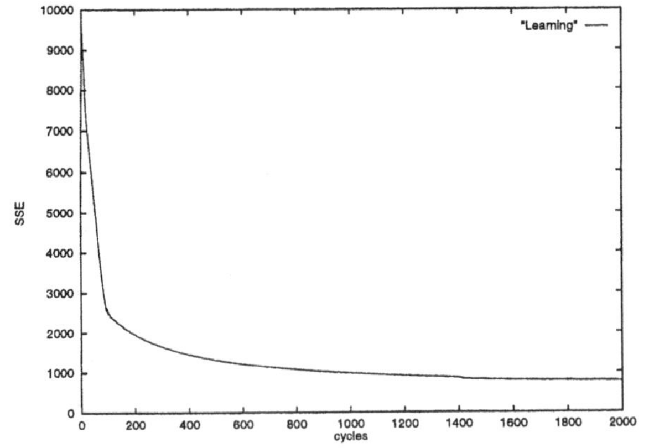

Fig. 5. Course of learning using ANFIS topology 3-3-5-3

considerable, it was able to solve the assigned task. The course of control for initial conditions $(x_v, \dot{x}_v, \varphi, \dot{\varphi}) = (0, 0, 3.14, 0)$ is on figure 6. The

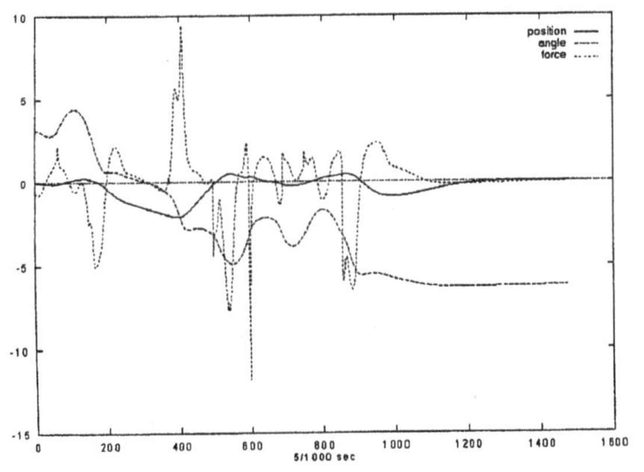

Fig. 6. Course of control using ANFIS topology 3-3-5-3

net had some problems as it used too much force to wobble the pendulum which resulted in it's inability to stabilize it on the top. But after a few trials the net succeeds to do so.

5.2 *Description of experiments on NARA using Automatic design*

During the process of learning 34 clusters were created using Kohonen neural network. After evaluation of the clusters four significant clusters were found for which a NARA neural network with corresponding topology (four NN_i subnets) was created. All these subnets were learned to a sufficient level.

After the learning phase the system was tested on different initial conditions. The regulating system was able to fulfill the assigned task only for initial conditions set to $(x_v, \dot{x}_v, \varphi, \dot{\varphi}) = (0, 0, 1.87, 0)$. **No other cases where the system would be able to fulfill the task were found.** The course of control for initial conditions set to $(x_v, \dot{x}_v, \varphi, \dot{\varphi}) = (0, 0, 1.87, 0)$ is shown on figure 7.

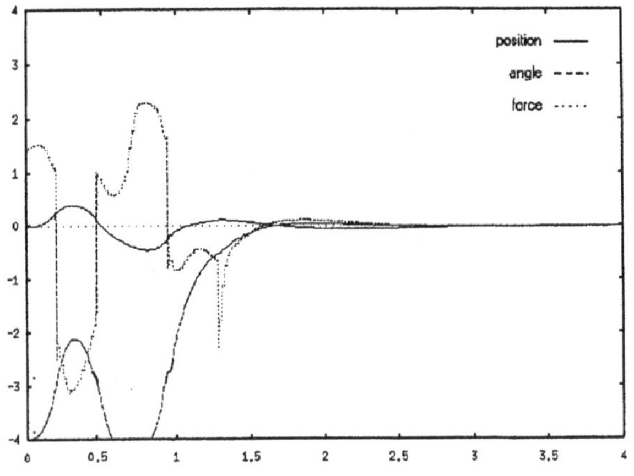

Fig. 7. Course of control for NARA with automatic design for initial parameters (0, 0, 1.87, 0)

5.3 *Description of experiments on NARA using Expert design*

The course of expert design was as follows. The control of a pendulum consists of two parts:

- To wobble the pendulum and get it as close to the needed position as possible (done by an NARA regulator).
- To stabilize the resulting position of pendulum (done by an optimum controller).

The two controllers were switched between one another based on the state, i.e. NARA controls the system as first and when the system reaches defined conditions the optimum regulator will be activated. In case the optimum regulator is not able to hold the position, NARA is activated again.

Then the system was tested on different initial conditions. The results are mixed. There were some initial conditions for which the system was able to fulfill the task. Initial conditions $(x_v, \dot{x}_v, \varphi, \dot{\varphi}) = (0, 0, 0, 0)$ and $(x_v, \dot{x}_v, \varphi, \dot{\varphi}) = (0.2, 0, -1.87, 0)$ belong among these cases. The course of control for initial conditions $(x_v, \dot{x}_v, \varphi, \dot{\varphi}) = (0.2, 0, -1.87, 0)$ is shown on figure 8.

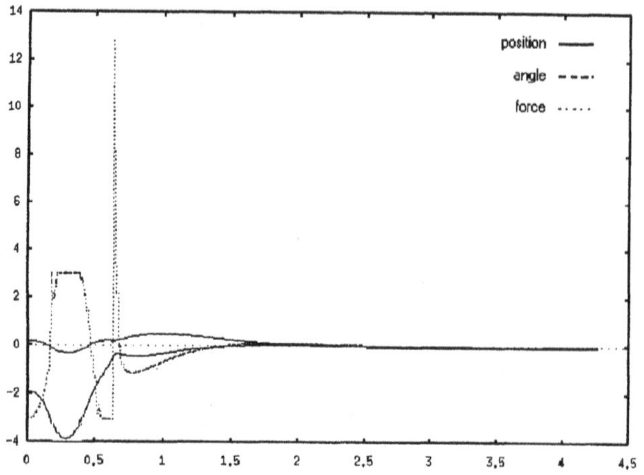

Fig. 8. Course of control using NARA with expert design for initial conditions (0.2, 0, -1.87, 0)

5.4 *Description of experiments on Feedforward NN using Backpropagation learning*

For this experiment classical feedforward neural network with Backpropagation learning was used. It was tested on different topologies in the learning phase and topology 4x20x10x1 showed to be the best topology for this task. This topology gained after 1000000 learning epochs with learning rate set to 0.02 mean error of 0.22164.

The trained system was tested on different initial conditions. The results obtained were equivocal. The ratio of successful tests to unsuccessful tests is 4 to 2. The course of control for initial conditions $(x_v, \dot{x}_v, \varphi, \dot{\varphi}) = (0, 0, 3.14, 0)$ is shown on figure 9.

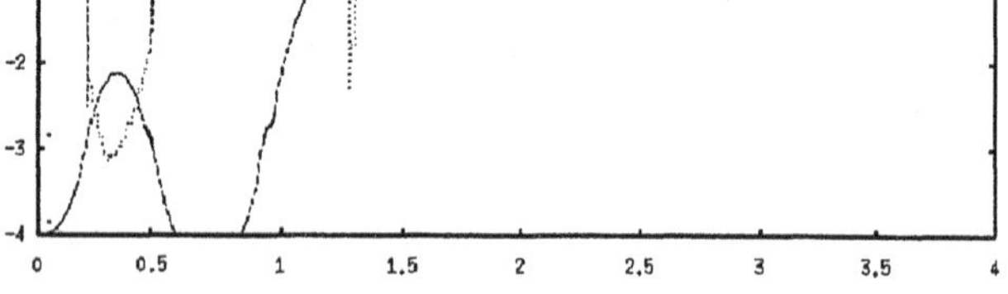

Fig. 9. Course of control using feedforward neural network for initial conditions (0, 0, 3.14, 0)

5.5 *Results and their evaluation*

The results show that ANFIS is the most suitable neuro-fuzzy controller for this task. A "classical" Feedforward NN using Backpropagation surprisingly showed to be also quite good regarding this task (It was expected that neural net of this kind should not cope with a task of such complexity). On the other hand the NARA system failed completely, both the automatic and the expert versions of it. **This result is especially interesting with regard to the work of (Han and Moraga, 1996), where it is shown that the NARA system is more suitable for modeling of functions. Our results show that for a control task ANFIS is more suitable neural network method.**

6. CONCLUSION

In this paper two neuro–fuzzy tools for non–linear system control were presented. Inverted pendulum was chosen as a typical non–linear model for control. The defined control task **to lift inverted pendulum from up–side-down position into standing position** by cart movement was the comparison criterion. Two different controllers were observed and their behavior was studied. ANFIS (**A**daptive **N**eural **I**nference **S**ystem) and NARA (**N**eural **A**pproximate **R**easoning **A**pproach) were methods of interest. Finally a simple neuro–controller based on feed–forward neural network with classical backpropagation adaptation was studied. Results showed that ANFIS is a powerful tool for this kind of task concerning non-linear control.

7. REFERENCES

Han, J. and C. Moraga (1996). Parametric networks based adaptive fuzzy model. In: *International ICSC Symposium on SOFT COMPUTING / SOCO'96.* 3rd ed. Reading, UK.

Jang, J. S. R. (1993). Anfis: Adaptive-network-based fuzzy inference system. In: *IEEE Transactions on systems, man, and cybernetics.* Vol. 23. pp. 665–685.

Jang, J. S. R. and Ch. T. Sun (1995). Neuro–fuzzy modeling and control. In: *The Proc. of the IEEE.* Vol. 83. pp. 378–406.

Kawaji, S., T. Maeda and N. Matsunaga (1992). Learning control of an inverted pendulum using neural networks. In: *Proceedings of the 31st Conference on Decision and Control.* Vol. 83. Tuscon, Arizona.

Kosko, B. (1992). *Neural Networks and Fuzzy Systems. A dynamical systems approach to machine intelligence.* Prentice-Hall. Singapore.

Riedmiller, M. (1993). Controlling an inverted pendulum by neural plant identification. In: *Proceedings of the IEEE International Conference on Systems, Man and Cybernetics.* Vol. 83. Le Touquet, France.

Vaščak, J. (1996). *Fuzzy Logic - Basic Principles.* Elfa Ltd. Košice, Slovakia.

THE NEURAL NETWORK MODEL OF A ROBOT JOINT

Jozef Suchý* and Jarmila Škrinárová**

**Institute of Control Theory and Robotics, Slovak Academy of Science*
Severná 5, 974 00 Banská Bystrica, Slovakia
***Department of Computer Science, Faculty of Natural Sciences*
Tajovského 40, 974 01 Banská Bystrica, Slovakia

Abstract: In the present paper the artificial neural network model of a robot joint is presented. The data used come from the real measurements on the first joint of Manutec R15 robot. The measured data correspond to the motor current and to the joint position, so that they include nonlinear phenomena like friction, backlash or stiffness present in the joint. For modelling of the robot joint based on the measured data the modification of CMAC neural network, the so called generalized CMAC (GCMAC), is used.

Keyword: Modelling, robot joint, neural-network models

1. INTRODUCTION

The research in the field of application of the artificial neural networks in robotics is according to the number of publications quite intensive. Among these publications there are already some overview works (Lee and Bekey, 1991, Sanchez and Hirzinger, 1992). This research concentrates on robot kinematics (Gardner, *et al.*, 1993, Yeung and Bekey, 1989, Yeung and Bekey, 1993), where attention is also paid to the learning of manipulator Jacobian matrix, robot dynamics (Eskandarian, *et al.*, 1994), control (Chen and Chang, 1996, Krijgsman, 1993, Larsen, *et al.*, 1995 and sensorial control of robots (Katic and Vukobratovic, 1993, Sanchez and Hirzinger, 1992). From among the latter works the number of those dealing with force control or impedance control of robots is significant. We have mentioned just one of them here. Recently, according to our knowledge, the first attempts to compensate for friction effects using neural networks have been performed (Larsen, *et al.*, 1995)

As for the types of artificial neural networks (ANN) used for research in robotics, most often multilayer feedforward networks are used (Gardner, *et al.*, 1993, Katic and Vukobratovic 1993, Lee and Bekey, 1991, Sanchez and Hirzinger, 1992, Yeung and Bekey, 1989, 1993). Especially in robotics and control the CMAC (Cerebellar Model Articulation Controller) neural networks become increasingly popular (Chen and Chang, 1996, Eskanderian *et al.*, 1996, Krijgsman, 1993, Larsen, *et al.* , 1995, Miller, 1990). With the multilayer artificial neural networks, the way are looked for how to effectively handle the high dimensional problems. One approach to solving this problem is the decomposition of the neural network (Katic and Vukobratovic, 1993) or construction of context-sensitive neural networks (Yeung and Bekey, 1989, 1993). Similar problems do also exist with CMAC algorithms, some authors (Eskanderian, *et al.*,1996) tried to solve them in context of robot dynamics by problem decomposition.

Not so much attention has been paid to the modelling of robot joints. Neural networks seem to be appropriate means for learning nonlinear mappings. Joints of geared robots are strongly nonlinear because of friction, backlash or stiffness. Besides, the dynamic nonlinear effects may become evident with rapid motion. They may, however, have minor influence while robot is performing fine motions. Besides nonlinear behaviour of a robot joint there is also linear one which can be described by a tranfer function. It could thus be useful to try to joint there is also linear one which can be described by a tranfer function. It could thus be useful to try to model these linear and nonlinear phenomena in robot joints by means of convenient ANN.

In this contribution the real data are used for modelling. These data have been taken while the robot Manutec R15 performed some predefined motions. During this motion the motor current of the first joint has been measured on the motor side of the joint and the position on the output side of the joint. From these data, also the speed of the joint could be approximately computed. It is well known that with DC electrical motors the motor current data are practically equivalent to torque data because there is linear relation between motor torque and motor current. This fact can be utilized in the modelling of torque/force relations of robot joint.

2. CEREBELLAR MODEL ARTICULATION CONTROLLER (CMAC)

For the purpose of robot joint modelling the CMAC neural network was selected in this work which was shown to be both useful and successful in robotics for its local approximative properties (Albus, 1975a, 1975b, Miller, 1990).

The CMAC neural network was first published in 1975 by J. S. Albus (Albus, 1975a). The CMAC was originally developped for controlling robotic systems. Since that time some authors worked out different modifications of this algorithm. One of the best known modifications is the so called GCMAC (generalized cerebellar model articulation controller) (Krijgsman, 1993). For GCMAC, there is the free simulation software also available. This sofware was also used in this work. All CMAC algorithms are in principle look-up tables with some additional properties. The basic configuration of the CMAC may be seen in Fig. 1. In the mapping S into A two properties have to be implemented: near vectors in

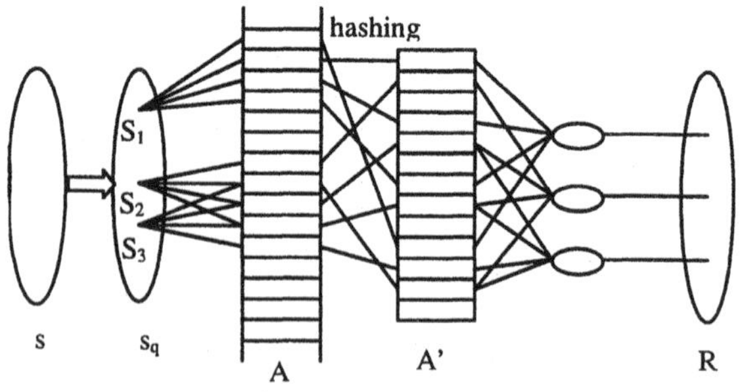

Fig. 1 The principle structure of CMAC neural network. S is the sensory input space, S_q is the quantized sensory input space, A and A' are the virtual memory and the actual memory of the association cells and R is the output space.

input space are mapped into near vectors in the association space and only a small number ñ of the cells will have nonzero value for a specific input. Number C defines the generalization property of CMAC and is called the generalization number. In case C=1 CMAC reduces to look-up table. After the generalization number has been selected, now another parameter has to be chosen, quantization or, equivalently, resolution r_i. The result is the number of intervals for each component s_i of the input vector. In this way the original input space S is mapped into quantized input space S_q. Hashing represents the random mapping from A into A'.

The GCMAC differs from the original CMAC by the explicit introduction of the so called kernel or basis function (Krijgsman, 1993). This function was implicitly present already in the original CMAC with just two values 0 or 1. According to the analogy with fuzzy sets, the function F was introduced as the kernel function with values from the interval [0,1]. The output of the GCMAC can now be calculated as

$$y_i = \frac{\sum_{j=1}^{C} F(S_q, C_{qj}) \cdot w_{ij}}{\sum_{j=1}^{C} F(S_q, C_{qj})} \tag{1}$$

where y_i from R is the i-th component of the output vector, C_{qj} is the center of the kernel function, S_q is the input vector and w_{ij} are synaptic weights associated with the associated cell j contributing to the i-th component of the output vector y.

The weight update in GCMAC is performed according to Widrow-Hoff rule as follows:

$$\Delta w_{ij} = b_j(y'_i - y_i)F(S_q, C_{qj}) \tag{2}$$

Here $(y_i'-y_i)$ is the error between the actual output and the desired output, b_j is the learning rate and is from [0,1].

3. MATHEMATICAL MODEL OF THE ROBOT JOINT

In this part the robot joint will be mathematically modelled. This model will correspond to the dependency of the joint position from the motor current. The model itself will consist of the linear and non-linear part. It will not be used with the measured data, the purpose of it is to gain some relevant information, like the order of the system, for the sake of non-parametric modeling with neural networks. This kind of information could, however, be found out also tentatively. The availability of parametric model may spare some effort for non-parametric modelling. According to (Williams and Khatib, 1996), the following equations describing the linear behaviour of the robot joint may be written:

$$\begin{aligned} &J_m\ddot{q}_m + b_m\dot{q}_m + k_t(q_m - Nq_l) = t_m \\ &J_l\ddot{q}_l + b_l\dot{q}_l + Nk_t(Nq_l - q_m) = 0 \\ &-Nk_t(Nq_l - q_m) = t_l \end{aligned} \qquad (3)$$

In these equations q_m and q_l are the positions of the motor shaft and the link respectively, J_m, J_l are the inertia of the motor and the inertia of the link, b_m and b_l are the coefficients of viscous friction of the motor and the link, k_t is the stiffness of the shaft. N is the gear ratio and t_m and t_l are the motor torque and the torque applied at the link. By taking the Laplace transform of the above equations and carrying out some manipulations, the desired transfer function may be obtained:

$$Q_l(s) = \frac{b_0}{s(s^3 + a_2 s^2 + a_1 s + a_0)} \cdot I_m(s) \qquad (4)$$

Here Q_l and I_m are the Laplace transforms of q_l and i_m, the motor current, b_0, a_i, i=0,1,2 are the coefficients dependent on the physical constants described in (3) If also the nonlinear effects are taken into account, then the last equation takes the following form:

$$Q_l(s) = \frac{b_0}{s(s^3 + a_2 s^2 + a_1 s + a_0)} \cdot I_m(s) + Q_n(s) \qquad (5)$$

where Q_a corresponds to the nonlinear part of the model This is caused by non-modelled nonlinear effects present in the joint, like friction, backlash or stiffness. The model is of the fourth order which gives some clue as for number of inputs to the neural network.

4. GCMAC MODEL OF THE ROBOT JOINT

The same relation as in foregoing section will now be modelled with neural network For the non-parametric modelling of the robot joint by means of GCMAC neural network the measured data are available. The data are measured on the first joint of manipulator Manutec r15 produced by the firm Siemens, Germany. Two sets of data have been available with approximatelly 6000 of pairs of data in every set. One set of the measured motor current and joint position data is in the Fig. 2. The strong noise is evident in the motor current signal and, therefore, it has to be filtered. After some simulation experiments, it was decided to use moving average filter of the order 50. The filtration effect was sufficient in this case while the signal was not distorted. Fig. 3 illustrates both the unfiltered and filtered motor current signals in one graph. These data have been used for training the GCMAC.

It goes out from the relations (4) and (5) that the

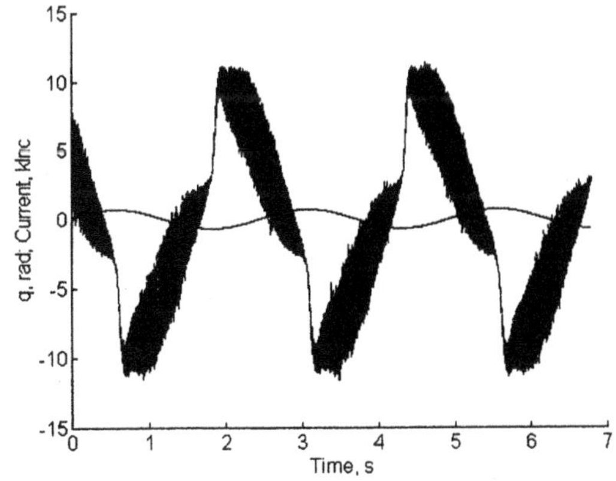

Fig. 2. The raw measured data of motor current and joint position at the first joint of Manutec r15 manipulator.

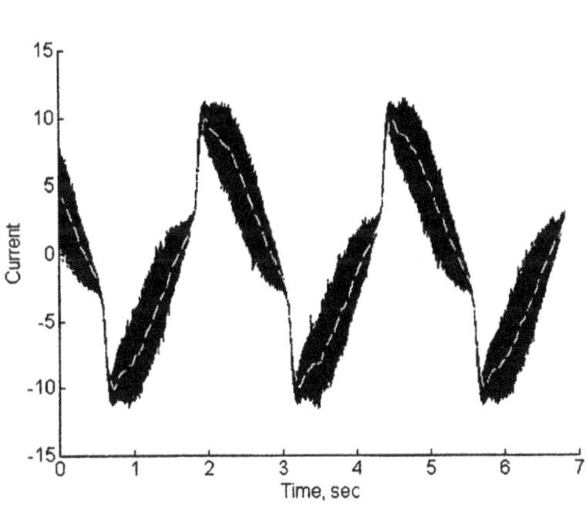

Fig. 3. Unfiltered (black) and filtered (white) motor current data.

maximum depth of history for both motor current and joint position, which should be taken into account, is four. It is, however, not obligatory to work with maximum possible number of samples. It was tried to use as few past samples as possible because with the number of inputs the complexity of the network increases considerably. Experiments have been performed in such a way that the depth of the past samples was continually increased. Sufficiently small error between the measured and the learned data was attained with the of depth of two for both current ,and position. This means that the number of inputs to the GCMAC is five: two samples of the joint position and three samples of the motor current. There is just one output corresponding to the joint position.

The GCMAC worked with the following parameters obtained experimentally:
generalization number C=4
the learning rate b_j=0.1 for all j
resolution r_i=0.025 for the samples of current and
r_i=0.0025 for the samples of position.

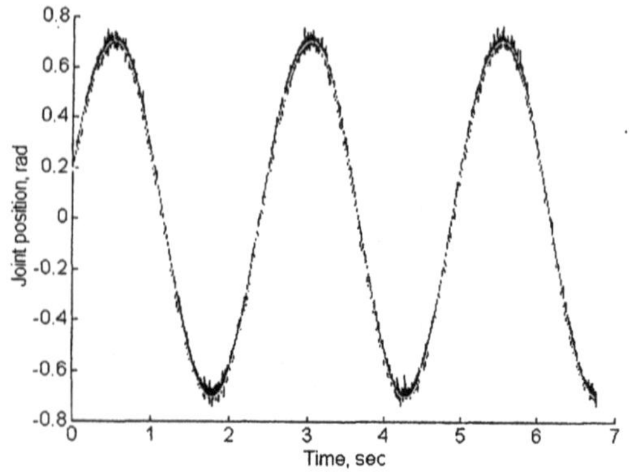

Fig. 4. Measured filtered (white) and learnt (black) joint position.

the output of the GCMAC neural network closely follow the measured and filtered data in straight line sections while in peaks of the graph the outputs of the GCMAC are slightly scattered around the measured data. Convergence of the learning proces characterized by the development of the sum of square errors is in Fig.5.

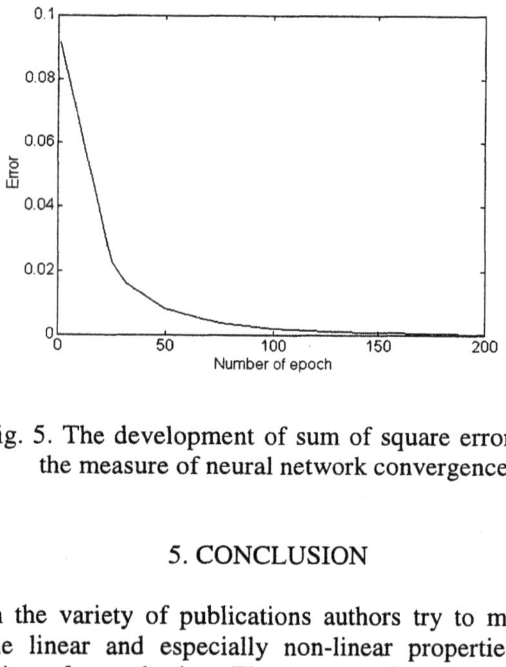

Fig. 5. The development of sum of square errors as the measure of neural network convergence

5. CONCLUSION

In the variety of publications authors try to model the linear and especially non-linear properties of joints of geared robots The purpose of these works is mainly to use the attained models for the compensation of these phenomena. Another goal may be the torque control of robots important, e.g., at contact tasks. Models of joints differ according to the purpose for which they are prepared. In difference to the model presented here, the model of the transfer of torque from the motor shaft to the output of the robot joint could be solved, e.g., in the same way as in this article.

In the present work the neural network model is built with the motor current as the input variable and position of the robot joint as the output variable. This model includes the linear as well as non-linear behaviour of the joint, the latter caused mainly by friction but possibly also by other effects. The non-parametric model corresponds with high measure of precision to the measured and at the same time filtered data. This work shows, that it is possible to model the properties of robot joints in the proposed way. The significance of this work consists mainly in introducing and proving the possibility of this kind of modelling.

Furher work in this direction will be concentrated on improving some steps in the proposed way of modelling, one goal being improved signal processing (filtration) of data, where the signal processing algorithms are more integrated with the algorithms of network learning. Another goal is building robot joint models for other purposes.

Acknowledgments The authors would like to express their gratitude to Mr. Martin Prüfer and Mr. Carsten Schmidt from the Institute for Robotics and Process Control (Prof. Friedrich Wahl), Technical University of Braunschweig, FRG for the preparation and rendering of data used in this work.
The authors do also appreciate the assistance of Mr. Rene Jager from TU Delft, Netherlands with the program GCMAC v. 3.0 developped by him and his willingness to agree with utilizing this program by the authors of the present article.
The first author prepared this work due to the grant No. 2/1279/97 financed by the Scientific Grant Agency (VEGA) and the second author due to grant No.1/4060/97 equally financed by VEGA.

REFERENCES

Albus, J.S. (1975a). A New Approach to Manipulator Control: The Cerebellar Model Articulation Controller (CMAC). *Transactions of the ASME, Journal of Dynamic Systems, Measurement, and Control,* **Vol. 97**, pp. 220-227.

Albus, J.S. (1975b). Data Storage in the Cerebellar Model Articulation Controller. *Transactions of the ASME, Journal of Dynamic Systems, Measurement, and Control,* **Vol. 97**, pp. 228-232.

Chen, F.-Ch. and Ch.-H. Chang (1996). Practical Stability Issues in CMAC Neural Network Control Systems. *IEEE Transactions on Control Systems Technology,* **Vol. 4**, pp. 86-91.

Eskanderian, A., N.E. Bedewi, B.M. Kramer, A.J. Barbera (1996). Dynamic Modeling of Robotic Manipulators Using Artificial Neural Network. *Journal of Robotic Systems,* **Vol. 11**, pp. 41-56.

Gardner, J.F., A. Brandt, G. Luecke (1993). Applications of Neural Networks for Coordinate Transformations in Robotics. *Journal of Intelligent and Robotic Systems,* **Vol. 8**, pp. 361-373

Journal of Intelligent and Robotic Systems, **Vol. 8**, pp. 361-373

Katic, D.M., M. Vukobratovic (1995). Highly Efficient Robot Dynamics Learning by Decomposed Connectionist Feedforward Control Structure. *IEEE Transactions on Systems, Man, and Cybernetics*, **Vol. 25**, pp. 145-158.

Krijgsman, A.J. (1993). *Artificial Intelligence in Real-Time Control. Ph.D. Thesis.* Delft University of Technology, Delft, Netherlands.

Larsen, G.A., S. Cetinkunt, A. Donmez (1995). CMAC Neural Control for High Precision Motion Control in the Presence of Large Friction. *Transactions of the ASME, Journal of Dynamic Systems, Measurement and Control*, **Vol. 117,** pp. 415-420.

Lee, S., G.A. Bekey, (1991). *Applications of Neural Networks to Robotics. In: Control and Dynamic Systems* (Leondes (Ed.)), **Vol. 39**, pp. 1-69. Academic Press, Inc.

Miller, W.T. (1990). Real-Time Application of Neural Networks for Sensor-Based Control of Robots. *IEEE Transactions on System, Man, and Cybernetics,* **Vol. 19**, pp.131-136.

Sanchez, V.D., G. Hirzinger (1992). The State of the Art of Robot Learning Control Using Artificial Neural Networks - An Overview. In: *The Robotics Review 2* (Khatib, O., J.J. Craig, T. Lozano-Perez (Ed.), pp. 261-283. The MIT Press, Cambridge, MA,.

Williams, D., O. Khatib (1996). Wrist Force Sensor Feedback for Improved Actuation Performance in Conventional Arms. In: *Proceedings of the IFAC 13th Triennal World Congress, pp.19-24,* San Francisco

Yeung, D.-Y., G.A. Bekey (1989). Using Context-Sensitive Learning Network for Robot Arm Control. In: *Proceedings of the Int. Conf. on Robotics and Automation,* IEEE Computer Society Press, pp. 142-148.

Yeung, D.-Y., G.A. Bekey (1993). On Reducing Learning Time in Context-Dependent Mappings. *IEEE Transactions on Neural Networks,* **Vol. 4**, pp. 31-42.

DEVELOPMENT OF INTELLIGENT SYSTEM DYNAMIC MODEL

K.A. Pupkov, V.N. Pilishkin.

Bauman Moscow State Technical University, Department of Automatic Control Systems,
5, 2nd Baumanskaya str., Moscow, Russia

Abstract: Synthesis of intelligent system (IS) dynamic model is considered. Intelligent transducer (IT) is introduced as the element of IS structure used for formation of the control signal. IT is considered as an interconnection of active and passive parts characterizing dynamics that is used for providing of control task. According to the operating concept, control task is presented in the form of phase restrictions that is formed on the base of structural-algorithmic operating approach. Three-level model of IT that consists of dynamic, forecasting and reproduction levels is offered. Synthesis of the levels and corresponding control laws proceeds simultaneously on the base of phase restriction technique. IS model is considered to be formed, if desired control for given task has been chosen and IT-model has been designed.

Keywords: Intelligent systems, intelligent control, control laws, dynamic models, phase restrictions

1. INTRODUCTION

One of the modern trends in automatic control theory is the theory of intellectual systems that includes formation of system structure and operating algorithms in general, developing of structural elements for IS and interaction techniques with environment, creating algorithms for information processing, ets. IS defined as systems that operate on the base of certain knowledge, data and information, make decisions and implement ones in the rational way, provided that target (motivation) is clearly defined (Pupkov, 1991). Problems of IS design and structure are considered by Pupkov (1994) and Pilishkin (1994). This paper is devoted to the problems of IS dynamic model design on the base of IS structural-algorithmic operating approach .

2. INTELLIGENT SYSTEM DESIGN

The IS synthesizing concept (Pupkov, 1991) is used for structural representation of IS (fig.1) termed as IS analytical structure (Pilishkin,1994). IT is used for task analysis and developing operations, demanded for gaining this task.

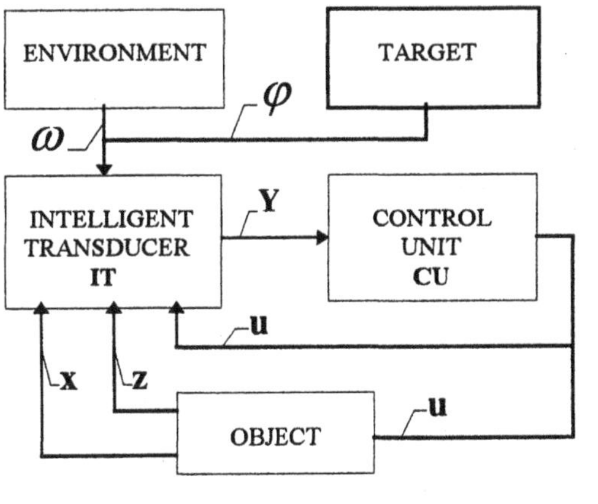

Fig. 1. Analytical structure of IS.
x, u, Y, w - n x 1, m x 1, n1 x 1, r x 1 object, control, IT, disturbance state vectors respectively; φ - target vector; z - object parameters vector.

Object state equations are the following:

$$x= f(x,u,w,z,t),\ x(t0)=x0;\ t>=t0, \tag{1}$$

f(*)- n x 1 given vector - function.

State equations of object could be presented as:

$$\dot{Y} = \tilde{f}(Y,x,u,\tilde{u},w,j,z,t),$$
$$Y(t_0) = Y_0, t\ t_0 \tag{2}$$

$\tilde{f}$ (*) - $n_1 \times 1$ apriori unknown vector - function defined during synthesis of control law;
$\tilde{u} - m_1 \times 1$ control vector;
IT is considered to be the unit, that consists of active and passive elements. Passive element is defined by dynamic characteristics of IT. Active element is used for forming additional signal, that modifies dynamic signal in order to gain certain control task. Therefore,

$$Y=F(Y_a, Y_p) \tag{3}$$

where Y_a - signal from active element, Y_p- signal from passive element, F(*) - given n_1 x 1 vector-function. These signals chose independently that can be described by corresponding equations. In this case above mentioned equation can be presented as (2)
Restrictions:

$$u \in U(t), w \in W(t), z \in Z(t), t \ge t_0 \tag{4}$$

U,W,Z - apriori known sets.
It is offered to build IT equations in the following form using (1),(2):

$$\dot{X} = F(X,U,w,\varphi,z,t),$$
$$X(t_0) = X_0, t \ge t_0 \tag{5}$$

F(*) -N x 1 vector - function formed in accordance with F(*)=Ô(f(*),f(*)),
Ô(*) - certain given vector - function N x 1 (probably operator in common form)
In particular $\Phi(f,\tilde{f}) = [f^{\mathrm{T}}, \tilde{f}^{\mathrm{T}}]^{\mathrm{T}}$.
X - N x 1 IS state vector, U - M x 1 IS control vector defined as

$$X = [x^{\mathrm{T}} Y^{\mathrm{T}}]^{\mathrm{T}},$$
$$U = [u^{\mathrm{T}} \tilde{u}^{\mathrm{T}}]^{\mathrm{T}}.$$

Certain target C must be provided by IS (5)on the base of corresponding model of IT (i.e. function f(*)) and acceptable control law. It is offered to realize formalization of target C in accordance with Functional - Set-Belonging (FSB) Concept introduced by authors.

3. FORMATION OF THE TASK FOR INTELLIGENT SYSTEM.

If dynamic system could be represented by state - space model then arbitrary target C can be formalized by introducing state - vector limitations that define the character of the state - vector belonging to the certain set during each time period. It is possible to consider that target C (5) could be formalized and represented as:

$$C{:}\Psi(y,t) \le 0,$$
$$y = \nu(X,t), \tag{6}$$
$$t \ge t_0$$

where $\Psi(*)$ – given scalar function, $\nu(*) - N_1 \times 1$ given vector - function (in particular if $N_1 = 1$, then (5) conform to $\Psi_1(X,t) \le 0$).
More common approach to target C formalization has been developed by authors on the base of introducing of sets ε-neibourhood. ε-neibourhood of Q considered as the following set:

$$Q_\varepsilon = \{X \in R^N{:}\rho(X,Q) \le \varepsilon\}$$

where unnegative scalar function $\rho(X,Q)$ treated as the distance between vector X and set Q. It is possible to set both targets in the form (5) and more common using the conception of ε-neibourhood.

For creating IS model according to fig.1 models of IT and CU required. Several models of above mentioned units are introduced. Functional and operating characteristics taken into account. In this way, if vector Y contains information about present target that govern to putting into effect target C, then CU must produce control for present target implementation.

IT dynamic model represented by (2) could be interpreted as:

$$\dot{Y} = f_t(Y,x,u,w,\varphi,z,t); Y(t_0) = Y_0; t \ge t_0 \tag{7}$$

$n_1 \times 1$ vector function $f_t(*)$ - time depend function.
Generally speaking, model (7) could be used instead of (2) in this approach.

CU dynamic model (limitations are not taken into consideration in this case)can be represented as the following:

$$\tilde{u} = \tilde{K}(Y,x) = K(X), \tag{8}$$

K-certain operator in the form given.
Then control U should be formed like

$$U=K(X,t), \quad (9)$$

where K(*) - M x 1 vector-function searched.

4. STRUCTURAL - ALGORITHMIC OPERATING APPROACH.

Within generalized IS synthesis concept (Pupkov, 1991) structural - algorithmic approach has been offered by authors. This approach involves introducing of so-called Intelligent Environment (IE) according to the state-space H of Intelligent System IS. IE is the set of elements S interrelated by certain conditions. In particular, IE is treated as some kind of entity (linear or unlinear) - IE can be considered as an hipersurface belonged to certain space or as certain unlinear space. Target C realization is analyzed on the base of IE elements.

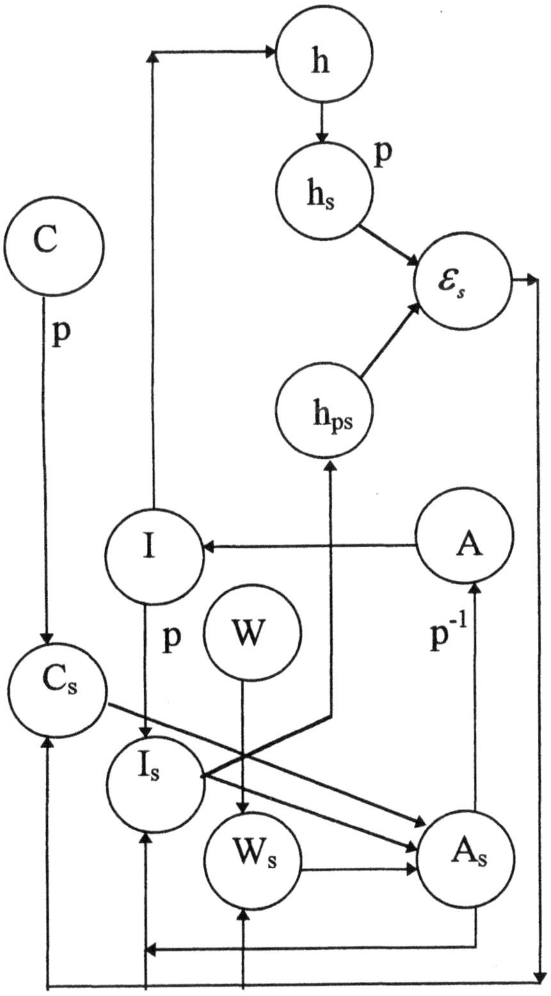

Fig.2. Structural-algorithmic operating approach.
A_s, A - control algorithms I_s, I;
h_{ps}, h, h_s - predicted, real, and mapping of real system dynamic.

It is offered to construct IT(2),(7) and CU (8) models and, therefore, total IS taking into consideration interrelation between state-space H and IE. Interacting of IE and H is realized on the base of the following technique. Certain operator P is used for mapping IS (I), target (C) and environment (W) models onto IE . The result is I_s, C_s, W_s that are used for forming IT_s and CU_s prototype models on the base of synthesis techniques (phase limiting technique(Pilishkin, 1994), (Q,ρ) - state -space subdivision technique) introduced by Pilishkin (1995). IE models and algorithms are mapped into H with the help of P^{-1} operator.
Above mentioned algorithm is illustrated by fig.2.

IS considered to be fixed (given) if target C has been fixed (formed)(in the form (5), for example), so IT(2), (7), and CU (8) models greatly depends on target C chosen. Therefore, IS must be considered as the "couple" <I,S>. In this case could be realized both synthesis of "couple" <I,C> and synthesis of I provided that C given. The later case will be considered.

Functions f(*), $\tilde{f}(*)$ or $f_t(*)$ will be treated as dynamic forms of object, IT, IS respectively. Then intelligent control of object is realized by forming fixed dynamic form $\tilde{f}(*)$ and control law (9), or by continuos forming of $f_t(*)$ and control law (9) for each time moment $t \geq t_0$.

5. THREE-LEVEL MODEL OF INTELLIGENT TRANSDUCER FOR SOLVING CONTROL TASKS

Intelligent control process is realized according to three-level IT model and structural-algorithmic approach offered by authors.

First level : Synthesis of dynamic form $f_t(*)$ or $\tilde{f}(*)$ and corresponding control law u(*)=K(X,t) provided that target C realization.
Is into IE is the following:

$$\begin{aligned} &\dot{s} = F^s(s,u^s,w^s,\varphi^s,z^s,t^s), \\ &s(t_0^s) = s_0; t^s \geq t_0^s \\ &u^s \in U_s, w^s \in W_s, z^s \in Z_s \end{aligned} \quad (10)$$

Generally, t_s of IE differs from t of H, (for example t_s - quick time).
Therefore, according to (10), synthesis of $u_s(*)$ and $\tilde{f}^s(*)$ or $f_{t^s}^s(*)$ forms is realized on the base of phase limitations technique (Pilishkin, 1994). Distinction feature of this method is that state vector Y must be formed according to:

$$P^{-1}[s(*)] = X(*) = [x^{\mathrm{T}}(*)Y^{\mathrm{T}}(*)]^{\mathrm{T}}, \quad (11)$$

i.e. after defining of trace s(*) into IE, vector Y(*) used in control algorithm is regenerated according to (11). If IE has not been defined (or fixed) for system (4), then first-level task conform to the conventional task of synthesis of automatic control system with the help of sequential dynamic correction unit.

Second level : Analysis of control law u(*) and dynamic forms $f(*)$ or $f_t(*)$ efficiency on the certain time period T_s. If tis time period contains time moments t_s not arrived yet, then on the second level the task of prediction IS behavior is also solved. On this level the model of IS must be designed for accurate approximation of IS real trace prototype. Real trace of IS with the help of us(*), $\tilde{f}^s(*)$ (or $f_t^s(*)$) chosen.

Approximation model is the following:

$$\dot{s}^a = F_a^s(s^a, u^s, u_a^s, w^s, \varphi^s, z^s, t^s);$$
$$s^a(t_0^s) = s_o^a; t^s \geq t_o^s \quad (12)$$

Where u_a^s - approximation impact and $F_u^s(*)$ - vector-function have chosen on the base of required approximation accuracy on the time period T_s.

Third level: Correction of control law $u^s(*)$ and dynamic form $f^s(*)$ is realized on this level in case of their inefficiency on T_s. For solving correction task IS model is formed on the base of (10),(12) in order to reproduce IS trace prototype (4) with required accuracy on certain time period T_s^b and for chosen $u^s(*)$, $u_a^s(*)$, $f^s(*)$.

Reproducing model

$$\dot{s}^b = F_b^s(s^b, u^s, u_a^s, u_b^s, w^s, \varphi^s, z^s, t^s)$$
$$s^b(t_0^s) = s_0^b; t^s \geq t_0^s, \quad (13)$$

u_b^s - reproducing influence F_b^s-vector-function according to given limitations. Intelligent control of object involves organization of interaction of IS levels. Synthesis of each level could be realized on the base of phase limitation technique. Generalized approach has been developed. According to this technique, IS synthesis problem will be solved when:

$$(\nabla_y \Psi, \nabla_X v \cdot F(*) + \frac{\partial v}{\partial t}) + \frac{\partial \Psi}{\partial t} \leq 0,$$
$$\forall y \in \Gamma Q(t), \forall X \in M(y,t); \quad (14)$$
$$t \geq t_0$$

where (*,*) -scalar multiplication in R^N; $\nabla_y \Psi, \nabla_X V$ - gradient of $\Psi(*)$ and Jacobean of function v(*).

$$\Gamma Q(t) = \{y \in R^{N_1} : \Psi(y,t) = 0\},$$
$$M(y,t) = \{X \in R^N : v(X,t) = y\}. \quad (15)$$

For solving (14) different analytical procedures can be used. It is effective to take into consideration control law restrictions: it is necessary to solve min-max mathematical programming task if above mentioned restrictions exist, and max-min task if there is no restrictions of control law.

6. CONCLUSION

Approach had been offered is used for solving wide range of practical tasks connected with synthesis of effective control laws for dynamic objects. Objects under consideration (like spacecraft and aircraft) operate under different conditions and information about their characteristics and environmental impact is usually ill-defined. For this class of objects intelligent system has been designed on the base of two-level (dynamic and forecasting) intelligent transducer. Furthermore, new approach connected with effective synthesis of three-level intelligent transducer has been offered. This approach provides effective interaction of levels and can be used for control of different types of objects

REFERENCES

Pilishkin, V.N. (1994) Phase restriction technique for intelligent system control. *University news, Electronic equipment design*, **No. 9-106**, pp.8-15.

Pilishkin, V.N. (1995). Intelligent system control on the base of functional set-belonging concept. *Problems of radioelectronics*, **No. 3**, pp. 16-21.

Pupkov, K.A. (1991) Scientific program: Intelligent systems. In: *Scientific technological and educational programs and projects in USSR for the period 1989-1994* pp. 198-214. USSR State Committee for Education, Moscow.

Pupkov, K. A. (1994) Theoretical and practical problems of intelligent systems. *Proc. First international symposia "Intelligent systems - 94"*, Russia, Dagestan.

APPLICATION OF GENETIC ALGORITHMS IN CONTROL

J. Hrabovec, I. Hantuch

STU, FEI, Department of Automation and Control, Bratislava

Abstract: Genetic algorithms represents new approach and programming techniques inspired by the natural evolution process that uses raw computer power for solving problems in field of static optimalisation and operational research. The algorithm is independent from solved function or problem definition and it's suitable for solving loosely defined problems, problems where algorithmical solution is unknown or in situations where other methods fails or became unstable from any reason. This work describes how genetic algorithm works and contains simple algorithm implementation for search of function's extremes.

Keywords: Genetic Algorithm, Optimisation, Operational Research, Application

1. GENETIC ALGORITHMS

There is a large number of problems in fields of mathematics, computing and in applied sciences like operational research and data processing, that couldn't be optimally solved in the past. The reasons were:

- algorithmical solution wasn't known at all
- the problem was defined too loosely to use mathematical methods
- known algorithms were not able to find solutions in reliable time or using accessible resources

Very often only a suboptimal solution was find but sufficient in the practise. For illustration there are problems from static optimisation and operational research:

1.1 Static optimisation

There is no known versatile algorithm that can find global extreme for all specified functions. For example all gradient methods fail when function has some absolute values inside, simplex methods can be 'catch' by a local extreme and some other methods became often numerically unstable or their results vary depending on initial conditions. The function can have non-continuous definition space as well, what brings some other difficulties. Different methods are suitable for different types of functions, where very often human intelligence and experiences are needed to decide which one is the best to use at the moment.

1.2 Operational Research

There are several algorithms for operational research problems, network analysis, production and resource management and so on. These methods work well on small and simple models, their complexity quickly grows on more complicated models and very often they are useless for highly complex models that correspond with real situations, especially when the structure of the model changes or is not known in the time of method implementation. Models from real world usually have a huge number of parameters of different types, many constraints that can not be broken, result mostly depends on many changing priorities. From this reasons exact structure of the model, it's parameters and constraints can't be part of the core of algorithm.

1.3 The Idea of Evolution

As it could be though, the name 'genetic algorithms' doesn't come from genetic laws, but from how genotype[1] changes through the **evolution** process. Two main principles of evolution are **natural selection** and **adaptability**. Very important condition for natural selection to work is that there are always new and new genotypes found to select from. The new genotypes are created from old ones in two main ways: **mutations** of some or all genes of one genotype and **crossings-over** of genes from two different genotypes (*Ferák and Sršeň, 1990*). The main differences between this two ways are (*Fig. 1*):

- Mutations of genes are spontaneous or caused by some external reason (e.g. radiation, some chemical agent) and are of different range. One or more genes from **one** original genotype changes, so in resulting genotypes we probably will find some new genes that can't be found in original genotype
- Crossing-over (crossing) of genes takes place in sexual replication process, what mean that **two** organisms must be involved in this process. In resulting genotype we can find only genes that can be found in original genotypes of this two organisms but **combined in different way**

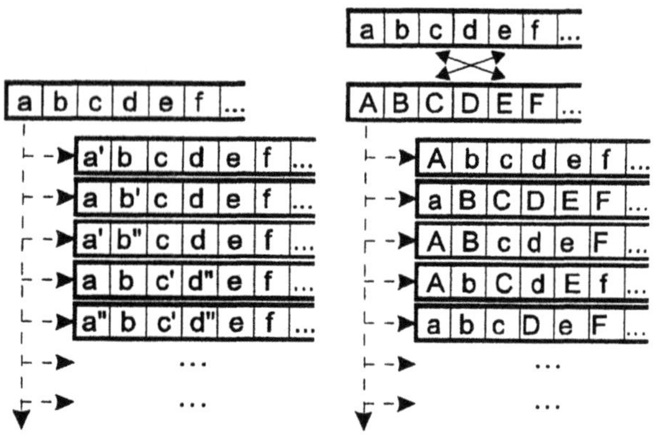

Fig. 1 Mutation and crossing-over

In each stage of evolution process only a couple of best[2] organisms (or organism types) **survives** the stage, or in other words, only genes from couple of best organisms (best genotypes) are involved in creation of organisms in next stage (natural selection).

This process runs continuously during the history of life on the Earth, but for better understanding only a discrete **generations** are taken.

1.4 Analogy with Evolution

The idea of genetic algorithms is based on analogy with evolution process.

- One organism is represented by one (any) solution (result). This can be a value on the function in mathematical problems, network throughput in network analysis, production expenses in production management, overall transporting distance in distribution management problems etc.
- The selection of 'good' organisms corresponds to selecting maximal or minimal values regarding on kind of searched extreme (maximal network throughput, minimal expenses and travelling distance etc.). Regardless of what kind of problem is solved there must be some criteria that can be useed for comparing two results
- Genotype of the organism corresponds with set of input data, for example function parameter vector, set of throughputs between nodes in network, set of assignments between supplying depot and customer etc.
- Particular genes of a genotype corresponds with particular items from set of input data, for example particular function parameters

1.5 Function Extremes

One of the areas where to use genetic algorithms is search of function maximum or minimum, especially of very complex non-linear functions, with absolute values inside and non-continuous definition space etc. Problem of searching for function roots can be transformed to problem of function minimum as well. Use of genetic algorithms in this case can be very often the only reliable method.

The basic application of the method is very simple. In the first step function value will be calculated in a large number of points of the function's definition space. The set of function values in these points is the initial generation and corresponds with different values of function parameters (parameter vectors). In the next step will be selected a couple of the best (the highest or the lowest) values and from these points will be created next generation by slightly modifying (mutating) some or all parameters and crossing some or all parameters together from two vectors[3]. Then the selection in this new generation will be repeated and so on.

If the function is to have non-continuous definition space it is enough for the algorithm to work well if function returns relatively low (if searching for

1 Genotype is set of genes of some type of organism. Organism qualities and characteristics are strongly tied to it's genotype

[2] 'best' in terms of organism ability to survive

[3] For example by averaging parameters from two parameter vectors

maximum) or relatively high values (if searching for minimum) in points where it is not defined. The rest of algorithm then treats these points as bad solutions.

The chance that this algorithmical 'evolution' process will be 'catch' by a local extreme is very low, although not equal to zero and it depends on given search parameters. As with some other methods the precision of the result grows and the process is slowing down with the number of times the function value is calculated.

1.6 Non-numerical problems

The great possibilities to employ genetic algorithms are in field of operational research. The implementation of the algorithm is more complicated as in numerical problems (see above), but for a large set of problems this method remains the only one that is useable.

As an example a process of scheduling fuel distribution is presented[4]:
The aim is to optimise schedule of fuel distribution from point of view of economical and business needs on day-by-day basis. Scheduling process itself is restricted by number of constraints that come from relationships in real world, business needs, legal rules etc. Some of them are obvious (one driver can drive only one tanker at a time). For illustration the constraints following can apply on described problem:

- Constraints applying to customer's site: presence of lights on site enabling night delivery, maximum tanker size that fits the site, maximum number of tankers that can serve the site, opening hours of the site, delivery hours of the site, delivery on the site can be controlled by the driver
- Constraints applying to supplying fuel depot: opening and closing hours, lights on the depot, automatic operation, number of tankers and drivers available, shifts configuration, load and ancillary time
- Constraints applying to driver: maximum shift length, necessary driver breaks (legal restriction)
- Constraints applying to tanker: weight and volume limits, safety constraints (mixing of some types of fuels on one tanker is not allowed), settling time, tanker pot configuration

Along with these constraints there are business preferences that influence scheduling process. Among other these can be: missing minimum desired drawings from a depot, changing a shift starting time, minimising driver's over-times, cutting down an order, using a night shift, using preferred type of tanker, achieving a time request specified on order etc.

This illustrates that scheduling such a process is really a serious task. Along with complexity of the problem itself we have to take into account size of data to be processed. System held information about tens of thousands of customers, hundreds of tankers and drivers and tens of depots and every day it processes more than thousand of orders. Only operational research conected with sophisticated optimising algorithms enabled to develop system that was fast enough to be used in day-by-day operation. As mentioned above this was done by **genetic algorithm**. Principal structure of integrating such a system into distribution process is shown on Fig. 2.

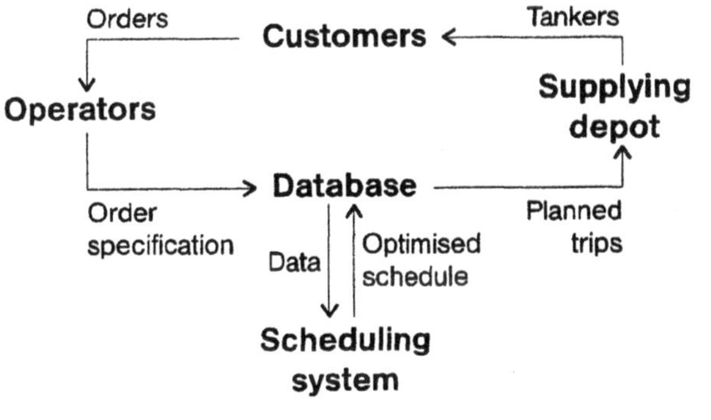

Fig. 2 Scheduling system

Implementation of scheduling system into described problem of optimising distribution process can have two main parts (*Fig. 3*):

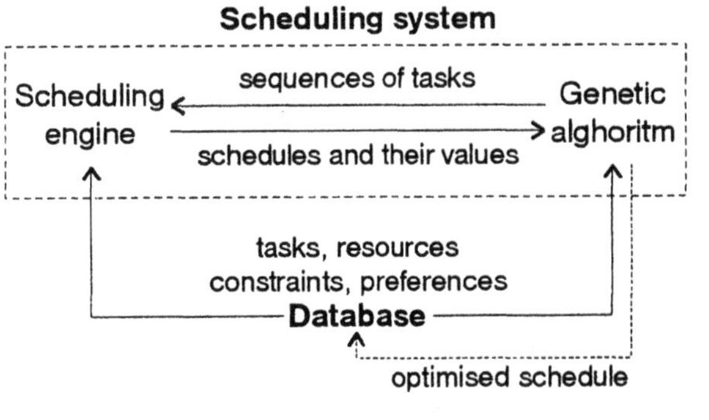

Fig. 3 Non-numerical problems

1. *Scheduler* - so called "scheduling engine", takes as input given set of available resources, required tasks (e.g. customers orders) and user defined

[4] System presented called 'TankerPOWER' was developed by International Computers Limited (Slough, England) for Texaco Limited (Swindon, England) (*International Computers Limited and Texaco Limited, 1993; International Computers Limited, 1994*)

constraints and allocates these resources to tasks while not breaking given constraints. Mostly scheduling engine uses recursive algorithms to do this and it starts with allocating the first resource to the first task. That effectively means that result depends on order in which tasks (or resources) are defined. Usually resources are fixed part of input data and order of tasks changes. Along with the allocation scheduler also evaluates each resulting schedule so it can be compared with others. The value of the schedule mostly correspond with expenses of the schedule but there can be some other priorities (e.g. business preferences) that scheduler takes into account when evaluating the schedule.

2. *Genetic algorithm* - produces a sets of tasks that sends to scheduling engine. These sets of tasks are different permutations of all defined tasks and from point of view of genetic algorithm they corresponds with different genotypes. As these sets are passed to the scheduler, different results are returned. Obviously just a small part of all possible permutations can be produced so it's the best to produce sets that are not similar to each other. Genetic algorithm then selects a couple of best results and produces next generation from these best genotypes. While genotype is represented not with a vector of numbers but with a sequence of non-numerical values (tasks), mutation and crossing of genotypes can not be applied in traditional way as in numerical problems. Instead of this genetic algorithm changes sequences by swapping consequent tasks, rotating sequences and so on. This new generation is then passed again to scheduling engine and process repeats until given time is consumed or results are not getting better from generation to generation. This part is usually simpler than implementation of scheduling engine. While the scheduling engine must be constructed for given type of problem, implementation of genetic algorithm for non-numeric problems is not that simple as for numerical problems and mostly requires different programs for different types of problems.

2. ALGORITHM IMPLEMENTATION - FUNCTION EXTREMES

Genetic algorithm was simply implemented for search of extremes of function of more variables in Pascal programming language and run on PC computer under Windows NT.

2.1 Evolution constants and parameters

Genotype dimension (number of parameters) *Dim*
Size of the generation created by mutations *MGen*
Number of selected 'best' genotypes for mutation $MBest\ (MBest \leq \frac{1}{2} MGen)$
Initial parameter mutation range *Range*
Number of selected 'best' genotypes for crossing *CBest* (>2)
Required precision (stop condition) *Eps*

2.2 Data structures

One of the most critical parts of the program is choosing only best cells (genotypes) from all computed cells. While there is a huge number of all computed cells from which only a small part is important for future calculations, it is best to have at each moment only heap of best cells held in memory. Each cell is storing vector of parameters *ParVect*, cell value (function value for ParVect input) *Value* and pointers to previous and next cells in heap. Heap of best cells is represented as bi-directionally chained dynamic linear list of records (*Fig.* 4). This heap is created at once during program initialisation from empty cells. Program held only pointers to the beginning and end of the list (FIRST, LAST). For moving around the list program needs to start at any end of the list and then following NEXT and PREV pointers it can move through it.

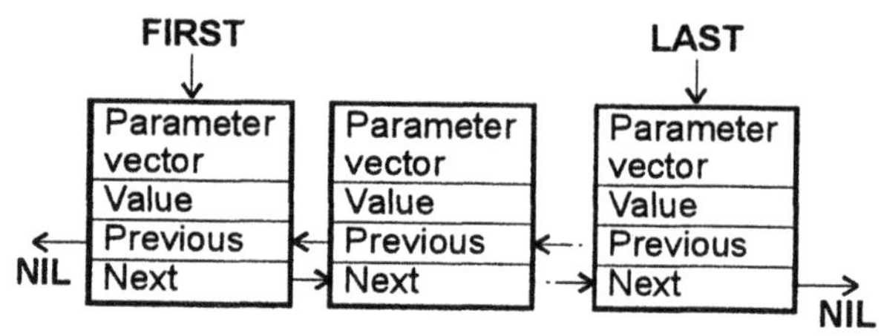

Fig. 4 Heap of best cells

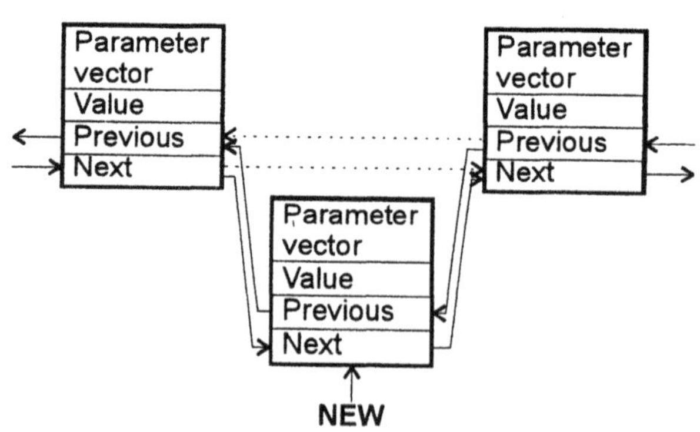

Fig. 5 New cell insertion

This heap is always sorted, so all program need to do is compare every new calculated cell with it's end. If this new cell is worse than end of the heap the cell can throw away, otherwise it is inserted into it's proper position in the heap. Insertion of the item into a dynamic linear list is very simple and requires only a couple of assignments (*Fig. 5*). To maintain constant heap size last cell from the heap must be then deleted.

2.3 Function representation

For maximum algorithm efficiency the solved function is implemented as part of the program and must be written into specified area in the source code as function with input array of numbers. Any parser analysing the function given as parameter would be more complicated than all rest of the algorithm and program in this case would spend much more then half of the time parsing the function.

If the function has non-continuous definition space the function should return any but relatively large (when searching for minimum) or relatively small number (when searching for maximum) in points where it is not defined.

2.4 How it works

Ideas of the genetic algorithms are implemented into the program in two main ways: mutations and crossings. Mutations can work separately and mostly they lead quickly to the result, but using crossings along them can help algorithm to "tear" from local extremes and start new branches of the search in between old ones.

In the first step program generates two empty heaps (see above) with very low (high) values for search of maximum (minimum). This two heaps are called *Target* and *Source* heaps. Let this call *initial generation*. Then program mutates zero cell, thus generating rectangular set of cells with length in both directions from zero of each dimension determined by parameter *Range* (*Fig. 6*). This is called *zero generation*. During creation of the zero generation cells are evaluated against target heap and the best ones are inserted into it. From this point starts cyclic process of mutations and crossings.

Mutations

At the beginning of the cycle target heap is copied into source heap to enable taking cells from one heap and inserting results into another one. The first *MBest* cells from the source heap are mutated. Mutation of the cell creates rectangular set of new cells around the old cell, covering area to it's nearest neighbours, because *Range* parameter is equal to *Step* value (mutation step) from previous generation. In this way *Range* value decreases exponentially from it's initial value thus enabling us to set initial search range to very large number. Generated cells are evaluated against target heap and when appropriate they are inserted into it.

Crossings

When crossing cells, all possible pair combinations from the first *CBest* cells from source heap are averaged together (*Fig. 6*). When switch *Whole* is specified, all parameters are averaged together otherwise all possible combinations of original and averaged parameters are created for each pair. New cells are again compared with target heap and inserted into it.

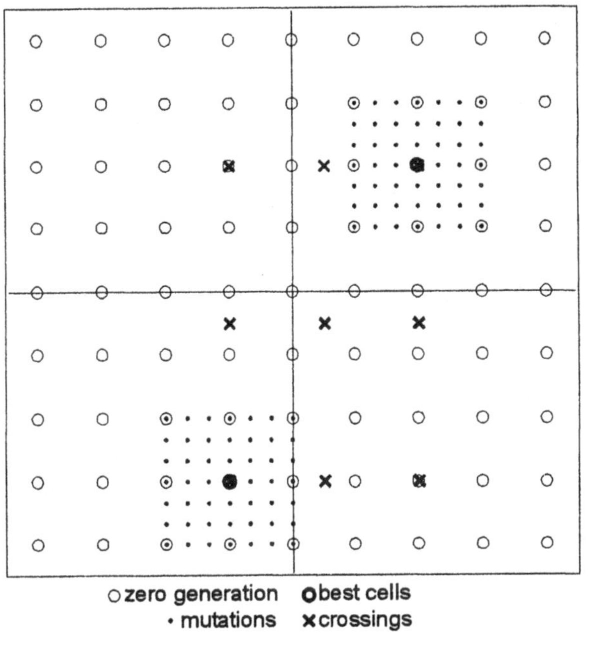

Fig. 6 Mutations and crossings for function of two variables

This process runs during stop condition is fulfilled. This condition can be that the new result differs from the last one less than *Eps* value but more than zero or *Range* is much less than *Eps*.

2.5 Results

To highlight advantages of the genetic algorithm, program was tested on functions that have many local extremes that are very similar to the global one. These functions use absolute values inside what brings difficulties with other methods. When searching for maximum, function of one variable can be for example of following type:

$$f(x) = sin\,(x) - \left|\frac{x-a}{b}\right| \tag{1}$$

Testing shows that program is quite independent from initial range and starting point, which allows to use one (any) starting point for all functions and very large initial range (e.g. 10^{10}). When local extremes are very close to each other, program sometimes can find only some local extreme, but this dramatically depends on the number of mutated and crossed cells. This generally but not always means that increasing this numbers reduces chances that algorithm is "catch" by some local extreme.

Except functions described above where local extremes are very close to each other algorithm works very well and mostly finds extreme in couple of seconds wherever the extreme is within it's initial range. Examining function of more variables noticeably slows down the process but still maintaining its versatility, robustness and efficiency (*Fig. 7, Fig. 8, Fig. 9*).

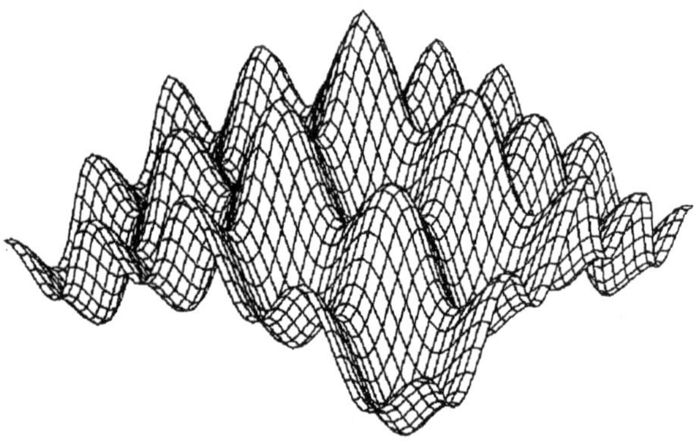

Fig. 7 Example of tested function of more variables with many local extremes

$$f(x,y) = a.cos(x) + b.cos(y) - |x/c| - |y/d|$$

Algorithm was previously successfuly used to determine parameters of thermal network model of inverter-fed induction motor (*Hrabovcová, 1992; Hrabovcová, 1994*). It was used to find the best fit between measured and calculated temperatures by changing six parameters of the network.

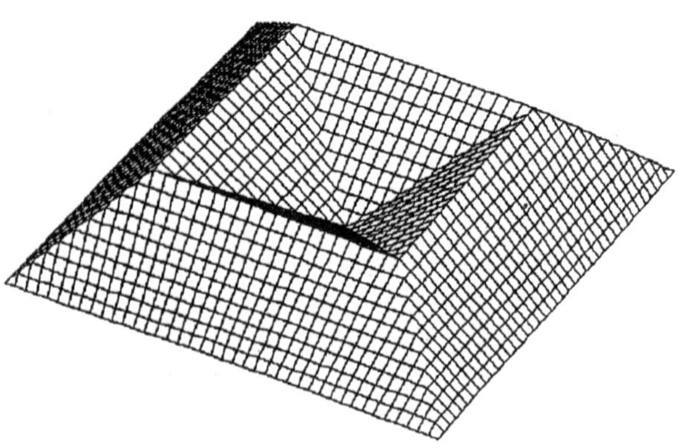

Fig. 8 Example of tested function of more variables with sharp edges $f(x,y) = \big|-|x+y|-|x+y|+a\big|$

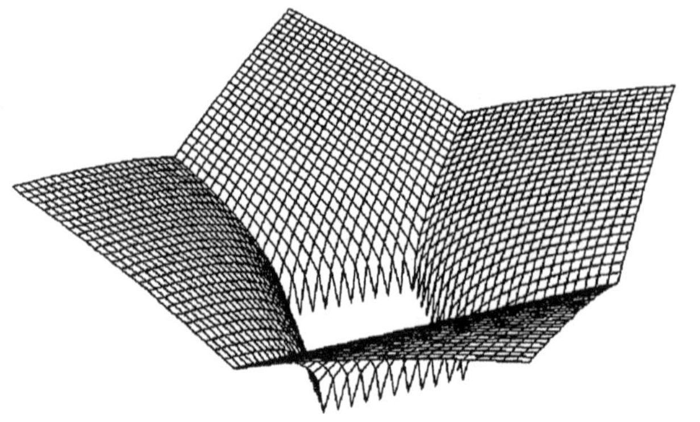

Fig. 9 Example of tested function of more variables with non-continuous definition space

$$f(x,y) = \sqrt{|x|+|y|-a}$$

3. CONCLUSION

A simple implementation of genetic algorithm for search of function's extremes of one or more variables was developed. Algorithm is intended for study and demonstration purposes although it can be used for searching for extreme of function of any type and complexity. Testing showed that algorithm mostly found the best or near to best solution in types of functions where other methods work incorrectly. The successfulness and performance of the method depends on given input parameters and some experimentation is necessary to gain the best results.

4. REFERENCES

Ferák, V., Sršeň, Š.: Genetika človeka. Slovenské pedagogické nakladateľstvo, Bratislava, 1990.

Hrabovcová,V.: Straty vo vinutiach asynchrónneho motora pri neharmonickom napájaní. Habilitačná práca, VŠDS Žilina, 1992

Hrabovcová, V.: Loss Distribution in an Inverter-fed Induction Motor. International Conference on Electric Drives and power Electronics (EDPE 94), October 18-20, 1994, the High Tatras, Slovakia, Proceedings, vol. 1, pp. 190-193

International Computers Limited, Texaco Limited: TankerPOWER system specification. International Computers Limited, Slough, England, 1993

International Computers Limited: TankerPOWER Release 1.0 User's Guide. International Computers Limited, Slough, England, 1994

ARTIFICIAL NEURAL NETWORK CONTROL DESIGN FOR CONTINUOUS FLOW STIRRED TANK REACTOR

Vladimír Popardovský

Institute of Control Theory and Robotics, Slovak Academy of Sciences
Dúbravska cesta 9, 842 37 Bratislava, Slovak Republic
E-mail: utrrpopa@nic.savba.sk

Abstract: This contribution shows artificial neural network (ANN) approach in system identification and control of the continuous flow stirred tank reactor (CSTR). The internal model control (IMC) strategy was used. Some simulation results are included.

Keywords: inverse model, forward model, backpropagation, IMC, CSTR.

1 INTRODUCTION

Chemical reactors are commonly used in the chemical industry. The reactor is a nonlinear process. In paper proposed approach to the CSTR control is based on using of ANN. The controlled object is a CSTR with first order exothermic reaction. Only the reactant temperature is measured and this variable represents the controlled output of the system.

2 PROCESS DESCRIPTION

We consider a CSTR in which an irreversible decomposition $A \xrightarrow{k} B$ proceeds. Assume that conversion is manipulated by adjusting the reactor temperature. The reaction temperature is controlled by the coolant temperature. Nonlinear model reactor is described by the following diferential equations of mass and heat balances.

$$V\frac{dc_A}{dt} = qc_{VA} - qc_A - Vkc_A$$

$$V\rho c_p \frac{dT}{dt} = q\rho c_p T_V + (-\Delta H_R)Vkc_A - q\rho c_p T - F\alpha(T - T_C)$$

$$k = k_0 e^{-\frac{E}{RT}}$$

Transfer function for linearizing model of CSTR is

$$F_{CSTR}(s) = \frac{b_1 s + b_0}{s^2 + a_1 s + a_0}$$

There was used following transfer function of the process (CSTR) for simulation experiments

$$F_{CSTR}(s) = \frac{0.09265s + 0.085839}{s^2 + 0.83982s + 0.1728988}$$

The reactor is assumed as SISO system with one input (T_C) and one output (T).

3 ARTIFICIAL NEURAL NETWORK CONTROLER DESIGN

3.1 Brief introduction to the theory of neural networks.

Neural networks, originaly inspirated by their biologicaly namesakes, are composed many intercomunicating elements (neurons), working a parallel to solve a problem.What makes them interesting is the fact that once a network has been set up, it can learn in a self-organizing way. In this sens, the internal behavior of a network is not designed - it grows as the network learns to operate properly based on a set of training data provided by the user. Once it has mastered the training data it can be turned loose on problems where the answers are not known. Because neural networks can be trained to respond in parallel to the inputs presented to them, they often can be much faster than more conventional methods. Basic functional unit of the artificial neural network is neuron. Neuron can be considered as a processor with n-inputs (synapses) and one output (axon), which performs relatively complicated mathematical operation. Input signals are multiplied by the positive or negative value (synaptic weight) in the synapses.. Then they are summed and the sum comes to a non-linearity (or linearity) which determines the intensity of the output signal. If the weighted sum of the inputs exceeds the threshold, neuron sends a signal which is close one. In the opposite case, output signal approaches to zero. Artificial neural networks are created via compiling the neurons into large integrates. Learning is performed on the so called training set, in which the input-otput relation for each element is known. Neural network can be understood as a function $y = T(x)$ adapted by weight and bias parameters for the training set. A neural network can have any number of layers, but networks with more than three are uncommon.

3.2 Control structure based on a model of the inverse dynamics.

A model of the inverse dynamics of the same process can be identified directly as shown in following figure.

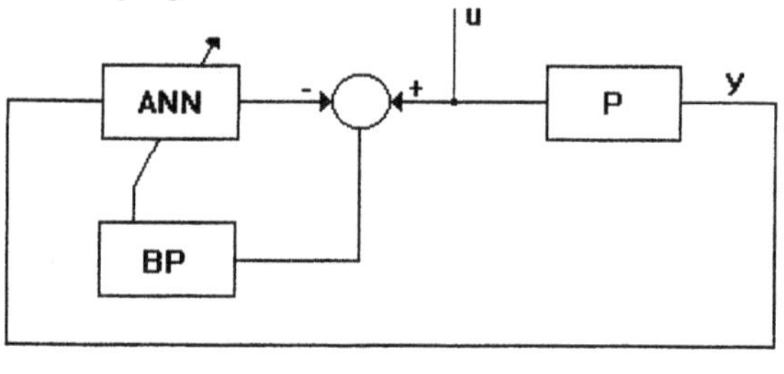

Fig.1. Direct identification of the inverse model

If the input to the ANN for the identification is now taken to be

$$I \cong [y(k+1); y(k);\ldots.; y(k-n); u(k-1);\ldots.; u(k-m)]^T$$

and the output is $u(k)$ then identified mapping F represents the inverse dynamics of the process.

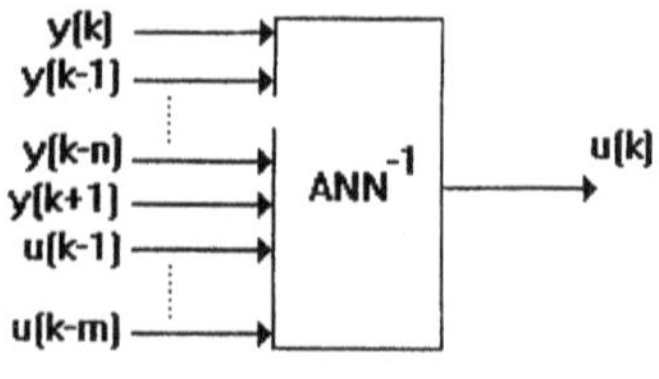

Fig.2. Inverse model

$$u(k) = F[y(k+1); y(k);\ldots.; y(k-n); u(k-1);\ldots.; u(k-m)]$$

The model obtained for the inverse dynamics will represent a generalized right inverse of the process.

3.3 Internal model control strategy.

The Internal Model Control (IMC) structure will be introduced as an alternative to the classic feedback structure. Because to the controller C it includes the plant model M explicitly, we refer to this feedback configuration as Internal Model Control.

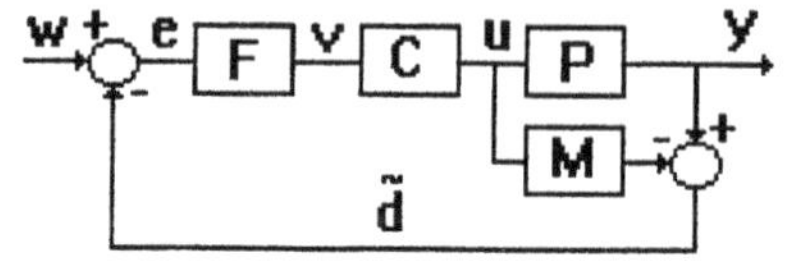

Fig.3. Internal model control

The feedback signal is $\tilde{d} = (P - M)u + d$

If the model is exact ($P = M$) and there are no disturbances ($d = 0$) then the model output $\tilde{y}$ and the process output y are the same and the feedback signal $\tilde{d}$ is zero. Thus, the control system is open-loop when there is no uncertainty - no model uncertainty and no unknown inputs d. If a process and all its inputs are known perfectly, there is no need for feedback control. The feedback signal $\tilde{d}$ expresses the uncertainty about the process. Control objective can be written in the dead-beat form as $y(k+1) = w(k)$. The resulting nonlinear IMC controller can be written $Q \sim CF$. Objective at each iteration is $y(k+1) = w(k)$. Control system output $y(k+1)$ can be expressed as $y(k+1) = PCw(k)$. We have $P = M$ and therefore can written $y(k+1) = MCw(k)$. Problem $y(k+1) = w(k)$ can be reformulated as choosing C such that $(I-MC)w(k)=0$. Eqn. $y(k+1) = PCw(k)$ is achieved by choosing the controller C to be the right inverse of the model $C = M_r^{-1}$. The feedback signal represents the error between the system output and the model output

$$d(k) = y(k) - y^*(k)$$
$$y^*(k+1) = w(k) - d(k)$$

We obtain the close-loop relation

$$y^*(k+1) - y^*(k) = w(k) - y(k)$$

3.4 Relationship with "classic feedback".

We transform IMC control structure in Fig.4. on the form in Fig.5.

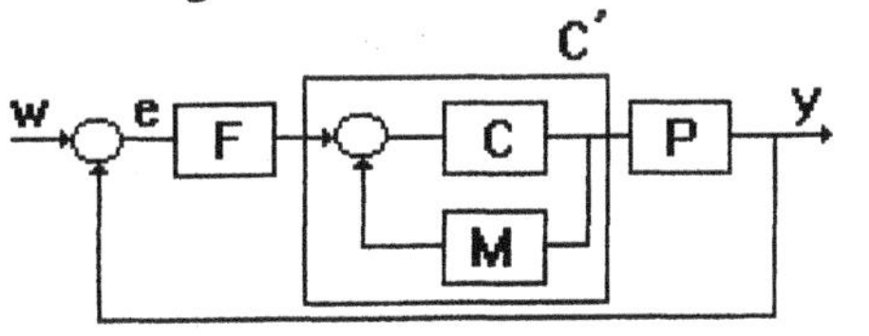

Fig.4. Equivalent feedback structure

$$u = Cv$$

$$u = C(e + \tilde{y})$$

$$u = Ce + C\tilde{y} = Ce + CMu$$

$$u(1 - CM) = Ce$$

$$F_{C'} = \frac{e}{u} = \frac{C}{1 - CM}$$

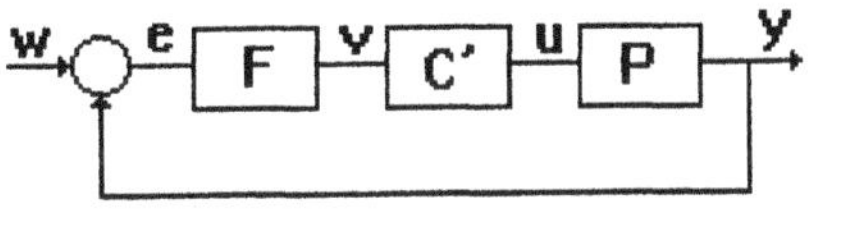

Fig.5. Alternative feedback scheme of IMC

Equivalent classic feedback (Fig.4,5) is stable because the internal signal u and y are unaffected by the transformation.

3.5 Filter design.

Inverse model as feedback controller can produce large control actions which can lead to a unstability. Therefore a robust filter is added as a precompensator to the ANN controller. A common choice for the filter is a exponential filter with following transfer function

$$F(z^{-1}) = \frac{1-\alpha}{1-\alpha z^{-1}}$$

where α is time constant of the filter. In this time domain the filter can be represented recursively as $v(k)=\alpha v(k-1)+(1-\alpha)[w(k)-d(k)]$. The pole of the filter α is used as tuning parameter. α is chosen such that $0<\alpha<1$.

4 SIMULATION RESULTS

Inverse and forward models of IMC controller was identified by the off-line direct identification. There was obtained 200 input-output pairs for each of the models with sampling time equal 5. For training phase of IMC controller was used adaptiv backpropagation algorithm. The linear transfer function of the inverse and forward models of IMC controller was used. Structure of inverse and forward models was 4-8-1.

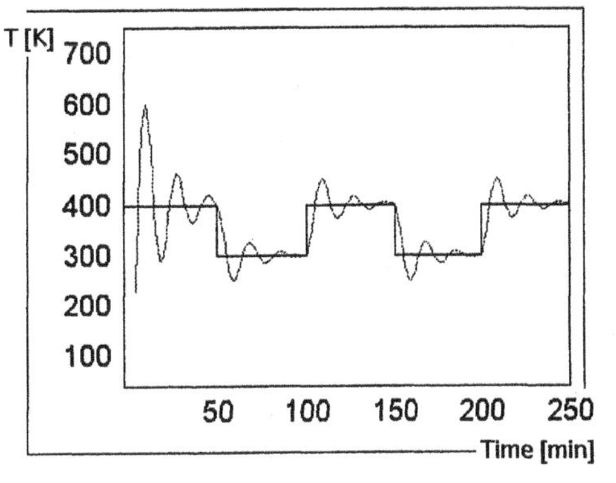

Fig.6. Controlled output of CSTR for $\alpha=0.7$

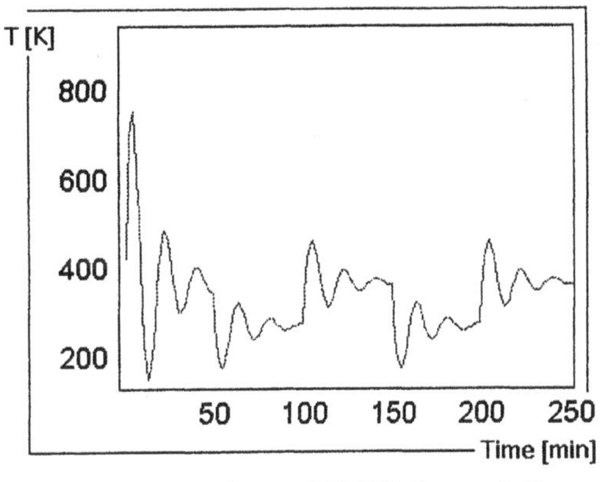

Fig.7. Control actions of CSTR for $\alpha=0.7$

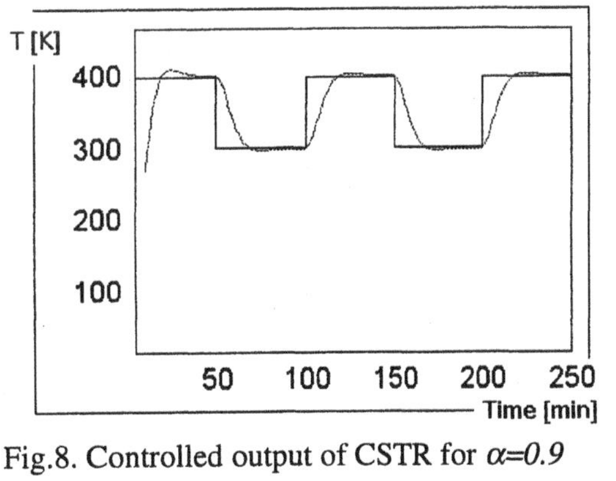

Fig.8. Controlled output of CSTR for $\alpha=0.9$

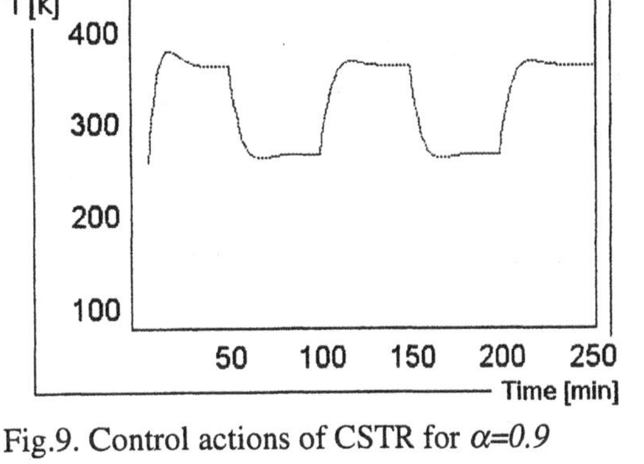

Fig.9. Control actions of CSTR for $\alpha=0.9$

5 CONCLUSION

An application of a IMC controller is given. The paper presents the ANN approach to chemical reactor control. The simulation results clearly show the advantages of this approach.

6 REFERENCES

Hertz J., A. Krogh, R. Palmer (1991). *Introduction to the theory of neural computation.* Addison-Wesley Publishing Company.

Henson M. A., Seborg D. E. (1991). An internal model control strategy for nonlinear systems. AIChE J., 37, 1065-1081.

Hunt K. J., Sbarbaro D., Zbikowski R. (1992). Neural Network for Control Systems-A survey. Automatica, Vol. 28, No. 6, 1083-1112.

Thibault J., Grandjean B. P. A. (1992). Process Control Using Feedforward Neural Networks. J. of Systems Engineering, 2, 199-212.

ADAPTIVE DECENTRALIZED CONTROL OF INTERCONNECTED NONLINEAR SYSTEMS WITH APPLICATION TO STABILIZATION OF POWER SYSTEMS

Druzhinina, M.V. *, Fradkov, A.L. * and Veselý, V.**

* *Institute for Problems of Mechanical Engineering 61, Bolshoy, VO, St.Petersburg, 199178, RUSSIA*

** *Slovak University of Technology, Department of Automatic Control systems Bratislava, SLOVAKIA*

Abstract. In this paper, an adaptive decentralized control scheme using the principle of dominant subsystems is presented for nonlinear large-scale interconnected system. The general stability theorem is obtained giving the applicability conditions for the designed algorithm. The proposed algorithm is applied to stabilization of a multimachine power system.

Keywords. Adaptive Nonlinear control; Decentralized control; Power systems.

1. INTRODUCTION

The stabilization of large-scale systems has been extensively studied over the last years (Ioannou and Kokotovic, 1983; Quiang Iu, 1989; Vesely and Murgas, 1993; Vesely, 1993; Simunek, 1996).

In the modelling of large-scale systems which are composed of interconnected subsystems, the significant uncertainty is represented by the interconnections among the subsystems. To receive realizable control of large-scale systems, decentralized controllers must be used. It is desirable to find the decentralized control which satisfies the following requirements: the structure and parameters of adaptive decentralized control have to be designed in terms of mathematical models of subsystems in such a way that the overall system asymptotic stability is guaranteed; the local controller must be robust and independent of either large-scale system local controllers. In particular, for nonlinear systems, these requirements demand new approaches to the design of decentralized control.

In this paper, we proposed an adaptive decentralized control algorithm using the principle of dominant subsystems for nonlinear large-scale interconnected system (Vesely, 1996). The general stability theorem is obtained giving the applicability conditions for the designed algorithm.

The proposed algorithm is applied to stabilization of a power system. We assume that the multimachine power system model can be split into N interconnected subsystems. According to the method developed in this paper an adaptive nonlinear steam valving and voltage control are designed using only a nonlinear mathematical model of the subsystem, in such a way that the stability of the multimachine power system is guaranteed. Adaptive decentralized controllers designed on the basis of the nonlinear model can provide proper effects for the power system, even during the violent swings caused by a large disturbances.

2. STABILIZATION OF LARGE-SCALE SYSTEMS

Consider a large-scale system described in general by nonlinear mathematical model consisting on N subsystems

$$\begin{aligned} \dot{x}_i &= f_i(x_i) + b_i(x_i, u_i) + h_i(x) \\ i &= 1 \dots N \end{aligned} \tag{1}$$

where $x_i \in \mathbb{R}^{n_i}$ is the local state vector of the i-th subsystem, $u_i \in \mathbb{R}^{m_i}$ is the control vector of the i-th subsystem, $n = \sum_{i=1}^{N} n_i$, $m = \sum_{i=1}^{N} m_i$; $f_i(\cdot), b_i(\cdot), h_i(\cdot)$ are continuously differentiable vector functions and the next condition holds:

$$\begin{aligned} f_i(0) = b_i(x_i, 0) = h_i(0) = 0, \\ ||h_i(x)|| \leq \xi_i ||x|| \end{aligned} \tag{2}$$

The numbers $\xi_i \leq 0$ are supposed to be bounded and unknown.

The problem of the local control agent is to find adaptive decentralized control of the form:

$$u_i = U_i(x_i, \theta_i) \tag{3}$$
$$\dot{\theta}_i = \Theta_i(x_i, \theta_i) \tag{4}$$

such that the closed-loop system consisting of the plant (1) and the adaptive decentralized controllers (ADC) (3),(4) is stable and robust against external input (interconnections between subsystems). In (3),(4) $\theta_i \in \mathbb{R}^{r_i}$ are the local controller parameters which are adapted by adaptation algorithm (4) and $\Theta_i(0, \theta_i) = 0$.

We will use the Lyapunov theory of stability to analyze individual subsystems and the whole system. If the controlled system consisted only of isolated subsystems without mutual interconnections, then the existence of a Lyapunov function $V_i(x_i)$ for each subsystem which satisfies quadratic type inequalities (Krasovsky, 1963):

$$\begin{aligned} \alpha_{i1}||x_i||^2 \le V_i(x_i) \le \alpha_{i2}||x_i||^2 \\ |gradV_i(x_i)| \le \beta_i||x_i|| \\ \dot{V}_i(x_i) \le -\rho_i||x_i||^2 \end{aligned} \tag{5}$$

would be the sufficient condition for stability of the whole system. Consideration of the interconnections among subsystems adds a new quality to the whole system and existence of Lyapunov functions for single subsystem doesn't mean the stability for the whole system.

For Lyapunov functions of subsystems $V_i(x_i)$ we find such numbers $\alpha_{i1}, \alpha_{i2}, \beta_i, \rho_i$ which satisfy the inequalities (5). These numbers characterize the properties of isolated subsystems from the stability point of wiew. Further, let ξ_{ij} characterize the interactions among subsystems in the following way:

$$h_i(x) \le \sum_{i=1}^{N} \xi_{ij}||x_j||. \tag{6}$$

Henceforth we will use the Lyapunov function of the whole system instead of Lyapunov functions of single subsystems.

According to the principle of dominant subsystems (Vesely, 1993) an appropriate quality of the whole system can be achieved by controlling only the isolated subsystems. The quality of a system is then determined by the stability measure $\alpha(x)$ defined as follows:

Definition 1 *Let $V(x)$ be a Lyapunov function of a system. The stability measure of this system is given by the following formula:*

$$\alpha(x) = -\frac{dV/dt}{V} \tag{7}$$

Choosing an adaptation algorithn from the principle of dominant subsystems (Vesely, 1993) to provide increasing the stability measure of subsystems, i.e. $\frac{\partial \alpha_i}{\partial \theta_i} > 0, i = 1 \ldots N$, we obtain the following form of adaptation algorithm:

$$\dot{\theta}_i = -\gamma_i \frac{\partial}{\partial \theta_i}\left((gradV_i)^T b_i)\right), \quad i = 1 \ldots N, \tag{8}$$

where $V_i(x_i)$ is the Lyapunov function of isolated i-th subsystem which satisfies sufficient conditions for stability. We can modify these Lyapunov functions for system (1) with adaptive control (3),(8) by adding an adaptive parameter to Lyapunov function. Then we choose

$$V_{ia}(x_i, \theta_i) = V_i(x_i) + (\theta_i - \theta_i^*)^T(\theta_i - \theta_i^*) \tag{9}$$

where θ_i^* are some constant parameter vector.

Now we are in position to formulate the theorem:

Theorem 1 *Let each i-th isolated subsystem of (1) possess smooth nonnegative and proper function $V_i(x_i)$ such that*

$$\begin{aligned} \alpha_{i1}||x_i||^2 \le V_i(x_i) \le \alpha_{i2}||x_i||^2 \\ |gradV_i(x_i)| \le \beta_i||x_i||, \quad i = 1 \ldots N, \end{aligned}$$

and the following assumptions hold:

1. There exist smooth functions $u_i^*(x_i, \theta_i)$, $i = 1 \ldots N$ and values of θ_i^* which satisfy the inequalities:

$$\dot{V}_i(x_i)\Big|_{u_i = u_i^*(x_i, \theta_i^*)} \le -\rho_i||x_i||^2 \tag{10}$$

(Stabilisability condition for the i-th isolated subsystem)

2. Time-derivative of the function $V_i(x_i)$ is convex in θ_i, i.e. any θ_i, θ_i', x_i satisfy the inequality:

$$\dot{V}_i(x_i, \theta_i') - \dot{V}_i(x_i, \theta_i) \ge (\theta_i - \theta_i')\frac{\partial}{\partial \theta_i}(\dot{V}_i(x_i, \theta_i)) \tag{11}$$

where

$$\dot{V}_i(x_i, \theta_i) = \frac{\partial V_i(x_i)}{\partial x_i}(f_i(x_i) + b_i(x_i, u_i(x_i, \theta_i)))$$

(convexity condition)

3. Interactions $h_i(x)$ satisfy (6) and the following inequalities hold:

$$\beta_i^2 + \sum_{j=1}^{N} \xi_{ij}^2 \le 2\rho_i/\sqrt{N}, \quad i = 1 \ldots N \tag{12}$$

Then the decentralized adaptive control law:

$$u_i = u_i^*(x_i, \theta_i) \tag{13}$$
$$\dot{\theta}_i = -\frac{1}{2}\gamma_i \frac{\partial}{\partial \theta_i}\left((gradV_i)^T b_i(x_i, u_i(x_i, \theta_i))\right), \quad i = 1 \ldots N,$$

where $\gamma_i > 0$ are constant parameters, provides the boundedness of the whole system (1),(13) trajectories and the achievement of the goal

$$x(t) \to 0, \quad when \quad t \to \infty \tag{14}$$

Proof: Define the Lyapunov function candidate for the large-scale system as follows:

$$\begin{aligned} V_a(x,\theta) &= \sum_{i=1}^{N} V_{ia}(x_i,\theta_i) = \\ &= \sum_{i=1}^{N} \{V_i(x_i) + \frac{1}{\gamma_i}(\theta_i - \theta_i^*)^T(\theta_i - \theta_i^*)\} \end{aligned}$$

Calculate its time-derivative with respect to (1):

$$\dot{V}_a \mid (1) = \sum_{i=1}^{N} \{(gradV_i)^T (f_i + b_i(x_i, u_i(x_i,\theta_i)) + \\ + \frac{2}{\gamma_i}(\theta_i - \theta_i^*)^T \dot{\theta}_i\}$$

Replacing the adaptive control law (13) and applying the convexity and stabilizability conditions gives:

$$\begin{aligned} \dot{V}_a &= \sum_{i=1}^{N} \{\dot{V}_i(x_i,\theta_i) - (\theta_i - \theta_i^*)^T \frac{\partial}{\partial \theta_i}(\dot{V}_i) + \\ &+ (gradV_i)^T h_i(x)\} \leq \\ &\leq \sum_{i=1}^{N} \{\dot{V}_i(x_i,\theta_i^*) + (gradV_i)^T h_i(x)\} \leq \\ &\leq -\sum_{i=1}^{N} \rho_i ||x_i||^2 + \sum_{i=1}^{N} \beta_i ||x_i|| \sum_{j=1}^{N} \xi_{ij} ||x_j||. \end{aligned}$$

Applying to the second term of the last expression the well-known inequality $2ab \leq a^2/c + b^2c$ we obtain:

$$\begin{aligned} &\sum_{i=1}^{N} \beta_i ||x_i|| \sum_{j=1}^{N} \xi_{ij} ||x_j|| \leq \\ &\leq \sum_i \sum_j \left(\frac{\beta_i^2 ||x_i||^2}{2\sqrt{N}} + \frac{\xi_{ij}^2 ||x_j||^2}{2} \sqrt{N} \right) = \\ &= \sum_i \frac{\sqrt{N}}{2} \beta_i^2 ||x_i||^2 + \sum_j \frac{\sqrt{N}}{2} ||x_j||^2 \sum_i \xi_{ij}^2 = \\ &= \frac{\sqrt{N}}{2} \left(\sum_i ||x_i||^2 (\beta_i^2 + \sum_j \xi_{ij}^2) \right) \end{aligned}$$

Therefore, the time-derivative of $\dot{V}_a$ becomes nonpositive while the inequality (12) holds.

We have proved that $\dot{V}_a(x,\theta) \leq 0$ and, therefore, all trajectories of system are bounded. Furthermore, LaSalle's invariance theorem guarantees that any bounded trajectory $x(t)$ converges to the largest invariant set of (1) on which $\dot{V}_a(x,\theta) = 0$. Hence, we have $x(t) \to 0$, *when* $t \to \infty$ that proves the theorem. ■

Remark 1. The adaptation algorithm (8) also can be obtained by the Speed-Gradient method (Fradkov, 1990). Stabilizability condition (10) means the existence of smooth feedback laws $u_i = u_i^*(x_i, \theta_i^*)$ exponentialy stabilizing the i-th isolated subsystem of (1). It corresponds to achievability condition of Speed-Gradient algorithm (Fradkov, 1990).

Remark 2. When the interactions among subsystems satisfy the weakened inequalities

$$h_i(x) \leq \sum_{i=1}^{N} \xi_{ij} ||x_j|| + d_i. \qquad (15)$$

where $d_i > 0$, the weakened goal is achieved. This control goal means convergence of the system trajectories to a some prespecified region which in fact is a neighbourhood of the origin. Additionally it is worth mentioning that parameter update algorithm should be robustyfied (Fradkov, 1990).

3. MATHEMATICAL MODEL OF A MULTIMACHINE POWER SYSTEM

In this section the mathematical model of a power system is formulated. The power system is a system of power stations connected to a large power network. Each power station is considered as an individual subsystem consisting of following aggregates:
- the generator of the input energy (steam aggregates, etc.);
- the turbine (transforms the input energy to the mechanical energy);
- the synchronous generator (transforms the mechanical energy to the electrical energy).

Each subsystem is connected to the common large power network via lines and loads.

Let us consider the model of a multimachine power system described by a nonlinear differential equations (Arrilaga, *et al.*, 1983; Venikov, 1978):

$$\dot{x}_i = f_i(x_i) + b_i(x_i, u_i) + h_i(x_i) \qquad (16)$$
$$i = 1 \dots N$$

where

$$x_i = \begin{pmatrix} x_{i1} \\ x_{i2} \\ x_{i3} \\ x_{i4} \\ x_{i5} \end{pmatrix}; \quad f_i(x_i) = \begin{pmatrix} f_{i1} \\ f_{i2} \\ f_{i3} \\ f_{i4} \\ f_{i5} \end{pmatrix};$$

$$\begin{aligned} f_{i1} &= x_{i2}; \\ f_{i2} &= (P_{T0i} + x_{i4} - D_{mi} x_{i2} - \\ &- e_i(E_{i0} + x_{i3}) \sin(\delta_{i0} + x_{i1}))/T_{ji}; \\ f_{i3} &= (x_{i5} - x_{i3}(1 + d_i/z_{ii}) + \\ &+ d_i e_i \cos(\delta_{i0} + x_{i1}) - d_i e_i \cos \delta_{i0})/T_{bi}; \\ f_{i4} &= -x_{i4}/T_{Ti}; \\ f_{i5} &= -x_{i5}/T_{Bi}; \\ d_i &= x_{di} - x'_{di}, \quad e_i = E_{N0}/z_{iN}; \end{aligned}$$

$$b_i(x_i, u_i) = \begin{pmatrix} 0 & 0 & 0 & \frac{k_{Ti}}{T_{Ti}} & 0 \\ 0 & 0 & 0 & 0 & \frac{k_{Bi}}{T_{Bi}} \end{pmatrix}^T \begin{pmatrix} u_{i1} \\ u_{i2} \end{pmatrix}$$

and the interactions among subsystems look as follows:

$$h_i(x) = \begin{pmatrix} h_{i1} \\ h_{i2} \\ h_{i3} \\ h_{i4} \\ h_{i5} \end{pmatrix} = \begin{pmatrix} 0 \\ h_{i2} \\ h_{i3} \\ 0 \\ 0 \end{pmatrix};$$

$$h_{i2} = -\sum_{k=1,\ k\neq i}^{N-1} \frac{(E_{i0} + x_{i3})(E_{k0} + x_{k3})}{z_{ik}} \sin(\delta_{i0} + \\ + x_{i1} - \delta_{k0} - x_{k1});$$

$$h_{i3} = d_i \sum_{k=1,\ k\neq i}^{N-1} \frac{E_{k0} + x_{k3}}{z_{ik}} \cos(\delta_{i0} + x_{i1} - \delta_{k0} - \\ - x_{k1}) - \frac{E_{k0}}{z_{ik}} \cos(\delta_{i0} - \delta_{k0});$$

N is number of synchronous machines;

$$x_i = \begin{pmatrix} x_{i1} \\ x_{i2} \\ x_{i3} \\ x_{i4} \\ x_{i5} \end{pmatrix} = \begin{pmatrix} \Delta\delta_i \\ \omega_i \\ \Delta E_i \\ \Delta P_{Ti} \\ \Delta U_{bi} \end{pmatrix};$$

Symbols in this model denote the i-th syncronous machine:

$\delta_i = \delta_{0i} + \Delta\delta_i$	load angle of i-th machine;
ω_i	rotor angular velocity;
$E_i = E_{0i} + \Delta E_i$	transient voltage;
$P_{Ti} = P_{T0i} + \Delta P_{Ti}$	turbine power;
$U_{bi} = U_{b0i} + \Delta Ubi$	exitation voltage;
$T_{ji} = T_{Mi}/\omega_0$	
T_{Mi}	inertia time constant;
D_{mi}	damping factor;
T_{bi}	d-axis time-constant of the synchronous machine;
$x_{di},\ x'_{di}$	d-axis and transient reactances;
T_{Bi}	time constant of the exciter;
T_{Ti}	turbine constant;
k_{Bi}	gain of the exciter;
k_{Ti}	governor constant;
u_{i1}	steam valve control signal;
u_{i2}	control signal for exciter;

The first three equations of (16) describe the generator model; the two last equations of (16) describe the turbine and excitation system respectively.

4. STABILIZATION OF THE POWER SYSTEM

The power system consist of any number of isolated subsystems which can be freely connected to or disconnected from the system. Because of the distribution of single subsystems, it is necessary to design a decentralized control algorithm for each subsystem. The controllers must simultaneously provide the stability for the whole system.

According to the theory of section 2, the control algorithm design for the power system is divided into two steps. In the first step we design the control algorithm for an isolated subsystem without considering the mutual interconnections which provides the stability of the subsystem. In the second step we modify this algorithm in order to provide the stability for the whole interconnected system.

Step 1.

The structure of the isolated subsystem is a special type of so-called cascaded system. Therefore, the procedure of integrator backstepping can be applied to design the control algorithm (Kanellakopoulos *et al.*, 1992; Druzhinina and Fradkov, 1994).

At first, equations for the generator can be considered as an independent subsystem with the fictive control input vector $(x_{i4},\ x_{i5})^T$. Then the control law $\alpha_i = [\alpha_{i1}\ \ \alpha_{i2}]^T$, where

$$\begin{aligned} \alpha_{i1} &= -x_{i1} - P_{T0i} + e_i(E_{i0} + x_{i1})\sin(\delta_{i0} + \\ &+ x_{i1}) - \gamma_1 x_{i2}; \\ \alpha_{i2} &= -\cos(\delta_{i0} + x_{i1}) + \cos\delta_{i0} - \gamma_2 x_{i3}; \\ &\gamma_1,\ \gamma_2 > 0, \end{aligned}$$

stabilizes the subsystem consisting on equations for the generator. Then, considering equations for the turbine and excitation system as integrator and taking into account that $(x_{i4},\ x_{i5})^T$ is not our actual control we apply the integrator backstepping algorithm to design the actual control law:

$$u_i = (u_{i1}\quad u_{i2})^T = \begin{pmatrix} (T_{Ti}\overline{u_{i1}} + x_{i4})/k_{Ti} \\ (T_{Bi}\overline{u_{i2}} + x_{i5})/k_{Bi} \end{pmatrix} \quad (17)$$

where

$$\begin{aligned} \overline{u_i} &= (\overline{u_{i1}}\quad \overline{u_{i2}})^T = \begin{pmatrix} -\lambda_{i1}(x_{i4} - \alpha_{i1}) \\ -\lambda_{i2}(x_{i5} - \alpha_{i2}) \end{pmatrix} - \\ &- \begin{pmatrix} x_{i2} \\ \frac{1}{d_i e_i} x_{i3} \end{pmatrix} + \begin{pmatrix} \frac{\partial \alpha_{i1}}{\partial x_i} \\ \frac{\partial \alpha_{i2}}{\partial x_i} \end{pmatrix} f_i(x_i) \end{aligned}$$

$$\frac{\partial \alpha_{i1}}{\partial x_i} = \begin{pmatrix} -1 - e_i(E_{i0} + x_{i3})\cos(\delta_{i0} + x_{i1}) \\ -\gamma_1 \\ e_i \sin(\delta_{i0} + x_{i1}) \end{pmatrix}^T$$

$$\frac{\partial \alpha_{i2}}{\partial x_i} = \begin{pmatrix} \sin(\delta_{i0} + x_{i1}) \\ 0 \\ -\gamma_2 \end{pmatrix}^T;$$

Step 2.

According to theorem of section 2, we introduce an adaptation into the obtained control algorithm (17) in order to achieve the stabilization of the whole system. The control algorithm $u_i(x_i)$ will be modified to $u_i(x_i, \theta_i)$, where $\theta_i = [\lambda_{i1} \ \lambda_{i2}]^T$ is a vector of adaptive parameters. Then, the adaptive control algorithm looks as follows:

$$\begin{aligned} \dot{\lambda}_{i1} &= \beta_{i1}(x_{i4} - \alpha_{i1})^2 \\ \dot{\lambda}_{i2} &= \beta_{i2}(x_{i5} - \alpha_{i2})^2 \end{aligned} \quad (18)$$

where $\beta_{i1}, \ \beta_{i2} > 0$
and provides the stabilization of interconnected subsystems in the large scale system.

The obtained adaptive control law (17-18) is valid both for one machine system and for multimachine one. We stress that this local feedback is based exclusively on local measurements. Another advantage of the supposed adaptive control law is that it is independent of the parameters and structure of the network and loads of the power system. Therefore, the proposed control scheme possesses a robustness property with respect to the network and load parameter alternation.

5. CONCLUSION

The adaptive control scheme has been developed using the dominant subsystem principle. Applying to stabilization of the multimachine power system model, the obtained control law is based exclusively on local measurements. Moreover, it is independent of the parameters and structure of the network and loads of the power system.

6. ACKNOWLEDGMENT

This research was partially supported by grant 1/4231/97 "Robust and adaptive control of dynamic systems".

REFERENCES

Arrilaga J., Arnold C.P., Harber B.J. (1983) *Computer modelling of electrical power systems.* N.Y.: Wiley.

Druzhinina M.V., Fradkov A.L. (1994) Speed- gradient and speed-difference algorithms for nonlinear control problems. *Diff.equations,* **N11,** pp.1861-1868.

Fradkov A.L. (1990) *Adaptive control in large-scale systems,* Nauka, Moscow (in Russian)

Ioannou P.A., Kokotovic P.V. (1983) *Adaptive systems which reduced models.* Berlin: Springer-Verlag.

Kanellakopoulos I., Kokotovic P.V., Morse A.S. (1992) A toolkit for nonlinear feedback design, *Syst.Control Letters,* **18**, pp.83-92

Krasovsky N.N. (1963) *Stability of motion,* Stanford.

Quiang Iu, Yuan Zhang-Sun (1989) Nonlinear stabilizing control of multimachine systems, *IEEE Trans. on Power Systems,* N4, pp.236-241

Simunek S. (1996) *Adaptive decentralized control of power system with disturbances attenuation,* Dissertation Thesis, Slovak Technical University

Venikov V.A. (1978) *Dynamic behavior of electrical power systems.* Moscow:Nauka (in Russian).

Vesely V., Murgas J. (1993) Decentralized adaptive stabilization of nonlinear system, *Int.J.Control,* **vol.58**, N6, pp.1445-1460

Vesely V. (1993) Large scale dynamic system stabilization using the principle of dominant subsystems approach, *Kybernetica,* **vol.29**, N1, pp.48-61

DECENTRALIZED ROBUST EXPONENTIAL STABILITY OF LINEAR SYSTEMS WITH OUPUT FEEDBACK

VOJTECH VESELÝ

Slovak University of Technology, Department of Automatic Control Systems, Faculty of Electrical Engineering and Information Technology, Bratislava, Slovakia, E-mail:vesely@kasr.elf.stuba.sk

Abstract. The paper addresses the problem of robust decentralized control design which guarantee exponential stability of uncertain linear complex systems with output feedback. No matching conditions are demanded for modelling the uncertain system, where the uncertainties, which are possibly nonlinear, may appear in the subsystems as well as in the interconnections .

Keywords Decentralized control, Robust control, Exponential stability, Lyapunov function, Linear continuous-time systems.

1. INTRODUCTION

The problem of designing decentralized control for complex systems has attracted a great amount of interest in recent years (Šiljak and Vukčevič, 1977; Singh,1981; Veselý et al, 1981; Bernussou and Titli, 1982; Šiljak, 1991). Robustness is frequently encountered in the design of decentralized controllers for many physical processes and its absence is very often the main cause for poor performance and instability of control systems. In view of this, the robustness is a topic of great practical importance which has attracted a lot of interest for several decades (Jury, 1990; Polla et al, 1990). If uncertainties arise in a complex system which consists of N interconnected subsystems, there an two classes of uncertainties considered. First, there may be internal uncertainties, subsystem and input ones, that is own to each subsystem. Secondly, external uncertainty may occur as well due to interconnections among subsytems. We assume that both the internal and external uncertainty upper bounds are partially known. It is difficult to estimate the bounds of uncertainties in real systems, especially in the interconnected uncertain systems where the knowledge of interactions between the subsystems is severely limited (Zhiming Gong et al, 1996; Halicka and Rosinova, 1994). Design of adaptive decentralized controller with uncertainties attenuation can be found in Ioannou (1986); Murgaš and Hejda (1990); Šiljak (1991); Veselý (1993). The other approaches to the robust control design can be found in Jury (1990), Polla et al (1990), Leitman (1996), Kozák (1995) and others. The paper deals with the robust decentralized control problem. The proposed control design strategy aims at achieving the closed-loop exponential stability with a prescribed decay rate. The robust exponential stability of systems plays a importance role in guaranteeing robustness and dynamical performance quality (Mao-Lin Ni and Ying Chen, 1996; Mao, 1996; Niculescu et al, 1995). Both the Lyapunov function method and Krasovskij (1959) definition of exponential stability of linear dynamic systems are of central importance in the design of static output feedback controllers which guarantee robust exponential stability of uncertain complex linear systems without matching conditions.

The paper is organized as follows. In the Section 2 the problem formulation with some preliminary results are given. The main results, and the design of procedure robust decentralized controller are given in Section 3 and the example is in Section 4.

2. Problem statement

Consider an uncertain complex system described by

$$\dot{x}_i = (A_i + \delta A_i)x_i + (B_i + \delta B_i)u_i + \sum_{j \neq i}^{N} h_{ij}(x_j) \quad (1)$$

$$y_i = C_i x_i, i \in \mathcal{N}$$

where

$$i \in \mathcal{N} = \{1, 2, ..., \mathrm{N}\}, \quad \mathrm{x_i} \in \mathrm{R^{n_i}}, \mathrm{u_i} \in \mathrm{R^{m_i}}, \mathrm{y_i} \in \mathrm{R^{l_i}}$$

are the state, control, input and output of the ith subsystem, respectively ; A_i, B_i, C_i are known constant matrices of appropriate dimensions which represent the nominal model of the ith subsystem; δA_i and δB_i are the matrices of appropriate dimensions which represent internal uncertainties own to each subsystem, the entries of the above uknown matrices are piecewise continuous and bounded on every time; the unknown vector function $h_{ij}(x_j)$ is continuous, sufficiently smooth in x_j and piecewise continuous in time and bounded, so that the system (1) has a unique continuous solutions for all initial conditions, $x_i(0) = x_{i0} \in R^{n_i}$, and all piecewise continuous inputs $u_i \in R^{m_i}$.

The isolated subsystem nominal model is

$$\dot{x}_i = A_i x_i + B_i u_i, \qquad y_i = C_i x_i, i \in \mathcal{N} \quad (2)$$

The following assumptions are made for the system (1).

Assumption 1. The internal uncertainties in each subsystem and the unknown vector function $h_i(x_j)$ which represents interconnections among the subsystems satisfy the following conditions:

$$\delta A_i^T \delta A_i \leq \gamma_{ai} Q_{0i} \quad (3)$$

$$\delta B_i^T \delta B_i \leq \gamma_{bi} R_i \quad (4)$$

$$||h_{ij}(x_j)|| \leq \xi_{ij} ||x_j|| \quad (5)$$

where $Q_{0i} = Q_{0i}^T \in R^{n_i \times n_i}, R_i^T = R_i \in R^{m_i \times m_i}$ are the known positive matrices, $\xi_{ij} \geq 0$ is the known positive constant and γ_{ai}, γ_{bi} are constants which will be determined below, $i, j \in \mathcal{N}$.

The system (1) can be rewritten in the following compact form

$$\dot{x} = (A + \delta A)x + (B + \delta B)u + h(x), \qquad y = Cx \quad (6)$$

where

$$x^T = [x_1^T, ..., x_N^T] \in R^n$$

$$u^T = [u_1^T, ..., u_N^T] \in R^m$$

$$y^T = [y_1^T, ..., y_N^T] \in R^l$$

are the state, input and output of the complex system; $A, \delta A, B, \delta B$ and C are the block diagonal matrices with the entries $A_i, \delta A_i, B_i, \delta B_i$ and C_i, respectively and $h(x)$ is a vector function with entries $h_{ij}(x_j)$.

The nominal model of the complex system is

$$\dot{x} = Ax + Bu + h(x) \quad (7)$$

The problem on the local level is to find the robust decentralized controller of the form

$$u = Ky = KCx, \qquad K = diag(K_i), K_i \in R^{m_i \times l_i} \quad (8)$$

such that the closed-loop uncertain system comprising the plant (6) and the decentralized controller (8)

$$\dot{x} = [A + \delta A + (B + \delta B)KC]x + h(x) \quad (9)$$

is robustly exponentially stable with a prescribed decay rate.

The necessary and sufficient conditions for the output feedback stabilizability of each subsystem are given in Kučera and Souza (1995).

Theorem 1. The nominal model of the ith subsystem (2) is output feedback stabilizable if and only if

- (A_i, B_i) is stabilizable and (A_i, C_i) is detectable and
- there exist real matrices K_i and G_i such that

  $$K_i C_i + B_i^T L_i = G_i$$

  where L_i is a real symmetric nonnegative definite solution of

  $$A_i^T L_i + L_i A_i - L_i B_i B_i^T L_i + G_i^T G_i = 0$$

Assumption 2. Assume that for each subsystem the conditions of Theorem 1 are met.

Throughout this paper the concept of robust exponential stability of complex systems will be used in the following sense.

Definition 1. The uncertain closed-loop system (9) is said to be robustly exponentially stable with a decay rate $\alpha > 0$ if for all initial conditions $x(t_0) = x_0$, and for all admissible uncertainties given by (3), (4) and (5) there exists a C^1 Lyapunov function $V(x) : R^n \rightarrow R_+$ for the nominal model (7) such that for its time derivative along the closed-loop system solution (9) the following inequalities hold

$$\frac{dV}{dt} \leq -\alpha V, \qquad V(x_0) = V_0 \quad (10)$$

where

$$\gamma_1 ||x|| \leq V \leq \gamma_2 ||x||$$

and

$$\frac{dV}{dt} \leq -\gamma_3 ||x||$$

where $\gamma_i, i = 1, 2, 3$ are positive constants.

Remark 1. The solution of the (10) is

$$V(x, t) = V(x_0, t_0)e^{-\alpha t} \quad (11)$$

Using the inequalities (10) we obtain

$$||x(t)|| \le \sqrt{\frac{\gamma_2}{\gamma_1}}||x_0||e^{-.5\alpha t} \qquad (12)$$

for all $x \in R^n$. Inequality (12) implies that if the closed-loop system (9) is robustly exponentially stable with decay rate $\alpha > 0$ then the exponential convergence of $||x||$ is to be by a decay rate $.5\alpha$ for all $x \in R^n$.

Definition 2. The system (9) is said to be *robustly exponentially stabilizable* if there exists K such that the closed-loop system (9) is robustly exponentially stable for all admissible uncertainties (3), (4) and (5).

The problem addressed in this paper can be stated as follows: Under the Assumptions 1, 2 with a given output matrix C find the matrix K (8) such that the closed-loop uncertain system (9) is exponentially stable with the prescribed decay rate α and the bounds $\gamma_{ai}, \gamma_{bi}, i \in \mathcal{N}$, in (3) and (4) achieve their maximum possible values for given $\xi_{ij}, i \neq j, i, j \in \mathcal{N}$.

3. Decentralized Robust Control Design with Output Feedback

In this section, we consider the problem of decentralized control design which guarantees the exponential stability of the uncertain complex system (1). Let choose the function

$$V_i = x_i^T P_i x_i, \qquad P_i = P_i^T \ge 0 \qquad (13)$$

as a candidate Lyapunov function of the nominal model of the ith isolated subsystem (2). The candidate Lyapunov function of the nominal model of the complex system (7) is given by

$$V(x) = \sum_{i=1}^{N} V_i(x) \qquad (14)$$

Sufficient conditions for the decentralized robust exponential stabilizability of a complex system with the output feedback are given by the following theorem.

Theorem 2. The uncertain closed-loop system (9) is decentralized output feedback robust exponentially stabilizable with a prescribe decay rate α if the following sufficient conditions are satisfied.

- Assumptions 1-2 hold.
- For the decay rate of the ith subsystem $\alpha_i > 0$, the following inequality holds

$$\alpha_i \ge \alpha, \qquad i \in \mathcal{N} \qquad (15)$$

- The matrix $M_i = M_i^T$ is nonnegative semidefinite for some K_i

$$M_i = (\alpha_i - \alpha)P_i + Q_{0i}(1 - \epsilon_{i1}) + Q_{1i} - \qquad (16)$$
$$P_i^2(\epsilon_{i1}^{-1}\gamma_{ai} + \epsilon_{i2}^{-1}\gamma_{bi}) +$$
$$(1 - \epsilon_{i2})C_i^T K_i^T R_i K_i C_i - \vartheta_i I_i$$

where $\epsilon_{i1}, \epsilon_{i2} > 0$ are positive constants; $Q_{1i} = Q_{1i}^T \in R^{n_i \times n_i}$ are positive definite matrix;

$$\vartheta_i = N\lambda_M^2(P_i)\sum_{j\neq i}^{N}\xi_{ij}^2 + \frac{N-1}{N}$$

$I_i \in R^{n_i \times n_i}$ is a unity matrix. The matrix P_i is calculated from the following matrix equation

$$(A_i + \frac{1}{2}\alpha_i I_i + B_i K_i C_i)^T P_i + P_i(A_i + \frac{1}{2}\alpha_i I_i + B_i K_i C_i) + Q_i + C_i^T K_i^T R_i K_i C_i = 0 \qquad (17)$$

where $Q_i = Q_{0i} + Q_{1i}$ and $K_i \in R^{m_i \times l_i} \in \mathcal{K}_i$ is the matrix of decentralized controller parameters; the set $\mathcal{K}_i, i \in \mathcal{N}$ is given as follows

$$\mathcal{K}_i = \{K_i : Re\lambda_M(A_i + \frac{1}{2}\alpha_i I_i + B_i K_i C_i) \le 0\}, \quad i \in \mathcal{N} \qquad (18)$$

$Re\lambda_M(.)$ denotes the maximum real part of all eigenvalues of the corresponding matrix.

Proof . For the time derivative of V defined by (14) along the solution of (9) we obtain

$$\frac{dV}{dt} = \sum_{i=1}^{N}(\dot{x}_i^T P_i x_i + x_i^T P_i \dot{x}_i) =$$

$$\sum_{i=1}^{N}\{x_i^T[(A_i + B_i K_i C_i)^T P_i + P_i(A_i + B_i K_i C_i)]x_i +$$
$$x_i^T[(\delta A_i + \delta B_i K_i C_i)^T P_i + P_i(\delta A_i + \delta B_i K_i C_i)]x_i + \qquad (19)$$
$$2x_i^T P_i \sum_{j\neq i}^{N} h_{ij}(x_j)\}$$

Owing to the Assumption 1 and to (17), the upper bound on (19) is

$$\frac{dV}{dt} \le -\alpha V - \sum_{i=1}^{N}[x_i^T\{(\alpha_i - \alpha)P_i + \delta G_i^T \delta G_i + Q_{1i} + \qquad (20)$$
$$\delta F_i^T \delta F_i - (\gamma_{ai} + \gamma_{bi})P_i^2\}x_i - 2||P_i||||x_i||\sum_{j\neq i}^{N}\xi_{ij}||x_j||]$$

where

$$\delta G_i = \frac{1}{\sqrt{\gamma_{ai}}}\delta A_i + P_i\sqrt{\gamma_{ai}}$$

$$\delta F_i = \frac{\delta B_i K_i C_i}{\sqrt{\gamma_{bi}}} + P_i\sqrt{\gamma_{bi}}$$

Using the following facts that

$$2ab \le ca^2 + \frac{b^2}{c}$$

$$\sum_{i=1}^{N}\{\frac{\|x_i\|^2}{N} + \sum_{j\neq i}^{N}\frac{\|x_j\|^2}{N}\} = \sum_{i=1}^{N}\|x_i\|^2 \qquad (21)$$

and

$$(F+G)^T(F+G) \le (1+\epsilon)F^TF + (1+\epsilon^{-1})G^TG, \epsilon > 0$$

the bound on the time derivative of the Lyapunov function (20) can be rewritten

$$\frac{dV}{dt} \le -\alpha V - \sum_{i=1}^{N}\{x_i^T[(\alpha_i-\alpha)P_i + (1-\epsilon_{i1})\frac{\delta A_i^T\delta A_i}{\gamma_{ai}} +$$

$$(1-\epsilon_{i1}^{-1})P_i\gamma_{ai} + Q_{1i} +$$

$$+(1-\epsilon_{i2})\frac{C_i^TK_i^T\delta B_i^T\delta B_iK_iC_i}{\gamma_{bi}} + (1-\epsilon_{i2}^{-1})P_i^2\gamma_{bi} -$$

$$-(\gamma_{ai}+\gamma_{bi})P_i^2]x_i - 2\|P_i\|\|x_i\|\sum_{j\neq i}^{N}\xi_{ij}\|x_j\|\}$$

or

$$\frac{dV}{dt} \le -\alpha V - \sum_{i=1}^{N}x_i^TM_ix_i \qquad (22)$$

where M_i is given by (16). If the third condition of Theorem 2 is met the uncertain system (9) is exponentially stable with a prescribed decay rate α. The proof of the theorem is completed.

For the next developments the following definition is introduced.

Definition 3. Consider the linear uncertain system (9) with $h(x) = 0$. Assume that the matrix $A+BKC$ is Hurwitz (stable). The robustness stability measure of the investigated system is defined as

$$m_r = \min_{\|\delta A+\delta BKC\|}\{\delta A + \delta BKC \in \Phi : A + \delta A + (B+\delta B)KC \quad \text{is not Hurwitz}\} \qquad (23)$$

where Φ is a set of all possible uncertainties of the system considered.

Assume that the matrices Q_{0i}, Q_{1i}, R_i and the constants $\gamma_{ai}, \gamma_{bi}, \alpha_i > \alpha$ and ξ_{ij} for $i,j \in \mathcal{N}$, are known. Then a detailed analysis of properties of M_i implies that the best way to guarantee nonnegative definiteness of M_i is to maximize the robustness stability measure for all uncertain isolated subsystems (3). It is known that when maximizing $m_{ri}, i \in \mathcal{N}$, the maximum eigenvalues $\lambda_M(P_i)$ should be minimized, where P_i is the solution of (17). To simplify the minimization task $\lambda_M(P_i)$ is replaced by the minimization of $Tr(P_i)$, where $Tr(X)$ denotes the trace of matrix X. The optimisation problem of the ith subsystem can be associated with the Lagrangian:

$$L_i = \min_{K_i,P_i}\max_{W_i} Tr\{P_i + W_i[(A_i + \frac{\alpha_i}{2}I_i + B_iK_iC_i)^TP_i$$

$$+ P_i(A_i + \frac{\alpha_i}{2}I_i + B_iK_iC_i) + Q_i + C_i^TK_i^TK_iC_i]\} \qquad (24)$$

$$i \in \mathcal{N}$$

The necessary optimality conditions are

$$I_i + A_{ci}W_i + W_iA_{ci}^T = 0 \qquad (25)$$

$$A_{ci}^TP_i + P_iA_{ci} + Q_i + C_i^TK_i^TK_iC_i = 0 \qquad (26)$$

$$K_i = -B_i^TP_iW_iC_i^T(C_iW_iC_i^T)^{-1} \qquad (27)$$

where

$$A_{ci} = A_i + \frac{\alpha_i}{2}I_i + B_iK_iC_i \qquad (28)$$

For finding the solution of the matrix equations (25), (26) and (27) we propose the following iterative algorithm.

Algorithm A.

1. Choose $K_i^{(1)}$ such that the matrix A_{ci} (28) is stable.
2. Calculate $P_i^{(1)}$ and $W_i^{(1)}$ from the well-known Lyapunov matrix equations (25) and (26).
3. Calculate the next approximation of $K_i^{(2)}$ using (27)

$$K_i^{(2)} = K_i^{(1)} - \beta_i(B_i^TP_i^{(1)}W_i^{(1)}C_i^T + K_i^{(1)}C_iW_i^{(1)}C_i^T)$$

   where $\beta_i > 0$, check the stability of the matrix $A_{ci}, i \in \mathcal{N}$. If any of them is unstable, for this i decrease β_i and calculate $K_i^{(2)}$ again.
4. Calculate $P_i^{(2)}$ and $W_i^{(2)}$ for $K_i^{(2)}$.
5. If $Tr(P_i^{(2)} - P_i^{(1)}) < 0$ then $P_i^{(1)} = P_i^{(2)}, W_i^{(1)} = W_i^{(2)}, K_i^{(1)} = K_i^{(2)}$, go to 3 else stop.

The proposed procedure for the design of decentralized robust controller that ensures the exponential stability of the uncertain system (1) with a prescribed decay rate α is summarized in the next algorithm.

Algorithm B.

1. Choose $\alpha_i > \alpha$, and $\gamma_{ai}, \gamma_{bi}, i \in \mathcal{N}$.
2. Use the Algorithm A for all $i \in \mathcal{N}$.
3. Check the nonnegative definiteness of the matrices $M_i, i \in \mathcal{N}$. The approximate values of the coefficients $\epsilon_{i1}, \epsilon_{i2}$ can be chosen using following relations

$$\epsilon_{i1} = (\frac{\lambda_M^2(P_i)\gamma_{ai}}{\lambda_M(Q_{0i})})^{\frac{1}{2}}$$

$$\epsilon_{i2} = (\frac{\lambda_M^2(P_i)\gamma_{bi}}{\lambda_M(C_i^T K_i^T R_i K_i C_i)})^{\frac{1}{2}}, \qquad i \in \mathcal{N}$$

If each M_i is nonnegative definite then stop, else increase α_i or the eigenvalues of Q_{1i} and/or decrease γ_{ai}, γ_{bi}.

Algorithms A and B yield for the given α optimal values of $\alpha_i, Q_{1i}, \gamma_{ai}$ and γ_{bi} which guarantee the nonnegative definiteness of M_i and the exponential stability of the complex uncertain system (9) with the prescribed decay rate α.

4. Example. Design of Robust Decentralized Controller.

Consider the 6th order system described by

$$\dot{x} = \begin{bmatrix} A_{11} & A_{12} \\ A_{21} & A_{22} \end{bmatrix} x + \begin{bmatrix} B_1 & 0 \\ 0 & B_2 \end{bmatrix} u +$$

$$\begin{bmatrix} \delta A_{11} & \delta A_{12} \\ \delta A_{21} & \delta A_{22} \end{bmatrix} x + \begin{bmatrix} \delta B_1 & 0 \\ 0 & \delta B_2 \end{bmatrix} u$$

$$y = \begin{bmatrix} C_1 & 0 \\ 0 & C_2 \end{bmatrix} x$$

where

$$u^T = [u_1^T u_2^T] \in R^4, \qquad x^T = [x_1^T x_2^T] \in R^6$$

$$y^T = [y_1^T y_2^T] \in R^4, \qquad n = 6, m = 4, l = 4$$

and

$$A_{11} = \begin{bmatrix} -6.1 & .23 & .5 \\ .3 & -5 & -3.5 \\ 1.5 & 2.3 & -5.3 \end{bmatrix}$$

$$\delta A_{11} = \begin{bmatrix} 1.5 & 1.5 & 1.5 \\ 1.1 & 1.23 & 1.1 \\ .6 & 1 & 1.5 \end{bmatrix}$$

$$A_{12} = \begin{bmatrix} .3 & .35 & .4 \\ .15 & -.32 & .3 \\ .32 & .43 & .22 \end{bmatrix}$$

$$A_{22} = \begin{bmatrix} -5.1 & .2 & -1 \\ .23 & -4.5 & -4 \\ 2 & 3.1 & -5.8 \end{bmatrix}$$

$$\delta A_{22} = \begin{bmatrix} 1 & 1 & 1 \\ 1.2 & 1.2 & 1.2 \\ .4 & .5 & .2 \end{bmatrix}$$

$$A_{21} = \begin{bmatrix} .23 & .35 & .2 \\ .2 & -.35 & .45 \\ .4 & .34 & .33 \end{bmatrix}$$

$$\delta A_{12} = 0, \qquad \delta A_{21} = 0$$

$$B_1 = \begin{bmatrix} 0 & 2 \\ .1 & .15 \\ 1.3 & 0 \end{bmatrix} \qquad \delta B_1 = \begin{bmatrix} .3 & .4 \\ .2 & .2 \\ .3 & .35 \end{bmatrix}$$

$$B_2 = \begin{bmatrix} 0 & 2 \\ .2 & .1 \\ 1 & 0 \end{bmatrix} \qquad \delta B_2 = \begin{bmatrix} .2 & .2 \\ .2 & .2 \\ .3 & .3 \end{bmatrix}$$

$$C_1 = C_2 = \begin{bmatrix} 0 & 1 & 0 \\ 0 & 0 & 1 \end{bmatrix}$$

Results of the decentralized procedure are the following.

1.subsystem. The matrix M_1 (16) is positive definite under the following conditions:

$$\alpha = .2, \quad \alpha_1 = 2.5, \quad \gamma_{a1} = .4 \quad \gamma_{b1} = .4$$

$$Q_{11} = q_1 diag\{1\}, \quad q_1 = 17, \quad Q_{01} = \delta A_{11}^T \delta A_{11}$$

$$R_1 = \delta B_1^T \delta B_1$$

The gain matrix K_1 (8) is computed using the Algorithm B

$$K_1 = \begin{bmatrix} -.1494 & -1.512 \\ -.5813 & -.6027 \end{bmatrix}$$

and the eigenvalues of the first nominal closed-loop subsystem are $-5.6234; -6.4222 \pm i2.7781$

2.subsystem. The matrix M_2 is positive definite for the following parameters:

$$\alpha_2 = 4.5, \quad \alpha = 1, \quad \gamma_{a2} = 1, \quad \gamma_{b2} = 1$$

$$Q_{12} = q_2 diag\{1\}, \quad q_2 = 8, \quad Q_{02} = \delta A_{22}^T \delta A_{22}$$

$$R_2 = \delta B_2^T \delta B_2$$

The gain matrix K_2

$$K_2 = \begin{bmatrix} -.0507 & -.6349 \\ -.3419 & -.2795 \end{bmatrix}$$

and the eigenvalues of the second nominal closed-loop system are $-4.8084, -5.6354 \pm i3.8965$.
According to the decentralized design procedure the uncertain closed-loop system (9) with the output static feedback controller is exponentially stable with the decay rate $\alpha = .2$ and the robustness properties are guaranteed by the following parameters: $\gamma_{a1} = .4, \gamma_{b1} = .4, \gamma_{a2} = 1, \gamma_{b2} = 1$.
For comparison we provide results of a centralized procedure as well.
The corresponding matrix M defined by (16) for the overall system is positive definite for the following matrix K

$$K = \begin{bmatrix} -.1956 & -.655 & -.1746 & .134 \\ -.8277 & -.4988 & .2264 & -.3265 \\ -.1551 & .0918 & .0546 & -.5019 \\ -.2012 & -.0945 & -.481 & -.3236 \end{bmatrix}$$

Robust exponential stability decay rate with centralized output feedback controller is $\alpha = 2$, the decay rate for the nominal model is $\alpha_1 = 5$. The eigenvalues of the nominal closed-loop system are $-5.7579 \pm i3.836; -5.9507 \pm i2.8218; -5.0881; -4.829$. The robustness properties of the closed-loop system are specified by

$$\delta A^T \delta A \le \gamma_a Q_0, \delta B^T \delta B \le \gamma_b R; \gamma_a = 1; \gamma_b = 1$$

$$Q_0 = \begin{bmatrix} Q_{01} & 0 \\ 0 & Q_{02} \end{bmatrix}, \qquad R = \begin{bmatrix} R_1 & 0 \\ 0 & R_2 \end{bmatrix}$$

5. CONCLUSION

The problem of the robust decentralized controllers design for complex linear systems with uncertainties and under the output feedback which guarantee the exponential stability with a specified decay rate is addressed. The uncertainties may be internal, these are inherent in each subsystem, and also external theseare due to interconnections among subsystems. The upper bounds of the internal and external uncertainties upper bounds are assumed to be known. In the proposed robust decentralized controller design procedure, in contrast with the references, no matching conditions are considered. The proposed algorithm is applied to a 6th order linear system with uncertainties.

6. ACKNOWLEDGMENT

This research was suported by grant 1/4231/97.

REFERENCES

Bernussou, J. and Titli,A. (1982), Interconnected dynamical systems stability, decomposition and decentralization, North Holland, Amsterdan.

Chen,Y.H. (1988), Decentralized robust control system design for large-scale uncertain systems, IJC, 47, 1195-1205.

Fradkov, A.L. (1990), Adaptive control of complex systems, Nauka, Moscow (in Russian)

Halicka, M. and Rosinova, D. (1994), Stability robustness bound estimates of discrete-time systems: Analysis and comparison, IJC, 60, 297-314.

Han, M.C. and Chen, Y.H. (1995), Decentralized control design: uncertain systems with strong interconnections, IJC, 61, 1363-1385.

Ioannou, P.I. (1986), Decentralized adaptive control of interconnected systems, IEEE Trans. on AC, AC-31, 290-298.

Jury, E.I. (1990), Robustness of discrete systems: A review. 11th IFAC World Congres, 1990, Tallin, USSR, 184-186.

Kozák,Š. (1995), Simple and robust PID controller, Applied Mathematics and Computation, 70, 2-8.

Kučera,V. and DeSouza,C.E. (1995), A necessary and sufficient condition for output feedback stabilizability, Automatica, 31, 1357-1359.

Krasovskij, N.N. (1959), Some tasks about theory of stability motion, Fizmatgiz, Moscow, 1959 (in Russian).

Leitman, G. (1996), One method for robust control of uncertain systems: Theory and practise, Kybernetika, 32, 1, 43-62.

Mao-Lin Ni and Ying Chen, (1996), Decentralized Stabilization and Output Tracking of Large-Scale Uncertain Systems, Automatica, Vol32, N2 , 1077-1088.

Mao, X., (1996), Robustness of exponential stability of stochastic differential delay systems, IEEE Trans. on AC, Vol.41, N3, 442-447.

Murgaš, J and Hejda, I. (1990), Decentralized adaptive stabilization with state regullators, Kybernetika, Vol26, N6, 496-504.

Niculescu, S.I., DeSouza, C.E. , Dugard, L. and Dion, J.M. (1995), Robust exponential stability of uncertain systems with time-varying delays, Proceedings of 3rd ECC'95, Italy, Sept. 1995, 1802-1807.

Poolla, K.S., Shamma, J.S., and Wise, K.A. (1990), Linear and nonlinear controller for robust stabilization problems: A survey, 11th IFAC World Congres, 1990, Talin, USSR, 176-183.

Singh, M.G. (1981), Decentralized control, North Holland, Amsterdan.

Šiljak,D.D. and Vukčevič, M.B. (1977), Decentrally stabilizable of linear and bilinear large-scale systems, IJC, 26, 289-305.

Šiljak,D.D. (1991), Decentralized control of complex systems, Academic Press, New York.

Zhiming Gong , Changyun Wen and Dinesh P.Mital, (1996), Decentralized robust controller design for a class of interconnected uncertain systems with unknown bound of uncertainty, IEEE Trans. on AC, Vol.41, N6, 850-854.

Veselý,V. , Bízik,J. Murgaš,J. and Gatial, J. (1981), Decentralized control of dynamic linear systems, Proceedings of VIII IFAC World Congres, Kyoto, Japan, IX39- IX43.

Veselý,V. (1993), Large scale dynamic system stabilization using the principle of dominant subsystems approach, Kybernetika, Vol29, N1, 48-61.

DECENTRALIZED DECISION AND CONTROL

J. Sarnovský* and J. Liguš*

* *Technical University of Košice, Department of Cybernetics and Artificial Intelligence, Košice, Letná 9, Slovak Republic*

Abstract. In this paper the methods of decentralized quantitative (statistical) and qualitative (fuzzy) decision and control algorithms are presented. The body of the paper is subdivided into two main sections. The first one describes decentralized statistical decision algorithms based on Bayesian conditional probabilities and the second one describes decentralized fuzzy control and decision algorithms. The derived mathematical apparatusies are described in detail. The principles of decentralization methods are designed and verified by simulations. The pendulum-car system and dynamic systems with transfer functions: $F(s) = \frac{1}{s(s+a)}$ are used as application models.

Key Words. decentralized statistical decision; decentralized fuzzy control

1. INTRODUCTION

Today there is a wide range of parallel and distributive systems all over the world. Parallel systems are normally used for scientific purposes, whereas Distributive Control Systems (DCS) represent typical distributive systems. So far tremendous work has been done in the field of fuzzy theory and applications. Very often in research, we naturally focus on simple systems in regard to the new methods in the field of fuzzy systems (Thiele, 1995; Zadeh, 1973).

From the previous paragraph it becomes evident, that are new chalanges arising in these areas of study: (1) creating of models for large-scale systems, namely from the viewpoint of modern approaches to control of such systems; (2) creating the decentralized decision and control methods and algorithms for use in large-scale systems; (3) implementation of these algorithms of control and decision in the hierarchical distributive systems.

II. HIERARCHY IN INTELLIGENT CONTROL SYSTEMS AND DISTRIBUTIVE CONTROL SYSTEMS

Next, it is important to explore the relationship between Intelligent Control Systems and Distributive Control Systems, even though these two systems were developed independently.

Decision processes in DCS allow create Intelligent Control Systems (ICS) (ICS is detaily presented in Saridis, 1985). When decision processes on different levels in the hierarchical structure of DCS are applied, adaptive, autonomous and intelligent control systems are created in the ICS (see Fig.1). Intelligent control is hierarchically distributed according to the principle Decreasing Precision with Increasing Intelligence (DPII) in ICS (Saridis, 1985; Filasová, 1994). Intelligent control is generally divided into three levels: (1) the organization level; (2) the coordination level (3) the execution level; where every level can be broken down into more layers. The form and representation of decision processes may differ from level to level. DCS for the practical application control of technological processes were developed independently from ICS (Liguš, 1995). Presently, DCS contains three main levels (Liguš, 1995; Sarnovský and Liguš, 1996b):

(1) **The information level** consists of very large database systems installed onto powerful servers and workstations. (2) **Visualization level** is considered functionally separated level of DCS. The vizualization level serves for interaction between the human operator and the technological process in hard real-time conditions; (3) **Controller level** is represented by a set of programmable logical controllers (PLC) and special elements of control systems.

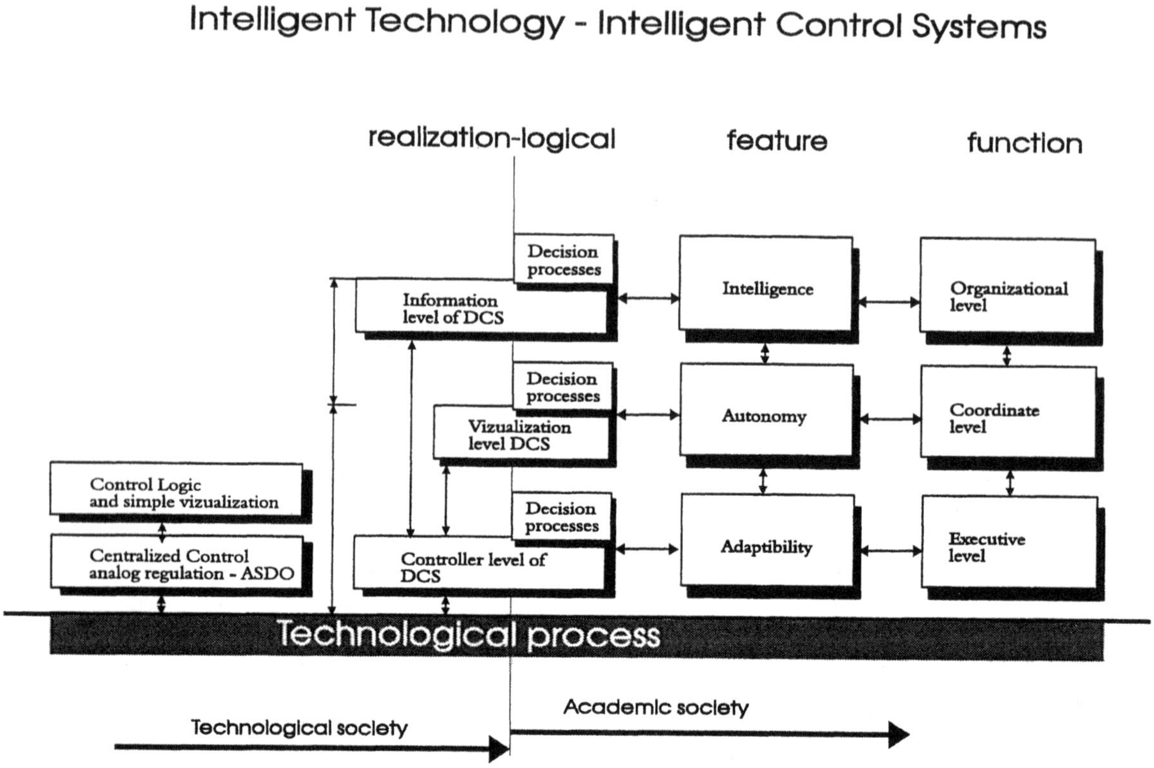

Fig.1 The relationship between DCS and ICS

By applying decision processes at the controller level a certain measure of adaptibility is achieved. By applying of decision processes at the vizualization level a certain measure of autonomy is achieved and likewise by applying decision process at the information level a certain measure of intelligence is achieved.

Therefore effort will be concentrated in this paper on the forms and representation of the control and decision algorithms at the different levels of DCS in order to reach ICS (Fig.1). This relationship between DCS and ICS is a new idea, which the authors of this paper want to emphasize. So far the decision algorithms have only been divided into two levels: (1) strategical decision; (2) operational decision. This division is not always sufficient. From the idea of three levels some conclusions regarding the relationship between DCS and ICS can be helpful in deciding the course for future research in the field of large-scale control systems. These conclusions can be summarized in the following way. For each of the mentioned levels it is worth considering the decentralized processes (Liguš, 1995; Sarnovský and Liguš, 1996a).

III. DECENTRALIZED STATISTICAL DECISION PROCESSES IN CONTROL SYSTEMS

This approach utilizes analytical (probabilistic) models to describe decision-making and control the various functions of the intelligent machine structured by the intuitively principle DPII. This section includes the decentralized quantitative statistic decision processes in the controller level of DCS and visualization level, as well.

In a basic controller level of DCS (e.g. closed-loop control) situations arise, which are characterized by changes of parameters of controlled systems (Sarnovský and Liguš, 1995; Sarnovský *et al.*, 1995). Based on measurement (valuation) of the parameters and according to the decision procedure a suitable structure of the feedback is assigned (Fig.2) (Liguš *et al.*, 1996; Valaštek and Liguš, 1996).

Consider the n such closed-loop systems (Fig.2) as blocks RO_i (Fig.3 and Fig.4) with transfer functions in the Laplace form $F_i(a,s)$, where parameters a_{ij} are changeable parameters, which are measured in an on-line way. In the place of the controllers, different control algorithms are used k_{ij}. The parameters a_{ij} can acquire a final number of discrete states. The parameters a^*_{ij} are measured under the influence of measurement noise or non-exact identification algorithm, when values of the parameters $a_{i,j}$ are actual. The conditional probabilities $p(a^*_{ij}|a_{ij})$ and apriori probabilities $p(a_{ij})$ are defined from the apriori knowledge of the systems. The respective values k_{ij} to a_{ij} are chosen to optimize the transient process. It can be considered, that criterion matrix (performance index) $U_i(a_{ij}, k_{ij})$, acquires minimal values for the correctly assigned values k_{ij} to a_{ij};

where block RO_i is i-th closed control loop; a_{ij} are parameters (variants); k_{ij} are alternatives for decision algorithms;
$i = 1, 2, \cdots, n$;
$j = 1, 2, \cdots, m$;
n – number of decentralized subsystems
m – number of variants and alternatives for each subsystem

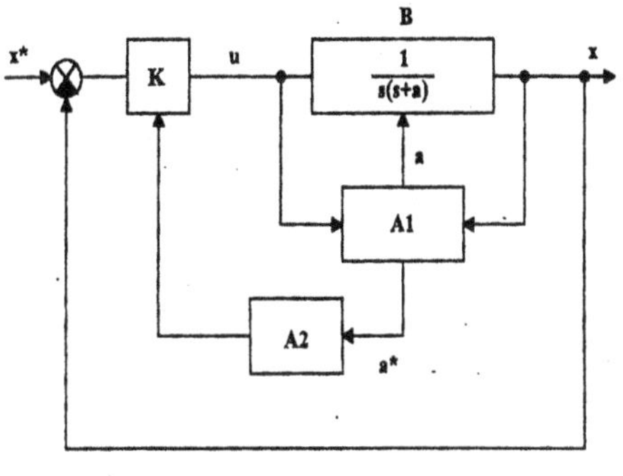

Fig.2 Typical example of decision-making in a closed-loop (A1-identification block, A2-block of decision algorithm).

According to the authors of this paper, decision algorithms in decentralized control systems can be divided as follows:

1. central decision-making in decentralized control systems - Fig.3.

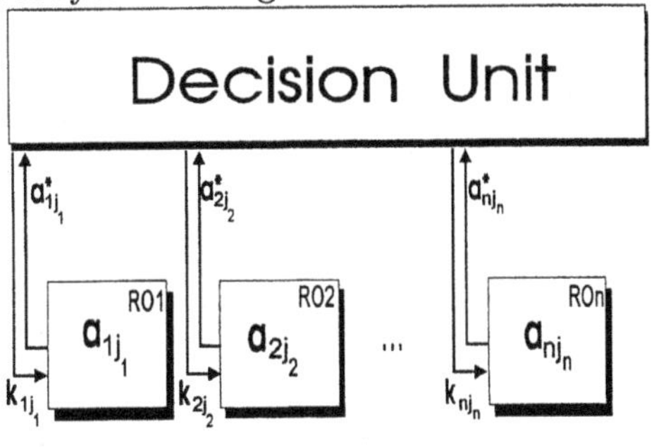

Fig.3 The scheme of central decision-making in decentralized control systems

The assignment of the "best" k_{ij} can be computed in one centralized calculation. This type of decision-making is frequently very difficult to realize because of complicated final calculation as well as criterion functions (matrices). Therefore they can be solved by:

2. partially decentralized decision making in decentralized control systems *eg* concerning criterion functions - Fig.4

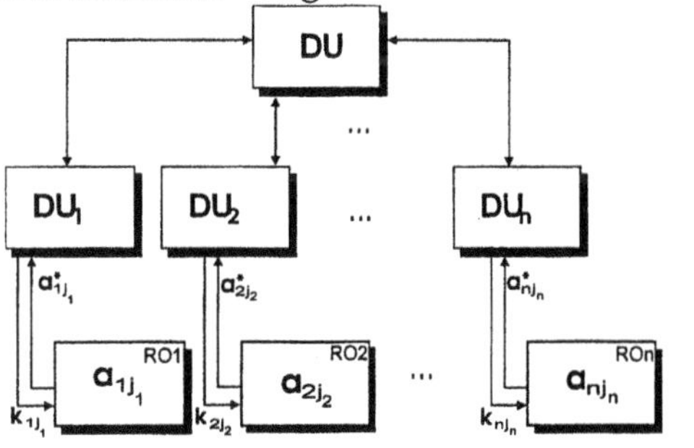

Fig.4 The scheme of partially decentralized decision making in decentralized control

We can consider two options regarding the partially decentralized decision:

- when parameters of each subsystems are independent,
- when parameters of each subsystem are mutually dependent.

We derived mathematical apparatusies for both options.

3.1 Independent parameters of subsystems

In the following sections n-subsystems will be discussed. When the measured parameters a_{ij} of each subsystems are independent, the risk of control algorithm assignment is computed as follows: (see relation 1):

$$r(\bigcap_{i=1}^{n}(a_{ij}^*, k_{iv})) = \\ = \frac{\prod_{i=1}^{n} R_i(a_{ij}^*, k_{iv})}{p(\bigcap_{i=1}^{n} a_{ij}^*)} \quad (1)$$

where $R_i(a_{ij}^*, k_{iv}) = \sum_{r=1}^{m} U_i(a_{ir}, k_{iv}).p(a_{ir}).$

$$.p(a_{ij}^*|a_{ir})\} \quad (2)$$

$i = 1, 2, \cdots, n,$
$j = 1, 2, \cdots, m,$
$v = 1, 2, \cdots, m.$

I-th decision unit DU_i of the *i-th* closed-loop subsystem (RO_i) computes relation $R_i(a_{ij}^*, k_{iv})$ for every alternative, when parameter a_{ij}^* is measured. Decision unit DU on the second level of hierarchy multiplies the results from every DU_i and finds minimal risk of assignment and provides setting of the control algorithm. In this kind of decentralization process is the same case when every DU_i computes relation $R_i(a_{ij}^*, k_{iv})$ for every alternative, when parameter a_{ij}^* is measured and finds minimal risk of assignment - $\min(R_i)$. It is because of validity:

$$\min(R_1.R_2.\cdots.R_n) = \min(R_1).\min(R_2). \cdots .\min(R_n). \quad (3)$$

Therefore the assignment also can be considered as total decentralized decision-making, i.e. without DU on the second level of decision-making hierarchy.

Decision algorithm (equation 1) **expressed in matrix form:**

$$\mathbf{r}_i = (\mathbf{U}_i^T.\mathbf{p}_{ai}.\mathbf{p}_{bi}^T)^T \quad (4)$$

where

$$\mathbf{U}_i = U_i(a_{ir}, k_{iv}) = \begin{bmatrix} U_{11}^i & U_{12}^i & \cdots \\ U_{21}^i & U_{22}^i & \cdots \\ \vdots & \vdots & \ddots \end{bmatrix} \quad (5)$$

$$\mathbf{p}_{ai} = p_{aij} = \begin{bmatrix} p(a_{i1}) & 0 & \cdots \\ 0 & p(a_{i2}) & \cdots \\ \vdots & \vdots & \ddots \end{bmatrix}$$

$$p(a_{ij}^*|a_{ir}) =$$

$$= \mathbf{p}_{bi} = \begin{bmatrix} p(a_{i1}^*|a_{i1}) & p(a_{i1}^*|a_{i2}) & \cdots \\ p(a_{i2}^*|a_{i1}) & p(a_{i2}^*|a_{i2}) & \cdots \\ \vdots & \vdots & \ddots \end{bmatrix} \quad (6)$$

Every DU_i finds minimum values according to raws of final matrix $\mathbf{r}_i$.

3.2 Mutually dependent parameters of subsystems

Furthermore when parameters of subystems are mutually dependent the final decision calculation is given:

$$r(\bigcap_{i=1}^n (a_{ij}^*, k_{iv})) =$$

$$= \frac{\prod_{i=1}^n \sum_{r=1}^m U_i(a_{ir}, k_{iv}).p(a_{ij}^*|a_{ir}).p(a_{ir}|a_{(i-1)j}^*)}{p(\bigcap_{i=1}^n a_{ij}^*)} \quad (7)$$

where

$$p(a_{ir}|a_{(i-1)j}^*) = p(a_{ir}) \text{for i=1}. \quad (8)$$

$i = 1, 2, \cdots, n,$
$j = 1, 2, \cdots, m,$
$v = 1, 2, \cdots, m.$

There are different modifications for the last relation 7. Modifications depend on the concrete Bayesian model *i.e* on the chosen order of the phenomenons which are arising. In equation 7 the order of the phenomenons are considered as follows:

$a_{1j}$of first subsystem $= A_1$
$a_{1j}^*$of first subsystem $= A_2$
$a_{2j}$of second subsystem $= A_3$
$a_{2j}^*$of second subsystem $= A_4$

The notion of the **adjoining phenomenons** can be introduced here in our case *eg* adjoining phenomenons of A_4 and A_3, are adjoining phenomenos when the arising of the phenomenon A_4 is dependent on the arising phenomenon A_3 and A_4 is not dependent on the arising phenomenons A_2 and A_1.

Decision algorithm in matrix form (mutually dependent parameters of subsystems - equation 7):

$$\mathbf{r}_i = (\mathbf{U}_i^T.\mathbf{p}_{ai}^T.\mathbf{p}_{bi}^T)^T \quad (9)$$

where

$\mathbf{U}_i$ and $\mathbf{p}_{bi}$ are the same matrices as in case, when parameters a_{ij} are independent (matrices 5, 6). Different is matrix $\mathbf{p}_{ai}$
$p(a_{ir}|a_{(i-1)j}^*) =$

$$\mathbf{p}_{ai} = \begin{bmatrix} p(a_{i1}|a_{(i-1)1}^*) & p(a_{i1}|a_{(i-1)2}^*) & \cdots \\ p(a_{i2}|a_{(i-1)1}^*) & p(a_{i2}|a_{(i-1)2}^*) & \cdots \\ \vdots & \vdots & \ddots \end{bmatrix}$$

for $i = 2, 3, \cdots, n$
for $i = 1$ the matrix $\mathbf{p}_{a1}$ is in the form:

$$\mathbf{p}_{a1} = p_{a1j} = \begin{bmatrix} p(a_{11}) & 0 & \cdots \\ 0 & p(a_{12}) & \cdots \\ \vdots & \vdots & \ddots \end{bmatrix}$$

The mutual dependence of parameters of subsystems in decision-making is analogous to interactions among subsystems in decentralized control. When dealing n subsystems, where for i-th subsystem x_i variants and y_i alternatives are given, then the dimension of criterion matrix is: $U(x_1 \text{ x } y_1 \text{ x } x_2 \text{ x } y_2 \text{ x } \cdots \text{ x } x_n \text{ x } y_n)$. Generally it is very difficult to design such a criterion matrix under real conditions. Under partial decentralization with mutually dependent parameters $a_{ij}, i = 1, \cdots, n$ the criterion matrices are designed independently for individual subsystems $U_i(x_i x y_i), i = 1, \cdots, n$ and they are gained from experts or they are cerated from the knowledge of each control loop.

The relations also were derived based on Bayesian Networks theory.

3. total decentralization of decision-making in decentralized control systems, when parameters a_{ij} are mutually dependent. This approach is under observation.

Simulation results

The decentralized decision algorithms were simulated for two closed-loop systems and power (energy) system of two controlled areas, as well. The results are quite promising. We will not present all the simulation results because of the limitation of this paper. Fig.5 shows the resulting trajectories of two closed-loop systems with transfer functions $F_1(s) = \frac{1}{s(s+a_{1j})}$ and $F_2(s) = \frac{1}{s(s+a_{2j})}$ as mentioned in the abstract.

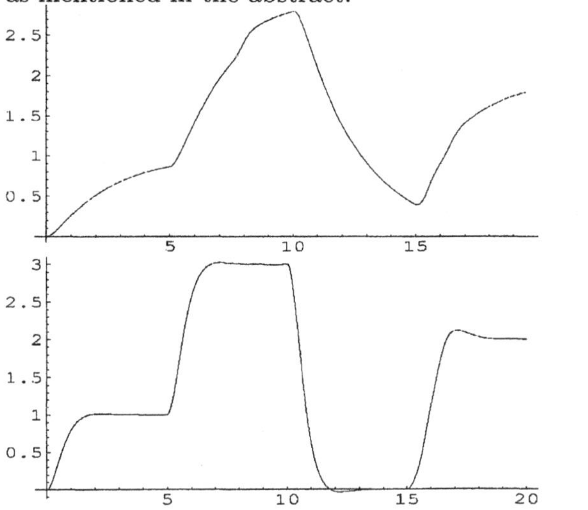

Fig.5 Simulation results without (1) and with (2) decision-making

IV. DECENTRALIZED FUZZY DECISION AND CONTROL ALGORITHMS

With respect to the numerous applications of IF-THEN rules to fuzzy control and fuzzy decision making and for the reason mentioned in section II, the importance of investigations concerning decentralized IF-THEN rule bases can not be overestimated (Thiele, 1995).

Regarding the decentralization of fuzzy rulebases can be considered:

1. cascade decentralization of rulebase
2. level decentralization rulebase
3. mixed decentralization rulebase

There are considered two kinds of decentralization processes in fuzzy systems:

4.1 Partiall decentralization

There is applied decentralization process to the three phases of fuzzy inference mechanism: (1) fuzzification, (2) inference, (3) composition phases in partiall decentralization. Here defuzification proces is executed only on the end of hierarchy.

The number of rules in a fuzzy systems grows overproportional with the number of linguistic variables and connected linguistic terms. For a complete rule base with input variables $\{X_i | i = 1, ..., n\}$ with terms $\{\Lambda_{ij} | j = 1, ..., q_i\}$ and output variables $\{Y_i | i = 1, ..., l\}$ with terms $\{B_{ij} | j = 1, ..., p_i\}$ one needs N fuzzy rules with

$$N = \prod_{i=1}^{n} q_i \tag{10}$$

The rules have the form:

$$if X_1 = \Lambda_1 \text{ and } \ldots \text{and } X_n = \Lambda_n \text{ then } Y_1 = B_1 \text{ and } \ldots \text{and } Y_l = B_l \text{ with } \Lambda_i \in \{\Lambda_{i_j}\}, B_i \in \{B_{ij}\}. \tag{11}$$

This large number of rules complicates the design of a rule base, because for each of the N different premises the expert must suggest a combination of terms for the output variables. It is possible to omit part of the rule base, if it could be guaranteed that certain combinations of input variables do not appear in the dynamic system during control. Such a proof requires an analytical model of the controlled system.

A hierarchical architecture of the rule base permits reduction of the number of rules. This could be done by introducing hidden fuzzy variables in analogy to hidden neurons in neural networks. The architecture of the hierarchical fuzzy controller is shown in Fig.6. In a first step the n linguistic input variables X_i are divided into disjunct sets $\{Q_m | m = 1, ..., r\}$ each with s_k input variables.

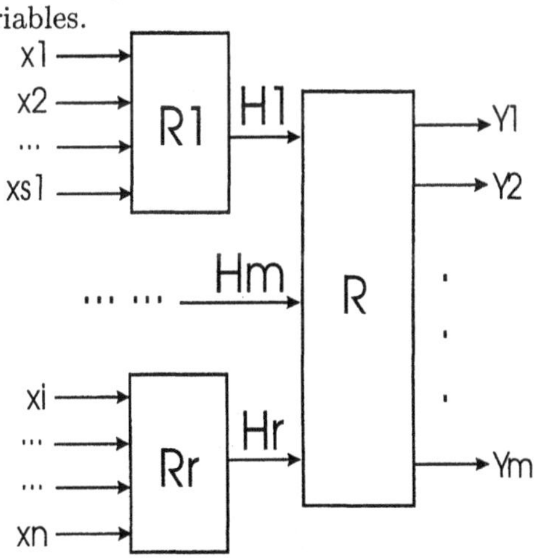

Fig.6 Architecture of the hierarchical fuzzy controller

With each set Q_m a complete rule base R_m of rules such as

$$if X_1 = \Lambda_1 \text{ and } \ldots \text{and } X_{s_m} = \Lambda_{s_m} \text{ then } H_m = C_m \tag{12}$$

is associated. There are no fuzzy sets associated to the hidden fuzzy variables H_m in the consequences. They are fully described by the membership values $\mu(C_{mj})$ of their fuzzy terms $C_m \in \{C_{mj} | j = 1, .., p_m\}$. The degrees of membership for fuzzy terms C_m are evaluated by inference of rule base R_m. Degree of membership $\mu(B_{mj})$ for output variables Y_i is obtained by inference of the overall rule base R with rules such as

if $H_1 = C_1$ and ...and $H_r = C_r$ then $Y_1 = B_1$ and ...and $Y_l = B_l$.

(13)

By defuzzification of fuzzy sets B_{ij} one obtains crisp values y_i for output variables. Grouping together different premises by associating them with the same hidden fuzzy term results in a fuzzy-OR of these premises. All premises of rule base R_m possessing the same hidden fuzzy term C_{mj} as consequence finally vote for the same consequence B_{ij} in rule base R. Let P_i^m be premises associated to the same term of hidden fuzzy variable H_m. The resulting overall rule base could be described by rules such as

if (P_1^1 or P_2^1 or ...) and (P_1^2 or P_2^2 or ...) and ... ...and (P_1^r or P_2^r or ...) then $Y_1 = B_1$ and ...and $Y_l = B_l$.

(14)

By mapping different premises to the same hidden fuzzy term the overall number of rules N is drastically reduced to:

$$N = \sum_m N_{n_m} + N_n \qquad (15)$$

where N_{n_m}, N_n are the number of rules for rule bases R_m, R, which could be obtained from the equation 10, but with smaller number of input variables n compared to a non-hierarchical fuzzy controller.

4.2 Total decentralization

In this case, every subsystem (decentralized subrulebase) includes all four phases of fuzzy inference mechanism. Defuzification methods are used in every subsystem output. This approach was also verified and simulated on the pendulum-car system.

4.3 Application of decentralized fuzzy controller for pendulum-car system

The methodology of decentralization of fuzzy rulebase is applied on chosen pendulum-car system (Fig.7)(Liguš, 1995). The three control systems: (1) swing-up controller; (2) balancing and centering controller and (3) decision-maker chooses between controller (1) and controller (2). The controller for balancing and centering uses two decentralized rulebases:
(1) rulebase includes θ $\dot{\theta}$;
(2) rulebase includes x and $\dot{x}$ (Fig.7).

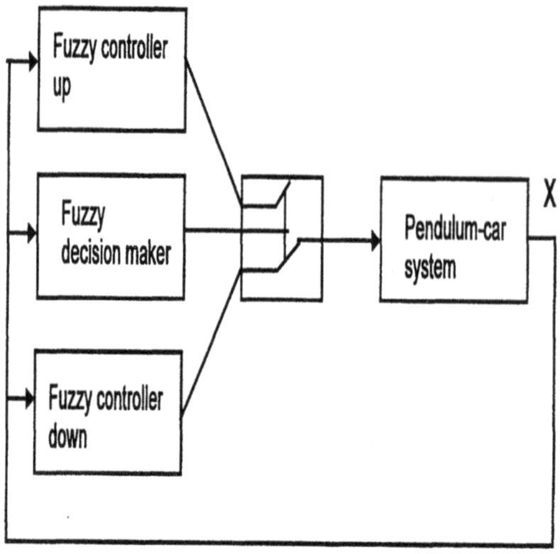

Fig.7 The scheme of pendulum-car control system

4.4 Results of the approach

Simulations were applied in Matlab/Simulink software environment. There were developed Matlab/Simulink infer functions **inferh** and **inferl** of decentralized approach for fuzzy decision and control. In the three controllers are used triangular membership functions. The values of the membership function parameters was acquired by experiments and tests.

Results of non-decentralized and decentralized approaches are the same (in Fig.8).

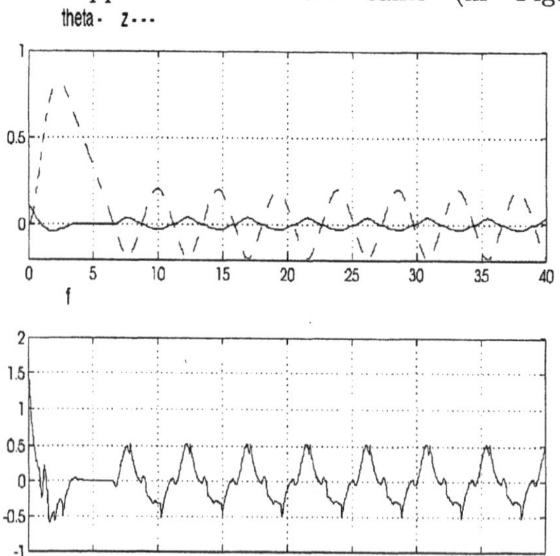

Fig. 8 Results of non-decentralized, partial decentralized and total decentralized rulebase for fuzzy control of the pendulum-car system.

V. CONCLUSION

The three main advantages of decentralized quantitative and most importantly qualitative decision and control algorithms is that, they are easier to design, easier to implement into *eg* Distributive Control Systems and easier to verify. Another advantage is increased computing speed of decomposed/decentralized systems.

Decentralization fuzzy control and decision is still fairly new and has not yet been widely used. Many different situations arise, when dealing with decentralization process especially with fuzzy decentralization process (investigation under different defuzification methods on every output of subsystems, investigation under different inference mechanism etc). The question is still important: "How to design a general decentralization process when deffuzification process would be applied in every subsystems?"

Applications of total decentralization are convenient for control systems, which have limited number of inputs and limited size of rulebase (*e.g.* fuzzy rulebase in PLC controllers).

At this point, regarding the fuzzy decision processes, authors want to emphasize, that so far was considered with fuzzy decision based on fuzzy rules/rulebases. Fuzzy decentralization is very suitable for system decentralization of fuzzy-neural control and decision *eg* for ANFIS fuzzy-neural systems.

REFERENCES

Cellier,F. and Mugica,F. (1992). Systematic Design of Fuzzy Controllers Using Inductive Reasoning, *Proceedings of the 1992 IEEE International Symposium on Intelligent Control*, 11.-13. August, Glasgow, Scotland, UK

Filasová, A., (1994). Synthesis of linear controller using pairwise autonomous subsystem approach, *Proc. 1th IFAC Workshop on New Trends in Design of control Systems*, Smolenice, 1994, pp. 185-189.

Jadlovský, J. and Liguš, J. (1996). The Principle of Direct Element Communication in Distributive Control Systems, *Conference on Intelligent Technologies*, Herľany, Slovakia, November 1996, (oral presentation).

Liguš, J. (1995). Decentralized Fuzzy Decision and Control, *Pre-work of dissertation thesis*, June 1995 (in Slovak)

Liguš, J., Marvan, I. and Štěpánková, O. (1996). Tools for Fuzzy Inductive Reasoning and Qualitative Simulation, *Proceedings of Symposium on Qualitative System Modelling, Qualitative Fault Diagnosis and Fuzzy Logic and Control*, April 17-20, 1996, Budapest and Balatonfured, Hungary, pp. 322-327.

Saridis G.N. (1985). An Integrated Theory of Intelligent Machines by Expressing the Control Performance as Entropy, *Control-Theory and Advanced Technology*, Vol.1, No.2, pp. 125-138, 1985

Sarnovský J. and Liguš, J. (1996a). Fuzzy-Bayesian Inference in Decision Algorithms, *Journal of Electrical Engineering 47*, No. 9-1, 1996, pp. 225-231

Sarnovský, J. and Liguš, J. (1996b). Intelligent Methods and Algorithms of Control and Decision in Large-Scale System, *Proceedings of the 2nd scientific-technical conference with international participation*, Process Control, Volume 1, University of Pardubice, 1996, pp. 297-302

Sarnovský, J. and Liguš, J. (1996). Fuzzy-Bayesian Inference in Decision Algorithms, *Proceedings of Symposium on Qualitative System Modeling*, Budapest, 1996, pp. 328-333

Sarnovský, J., Liguš, J. and Jadlovský, J. (1993). Decision Processes in Automatic Control Systems and in Decentralized Control, *Bulletin for Applied Mathematics, Techn. University Budapest*, 971/94, pp. 117-122

Thiele, H. (1995). On the Calculus of Fuzzy if-then Rules, *Intelligent Technologies and Soft Computing*, Mangalia, Romania, September 10-16, 1995

Valaštek, M. and Liguš, J. (1996). Tool for Fuzzy Inductive Reasoning, In: *Proceedings of Conference on Intelligent Technologies*, Vol.II, Herľany, Slovakia, November 1996, pp. 133-144.

Zadeh, L. (1973). Outline of a New Approach to the Analysis of Complex Systems and Decision Processes, *IEEE Trans. on Systems, Man and Cybernetics*, 3(1):28-44

ROBUST DECENTRALIZED CONTROL OF MIMO SYSTEMS IN THE FREQUENCY DOMAIN

Alena Kozáková [1]

Dept. of Automatic Control Systems,
Faculty of Electrical Engineering and Information Technology,
Slovak University of Technology, Bratislava, Slovak Republic.
E-mail: kozakova@kasr.elf.stuba.sk

Abstract: The paper presents an application-oriented robust decentralized control design technique based upon the sufficient condition for robust stability (SCRS) of systems under decentralized control (DC) which is fulfilled using parameter tuning of fixed-structure local controllers. Graphical interpretation of the SCRS provides two useful stability criteria used to test stability of the overall system as well as stability of subsystems. Theoretical results are illustrated in a case study.

Keywords: Robust control, Decentralized control, Stability robustness, Frequency domain

1. INTRODUCTION

Real plants usually have a complex structure. However, it is often possible to decompose them into smaller subsystems w.r.t. the specific functions and goals defined by technological specifications. Then, the mathematical model of the overall system is an interconnection of the subsystem models and interactions among them. In such cases it is beneficial to apply the decentralized control (DC) with following properties (Veselý and Murgaš, 1992): The DC should 1. be designed using information from subsystems only; 2. guarantee the overall system stability and satisfy given performance requirements; 3. be robust against possible changes (a) in interconnections among subsystems, (b) in structure and parameters of individual subsystems; 4. the DC laws have to be obtained in a decentralized design procedure.

In this paper an individual design procedure has been developed to guarantee the fulfillment of these requirements.

[1] The work has been partially supported by the Slovak Scientific Grant Agency, Grant No. 95/5195/198

2. PROBLEM STATEMENT

Consider a square transfer function matrix $\tilde{G}(s)$ to be the mathematical model of an uncertain MIMO system with n subsystems

$$\tilde{G}(s) = G(s) + \Delta(s) \tag{1}$$

The uncertainty is described by the perturbation term $\Delta(s) \in \{\mathcal{D}_U, \mathcal{D}_S\}, s = j\omega$, where $\mathcal{D}_U$ denotes the class of unstructured and $\mathcal{D}_S$ the class of structured perturbations. After choosing an appropriate input/output variables pairing, the plant transfer matrix $G(s)$ diagonal entries correspond to the paired variables (individual SISO subsystems) and are collected in the diagonal transfer-function matrix $G_D(s) = diag\{G_{ii}(s)\}_{i=1,\dots,n}$. The interactions among subsystems are in the matrix $G_N(s) = G(s) - G_D(s)$. Thus (1) can be rewritten as follows

$$\tilde{G}(s) = G_D(s) + G_N(s) + \Delta(s) \tag{2}$$

Throughout this paper, all these three transfer matrices are supposed to have equal number of RHP poles i.e. $\eta[G(s)] = \eta[\tilde{G}(s)] = \eta[G_D(s)]$. In the decentralized control $G_D(s)$ is considered to be the nominal model and the term $[G_N(s) + \Delta(s)] = [\tilde{G}(s) - G_D(s)]$ to represent the new defined additive perturbation. The decentralized controller transfer function matrix is $R(s) = diag\{R_i(s)\}_{i=1,\dots,n}$ where $R_i(s)$ represents the $i-th$ local controller transfer function. $R(s)$ is supposed to guarantee stability of the nominal (diagonal) closed-loop system $H_D(s) = G_D(s)R(s)[I + G_D(s)R(s)]^{-1} = diag\{H_i(s)\}_{i=1,\dots,n}$. The point at issue is this: *If the nominal (diagonal) closed-loop system is stable and meets certain performance requirements, what constraints have to be placed on individual designs in order to guarantee robust stability of the overall closed-loop system* $\tilde{H}(s) = \tilde{G}(s)R(s)[I + \tilde{G}(s)R(s)]$ *?* The solution to this issue evolves from the SCRS for systems under DC (Grosdidier and Morari, 1986; Morari and Zafiriou, 1989; Zietz and Schufft, 1989) which has been extended to consider uncertainties not only interactions among subsystems but in subsystem models as well.

3. MAIN RESULTS

3.1 *Necessary and Sufficient Stability Robustness Conditions*

According to Grosdidier and Morari (1986), under the assumption that $H_D(s)$ is stable, the decentralized closed-loop feedback system $H(s) = G(s)R(s)[I + G(s)R(s)]^{-1}$ remains stable iff the encirclement condition of the Generalized Nyquist criterion $N[0, \det(I + L_H(s)H_D(s)] = 0$ is met. This provides a condition for $H_D(s)$ to remain stable under the multiplicative perturbation $L_H(s) = [G(s) - G_D(s)]G_D^{-1}(s)$. This idea has been extended to derive necessary and sufficient condition for robust stability of decentralized feedback systems in the following theorem by defining

$$L_R(s) = [\tilde{G}(s) - G_D(s)]G_D(s)^{-1} \qquad (3)$$

to be a new *(generalized) multiplicative perturbation matrix.* Besides the uncertainty inherent in decentralized feedback systems represented by interactions, the matrix $L_R(s)$ involves uncertainties of subsystems as well.

Theorem 1. Assume that $\eta[\tilde{G}(s)] = \eta[G_D(s)] = \eta[G(s)]$ and that $H_D(s)$ is stable. Then the perturbed closed-loop system $\tilde{H}(s)$ is stable iff

$$N[0, \det(I + L_R(s)H_D(s)] = 0 \qquad (4)$$

Proof 1. Applying the idea from Zietz and Shufft (1989) we have factorized the return-difference matrix of the overall perturbed system under the diagonal controller $\tilde{F}(s)$ in terms of the decomposed system (2). It is assumed that $\det G_D(s) \neq 0$ and $G_D^{-1}(s)$ exists.

$$\begin{aligned}\tilde{F}(s) &= [I + \tilde{Q}(s)] = [I + \tilde{G}(s)R(s)] = \\ &= [I + L_R(s)H_D(s)][I + G_D(s)R(s)]\end{aligned} \qquad (5)$$

Denote the number of open-loop unstable poles of $\tilde{Q}(s)$ and $Q_D(s)$ by p_{QRHP}. Using the Generalized Nyquist stability criterion (counting the clockwise encirclements), $\tilde{H}(s)$ is stable iff

$$\begin{aligned}&N\{0, \det[I + \tilde{G}(s)R(s)]\} = \\ &N\{0, \det[I + G_D(s)R(s)]\} + \\ &+N\{0, \det[I + L_R(s)H_D(s)]\}\end{aligned} \qquad (6)$$

As $H_D(s)$ is stable by assumption, i.e.

$$N\{0, \det[I + G_D(s)R(s)]\} = -p_{QRHP} \qquad (7)$$

after substituting (7) into (6), (4) follows immediately.

3.2 *Sufficient Condition for Robust Stability (SCRS)*

The following theorem is an extension of the results of Grosdidier and Morari (1986) in that instead of L_H it uses the new-defined matrix L_R. The proof is similar as the one in the cited reference and therefore omitted.

Theorem 2. Assume $H_D(s)$ to be stable. U.t.c. the closed-loop system $\tilde{H}(s)$ is stable if

$$\|L_R(j\omega)H_D(j\omega)\| < 1 \qquad \forall\omega \qquad (8)$$

where $\|\cdot\|$ denotes any compatible norm.

Modification of (8) using norm properties yields

$$\|H_D(j\omega)\| < \|L_R(j\omega)\|^{-1} \qquad \forall\omega \qquad (9)$$

Substituting (3) into (9) yields

$$|H_i(j\omega)| < \frac{1}{\left\|[\tilde{G}(j\omega) - G_D(j\omega)]G_D^{-1}(j\omega)\right\|} \qquad (10)$$

$$\forall i, \qquad \forall\omega$$

The last relation expresses the *sufficient condition for stability robustness of a system under decentralized control.* The frequency-dependent function on the r.h.s. is independent of any information from subsystems. It assesses the uncertainties and interactions in the overall system and expresses a constraint imposed on the moduli of subsystem closed-loop transfer functions. It will be denoted $M_0(\omega)$. Thus

$$\left|\frac{G_{ii}(j\omega)R_i(j\omega)}{1 + G_{ii}(j\omega)R_i(j\omega)}\right| < M_0(\omega) \qquad \forall\omega \quad (11)$$

and

$$M_0(\omega) = \frac{1}{\left\| [\tilde{G}(j\omega) - G_D(j\omega)]G_D^{-1}(j\omega) \right\|} \quad (12)$$

The main idea of the individual design philosophy developed on (11) and (12) is summarized in the following corollary.

Corollary 1. Let $\eta[G(s)] = \eta[\tilde{G}(s)] = \eta[G_D(s)]$ and the nominal (diagonal) closed-loop system under the diagonal controller $H_D(s) = G_D(s)R(s)[I + G_D(s)R(s)]^{-1} = diag\{H_i(s)\}_{i=1,...,n}$ be stable. U.t.c. the overall closed-loop perturbed system $\tilde{H}(s) = \tilde{G}(s)R(s)[I + \tilde{G}(s)R(s)]^{-1}$ is guaranteed to be stable as long as there are local controllers $R_i(s)$ which guarantee the fulfillment of (11), (12).

Fulfillment of (11) can be ensured in several ways. The proposed graphical-analytical approach results in tuning parameters of local fixed-structure controllers. The graphical interpretation of (11) in the complex plane provides a stability criterion for testing stability of the overall system and of the local subsystems as well.

3.3 *Graphical Interpretation of the Sufficient Condition for Stability Robustness*

The graphical-analytical approach is based upon the relation between the closed-loop- and open-loop transfer functions of subsystems. Denote the real and imaginary parts of the open-loop transfer functions $Q(j\omega)$, respectively, $Re[Q_i(j\omega)] = U(\omega)$, $Im[Q(j\omega)] = V(\omega), \forall\omega$. After substituting in the (11) we obtain

$$|H_i(j\omega)| = \left| \frac{Q_i(j\omega)}{1 + Q_i(j\omega)} \right| = \left| \frac{U_i(\omega) + jV_i(\omega)}{1 + U_i(\omega) + jV_i(\omega)} \right| \quad \forall i, \forall\omega \quad (13)$$

Squaring both sides of (11) yields

$$\left| \frac{U_i^2(\omega) + V_i^2(\omega)}{1 + 2U_i(\omega) + U_i^2(\omega) + V_i^2(\omega)} \right|^2 < M_0^2(\omega) \quad \forall i, \forall\omega \quad (14)$$

Re-arranging (14) yields the SCRS condition in terms of $U(\omega), V(\omega)$

$$\left[U_i(\omega) + \frac{M_0^2(\omega)}{M_0^2(\omega) - 1} \right]^2 + [V_i(\omega)]^2 > \left[\frac{M_0(\omega)}{M_0^2(\omega) - 1} \right]^2 \quad \forall i, \forall\omega \quad (15)$$

which, depending upon the value of $M_0(\omega)$, describes the inside and/or outside of a circle centered at $(-\frac{M_0^2(\omega)}{M_0^2(\omega)-1}, 0)$ with a radius $r = \left|\frac{M_0(\omega)}{M_0^2(\omega)-1}\right|$. Evaluation of (15) for different constant M_0 provides a family of circles in the complex plane $[U(\omega), jV(\omega)]$ called the constant M_0-circles. For $M_0 = 1$, (14) reduces to the well-known condition $U(\omega) > -0.5$ which underlies the Optimum magnitude tuning technique. For $M_0 \neq 1$, (15) defines the the region to be avoided by the Nyquist plot of the open-loop subsystem w.r.t. stability of the global system. The 'M_0-versus-ω plot' depicts the upper bound for the closed-loop magnitude $|H_i(j\omega)|$ for each subsystem. The foregoing is summarized in the following corollary.

Corollary 2. Let $\eta[G(s)] = \eta[\tilde{G}(s)] = \eta[G_D(s)]$. The overall perturbed closed-loop system under a diagonal controller $\tilde{H}(s) = \tilde{G}(s)R(s)[I + \tilde{G}(s)R(s)]^{-1}$ is guaranteed to be stable in face of uncertainty specified by (12) as long as there are local controllers $R_i(s)$, $i = 1, ..., n$, which stabilize corresponding subsystems and ensure that

a. the Nyquist plot of each subsystem open-loop $Q_i(j\omega) = G_{ii}(j\omega)R_i(j\omega), \forall i, \forall\omega$:
   (i) *neither crosses nor is tangent to the corresponding M_0-circle if $M_0 > 1$ and*
   (ii) *lies within the corresponding M_0-circle if $M_0 < 1$.*
b. over the given range of ω, the $|H_i(j\omega)|$-versus-ω plot for $i = 1, ..., n$, is lying 'under' the M_0-versus-ω plot specified by (12).

Cor.(2a) will be termed the '*M_0-circle/Nyquist plot stability criterion*' and Cor.(2b) the '*Magnitude stability criterion*'. Both criteria are useful graphical tools for testing stability of the overall system in face of a prespecified uncertainty. The '*M_0-circle/Nyquist plot stability criterion*' assesses stability not only from the overall system point of view (situation of the Nyquist plot $Q_i(j\omega)$ w.r.t. the specific $M_0(\omega)$-circle for every specific ω) but from the viewpoint of isolated subsystems as well (classical Nyquist criterion) at one glance. Moreover, the M_0-circles together with the Nyquist plots of subsystems provide information on closed-loop performances in terms of relative stability measures (bandwidth frequency, decay rate, resonance peak and stability margins). If a computer-aided graphical facility for carrying out the above stability tests is available, they can be used for an additional open-loop shaping in subsystems as well as for a heuristic individual design, in which the designer's skills, experience and intuition play a considerable role.

4. INDIVIDUAL DESIGN

The task of finding local controllers which stabilize the corresponding subsystems w.r.t. the SCRS (11) and (12) has been solved via *parameter tuning of local fixed - structure controllers.*

4.1 *Parameter Tuning of Local Fixed-Structure Controllers*

Interpreting of (11) within the frequency-domain framework provides a design strategy for computing parameters of local controllers with an arbitrary but fixed in advance structure. The main idea is as follows. If $M_0(\omega)$ determines the closed-loop magnitude bound then it is possible to guarantee the fulfillment of (11) by tuning parameters of local controllers.

Consider a square of (11). The l.h.s. expresses the subsystem closed-loop magnitudes.

$$\left|\frac{G_{ii}(j\omega)R_i(j\omega)}{1+G_{ii}(j\omega)R_i(j\omega)}\right|^2 < M_0^2(\omega), \forall\omega, \forall i \quad (16)$$

Express $H_i(j\omega)$ as a complex rational function in $j\omega$ (the subscript i is dropped in further considerations).

$$H(j\omega) = \frac{B_n(j\omega)^n + ... + B_1(j\omega) + B_0}{A_n(j\omega)^n + ... + A_1(j\omega) + A_0} \quad (17)$$

The closed-loop magnitude square is then

$$|H(j\omega)|^2 = \frac{Q_n\omega^{2n} + ... + Q_1\omega^2 + Q_0}{P_n\omega^{2n} + ... + P_1\omega^2 + P_0} \quad (18)$$

where the coefficients Q_i, P_i are determined by B_i, A_i using the known formulas. Condition (16) can be then rewritten as

$$\frac{Q_n\omega^{2n} + ... + Q_1\omega^2 + Q_0}{P_n\omega^{2n} + ... + P_1\omega^2 + P_0} < M_0^2(\omega) \quad \forall\omega (19)$$

For a fixed ω (19) can be solved via the modified Optimum magnitude technique, according to which (19) has to hold for every ratio of numerator and denominator coefficients standing at equal powers of ω.

$$\frac{Q_n}{P_n} < M_0^2; \quad \dots \quad \frac{Q_1}{P_2} < M_0^2; \quad \frac{Q_0}{P_0} = 1 \quad (20)$$

Now, the problem reduces to the solution of (20). The proposed the graphical interpretation of each inequality uses a standard form of subsystem- and controller (here the PID) transfer functions

$$G_{ii}(s) = K_i\frac{1 + \tau_{1i}s + \tau_{2i}s^2 + ...\tau_{mi}s^m}{1 + T_{1i}s + T_{2i}s^2 + ...T_{ni}s^n} \quad (21)$$

$$m < n, \forall i$$

$$R_i(s) = r_{pi} + \frac{r_{ii}}{s} + r_{di}s \quad \forall i \quad (22)$$

Coefficients B_i, A_i and Q_i, P_i have been expressed up to the 4th order of the subsystem transfer function and the related set of inequalities has been expressed as well. As r_{di} occurs only in the last inequality as a single unknown variable, it can be computed first and used as a parameter in the remaining two inequalities. Now, only the set of nonlinear inequalities in two unknown variables (r_{pi} and r_{ii}) is to be solved.

For each subsystem (21) under a PID controller (22) the inequalities (20) represent regions in the space of controller parameters (r_{pi}, r_{ii}) determined by pertinent conic section equations. The PID controller parameters found in the inter of particular regions ensure fulfillment of the SCRS (16). However, they may happen to be either unfeasible and/or may not guarantee stability of the given isolated subsystem. Therefore it is inevitable to superimpose onto the obtained inter the region of controller parameters stabilizing the isolated subsystems, e.g. using the Neymark method of D-contours (Kozáková and Veselý, 1994).

5. CASE STUDY

The proposed robust decentralized design technique has been applied to the mathematical model of a laboratory process - a power supply unit (PSU) with two subsystems. The process was identified in two operating points by two transfer function matrices $\tilde{G}_1$ and $\tilde{G}_2$. considered as perturbed models and defining the operating range of the system. The nominal model $G(s)$ and the perturbation matrix Δ have been obtained using the procedure described in Kozáková and Tuchyňa (1996).

$$G(s) = \begin{pmatrix} G_{11} & G_{12} \\ G_{21} & G_{22} \end{pmatrix} \quad (23)$$

where

$$\begin{aligned} G_{11} &= 0.369\frac{0.487s + 1}{0.622s^2 + 1.515s + 1} \\ G_{12} &= -0.133\frac{0.206s + 1}{0.501s^2 + 2.151s + 1} \\ G_{21} &= 0.341\frac{0.117s + 1}{0.145s^2 + 0.963s + 1} \\ G_{22} &= 0.857\frac{0.648s + 1}{0.275s^2 + 1.322s + 1} \end{aligned} \quad (24)$$

$$\Delta = \begin{pmatrix} 0.122 & 0.050 \\ 0.110 & 0.067 \end{pmatrix} \quad (25)$$

Local controllers have to stabilize the process within the operating range given by $G(s)$ and Δ.

Prior to begin the design, the correct pairing was verified using the RGA array (Morari and Zafiriou, 1989) followed by the computation of $M_0(\omega)$ (the most upper curve in Fig.3) and of the bandwidth frequency ω_{BW} given by definition $M_0(\omega_{BW}) = 0.707 M_0(0)$. In our case $M_0(0) = 1.948$ and $M_0(\omega_{BW}) = 1.3776$. Consequently, $\omega_{BW} = 7 s^{-1}$.

In each subsystem a PID controller (22) is considered. The subsystems are of 2-nd order and have a standard form (21).

$$G_{ii}(s) = K \frac{\tau_1 s + 1}{T_2 s^2 + T_1 s + 1} \qquad \forall i \qquad (26)$$

The numerator and denominator coefficients B_i, A_i of the closed-loop transfer function have been expressed and used to determine the coefficients Q_i, P_i of the closed-loop magnitude square $|M_0(j\omega)|^2$.

$$\begin{aligned}
Q_0 &= K^2 r_i^2 \\
Q_1 &= K^2(r_p^2 - 2 r_d r_i + \tau_1^2 r_i^2) \\
Q_2 &= K^2(r_d^2 + \tau_1^2 r_d^2 - 2\tau_1^2 r_d r_i) \\
Q_3 &= K^2 \tau_1^2 r_d^2 \\
\\
P_0 &= K^2 r_i^2 \\
P_1 &= K^2 r_p^2 + 2K r_p - + K^2 \tau_1^2 r_i^2 - \\
&\quad -2K r_i (K r_d + T_1 - \tau_1) + 1 \\
P_2 &= K^2 r_d^2 + K^2 \tau_1^2 r_p^2 + 2K(\tau_1 T_1 - T_2) - \\
&\quad -2K r_i (\tau_1^2 r_d - \tau_1 T_2) + \\
&\quad +2K r_d (T_1 - \tau_1) + (T_1^2 - 2T_2) \\
P_3 &= K^2 \tau_1^2 r_d^2 + 2K T_2 \tau_1 r_d + T_2^2
\end{aligned} \qquad (27)$$

The final set of three inequalities is

$$\begin{aligned}
&\frac{Q_1}{P_1}: \\
&K^2 \frac{M_0^2 - 1}{M_0^2} r_p^2 + 2K r_p + \\
&+K^2 \frac{M_0^2 - 1}{M_0^2} \tau_1^2 r_i^2 - \\
&-2K r_i \left(K \frac{M_0^2 - 1}{M_0^2} r_d + T_1 - \tau_1\right) + 1 > 0 \\
\\
&\frac{Q_2}{P_2}: \\
&K^2 \frac{M_0^2 - 1}{M_0^2} \tau_1^2 r_p^2 + 2K r_p (\tau_1 T_1 - T_2) - \\
&-2K r_i \left(K r_d \tau_1^2 \frac{M_0^2 - 1}{M_0^2} + \tau_1 T_2\right) + \\
&+2K r_d (T_1 - \tau_1) + K^2 \frac{M_0^2 - 1}{M_0^2} r_d^2 + \\
&+(T_1^2 - 2T_2) > 0 \\
\\
&\frac{Q_3}{P_3}: \\
&K^2 \frac{M_0^2 - 1}{M_0^2} \tau_1^2 r_d^2 + 2K T_2 \tau_1 r_d + T_2^2 > 0
\end{aligned} \qquad (28)$$

For $M_0(\omega_{BW}) = 1.3776$ the intervals for r_{di} have been computed from the last inequality. Since for subsystems of order $n = 2$, the PI-controller is sufficient to achieve the offset-free performance, $r_{di} = 0$ have been chosen $\forall i$. The stability regions in the (r_p, r_i) plane are determined independently for each subsystem by the inter of

(1) *the outside of an ellipse (corresponding to the inequality* $\frac{Q_1}{P_1} < M_0^2$*),*
(2) *the outside of a parabola (corresponding to the inequality* $\frac{Q_2}{P_2} < M_0^2$*) and of*
(3) *the stable region in the* (r_p, r_i) *plane specified by the Neymark method of D-contours.*

The resulting subsystem stability regions are depicted in Fig.1.

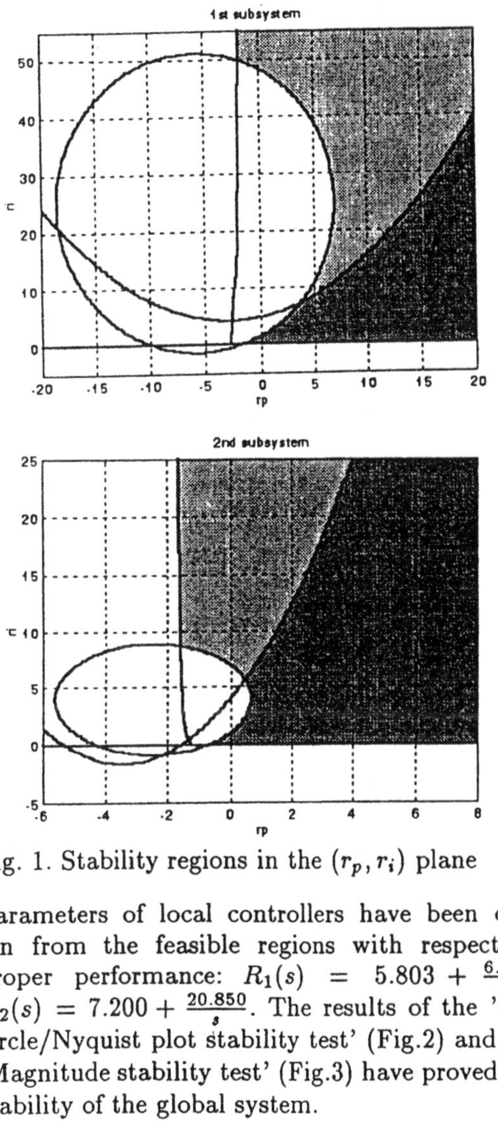

Fig. 1. Stability regions in the (r_p, r_i) plane

Parameters of local controllers have been chosen from the feasible regions with respect to proper performance: $R_1(s) = 5.803 + \frac{6.946}{s}$, $R_2(s) = 7.200 + \frac{20.850}{s}$. The results of the '$M_0$-circle/Nyquist plot stability test' (Fig.2) and the 'Magnitude stability test' (Fig.3) have proved the stability of the global system.

Figs.4, 5 show the the closed-loop system step responses under the decentralized PID controller for the nominal ('nom') and the perturbed models (working points 'wp1, wp2').

6. CONCLUSION

The paper describes a robust decentralized control design strategy in the frequency domain which

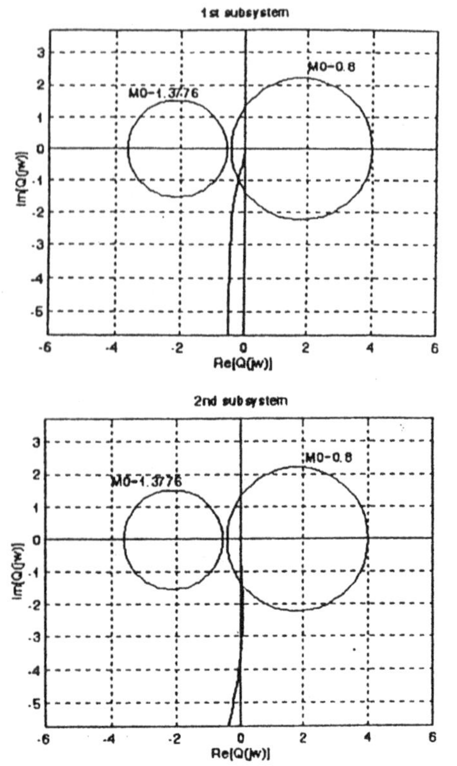

Fig. 2. 'M_0-circle/Nyquist plot stability test'

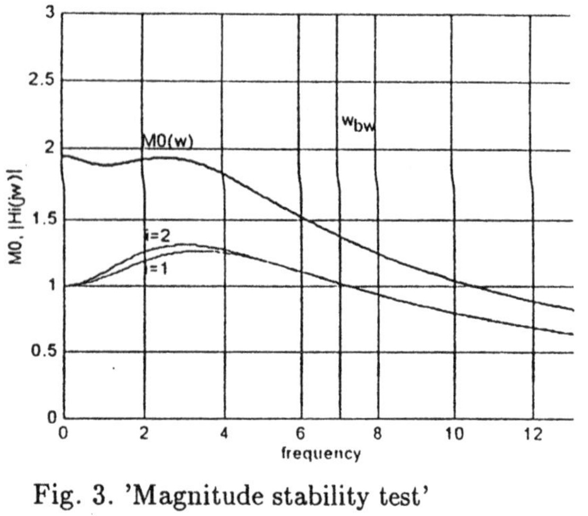

Fig. 3. 'Magnitude stability test'

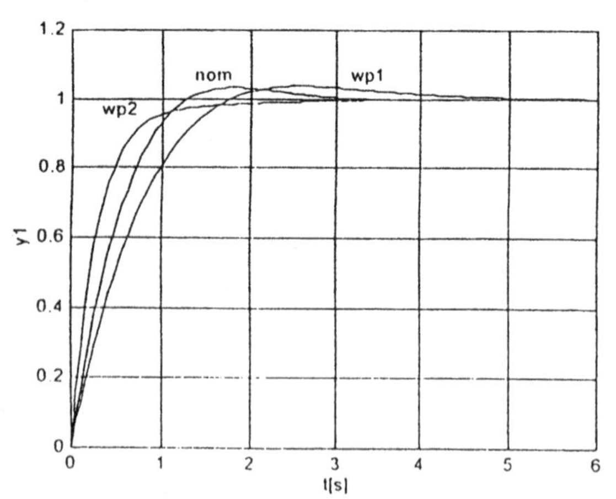

Fig. 4. Closed-loop step response: 1st subsystem

evolves on the fulfillment of the sufficient condition for robust stability (SCSR) of systems under decentralized control fulfilled via parameter tun-

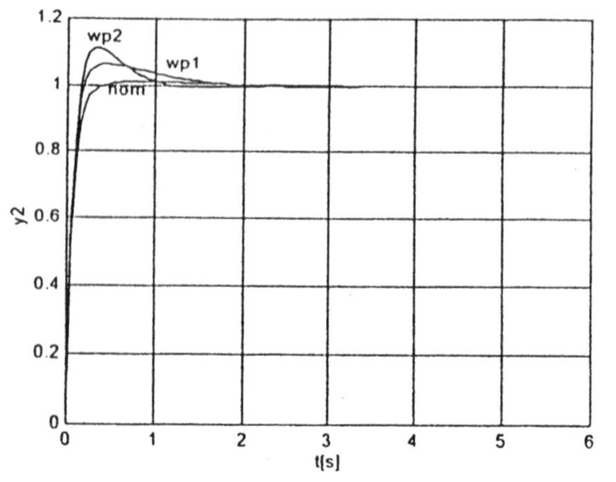

Fig. 5. Closed-loop step response: 2nd subsystem

ing of fixed-structure local controllers. The graphical interpretation of SCSR provides two stability tests for a visual stability testing of the overall system as well as of the subsystems at one glance. The results have proved the practicality of the proposed approach.

7. REFERENCES

Grosdidier, P. and M. Morari (1986). Interaction measures for systems under decentralized control. *Automatica*, **22**, 309-319.

Kozáková, A. and V. Veselý (1994). Decentralized control of linear systems. In *Preprints of the Conference 'Process Control ŘIP'94'*, Horní Bečva, Czech Republic, 437-442.

Kozáková, A. and D. Tuchyňa (1996). Modelling of uncertain systems. In *Proceedings of the 12th Int. Conference on Process Control and Simulation ASR TP '96*, Košice, Slovak Republic, 119-124.

Morari, M. and E. Zafiriou (1989). *Robust Process Control.* Prentice Hall, London.

Veselý, V. and J. Murgaš (1992). Design of decentralized adaptive controllers using the principle of dominant subsystems. In *Preprints of International Symposium on Adaptive Systems in Control and Signal Processing.* Grenoble, France, 651-655.

Zietz, H. and M. Schufft (1989). Decomposition and design of decentralized robust control systems in the frequency domain. *LSS '89 Theory and Application*, **1**, 188-193.

DECENTRALIZED ADAPTIVE APPROACH TO CONTROL OF CHEMICAL REACTORS BASED ON DELTA MODEL REPRESENTATION

Monika Bakošová*, Roman Prokop, Mária Karšaiová***

** Slovak University of Technology, Faculty of Chemical Technology, Department of Process Control, Radlinského 9, 812 37 Bratislava, Slovak Republic*
*** Technical University of Brno, Faculty of Technology, Department of Process Control, 762 72 Zlín, Czech Republic*

Abstract: The contribution presents decentralized adaptive approach to control of multi input-multi output (MIMO) continuous-time dynamic systems. The controlled plant is represented by two serially connected chemical reactors with a recycle. The dynamics of this system is described by fourteen state-space differential equations. The external model of the controlled plant as well as the control design are formulated in a delta operator form. The decentralization of a MIMO plant is performed in the matrix fraction description as a two input - two output system. Control loop consists of local and global control loops. The local controllers are adaptive ones obtained via Diophantine equations in delta polynomials. The global control strategy ensures asymptotic stability of the entire control system. The adaptive property is based on a recursive identification procedure.

Key words: Adaptive control, decentralized control, local control, recursive estimation, pole assignment, chemical reactors control.

1. INTRODUCTION

Chemical reactors present a wide class of systems which are characterised by a non-linear behaviour. A traditional approach to design a control system for such plants includes open-loop strategies, see e.g. (Rippin, 1983), the state-space utilization (Hoo and Kantor, 1985), adaptive regulators (Chidambaram and Malleswararao, 1992), etc. Continuous-time (CT) dynamic systems are usually modelled by differential equations. However, digital processing of these models in the time domain is impossible. All signals have to be sampled and discrete-time (DT) approximations of CT operators involved. The traditional DT approximation utilised the well-known Z-transform technique yielding difference equations and Z-transform functions. However, these models suffer from several drawbacks, namely that they do not converge to their CT counterparts when the sampling period tends to zero. Moreover, the poles and zeros of Z-transfer functions become unstable, see (Mukhopadhyay, *et al.*, 1992) for details. Some of these problems can be circumvented by involving a class of delta operators (Middleton and Goodwin, 1986, 1990). Models expressed in the delta representation converge to the CT ones. The control synthesis can be easily formulated by Diophantine equations.

The adaptive control strategy, proposed herein, consists of several steps. The parameters of an external model of a MIMO system are recursively estimated from sampled data (Prokop, *et al.*, 1992; (Dostál, *et al.*, 1994). Then, the principles of decomposition are applied and the whole system is split into subsystems, see e.g. (Sundareshan and Elbanna, 1990; Iwai, *et al.*, 1993; Iftar, *et al.*, 1993; Filasová, 1994). The local control design, based on delta model algebra, is performed and local control variables are generated. The global control variables compensate the undesirable influence of interactions (Bakošová, *et al.*, 1994).

The resultant adaptive control scheme is applied to control of two serially connected non-linear continuous-time stirred tank reactors.

2. CONTROLLER DESIGN

2.1 Problem statement

Consider the multivariable large-scale system S described by the matrix equation

$$\mathbf{A}(\delta)\,\mathbf{y}(t) = \mathbf{B}(\delta)\,\mathbf{u}(t) \qquad (1)$$

where δ is one of the discrete operator defined by Mukhopadhyay, *et al.* (1992),

$$\delta = \frac{z-1}{T_S} \qquad (2)$$

T_s is the sampling period, z is the shifting operator, y(t) is the output vector: $\mathbf{y}(t) \in R^r$, $\mathbf{u}(t)$ is the input vector: $\mathbf{u}(t) \in R^m$, **A**, **B** are (r × r) respectively (r × m) polynomial matrices. The condition m >= r must be satisfied. **A**, **B** can be written as

$$\mathbf{A}(\delta) = \mathbf{I}\delta^n + \mathbf{A}_{n-1}\delta^{n-1} + \dots + \mathbf{A}_0$$
$$\mathbf{B}(\delta) = \mathbf{B}_{n-1}\delta^{n-1} + \dots + \mathbf{B}_0 \qquad (3)$$

Consider further that it is possible to decompose the system (1) into N subsystems S_i, described by equations

$$\mathbf{A}_i(\delta)\,\mathbf{y}_i(t) = \mathbf{B}_i(\delta)\,\mathbf{u}_i(t) + \mathbf{h}_i(t) \qquad (4)$$

where i=1,...,N, $\mathbf{y}_i(t)$ is the output vector of the ith subsystem : $\mathbf{y}_i \in R^{r_i}$ and $\mathbf{y}^T = (\mathbf{y}_1, \dots, \mathbf{y}_N)$, $\mathbf{u}_i(t)$ is the input vector of the ith subsystem: $\mathbf{u}_i \in R^{m_i}$ and $\mathbf{u}^T = (\mathbf{u}_1, \dots, \mathbf{u}_N)$. The conditions

$$\sum_{i=1}^{N} r_i = r, \qquad \sum_{i=1}^{N} m_i = m \qquad (5)$$

also hold. The interaction vector of the ith subsystem $\mathbf{h}_i$ is described by

$$\mathbf{h}_i(t) = -\sum_{\substack{j=1\\ j\neq i}}^{N} \mathbf{A}_{ij}(\delta)\mathbf{y}_j(t) + \sum_{\substack{j=1\\ j\neq i}}^{N} \mathbf{B}_{ij}(\delta)\mathbf{u}_j(t) \qquad (6)$$

The task consists in finding a control vector $\mathbf{u}(t)^T = (\mathbf{u}_1,\dots,\mathbf{u}_N)$ with its components in the form

$$\mathbf{u}_i(t) = \mathbf{u}_i^{l}(t) + \mathbf{u}_i^{g}(t) \qquad (7)$$

where $\mathbf{u}_i^l(t)$ is called the local part of control $\mathbf{u}_i(t)$ and $\mathbf{u}_i^g(t)$ is called the global one. The closed-loop control is desired to be asymptotic stabile and the characteristic polynomials of local loops must have the prescribed poles.

2.2 Local subsystem control

In the stabilisation of large scale or multivariable systems, the individual subsystems play the important role. The design of local control is based on polynomial approach, see e.g. (Kučera, 1983).

Let each of decomposed isolated subsystems be modelled

$$\mathbf{A}_i(\delta)\,\mathbf{y}_i(t) = \mathbf{B}_i(\delta)\,\mathbf{u}_i^{l}(t) \qquad (8)$$

The reference $\mathbf{w}_i$ for each local subsystem is governed by the equation

$$\mathbf{F}_i(\delta)\,\mathbf{w}_i(t) = \mathbf{G}_i(\delta) \qquad (9)$$

For the local control loops with feed-back and feed-forward controllers, the control law is in the form

$$\mathbf{P}_i(\delta)\,\mathbf{u}_i^{l}(t) = \mathbf{R}_i(\delta)\,\mathbf{w}_i(t) - \mathbf{Q}_i(\delta)\,\mathbf{y}_i(t) \qquad (10)$$

The output regulation problem in the pole-placement sense consists in finding admissible control law (10) so that the appropriate response to reference (9) is obtained. The appropriate matrices $\mathbf{P}_i(\delta)$, $\mathbf{R}_i(\delta)$, $\mathbf{Q}_i(\delta)$ are obtained via solution of the couple of Diophantine equations:

$$\mathbf{A}_i(\delta)\mathbf{P}_i(\delta) + \mathbf{B}_i(\delta)\mathbf{Q}_i(\delta) = \mathbf{D}_i(\delta)$$
$$\mathbf{T}_i(\delta)\mathbf{F}_i(\delta) + \mathbf{B}_{ri}(\delta)\mathbf{R}_i(\delta) = \mathbf{D}_i(\delta) \qquad (11)$$

where $\mathbf{D}_i(\delta)$ is a given polynomial matrix with desired invariant polynomials, see (Prokop, e*t al.*, 1992) for further details.

2.3 Global Control

The task of global parts of control inputs is to compensate the undesirable influence of interactions. From (4), (6) and (7), it follows

$$\mathbf{A}_i(\delta)\mathbf{y}_i(t) = \mathbf{B}_i(\delta)\mathbf{u}_i^{l}(t) + \mathbf{B}_i(\delta)\mathbf{u}_i^{g}(t) - \sum_{\substack{j=1\\ j\neq i}}^{N} \mathbf{A}_{ij}(\delta)\mathbf{y}_j(t)$$
$$+ \sum_{\substack{j=1\\ j\neq i}}^{N} \mathbf{B}_{ij}(\delta)\mathbf{u}_j^{l}(t) + \sum_{\substack{j=1\\ j\neq i}}^{N} \mathbf{B}_{ij}(\delta)\mathbf{u}_j^{g}(t) \qquad (12)$$

The local control ensures the optimal behaviour of each subsystem (8). So the term $\mathbf{B}_i\mathbf{u}_i^g$ in (12) has to compensate the last three terms on the right side of (12). This condition leads to

$$\mathbf{B}(\delta)\,\mathbf{u}^{g}(t) = \mathbf{A}_{ND}(\delta)\,\mathbf{y}(t) - \mathbf{B}_{ND}(\delta)\,\mathbf{u}^{l}(t) \qquad (13)$$

where $\mathbf{A}_{ND}(\delta)$, $\mathbf{B}_{ND}(\delta)$ represent the non-diagonal blocks of matrices $\mathbf{A}$, $\mathbf{B}$:

$$\mathbf{A}_{ND}(\delta) = \mathbf{A}(\delta) - \mathrm{diag}\,\mathbf{A}(\delta)$$
$$\mathbf{B}_{ND}(\delta) = \mathbf{B}(\delta) - \mathrm{diag}\,\mathbf{B}(\delta) \qquad (14)$$

Equation (13) represents a system of difference equations which must be solved along with the control law.

3. CONTROLLED SYSTEM

3.1 Dynamic Model

Consider a controlled system consisting of two serially connected non-isothermal continuous-time stirred tank reactors (CSTRs) with a recycle, as it is shown in Fig. 1. The first reactor is the CSTR with the first-order irreversible parallel exothermic reactions according to the scheme $A \xrightarrow{k_1} B$, $A \xrightarrow{k_2} C$. The second reactor is also a CSTR with the first order consecutive exothermic reactions according to the scheme $B \xrightarrow{k_3} D \xrightarrow{k_4} E$. k_1, k_2, k_3, k_4 are reaction rate coefficients. Under the condition of perfect mixing, the dynamic mathematical model of the entire system was obtained by mass balances of reactants, energy balance of the reactant mixture and energy balance of the coolant. Many simplifications were accepted by modelling of considered reactors. Some of these simplifications are e.g. perfect mixing in the reactors, constant mass densities, volumes of reaction mixture, specific heat capacities, etc.

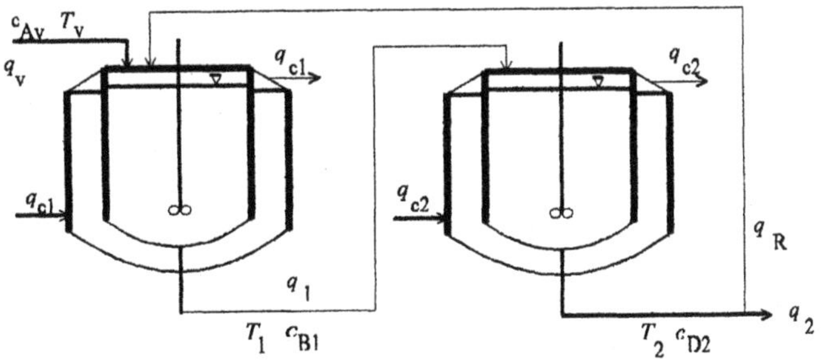

Fig. 1. Scheme of two serial chemical reactors with a recycle

The non linear model of two serially connected chemical reactors with a recycle is described by fourteen differential equations in the form:

1st reactor

$$\frac{dc_{A1}}{dt} = -(1/\tau_1 + k_1 + k_2)c_{A1} + (q_v c_{Av} + q_R c_{A2})/(\tau_1 q_1) \qquad (15)$$

$$\frac{dc_{B1}}{dt} = -c_{B1}/\tau_1 + k_1 c_{A1} + q_R c_{B2}/(\tau_1 q_1) \qquad (16)$$

$$\frac{dc_{C1}}{dt} = -c_{C1}/\tau_1 + k_2 c_{A1} + q_R c_{C2}/(\tau_1 q_1) \qquad (17)$$

$$\frac{dc_{D1}}{dt} = -c_{D1}/\tau_1 + q_R c_{D2}/(\tau_1 q_1) \qquad (18)$$

$$\frac{dc_{E1}}{dt} = -c_{E1}/\tau_1 + q_R c_{E2}/(\tau_1 q_1) \qquad (19)$$

$$\frac{dT_1}{dt} = -T_1/\tau_1 + q_v T_v/(\tau_1 q_1) + \frac{Q_{R1} - Q_{C1}}{\tau_1 q_1 \rho_1 c_{p1}} + \frac{q_R \rho_2 c_{p2}}{\tau_1 q_1 \rho_1 c_{p1}} T_2 \qquad (20)$$

$$\frac{dQ_{c1}}{dt} = -Q_{c1}/\tau_{c1} + \varphi_1(T_1 - T_{cv})/\tau_{c1} \qquad (21)$$

2nd reactor

$$\frac{dc_{A2}}{dt} = -c_{A2}/\tau_2 + c_{A1}\, q_1/((q_2 + q_R)\tau_2) \qquad (22)$$

$$\frac{dc_{B2}}{dt} = -(1/\tau_2 + k_3)c_{B2} + c_{B1} q_1/((q_2 + q_R)\tau_2) \qquad (23)$$

$$\frac{dc_{C2}}{dt} = -c_{C2}/\tau_2 + c_{C1}\, q_1/((q_2 + q_R)\tau_2) \qquad (24)$$

$$\frac{dc_{D2}}{dt} = -(1/\tau_2 + k_4)c_{D2} + c_{D1} q_1/((q_2 + q_R)\tau_2) + k_3 c_{B2} \qquad (25)$$

$$\frac{dc_{E2}}{dt} = -c_{E2}/\tau_2 + k_4 c_{D2} + c_{E1} q_1/((q_2 + q_R)\tau_2) \qquad (26)$$

$$\frac{dT_2}{dt} = -T_2/\tau_2 + \frac{Q_{R2} - Q_{C2}}{\tau_2 (q_2 + q_R)\rho_2 c_{p2}} + \frac{q_1 \rho_1 c_{p1}}{\tau_2 (q_2 + q_R)\rho_2 c_{p2}} T_1 \qquad (27)$$

$$\frac{dQ_{c2}}{dt} = -Q_{c2}/\tau_{c2} + \varphi_2(T_2 - T_{cv})/\tau_{c2} \qquad (28)$$

where

$$\tau_1 = V_1/q_1,\ \tau_{c1} = \frac{V_{c1}\rho_c c_{pc}}{q_{c1}\rho_c c_{pc} + F_1\alpha_1},$$
$$\tau_2 = V_2/(q_2 + q_R),\ \tau_{c2} = \frac{V_{c2}\rho_c c_{pc}}{q_{c2}\rho_c c_{pc} + F_2\alpha_2} \qquad (29)$$

$$\varphi_1 = \frac{q_{c1}\rho_c c_{pc} F_1\alpha_1}{q_{c1}\rho_c c_{pc} + F_1\alpha_1}, \quad \varphi_2 = \frac{q_{c2}\rho_c c_{pc} F_2\alpha_2}{q_{c2}\rho_c c_{pc} + F_2\alpha_2} \tag{30}$$

$$Q_{R1} = \left[(-\Delta H_1)k_1 + (-\Delta H_2)k_2\right] V_1 c_{A1},$$

$$Q_{R2} = \left[(-\Delta H_3)k_3\, c_{B2} + (-\Delta H_4)k_4\, c_{D2}\right] V_2 \tag{31}$$

The rates of reactions per volume unit are

$$r_1 = k_1\, c_{A1}, \quad r_2 = k_2\, c_{A1} \tag{32}$$

$$r_3 = k_3\, c_{B2}, \quad r_4 = k_4\, c_{D2} \tag{33}$$

The rate coefficients k_i are nonlinear functions of reaction temperatures being described by the Arrhenius relation

$$k_i = k_{i0}\, e^{\left(-\frac{E_i}{RT_i}\right)}, \quad i = 1, 2, 3, 4 \tag{34}$$

The mathematical state-space model of the two reactors described above is then a system of fourteen non-linear differential equations and the controller design of such a system by classical methods of the control theory would be very complicated.

3.2 Control of serially connected chemical reactors

The approach to control of reactors described by (15) - (28) is based on having the external model of the controlled process with several inputs and several outputs with predefined order of dynamics. In this case, it is necessary to predefine the number of inputs and outputs and also the sufficient order of dynamics. From the chemical engineering studies of chemical reactors follows that the most important internal independent variable is the temperature of the reaction mixture, while concentrations of reactants and products are functions of it according to the eq. (34). The temperature of the reaction mixture is in practice most often manipulated by the coolant flow rate. So for the model (15) - (28), it is supposed that controlled reactors are well described by the external model with two inputs $u_1(k)$, $u_2(k)$ and two outputs $y_1(k)$, $y_2(k)$, where coolant flow rates q_1, q_2 in reactors were chosen as the inputs $u_1(k)$, $u_2(k)$ and temperatures of reaction mixtures in the individual reactors T_1, T_2 as the outputs $y_1(k)$, $y_2(k)$. After the step response analysis of the model (15) - (28), it was decided that the second - order dynamics of the external model will be sufficient.

The external mathematical model of the system with two inputs, two outputs and the second-order dynamics is described by

$$\begin{pmatrix} \delta^2 + a_1\delta + a_0 & a_3\delta + a_2 \\ a_5\delta + a_4 & \delta^2 + a_7\delta + a_6 \end{pmatrix} \begin{pmatrix} y_1(k) \\ y_2(k) \end{pmatrix} =$$

$$\begin{pmatrix} b_1\delta + b_0 & 0 \\ 0 & b_5\delta + b_4 \end{pmatrix} \begin{pmatrix} u_1(k) \\ u_2(k) \end{pmatrix} \tag{35}$$

The inherent property of the δ operator is that it converges to the continuous derivative operator as the sampling period tends to zero. Thus, the model (35) represents two second-order differential equations, when $T_s \to 0^+$. The model (35) is much simpler than that described by eqs. (15) - (28) because of having two difference equations instead of fourteen differential ones. The other advantage of this model consists in a possibility to decompose the system (35) in two subsystems. The first of them represents the first reactor and is described by

$$(\delta^2 + a_1\delta + a_0)y_1(k) + h_1(k) = (b_1\delta + b_0)u_1(k) \tag{36}$$

and the second one represents the second reactor

$$(\delta^2 + a_7\delta + a_6)y_2(k) + h_2(k) = (b_5\delta + b_4)u_2(k) \tag{37}$$

Variables $h_1(k)$, $h_2(k)$ are the interconnections expressing the mutual influence of reactors defined as

$$h_1(k) = (a_3\delta + a_2)y_2(k) \tag{38}$$

$$h_2(k) = (a_5\delta + a_4)y_1(k) \tag{39}$$

The local control synthesis equations for the first subsystem have the form

$$(\delta^2 + a_1\delta + a_0)(p_{11}\delta + p_{01}) + (b_1\delta + b_0)(q_{11}\delta + q_{01}) = \delta^3 + d_2\delta^2 + d_1\delta + d_0 \tag{40}$$

$$\delta\,(t_{21}\delta^2 + t_{11}\delta + t_{01}) + (b_1\delta + b_0)\, r_{01} = \delta^3 + d_2\delta^2 + d_1\delta + d_0 \tag{41}$$

Then, the local control law of the first subsystem can be described by

$$u_1^l(k) = [(p_{11} - p_{01}T_s)u_1^l(k-1) + r_{01}T_s w_1(k-1) - q_{11}y_1(k) + (q_{11} - q_{01}T_s)y_1(k-1)]/p_{11} \tag{42}$$

Similarly, the local control law of the second subsystem is

$$u_2^l(k) = [(p_{12} - p_{02}T_s)u_2^l(k-1) + r_{02}T_s w_2(k-1) - q_{12}y_1(k) + (q_{12} - q_{02}T_s)y_2(k-1)]/p_{12} \tag{43}$$

The global parts of subsystem control were determined so that they are able to compensate the influence of interconnections between subsystems. The following equations hold for the first and for the second subsystem

$$(a_3\delta + a_2)y_2(k) = (b_1\delta + b_0)u_1^g(k) \tag{44}$$

$$(a_5\delta + a_4)y_1(k) = (b_5\delta + b_4)u_2^g(k) \tag{45}$$

Eqs. (44) and (45) represent generally a system of difference equations which must be solved along with the control law. These equations can be transformed in algebraic equations at the conditions $\delta u(k) = 0$ and $\delta y(k) = 0$ in the steady state. Under this simplifying condition, it is possible to obtain the global parts of control in the form

$$\begin{pmatrix} u_1^g(k) \\ u_2^g(k) \end{pmatrix} = B^{-1}A \begin{pmatrix} y_2(k) \\ y_1(k) \end{pmatrix} \tag{46}$$

where $A = \begin{pmatrix} 0 & a_2 \\ a_4 & 0 \end{pmatrix}$, $B = \begin{pmatrix} b_0 & 0 \\ 0 & b_4 \end{pmatrix}$.

4. RESULTS

The algorithm described above was implemented for simulation of control in the adaptive form. The parameters used for simulation are in Table 1. The recursive identification method LDDIF based on the least-squares method was used. The recursive identification enables to take changes of output and input variables into account. After the parameters of external model are calculated, the parameters of controllers can be evaluated and control laws can be calculated by eqs. (42), (43), (46) and so the PC serves as a self-tuner, which performs the control laws. To compare the dynamic behaviour of the uncontrolled and controlled systems, the step responses of the CSTRs were obtained at first and they are shown in Fig. 2. These step responses exhibited overdamped dynamics and it took about 40min to reach a new steady state. The control responses of the CSTRs are shown in Figs. 3,4,5. The controlled outputs reached their set points w_1, w_2 after less than 20min. For the reason of occurring many industrial boundaries the inputs in the controlled system were limited into the interval $\langle 0; 0.4\,m^3 min^{-1} \rangle$.

Table 1 Parameter values used for simulation

$q_v = 0.32\ m^3 min^{-1}$	$\rho_c = 998\ kg\ m^{-3}$
$q_R = 0.03\ m^3 min^{-1}$	$c_{pc} = 4.182\ kJ\ kg^{-1}K^{-1}$
$q_1 = 0.35\ m^3 min^{-1}$	$V_{c1} = V_{c2} = 0.64\ m^3$
$q_2 = 0.32\ m^3 min^{-1}$	$T_{cv} = 293\ K$
$c_{Av} = 9.82\ kmol\ m^{-3}$	$T_s = 0.5\ min$
$T_v = 328\ K$	
$V_1 = 1.2\ m^3$	$V_2 = 1.2\ m^3$
$F_1 = 5.6\ m^2$	$F_2 = 5.6\ m^2$
$\rho_1 = 802\ kg\ m^{-3}$	$\rho_2 = 796\ kg\ m^{-3}$
$c_{p1} = 3.97\ kJ\ kg^{-1}K^{-1}$	$c_{p2} = 3.94\ kJ\ kg^{-1}K^{-1}$
$\alpha_1 = 38.9\ kJ\ m^{-2}min^{-1}K^{-1}$	$\alpha_2 = 41.2\ kJ\ m^{-2}min^{-1}K^{-1}$
$k_{10} = 6.219 x 10^{12}\ min^{-1}$	$k_{30} = 2.703 x 10^{8}\ min^{-1}$
$k_{20} = 1.262 x 10^{26}\ min^{-1}$	$k_{40} = 2.223 x 10^{27}\ min^{-1}$
$E_1/R = 10\ 032\ K$	$E_3/R = 6896\ K$
$E_2/R = 20\ 672\ K$	$E_4/R = 23453\ K$
$(-\Delta H_1) = 1\ 750\ kJ\ kmol^{-1}$	$(-\Delta H_3) = 10\ 500\ kJ\ kmol^{-1}$
$(-\Delta H_2) = 12\ 800\ kJ\ kmol^{-1}$	$(-\Delta H_4) = 7\ 500\ kJ\ kmol^{-1}$

5. CONCLUSION

It is very important from the viewpoint of its industrial utilisation to have a good control algorithm for chemical processing plants. The control algorithm designed in the presented paper has several advantages. It does not require to have an exact mathematical model of the controlled process. The external input-output model can be

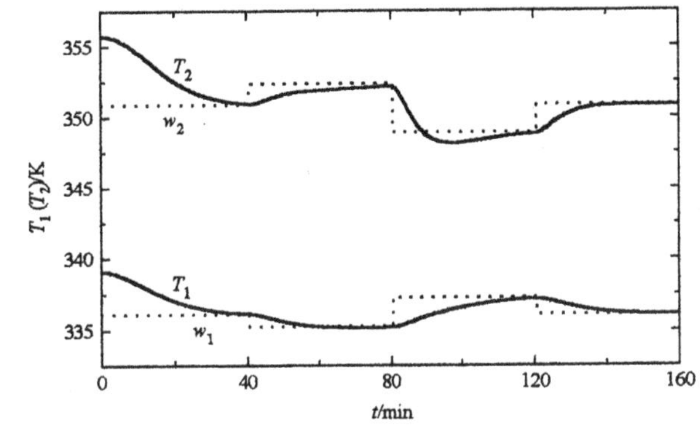

Fig. 2. Step responses of the CSTRs: output variables

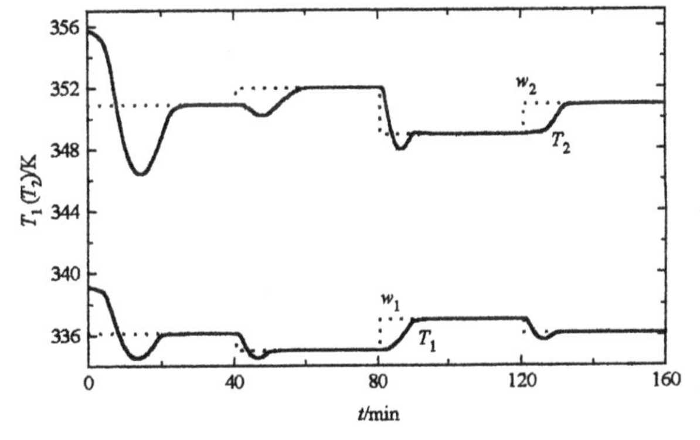

Fig. 3. Control responses of the CSTRs: output variables and references

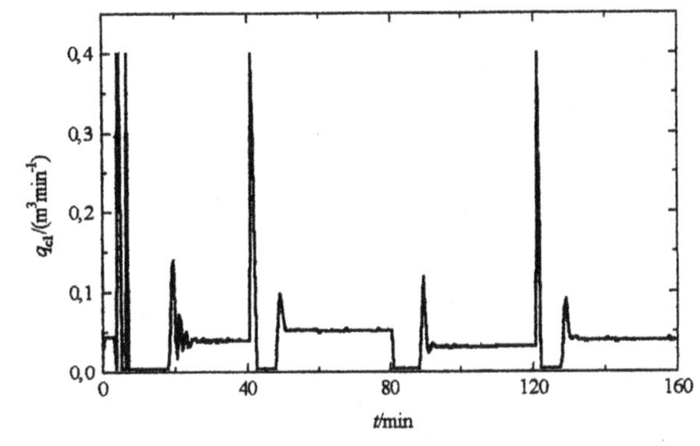

Fig. 4. Control of the CSTRs: input variable in the first reactor

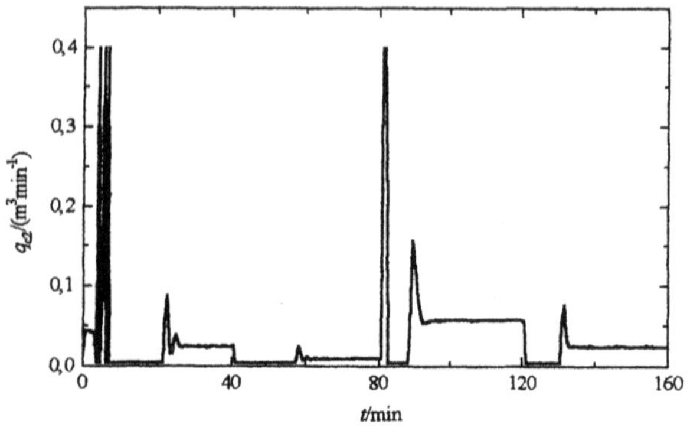

Fig. 5. Control of the CSTRs - input variable in the second reactor

obtained by the measurement of the input and output variables and the use of some of the recursive identification methods. Having the real process, the inputs and outputs can be measured on it. For these purposes, it is not necessary to know the reaction kinetics exactly. The investigated control algorithm is adaptive and the controller parameters vary as some properties of the controlled system change. Presented control algorithm is also very simple as it is based on the decomposition and decenralization approaches. The local and global controls are very simple and the entire control is only the sum of two previous ones. The convergence properties of the proposed algorithm are very good, too, and realistic simulations demonstrate convenience of using delta operators. The formulation and implementation of delta model self-tuning controllers do not bring any additional problems and difficulties.

ACKNOWLEDGEMENT

The authors are pleased to acknowledge the financial support of the Scientific Grant Agency of Ministry of Slovak Republic and Slovak Academy of sciences under grants N° 95/5195/198 and N° 1/4200/97.

SYMBOLS

c_A, c_B, c_C, c_D	reactant concentrations	kmol m^{-3}
t	time	min
k_1, k_2, k_3, k_4	reaction rate coefficients	min^{-1}
c_{Av}	feed concentration	kmol m^{-3}
q, q_c, q_R	volumetric flow rates	m^3 min^{-1}
q_v	feed volumetric flow rate	$m^3 min^{-1}$
Q_R	reaction heat	kJ min^{-1}
Q_c	heat accepted by coolant	kJ min^{-1}
ρ, $ρ_c$	reactant and coolant densities	kg m^{-3}
c_p, c_{pc}	specific heat capacities	kJ kg^{-1} K^{-1}
T, T_c	reactant and coolant temperatures	K
T_v, T_{cv}	feed values for T and T_c	K
V, V_c	reactant and coolant volumes	m^3
F	heat exchange surface area	m^2
α	heat transfer coefficient	kJ $m^{-2}min^{-1}K^{-1}$
r_i, i=1,...,4	reaction rates	kmol m^{-3} min^{-1}
E_i	activation energies	kJ $kmol^{-1}$
k_{i0}	frequency factors	min^{-1}
$(-ΔH_i)$	specific reaction heats	kJ $kmol^{-1}$
R	gas constant	kJ $kmol^{-1}K^{-1}$
k	discrete time	
u	control input	m^3min^{-1}
y	controlled output	K
δ	discrete time operator	
T_s	sampling period	min
w	reference signal	

REFERENCES

Bakošová, M., R. Prokop and P. Dostál (1994). Decentralized adaptive control based on the delta model representation. In: *Prep. IFAC Workshop NTDCS'94*, pp. 272 —277. Smolenice.

Dostál, P., A. Mészáros, R. Prokop and M. Bakošová (1994). Self-tuning control of a continuous stirred tank reactor. *Hung. J. of Ind. Chem.*, **22**, 315-320.

Filasová, A. (1994). Decentralized control of interconnected systems with unmodelled nonlinearity and interaction. In: *Prep. Int. Conf. ASRTP '94*, pp. 484 - 492. Košice.

Hoo, K. and J.C. Kantor, (1985). *Chem. Eng. Comm.*, **37**, 1.

Chidambaram, M. and Y.S.N. Malleswararao, (1992). *J. Proc. Cont.*, **2**, 1-9.

Iftar, A. (1993). Overlapping decentralized dynamic optimal control. *Int. J. Control*, **58**, 187-209.

Iwai, Z., I. Mizamoto. and H. Douzono. (1993). Decentralized simple adaptive control. In: *Prep. World Congress IFAC*, Vol. 3, pp. 9-16. Sydney.

Kučera, V. (1983). Linear quadratic control (St. space v. pol. eq.). *Kybernetika*, **19**, 185-195.

Middleton, R.H. and G.C. Goodwin (1986). Improved finite word length characteristics in digital control using delta operators. *IEEE Trans. on AC*, **31**, 1015 - 1023.

Middleton, R.H. and G.C. Goodwin (1990). *Digital control and estimation - A unified approach.* Prentice Hall, New York.

Mukhopadhyay, S. et al. (1992): Discrete-time models for continuous-time systems. *Int. J. Control*, **55**, 1161-87.

Prokop, R., M. Bakošová. and Z. Turisová (1992). Decentralized hybrid adaptive control based on polynomial approach. In: *Prep. IFAC Workshop "MICC'92"*, pp. 215-220. Prague.

Rippin, D.W.T. (1983). *Comp. Chem. Eng.*, **7**, 137.

Sundareshan, M.K. and R.M. Elbanna (1990). Design of decentralized observation schemes for large-scale interconnected systems. some new results. *Automatica*, **26**, 789-796.

ROBUST CONTROLLER DESIGN FOR LARGE-SCALE UNCERTAIN DYNAMIC SYSTEMS

Anna Filasová

Technical University of Košice
Faculty of Electrical Engineering and Informatics
Department of Cybernetics and Artificial Intelligence
Letná 9/B, 042 00 Košice, Slovak Republic
e-mail: filasova@ccsun.tuke.sk

Abstract: A useful technique for determining a linear feedback control law which stabilizes an uncertain systems is the Riccati equation approach of Schmitendorf (1988) He consider systems with time-varying uncertainty in the system matrix and the input connection matrix. We extend this technique to include problems with the control in the time range and the concept of 'partial decentralization'.

Keywords: Linear quadratic control, large-scale system, decentralized control.

1. INTRODUCTION

An industrial plant always contains parametric uncertainties. In the design of a controller, these uncertainties have to be considered. Otherwise, if the real plant differs from the assumed plant model , a controller design based on classical approaches may not ensure the stability of the overall system. One of the principal reasons for introducing feedback into a control system is to obtain relative insensivity to changes in plant parameters and to perturbations.

Petersen and Hollot (1986) proposed an linear feedback control law with stabilizes an uncertain systems with time-varying uncertainty in the system matrix and Schmitendorf extended this technique to include problems with time-varying uncertainty in the input connection matrix. They give a procedure for determining a linear feedback control law where the gains are obtained directly from the solution of a algebraic Riccati equation.

Here, we generalize their procedure for computing the gains in the linear controller so that uncertain large-scale systems and the control in the time range can be analyzed.

This approach is based on the concept of 'partial decentralization' (Filasová, 1994). The resulting controller is feedback with the features of utilizing with complete weighting in the objective function the subsystem interconnections, parallel processing for on-line implementation, and structure flexibility in the sense that the addition and/or deletion of a subsystem does not require the design of the control law of the overall large-scale system from the beginning again.

The others applicable publications which have dealt with the mentioned problem are (Filasová, 1996), (Krokavec, 1995), (Schmitendorf, 1988).

2. PROBLEM FORMULATION

In general, a continuous-time multivariable system can be considered in the form of differential equation

$$\dot{\mathbf{x}}(t) = \mathbf{A}\mathbf{x}(t) + \mathbf{B}\mathbf{u}(t) \qquad (1)$$

and the measurements can be governed by

$$\mathbf{y}(t) = \mathbf{C}\mathbf{x}(t) \qquad (2)$$

where

$\mathbf{x}(t)$	nx1 state vector of the system
$\mathbf{u}(t)$	qx1 input vector
$\mathbf{y}(t)$	mx1 output measurement vector
$\mathbf{A}$	nxn system state transition matrix
$\mathbf{B}$	nxr constant input matrix
$\mathbf{C}$	mxn constant output matrix

In order to develop the equations for the overall large-scale systems comprising M interconnected subsystems, it is assumed this form of matrices $\mathbf{A}$ and $\mathbf{B}$

$$\mathbf{A} = \begin{bmatrix} \mathbf{A}_{11} & \mathbf{A}_{12} & \dots & \mathbf{A}_{1M} \\ \mathbf{A}_{21} & \mathbf{A}_{22} & \dots & \mathbf{A}_{2M} \\ & & \vdots & \\ \mathbf{A}_{M1} & \mathbf{A}_{M2} & \dots & \mathbf{A}_{MM} \end{bmatrix} \tag{3}$$

$$\mathbf{B} = \begin{bmatrix} \mathbf{B}_1 & 0 & \dots & 0 \\ 0 & \mathbf{B}_2 & \dots & 0 \\ & & \vdots & \\ 0 & 0 & \dots & \mathbf{B}_M \end{bmatrix} \tag{4}$$

$$\begin{aligned} \mathbf{A}_{ij} &= \mathbf{A}_{0ij} + \Delta\mathbf{A}_{ij}(r(t)) \\ \mathbf{B}_i &= \mathbf{B}_{0i} + \Delta\mathbf{B}_i(s(t)) \end{aligned} \tag{5}$$

where $\mathbf{A}_{0ij}, \mathbf{B}_{0i}$ are the nominal system matrices. Uncertainty enters the system through the vectors $r(t)$ and $s(t)$. The only information available about the uncertainty is that $r(t) \in R$ and $s(t) \in S$ and R and S are compact sets.

The control design problem may be state as follows:

For a continuous-time uncertain multivariable linear system to determine the control

$$\mathbf{u}(t) = -\mathbf{K}(t)\mathbf{x}(t) \tag{6}$$

where $\mathbf{K}(t)$ is the control gain matrix of the form

$$\mathbf{K}(t) = \begin{bmatrix} \mathbf{K}_{11}(t) & \mathbf{K}_{12}(t) & \dots & \mathbf{K}_{1M}(t) \\ \mathbf{K}_{21}(t) & \mathbf{K}_{22}(t) & \dots & \mathbf{K}_{2M}(t) \\ & & \vdots & \\ \mathbf{K}_{M1}(t) & \mathbf{K}_{M2}(t) & \dots & \mathbf{K}_{MM}(t) \end{bmatrix} \tag{7}$$

$$\mathbf{K}_{ij}(t) = \mathbf{K}_{0ij}(t) + \Delta\mathbf{K}_{ij}(t) \tag{8}$$

$$\mathbf{K}_{ii}(t) = \sum_{\substack{j=1 \\ j\neq i}}^{M} (\mathbf{K}_{oi}^{j}(t) + \Delta\mathbf{K}_{i}^{j}(t)) \tag{9}$$

which verify the asymptotic stability of the closed loop system.

3. ASYMPTOTIC STABILITY

The method for constructing Lyapunov functions can by used for investigation of asymptotic stability in the control system. The best obtainable Lyapunov function is

$$V(\mathbf{x}(t), t) = \mathbf{x}^T(t)\mathbf{P}(t)\mathbf{x}(t) \tag{10}$$

where $\mathbf{P}(t)$ is a symmetric positive definite matrix

$$\mathbf{P}(t) = \begin{bmatrix} \mathbf{P}_{11}(t) & \mathbf{P}_{12}(t) & \dots & \mathbf{P}_{1M}(t) \\ \mathbf{P}_{21}(t) & \mathbf{P}_{22}(t) & \dots & \mathbf{P}_{2M}(t) \\ & & \vdots & \\ \mathbf{P}_{M1}(t) & \mathbf{P}_{M2}(t) & \dots & \mathbf{P}_{MM}(t) \end{bmatrix} \tag{11}$$

If it is assumed that the diagonal elements of matrix $\mathbf{P}(t)$ are defined by

$$\mathbf{P}_{ii}(t) = \sum_{\substack{j=1 \\ j\neq i}}^{M} \mathbf{P}_{i}^{j}(t); \tag{12}$$

the resulting structure of the matrix $\mathbf{P}(t)$ will be

$$\begin{aligned} \mathbf{P} = \Bigg(\dots \Bigg(& \begin{bmatrix} \mathbf{P}_1^2 & \mathbf{P}_{12} & 0 & \dots & 0 \\ \mathbf{P}_{21} & \mathbf{P}_2^1 & 0 & \dots & 0 \\ 0 & 0 & 0 & \dots & 0 \\ & & & \vdots & \\ 0 & 0 & 0 & \dots & 0 \end{bmatrix} + \\ + & \begin{bmatrix} \mathbf{P}_1^3 & 0 & \mathbf{P}_{13} & \dots & 0 \\ 0 & 0 & 0 & \dots & 0 \\ \mathbf{P}_{31} & 0 & \mathbf{P}_3^1 & \dots & 0 \\ & & & \vdots & \\ 0 & 0 & 0 & \dots & 0 \end{bmatrix} + \dots + \\ + & \begin{bmatrix} \mathbf{P}_1^M & 0 & \dots & \mathbf{P}_{1M} \\ 0 & 0 & \dots & 0 \\ 0 & 0 & \dots & 0 \\ & & \vdots & \\ \mathbf{P}_{M1} & 0 & \dots & \mathbf{P}_M^1 \end{bmatrix} \Bigg) + \dots + \\ + & \begin{bmatrix} 0 & 0 & \dots & 0 & 0 \\ 0 & 0 & \dots & 0 & 0 \\ & & \vdots & & \\ 0 & 0 & \dots & \mathbf{P}_{M-1}^M & \mathbf{P}_{M-1,M} \\ 0 & 0 & \dots & \mathbf{P}_{M,M-1} & \mathbf{P}_M^{M-1} \end{bmatrix} \Bigg) \end{aligned} \tag{13}$$

and the Lyapunov function can be transformed into the next form

$$V(\mathbf{x}(t), t) = \sum_{h=1}^{M-1} \sum_{k=h+1}^{M} \begin{bmatrix} \mathbf{x}_h(t) \\ \mathbf{x}_k(t) \end{bmatrix}^T \begin{bmatrix} \mathbf{P}_h^k(t) & \mathbf{P}_{hk}(t) \\ \mathbf{P}_{kh}(t) & \mathbf{P}_k^h(t) \end{bmatrix} \begin{bmatrix} \mathbf{x}_h(t) \\ \mathbf{x}_k(t) \end{bmatrix} \tag{14}$$

The condition for the investigation of asymptotic stability of the control system is negative definite of the derivative of the Lyapunov function

$$\dot{V}(\mathbf{x}(t),t)=\sum_{h=1}^{M-1}\sum_{k=h+1}^{M}$$

$$\left(\begin{bmatrix}\mathbf{x}_h(t)\\ \mathbf{x}_k(t)\end{bmatrix}^T\begin{bmatrix}\dot{\mathbf{P}}_h^k(t) & \dot{\mathbf{P}}_{hk}(t)\\ \dot{\mathbf{P}}_{kh}(t) & \dot{\mathbf{P}}_k^h(t)\end{bmatrix}\begin{bmatrix}\mathbf{x}_h(t)\\ \mathbf{x}_k(t)\end{bmatrix}+\right.$$
$$+\begin{bmatrix}\dot{\mathbf{x}}_h(t)\\ \dot{\mathbf{x}}_k(t)\end{bmatrix}^T\begin{bmatrix}\mathbf{P}_h^k(t) & \mathbf{P}_{hk}(t)\\ \mathbf{P}_{kh}(t) & \mathbf{P}_k^h(t)\end{bmatrix}\begin{bmatrix}\mathbf{x}_h(t)\\ \mathbf{x}_k(t)\end{bmatrix}+$$
$$\left.+\begin{bmatrix}\mathbf{x}_h(t)\\ \mathbf{x}_k(t)\end{bmatrix}^T\begin{bmatrix}\mathbf{P}_h^k(t) & \mathbf{P}_{hk}(t)\\ \mathbf{P}_{kh}(t) & \mathbf{P}_k^h(t)\end{bmatrix}\begin{bmatrix}\dot{\mathbf{x}}_h(t)\\ \dot{\mathbf{x}}_k(t)\end{bmatrix}\right)$$

Assuming a separable model where the subsystem pair is defined by

$$\begin{bmatrix}\mathbf{P}_h^k(t) & \mathbf{P}_{hk}(t)\\ \mathbf{P}_{kh}(t) & \mathbf{P}_k^h(t)\end{bmatrix}\begin{bmatrix}\dot{\mathbf{x}}_h(t)\\ \dot{\mathbf{x}}_k(t)\end{bmatrix}=$$
$$=\begin{bmatrix}\mathbf{P}_h^k(t) & \mathbf{P}_{hk}(t)\\ \mathbf{P}_{kh}(t) & \mathbf{P}_k^h(t)\end{bmatrix}\begin{bmatrix}\mathbf{A}_{hh} & \mathbf{A}_{hk}\\ \mathbf{A}_{kh} & \mathbf{A}_{kk}\end{bmatrix}\begin{bmatrix}\mathbf{x}_h(t)\\ \mathbf{x}_k(t)\end{bmatrix}+$$
$$+\begin{bmatrix}\mathbf{P}_h^k(t) & \mathbf{P}_{hk}(t)\\ \mathbf{P}_{kh}(t) & \mathbf{P}_k^h(t)\end{bmatrix}\begin{bmatrix}\mathbf{B}_h & 0\\ 0 & \mathbf{B}_k\end{bmatrix}\begin{bmatrix}\mathbf{u}_h^k(t)\\ \mathbf{u}_k^h(t)\end{bmatrix}$$

for

$$\begin{bmatrix}\mathbf{u}_h^k(t)\\ \mathbf{u}_k^h(t)\end{bmatrix}=-\begin{bmatrix}\mathbf{K}_h^k(t) & \mathbf{K}_{hk}(t)\\ \mathbf{K}_{kh}(t) & \mathbf{K}_k^h(t)\end{bmatrix}\begin{bmatrix}\mathbf{x}_h(t)\\ \mathbf{x}_k(t)\end{bmatrix} \tag{15}$$

the above mentioned derivative of Lyapunov function takes the form

$$\dot{V}(\mathbf{x}(t),t)=\sum_{h=1}^{M-1}\sum_{k=h+1}^{M} \begin{bmatrix}\mathbf{x}_h(t)\\ \mathbf{x}_k(t)\end{bmatrix}^T\mathbf{P}_{hk}^{**}(t)\begin{bmatrix}\mathbf{x}_h(t)\\ \mathbf{x}_k(t)\end{bmatrix} \tag{16}$$

$$\begin{aligned}\mathbf{P}_{hk}^{**}(t)=&\begin{bmatrix}\dot{\mathbf{P}}_h^k(t) & \dot{\mathbf{P}}_{hk}(t)\\ \dot{\mathbf{P}}_{kh}(t) & \dot{\mathbf{P}}_k^h(t)\end{bmatrix}+\\ &+\begin{bmatrix}\mathbf{A}_{hh} & \mathbf{A}_{hk}\\ \mathbf{A}_{kh} & \mathbf{A}_{kk}\end{bmatrix}^T\begin{bmatrix}\mathbf{P}_h^k(t) & \mathbf{P}_{hk}(t)\\ \mathbf{P}_{kh}(t) & \mathbf{P}_k^h(t)\end{bmatrix}+\\ &+\begin{bmatrix}\mathbf{P}_h^k(t) & \mathbf{P}_{hk}(t)\\ \mathbf{P}_{kh}(t) & \mathbf{P}_k^h(t)\end{bmatrix}\begin{bmatrix}\mathbf{A}_{hh} & \mathbf{A}_{hk}\\ \mathbf{A}_{kh} & \mathbf{A}_{kk}\end{bmatrix}+\\ &-\begin{bmatrix}\mathbf{K}_h^k(t) & \mathbf{K}_{hk}(t)\\ \mathbf{K}_{kh}(t) & \mathbf{K}_k^h(t)\end{bmatrix}^T\begin{bmatrix}\mathbf{B}_h & 0\\ 0 & \mathbf{B}_k\end{bmatrix}^T\cdot\\ &\cdot\begin{bmatrix}\mathbf{P}_h^k(t) & \mathbf{P}_{hk}(t)\\ \mathbf{P}_{kh}(t) & \mathbf{P}_k^h(t)\end{bmatrix}-\begin{bmatrix}\mathbf{P}_h^k(t) & \mathbf{P}_{hk}(t)\\ \mathbf{P}_{kh}(t) & \mathbf{P}_k^h(t)\end{bmatrix}\cdot\\ &\cdot\begin{bmatrix}\mathbf{B}_h & 0\\ 0 & \mathbf{B}_k\end{bmatrix}\begin{bmatrix}\mathbf{K}_h^k(t) & \mathbf{K}_{hk}(t)\\ \mathbf{K}_{kh}(t) & \mathbf{K}_k^h(t)\end{bmatrix}\end{aligned} \tag{17}$$

To guarantee that the control law is stabilizing, the summation of the matrices $\mathbf{P}_{hk}^{**}(t)$ must be negative definite.

4. CONTROL DESIGN

To guarantee that the control is stabilizing the set of matrix

$$\mathbf{P}_{hk}^{*}(t)=\begin{bmatrix}\mathbf{P}_h^k(t) & \mathbf{P}_{hk}(t)\\ \mathbf{P}_{kh}(t) & \mathbf{P}_k^h(t)\end{bmatrix} \tag{18}$$

must be chosen such that $\dot{V}(\mathbf{x}(t),t)$ is negative definite for all admissible values of the uncertainty.

For the positive definite symmetric matrix

$$\mathbf{Q}_{hk}^{*}=\begin{bmatrix}\mathbf{Q}_h^k & \mathbf{Q}_{hk}\\ \mathbf{Q}_{kh} & \mathbf{Q}_k^h\end{bmatrix} \tag{19}$$

from (5) and (17) implies that

$$\begin{aligned}-&\begin{bmatrix}\dot{\mathbf{P}}_h^k(t) & \dot{\mathbf{P}}_{hk}(t)\\ \dot{\mathbf{P}}_{kh}(t) & \dot{\mathbf{P}}_k^h(t)\end{bmatrix}=\begin{bmatrix}\mathbf{Q}_h^k & \mathbf{Q}_{hk}\\ \mathbf{Q}_{kh} & \mathbf{Q}_k^h\end{bmatrix}+\\ +&\begin{bmatrix}\mathbf{P}_h^k(t) & \mathbf{P}_{hk}(t)\\ \mathbf{P}_{kh}(t) & \mathbf{P}_k^h(t)\end{bmatrix}\left(\begin{bmatrix}\mathbf{A}_{hh0} & \mathbf{A}_{hk0}\\ \mathbf{A}_{kh0} & \mathbf{A}_{kk0}\end{bmatrix}-\right.\\ -&\left.\begin{bmatrix}\mathbf{B}_{h0} & 0\\ 0 & \mathbf{B}_{k0}\end{bmatrix}\begin{bmatrix}\mathbf{K}_{h0}^k(t) & \mathbf{K}_{hk0}(t)\\ \mathbf{K}_{kh0}(t) & \mathbf{K}_{k0}^h(t)\end{bmatrix}\right)+\\ &\left(-\begin{bmatrix}\mathbf{K}_{h0}^k(t) & \mathbf{K}_{hk0}(t)\\ \mathbf{K}_{kh0}(t) & \mathbf{K}_{k0}^h(t)\end{bmatrix}^T\begin{bmatrix}\mathbf{B}_{h0} & 0\\ 0 & \mathbf{B}_{k0}\end{bmatrix}^T+\right.\\ +&\left.\begin{bmatrix}\mathbf{A}_{hh0} & \mathbf{A}_{hk0}\\ \mathbf{A}_{kh0} & \mathbf{A}_{kk0}\end{bmatrix}^T\right)\begin{bmatrix}\mathbf{P}_h^k(t) & \mathbf{P}_{hk}(t)\\ \mathbf{P}_{kh}(t) & \mathbf{P}_k^h(t)\end{bmatrix}\end{aligned} \tag{20}$$

and

$$\mathbf{P}_{hk}^{*}(\tau)=\begin{bmatrix}\mathbf{P}_h^k(\tau) & \mathbf{P}_{hk}(\tau)\\ \mathbf{P}_{kh}(\tau) & \mathbf{P}_k^h(\tau)\end{bmatrix}=\begin{bmatrix}0 & 0\\ 0 & 0\end{bmatrix} \tag{21}$$

where τ is the terminal time. The second condition is, that the summation of matrices

$$\begin{aligned}&-\mathbf{Q}_{hk}^{*}+\mathbf{P}_{hk}^{*}(t)\Delta\mathbf{A}_{hk}^{*}(r(t))+\\ &+\Delta\mathbf{A}_{hk}^{*T}(r(t))\mathbf{P}_{hk}^{*}(t)-\\ &-\Delta\mathbf{K}_{hk}^{*T}(t)\mathbf{B}_{hk0}^{*T}\mathbf{P}_{hk}^{*}(t)-\\ &-\Delta\mathbf{K}_{hk}^{*T}(t)\Delta\mathbf{B}_{hk}^{*T}(s(t))\mathbf{P}_{hk}^{*}(t)-\\ &-\mathbf{K}_{hk0}^{*T}(t)\Delta\mathbf{B}_{hk}^{*T}(s(t))\mathbf{P}_{hk}^{*}(t)-\\ &-\mathbf{P}_{hk}^{*}(t)\mathbf{B}_{hk0}^{*}\Delta\mathbf{K}_{hk}^{*}(t)-\\ &-\mathbf{P}_{hk}^{*}(t)\Delta\mathbf{B}_{hk}^{*}(s(t))\Delta\mathbf{K}_{hk}^{*}(t)-\\ &-\mathbf{P}_{hk}^{*}(t)\Delta\mathbf{B}_{hk}^{*}(s(t))\mathbf{K}_{hk0}^{*}(t)\end{aligned} \tag{22}$$

must be negative definite for

$$\begin{aligned}\mathbf{K}_{hk}^{*}(t)=&\mathbf{K}_{hk0}^{*}(t)+\Delta\mathbf{K}_{hk}^{*}(t)=\\ =&\begin{bmatrix}\mathbf{K}_{h0}^k(t) & \mathbf{K}_{hk0}(t)\\ \mathbf{K}_{kh0}(t) & \mathbf{K}_{k0}^h(t)\end{bmatrix}+\\ +&\begin{bmatrix}\Delta\mathbf{K}_h^k(t) & \Delta\mathbf{K}_{hk}(t)\\ \Delta\mathbf{K}_{kh}(t) & \Delta\mathbf{K}_k^h(t)\end{bmatrix}\end{aligned} \tag{23}$$

We assume that the matched uncertainty is defined by the existence of matrices $\mathbf{D}_{hk}^{*}(r(t))$ and $\mathbf{E}_{hk}^{*}(s(t))$ such that

$$\begin{aligned}\Delta\mathbf{A}_{hk}^{*}(r(t))=&\begin{bmatrix}\Delta\mathbf{A}_{hh}(r(t)) & \Delta\mathbf{A}_{hk}(r(t))\\ \Delta\mathbf{A}_{kh}(r(t)) & \Delta\mathbf{A}_{kk}(r(t))\end{bmatrix}=\\ =&\begin{bmatrix}\mathbf{B}_{h0} & 0\\ 0 & \mathbf{B}_{k0}\end{bmatrix}\begin{bmatrix}\mathbf{D}_{hh}(r(t)) & \mathbf{D}_{hk}(r(t))\\ \mathbf{D}_{kh}(r(t)) & \mathbf{D}_{kk}(r(t))\end{bmatrix}=\\ =&\mathbf{B}_{hk0}^{*}\mathbf{D}_{hk}^{*}(r(t))\end{aligned} \tag{24}$$

$$\Delta\mathbf{B}^*_{hk}(s(t)) = \begin{bmatrix} \Delta\mathbf{B}_h(s(t)) & 0 \\ 0 & \Delta\mathbf{B}_k(s(t)) \end{bmatrix} = \\ = \begin{bmatrix} \mathbf{B}_{h0} & \mathbf{0} \\ \mathbf{0} & \mathbf{B}_{k0} \end{bmatrix} \begin{bmatrix} \mathbf{E}_h(s(t)) & 0 \\ 0 & \mathbf{E}_k(s(t)) \end{bmatrix} = \\ = \mathbf{B}^*_{hk0}\mathbf{E}^*_{hk}(s(t)) \tag{25}$$

where $\| \mathbf{E}^*_{hk}(s(t)) \| < 1$ and the matrix norm is defined as the maximum singular value of the matrix. Using property

$$\mathbf{XX}^T + \mathbf{YY}^T \geq \mathbf{XY}^T + \mathbf{YX}^T \tag{26}$$

one can verify that

$$\mathbf{P}^*_{hk}(t)\Delta\mathbf{A}^*_{hk}(t) + \Delta\mathbf{A}^{*T}_{hk}(t)\mathbf{P}^*_{hk}(t) \leq \\ \leq \mathbf{D}^{*T}_{hk}(r(t))\mathbf{D}^*_{hk}(r(t)) + \\ +\mathbf{P}^*_{hk}(t)\mathbf{B}^*_{hk0}\mathbf{B}^{*T}_{hk0}\mathbf{P}^*_{hk}(t) \tag{27}$$

$$-\mathbf{K}^{*T}_{hk0}(t)\Delta\mathbf{B}^*_{hk}(s(t))\mathbf{P}^*_{hk}(t) - \\ -\mathbf{P}^*_{hk}(t)\Delta\mathbf{B}^*_{hk}(s(t))\mathbf{K}^*_{hk0}(t) \leq \\ \leq \mathbf{P}^*_{hk}(t)\mathbf{B}^*_{hk0}\mathbf{B}^{*T}_{hk0}\mathbf{P}^*_{hk}(t) + \\ +\mathbf{K}^{*T}_{hk0}(t)\mathbf{E}^{*T}_{hk}(s(t))\mathbf{E}^*_{hk}(s(t))\mathbf{K}^*_{hk0}(t) \tag{28}$$

and

$$-\Delta\mathbf{K}^{*T}_{hk}(t)\mathbf{B}^{*T}_{hk0}\mathbf{P}^*_{hk}(t) - \\ -\mathbf{P}^*_{hk}(t)\mathbf{B}^*_{hk0}\Delta\mathbf{K}^*_{hk}(t) - \\ -\Delta\mathbf{K}^{*T}_{hk}(t)\Delta\mathbf{B}^{*T}_{hk}(s(t))\mathbf{P}^*_{hk}(t) - \\ -\mathbf{P}^*_{hk}(t)\Delta\mathbf{B}^*_{hk}(s(t))\Delta\mathbf{K}^*_{hk}(t) = \\ = -\Delta\mathbf{K}^{*T}_{hk}(t)\mathbf{B}^{*T}_{hk0}\mathbf{P}^*_{hk}(t) - \\ -\mathbf{P}^*_{hk}(t)\mathbf{B}^*_{hk0}\Delta\mathbf{K}^*_{hk}(t) - \\ -\Delta\mathbf{K}^{*T}_{hk}(t)\mathbf{E}^{*T}_{hk}(s(t))\mathbf{B}^{*T}_{hk0}\mathbf{P}^*_{hk}(t) - \\ -\mathbf{P}^*_{hk}(t)\mathbf{B}^*_{hk0}\mathbf{E}^*_{hk}(s(t))\Delta\mathbf{K}^*_{hk}(t) \tag{29}$$

Let

$$\Delta\mathbf{K}^*_{hk}(t) = \mathbf{\Gamma}^*_{1hk}\mathbf{B}^{*T}_{hk0}\mathbf{P}^*_{hk}(t) \tag{30}$$

where

$$\mathbf{\Gamma}^*_{1hk} = \begin{bmatrix} \gamma_h\mathbf{I} & \mathbf{0} \\ \mathbf{0} & \gamma_k\mathbf{I} \end{bmatrix} \tag{31}$$

From (27)-(30) it follows, that summation of the matrices

$$-\mathbf{Q}^*_{hk} + \mathbf{D}^{*T}_{hk}(r(t))\mathbf{D}^*_{hk}(r(t)) + \\ +\mathbf{K}^{*T}_{hk0}(t)\mathbf{E}^{*T}_{hk}(s(t))\mathbf{E}^*_{hk}(s(t))\mathbf{K}^*_{hk0}(t) - \\ -\mathbf{\Gamma}^*_{1hk}\mathbf{P}^*_{hk}(t)\mathbf{B}^*_{hk0} \cdot \\ \cdot[2\mathbf{I} + \mathbf{E}^{*T}_{hk}(s(t)) + \mathbf{E}^*_{hk}(s(t)) - 2\mathbf{\Gamma}^*_{2hk}] \cdot \\ \cdot\mathbf{B}^{*T}_{hk0}\mathbf{P}^*_{hk}(t) \tag{32}$$

must be negative definite, where

$$\mathbf{\Gamma}^*_{2hk} = \begin{bmatrix} \frac{1}{\gamma_h}\mathbf{I} & \mathbf{0} \\ \mathbf{0} & \frac{1}{\gamma_k}\mathbf{I} \end{bmatrix} \tag{33}$$

It is obvious if for some $\theta_k > 0$

$$2\mathbf{I} + \mathbf{E}^{*T}_{hk}(s(t)) + \mathbf{E}^*_{hk}(s(t)) \geq \mathbf{\Theta}^*_{1hk} \tag{34}$$

where

$$\mathbf{\Theta}^*_{1hk} = \begin{bmatrix} \theta_h\mathbf{I} & \mathbf{0} \\ \mathbf{0} & \theta_k\mathbf{I} \end{bmatrix} \tag{35}$$

and $\mathbf{Q}^*_{hk}$ and $\mathbf{\Gamma}^*_{1hk}$ are chosen to satisfy

$$\mathbf{Q}^*_{hk} > \mathbf{D}^{*T}_{hk}(r(t))\mathbf{D}^*_{hk}(r(t)) + \\ +\mathbf{K}^{*T}_{hk0}(t)\mathbf{E}^{*T}_{hk}(s(t))\mathbf{E}^*_{hk}(s(t))\mathbf{K}^*_{hk0}(t) \tag{36}$$

and

$$\mathbf{\Gamma}^*_{1hk} \geq 2\mathbf{\Theta}^*_{2hk} \tag{37}$$

where

$$\mathbf{\Theta}^*_{2hk} = \begin{bmatrix} \frac{1}{\theta_h}\mathbf{I} & \mathbf{0} \\ \mathbf{0} & \frac{1}{\theta_k}\mathbf{I} \end{bmatrix} \tag{38}$$

then the derivative of Lyapunov function is negative definite.

5. A DESIGN EXAMPLE

The properties of the algorithms are illustrated here with a simplified power system design example for three control areas of the system. A stick-fixed linearisation established by considering small changes in load about a chosen state trajectory of the nonlinear system gives the system linear model (1), (2), where

$$\mathbf{A} = \begin{bmatrix} 0 & 1 & 0 & 0 & 0 & 0 & 0 & 0 & 0 & 0 & 0 & 0 \\ a_{21} & a_{22} & a_{23} & 0 & a_{25} & 0 & 0 & 0 & a_{29} & 0 & 0 & 0 \\ 0 & 0 & a_{33} & a_{34} & 0 & 0 & 0 & 0 & 0 & 0 & 0 & 0 \\ 0 & 0a_{42} & 0 & a_{44} & 0 & 0 & 0 & 0 & 0 & 0 & 0 & 0 \\ 0 & 0 & 0 & 0 & 0 & 1 & 0 & 0 & 0 & 0 & 0 & 0 \\ a_{61} & 0 & 0 & 0 & a_{65} & a_{66} & a_{67} & 0 & a_{69} & 0 & 0 & 0 \\ 0 & 0 & 0 & 0 & 0 & 0 & a_{77} & a_{78} & 0 & 0 & 0 & 0 \\ 0 & 0 & 0 & 0 & 0 & a_{86} & 0 & a_{88} & 0 & 0 & 0 & 0 \\ 0 & 0 & 0 & 0 & 0 & 0 & 0 & 0 & 0 & 1 & 0 & 0 \\ a_{101} & 0 & 0 & 0 & a_{105} & 0 & 0 & 0 & a_{109} & a_{1010} & a_{1011} & 0 \\ 0 & 0 & 0 & 0 & 0 & 0 & 0 & 0 & 0 & 0 & a_{1111} & a_{1112} \\ 0 & 0 & 0 & 0 & 0 & 0 & 0 & 0 & 0 & a_{1210} & 0 & a_{1212} \end{bmatrix}$$

$$\mathbf{B} = \begin{bmatrix} 0 & 0 & 0 \\ 0 & 0 & 0 \\ 0 & 0 & 0 \\ \frac{1}{T_{g1}} & 0 & 0 \\ 0 & 0 & 0 \\ 0 & 0 & 0 \\ 0 & 0 & 0 \\ 0 & \frac{1}{T_{g2}} & 0 \\ 0 & 0 & 0 \\ 0 & 0 & 0 \\ 0 & 0 & 0 \\ 0 & 0 & \frac{1}{T_{g3}} \end{bmatrix}$$

where

$$a_{22} = -\frac{1}{T_{p1}} \qquad a_{23} = \frac{K_{p1}}{T_{p1}}$$
$$a_{25} = \frac{K_{p1}}{T_{p1}} T_{12}^0 \qquad a_{29} = \frac{K_{p1}}{T_{p1}} T_{13}^0$$
$$a_{33} = -\frac{1}{T_{t1}} \qquad a_{34} = \frac{1}{T_{t1}}$$
$$a_{42} = -\frac{1}{T_{g1} R_1} \qquad a_{44} = -\frac{1}{T_{g1}}$$
$$a_{61} = -\frac{K_{p2}}{T_{p2}} a_{12} T_{12}^0 \qquad a_{66} = -\frac{1}{T_{p2}}$$
$$a_{67} = \frac{K_{p2}}{T_{p2}} \qquad a_{69} = \frac{K_{p3}}{T_{p3}} T_{23}^0$$
$$a_{77} = -\frac{1}{T_{t2}} \qquad a_{78} = \frac{1}{T_{t2}}$$
$$a_{86} = -\frac{1}{T_{g2} R_2} \qquad a_{88} = -\frac{1}{T_{g2}}$$
$$a_{101} = -\frac{K_{p3}}{T_{p3}} a_{13} T_{13}^0 \qquad a_{105} = -\frac{K_{p3}}{T_{p3}} a_{23} T_{23}^0$$
$$a_{1010} = -\frac{1}{T_{p3}} \qquad a_{1011} = \frac{K_{p3}}{T_{p3}}$$
$$a_{1111} = -\frac{1}{T_{t3}} \qquad a_{1112} = \frac{1}{T_{t3}}$$
$$a_{1210} = -\frac{1}{T_{g3} R_3} \qquad a_{1212} = -\frac{1}{T_{g3}}$$
$$a_{21} = -\frac{K_{p1}}{T_{p1}} (T_{12}^0 + T_{13}^0)$$
$$a_{65} = -\frac{K_{p2}}{T_{p2}} (T_{23}^0 - a_{12} T_{12}^0)$$
$$a_{109} = \frac{K_{p3}}{T_{p3}} (a_{23} T_{23}^0 + a_{13} T_{13}^0)$$

T_{gi}	i-th governor time constant
T_{ti}	i-th turbine time constant
T_{pi}	i-th plant model time constant
K_{pi}	i-th plant gain
R_i	i-th generator self-regulation

For the simulation the following parameter values were used

$$\begin{aligned} T_{g1} &= T_{g2} = T_{g3} = 0.08 \text{ s} \\ T_{t1} &= T_{t2} = T_{t3} = 0.30 \text{ s} \\ T_{p1} &= T_{p2} = T_{p3} = 30 \text{ s} \\ T_{12}^0 &= T_{23}^0 = T_{13}^0 = 0.545 \text{ pu MW/Hz} \\ K_{p1} &= K_{p2} = K_{p3} = 150 \end{aligned}$$

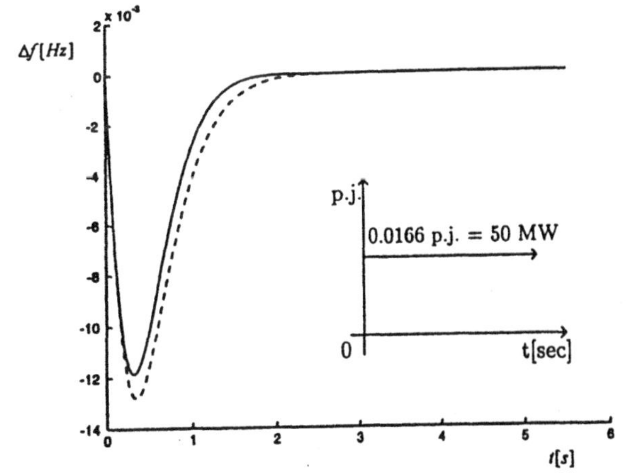

Fig. 1. $\Delta f_1(t)$ response

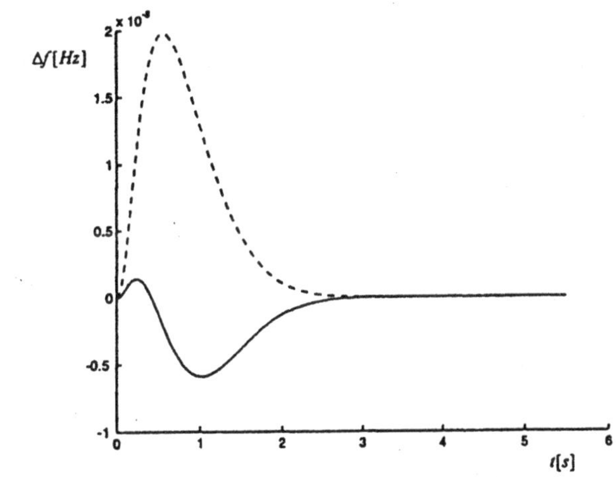

Fig. 2. $\Delta f_2(t)$, $\Delta f_3(t)$ responses

$$R_1 = R_2 = R_3 = 1.60 \text{ Hz/pu MW}$$

The provided range of the system parameters was

$$\begin{aligned} T_{g1} &= T_{g2} = T_{g3} = (0.06 \text{ - } 0.10) \text{ s} \\ R_1 &= R_2 = R_3 = (1.40 \text{ - } 1.80) \text{ Hz/pu MW} \end{aligned}$$

For the load disturbance affecting the first subsystem the simulation results of the incremental frequency deviation (Hz) $\Delta f_i(t)$ and the incremental generator output change (pu MW) $\Delta P_{gi}(t)$ for the robust centralized regulator (- - -) as well as for the robust pairwise autonomous regulator (—) are shown in Figs. 1-4.

6. CONCLUDING REMARKS

In the paper is presented the technique for determining a linear feedback control law which stabilizes an uncertain dynamic large-scale systems consisting of connection of the subsystems.The design procedure involves solving Riccati equations and is based on the concept of partial decentralization. Obtained sets of Riccati equations are functions of the interconnections, and

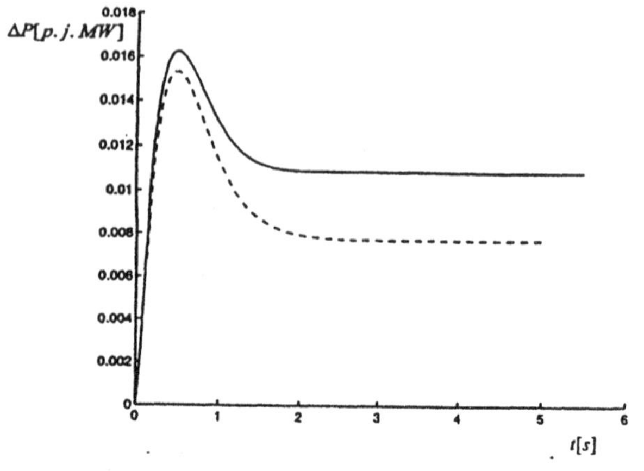

Fig. 3. $\Delta P_{g1}(t)$ response

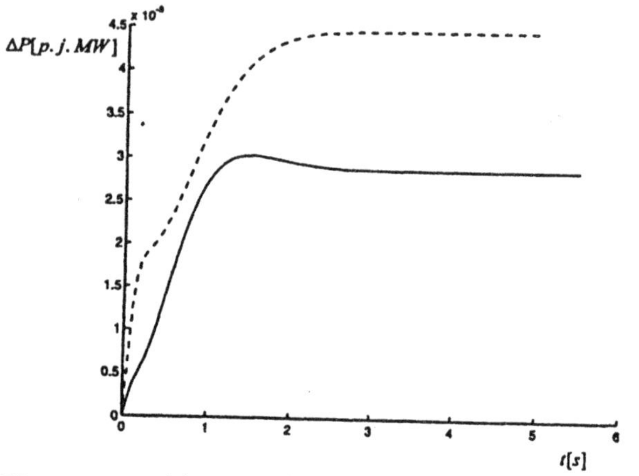

Fig. 4. $\Delta P_{g2}(t)$, $\Delta P_{g3}(t)$ responses

separate and independent, afterwards the resulting controller structures are pairwise autonomous (implying parallel processing for on-line implementation), decentralized (implying interconnection utilization) and separable. Priori information required for the plant simulation are only the bounds of time-varying parameters. Presented simulation results illustrate the control process quality.

ACKNOWLEDGEMENTS

The work presented in this paper was supported by the Grant Agency of Ministry of Education and Academy of Science of Slovak Republic under Grant No. 1/3206/96.

7. REFERENCES

Filasová, A. (1994). Synthesis of linear controllers using pairwise autonomous subsystem approach. In: *Proc. 1th IFAC Workshop on New Trends in Design of Control Systems*, Smolenice, pp. 185-189.

Filasová, A. (1996). Control systems for the plants with the parametric uncertainties. In: *ELMAT '96*, Ferrost a.s., Košice, Slovakia, pp. 127-132. (in Slovak)

Fosha, C.E., Elgerd, O.I. (1970). The megawatt-frequency control problem: A new approach via optimal control theory. *IEEE Transactions on Power Apparatus and Systems*, **PAS-89**, pp. 563-571.

Krokavec, D. (1995). Partially decentralized algorithms for the Kalman estimator. In: *New Trends in Power System Automation*, Academia Centrum, Zlin, Czech Republic, pp. 196-200.

Petersen, I.R., Hollot, C.V. (1986). A Riccati equation approach to the stabilization of uncertain linear systems. *Automatica*, **22**, pp. 397-411.

Sarnovský, J. (1982). Technological process control. Alfa, Bratislava. (In slovak)

Schmitendorf, W.E. (1988). Designing stabilizing controllers for uncertain systems using the Riccati equation approach, *IEEE Transaction on Automatic Control*, **AC-33**, pp. 376-378.

HIERARCHICAL STEADY STATE CONTROL OF COMPLEX SYSTEMS

Ladislav Harsányi*, Zdenka Králová*, Vojtech Veselý*, Khalil S. Hindi**

**Department of Automatic Control Systems, Slovak University of Technology in Bratislava, Ilkovicova 3, 812 19 Bratislava, Slovak Republic*[1]
***Department of Manufacturing Systems, Brunel University, Uxbridge, Middlesex, UK*

Abstract: This paper presents the basic principles and main properties of new modified feasible methods that can be applied in steady-state optimization of complex systems. The essential features of the proposed structure are both using the feedback from the controlled system and the linearity of a mathematical model for finding the optimal solution on the coordinator level and on the subsystem one. These properties are important from a practical point of view. An example of optimal steady state control of a complex system is given.

Keywords: hierarchical systems, steady states, optimization

1. INTRODUCTION

The problem of complex system optimization for industrial processes is becoming more and more important because of increasing raw material and energy costs. To handle a steady-state optimization problem of a complex system which consists of N subsystems, the hierarchical control structure, composed of local decision making units, local control systems and a coordinator can be used (Mesarovic, *et al.*, 1970, Findeisen, *et al.*, 1978, Singh and Titli, 1978). There are two principal modes by which the coordinator could intervene into the local problem:

-action on the objective functions of the subproblems (goal cordination),
-interaction prediction among subproblems (model coordination).

Generally, the steady-state optimising technique can be divided as follows.

1) Methods that do not use the feedback from real processes,
- goal coordination or non-feasible method,
- model coordination or feasible method,
- mixed method.

2) Techniques where the feedback from real processes is used
- interaction balance method with feedback, IBMF,
- modified feasible method, MFM, proposed in this paper.

3) Integrated system optimization and parameter estimation, ISOPE.

Later modifications have been made to these methods which extend their utility and reduce some of their disadvantages by incorporating several types of approaches according to their availability (Tatjewski, 1985, Sidaou, *et al.*, 1991, Tatjewski, *et al.*, 1990, Roberts and Lin, 1991, Brdiš, *et al.*, 1987, Veselý, *et al.*, 1992, etc.).

In this paper the modified feasible method, MFM is presented. Its main advantage is that the linear algebraic system equations carry out on the coordinator level.

The work is organised as follows. First, the problem statement and some preliminary results are given in Section 2. The main results are given in Section 3 and an example in Section 4.

[1] The research work has been partially supported by the Slovak Scientific Grant Agency, grant No. 1/4231/97.

2. PROBLEM STATEMENT

Consider an industrial process consisting of N interconnected subsystems. Each subsystem is represented by the following model (Fig. 1).

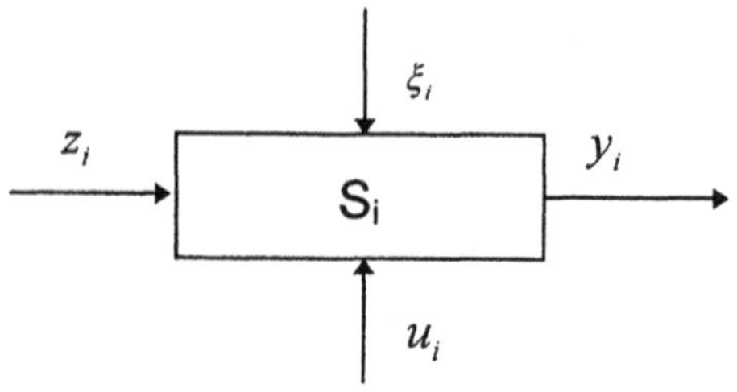

Fig. 1. Representation of the ith subsystem

where $u_i \in R^{m_i}$, $y_i \in R^{l_i}$, $z_i \in R^{p_i}$ and ξ_i are the control input, output (controller set points in many cases), input and disturbances, respectively. It is assumed that external disturbances ξ_i affecting the ith subsystem are constant over the considered time interval and therefore disturbances can be omitted from the system equations or the system equations described by the mean value of process variables. The input-output steady-state behaviour of the ith subsystem is described by the following functions

$$y_i = T_i(u_i, z_i), \quad i \in \mathbf{N} = \{1,2,\ldots,N\} \tag{1}$$

The interconnections between the subsystems are identified as the physical connection between the inputs of a subsystem and the outputs of the other subsystems. This can be described by N linear coupling equations

$$z_i = \sum_{j=i}^{N} C_{ij} y_i \tag{2}$$

where $C_{ij} \in R^{p_i \times l_j}$ is the interconnection matrix.

Generally, the purpose of process operations optimization is to achieve the optimum performance of the process in the sense that the production process yields the maximum profit for the minimum cost, Findeisen et al. (1978), Roberts and Lin (1991).

The global performance index in hierarchical situation is specified as a summation of the performed indices of each individual subsystem.

$$J = \sum_{i=1}^{N} Q_i(u_i, z_i) \tag{3}$$

In order to ensure that the process always operates within the region specified by standards and capacity limits it is necessary to impose constraints on the process optimization. Such constraints are usually identified as the local constraints. The local constraints are

$$g_i(u_i, z_i) \le 0, \quad i \in \mathbf{N} \tag{4}$$

Generally, we assume that the mappings $T_i(u_i, z_i)$, $Q_i(u_i, z_i)$ and $g_i(u_i, z_i): R^{m_i \times p_i}$ are C^2 functions, continuously differentiable and $Q_i(u_i, z_i)$ is convex for $i \in \mathbf{N}$.

Based on the above, a process optimising control problem may be formulated as follows:

$$\begin{aligned} &\min J(u, z) \\ &\text{subject to } z = Cy \\ &\qquad y = T(u, z) \\ &\qquad g(u, z) \le 0 \end{aligned} \tag{5}$$

where

$$u^T = \left[u_1^T, \ldots, u_N^T\right] \in R^m,$$
$$z^T = \left[z_1^T, \ldots, z_N^T\right] \in R^p,$$
$$y^T = \left[y_1^T, \ldots, y_N^T\right] \in R^l.$$

are control input, input and output of the complex system. Different coordination methods could be used for finding out the solution of (5) (Findeisen, 1978, Singh and Titli, 1978, Roberts, 1978, Tatjewski, 1985, and references therein).

With this optimization problem we can associate the Lagrangian:

$$L = \sum_{i=1}^{N} \left\{ Q_i(u_i, z_i) + \lambda_i^T \left[T_i(u_i, z_i) - y_i\right] + \right.$$

$$+ \vartheta_i^T \left(\sum_{j=1}^{N} C_{ij} y_j - z_i \right) + \rho_i^T g_i(u_i, z_i) \tag{6}$$

where λ_i, ϑ_i are Lagrange multiplier vectors and ρ_i is the Kuhn-Tucker multiplier which have been introduced to take into account the equalities (1), (2) and inequality (4) as constraints. In the feasible or model coordination method (Singh and Titli, 1978), the coordinator level fixes y_j and provides it for the lower level. At the subsystem level, the Lagrangian (6) can be written in the following separable form

$$L = \sum_{i=1}^{N} L_i \tag{7}$$

where $L_i = Q_i(u_i, z_i) + \lambda_i^T \left[T_i(u_i, z_i) - y_i\right] +$

$$+ \sum_{j=1}^{N} C_{ij} y_j - z_i + \rho_i^T g_i(u_i, z_i)$$

The ith subproblem is

$\min \; Q_i(u_i, z_i)$

subject to $y_i = T_i(u_i, z_i)$

$$z_i = \sum_{j=1}^{N} C_{ij} y_j \qquad (8)$$

$$g_i(u_i, z_i) \le 0$$

for a given y_j, $j \in \mathbf{N}$. For fixed y_j, $j \in \mathbf{N}$ the model constraints (4) and interconnections constraints are always satisfied. The gradient type coordination algorithm for model coordination is obtained as follows

$$y^{(n)} = y^{(n-1)} \pm \alpha \frac{\partial L}{\partial y} \qquad (9)$$

where $\alpha > 0$, + for the maximum and - for the minimum.

Thus the coordination algorithm will improve the value of the performance index (3) at each iteration and the results could be applied in "real time" since all the constraints are satisfied at each iteration. Generally, in the subsystem level, the nonlinear equations (8) have to be solved. On the gradient coordinator level the calculations are simpler. For the solution of complex optimization problems the designer needs N subsystems with complicated microprocessors, and for the coordinator just a simple one. This approach may not be economical from a practical point of view, which is one of important drawbacks of the proposed coordination method (Findeisen, *et al.,* 1978, Singh and Titli, 1978, Roberts, 1978, Tatjewski, 1985).

3. MODIFIED FEASIBLE METHOD

Hence, in the feasible method the constraints are satisfied at each iteration. If $g_i(u_i, z_i)$ exists in (7) then its constraints should be included in the cost function (3) as a penalty function, Singh and Titli (1978). Let us assume that output variables of all subsystems y_i, $i \in \mathbf{N}$ (1) can be measured.

The optimal solution of (8) must satisfy the stationary conditions of the Lagrangian

$$L_{u_i} = \frac{\partial L}{\partial u_i} = \frac{\partial Q_i}{\partial u_i} + \left[\frac{\partial T_i}{\partial u_i}\right]^T \lambda_i = 0$$

$$L_{z_i} = \frac{\partial L}{\partial z_i} = \frac{\partial Q_i}{\partial z_i} + \left[\frac{\partial T_i}{\partial u_i}\right]^T \lambda_i - \vartheta_i = 0$$

$$L_{y_i} = \frac{\partial L}{\partial y_i} = -\lambda_i + \sum_{j=1}^{N} C_{ji}{}^T \vartheta_j = 0 \qquad (10)$$

$$L_{\lambda_i} = \frac{\partial L}{\partial \lambda_i} = T_i(u_i, z_i) - y_i = 0$$

$$L_{\vartheta_i} = \frac{\partial L}{\partial \vartheta_i} = \sum_{i=1}^{N} C_{ij} y_j - z_i = 0 \qquad i \in \mathbf{N}$$

where

$$\frac{\partial T_i}{\partial z_i} = \begin{bmatrix} \frac{\partial T_{i1}}{\partial z_{i1}} & \dots & \frac{\partial T_{i1}}{\partial z_{ip_i}} \\ \dots & \dots & \dots \\ \frac{\partial T_{il_i}}{\partial z_{i1}} & \dots & \frac{\partial T_{il_i}}{\partial z_{ip_i}} \end{bmatrix} \in R^{l_i \times p_i}, \quad i \in \mathbf{N} \text{ etc.}$$

For the ith subsystem the number of equations and the number of searched parameters are equal to $m_i + 2(p_i + l_i)$. In the modified feasible method the following procedure is proposed:

1. On the local level for the each subsystem the input control variables u_i, $i \in \mathbf{N}$ have to be chosen. The chosen $u_i{}^T$ would be adjusted to the real process and the output variables y_i of all subsystems will have to be measured.

2. The input variables of z_i, $i \in \mathbf{N}$ are determined on the coordinator level from the last equation of (10).

3. Substituting the third equation for the second one of (10) we obtain $N.\sum_{j=1}^{N} l_j$ linear equations in the form

$$-\lambda_i + \sum_{j=1}^{N} C_{ji}{}^T \left[\frac{\partial Q_j}{\partial z_j} + \left[\frac{\partial T_j}{\partial z_j}\right]^T \lambda_j\right] = 0, \qquad i \in \mathbf{N} \qquad (11)$$

with $N.\sum_{j=1}^{N} l_j$ unknown Lagrangian multipliers λ_i, $i \in \mathbf{N}$. The coordinator carries out the solution for the system of linear algebraic equations (11) with respect to λ_i, and the results would be sent to the local level.

4. Local control systems carry out the solution

$$L_{u_i} = \frac{\partial Q_i}{\partial u_i} + \left[\frac{\partial T_i}{\partial u_i}\right]^T \lambda_i, \ i \in \mathbf{N} \qquad (12)$$

If for some $i \in \mathbf{N}_1$ and $\varepsilon_i > 0$ is $|Lu_i| < \varepsilon_i$, then stop the operation for those control systems, else

$$u_i{}^{(k+1)} = u_i{}^k \pm \alpha L u_i, \ i \in \mathbf{N}_2 \qquad (13)$$

where $N = N_1 \cup N_2$, sign +(-) valid for maximum (minimum) of L with respect to u_i, $i \in \mathbf{N}$.

Remark 1. Substituting λ_i, $i \in \mathbf{N}$ from the third equation of (10) to the second one we can obtain a system of linear algebraic equations with respect to $\vartheta_i, i \in \mathbf{N}$.

Remark 2. The most practical situations in a real process: the input and output variables z_i and y_i, $i \in \mathbf{N}$ can be measured. In this case the second step of the above proposed procedure can be omitted.

Remark 3. If the process has disturbances, thus it is not noise free, the input-output mappings of the ith subsystem (1) for the real process

$$y_i = T_{iR}(u_i, z_i) \tag{14}$$

may be slightly different from those given by (1).

The sufficient conditions for u_i, $i \in \mathbf{N}$ to be the solution of the system optimization problem (10) are given by

$$y_i|_{u=u^*} = y_{iR}|_{u=u^*}$$

$$\frac{\partial T_i}{\partial u_i}|_{u=u^*} = \frac{\partial T_{iR}}{\partial u_i}|_{u=u^*}, \quad i = 1,2,\ldots,N \tag{15}$$

where T_{iR} are the input-output mappings of the real systems,

y_{iR} is the output of the real system.

Thus, it is sufficient that both the respective outputs and the derivatives with respect to the input controls of the model and system should match the solution to the model based problem at the optimum. Convergence of the proposed coordination algorithm is given by the following theorem.

Theorem 1.

The gradient type coordinator with proposed steps 1-4 is asymptotically stable with respect to the optimal solution of the system (10) if there exists a positive coefficient $\alpha < \alpha^*$ such that the following inequality holds

$$L_u^T L_u > \frac{\rho(\Delta u(t), \Delta q(t))}{\alpha^*} \tag{16}$$

where $\rho(\Delta u(t), \Delta q(t))$ is given later.

Proof.

Let q be the set of variables given as follows:

$$q^T = \left[z^T, y^T, \lambda^T, \vartheta^T\right],$$

where $\lambda^T = \left[\lambda_1^T, \ldots, \lambda_N^T\right], \vartheta^T = \left[\vartheta_1^T, \ldots, \vartheta_N^T\right]$

The local control systems equation can be written in the form of difference equation (13)

$$u(t+1) = u(t) - \alpha L_u \tag{17}$$

where $t = 1,2\ldots,n$ is the discrete sample time.

We seek the minimum of L by a gradient procedure. Let $\hat{L}$ be the optimal solution, and let the Lyapunov function be chosen as (Singh and Titli,1978)

$$V(t) = L(t) - \hat{L} \tag{18}$$

The first difference of Lyapunov function is given

$$\Delta V(t) = L(t+1) - L(t) = \Delta L(t)$$

Using the Taylor series expansion of Lagrangian L (7)

$$\Delta L = L_u^T(t)\Delta u(t) + L_q^T \Delta q(t) + \rho(\Delta u(t), \Delta q(t)) \tag{19}$$

where $\lim\limits_{\Delta u(t), \Delta q(t) \to 0} \rho(\Delta u(t), \Delta q(t)) = 0$ is the remaining part of the Taylor series expansion.

Because of (10), $L_q = 0$ for each iteration process, then using (17) we obtain

$$\Delta V(t) = -\alpha L u^T(t) L u(t) + \rho(\Delta u(t), \Delta q(t)) \tag{20}$$

Equation (20) implies the sufficient stability condition given by (16). This completes the proof.

Summarizing the main properties of the proposed modified feasible method we can conclude that the situation for carrying out the optimal solution is easier than in Findeisen, *et al.* (1978), Singh and Titli (1978). On the local "coordinator" level the proposed method uses the very simple equation (13). On the above level the control system carries out the solution of the system of linear algebraic equations, which is independent on linearity or nonlinearity of the mathematical model of each subsystems. The above mentioned properties are the main advantage of the proposed method.

In general, a process model may not be a true representation of the real system and this usually results in a non-optimum solution. To overcome this disadvantage an Integrated System Optimization and Parameter Estimation method has been proposed by Roberts (1979). An important feature of the ISOPE method is that it is capable of producing a true optimum in the presence of model reality differences. This is achieved by introducing a modifier into the model based optimization problem so that the error caused by model reality differences is compensated for when the algorithm converges. Using the ISOPE techniques the following Lagrangian forms:

$$L_S = \sum_{i=1}^{N} \Big\{ Q_i(u_i, z_i, a_i) - \vartheta_i^T z_i +$$

$$+ \sum_{j=1}^{N} \vartheta_j^T C_{ji} T_i(u_i, z_i, a_i) + \gamma_i^T (w_i - z_i) +$$

$$+ \lambda_i^T (v_i - u_i) + \eta_i^T \left[y_i'(v) - T_i(v_i, w_i, Q_i) \right] \tag{21}$$

where $y_i'(v) \in R^{l_i}$ is the measured output variable of ith subsystem

$\vartheta_i, \gamma_i, \lambda_i, \eta_i$ are the Lagrangean multiplier vectors

v_i, w_i are the vector variables which associate with identification process of process model

a_i are the unknown parameters of the model process

$v^T = \left[v_1^T, \ldots, v_N^T\right]$

The necessity conditions for optimal solution of (21) are as follows.

a) $L_{ui} = \frac{\partial}{\partial u_i}\left[Q_i + \sum_{j=1}^{N} T_i^T C_{ji}^T \vartheta_j\right] - \lambda_i = 0$

b) $L_{zi} = \frac{\partial}{\partial z_i}\left[Q_i + \sum_{j=1}^{N} T_i^T C_{ji}^T \vartheta_j\right] - \vartheta_i - \gamma_i = 0$

c) $L_{v_i} = \lambda_i + \left[\frac{\partial y_i'(v)}{\partial v_i}\right]\eta_i - \frac{\partial}{\partial v_i}\left[T_i^T\right]\eta_i = 0$

d) $L_{w_i} = \gamma_i - \frac{\partial}{\partial w_i}\left(T_i^T\right)\eta_i = 0$ (22)

e) $L_{ai} = \frac{\partial}{\partial a_i}\left[Q_i + \sum_{j=1}^{N} T_i^T C_{ji}^T \vartheta_j\right] - \frac{\partial}{\partial a_i}(T_i)\eta_i = 0$

f) $L_{\eta_i} = y_i'(v) - T(v_i, w_i, a_i) = 0$

g) $L_{\vartheta_i} = \sum_{j=1}^{N} C_{ij} T_i - z_i = 0$

h) $L_{\gamma_i} = w_i - u_i,\ L_{\lambda_i} = v_i - u_i = 0$

The following procedure for finding the solution of (12) is proposed.

1. On the local level for each subsystem the input control variables $v_i \in R^{m_i}$, $i \in \mathbf{N}$ have to be chosen. The chosen v_i would be adjusted on the real process and the output variables of the ith subsystem $y_i'(v)$ (and z_i) have to be measured. If z_i could not be measured on the subsystem level, the output variables $y_i'(v)$ have to be sent to coordinator level using the equations (22g) and variables z_i, $i \in \mathbf{N}$ can be calculated. From (22f) the unknown parameter a_i, $i \in \mathbf{N}$ is calculated. The derivatives of $\partial y_i'(v)/\partial v_i$, $i \in \mathbf{N}$, could be determined as in Roberts and Lin (1991).

2. Substituting (22d) and (22e) to (22b) we obtain

$$\frac{\partial}{\partial z_i}\left[Q_i + \sum_{j=1}^{N} T_i^T C_{ji}^T \vartheta_j\right] - \vartheta_i + \frac{\partial}{\partial w_i}(T_i^T)\left[\frac{\partial}{\partial a_i}\left(T_i^T\right)\right]^{-1} \times$$

$$\times \frac{\partial}{\partial a_i}\left[Q_i + \sum_{j=1}^{N} T_i^T C_{ji}^T \vartheta_j\right] = 0, \quad i \in \mathbf{N} \qquad (23)$$

Equations (23) are a system of linear algebraic equations with respect to ϑ_i, $i \in \mathbf{N}$ which can be discovered on the coordinator level. The solutions of ϑ_i, $i \in \mathbf{N}$ have to be sent to the local level.

3. On the local level the vectors η_i, λ_i are calculated from (22e) and (22c). The local level carries out the solution of L_{u_i}, $i \in \mathbf{N}$ and if for some $i \in \mathbf{N}$ and for a positive small number ε_i $\left|L_{ui}\right| < \varepsilon_i$ then stop the operation of those local control systems, else

$u_i^{(k+1)} = u_i(k) \pm \alpha L u_i$

4. STEADY STATE OPTIMIZATION OF CHEMICAL TECHNOLOGY PROCESSES - SIMULATION EXAMPLE

Let us consider the following nonlinear three-subsystem example. The subsystem output mappings are

$$y_i = a_i u_i^2 + b_i u_i + \eta_i z_i, \quad i = 1,2,3 \qquad (25)$$

with interconnection constraints (2)

$z_i = y_i - 1$ for $i = 1,2,3$

where $y_0 = const > 0$.

The local constraints set is

$y_{i\,\min} \le y_i \le y_{i\,\max}$ for $i = 1,2,3$

and local performance indices are (3)

$$Q_i = q_i u_i^2 + p_i u_i \to \min \quad i = 1,2,3 \qquad (26)$$

where $a_i, b_i, \eta_i, q_i, p_i$ for $i = 1,2,3$ are supposed to be known parameters.

With this optimization problem we can associate the Lagrangian (6):

$$L = \sum_{i=1}^{3}(q_i u_i^2 + p_i u_i + \lambda_i\left[y_i - \left(a_i u_i^2 + b_i u_i + \eta_i z_i\right)\right] + \rho_i\left(y_i - y_{i2}\right)^2) + \sum_{i=1}^{2}\vartheta_i\left(y_i - z_{i+1}\right) \qquad (27)$$

The optimal solution of (27) must satisfy the following conditions:

$$L_{u_i} = \frac{\partial L}{\partial u_i} = 2q_i u_i + \rho_i - \lambda_i(2a_i u_i + b_i) = 0$$

$$L_{y_i} = \lambda_i + \gamma_i + 2\rho_i(y_i - y_{iz}) = 0 \quad (28)$$

$$L_{\lambda_i} = y_i - a_i u_i^2 - b_i u_i - \eta_i z_i = 0 \quad i = 1,2,3$$

$$L_{v_j} = y_j - z_{j+1} = 0 \quad j = 1,2$$

$$L_{z_k} = -\lambda_k \eta_{ki} - \vartheta_{k-1} = 0 \quad k = 2,3$$

The parameters of both models with constraints (25) and performances indices (26) are as follows.

i	a_i	b_i	η_i	q_i	p_i
1	-1.4958	19.6062	1	1.1626	6.6759
2	0.2467	13.9693	0.8	4.3525	-3.8628
3	0.2739	-8.0270	0.8	0.0085	14.7900

$$43 \le y_1 \le 47 \quad 73 \le y_2 \le 77 \quad 20 \le y_3 \le 26$$

and $\rho_1 = 0.2,\ \rho_2 = 1.5,\ \rho_3 = 1.$

For the first step the following input control variables have been chosen:

$$u_1 = 1.4,\ u_2 = 2.7,\ u_3 = 5,\ y_0 = 20$$

On the coordinator level the following linear algebraic equations have been obtained (11):

$$\lambda_1 - \lambda_2\eta_2 + 2\rho_1(y_1 - y_{1z}) = 0$$

$$\lambda_2 - \lambda_3\eta_3 + 2\rho_2(y_2 - y_{2z}) = 0$$

$$\lambda_3 + 2\rho_3(y_3 - y_{3z}) = 0$$

The optimal solution of (28) is given as follows:

$$u_1 = 1.5638, \quad y_1 = 46.9863, \ \lambda_1 = 0.69$$

$$u_2 = 2.4662, \quad y_2 = 73.6084, \ \lambda_2 = 1.8539$$

$$u_3 = 5.2266, \quad y_3 = 24.4501, \ \lambda_3 = -2.9005$$

and $\vartheta_1 = -1.4831,\ \vartheta_2 = 2.3204$

with $Q = \sum_{i=1}^{n} Q_i = 133.7053$

5. CONCLUSION

The paper presents the main properties of a new modified feasible method. The essential features of the proposed method are both in the procedure for finding the optimal solution for the steady state optimization problems on the coordinator level and on the local one, only the linear algebraic equations are used. In the process of calculation of optimal solution the feedback variables from the real process are needed. The above mentioned properties of the proposed method are important from a practical implementation point of view. A nonlinear three-subsystem example is given.

REFERENCES

Brdiš, M., N. Abdullah and P.D. Roberts (1987). Augmented model based double iterative loop techniques for integrated system optimisation and parameter estimation of large-scale industrial processes. *Proceedings of 10th IFAC World Congress*, 7, 97-102.

Findeisen, W., M. Brdiš, K. Malinowski, P. Tatjewski and A. Wozniak (1978). On-Line Hierarchical Control for Steady-State Systems. *IEEE Trans. on AC*, **13**, No. 3, 189-208.

Mesarovic, M.D., D. Macko and Y.Takahara (1970). *Theory of Hierarchical Multilevel Systems.* Academic Press, New York.

Roberts, P.D., (1979). An Algorithm for Steady-State System Optimisation and Parameter Estimation, *Int. Journal of System Science*, **10**, 719-734.

Roberts, P.D. and J. Lin (1991). Potential for Hierarchical Optimising Control in the Process Industry, *Proceedings ECC 91*, Grenoble, France, 213-217.

Sidaoui, H., Z.Binder and R. Perret (1991). Tracking Approach in Hierarchical Optimization of Large Scale Systems. *Proceedings ECC 91*, Grenoble, 2376-2379.

Singh, M.G. and A. Titli (1978). *Decomposition, Optimisation and Control.* Pergamon Press, Oxford.

Tatjewski, P., (1985). On Hierachical Control of Steady-State System Using IBMF. *Large Scale Systems*, **1**, No. 1.

Tatjewski, P., N. Abdullah and P.D. Roberts (1990). Comparative Study of Development of Integrated Optimisation and Parameter Estimation Algorithms for Hierarchical Steady-State Control. *IJC*, **51**, No. 2, 421-443.

Veselý, V. *et al.* (1992). *Design of Complex Systems Control.* ALFA, Bratislava (in Slovak).

DECENTRALIZED OPTIMAL FILTERING OF THE LARGE-SCALE SYSTEM STATE

Dušan Krokavec

Technical University of Košice,
Faculty of Electrical Engineering and Informatics
Department of Cybernetics and Artificial Intelligence
Letná 9/B, 042 00 Košice, Slovak Republic
e-mail: krokavec@hron.fei.tuke.sk

Abstract: The purpose of the paper is to present some decentralized algorithms based on successive approximation in the correction stage for the stochastic state estimation in large-scale linear interconnected sampled-data dynamical system with well-defined interactions where, each output of the subsystems is decoupled from one another.

Keywords: Stochastic estimation, Kalman filters, large-scale systems, decentralized control.

1. INTRODUCTION

All real systems operate in a stochastic environment where they are subject to unknown disturbances and in addition, the controller has to rely, in practice, on imperfect measurements. This difficulties can be overcome, at least in principle, by the optimal stochastic state estimation.

Stochastic estimation is the operation of assigning a value to an unknown system state based on noise-corrupted observations involving some functions of the state. The noise is assumed to have known statistical properties and usually Gaussian noise models provide an adequate description. The value assigned is called the estimate and the system of the observations yielding the estimate is called the estimator. The estimation operation will be optimal if the assignment of an estimate is in accordance with minimization of some estimation criterion (minimum mean squared error, minimum least absolute error value, etc.)

In recent years many applications have spurred interest in recursive estimation based on the results of Kalman for linear dynamic systems, where the optimal steady-state filter gain can be obtained by an iterative procedure. The disadvantage of applying the single-level Kalman filter, however, is that the computational burden associated with the implementation of the filter is proportional to the cube of system order. In order to overcome this, there is a need for finding ways of decomposing the optimal global single-level Kalman filter.

A new approach is outlined to obtain partially decentralized algorithms for optimal state estimation in large-scale linear discrete-time dynamical system where each output of the system is decoupled from one another. Application of these ideas for the purpose of decentralization and reduction of computational requirements as well as the conditions under which the filter gives near optimal performance is the subject of the presented paper. The most applicable publications which have dealt with the above mentioned problem are (Chui and Cheng, 1991), (Hassan, Salut, Singh and Titli, 1978), (Verhaegen and Dooren, 1986) and in (Filasová, 1994), (Krokavec, 1995), (Myers and Tapley, 1976) are proposed the basic techniques for dealing with them.

2. KALMAN FILTER

In general, a discrete-time stochastic multivariable linear system can be considered in the form of linear difference equations

$$\mathbf{x}(i+1) = \mathbf{F}\mathbf{x}(i) + \mathbf{G}\mathbf{u}(i) + \mathbf{v}(i) \qquad (1)$$

and the measurements can be governed by

$$\mathbf{y}(i) = \mathbf{H}\mathbf{x}(i) + \mathbf{o}(i) \qquad (2)$$

where

- $\mathbf{x}(i)$ nx1 state vector of the system,
- $\mathbf{u}(i)$ rx1 input vector of the system,
- $\mathbf{y}(i)$ mx1 output vector of the system,
- $\mathbf{v}(i)$ nx1 random vector of process noise,
- $\mathbf{o}(i)$ mx1 random vector of output noise,
- $\mathbf{F}$ nxn nonsingular state transition matrix,
- $\mathbf{G}$ nxr constant input matrix,
- $\mathbf{H}$ mxn constant output matrix.

The Gaussian distributed process noise and measurement output noise random sequences $\{\mathbf{v}(i) : i = 0,1,2,\ldots\}$ and $\{\mathbf{o}(i) : i = 0,1,2,\ldots\}$, as well as the initial state $\mathbf{x}_0$ are mutually independent with means and covariances as follows

$$\begin{array}{l} E\{\mathbf{v}(i)\} = 0; \quad E\{\mathbf{v}(i)\mathbf{v}^T(j)\} = \mathbf{Q}\delta_{ij} \\ E\{\mathbf{o}(i)\} = 0; \quad E\{\mathbf{o}(i)\mathbf{o}^T(j)\} = \mathbf{R}\delta_{ij} \\ E\{\mathbf{x}(0)\} = \mathbf{x}_0; \ E\{\mathbf{v}(i)\mathbf{o}^T(j)\} = \mathbf{0} \\ E\{(\mathbf{x}(0) - \mathbf{x}_0)\mathbf{v}^T(j)\} = \mathbf{0} \\ E\{(\mathbf{x}(0) - \mathbf{x}_0)\mathbf{o}^T(j)\} = \mathbf{0} \\ E\{(\mathbf{x}(0) - \mathbf{x}_0)(\mathbf{x}(0) - \mathbf{x}_0)^T\} = \mathbf{R}_0 \end{array} \qquad (3)$$

where $E\{.\}$ denotes the expectation, and δ_{ij} denotes the Kronecker delta function and matrices

- $\mathbf{R}_0$ nxn initial state covariance matrix,
- $\mathbf{Q}$ nxn system noise covariance matrix,
- $\mathbf{R}$ mxm output noise covariance matrix

are bounded positive definite ones. The goal is to compute the minimum variance estimate $\mathbf{x}_e(i)$ of the variable $\mathbf{x}(i)$ of the form

$$\mathbf{x}_e(i|\mathbf{y}(j), \mathbf{y}(j-1), \ldots, \mathbf{y}(1)) = \mathbf{x}_e(i|j) \quad (4)$$

This estimate is called the filtered if j=i and the one-step predicted if j=i-1.

For such an above mentioned system (1), (2), given an initial estimate of the state $\mathbf{x}_e(0|0) = E\{\mathbf{x}(0)\} = \mathbf{x}_0$ the stochastic state filter is defined on the base of orthogonal projection as

$$\mathbf{x}_e(i|i-1) = \mathbf{F}\mathbf{x}_e(i-1|i-1) + \mathbf{G}\mathbf{u}(i-1) \quad (5)$$

$$\mathbf{x}_e(i|i) = \mathbf{x}_e(i|i-1) + \mathbf{K}(i)[\mathbf{y}(i) - \mathbf{y}_e(i)] \quad (6)$$

$$\mathbf{y}_e(i) = \mathbf{H}\mathbf{x}_e(i|i-1) \qquad (7)$$

where $\mathbf{x}_e(i|i)$ is introduced as a state-variable vector $\mathbf{x}(i)$ estimate at time i in dependence on all the measurements $\{\mathbf{y}(j) : j = 0,1,2,\ldots i\}$ of signal and noise up to time i, and gain sequence $\{\mathbf{K}(i) : i = 0,1,2,\ldots\}$ is the filter gain matrix sequence that depends upon past data.

The actual errors in the estimates of the filter can be written by using (1) - (7) as

$$\begin{array}{c} \mathbf{e}(i|i) = \mathbf{x}(i) - \mathbf{x}_e(i|i) = \\ = (\mathbf{I} - \mathbf{K}(i)\mathbf{H})\mathbf{e}(i-1|i-1) + \\ +(\mathbf{I} - \mathbf{K}(i)\mathbf{H})\mathbf{v}(i-1) - \mathbf{K}(i)\mathbf{o}(i) \end{array} \qquad (8)$$

The error expectation is $E\{\mathbf{e}(i|i)\} = \mathbf{0}$ and covariance of error is given by the equation

$$\begin{array}{l} E\{\mathbf{e}(i|i)\mathbf{e}^T(i|i)\} = \mathbf{P}(i|i) = \\ (\mathbf{I} - \mathbf{K}(i)\mathbf{H})\mathbf{F}\mathbf{P}(i-1|i-1)\mathbf{F}^T(\mathbf{I} - \mathbf{K}(i)\mathbf{H})^T + \\ +(\mathbf{I} - \mathbf{K}(i)\mathbf{H}\mathbf{Q}(\mathbf{I} - \mathbf{K}(i)\mathbf{H})^T + \mathbf{K}(i)\mathbf{R}\mathbf{K}^T(i) \end{array} \qquad (9)$$

which can be minimized, if

$$\mathbf{P}(i|i-1) = \mathbf{F}\mathbf{P}(i-1|i-1)\mathbf{F}^T + \mathbf{Q} \quad (10)$$

$$\mathbf{K}(i) = \mathbf{P}(i|i-1)\mathbf{H}^T(\mathbf{R} + \mathbf{H}\mathbf{P}(i|i-1)\mathbf{H}^T)^{-1} \qquad (11)$$

The predicted error covariance matrix is a solution of the Riccati equation (10), where $\mathbf{P}(0|0) = \mathbf{R}_0$ is the initial predicted error covariance, and the covariance matrix of actual error $\mathbf{e}(i|i)$ is given as

$$\mathbf{P}(i|i) = (\mathbf{I} - \mathbf{K}(i)\mathbf{H})\mathbf{P}(i|i-1) \qquad (12)$$

The main part of the computational burden is in computing above Riccati equation since this is an nxn matrix where n is the order of the overall state vector $\mathbf{x}(i)$. Clearly, if $\mathbf{x}(i)$ could be portioned into vectors of lower order, much computational saving may result. This is essentially ta basic principle of several modified implementation in order to reduce the amount of computation.

3. SINGLY DECENTRALIZED ALGORITHM

In order to develop the filter equations for the overall large-scale systems (1), (2), comprising M interconnected subsystems with well-defined interaction structure, it is assumed this form of decomposition

$$\begin{array}{c} \mathbf{x}_h(i+1) = \\ = \sum_{l=1}^{M}(\mathbf{F}_{hl}\mathbf{x}_l(i) + \mathbf{G}_{hl}\mathbf{u}_l(i)) + \mathbf{v}_h(i) \end{array} \qquad (13)$$

$$\mathbf{y}_h(i) = \mathbf{H}_h\mathbf{x}_h(i) + \mathbf{o}_h(i) \qquad (14)$$

where

$\mathbf{x}_h(i)$ n_hx1 state vector of the subsystem,
$\mathbf{u}_h(i)$ r_hx1 input vector of the subsystem,
$\mathbf{y}_h(i)$ m_hx1 output vector of the subsystem,
$\mathbf{v}_h(i)$ n_hx1 random vector of hth process noise,
$\mathbf{o}_h(i)$ m_hx1 random vector of hth output noise,
$\mathbf{F}_{hk}$ n_hxn_k interaction state transition matrix,
$\mathbf{G}_{hk}$ n_hxr_k interaction input matrix,
$\mathbf{H}_h$ m_hxn_h hth output matrix.

It is assumed the random sequences $\{\mathbf{v}_h(i) : i = 0,1,2,\ldots\}$ and $\{\mathbf{o}_h(i) : i = 0,1,2,\ldots\}$ are independent of the initial state $\mathbf{x}_0$ for h = 1,2,...,M, and are mutually independent zero-mean Gaussian white noise sequence with covariance

$$\begin{aligned} &E\{\mathbf{w}(i)\mathbf{w}(j)^T\} = \\ &= E\{[\mathbf{v}_1(i)\ldots\mathbf{v}_M(i)\mathbf{o}_1(i)\ldots\mathbf{o}_M(i)]\mathbf{w}(j)^T\} = \\ &= diag[\mathbf{Q}_1\ldots\mathbf{Q}_M\mathbf{R}_1\ldots\mathbf{R}_M]\delta_{ij} \end{aligned} \tag{15}$$

what is a realistic assumption for practical systems.

For hth subsystem the state prediction is given as

$$\begin{aligned} &\mathbf{x}_{eh}(i|i-1) = \\ &= \sum_{l=1}^{M}(\mathbf{F}_{hl}\mathbf{x}_{el}(i-1|i-1) + \mathbf{G}_{hl}\mathbf{u}_l(i-1)) \end{aligned} \tag{16}$$

where $\mathbf{x}_{eh}(i|i-1)$ is introduced as a prediction of the hth subsystem state-variable vector $\mathbf{x}_h(i)$ at time i in dependence on all the measurements $\{\mathbf{y}(j) : j = 0,1,\ldots,i-1\}$ of signal and noise up to time i-1.

The error of the prediction $\mathbf{x}_{eh}(i|i-1)$ related to the state $\mathbf{x}$(i) for $\mathbf{e}_h(i|i) = \mathbf{x}_h(i) - \mathbf{x}_{eh}(i|i)$ is

$$\begin{aligned} &\mathbf{e}_h(i|i-1) = \mathbf{x}_h(i) - \mathbf{x}_{eh}(i|i-1) = \\ &\sum_{l=1}^{M}\mathbf{F}_{hl}\mathbf{e}_l(i-1|i-1) + \mathbf{v}_h(i) \end{aligned} \tag{17}$$

and the covariance of the error $\mathbf{e}_h(i|i-1)$ is given by the following equation

$$\begin{aligned} &\mathbf{X}_{hk}(i|i-1) = \mathbf{X}_{hk}(i|i)_0 = \\ &= \sum_{l=1}^{M}\sum_{j=1}^{M}\mathbf{F}_{hl}\mathbf{X}_{lj}(i-1|i-1)\mathbf{F}_{kj}^T + \\ &+\mathbf{Q}_{kh}\delta_{hk} \end{aligned} \tag{18}$$

The basis of the method is to convert the problem of finding the best estimate $\mathbf{x}_e(i|i)$ of $\mathbf{x}(i)$ to a functional minimization to solve this using the decentralized principles and successive approximation. To obtain successive approximation in the correction stage it is considered, that

$$\begin{aligned} &\mathbf{x}_{eh}(i|i)_j = \\ &= \mathbf{x}_{eh}(i|i)_{j-1} + \mathbf{K}_{hl}(i)_{j-1}\mathbf{e}_{yl}(i|i)_{j-1} \\ &j = 1,2,\ldots,M\ ;\ l = (h+j-1)_{modM} \end{aligned} \tag{19}$$

where

$$\begin{aligned} \mathbf{x}_{eh}(i|i)_0 &= \mathbf{x}_{eh}(i|i-1) \\ \mathbf{e}_{yl}(i|i)_{j-1} &= \mathbf{y}_l(i) - \mathbf{H}_l\mathbf{x}_{el}(i|i)_{j-1} \end{aligned} \tag{20}$$

The above recursive solution of the actual error $\mathbf{e}_h(i|i)_j = \mathbf{x}_h(i) - \mathbf{x}_{eh}(i|i)_j$ has the covariance

$$\begin{aligned} &\mathbf{X}_{hk}(i|i)_j = E\{\mathbf{e}_h(i|i)_j\mathbf{e}_k^T(i|i)_j\} = \\ &= E\{\mathbf{e}_h(i|i)_{j-1}\mathbf{e}_k^T(i|i)_{j-1}\} - \\ &-\mathbf{K}_{hl}(i)_{j-1}E\{\mathbf{e}_{yl}(i|i)_{j-1}\mathbf{e}_k^T(i|i)_{j-1}\} - \\ &-(E\{\mathbf{e}_h(i|i)_{j-1}\mathbf{e}_{yp}^T(i|i)_{j-1}\}\mathbf{K}_{kp}^T(i)_{j-1} - \\ &-\mathbf{K}_{hl}(i)_{j-1}E\{\mathbf{e}_{yl}(i|i)_{j-1}\mathbf{e}_{yp}^T(i|i)_{j-1}\})\mathbf{K}_{kp}^T(i)_{j-1} \end{aligned} \tag{21}$$

and the minimal value of the error covariance, i.e.

$$\begin{aligned} &\mathbf{X}_{hk}(i|i)_j = \\ &E\{\mathbf{e}_h(i|i)_{j-1}\mathbf{e}_k^T(i|i)_{j-1}\} - \\ &-\mathbf{K}_{hl}(i)_{j-1}E\{\mathbf{e}_{yl}(i|i)_{j-1}\mathbf{e}_k^T(i|i)_{j-1}\} = \\ &= \mathbf{X}_{hk}(i|i)_{j-1} - \mathbf{K}_{hl}(i)_{j-1}\mathbf{Z}_{lk}(i|i)_{j-1} \end{aligned} \tag{22}$$

is obtained if the gain matrix $\mathbf{K}_{hl}(i)_{j-1}$ is

$$\begin{aligned} &\mathbf{K}_{hl}(i)_{j-1} = \\ &= E\{\mathbf{e}_h(i|i)_{j-1}\mathbf{e}_{yp}^T(i|i)_{j-1}\}\cdot \\ &\cdot[E\{\mathbf{e}_{yl}(i|i)_{j-1}\mathbf{e}_{yp}^T(i|i)_{j-1}\}]^{-1} = \\ &= \mathbf{Z}_{hp}(i|i)_{j-1}\mathbf{Y}_{lp}^{-1}(i|i)_{j-1} \end{aligned} \tag{23}$$

As a consequence of these last two results it is easy to verify that some covariances can be written in terms of the subcomponents, i.e.

$$\begin{aligned} &\mathbf{Y}_{hk}(i|i)_{j-1} = \\ &= E\{\mathbf{e}_{yh}(i|i)_{j-1}\mathbf{e}_{yk}^T(i|i)_{j-1}\} = \\ &= \mathbf{H}_hE\{\mathbf{e}_h(i|i)_{j-1}\mathbf{e}_k^T(i|i)_{j-1}\}\mathbf{H}_k^T + \\ &+E\{\mathbf{o}_h\mathbf{o}_k^T\} = \\ &= \mathbf{H}_h\mathbf{X}_{hk}(i|i)_{j-1}\mathbf{H}_k^T + \mathbf{R}_{hk}\delta_{hk} \end{aligned} \tag{24}$$

$$\begin{aligned} &\mathbf{Z}_{hk}(i|i)_{j-1} = \\ &= E\{\mathbf{e}_h(i|i)_{j-1}\mathbf{e}_{yk}^T(i|i)_{j-1}\} = \\ &= E\{\mathbf{e}_h(i|i)_{j-1}\mathbf{e}_k^T(i|i)_{j-1}\}\mathbf{H}_k^T = \\ &= \mathbf{X}_{hk}(i|i)_{j-1}\mathbf{H}_k^T \end{aligned} \tag{25}$$

This set of equations can be implemented by an efficient principle that exploits the prediction estimate as well as its covariance matrix and an iterative solution of the filtering estimate and the corresponding covariance matrix, where the resulting estimate is the optimal Kalman estimate.

4. PAIRWISE DECENTRALIZED ALGORITHM

To obtain successive approximation in the correction stage it is considered, that the filtered estimate of a pair of subsystems is

$$\begin{bmatrix} \mathbf{x}_{eh}(i|i)_j \\ \mathbf{x}_{ek}(i|i)_j \end{bmatrix} = \begin{bmatrix} \mathbf{x}_{eh}(i|i)_{j-1} \\ \mathbf{x}_{ek}(i|i)_{j-1} \end{bmatrix} + \\ + \begin{bmatrix} \mathbf{K}_{hl}(i)_{j-1} & \mathbf{K}_{hf}(i)_{j-1} \\ \mathbf{K}_{kl}(i)_{j-1} & \mathbf{K}_{kf}(i)_{j-1} \end{bmatrix} \begin{bmatrix} \mathbf{e}_{yl}(i|i)_{j-1} \\ \mathbf{e}_{yf}(i|i)_{j-1} \end{bmatrix} \quad (26)$$

for $j = 1, 2, \ldots, M$; $l = (h+j-1)_{modM}$; $f = (k+j-1)_{modM}$ where

$$\mathbf{x}_{ep}(i|i)_0 = \mathbf{x}_{ep}(i|i-1) \\ \mathbf{e}_{yp}(i|i)_{j-1} = \mathbf{y}_p(i) - \mathbf{H}_p \mathbf{x}_{ep}(i|i)_{j-1} \quad (27)$$

If the covariance matrix of the actual error takes the form

$$\mathbf{X}_{hkps}(i|i)_j = \\ = E\{\begin{bmatrix} \mathbf{e}_h(i|i)_j \\ \mathbf{e}_k(i|i)_j \end{bmatrix} [\mathbf{e}_p^T(i|i)_j \mathbf{e}_s^T(i|i)_j]\} = \\ = E\{\begin{bmatrix} \mathbf{e}_h(i|i)_{j-1} \\ \mathbf{e}_k(i|i)_{j-1} \end{bmatrix} [\mathbf{e}_p^T(i|i)_{j-1} \mathbf{e}_s^T(i|i)_{j-1}]\} - \\ - \begin{bmatrix} \mathbf{K}_{hl}(i)_{j-1} & \mathbf{K}_{hf}(i)_{j-1} \\ \mathbf{K}_{kl}(i)_{j-1} & \mathbf{K}_{kf}(i)_{j-1} \end{bmatrix} . \\ E\{\begin{bmatrix} \mathbf{e}_{yl}(i|i)_{j-1} \\ \mathbf{e}_{yf}(i|i)_{j-1} \end{bmatrix} [\mathbf{e}_p^T(i|i)_{j-1} \mathbf{e}_s^T(i|i)_{j-1}]\} - \\ - E\{\begin{bmatrix} \mathbf{e}_{yl}(i|i)_{j-1} \\ \mathbf{e}_{yf}(i|i)_{j-1} \end{bmatrix} [\mathbf{e}_p^T(i|i)_{j-1} \mathbf{e}_s^T(i|i)_{j-1}]\}^T . \\ \begin{bmatrix} \mathbf{K}_{pv}(i)_{j-1} & \mathbf{K}_{pw}(i)_{j-1} \\ \mathbf{K}_{sv}(i)_{j-1} & \mathbf{K}_{sw}(i)_{j-1} \end{bmatrix}^T + \\ + \begin{bmatrix} \mathbf{K}_{hl}(i)_{j-1} & \mathbf{K}_{hf}(i)_{j-1} \\ \mathbf{K}_{kl}(i)_{j-1} & \mathbf{K}_{kf}(i)_{j-1} \end{bmatrix} . \\ E\{\begin{bmatrix} \mathbf{e}_{yl}(i|i)_{j-1} \\ \mathbf{e}_{yf}(i|i)_{j-1} \end{bmatrix} [\mathbf{e}_{yv}^T(i|i)_{j-1} \mathbf{e}_{yw}^T(i|i)_{j-1}]\} . \\ \begin{bmatrix} \mathbf{K}_{pv}(i)_{j-1} & \mathbf{K}_{pw}(i)_{j-1} \\ \mathbf{K}_{sv}(i)_{j-1} & \mathbf{K}_{sw}(i)_{j-1} \end{bmatrix}^T \quad (28)$$

then the minimal value of the actual error covariance exists, is unique, ad is given by the equation

$$\mathbf{X}_{hkps}(i|i)_j = \\ = E\{\begin{bmatrix} \mathbf{e}_h(i|i)_{j-1} \\ \mathbf{e}_k(i|i)_{j-1} \end{bmatrix} [\mathbf{e}_p^T(i|i)_{j-1} \mathbf{e}_s^T(i|i)_{j-1}]\} - \\ - \begin{bmatrix} \mathbf{K}_{hl}(i)_{j-1} & \mathbf{K}_{hf}(i)_{j-1} \\ \mathbf{K}_{kl}(i)_{j-1} & \mathbf{K}_{kf}(i)_{j-1} \end{bmatrix} . \\ E\{\begin{bmatrix} \mathbf{e}_{yl}(i|i)_{j-1} \\ \mathbf{e}_{yf}(i|i)_{j-1} \end{bmatrix} [\mathbf{e}_p^T(i|i)_{j-1} \mathbf{e}_s^T(i|i)_{j-1}]\} = \\ = \mathbf{X}_{hkps}(i|i)_{j-1} - \mathbf{K}_{hklf}(i)_{j-1} \mathbf{Z}_{lfps}(i|i)_{j-1} \quad (29)$$

if the gain matrix $\mathbf{K}_{hklf}(i)_{j-1}$ is the solution of matrix equation

$$\mathbf{K}_{hklf}(i)_{j-1} = \mathbf{Z}_{hkvw}(i|i)_{j-1} \mathbf{Y}_{vwlf}^{-1}(i|i)_{j-1} = \\ = E\{\begin{bmatrix} \mathbf{e}_{yl}(i|i)_{j-1} \\ \mathbf{e}_{yf}(i|i)_{j-1} \end{bmatrix} [\mathbf{e}_p^T(i|i)_{j-1} \mathbf{e}_s^T(i|i)_{j-1}]\}^T . \\ E\{\begin{bmatrix} \mathbf{e}_{yl}(i|i)_{j-1} \\ \mathbf{e}_{yf}(i|i)_{j-1} \end{bmatrix} [\mathbf{e}_{yv}^T(i|i)_{j-1} \mathbf{e}_{yw}^T(i|i)_{j-1}]\}^{-1} \quad (30)$$

In this case some covariances can be written in terms of the subcomponents, i.e.

$$E\{\begin{bmatrix} \mathbf{e}_{yl}(i|i)_{j-1} \\ \mathbf{e}_{yf}(i|i)_{j-1} \end{bmatrix} [\mathbf{e}_{yv}^T(i|i)_{j-1} \mathbf{e}_{yw}^T(i|i)_{j-1}]\} = \\ = E\{\begin{bmatrix} \mathbf{H}_l \mathbf{e}_l(i|i)_{j-1} + \mathbf{o}_l(i) \\ \mathbf{H}_f \mathbf{e}_f(i|i)_{j-1} + \mathbf{o}_f(i) \end{bmatrix} . \\ [\mathbf{e}_{yv}^T(i|i)_{j-1} \mathbf{e}_{yw}^T(i|i)_{j-1}]\} = \\ = \begin{bmatrix} \mathbf{H}_l & 0 \\ 0 & \mathbf{H}_f \end{bmatrix} \mathbf{X}_{lfvw}(i|i)_{j-1} \begin{bmatrix} \mathbf{H}_l^T & 0 \\ 0 & \mathbf{H}_f^T \end{bmatrix} + \\ + \begin{bmatrix} \mathbf{R}_{lv}\delta_{lv} & \mathbf{R}_{lw}\delta_{lw} \\ \mathbf{R}_{fv}\delta_{fv} & \mathbf{R}_{fw}\delta_{fw} \end{bmatrix} = \mathbf{Y}_{lfvw}(i|i)_{j-1} \quad (31)$$

$$E\{\begin{bmatrix} \mathbf{e}_{yl}(i|i)_{j-1} \\ \mathbf{e}_{yf}(i|i)_{j-1} \end{bmatrix} [\mathbf{e}_p^T(i|i)_{j-1} \mathbf{e}_s^T(i|i)_{j-1}]\} = \\ = E\{\begin{bmatrix} \mathbf{H}_l \mathbf{e}_l(i|i)_{j-1} + \mathbf{o}_l(i) \\ \mathbf{H}_f \mathbf{e}_f(i|i)_{j-1} + \mathbf{o}_f(i) \end{bmatrix} . \\ [\mathbf{e}_p^T(i|i)_{j-1} + \mathbf{o}_p^T(i), \mathbf{e}_s^T(i|i)_{j-1} + \mathbf{o}_s^T(i)]\} = \\ = \begin{bmatrix} \mathbf{H}_l & 0 \\ 0 & \mathbf{H}_f \end{bmatrix} \mathbf{X}_{lfps}(i|i)_{j-1} + \\ + \begin{bmatrix} \mathbf{R}_{lv}\delta_{lv} & \mathbf{R}_{lp}\delta_{ls} \\ \mathbf{R}_{fv}\delta_{fv} & \mathbf{R}_{fp}\delta_{fs} \end{bmatrix} = \mathbf{Z}_{hkvw}(i|i)_{j-1} \quad (32)$$

Starting from predicted value the successive approximation can repeat for h=2,3,...,M, and set of equations can be implemented by an efficient principle that exploits an iterative solution of the filtering estimate and the corresponding covariance matrix.

5. ILLUSTRATIVE EXAMPLE

To demonstrate the properties of the iterative solution of the filtering estimate it was assumed that measured inputs and ouputs are generated by the controlable and observable discrete-time linear system (1), (2), where

$$\mathbf{F} = \begin{bmatrix} 1.0464 & -0.0040 & 0.0197 & -0.1546 & 0.3584 \\ -0.0086 & 1.0031 & -0.0005 & -0.0055 & -0.0888 \\ -0.0101 & 0.0040 & 0.9949 & 0.0123 & -0.0884 \\ -0.0174 & 0.0182 & -0.0036 & 0.9708 & -0.2236 \\ -0.0214 & 0.0085 & -0.0070 & 0.0346 & 0.8062 \end{bmatrix}$$

$$\mathbf{G} = \begin{bmatrix} 0.0400 & 0.0500 & 0.0600 \\ 0.0028 & 0.0040 & 0.0053 \\ 0.0230 & 0.0076 & 0.0021 \\ 0.0340 & 0.0123 & 0.0060 \\ 0.0037 & 0.0030 & 0.0021 \end{bmatrix}$$

$$\mathbf{H} = \begin{bmatrix} 1 & 0 & 0 & 0 & 0 \\ 0 & 0 & 1 & 0 & 0 \\ 0 & 0 & 0 & 0 & 1 \end{bmatrix}$$

The process noise and measurement output noise random sequences $\{\mathbf{v}(i) : i = 0, 1, 2, \ldots\}$ and $\{\mathbf{o}(i) : i = 0, 1, 2, \ldots\}$, as well as the initial state $\mathbf{x}_0$ are mutually independent Gaussian distributed random processes and are assumed to have the following statics:

$$\begin{aligned} &E\{\mathbf{v}(i)\} = \mathbf{0} \\ &E\{\mathbf{o}(i)\} = \mathbf{0} \\ &E\{\mathbf{x}(0)\} = \mathbf{x}_0 \\ &E\{\mathbf{v}(i)\mathbf{v}^T(j)\} = \mathbf{Q}\delta_{ij} \\ &E\{\mathbf{o}(i)\mathbf{o}^T(j)\} = \mathbf{R}\delta_{ij} \\ &E\{(\mathbf{x}(0) - \mathbf{x}_0)(\mathbf{x}(0) - \mathbf{x}_0)^T\} = \mathbf{R}_0 \end{aligned}$$

$$\mathbf{R}_0 = \begin{bmatrix} 0.45 & 0 & 0 & 0 & 0 \\ 0 & 37.5 & 0 & 0 & 0 \\ 0 & 0 & 0.365 & 0 & 0 \\ 0 & 0 & 0 & 3.65 & 0 \\ 0 & 0 & 0 & 0 & 0.185 \end{bmatrix}$$

$$\mathbf{Q} = \begin{bmatrix} 0.825 & 0 & 0 & 0 & 0 \\ 0 & 0.825 & 0 & 0 & 0 \\ 0 & 0 & 0.475 & 0 & 0 \\ 0 & 0 & 0 & 0.475 & 0 \\ 0 & 0 & 0 & 0 & 0.25 \end{bmatrix}$$

$$\mathbf{R} = \begin{bmatrix} 0.65 & 0 & 0 \\ 0 & 0.65 & 0 \\ 0 & 0 & 0.35 \end{bmatrix}$$

The Kalman fiter obtained estimates of the states and the unknown Kalman gain matrix parameters. The results of the typical run for given $\mathbf{R}_0$ are the steady-state error covariance matrix $\mathbf{P}(i|i)$ and the stedy-state Kalman gain matrix $\mathbf{K}(i)$.

$$\mathbf{P}(i|i) = \begin{bmatrix} 0.4539 & -0.2350 & -0.0034 & \vdots \\ -0.2350 & 37.2703 & 0.1262 & \vdots \\ -0.0034 & 0.1262 & 0.3669 & \vdots \\ -0.2420 & 3.2726 & 0.0383 & \vdots \\ -0.0009 & 0.3224 & 0.0019 & \vdots \end{bmatrix}$$

$$\begin{bmatrix} \vdots & -0.2420 & -0.0009 \\ \vdots & 3.2726 & 0.3224 \\ \vdots & 0.0383 & -0.0019 \\ \vdots & 3.6521 & 0.0986 \\ \vdots & 0.0986 & 0.1843 \end{bmatrix}$$

$$\mathbf{K}(i) = \begin{bmatrix} 0.6983 & -0.0052 & -0.0025 \\ -0.3616 & 0.1941 & 0.9211 \\ -0.0052 & 0.5646 & -0.0056 \\ -0.3724 & 0.0589 & 0.2818 \\ -0.0013 & -0.0030 & 0.5266 \end{bmatrix}$$

Using the same initial conditions decentralized algorithms give the comparable results, where the steady-state error covariance matrix components are, e.g.

$$\mathbf{X}_{11}(i|i) = \begin{bmatrix} 0.4546 & -0.2328 \\ -0.2328 & 37.3104 \end{bmatrix}$$

$$\mathbf{X}_{12}(i|i) = \mathbf{X}_{21}^T(i|i) = \begin{bmatrix} -0.0034 & -0.2388 \\ 0.1274 & 3.2846 \end{bmatrix}$$

$$\mathbf{X}_{13}(i|i) = \mathbf{X}_{31}^T(i|i) = \begin{bmatrix} -0.0014 \\ 0.3198 \end{bmatrix}$$

$$\mathbf{X}_{22}(i|i) = \begin{bmatrix} 0.3721 & 0.0366 \\ 0.0366 & 3.6599 \end{bmatrix}$$

$$\mathbf{X}_{23}(i|i) = \mathbf{X}_{32}^T(i|i) = \begin{bmatrix} -0.0019 \\ 0.0964 \end{bmatrix}$$

$$\mathbf{X}_{22}(i|i) = \begin{bmatrix} 0.1799 \end{bmatrix}$$

The performances of both filters are quite satisfactory and demonstrate robustness and economy of the decentralized algorithms.

6. CONCLUDING REMARKS

In this paper is presented a decentralized algorithm of the Kalman filter with decentralization in the prediction and the correction stages, where the successive approximation in the correction stage is supposed. The Kalman filter is used for optimal state estimation in large-scale linear interconnected discrete-time dynamical systems, where the well-defined interaction are considered. The interconnected outputs are assumed to be completely measurable through certain corrupted measurements. Considerations for the design procedures are given and the more informal approach is outlined. As pointed out in the body of paper, there is no need to solve Riccati equation for the discrete Kalman filters of the defined overall large-scale systems, what ensures substantial savings in computational time. It was obtained, the estimates using global Kalman filter and decentralized algorithms are corresponding and the error covariances provide enough information to estimate the state of the system.

7. REFERENCES

Chui, C.K., Chen, G. (1991). *Kalman Filtering with Real-time Applications.* Springer-Verlag, Berlin.

Filasová, A. (1994). Synthesis of linear controllers using pairwise autonomous subsystem approach. In: *Proceedings of the 1th IFAC Workshop ob New Trends in Design of Control Systems.* Smolenice, Slovak Republic, pp. 185-189.

Hassan, M.F.A., Salut, G., Singh, M.G., Titli, A. (1978). A decentralized computational algorithm for the global Kalman filter. *IEEE Trans. on Aut. Cont.*, **AC-23**, pp. 262-268.

Jover, J.M., Kailath, T. (1986). A parallel architecture for Kalman filter measurement update and parameter estimation. *Automatica*, **22**, pp. 43-57.

Krokavec, D. (1995). Partially decentralized algorithms for the Kalman estimator. In: *New Trends in Power System Automation*, Academia Centrum, Zlin, Czech Republic, pp. 196-200.

Myers, K.A., Tapley, B.D. (1976). Adaptive sequential estimation with unknown noise statistics. *IEEE Trans. on Aut. Cont.*, **AC-21**, pp. 520-523.

Verhaegen, M., Dooren, P. (1986). Numerical aspects of different Kalman filter implementations, *IEEE Trans. on Aut. Cont.*, **AC-31**, pp. 907-917.

ROBUST CONTROL DESIGN FOR HYBRID SYSTEMS BASED ON MARKOVIAN JUMPING MODELS

P. V. Pakshin and D.M. Fomin*

**Nizhny Novgorod State Technical University at Arzamas, 19, Kalinina Str., Arzamas, 607220, Russia. E-mail: pakshin@afngtu.nnov.ru*

Abstract. A class of hybrid systems, which represented as a finite family of stochastic differential models is considered. Each model of this family describes the individual regime of the system. The transitions between these regimes are modelled by a homogeneous Markov chain. Both state and output feedback control laws are obtained, which guarantee stablity in the mean square of closed-loop hybrid system for arbitrary transitions intensity and at the same time minimize a quadratic cost functional. An application to autopilot synthesis for high maneuvering aircraft is also given.

Key Words. Stochastic systems, jump process, Markov models, robust stability, robust control, optimal control, linear control, state feedback, output feedback, adaptive control

1. INTRODUCTION

Many real systems, such as failure prone systems exibit variations in their stuctures or abrupt changes in their inputs or internal variables and other system parameters. Standard state space system models cannot adequately represent these systems. Consequently, a new class of models has been proposed to describe such systems (Kazakov and Artem'ev, 1980; Mariton, 1990). This class is called random structure systems (Kazakov and Artem'ev, 1980), jump systems or hybrid systems due the existence both discrete and continuous variables in their state space (Mariton, 1990). We indicate some typical situations where hybrid jump models arise naturally: a manufacturing process, a solar thermal receiver, fault-tolerant control and target tracking problem.

The control theory of jumping systems began to developed since the pioneering works by Krasovskii and Lidskii (1961). The linear quadratic (LQ) control problem was solved by Sworder (1969) using stochastic maximum principle for state feedback in finite horizon case. Wonham (1970) also obtained the same results using dynamic programming for both finite and infinite horizon cases. Mariton and Bertrand (1985) provided an approach to output feedback jump LQ control. An investigation of discrete-time version of jump LQ control problem was given by Chizeck *et.al.* (1986), Pakshin (1982). Now many important results concerning the stochastic stability, stabilization and control of jump systems have been obtained, see (Ji and Chizeck, 1990; Mariton, 1990; Pakshin, 1994) and references therein. Recently interest increased appreciably to application of the jumping systems theory to research of complex dynamical systems (Siljak, 1991).

The present paper considers the problem of robust stabilizing control design for jumping stochastic systems based on robust stochastic stability concept. A stochastic versions of robust stability and control problems has been studied considerably less than deterministic ones, see (Bernstein, 1987; Krasovskii, 1965; Shi,1996; Willems and Willems, 1976,1978; Yaz, 1995) and references therein. These problems for jumping stochastic systems has been studied in recent papers see (Shi,1996; Yaz, 1995) and series of papers on subject in *Proc. 13-th IFAC World Congress*, but we use the concept of robustness absolutely different from adopted in these papers. We define the robust stability as a general property of the considered family of stochastic models, more exactly, as stochastic stability (in suitable sense) of the hybrid system at arbitrary transition intensities between the individual models.

The paper is organized as follows. In section 2 we formulate and solve the problems of robust stochastic stability of the jumping linear systems. In section 3 the problem of linear robust stabilizing control design is considered. We obtain both the state feedback control and the output feedback control, which guarantee robust stability of the closed-loop jumping system and at the same time minimize a quadratic cost functional. Comparison of the obtained robust stabilizing control with adaptive one is given. Application of presented mathematical results to autopilot design problem of a high manoeuvring aircraft is considered in section 4. In section 5 the computational and analytical aspects of arising nonlinear algebraic problems are discussed.

2. BASIC ROBUST STABILITY RESULTS

We consider a control system described by the family of stochastic differential equations

$$d\mathbf{x}(t) = \mathbf{A}(i_t)\mathbf{x}(t)dt + \\ + \sum_{l=1}^{N} \sigma_l \mathbf{A}_l(i_t)\mathbf{x}(t)dw_l(t), \quad (1)$$

where $\mathbf{x}(t)$ is n-dimensional state vector; $w_l(t)$ $(l = 1, \ldots, N)$ are independent standard Wiener processes; $\mathbf{A}(i)$, $\mathbf{A}_l(i)$ $(l = 1, \ldots, N)$ are $n \times n$ matrices; σ_l are positive scalars, which characterize the intensity of components w_l $(l = 1, \ldots, N)$; i_t is Markov chain with discrete set of states $\mathbb{N} = \{1, \ldots, \nu\}$ and with matrix of transition probabilities $\mathbf{P}(\tau) = [P_{ij}(\tau)]_1^\nu = \exp(\mathbf{Q}\tau)$, $P_{ij}(\tau) = \mathrm{P}\{i(t+\tau) = j \mid i(\tau) = i\}$ $(i, j \in \mathbb{N})$; $\mathbf{Q} = [q_{ij}]_1^\nu$, $q_{ij} \geq 0$ $(i \neq j)$, $q_{ii} = -\sum_{j \neq i}^{\nu} q_{ij}$; this chain describes regime (or mode) changes.

We shall assume that the initial state $(\mathbf{x}_0, i_0)$ of system (1) does not depend on the augmented Wiener process $\mathbf{w}(t) = [w_1(t), \ldots, w_n(t)]^T$. Then it is possible to construct a stochastic process as the solution of equation (1) in the sense of Ito (Khas'minskii, 1980) at given realization i_t. The set of all possible pairs $\{\mathbf{x}(t), i(t), t \geq 0\}$ will define Markov process $\{\mathrm{X}, \mathrm{I}\} = \{\mathbf{x}(t), i(t), t \geq 0\}$. We notice, incidentally, that separately considered process $\{\mathrm{X}\} = \{\mathbf{x}(t), t \geq 0\}$ is not Markov.

In this section we investigate the following robust stability problems.

Problem 1. Determine conditions of exponential stability in the mean square (ESMS) of system (1) (Khas'minskii, 1980) for given σ_l $(l = 1, \ldots, N)$ and arbitrary transition intensities q_{ij} $(i, j \in \mathbb{N})$.

Problem 2. Determine conditions of ESMS of system (1) for arbitrary σ_l $(l = 1, \ldots, N)$ and arbitrary transition intensities q_{ij} $(i, j \in \mathbb{N})$.

Let' s consider matrices

$$\mathbf{F}_{ii} = \mathbf{A}^T(i) \otimes \mathbf{I} + \mathbf{I} \otimes \mathbf{A}^T(i) + \\ + \sum_{l=1}^{N} \sigma_l^2 \mathbf{A}_l^T(i) \otimes \mathbf{A}_l^T(i), \; i \in \mathbb{N}, \quad (2)$$

where T and $\otimes$ denote the transpose symbol and the Kronecker product symbol respectively.

Assume that all these matrices are Hurwitz. Then according to Mariton (1990) for all positive definite matrices $\mathbf{M}(i) = \mathbf{M}^T(i)$ $(i \in \mathbb{N})$ there exists a positive definite solution $\mathbf{H}(i) = \mathbf{H}^T(i)$ of the system of matrix equations

$$\mathbf{A}^T(i)\mathbf{H}(i) + \mathbf{H}(i)\mathbf{A}(i) + \Delta_i(\mathbf{H}(i)) + \\ + \mathbf{M}(i) = 0, \; i \in \mathbb{N}, \quad (3)$$

where

$$\Delta_i(\mathbf{H}(j)) = \sum_{l=1}^{N} \sigma_l^2 \mathbf{A}_l^T(i)\mathbf{H}(j)\mathbf{A}_l(i).$$

Define for some fixed $k \in \mathbb{N}$ $\mathbf{M}(i)$ matrices according to formula

$$\mathbf{M}(i) = -(\mathbf{A}^T(i)\mathbf{H}(k) + \mathbf{H}(k)\mathbf{A}(i) + \\ + \Delta_i(\mathbf{H}(k)), \; i \in \mathbb{N}, i \neq k, \quad (4)$$

where $\mathbf{H}(k)$ is a solution of k-th equation from (3). If matrices constructed by this formula are positive definite, then equations (4) will have the constant solution $\mathbf{H}(i) = \mathbf{H}(k) = \mathbf{H}$, for all $i \in \mathbb{N}$. It is evident that this solution simultaneously satisfies the equations system

$$\mathbf{A}^T(i)\mathbf{H}(i) + \mathbf{H}(i)\mathbf{A}(i) + \Delta_i(\mathbf{H}(i)) + \\ + q_{ii}\mathbf{H}(i) + \sum_{j \neq i}^{\nu} q_{ij}\mathbf{H}(j) + \mathbf{M}(i) = 0, \quad (5)$$

for arbitrary q_{ij}, which according to Mariton (1990) means ESMS of system (1) for given σ_l $(l = 1, \ldots, N)$ and arbitrary q_{ij}.

So, we have proved the following theorem.

Theorem 1. Let all matrices (2) be Hurwitz and there exists at least one index $k \in \mathbb{N}$ and positive definite matrix $\mathbf{M}(k) = \mathbf{M}^T(k)$ such that all matrices $\mathbf{M}(i)$ $(i \in \mathbb{N})$, defined by formula (4) be positive definite. Then system (1) is ESMS for given σ_l $(l = 1, \ldots, N)$ and arbitrary q_{ij}.

Now we consider Problem 2. Suppose that for all $i \in \mathbb{N}$, $\mathbf{A}_l(i) = \mathbf{A}_l$ $(l = 1, \ldots, N)$, then $\Delta_i(\mathbf{H})$ not depend on i and we shall write $\Delta(\mathbf{H})$ instead of $\Delta_i(\mathbf{H})$, $i \in \mathbb{N}$. In the following we use some ideas by Willems and Willems (1976). Let $\mathbf{H}_0(k)$ denotes the solution of Lyapunov equation

$$\mathbf{A}^T(k)\mathbf{H}(k) + \mathbf{H}(k)\mathbf{A}(k) + \mathbf{M}(k) = 0 \quad (6)$$

with $\mathbf{M}(k) = \mathbf{M}_0(k)$ where $\mathbf{M}_0(k) \geq 0$, but $\mathbf{x}^T\mathbf{M}_0(k)\mathbf{x} > 0$ for all $\mathbf{x} \notin \Omega$, $\Omega = \{\mathbf{x} : \mathbf{A}_l\mathbf{x} = 0, \; l = 1, \ldots, N\}$; $\mathbf{H}_\epsilon(k)$ denotes the solution of equation (6) with $\mathbf{M}(k) = \mathbf{M}_\epsilon(k) = \mathbf{M}_0(k) + \epsilon\mathbf{M}_1$, $\epsilon > 0$, $\mathbf{M}_1$ a positive definite matrix and $\mathbf{M}_\epsilon(i) = -\mathbf{A}^T(i)\mathbf{H}_\epsilon(k) - \mathbf{H}_\epsilon(k)\mathbf{A}(i)$ $(i \neq k)$.

Theorem 2. Let matrices $\mathbf{A}(i)$ $(i \in \mathbb{N})$ be Hurwitz and there exists at least one number $k \in \mathbb{N}$ such that $\Delta(\mathbf{H}_0(k)) = 0$ and for $\epsilon > 0$ sufficiently small

$$\mathbf{x}^T\mathbf{M}_\epsilon(i)\mathbf{x} > 0, \ \mathbf{x} \in \Omega,$$
$$\mathbf{x}^T\mathbf{M}_\epsilon(i)\mathbf{x} \geq \mathbf{x}^T\mathbf{M}_\epsilon(k)\mathbf{x}, \ \mathbf{x} \notin \Omega,$$

Then system (1) is exponentially stable in the mean square for arbitrary $\sigma_l(l = 1, \ldots, N)$ and arbitrary q_{ij} $(i, j \in \mathbb{N})$.

Proof: If $\mathbf{x} \in \Omega$, then $\mathbf{x}^T\Delta(\mathbf{H})\mathbf{x} = 0$ and from this for $i = k$ we have

$$0 = \mathbf{x}^T(\mathbf{A}^T(k)\mathbf{H}_\epsilon(k) + \mathbf{H}_\epsilon(k)\mathbf{A}(k) + \\ +\mathbf{M}_\epsilon(k))\mathbf{x} = \mathbf{x}^T(\mathbf{A}^T(k)\mathbf{H}_\epsilon(k) + \mathbf{H}_\epsilon(k)\mathbf{A}(k) + \\ +\Delta(\mathbf{H}_\epsilon(k)) + \mathbf{M}_\epsilon(k))\mathbf{x}.$$

For $i \neq k$

$$\mathbf{x}^T(\mathbf{A}^T(i)\mathbf{H}_\epsilon(i) + \mathbf{H}_\epsilon(i)\mathbf{A}(i) + \Delta(\mathbf{H}\epsilon(i)) - \\ -(\mathbf{A}^T(i)\mathbf{H}_\epsilon(k) + \mathbf{H}_\epsilon(k)\mathbf{A}(i)))\mathbf{x} = 0$$

and it is easily seen that equations of the type (3) are valid on Ω and $\mathbf{H}_\epsilon(i) = \mathbf{H}_\epsilon(k) = \mathbf{H}$ $(i \in \mathbb{N})$ is a solution of these equations. Consequently equations of the type (5) are also valid on Ω.

If $\mathbf{x} \notin \Omega$ then for $i = k$ and $\epsilon > 0$ sufficiently small $\mathbf{x}^T\mathbf{M}_0(k)\mathbf{x} \geq \mathbf{x}^T\Delta(\mathbf{H}_\epsilon(k))\mathbf{x}$, and we have

$$0 = \mathbf{x}^T(\mathbf{A}^T(k)\mathbf{H}_\epsilon(k) + \mathbf{H}_\epsilon(k)\mathbf{A}(k) + \\ +\mathbf{M}_\epsilon(k))\mathbf{x} = \mathbf{x}^T(\mathbf{A}^T(k)\mathbf{H}_\epsilon(k) + \mathbf{H}_\epsilon(k)\mathbf{A}(k) + \\ +\Delta(\mathbf{H}_\epsilon(k)) + \mathbf{M}_\Delta(k))\mathbf{x},$$

where $\mathbf{M}_\Delta(k) = \mathbf{M}_\epsilon(k) - \Delta(\mathbf{H}_\epsilon(k)) > 0$.

For $i \neq k$

$$\mathbf{x}^T(\mathbf{A}^T(i)\mathbf{H}_\epsilon(i) + \mathbf{H}_\epsilon(i)\mathbf{A}(i) - \\ -(\mathbf{A}^T(i)\mathbf{H}_\epsilon(k) + \mathbf{H}_\epsilon(k)\mathbf{A}(i)))\mathbf{x} = 0.$$

Obviously that $\mathbf{H}(k)$ matrix is a solution of these equations and we obtain

$$\mathbf{x}^T\mathbf{M}_\Delta(i)\mathbf{x} = -\mathbf{x}^T(\mathbf{A}^T(i)\mathbf{H}_\epsilon(k) + \\ +\mathbf{H}_\epsilon(k)\mathbf{A}(i)) - \Delta(\mathbf{H}_\epsilon(i))\mathbf{x} = \\ = \mathbf{x}^T(\mathbf{M}_\epsilon(i) - \Delta(\mathbf{H}_\epsilon(k)))\mathbf{x} \geq \mathbf{x}^T(\mathbf{M}_\epsilon(k) - \\ -\Delta(\mathbf{H}_\epsilon(k)))\mathbf{x} = \mathbf{x}^T(\mathbf{M}_\Delta(k))\mathbf{x} > 0.$$

Then for $i \neq k$ we can write

$$\mathbf{x}^T(\mathbf{A}^T(i)\mathbf{H}_\epsilon(i) + \mathbf{H}_\epsilon(i)\mathbf{A}(i) + \\ +\Delta(\mathbf{H}_\epsilon(i)) + \mathbf{M}_\Delta(i))\mathbf{x} = 0.$$

Thus equations of the type (5) with $\mathbf{M}(i) = \mathbf{M}_\Delta(i) > 0$ are valid for all $i \in \mathbb{N}, \mathbf{x} \in \mathbb{R}^n$ and according to Mariton (1980) system (1) is ESMS for arbitrary σ_l $(l = 1, \ldots, N)$ and arbitrary q_{ij} $(i, j \in \mathbb{N})$. ∎

The following remark will be useful: since $\mathbf{A}(k)$ matrix is Hurwitz, the solution of Lyapunov equation (6) is given by formula

$$\mathbf{H}_0(k) = \int_0^\infty \exp(\mathbf{A}^T(k)t)\mathbf{M}_0(k)\exp(\mathbf{A}(k)t)dt.$$

Then it is easily seen that the condition $\Delta(\mathbf{H}_0(k)) = 0$ is equivalent to

$$\begin{bmatrix} \mathbf{M}_0(k) \\ \mathbf{M}_0(k)\mathbf{A}(k) \\ \cdots \\ \mathbf{M}_0(k)\mathbf{A}^{n-1}(k) \end{bmatrix} \mathbf{A}_l = 0, \ l = 1, \ldots, N.$$

It follows from this relation, that the columns of $\mathbf{A}_l$ matrices $(l = 1, \ldots, N)$ belong to the null space of the transpose observability matrix on the left hand side.

3. ROBUST STATE AND OUTPUT FEEDBACK CONTROL DESIGN

The considered system in linear case with control action is described by the family of stochastic differential equations

$$d\mathbf{x}(t) = [\mathbf{A}(i_t)\mathbf{x}(t) + \mathbf{B}(i_t)\mathbf{u}(t)]dt + \\ +\sum_{l=1}^{N}\sigma_l\mathbf{A}_l(i_t)\mathbf{x}(t)dw_l(t), \quad (7)$$

$$\mathbf{y}(t) = \mathbf{C}(i_t)\mathbf{x}(t), \quad (8)$$

where $\mathbf{u}(t)$ is k-dimensional control vector, $\mathbf{y}(t)$ is m-dimensional output vector, $\mathbf{B}(i)$ and $\mathbf{C}(i)$ are $n \times k$ and $m \times n$ matrices.

In this section we study the following robust stabilizing control problem.

Problem 3. Determine the state feedback control law

$$\mathbf{u}(t) = -\mathbf{K}\mathbf{x}(t), \quad (9)$$

which provides robust stability in the sence of Problem 1 and at the same time minimizes cost functional

$$J = \mathcal{E}_x\{\int_0^\infty [\mathbf{x}^T(t)\mathbf{M}(i_t)\mathbf{x}(t) + \\ +\mathbf{u}^T(t)\mathbf{Q}(i_t)\mathbf{u}(t)]dt\}, \quad (10)$$

where $\mathcal{E}_x-$ expectation operator at $\mathbf{x}_0 = \mathbf{x}$, $\mathbf{M}(i)$ and $\mathbf{Q}(i)$ $(i \in \mathbb{N})$ are $n \times n$ and $k \times k$ positive definite weight matrices.

Problem 4. Determine the output feedback control law

$$\mathbf{u}(t) = -\mathbf{K}\mathbf{y}(t), \quad (11)$$

which provides robust stability in the sence above and at the same time minimizes cost functional (10) Note that a similar problems for discrete time systems was considered by Pakshin (1995), Frangos and Yavin (1996).

Substituting (9) in (7) we obtain

$$d\mathbf{x}(t) = [\mathbf{A}(i_t) - \mathbf{B}(i_t)\mathbf{K}]\mathbf{x}(t)dt + \\ + \sum_{l=1}^{N} \sigma_l \mathbf{A}_l(i_t)\mathbf{x}(t)dw_l(t), \quad (12)$$

Let function $V(\mathbf{x}, i)$ satisfies condition

$$\mathcal{L}V(\mathbf{x}, i) = -\mathbf{x}^T[\mathbf{M}(i) + \mathbf{K}^T\mathbf{Q}(i)\mathbf{K}]\mathbf{x}.$$

Suppose that this function guarantees ESMS of the system (12). Then the cost functional (10) can be rewritten as

$$J = \sum_{i=1}^{\nu} V(\mathbf{x}, i)\mathrm{P}(i_0 = i).$$

For $V(\mathbf{x}, i) = V(\mathbf{x}) = \mathbf{x}^T\mathbf{H}\mathbf{x}$ we have

$$J = V(\mathbf{x}) = \mathrm{tr}[\mathbf{H}\mathbf{R}_0], \quad (13)$$

where $\mathbf{R}_0 = \mathbf{x}\mathbf{x}^T$ and $\mathbf{H}$ matrix satisfies the following equations

$$(\mathbf{A}(i) - \mathbf{B}(i)\mathbf{K})^T\mathbf{H} + \\ +\mathbf{H}(\mathbf{A}(i) - \mathbf{B}(i)\mathbf{K}) + \Delta_i(\mathbf{H}) + \\ +\mathbf{K}^T\mathbf{Q}(i)\mathbf{K} + \mathbf{M}(i) = 0, \; i \in \mathbb{N}. \quad (14)$$

Thus we come to the problem of minimization of function (13) with restrictions in the form of equations (14) and in order to find matrix $\mathbf{K}$ of control law (9) it is possible to use Lagrange's multipliers method. Let's write Lagrangian function

$$\Phi = \mathrm{tr}[\mathbf{H}\mathbf{R}_0] + \sum_{i=1}^{\nu} \mathrm{tr}[\mathbf{R}(i)[(\mathbf{A}(i) - \\ -\mathbf{B}(i)\mathbf{K})^T\mathbf{H} + +\mathbf{H}(\mathbf{A}(i) - \mathbf{B}(i)\mathbf{K}) + \\ +\Delta_i(\mathbf{H}) + \mathbf{K}^T\mathbf{Q}(i)\mathbf{K} + \mathbf{M}(i)],$$

where $\mathbf{R}(i)$ is Lagrange's multipliers matrix corresponding to $i-$ th equation (14).

Necessary conditions of minimum are written in the following form

$$\frac{\partial\Phi}{\partial\mathbf{R}(i)} = 0, \; \frac{\partial\Phi}{\partial\mathbf{H}} = 0, \; \frac{\partial\Phi}{\partial\mathbf{K}} = 0.$$

Calculating the derivatives we obtain

$$(\mathbf{A}(i) - \mathbf{B}(i)\mathbf{K})^T\mathbf{H} + \mathbf{H}(\mathbf{A}(i) - \mathbf{B}(i)\mathbf{K}) + \\ +\Delta_i(\mathbf{H}) + \mathbf{K}^T\mathbf{Q}(i)\mathbf{K} + \mathbf{M}(i) = 0, \; i \in \mathbb{N}.$$

$$\sum_{i=1}^{\nu}(\mathbf{A}(i) - \mathbf{B}(i)\mathbf{K})\mathbf{R}(i) + \mathbf{R}(i)(\mathbf{A}(i) - \\ -\mathbf{B}(i)\mathbf{K})^T + \Omega_i(\mathbf{R}(i)) + \mathbf{R}_0 = 0, \quad (15)$$

$$\sum_{i=1}^{\nu} \mathbf{Q}(i)\mathbf{K}\mathbf{R}(i) = \sum_{i=1}^{\nu} \mathbf{B}^T(i)\mathbf{H}\mathbf{R}(i). \quad (16)$$

$$\Omega_i(\mathbf{R}(i)) = \sum_{l=1}^{N} \sigma_l^2 \mathbf{A}_l(i)\mathbf{R}(i)\mathbf{A}_l^T(i)$$

We show that if there exists a constant positive definite matrix $\mathbf{H} = \mathbf{H}^T$, satisfying equations (14), then the equation (15) has also a constant solution $\mathbf{R}(i) = \mathbf{R}$. Then $\mathbf{K}$ matrix can be obtained from the equation (16) in its explicit form:

$$\mathbf{K} = [\sum_{i=1}^{\nu} \mathbf{Q}(i)]^{-1} \sum_{i=1}^{\nu} \mathbf{B}^T(i)\mathbf{H}. \quad (17)$$

Let's introduce the matrix differential equations

$$\frac{d\mathbf{H}(t)}{dt} + \sum_{i=1}^{\nu}(\mathbf{A}(i) - \mathbf{B}(i)\mathbf{K})^T\mathbf{H}(t) + \\ +\mathbf{H}(t)(\mathbf{A}(i) - \mathbf{B}(i)\mathbf{K}) + \\ +\Delta_i(\mathbf{H}(t)) + \mathbf{K}^T\mathbf{Q}(i)\mathbf{K} + \\ +\mathbf{M}(i) = 0, \mathbf{H}(0) = \mathbf{H}_0, \; t \leq 0. \quad (18)$$

According to Wonham (1970) $\mathbf{H}(t) \to \mathbf{H}$ as $t \to -\infty$.

We write equation (18) in the equivalent vector form:

$$\dot{\mathbf{h}}(t) + \mathbf{F}\mathbf{h}(t) + \mathbf{m} = 0, \quad (19)$$

where $\mathbf{h}(t)$, $\mathbf{m}$ are n^2 - dimensional vectors, formed out of matrix rows $\mathbf{H}(t)$ and $\mathbf{K}^T\mathbf{Q}(i)\mathbf{K} + \mathbf{M}(i)$ correspondingly,

$$\mathbf{F} = \sum_{i=1}^{\nu}(\mathbf{A}(i) - \mathbf{B}(i)\mathbf{K})^T \otimes \mathbf{I} + \\ +\mathbf{I} \otimes (\mathbf{A}(i) - \mathbf{B}(i)\mathbf{K})^T + \sum_{l=1}^{N} \sigma_l^2 \mathbf{A}_l^T(i) \otimes \mathbf{A}_l^T(i).$$

Since $\mathbf{H}(t) \to \mathbf{H}$ as $t \to -\infty$, then $\mathbf{h}(t) \to \mathbf{h}$ as $t \to -\infty$, i.e. the solution of differential equation (19) is asymptotically stable. This is possible if and only if $\mathbf{F}$ matrix be Hurwitz. It follows that $\mathbf{F}$ matrix is nonsingular. Assuming in (15) $\mathbf{R}(i) = \mathbf{R}$ ($i \in \mathbb{N}$) and rewriting this equation in vector form, we have

$$\mathbf{F}^T\mathbf{r} + \mathbf{r}_0 = 0.$$

Since $\mathbf{F}$ matrix is nonsingular, $\mathbf{r}$ vector is unique solution of this equation defined by $\mathbf{r} =$

$-[\mathbf{F}^T]^{-1}\mathbf{r}_0$. Hence, constant matrix $\mathbf{R}$ is also unique and satisfies equation (15).

According to the results of section 2 the system (7)-(8) is ESMS for arbitrary q_{ij} $(i,j \in \mathbb{N})$ and given σ_l $(l = 1,\ldots,N)$. If such matrices are unique solution of system (14), (17) then the control law (9) minimizes cost functional (12). Indeed, this control is obtained from necessary condition of minimum. Besides, it is easy to see, that if $\bar{\mathbf{H}} > 0$ is a solution of equation system (14) at any other $\mathbf{K}$, different from (17), then $\mathbf{H} - \bar{\mathbf{H}} > 0$.

So we have the following result

Theorem 3. If there exist the constant matrices $\mathbf{H} > 0$ and $\mathbf{K}$, satisfying the system of equation (14), (17), then the system (12) is ESMS for arbitrary $q_{ij}(i,j \in \mathbb{N})$ and given σ_l $(l = 1,\ldots,N)$. If such matrices are unique solution of system (14), (17) then the control law (9) minimizes cost functional (10).

Let us dwell upon the comparison of robust control with adaptive one. The adaptive control includes regime changes and has the form of

$$\mathbf{u}(t) = -\mathbf{K}(i_t)\mathbf{x}(t), \tag{20}$$

where matrix $\mathbf{K}(i)$ is defined by equations see (Wonham, 1970)

$$\begin{gathered}\mathbf{K}(i) = [\mathbf{Q}(i)]^{-1}\mathbf{B}^T(i)\mathbf{H}(i),\\ (\mathbf{A}(i) - \mathbf{B}(i)\mathbf{K})^T\mathbf{H} + \\ +\mathbf{H}(\mathbf{A}(i) - \mathbf{B}(i)\mathbf{K}) + \Delta_i(\mathbf{H}(i)) + \\ +\textstyle\sum_{j=1}^{\nu} q_{ij}\mathbf{H}(j) + \mathbf{M}(i) + \\ +\mathbf{K}^T\mathbf{Q}(i)\mathbf{K} = 0,\ i \in \mathbb{N},\end{gathered} \tag{21}$$

It follows from the minimality property, established by (Wonham, 1970) that $\mathbf{H}(i) \leq \mathbf{H}(i \in \mathbb{N})$. So the adaptive control (20) provides better quality in the sence of functional (10).

Now we consider problem 4. In this case we obtain

$$\begin{gathered}\sum_{i=1}^{\nu}(\mathbf{A}(i) - \mathbf{B}(i)\mathbf{K}\mathbf{C}(i))\mathbf{R} + \mathbf{R}(\mathbf{A}(i) - \\ -\mathbf{B}(i)\mathbf{K}\mathbf{C}(i))^T + \Omega_i(\mathbf{R}) + \mathbf{R}_0 = 0, \qquad (22)\\ (\mathbf{A}(i) - \mathbf{B}(i)\mathbf{K}\mathbf{C}(i))^T\mathbf{H} + \\ +\mathbf{H}(\mathbf{A}(i) - \mathbf{B}(i)\mathbf{K}\mathbf{C}(i)) + \Delta_i(\mathbf{H}) + \\ +\mathbf{C}^T(i)\mathbf{K}^T\mathbf{Q}(i)\mathbf{K}\mathbf{C}(i) + \mathbf{M}(i) = 0, \qquad (23)\\ \sum_{i=1}^{\nu}\mathbf{Q}(i)\mathbf{K}\mathbf{C}(i)\mathbf{R}\mathbf{C}^T(i) = \\ = \sum_{i=1}^{\nu}\mathbf{B}^T(i)\mathbf{H}\mathbf{R}\mathbf{C}^T(i),\ i \in \mathbb{N}. \qquad (24)\end{gathered}$$

Theorem 4. If there exist the constant matrices $\mathbf{R} > 0$, $\mathbf{H} > 0$ and $\mathbf{K}$, satisfying the system of equation (22) - (24), then the system (12) is ESMS for arbitrary $q_{ij}(i,j \in \mathbb{N})$ and given σ_l $(l = 1,\ldots,N)$. If such matrices are unique solution of system (22) - (24) then the control law (11) minimizes cost functional (10).

Note that if $\mathbf{C}(i) = \mathbf{C}$, then matrix $\mathbf{K}$ can be obtained from the equation (24) in its explicit form:

$$\mathbf{K} = [\sum_{i=1}^{\nu}\mathbf{Q}(i)]^{-1}\sum_{i=1}^{\nu}\mathbf{B}^T(i)\mathbf{H}\mathbf{R}\mathbf{C}^T[\mathbf{C}\mathbf{R}\mathbf{C}^T]^{-1}.$$

4. APPLICATION TO MULTIREGIME AUTOPILOT SYNTHESIS

We made a MATLAB toolbox based on proposed method and applied it to autopilot synthesis problem below. This toolbox includes a weight matrices obtaining procedure based on eigenvalue assignment, a procedure for solving of equations (22) - (24) and friendly interface. The longitudinal linear equations of angular aircraft motion can be expressed in following form (McRuer *et. al.*, 1973)

$$\begin{bmatrix}\dot{\alpha}\\ \dot{q}\\ \dot{\theta}\end{bmatrix} = \mathbf{A}(r_t)\begin{bmatrix}\alpha\\ q\\ \theta\end{bmatrix} + \begin{bmatrix}0\\ M_\delta(r_t)\\ 0\end{bmatrix}\delta, \tag{25}$$

$$\mathbf{A}(r_t) = \begin{bmatrix} Z_\alpha(r_t) & Z_q(r_t) & g\frac{\sin\theta_0(r_t)}{U_0(r_t)}\\ M_\alpha(r_t) & M_q(r_t) & 0\\ 0 & 1 & 0\end{bmatrix},$$

where α is angle of attack in rad, θ is pitch angle in rad, q is pitch rate in rad s^{-1}, δ is elevator angle (elevator servo command in rad), g is gravity, $Z_\delta, Z_q, M_\alpha, M_q$ are longitudinal stability derivatives, M_δ is longitudinal control derivative; all these derivatives as so as equlibrium pitch angle θ_0 and equlibrium velocity U_0 are depend on the flight regime r_t.

The problem of autopilot synthesis is to obtain an output feedback control law

$$\delta = K_q q + K_\theta\theta \tag{26}$$

providing satisfactory time responses for closed loop system (25), (26) with regime independent K_q and K_θ.

There is not a regular procedure for solving of this (or another similar) problems. In engineering practice in this case some direct modelling based approaches are applied.

The regime changes of considered high maneuring aircraft depends on fligh conditions, such that attitude is changes from 1000 m. to 18 000 m. and Mach number is changes from 0,3 to 1,95. Using the toolbox above we have obtained that in this case control law (26), guarantees good time responses for all possible fligh conditions (Fig.1).

5. CONCLUSION

The problems, of computing of state feedback control (9) or output feedback control (11) can be reduced to a special LMI algorithms (Boyd *et.al.*, 1994) We also obtained another effective algorithms of computing these controls based on Theorem 1. At the same time both these algorithms permit to calculate **H** matrix, but the condition of their applications remain unknown. So a new interesting algebraic problem of solvability of matrix equations (14), (17) or (22)-(24) is arised.

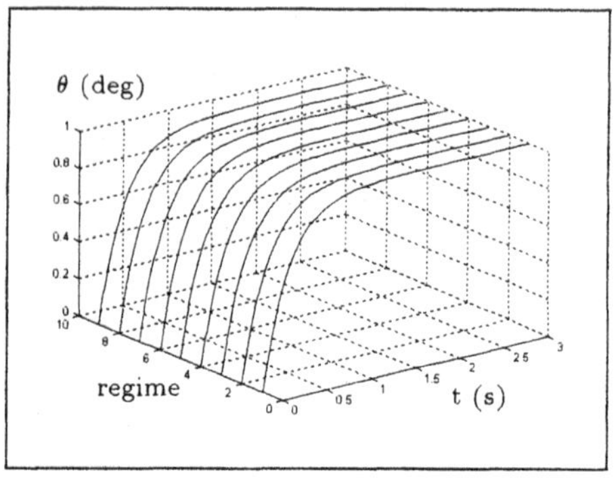

Fig. 1. Unit step responses for selected regimes

Acknowledgements. Authors thank the Russian Foundation for Basic Research and the State Committee of Russian Federation on Higher Education for financial support. Authors also thank professor Dragoslav D. Siljak from Santa Clara University (Santa Clara, California) for valuable remarks and suggestions.

6. REFERENCES

Bernstein, D.S. (1987). Robust static and dynamic output stabilization: deterministic and stochastic perspectives. *IEEE Trans. Automat. Control*, **32**, 1076-1084.

Boyd, S., E. El Ghaoui, E. Feron and V. Balakrishnan (1994). *Linear matrix inequalities in control and system theory*. SIAM, PA.

Chizeck, H.J., A.S. Willsky and D. Castanon (1986). Discrete-time, markovian-jump linear quadratic optimal control. *Int. J. Control*, **43**, 213-231.

Frangos, C. and Y. Yavin (1996). Feasible control design for plants with discrete-time Markov jump parameters. *Automatica*, **32**, 95-100.

Ji, Y. and H.J. Chizeck (1990). Jump linear quadratic Gaussian control: Steady state solution and testable conditions. *Control-Theory and Advanced Tech.*, **6**, 289-319.

Kazakov, I.Ye. and V.M. Artem'ev (1980). *Optimization of dynamical systems with random structure*. Nauka, Moscow (In Russian).

Khas'minskii, R.Z. (1980). *Stochastic stability of differential equations*. Sijthoff & Nordhoff. Netherlands.

Krasovskii, N.N. and E.A. Lidskii (1961). Analytic design of controller in systems with random attributes, I, II, III. *Aut. Remote Control*, **22**, 1021-1025, 1141-1146, 1289-1294.

Krasovskii, N.N. (1965). On stabilization of systems with control dependent noises. *Izv. AN SSSR. Techn. kybernetika*, **2**, 102-109 (In Russian).

Mariton, M. and P. Bertrand (1985). Output feedback for a class of linear systems with stochastic jump parameters. *IEEE Trans. Automat. Control*, **30**, 898-900.

McRuer, D., I. Ashkenas and D. M. Graham (1973) *Aircraft Dynamics and Automatic Control*. Princeton University Press, Princeton, NJ.

Mariton, M. (1990). *Jump linear systems in automatic control*. Marcel Dekker, Inc., New York.

Pakshin, P.V. (1982). Optimal linear control of discrete-time systems with random jump parameters. *Problems of Control and Information Theory*, **11**, 179-193.

Pakshin, P.V. (1994). *Discrete-time systems with random parameters and structure*. Fizmatlit, Moscow (in Russian).

Pakshin, P.V. (1995). Robust stability and control of discrete-time systems with random jump parameters and state dependent noise. In: *Proc. 5th IFAC Symposium on adaptive systems in control and signal processing*, pp. 545-550.

Shi, P. (1996). Robust control for linear systems with markovian jumping parameters. In: *Proc. 13th IFAC World Congress*, pp. 433-438.

Siljak, D.D. (1991). *Dezentralized control of complex systems*. Academic Press, Cambridge, MA.

Sworder, D.D. (1969). Feedback control of a class of linear systems with jump parameters. *IEEE Trans. Automat. Control*, **14**, 9-14.

Willems, J.L. and J.C. Willems (1976). Feedback stabilizability for stochastic systems with state and control dependent noise. *Automatica*, **12**, 277-283.

Willems, J.L. and J.C. Willems (1978). Robust stabilization of uncertain systems. *SIAM J. Control and Optimization*, **21**, 352-374.

Wonham, W.M. (1970). Random differential equations in control theory. In: *Probabilistic Methods in Applied Mathematics* (A. T. Bharucha-Reid, Ed.), **2**, pp. 131-212. Academic Press, New York.

Yaz, E. (1995). Robust stability of discrete time randomly perturbed systems. In: *Control and Dynamic Systems* (C. T. Leondes, Ed.), **73**, pp. 89-113. Academic Press, New York.

APPLICATION OF AGGREGATION METHODS IN CONTROL OF POWER SYSTEMS

Juraj Bízik, Štefan Kozák

Dept. of Automatic Control Systems, Faculty of Electrical Engineering and Information Technology, Slovak University of Technology, Ilkovičova 3, 812 19 Bratislava, Slovak Republic. E-mail: bizik@kasr.elf.stuba.sk

Abstract: Model reduction has significant importance in Large-Scale Systems. In this paper one of the main classes of model reduction-aggregation is used and the results of the two different methods are discused and compared for application in Power Systems.

Keywords: Modelling and aggregation, state-space model, LQ controller design, power and excitation control, synchronous generator, transformer, line.

1. INTRODUCTION

A high order dynamic system is often simplified to a low-order one, thus the simulation and control algorithms can be easily verified. To simplify the model of Large-scale dynamic systems aggregation methods are used (Aoki 1967). Among the various reasons in going for low order models, the view point of designers is that the reduced order models should approximately replace the high order controlled process model and a controller designed on the basis of reduced order model should effectively control the original high order system. The paper deals with the application of aggregation methods in power industry for modelling the cooperation of a synchronous generator with a large-power network as well as modelling excitation and power control. Such complex models represent multi-input multi- output systems.

A complete model which describes a synchronous generator is based on a set of differential equations of 7th order expressing electromagnetic and electromechanical transient processes. This model is linked to the model of transformer and connection lines up to the to the network which power is much greater than the power of the synchronous generator. In the proposed paper, the synchronous generator model is created using the matrix form of the differential equations set using the known matrices P,Q and R such that it is possible to express the model in the state space.

The transformer and the line are substituted by an appropriate reactance. To reduce the order of the global model, aggregation methods have been used based on modal matrix (Davison and Rao 1975). The original model solution has been compared with the aggregated models of the different orders using responses to step change in excitation and power, respectively. The control design has been verified on the base of step responses of the original and the aggregated model under given LQ controllers.

2. PROBLEM STATEMENT

The basic idea underlying building of an aggregation model is to combine appropriate system variables involving weighted averaging of the state vectors and to find an approximate model for the investigated large -scale system. The linear aggregation of the linear time-invariant systems has been considered..

Consider the following linear time-invariant dynamic system

$$\dot{x}(t) = Ax(t) + Bu(t) \qquad x(0) = x_0$$

$$y(t) = Cx(t) \tag{1}$$

where $x(t)$ is n x 1 state vector,

$u(t)$ m x 1 input vector,

$y(t)$ r x 1 output vector,

A is n x n system matrix,

B n x m control matrix,

C nxn output matrix

It is assumed that the pair (A,B) is controllable and the pair (A,C) is observable. The aim of the aggregation is to find a reduced model in the form

$$\dot{z}(t) = Fz(t) + Gu(t), \quad z(0)=z_0 \tag{2}$$
$$v(t) = Kz(t) \tag{3}$$

and $z(t) = C_1\, x(t)$ (4)

where $\mathbf{C_1}$ is s x n constant aggregation matrix,

and $s<n$, $rank(\mathbf{C_1}) = s$

where F, G, H are appropriate matrices of an aggregated model with dimensions s x s, s x m, s x s, respectively,

$v(t)$ is output reduced vector (r x 1),

$z(t)$ is aggregated vector (s x 1).

There are several aggregation methods for time domain models. Two of them (Davison and Rao methods) are investigated and their results are discussed and compared. The exact aggregation method is complicated from numerical aspect and the continued fraction aggregation method is beneficial mainly for single input and single output systems. Both mentioned methods are very suitable for practical usage.

The modal aggregation method is based upon the eigenvalues of the original system and the so-called modal matrix (matrix constructed from eigenvectors of A in descending order):

$$M = \begin{pmatrix} M_r & M_{12} \\ M_{21} & M_2 \end{pmatrix} \tag{6}$$

where M_r is a square (s x s) matrix

To compute the aggregated model matrices F, G, C_1, K the matrices P, L have to be constructed in following form :

$$P = [I_s \quad 0] \tag{7}$$

with dimension s x n

where I_s is {s x s) identity matrix

O is s x (n-s) zero matrix

Let denote $L=M^{-1}A\,M$ (8)

L is a (s x n) matrix and M^{-1} represents the inverse (n x n) modal matrix.

The aggregated matrices F, G, C_1, K are then computed using following formulas

$$F = M_r P L P^T M_r^{-1} \qquad (s \times s) \tag{9}$$

$$G = M_r P M^{-1} \qquad (s \times m) \tag{10}$$

$$C_1 = M_r P M^{-1} \qquad (s \times n) \tag{11}$$

$$K = DC^{+} \qquad (r \times s) \tag{12}$$

where C^+ is a pseudoinverse of C.

Modal aggregation methods by Lambda, Rao use the following structure of matrices:

$$A = \begin{pmatrix} A_1 & A_{12} \\ A_{21} & A2 \end{pmatrix} \tag{13}$$

$$B = \begin{pmatrix} B_1 \\ B_2 \end{pmatrix} \tag{14}$$

where the submatrices in (13) and (14) have dimensions:

A_1 is	(s x s),	A_{12} is	s x (n-s)
A_{21}	(n-s) x s,	A_2	(n-s) x (n-s)
B_1	is s x m	B_2	(n-s) x m

The resulting matrices F and G are computed using auxilliary matrices:

$$N=M_{21}M_r^{-1} \tag{15}$$

$$G_a=M^{-1}B = (G_{a1} \quad G_{a2})^T \tag{16}$$

$$L=M^{-1}AM=diag(L_1,L_2) \tag{17}$$

$$L_a=-L_2^{-1}G_{a2} \tag{18}$$

$$E=(M_2-NM_{12})L_a \tag{19}$$

and finally $F=A_1 + A_{12}N$ (20)

$$G=B_1 + A_{12}E \tag{21}$$

The output matrix C is determined from the relations (11) and (12).

The aggregation method based on the continuted fraction uses the canonical form of the original system. Consider the original dynamic system in the form

$$\dot{x}(t) = Ax(t) + Bu(t), \quad x(t_0) = x_0 \tag{22}$$

$$y(t) = Cx(t) \tag{23}$$

where

$$A = \begin{pmatrix} 0 & 1 & 0 & \dots & 0 \\ 0 & 0 & 1 & \dots & 0 \\ \dots & \dots & \dots & \dots & \dots \\ -a_{11} & -a_{12} & -a_{13} & \dots & -a_{1n} \end{pmatrix}, \quad B = \begin{pmatrix} 0 \\ 0 \\ \dots \\ 1 \end{pmatrix}$$

The original model can be aggregated through the transformation matrix P

$$q(t) = Px(t) \tag{24}$$

$$\dot{q}(t) = Vq(t) + Ku(t) \tag{25}$$

$$v(t) = Cq(t) \tag{26}$$

Matrix P is formed from the elements of A:

$$P = \begin{pmatrix} a_{31} & a_{32} & a_{33} & \dots & a_{3n-1} & 1 \\ 0 & a_{51} & a_{52} & \dots & a_{5,n-2} & 1 \\ 0 & 0 & a_{71} & \dots & a_{7,n-3} & 1 \\ \dots & \dots & \dots & \dots & \dots & \dots \\ 0 & 0 & 0 & \dots & a_{2n-1,1} & 1 \\ 0 & 0 & 0 & \dots & 0 & 1 \end{pmatrix}$$

Introducing P into (25)

$$\dot{q}(t) = PAP^{-1}q(t) + PBu(t) = Vq(t) + ku(t) \tag{27}$$

the desired aggregated model is given by

$$\dot{z}(t) = Fz(t) + Gu(t) \tag{28}$$

where

$$F = RPAP^{-1}R^T(RR^T)^{-1}, \quad G = RPB \tag{29|}$$

and $M=RP$, $H = CM^T(MM^T)^{-1}$

The design of the state space controller is carried out using the LQ approach for both the original and the aggregated model.

3. THE LQ REGULATION PROBLEM

Consider the linear time invariant dynamic system

$$\dot{x}(t) = Ax(t) + Bu(t), \quad x \in R^n, \quad u \in R$$

$$y(t)=Cx(t)$$

where (A,B) is stabilizable and (A,C) is observable.

The problem of finding the linear state feedback which minimizes the performance index: $\lim_{t\to\infty} \int_{t_0}^{t_f} \|y(t)\|_Q + \|u(t)\|_R dt$

where Q and R are suitable weighting matrices. From the classical LQ theory it is known that the optimal solution provides the control law

$$u(t) = -K_c x(t)$$

and the closed loop dynamical matrix is

$$K_c = R^{-1}B^TK$$

$$G_c = A - BK_c$$

where K is the solution of nonlinerar algebraic Riccati equation

$$A^{TK}P + KA + KB^TR^{-1}B^TK + C^TQC^T = 0$$

The overal system is

$$\dot{x}(t) = (A + BK_c)x(t)$$

4. CASE STUDY

Consider the model of synchronouos generator working through transformer to an infinite bus

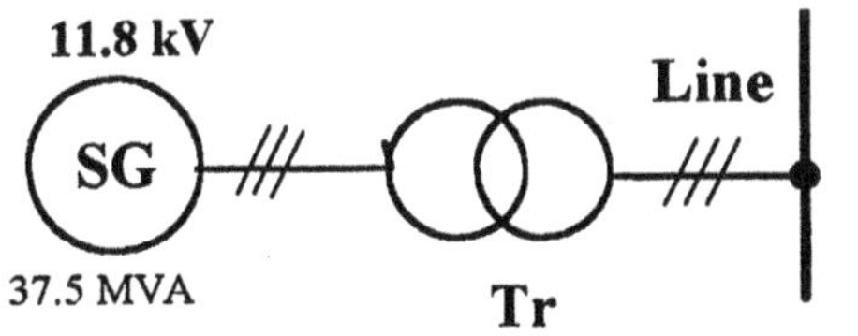

Fig.1 Block scheme of the synchronous generator

Generator rating :

37.5 MVA , 30 MW, 3000 r/min, 11.8 kV

$\cos\varphi=0.8$

Parameters obtained from the real process are

$x_d=x_q=x_{ff}=2.0$ p.u.

$x_{af}=1.86$ p.u. $r=0.002$ p.u.

$r_f=0.00107$ p.u.

$r_t=0.0223$ p.u. (resistance of the line)

H=0.625 s (inertial constant)

K_d=1.5 (damping coefficient)

Base stator voltage: 9.68 kV peak per phase

Base stator current: 2.59 kA peak per phase

The original model is expressed :

$$\dot{x}(t) = A_{or}x(t) + B_{or}u(t$$

$$y(t)=C_{or}x(t)$$

where A_{or} is a (7x7) square matrix , and B_{or} is a (7x3) column matrix ,

C_{or} is a (12 x 7) rectangular matrix.

From the original model the aggregated models of fifth and fourth order are obtained by Davison and the fifth-order model by Lambda-Rao.

The original model:

A3 =

0	1.0000	0	0	0	0	0
0	−0.0700	0.0003	−0.0006	−0.0001	0.0000	0.0000
−1.5754	1.1465	−0.0480	4.3092	−0.0018	0.0000	0.0198
0.2707	0.3794	−1.0000	−0.0111	0.8991	0.0046	0.0000
−1.4651	1.0662	−0.0447	4.0075	−0.0022	0.0000	0.0184
−0.0158	0.0115	−0.0005	0.0431	0.0000	−0.0010	0.0002
0.0027	0.0038	−0.0100	−0.0001	0.0090	0.0000	0.0010

$$B3 = \begin{pmatrix} 0 & 0 & 0 \\ 0 & 0.0002 & 0 \\ 7.1429 & 0 & -5.7143 \\ 0 & 0 & 0 \\ 7.1429 & 0 & -5.7143 \\ 271.4286 & 0 & -157.1429 \\ 0 & 0 & 80 \end{pmatrix}$$

C3 =

0.5134	0.1564	0.0137	0.5956	-0.0003	0.0000	-0.0036
-0.5413	0.2413	0.0000	0.0203	0.1618	0.0008	0.0000
0.7970	0	0	0	0	0	0
-0.5900	0	0	0	0	0	0
0.3440	0.2165	-0.5469	1.5474	0.0442	0.0002	-0.0034
-1.3259	1.7853	0.0423	1.9559	0.8713	0.0044	-0.0110
0	0	-1.3500	2.5200	0.5115	0.0028	-0.0096
0	0	-2.0000	0	1.8600	0.0100	0
0	0	0	-2.0000	0	0	0.0100
0	0	1.8600	0	2.0000	0	0
0	0	-0.0100	0	0	1.0000	0
0	0	0	-0.010	0	0	1.0000

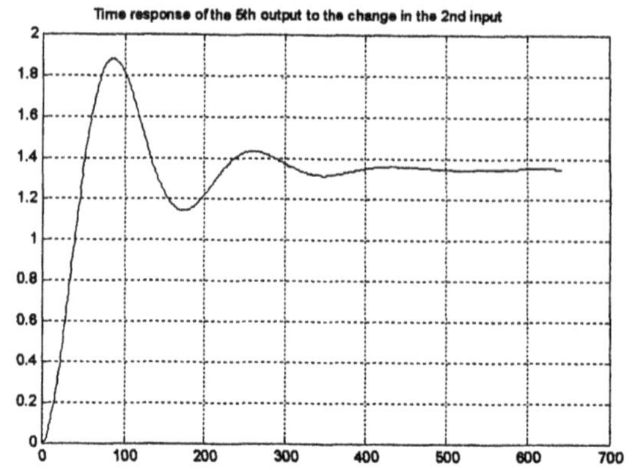

Fig.2 Time response of the original model (5th output under 2th input step change)

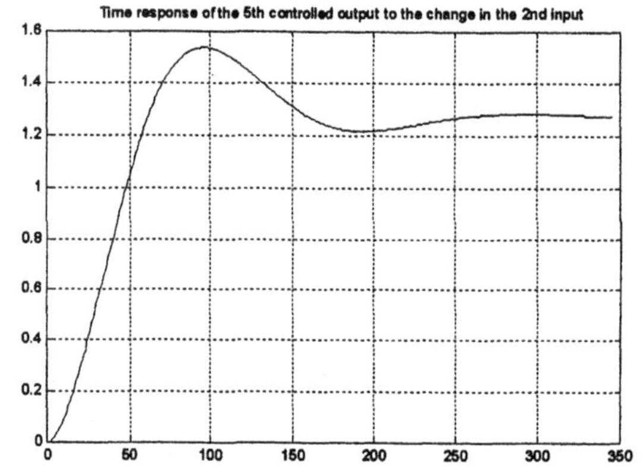

Fig. 3 Time response of the original controlled model (5th output)

Aggregated model (by Davison):

F=

0.0000	1.0000	0.0000	0.0000
0.0000	−0.0697	0.0002	−0.0006
−1.5751	1.1508	−0.0497	4.3090
0.1274	−1.7628	−0.1641	−0.0113

G =

1.1063	−0.0016	0.3959
−0.0142	0.0002	0.0107
2.8545	−0.0028	−3.4293
0.3565	−0.0006	−0.1963

K=

0.3590	0.8762	−0.2410	0.5201
−0.4703	−0.5268	0.2214	0.0654
0.6315	0.7628	−0.2626	−0.0807
−0.4675	−0.5647	0.1944	0.0598
0.2153	0.5044	−0.4801	1.4934
−1.4221	−0.1231	0.4463	1.9647
−0.1404	−1.2108	−0.6245	2.4996
−0.1911	−4.8925	−0.1352	0.0503
0.1547	−0.7210	0.2557	−1.9242
−0.2443	−4.9796	−0.0491	0.0319
0.0040	−0.0105	−0.0023	0.0018
0.0282	0.0456	−0.0176	−0.0054

Optimal gain for the original model using the LQ controller:

k =

−0.0171	−0.0001	−0.1327	0.1750	0.1220	0.1671	0.0743
0.2351	3.3544	−0.7564	0.0015	0.8138	−0.0015	0.0011
0.0246	−0.4830	−0.2145	0.0532	0.1693	−0. 0734	0.1660

Eigenvalues of the closed-loop system for the original model

e =[-57.7984 -12.5365 -0.0717 -0.0336 + 0.8400i
-0.0336 - 0.8400i
-0.0011 + 0.0022i
-0.0011 - 0.0022i]

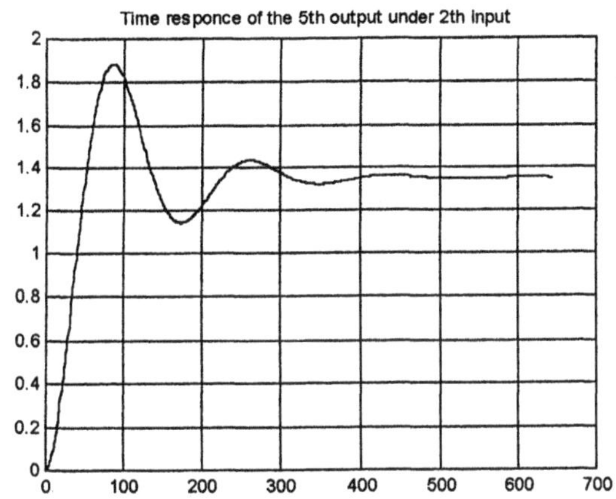

Fig. 4 Time response of the aggregated model (5th order by Davison)

Optimal gain of the aggregated model:

k =

0.1025	0.3530	-0.0986	0.2442	0.2551
0.0000	0.0004	0.0000	0.0000	0.0000
0.2637	0.2305	-0.1236	-0.3222	-0.0637]

Eigenvalues of the closed-loop system

e =[-3.0992 -0.3891 -0.0947 + 0.9289i
-0.0947 - 0.9289i
-0.0689]

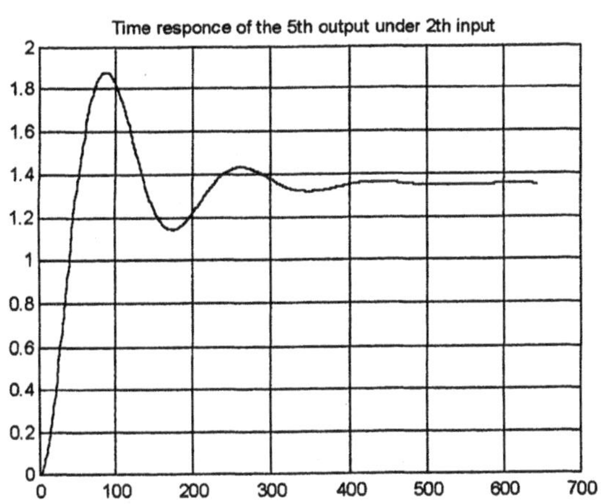

Fig. 5 Time response of the aggregated model (5th order by Rao)

Optimal gain of the aggregated controlled model (by Rao)

k =

0.2210	-2.8164	-0.0467	0.1770	-0.0029
0.0087	0.1110	0.0014	0.0000	-0.0015
0.0165	0.1957	-0.1521	-0.0390	-0.0992

Eigenvalues of the closed-loop system

e =[-424.6997 -211.2520 -0.0714 -0.0160 -0.0007]

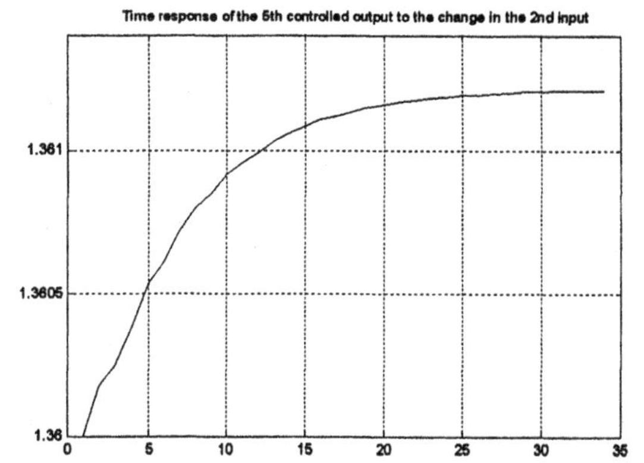

Fig. 6 Time response of the aggregated controlled model (5th output by Rao)

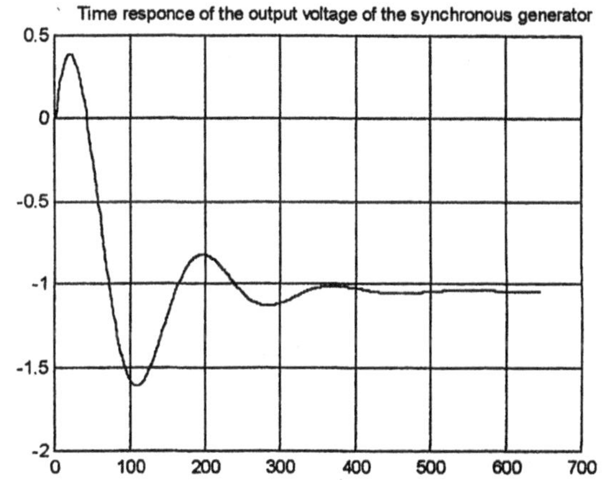

Fig.7 Time response of the output voltage of the aggregated model (without controller)

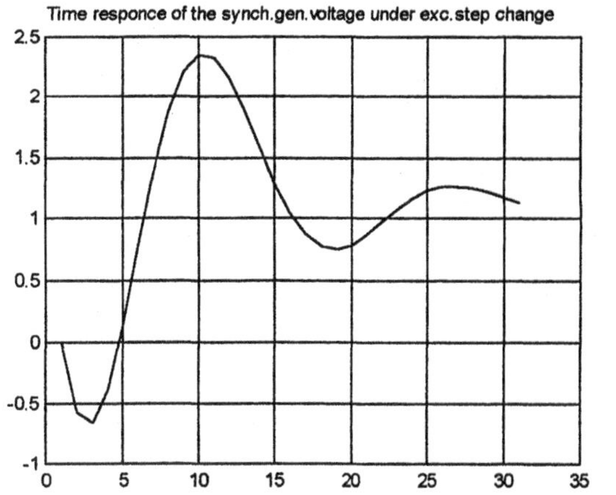

Fig.8 Time response of the synchronous generator to the excitation step change

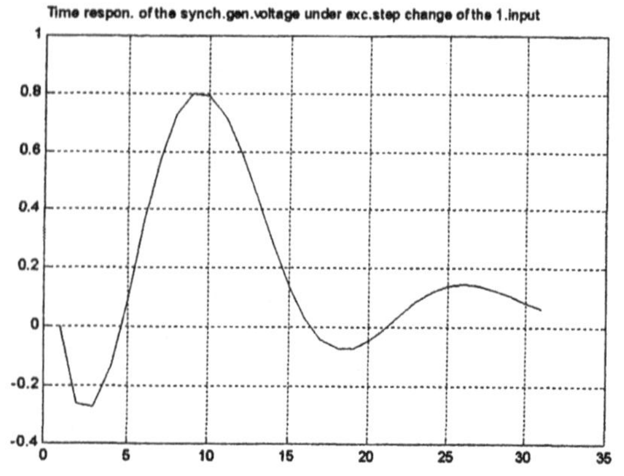

Fig.9 Time response of the synchronous generator voltage to the power and excitation step change

5. CONCLUSIONS

Simulation results of the power system modelling, aggregation and control indicate a good applicability of all used methods. The proposed aggregation methods are useful for the reduction of linear dynamical systems. The reduced dynamical input-output model is more suitable for linear controller design. By using an appropriate aggregation method a suitable model describing the dynamical behaviour with the desired accuracy can be found.

6. REFERENCES

Wismer,D.A.,(1971). *OptimizationMethods for Large - Scale Systems*, Mc. Graw - Hill, New York, 191-232.

Jamshidi, M.,(1981). An Overwiew on the Aggregation of Large-Scale Systems, IFAC KYOTO, 44.1, IX-86-IX-91.

Aoki, M., (1968). Control of Large - Scale Dynamic Systems by Aggregation,*IEEE Trans. Autom. Control AC-13 245-253*

Anderson, J.H.(1969). Matrix Methods for the Study of a Regulated Synchronouos Machines, in Proceedings of the IEEE, **Vol.57**, No. 12, 2122-2136.

Bízik, J., Kozák, Š.,. (1992). Comparison of several Aggregation Methods, Electrotechnical Journal , **Vol.43** , **No.** 7, 193- 196.

A NEW SIMPLE AUTO-TUNING METHOD FOR PID CONTROLLERS

Damir Vrančić, Youbin Peng*, and Janko Petrovčič

J. Stefan Institute
Jamova 39, SI-1000 Ljubljana, Slovenia
e-mail: damir.vrancic@ijs.si
home page: http://www-e2.ijs.si/Damir.Vrancic

**Free University of Brussels*
F. D. Roosevelt 50, CP. 165, B-1050 Brussels, Belgium
e-mail: peng@labauto.ulb.ac.be

Abstract: A simple procedure of tuning the PID controllers, based on the so-called magnitude optimum technique, is given. The tuning procedure requires only the process open-loop step response in order to calculate the PID controller parameters.

Keywords: PID controllers, amplitude optimum, moment method, step response, self-tuning regulators

1. INTRODUCTION

During the past 50 years, many different approaches for tuning the PID types of controllers have been developed. Each tuning method has its advantages and drawbacks. In general, the more information that can be derived from the process, the better are the tuning results that can be obtained. However, getting more process information requires more extensive process identification and extensive computations, with longer and more complicated experiments on the testing plant. Moreover, it has been recognised (see Peterka and Åström, 1980) that developing *accurate models* for the process industry, and identifying the parameters within them, is often not worthwhile; the problem is, rather, how to identify the *controller* of the process.

On the other hand, tuning rules which gather less process information (e.g. process gain, lag and rise times, or the process ultimate point) generally do not give the optimal response, since the available information is based only on few identified process parameters. For instance, there are some quite different processes with the same pair of lag and rise times (or ultimate points); clearly, these processes require different controller parameters. Several problems, which arise when using the Ziegler-Nichols tuning rules, are reported in Hang and Cao, (1993); Hang and Sin, (1991); Voda and Landau, (1995). Some authors have suggested the introduction of so-called set-point weighting approach (Hang and Cao, 1993) and of the additional change of the tuning rules (Åström and Hägglund, 1995). The improved closed-loop response has shown that they were successful for some types of processes, but the rules were still based on measurements of the process lag and rise time (or ultimate point), so their applicability to a wide spectrum of processes still remains an open problem.

Our approach was to find a simple tuning procedure which could easily be executed by the real-time auto-tuning algorithm, and which would give the optimal controller settings for a large class of process models. It has been decided that the tuning algorithm should be based on the process step response, because it can be simply obtained on the real plant. Instead of detecting the lag and rise times (which can be difficult in a noisy environment), we tried to extract as much information as possible from the process step response by using the function which is quite inert to the high-frequency process noise and can be simply performed by the real-time digital algorithm: an integration (Nishikawa et al., 1984; Vrančić et al., 1996). The multiple integration approach is applied, originally used as one of the identification techniques (see e.g. Rake, 1987; Strejc, 1959).

It was also decided to use the "magnitude" ("modulus", "Betrags") optimum (MO) tuning approach (Åström and Hägglund, 1995; Hanus, 1975; Kessler, 1955; Umland and Safiuddin, 1990). This technique guarantees fast, stable and non-oscillatory closed-loop behaviour for a large class of industrial processes.

In our research we found the relation between the five integrals (areas), obtained directly from the process step response (using the multiple integration approach), and the MO technique. This allows for the optimal PID controller parameters to be calculated merely from the integrals of the process open-loop step response, regardless of the process order.

2. THEORETICAL EVALUATION

The tuning procedure for the PID controller is given for processes which can be approximated by the following transfer function:

$$G_P(s) = K_{PR}\frac{1+b_1s+b_2s^2+\cdots+b_ms^m}{1+a_1s+a_2s^2+\cdots+a_ns^n}e^{-sT_{del}}, \quad (1)$$

where K_{PR} denotes the process steady-state gain and a_1 to a_n and b_1 to b_m are the corresponding parameters ($m \leq n$) of the process transfer function. A parameter T_{del} represents the process pure time delay.

The PID controller is given by the following transfer function:

$$G_C(s) = \frac{U(s)}{E(s)} = K\left(1+\frac{1}{sT_i}+\frac{sT_d}{1+sT_f}\right), \quad (2)$$

where U and E denote the Laplace transforms of the controller output and the control error ($e=w-y$), respectively. The controller parameters K, T_i, T_d and T_f denote proportional gain, integral time constant, differential time constant and filter time constant, respectively.

The PID controller in a closed-loop configuration with a process is shown in Fig. 1, where d denotes a load disturbance.

A tuning goal was to find such a controller that makes the closed-loop amplitude (magnitude) frequency response from the set-point to the plant output as flat and as close to unity as possible for a large bandwidth for a given plant and controller structure. This technique is called the magnitude optimum (Umland and Safiuddin, 1990) or modulus optimum technique (Åström and Hägglund, 1995). Such design objective results in a fast and non-oscillatory closed-loop time response for a large class of process models.

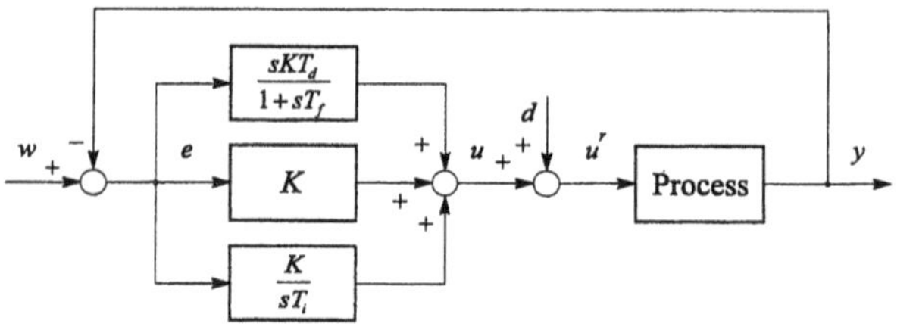

Fig. 1. The closed loop system with PID controller.

Following the procedure given by Hanus, (1975), such tuning goal can be achieved by moving the zeros of the function $\mathbf{Re}\{G_P(j\omega)G_C(j\omega)\}+1/2$ toward $\omega=0$.

When developing the pure time delay in (1), into the Taylor series:

$$e^{-sT_{del}} = 1 - sT_{del} + \frac{(sT_{del})^2}{2!} - \frac{(sT_{del})^3}{3!} + \cdots, \quad (3)$$

the open-loop system transfer function can be expressed in the following way:

$$G_P(s)G_C(s) = \frac{d_0+d_1s+d_2s^2+d_3s^3+\cdots}{c_0s+c_1s^2+c_2s^3+c_3s^4+\cdots}. \quad (4)$$

Parameters c_i and d_i can be calculated by inserting (1) to (3) into (4):

$$c_0 = T_i,\ c_1 = a_1T_i,\ c_2 = a_2T_i,\ c_3 = a_3T_i,\ \cdots \quad (5a)$$

$$d_0 = K_{PR}K,\ d_1 = K_{PR}K\left[T_i+b_1-T_{del}\right],$$
$$d_2 = K_{PR}K\left[T_iT_d+b_1T_i+b_2-T_{del}(T_i+b_1)+\frac{T_{del}^2}{2}\right], \quad (5b)$$

$$d_3 = K_{PR}K\begin{bmatrix} b_1T_iT_d+b_2T_i+b_3- \\ -T_{del}(T_iT_d+b_1T_i+b_2)+ \\ +\frac{T_{del}^2}{2}(T_i+b_1)-\frac{T_{del}^3}{6} \end{bmatrix}, \quad (5c)$$

$$d_4 = K_{PR}K\begin{bmatrix} b_2T_iT_d+b_3T_i+b_4- \\ -T_{del}(b_1T_iT_d+b_2T_i+b_3)+ \\ +\frac{T_{del}^2}{2}(T_iT_d+b_1T_i+b_2)- \\ -\frac{T_{del}^3}{6}(T_i+b_1)+\frac{T_{del}^4}{24} \end{bmatrix}, \quad (5d)$$

$$d_5 = K_{PR}K\begin{bmatrix} b_3T_iT_d + b_4T_i + b_5 - \\ -T_{del}\left(b_2T_iT_d + b_3T_i + b_4\right) + \\ +\frac{T_{del}^2}{2}\left(b_1T_iT_d + b_2T_i + b_3\right) - \\ -\frac{T_{del}^3}{6}\left(T_iT_d + b_1T_i + b_2\right) + \\ +\frac{T_{del}^4}{24}\left(T_i + b_1\right) - \frac{T_{del}^5}{120} \end{bmatrix} \quad (5e)$$

Note that T_f is fixed to zero so as to simplify a derivation of the PID controller parameters.

In order for the PID controller parameters to be found, as required by the presented magnitude optimum criterion, the first three (n=0..2) equations from the following set of equations are to hold (Hanus, 1975):

$$\sum_{i=0}^{2n+1}(-1)^i d_i c_{2n+1-i} = \frac{1}{2}\sum_{i=0}^{2n}(-1)^i c_i c_{2n-i} \quad (6)$$

When inserting (5) into (6), by setting n=0..2, the following PID controller parameters can be expressed by the unknown process parameters:

$$K = \frac{\begin{matrix} a_1^3 - a_1^2b_1 + a_1b_2 - 2a_1a_2 + a_2b_1 + a_3 - b_3 + \\ +T_{del}\left(a_1^2 - a_1b_1 - a_2 + b_2\right) + \frac{T_{del}^2}{2}\left(a_1 - b_1\right) + \frac{T_{del}^3}{6} \end{matrix}}{2K_{PR}\begin{bmatrix} -a_1^2b_1 + a_1a_2 + a_1b_1^2 - a_3 - b_1b_2 + b_3 + \\ +T_{del}\left(a_1 - b_1\right)^2 + T_{del}^2\left(a_1 - b_1\right) + \frac{T_{del}^3}{3} - \\ -T_d\left(a_1 - b_1 + T_{del}\right)^2 \end{bmatrix}} \quad (7)$$

$$T_i = \frac{\begin{matrix} a_1^3 - a_1^2b_1 + a_1b_2 - 2a_1a_2 + a_2b_1 + a_3 - b_3 + \\ +T_{del}\left(a_1^2 - a_1b_1 - a_2 + b_2\right) + \frac{T_{del}^2}{2}\left(a_1 - b_1\right) + \frac{T_{del}^3}{6} \end{matrix}}{\begin{matrix} a_1^2 - a_1b_1 - a_2 + b_2 + T_{del}\left(a_1 - b_1\right) + \frac{T_{del}^2}{2} - \\ -T_d\left(a_1 - b_1 + T_{del}\right) \end{matrix}} \quad (8)$$

$$T_d = f\left(a_1 \cdots a_5, b_1 \cdots b_5, T_{del}\right) \quad (9)$$

Note that the explicit result for the derivative time constant is not given due to the limitation of the space.

In order for the method to be applied, the real process must be approximated by the transfer function (1), which requires an explicit identification of the parameters K_{PR}, a_1 ... a_5, b_1 ... b_5, and T_{del}. Note that the identified model of the process must be of the same or a higher order than the real process so as to avoid the modelling error which could appear due to the insufficient order of the model. Therefore, calculation of (7) to (9), by using the process identification, is frequently impossible in practice.

However, it will be shown that the explicit identification of the process model can be avoided by using the concept of multiple integration (Strejc, 1959).

When applying the step-change ΔU at the process input, the following five areas can be expressed by integrating the process step response (y):

$$A_1 = y_1(\infty) = K_{PR}\left(a_1 - b_1 + T_{delay}\right), \quad (10)$$

$$A_2 = y_2(\infty) = K_{PR}\left[b_2 - a_2 + \frac{A_1a_1}{K_{PR}} - T_{delay}b_1 + \frac{T_{delay}^2}{2}\right] \quad (11)$$

$$A_3 = y_3(\infty) = K_{PR}\begin{bmatrix} a_3 - b_3 + \frac{A_2a_1}{K_{PR}} - \frac{A_1a_2}{K_{PR}} + \\ +T_{delay}b_2 - \frac{T_{delay}^2b_1}{2} + \frac{T_{delay}^3}{6} \end{bmatrix}, \quad (12)$$

$$A_4 = y_4(\infty) = K_{PR}\begin{bmatrix} b_4 - a_4 + \frac{A_3a_1}{K_{PR}} - \frac{A_2a_2}{K_{PR}} + \\ +\frac{A_1a_3}{K_{PR}} - T_{delay}b_3 + \frac{T_{delay}^2b_2}{2} - \\ -\frac{T_{delay}^3b_1}{6} + \frac{T_{delay}^4}{24} \end{bmatrix}, \quad (13)$$

$$A_5 = y_5(\infty) = K_{PR}\begin{bmatrix} a_5 - b_5 + \frac{A_4a_1}{K_{PR}} - \frac{A_3a_2}{K_{PR}} + \\ +\frac{A_2a_3}{K_{PR}} - \frac{A_1a_4}{K_{PR}} + T_{delay}b_4 - \\ -\frac{T_{delay}^2b_3}{2} + \frac{T_{delay}^3b_2}{6} - \\ -\frac{T_{delay}^4b_1}{24} + \frac{T_{delay}^5}{120} \end{bmatrix}, \quad (14)$$

where

$$\begin{aligned} y_1(t) &= \int_0^t \left(K_{PR} - \frac{y(\tau)}{\Delta U}\right)d\tau, \\ y_2(t) &= \int_0^t \left(A_1 - y_1(\tau)\right)d\tau, \\ y_3(t) &= \int_0^t \left(A_2 - y_2(\tau)\right)d\tau, \\ y_4(t) &= \int_0^t \left(A_3 - y_3(\tau)\right)d\tau, \\ y_5(t) &= \int_0^t \left(A_4 - y_4(\tau)\right)d\tau \end{aligned} \quad (15)$$

In order to clarify the mathematical derivation, the graphical representations of the first two areas (A_1 and A_2) are shown in Figures 2 and 3.

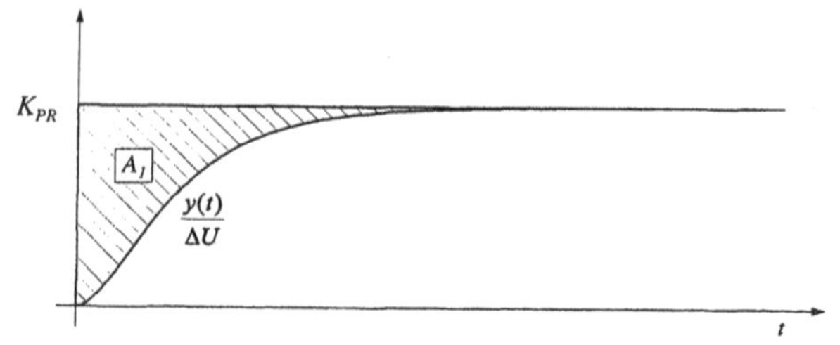

Fig. 2. The graphical representation of area A_1.

Fig. 3. The graphical representation of area A_2.

When inserting the calculated areas (10) to (14), obtained from the process step response, into equations (7) to (9), the following result is obtained:

$$K = \frac{1}{2K_{PR}\alpha} \tag{16}$$

$$T_i = \frac{A_1}{K_{PR}(1+\alpha)} \tag{17}$$

$$T_d = \frac{A_3 A_4 - A_2 A_5}{A_3 A_3 - A_1 A_5} \tag{18}$$

where

$$\alpha = \frac{A_1 A_2 - T_d A_1^2}{K_{PR} A_3} - 1 \tag{19}$$

Now obviously only areas A_1 to A_5, and the process gain K_{PR} are needed to calculate the unknown controller parameters.

As can be seen from equations (10) to (15), the areas can be calculated from the process open-loop step response by simple numerical integration while the gain K_{PR} can be determined from the steady-state value of the step response in the usual way.

The PID controller tuning procedure can therefore go on as follows:

- measure a process step response
- find a process steady-state gain K_{PR} and areas A_1, to A_5 (by numerical integration (summation) from the start to the end of the process step response).
- calculate PID controller parameters by using expressions (16) to (19).

The main point is, however, that the PID controller parameters can be tuned exactly, according to the amplitude optimum criterion (6), for a wide spectrum of process models (1), only by measuring the process open-loop step response.

3. REAL-TIME EXPERIMENTS WITH AUTO-TUNING ALGORITHM

The real-time experiments were performed by using the auto-tuning algorithm written in the program language PASCAL (some details are given in Vrančić et al., 1997a) with the Burr-Brown acquisition system PCI-20000.

Two real-time experiments were performed on the laboratory set-ups. The first experiment was made on a pneumatic set-up (process), given by Fig. 4. The input of the process is the current reference i_{in} (4/20 mA) on the servo-driven valve V_1 and the output is the pressure p_1 between valves V_1 and V_2 (transferred to the voltage u_{out} by using the pressure-to-voltage transmitter in the range from 0 to 10V).

Fig. 5 shows the results of the closed-loop experiment on the pneumatic set-up, when using the auto-tuning algorithm. After the process open-loop step response is obtained (at t=1.2sec), the following values of the process gain K_{PR}, and areas A_1 to A_5 are measured: K_{PR}=-0.0782, A_1=-0.0248, A_2=-4.919·10^{-3}, A_3=-7.746·10^{-4}, A_4=-1.039·10^{-4}, A_5=-1.219·10^{-5}.

The following PI and PID controller parameters and factors α are obtained from (16) to (19):

	α	K	T_i	T_d
PI	1.014	-6.31	0.158	
PID	0.313	-20.43	0.241	0.069

The closed-loop process responses (see Fig. 6) are quite good. It is obvious that the closed-loop time response is faster when using the PID controller, without significant increase of the process overshoot. It can be observed that the closed-loop responses are quite different when comparing the process responses on positive and on negative reference changes. The reason is that the process is non-linear.

The second experiment was made on a motor-generator laboratory set-up which is shown in Fig. 7.

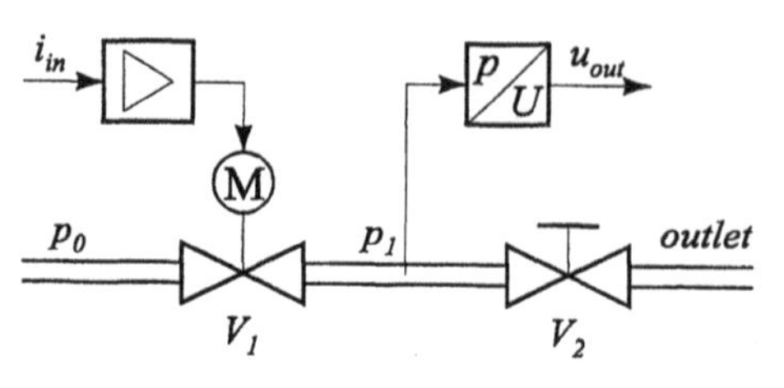

Fig. 4. Pneumatic set-up.

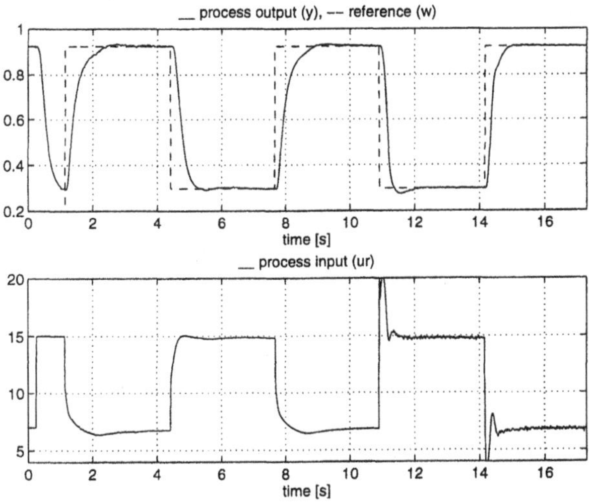

Fig. 5. The system responses under auto-tuning algorithm for the pneumatic set-up.

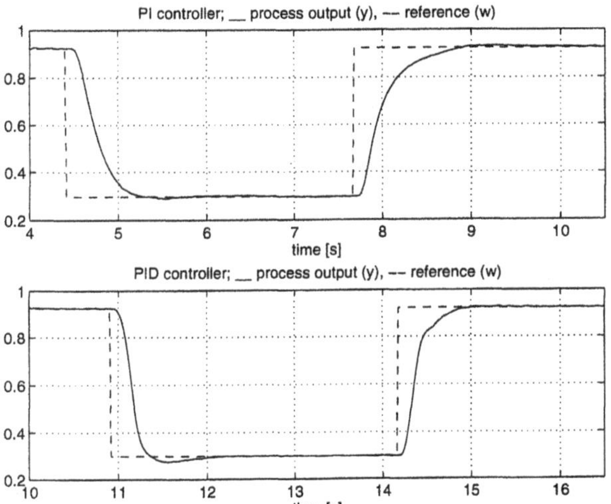

Fig. 6. The closed-loop process responses under PI and PID controller for the pneumatic set-up (the detailed view of the Fig. 5).

The input of the process is the voltage on the input of the amplifier (u_{in}) which drives the motor, and the output is the speed of the motor-generator system, obtained by the speed-to-voltage converter (u_{out}). Both, input and output signals, are in the range from 0 to 10V.

The system time responses, when driven by the auto-tuning algorithm, are shown in Fig. 8.

The following values of the process gain, and areas are obtained: K_{PR}=0.7144, A_1=0.187, A_2=3.198$\cdot 10^{-2}$, A_3=4.357$\cdot 10^{-3}$, A_4=4.989$\cdot 10^{-4}$, A_5=4.881$\cdot 10^{-5}$.

The following PI and PID controller parameters and factors α are obtained from (16) to (19):

	α	K	T_i	T_d
PI	0.921	0.759	0.136	
PID	0.223	3.14	0.214	0.062

Both closed-loop time responses (when using PI and PID controller) are quite good (see Fig. 9). It is obvious that the closed-loop time response is faster when using the PID controller, without significant increase of the process overshoot. The process non-linear behaviour can be clearly observed from the different closed-loop responses at higher and at lower reference values.

4. CONCLUSIONS

The purpose of this paper was to give a simple tuning method for the PI(D) controller, suitable for a large class of processes. A frequency domain tuning criterion (magnitude optimum) was chosen. It was shown that the parameters of the PID controller can be calculated from the step response by using the multiple integration method in a very simple way.

Tests on the laboratory plants showed that the method is quite robust to the process high-frequency noise and non-linearity. The main point, however, is that in spite of the quite demanding frequency criterion - and the fact that the calculation of the controller parameters is based on a quite complex process model - the implementation, using the time domain approach, is very simple and straightforward.

The advantages of the new tuning approach are:

- There is no need to detect the process inflection point and the slope from the process step response (difficult to do for noisy processes) as required by some other tuning methods.
- The method is based on integrations (summations) and is therefore suitable for use in the auto-tuning algorithms (as shown in examples).
- The controller parameters are calculated exactly, according to the given MO criterion, for a wide spectrum of process models.

The drawback of such an approach is that the method requires a stable open-loop process response in order to determine the appropriate controller parameters. Moreover, the closed-loop stability is not guaranteed.

However, the new tuning method has been tested by Vrančić, (1995) and Vrančić et al., (1995) and the experimental results showed very good closed-loop time responses for nine different process models.

Besides the given two examples, the new tuning approach has shown excellent closed-loop responses also for the high-order processes, the highly non-minimal phase processes and processes with larger time delays (Vrančić, 1995).

The new method can also be successfully used for improving the classical tuning rules (see Vrančić, 1995). The extension of the method (the areas) to other types of controllers so as PID controllers with D or/and P part only partially connected to the reference (Åström and Hägglund, 1995; Hang and

Cao, 1993) or multivariable controllers (Vrančić et al., 1997b) is also straightforward.

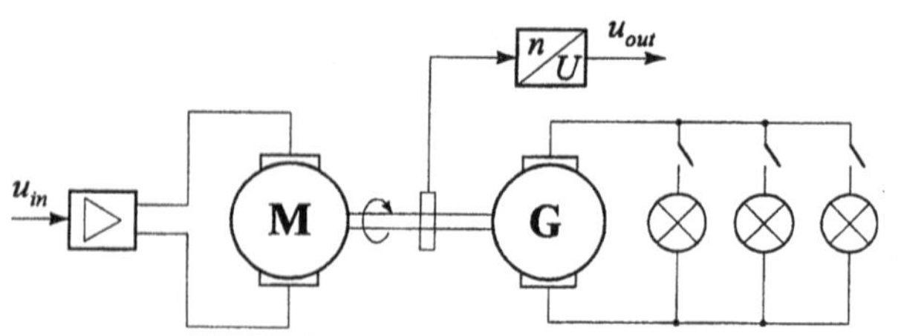

Fig. 7. Motor-generator laboratory set-up.

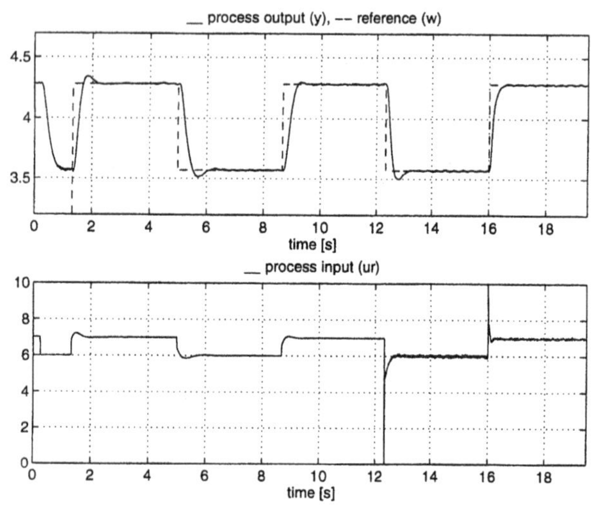

Fig. 8. The system responses under auto-tuning algorithm for the motor-generator set-up.

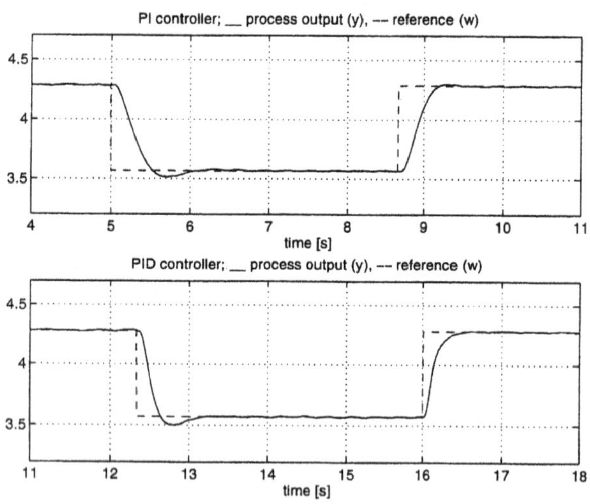

Fig. 9. The closed-loop process responses under PI and PID controller for the motor-generator set-up (the detailed view of the Fig. 8).

ACKNOWLEDGEMENT

The authors would like to thank three anonymous reviewers for providing many useful comments.

REFERENCES

Åström, K. J., and T. Hägglund (1995). PID Controllers: Theory, Design, and Tuning. *Instrument Society of America*, 2nd edition.

Hang, C. C., and L. S. Cao (1993). Improvement of Transient Response by Means of Setpoint Weighting. *12th World Congress IFAC*, Sydney, Proceedings, **3**, pp. 69-72.

Hang, C. C., and K. K. Sin (1991). On-line Auto-tuning of PID Controllers Based on the Cross-correlation Technique. National University of Singapore, Department of Electrical Engineering, *report* **CI-91-4**.

Hanus, R. (1975). Determination of controllers parameters in the frequency domain. Journal A, **XVI** (3).

Kessler, C. (1955). Über die Vorausberechnung optimal abgestimmter Regelkreise Teil III. Die optimale Einstellung des Reglers nach dem Betragsoptimum. Regelungstechnik, Jahrg. **3**, pp. 40-49.

Nishikawa, Y., N. Sannomiya, T. Ohta, and H. Tanaka (1984). A Method for Auto-tuning of PID Control Parameters. Automatica, **20** (3), pp. 321-332.

Peterka, V., and K. J. Åström (1980). Control of multivariable systems with unknown but constrained parameters. Automatica, **16**, pp. 393-398.

Rake, H. (1987). Identification: Transient- and Frequency-Response Methods. Systems & Control Encyclopedia; Theory, Technology, Applications, Madan G Singh, ed., *Pergamon Press*.

Strejc, V. (1959). Näherungsverfrahren für aperiodische Übergangscharakteristiken. *Regelungstechnik 7*, pp. 124-128.

Umland, J. W., and M. Safiuddin (1990). Magnitude and Symmetric Optimum Criterion for the Design of Linear Control Systems: What Is It and How Does It Compare with the Others?. IEEE Transactions on Industry Applications, **26** (3), pp. 489-497.

Voda, A. A., and I. D. Landau (1995). A Method for the Auto-calibration of PID Controllers. Automatica, **31** (1), pp. 41-53.

Vrančić, D. (1995). A new PI(D) tuning method based on the process reaction curve. *J. Stefan Institute, Report DP-7298*.

Vrančić, D., Y. Peng, C. Danz (1995). A comparison between different PI controller tuning methods. *J. Stefan Institute, Report DP-7286*.

Vrančić, D., Y. Peng, S. Strmčnik, R. Hanus (1996). A new tuning method for PI controllers based on a process step response. *Pre-prints of the CESA'96 IMACS Multiconference, Symposium on Control, Optimization and Supervision*, Lille, **2**, pp. 790-794.

Vrančić, D., Y. Peng, J. Petrovčič, and R. Hanus (1997a). A new tuning method for PID controllers. Accepted by the 4th IFAC Conference on System Structure and Control, Bucharest, October 23-25.

Vrančić, D., Y. Peng, J. Lieslehto, S. Strmčnik, and R. Hanus (1997b). Design of MIMO PI Controllers Using the Multiple Integration Approach. Accepted by the IMACS World Congress, Berlin, August 24-29.

ADAPTIVE APPROACH TO THE OPTIMAL PORTFOLIO SELECTION

Miroslav Šimandl and Karel Novák

University of West Bohemia, Faculty of Applied Sciences,
Department of Cybernetics
Univerzitní 22, 306 14 Plzeň, Czech Republic,
Tel.: +420-19-274252, Fax.: +420-19-279050
e-mail: simandl@kky.zcu.cz, novakka@kky.zcu.cz

Abstract: An adaptive approach to the synthesis of optimal portfolio is presented. The approach is based on tracking of time-varying parameters of the security characteristic lines by techniques of exponential or linear forgetting. The estimated parameters are used for finding returns and risks of assets. Finally the efficient frontier is finding by quadratic programming. The new adaptive optimal portfolio design, Markowitz and capital asset pricing design are applied on collection of assets and results are compared.

Keywords: adaptation, time-varying systems, financial systems, estimation parameters, quadratic programming, economic systems, Kalman filters

1. INTRODUCTION

One of the most common and difficult problem of investment and share funds managers is an efficient diversification of investment, i.e. finding optimal collection of assets - optimal portfolio. Portfolio theory deals with the question of how to invest a certain amount of wealth among a collection of assets.

The modern portfolio theory (Sharpe and Gordon, 1990) was based on statistical analysis of asset values in certain period. Each asset is characterized by its expected return and risk. The efficient combination of returns and risks is called Markowitz efficient frontier which is necessary to the synthesis of optimal portfolio. One of the disadvantages of the Markowitz approach is a great numerical complexity of the calculation. A significant simplification of optimal portfolio selection is based on the security characteristic line (SCL) (Sharpe, 1964). The static regression model and simple least squares method are used for the estimation of constant parameters of the SCL representing time invariant economic system and then quadratic programming is used for the portfolio design.

Some economists showed later that the static regression model and basic least square estimation technique are not suitable for estimates of the SCL's parameters. They recommended to consider time variable parameters of the regression models and designed several simple algorithms for corrections of parameter estimates of the static models (Rosenberg, 1985; Statman, 1981).

The presented paper shows that the parameters of the SCL can be seen as a random walk and described by a state-space model or as time-varying parameters without an exactly defined behaviour. A vague model of parameter variations is often and successfully considered in design of adaptive systems in technical fields too. In such cases various techniques of forgetting of obsolete information can be commonly used to make parameter estimation capable of tracking slow and/or abrupt parameter changes (Kulhavý and Kárný, 1984; Parkum *et al.*, 1991; Ljung, 1991; Fortescue *et al.*, 1981; Šimandl, 1996). Using the assumption that slow parameter changes better represent stock market, it is possible to use various techniques of parameter estimation as exponential forgetting motivated by time weighted least squares (Ljung, 1991) and linear forgetting justified on regression models by Kalman filtering view of estimation of time-varying parameters (Anderson and Moore, 1979). The estimated parameters then will be used for finding the efficient frontier by quadratic programming and for optimal portfolio design. The new adaptive optimal portfolio design, Markowitz and capital asset pricing design will be applied to a collection of stocks from Prague Stock Exchange and their properties will be compared.

2. MODERN PORTFOLIO THEORY

The section deals with problem of modern optimal portfolio design. Two approaches will be described: first Markowitz approach, and then approach, which is based on the capital asset pricing model.

2.1 *Markowitz approach*

Modern portfolio theory proceeds from Markowitz approach (Markowitz, 1952; Markowitz, 1959). The main assumptions of the theory are: (1) Portfolios are evaluated according to expected returns and standard deviations. (2) Investors are never satisfied. If they can choose between two structurally identical portfolios, they choose just that one which has higher expected return. (3) Investors have risk aversion. If they can choose between two structurally identical portfolios, they choose just that one which has smaller standard deviation. (4) Every asset is infinitely separable. It means that investors can buy any fragment of an asset.

Let us consider investments z_i, $i = 1, 2, ..., n$ to n risky assets. The whole amount of invested wealth z is given $z = \sum_{i=1}^{n} z_i$ and the relative amount of the ith investment regarding z is $x_i = z_i/z$, where $x_i \in \langle 0, 1\rangle$ and $\sum_{i=1}^{n} x_i = 1$. Hence the vector $\mathbf{x} \triangleq [x_1, x_2, ..., x_i, ..., x_n]^T$ can represent a potential portfolio. From assumption (4) it is clear that for a finite number of assets it is possible to obtain an infinite number of possible portfolios.

The set of all possible portfolios is called the admissible set. In general case the set has a shape as in the Fig. 1 (the shaded area). Symbols $E(r)$ and σ represent expected value of rate of return and risk of an asset or a portfolio respectively. The portfolios E, G, H, S have special properties: G is a portfolio with the lowest return, S is a portfolio with the highest return, E is a portfolio with the lowest risk, H is a portfolio with the highest risk. Another important term is the efficient frontier which lies on bold frontier between E and S in Fig. 1. It is a set of portfolios which have maximal expected return for a given risk level and minimal risk for a given expected return level. The optimal portfolio (O) must lie on the efficient frontier. It can be found by means of indifference curves (IC) (Swalm, 1966; Alexander and Francis, 1986) which can have different shapes for different investors and the curves characterize investor's attitude toward risk. An example of such a curve is in Fig. 1.

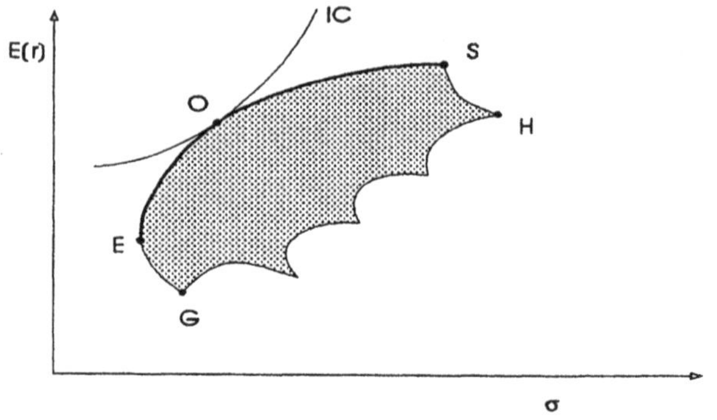

Fig. 1. The efficient frontier and the optimal portfolio

Finding of the efficient frontier is based on the theory of quadratic programming. Variables $x_1, ..., x_n$ fulfil linear constraints:

$$\sum_{i=1}^{n} x_i = 1 \tag{1}$$

$$x_i \geq 0 \quad i = 1, 2, \ldots, n \tag{2}$$

The payoff r_P (portfolio return) is defined by

$$r_P = \sum_{i=1}^{n} r_i x_i \tag{3}$$

where coefficients r_i (return of the ith asset) are unknown at the time of choosing x_i. The r_i are random variables with expected values $\bar{r}_i$ (expected returns) and covariances σ_{ij} (covariances of returns). The expected value of r_P is given by

$$\bar{r}_P = \sum_{i=1}^{n} \bar{r}_i x_i \tag{4}$$

The variance of r_p (risk of portfolio) is then

$$\sigma_P^2 = \sum_{j=1}^{n} \sum_{i=1}^{n} \sigma_{ij} x_i x_j \tag{5}$$

The main problem is to compute: (a) the efficient combinations of $\bar{r}_P$ and σ_P^2, i.e. such combinations of $[\bar{r}_P, \sigma_P^2]$ from admissible set which give minimum $\bar{r}_P$ for given σ_P^2 and maximum σ_P^2 for given $\bar{r}_P$; and (b) the portfolios $\mathbf{x}$ associated with the efficient combinations $[\bar{r}_P, \sigma_P^2]$. Quadratic programming (Sharpe and Gordon, 1990; Markowitz, 1959; Markowitz, 1956) is a computational technique of determination of the set of efficient combinations of $\bar{r}_P$ and σ_P^2, i.e. the efficient frontier.

2.2 *The Capital Asset Pricing Model*

One of the disadvantages of the Markowitz approach is a great numerical complexity of the calculation. The capital asset pricing model (CAPM) tries to solve this problem. However in contrast to Markowitz model which takes interest in only one investor, the capital asset pricing model is interested in the whole capital market with all its investors. The model presented here considers all investors as rational mean-variance optimizers, it means they all use the Markowitz portfolio selection model. Main consequence of the approach is the connection of expected return of an asset with a measure of this asset risk. Exact way of the connection is determined by the capital asset pricing model.

In addition to assumptions used in Markowitz theory this theory states following ones: (1) All investors will keep the portfolio for the same period. (2) All investors may borrow or lend any amount at a fixed risk free rate. (3) Information is free and instantly reachable for all investors. (4) Investors have homogeneous expectation, i.e. they use the same expected returns and the same covariance matrix of asset returns to generate the efficient frontier. (5) Investors pay no taxes on returns and no transaction costs on trades in assets.

In this paper the Security Market Line (SML) will be used as one kind of the CAPM. Connection between total risk of market portfolio σ_M^2 and mutual covariance of a single asset and market portfolio σ_{iM} issues from

$$\bar{r}_i = r_F + \left[\frac{\bar{r}_M - r_F}{\sigma_M^2}\right]\sigma_{iM} \tag{6}$$

where r_F is unknown risk free rate and $\bar{r}_M$ is expected value of random variable r_M. The variable r_M describes rate of return of market portfolio with variance σ_M^2 (square of total risk of market portfolio return change) defined by $\sigma_M = \sqrt{\sum_{i,j=1}^{n} x_{iM} x_{jM} \sigma_{ij}}$, where x_{iM} is relative amount of the ith asset invested in the market portfolio and σ_{ij} is covariance between the ith and the jth assets. The variable σ_{iM} is covariance between market portfolio and asset i. The market portfolio includes all traded assets. Ratio of each asset in the market portfolio equals the market value of the asset divided by total market value of all assets. Equation (6) thus represents the SML. Another form of the SML expression is $\bar{r}_i = r_F + [\bar{r}_M - r_F]\beta_i$, where $\beta_i = \frac{\sigma_{iM}}{\sigma_M^2}$. The model given by equation of the SML may not match the real situation of the market, therefore an error $v_i(t)$ with $E(v_i(t)) = \bar{v}_i$ is usually involved in the model: $r_i(t) = r_F + [r_M(t) - r_F]\beta_i + \bar{v}_i + e_i(t)$ where $e_i(t) = v_i(t) - \bar{v}_i$. For the sake of simplicity a substitution $\alpha_i \triangleq r_F - \beta_i r_F + \bar{v}_i$ is performed and the model has the following form

$$r_i(t) = \alpha_i + \beta_i r_M(t) + e_i(t) \tag{7}$$

Equation (7) represents the SCL and static regression model with unknown parameters α_i and β_i from mathematical point of view. If $\{e_i\}$ is white noise then an important equation

$$\sigma_i^2 = \beta_i^2 \sigma_M^2 + \sigma_{ei}^2 \tag{8}$$

can be written, where σ_i^2 is variance of the ith asset return change (square of total risk of the ith asset), $\beta_i^2\sigma_M^2$ is a square of systematic risk of the ith asset return change in the market, σ_{ei}^2 is variance of the ith asset random error (square of nonsystematic risk of the ith asset). The coefficients β_i known as beta coefficients of the ith asset are an alternative way of covariance risk expression of the ith asset. The difference between expected value of r_i from (6) and expected value of actual (measured) rate of return r_i^a, which is defined by

$$r_i^a(t) = \frac{p_i(t) - p_i(t-T)}{p_i(t-T)} \tag{9}$$

where $p_i(t)$ is price of the ith asset at time instant t, $t \in (1, 2, \ldots, t_L)$, t_L is a number of observed market sessions, T is a period on which the rate of return is defined, is equal $\bar{v}_i$. The actual rate of market portfolio return r_M^a can be given by

$$r_M^a(t) = \sum_{i=1}^{n} x_{iM} r_i^a(t) \tag{10}$$

Now the efficient frontier can be find by using the CAPM. The design procedure is analogical with Markowitz approach, but only input parameters of quadratic programming algorithm are different. The expected returns for each asset are then determined by equation

$$\bar{r}_i = \alpha_i + \beta_i \bar{r}_M \tag{11}$$

and the risks of each asset is given by (8). Finally covariances between the ith asset and the jth asset can be find by equation (7) and are given

$$\sigma_{ij} = \beta_i \beta_j \sigma_M^2 \tag{12}$$

All values of variables in equation (8), (11), (12) are unknown, therefore they must be substituted by their estimates. Parameters α_i, β_i and σ_{ei}^2 are estimated by least squares methods in equation (7). Characteristics of market portfolio $\bar{r}_M$ and σ_M^2 are substituted by expected value and variance of r_M^a (10). Finally the efficient frontier is determined by Markowitz approach.

3. PROBLEM STATEMENT

Some economists have shown that the static regression model (7) representing the CAPM by the SCL and basic least squares estimation technique are not suitable to estimate the SCL's parameters. These methods were modified and several simple algorithms for corrections of parameter estimates of static model were designed (Sharpe and Gordon, 1990; Rosenberg, 1985; Statman, 1981). The ad hoc corrections done after getting the SCL were based on the principles of application of prior subjective information.

Presented approach proposes not only to correct parameter estimates of the static model but to change the CAPM (7) to a model with time-varying parameters which will be called the dynamic capital asset pricing model (DCAPM). It means that the parameters of the DCAPM can be seen as time-varying parameters without exactly defined behaviour or as random walk and described by a state-space model. The DCAPM in the variant the SCL supposes form

$$r_i(t) = \alpha_i(t) + \beta_i(t) r_M(t) + e_i(t) \tag{13}$$

where $\alpha_i(t)$ and $\beta_i(t)$ are time-varying parameters. Equation (13) rewritten to standard regression model has a form

$$r_i(t) = \phi(t)^T \theta_i(t) + e_i(t) \tag{14}$$

where $\phi(t) = [1 \quad r_M(t)]^T$ is a vector of the regressors and $\theta_i(t) = [\alpha_i(t) \quad \beta_i(t)]^T$ is a vector of the parameters.

4. PARAMETER ESTIMATIONS OF THE DCAPM

Parameters of the DCAPM (13) are time-varying, therefore methods with exponential or linear forgetting for a estimation of those parameters were used. Linear forgetting was chosen for a presentation in this paper. Development of the DCAPM parameters is modelled as random walk.

4.1 *Dynamic capital asset pricing model*

The following state space model can be used for explicit description of parameter evolution and rate of return

$$\theta_i(t) = \theta_i(t-1) + w_{\theta i}(t) \tag{15}$$

$$r_i(t) = \phi^T(t)\theta_i(t) + e_i(t) \tag{16}$$

where $\{e_i(t)\}$ and $\{w_{\theta i}(t)\}$ denote the mutually independent white noise sequences with variance R_i and covariance matrix Q_i respectively and $E(w_{\theta i}(t)) = 0$. Now the DCAPM is described by (15), (16).

4.2 *Parameter estimation by Kalman filter*

The time-varying parameters of the DCAPM can then be estimated using the Kalman filter (Anderson and Moore, 1979)

$$L_i(t) = \frac{P_i(t)\phi(t)}{R_i(t) + \phi^T P_i(t-1)\phi(t)} \tag{17}$$

$$\hat{\theta}_i(t) = \hat{\theta}_i(t-1) + L_i(t)\varepsilon_i(t) \tag{18}$$

$$\varepsilon_i(t) = r_i(t) - \phi(t)\hat{\theta}_i(t) \tag{19}$$

$$P_i(t) = P_i(t-1) - L_i(t)\phi^T(t)P_i(t-1) + Q_i(t) \tag{20}$$

Variance R_i is selected by an analysis of return changes and covariance matrix Q_i is selected by minimizing of criterion

$$V(\hat{\theta}_i, Q_i) = \frac{1}{t_L}\sum_{t=1}^{t_L} \varepsilon_i^2(t) \tag{21}$$

Adaptive Kalman filter (Mehra, 1970) can be also used for estimation θ_i, variance R_i and covariance matrix Q_i.

5. A NEW ALGORITHM FOR ADAPTIVE SELECTION OF OPTIMAL PORTFOLIO

The new adaptive approach to optimal portfolio synthesis was explained in previous sections. Now it is possible to summarize achieved results in following algorithm:

(1) find the estimate of expected value (expected rate of return) $\bar{r}_M$ and variance (square of the risk) σ_M^2 of random variable r_M (rate of return of the market portfolio) (10)
(2) find the estimates of parameters $\alpha_i(t_L)$, $\beta_i(t_L)$ and variance σ_{ei}^2 in the model (14), where $r_i(t)$ and $r_M(t)$ are substituted by $r_i^a(t)$ and $r_M^a(t)$ respectively, for each asset by some of the estimation technique

(3) find the estimates of an expected value (expected rate of return) and covariance matrix of the vector random variable $\mathbf{r} \triangleq [r_1, r_2, \ldots, r_n]^T$ (8), (11), (12)
(4) find the efficient frontier, i.e. solve the problem described in subsection 2.1 by using characteristics obtained in step (3) and quadratic programming
(5) determine the indifferent curves (IC) for an investor (Swalm, 1966; Alexander and Francis, 1986)
(6) find the point of a contact between indifferent curve (IC) and the efficient frontier and determine optimal portfolio for the investor

6. EXPERIMENTAL RESULTS

Markowitz approach and the approaches based on the CAPM and newly designed the DCAPM were applied to selected stocks from Prague Stock Exchange (PSE) and the results are presented by tables and figures. The stocks used for optimal portfolio synthesis are Komerční banka(KB), RIF(RIF) and Škoda Plzeň(SKODA) from 11/01/94 to 11/14/96. Number of observation is 481 days. The portfolio returns are shown on time series (17 days) from 11/15/96 to 12/06/96. Graphical development of the stocks and PX-GLOB index representing market portfolio are shown in Fig. 2.

In Table 1. the SCL's parameters α, β and variance σ_e are presented for two models, the CAPM and the DCAPM. The CAPM parameters are estimated by standard least squares method. The DCAPM parameters are estimated by Kalman filter. Illustrations of parameter evolution which were estimated are shown in Fig. 3. Expected returns of each asset and covariance matrix between all assets were computed by standard mathematical procedures, using the CAPM parameters and the DCAPM parameters presented in Table 1.

Table 1. Estimated stock parameters

	PAR.	CAPM	DCAPM
	$\alpha(t_L)$	0.1185	0.4740
KB	$\beta(t_L)$	1.374	2.094
	σ_e	1.705	1.466
	$\alpha(t_L)$	0.1038	0.1038
RIF	$\beta(t_L)$	0.8284	0.8284
	σ_e	2.068	2.066
	$\alpha(t_L)$	0.1983	0.2535
SKODA	$\beta(t_L)$	0.841	1.437
	σ_e	1.797	1.743

Table 2. Designed portfolios and their rates of returns on 12/06/96

	Mark.	CAPM	DCAPM
KB	0.3550	0.3094	0.2785
RIF	0.2773	0.2957	0.3626
SKODA	0.3677	0.3949	0.3590
Returns	0.0794	0.0832	0.0896

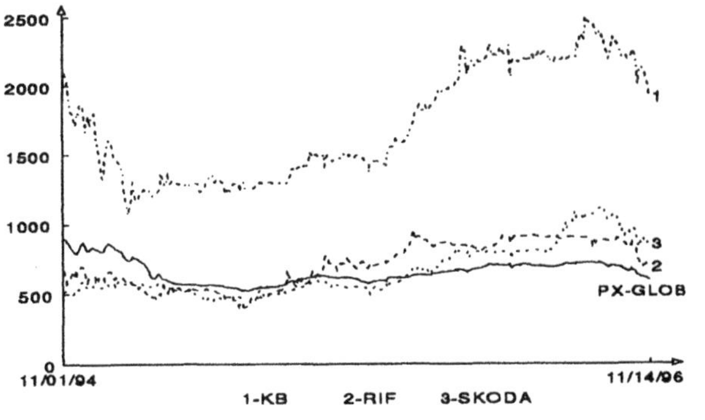

Fig. 2. Stocks and index development

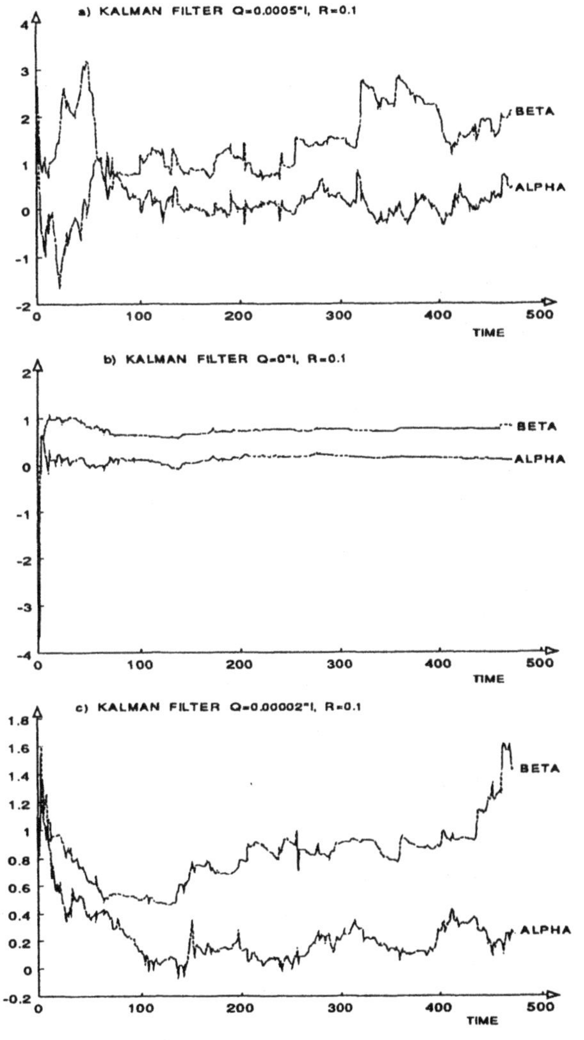

Fig. 3. The development of parameter Alpha and Beta: a)KB b)RIF c)SKODA

These parameters were used in equations (8), (11), (12) for determination of expected returns and covariance matrices and finally the efficient frontiers were found by quadratic programming (Financial Toolbox in MATLAB v. 4.21). All designed portfolios are presented in Table 2. Actual rate of return after 17 days is presented for each portfolio too.

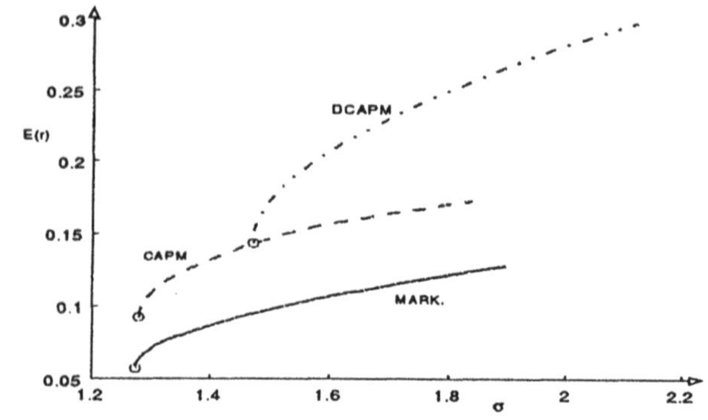

Fig. 4. The efficient frontiers

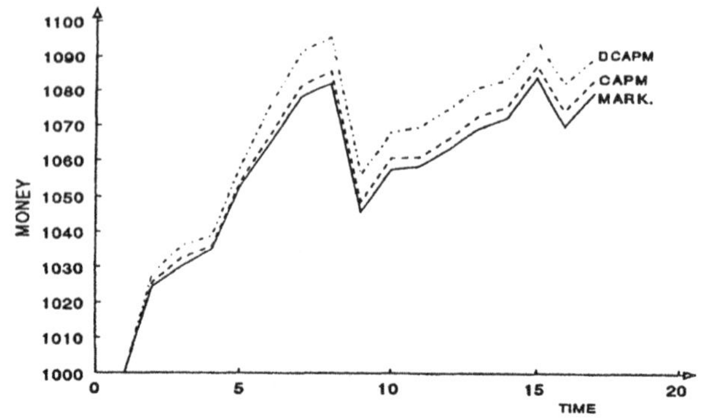

Fig. 5. The development of portfolio values

Daily wealth obtained by holding of portfolio designed by Markowitz approach, approach based on the CAPM and the best portfolio designed by approach based on the DCAPM are illustrated in Fig.5. Daily wealth represents values of the selected portfolios bought for 1000 Kč on portfolios holding period. The portfolios were selected from the efficient frontiers which are illustrated in Fig. 4.

The new approach based on the DCAPM has higher returns of portfolios than approach by the CAPM and Markowitz approach in the majority of experiments. This property of the DCAPM is more apparent on nonstacionary security market which is sometimes called "market without history".

7. CONCLUSION

The new approach to optimal portfolio selection based on adaptive viewpoint was presented. The dynamic capital asset pricing model was defined and its time-varying parameters were estimated by tracking algorithms. Adaptive approach contrary to modern portfolio selection theory using the static capital asset pricing model respects a development of asset characteristics. This fact affects the security characteristic line and finally the portfolio return. The new adaptive portfolio selection algorithm was applied to several assets from PSE and the results were presented. The designed portfolios and namely their portfolio returns were successfully confronted with the results representing Markowitz and the CAPM approach to portfolio selection.

8. ACKNOWLEDGMENTS

The research was supported by the Grant Agency of Czech Republic under grant GA ČR 102/97/0466.

9. REFERENCES

Alexander, G. J. and J. C. Francis (1986). *Portfolio Analysis.* Englewood Cliffs.

Anderson, B. D. O. and J. B. Moore (1979). *Optimal Filtering.* Prentice Hall.

Fortescue, T. R., L. S. Kershenbaum and B. F. Ydstie (1981). Implementation of self-tuning regulators with variable forgetting factors. *Automatica* **17**, 831–835.

Kulhavý, P. and M. Kárný (1984). Tracking of slowly varying parameters by directional forgetting. In: *Preprints of 9th triennial World Congress IFAC.* Vol. 2. IFAC. pp. 687–692.

Ljung, L. (1991). Optimal and ad hoc adaptation mechanism. In: *Preprints of European Control Control Conference ECC91.* ECC91.

Markowitz, H. M. (1956). The optimization of a quadratic function subject to linear constraints. *Naval Research Logistics Quartely.*

Markowitz, H. M. (1959). *Portfolio Selection - Efficient Diversification of Investment.* John Wiley & Sons, Inc.

Markowitz, H. M. (1952). Portfolio selection. *Journal of Finance* **7**, 77–91.

Mehra, R. M. (1970). On the identification of variances and adaptive kalman filtering. *IEEE Transactions on Automatic Control* **AC-15**, 175–184.

Parkum, J. E., N. K. Poulsen and J. Holst (1991). Selective forgetting in adaptive procedures. In: *Preprint of 11th World Congress IFAC.* IFAC. pp. 180–186.

Rosenberg, B. (1985). Prediction of common stock betas. *Journal of Portfolio Management* **11**, 5–14.

Sharpe, W. F. (1964). Capital asset prices: a theory market equilibrium under conditions of risk. *Journal of Finance.*

Sharpe, W. F. and J. A. Gordon (1990). *Investments.* Prentice-Hall, Inc.

Statman, M. (1981). Betas compared: Merril lynch vs. value line. *Journal of Portfolio Management* **7**, 41–44.

Swalm, R. O. (1966). Utility theory - insights into risk taking. *Harvard Business Review* **44**, 123–36.

Šimandl, M. (1996). State estimation for non gaussian models. In: *Preprints of 13th World Congress of IFAC.* Vol. H. IFAC. pp. 463–468.

MODEL REFERENCE ADAPTIVE CONTROL OF NONLINEAR SYSTEMS WITH CANONICAL STRUCTURE

Ján Murgaš and Eva Miklovičová[1]

Department of Automatic Control Systems, Faculty of Electrical Engineering and Information Technology, Slovak Technical University, Ilkovičova 3, 812 19 Bratislava, Slovakia. e-mail:murgas@kasr.elf.stuba.sk, miklovic@kasr.elf.stuba.sk.

Abstract: **M**odel reference adaptive control (MRAC) has been developed in many modifications. The stability proofs have been given under the assumption that the plant model is linear. However, MRAC approaches give often convergent solutions even for nonlinear systems. The aim of this paper is to show that for nonlinear systems with the model in canonical form, the standard MRAC with state feedback structure can be used to obtain the perfect model matching. The stability proof is derived using the general Laypunov stability theory. At the end of paper, an example of adaptive control of third order nonlinear system is presented, where a considerable improvement of control system dynamical behavior has been obtained.

Keywords: nonlinear system, model reference adaptive control

1. INTRODUCTION

Even though MRAC principle is mostly employed in tracking problems, it can also be effectively used for adaptive regulation, where the reference model serves as an etalon of performance and the perfect model tracking is not required. If the plant to be controlled is linearizable, then standard methods can be used, either with linear state controller (Landau, 1979) if all state space variables are available, or with output feedback (Narendra and Annaswamy, 1989; Sastry and Bodson, 1989) otherwise.

Up to now, several MRAC methods for linear systems have been proposed, which are convergent provided that some apriori assumptions are satisfied. The main problems are represented by low robustness and high complexity of resulting algorithms. Several authors aimed at showing the effectiveness of MRAC algorithms also in the control of nonlinear plants. In some papers, the problem is solved for the nonlinear system of the 1st order. For example, in (Lozano and Brogliato, 1992) a linear reference model is considered and the apriori knowledge of the system gain is not required. The control law is chosen such that it compensates the plant nonlinearity and the linear dynamics of reference model is imposed to the closed loop system. The adaptation law is based on the normalized least-squares method. In (Papadakis and Thomopoulos, 1994), the modified standard input/output algorithm is investigated for nonlinear systems with exponentially stable zero dynamics. A modifying term is included into the control law, in order to improve the performance of the standard MRAC scheme. MRAC scheme for general nonlinear infinite dimensional system is dealt with in (Bohm, *et al.*, 1996). A general form of adaptation law for adaptive control of nonlinear systems which guarantees the negative definiteness of Lyapunov function is derived in (Murgaš, *et al.*, 1992). The mostly used application domain of MRAC algorithms are the servosystems for their fast dynamics.

All adaptive control methods require the satisfaction of some apriori assumptions concerning the plant

[1] This work has been partially supported by the Slovak Scientific Grant Agency, grant No. 1/4231/97.

model. In practice, some information about the structure of an acceptable plant model can be available, often together with the structure of dominant nonlinearities, while the parameters are unknown. Another frequently appearing property of these models is its description with separable linear and nonlinear part. The models of power systems, robotic systems, etc. can be noted as an example.

The aim of the work is to show that MRAC schemes proved for linear systems can also be used for adaptive regulation of nonlinear systems. The paper consists of 4 sections. In Section 2 we describe the standard MRAC schemes for linear systems, which constitute a basis for nonlinear systems control as well. The state space approach with an assumption of measurability (direct or indirect) of all state variables is presented together with the input/output method with augmented error. In Section 3, MRAC algorithm for nonlinear systems with canonical structure is proposed. The effectiveness of the proposed method is illustrated by an example in Section 4.

2. STANDARD MRAC SCHEMES

Intensive research of MRAC schemes over the last decades has brought the following standard MRAC algorithms

- the algorithm with state feedback structure (Bohm, *et al.*, 1996; Murgaš, 1994; Murgaš, *et al.*, 1995), when all the state space variables are available
- the augmented error method (Datta and Ioannou, 1994; Narendra and Annaswamy, 1989; Sastry and Bodson, 1989), when only the input and output information can be used

Many other methods have been proposed, like CGT, gradient method, etc., but they have not attracted much attention in the adaptive control community.

2.1 MRAC with state space structure

Consider that all state space variables are available. This represents a relatively severe requirement, which can be solved by calculation of state variables from the measurable signals.

Then for SISO systems, the control law can be expressed by linear form

$$u = k_x^T(x,e)x + k_r(r,e)r = \Theta^T\omega \tag{1}$$

where $\Theta = \left[k_x^T, k_r\right]^T$ is the adaptive controller parameter vector, $\omega = \left[x^T, r\right]^T$ is the signal vector, and x, r are the state vector and the reference input signal, respectively.

The adaptation law can be designed using the general Lyapunov stability theory

$$\dot{\Theta} = -\alpha\varepsilon\omega, \quad \alpha > 0, \quad \varepsilon = b^T Pe \tag{2}$$

where $e = x - x_m$ is the adaptation error vector, b is the system input vector and P is the solution of the Lyapunov's matrix equation $A_m^T P + PA_m = -Q$, P and Q being the positive symmetric matrices.

Equations (1) and (2) represent the basic relations of MRAC with state structure. Note that if the plant transfer function contains stable zeros, the above algorithm has to be completed with an adaptive prefilter (Mugaš, *et al.*, 1995).

2.2 MRAC structure with augmented error

MRAC methods using only the measurable input/output information are often based on the approach originally proposed by Monopoli (1979) and consequently developed and improved by Narendra, *et al* (1989), Datta and Ioannou (1994) and Tao (1996). To implement the algorithm, the satisfaction of some necessary assumptions is required

- the order of the plant transfer function $G_p(s)$ is known
- the relative degree of the $G_p(s)$ is known and the reference model transfer function $G_m(s)$ has the same degree
- the plant and model are both minimum phase systems
- the high frequency gain is known.

We consider MRAC scheme which taking into account its large popularity in adaptive control literature can be considered as the standard one:

Control law

$$u = \Theta^T\omega + c_0 r, \quad c_0 = \frac{k_m}{k_p}$$

where
$$\Theta = \left[\Theta_1^T, \Theta_2^T, \Theta_3\right]^T$$

$$\omega = \left[\omega_1^T, \omega_2^T, y\right]^T$$

$$\omega_1(s) = \frac{\alpha(s)}{\Lambda(s)}u(s) \qquad \omega_2(s) = \frac{\alpha(s)}{\Lambda(s)}y(s)$$

$$\alpha(s) = \left[s^{n-2}, s^{n-1}, \ldots, s, 1\right]^T$$

$$\Lambda(s) = G_m(s)\Lambda_1(s)$$

$\Lambda_1(s)$ is a monic Hurwitz polynomial of degree $n - 1 - n_m$.

Adaptation law $\dot{\Theta} = -\Gamma \dfrac{\varepsilon\xi}{m^2}$

$$\xi(s) = G_m(s)\omega(s)$$

where $\Gamma = \Gamma^T > 0$ is the adaptation gain matrix, m is the normalizing signal and

$$\varepsilon(s) = \Theta^T\xi(s) + c_O y(s) + G_m(s)u(s)$$

is an augmented error.

3. MRAC FOR THE CLASS OF NONLINEAR SYSTEMS

Consider the class of nonlinear dynamical systems which can be described as follows

$$\begin{aligned} \dot{x}_1 &= x_2 \\ \dot{x}_2 &= x_3 \\ &\vdots \\ \dot{x}_n &= f(x) + bu \end{aligned} \tag{3}$$

where the nonlinear function $f(x)$ is separable, i.e. $f(x) = \sum_{i=1}^{n} f_i(x_i)$.

The reference model is assumed to be linear and described by the canonical form

$$\begin{aligned} \dot{x}_{m1} &= x_{m2} \\ \dot{x}_{m2} &= x_{m3} \\ &\vdots \\ \dot{x}_{mn} &= \sum_{i=1}^{n} a_{mi}x_{mi} + b_m r \end{aligned} \tag{4}$$

where the constants a_{mi} $(i = 1,\ldots,n)$ and b_m are determined taking into account the qualitative requirements for the controlled system and r is the reference input signal.

The control law is of the form

$$u = \sum_{i=1}^{n} g_i(x_i)x_i + k_r r \tag{5}$$

where $g_i(x_i)$ are for the present not specified functions and k_r is the adaptive controller parameter which is introduced to adjust the closed loop gain to the gain of the reference model.

System (3) with the controller (5) is described by

$$\begin{aligned} \dot{x}_1 &= x_2 \\ \dot{x}_2 &= x_3 \\ &\vdots \\ \dot{x}_n &= \sum_{i=1}^{n}\left[f_i(x_i) + bg_i(x_i)x_i\right] + bk_r r \end{aligned} \tag{6}$$

The model matching conditions are expressed by equations

$$\sum_{i=1}^{n}\left[f_i(x_i) + bg_i(x_i)x_i\right] + bk_r r = \sum_{i=1}^{n} a_{mi}x_i + b_m r \tag{7}$$

or

$$b_m = bk_r^* \quad \Rightarrow \quad k_r^* = \frac{b_m}{b}$$

$$a_{m1}x_1 = f_1(x_1) + bg_1^*(x_1)x_1 \quad \Rightarrow \quad g_1^* = \frac{a_{m1}}{b} - \frac{f_1(x_1)}{bx_1}$$

$$\vdots$$

$$a_{mn}x_n = f_n(x_n) + bg_n^*(x_n)x_n \quad \Rightarrow \quad g_n^* = \frac{a_{mn}}{b} - \frac{f_n(x_n)}{bx_n} \tag{8}$$

Using the model matching conditions (8) the closed loop system equations (6) can be rewritten as follows

$$\begin{aligned} \dot{x}_1 &= x_2 \\ \dot{x}_2 &= x_3 \\ &\vdots \\ \dot{x}_n &= \sum_{i=1}^{n} a_{mi}x_i + b\Theta^T\omega \end{aligned} \tag{9}$$

where

$$\Theta = \left[g_1 - g_1^*, g_2 - g_2^*, \ldots, g_n - g_n^*, k_r - k_r^*\right]^T$$

$$\omega = \left[x^T, r\right]^T$$

The dynamics of the adaptation error $e = x - x_m$ can then be described by

$$\begin{aligned} \dot{e}_1 &= e_2 \\ \dot{e}_2 &= e_3 \\ &\vdots \\ \dot{e}_n &= \sum_{i=1}^{n} a_{mi}e_i + b\Theta^T\omega \end{aligned} \tag{10}$$

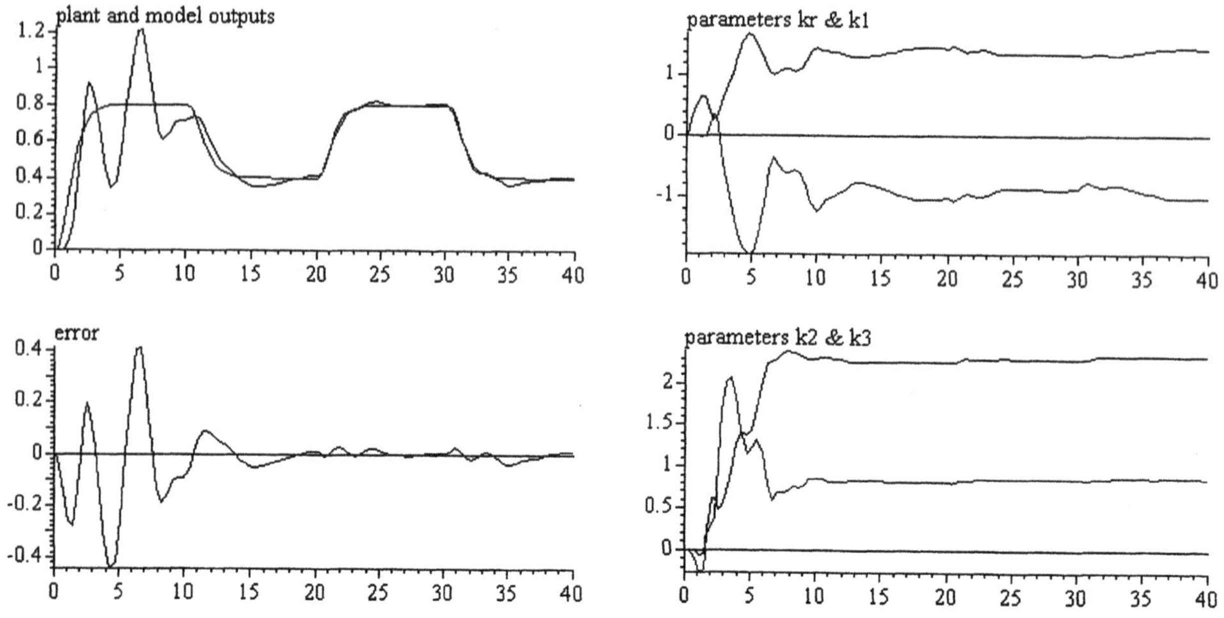

Fig. 1. Simulation results.

The stability of the equilibrium state $\left[e_1 = 0,\ldots,e_n = 0, \Theta_1 = 0,\ldots,\Theta_n = 0\right]$ of the system (10) can be proved using the standard function

$$V(e,\Theta) = e^T Pe + \alpha^{-1}\Theta^T\Theta \qquad e = \left[e_1,\ldots,e_n\right]^T \tag{11}$$

where $P = P^T > 0,\quad \alpha > 0$.

The function (11) is the Lyapunov function candidate because the following conditions are satisfied

i. $V(0,0) = 0$, $V(e,\Theta)$ is continuous

ii. $V(e,\Theta)$ is positive definite for all (e,Θ)

iii. $\dot{V}(e,\Theta) \le 0$ if the adaptation law is chosen as

$$\dot{\Theta} = -\alpha\varepsilon\omega, \qquad \varepsilon = b^T Pe \tag{12}$$

The satisfaction of conditions *i* and *ii* is evident. Now it is necessary to show that also the condition *iii* is satisfied. The derivative of the function (11) along the system trajectory is

$$\dot{V}(e,\Theta) = e^T\left(A_m^T P + PA_m\right)e + 2b^T Pe\Theta^T\omega + 2\alpha^{-1}\Theta^T\dot{\Theta} \tag{13}$$

If the adaptation law is of the form (12) then

$$\dot{V}(e,\Theta) = -e^T Qe \tag{14}$$

where $A_m^T P + PA_m = -Q,\quad Q = Q^T > 0$.

Taking into account (14) it can be concluded that the function (11) is negative semidefinite and negative definite with respect to e which means that the equilibrium state of system (10) is asymptotically stable with respect to e and stable with respect to Θ.

4. EXAMPLE

Consider the 3rd order system described by equations

$$\dot{x}_1 = x_2$$
$$\dot{x}_2 = x_3$$
$$\dot{x}_3 = bu - a_1 \sin x_1 - a_2 \sin x_2 + a_3 x_3^2$$

and the reference model

$$\dot{x}_{m1} = x_{m2}$$
$$\dot{x}_{m2} = x_{m3}$$
$$\dot{x}_{m3} = 8r - 8x_1 - 12x_2 - 6x_3$$

The control law is according to (3.3)

$$u = -k_1x_1 - k_2x_2 - k_3x_3 + k_r r$$

and the adaptation laws are as follows

$$\dot{k}_1 = \varepsilon x_1$$
$$\dot{k}_2 = \varepsilon x_2$$
$$\dot{k}_3 = \varepsilon x_3$$
$$\dot{k}_r = -\varepsilon r$$

where $\varepsilon = 2e_1 + e_2 + e_3$.

The adaptive controller adjusting is depicted in Fig. 1.

CONCLUSION

This paper has shown that for nonlinear systems with canonical structure the standard state space MRAC scheme can result in stable adaptive system. This fact has been illustrated by the example of the 3rd order nonlinear system.

REFERENCES

Bohm, M., M.A. Demetriou, S. Reich and I.G. Rosen (1996). A Model Reference Adaptive Control Scheme for Nonlinear Infinite Dimensional Systems. *13th Triennial World Congres*, San Francisco, 205-210.

Datta, A. and P.A. Ioannou (1994). Performance Analysis and Improvement in Model Reference Adaptive Control. *IEEE Trans. on Automatic Control*, **39**, 2370-2387.

Druzhinina, M.V., V.O. Nikiforov and A.L. Fradkov (1996). Adaptive Control Methods for Nonlinear Systems with Input Nonlinearities (in Russian). *Automatica and Remonte*, 3-32.

Landau, I.D (1979). Adaptive Control - The Model Reference Approach. Marcel Decker, New York.

Lozano, R. and B. Brogliato (1992). Adaptive Control of a Simple Nonlinear System without a priori Information on the Plant Parameters. *IEEE Trans. Automat. Contr.*, **37**, 30-37.

Monopoli, R.V. (1979). Model Reference Adaptive Control with an Augmented Error Signal. *IEEE Trans. Automat. Contr.*, **19**, 474-484.

Murgaš, J. (1994). Present View on a Class of Adaptive Control with Reference Model. *Journal of Electrical Eng.*, **45**, 383-388.

Murgaš, J. and I. Hejda (1990). Decentralized Adaptive Stabilization with State Regulators. *Kybernetika*, **26**, 496-504.

Murgaš, J., M. Matejovič and V. Červenka (1995). MRAC Algorithm for High Relative Degree Plants. *Proc. of the 5th IFAC Symposium ACASP'95*, Budapest, 276-281.

Murgaš, J., V. Veselý and I. Hejda (1992). State Space Structures in MRAC. *IFAC Workshop System Structure and Control*, Prague.

Narendra, K.S. and A.M. Annaswamy (1989). *Stable Adaptive Systems*, Prentice-Hall, New Jersey.

Papadakis, I.N.M. and S.C.A. Thomopoulos (1994). Improved Performance MRAC of Linear and Nonlinear Systems. *Int. J. of Adaptive Control and Signal Proc.*, **8**, 337-357.

Sastry, S. and M. Bodson (1989). *Adaptive Control, Stability, Convergence and Robustness*. Prentice-Hall, New Jersey.

Tao, G. (1996). Model Reference Adaptive Control with L Tracking. *Int. J. of Control*, **64**, 859-870.

Veselý, V. and J. Murgaš (1993). Decentralized Adaptive Stabilization of Nonlinear Systems. *Int. J. of Control*, **58**, 1445-1460.

ADAPTIVE SELF-TUNING PID ALGORITHM BASED ON CONTINUOUS SYNTHESIS

Mikuláš Alexík

University of Žilina, dept. of Technical cybernetics, Slovak republic

Abstract. This paper describes the analytical PID algorithm which has the adaptive self-tuning performance. It's synthesis is made in a continuous domain. This means that identification is followed by recalculation into the continuous domain and the final gain of the algorithm is compensated due to sampling interval. Simulation experiments performed on analogue models as hybrid simulation in real time are also shown. The suggested design of the control algorithm is robust against the changes of the plant parameters and change of sampling interval. The simplicity of the control algorithm makes it suitable for initial experiments with adaptive algorithms in the teaching process and also in real control of heat exchangers and energetic plants

Keywords: PID algorithm, quality criterion, sampling interval, on-line identification, analytical synthesis, hybrid simulation.

1. INTRODUCTION

Design of the described control algorithm is based on classical advance. This is the design of feedback for demanded quality of control. In principle the on-line mathematical model of controlled plant and on-line synthesis of control algorithm for demanded quality of control process are created. This advance is called Self-Tuning Control (STC) (Ästrom and Wittenmark 1989) or indirect adaptive control because the parameters of the control systems are adjusted after the calculation of plant parameters. Fig.1 shows the basic block scheme of the adaptive control system.

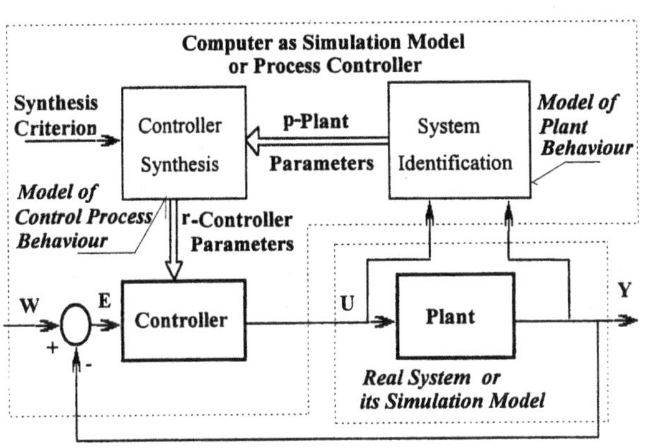

Fig. 1. Block scheme of the adaptive control system based on on-line identification.

This scheme corresponds to the classical off-line computation of controller parameters, which is still very important in the process of evaluation of the control algorithm. A modification of STC algorithm based on continuous synthesis is shown in Fig. 2

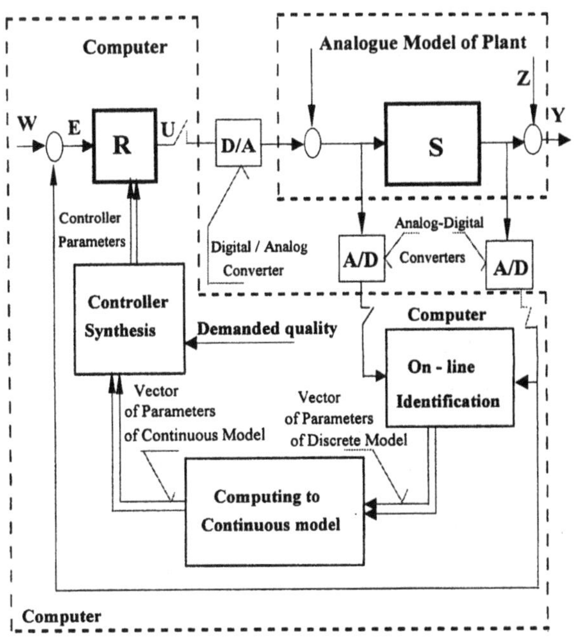

Fig. 2. Block scheme of the adaptive analytical PID algorithm.

The basic design of STC from Fig. 1 can have many modifications which differ from one another by principles of minimalization of adaptive criteria, complex of computation in real time, by the identifi cation algorithm (Kozák, 1994), by applied advance during control algorithm synthesis (Bobál, 1994) and with the quality of final behavior of controlled plant

One modification which is shown in Figure 2 was designed based on (Alexík, 1994) and is described including its verification with the simulation experiments For experimental evaluation of the control algorithm in real time the hybrid simulation which replaces real plant with its analogue model and with A/D and D/A converters was used

2. ANALYTICAL PID ALGORITHM

The described algorithm of the adaptive control has PID structure and differs from the algorithm known in this area for its synthesis which is made in a continuous domain. There is a possibility to design the analytical algorithm from P to PID to meet the demands of the control system second or third order but without transportation delay. These models represent a wide scale of the control system including systems of higher order, which are shown in the third section of this paper. The individual steps of the control algorithm design are shown in Fig. 2 and they are explained further respectively: recalculation of parameters of the discrete model into parameters of the continuous system, the synthesis of the analytical PID algorithm and the recalculation of parameters of the control algorithm into a discrete domain taking into account the sampling interval and demanded quality of control.

Relations for the recalculation of the discrete model of the second or third order parameters into parameters of continuous system were derived for possibility of the control algorithm synthesis in the continuous domain after the identification of its parameters in the discrete domain. If the proportional control system of the second order is assumed (1) then after the identification it has form (2) in Z-transform

$$S(p)=\frac{K}{T^2 * p^2 + 2aT * p + 1} \qquad (1)$$

$$S(z)=\frac{b_0 + b_1 * z^{-1} + b_2 * z^{-2}}{1 + a_1 * z^{-1} + a_2 * z^{-2}} \qquad (2)$$

Parameters of the discrete transfer function (2) can be rewritten for real poles of transfer function (1) or for time constants T_1 and T_2.

Parameters of the discrete model (2) a_1, a_2, b_1, b_2 are gained by the identification. Parameter b_0 depends on initial conditions of the taking of the discrete samples of a continuous model. If the transfer function with real poles (2) is rearranged then relations (3), (4) and (5) for calculation of the transfer function S(p) parameters are derived from (1)

$$z_1 = \frac{-a_1 + \sqrt{a_1^2 - 4a_2}}{2}; \; z_2 = \frac{-a_1 - \sqrt{a_1^2 - 4a_2}}{2}; \qquad (3)$$

$$T_1 = \frac{-T_0}{\ln(z_1)}; \; T_2 = \frac{-T_0}{\ln(z_2)}; \; K = \frac{b_0 + b_1 + b_2}{1 + a_1 + a_2} \qquad (4)$$

$$T^2 = T_1 T_2 ; \qquad 2aT = (T_1 + T_2) \qquad (5)$$

In the case of complex poles in the transfer function (2) a similar advance can be applied and for the calculation of coefficients of transfer function (1) relations (6), (7) and (8) are valid. The same advance is realized for the calculation of parameters for the system of the third order and relations (6), (7) and (8), which were derived previously are applied again.

$$z_1 = (1/T_0) * \mathrm{arctg}\left(\frac{\sqrt{\left\{1 - \left[(a_1)/(2\sqrt{a_2})\right]^2\right\}}}{(a_1)/(2\sqrt{a_2})}\right) \qquad (6)$$

$$z_2 = \frac{\ln\left(1/\sqrt{a_2}\right)}{T_0}; \qquad (7)$$

$$K = \frac{b_0 + b_1 + b_2}{1 + a_1 + a_2}; \quad T^2 = \left\{1/\left[(z_1)^2 + (z_2)^2\right]\right\}$$

$$2aT = \frac{2z_2}{\left[(z_1)^2 + (z_2)^2\right]}; \qquad (8)$$

The synthesis of the analytical PID controller in the continuous domain is based on the demand of the best quality of the control which was defined in the theory of optimal control. For the known system description and the known quality criterion it is possible to design software or feedback control also in the case of saturated values of the manipulated variable. Very frequent criterion which are time-optimal control and quadratic criteria are used for the design of the servomechanism. Another possibility introduces an integral criterion for a linear system where non-linearity as saturation of the manipulated variable is solved by means of an antiwindup (Alexík,1994).

The third possibility is to prescribe the demanded course of the control process in the form of the Laplace transform. For the linear system and for the compensation of saturation it means the use of antiwindup correction again. Essential for this method is the placement of poles of the transfer function in a complex plane, which has is called "pole placement method". A suitable integral criterion formula (9) is used in (Vindincev, 84) which resulted in an aperiodical course of the control process.

$$Q(e)=\int_0^{\infty}\left[e(t)+\frac{d}{dt}e(t)\right]^2 dt \qquad (9)$$

The method which is described further gives the same results as the criterion (9), but is less time consuming during the calculation and can be applied also on systems with time delay. The synthesis for the second order system with time delay "D" (10) which is described further and then the results are applied for the system from the second to the fourth order and verified experimentally by the simulation.

$$S_1(p)=\frac{K*\exp(-D*p)}{T^2*p^2+2aT*p+1}=\frac{K*\exp(-D*p)}{A_1(p)} \qquad (10)$$

The demanded courses of the control variables during the step change of the manipulated variable are shown in Fig. 3. An ideal response which is introduced by curve "0" cannot be realized because it needs non-realistic (equal to infinity) parameters of the controller. The closest to the ideal case is the characteristic of the first order marked as "1". The time constant Tr completely characterizes the time of control "tr" which can be chosen as the multiple of Tr The transfer function of the closed loop control circuit can be expressed by (11):

$$Fw(p)=\frac{S_1(p)*R(p)}{1+S_1(p)*R(p)}=\frac{\frac{K*\exp(-D)}{A_1(p)}*R(p)}{1+\frac{K*\exp(-D)}{A_1(p)}*R(p)}$$

$$=\frac{Y(p)}{W(p)}=\frac{1}{Tr*p+1}*\exp(-D) \qquad (11)$$

After the rearrangement and calculation with only the first two members of the Taylor series

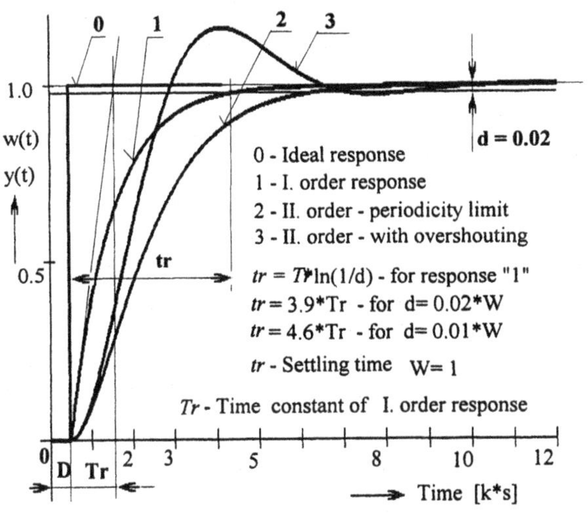

Fig. 3. Demanded quality of control process

$\exp(-D*p)=1-D*p$ for parameters of controllers, relations (12) and (13) can be derived as follows

$$R(p)=\frac{\left(1+2aT*p+T^2*p^2\right)}{K\left[Tr*p+1-\exp(-D*p)\right]} \qquad (12)$$

$$R(p)=\frac{2aT}{K*(Tr+D)}*\left(1+\frac{1}{2aT*p}+\frac{T^2}{2aT}\right)=$$

$$=rp*\left(1+\frac{1}{Ti*p}+Td*p\right) \qquad (13)$$

In spite of the fact that the structure of the control algorithm was not assumed, the best quality of control can be achieved for the proportional system with the second order delay (with time delay or without) with the PID control algorithm. Constants of the continuous PID algorithm clearly depend on plant parameters and the demanded time of control as described in the equation (11). This conclusion is valid also for the elimination of deterministic disturbances. The same is valid for the saturation of the manipulated variable because it was shown in [8], [9] that the application of the antiwind-up algorithm keeps global quality of control in this case it is an aperiodical course. The correction of the expression for proportional gain will be described later.

The same advance can be applied for the structure and parameters of the controller in the case of the proportional system with the third order delay (14). As a result there is a resulting $PIDD^2$ structure of the controller. Taking into account that the ideal derivation of the second order cannot be realized continuously, it means that the assumed courses 2 or 3 from the Fig. 3 shall be applied. It is clear from relations (15) and (16) that the settling time cannot be computed from Tr because it is a function of T_1 in the transfer function (14) but on the other hand a damping coefficient of control process a_r is eligible.

$$S_2(p)=\frac{K\ \exp(-D*p)}{(T_1*p+1)(T^2*p^2+2aT*p+1)} \qquad (14)$$

$$Fw(p)=\frac{S_2(p)*R(p)}{1+S_2(p)*R(p)}=\frac{K\ \exp(-D*p)\ R(p)}{A_2(p)+K\exp(-D*p)R(p)}$$

$$=\frac{Y(p)}{W(p)}=\frac{K\ \exp(-D*p)}{Tr^2*p^2+2arTr*p+1}$$

$$R(p)=\frac{(T_1*p+1)\left(T^2*p^2+2aT*p+1\right)}{K(2arTr+D)*p\left[\frac{Tr^2}{(2arTr+D)}*p+1\right]} \qquad (15)$$

$$R(p)=\frac{2aT}{K*(2arTr+D)}\left(1+\frac{1}{2aT*p}+\frac{T^2}{2aT}*p\right) \qquad (16)$$

Time constant Tr and modification of the controller parameters based on the sampling interval is chosen as it is described in the next section. Based on (11) it is the demanded behaviour with the first order delay but for the reason that controller is not continuous but discrete, the control cycle has course of second order with one time constant which characterizes the sampling interval of the control variable. After the on-line identification and the recalculation into the continuous domain the continuous transfer function of the second order for control system is at disposal. The control process should last a shorter time interval than the transient of the system. The settling time of the system (1) is possible to estimate from the dominant roots. In the case of overshoot or complex poles of the system the assumption is valid. The same is true for the aperiodical very damped systems. It is not valid for the aperiodical and periodical systems very close to the periodicity limit when the relative damping "a" is equal to 1. In this case it is possible to estimate the settling time from the time constant "T" from the transfer function (1). In both cases for settling time it is demanded to be less than 2.5 times shorter in comparison with the time while the controlled system reaches a steady state.

For the aperiodical very damped system, relation (4) can be used and the demanded time constant is (17), but for the aperiodical system close to the periodicity limit, the relation (18) is more suitable. For the oscillating system should by valid (19), but for the oscillating system with relative damping $0.707 \langle \mathbf{a} \langle 1$ (18) is more suitable. For any relative damping of the second order system it is then suitable to compute the demanded settling time from (20).

$$Tr = 0.4(1/\lambda) = 0.4T1 \qquad (17)$$

$$Tr = 0.8\sqrt{T^2} \qquad (18)$$

$$Tr = 0.4(1/\lambda) = (0.4/z2) \qquad (19)$$

$$Tr = 0.5\left[\,0.4(1/\lambda) + 0.8\sqrt{T^2}\,\right] \qquad (20)$$

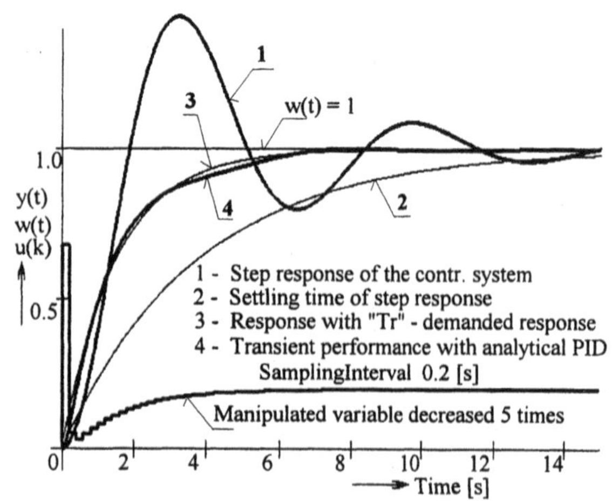

Fig. 4. Selection of the time constant Tr.

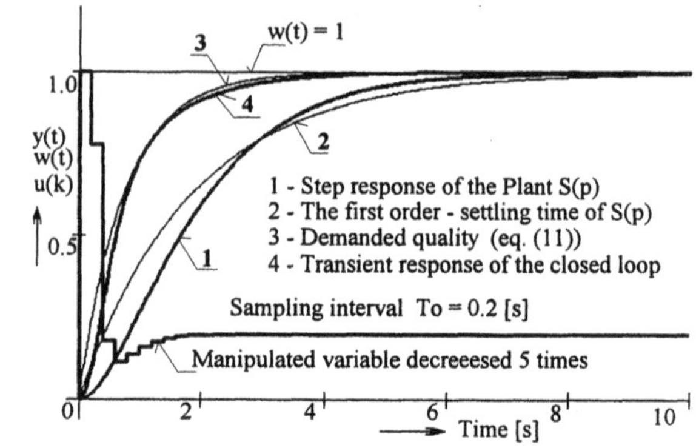

Fig. 5. Transient characteristic and control process for the system close to the periodicity limit.

It can be clearly seen that synthesis under the PID algorithm ensures control process without overshooting also for oscillating systems where assumption (11), which characterizes control process, is valid. It is evident from Fig. 4 and Fig. 5 that relation (20) precisely enough characterizes time for the control system to reache the steady state. Based on simulation experiments in (Alexík, 1994) it is possible to conclude that if PID control algorithm (13) and (16) and their prescription into discrete domain are applied, it is necessary with increasing interval of sampling of action variable T_0 to decrease proportional gain of R(p) in algorithm (13) or (16) to obtain the same quality of control. The input of the control system is usually a D/A converter which behaves as forming element of zero order with transfer function $H(p) = \{[1-\exp(p)]/p\}$. In both algorithm (13) and (16) it is necessary to calculate with transportation delay $Ds = D + 0.707{*}T_0$ or with $Ds = 0.707T_0$ if control systems (10) and (14) are without transportation delay. Fig. 6 and Fig. 7 show the influence of described synthesis including compensation of sampling interval on the control process of the second order system with dynamics corresponding to Fig. 4. This problem is the best solved in (Vitečková, 1995)

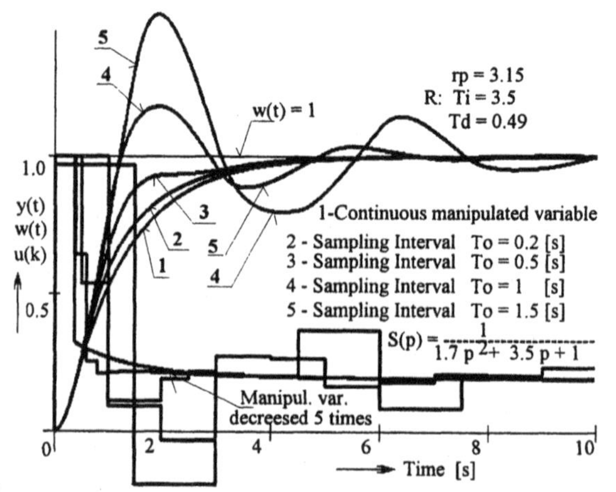

Fig. 6. Analytical PID algorithm without compensa ting of influence of sampling interval

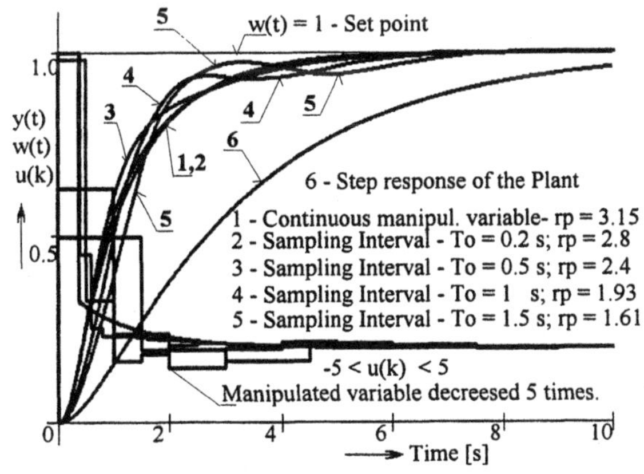

Fig. 7. Analytical PID controller with compensation of sampling interval.

3. SIMULATION EXPERIMENTS WITH ANALOGUE MODEL OF THE SYSTEM.

The adaptive algorithms with on-line identification are very demanding as far as the computing time is concerned. For their evaluation the hybrid simulation in comparison with the digital one is more suitable. In our case it is the continuous process modelled with the continuous model working in real time. The technical design of the continuous model enables us to model the proportional linear system of the first to the sixth order and also to change the parameters and the structure of the controlled system which is suitable for the evaluation of robustness of the control algorithm. Fig. 8 shows and describes the model.

The dynamics of the process realized with the model under Fig. 8 is described with transfer functions of the second and first orders and during the transient state, it is possible to change the relative damping for the second order system or time constant of the second or first order and the total gain or order of the system from the first to sixth orders

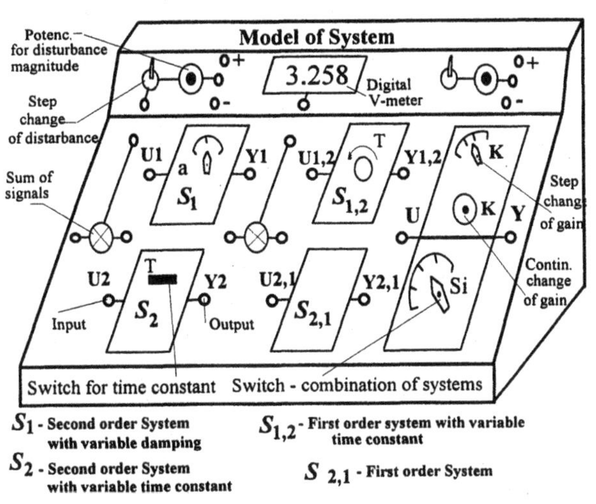

Fig. 8. Analogue model of the controlled system.

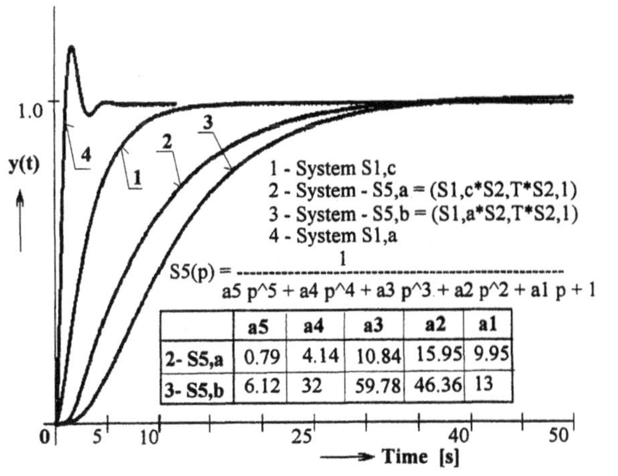

Fig. 9. Transient characteristics of the systems of the second to fifth order.

The individual models can be connected to the input of the gain block which is marked in Fig. 8 as U and then a change of the system gain during the transient state is possible in a continuous or step manner. The switch S1 enables to combine individual transfer functions in such a way that the Y output has dynamic characteristics of the proportional system till the sixth order. This kind of model of the system enables us to change individual time constants, gains and structure of the controlled system in a continuous or step by step manner.

Some results of the simulation experiments with the described adaptive algorithm are shown in Fig. 10 and Fig. 11. All experiments were performed in real time with the application of A/D and D/A converters and the simulations can be repeated when it is demanded.

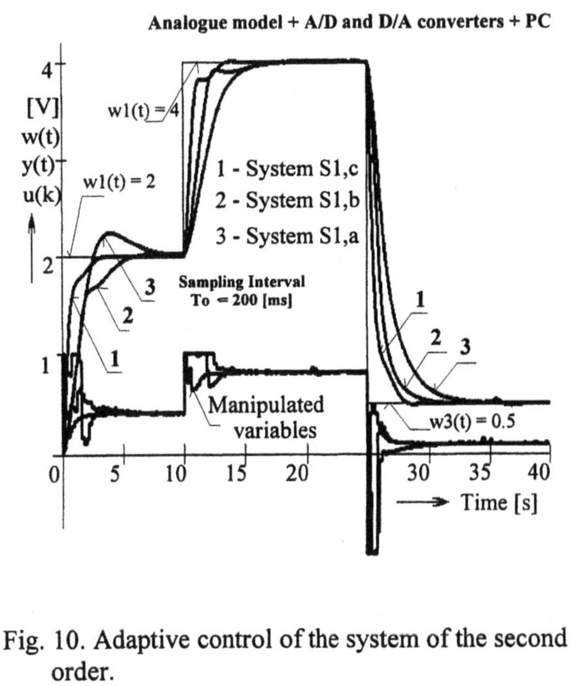

Fig. 10. Adaptive control of the system of the second order.

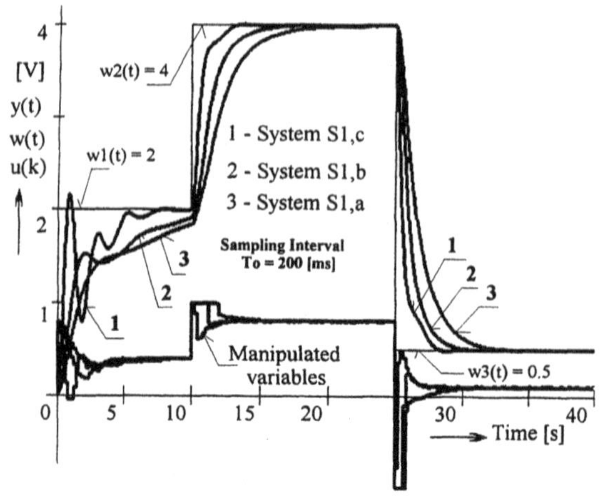

Fig. 11. Adaptive control of the system of the second order which starts with P controller.

It can be seen from both simulation experiments that the control process in steady states satisfies assumptions which created the base for the design of the control algorithm. There is a problem to ensure that the demanded quality of control during starting conditions for the adaptive control which can be followed from both figures.

Fig.12 shows the assumed behaviour of the algorithm also during step changes of the parameters of the controlled system, which is pretentious disturbance normally not supposed for continuous systems.

The influence of the sampling interval on the quality of the control process is shown in Fig. 13. The presented results show that the assumptions for the gain correction which were shown in Fig. 7 are correct. The figure again shows the necessity to solve the problem of the start of adaptation process

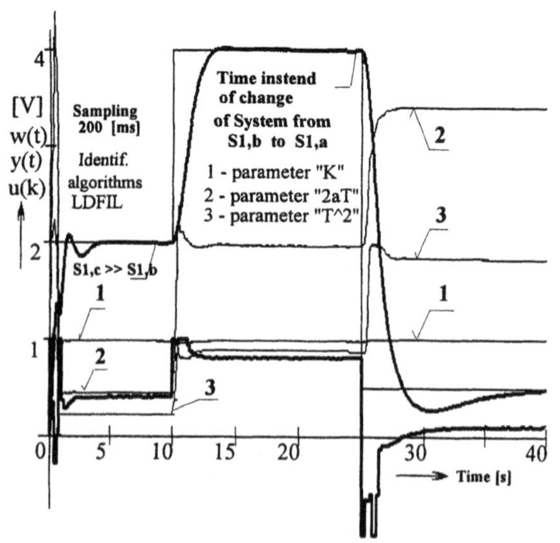

Fig. 12. Adaptation process of tuning identified parameters during the controlled system parameters changes

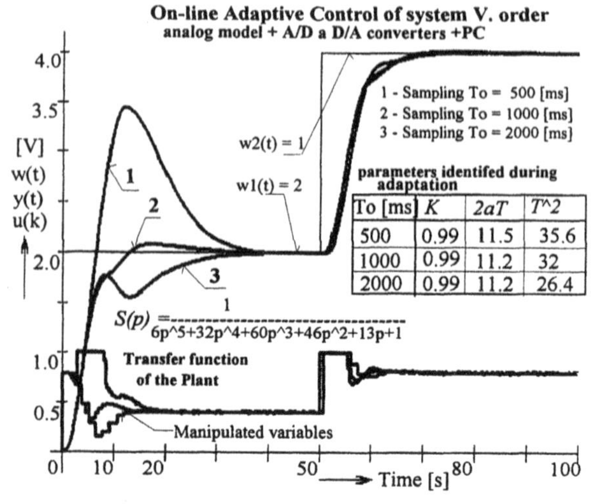

Fig.13. Adap. control of the system of the fifth order.

Simulation experiments clearly emphasize the advantage of hybrid simulation which enables the process of identification of the plant parameters and the adaptation of the controller parameters in a more sophisticated way compared with only a discrete model of the plant and controller.

CONCLUSIONS.

Based on simulation results the described analytical adaptive PID algorithm has a good dynamics performance which was verified in the wide scale of the controlled systems. It is supposed to solve the adaptive control problems also in the future as combination of individual methods and individual control algorithm and the hybrid model built including the simulation experiments described create a good base for the further evaluation of the analytical adaptive algorithm.

REFERENCES

Alexík M. (1994): Adaptive control of systems and process. *Habilitation work* (in Slovak), VŠDS Žilina, dec. 1994

Aström, K. J. and B. Wittenmark:(1989). *Adaptive Control.* New York, Addison-Wesley.

Bobál V.(1994): Robust Self-Tuning PID Controller. *IFAC Workshop: New Trends in Design of Control Systems.* Sept. 1994, Smolenice, pp. 312- 317.

Kozák, Š. (1994): Non-linear Selftuning Controller Based upon Laguerre Series Representation. *IFAC Workshop: New Trends in Design of Control Systems.* Sept. 1994, pp. 432-438.

Vindinčev P. (1984): Analytical synthesis of one class of control systems third order (in Czech). *Automatizace* **7/84**, pp. 111-115.

Vitečková M. (1995).: New Approach to Controller Tuning. *International Conference "micro CAD'95"* Miskolc, Feb. 23, 1995, pp. 18 - 21.

ADAPTIVE PREDICTIVE CONTROL OF DISTRIBUTED PARAMETER SYSTEMS

Szuda, J., Antoniová, M., Rohal'-Il'kiv, B., Hulkó, G.

Department. of Automatic Control and Measurement, Faculty of Mechanical Engineering, Slovak Technical University Nám. Slobody 17, 812 31 Bratislava, Slovak Republic

Abstract: In the paper a predictive control scheme of distributed parameter systems is outlined. The scheme of predictive control is based on spatial and time decomposition of system dynamics. Distributed parameter system is considered as a system with lumped input and distributed output. GPC and LQG synthesis are designed for this system in the paper. Result algorithms are verified on physical model of distributed parameter system.

Keywords: Distributed parameter systems, adaptive control, LQG synthesis, GPC.

1. INTRODUCTION

A large attention today is devoted to the problems of predictive control of technical systems. Many theories and many algorithms were developed, mainly for SISO and MIMO systems. In this paper we will show some possibilities how to use these results in distributed parameter systems (DPS) area. Our approach to the predictive control is based on the fact, that the real distributed parameter systems are mostly found as lumped input and distributed output systems (LDS) in engineering practice (M. Kárny *et all.*,1985).

In the first part of the paper we will introduce the models of LDS. The controlled system response in each step will be decomposed in the time and space parts. Control synthesis in space domain will be realized by approximation method. Control synthesis in time domain will be realized using LQG synthesis, respectively GPC for SISO systems.

2. LUMPED INPUT AND DISTRIBUTED OUTPUT MODELING

Let's suppose, we have LDS with first order holder H: HLDS. Let this system is distributed on interval <0, L>. We can consider measuring points set-up in interval points $\mathbf{x} = [x_1,\dots x_i,\dots x_j,\dots x_m]$. On the figure 1.we illustrate measuring points set-up and partial response of distributed parameter system to i-th input signal for the time instant k.

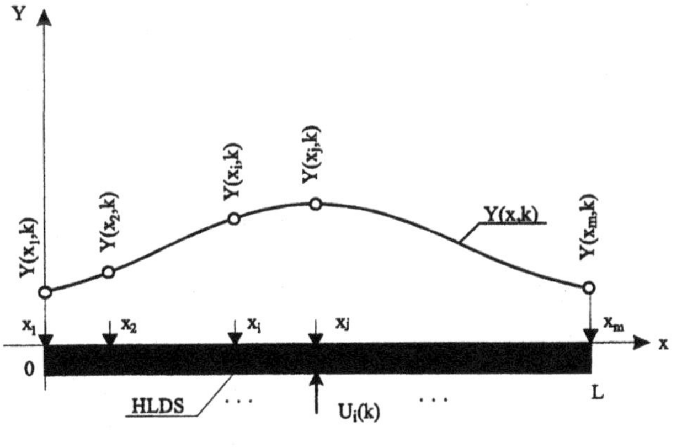

Fig. 1. Measuring points set-up and system response at time k
HLDS - lumped input distributed output system with first order holder
$U_i(k)$ - i-th lumped input quantity
$Y(x,k)$ - distributed output quantity at time k
$Y(x_j,k)$ - value of distributed output quantity on measurement point x_j at time k

Let us define regression models between sequences $\mathbf{U}(k)=[U_1(k),\dots,U_n(k)]^T$, $\mathbf{Y}(\mathbf{x},k)=[Y(x_1,k),\dots,Y(x_m,k)]^T$:

$$Y(x,k)=\sum_{i=1}^{na}A_i Y(x,k-i)+\sum_{i=0}^{nb}B_i U(k-i)+ \\ +\sum_{i=1}^{nc}C_i e(x,k-i)+K+e(x,k) \quad (1)$$

where na, nb, nc are orders of the model and **K** is a vector of absolute elements of the model, **e**(**x**, k), is non-correlated random sequence with zero mean.

Further, we denote a set of unknown recursively estimated parameters as $\hat{\mathbf{P}}$: $\hat{\mathbf{P}}=[\mathbf{A}_1,\mathbf{A}_2,\ldots,\mathbf{A}_{na},\mathbf{B}_0,\ldots,\mathbf{B}_{nb},\mathbf{C}_1,\ldots\mathbf{C}_{nc},\mathbf{K}]$
Then we can express distributed discrete output quantities at the time point k+r as:

$$\mathbf{Y}(\mathbf{x},k+r)=\hat{\mathbf{Y}}(\mathbf{x},k+r)\big|_k+\mathbf{e}(\mathbf{x},k+r) \quad (2)$$

where $\hat{\mathbf{Y}}(\mathbf{x},k+r)$ is predicted conditional mean value of discrete distributed output quantities at the time instant k+r, where r is a natural number, **e**(**x**,k+r) is non-correlated random sequence with zero mean.

Base on these vectors we can approximate the continuous distributed variable. A lot of interpolation methods can be used for this purpose. Further, we will not distinguish between approximated variables and aproximants. Because the output variable can be measured in arbitrary points we can generally write (for example as in case of numerical solving of partial differential equation or finite element method) the course of distributed output quantity as a continuous function of x:

$$Y(x,k+r)=\hat{Y}(x,k+r)\big|_k+e(x,k+r) \quad (3)$$

3. DECOMPOSITION OF SYSTEM DYNAMICS

For control synthesis purposes we will introduce system dynamics decomposition into space and time components. Space components will be used at the synthesis in space domain, the time parts will be utilized at the synthesis in time domain.

Further, we will consider at the time k all input, output quantities in deviation from given reference state due to linearization.

Reference state at the time instant k is defined with sequences of input variables **U**(k) and sequences of distributed output variables **Y**(**x**, k-1). At time after k we consider input reference variable values to be equal to the values of input variables at the time k, **U**(k+r) = **U**(k), where r is a natural number. This is the state, in which system would operate if the input signal will be not changed..

Let $\hat{\mathbf{Y}}\mathbf{P}(\mathbf{x},k+r)$ denotes discrete output variable, which is predicted to the time instant k+r as a response to the vector **Y**(**x**, k-1) and to the actuating variables sequence **U**(k+r) = **U**(k). This vector is predicted reference state. In the continuous case $\hat{Y}P(x,k+r)$.

Let us consider the appropriate test signal at the given reference state:

$$\mathbf{UT}_i(k+r) = [U_1(k+r),\ U_2(k+r),\ \ldots,\ U_i(k+r)+ \\ +\ UT_i(k+r),\ \ldots,\ U_n(k+r)]^T \quad (4)$$

The HLDS response is modeled by the regression model (1):

$$\mathbf{YT}(\mathbf{x},k+r)=\hat{\mathbf{Y}}\mathbf{T}(\mathbf{x},k+r)+\mathbf{e}(\mathbf{x},k+r)$$

The deviation of $\mathbf{YT}_i$(**x**, k+r) and $\hat{\mathbf{Y}}\mathbf{P}(\mathbf{x},k+r)$ is:

$$\Delta\hat{\mathbf{Y}}(\mathbf{x},k+r)=\hat{\mathbf{Y}}\mathbf{T}(\mathbf{x},k+r)-\hat{\mathbf{Y}}\mathbf{P}(\mathbf{x},k+r) \quad (5)$$

Also, the regression model between sequences $\Delta Y_i(x_j$, k+r) and UT_i(k+r) is valid:

$$\Delta Y_i(x_j,k+r)=\sum_{m=1}^{na}a^i_m\Delta Y_i(x_j,k+r-m)+ \\ +\sum_{m=1}^{nb}b^i_m UT_i(k+r-m)+\sum_{m=1}^{nc}c^i_m e(k+r-m)+e(k+r)= \\ =\mathbf{a}^i\Delta Y_i(x_j,\overline{k+r})+\mathbf{b}^i UT_i(\overline{k+r})+\mathbf{c}^i e(\overline{k+r})+e(k+r) \quad (6)$$

where $\mathbf{a}^i$, $\mathbf{b}^i$ and $\mathbf{c}^i$ are the estimated parameters $\hat{\mathbf{P}}_i$.

If the test signal UT_i(k+r) is unite step function the deviation (5) gives the conditional mean value of distributed discrete step response $\mathbf{H}_i$(**x**,k+r) at the time instant k+r.

If we apply this procedure to all input variables, we will obtain parameter estimates $\{\hat{\mathbf{P}}_i\} i=1,n$ for all inputs and matrix of conditional mean values of distributed step responses:

$$\mathbf{H}(\mathbf{x},k+r) = [\mathbf{H}_1(\mathbf{x},k+r),\ \ldots,\ \mathbf{H}_n(\mathbf{x},k+r)]. \quad (7)$$

Due to (3) we can write the distributed step responses as a continuous function too:

$\{H_1(x,k+r), \ldots, H_n(x,k+r)\}$.(8)

Let us introduce a reduced distributed step responses by:

$$HR_i(x,k+r) = \frac{H_i(x,k+r)}{H_i(x_j,k+r)} \qquad (9)$$

for i = 1,n

These reduced step responses we use on the system dynamics decomposition in space domain.

For unit step function on all inputs by reference state it holds:

$$\hat{Y}(x,k+r)\big|_k = \hat{Y}P(x,k+r)\big|_k + \sum_{i=1}^{n} H_i(x,k+r)\big|_k \qquad (10)$$

We can rewrite this expression as:

$$\hat{Y}(x,k+r)\big|_k = \hat{Y}P(x,k+r)\big|_k + \\ + \sum_{i=1}^{n} \Delta\hat{Y}_i(x_j,k+r)\big|_k HR_i(x,k+r)\big|_k \qquad (11)$$

where $\{\Delta\hat{Y}_i(x_j,k+r)\}i = 1,n$ are conditional mean values predicted by (6) at the unite step functions.

4. SPACE SYNTHESIS

In the block space synthesis (SS) will be entered the reduced step responses $\{HR_i(x,k+r)\}i = 1,n$ and control variable W(x,k+r), fig. 2. Let us introduce the following control criterion and analyze at first generally its minimization possibilities:

$$E\|W(x,k+r) - Y(x,k+r)\| \qquad (12)$$

where E(.) is expected value and ||.|| is a quadratic norm.

At the step input quantities next relations hold:

$$\min E\|W(x,k+r) - Y(x,k+r)\| =$$
$$= \min E\left\|W(x,k+r) - \hat{Y}(x,k+r)\big|_k + e(x,k+r)\right\| \le$$
$$\le \min E\left\|W(x,k+r) - \hat{Y}(x,k+r)\big|_k\right\| + E\|e(x,k+r)\| =$$
$$= \min E\Big\|W(x,k+r) - \hat{Y}P(x,k+r)\big|_k -$$
$$- \sum_{i=1}^{n} \Delta\hat{Y}_i(x_j,k+r)\big|_k HR_i(x,k+r)\big|_k\Big\| + E\|e(x,k+r)\| =$$
$$= \min E\Big\|WP(x,k+r)\big|_k -$$
$$- \sum_{i=1}^{n} \Delta\hat{Y}_i(x_j,k+r)\big|_k HR_i(x,k+r)\big|_k\Big\| + E\|e(x,k+r)\| \qquad (13)$$

where

$$WP(x,k+r)\big|_k = W(x,k+r) - \hat{Y}P(x,k+r)\big|_k$$

Then we can solve at the time k only optimization problem

$$\min\left\|WP(x,k+r)\big|_k - \sum_{i=1}^{n} \Delta\hat{Y}_i(x_j,k+r)\big|_k HR_i(x,k+r)\big|_k\right\| =$$
$$= \left\|WP(x,k+r)\big|_k - \sum_{i=1}^{n} \breve{W}S_i(k+r)\big|_k HR_i(x,k+r)\big|_k\right\| \qquad (14)$$

because E||e(x, k+r)|| is constrained.

Where $\{\breve{W}S_i(k+r)\}i = 1,n$ is the optimal solution of the approximation problem. In block SS is solved optimization problem (14). The result of SS is a sequence of values of $\breve{W}S_i(k+r)i = 1,n$, control values entering to single parameter control loops in block TS, fig. 2.

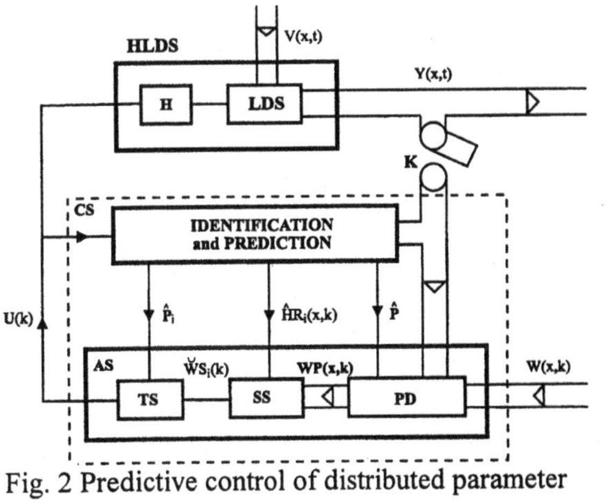

Fig. 2 Predictive control of distributed parameter system
HLDS - controlled lumped input distributed output system (LDS) with first order holders (H)
K - sampling
SS - space synthesis
TS - time synthesis
AS - synthesis of actuating quantities

PD - prediction, signal subtracting
CS - control system
$\mathbf{P}_i, \hat{\mathbf{P}}$ - estimated parameters
$\hat{HR}_i(x,k)$ - i-th reduced conditional distributed step response
Y(x,t) - controlled quantity
W(x,k) - control quantity
V(x,t) - disturbance
$\breve{WS}_i(k)$ - i-th approximation problem solution
U(k) - actuating quantities

5. TIME CONTROL SYNTHESIS

The block time synthesis (TS) consists of n control loops. Each control loop belong to one lumped input. The desired values from the block SS will be entered into control loops. From these loops we can obtain the actuating quantities to HLDS. Procedure of time control synthesis we can differentiate for LQG control and for GPC.

LQG synthesis

The i-th control loop is designed to minimize the quadratic criterion:

$$\lim_{N\to\infty} E\left[\frac{1}{N}\sum_{k}^{N} e_y^T(k)^i Q_y^i e_y(k)^i + e_u^T(k)^i Q_u^i e_u(k)^i\right] \tag{15}$$

where expected value E(.) is taken of an average loss per control step, weighting by positive semidefinite matrix Q_y, Q_u, the output error e_y and the input control effort e_u.

$$\begin{aligned} e_y(k)^i &= Y_i(x_j,k) - \breve{WS}_i(k) \\ e_u(k)^i &= U_i(k) - UR_i(k) \end{aligned} \tag{16}$$

where $UR_i(k)$ is the reference value (J. Richalet,1993)

GPC synthesis

We can write the model obtained by predictor (6) for i-th input in CARIMA form (D. W. Clarke, C. Mohtadi, P. S. Tuffs,1987):

$$A_i(q^{-1})Y_i(x_j,k) = B_i(q^{-1})U_i(k-1) + C_i(q^{-1})e(k)/\Delta \tag{17}$$

where A_i, B_i and C_i are polynomials in the backward shift operator q^{-1}:

$$A_i(q^{-1}) = 1 + a_1^i q^{-1} + \ldots + a_{na}^i q^{-na}$$
$$B_i(q^{-1}) = b_0^i + b_1^i q^{-1} + \ldots + b_{nb}^i q^{-nb}$$
$$C_i(q^{-1}) = 1 + c_1^i q^{-1} + \ldots + c_{nc}^i q^{-nc} \tag{18}$$

and e(k) is the zero mean white noise and Δ is the differencig operator $\Delta = 1-q^{-1}$.

The GPC algorithm consists of applying a control sequence that minimize a multistage cost function of the form

$$J = E\left\{\sum_{r=N_1}^{N_2} \delta(r)\left[Y_i(x_j,k+r)\Big|_k - \breve{WS}_i(k+r)\Big|_k\right]^2 + \sum_{r=1}^{N_2} \lambda(r)\left[\Delta U_i(k+1-r)\Big|_k\right]^2\right\} \tag{19}$$

N_1 and N_2 are the minimum and maximum costing horizons, $\delta(r)$, $\lambda(r)$ are weighting sequences, $\delta(r)$ is considered to be 1 and $\lambda(r)$ is considered to be constant, $\breve{WS}_i(k+r)\big|_k$ is the future reference trajectory, which can be calculated by (14).

To solve the GPC problem the set of control signals $U_i(k)$, $U_i(k+1)$, ...$U_i(k+N)$ has to be obtained in order to optimize the equation (19). The values N_1 and N_2 defining the horizon can be defined $N_1 = 1$, $N_2 = N$ (if the considered system has dead time zero).

Now consider the following set of r ahead optimal predictions:

$$\hat{\mathbf{Y}}_i = \mathbf{G}_i\tilde{\mathbf{U}}_i + \mathbf{f}_i \tag{20}$$

where the vectors are all $N\times 1$:

$$\hat{\mathbf{Y}}_i = \left[\hat{Y}_i(x_j,k+1), \hat{Y}_i(x_j,k+2), \ldots, \hat{Y}_i(x_j,k+N)\right]^T$$
$$\tilde{\mathbf{U}}_i = \left[\Delta U_i(k+1), \Delta U_i(k+2), \ldots, \Delta U_i(k+N)\right]^T$$
$$\mathbf{f}_i = \left[f_i(k+1), f_i(k+2), \ldots, f_i(k+N)\right]^T$$

The matrix **G** is lower-triangular of dimension $N\times N$:

$$\mathbf{G}_i = \begin{bmatrix} g_0^i & 0 & \cdots & 0 \\ g_1^i & g_0^i & \cdots & 0 \\ \vdots & & & \vdots \\ g_{N-1}^i & g_{N-2}^i & \cdots & g_0^i \end{bmatrix}$$

The vector $\mathbf{f}_i$ and the matrix $\mathbf{G}_i$ we obtain by method described in D. W. Clarke, C. Mohtadi, P. S. Tuffs,1987.

Equation (19) can be written as:

$$J = \left\{ \left(\mathbf{G}_i \widetilde{\mathbf{U}}_i + \mathbf{f}_i - \mathbf{ws}_i\right)^T \left(\mathbf{G}_i \widetilde{\mathbf{U}}_i + \mathbf{f}_i - \mathbf{ws}_i\right) + \lambda \widetilde{\mathbf{U}}_i^T \widetilde{\mathbf{U}}_i \right\} \tag{21}$$

where

$$\mathbf{ws}_i = \left[\breve{W}S_i(k+1), \breve{W}S_i(k+2), \ldots, \breve{W}S_i(k+N)\right]^T$$

The cost function (21) we can write as:

$$J = \frac{1}{2} \widetilde{\mathbf{U}}_i^T \mathbf{N}_i \widetilde{\mathbf{U}}_i + \mathbf{b}_i^T \widetilde{\mathbf{U}}_i + \mathbf{f}_0^i \tag{22}$$

where

$$\mathbf{N}_i = 2\left(\mathbf{G}_i^T \mathbf{G}_i + \lambda \mathbf{I}\right)$$
$$\mathbf{b}_i^T = 2\left(\mathbf{f}_i - \mathbf{ws}_i\right)^T \mathbf{G}_i$$
$$\mathbf{f}_0^i = \left(\mathbf{f}_i - \mathbf{ws}_i\right)^T \left(\mathbf{f}_i - \mathbf{ws}_i\right)$$

The equation (22) is solved by quadratic programming. Notice that only the first element of $\widetilde{\mathbf{U}}_i$ is applied and the procedure is repeated at next sampling time.

6. DEMONSTRATION

The performance of adaptive predictive control designed was verified successfully by the control of physical model with distributed parameter. Used physical model is a model of distributed parameter continual thermal process, fig. 3.

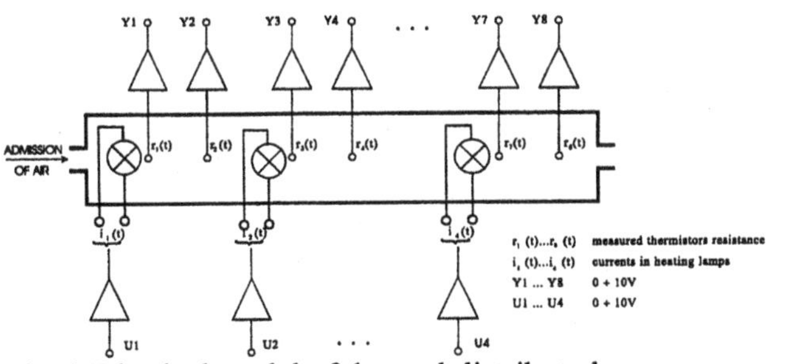

Fig. 3. Physical model of thermal distributed parameter system

The physical basis of the apparatus is the flow of air trough a long tube. The flowing air is heated by four electrical heaters (lumped inputs) which are distributed in the length of the tube, fig. 3. The active length L of the apparatus is L = 600 mm. The temperature of the hot air is measured by eight bead thermistors. All elements of the model are constructed as simple exchangeable modules which can be placed anywhere in the direction of the apparatus length. The system inputs consists of heating inputs (electrical heaters with maximum power 25 W, the manipulated variable is voltage $U(t) \in [0\ V \div 10\ V]$) and cooling disturbing inputs (electrical fans). The system output is a spatial temperature distribution $Y(x,t)$, $x \in <0, L>$, $t \geq 0$, in the tube, which is measured by thermistors. For data conversion we use the A/D and D/A serial data converter CTRL, which has four inputs and eight outputs. Its minimal sampling time is 20 ms and it works in scale range 0-10 V. The converter is the product of UTIA Praha.

We considered the operating state whit input signal 5 V on all inputs. We obtained the step responses over the lumped inputs (fig. 4., 5.), the steady states and the reduced steady states (fig. 6.). The system dynamics is evident from this step responses and we can joint it on the fast and slow parts. The fast part of dynamics is given by heating of the air in the tube and the slow part is given by the heating of apparatus body.

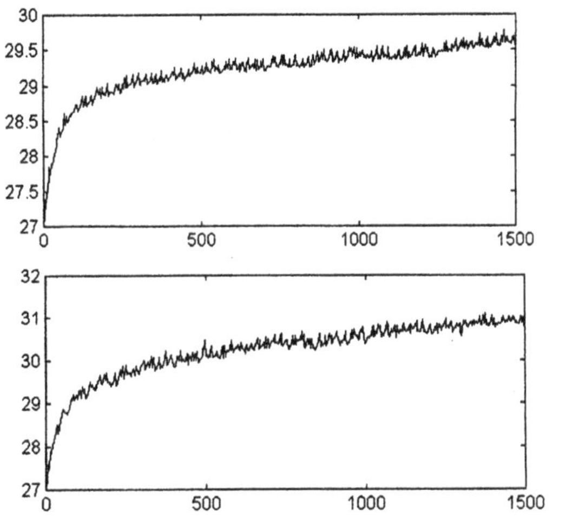

Fig. 4. The partial system step responses over the inputs U_1 and U_2

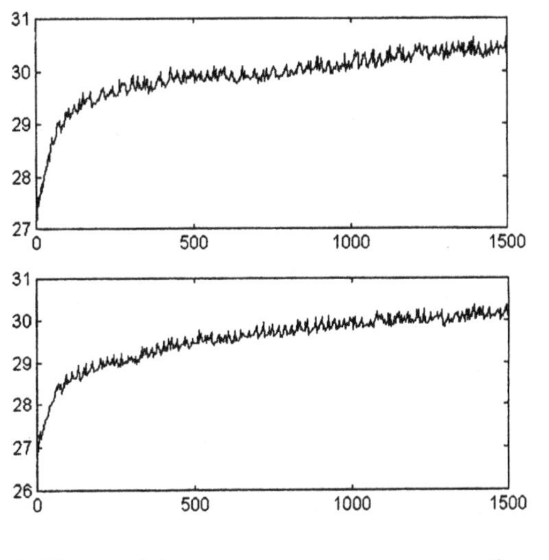

Fig. 5. The partial system step responses over the inputs U_3 and U_4

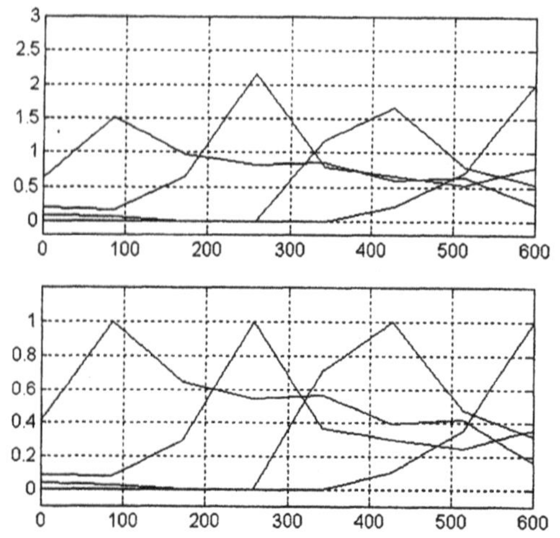

Fig. 6. The steady states of system and their reduction

The control algorithms for experiment was made in MATLAB software environment. For the identification of the system was used the recursive algorithm least-square method with exponential forgetting. We obtained the regression models by program algorithms md01 and md32, products ÚTIA Prague. We considered the second order regression models. The sampling time was T = 2,7 s. The space synthesis is solved by MATLAB functions from Optimization toolbox. LQG algorithm for time synthesis is described in program module rb77, product UTIA Prague. The GPC algorithm for single parameter loops was prepared at our department. The control horizon in GPC was considered N = 3.

The distributed control quantity in our experiments is on the fig. 7.

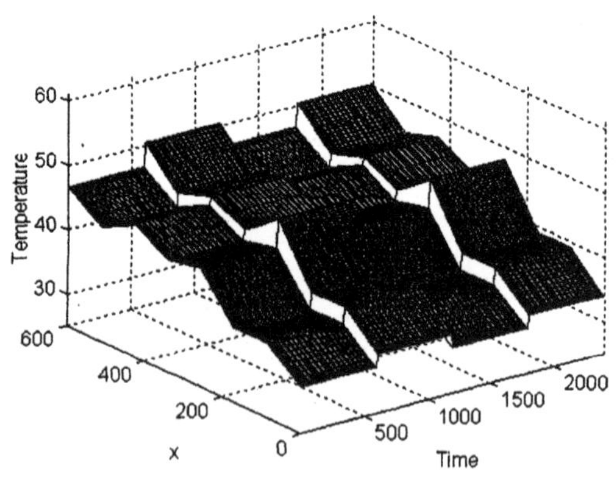

Fig. 7. The distributed control quantity

The results of the experiments are on pictures 8. -15. The measured data and computed control data with LQG algorithm are on the figures 8. - 11. and data with GPC algorithm are on the figures 12. - 15.

The trajectories of desired values for single parameter control loops are on the pictures 9. and 11. These values are the result of the space synthesis. Actuating quantities, the results of the time synthesis, is on the figure 10. for the LQG algorithm and on the figure 14. for the GPC algorithm. The control error is calculated by relation $e(k) = (\mathbf{W}(\mathbf{x},k) - \mathbf{Y}(\mathbf{x},k))^T (\mathbf{W}(\mathbf{x},k) - \mathbf{Y}(\mathbf{x},k))$. Its trajectory for LQG synthesis is on the figure 11. and control error trajectory for GPC is on the figure 15.

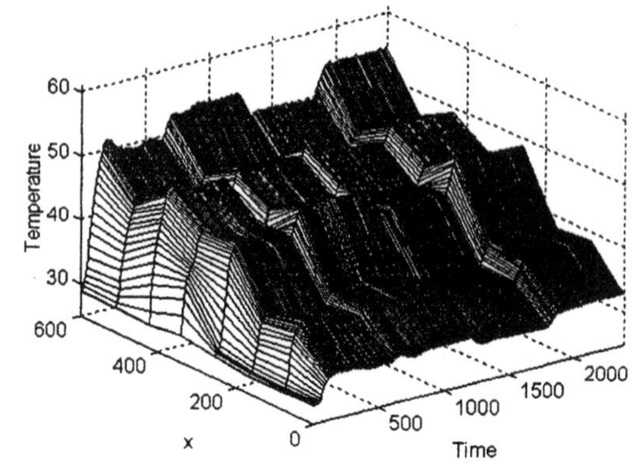

Fig 8. Controlled quantity at LQG algorithm in single parameter control loops

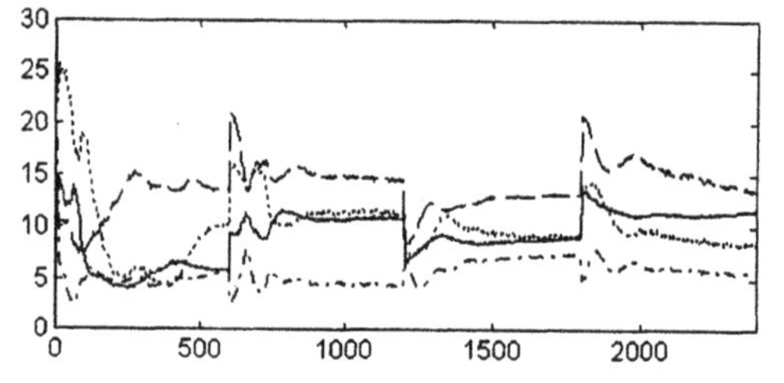

Fig. 9. Input quantities to single parameter control loops
—- $\breve{WS}_1$, --- - $\breve{WS}_2$, -.-- $\breve{WS}_3$, ... - $\breve{WS}_4$

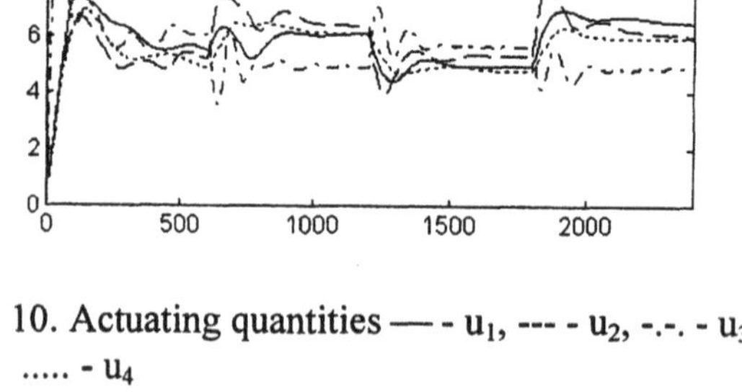

Fig. 10. Actuating quantities —- u_1, --- - u_2, -.-. - u_3, ... - u_4

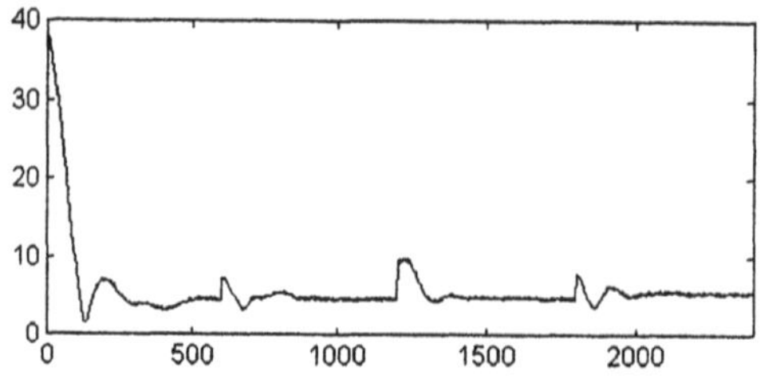

Fig. 11. Quadratic norm of control error

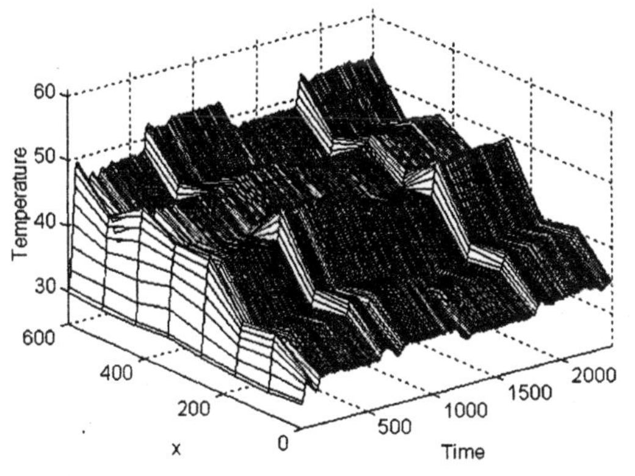

Fig 12. Controlled quantity at GPC algorithm in single parameter control loops

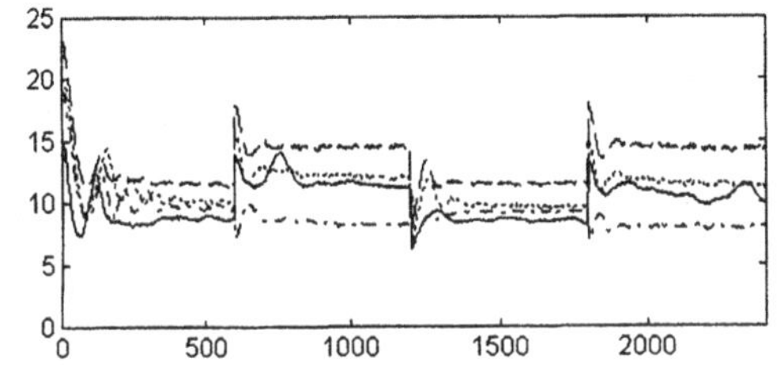

Fig. 13. Input quantities to single parameter control loops

—- $\breve{W}S$, --- - $\breve{W}S_2$, -.-- $\breve{W}S_3$, ... - $\breve{W}S_4$

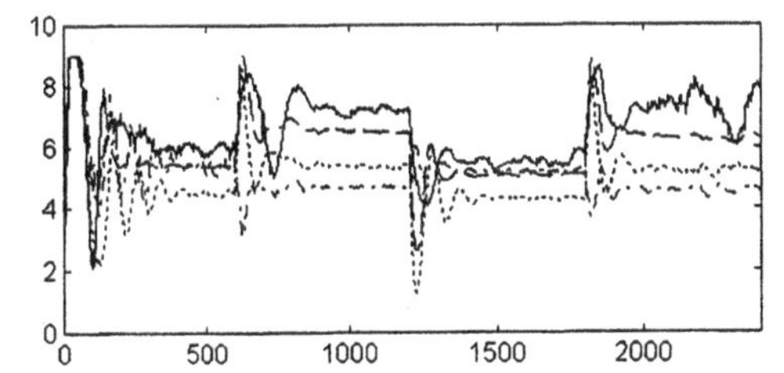

Fig. 14. Actuating quantities

—- u_1, --- - u_2, -.-. - u_3, ... - u_4

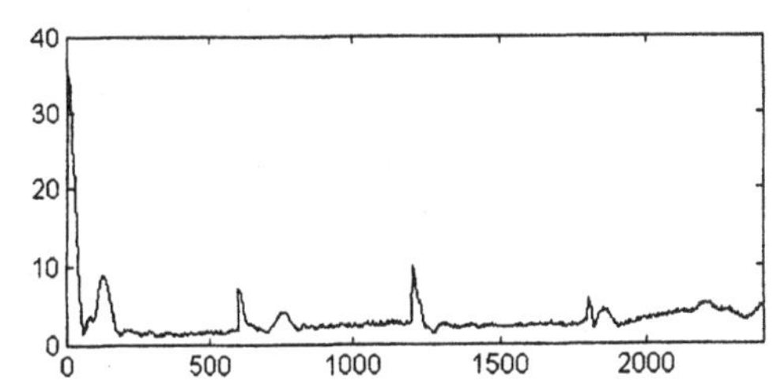

Fig. 15. Quadratic norm of control error

7. CONCLUSION

In the paper was outlined the predictive control for distributed parameter systems. In the first part we introduced the model of LDS by regression model. Then we made the decomposition of system dynamics into space and time components. The space synthesis was solved as approximation problem of given control criterion. The time control synthesis for n control loops was differentiated for LQG control and GPC. The algorithms was successfully verified by physical model of continual thermal process.

ACKNOWLEDGMENT

This work has been carried out under the financial support of the VEGA project „Development of automatic control methods“ (grant 95/5196/198) and the EC granted COPERNICUS project „Adaptive and predictive control with physical constraints“ (grant CP941174).

REFERENCES

E. F. Camacho, C. Bordons (1995). *Model Predictive Control in the Crocess Industry*, Springer-Verlag.

D. W. Clarke, C. Mohtadi, P. S. Tuffs (1987). *Generalized Predictive Control*, Automatica, **Vol. 23, No. 2.**

D. W. Clarke (1988). *Application of Generalized Predictive Control to Industrial Processes*, 122:49-55, IEEE Control Systems Magazine.

M. Kárny *et all.* (1985). *Design of Linear Quadratic Adaptive Control: Theory and Algorithms for Practice*, Kybernetika, **Vol. 21**, Academia Praha.

J. Richalet (1993). *Industrial Applications of Model Base Predictive Control*, Automatica, **Vol. 23, No. 5**,.

Rohal-Ilkiv, B. (1995). *Identification of Distributed Parameter Systems*. Preprints of the summer school course on identification and optimization, Prague.

Rohal-Ilkiv, B., Zelinka, P., Richter, R. (1994). *Experimental Systems for Distributed Process Control Education*, The 3rd symposium on advances in education, Tokyo.

R. Söeterboerk (1992). *Predictive Control. A Unified Approach.* Prentice-Hall.

Hulko, G., Belansky, J., Belavy, C., Kovalcik, J., Antoniova, M., Szuda, J., Vegh, P. (1995). *Self-Tuning Control of Lumped Input and Distributed Output Systems.* Preprints of 5th IFAC Symposium on Adaptive Systems in Control and Signal Processing, Budapest.

Hulko, G., Belansky, J., Belavy, C., Kovalcik, J., Antoniova, M., Szuda, J., Vegh, P. (1995). *Identification and Modelling of Lumped Input and Distributed Output Systems of Accuracy Requested.* Preprint of European Control Conference ECC 95, Roma.

Hulko, G., Belansky, J., Belavy, C., Kovalcik, J., Antoniova, M., Szuda, J., Vegh, P. (1995). *Control of Thermal Processes as Lumped Input and Distributed Output Systems.* Preprints of THERMO International Symposium, Budapest.

COMPARISON OF TRACKING ABILITIES OF SELECTED FORGETTING METHODS

Holička V. – Benčurik B. – Rohal̆-Ilkiv B. *,[1]

* *Department of Automatic Control and Measurement*
Faculty of Mechanical Engineering, Slovak University of Technology
Nám. Slobody č. 17, 812 31 Bratislava, Slovak Republic

Abstract:
This paper presents a comparison of some selected regularized identification methods whose time updates are based on the weighted sum of matrices to the standard Directional Forgetting. Procedure of Dyadic reduction is used as a basic tool to perform this sum on the L'DL factors of the covariance matrices. Their abilities to track time varying parameters are compared under different types of excitation.

Keywords: Robust identification, Alternative covariance matrix, Regularized forgetting methods

1. INTRODUCTION

In connection with adaptive control it is often unrealistic to assume that the parameters of the process to be controlled are constant. The dynamics of the process may vary in time, perhaps due to the age or deformation. Frequently the model may be a local approximation, valid only near a particular operating point. If this operating point drifts in time, then the same does the parameter vector. *The standard recursive least squares method* for parameter estimation is not directly applicable in such case. Due to the persistent flow of information, the algorithm gain tends to zero, making it impossible to track time variations. To cope with this situation we can use various techniques of forgetting of obsolete information to make parameter estimation capable of tracking slow parameter changes. *Forgetting* is essentially a *heuristic technique* owing to the fact that *no or just vague model* of parameter variations is typically considered in design of adaptive systems. That makes a good sense. The assumption of a precise model of nonstationarities would go against the philosophy of adaptive systems which prefers rather simple models to be adopted to changing working conditions. In the literature a large number of methods for estimations of timevarying parameters have been proposed. The first attempt to tracking of timevarying parameters known as *Exponential Forgetting, (EF)* was motivated by time weighted (discounted) least squares in Jazwinsky (1970) or more generally, by the idea of flattening of likelihood or posterior density function in (Peterka, 1981). A number of ways to defeat the well-known *blow up* problem have been proposed with *variable forgetting factors* as in (Fortescue *et al.*, 1981; Kulhavý, 1987). (Kulhavý and Kárny, 1984) suggested an algorithm forgetting only on the part of parameter space for which the new information is restricted called *Directional Forgetting, (DF)*. In addition to data forgetting, one can modify the *covariance matrix* obtained from the ordinary LS estimator so as to retain alertness. Typical example includes *LS with constant trace* in (Lozano-Leal and Goodwin, 1985). Another different ways of regularizing or stabilizing the covariance update formula have

[1] Supported by ECC Commission of the European Communities (COPERNICUS grant CP941174)

been proposed including *Regularized Exponential Forgetting, (REF)* in (Kulhavý and Zarrop, 1993) and *Stabilized Linear Forgetting, (SLF)* in (Milek and Kraus, 1991). The last two techniques were shown to be *dual each other* in (Kulhavý and Kraus, 1996).

The paper is organized as follows: In section 2 we will introduce the problem of collecting the data for recursive identification in a sufficient statistic from Bayesian point of view using well-known *Bayes Rule.* In section 3 we will choose some of identification methods previously mentioned in this section for a slight deeper analysis. In section 4 we will have a closer look on *REF* and *SLF* which were both derived by solving basically the same Bayesian decision-making problem using *Kullback-Leibler divergence* as a measure (quasi)distance between probability distribution. In section 5 we will mention some notes to algorithmical aspect of selected methods. Then in section 6, a comparison of the performance of some selected recursive identification algorithms is carried out via computer simulations. Finally section 7 concludes this paper.

2. PROBLEM STATEMENT

We shall consider a stochastic system on which observations are made at discrete time instants $k = 1, 2, \ldots$. A directly manipulated input u_k and an indirectly affected output y_k (booth possibly multivariate) can be distinguished in the data pair $d_k = (u_k, y_k)$. The collection of all data observed on the system up to time t is denoted by

$$D_t = (d_1, d_2, \ldots, d_t).$$

The dependence of a new pair of data (u_k, y_k) on previous observations D_{k-1} can be described by a conditional probability density function (p.d.f) with the following structure

$$p(y_k, u_k|D_{k-1}, \theta_k) = p(y_k|u_k, D_{k-1}, \theta_k)\, p(u_k|D_{k-1}) \qquad (1)$$

Incomplete knowledge of the system behaviour is expressed through a vector of unknown, time varying parameters $\theta_k \in \theta$. Note that *the input generator* described by the second term does not depend on these parameters directly, it is expected to utilize only prior information and information contained in observed data. The first term actually characterizes *the system.*

BAYES PARAMETRIC INFERENCE: When the unknown parameter θ is interpreted as a random variable, the uncertainty of θ given the observed data D_t is naturally described by the posterior p.d.f $p(\theta|D_t)$ conditional on D_t. This is generally determined by *Bayes' theorem.* Provided the input generator employs no other information about θ than the observed data *(see the natural conditions of control in (Peterka, 1981))*

$$p(u_k|D_{k-1}, \theta_k) = p(u_k|D_{k-1}) \qquad k = 1, \ldots, t \qquad (2)$$

Bayes' rule simplifies to the formula

$$p(\theta|D_t) \propto p(\theta) \prod_{k=1}^{t} p(y_k|u_k, D_{k-1}, \theta) \qquad (3)$$

Where $\propto$ stands for proportionality, i.e. equality up to a normalizing factor.

RLS: An important special case of the first term of equation (1) (which represents the system) arises when the output y_k depends on previous data u_k, D_{k-1} through a known finite-dimensional vector function $\phi(u_k, D_{k-1}) = \phi_k$ in the following special way

$$y_k = \theta' \phi_k + e_k, \quad e_k \sim N(0, \sigma^2) \qquad (4)$$

Here θ' denotes transposition of θ. The dependence is thus linear in the unknown parameters. The stochastic component e_k is supposed to be a discrete white noise, normally distributed with known (for simplicity) variance σ^2. The model (4) induces a conditional p.d.f. of the form

$$p(y_k|u_k, D_{k-1}, \theta) = \frac{1}{\sqrt{2\pi\sigma^2}} exp(-\frac{1}{2\sigma^2}(y_k - \theta'\phi_k)^2) \qquad (5)$$

Suppose the prior p.d.f. $p(\theta)$ is normal $N(\hat{\theta}_0, P_0)$. Then one can easily prove that the posterior p.d.f. $p(\theta|D_t)$ produced by Bayes' rule (3) is normal $N(\hat{\theta}_t, P_t)$ with modified statistics $\hat{\theta}_t$ and P_t. The mean $\hat{\theta}_k$ and covariance P_k update recursively according to the formulae

$$\hat{\theta}_k = \hat{\theta}_{k-1} + \frac{P_{k-1}\phi_k}{\sigma^2 + \phi'_k P_{k-1}\phi_k} \hat{e}_k \qquad (6)$$

$$P_k^{-1} = P_{k-1}^{-1} + \frac{\phi_k \phi'_k}{\sigma^2} \qquad (7)$$

where

$$\hat{e}_k = y_k - \hat{\theta}'_{k-1}\phi_k \qquad (8)$$

denotes prediction error. We can see that the recursive formulae (6)-(8) coincide with well-known algorithm of recursive least squares. From (7) it can be seen that if ϕ is sufficiently exciting, the elements of P converge to zero, and $\hat{\theta}_k$ eventually becomes constant. Hence, in the case of timevarying parameters, this basic method is unsuited due to the loss of tracking capability. One has to agree that the assumption of constant parameters in model (4) is not realistic in typical application and need to be released.

3. A GENERAL RECURSIVE FORGETTING ALGORITHM

In the previous section was shown that *standard RLS method* is in the case of time-varying parameters $\theta = \theta_k$ unsuitable owing to the loss of *tracking capability*. Most of methods designed for the time varying case can be shown as modification of the RLS method, typicaly introducing some kind of forgetting of old data in order to prevent the estimator gain from becoming zero. Every method can be seen as a special case of the following general forgetting algorithm.
MEASUREMENT UPDATE :

$$\hat{\theta}_{k|k} = \hat{\theta}_{k|k-1} + \frac{P_{k|k-1}\phi_k}{\sigma^2 + \phi_k' P_{k|k-1}\phi_k}\hat{e}_k \quad (9)$$

$$P_{k|k}^{-1} = P_{k|k-1}^{-1} + \frac{\phi_k \phi_k'}{\sigma^2} \quad (10)$$

TIME UPDATE :

$$\hat{\theta}_{k+1|k} = \hat{\theta}_{k|k} \quad (11)$$

$$P_{k+1|k} = F\{P_{k|k}\}, \quad (\geq P_{k|k}) \quad (12)$$

The measurement update is not affected by the time variations of the parameters. The time update (11)-(12) represents the change of parameter vector between two measurements. The time update equations can be derived analytically if we have an explicit model of parameter changes. Such a case is extremely rare in practice situations. So we are forced to use some heuristic procedure. Usually it is simply assumed that θ is approximately constant within a time horizon $N \gg 0$, and some kind of *ad hoc* forgetting is introduced via the choice of forgetting operator $F\{.\}$. This gives a raise to many possibilities, and explains the abundance of techniques that have appeared. The only difference between all currently used forgetting methods lies in the choice of $F\{.\}$, i.e. in the time update of P. The following short review of existing methods illustrates this point. The unmodified RLS method corresponds to the time update

$$P_{k+1|k} = P_{k|k} \quad (13)$$

We can easily derive the same equation by application of Kalman filter to the model

$$\theta_{k+1} = \theta_k \quad (14)$$

$$y_k = \theta_k' \phi_k \quad (15)$$

In case when little information is available about parameter changes, but we are expecting their slow variations, the model of a *random-walk* discrete process is often used.

$$\theta_k = \theta_k + w_k \quad (16)$$

where w_k is a white noise sequence, $w_k \in N(0, \sigma^2 R)$. The Kalman filter aproach then leads to the time update.

$$P_{k+1|k} = P_{k|k} + R \quad (17)$$

If the a priori knowledge does not justify the formal statistical assumptions connected with (16), the update can still be used for *ad hoc* estimation. In this case, the elements in R can be regarded, as tuning parameters. Well-known *Exponential Forgetting* method with the following choice of forgetting operator $F\{.\}$.

$$P_{k+1|k} = (P_{k|k})/\lambda \quad (18)$$

leads to the following recursive update of the covariance matrix.

$$P_{k+1|k} = (P_{k|k-1} - \frac{P_{k|k-1}\phi_k\phi_k' P_{k|k-1}}{1 + \phi_k' P_{k|k-1}\phi_k})\frac{1}{\lambda} \quad (19)$$

or in the inverse form

$$P_{k+1|k}^{-1} = \lambda(P_{k|k-1}^{-1} + \phi_k\phi_k') \quad (20)$$

The last two equations (19),(20) joins together measurement and time update which is helpful to reduce computational complexity in real time situations.
The Kalman filter method and the EF method have been widely used, and they perform well in many situations. However in cases when the data are not sufficiently informative for a long time (the sequence of regression vectors $\{\phi\}$ fails to excite part of parameter space, the corresponding eigenvalues of P will grow unbounded-blow up). Even a small excitation in a direction where the corresponding P-elements are large, may result in a dramatic change of the estimates. One possibility to defeat the blow up problem is to keep the eigenvalues of $P_{k+1|k}$ in preliminary specified interval $[\lambda_0, \lambda_1]$ as in (Praly, 1983). Then the following time update is used.

$$P_{k+1|k} = (1 - \frac{\lambda_0}{\lambda_1})P_{k|k} + \lambda_0 I \quad (21)$$

where $(P_{1|0} = \lambda_1 I)$ and I is the identity matrix of appropriate dimension, $\lambda_0 (\geq 0)$ and $\lambda_1 (y > 0)$.

Even though a non-zero value of λ_0 affects the direction of the estimates, it can help to track time varying parameters in a simple but efficient way, see (Yoon and Clarke, 1994a). Note that the specification of λ_0 and λ_1 allows a compromise between the *LS* and *projection* ($\lambda_0 = \lambda_1$) algorithms. The covariance modification (21) was employed in order to prevent $P_{k+1|k}$ from tending to zero (its eigenvalues). In a general form the matrix $P_{k+1|k}$ may be selected as

$$P_{k|k} \leq P_{k+1|k} < \infty I \qquad (22)$$

So it's bounded from above and below. In this case exponencial convergence of the above algorithms is ensured, see (Parkum *et al.*, 1992). Another simple example satisfying the condition (22) is *the constant trace algorithm* used in (Goodwin, 1986; DeSouza, 1988) in which the time update is given by

$$P_{k+1|k} = \frac{k_0}{trace[P_{k|k}]} P_{k|k} + k_1 I \qquad (23)$$

The estimator suggested in (Lozano-Leal and Goodwin, 1985) also results from (23) by setting the constant k_1 to zero.

In contrast with *EF* time update (19) in which old information is discounted uniformly in space , (Kulhavý and Kárny, 1984) suggested an algorithm forgetting only on the part of parameter space, which was affected by last measured data. In such a case the time update is given by

$$P^{-1}_{k+1|k} = P^{-1}_{k|k} - k_k \phi_k \phi'_k \qquad (24)$$

where

$$k_k = (1 - \lambda)[1 + \frac{1}{\phi'_k P_{k|k-1} \phi_k}] \qquad (25)$$

for $\phi'_k P_{k|k-1} \phi_k > 0$. From (25) can be easily seen that at each time update, information is subtracted from $P_{k|k}$ (covariance matrix after measur. update) in the direction determined by ϕ_k, i.e in the direction where new information has been received.

At the end of this section we may mention the following. The crucial point in all above mentioned forgetting methods was choosing the time update equation for covariance matrix $P_{k+1|k}$ (choosing of forgetting operator $F\{.\}$) which must be done very carefully because of its high importance in *RLS-like* forgetting algorithms. All forgetting methods which appeared in the literature have their own particular advantages and drawbacks. They perform well in many situations.

In general it can be said that by recursive update of covariance matrix we must avoid loss of tracking capability on the one hand and unbounded growth of $P_{k+1|k}$ on the other as was previously mentioned in equation (22).

Forgetting techniques from this section didn't give us opportunity sufficiently use the prior information about model in the form of covariance matrix. Prior information in this form (if exist) is disappearing step by step and after a few steps is completely forced by information from measured data. *REF* introduced by (Kulhavý and Zarrop, 1993), and *SLF* introduced by (Milek and Kraus, 1991) are able to avoid this loss of prior knowledge.

4. REF AND SLF TECHNIQUES

Suppose that no explicit model of parameter changes is known. Yet, we can quantify our prior information (and possibly information taken from data already available) by introducing an alternative *p.d.f.* $p^*(\theta_{k+1}|D_k)$. The problem is then to construct a *p.d.f.* $p(\theta_{k+1}|Dk)$ based on two hypotheses described by the posterior *p.d.f.* $p(\theta_k|D_k)$ (the case of no parameter changes) and the alternative *p.d.f.* $p^*(\theta_{k+1}|D_k)$ (the case of worst expected changes). For simplicity in this section we use the notation $p_0(\theta), p_1(\theta)$ and $p^*(\theta)$ for the posterior, alternative and resulting *p.d.f.'s*, respectively.

Kulhavy and Kraus formulated the task of choosing p^* given p_0 and p_1 as a Bayesian decision making problem. In the next we will make a short review of their solutions.

After assigning probabilities λ and $1 - \lambda$ to the hypotheses p_0 and p_1 it is suitable to define the best decision p^* as the *p.d.f.* minimizing expectation of a distance or divergence $d(p, p_i)$ between p and $p_i, i = 0, 1$.

$$min[\lambda d(p, p_0) + (1 - \lambda) d(p, p_1)] \qquad (26)$$

As a distance can be taken *Kullback-Leibler divergence* (also known as relative entropy).

$$D(p||q) = \int p(\theta) log \frac{p(\theta)}{q\theta} d\theta \qquad (27)$$

The general solution (26) in this case take on two form which were shown to be dual to each other.

PROBLEM EF:

$$min[\lambda D(p||p_0) + (1 - \lambda) D(p||p_1)] \qquad (28)$$

PROBLEM LF:

$$min[\lambda D(p_0||p) + (1 - \lambda) d(p_1||p)] \qquad (29)$$

It was shown that if the *p.d.f's* p_0 and p_1 aren't mutually orthogonal, i.e. their product wasn't zero almost everywhere then the optimization problems (28),(29) have a unique solutions

$$p^*(\theta) \propto p_0(\theta)^\lambda p_1(\theta)^{1-\lambda} \qquad (30)$$

$$p^*(\theta) = \lambda p_0(\theta) + (1-\lambda)p_1(\theta) \qquad (31)$$

for problems EF, LF respectively. The general solutions (30) and (31) are functional expressions that are difficult to implement if nothing more is known about the *p.d.f.'s* p_0 and p_1. If p_0, p_1 belongs to the normal family N composed of *p.d.f.'s* of the following form

$$p_{\hat{\theta},P} = \frac{1}{\sqrt{2\Pi}}|P|^{-\frac{1}{2}} \\ exp(-\frac{1}{2}(\theta-\hat{\theta})'P^{-1}(\theta-\hat{\theta})) \qquad (32)$$

where $\hat{\theta}$ and P denote the mean and covariance of a particular *p.d.f.* then the following solutions were shown.
EF:

$$P^{*-1}\hat{\theta}^* = \lambda P_0^{-1}\hat{\theta}_0 + (1-\lambda)P_1^{-1}\hat{\theta}_1 \qquad (33)$$

$$P^{*-1} = \lambda P_0^{-1} + (1-\lambda)P_1^{-1} \qquad (34)$$

LF:

$$\hat{\theta}^* = \lambda\hat{\theta}_0 + (1-\lambda)\hat{\theta}_1 \qquad (35)$$

$$P^* = \lambda P_0 + (1-\lambda)P_1 + \\ \lambda(1-\lambda)(\hat{\theta}_0 - \hat{\theta}_1)(\hat{\theta}_0 - \hat{\theta}_1)' \qquad (36)$$

By considering $\hat{\theta_1} = \hat{\theta_0}$ the formulae (33),(34) simplify to

$$\hat{\theta}^* = \hat{\theta}_0 \qquad (37)$$

$$P^{*-1} = \lambda P_0^{-1} + (1-\lambda)P_1^{-1} \qquad (38)$$

and the formulae (35),(36) simplify to

$$\hat{\theta}^* = \hat{\theta}_0 \qquad (39)$$

$$P^* = \lambda P_0 + (1-\lambda)P_1 \qquad (40)$$

Let's consider the model of system with time varying parameters θ_k in the form (4).
In order to be able to track parameter variations we complement the *standard RLS algorithm* (9)-(10) by exponential or linear forgetting according to (37)-(38) or (39)-(40) respectively. In addition the alternative mean is set equal to posterior mean $\hat{\theta}^{alt}_{k+1|k} = \hat{\theta}_{k|k}$ and for simplicity the alternative covariance is set equal to the prior covariance $P^{alt}_{k+1|k} = P_{1|0} = Q$.
With this choice we can use general forgetting algorithm with the following choice of forgetting operator

$$F\{P_{k|k}, Q\} = [\lambda P_{k|k}^{-1} + (1-\lambda)Q^{-1}]^{-1} \quad (41)$$

which construct *harmonic mean* for *REF* and

$$F\{P_{k|k}, Q\} = \lambda P_{k|k} + (1-\lambda)Q \qquad (42)$$

which construct *arithmetic mean* for *SLF*.
In both cases the prior covariance matrix Q isn't forgotten and is repetively taking into account in every step k.

5. NOTES TO NUMERICAL ASPECTS

Measurement update of the general forgetting algorithm is given by (9)-(10). It can be seen that it requires explicitly inversion of a full rank covariance matrix P (indexes are omitted for simplicity). Using the well-known *matrix inversion lemma* we can rewrite it to the following

$$P_{k|k} = (P_{k|k-1} - \frac{P_{k|k-1}\phi_k\phi_k' P_{k|k-1}}{\sigma^2 + \phi_k' P_{k|k-1}\phi_k}) \quad (43)$$

In this alternative of the recursive updating of the covariance matrix the inversion is performed implicitly. Moreover covariance matrix P must be positive definite to give the correct sense due to its nature. Let's remind the fact that in the case of ill-excited system or redundant parameters of the model it may lose this very important property. Reduced *word length* of the digital computers which are used to perform this evaluation of the matrix makes the situation more difficult. Simple way to avoid troubles is to make recursive updating of the covariance matrix in the factorized forms, which ensures its positive definiteness. In this case we may consider *Cholesky(square root)* factorization of the covariance matrix

$$P = G'G \qquad (44)$$

the result of which is the lower triangular matrix with positive diagonal entries or the *L-D factorization* of the same matrix

$$P = L'DL \qquad (45)$$

the results of which are *the monic LT matrix* (lower triangular matrix with units on the main diagonal) and the diagonal matrix D. Because of the slightly more effectiveness of the second factorization (it requires less computations) we turn our look to this one ($L'DL$). Now we can rewrite equation (43) in the form

$$L'_{k|k}D_{k|k}L_{k|k} = L'_{k|k-1}D_{k|k-1}L_{k|k-1} - \quad (46) \\ \frac{L'_{k|k-1}D_{k|k-1}L_{k|k-1}\phi_k\phi_k' L'_{k|k-1}D_{k|k-1}L_{k|k-1}}{\sigma^2 + \phi_k' L'_{k|k-1}D_{k|k-1}L_{k|k-1}\phi_k}$$

So instead of updating of the whole covariance matrix it is possible to update only its factors what ensures the possitive definiteness. Algorithm which performs above mentioned recursive updating (46) is *LDFIL* which was derived on UTIA

Prague. Now it's time to turn our look to the second part of the general forgetting algorithm - *time update.* Implementation of regularization of the covariance matrix by the alternative covariance matrix calls for more computational efforts. In both cases (38),(40) we require weighted addition of two full rank matrices P_0^{-1}, P_1^{-1} or P_0, P_1 respectively. There are many ways how to do it but we have chosen *dyadic reduction*, see (Peterka, 1986) which is a basic tool for solving such problems.

DYADIC REDUCTION: Consider a symmetric nonnegative definite matrix M of rank at most two expressed as a weighted sum of two symmetric dyads

$$M = f'D_f f + r'D_r r \tag{47}$$

where f and r are row-vectors

$$f = [1, f_1, f_2 \ldots, f_n], r = [r_0, r_1, r_2 \ldots, r_n]$$

while D_f and D_r are nonnegative scalar weights. Note that the element f_0 is 1. Such a representation of the matrix M is not unique and can be modified.

$$M = f'D_f f + r'D_r r = \tilde{f}'\tilde{D}_f \tilde{f} + \tilde{r}'\tilde{D}_r \tilde{r} \tag{48}$$

Algorithm of dyadic reduction performs the modification by which the element $\tilde{f}_0$ remains to be equal to 1 but the element in the same position in $\tilde{r}$ is zeroed.

$$\tilde{f}_0 = 1, \tilde{r}_0 = 1$$

Why we decided to use just dyadic reduction is easy to understand from the following lines. Using updating of the covariance matrix in the factorized form (45) then it can be expressed as a sum of weighted dyads .

$$P = \sum_{i=1}^{dim_L} L_i' D_i L_i \tag{49}$$

where L_i are the row-vectors introduced as the ith row of the matrix L. If all weights D_i are nonnegative then also the covariance matrix P is positive definite. Now we will show how can be used dyadic reduction in SLF to perform weighted sum of two full rank matrices considered in factorized form ($L'DL$). Equation (40) representing the time update of the covariance matrix can be written

$$P = \lambda \sum_{i=1}^{dim_{L_0}} L_{0,i}' D_{0,i} L_{0,i} + (1-\lambda) \sum_{i=1}^{dim_{L_1}} L_{1,i}' D_{1,i} L_{1,i} \tag{50}$$

where $dim_{L_0} = dim_{L_1}$. Diagonal elements are units for the both triangular matrices which is the crucial point here, because the algorithm of dyadic reduction assumes that the reduction row, called f has the element $f_0 = 1$ in the proper place.

ALGORITHM USING DYADIC REDUCTION: Usual output of the recursive updating of the covariance matrix in the *L'DL factorization* (algorithm LDFIL) is one matrix which is constructed as an addition of the *monic LT matrix* and the *diagonal matrix* with weights corresponding to single rows of LT monic matrix. So instead of two matrices (LT monic and diagonal) we have just one constructed from both ones. The alternative måtrix (for regularization P_1) is considered to be constructed in the same way. Our aim is to make weighted addition of these both matrices (P_0-matrix after data update, P_1-alternative matrix) considered in factorized forms. *The algorithm can be divided in three steps:*

1: Separating of the diagonal factors and multiplying them by the weight λ or $1-\lambda$ (for alternative matrix). After it we have both matrices in the following form.

$$L_i = \begin{pmatrix} 1 & & & & \\ L_i(2,1) & 1 & & & \\ L_i(3,1) & L_i(3,2) & 1 & & \\ \cdot & \cdot & \cdot & & \\ \cdot & \cdot & & \cdot\ \cdot & \\ L_i(k,1) & \cdot & & \cdot\ L_i(k,k-1) & 1 \end{pmatrix} \tag{51}$$

where $k = dim_{L_i}$ and diagonal factors in the vector

$$D_i' = [D_i(1), D_i(2), \ldots D_i(k)],\ i = 0,1 \tag{52}$$

2: Now it is possible to use *dyadic reduction* so that the last row of the matrix L_0 is used to reduce to zero the last column of the matrix L_1. Using the last but one row the last but one column of the matrix L_1 is zeroed and so on until the matrix is completely zeroed.

3: In the last step when the matrix L_1 is completely zeroed and matrices L_0 and D_0 are updated to L_0^* and D_0^* so that *regularized covariance matrix* may be written

$$P^* = L_0^{*'} D_0^* L_0^* \tag{53}$$

we must put the weighting factors from the diagonal matrix D_0^* back on the main diagonal of the *LT monic matrix* L_0^*. We must do it because *LDFIL* needs this form for recursive measurement update. After this step time updating (regularizing) is finished.

POSSIBLE MODIFICATION: It's very easy to make a little modification of the above algorithm to make it construct the sum of factorized matrices in inverse form *(REF)*. In this case we must make a simple extension of the above algorithm, namely we must have another algorithm for computing the inverse of the matrix. Because of triangular matrix such an algorithm is very simple. It's really possible to make many extensions to cope with all regularized methods considering regularization as a weighted sum of matrices in factorized forms.

6. SIMULATION EXAMPLES

From all mentioned forgetting methods we selected *Directional Forgetting DF*, and three regularized methods namely *REF*, *SLF*, and *PF* with time update of the covariance matrix (21). We compared their abilities to track time varying parameters of *two different* simulated models with different types of *excitation*.

DESCRIPTION OF THE MODEL 1:
Second order model (4) with external disturbance v_t.

$$y_k = \sum_{i=1}^{2} a_i y_{k-i} + \sum_{i=0}^{2} b_i u_{k-i} + \sum_{i=1}^{2} d_i v_{k-i} + e_t \quad (54)$$

where $e_k \sim N(0, \sigma^2)$. Constant parameter values were $a_2 = -0.9$, $b_0 = 0.5$, $b_1 = -0.25$, $b_2 = 0.1$, $d_1 = 0.8$, $d_2 = 0.2$ and $\sigma = 0.1$
Time varying parameter was $a_1 = 0.98$ which had the same value till the simulation took 300 steps and at the time $t = 300$ this parameter was changed to -0.98. The external disturbance was simulated as a rectangular signal alternating periodically between the values $+1$ and -1 at $k = 100, 200, \ldots$ Identification was complicated mostly by a rarely changing disturbance which carries little information about the parameters d_i. We considered two different choices of the input generator:

- **A:** The input was generated as a normally distributed white noise.
- **B:** The input was generated in the following way

$$u_k = 0.8u_{k-1} + 0.2u_k \quad (55)$$

where u_k was normally distributed white noise and u_{k-1} was a previous value of the input.

DESCRIPTION OF THE MODEL 2:
The only difference comparing with the model 1 was the time varying parameter: $a_{1,k} = 0.98cos(2\pi k/250)$. We also considered two *different choices* of the input generator.

- **A:** The input was generated as a normally distributed white noise.
- **B:** The input was generated in the same way as in the *model 1* with the small difference $u_k \sim (0.5, 1)$.

DESIGN PARAMETERS :

REF alternative matrix is $P_1 = 50 * I$ for **1A,2A,1B** the only dif. is $P_1(1,1) = 1000$ in **2B**, (alternative=prior) $\lambda = 0.8$, $\hat{\theta}_{1|0} = 0$ for all cases

SLF alternative matrix is $P_1 = 50 * I$,the only dif. is $P_1(1,1) = 1000$ for all cases, (alternative=prior) $\lambda = 1 - e - 4$, $\hat{\theta}_{1|0} = 0$ for all cases

PF $\lambda_0 = 0.1$ $\lambda_1 = 100$ prior matrix is $P_{1|0} = \lambda_1 I$, $\hat{\theta}_{1|0} = 0$ for all cases

DF alternative matrix is $P_1 = 50 * I$ for **1A,2A,1B** the only dif. is $P_1(1,1) = 1000$ in **2B**, (alternative=prior) $\lambda = 0.2$, $\hat{\theta}_{1|0} = 0$ for all cases

SUMMARY OF IDENTIFICATION RESULTS :

	1A	**2A**	**1B**	**2B**
REF				
	IS=57.8	IS=65.5	IS=110.6	IS=99
	PEOR=20	PEOR=27	PEOR=10	PEOR=9
SLF				
	IS=75.3	IS=94.8	IS=105.6	IS=103.4
	PEOR=20	PEOR=29	PEOR=17	PEOR=9
PF				
	IS=70.5	IS=92.1	IS=74.8	IS=143
	PEOR=17	PEOR=32	PEOR=15	PEOR=19
DF				
	IS=167.8	IS=221	IS=217	IS=311
	PEOR=100	PEOR=170	PEOR=46	PEOR=99

IS-means the integral sum of the *Euclidean norm* of a deviation of the parameter estimates $\hat{\theta}_{k+1|k}$ from their true values.
PEOR-indicates how many times the prediction error $\hat{e}_{k|k-1}$ was outside the $\pm 3\sigma -$ *band* during 600 steps of simulation.

On all four figures (1-4) in the left part of the figure the graphs plot the Euclidean norm of a

deviation of the parameter estimates $\hat{\theta}_{k+1|k}$ from their true values.
In the right part of the page the graphs plot the eigenvalues of the covariance matrix $P_{k+1|k}$. From our experiments we can see the following connections:

- **1:** Decrease of excitation (B) makes the increase of the Euclidean norm (increase of the time of adaptation).
- **2:** Decrease of excitation (B) makes also the decrease of the number of prediction errors $\hat{e}_{k|k-1}$ outside the $\pm 3\sigma$ – *band*.

7. CONCLUSION REMARKS

Identification procedures for all compared regularized methods were written in Fortran in a numerically safe and effective ways. In all three methods both updates (measurement and time) of the covariance matrices were performed on L'DL factors to get numerical stability of the computation. For comparing to these methods was selected procedure performing Directional forgetting (md01) which was a standard identification procedure at our department for many years. All simulation experiments were performed in the system MATLAB on PC 586/100. The presented identification method are composed of Fortran - coded MEX and M files. The experiments confirm better tracking abilities of selected regularized forgetting methods comparing to the standard Directional Forgetting.

8. ACKNOWLEDGEMENT

The authors acknowledge the current financial support of the Commission of the European Communities (COPERNICUS grant CP941174).

9. REFERENCES

DeSouza, C.E. (1988). On adaptive control algorithm for linear system having unknown time delay. *Automatica* **24(3)**, 327–341.

Fortescue, T.R., L.S. Kershenbaum and B.E. Ydstie (1981). Implementation of selftuning regulators with variable forgetting factors. *Automatica* **17(6)**, 831–835.

Goodwin, G.C. (1986). Rapprochement between continuous and discrete model reference adaptive control. *Automatica* **22(2)**, 199–207.

Kulhavý, R. (1987). Restricted exponential forgetting in real time identification. *Automatica* **23**, 598–600.

Kulhavý, R. and F.J. Kraus (1996). On duality of regularized exponential and linear forgetting. *Automatica* **32(10)**, 1403–1415.

Kulhavý, R. and M. B. Zarrop (1993). On a general concept of forgetting in real time identification. *Internat.J.Control* **58**, 905–924.

Kulhavý, R. and M. Kárny (1984). Tracking of slowly varying parameters by directional forgetting. In: *9th IFAC World Congress Budapest.* Budapest.

Lozano-Leal, R. and G.C. Goodwin (1985). A globally convergent adaptive pole placement algorithm without a persistancy of excitation. *IEEE Trans. on Auto.Control* **AC30(8)**, 795–798.

Milek, J.J and F.J. Kraus (1991). Stabilized least squaresestimators for time variant processes. In: *In:Proceedings of the 1st IFAC Symposium on Design Methods of Control Systems.* Zurich, Switzerland.

Parkum, J.E., N.K. Poulsen and J. Holst (1992). Recursive forgetting algorithms. *Internat.J.Control* **55**, 109–128.

Peterka, V. (1981). Bayesian approach to system identification. In: *Trends and Progress in System Identification ,Chap. 8.* pp. 239–340. Pergamon. New York.

Peterka, V. (1986). Applied theory and algorithms. *Supplement to the Journal Kybernetika.*

Praly, L. (1983). Robustness of model reference adaptive control. In: *In 3rd Yale workshop on application of adaptive system theory.* Yale University.

Yoon, T.W. and D.W. Clarke (1994a). Adaptive predictive control of the benchmark plant. *Automatica* **30(4)**, 621–628.

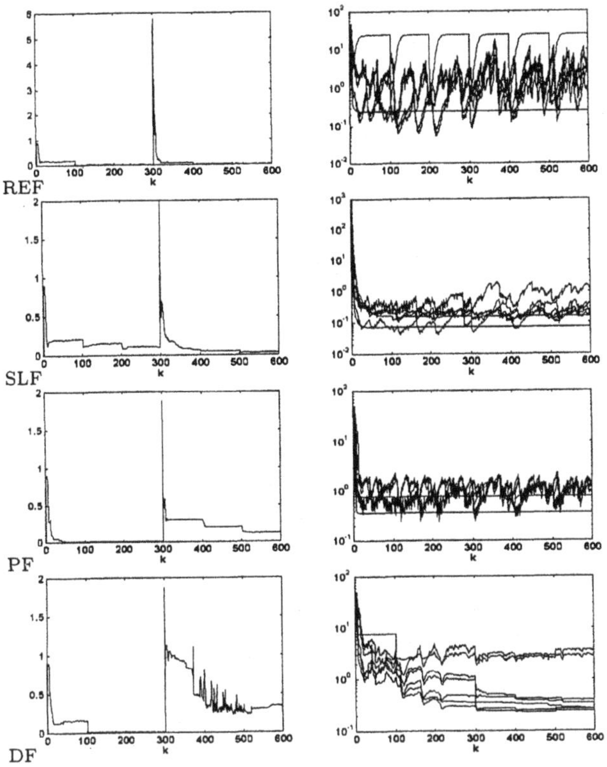

Fig. 1. 1A - results, model 1 with excitation A

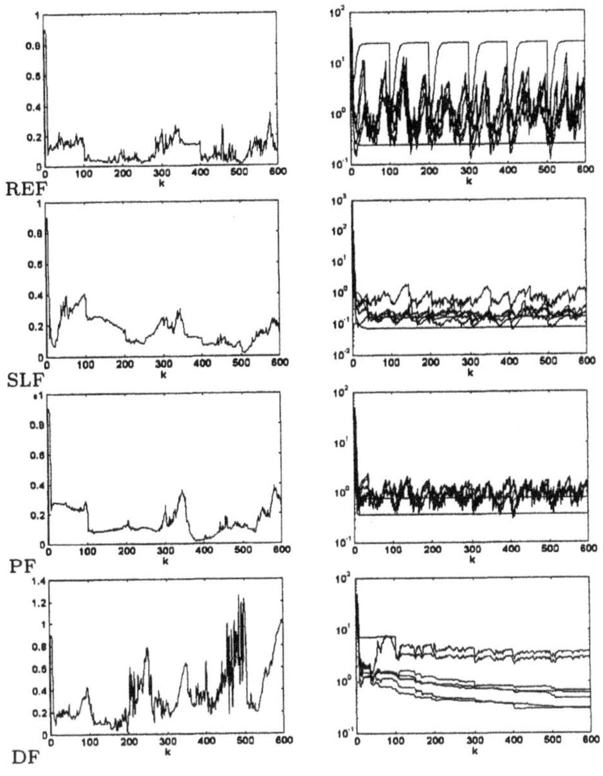

Fig. 3. 2A - results, model 2 with excitation A

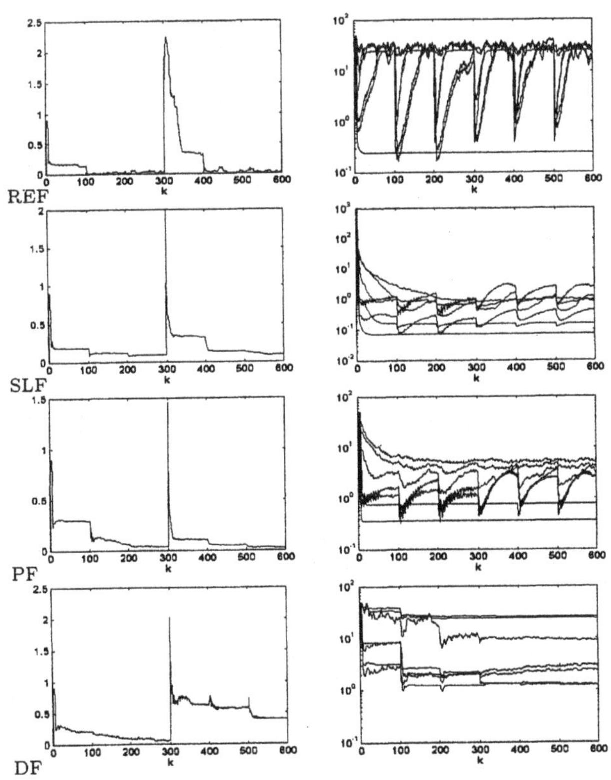

Fig. 2. 1B - results, model 1 with excitation B

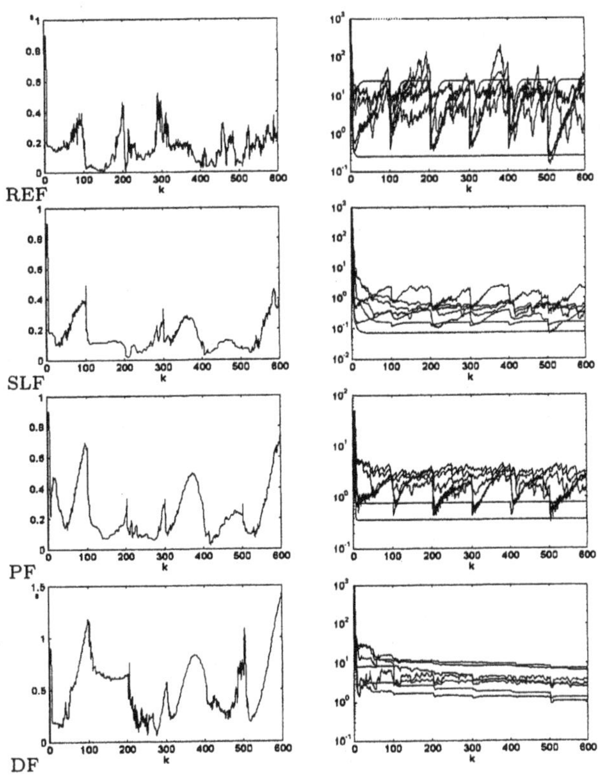

Fig. 4. 2B - results, model 2 with excitation B

MODEL REFERENCE ADAPTIVE CONTROL-
SOME DEVELOPMENT TRENDS

Ján Murgaš and Eva Miklovičová[1]

Department of Automatic Control Systems, Faculty of Electrical Engineering and Information Technology, Slovak Technical University, Ilkovičova 3, 812 19 Bratislava, Slovakia. E-mail:murgas@kasr.elf.stuba.sk, miklovic@kasr.elf.stuba.sk.

Abstract. The aim of the work is to present the state-of-the-art in this filed of adaptive control. Some development trends based especially on the general stability theory are analyzed. The characteristics of basic model reference adaptive control (MRAC) structures and algorithms are given in order to provide an insight into the development of the theory and to indicate the possibilities of their practical application. As the paper makes only a survey of the subject area, the stability proofs of the presented algorithms are omitted. The plant to be controlled is assumed to be described by a linear model.

Keywords: model reference adaptive control

1. INTRODUCTION

The adaptive control algorithms have been proposed especially for control of systems with unknown or time-varying parameters. The model reference adaptive control has evolved during the last three decades. First MRAC algorithms had the gradient character and the overall stability of the adaptive system had not been guaranteed. The stability proofs were given later, using the passivity theory (Landau, 1993) and the general stability theory (Narendra and Annaswamy, 1989). There is a variety of MRAC methods, which can be generally divided into the algorithms using complete state space information and those using the input/output information only. First group of algorithms assume that all state space variables can be measured or calculated (Butler, 1992; Ioannou and Kokotovic, 1983; Landau, 1973; Murgaš, *et al.*, 1992) and results in the simple structure of control system even when the transfer function relative degree is high (Murgaš, *et al.*, 1995). Much more attention has been devoted to the design of algorithms when only process input and output is available (Aström and Wittenmark, 1989; Goodwin and Sin, 1984; Iserman, *et al.*, 1992; Ioannou and Kokotovic, 1983; Narendra and Annaswamy, 1989; Rohrs, *et al.*, 1985; Sastry and Bodson, 1989). The standard MRAC algorithm, referred to as augmented error method, was first introduced by Monopoli (1974) and subsequently developed and modified in many papers (Butler, 1992; Narendra and Annaswamy, 1989; Sastry and Bodson, 1989; Slotine and Li, 1991; Tao, 1996).

The continuous-time approaches give good results in the ideal case when apriori defined requirements are satisfied. However, their practical applicability is limited by their low robustness, even though this problem has been addressed by many authors (Chai and Tao, 1994; Datta and Ioannou, 1994; Ioannou and Datta, 1991; Ortega and Tang, 1989; Rohrs, *et al.*, 1985; Sastry and Bodson, 1989; Slotine and Li, 1991). Another important problem is the discrete-time realization of MRAC algorithms. As it is well known, when discretize a system with relative degree greater then two, the unstable zeros can occur in the discrete-time transfer function, which cause the problems in solving the tracking problem.

2. PROBLEM STATEMENT

Consider the class of linear time-invariant SISO

[1] This work has been partially supported by the Slovak Scientific Grant Agency, grant No. 95/5195/198.

systems described by the transfer function

$$kW(s) = k\frac{B(s)}{A(s)} = k\frac{s^q + b_{q-1}s^{q-1} + \cdots + b_1 s + b_0}{s^n + a_{n-1}s^{n-1} + \cdots + a_1 s + a_0} \quad (1)$$

or in the state space representation by equations

$$\dot{x} = Ax + bu, \quad y = c^T x, \quad x \in R^n, \quad u \in R \quad (2)$$

The desired closed loop dynamical behavior is represented by the reference model

$$\dot{x}_m = A_m x_m + b_m r, \quad y_m = c_m^T x_m, \quad x_m \in R^n, \quad r \in R \quad (3)$$

where A_m is an $(n \times n)$ matrix of known constants similarly as vectors b_m and c_m, r is the reference input signal. The equivalent transfer function is

$$k_m W_m(s) = k_m \frac{B_m(s)}{A_m(s)} = c_m^T (sI - A_m)^{-1} b_m \quad (4)$$

The control objective is to choose the control input

$$u = f(\Theta, \omega)$$

where Θ is the controller parameter vector and ω is the vector of signals used in the adaptation mechanism, so that the closed loop transfer function matches the reference model one as close as possible. To meet this control objective, the controller parameters are adapted in such the way that the tracking error

$$e = x - x_m \quad \text{or} \quad e_1 = y - y_m \quad (5)$$

tends to zero.

The derivation of the adaptation law

$$\dot{\Theta} = g(\omega, e, \Theta)$$

is a key issue in MRAC design and can be carried out using a variety of techniques. First adaptation laws have been designed using gradient methods. The main disadvantage is the absence of the stability proof, which has led to the use of the stability theory of Lyapunov and Popov's hyperstability theory. These approaches serve as standard design techniques, yielding the guaranteed stable adaptive system.

3. CONTINUOUS-TIME MRAC

Development of MRAC started with the continuous-time analysis methods and this technique is still the basis of most MRAC designs. Generally, the MRAC algorithms can be divided into algorithms using all state space variables and those using the input and output information only.

3.1 MRAC with complete state space information

Assume that the system (2) is completely controllable with A and b unknown but bounded and it has no zeros. If the complete state vector x of the plant (2) is available, the control law can be chosen in the form of state feedback with a feedforward term to ensure the required closed loop gain

$$u = k_x^T(x,e)x + k_r(r,e)r = \Theta^T \omega \quad (6)$$

The adaptation law ensuring the overall stability of the adaptive system is designed using Lyapunov's direct method

$$\dot{\Theta} = \dot{\varphi} = -\Gamma \varepsilon \omega, \quad \Gamma > 0, \quad \varepsilon = p^T e \quad (7)$$

where $\Theta = [k_x^T, k_r]^T$ the adaptive controller parameter vector, $\omega = [x^T, r]^T$ is the signal vector and $\varphi = \Theta - \Theta^*$ is the error of instantaneous adjustment of adaptive controller parameters, where Θ^* denotes the value for which the closed loop behavior matches the reference model. Γ is the adaptation gain matrix, which must be diagonal and positive definite, $p = Pb$ and P is a solution of Lyapunov's equation $A_m^T P + PA_m = -Q$, P and Q being the positive symmetric matrices.

3.2 MRAC with input/output information only

In practice, limited state information or only the process output may be measurable. The parallel state reconstruction is difficult because of the unknown process parameters. In this subsection, several algorithms developed for model reference adaptive control when using output feedback are presented, namely the algorithms with augmented error, the algorithms using optimization methods (including gradient methods) and CGT algorithms as well.

The augmented error method

The augmented error method represents the standard scheme for MRAC when only process output is available. The plant to be controlled (1) is assumed to satisfy certain conditions and some prior information is required

i. the plant is minimum phase
ii. the upper bound on the plant order n is known
iii. the relative degree $n^* = n - q$ is known
iv. the sign of the high frequency gain k is known

This approach makes use of linear filters called auxiliary signal generators to generate $2(n-1)$ variables ν_1 and ν_2 from the process input and output

$$\dot{\nu}_1 = \Lambda \nu_1 + hu, \quad \dot{\nu}_2 = \Lambda \nu_2 + hy, \quad \nu_1, \nu_2 \in R^n$$
$$y_1 = c^T \nu_1 \qquad y_2 = d^T \nu_2 \quad (8)$$

$h^T = [0,0,\ldots,1]$, Λ is an $[(n-1)\times(n-1)]$ stable matrix

.

The transfer functions of the signal generators (8) can be written as follows

$$\frac{y_1(s)}{u(s)} = c^T(sI-\Lambda)^{-1}h = \frac{C(s)}{G(s)}$$

$$\frac{y_2(s)}{y(s)} = d^T(sI-\Lambda)^{-1}h = \frac{D(s)}{G(s)}$$

where C, D and G are the polynomials

$$C(s) = c_{n-2}s^{n-2} + \cdots + c_1 s + c_0$$

$$D(s) = d_{n-2}s^{n-2} + \cdots + d_1 s + d_0$$

$$G(s) = s^{n-1} + g_{n-2}s^{n-2} + \cdots + g_1 s + g_0$$

The control law can be given in the form

$$u = \Theta^T\omega = k_r r + c^T\nu_1 + k_y y + d^T\nu_2 \tag{9}$$

with the signal vector $\omega = \left[r, \nu_1^T, y, \nu_2^T\right]^T$ and the parameter vector $\Theta = \left[k_r, c^T, k_y, d^T\right]^T$.

Design of MRAC can be divided into two parts, the algebraic and the analytic one. First, we deal with the algebraic part, i.e. the solvability of the MRAC problem with a chosen structure is investigated. Then, in the analytic part of design, the adaptation law is chosen and the necessary conditions to guarantee stable solutions are stated.

The perfect model-matching condition requires the closed loop transfer function of the plant (1) together with the controller (8)-(9)

$$W_c(s) = \frac{y(s)}{r(s)} = \frac{kk_r B(s)G(s)}{[G(s)-C(s)]A(s) - kB(s)[k_y G(s) + D(s)]}$$

to be identical to the model reference transfer function (4). One solution can be found in (Lozano and Landau, 1981)

$$kk_r = k_m \qquad G(s) = B_m(s)G_1(s)$$

$$[G(s)-C(s)]A(s) - kB(s)[k_y G(s) + D(s)] = $$
$$= A_m(s)B(s)G_1(s)$$

where $G_1(s)$ is a Hurwitz polynomial of degree $(n-q-1)$. Then the closed loop transfer function is

$$W_c(s) = k_m\frac{B(s)B_m(s)G_1(s)}{A_m(s)B(s)G_1(s)} = k_m\frac{B_m(s)}{A_m(s)} \tag{10}$$

The process zeros $B(s)$ must be canceled by poles of the transfer function (10), which explains the minimum phase requirement on the plant. If $q = n-1$, $G(s)$ can be chosen equal to $B_m(s)$.

The adaptation law is derived using Lyapunov's second method and depends on whether the reference model transfer function $W_m(s)$ satisfies the strictly positive real (SPR) property or not.

i. $W_m(s)$ is SPR: $\dot{\Theta} = \dot{\varphi} = -\Gamma e_1\omega$ (11)

ii. $W_m(s)$ is not SPR: $\dot{\Theta} = \dot{\varphi} = -\Gamma e_d\xi$ (12)

where $\Gamma^T = \Gamma > 0$ is the adaptation gain matrix, e_d is the augmented error, ξ is the filtered signal vector

$$e_d(s) = e_1(s) + \varepsilon(s) = $$
$$= e_1(s) + W_m(s)L(s)\left(\Theta^T(s)\xi(s) - L^{-1}(s)u(s)\right) \tag{13}$$

$$\xi(s) = L^{-1}(s)\omega(s)$$

and $L(s)$ is the filter which is chosen in such the way that the transfer function $W_m(s)L(s)$ is SPR.

Gradient methods

These are very popular in spite of the fact that they do not ensure adaptive control scheme stability. First MRAC method was based on the use of sensitivity models to update the controller parameters in the correct direction. The control law has various structures, the most frequently it takes the general form

$$F(s)u(s) = H(s)r(s) - G(s)y(s)$$

where F, G, H are the polynomials of orders n_f, n_g, n_h. The adaptation law for this case takes the following form

$$\dot{\Theta} = -\Gamma e_1(t)\nabla_\Theta e_1(t)$$

with the parameter vector $\Theta = \left[f^T, g^T, h^T\right]^T$, where f, g, h are vectors of coefficients of F, G, H polynomials. The sensitivity functions derivations $\frac{\partial e_1(t)}{\partial\Theta} = \nabla_\Theta e_1(t)$ depending on known plant model constants only is the main problem. Some approximations can be used and only the existence of stable zeros of the plant transfer function is assumed (Aström and Wittenmark, 1989).

CGT (command generator tracker) approaches.

Assuming that the plant state information is not available for the control design purposes, not only input/output but also the information on the reference model is used. The preliminary condition for methods of this category is that it is possible to stabilize the plant from the output. The controller realization takes the following form (Bar-Kana and Kaufman, 1985)

$$u = k_e e_1 + I k_x^T x_m + k_r r = \Theta^T \omega$$

$$\Theta = \Theta_p + \Theta_I, \quad \Theta_p = \alpha_1 e_1 \omega^T, \quad \dot{\Theta}_I = \alpha_2 e_1 \omega^T$$

where the parameter vector $\Theta = \left[k_e, k_x^T, k_r\right]^T$, $\Theta \in R^{n+2}$, the signal vector $\omega = \left[e_1, x_m, r\right]^T$ and α_1, α_2 are the weighting constants.

4. DISCRETE-TIME MRAC

First steps towards the discrete-time MRAC were based on the existing continuous-time designs which was possible due to the availability of discrete version of the stability theories. Later, the MRAC schemes were introduced which were completely discrete-time oriented and did not have a continuous-time predecessor.

4.1 Augmented error method

The continuous-time augmented error method can be transformed into a discrete version without much difficulty. The controller structure remains unaltered

$$u(k) = \Theta^T(k)\omega(k)$$

and the order of the auxiliary generators is increased to the process order n, so that the perfect model matching condition is satisfied if $q \le n$. The signal vector and the parameter vector become

$$\omega(k) = \left(r(k), v_1^T(k), v_2^T(k)\right)^T$$

$$\Theta(k) = \left(k_r(k), c^T(k), d^T(k)\right)^T$$

The signal vector contains one element more than in the continuous case $(2n+1$ total$)$ but the process output $y(k)$ is excluded. The adaptation law is derived using the Lyapunov's stability method

$$\Theta(k) = \Theta(k-1) - \Gamma\varepsilon(k)\xi(k), \qquad \xi(k) = L^{-1}(z)\omega(k)$$

where $\varepsilon(k)$ is the augmented error signal in order to obtain an SPR transfer function $W_m L$

$$\varepsilon(k) = e_1(k) - W_m(z)L(z)\left[W_m(z)u(k) - \Theta^T(k)\xi(k)\right] \tag{14}$$

$$e_1(k) = y(k) - y_m(k)$$

The presence of a one-step delay in the adaptation loop needs a correction factor for the output error, linking the apriori error (which is only available)

$$e_1^0(k) = y^0(k) - y_m(k), \quad y^0(k) = W_c\left[\Theta(k-1)\right]u(k)$$

to the aposteriori error $e_1(k)$. The parameters are then updated using the adaptive law

$$\Theta(k) = \Theta(k-1) - \Gamma\frac{\varepsilon^0(k)\xi(k)}{1 + W_m(z)L(z)\left[\xi^T(k)\Gamma\xi(k)\right]} \tag{15}$$

Note that the auxiliary signal generator orders can be reduced from n to $n-1$ which can lead to the improving of the convergence properties (Butler, 1992).

In the equation (15), the fixed adaptation gain has been used. Applying the least-squares-type algorithm, this gain can be computed recursively

$$\Gamma(k+1) = \frac{1}{\lambda_1}\left[\Gamma(k) - \frac{\Gamma(k)\omega(k-d)\omega^T(k-d)\Gamma(k)}{(\lambda_1/\lambda_2) + \omega^T(k-d)\Gamma(k)\omega(k-d)}\right] \tag{16}$$

where $0 < \lambda_1 \le 1, 0 \le \lambda_2 < 2, \Gamma(0) > 0, d$ is the plant time delay, λ_1 corresponds to the forgetting factor in the least squares estimation and λ_2 is included for generalization purposes (in the case of $\lambda_1 = 1$ and $\lambda_2 = 0$ the fixed-gain adaptation appears). Least-squares adaptation allows user to avoid choosing Γ himself, but on the other hand it introduces other tuning factors such as λ_1, λ_2 and $\Gamma(0)$ and it brings on all the common problems of least-squares methods.

4.2 Independent tracking and regulation (ITR)

An example of the complete discrete-time design is MRAC scheme which makes it possible to specify the desired tracking and regulation properties separately (Landau and Lozano, 1981; Lozano and Landau, 1981), whereas existing MRAC schemes only dealt with tracking requirements. This control structures resemble some self-tuning regulators.

Assuming that the plant can be described by the transfer function

$$A\left(z^{-1}\right)y(k) = z^{-d}B\left(z^{-1}\right)u(k)$$

where $A\left(z^{-1}\right) = 1 + a_1 z^{-1} + \cdots + a_n z^n$

$$B\left(z^{-1}\right) = b_0 + b_1 z^{-1} + \cdots + b_m z^m$$

and the plant numerator and denominator orders m, n and the time delay d are assumed to be known.

The control objectives are twofold. First, in tracking, $\frac{y(k)}{r(k)}$ should be equal to the reference model

$$A_m\left(z^{-1}\right)y_m(k) = z^{-d} B_m\left(z^{-1}\right)r(k)$$

and second, in regulation, an initial disturbance of the output error $e_1(k) = y(k) - y_m(k)$ should be eliminated with dynamics specified by

$$C\left(z^{-1}\right)e(k+d) = 0 \quad \forall k \geq 0$$

The control signal is calculated as

$$u(k) = \frac{1}{b_0}\left[C\left(z^{-1}\right)y_m(k+d) - Q\left(z^{-1}\right)u(k) - R\left(z^{-1}\right)y(k)\right]$$
$$= \frac{1}{b_0}\left[C\left(z^{-1}\right)y_m(k+d) - \Theta_0^T(k)\omega_0(k)\right]$$

where $\quad Q\left(z^{-1}\right) = q_1 z^{-1} + \cdots + b_{n_Q} z^{n_Q}$

$$R\left(z^{-1}\right) = r_0 + r_1 z^{-1} + \cdots + r_{n_R} z^{n_R}$$

The parameter and signal vectors are given by

$$\Theta_0^T(k) = \left[q_1(k), \ldots q_{n_Q}(k), r_0(k), \ldots, r_{n_R}(k)\right]$$
$$\Theta^T(k) = \left[b_0(k), \Theta_0^T(k)\right]$$
$$\omega_0^T(k) = \left[u(k-1), \ldots, u(k-n_Q), y(k), \ldots y(k-n_R)\right]$$
$$\omega^T(k) = \left[u(k), \omega_0^T(k)\right]$$

The adaptation mechanism consists of the direct least-squares parameter estimator

$$\Theta(k) = \Theta(k-1) + \Gamma(k)\omega(k-d)\varepsilon(k)$$

where the adaptation gain matrix is updated according to the equation (16) and the error equation is calculated by using known signals up to $t = k-1$

$$\varepsilon(k) = \frac{C\left(z^{-1}\right)y(k) - \Theta^T(k-1)\omega(k-d)}{1 + \omega^T(k-d)\Gamma(k)\omega(k-d)}$$

4.3 Deadbeat MRAC

The above discussed ITR schemes require that the process have all its zeros inside the unit circle to avoid instability and simulations have shown its considerable ringing-like behavior. To avoid these problems, MRAC scheme with deadbeat control strategy has been developed (Butler, 1992). This gives rise to a new definition of the parameter and signal vectors

$$\Theta^T(k) = \left[\hat{b}_0(k), \ldots \hat{b}_m(k), -\hat{a}_1(k), \ldots, \hat{a}_n(k)\right]$$

$$\omega^T(k-d) =$$
$$\left[u(k-d), \ldots u(k-d-m), y(k-1), \ldots, y(k-n)\right]$$

i.e. the controller parameters are replaced by the estimates of the plant parameters in the parameter vector. The control signal is calculated as follows

$$u(k) = \frac{\hat{A}}{\hat{b}C - z^{-d}\hat{B}}\left[Cy_m(k+d) - y(k)\right]$$

where $\hat{b} = \sum_{i=0}^{m} \hat{b}_i$ is a compensation for the DC gain of $\hat{B}\left(z^{-1}\right)$. The error signal is similar to that of the ITR scheme

$$\varepsilon(k) = \frac{y(k) - \Theta^T(k-1)\omega(k-d)}{1 + \omega^T(k-d)\Gamma(k)\omega(k-d)} \qquad (17)$$

In the ITR case the actual meaning of the error signal is

$$\varepsilon(k) = C\left(z^{-1}\right)\left[y(k) - y_m(k)\right] + \bar{\varepsilon}(k)$$

in which $\bar{\varepsilon}(k)$ vanishes as adaptation converges. In the deadbeat MRAC controller, the error equation (17) can be reformulated as follows

$$\varepsilon(k) = \frac{C\hat{A}}{\hat{b}C - z^{-d}\hat{B}}\left[\hat{b}y(k) - \hat{B}y_m(k)\right]$$

Hence, where $\varepsilon(k) \to 0$ involved $y(k) \to y_m(k)$ in ITR, in the deadbeat MRAC $y(k) \to \frac{B\left(z^{-1}\right)}{b} y_m(k)$. The controller does not attempt to compensate the existing process zeros, so it allows controlling nonminimum-phase processes.

5. IMPROVING QUALITIES OF MRAC

First stability results for direct adaptive controllers were obtained under ideal conditions:

i. the plant is free of noise disturbances, unmodelled dynamics and unknown nonlinearities
ii. the order of the plant is known exactly
iii. the unknown parameters are constant for all time

Subsequently it was demonstrated (Ioannou and Datta, 1991; Rohrs, *et al.*, 1985) that in the presence of bounded disturbances, time-varying parameters and modeling errors the resulting adaptive systems could be made unstable. This triggered in the 1980s the interest on robustness issues, that is, on the preservation of stability or boundedness properties when the ideal conditions are not met (Ortega and Tang, 1989).

Several modified parameter update laws were proposed to improve the robustness against external disturbances (Narendra and Annaswamy, 1989)

Dead-zone:

$$\dot{\Theta} = -\Gamma\frac{\varepsilon_d \xi}{1+\xi^T\xi} \quad \text{if} \quad \|\varepsilon_d\| \geq \nu_0$$

$$\dot{\Theta} = 0 \quad \text{if} \quad \|\varepsilon_d\| < \nu_0$$

Gain limitation:

$$\dot{\Theta} = -\Gamma\frac{\varepsilon_d \xi}{1+\xi^T\xi} - \Theta f(\Theta)$$

$$f(\Theta) = \begin{cases} \left(\dfrac{\|\Theta\|}{\|\Theta^*\|_{max}} - 1\right)^2 & \text{if} \quad \|\Theta\| \geq \|\Theta^*\|_{max} \\ 0 & \text{if} \quad \|\Theta\| < \|\Theta^*\|_{max} \end{cases}$$

Sigma modification:

$$\dot{\Theta} = -\Gamma\frac{\varepsilon_d \xi}{1+\xi^T\xi} - \frac{\sigma}{1+\xi^T\xi}\Theta, \qquad \sigma > 0$$

Gamma modification:

$$\dot{\Theta} = -\Gamma\frac{\varepsilon_d \xi}{1+\xi^T\xi} - \frac{\gamma|\varepsilon_d|}{1+\xi^T\xi}\Theta, \qquad \gamma > 0$$

The basic idea of the modifications is to prevent the instability by limiting the gain of the adaptation loop and to counteract the parameter drift by eliminating the integral action. The implementation of the above mentioned modifications could be difficult, because the knowledge may be needed about an upper bound on the disturbance ν_0 or on the norm of the parameter vector $\|\Theta^*\|_{max}$. An alternative adaptation law based on the σ-modification that requires no prior knowledge of $\|\Theta^*\|_{max}$ and allows to avoid the bursting phenomena appearing in the use of σ-modification is proposed in (Chai and Tao, 1994)

$$\dot{\Theta} = -\Gamma\frac{\varepsilon_d \xi}{m^2} - -\Gamma|\delta|\,\Theta, \qquad \dot{\delta} = -a_1\delta - a_0\frac{\varepsilon_d}{m}$$

$$\dot{m} = -\delta_0 m + \delta_1(|u| + |y| + 1), \quad m(0) \geq \frac{\delta_1}{\delta_0}$$

where $a_0, a_1, \delta_0, \delta_1$ are positive constants. In all considered adaptation law modifications the normalization factor is used in order to obtain the speed selection effect on parameter convergence.

Improving the performances of MRAC subjected to unmodelled dynamics can also be obtained by certain modifications of the control law (Murgaš, 1994), using the standard adaptation law (7). Consider that the transfer function characterizing the plant exactly takes the form (1) and that the system can be expressed by the equivalent state space description

$$\dot{x}_1 = A_1 x_1 + b_1 u + A_{12} x_2, \quad x_1 \in R^{n1} \qquad (18)$$

$$\dot{x}_2 = A_2 x_2 + A_2 x_1, \qquad x_2 \in R^{n-n1} \qquad (19)$$

$$y = [1, 0, \ldots, 0] x_1 \qquad (20)$$

where A_2 is a stable matrix and A_{12}, A_{21} are bounded matrices.

Decomposition of (1) to (18)-(20) means that there are $n1$ dominant poles and $n - n1$ so called fast poles in the original system, which are placed at a further distance on the left from the imaginary axis in the complex plane. Similar decomposition is also used in (Slotine, 1991) for nonlinear systems.

In MRAS only the dominant part of original system will be modelled using the reference model

$$\dot{x}_m = A_m x_m + b_m r, \qquad x_m \in R^{n1}$$

Consider the control law in its modified form

$$u = k_x^T(x_1, e)x_1 + k_r(r, e)r - f(e)$$

with

$$e = x_1 - x_m, \qquad f(e) = \varepsilon\|e\|^2$$

The adaptation law is used in its standard form (7) with $\Theta = \left[k_x^T, k_r\right]^T$ and $\omega = \left[x_1^T, r\right]^T$.

Recently, MRAC using a non-quadratic Lyapunov function to derive the parameter update law has been proposed for plants with relative degree one (Tao, 1996). This redesign allows to achieve a generalized tracking performance

$$\int_0^\infty |e_1(t)|^{1+\alpha}\, dt < \infty \quad \alpha > 0$$

An adaptation law

$$\dot{\Theta} = -\Gamma\omega e_1^\alpha$$

with the choice of $0 < \alpha < 1$ results in the tracking error with less energy and the parameter derivative with more energy than those with the standard scheme with $\alpha = 1$, which is favorable for achieving the better tracking performance. A projection modification could be used to make the system robust with respect to the external disturbances.

Since the problems with the stability and robustness of MRAC were resolved, researchers working in this area have begun to address the performance issue. The criteria like boundedness of all closed loop signals and asymptotic convergence of the tracking error, which have been commonly used to characterize the performance of MRAC schemes, do not reveal large transient oscillations often observed in simulations and do not exclude the possibility of bursting at steady state in the presence of small modeling errors. To study the performance of the standard MRAC schemes, the L_2 and L_∞ norms of the tracking error have been used, which provide true transient performance measures instead of the usual asymptotic performance analysis.

A typical approach proposed to address the performance issue is using persistent excitation signals to enhance the parameter adaptation. However, its beneficial effect on the closed loop behavior during the transient phase is not so obvious and the exogenous excitation signal may destroy the tracking properties of the adaptive scheme, which renders the persistent excitation condition inapplicable in many cases. Consequently, the modified MRAC schemes have been proposed which improve the transient performance without destroying ideal properties of the standard MRAC system (Datta and Ioannou, 1994; Sun, 1993). The algorithm used for parameter updating is kept the same as that of the standard MRAC, while the control law is modified with an additional feedback signal, which is introduced to attenuate the effect of large parameter mismatch on the tracking error. The performance of the modified scheme can be arbitrarily improved in the presence of large parametric uncertainties and bounded input disturbances simply by choosing a control design parameter.

6. CONCLUSION

The model reference adaptive control schemes give very good results in tracking and regulation even for fast plants, providing that some apriori conditions are satisfied. However, it is important to increase the robustness of these methods in the presence of unmodelled dynamics as well as to work out the possibility of controlling nonminimum phase systems. For the above reasons, our article is intended for research workers who want to get into touch with this research area and also for control engineers who would like to realize this form of adaptive control in practice.

REFERENCES

Aström, K.J. and B. Wittenmark (1989). *Adaptive Control*, Addison-Wesley, New York.

Bar-Kana, I. and H. Kaufman (1985). Global Stability and Performance of a Simplified Adaptive Control Algorithm. *Int. J. Control*, **42**, 1491-1505.

Butler, H (1992). Model Reference Adaptive Control: From Theory to Practice, Prentice-Hall, New York.

Chai, T.Y. and Z. Tao (1994). A New Model Reference Robust Adaptive Controller in the Presence of Unmodelled Dynamics and Bounded Disturbances. *Automatica*, **30**, 865-869.

Datta, A. and P. Ioannou (1994). Performance Analysis and Improvement in Model Reference Adaptive Control. *IEEE Trans. on Aut. Control*, **39**, 2370-2387.

Egart, B. (1979). *Stability of Adaptive Controllers*, Springer Verlag, New York.

Goodwin, G.C. and K.S. Sin (1984). *Adaptive Filtering, Prediction and Control*, Prentice-Hall, New Jersey.

Isermann, R., K.H. Lachmann and D. Matko (1992). *Adaptive Control Systems*. Prentice-Hall, New York.

Ioannou, P.A, and A. Datta (1991). Robust Adaptive Control: A Unified Approach. *Proc. of the IEEE*, **79**, 1736-1767.

Ioannou, P.A. and P.V. Kokotovic (1983). *Adaptive System with Reduced Models*, Springer Verlag, New York.

Landau, I.D. (1973). *Adaptive Control - The model Reference Approach*, Marcel Dekker, New York.

Landau, I.D. and R. Lozano (1981). Unification of Discrete-Time Explicit Model Reference Adaptive Control Designs. *Automatica*, **17**, 593-611.

Lozano, R. and I.D. Landau (1981). Redesign of Explicit and Implicit Discrete-Time Model Reference Adaptive Control Schemes. *Int. J. of Control*, **33**, 247-268.

Monopoli, R.V. (1974). Model Reference Adaptive Control with an Augmented Error Signal. *IEEE Trans. on Aut. Control*, **19**, 474-484.

Murgaš, J. (1994). Present View on a Class of Adaptive Control with Reference Model. *J. of Electrical. Egineering*, **45**, 383-388.

Murgaš, J., M. Matejovič and V. Červenka (1995). MRAC Algorithm for High Relative Degree Plants. *Proc. of the 5th IFAC Symposium Adaptive System in Control and Signal Processing*, Budapest, 276-281.

Murgaš, J., V. Veselý and I. Hejda (1992). State Space Structures in MRAC. *Proc. of the IFAC Workshop on Structures and Control*, Prague, 80-83.

Narendra, K.S. and A.M. Annaswamy (1989). *Stable Adaptive Systems*, Prentice-Hall, New Jersey.

Ortega, R. and Y. Tang (1989). Robustness of Adaptive Controllers - A Survey. *Automatica*, **25**, 651-677.

Rohrs, C.E., L. Valavani, M. Athans and G. Stein (1985). Robustness of Continuous-Time Adaptive Control Algorithms in the Presence of Unmodelled Dynamics. *IEEE Trans. on Aut. Control*, **30**, 881-889.

Sastry, S. and M. Bodson (1989). *Adaptive Control. Stability, Convergence and Robustness*, Prentice-Hall, New Jersey.

Slotine, J.J.E. and W. Li (1991). *Applied Nonlinear Control*, Prentice-Hall, New Jersey.

Sun, J. (1993). A Modified Model Reference Adaptive Control Scheme for Improved Transient Performance. *IEEE Trans. on Aut. Control*, **38**, 1255-1259.

Tao, G. (1996). Model Reference Adaptive Control with $L^{1+\alpha}$ Tracking. *Int. J. Control*, **64**, 859-870.

REALISATION PROBLEM FOR 2-D POSITIVE SYSTEMS

Tadeusz Kaczorek

Warsaw University of Technology, Faculty of Electrical Engineering
Institute of Control and Industrial Electronics
00-662 Warszawa, Koszykowa 75, Poland
Tel. (+48 22) 6256278, 6280665, fax (+48 22) 6256633
e mail: kaczorek @ nov.isep.pw.edu.pl

Abstract: Sufficient conditions for the existence of a realisation being a positive 2-D Roesser type model of a given 2-D proper transfer matrix are established. A procedure for calculating a positive realization of a transfer matrix is given and illustrated by an example.

Keywords: 2-D positive system, realisation, procedure.

1. INTRODUCTION

The most popular models of two-dimensional (2-D) systems are the models introduced by Roesser (1975), Fornasini and Marchesini (1976 and 1978).The realization problem for 2-D linear systems has been considered in many books (Eising 1979, Kaczorek 1993, Kaczorek 1985) and papers (Eising 1978, Fonasini and Marchesini 1976, Himamoto and Fairman 1978, Roesser 1975, Kaczorek 1987). Recently Valcher and Fornasini in (Valcher and Fornasini 1995) have investigated some interesting properties of homogeneous 2-D positive systems described by the second Fornasini-Marchesini type models. In (Kaczorek 1996, Kaczorek 1997b) necessary and sufficient conditions have been established for the reachability of a positive (nonnegative) 2-D Roesser type model for zero boundary conditions and the minimum energy control problem has been solved. In this paper sufficient conditions are established for the existence of a realization being a positive 2-D Roesser type model of a given 2-D transfer matrix. A procedure for calculating a positive realisation of a proper transfer matrix will be presented.

2.PRELIMINARIES AND PROBLEM FORMULATION

Let $R_+ := [0,+\infty)$ be the set of the nonnegative numbers and let $Z_+ := \{0,1,2,\ldots\}$ be the set of nonnegative integers. Denote by R_+^n the set of n-tuples of the nonnegative numbers. The set of nonnegative matrices of size n by m will be denoted by $R_+^{n\times m}$.

Let $R[z_1,z_2]$ be the set of polynomials in z_1 and z_2 with real coefficients. The set of $p\times m$ polynomial matrices with entries from $R[z_1,z_2]$ will be denoted by $R^{p\times m}[z_1,z_2]$. The field of rational functions in z_1 and z_2 with real coefficients will be denoted by $R(z_1,z_2)$ and the set of $p\times m$ rational matrices with entries from $R(z_1,z_2)$ will be denoted by $R^{p\times m}(z_1,z_2)$.

Consider the 2-D Roesser model (Roesser 1985, Klamka 1991)

$$x_{ij}^{(1)} = Ax_{ij} + Bu_{ij} \tag{1}$$

$$y_{ij} = Cx_{ij} + Du_{ij} \tag{2}$$

$$x_{ij}^{(1)} = \begin{bmatrix} x_{i+1,j}^h \\ x_{i,j+1}^v \end{bmatrix}, \quad x_{ij} = \begin{bmatrix} x_{ij}^h \\ x_{ij}^v \end{bmatrix} \quad i,j \in Z_+$$

where $x_{ij}^h \in R^{n_1}$ is the horizontal state vector, $x_{ij}^v \in R^{n_2}$ is the vertical state vector, $u_{ij} \in R^m$ is the input vector, $y_{ij} \in R^p$ is the output vector,

$$A = \begin{bmatrix} A_1 & A_2 \\ A_3 & A_4 \end{bmatrix}, B = \begin{bmatrix} B_1 \\ B_2 \end{bmatrix}, C = [C_1 \quad C_2]$$

$$A_1 \in R^{n_1 \times n_1}, B_1 \in R^{n_1 \times m}, C_1 \in R^{p \times n_1},$$

$$A_4 \in R^{n_2 \times n_2}, B_2 \in R^{n_2 \times m}, C_2 \in R^{p \times n_2}$$

Definition 1. (Kaczorek 1996) The 2-D Roesser model (1), (2) is called nonnegative 2-D Roesser type model if for all boundary conditions

$$x_{0j}^h \in R_+^{n_1}, j \in Z_+, \; x_{i0}^v \in R_+^{n_2}, i \in Z_+ \tag{3}$$

and for all $u_{ij} \in R_+^m, i,j \in Z_+$ we have $x_{ij} \in R_+^n, n = n_1 + n_2$ and $y_{ij} \in R_+^p$ for $i,j \in Z_+$.

Proposition 1. The 2-D Roesser model (1), (2) is nonnegative if and only if

$$A \in R_+^{n \times n}, B \in R_+^{n \times m}, C \in R_+^{p \times n}, D \in R_+^{p \times m}$$

(4)

Proof. From (1) and (2) it follows that $x_{ij} \in R_+^n$ *and* $y_{ij} \in R_+^p, i,j \in Z_+$ for all boundary conditions (3) and $u_{ij} \in R_+^m, i,j \in Z_+$ if and only if (4) holds. □

The transfer matrix of (1)-(2) is given by (Roesser 1975, Kaczorek 1985)

$$T(z_1,z_2) = C \begin{bmatrix} I_{n_1} z_1 - A_1, & -A_2 \\ -A_3 & , I_{n_2} z_2 - A_4 \end{bmatrix}^{-1} B + D \in R^{p \times m}(z_1,z_2)$$

(5)

Definition 2 (Eising 1978 and Kaczorek 1985) The matrices A,B,C,D are called a realization of $T(z_1,z_2)$ if they satisfy (5). A realization (A,B,C,D) is called a minimal one if the matrix A has minimal dimension among all realizations of $T(z_1,z_2)$.

The realization problem can be stated as follows: Given a proper 2-D transfer matrix $T(z_1,z_2) \in R^{p \times m}(z_1,z_2)$ find matrices (4) such that (5) holds.

The main goal of this paper is:

① to establish conditions under which a proper transfer matrix $T(z_1,z_2)$ has a realization (4)

② to give a procedure for calculating a realization (4) for a given proper

$$T(z_1,z_2) \in R^{p \times m}(z_1,z_2)$$

3. THE EXISTENCE OF A REALIZATION

We shall use the following lemmas (compare with Eising 1978 and 1979)

Lemma 1. The transfer matrix (5) may be written in the form

$$T(z_1,z_2) = C(z_1)[I_{n_2} z_2 - A(z_1)]^{-1} B(z_1) + D(z_1) \tag{6a}$$

or

$$T(z_1,z_2) = C(z_2)[I_{n_1} z_1 - A(z_2)]^{-1} B(z_2) + D(z_2) \tag{6b}$$

where:

$$A(z_1) := A_3[I_{n_1} z_1 - A_1]^{-1} A_2 + A_4,$$
$$B(z_1) := A_3[I_{n_1} z_1 - A_1]^{-1} B_1 + B_2$$
$$C(z_1) := C_1[I_{n_1} z_1 - A_1]^{-1} A_2 + C_2,$$
$$D(z_1) := C_1[I_{n_1} z_1 - A_1]^{-1} B_1 + D \tag{7a}$$

$$A(z_2) := A_2[I_{n_2} z_2 - A_4]^{-1} A_3 + A_1,$$
$$B(z_2) := A_2[I_{n_2} z_2 - A_4]^{-1} B_2 + B_1$$
$$C(z_2) := C_2[I_{n_2} z_2 - A_4]^{-1} A_3 + C_1,$$
$$D(z_2) := C_2[I_{n_2} z_2 - A_4]^{-1} B_2 + D \tag{7b}$$

Proof. It is easy to show that

$$\begin{bmatrix} I_{n_1} z_1 - A_1, & -A_2 \\ -A_3 & , I_{n_2} z_2 - A_4 \end{bmatrix}^{-1} =$$
$$= \begin{bmatrix} [I_{n_1} z_1 - A_1]^{-1}, & [I_{n_1} z_1 - A_1]^{-1} A_2 \\ 0 & , I_{n_2} \end{bmatrix} \tag{8}$$
$$\begin{bmatrix} I_{n_1} & , 0 \\ [I_{n_2} z_2 - A(z_1)]^{-1} A_3 [I_{n_1} z_1 - A_1]^{-1} & , [I_{n_2} z_2 - A(z_1)]^{-1} \end{bmatrix}$$

Substitution of (8) into (5) yields

$$T(z_1,z_2) = [C_1 \quad C_2] \begin{bmatrix} I_{n_1} z_1 - A_1, & -A_2 \\ -A_3 & , I_{n_2} z_2 - A_4 \end{bmatrix}^{-1} \begin{bmatrix} B_1 \\ B_2 \end{bmatrix} + D =$$

$$= [C_1 \quad C_2] \begin{bmatrix} [I_{n_1} z_1 - A_1]^{-1}, & [I_{n_1} z_1 - A_1]^{-1} A_2 \\ 0 & , I_{n_2} \end{bmatrix}$$

$$\begin{bmatrix} I_{n_1} & , & 0 \\ [I_{n_2}z_2 - A(z_1)]^{-1}A_3[I_{n_1}z_1 - A_1]^{-1} & , & [I_{n_2}z_2 - A(z_1)]^{-1} \end{bmatrix}$$

$$\begin{bmatrix} B_1 \\ B_2 \end{bmatrix} + D = C(z_1)[I_{n_2}z_2 - A(z_1)]^{-1}B(z_1) + D(z_1)$$

The proof of (6b) is similar. □

Lemma 2.
For $T(z_1,z_2)$ the following relations hold

$$\begin{bmatrix} A_1 & B_1 \\ C_1 & D \end{bmatrix} = \lim_{z_2\to\infty}\begin{bmatrix} A(z_2) & B(z_2) \\ C(z_2) & D(z_2) \end{bmatrix} \tag{9a}$$

$$\begin{bmatrix} A_4 & B_2 \\ C_2 & D \end{bmatrix} = \lim_{z_1\to\infty}\begin{bmatrix} A(z_1) & B(z_1) \\ C(z_1) & D(z_1) \end{bmatrix} \tag{9b}$$

Proof. From (7a) we have

$$\begin{bmatrix} A(z_1) & B(z_1) \\ C(z_1) & D(z_1) \end{bmatrix} = \begin{bmatrix} A_3 \\ C_1 \end{bmatrix}[I_{n_1}z_1 - A_1]^{-1}[A_2 \ B_1] + \begin{bmatrix} A_4 & B_2 \\ C_2 & D \end{bmatrix} \tag{10}$$

Taking into account that $\lim_{z_1\to\infty}[I_{n_1}z_1 - A_1]^{-1} = 0$ from (10) we obtain (9b).
The proof of (9a) is similar. □
The given transfer matrix $T(z_1,z_2)$ may be always written in the form

$$T(z_1,z_2) = \frac{N(z_1,z_2)}{d(z_1,z_2)}$$

$$= \frac{N_{q_1}(z_2)}{d_{q_1}(z_2)} + \frac{\overline{N}_{q_1-1}(z_2)z_1^{q_1-1} + \cdots + \overline{N}_1(z_2)z_1 + \overline{N}_0(z_2)}{z_1^{q_1} + \overline{d}_{q_1-1}(z_2)z_1^{q_1-1} + \cdots + \overline{d}_1(z_2)z_1 + \overline{d}_0(z_2)} \tag{11a}$$

or

$$T(z_1,z_2) =$$

$$\frac{N_{q_2}(z_1)}{d_{q_2}(z_1)} + \frac{\overline{N}_{q_2-1}(z_1)z_2^{q_2-1} + \cdots + \overline{N}_1(z_1)z_2 + \overline{N}_0(z_1)}{z_2^{q_2} + \overline{d}_{q_2-1}(z_1)z_2^{q_2-1} + \cdots + \overline{d}_1(z_1)z_2 + \overline{d}_0(z_1)} \tag{11b}$$

where

$N(z_1,z_2) \in R^{p\times m}[z_1,z_2], d(z_1,z_2) \in R[z_1,z_2]$,
$N_{q_1}(z_2) \in R^{p\times m}[z_2]$
$N_{q_2}(z_1) \in R^{p\times m}[z_1], d_{q_1}(z_2) \in R[z_2], d_{q_2}(z_1) \in R[z_1]$

$\overline{N}_i(z_2) \in R^{p\times m}(z_2), \overline{d}_i(z_2) \in R(z_2), i = 0.1;\dots,q_1 - 1$,
$\overline{N}_j(z_1) \in R^{p\times m}(z_1), \overline{d}_j(z_1) \in R(z_1), j = 0,1,\dots,q_2 - 1$

Let

$$T(z_2) := \begin{bmatrix} A(z_2) - A_1, B(z_2) - B_1 \\ C(z_2) - C_1, D(z_2) - D \end{bmatrix} = \frac{N'(z_2)}{d(z_2)} \tag{12a}$$

and

$$T(z_1) := \begin{bmatrix} A(z_1) - A_4, B(z_1) - B_2 \\ C(z_1) - C_2, D(z_1) - D \end{bmatrix} = \frac{N'(z_1)}{d(z_1)} \tag{12b}$$

If the equation $d(z_2) = 0$ has roots $z_{21}, z_{22}, \dots, z_{2q}$ with multiplicities $m_1, m_2, \dots, m_q$, respectively, then the strictly proper transfer matrix (12a) can be written in the form (Kaczorek 1997a)

$$T(z_2) = \sum_{k=1}^{q}\sum_{i=1}^{m_k}\frac{T_{ki}}{(z_2 - z_{2k})^{m_k - i + 1}} \tag{13}$$

where

$$T_{ki} = \frac{1}{(i-1)!}\frac{d^{i-1}}{dz_2^{i-1}}\left[(z_2 - z_{2k})^{m_k}T(z_2)\right]_{|z_2 = z_{2k}} \tag{14}$$

Theorem. Let

$d_{i\infty} := -\lim_{z_2\to\infty}\overline{d}_i(z_2)$, $\ \overline{d}_{i\infty} := \lim_{z_2\to\infty} z_2[\overline{d}_i(z_2) - d_{i\infty}]$,

$N_{i\infty} := \lim_{z_2\to\infty}\overline{N}_i(z_2), i = 0,1,\dots,q_1 - 1$

$d_{j\infty} := -\lim_{z_1\to\infty}\overline{d}_j(z_1)$, $\ \overline{d}_{j\infty} := \lim_{z_1\to\infty} z_1[\overline{d}_j(z_1) - d_{j\infty}]$

$N_{j\infty} := \lim_{z_1\to\infty}\overline{N}_j(z_1), j = 0,1,\dots,q_2 - 1$

and

$$T_\infty := \lim_{z_1,z_2\to\infty} T(z_1,z_2) = \lim_{z_1\to\infty}\frac{N_{q_2}(z_1)}{d_{q_2}(z_1)} = \lim_{z_2\to\infty}\frac{N_{q_1}(z_2)}{d_{q_1}(z_2)}$$

Then there exists a realization (4) of $T(z_z,z_2)$ if

$T_\infty \in R_+^{p\times m}$ (15a)

$d_{i\infty} \in R_+, N_{i\infty} \in R_+^{p\times m}$ for $i = 0,1,\dots,q_1 - 1$ (15b)

$d_{j\infty} \in R_+, N_{j\infty} \in R_+^{p\times m}$ for $j = 0,1,\dots,q_2 - 1$ (15c)

$\overline{d}_{i\infty} \in R_+$ for $i = 0,1,\dots,q_1 - 1$

(or $\overline{d}_{j\infty} \in R_+$ for $j = 0,1,\dots,q_2 - 1$) (15d)

or (15a) and (15b) hold and

$z_{2k} \in R_+$ for $k = 1,\dots,q$ (16a)

$T_{ki} \in R_+^{(n_2+p)\times(n_2+m)}$ for $k = 1,\dots,q; i = 1,\dots,m_k$ (16b)

where T_{ki} are defined by (14).

Proof.
It is easy to check that the matrices

$$A(z_2) = \begin{bmatrix} 0 & 0 & \cdots & 0 & -\overline{d}_0(z_2)I_p \\ I_p & 0 & \cdots & 0 & -\overline{d}_1(z_2)I_p \\ 0 & I_p & \cdots & 0 & -\overline{d}_2(z_2)I_p \\ \vdots & & & & \vdots \\ 0 & 0 & \cdots & 0 & -\overline{d}_{q_1-2}(z_2)I_p \\ 0 & 0 & \cdots & I_p & -\overline{d}_{q_1-1}(z_2)I_p \end{bmatrix},$$

$$B(z_2) = \begin{bmatrix} \overline{N}_0(z_2) \\ \overline{N}_1(z_2) \\ \vdots \\ \overline{N}_{q_1-1}(z_2) \end{bmatrix} \tag{17a}$$

$$C(z_2) = [0 \ \ 0 \cdots 0 \ \ I_p], D(z_2) = \frac{N_{q_1}(z_2)}{d_{q_1}(z_2)}$$

are a realization of (11a) and the matrices

$$A(z_1)=\begin{bmatrix} 0 & 0 & \cdots & 0 & -\bar{d}_0(z_1)I_p \\ I_p & 0 & \cdots & 0 & -\bar{d}_1(z_1)I_p \\ 0 & I_p & \cdots & 0 & -\bar{d}_2(z_1)I_p \\ \hdashline 0 & 0 & \cdots & 0 & -\bar{d}_{q_2-2}(z_1)I_p \\ 0 & 0 & \cdots & I_p & -\bar{d}_{q_2-1}(z_1)I_p \end{bmatrix}$$

$$B(z_1)=\begin{bmatrix} \bar{N}_0(z_1) \\ \bar{N}_1(z_1) \\ \vdots \\ \bar{N}_{q_2-1}(z_1) \end{bmatrix} \qquad (17b)$$

$$C(z_1)=\begin{bmatrix} 0 & 0 \cdots 0 & I_p \end{bmatrix},\ D(z_1)=\frac{N_{q_2}(z_1)}{d_{q_2}(z_1)}$$

are a realization of (11b).

From (9) and (17) it follows that if (15a), (15b) and (15c) hold then $A_1 \in R_+^{n_1\times n_1}$, $B_1 \in R_+^{n_1\times m}$, $C_1 \in R_+^{p\times n_1}$, $D \in R_+^{p\times m}$, $A_4 \in R_+^{n_2\times n_2}$, $B_2 \in R_+^{n_2\times m}$, $C_2 \in R_+^{p\times n_2}$

From (10) and (9a) we have

$$\lim_{z_1\to\infty} z_1\begin{bmatrix} A(z_1)-A_4, & B(z_1)-B_2 \\ C(z_1)-C_2, & D(z_1)-D \end{bmatrix}=$$

$$\lim_{z_1\to\infty} z_1\begin{bmatrix} A_3 \\ C_1 \end{bmatrix}\left[I_{n_1}z_1-A_1\right]^{-1}\left[A_2\ \ B_1\right]=\begin{bmatrix} A_3 \\ C_1 \end{bmatrix}\left[A_2\ \ B_1\right] \qquad (18)$$

since

$$\lim_{z_1\to\infty} z_1\left[I_{n_1}z_1-A_1\right]^{-1}=I_{n_1}$$

From (18) and $\bar{d}_{j\infty} \in R_+$, $j=0,1,\ldots,q_2-1$ it follows that $A_2 \in R_+^{n_1\times n_2}$ and $A_3 \in R_+^{n_2\times n_1}$.

Similarly it can be shown that $\bar{d}_{i\infty} \in R_+$, $i=0,1,\ldots,q_1-1$ implies $A_2 \in R_+^{n_1\times n_2}$ and $A_3 \in R_+^{n_2\times n_1}$.

To simplify the notation it is assumed that $rank_+ T_{ki} \le i$. Then by Lemma 3 (given in Appendix) there exist nonnegative column matrices $G_{k1},\ldots,G_{ki}$ and nonnegative row matrices $H_{k,m_k-i+1},\ldots,H_{k,m_k}$ such that

$$T_{ki}=\sum_{j=1}^{i} G_{kj}H_{k,m_k+j-i} \quad \text{for } k=1,\ldots,q; i=1,\ldots,m_k \qquad (19)$$

We shall show that a realization (of type (4)) of $T(z_2)$ has the form

$$A_4:=diag\left[J_1 J_2,\ldots,J_q\right],\ \left[A_3\ \ B_2\right]=\begin{bmatrix} H_1 \\ H_2 \\ \vdots \\ H_q \end{bmatrix},\ \begin{bmatrix} A_2 \\ C_2 \end{bmatrix}=\left[G_1\ G_2 \ldots G_q\right]$$

(20)

where

$$J_k:=\begin{bmatrix} z_{2k} & 1 & 0 & \cdots & 0 & 0 \\ 0 & z_{2k} & 1 & \cdots & 0 & 0 \\ \hdashline 0 & 0 & 0 & \cdots & z_{2k} & 1 \\ 0 & 0 & 0 & \cdots & 0 & z_{2k} \end{bmatrix} \in R_+^{m_k\times m_k},$$

$$H_k:=\begin{bmatrix} H_{k1} \\ H_{k2} \\ \vdots \\ H_{km_k} \end{bmatrix}$$

$$k=1,\ldots,q \qquad (21)$$

$$G_k:=\left[G_{k1}\ G_{k2}\ldots G_{km_k}\right]$$

Taking into account that

$$\left[I_{m_k}z_2-J_k\right]^{-1}=$$

$$\begin{bmatrix} \frac{1}{z_2-z_{2k}}, & \frac{1}{(z_2-z_{2k})^2}, & \cdots, & \frac{1}{(z_2-z_{2k})^{m_k}} \\ 0\ , & \frac{1}{z_2-z_{2k}}, & \cdots, & \frac{1}{(z_2-z_2)^{m_k-1}} \\ \hdashline 0\ , & 0\ , & \cdots, & \frac{1}{z_2-z_{2k}} \end{bmatrix}$$

we obtain

$$G_k\left[I_{m_k}z_2-J_k\right]^{-1}H_k=$$

$$\frac{1}{z_2-z_{2k}}\sum_{i=1}^{m_k}G_{ki}H_{ki}+\frac{1}{(z_2-z_{2k})^2}\sum_{i=1}^{m_k-1}G_{ki}H_{k,i+1}+\cdots$$

$$+\frac{1}{(z_2-z_{2k})^{m_k}}G_{k1}H_{km_k}=\sum_{j=1}^{m_k}\frac{T_{kj}}{(z_2-z_{2k})^{m_k-j+1}}$$

Hence from (20), (19) and (13) we have

$$\begin{bmatrix} A_2 \\ C_2 \end{bmatrix}\left[I_{n_2}z_2-A_4\right]^{-1}\left[A_3\ B_2\right]=\sum_{k=1}^{q}G_k\left[I_{m_k}z_2-J_k\right]^{-1}H_k=$$

$$\sum_{k=1}^{q}\sum_{j=1}^{m_k}\frac{T_{kj}}{(z_2-z_{2k})^{m_k-j+1}}=T(z_2)$$

Thus, the matrices (20) are a realization of $T(z_2)$. Similar (dual) results can be obtained for $T(z_1)$. □

4. CALCULATION OF A REALIZATION.

If the conditions of Theorem are satisfied the following procedure can be used for calculation of a realization (4) of a given $T(z_1,z_2) \in R^{p\times m}(z_1,z_2)$.

Procedure

Step 1. Write $T(z_1,z_2)$ in the form (11a) (or equivalently (11b)).

Step 2. Find $A(z_2), B(z_2), C(z_2)$ and $D(z_2)$ given by (17a) (or equivalently $A(z_1), B(z_1), C(z_1)$ and $D(z_1)$ given by (17b)).

Step 3. Using (9a) find A_1, B_1, C_1 and D (or equivalently using (9b) find A_4, B_2, C_2 and D).

Step 4. Find (12a) (or (12b)) and using (13), (14), (19), (20) and (21) calculate matrices A_2, A_3, A_4, B_2, C_2 (or A_1, A_2, A_3, B_1, C_1).

Remark. If the equation $d(z_2)=0$ has only distinct roots $z_{21}, z_{22}, \ldots, z_{2q}$ then $T(z_2)$ can be written as

$$T(z_2)=\sum_{k=1}^{q}\frac{T_k}{z_2-z_{2k}}$$

where

$$T_k=\lim_{z_2\to z_{2k}}(z_2-z_{2k})T(z_2)=G_kH_k$$

and

$$A_4=diag\left[I_{r_1}z_{21}, I_{r_2}z_{22}, \ldots, I_{r_q}z_{2q}\right],\ \begin{bmatrix}A_3 & B_2\end{bmatrix}=\begin{bmatrix}H_1\\H_2\\ \vdots \\ H_q\end{bmatrix}$$

$$\begin{bmatrix}A_2\\C_2\end{bmatrix}=\left[G_1, G_2, \ldots, G_q\right]$$

Example. Find a realization (4) of the transfer matrix

$$T(z_1,z_2)=\frac{1}{d(z_1,z_2)}\begin{bmatrix}(z_1-1)[(z_2-2)^2+1]+(z_2-2)(z_2-1), & (z_1-1)(z_2-2)(2z_2-3)\\(z_1-1)(z_2-2)(z_2-1) & , \quad (z_1-1)(z_2-2)^2\end{bmatrix} \tag{22}$$

where $d(z_1,z_2)=(z_1-1)(z_2-2)^2$

In this case $m=p=2$ and (22) can be written as

$$T(z_1,z_2)=\frac{1}{(z_2-2)^2}\begin{bmatrix}(z_2-2)^2+1 & ,(z_2-2)(2z_2-3)\\(z_2-1)(z_2-2) & ,(z_2-2)^2\end{bmatrix}$$

$$+\frac{1}{z_1-1}\begin{bmatrix}\frac{z_2-1}{z_2-2}, & 0\\0 \ , & 0\end{bmatrix}$$

Hence $d_1(z_2)=(z_2-2)^2$,

$$N_1(z_2)=\begin{bmatrix}(z_2-2)^2+1 & ,(z_2-2)(2z_2-3)\\(z_2-1)(z_2-2) & ,(z_2-2)^2\end{bmatrix},$$

$$\bar{d}_0(z_2)=-1,\ \bar{N}_0(z_2)=\begin{bmatrix}\frac{z_2-1}{z_2-2}, & 0\\0 \ , & 0\end{bmatrix}$$

and from (17a) we have

$$A(z_2)=\left[-I_2\bar{d}_0(z_2)\right]=\begin{bmatrix}1 & 0\\0 & 1\end{bmatrix},$$

$$B(z_2)=\bar{N}_0(z_2)=\begin{bmatrix}\frac{z_2-1}{z_2-2}, & 0\\0 \ , & 0\end{bmatrix}$$

$$C_2(z_2)=\left[I_2\right],\ D(z_2)=\frac{N_1(z_2)}{d_1(z_2)}$$

Using (9a) we obtain

$$\begin{bmatrix}A_1 & B_1\\C_1 & D\end{bmatrix}=\lim_{z_2\to\infty}\begin{bmatrix}A(z_2) & B(z_2)\\C(z_2) & D(z_2)\end{bmatrix}=\left[\begin{array}{cc:cc}1 & 0 & 1 & 0\\0 & 1 & 0 & 0\\ \hdashline 1 & 0 & 1 & 2\\0 & 1 & 1 & 1\end{array}\right]$$

and

$$T(z_2)=\begin{bmatrix}A(z_2)-A_1\ , & B(z_2)-B_1\\C(z_2)-C_1\ , & D(z_2)-D\end{bmatrix}=$$

$$\frac{1}{(z_2-2)^2}\left[\begin{array}{cc:cc}0 & 0 & z_2-2, & 0\\0 & 0 & 0\ , & 0\\ \hdashline 0 & 0 & 1\ , & z_2-2\\0 & 0 & z_2-2, & 0\end{array}\right]$$

In this case the equation $d(z_2)=(z_2-2)^2=0$ has only one double root $z_{21}=2\ (m_1=2)$. Then using (13), (14) and (19) we obtain

$$T(z_2)=\frac{T_{11}}{(z_2-2)^2}+\frac{T_{12}}{z_2-2}$$

where

$$T_{11}=\begin{bmatrix}0 & 0 & 0 & 0\\0 & 0 & 0 & 0\\0 & 0 & 1 & 0\\0 & 0 & 0 & 0\end{bmatrix}=G_{11}H_{12},$$

$$T_{12}=\begin{bmatrix}0 & 0 & 1 & 0\\0 & 0 & 0 & 0\\0 & 0 & 0 & 1\\0 & 0 & 1 & 0\end{bmatrix}=G_{11}H_{11}+G_{12}H_{12}$$

$$G_{11}=\begin{bmatrix}0\\0\\1\\0\end{bmatrix}, G_{12}=\begin{bmatrix}1\\0\\0\\1\end{bmatrix}, H_{11}=\begin{bmatrix}0 & 0 & 0 & 1\end{bmatrix}, H_{12}=\begin{bmatrix}0 & 0 & 1 & 0\end{bmatrix}$$

From (20) and (21) we have

$$A_4=J_1=\begin{bmatrix}z_{21} & 1\\0 & z_{21}\end{bmatrix}=\begin{bmatrix}2 & 1\\0 & 2\end{bmatrix}$$

$$[A_3, B_2] = H_1 = \begin{bmatrix} H_{11} \\ H_{12} \end{bmatrix} = \begin{bmatrix} 0 & 0 & 0 & 1 \\ 0 & 0 & 1 & 0 \end{bmatrix}$$

$$\begin{bmatrix} A_2 \\ C_2 \end{bmatrix} = G_1 = [G_{11} \;\; G_{12}] = \begin{bmatrix} 0 & 1 \\ 0 & 0 \\ 1 & 0 \\ 0 & 1 \end{bmatrix}$$

Therefore, a desired realization of (22) has the form

$$A = \begin{bmatrix} A_1 & A_2 \\ A_3 & A_4 \end{bmatrix} = \begin{bmatrix} 1 & 0 & 0 & 1 \\ 0 & 1 & 0 & 0 \\ 0 & 0 & 2 & 1 \\ 0 & 0 & 0 & 2 \end{bmatrix}, B = \begin{bmatrix} B_1 \\ B_2 \end{bmatrix} = \begin{bmatrix} 1 & 0 \\ 0 & 0 \\ 0 & 1 \\ 1 & 0 \end{bmatrix} \quad (23)$$

$$C = [C_1, C_2] = \begin{bmatrix} 1 & 0 & 1 & 0 \\ 0 & 1 & 0 & 1 \end{bmatrix}, D = \begin{bmatrix} 1 & 2 \\ 1 & 1 \end{bmatrix}$$

The realization (23) is not a minimal one. It can be easily checked that the matrices

$$A = \begin{bmatrix} 1 & 0 & 1 \\ 0 & 2 & 1 \\ 0 & 0 & 2 \end{bmatrix}, B = \begin{bmatrix} 1 & 0 \\ 0 & 1 \\ 1 & 0 \end{bmatrix}, C = \begin{bmatrix} 1 & 1 & 0 \\ 0 & 0 & 1 \end{bmatrix}, D = \begin{bmatrix} 1 & 2 \\ 1 & 1 \end{bmatrix}$$

are also a realization of (22).

5. CONCLUDING REMARKS

Sufficient conditions for the existence of a realization being a positive 2-D Roesser type model of a given proper transfer matrix have been established. A procedure for calculating a positive realization of 2-D transfer matrix has been given and illustrated by an example. The relationship between the minimality of a realization obtained by the procedure and its controllability and observability will be considered in a sequential paper. An open problem is also necessary and sufficient conditions for the existence of a positive stable 2-D Roesser type model for a given 2-D proper transfer matrix.

REFERENCES

Cohen, J.E. (1993), *Nonnegative Ranks, Decompositions, and Factorizations of Nonnegative Matrices,* Linear Algebra and Its Applications 190, 149-168.

Eising, R. (1979), 2-D *Systems: An Algebraic Approach,* Mathematical Centrum, Amsterdam

Eising, R. (1978), *Realization and stabilization of 2-D systems,* IEEE Trans. Autom. Contr., vol. **AC-23**, No 5, 795-799.

Fornasini, E. and Marchesini, G. (1976), *State space realization of two-dimensional filters,* IEEE Trans. Autom.Control, **AC-21**, 484-491.

Fornasini, E. and Marchesini, G. (1978), *Doubly indexed dynamical systems: State space models and structural properties, Math. Syst.Theory* **12**

Hinamoto, T.and Fairman, F.W. (1984), *Realisation of the Attasi state space model for 2-D filters.* Int. J. Systems Sci., vol. 15, No 2, 215-228.

Kaczorek, T. (1987), *Realization problem for general model of two-dimensional linear systems,* Bull. Acad. Pol. Sci. Techn. Sci., vol. 35, No 11-12, 633-637.

Kaczorek, T. (1996), *Reachability and controllability of nonnegative 2-D Roesser type models,* Bull. Pol.Acad. Techn. Sci., vol . 44, No 4, 405-410

Kaczorek, T. (1997) a): *Realisation problem for discrete-time positive linear systems,* Appl. Math. and Comp. Sc., vol. 7, No 1, 117-124; b): *Reachability and minimum energy control of nonnegative 2-D Roesser type models,* Bull. Pol. Acad. Techn. Sci., vol .45 (in press)

Kaczorek, T. (1993), *Linear Control Systems,* vol. 2, Research Studies Press and J. Wiley, New York.

Kaczorek, T. (1985), *Two-Dimensional Linear Systems,* Springer-Verlag, New York.

Klamka, J. (1991), *Controllability of Dynamical Systems,* Kluwer Academic Publ., Dordrecht.

Roesser, P.R. (1975), *A discrete state-space model for linear image processing,* IEEE Trans. Autom. Contr., vol. AC-20, No. 1, 1-10.

Valcher, M.E. and Fornasini, E. (1995), *State Models and Asymptotic Behaviour of 2D Positive Systems,* IMA Journal of Mathematical Control & Information, No 12, 17-36.

APPENDIX

Let $A \in R_+^{m\times n}$ and r be its positive column *rank* defined as follows (Cohen 1993). The positive column *rank* of A (denoted by $rank_+ A$) is the smallest nonnegative integer r for which there exist linearly independent vectors $v_i \in R_+^m$, $i = 1,\dots,r$ such that each column of A is a linear combination of the vectors with nonnegative coefficients.
It can be shown (Cohen 1993) that

$$rank \le rank_+ A \le \min(m,n)$$

and if for $A \in R_+^{m\times n}$, $m < 4$ and $n < 4$ then $rank_+ A = rank\, A$.

Lemma 3. Let $A \in R_+^{m\times n}$ and $r = rank_+ A$. Then there exists a pair of matrices $B \in R_+^{r\times m}, C \in R_+^{p\times r}$ such that $A = CB$.
The proof is given in (Cohen 1993).

MULTITIME SCALE SYSTEMS - THE ND APPROACH

Krzysztof Gałkowski
Artur Gramacki

Technical University of Zielona Góra, Institute of Robotics and Software Engineering, Podgórna Str. 50, 65-246 Zielona Góra, Poland, Fax: 48 68 253944
email: galko@irio.pz.zgora.pl

Technical University of Zielona Góra, Institute of Computer Engineering and Electronics, Podgórna Str. 50, 65-246 Zielona Góra, Poland, Fax: 48 68 253944
mail: A.Gramacki@iie.pz.zgora.pl

Abstract: In this paper so-called multitime scale systems are described in terms of discrete repetitive processes which belong to the wide family of so-called 2D systems.

Keywords: Multidimensional systems, the Roesser model, the Fornasini-Marchesini model, discrete repetitive processes, multitime scale systems.

1. INTRODUCTION

Many physical systems, data analysis procedures, computational algorithms and (more recently) learning algorithms have a natural (and underexploited) two-dimensional (2D) structure due to the presence of more than one spatial variable, the combined effect of space and time or the combined effect of a spatial/time variable and an integer index representing iteration, pass or trial number. Physical examples of such systems include bench mining systems, metal rolling, automatic ploughing aids and vehicle convoy co-ordination on motorways whilst algorithmic examples include image processing, discrete models of spatial behaviour, point mapping algorithms and recursive learning schemes as illustrated by trajectory learning in iterative learning control or recursive/feedback neural networks. The structure of all such problems is strongly related and there is a need to exploit current knowledge, develop new techniques and develop new concepts with application to industrially related benchmark examples. The required techniques are those of modelling, computational support, systems analysis tools, control design tools and software development and the formulation of applications problems in the format of the theory. There is a rich literature dealing with such problems. The basic choice is, for example, Bose (1985); Fornasini (1978); Kaczorek (1992); Roesser (1975); Rogers (1992).

On the other hand, following Peponides, Kokotovic and Hong Chow (1982) „Singular perturbations, a traditional tool of fluid dynamics and non-linear mechanics, have recently entered the system theory field and are evolving into a systematic approach to multitime scale modelling, analysis and control of dynamic systems with widely different speeds. The approach is asymptotic, that is exact in the limit as the ratio ε of the speeds of the slow versus the fast dynamics tends to zero. When ε is small approximations are obtained from reduced order models in separate time scales." Such a situation, when a system behaviour/dynamics develops with two or more different speeds, seems to be interesting in many physical, biological or economic systems.

The continual developments in digital techniques for the design/implementation of filters/controllers is, of course, of direct relevance to the control of continuous time linear (and non-linear) processes. By analogy with other areas, a common approach is to convert filters/controllers designed in the continuous domain to a discrete equivalent for implementation. This could be done in either (i) the time domain or (ii) the generalised frequency domain which requires appropriate frequency domain mapping tools. A key technique is Euler, or trapezoidal, numerical integration (Schneider, 1991) which is isomorphic to the well known bilinear transform (frequency domain).

The nD systems approach, especially its particular case, the so-called repetitive processes seems to provide the effective tool to describe multitime scale problems. In the paper, some initial work on presentation of multitime scale systems in 2D frameworks along with first analysis is reported.

2. BACKGROUND

The state-space model of a discrete linear repetitive process has the structure (Rogers, 1992)

$$x(k+1,p+1) = \Phi x(k+1,p) + \Delta_0 y(k,p) + \Delta u(k+1,p) , \tag{1}$$

$$y(k+1,p) = Cx(k+1,p) + D_1 y(k,p) + D_0 u(k+1,p) . \tag{2}$$

Here, on pass k, $x(k,p)$ is the $n\times 1$ state vector, $y(k,p)$ is the $m\times 1$ vector pass profile and $u(k,p)$ denotes the $t\times 1$ vector of control inputs and Φ, Δ_0, Δ, C, D_1, D_0 are matrices of appropriate dimensions. This model has the so-called unit memory property, i.e. it is only the pass profile on pass k which (explicitly) contributes to its counterpart on pass $k+1$, $k\geq 0$. In the more general case, it is the previous $M>1$ pass profiles which (explicitly) contribute to the current one. These so-called non-unit memory processes are not considered further here since the results developed generalise in a natural manner. The intrinsic feature of repetitive process is that all passes have the finite and fixed length α, i.e. $p = 0,1,\ldots,\alpha-1$. Consider a 2D linear system with time invariant dynamics and let i and j denote the co-ordinates in the horizontal and vertical directions of information propagation respectively. Then for systems recursive in the positive quadrant $((i,j)\geq 0)$ the Roesser state-space model has the structure (Roesser, 1975)

$$\begin{bmatrix} x^h(i+1,j) \\ x^v(i,j+1) \end{bmatrix} = A\begin{bmatrix} x^h(i,j) \\ x^v(i,j) \end{bmatrix} + Bu(i,j) , \tag{3}$$

$$y(i,j) = C\begin{bmatrix} x^h(i,j) \\ x^v(i,j) \end{bmatrix} + Du(i,j) . \tag{4}$$

Here $x^h(i,j)$ is the $n_1\times 1$ partial horizontal state vector, $x^v(i,j)$ is the $n_2\times 1$ partial vertical state vector, $y(i,j)$ is the $m\times 1$ output vector and $u(i,j)$ is the $t\times 1$ input vector. The matrices A, B, C, D are of appropriate dimensions, and constitute the so-called state-space realisation of the Roesser form.

Alternatively, let $x(i,j)$ denotes the $n\times 1$ partial state and define $y(i,j)$ and $u(i,j)$ as before. In particular, set $n = n_1 + n_2$, and hence the state vector $x(i,j)$ is the collection of the horizontal and vertical subvectors, defined previously, i.e. $x(i,j) \triangleq \left[x^h(i,j)^T \;\; x^v(i,j)^T\right]^T$. Then, the Fornasini-Marchesini model generalised by Kurek has the structure (Fornasini, 1978; Kurek, 1985)

$$x(i+1,j+1) = A_1 x(i+1,j) + A_2 x(i,j+1) + B_1 u(i+1,j) + B_2 u(i,j+1) , \tag{5}$$

$$y(i,j) = Cx(i,j) + Du(i,j) . \tag{6}$$

Now, following Sueur and Dauphin-Tanguy (Sueur, 1991) refer to the linear continuous multitime scale linear system

$$\dot{X}_1 = A_{11}X_1 + A_{12}X_2 + B_1 U$$

$$\varepsilon\dot{X}_2 = A_{21}X_1 + A_{22}X_2 + B_2 U , \tag{7}$$

$$Y = C_1 X_1 + C_2 X_2 + DU , \tag{8}$$

where

$$X_1 \in \Re^n,\;\; X_2 \in \Re^m,\;\; U \in \Re^p,\;\; Y \in \Re^q ,$$

$$X_1(t_0) \triangleq X_{10},\;\; X_{21}(t_0) \triangleq X_{20} .$$

The intrinsic feature of these systems is presence of the small scalar ε in the second equation of (7). This is interpreted as that there are two time scales: the slow $t - t_0$ and the fast $\tau = (t-t_0)/\varepsilon$. Hence, the dynamics of the system represented by the variables X_2 and U have two parts: the slow and the fast, when, due to the lack of the small scalar, the variable X_1 has only the slow dynamics:

$$X_1 \cong X_{1s},\;\; X_2 \cong X_{2s} + X_{2f} ,$$

$$U \cong U_s + U_f ,\;\; Y \cong Y_s + Y_f , \tag{9}$$

where the slow dynamics is obtained when substituting $\varepsilon = 0$ to the second equation of (7). It is straightforward task to show that (Sueur, 1991) both the slow and the fast dynamics are governed by the following equations

$$\dot{X}_{1s} = \left(A_{11} - A_{12}A_{22}^{-1}A_{21}\right)X_{1s} + \left(B_1 - A_{12}A_{22}^{-1}B_2\right)U_s,$$

$$X_{2s} = -A_{22}^{-1}A_{21}X_{1s} - A_{22}^{-1}B_2U_s,$$

$$Y_s = \left(C_1 - C_2A_{22}^{-1}A_{21}\right)X_{1s} + \left(D - C_2A_{22}^{-1}B_2\right)U_s, \quad (10)$$

and

$$\varepsilon\dot{X}_{2f} = A_{22}X_{2f} + B_2U_f,$$

$$X_{1f} = 0, \qquad Y_f = C_2X_{2f}, \quad (11)$$

where, the initial conditions clearly satisfy

$$X_{1s}(t_0) = X_1(t_0),$$

$$X_{2f}(t_0) = X_2(t_0) - X_{2s}(t_0) = X_2(t_0) + A_{22}^{-1}A_{21}X_1(t_0). \quad (12)$$

In the remainder of this paper the representation of the multitime scale systems in terms of 2D systems is provided.

3. ND SYSTEMS APPROACH

The continuous version of the Roesser model presented in the equations of (3)-(4), called also the Darboux equations, can be clearly presented as

$$\begin{bmatrix} \frac{\partial}{\partial t}x^h(t,\tau) \\ \frac{\partial}{\partial \tau}x^v(t,\tau) \end{bmatrix} = A\begin{bmatrix} x^h(t,\tau) \\ x^v(t,\tau) \end{bmatrix} + Bu(t,\tau), \quad (13)$$

$$y(t,\tau) = C\begin{bmatrix} x^h(t,\tau) \\ x^v(t,\tau) \end{bmatrix} + Du(t,\tau). \quad (14)$$

Observe, here, the interesting feature of nD systems, unknown in the classic 1D case, that system may be continuous in the one direction but discrete in the other (Rogers, 1992; Kaczorek, 1994).

It is easy to see that even the initial state-space model of multitime scale systems may be presented in the form of (13)-(14). Hence, assume that

$$x^h(t,\tau) \triangleq X_{1s}(t), \quad x^v(t,\tau) \triangleq X_{2s}(t) + X_{2f}(\tau), \quad (15)$$

where

$$\tau = t/\varepsilon, \quad (16)$$

here, for simplicity $t_0 = 0$ is taken. Note, that due to (9) the 'horizontal' sub vector does not depend on the fast time. In this framework, horizontal means the slow dynamics and the vertical - the fast.

The most interesting models are, however, obtained when discretizing the equations of (7)-(8).
First, use the well-known trapezoidal (Euler) method of discretization (Schneider, 1991) to this model

$$X_1(i+1) = X_1(i) + (H/2)\left[\dot{X}_1(i+1) + \dot{X}_1(i)\right], \quad (17)$$

$$X_2(j+1) = X_2(j) + (h/2)\left[\dot{X}_2(j+1) + \dot{X}_2(j)\right]. \quad (18)$$

It is seen that, due to the fact that the state sub-vector X_1 does not have the fast dynamics, it is discretized 'slowly' when X_2 'quickly'. Hence, the index 'i' concerns the slow dynamics and 'j' the fast and 'H' denotes a 'long period' when 'h' denotes the short. Moreover, both of them satisfy

$$h = H\varepsilon. \quad (19)$$

Now, introduce the following notation

$$\varphi(i,k) \triangleq \varphi(iH + jh), \; i=0,1,...; j=0,1,..., N-1, \quad (20)$$

where

$$N = \frac{1}{\varepsilon}. \quad (21)$$

It is the straightforward task to show from (19) and (21) that

$$H = Nh. \quad (22)$$

Remind that N has to be natural number which introduces the limitation for ε which, however, seems not to be essential in practice.

Finally, one may obtain the discrete version of the multitime scale system described by the equations of (7)-(8). Hence, insert the first equation of (7) into that of (17) to achieve

$$\begin{aligned} X_1(i+1) = \mathbf{A}_1X_1(i) + \mathbf{A}_2X_2(i) + \mathbf{A}_2X_2(i+1) \\ + \mathbf{B}_1U(i) + \mathbf{B}_1U(i+1) \end{aligned}, \quad (23)$$

where i=0,1,... and

$$\mathbf{A}_1 = \left[I - \frac{H}{2}A_{11}\right]^{-1}\left[I + \frac{H}{2}A_{11}\right],$$

$$\mathbf{A}_2 = \left[I - \frac{H}{2}A_{11}\right]^{-1}\frac{H}{2}A_{12},$$

$$\mathbf{B}_1 = \left[I - \frac{H}{2}A_{11}\right]^{-1}\frac{H}{2}B_1. \quad (24)$$

Here, due to the fact that the subvector X_2 has simultaneously the slow and the fast dynamics, the usual assumption in such cases that the function is step wise, could not hold, i.e. one must assume that

$$X_2(i) \neq X_2(i+1), \quad U(i) \neq U(i+1). \quad (25)$$

As will be shown, the further value $X_2(i+1)$ may be calculated from the equation for the second subvector, i.e. X_2. Applying the same approach to the second equation of (7) and (18) yields

$$X_2(i,j+1) = \mathbf{A}_3X_1(i) + \mathbf{A}_4X_2(i,j) + \mathbf{B}_2U(i,j), \quad (26)$$

where i=0,1,...; j=0,1,..., N-1, and

$$\mathbf{A}_3 = \left[I - \frac{h}{2}A_{22}\right]^{-1}hA_{21},$$

$$\mathbf{A}_4 = \left[I - \frac{h}{2}A_{22}\right]^{-1}\frac{h}{2}A_{22},$$

$$\mathbf{B}_2 = \left[I - \frac{h}{2}A_{22}\right]^{-1}hB_2. \quad (27)$$

Here, the fact that the subvector X_1 has only the slow dynamics, i.e.

$$X_1(i,j) = X_1(i), \quad j = 0,1,\ldots,N-1, \quad (28)$$

implies that this is step-wise along the fast 'direction', i.e.

$$X_1(i,j) = X_1(i,j+1), \quad j = 0,1,\ldots,N-2. \quad (29)$$

Similarly, the input may be assumed to be step-wise along 'j' too

$$U(i,j) = U(i,j+1), \quad j = 0,1,\ldots,N-2. \quad (30)$$

In order to avoid any misunderstanding when comparing the equation of (23) to that of (26), note that

$$U(i) \triangleq U(i,j)_{j=0}. \quad (31)$$

It is clear that the equations of (23) and (26) have very similar structure to those for a linear discrete repetitive process, i.e. (1)-(2). The insignificant differences between indexing in both the pairs of equations are not important. There is simple approach to transform the one form to the other. The intrinsic, common feature of both the models, (1)-(2) and (23), (26) is the fact that the 'pass' is limited. Hence the functions have the limited region of support. In fact, 'N' in (23), (26) plays the same role as α in (1)-(2), and the long period H plays a role of the pass in (1)-(2). The model of (23), (26) may be considered as the particular case of the standard repetitive process. The main difference is that there is no, here, interactions between passes/ long periods, i.e. there is no recurrence along 'j' for the given 'i'. If such a recurrence exists, it means that there are constant delays of the pass/long period length in the process equations.

To have the clear 'repetitive' or 'nD' structure in the equation of (23) it is necessary to calculate the value of $X_2(i+1)$ from the equation of (26). Note that

$$X_2(i+1) \triangleq X_2(i,N). \quad (32)$$

Hence

$$X_2(i+1) = \left[\sum_{j=0}^{N-1}\mathbf{A}_4^j\mathbf{A}_3\right]X_1(i) + \mathbf{A}_4^NX_2(i) + \sum_{j=0}^{N-1}\mathbf{A}_4^{N-j-1}\mathbf{B}_2U(i,j) \quad (33)$$

Hence, the final form of the 'slow' equation can be presented as

$$X_1(i+1) = \left\{\mathbf{A}_1 + \mathbf{A}_2\left[\sum_{j=0}^{N-1}\mathbf{A}_4^j\mathbf{A}_3\right]\right\}X_1(i) + \mathbf{A}_2\left[I + \mathbf{A}_4^N\right]X_2(i) + \mathbf{A}_2\sum_{j=1}^{N-1}\mathbf{A}_4^{N-j-1}\mathbf{B}_2U(i,j) + \left[\mathbf{B}_1 + \mathbf{A}_2\mathbf{A}_4^{N-1}\mathbf{B}_2\right]U(i) + \mathbf{B}_1U(i+1),$$

$$i=0,1,\ldots;\ j=0,1,\ldots,N-1. \quad (34)$$

Now, it is seen that due to (25) the slow subvector X_1 has in fact also the fast fraction which has been aggregated in the equation of (34).

To transform the equation of (34) to the standard form used in dynamic systems description, apply the substitution

$$Z_1(i) \triangleq X_1(i) - \mathbf{B}_1U(i). \quad (35)$$

Then, this equation may be rewritten as

$$Z_1(i+1) = \left\{ \mathbf{A}_1 + \mathbf{A}_2 \left[\sum_{j=0}^{N-1} \mathbf{A}_4^j \mathbf{A}_3 \right] \right\} Z_1(i)$$

$$+\mathbf{A}_2 \left[I + \mathbf{A}_4^N \right] X_2(i) + \mathbf{A}_2 \sum_{j=1}^{N-1} \mathbf{A}_4^{N-j-1} \mathbf{B}_2 U(i,j)$$

$$+\left\{ (I+\mathbf{A}_1)\mathbf{B}_1 + \mathbf{A}_2 \left[\sum_{j=0}^{N-1} \mathbf{A}_4^j \mathbf{A}_3 \right] \mathbf{B}_1 + \mathbf{A}_2 \mathbf{A}_4^{N-1} \mathbf{B}_2 \right\} U(i),$$

$$i=0,1,...; j=0,1,..., N-1. \qquad (36)$$

Introducing the new variable Z_1 yields also the change in the fast equation of (26)

$$\begin{aligned} X_2(i,j+1) &= \mathbf{A}_3 Z_1(i) + \mathbf{A}_4 X_2(i,j) \\ &+ \mathbf{A}_3 \mathbf{B}_1 U(i) + \mathbf{B}_2 U(i,j) \end{aligned} , \qquad (37)$$

where i=0,1,...; j=0,1,..., N-1. In what follows, the output equation is

$$Y(i,j) = C_1 Z_1(i) + C_2 X_2(i,j) + DU(i,j) + C_1 B_1 U(i), \qquad (38)$$

where i=0,1,...; j=0,1,..., N-1.

The equations of (36)-(38) present a discrete multitime linear system described in the repetitive/2D form.

4. COMPUTER SIMULATIONS

Below a simple computer simulation of a sample multitime scale system (the ND approach) is presented. To make the obtained results more readable and more simple to analyse the scalars instead matrices were used. Such simplification does not change the general features of multitime scale systems.
Moreover N is relatively small so the main features of the system are easier to notice (In the examples below N=6). In practical systems N should be much more bigger (for example if short periods range from intervals of seconds the corresponding long periods range from several minutes). The computations were made with respect to the assumptions (28)-(31).
Fig. 1 shows the impulse response while Fig. 2 shows the step response of the system. In both cases state variables (slow and fast) and outputs are stable.

The matrices (scalars) of the model (7)-(8) are as follows:

A_{11} = -3, A_{22} = -2, A_{12} = -0.5, A_{21} = 2,
B_1 = 5, B_2 = -2, C_1 = 1, C_2 = -1, D = 1,

and the corresponding matrices (scalars) of the model after discretization (24), (27) are as follows:

$\mathbf{A}_1$ = -0.8, $\mathbf{A}_2$ = -0.15, $\mathbf{A}_3$ = 1,
$\mathbf{A}_4$ = -0.5, $\mathbf{B}_1$ = -0.5, $\mathbf{B}_2$ = -1,

long period (H)=6, short period (h)=1,
initial conditions: $Z_1(0)$=0, $X_2(0,0)$=0,

discrete impulse input:

$$U(i,j) = \begin{cases} 1 & i=0, j=0 \\ 0 & i \neq 0, j \neq 0 \end{cases}, \quad i,j \geq 0,$$

discrete step input:

$$U(i,j) = \begin{cases} 0 & i=0, j=0 \\ 1 & i \neq 0, j \neq 0 \end{cases}, \quad i,j \geq 0.$$

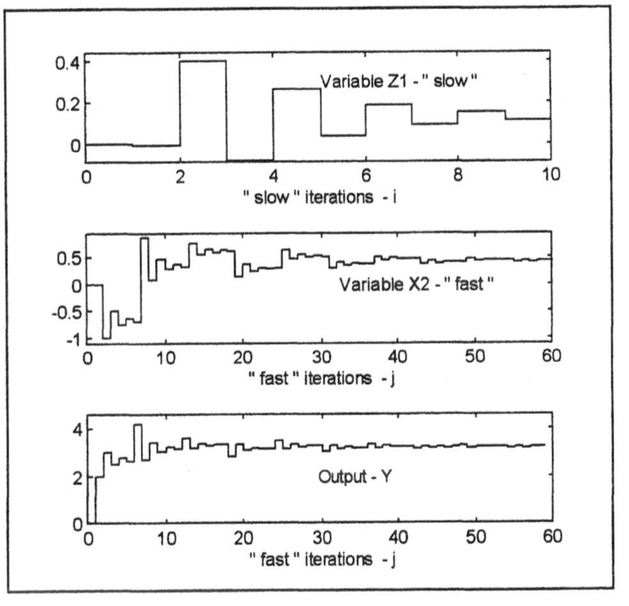

Fig. 1. Multitime scale system (the ND approach) example - impulse response.

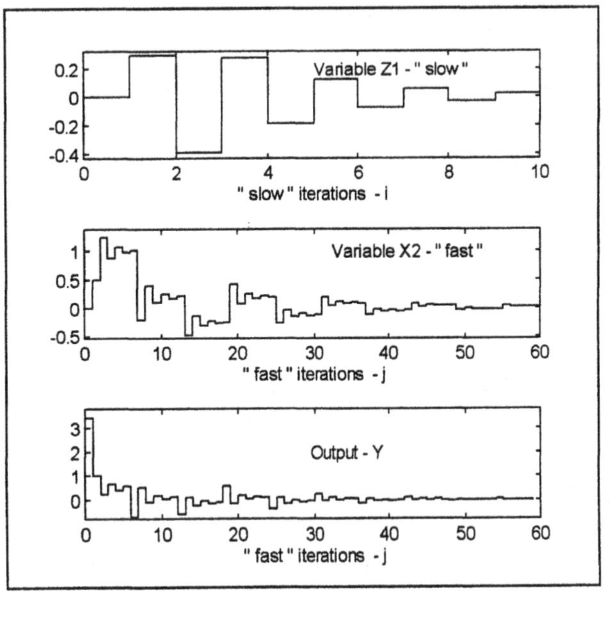

Fig. 2. Multitime scale system (the ND approach) example - step response.

5. CONCLUDING REMARKS

The article presents a new application of ND systems theory. The standard theory (1D approach) of the multitime scale systems (Peponides, 1982) was expanded for the ND approach. The obtained results show that the newly developed model is very similar to the special case of the ND systems called discrete repetitive processes.

There are a lot of physical examples were the multitime scale system's approach can be used (for examples Power Systems and even Economics).

The problem is developing and obtained results will be reported in due course.

REFERENCES

Bose N.K.B. (1985). Multidimensional Systems Theory. *Progress, Directions and Open Problems in Multidimensional Systems, D. Reidel, Dortrecht, Boston, Lancaster.*

Fornasini E., G. Marchesini (1978). Doubly indexed dynamical systems: state-space models and structural properties. *Math. Systems Theory,* **12**, 59-72.

Kaczorek T. (1992). Linear control systems. *Research Studies Press LTD, Taunton, Somerset, England & John Wiley & Sons INC., New York.*

Kaczorek T. (1994). 2-D continuous-discrete linear systems, *Proc. Tenth Intern. Conference. on System Engineering ICSE'94, Coventry, 6-8 Sept. 1994,* **Vol. 1**, 550-557.

Kurek J.E. (1985). The general state-space-model for a two-dimensional linear digital systems. *IEEE Trans. on Automatic Control,* **Vol. AC-30**, 600-602.

Peponides G., P.V. Kokotovic, J. Hong Chow (1982). Singular perturbations and time scales in nonlinear models of power systems. *IEEE Trans. on Circuits and Systems,* **Vol. CAS-29**, 758-767.

Roesser R. P., (1975), A discrete state-space model for linear image processing, *IEEE Trans. on Automatic Control,* **Vol. AC-20**, 1-10.

Rogers E., D.H. Owens (1992). Stability analysis for linear repetitive processes. *Lecture Notes in Control and Information Sciences,* **175**, Springer Verlag, Berlin, 1992.

Schneider A.M., J.T. Kanesige, F.D. Groutage (1991). Higher order s-to-z mapping functions and their applications in digitizing continuous-time filters. *Proc. of The IEEE,* **Vol. 79, no. 11,** 1661-1674.

Sueur C., G. Dauphin-Tanguy (1991). *Bond graph approach to multi-time scale systems analysis.* J. of the Franklin Inst., **Vol. 328, no. 5/6**, 1005-1026.

AUTHOR INDEX

www.ingramcontent.com/pod-product-compliance
Ingram Content Group UK Ltd.
Pitfield, Milton Keynes, MK11 3LW, UK
UKHW052301280926
15471UKWH00006B/1044

9 780080 429359